ALGEBRA

Arithmetic Operations

$a(b + c) = ab + ac$

$\dfrac{a}{b} + \dfrac{c}{d} = \dfrac{ad + bc}{bd}$

$\dfrac{a + c}{b} = \dfrac{a}{b} + \dfrac{c}{b}$

$\dfrac{\frac{a}{b}}{\frac{c}{d}} = \dfrac{a}{b} \times \dfrac{d}{c} = \dfrac{ad}{bc}$

Exponents and Radicals

$x^m x^n = x^{m+n}$

$\dfrac{x^m}{x^n} = x^{m-n}$

$(x^m)^n = x^{mn}$

$x^{-n} = \dfrac{1}{x^n}$

$(xy)^n = x^n y^n$

$\left(\dfrac{x}{y}\right)^n = \dfrac{x^n}{y^n}$

$x^{1/n} = \sqrt[n]{x}$

$x^{m/n} = \sqrt[n]{x^m} = \left(\sqrt[n]{x}\right)^m$

$\sqrt[n]{xy} = \sqrt[n]{x}\sqrt[n]{y}$

$\sqrt[n]{\dfrac{x}{y}} = \dfrac{\sqrt[n]{x}}{\sqrt[n]{y}}$

Factoring Special Polynomials

$x^2 - y^2 = (x + y)(x - y)$

$x^3 + y^3 = (x + y)(x^2 - xy + y^2)$

$x^3 - y^3 = (x - y)(x^2 + xy + y^2)$

Binomial Theorem

$(x + y)^2 = x^2 + 2xy + y^2$ $\qquad (x - y)^2 = x^2 - 2xy + y^2$

$(x + y)^3 = x^3 + 3x^2y + 3xy^2 + y^3$

$(x - y)^3 = x^3 - 3x^2y + 3xy^2 - y^3$

$$(x + y)^n = x^n + nx^{n-1}y + \frac{n(n-1)}{2}x^{n-2}y^2 + \cdots + \binom{n}{k}x^{n-k}y^k + \cdots + nxy^{n-1} + y^n$$

where $\dbinom{n}{k} = \dfrac{n(n-1)\cdots(n-k+1)}{1 \cdot 2 \cdot 3 \cdot \cdots \cdot k}$

Quadratic Formula

If $ax^2 + bx + c = 0$, then $x = \dfrac{-b \pm \sqrt{b^2 - 4ac}}{2a}$.

Inequalities and Absolute Value

If $a < b$ and $b < c$, then $a < c$.

If $a < b$, then $a + c < b + c$.

If $a < b$ and $c > 0$, then $ca < cb$.

If $a < b$ and $c < 0$, then $ca > cb$.

If $a > 0$, then

$|x| = a$ means $x = a$ or $x = -a$

$|x| < a$ means $-a < x < a$

$|x| > a$ means $x > a$ or $x < -a$

GE(

Geon

Formu

Triangle

$A = \frac{1}{2}bh$

$= \frac{1}{2}ab \sin \theta$

Circle

$A = \pi r^2$

$C = 2\pi r$

Sector of Circle

$A = \frac{1}{2}r^2\theta$

$s = r\theta$ (θ in radians)

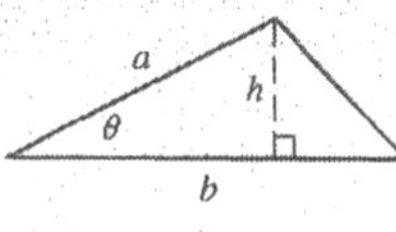

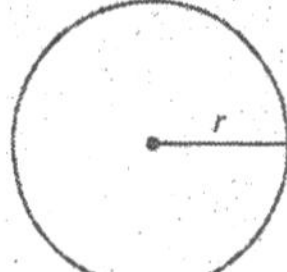

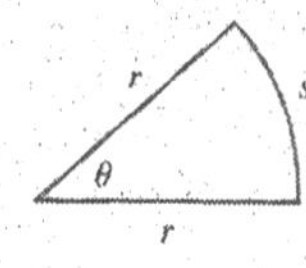

Sphere

$V = \frac{4}{3}\pi r^3$

$A = 4\pi r^2$

Cylinder

$V = \pi r^2 h$

Cone

$V = \frac{1}{3}\pi r^2 h$

$A = \pi r\sqrt{r^2 + h^2}$

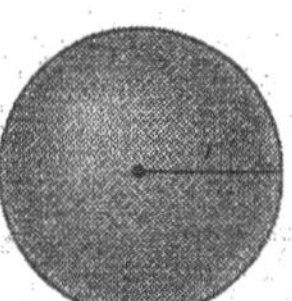

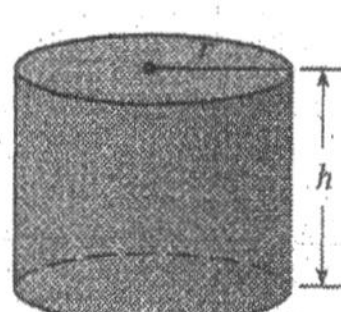

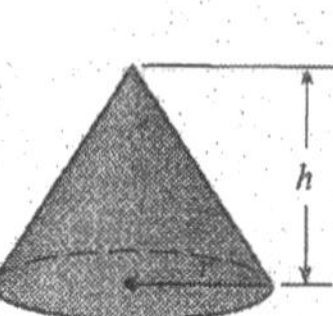

Distance and Midpoint Formulas

Distance between $P_1(x_1, y_1)$ and $P_2(x_2, y_2)$:

$$d = \sqrt{(x_2 - x_1)^2 + (y_2 - y_1)^2}$$

Midpoint of $\overline{P_1P_2}$: $\left(\dfrac{x_1 + x_2}{2}, \dfrac{y_1 + y_2}{2}\right)$

Lines

Slope of line through $P_1(x_1, y_1)$ and $P_2(x_2, y_2)$:

$$m = \frac{y_2 - y_1}{x_2 - x_1}$$

Point-slope equation of line through $P_1(x_1, y_1)$ with slope m:

$$y - y_1 = m(x - x_1)$$

Slope-intercept equation of line with slope m and y-intercept b:

$$y = mx + b$$

Circles

Equation of the circle with center (h, k) and radius r:

$$(x - h)^2 + (y - k)^2 = r^2$$

TRIGONOMETRY

Angle Measurement

π radians $= 180°$

$1° = \frac{\pi}{180}$ rad 1 rad $= \frac{180°}{\pi}$

$s = r\theta$

(θ in radians)

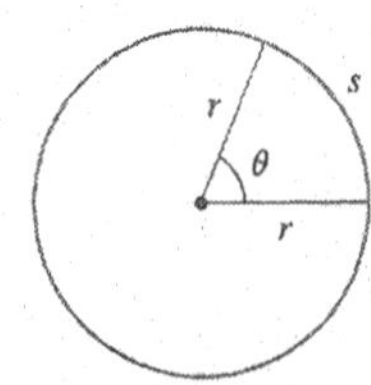

Right Angle Trigonometry

$\sin\theta = \frac{\text{opp}}{\text{hyp}}$ $\csc\theta = \frac{\text{hyp}}{\text{opp}}$

$\cos\theta = \frac{\text{adj}}{\text{hyp}}$ $\sec\theta = \frac{\text{hyp}}{\text{adj}}$

$\tan\theta = \frac{\text{opp}}{\text{adj}}$ $\cot\theta = \frac{\text{adj}}{\text{opp}}$

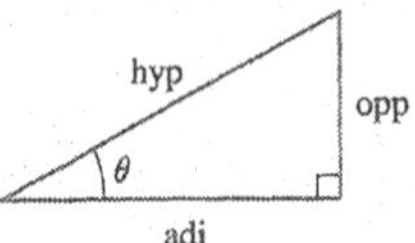

Trigonometric Functions

$\sin\theta = \frac{y}{r}$ $\csc\theta = \frac{r}{y}$

$\cos\theta = \frac{x}{r}$ $\sec\theta = \frac{r}{x}$

$\tan\theta = \frac{y}{x}$ $\cot\theta = \frac{x}{y}$

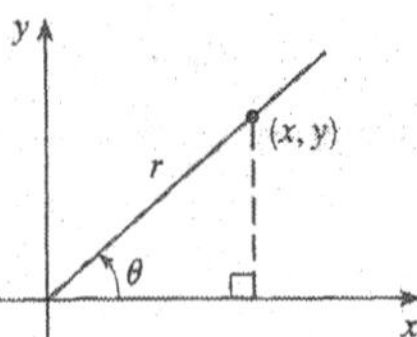

Graphs of Trigonometric Functions

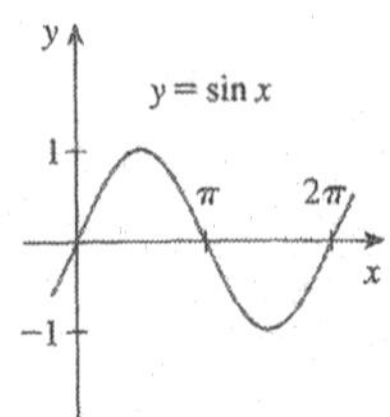

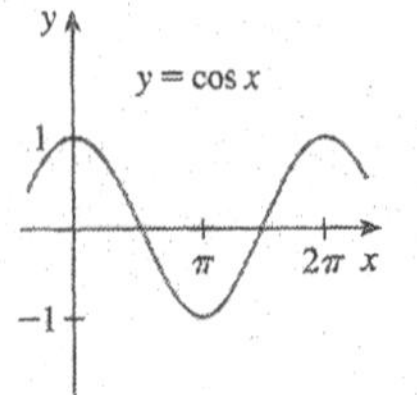

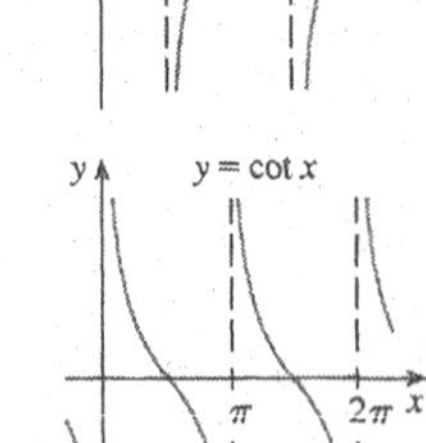

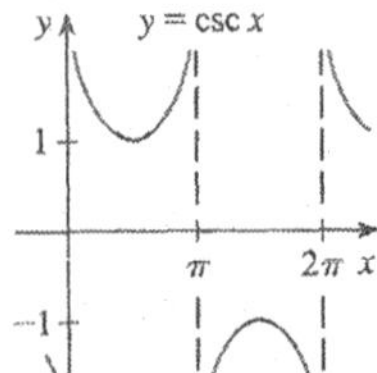

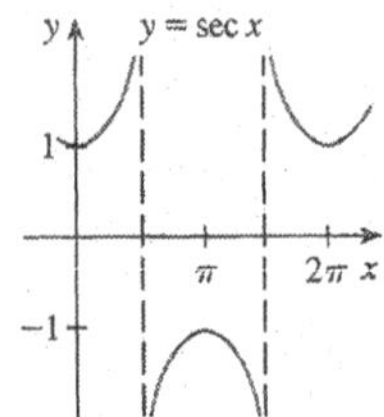

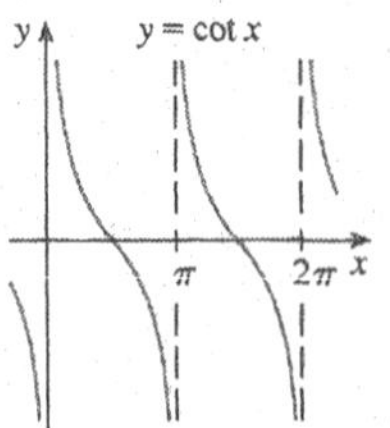

Trigonometric Functions of Important Angles

θ	radians	$\sin\theta$	$\cos\theta$	$\tan\theta$
0°	0	0	1	0
30°	$\pi/6$	$1/2$	$\sqrt{3}/2$	$\sqrt{3}/3$
45°	$\pi/4$	$\sqrt{2}/2$	$\sqrt{2}/2$	1
60°	$\pi/3$	$\sqrt{3}/2$	$1/2$	$\sqrt{3}$
90°	$\pi/2$	1	0	—

Fundamental Identities

$\csc\theta = \frac{1}{\sin\theta}$ $\sec\theta = \frac{1}{\cos\theta}$

$\tan\theta = \frac{\sin\theta}{\cos\theta}$ $\cot\theta = \frac{\cos\theta}{\sin\theta}$

$\cot\theta = \frac{1}{\tan\theta}$ $\sin^2\theta + \cos^2\theta = 1$

$1 + \tan^2\theta = \sec^2\theta$ $1 + \cot^2\theta = \csc^2\theta$

$\sin(-\theta) = -\sin\theta$ $\cos(-\theta) = \cos\theta$

$\tan(-\theta) = -\tan\theta$ $\sin\left(\frac{\pi}{2} - \theta\right) = \cos\theta$

$\cos\left(\frac{\pi}{2} - \theta\right) = \sin\theta$ $\tan\left(\frac{\pi}{2} - \theta\right) = \cot\theta$

The Law of Sines

$$\frac{\sin A}{a} = \frac{\sin B}{b} = \frac{\sin C}{c}$$

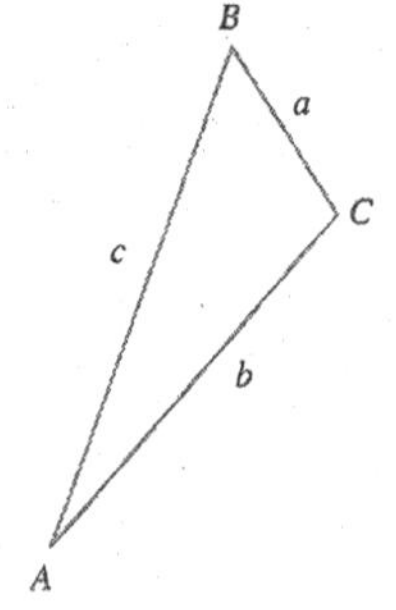

The Law of Cosines

$a^2 = b^2 + c^2 - 2bc\cos A$

$b^2 = a^2 + c^2 - 2ac\cos B$

$c^2 = a^2 + b^2 - 2ab\cos C$

Addition and Subtraction Formulas

$\sin(x + y) = \sin x\cos y + \cos x\sin y$

$\sin(x - y) = \sin x\cos y - \cos x\sin y$

$\cos(x + y) = \cos x\cos y - \sin x\sin y$

$\cos(x - y) = \cos x\cos y + \sin x\sin y$

$\tan(x + y) = \frac{\tan x + \tan y}{1 - \tan x\tan y}$

$\tan(x - y) = \frac{\tan x - \tan y}{1 + \tan x\tan y}$

Double-Angle Formulas

$\sin 2x = 2\sin x\cos x$

$\cos 2x = \cos^2 x - \sin^2 x = 2\cos^2 x - 1 = 1 - 2\sin^2 x$

$\tan 2x = \frac{2\tan x}{1 - \tan^2 x}$

Half-Angle Formulas

$\sin^2 x = \frac{1 - \cos 2x}{2}$ $\cos^2 x = \frac{1 + \cos 2x}{2}$

Calculus

Early Transcendentals
Volume 2

7E

James Stewart

Australia • Brazil • Japan • Korea • Mexico • Singapore • Spain • United Kingdom • United States

Calculus: Early Transcendentals: Volume 2, 7E

Executive Editors:
Maureen Staudt
Michael Stranz

Senior Project Development Manager:
Linda deStefano

Marketing Specialist:
Courtney Sheldon

Senior Production/Manufacturing Manager:
Donna M. Brown

PreMedia Manager:
Joel Brennecke

Sr. Rights Acquisition Account Manager:
Todd Osborne

Cover Image:
Getty Images*

*Unless otherwise noted, all cover images used by Custom Solutions, a part of Cengage Learning, have been supplied courtesy of Getty Images with the exception of the Earthview cover image, which has been supplied by the National Aeronautics and Space Administration (NASA).

Printed in the United States of America

Source:

Calculus: Early Transcendentals, 7th Edition
James Stewart

For product information and technology assistance, contact us at
Cengage Learning Customer & Sales Support, 1-800-354-9706
For permission to use material from this text or product,
submit all requests online at **cengage.com/permissions**
Further permissions questions can be emailed to
permissionrequest@cengage.com

This book contains select works from existing Cengage Learning resources and was produced by Cengage Learning Custom Solutions for collegiate use. As such, those adopting and/or contributing to this work are responsible for editorial content accuracy, continuity and completeness.

ISBN-13: 978-1-133-27288-5

ISBN-10: 1-133-27288-6

Cengage Learning
5191 Natorp Boulevard
Mason, Ohio 45040
USA

Cengage Learning is a leading provider of customized learning solutions with office locations around the globe, including Singapore, the United Kingdom, Australia, Mexico, Brazil, and Japan. Locate your local office at:
international.cengage.com/region.

Cengage Learning products are represented in Canada by Nelson Education, Ltd.
For your lifelong learning solutions, visit **www.cengage.com/custom.**
Visit our corporate website at **www.cengage.com.**

Table of Contents

Chapter 12 Vectors and the Geometry of Space 785

Chapter 13 Vector Functions 839

Chapter 14 Partial Derivatives 877

Chapter 15 Multiple Integrals 973

Chapter 16 Vector Calculus 1055

Chapter 17 Second-Order Differential Equations 1141

Appendixes A1

Index A135

12 Vectors and the Geometry of Space 785

12.1 Three-Dimensional Coordinate Systems 786

12.2 Vectors 791

12.3 The Dot Product 800

12.4 The Cross Product 808

Discovery Project ▪ The Geometry of a Tetrahedron 816

12.5 Equations of Lines and Planes 816

Laboratory Project ▪ Putting 3D in Perspective 826

12.6 Cylinders and Quadric Surfaces 827

Review 834

Problems Plus 837

13 Vector Functions 839

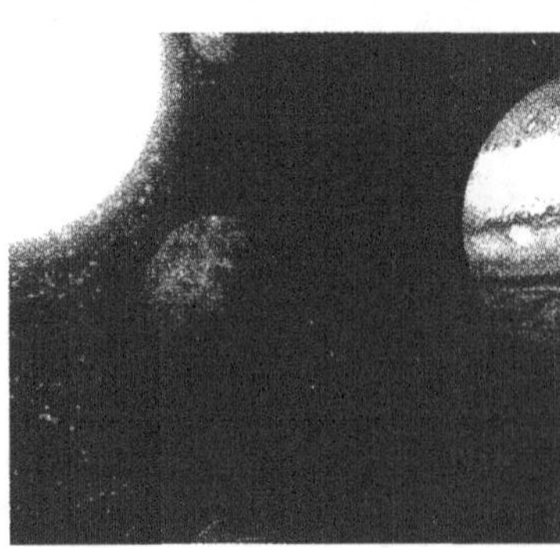

13.1 Vector Functions and Space Curves 840

13.2 Derivatives and Integrals of Vector Functions 847

13.3 Arc Length and Curvature 853

13.4 Motion in Space: Velocity and Acceleration 862

Applied Project ▪ Kepler's Laws 872

Review 873

Problems Plus 876

14 Partial Derivatives 877

14.1 Functions of Several Variables 878

14.2 Limits and Continuity 892

14.3 Partial Derivatives 900

14.4 Tangent Planes and Linear Approximations 915

14.5 The Chain Rule 924

14.6 Directional Derivatives and the Gradient Vector 933

14.7 Maximum and Minimum Values 946

Applied Project ▪ Designing a Dumpster 956

Discovery Project ▪ Quadratic Approximations and Critical Points 956

14.8 Lagrange Multipliers 957

Applied Project ▪ Rocket Science 964

Applied Project ▪ Hydro-Turbine Optimization 966

Review 967

Problems Plus 971

15 Multiple Integrals 973

15.1 Double Integrals over Rectangles 974

15.2 Iterated Integrals 982

15.3 Double Integrals over General Regions 988

15.4 Double Integrals in Polar Coordinates 997

15.5 Applications of Double Integrals 1003

15.6 Surface Area 1013

15.7 Triple Integrals 1017

Discovery Project ▪ Volumes of Hyperspheres 1027

15.8 Triple Integrals in Cylindrical Coordinates 1027

Discovery Project ▪ The Intersection of Three Cylinders 1032

15.9 Triple Integrals in Spherical Coordinates 1033

Applied Project ▪ Roller Derby 1039

15.10 Change of Variables in Multiple Integrals 1040

Review 1049

Problems Plus 1053

16 Vector Calculus 1055

16.1 Vector Fields 1056

16.2 Line Integrals 1063

16.3 The Fundamental Theorem for Line Integrals 1075

16.4 Green's Theorem 1084

16.5 Curl and Divergence 1091

16.6 Parametric Surfaces and Their Areas 1099

16.7 Surface Integrals 1110

16.8 Stokes' Theorem 1122

Writing Project ▪ Three Men and Two Theorems 1128

16.9 The Divergence Theorem 1128
16.10 Summary 1135
Review 1136

Problems Plus 1139

17 Second-Order Differential Equations 1141

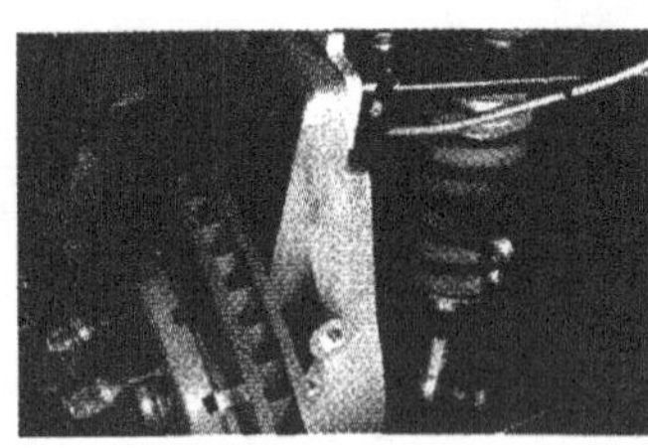

17.1 Second-Order Linear Equations 1142
17.2 Nonhomogeneous Linear Equations 1148
17.3 Applications of Second-Order Differential Equations 1156
17.4 Series Solutions 1164
Review 1169

Appendixes A1

A Numbers, Inequalities, and Absolute Values A2
B Coordinate Geometry and Lines A10
C Graphs of Second-Degree Equations A16
D Trigonometry A24
E Sigma Notation A34
F Proofs of Theorems A39
G The Logarithm Defined as an Integral A50
H Complex Numbers A57
I Answers to Odd-Numbered Exercises A65

Index A135

12 Vectors and the Geometry of Space

Examples of the surfaces and solids we study in this chapter are paraboloids (used for satellite dishes) and hyperboloids (used for cooling towers of nuclear reactors).

In this chapter we introduce vectors and coordinate systems for three-dimensional space. This will be the setting for our study of the calculus of functions of two variables in Chapter 14 because the graph of such a function is a surface in space. In this chapter we will see that vectors provide particularly simple descriptions of lines and planes in space.

12.1 Three-Dimensional Coordinate Systems

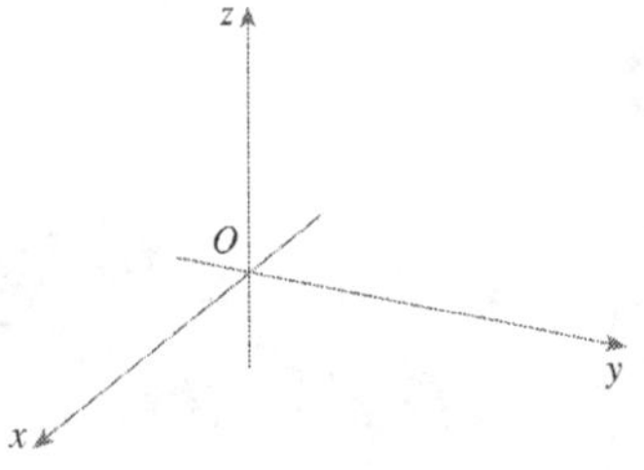

FIGURE 1
Coordinate axes

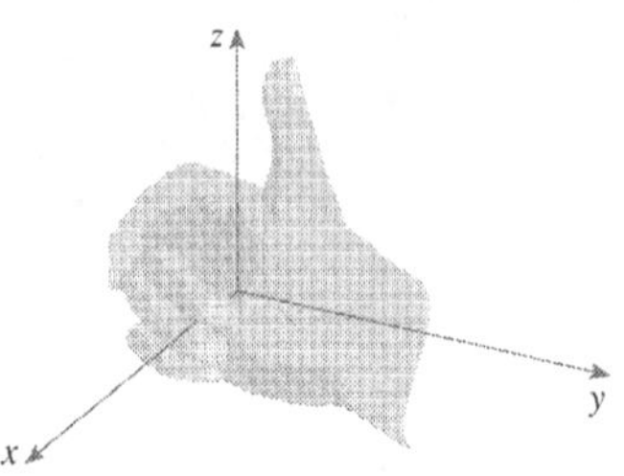

FIGURE 2
Right-hand rule

To locate a point in a plane, two numbers are necessary. We know that any point in the plane can be represented as an ordered pair (a, b) of real numbers, where a is the x-coordinate and b is the y-coordinate. For this reason, a plane is called two-dimensional. To locate a point in space, three numbers are required. We represent any point in space by an ordered triple (a, b, c) of real numbers.

In order to represent points in space, we first choose a fixed point O (the origin) and three directed lines through O that are perpendicular to each other, called the **coordinate axes** and labeled the x-axis, y-axis, and z-axis. Usually we think of the x- and y-axes as being horizontal and the z-axis as being vertical, and we draw the orientation of the axes as in Figure 1. The direction of the z-axis is determined by the **right-hand rule** as illustrated in Figure 2: If you curl the fingers of your right hand around the z-axis in the direction of a 90° counterclockwise rotation from the positive x-axis to the positive y-axis, then your thumb points in the positive direction of the z-axis.

The three coordinate axes determine the three **coordinate planes** illustrated in Figure 3(a). The xy-plane is the plane that contains the x- and y-axes; the yz-plane contains the y- and z-axes; the xz-plane contains the x- and z-axes. These three coordinate planes divide space into eight parts, called **octants**. The **first octant**, in the foreground, is determined by the positive axes.

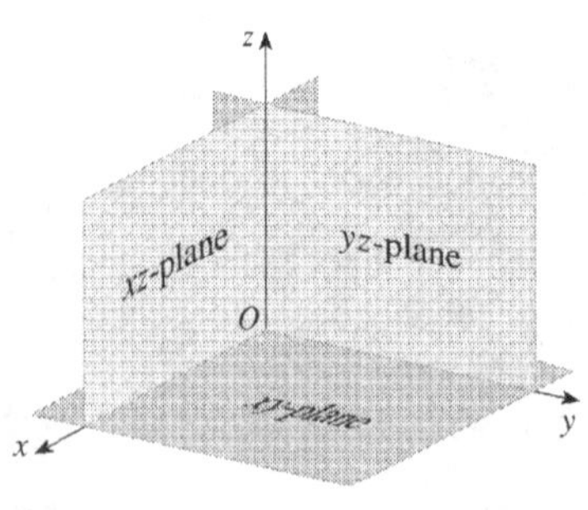

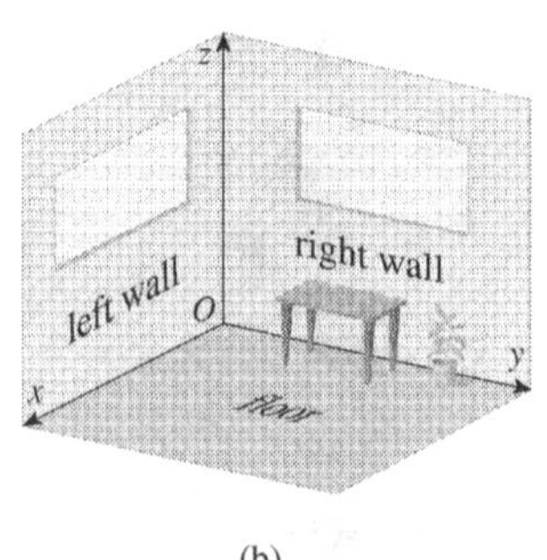

FIGURE 3 (a) Coordinate planes (b)

Because many people have some difficulty visualizing diagrams of three-dimensional figures, you may find it helpful to do the following [see Figure 3(b)]. Look at any bottom corner of a room and call the corner the origin. The wall on your left is in the xz-plane, the wall on your right is in the yz-plane, and the floor is in the xy-plane. The x-axis runs along the intersection of the floor and the left wall. The y-axis runs along the intersection of the floor and the right wall. The z-axis runs up from the floor toward the ceiling along the intersection of the two walls. You are situated in the first octant, and you can now imagine seven other rooms situated in the other seven octants (three on the same floor and four on the floor below), all connected by the common corner point O.

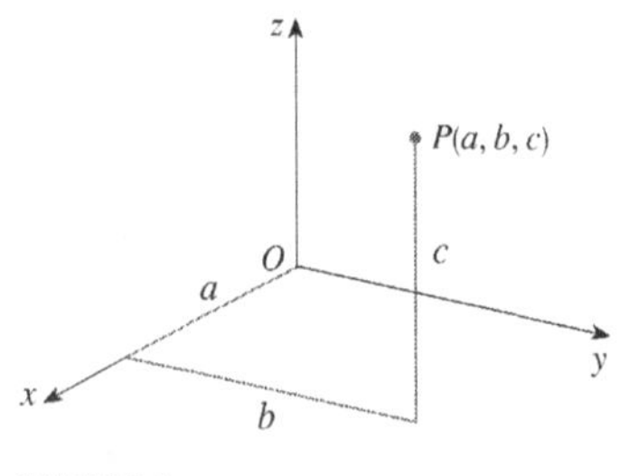

FIGURE 4

Now if P is any point in space, let a be the (directed) distance from the yz-plane to P, let b be the distance from the xz-plane to P, and let c be the distance from the xy-plane to P. We represent the point P by the ordered triple (a, b, c) of real numbers and we call a, b, and c the **coordinates** of P; a is the x-coordinate, b is the y-coordinate, and c is the z-coordinate. Thus, to locate the point (a, b, c), we can start at the origin O and move a units along the x-axis, then b units parallel to the y-axis, and then c units parallel to the z-axis as in Figure 4.

The point $P(a, b, c)$ determines a rectangular box as in Figure 5. If we drop a perpendicular from P to the xy-plane, we get a point Q with coordinates $(a, b, 0)$ called the **projection** of P onto the xy-plane. Similarly, $R(0, b, c)$ and $S(a, 0, c)$ are the projections of P onto the yz-plane and xz-plane, respectively.

As numerical illustrations, the points $(-4, 3, -5)$ and $(3, -2, -6)$ are plotted in Figure 6.

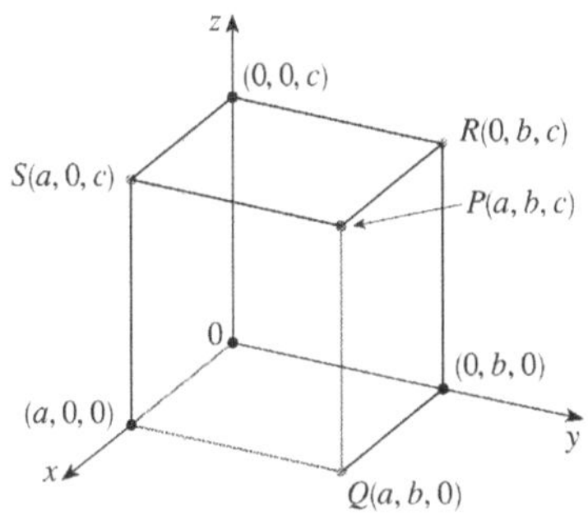

FIGURE 5

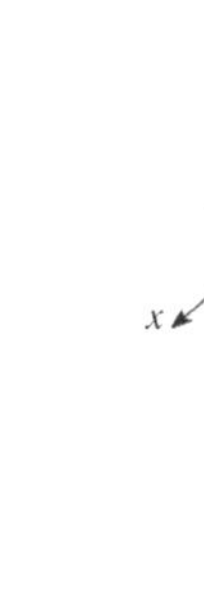

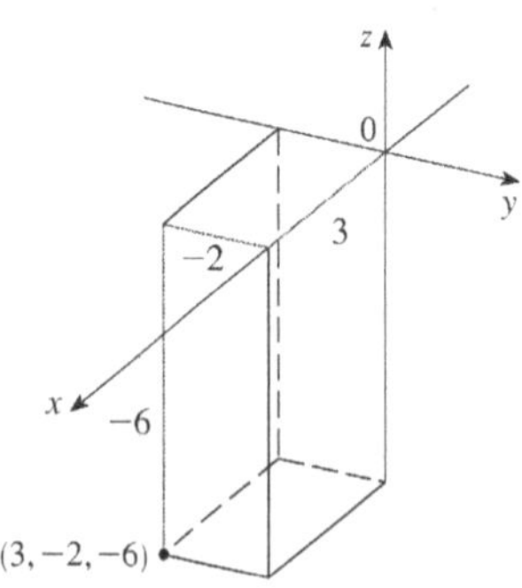

FIGURE 6

The Cartesian product $\mathbb{R} \times \mathbb{R} \times \mathbb{R} = \{(x, y, z) \mid x, y, z \in \mathbb{R}\}$ is the set of all ordered triples of real numbers and is denoted by $\mathbb{R}^3$. We have given a one-to-one correspondence between points P in space and ordered triples (a, b, c) in $\mathbb{R}^3$. It is called a **three-dimensional rectangular coordinate system**. Notice that, in terms of coordinates, the first octant can be described as the set of points whose coordinates are all positive.

In two-dimensional analytic geometry, the graph of an equation involving x and y is a curve in $\mathbb{R}^2$. In three-dimensional analytic geometry, an equation in x, y, and z represents a *surface* in $\mathbb{R}^3$.

V EXAMPLE 1 What surfaces in $\mathbb{R}^3$ are represented by the following equations?

(a) $z = 3$ (b) $y = 5$

SOLUTION

(a) The equation $z = 3$ represents the set $\{(x, y, z) \mid z = 3\}$, which is the set of all points in $\mathbb{R}^3$ whose z-coordinate is 3. This is the horizontal plane that is parallel to the xy-plane and three units above it as in Figure 7(a).

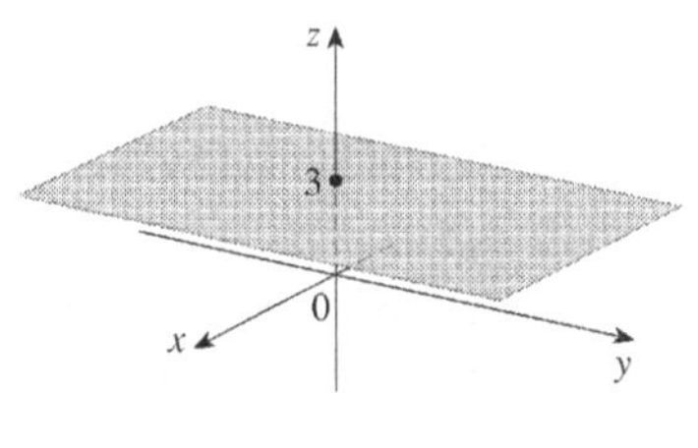

(a) $z = 3$, a plane in $\mathbb{R}^3$

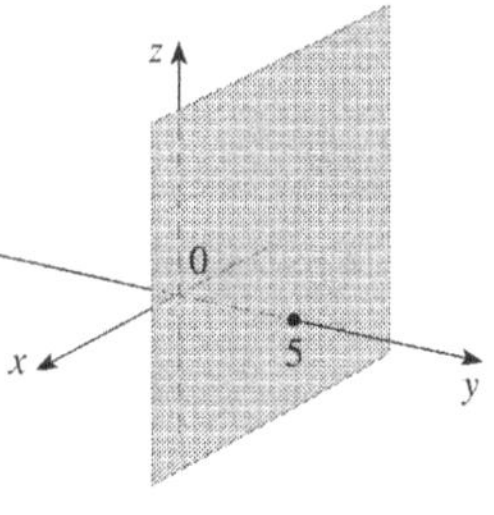

(b) $y = 5$, a plane in $\mathbb{R}^3$

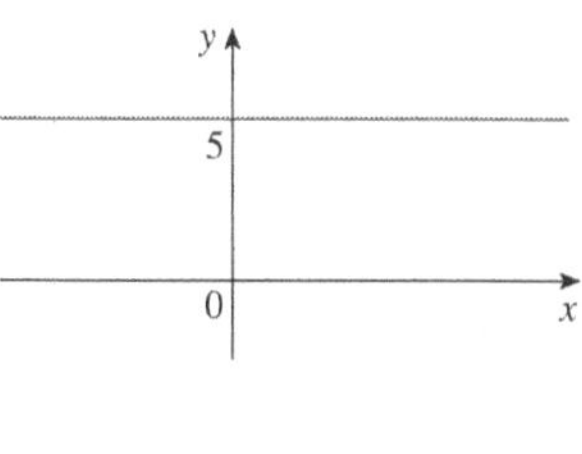

(c) $y = 5$, a line in $\mathbb{R}^2$

FIGURE 7

(b) The equation $y = 5$ represents the set of all points in $\mathbb{R}^3$ whose y-coordinate is 5. This is the vertical plane that is parallel to the xz-plane and five units to the right of it as in Figure 7(b). ■

NOTE When an equation is given, we must understand from the context whether it represents a curve in $\mathbb{R}^2$ or a surface in $\mathbb{R}^3$. In Example 1, $y = 5$ represents a plane in $\mathbb{R}^3$, but of course $y = 5$ can also represent a line in $\mathbb{R}^2$ if we are dealing with two-dimensional analytic geometry. See Figure 7(b) and (c).

In general, if k is a constant, then $x = k$ represents a plane parallel to the yz-plane, $y = k$ is a plane parallel to the xz-plane, and $z = k$ is a plane parallel to the xy-plane. In Figure 5, the faces of the rectangular box are formed by the three coordinate planes $x = 0$ (the yz-plane), $y = 0$ (the xz-plane), and $z = 0$ (the xy-plane), and the planes $x = a$, $y = b$, and $z = c$.

EXAMPLE 2

(a) Which points (x, y, z) satisfy the equations

$$x^2 + y^2 = 1 \qquad \text{and} \qquad z = 3$$

(b) What does the equation $x^2 + y^2 = 1$ represent as a surface in $\mathbb{R}^3$?

SOLUTION

(a) Because $z = 3$, the points lie in the horizontal plane $z = 3$ from Example 1(a). Because $x^2 + y^2 = 1$, the points lie on the circle with radius 1 and center on the z-axis. See Figure 8.

(b) Given that $x^2 + y^2 = 1$, with no restrictions on z, we see that the point (x, y, z) could lie on a circle in any horizontal plane $z = k$. So the surface $x^2 + y^2 = 1$ in $\mathbb{R}^3$ consists of all possible horizontal circles $x^2 + y^2 = 1$, $z = k$, and is therefore the circular cylinder with radius 1 whose axis is the z-axis. See Figure 9.

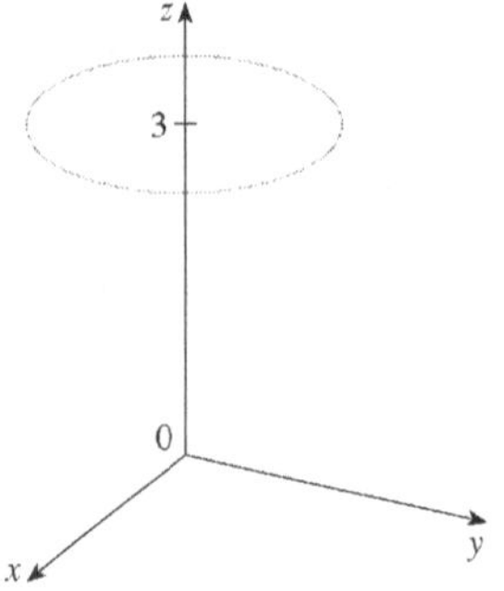

FIGURE 8
The circle $x^2 + y^2 = 1$, $z = 3$

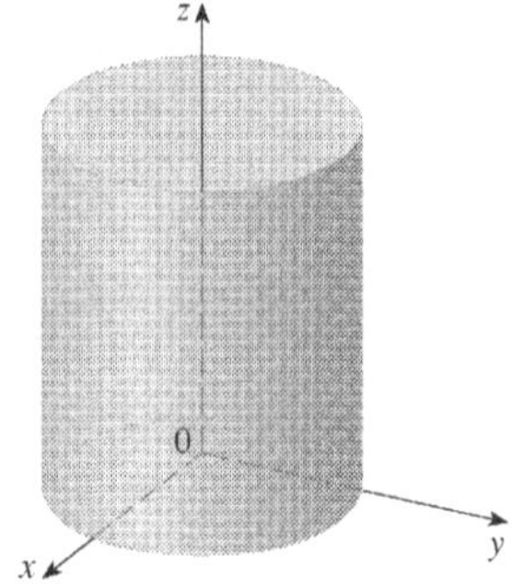

FIGURE 9
The cylinder $x^2 + y^2 = 1$

V EXAMPLE 3 Describe and sketch the surface in $\mathbb{R}^3$ represented by the equation $y = x$.

SOLUTION The equation represents the set of all points in $\mathbb{R}^3$ whose x- and y-coordinates are equal, that is, $\{(x, x, z) \mid x \in \mathbb{R}, z \in \mathbb{R}\}$. This is a vertical plane that intersects the xy-plane in the line $y = x$, $z = 0$. The portion of this plane that lies in the first octant is sketched in Figure 10.

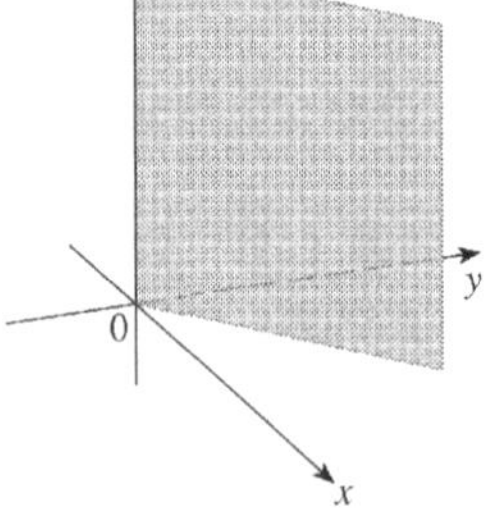

FIGURE 10
The plane $y = x$

The familiar formula for the distance between two points in a plane is easily extended to the following three-dimensional formula.

Distance Formula in Three Dimensions The distance $|P_1P_2|$ between the points $P_1(x_1, y_1, z_1)$ and $P_2(x_2, y_2, z_2)$ is

$$|P_1P_2| = \sqrt{(x_2 - x_1)^2 + (y_2 - y_1)^2 + (z_2 - z_1)^2}$$

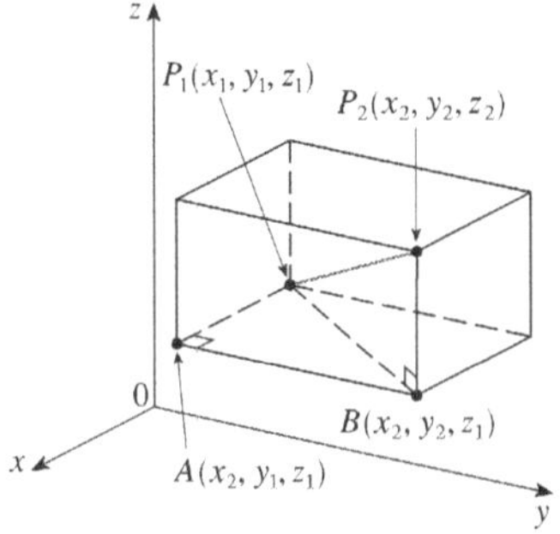

FIGURE 11

To see why this formula is true, we construct a rectangular box as in Figure 11, where P_1 and P_2 are opposite vertices and the faces of the box are parallel to the coordinate planes. If $A(x_2, y_1, z_1)$ and $B(x_2, y_2, z_1)$ are the vertices of the box indicated in the figure, then

$$|P_1A| = |x_2 - x_1| \qquad |AB| = |y_2 - y_1| \qquad |BP_2| = |z_2 - z_1|$$

Because triangles P_1BP_2 and P_1AB are both right-angled, two applications of the Pythagorean Theorem give

$$|P_1P_2|^2 = |P_1B|^2 + |BP_2|^2$$

and

$$|P_1B|^2 = |P_1A|^2 + |AB|^2$$

Combining these equations, we get

$$\begin{aligned} |P_1P_2|^2 &= |P_1A|^2 + |AB|^2 + |BP_2|^2 \\ &= |x_2 - x_1|^2 + |y_2 - y_1|^2 + |z_2 - z_1|^2 \\ &= (x_2 - x_1)^2 + (y_2 - y_1)^2 + (z_2 - z_1)^2 \end{aligned}$$

Therefore

$$|P_1P_2| = \sqrt{(x_2 - x_1)^2 + (y_2 - y_1)^2 + (z_2 - z_1)^2}$$

EXAMPLE 4 The distance from the point $P(2, -1, 7)$ to the point $Q(1, -3, 5)$ is

$$|PQ| = \sqrt{(1 - 2)^2 + (-3 + 1)^2 + (5 - 7)^2} = \sqrt{1 + 4 + 4} = 3$$

V EXAMPLE 5 Find an equation of a sphere with radius r and center $C(h, k, l)$.

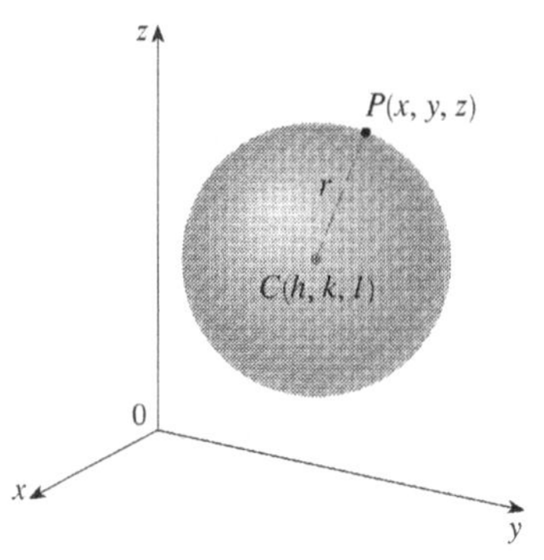

FIGURE 12

SOLUTION By definition, a sphere is the set of all points $P(x, y, z)$ whose distance from C is r. (See Figure 12.) Thus P is on the sphere if and only if $|PC| = r$. Squaring both sides, we have $|PC|^2 = r^2$ or

$$(x - h)^2 + (y - k)^2 + (z - l)^2 = r^2$$

The result of Example 5 is worth remembering.

Equation of a Sphere An equation of a sphere with center $C(h, k, l)$ and radius r is

$$(x - h)^2 + (y - k)^2 + (z - l)^2 = r^2$$

In particular, if the center is the origin O, then an equation of the sphere is

$$x^2 + y^2 + z^2 = r^2$$

EXAMPLE 6 Show that $x^2 + y^2 + z^2 + 4x - 6y + 2z + 6 = 0$ is the equation of a sphere, and find its center and radius.

SOLUTION We can rewrite the given equation in the form of an equation of a sphere if we complete squares:

$$(x^2 + 4x + 4) + (y^2 - 6y + 9) + (z^2 + 2z + 1) = -6 + 4 + 9 + 1$$

$$(x + 2)^2 + (y - 3)^2 + (z + 1)^2 = 8$$

Comparing this equation with the standard form, we see that it is the equation of a sphere with center $(-2, 3, -1)$ and radius $\sqrt{8} = 2\sqrt{2}$.

EXAMPLE 7 What region in $\mathbb{R}^3$ is represented by the following inequalities?

$$1 \leqslant x^2 + y^2 + z^2 \leqslant 4 \qquad z \leqslant 0$$

SOLUTION The inequalities

$$1 \leqslant x^2 + y^2 + z^2 \leqslant 4$$

can be rewritten as

$$1 \leqslant \sqrt{x^2 + y^2 + z^2} \leqslant 2$$

so they represent the points (x, y, z) whose distance from the origin is at least 1 and at most 2. But we are also given that $z \leqslant 0$, so the points lie on or below the xy-plane. Thus the given inequalities represent the region that lies between (or on) the spheres $x^2 + y^2 + z^2 = 1$ and $x^2 + y^2 + z^2 = 4$ and beneath (or on) the xy-plane. It is sketched in Figure 13.

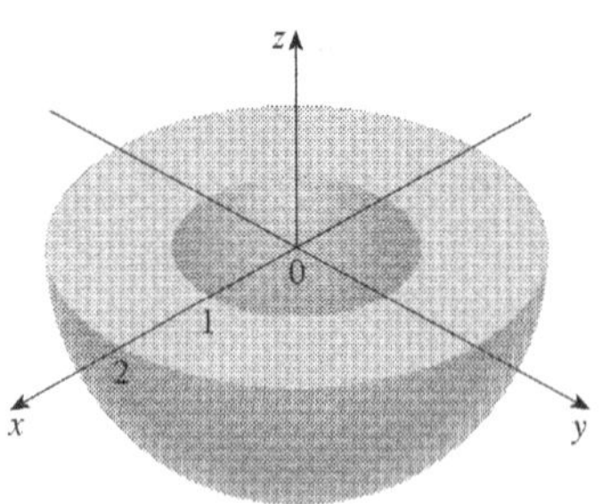

FIGURE 13

12.1 Exercises

1. Suppose you start at the origin, move along the x-axis a distance of 4 units in the positive direction, and then move downward a distance of 3 units. What are the coordinates of your position?

2. Sketch the points $(0, 5, 2)$, $(4, 0, -1)$, $(2, 4, 6)$, and $(1, -1, 2)$ on a single set of coordinate axes.

3. Which of the points $A(-4, 0, -1)$, $B(3, 1, -5)$, and $C(2, 4, 6)$ is closest to the yz-plane? Which point lies in the xz-plane?

4. What are the projections of the point $(2, 3, 5)$ on the xy-, yz-, and xz-planes? Draw a rectangular box with the origin and $(2, 3, 5)$ as opposite vertices and with its faces parallel to the coordinate planes. Label all vertices of the box. Find the length of the diagonal of the box.

5. Describe and sketch the surface in $\mathbb{R}^3$ represented by the equation $x + y = 2$.

6. (a) What does the equation $x = 4$ represent in $\mathbb{R}^2$? What does it represent in $\mathbb{R}^3$? Illustrate with sketches.
   (b) What does the equation $y = 3$ represent in $\mathbb{R}^3$? What does $z = 5$ represent? What does the pair of equations $y = 3$, $z = 5$ represent? In other words, describe the set of points (x, y, z) such that $y = 3$ and $z = 5$. Illustrate with a sketch.

7–8 Find the lengths of the sides of the triangle PQR. Is it a right triangle? Is it an isosceles triangle?

7. $P(3, -2, -3)$, $Q(7, 0, 1)$, $R(1, 2, 1)$

8. $P(2, -1, 0)$, $Q(4, 1, 1)$, $R(4, -5, 4)$

9. Determine whether the points lie on straight line.
   (a) $A(2, 4, 2)$, $B(3, 7, -2)$, $C(1, 3, 3)$
   (b) $D(0, -5, 5)$, $E(1, -2, 4)$, $F(3, 4, 2)$

10. Find the distance from $(4, -2, 6)$ to each of the following.
    (a) The xy-plane (b) The yz-plane
    (c) The xz-plane (d) The x-axis
    (e) The y-axis (f) The z-axis

11. Find an equation of the sphere with center $(-3, 2, 5)$ and radius 4. What is the intersection of this sphere with the yz-plane?

12. Find an equation of the sphere with center $(2, -6, 4)$ and radius 5. Describe its intersection with each of the coordinate planes.

13. Find an equation of the sphere that passes through the point $(4, 3, -1)$ and has center $(3, 8, 1)$.

14. Find an equation of the sphere that passes through the origin and whose center is $(1, 2, 3)$.

15–18 Show that the equation represents a sphere, and find its center and radius.

15. $x^2 + y^2 + z^2 - 2x - 4y + 8z = 15$

16. $x^2 + y^2 + z^2 + 8x - 6y + 2z + 17 = 0$

17. $2x^2 + 2y^2 + 2z^2 = 8x - 24z + 1$

18. $3x^2 + 3y^2 + 3z^2 = 10 + 6y + 12z$

1. Homework Hints available at stewartcalculus.com

19. (a) Prove that the midpoint of the line segment from $P_1(x_1, y_1, z_1)$ to $P_2(x_2, y_2, z_2)$ is

$$\left(\frac{x_1 + x_2}{2}, \frac{y_1 + y_2}{2}, \frac{z_1 + z_2}{2}\right)$$

(b) Find the lengths of the medians of the triangle with vertices $A(1, 2, 3)$, $B(-2, 0, 5)$, and $C(4, 1, 5)$.

20. Find an equation of a sphere if one of its diameters has endpoints $(2, 1, 4)$ and $(4, 3, 10)$.

21. Find equations of the spheres with center $(2, -3, 6)$ that touch (a) the xy-plane, (b) the yz-plane, (c) the xz-plane.

22. Find an equation of the largest sphere with center $(5, 4, 9)$ that is contained in the first octant.

23–34 Describe in words the region of $\mathbb{R}^3$ represented by the equations or inequalities.

23. $x = 5$

24. $y = -2$

25. $y < 8$

26. $x \geqslant -3$

27. $0 \leqslant z \leqslant 6$

28. $z^2 = 1$

29. $x^2 + y^2 = 4, \quad z = -1$

30. $y^2 + z^2 = 16$

31. $x^2 + y^2 + z^2 \leqslant 3$

32. $x = z$

33. $x^2 + z^2 \leqslant 9$

34. $x^2 + y^2 + z^2 > 2z$

35–38 Write inequalities to describe the region.

35. The region between the yz-plane and the vertical plane $x = 5$

36. The solid cylinder that lies on or below the plane $z = 8$ and on or above the disk in the xy-plane with center the origin and radius 2

37. The region consisting of all points between (but not on) the spheres of radius r and R centered at the origin, where $r < R$

38. The solid upper hemisphere of the sphere of radius 2 centered at the origin

39. The figure shows a line L_1 in space and a second line L_2, which is the projection of L_1 on the xy-plane. (In other words, the points on L_2 are directly beneath, or above, the points on L_1.)
(a) Find the coordinates of the point P on the line L_1.
(b) Locate on the diagram the points A, B, and C, where the line L_1 intersects the xy-plane, the yz-plane, and the xz-plane, respectively.

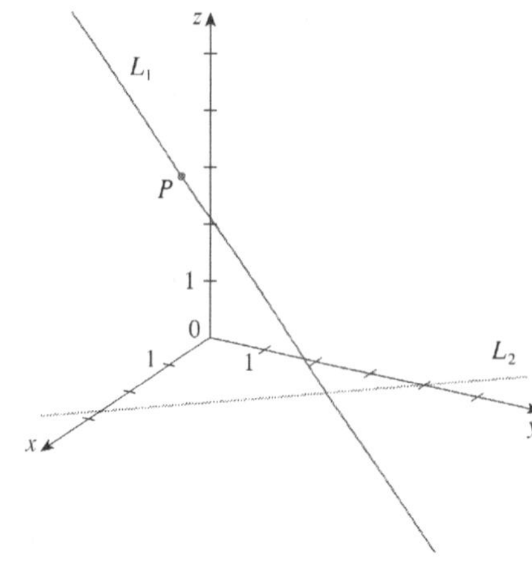

40. Consider the points P such that the distance from P to $A(-1, 5, 3)$ is twice the distance from P to $B(6, 2, -2)$. Show that the set of all such points is a sphere, and find its center and radius.

41. Find an equation of the set of all points equidistant from the points $A(-1, 5, 3)$ and $B(6, 2, -2)$. Describe the set.

42. Find the volume of the solid that lies inside both of the spheres

$$x^2 + y^2 + z^2 + 4x - 2y + 4z + 5 = 0$$

and
$$x^2 + y^2 + z^2 = 4$$

43. Find the distance between the spheres $x^2 + y^2 + z^2 = 4$ and $x^2 + y^2 + z^2 = 4x + 4y + 4z - 11$.

44. Describe and sketch a solid with the following properties. When illuminated by rays parallel to the z-axis, its shadow is a circular disk. If the rays are parallel to the y-axis, its shadow is a square. If the rays are parallel to the x-axis, its shadow is an isosceles triangle.

12.2 Vectors

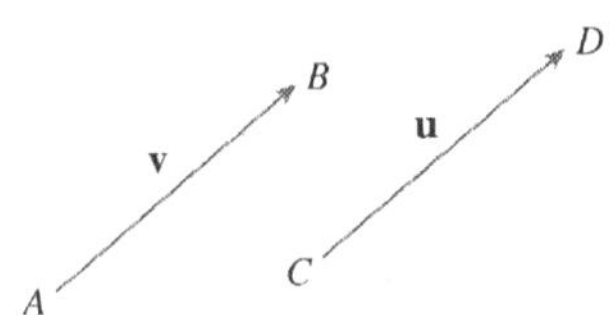

FIGURE 1
Equivalent vectors

The term **vector** is used by scientists to indicate a quantity (such as displacement or velocity or force) that has both magnitude and direction. A vector is often represented by an arrow or a directed line segment. The length of the arrow represents the magnitude of the vector and the arrow points in the direction of the vector. We denote a vector by printing a letter in boldface (**v**) or by putting an arrow above the letter ($\vec{v}$).

For instance, suppose a particle moves along a line segment from point A to point B. The corresponding **displacement vector** **v**, shown in Figure 1, has **initial point** A (the tail) and **terminal point** B (the tip) and we indicate this by writing $\mathbf{v} = \overrightarrow{AB}$. Notice that the vector $\mathbf{u} = \overrightarrow{CD}$ has the same length and the same direction as $\mathbf{v}$ even though it is in a different position. We say that $\mathbf{u}$ and $\mathbf{v}$ are **equivalent** (or **equal**) and we write $\mathbf{u} = \mathbf{v}$. The **zero vector**, denoted by $\mathbf{0}$, has length 0. It is the only vector with no specific direction.

Combining Vectors

C
B
A

FIGURE 2

Suppose a particle moves from A to B, so its displacement vector is $\overrightarrow{AB}$. Then the particle changes direction and moves from B to C, with displacement vector $\overrightarrow{BC}$ as in Figure 2. The combined effect of these displacements is that the particle has moved from A to C. The resulting displacement vector $\overrightarrow{AC}$ is called the *sum* of $\overrightarrow{AB}$ and $\overrightarrow{BC}$ and we write

$$\overrightarrow{AC} = \overrightarrow{AB} + \overrightarrow{BC}$$

In general, if we start with vectors $\mathbf{u}$ and $\mathbf{v}$, we first move $\mathbf{v}$ so that its tail coincides with the tip of $\mathbf{u}$ and define the sum of $\mathbf{u}$ and $\mathbf{v}$ as follows.

> **Definition of Vector Addition** If $\mathbf{u}$ and $\mathbf{v}$ are vectors positioned so the initial point of $\mathbf{v}$ is at the terminal point of $\mathbf{u}$, then the **sum** $\mathbf{u} + \mathbf{v}$ is the vector from the initial point of $\mathbf{u}$ to the terminal point of $\mathbf{v}$.

The definition of vector addition is illustrated in Figure 3. You can see why this definition is sometimes called the **Triangle Law**.

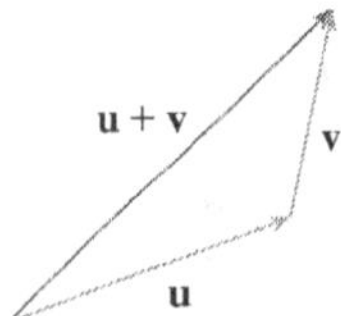

FIGURE 3 The Triangle Law

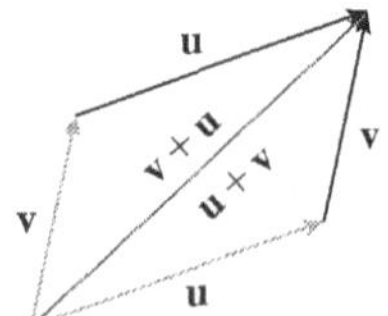

FIGURE 4 The Parallelogram Law

In Figure 4 we start with the same vectors $\mathbf{u}$ and $\mathbf{v}$ as in Figure 3 and draw another copy of $\mathbf{v}$ with the same initial point as $\mathbf{u}$. Completing the parallelogram, we see that $\mathbf{u} + \mathbf{v} = \mathbf{v} + \mathbf{u}$. This also gives another way to construct the sum: If we place $\mathbf{u}$ and $\mathbf{v}$ so they start at the same point, then $\mathbf{u} + \mathbf{v}$ lies along the diagonal of the parallelogram with $\mathbf{u}$ and $\mathbf{v}$ as sides. (This is called the **Parallelogram Law**.)

FIGURE 5

V EXAMPLE 1 Draw the sum of the vectors **a** and **b** shown in Figure 5.

SOLUTION First we translate **b** and place its tail at the tip of **a**, being careful to draw a copy of **b** that has the same length and direction. Then we draw the vector $\mathbf{a} + \mathbf{b}$ [see Figure 6(a)] starting at the initial point of **a** and ending at the terminal point of the copy of **b**.

Alternatively, we could place **b** so it starts where **a** starts and construct $\mathbf{a} + \mathbf{b}$ by the Parallelogram Law as in Figure 6(b).

TEC Visual 12.2 shows how the Triangle and Parallelogram Laws work for various vectors **a** and **b**.

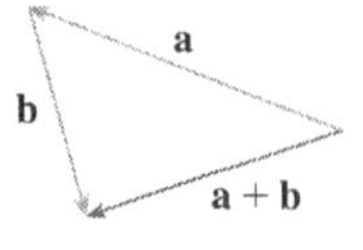

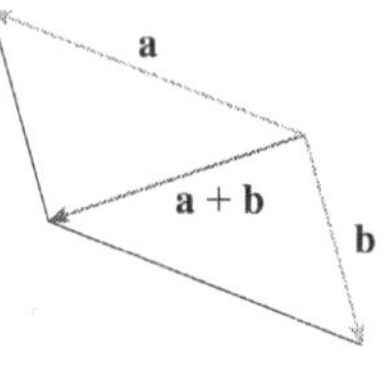

FIGURE 6 (a) (b)

It is possible to multiply a vector by a real number c. (In this context we call the real number c a **scalar** to distinguish it from a vector.) For instance, we want $2\mathbf{v}$ to be the same vector as $\mathbf{v} + \mathbf{v}$, which has the same direction as $\mathbf{v}$ but is twice as long. In general, we multiply a vector by a scalar as follows.

> **Definition of Scalar Multiplication** If c is a scalar and $\mathbf{v}$ is a vector, then the **scalar multiple** $c\mathbf{v}$ is the vector whose length is $|c|$ times the length of $\mathbf{v}$ and whose direction is the same as $\mathbf{v}$ if $c > 0$ and is opposite to $\mathbf{v}$ if $c < 0$. If $c = 0$ or $\mathbf{v} = \mathbf{0}$, then $c\mathbf{v} = \mathbf{0}$.

FIGURE 7
Scalar multiples of $\mathbf{v}$

This definition is illustrated in Figure 7. We see that real numbers work like scaling factors here; that's why we call them scalars. Notice that two nonzero vectors are **parallel** if they are scalar multiples of one another. In particular, the vector $-\mathbf{v} = (-1)\mathbf{v}$ has the same length as $\mathbf{v}$ but points in the opposite direction. We call it the **negative** of $\mathbf{v}$.

By the **difference** $\mathbf{u} - \mathbf{v}$ of two vectors we mean

$$\mathbf{u} - \mathbf{v} = \mathbf{u} + (-\mathbf{v})$$

So we can construct $\mathbf{u} - \mathbf{v}$ by first drawing the negative of $\mathbf{v}$, $-\mathbf{v}$, and then adding it to $\mathbf{u}$ by the Parallelogram Law as in Figure 8(a). Alternatively, since $\mathbf{v} + (\mathbf{u} - \mathbf{v}) = \mathbf{u}$, the vector $\mathbf{u} - \mathbf{v}$, when added to $\mathbf{v}$, gives $\mathbf{u}$. So we could construct $\mathbf{u} - \mathbf{v}$ as in Figure 8(b) by means of the Triangle Law.

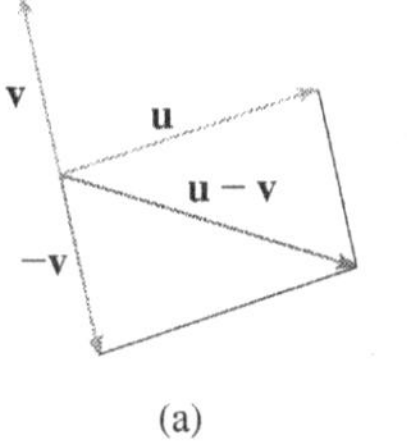

(a)

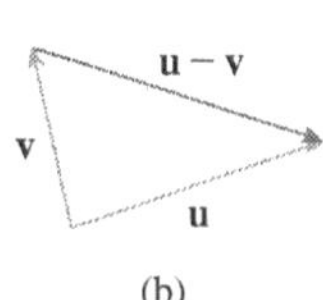

(b)

FIGURE 8
Drawing $\mathbf{u} - \mathbf{v}$

EXAMPLE 2 If $\mathbf{a}$ and $\mathbf{b}$ are the vectors shown in Figure 9, draw $\mathbf{a} - 2\mathbf{b}$.

FIGURE 9

SOLUTION We first draw the vector $-2\mathbf{b}$ pointing in the direction opposite to $\mathbf{b}$ and twice as long. We place it with its tail at the tip of $\mathbf{a}$ and then use the Triangle Law to draw $\mathbf{a} + (-2\mathbf{b})$ as in Figure 10.

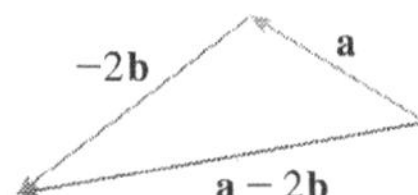

FIGURE 10

Components

For some purposes it's best to introduce a coordinate system and treat vectors algebraically. If we place the initial point of a vector $\mathbf{a}$ at the origin of a rectangular coordinate system, then the terminal point of $\mathbf{a}$ has coordinates of the form (a_1, a_2) or (a_1, a_2, a_3), depending on whether our coordinate system is two- or three-dimensional (see Figure 11).

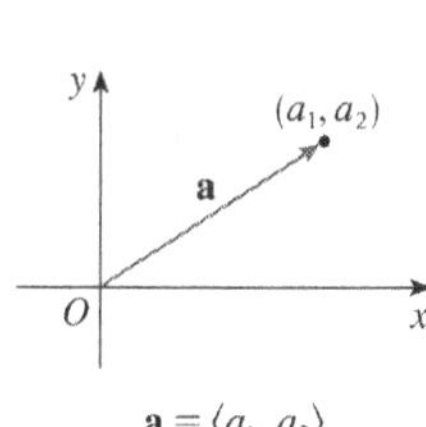

$\mathbf{a} = \langle a_1, a_2 \rangle$

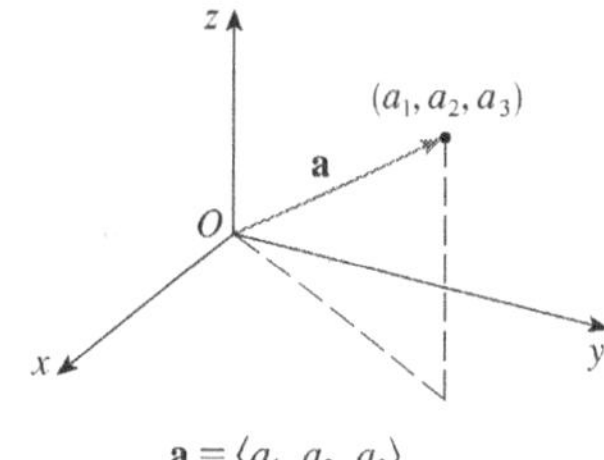

$\mathbf{a} = \langle a_1, a_2, a_3 \rangle$

FIGURE 11

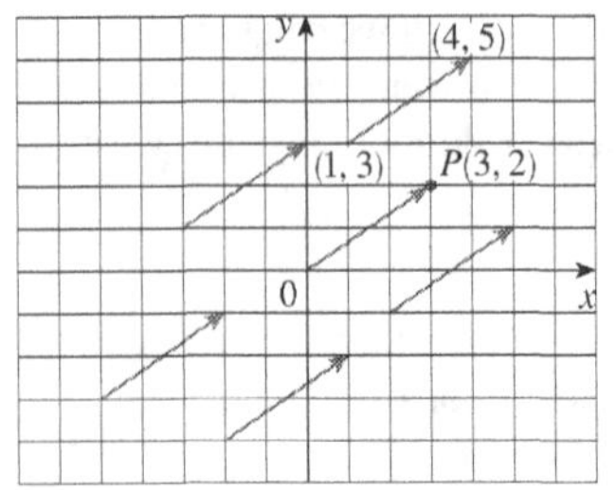

FIGURE 12
Representations of the vector $\mathbf{a} = \langle 3, 2 \rangle$

These coordinates are called the **components** of **a** and we write

$$\mathbf{a} = \langle a_1, a_2 \rangle \qquad \text{or} \qquad \mathbf{a} = \langle a_1, a_2, a_3 \rangle$$

We use the notation $\langle a_1, a_2 \rangle$ for the ordered pair that refers to a vector so as not to confuse it with the ordered pair (a_1, a_2) that refers to a point in the plane.

For instance, the vectors shown in Figure 12 are all equivalent to the vector $\overrightarrow{OP} = \langle 3, 2 \rangle$ whose terminal point is $P(3, 2)$. What they have in common is that the terminal point is reached from the initial point by a displacement of three units to the right and two upward. We can think of all these geometric vectors as **representations** of the algebraic vector $\mathbf{a} = \langle 3, 2 \rangle$. The particular representation $\overrightarrow{OP}$ from the origin to the point $P(3, 2)$ is called the **position vector** of the point P.

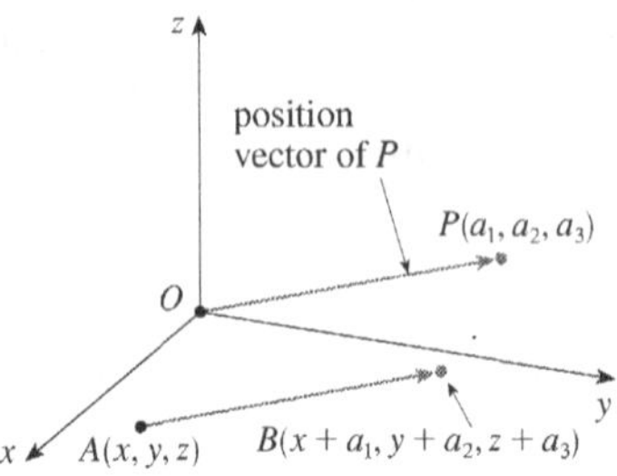

FIGURE 13
Representations of $\mathbf{a} = \langle a_1, a_2, a_3 \rangle$

In three dimensions, the vector $\mathbf{a} = \overrightarrow{OP} = \langle a_1, a_2, a_3 \rangle$ is the **position vector** of the point $P(a_1, a_2, a_3)$. (See Figure 13.) Let's consider any other representation $\overrightarrow{AB}$ of **a**, where the initial point is $A(x_1, y_1, z_1)$ and the terminal point is $B(x_2, y_2, z_2)$. Then we must have $x_1 + a_1 = x_2$, $y_1 + a_2 = y_2$, and $z_1 + a_3 = z_2$ and so $a_1 = x_2 - x_1$, $a_2 = y_2 - y_1$, and $a_3 = z_2 - z_1$. Thus we have the following result.

1 Given the points $A(x_1, y_1, z_1)$ and $B(x_2, y_2, z_2)$, the vector **a** with representation $\overrightarrow{AB}$ is

$$\mathbf{a} = \langle x_2 - x_1, y_2 - y_1, z_2 - z_1 \rangle$$

V EXAMPLE 3 Find the vector represented by the directed line segment with initial point $A(2, -3, 4)$ and terminal point $B(-2, 1, 1)$.

SOLUTION By **1**, the vector corresponding to $\overrightarrow{AB}$ is

$$\mathbf{a} = \langle -2 - 2, 1 - (-3), 1 - 4 \rangle = \langle -4, 4, -3 \rangle$$

The **magnitude** or **length** of the vector **v** is the length of any of its representations and is denoted by the symbol $|\mathbf{v}|$ or $\|\mathbf{v}\|$. By using the distance formula to compute the length of a segment OP, we obtain the following formulas.

The length of the two-dimensional vector $\mathbf{a} = \langle a_1, a_2 \rangle$ is

$$|\mathbf{a}| = \sqrt{a_1^2 + a_2^2}$$

The length of the three-dimensional vector $\mathbf{a} = \langle a_1, a_2, a_3 \rangle$ is

$$|\mathbf{a}| = \sqrt{a_1^2 + a_2^2 + a_3^2}$$

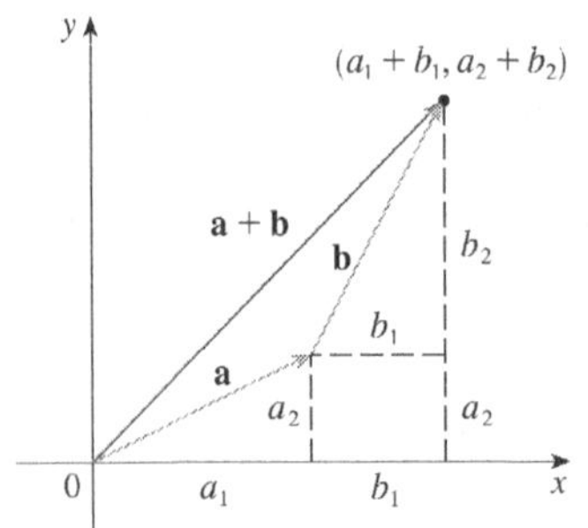

FIGURE 14

How do we add vectors algebraically? Figure 14 shows that if $\mathbf{a} = \langle a_1, a_2 \rangle$ and $\mathbf{b} = \langle b_1, b_2 \rangle$, then the sum is $\mathbf{a} + \mathbf{b} = \langle a_1 + b_1, a_2 + b_2 \rangle$, at least for the case where the components are positive. In other words, *to add algebraic vectors we add their components.* Similarly, *to subtract vectors we subtract components.* From the similar triangles in

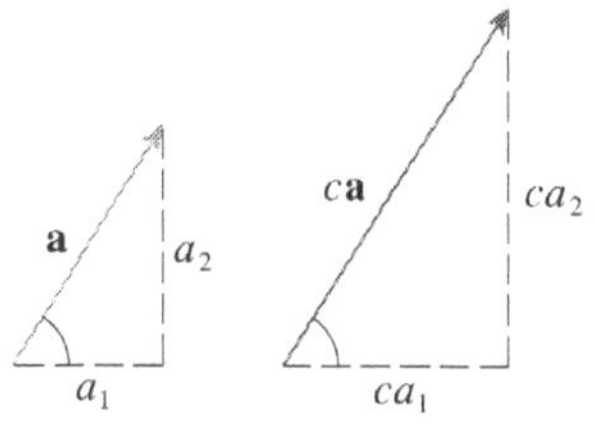

FIGURE 15

Figure 15 we see that the components of $c\mathbf{a}$ are ca_1 and ca_2. So *to multiply a vector by a scalar we multiply each component by that scalar.*

> If $\mathbf{a} = \langle a_1, a_2 \rangle$ and $\mathbf{b} = \langle b_1, b_2 \rangle$, then
>
> $$\mathbf{a} + \mathbf{b} = \langle a_1 + b_1, a_2 + b_2 \rangle \qquad \mathbf{a} - \mathbf{b} = \langle a_1 - b_1, a_2 - b_2 \rangle$$
>
> $$c\mathbf{a} = \langle ca_1, ca_2 \rangle$$
>
> Similarly, for three-dimensional vectors,
>
> $$\langle a_1, a_2, a_3 \rangle + \langle b_1, b_2, b_3 \rangle = \langle a_1 + b_1, a_2 + b_2, a_3 + b_3 \rangle$$
>
> $$\langle a_1, a_2, a_3 \rangle - \langle b_1, b_2, b_3 \rangle = \langle a_1 - b_1, a_2 - b_2, a_3 - b_3 \rangle$$
>
> $$c\langle a_1, a_2, a_3 \rangle = \langle ca_1, ca_2, ca_3 \rangle$$

V EXAMPLE 4 If $\mathbf{a} = \langle 4, 0, 3 \rangle$ and $\mathbf{b} = \langle -2, 1, 5 \rangle$, find $|\mathbf{a}|$ and the vectors $\mathbf{a} + \mathbf{b}$, $\mathbf{a} - \mathbf{b}$, $3\mathbf{b}$, and $2\mathbf{a} + 5\mathbf{b}$.

SOLUTION

$$|\mathbf{a}| = \sqrt{4^2 + 0^2 + 3^2} = \sqrt{25} = 5$$

$$\begin{aligned}
\mathbf{a} + \mathbf{b} &= \langle 4, 0, 3 \rangle + \langle -2, 1, 5 \rangle \\
&= \langle 4 + (-2), 0 + 1, 3 + 5 \rangle = \langle 2, 1, 8 \rangle \\
\mathbf{a} - \mathbf{b} &= \langle 4, 0, 3 \rangle - \langle -2, 1, 5 \rangle \\
&= \langle 4 - (-2), 0 - 1, 3 - 5 \rangle = \langle 6, -1, -2 \rangle \\
3\mathbf{b} &= 3\langle -2, 1, 5 \rangle = \langle 3(-2), 3(1), 3(5) \rangle = \langle -6, 3, 15 \rangle \\
2\mathbf{a} + 5\mathbf{b} &= 2\langle 4, 0, 3 \rangle + 5\langle -2, 1, 5 \rangle \\
&= \langle 8, 0, 6 \rangle + \langle -10, 5, 25 \rangle = \langle -2, 5, 31 \rangle
\end{aligned}$$

We denote by V_2 the set of all two-dimensional vectors and by V_3 the set of all three-dimensional vectors. More generally, we will later need to consider the set V_n of all n-dimensional vectors. An n-dimensional vector is an ordered n-tuple:

$$\mathbf{a} = \langle a_1, a_2, \ldots, a_n \rangle$$

where $a_1, a_2, \ldots, a_n$ are real numbers that are called the components of $\mathbf{a}$. Addition and scalar multiplication are defined in terms of components just as for the cases $n = 2$ and $n = 3$.

Vectors in n dimensions are used to list various quantities in an organized way. For instance, the components of a six-dimensional vector

$$\mathbf{p} = \langle p_1, p_2, p_3, p_4, p_5, p_6 \rangle$$

might represent the prices of six different ingredients required to make a particular product. Four-dimensional vectors $\langle x, y, z, t \rangle$ are used in relativity theory, where the first three components specify a position in space and the fourth represents time.

> **Properties of Vectors** If $\mathbf{a}$, $\mathbf{b}$, and $\mathbf{c}$ are vectors in V_n and c and d are scalars, then
>
> 1. $\mathbf{a} + \mathbf{b} = \mathbf{b} + \mathbf{a}$
> 2. $\mathbf{a} + (\mathbf{b} + \mathbf{c}) = (\mathbf{a} + \mathbf{b}) + \mathbf{c}$
> 3. $\mathbf{a} + \mathbf{0} = \mathbf{a}$
> 4. $\mathbf{a} + (-\mathbf{a}) = \mathbf{0}$
> 5. $c(\mathbf{a} + \mathbf{b}) = c\mathbf{a} + c\mathbf{b}$
> 6. $(c + d)\mathbf{a} = c\mathbf{a} + d\mathbf{a}$
> 7. $(cd)\mathbf{a} = c(d\mathbf{a})$
> 8. $1\mathbf{a} = \mathbf{a}$

These eight properties of vectors can be readily verified either geometrically or algebraically. For instance, Property 1 can be seen from Figure 4 (it's equivalent to the Parallelogram Law) or as follows for the case $n = 2$:

$$\begin{aligned} \mathbf{a} + \mathbf{b} &= \langle a_1, a_2 \rangle + \langle b_1, b_2 \rangle = \langle a_1 + b_1, a_2 + b_2 \rangle \\ &= \langle b_1 + a_1, b_2 + a_2 \rangle = \langle b_1, b_2 \rangle + \langle a_1, a_2 \rangle \\ &= \mathbf{b} + \mathbf{a} \end{aligned}$$

FIGURE 16

We can see why Property 2 (the associative law) is true by looking at Figure 16 and applying the Triangle Law several times: The vector $\overrightarrow{PQ}$ is obtained either by first constructing $\mathbf{a} + \mathbf{b}$ and then adding $\mathbf{c}$ or by adding $\mathbf{a}$ to the vector $\mathbf{b} + \mathbf{c}$.

Three vectors in V_3 play a special role. Let

$$\mathbf{i} = \langle 1, 0, 0 \rangle \qquad \mathbf{j} = \langle 0, 1, 0 \rangle \qquad \mathbf{k} = \langle 0, 0, 1 \rangle$$

These vectors $\mathbf{i}$, $\mathbf{j}$, and $\mathbf{k}$ are called the **standard basis vectors**. They have length 1 and point in the directions of the positive x-, y-, and z-axes. Similarly, in two dimensions we define $\mathbf{i} = \langle 1, 0 \rangle$ and $\mathbf{j} = \langle 0, 1 \rangle$. (See Figure 17.)

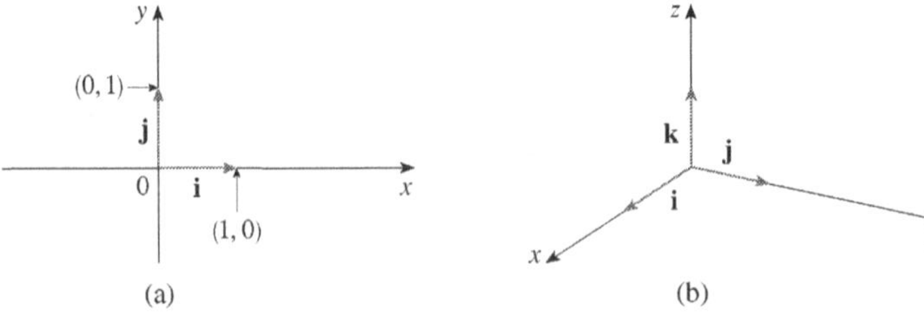

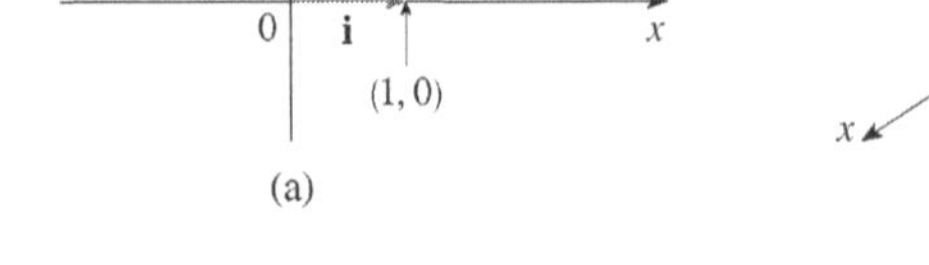
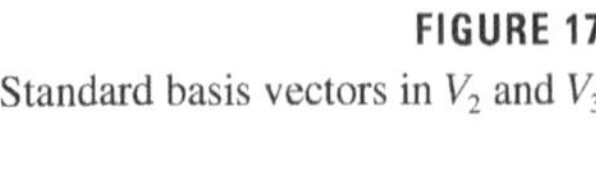

FIGURE 17
Standard basis vectors in V_2 and V_3

If $\mathbf{a} = \langle a_1, a_2, a_3 \rangle$, then we can write

$$\begin{aligned} \mathbf{a} &= \langle a_1, a_2, a_3 \rangle = \langle a_1, 0, 0 \rangle + \langle 0, a_2, 0 \rangle + \langle 0, 0, a_3 \rangle \\ &= a_1 \langle 1, 0, 0 \rangle + a_2 \langle 0, 1, 0 \rangle + a_3 \langle 0, 0, 1 \rangle \end{aligned}$$

2 $$\mathbf{a} = a_1 \mathbf{i} + a_2 \mathbf{j} + a_3 \mathbf{k}$$

Thus any vector in V_3 can be expressed in terms of $\mathbf{i}$, $\mathbf{j}$, and $\mathbf{k}$. For instance,

$$\langle 1, -2, 6 \rangle = \mathbf{i} - 2\mathbf{j} + 6\mathbf{k}$$

Similarly, in two dimensions, we can write

3 $$\mathbf{a} = \langle a_1, a_2 \rangle = a_1 \mathbf{i} + a_2 \mathbf{j}$$

See Figure 18 for the geometric interpretation of Equations 3 and 2 and compare with Figure 17.

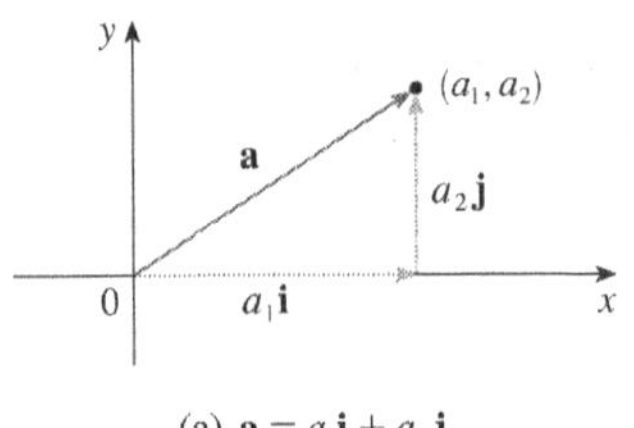

(a) $\mathbf{a} = a_1\mathbf{i} + a_2\mathbf{j}$

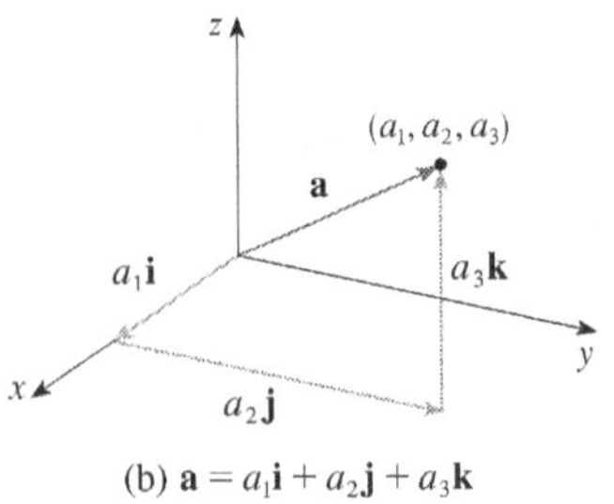

(b) $\mathbf{a} = a_1\mathbf{i} + a_2\mathbf{j} + a_3\mathbf{k}$

FIGURE 18

EXAMPLE 5 If $\mathbf{a} = \mathbf{i} + 2\mathbf{j} - 3\mathbf{k}$ and $\mathbf{b} = 4\mathbf{i} + 7\mathbf{k}$, express the vector $2\mathbf{a} + 3\mathbf{b}$ in terms of $\mathbf{i}$, $\mathbf{j}$, and $\mathbf{k}$.

SOLUTION Using Properties 1, 2, 5, 6, and 7 of vectors, we have

$$\begin{aligned} 2\mathbf{a} + 3\mathbf{b} &= 2(\mathbf{i} + 2\mathbf{j} - 3\mathbf{k}) + 3(4\mathbf{i} + 7\mathbf{k}) \\ &= 2\mathbf{i} + 4\mathbf{j} - 6\mathbf{k} + 12\mathbf{i} + 21\mathbf{k} = 14\mathbf{i} + 4\mathbf{j} + 15\mathbf{k} \end{aligned}$$

Gibbs

Josiah Willard Gibbs (1839–1903), a professor of mathematical physics at Yale College, published the first book on vectors, *Vector Analysis*, in 1881. More complicated objects, called quaternions, had earlier been invented by Hamilton as mathematical tools for describing space, but they weren't easy for scientists to use. Quaternions have a scalar part and a vector part. Gibb's idea was to use the vector part separately. Maxwell and Heaviside had similar ideas, but Gibb's approach has proved to be the most convenient way to study space.

A **unit vector** is a vector whose length is 1. For instance, $\mathbf{i}$, $\mathbf{j}$, and $\mathbf{k}$ are all unit vectors. In general, if $\mathbf{a} \neq \mathbf{0}$, then the unit vector that has the same direction as $\mathbf{a}$ is

4 $$\mathbf{u} = \frac{1}{|\mathbf{a}|}\mathbf{a} = \frac{\mathbf{a}}{|\mathbf{a}|}$$

In order to verify this, we let $c = 1/|\mathbf{a}|$. Then $\mathbf{u} = c\mathbf{a}$ and c is a positive scalar, so $\mathbf{u}$ has the same direction as $\mathbf{a}$. Also

$$|\mathbf{u}| = |c\mathbf{a}| = |c||\mathbf{a}| = \frac{1}{|\mathbf{a}|}|\mathbf{a}| = 1$$

EXAMPLE 6 Find the unit vector in the direction of the vector $2\mathbf{i} - \mathbf{j} - 2\mathbf{k}$.

SOLUTION The given vector has length

$$|2\mathbf{i} - \mathbf{j} - 2\mathbf{k}| = \sqrt{2^2 + (-1)^2 + (-2)^2} = \sqrt{9} = 3$$

so, by Equation 4, the unit vector with the same direction is

$$\tfrac{1}{3}(2\mathbf{i} - \mathbf{j} - 2\mathbf{k}) = \tfrac{2}{3}\mathbf{i} - \tfrac{1}{3}\mathbf{j} - \tfrac{2}{3}\mathbf{k}$$

Applications

Vectors are useful in many aspects of physics and engineering. In Chapter 13 we will see how they describe the velocity and acceleration of objects moving in space. Here we look at forces.

A force is represented by a vector because it has both a magnitude (measured in pounds or newtons) and a direction. If several forces are acting on an object, the **resultant force** experienced by the object is the vector sum of these forces.

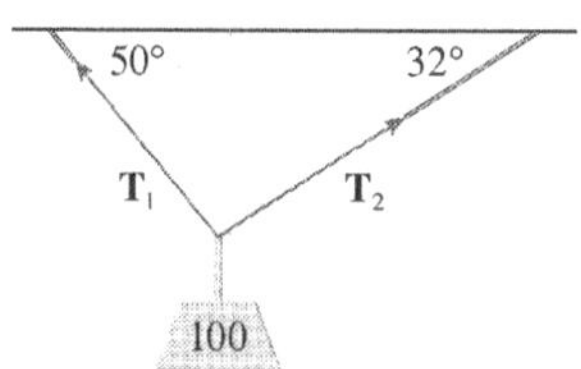

FIGURE 19

EXAMPLE 7 A 100-lb weight hangs from two wires as shown in Figure 19. Find the tensions (forces) $\mathbf{T}_1$ and $\mathbf{T}_2$ in both wires and the magnitudes of the tensions.

SOLUTION We first express $\mathbf{T}_1$ and $\mathbf{T}_2$ in terms of their horizontal and vertical components. From Figure 20 we see that

5 $$\mathbf{T}_1 = -|\mathbf{T}_1|\cos 50^\circ\,\mathbf{i} + |\mathbf{T}_1|\sin 50^\circ\,\mathbf{j}$$

6 $$\mathbf{T}_2 = |\mathbf{T}_2|\cos 32^\circ\,\mathbf{i} + |\mathbf{T}_2|\sin 32^\circ\,\mathbf{j}$$

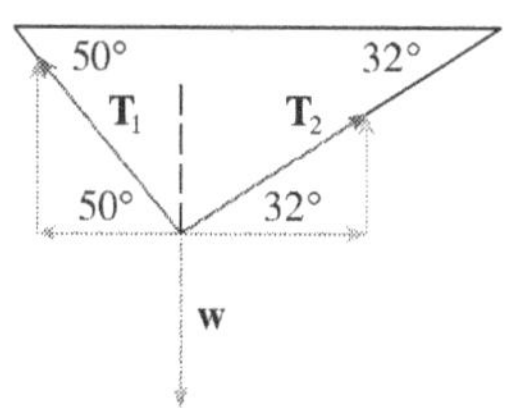

FIGURE 20

The resultant $\mathbf{T}_1 + \mathbf{T}_2$ of the tensions counterbalances the weight $\mathbf{w}$ and so we must have

$$\mathbf{T}_1 + \mathbf{T}_2 = -\mathbf{w} = 100\mathbf{j}$$

Thus

$$\left(-|\mathbf{T}_1|\cos 50^\circ + |\mathbf{T}_2|\cos 32^\circ\right)\mathbf{i} + \left(|\mathbf{T}_1|\sin 50^\circ + |\mathbf{T}_2|\sin 32^\circ\right)\mathbf{j} = 100\mathbf{j}$$

Equating components, we get

$$-|\mathbf{T}_1|\cos 50° + |\mathbf{T}_2|\cos 32° = 0$$
$$|\mathbf{T}_1|\sin 50° + |\mathbf{T}_2|\sin 32° = 100$$

Solving the first of these equations for $|\mathbf{T}_2|$ and substituting into the second, we get

$$|\mathbf{T}_1|\sin 50° + \frac{|\mathbf{T}_1|\cos 50°}{\cos 32°}\sin 32° = 100$$

So the magnitudes of the tensions are

$$|\mathbf{T}_1| = \frac{100}{\sin 50° + \tan 32° \cos 50°} \approx 85.64 \text{ lb}$$

and

$$|\mathbf{T}_2| = \frac{|\mathbf{T}_1|\cos 50°}{\cos 32°} \approx 64.91 \text{ lb}$$

Substituting these values in $\boxed{5}$ and $\boxed{6}$, we obtain the tension vectors

$$\mathbf{T}_1 \approx -55.05\,\mathbf{i} + 65.60\,\mathbf{j} \qquad \mathbf{T}_2 \approx 55.05\,\mathbf{i} + 34.40\,\mathbf{j}$$

12.2 Exercises

1. Are the following quantities vectors or scalars? Explain.
   (a) The cost of a theater ticket
   (b) The current in a river
   (c) The initial flight path from Houston to Dallas
   (d) The population of the world

2. What is the relationship between the point (4, 7) and the vector $\langle 4, 7 \rangle$? Illustrate with a sketch.

3. Name all the equal vectors in the parallelogram shown.

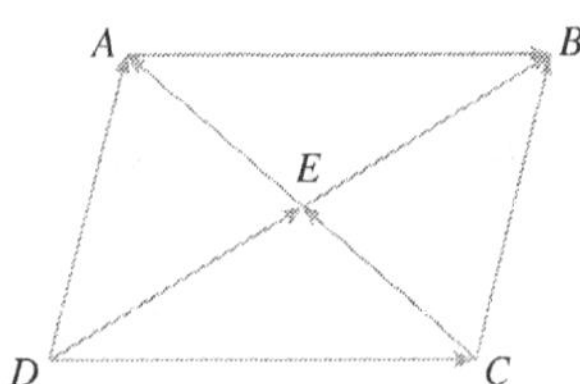

4. Write each combination of vectors as a single vector.
   (a) $\overrightarrow{AB} + \overrightarrow{BC}$ (b) $\overrightarrow{CD} + \overrightarrow{DB}$
   (c) $\overrightarrow{DB} - \overrightarrow{AB}$ (d) $\overrightarrow{DC} + \overrightarrow{CA} + \overrightarrow{AB}$

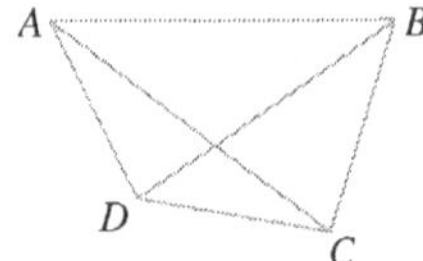

5. Copy the vectors in the figure and use them to draw the following vectors.
   (a) $\mathbf{u} + \mathbf{v}$ (b) $\mathbf{u} + \mathbf{w}$
   (c) $\mathbf{v} + \mathbf{w}$ (d) $\mathbf{u} - \mathbf{v}$
   (e) $\mathbf{v} + \mathbf{u} + \mathbf{w}$ (f) $\mathbf{u} - \mathbf{w} - \mathbf{v}$

6. Copy the vectors in the figure and use them to draw the following vectors.
   (a) $\mathbf{a} + \mathbf{b}$ (b) $\mathbf{a} - \mathbf{b}$
   (c) $\frac{1}{2}\mathbf{a}$ (d) $-3\mathbf{b}$
   (e) $\mathbf{a} + 2\mathbf{b}$ (f) $2\mathbf{b} - \mathbf{a}$

7. In the figure, the tip of $\mathbf{c}$ and the tail of $\mathbf{d}$ are both the midpoint of QR. Express $\mathbf{c}$ and $\mathbf{d}$ in terms of $\mathbf{a}$ and $\mathbf{b}$.

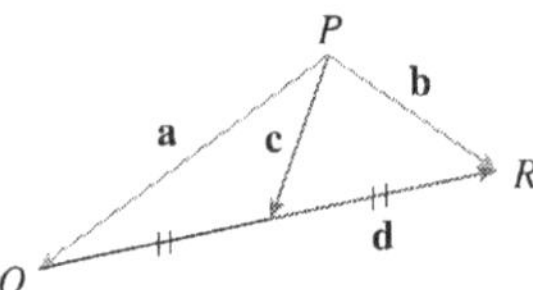

8. If the vectors in the figure satisfy $|\mathbf{u}| = |\mathbf{v}| = 1$ and $\mathbf{u} + \mathbf{v} + \mathbf{w} = \mathbf{0}$, what is $|\mathbf{w}|$?

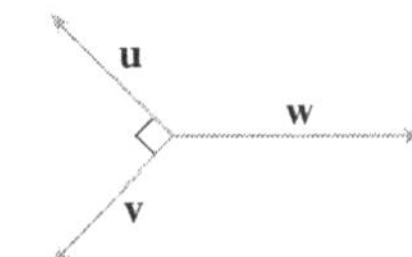

9–14 Find a vector $\mathbf{a}$ with representation given by the directed line segment $\overrightarrow{AB}$. Draw $\overrightarrow{AB}$ and the equivalent representation starting at the origin.

9. $A(-1, 1)$, $B(3, 2)$ **10.** $A(-4, -1)$, $B(1, 2)$

11. $A(-1, 3)$, $B(2, 2)$ **12.** $A(2, 1)$, $B(0, 6)$

13. $A(0, 3, 1)$, $B(2, 3, -1)$ **14.** $A(4, 0, -2)$, $B(4, 2, 1)$

15–18 Find the sum of the given vectors and illustrate geometrically.

15. $\langle -1, 4 \rangle$, $\langle 6, -2 \rangle$ **16.** $\langle 3, -1 \rangle$, $\langle -1, 5 \rangle$

17. $\langle 3, 0, 1 \rangle$, $\langle 0, 8, 0 \rangle$ **18.** $\langle 1, 3, -2 \rangle$, $\langle 0, 0, 6 \rangle$

19–22 Find $\mathbf{a} + \mathbf{b}$, $2\mathbf{a} + 3\mathbf{b}$, $|\mathbf{a}|$, and $|\mathbf{a} - \mathbf{b}|$.

19. $\mathbf{a} = \langle 5, -12 \rangle$, $\mathbf{b} = \langle -3, -6 \rangle$

20. $\mathbf{a} = 4\mathbf{i} + \mathbf{j}$, $\mathbf{b} = \mathbf{i} - 2\mathbf{j}$

21. $\mathbf{a} = \mathbf{i} + 2\mathbf{j} - 3\mathbf{k}$, $\mathbf{b} = -2\mathbf{i} - \mathbf{j} + 5\mathbf{k}$

22. $\mathbf{a} = 2\mathbf{i} - 4\mathbf{j} + 4\mathbf{k}$, $\mathbf{b} = 2\mathbf{j} - \mathbf{k}$

23–25 Find a unit vector that has the same direction as the given vector.

23. $-3\mathbf{i} + 7\mathbf{j}$ **24.** $\langle -4, 2, 4 \rangle$

25. $8\mathbf{i} - \mathbf{j} + 4\mathbf{k}$

26. Find a vector that has the same direction as $\langle -2, 4, 2 \rangle$ but has length 6.

27–28 What is the angle between the given vector and the positive direction of the x-axis?

27. $\mathbf{i} + \sqrt{3}\,\mathbf{j}$ **28.** $8\mathbf{i} + 6\mathbf{j}$

29. If $\mathbf{v}$ lies in the first quadrant and makes an angle $\pi/3$ with the positive x-axis and $|\mathbf{v}| = 4$, find $\mathbf{v}$ in component form.

30. If a child pulls a sled through the snow on a level path with a force of 50 N exerted at an angle of 38° above the horizontal, find the horizontal and vertical components of the force.

31. A quarterback throws a football with angle of elevation 40° and speed 60 ft/s. Find the horizontal and vertical components of the velocity vector.

32–33 Find the magnitude of the resultant force and the angle it makes with the positive x-axis.

32.

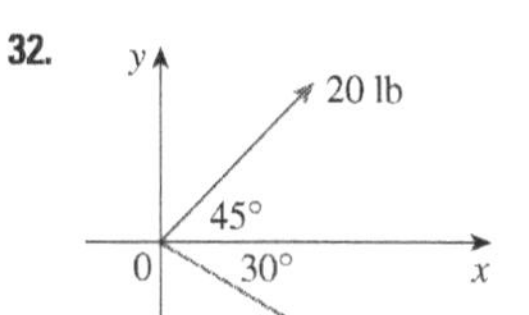

33.

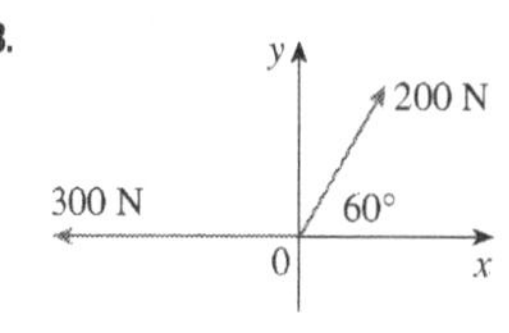

34. The magnitude of a velocity vector is called *speed*. Suppose that a wind is blowing from the direction N45°W at a speed of 50 km/h. (This means that the direction from which the wind blows is 45° west of the northerly direction.) A pilot is steering a plane in the direction N60°E at an airspeed (speed in still air) of 250 km/h. The *true course*, or *track*, of the plane is the direction of the resultant of the velocity vectors of the plane and the wind. The *ground speed* of the plane is the magnitude of the resultant. Find the true course and the ground speed of the plane.

35. A woman walks due west on the deck of a ship at 3 mi/h. The ship is moving north at a speed of 22 mi/h. Find the speed and direction of the woman relative to the surface of the water.

36. Ropes 3 m and 5 m in length are fastened to a holiday decoration that is suspended over a town square. The decoration has a mass of 5 kg. The ropes, fastened at different heights, make angles of 52° and 40° with the horizontal. Find the tension in each wire and the magnitude of each tension.

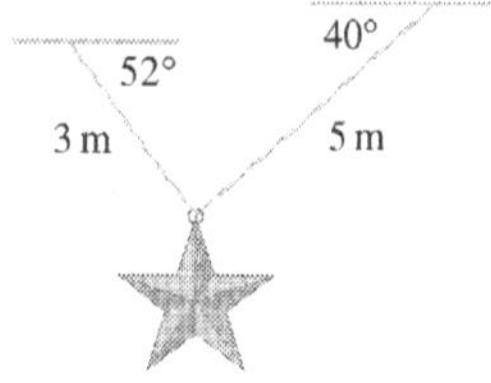

37. A clothesline is tied between two poles, 8 m apart. The line is quite taut and has negligible sag. When a wet shirt with a mass of 0.8 kg is hung at the middle of the line, the midpoint is pulled down 8 cm. Find the tension in each half of the clothesline.

38. The tension $\mathbf{T}$ at each end of the chain has magnitude 25 N (see the figure). What is the weight of the chain?

39. A boatman wants to cross a canal that is 3 km wide and wants to land at a point 2 km upstream from his starting point. The current in the canal flows at 3.5 km/h and the speed of his boat is 13 km/h.

(a) In what direction should he steer?

(b) How long will the trip take?

40. Three forces act on an object. Two of the forces are at an angle of 100° to each other and have magnitudes 25 N and 12 N. The third is perpendicular to the plane of these two forces and has magnitude 4 N. Calculate the magnitude of the force that would exactly counterbalance these three forces.

41. Find the unit vectors that are parallel to the tangent line to the parabola $y = x^2$ at the point $(2, 4)$.

42. (a) Find the unit vectors that are parallel to the tangent line to the curve $y = 2 \sin x$ at the point $(\pi/6, 1)$.
(b) Find the unit vectors that are perpendicular to the tangent line.
(c) Sketch the curve $y = 2 \sin x$ and the vectors in parts (a) and (b), all starting at $(\pi/6, 1)$.

43. If A, B, and C are the vertices of a triangle, find $\overrightarrow{AB} + \overrightarrow{BC} + \overrightarrow{CA}$.

44. Let C be the point on the line segment AB that is twice as far from B as it is from A. If $\mathbf{a} = \overrightarrow{OA}$, $\mathbf{b} = \overrightarrow{OB}$, and $\mathbf{c} = \overrightarrow{OC}$, show that $\mathbf{c} = \frac{2}{3}\mathbf{a} + \frac{1}{3}\mathbf{b}$.

45. (a) Draw the vectors $\mathbf{a} = \langle 3, 2 \rangle$, $\mathbf{b} = \langle 2, -1 \rangle$, and $\mathbf{c} = \langle 7, 1 \rangle$.
(b) Show, by means of a sketch, that there are scalars s and t such that $\mathbf{c} = s\mathbf{a} + t\mathbf{b}$.
(c) Use the sketch to estimate the values of s and t.
(d) Find the exact values of s and t.

46. Suppose that $\mathbf{a}$ and $\mathbf{b}$ are nonzero vectors that are not parallel and $\mathbf{c}$ is any vector in the plane determined by $\mathbf{a}$ and $\mathbf{b}$. Give a geometric argument to show that $\mathbf{c}$ can be written as $\mathbf{c} = s\mathbf{a} + t\mathbf{b}$ for suitable scalars s and t. Then give an argument using components.

47. If $\mathbf{r} = \langle x, y, z \rangle$ and $\mathbf{r}_0 = \langle x_0, y_0, z_0 \rangle$, describe the set of all points (x, y, z) such that $|\mathbf{r} - \mathbf{r}_0| = 1$.

48. If $\mathbf{r} = \langle x, y \rangle$, $\mathbf{r}_1 = \langle x_1, y_1 \rangle$, and $\mathbf{r}_2 = \langle x_2, y_2 \rangle$, describe the set of all points (x, y) such that $|\mathbf{r} - \mathbf{r}_1| + |\mathbf{r} - \mathbf{r}_2| = k$, where $k > |\mathbf{r}_1 - \mathbf{r}_2|$.

49. Figure 16 gives a geometric demonstration of Property 2 of vectors. Use components to give an algebraic proof of this fact for the case $n = 2$.

50. Prove Property 5 of vectors algebraically for the case $n = 3$. Then use similar triangles to give a geometric proof.

51. Use vectors to prove that the line joining the midpoints of two sides of a triangle is parallel to the third side and half its length.

52. Suppose the three coordinate planes are all mirrored and a light ray given by the vector $\mathbf{a} = \langle a_1, a_2, a_3 \rangle$ first strikes the xz-plane, as shown in the figure. Use the fact that the angle of incidence equals the angle of reflection to show that the direction of the reflected ray is given by $\mathbf{b} = \langle a_1, -a_2, a_3 \rangle$. Deduce that, after being reflected by all three mutually perpendicular mirrors, the resulting ray is parallel to the initial ray. (American space scientists used this principle, together with laser beams and an array of corner mirrors on the moon, to calculate very precisely the distance from the earth to the moon.)

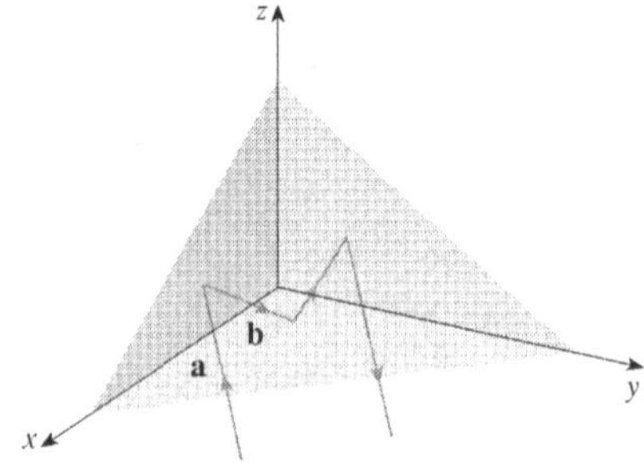

12.3 The Dot Product

So far we have added two vectors and multiplied a vector by a scalar. The question arises: Is it possible to multiply two vectors so that their product is a useful quantity? One such product is the dot product, whose definition follows. Another is the cross product, which is discussed in the next section.

> **1 Definition** If $\mathbf{a} = \langle a_1, a_2, a_3 \rangle$ and $\mathbf{b} = \langle b_1, b_2, b_3 \rangle$, then the **dot product** of $\mathbf{a}$ and $\mathbf{b}$ is the number $\mathbf{a} \cdot \mathbf{b}$ given by
>
> $$\mathbf{a} \cdot \mathbf{b} = a_1 b_1 + a_2 b_2 + a_3 b_3$$

Thus, to find the dot product of $\mathbf{a}$ and $\mathbf{b}$, we multiply corresponding components and add. The result is not a vector. It is a real number, that is, a scalar. For this reason, the dot product is sometimes called the **scalar product** (or **inner product**). Although Definition 1 is given for three-dimensional vectors, the dot product of two-dimensional vectors is defined in a similar fashion:

$$\langle a_1, a_2 \rangle \cdot \langle b_1, b_2 \rangle = a_1 b_1 + a_2 b_2$$

V EXAMPLE 1

$$\langle 2, 4\rangle \cdot \langle 3, -1\rangle = 2(3) + 4(-1) = 2$$

$$\langle -1, 7, 4\rangle \cdot \langle 6, 2, -\tfrac{1}{2}\rangle = (-1)(6) + 7(2) + 4(-\tfrac{1}{2}) = 6$$

$$(\mathbf{i} + 2\mathbf{j} - 3\mathbf{k}) \cdot (2\mathbf{j} - \mathbf{k}) = 1(0) + 2(2) + (-3)(-1) = 7$$

The dot product obeys many of the laws that hold for ordinary products of real numbers. These are stated in the following theorem.

> **2 Properties of the Dot Product** If **a**, **b**, and **c** are vectors in V_3 and c is a scalar, then
>
> 1. $\mathbf{a} \cdot \mathbf{a} = |\mathbf{a}|^2$
> 2. $\mathbf{a} \cdot \mathbf{b} = \mathbf{b} \cdot \mathbf{a}$
> 3. $\mathbf{a} \cdot (\mathbf{b} + \mathbf{c}) = \mathbf{a} \cdot \mathbf{b} + \mathbf{a} \cdot \mathbf{c}$
> 4. $(c\mathbf{a}) \cdot \mathbf{b} = c(\mathbf{a} \cdot \mathbf{b}) = \mathbf{a} \cdot (c\mathbf{b})$
> 5. $\mathbf{0} \cdot \mathbf{a} = 0$

These properties are easily proved using Definition 1. For instance, here are the proofs of Properties 1 and 3:

1. $\mathbf{a} \cdot \mathbf{a} = a_1^2 + a_2^2 + a_3^2 = |\mathbf{a}|^2$

3. $$\begin{aligned} \mathbf{a} \cdot (\mathbf{b} + \mathbf{c}) &= \langle a_1, a_2, a_3\rangle \cdot \langle b_1 + c_1, b_2 + c_2, b_3 + c_3\rangle \\ &= a_1(b_1 + c_1) + a_2(b_2 + c_2) + a_3(b_3 + c_3) \\ &= a_1b_1 + a_1c_1 + a_2b_2 + a_2c_2 + a_3b_3 + a_3c_3 \\ &= (a_1b_1 + a_2b_2 + a_3b_3) + (a_1c_1 + a_2c_2 + a_3c_3) \\ &= \mathbf{a} \cdot \mathbf{b} + \mathbf{a} \cdot \mathbf{c} \end{aligned}$$

The proofs of the remaining properties are left as exercises.

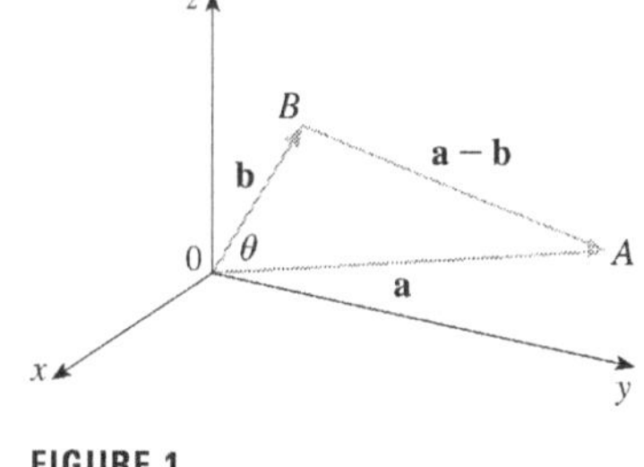

FIGURE 1

The dot product $\mathbf{a} \cdot \mathbf{b}$ can be given a geometric interpretation in terms of the **angle θ between a and b**, which is defined to be the angle between the representations of **a** and **b** that start at the origin, where $0 \leqslant \theta \leqslant \pi$. In other words, θ is the angle between the line segments $\overrightarrow{OA}$ and $\overrightarrow{OB}$ in Figure 1. Note that if **a** and **b** are parallel vectors, then $\theta = 0$ or $\theta = \pi$.

The formula in the following theorem is used by physicists as the *definition* of the dot product.

> **3 Theorem** If θ is the angle between the vectors **a** and **b**, then
>
> $$\mathbf{a} \cdot \mathbf{b} = |\mathbf{a}||\mathbf{b}| \cos \theta$$

PROOF If we apply the Law of Cosines to triangle OAB in Figure 1, we get

4 $$|AB|^2 = |OA|^2 + |OB|^2 - 2|OA||OB| \cos \theta$$

(Observe that the Law of Cosines still applies in the limiting cases when $\theta = 0$ or π, or $\mathbf{a} = \mathbf{0}$ or $\mathbf{b} = \mathbf{0}$.) But $|OA| = |\mathbf{a}|$, $|OB| = |\mathbf{b}|$, and $|AB| = |\mathbf{a} - \mathbf{b}|$, so Equation 4 becomes

5 $$|\mathbf{a} - \mathbf{b}|^2 = |\mathbf{a}|^2 + |\mathbf{b}|^2 - 2|\mathbf{a}||\mathbf{b}| \cos \theta$$

Using Properties 1, 2, and 3 of the dot product, we can rewrite the left side of this equation as follows:

$$\begin{aligned} |\mathbf{a} - \mathbf{b}|^2 &= (\mathbf{a} - \mathbf{b}) \cdot (\mathbf{a} - \mathbf{b}) \\ &= \mathbf{a} \cdot \mathbf{a} - \mathbf{a} \cdot \mathbf{b} - \mathbf{b} \cdot \mathbf{a} + \mathbf{b} \cdot \mathbf{b} \\ &= |\mathbf{a}|^2 - 2\mathbf{a} \cdot \mathbf{b} + |\mathbf{b}|^2 \end{aligned}$$

Therefore Equation 5 gives

$$|\mathbf{a}|^2 - 2\mathbf{a} \cdot \mathbf{b} + |\mathbf{b}|^2 = |\mathbf{a}|^2 + |\mathbf{b}|^2 - 2|\mathbf{a}||\mathbf{b}| \cos \theta$$

Thus

$$-2\mathbf{a} \cdot \mathbf{b} = -2|\mathbf{a}||\mathbf{b}| \cos \theta$$

or

$$\mathbf{a} \cdot \mathbf{b} = |\mathbf{a}||\mathbf{b}| \cos \theta$$

EXAMPLE 2 If the vectors **a** and **b** have lengths 4 and 6, and the angle between them is $\pi/3$, find $\mathbf{a} \cdot \mathbf{b}$.

SOLUTION Using Theorem 3, we have

$$\mathbf{a} \cdot \mathbf{b} = |\mathbf{a}||\mathbf{b}| \cos(\pi/3) = 4 \cdot 6 \cdot \tfrac{1}{2} = 12$$

The formula in Theorem 3 also enables us to find the angle between two vectors.

> **6** **Corollary** If θ is the angle between the nonzero vectors **a** and **b**, then
>
> $$\cos \theta = \frac{\mathbf{a} \cdot \mathbf{b}}{|\mathbf{a}||\mathbf{b}|}$$

V **EXAMPLE 3** Find the angle between the vectors $\mathbf{a} = \langle 2, 2, -1 \rangle$ and $\mathbf{b} = \langle 5, -3, 2 \rangle$.

SOLUTION Since

$$|\mathbf{a}| = \sqrt{2^2 + 2^2 + (-1)^2} = 3 \qquad \text{and} \qquad |\mathbf{b}| = \sqrt{5^2 + (-3)^2 + 2^2} = \sqrt{38}$$

and since

$$\mathbf{a} \cdot \mathbf{b} = 2(5) + 2(-3) + (-1)(2) = 2$$

we have, from Corollary 6,

$$\cos \theta = \frac{\mathbf{a} \cdot \mathbf{b}}{|\mathbf{a}||\mathbf{b}|} = \frac{2}{3\sqrt{38}}$$

So the angle between **a** and **b** is

$$\theta = \cos^{-1}\left(\frac{2}{3\sqrt{38}}\right) \approx 1.46 \quad (\text{or } 84°)$$

Two nonzero vectors **a** and **b** are called **perpendicular** or **orthogonal** if the angle between them is $\theta = \pi/2$. Then Theorem 3 gives

$$\mathbf{a} \cdot \mathbf{b} = |\mathbf{a}||\mathbf{b}| \cos(\pi/2) = 0$$

and conversely if $\mathbf{a} \cdot \mathbf{b} = 0$, then $\cos\theta = 0$, so $\theta = \pi/2$. The zero vector $\mathbf{0}$ is considered to be perpendicular to all vectors. Therefore we have the following method for determining whether two vectors are orthogonal.

> **7** Two vectors $\mathbf{a}$ and $\mathbf{b}$ are orthogonal if and only if $\mathbf{a} \cdot \mathbf{b} = 0$.

EXAMPLE 4 Show that $2\mathbf{i} + 2\mathbf{j} - \mathbf{k}$ is perpendicular to $5\mathbf{i} - 4\mathbf{j} + 2\mathbf{k}$.

SOLUTION Since

$$(2\mathbf{i} + 2\mathbf{j} - \mathbf{k}) \cdot (5\mathbf{i} - 4\mathbf{j} + 2\mathbf{k}) = 2(5) + 2(-4) + (-1)(2) = 0$$

these vectors are perpendicular by 7.

a, b, θ — $\mathbf{a} \cdot \mathbf{b} > 0$, θ acute

a, b — $\mathbf{a} \cdot \mathbf{b} = 0$, $\theta = \pi/2$

a, b, θ — $\mathbf{a} \cdot \mathbf{b} < 0$, θ obtuse

FIGURE 2

Visual 12.3A shows an animation of Figure 2.

Because $\cos\theta > 0$ if $0 \leqslant \theta < \pi/2$ and $\cos\theta < 0$ if $\pi/2 < \theta \leqslant \pi$, we see that $\mathbf{a} \cdot \mathbf{b}$ is positive for $\theta < \pi/2$ and negative for $\theta > \pi/2$. We can think of $\mathbf{a} \cdot \mathbf{b}$ as measuring the extent to which $\mathbf{a}$ and $\mathbf{b}$ point in the same direction. The dot product $\mathbf{a} \cdot \mathbf{b}$ is positive if $\mathbf{a}$ and $\mathbf{b}$ point in the same general direction, 0 if they are perpendicular, and negative if they point in generally opposite directions (see Figure 2). In the extreme case where $\mathbf{a}$ and $\mathbf{b}$ point in exactly the same direction, we have $\theta = 0$, so $\cos\theta = 1$ and

$$\mathbf{a} \cdot \mathbf{b} = |\mathbf{a}||\mathbf{b}|$$

If $\mathbf{a}$ and $\mathbf{b}$ point in exactly opposite directions, then $\theta = \pi$ and so $\cos\theta = -1$ and $\mathbf{a} \cdot \mathbf{b} = -|\mathbf{a}||\mathbf{b}|$.

Direction Angles and Direction Cosines

The **direction angles** of a nonzero vector $\mathbf{a}$ are the angles α, β, and γ (in the interval $[0, \pi]$) that $\mathbf{a}$ makes with the positive x-, y-, and z-axes. (See Figure 3.)

The cosines of these direction angles, $\cos\alpha$, $\cos\beta$, and $\cos\gamma$, are called the **direction cosines** of the vector $\mathbf{a}$. Using Corollary 6 with $\mathbf{b}$ replaced by $\mathbf{i}$, we obtain

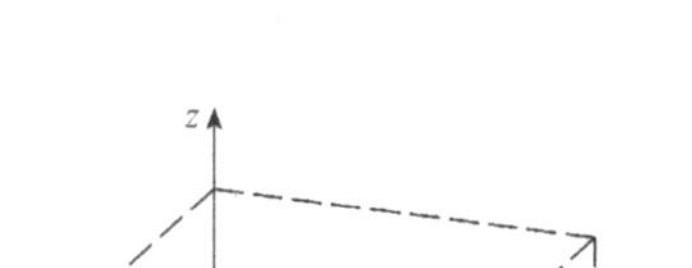

FIGURE 3

$$\textbf{8} \qquad \cos\alpha = \frac{\mathbf{a} \cdot \mathbf{i}}{|\mathbf{a}||\mathbf{i}|} = \frac{a_1}{|\mathbf{a}|}$$

(This can also be seen directly from Figure 3.)

Similarly, we also have

$$\textbf{9} \qquad \cos\beta = \frac{a_2}{|\mathbf{a}|} \qquad \cos\gamma = \frac{a_3}{|\mathbf{a}|}$$

By squaring the expressions in Equations 8 and 9 and adding, we see that

$$\textbf{10} \qquad \cos^2\alpha + \cos^2\beta + \cos^2\gamma = 1$$

We can also use Equations 8 and 9 to write

$$\mathbf{a} = \langle a_1, a_2, a_3 \rangle = \langle |\mathbf{a}|\cos\alpha, |\mathbf{a}|\cos\beta, |\mathbf{a}|\cos\gamma \rangle$$
$$= |\mathbf{a}|\langle \cos\alpha, \cos\beta, \cos\gamma \rangle$$

Therefore

$$\boxed{11} \qquad \frac{1}{|\mathbf{a}|}\mathbf{a} = \langle \cos\alpha, \cos\beta, \cos\gamma \rangle$$

which says that the direction cosines of **a** are the components of the unit vector in the direction of **a**.

EXAMPLE 5 Find the direction angles of the vector $\mathbf{a} = \langle 1, 2, 3 \rangle$.

SOLUTION Since $|\mathbf{a}| = \sqrt{1^2 + 2^2 + 3^2} = \sqrt{14}$, Equations 8 and 9 give

$$\cos\alpha = \frac{1}{\sqrt{14}} \qquad \cos\beta = \frac{2}{\sqrt{14}} \qquad \cos\gamma = \frac{3}{\sqrt{14}}$$

and so

$$\alpha = \cos^{-1}\left(\frac{1}{\sqrt{14}}\right) \approx 74° \qquad \beta = \cos^{-1}\left(\frac{2}{\sqrt{14}}\right) \approx 58° \qquad \gamma = \cos^{-1}\left(\frac{3}{\sqrt{14}}\right) \approx 37°$$

Projections

TEC Visual 12.3B shows how Figure 4 changes when we vary **a** and **b**.

Figure 4 shows representations $\overrightarrow{PQ}$ and $\overrightarrow{PR}$ of two vectors **a** and **b** with the same initial point P. If S is the foot of the perpendicular from R to the line containing $\overrightarrow{PQ}$, then the vector with representation $\overrightarrow{PS}$ is called the **vector projection** of **b** onto **a** and is denoted by $\text{proj}_\mathbf{a}\,\mathbf{b}$. (You can think of it as a shadow of **b**).

The **scalar projection** of **b** onto **a** (also called the **component of b along a**) is defined to be the signed magnitude of the vector projection, which is the number $|\mathbf{b}|\cos\theta$, where θ is the angle between **a** and **b**. (See Figure 5.) This is denoted by $\text{comp}_\mathbf{a}\,\mathbf{b}$. Observe that it is negative if $\pi/2 < \theta \leq \pi$. The equation

$$\mathbf{a}\cdot\mathbf{b} = |\mathbf{a}||\mathbf{b}|\cos\theta = |\mathbf{a}|(|\mathbf{b}|\cos\theta)$$

shows that the dot product of **a** and **b** can be interpreted as the length of **a** times the scalar projection of **b** onto **a**. Since

$$|\mathbf{b}|\cos\theta = \frac{\mathbf{a}\cdot\mathbf{b}}{|\mathbf{a}|} = \frac{\mathbf{a}}{|\mathbf{a}|}\cdot\mathbf{b}$$

the component of **b** along **a** can be computed by taking the dot product of **b** with the unit vector in the direction of **a**. We summarize these ideas as follows.

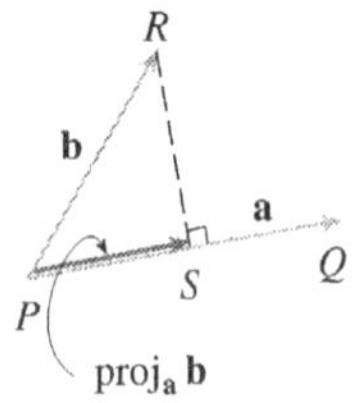

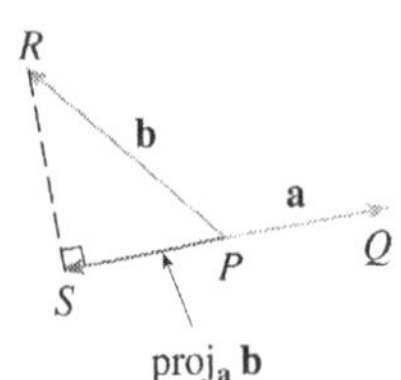

FIGURE 4
Vector projections

Scalar projection of **b** onto **a**:	$\text{comp}_\mathbf{a}\,\mathbf{b} = \dfrac{\mathbf{a}\cdot\mathbf{b}}{|\mathbf{a}|}$
Vector projection of **b** onto **a**:	$\text{proj}_\mathbf{a}\,\mathbf{b} = \left(\dfrac{\mathbf{a}\cdot\mathbf{b}}{|\mathbf{a}|}\right)\dfrac{\mathbf{a}}{|\mathbf{a}|} = \dfrac{\mathbf{a}\cdot\mathbf{b}}{|\mathbf{a}|^2}\mathbf{a}$

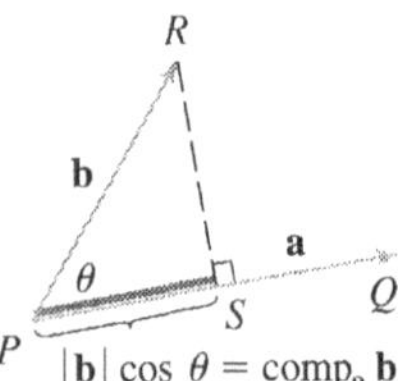

FIGURE 5
Scalar projection

Notice that the vector projection is the scalar projection times the unit vector in the direction of **a**.

V EXAMPLE 6 Find the scalar projection and vector projection of $\mathbf{b} = \langle 1, 1, 2 \rangle$ onto $\mathbf{a} = \langle -2, 3, 1 \rangle$.

SOLUTION Since $|\mathbf{a}| = \sqrt{(-2)^2 + 3^2 + 1^2} = \sqrt{14}$, the scalar projection of $\mathbf{b}$ onto $\mathbf{a}$ is

$$\text{comp}_{\mathbf{a}}\, \mathbf{b} = \frac{\mathbf{a} \cdot \mathbf{b}}{|\mathbf{a}|} = \frac{(-2)(1) + 3(1) + 1(2)}{\sqrt{14}} = \frac{3}{\sqrt{14}}$$

The vector projection is this scalar projection times the unit vector in the direction of $\mathbf{a}$:

$$\text{proj}_{\mathbf{a}}\, \mathbf{b} = \frac{3}{\sqrt{14}} \frac{\mathbf{a}}{|\mathbf{a}|} = \frac{3}{14}\mathbf{a} = \left\langle -\frac{3}{7}, \frac{9}{14}, \frac{3}{14} \right\rangle$$

One use of projections occurs in physics in calculating work. In Section 6.4 we defined the work done by a constant force F in moving an object through a distance d as $W = Fd$, but this applies only when the force is directed along the line of motion of the object. Suppose, however, that the constant force is a vector $\mathbf{F} = \overrightarrow{PR}$ pointing in some other direction, as in Figure 6. If the force moves the object from P to Q, then the **displacement vector** is $\mathbf{D} = \overrightarrow{PQ}$. The **work** done by this force is defined to be the product of the component of the force along $\mathbf{D}$ and the distance moved:

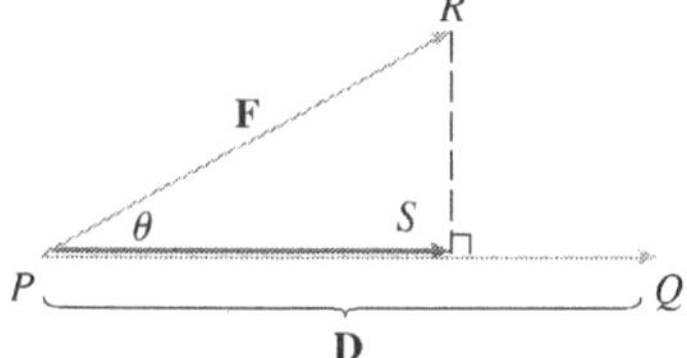

FIGURE 6

$$W = (|\mathbf{F}| \cos \theta)\,|\mathbf{D}|$$

But then, from Theorem 3, we have

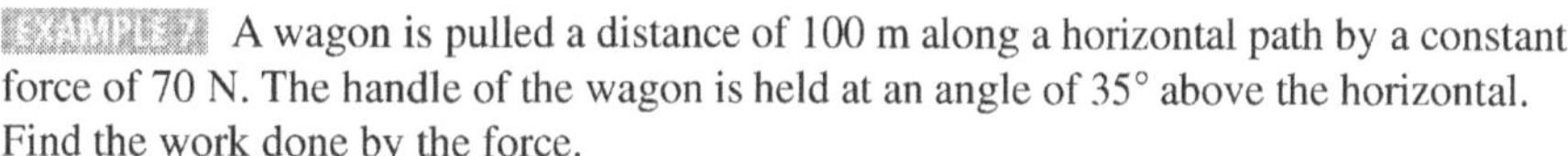

12 $$W = |\mathbf{F}|\,|\mathbf{D}| \cos \theta = \mathbf{F} \cdot \mathbf{D}$$

Thus the work done by a constant force $\mathbf{F}$ is the dot product $\mathbf{F} \cdot \mathbf{D}$, where $\mathbf{D}$ is the displacement vector.

EXAMPLE 7 A wagon is pulled a distance of 100 m along a horizontal path by a constant force of 70 N. The handle of the wagon is held at an angle of 35° above the horizontal. Find the work done by the force.

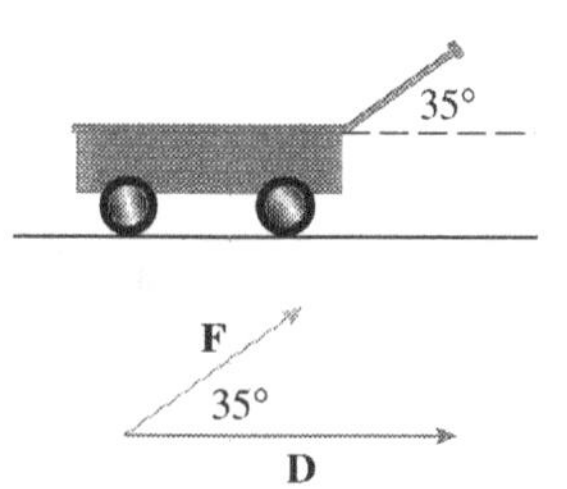

FIGURE 7

SOLUTION If $\mathbf{F}$ and $\mathbf{D}$ are the force and displacement vectors, as pictured in Figure 7, then the work done is

$$W = \mathbf{F} \cdot \mathbf{D} = |\mathbf{F}|\,|\mathbf{D}| \cos 35°$$
$$= (70)(100) \cos 35° \approx 5734 \text{ N}\cdot\text{m} = 5734 \text{ J}$$

EXAMPLE 8 A force is given by a vector $\mathbf{F} = 3\mathbf{i} + 4\mathbf{j} + 5\mathbf{k}$ and moves a particle from the point $P(2, 1, 0)$ to the point $Q(4, 6, 2)$. Find the work done.

SOLUTION The displacement vector is $\mathbf{D} = \overrightarrow{PQ} = \langle 2, 5, 2 \rangle$, so by Equation 12, the work done is

$$W = \mathbf{F} \cdot \mathbf{D} = \langle 3, 4, 5 \rangle \cdot \langle 2, 5, 2 \rangle$$
$$= 6 + 20 + 10 = 36$$

If the unit of length is meters and the magnitude of the force is measured in newtons, then the work done is 36 J.

12.3 Exercises

1. Which of the following expressions are meaningful? Which are meaningless? Explain.
(a) $(\mathbf{a} \cdot \mathbf{b}) \cdot \mathbf{c}$ (b) $(\mathbf{a} \cdot \mathbf{b})\mathbf{c}$
(c) $|\mathbf{a}|(\mathbf{b} \cdot \mathbf{c})$ (d) $\mathbf{a} \cdot (\mathbf{b} + \mathbf{c})$
(e) $\mathbf{a} \cdot \mathbf{b} + \mathbf{c}$ (f) $|\mathbf{a}| \cdot (\mathbf{b} + \mathbf{c})$

2–10 Find $\mathbf{a} \cdot \mathbf{b}$.

2. $\mathbf{a} = \langle -2, 3 \rangle$, $\mathbf{b} = \langle 0.7, 1.2 \rangle$

3. $\mathbf{a} = \langle -2, \frac{1}{3} \rangle$, $\mathbf{b} = \langle -5, 12 \rangle$

4. $\mathbf{a} = \langle 6, -2, 3 \rangle$, $\mathbf{b} = \langle 2, 5, -1 \rangle$

5. $\mathbf{a} = \langle 4, 1, \frac{1}{4} \rangle$, $\mathbf{b} = \langle 6, -3, -8 \rangle$

6. $\mathbf{a} = \langle p, -p, 2p \rangle$, $\mathbf{b} = \langle 2q, q, -q \rangle$

7. $\mathbf{a} = 2\mathbf{i} + \mathbf{j}$, $\mathbf{b} = \mathbf{i} - \mathbf{j} + \mathbf{k}$

8. $\mathbf{a} = 3\mathbf{i} + 2\mathbf{j} - \mathbf{k}$, $\mathbf{b} = 4\mathbf{i} + 5\mathbf{k}$

9. $|\mathbf{a}| = 6$, $|\mathbf{b}| = 5$, the angle between $\mathbf{a}$ and $\mathbf{b}$ is $2\pi/3$

10. $|\mathbf{a}| = 3$, $|\mathbf{b}| = \sqrt{6}$, the angle between $\mathbf{a}$ and $\mathbf{b}$ is 45°

11–12 If $\mathbf{u}$ is a unit vector, find $\mathbf{u} \cdot \mathbf{v}$ and $\mathbf{u} \cdot \mathbf{w}$.

11.
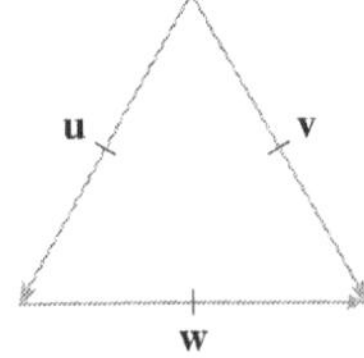

12.
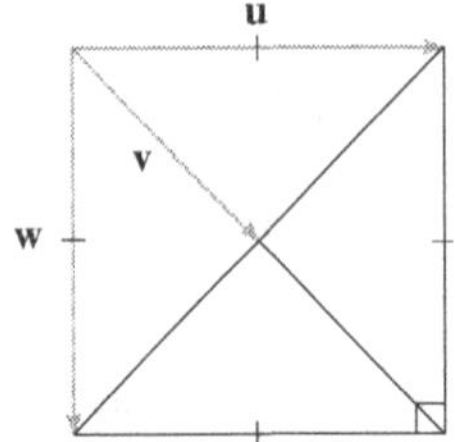

13. (a) Show that $\mathbf{i} \cdot \mathbf{j} = \mathbf{j} \cdot \mathbf{k} = \mathbf{k} \cdot \mathbf{i} = 0$.
(b) Show that $\mathbf{i} \cdot \mathbf{i} = \mathbf{j} \cdot \mathbf{j} = \mathbf{k} \cdot \mathbf{k} = 1$.

14. A street vendor sells a hamburgers, b hot dogs, and c soft drinks on a given day. He charges \$2 for a hamburger, \$1.50 for a hot dog, and \$1 for a soft drink. If $\mathbf{A} = \langle a, b, c \rangle$ and $\mathbf{P} = \langle 2, 1.5, 1 \rangle$, what is the meaning of the dot product $\mathbf{A} \cdot \mathbf{P}$?

15–20 Find the angle between the vectors. (First find an exact expression and then approximate to the nearest degree.)

15. $\mathbf{a} = \langle 4, 3 \rangle$, $\mathbf{b} = \langle 2, -1 \rangle$

16. $\mathbf{a} = \langle -2, 5 \rangle$, $\mathbf{b} = \langle 5, 12 \rangle$

17. $\mathbf{a} = \langle 3, -1, 5 \rangle$, $\mathbf{b} = \langle -2, 4, 3 \rangle$

18. $\mathbf{a} = \langle 4, 0, 2 \rangle$, $\mathbf{b} = \langle 2, -1, 0 \rangle$

19. $\mathbf{a} = 4\mathbf{i} - 3\mathbf{j} + \mathbf{k}$, $\mathbf{b} = 2\mathbf{i} - \mathbf{k}$

20. $\mathbf{a} = \mathbf{i} + 2\mathbf{j} - 2\mathbf{k}$, $\mathbf{b} = 4\mathbf{i} - 3\mathbf{k}$

21–22 Find, correct to the nearest degree, the three angles of the triangle with the given vertices.

21. $P(2, 0)$, $Q(0, 3)$, $R(3, 4)$

22. $A(1, 0, -1)$, $B(3, -2, 0)$, $C(1, 3, 3)$

23–24 Determine whether the given vectors are orthogonal, parallel, or neither.

23. (a) $\mathbf{a} = \langle -5, 3, 7 \rangle$, $\mathbf{b} = \langle 6, -8, 2 \rangle$
(b) $\mathbf{a} = \langle 4, 6 \rangle$, $\mathbf{b} = \langle -3, 2 \rangle$
(c) $\mathbf{a} = -\mathbf{i} + 2\mathbf{j} + 5\mathbf{k}$, $\mathbf{b} = 3\mathbf{i} + 4\mathbf{j} - \mathbf{k}$
(d) $\mathbf{a} = 2\mathbf{i} + 6\mathbf{j} - 4\mathbf{k}$, $\mathbf{b} = -3\mathbf{i} - 9\mathbf{j} + 6\mathbf{k}$

24. (a) $\mathbf{u} = \langle -3, 9, 6 \rangle$, $\mathbf{v} = \langle 4, -12, -8 \rangle$
(b) $\mathbf{u} = \mathbf{i} - \mathbf{j} + 2\mathbf{k}$, $\mathbf{v} = 2\mathbf{i} - \mathbf{j} + \mathbf{k}$
(c) $\mathbf{u} = \langle a, b, c \rangle$, $\mathbf{v} = \langle -b, a, 0 \rangle$

25. Use vectors to decide whether the triangle with vertices $P(1, -3, -2)$, $Q(2, 0, -4)$, and $R(6, -2, -5)$ is right-angled.

26. Find the values of x such that the angle between the vectors $\langle 2, 1, -1 \rangle$, and $\langle 1, x, 0 \rangle$ is 45°.

27. Find a unit vector that is orthogonal to both $\mathbf{i} + \mathbf{j}$ and $\mathbf{i} + \mathbf{k}$.

28. Find two unit vectors that make an angle of 60° with $\mathbf{v} = \langle 3, 4 \rangle$.

29–30 Find the acute angle between the lines.

29. $2x - y = 3$, $3x + y = 7$

30. $x + 2y = 7$, $5x - y = 2$

31–32 Find the acute angles between the curves at their points of intersection. (The angle between two curves is the angle between their tangent lines at the point of intersection.)

31. $y = x^2$, $y = x^3$

32. $y = \sin x$, $y = \cos x$, $0 \leqslant x \leqslant \pi/2$

33–37 Find the direction cosines and direction angles of the vector. (Give the direction angles correct to the nearest degree.)

33. $\langle 2, 1, 2 \rangle$

34. $\langle 6, 3, -2 \rangle$

35. $\mathbf{i} - 2\mathbf{j} - 3\mathbf{k}$

36. $\frac{1}{2}\mathbf{i} + \mathbf{j} + \mathbf{k}$

37. $\langle c, c, c \rangle$, where $c > 0$

38. If a vector has direction angles $\alpha = \pi/4$ and $\beta = \pi/3$, find the third direction angle γ.

39–44 Find the scalar and vector projections of **b** onto **a**.

39. $\mathbf{a} = \langle -5, 12 \rangle$, $\mathbf{b} = \langle 4, 6 \rangle$

40. $\mathbf{a} = \langle 1, 4 \rangle$, $\mathbf{b} = \langle 2, 3 \rangle$

41. $\mathbf{a} = \langle 3, 6, -2 \rangle$, $\mathbf{b} = \langle 1, 2, 3 \rangle$

42. $\mathbf{a} = \langle -2, 3, -6 \rangle$, $\mathbf{b} = \langle 5, -1, 4 \rangle$

43. $\mathbf{a} = 2\mathbf{i} - \mathbf{j} + 4\mathbf{k}$, $\mathbf{b} = \mathbf{j} + \frac{1}{2}\mathbf{k}$

44. $\mathbf{a} = \mathbf{i} + \mathbf{j} + \mathbf{k}$, $\mathbf{b} = \mathbf{i} - \mathbf{j} + \mathbf{k}$

---

45. Show that the vector $\text{orth}_{\mathbf{a}}\,\mathbf{b} = \mathbf{b} - \text{proj}_{\mathbf{a}}\,\mathbf{b}$ is orthogonal to **a**. (It is called an **orthogonal projection** of **b**.)

46. For the vectors in Exercise 40, find $\text{orth}_{\mathbf{a}}\,\mathbf{b}$ and illustrate by drawing the vectors **a**, **b**, $\text{proj}_{\mathbf{a}}\,\mathbf{b}$, and $\text{orth}_{\mathbf{a}}\,\mathbf{b}$.

47. If $\mathbf{a} = \langle 3, 0, -1 \rangle$, find a vector **b** such that $\text{comp}_{\mathbf{a}}\,\mathbf{b} = 2$.

48. Suppose that **a** and **b** are nonzero vectors.
(a) Under what circumstances is $\text{comp}_{\mathbf{a}}\,\mathbf{b} = \text{comp}_{\mathbf{b}}\,\mathbf{a}$?
(b) Under what circumstances is $\text{proj}_{\mathbf{a}}\,\mathbf{b} = \text{proj}_{\mathbf{b}}\,\mathbf{a}$?

49. Find the work done by a force $\mathbf{F} = 8\mathbf{i} - 6\mathbf{j} + 9\mathbf{k}$ that moves an object from the point $(0, 10, 8)$ to the point $(6, 12, 20)$ along a straight line. The distance is measured in meters and the force in newtons.

50. A tow truck drags a stalled car along a road. The chain makes an angle of 30° with the road and the tension in the chain is 1500 N. How much work is done by the truck in pulling the car 1 km?

51. A sled is pulled along a level path through snow by a rope. A 30-lb force acting at an angle of 40° above the horizontal moves the sled 80 ft. Find the work done by the force.

52. A boat sails south with the help of a wind blowing in the direction S36°E with magnitude 400 lb. Find the work done by the wind as the boat moves 120 ft.

53. Use a scalar projection to show that the distance from a point $P_1(x_1, y_1)$ to the line $ax + by + c = 0$ is

$$\frac{|ax_1 + by_1 + c|}{\sqrt{a^2 + b^2}}$$

Use this formula to find the distance from the point $(-2, 3)$ to the line $3x - 4y + 5 = 0$.

54. If $\mathbf{r} = \langle x, y, z \rangle$, $\mathbf{a} = \langle a_1, a_2, a_3 \rangle$, and $\mathbf{b} = \langle b_1, b_2, b_3 \rangle$, show that the vector equation $(\mathbf{r} - \mathbf{a}) \cdot (\mathbf{r} - \mathbf{b}) = 0$ represents a sphere, and find its center and radius.

55. Find the angle between a diagonal of a cube and one of its edges.

56. Find the angle between a diagonal of a cube and a diagonal of one of its faces.

57. A molecule of methane, CH_4, is structured with the four hydrogen atoms at the vertices of a regular tetrahedron and the carbon atom at the centroid. The *bond angle* is the angle formed by the H—C—H combination; it is the angle between the lines that join the carbon atom to two of the hydrogen atoms. Show that the bond angle is about 109.5°. [*Hint:* Take the vertices of the tetrahedron to be the points $(1, 0, 0)$, $(0, 1, 0)$, $(0, 0, 1)$, and $(1, 1, 1)$, as shown in the figure. Then the centroid is $\left(\frac{1}{2}, \frac{1}{2}, \frac{1}{2}\right)$.]

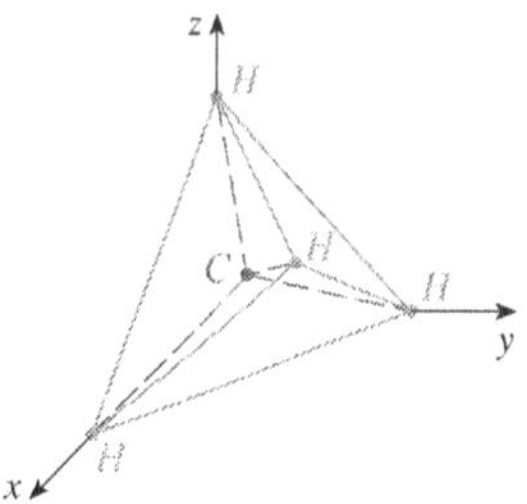

58. If $\mathbf{c} = |\mathbf{a}|\mathbf{b} + |\mathbf{b}|\mathbf{a}$, where **a**, **b**, and **c** are all nonzero vectors, show that **c** bisects the angle between **a** and **b**.

59. Prove Properties 2, 4, and 5 of the dot product (Theorem 2).

60. Suppose that all sides of a quadrilateral are equal in length and opposite sides are parallel. Use vector methods to show that the diagonals are perpendicular.

61. Use Theorem 3 to prove the Cauchy-Schwarz Inequality:

$$|\mathbf{a} \cdot \mathbf{b}| \leqslant |\mathbf{a}|\,|\mathbf{b}|$$

62. The Triangle Inequality for vectors is

$$|\mathbf{a} + \mathbf{b}| \leqslant |\mathbf{a}| + |\mathbf{b}|$$

(a) Give a geometric interpretation of the Triangle Inequality.
(b) Use the Cauchy-Schwarz Inequality from Exercise 61 to prove the Triangle Inequality. [*Hint:* Use the fact that $|\mathbf{a} + \mathbf{b}|^2 = (\mathbf{a} + \mathbf{b}) \cdot (\mathbf{a} + \mathbf{b})$ and use Property 3 of the dot product.]

63. The Parallelogram Law states that

$$|\mathbf{a} + \mathbf{b}|^2 + |\mathbf{a} - \mathbf{b}|^2 = 2|\mathbf{a}|^2 + 2|\mathbf{b}|^2$$

(a) Give a geometric interpretation of the Parallelogram Law.
(b) Prove the Parallelogram Law. (See the hint in Exercise 62.)

64. Show that if $\mathbf{u} + \mathbf{v}$ and $\mathbf{u} - \mathbf{v}$ are orthogonal, then the vectors **u** and **v** must have the same length.

12.4 The Cross Product

Given two nonzero vectors $\mathbf{a} = \langle a_1, a_2, a_3 \rangle$ and $b = \langle b_1, b_2, b_3 \rangle$, it is very useful to be able to find a nonzero vector $\mathbf{c}$ that is perpendicular to both $\mathbf{a}$ and $\mathbf{b}$, as we will see in the next section and in Chapters 13 and 14. If $\mathbf{c} = \langle c_1, c_2, c_3 \rangle$ is such a vector, then $\mathbf{a} \cdot \mathbf{c} = 0$ and $\mathbf{b} \cdot \mathbf{c} = 0$ and so

1 $$a_1c_1 + a_2c_2 + a_3c_3 = 0$$

2 $$b_1c_1 + b_2c_2 + b_3c_3 = 0$$

To eliminate c_3 we multiply **1** by b_3 and **2** by a_3 and subtract:

3 $$(a_1b_3 - a_3b_1)c_1 + (a_2b_3 - a_3b_2)c_2 = 0$$

Equation 3 has the form $pc_1 + qc_2 = 0$, for which an obvious solution is $c_1 = q$ and $c_2 = -p$. So a solution of **3** is

$$c_1 = a_2b_3 - a_3b_2 \qquad\qquad c_2 = a_3b_1 - a_1b_3$$

Substituting these values into **1** and **2**, we then get

$$c_3 = a_1b_2 - a_2b_1$$

This means that a vector perpendicular to both $\mathbf{a}$ and $\mathbf{b}$ is

$$\langle c_1, c_2, c_3 \rangle = \langle a_2b_3 - a_3b_2, a_3b_1 - a_1b_3, a_1b_2 - a_2b_1 \rangle$$

The resulting vector is called the *cross product* of $\mathbf{a}$ and $\mathbf{b}$ and is denoted by $\mathbf{a} \times \mathbf{b}$.

Hamilton

The cross product was invented by the Irish mathematician Sir William Rowan Hamilton (1805–1865), who had created a precursor of vectors, called quaternions. When he was five years old Hamilton could read Latin, Greek, and Hebrew. At age eight he added French and Italian and when ten he could read Arabic and Sanskrit. At the age of 21, while still an undergraduate at Trinity College in Dublin, Hamilton was appointed Professor of Astronomy at the university and Royal Astronomer of Ireland!

4 Definition If $\mathbf{a} = \langle a_1, a_2, a_3 \rangle$ and $\mathbf{b} = \langle b_1, b_2, b_3 \rangle$, then the **cross product** of $\mathbf{a}$ and $\mathbf{b}$ is the vector

$$\mathbf{a} \times \mathbf{b} = \langle a_2b_3 - a_3b_2, a_3b_1 - a_1b_3, a_1b_2 - a_2b_1 \rangle$$

Notice that the **cross product** $\mathbf{a} \times \mathbf{b}$ of two vectors $\mathbf{a}$ and $\mathbf{b}$, unlike the dot product, is a vector. For this reason it is also called the **vector product**. Note that $\mathbf{a} \times \mathbf{b}$ is defined only when $\mathbf{a}$ and $\mathbf{b}$ are *three-dimensional* vectors.

In order to make Definition 4 easier to remember, we use the notation of determinants. A **determinant of order 2** is defined by

$$\begin{vmatrix} a & b \\ c & d \end{vmatrix} = ad - bc$$

For example,

$$\begin{vmatrix} 2 & 1 \\ -6 & 4 \end{vmatrix} = 2(4) - 1(-6) = 14$$

A **determinant of order 3** can be defined in terms of second-order determinants as follows:

5 $$\begin{vmatrix} a_1 & a_2 & a_3 \\ b_1 & b_2 & b_3 \\ c_1 & c_2 & c_3 \end{vmatrix} = a_1 \begin{vmatrix} b_2 & b_3 \\ c_2 & c_3 \end{vmatrix} - a_2 \begin{vmatrix} b_1 & b_3 \\ c_1 & c_3 \end{vmatrix} + a_3 \begin{vmatrix} b_1 & b_2 \\ c_1 & c_2 \end{vmatrix}$$

Observe that each term on the right side of Equation 5 involves a number a_i in the first row of the determinant, and a_i is multiplied by the second-order determinant obtained from the left side by deleting the row and column in which a_i appears. Notice also the minus sign in the second term. For example,

$$\begin{vmatrix} 1 & 2 & -1 \\ 3 & 0 & 1 \\ -5 & 4 & 2 \end{vmatrix} = 1\begin{vmatrix} 0 & 1 \\ 4 & 2 \end{vmatrix} - 2\begin{vmatrix} 3 & 1 \\ -5 & 2 \end{vmatrix} + (-1)\begin{vmatrix} 3 & 0 \\ -5 & 4 \end{vmatrix}$$

$$= 1(0 - 4) - 2(6 + 5) + (-1)(12 - 0) = -38$$

If we now rewrite Definition 4 using second-order determinants and the standard basis vectors **i**, **j**, and **k**, we see that the cross product of the vectors $\mathbf{a} = a_1\mathbf{i} + a_2\mathbf{j} + a_3\mathbf{k}$ and $\mathbf{b} = b_1\mathbf{i} + b_2\mathbf{j} + b_3\mathbf{k}$ is

6 $$\mathbf{a} \times \mathbf{b} = \begin{vmatrix} a_2 & a_3 \\ b_2 & b_3 \end{vmatrix}\mathbf{i} - \begin{vmatrix} a_1 & a_3 \\ b_1 & b_3 \end{vmatrix}\mathbf{j} + \begin{vmatrix} a_1 & a_2 \\ b_1 & b_2 \end{vmatrix}\mathbf{k}$$

In view of the similarity between Equations 5 and 6, we often write

7 $$\mathbf{a} \times \mathbf{b} = \begin{vmatrix} \mathbf{i} & \mathbf{j} & \mathbf{k} \\ a_1 & a_2 & a_3 \\ b_1 & b_2 & b_3 \end{vmatrix}$$

Although the first row of the symbolic determinant in Equation 7 consists of vectors, if we expand it as if it were an ordinary determinant using the rule in Equation 5, we obtain Equation 6. The symbolic formula in Equation 7 is probably the easiest way of remembering and computing cross products.

V EXAMPLE 1 If $\mathbf{a} = \langle 1, 3, 4 \rangle$ and $\mathbf{b} = \langle 2, 7, -5 \rangle$, then

$$\mathbf{a} \times \mathbf{b} = \begin{vmatrix} \mathbf{i} & \mathbf{j} & \mathbf{k} \\ 1 & 3 & 4 \\ 2 & 7 & -5 \end{vmatrix}$$

$$= \begin{vmatrix} 3 & 4 \\ 7 & -5 \end{vmatrix}\mathbf{i} - \begin{vmatrix} 1 & 4 \\ 2 & -5 \end{vmatrix}\mathbf{j} + \begin{vmatrix} 1 & 3 \\ 2 & 7 \end{vmatrix}\mathbf{k}$$

$$= (-15 - 28)\,\mathbf{i} - (-5 - 8)\,\mathbf{j} + (7 - 6)\,\mathbf{k} = -43\mathbf{i} + 13\mathbf{j} + \mathbf{k}$$

V EXAMPLE 2 Show that $\mathbf{a} \times \mathbf{a} = \mathbf{0}$ for any vector **a** in V_3.

SOLUTION If $\mathbf{a} = \langle a_1, a_2, a_3 \rangle$, then

$$\mathbf{a} \times \mathbf{a} = \begin{vmatrix} \mathbf{i} & \mathbf{j} & \mathbf{k} \\ a_1 & a_2 & a_3 \\ a_1 & a_2 & a_3 \end{vmatrix}$$

$$= (a_2a_3 - a_3a_2)\,\mathbf{i} - (a_1a_3 - a_3a_1)\,\mathbf{j} + (a_1a_2 - a_2a_1)\,\mathbf{k}$$

$$= 0\mathbf{i} - 0\mathbf{j} + 0\mathbf{k} = \mathbf{0}$$

We constructed the cross product $\mathbf{a} \times \mathbf{b}$ so that it would be perpendicular to both $\mathbf{a}$ and $\mathbf{b}$. This is one of the most important properties of a cross product, so let's emphasize and verify it in the following theorem and give a formal proof.

8 Theorem The vector $\mathbf{a} \times \mathbf{b}$ is orthogonal to both $\mathbf{a}$ and $\mathbf{b}$.

PROOF In order to show that $\mathbf{a} \times \mathbf{b}$ is orthogonal to $\mathbf{a}$, we compute their dot product as follows:

$$\begin{aligned}(\mathbf{a} \times \mathbf{b}) \cdot \mathbf{a} &= \begin{vmatrix} a_2 & a_3 \\ b_2 & b_3 \end{vmatrix} a_1 - \begin{vmatrix} a_1 & a_3 \\ b_1 & b_3 \end{vmatrix} a_2 + \begin{vmatrix} a_1 & a_2 \\ b_1 & b_2 \end{vmatrix} a_3 \\ &= a_1(a_2b_3 - a_3b_2) - a_2(a_1b_3 - a_3b_1) + a_3(a_1b_2 - a_2b_1) \\ &= a_1a_2b_3 - a_1b_2a_3 - a_1a_2b_3 + b_1a_2a_3 + a_1b_2a_3 - b_1a_2a_3 \\ &= 0\end{aligned}$$

A similar computation shows that $(\mathbf{a} \times \mathbf{b}) \cdot \mathbf{b} = 0$. Therefore $\mathbf{a} \times \mathbf{b}$ is orthogonal to both $\mathbf{a}$ and $\mathbf{b}$. ■

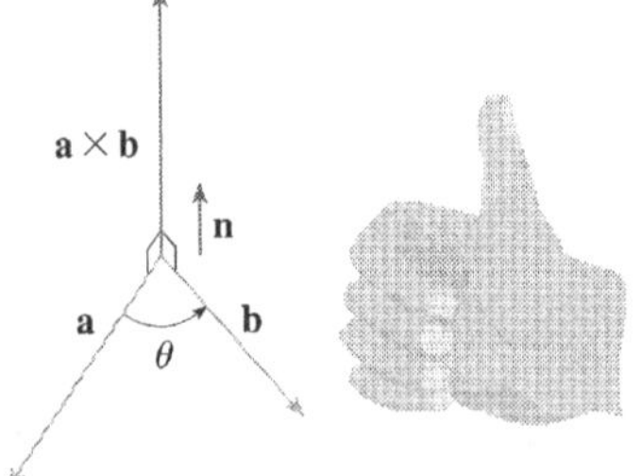

FIGURE 1
The right-hand rule gives the direction of $\mathbf{a} \times \mathbf{b}$.

TEC Visual 12.4 shows how $\mathbf{a} \times \mathbf{b}$ changes as $\mathbf{b}$ changes.

If $\mathbf{a}$ and $\mathbf{b}$ are represented by directed line segments with the same initial point (as in Figure 1), then Theorem 8 says that the cross product $\mathbf{a} \times \mathbf{b}$ points in a direction perpendicular to the plane through $\mathbf{a}$ and $\mathbf{b}$. It turns out that the direction of $\mathbf{a} \times \mathbf{b}$ is given by the *right-hand rule:* If the fingers of your right hand curl in the direction of a rotation (through an angle less than 180°) from $\mathbf{a}$ to $\mathbf{b}$, then your thumb points in the direction of $\mathbf{a} \times \mathbf{b}$.

Now that we know the direction of the vector $\mathbf{a} \times \mathbf{b}$, the remaining thing we need to complete its geometric description is its length $|\mathbf{a} \times \mathbf{b}|$. This is given by the following theorem.

9 Theorem If θ is the angle between $\mathbf{a}$ and $\mathbf{b}$ (so $0 \leqslant \theta \leqslant \pi$), then

$$|\mathbf{a} \times \mathbf{b}| = |\mathbf{a}||\mathbf{b}| \sin \theta$$

PROOF From the definitions of the cross product and length of a vector, we have

$$\begin{aligned}|\mathbf{a} \times \mathbf{b}|^2 &= (a_2b_3 - a_3b_2)^2 + (a_3b_1 - a_1b_3)^2 + (a_1b_2 - a_2b_1)^2 \\ &= a_2^2b_3^2 - 2a_2a_3b_2b_3 + a_3^2b_2^2 + a_3^2b_1^2 - 2a_1a_3b_1b_3 + a_1^2b_3^2 \\ &\qquad + a_1^2b_2^2 - 2a_1a_2b_1b_2 + a_2^2b_1^2 \\ &= (a_1^2 + a_2^2 + a_3^2)(b_1^2 + b_2^2 + b_3^2) - (a_1b_1 + a_2b_2 + a_3b_3)^2 \\ &= |\mathbf{a}|^2|\mathbf{b}|^2 - (\mathbf{a} \cdot \mathbf{b})^2 \\ &= |\mathbf{a}|^2|\mathbf{b}|^2 - |\mathbf{a}|^2|\mathbf{b}|^2\cos^2\theta \qquad \text{(by Theorem 12.3.3)} \\ &= |\mathbf{a}|^2|\mathbf{b}|^2(1 - \cos^2\theta) \\ &= |\mathbf{a}|^2|\mathbf{b}|^2\sin^2\theta\end{aligned}$$

Taking square roots and observing that $\sqrt{\sin^2\theta} = \sin\theta$ because $\sin\theta \geqslant 0$ when $0 \leqslant \theta \leqslant \pi$, we have

$$|\mathbf{a} \times \mathbf{b}| = |\mathbf{a}||\mathbf{b}| \sin \theta$$ ■

Geometric characterization of $\mathbf{a} \times \mathbf{b}$

Since a vector is completely determined by its magnitude and direction, we can now say that $\mathbf{a} \times \mathbf{b}$ is the vector that is perpendicular to both $\mathbf{a}$ and $\mathbf{b}$, whose orientation is determined by the right-hand rule, and whose length is $|\mathbf{a}||\mathbf{b}|\sin\theta$. In fact, that is exactly how physicists *define* $\mathbf{a} \times \mathbf{b}$.

10 Corollary Two nonzero vectors $\mathbf{a}$ and $\mathbf{b}$ are parallel if and only if

$$\mathbf{a} \times \mathbf{b} = \mathbf{0}$$

PROOF Two nonzero vectors $\mathbf{a}$ and $\mathbf{b}$ are parallel if and only if $\theta = 0$ or π. In either case $\sin\theta = 0$, so $|\mathbf{a} \times \mathbf{b}| = 0$ and therefore $\mathbf{a} \times \mathbf{b} = \mathbf{0}$.

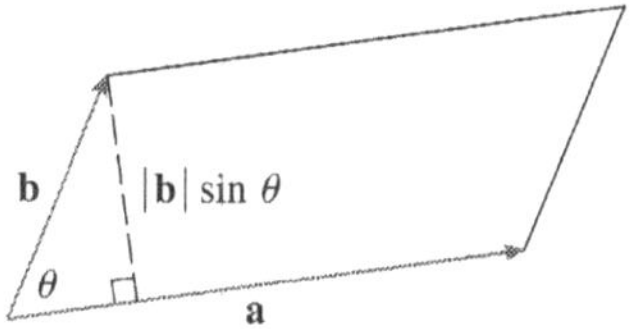

FIGURE 2

The geometric interpretation of Theorem 9 can be seen by looking at Figure 2. If $\mathbf{a}$ and $\mathbf{b}$ are represented by directed line segments with the same initial point, then they determine a parallelogram with base $|\mathbf{a}|$, altitude $|\mathbf{b}|\sin\theta$, and area

$$A = |\mathbf{a}|(|\mathbf{b}|\sin\theta) = |\mathbf{a} \times \mathbf{b}|$$

Thus we have the following way of interpreting the magnitude of a cross product.

The length of the cross product $\mathbf{a} \times \mathbf{b}$ is equal to the area of the parallelogram determined by $\mathbf{a}$ and $\mathbf{b}$.

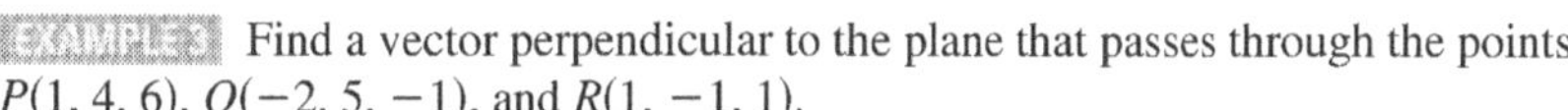

EXAMPLE 3 Find a vector perpendicular to the plane that passes through the points $P(1, 4, 6)$, $Q(-2, 5, -1)$, and $R(1, -1, 1)$.

SOLUTION The vector $\overrightarrow{PQ} \times \overrightarrow{PR}$ is perpendicular to both $\overrightarrow{PQ}$ and $\overrightarrow{PR}$ and is therefore perpendicular to the plane through P, Q, and R. We know from (12.2.1) that

$$\overrightarrow{PQ} = (-2 - 1)\,\mathbf{i} + (5 - 4)\,\mathbf{j} + (-1 - 6)\,\mathbf{k} = -3\mathbf{i} + \mathbf{j} - 7\mathbf{k}$$

$$\overrightarrow{PR} = (1 - 1)\,\mathbf{i} + (-1 - 4)\,\mathbf{j} + (1 - 6)\,\mathbf{k} = -5\mathbf{j} - 5\mathbf{k}$$

We compute the cross product of these vectors:

$$\overrightarrow{PQ} \times \overrightarrow{PR} = \begin{vmatrix} \mathbf{i} & \mathbf{j} & \mathbf{k} \\ -3 & 1 & -7 \\ 0 & -5 & -5 \end{vmatrix}$$

$$= (-5 - 35)\,\mathbf{i} - (15 - 0)\,\mathbf{j} + (15 - 0)\,\mathbf{k} = -40\mathbf{i} - 15\mathbf{j} + 15\mathbf{k}$$

So the vector $\langle -40, -15, 15 \rangle$ is perpendicular to the given plane. Any nonzero scalar multiple of this vector, such as $\langle -8, -3, 3 \rangle$, is also perpendicular to the plane.

EXAMPLE 4 Find the area of the triangle with vertices $P(1, 4, 6)$, $Q(-2, 5, -1)$, and $R(1, -1, 1)$.

SOLUTION In Example 3 we computed that $\overrightarrow{PQ} \times \overrightarrow{PR} = \langle -40, -15, 15 \rangle$. The area of the parallelogram with adjacent sides PQ and PR is the length of this cross product:

$$|\overrightarrow{PQ} \times \overrightarrow{PR}| = \sqrt{(-40)^2 + (-15)^2 + 15^2} = 5\sqrt{82}$$

The area A of the triangle PQR is half the area of this parallelogram, that is, $\frac{5}{2}\sqrt{82}$.

If we apply Theorems 8 and 9 to the standard basis vectors **i**, **j**, and **k** using $\theta = \pi/2$, we obtain

$$\mathbf{i} \times \mathbf{j} = \mathbf{k} \qquad \mathbf{j} \times \mathbf{k} = \mathbf{i} \qquad \mathbf{k} \times \mathbf{i} = \mathbf{j}$$

$$\mathbf{j} \times \mathbf{i} = -\mathbf{k} \qquad \mathbf{k} \times \mathbf{j} = -\mathbf{i} \qquad \mathbf{i} \times \mathbf{k} = -\mathbf{j}$$

Observe that

$$\mathbf{i} \times \mathbf{j} \neq \mathbf{j} \times \mathbf{i}$$

Thus the cross product is not commutative. Also

$$\mathbf{i} \times (\mathbf{i} \times \mathbf{j}) = \mathbf{i} \times \mathbf{k} = -\mathbf{j}$$

whereas

$$(\mathbf{i} \times \mathbf{i}) \times \mathbf{j} = \mathbf{0} \times \mathbf{j} = \mathbf{0}$$

So the associative law for multiplication does not usually hold; that is, in general,

$$(\mathbf{a} \times \mathbf{b}) \times \mathbf{c} \neq \mathbf{a} \times (\mathbf{b} \times \mathbf{c})$$

However, some of the usual laws of algebra *do* hold for cross products. The following theorem summarizes the properties of vector products.

11 Theorem If **a**, **b**, and **c** are vectors and c is a scalar, then

1. $\mathbf{a} \times \mathbf{b} = -\mathbf{b} \times \mathbf{a}$
2. $(c\mathbf{a}) \times \mathbf{b} = c(\mathbf{a} \times \mathbf{b}) = \mathbf{a} \times (c\mathbf{b})$
3. $\mathbf{a} \times (\mathbf{b} + \mathbf{c}) = \mathbf{a} \times \mathbf{b} + \mathbf{a} \times \mathbf{c}$
4. $(\mathbf{a} + \mathbf{b}) \times \mathbf{c} = \mathbf{a} \times \mathbf{c} + \mathbf{b} \times \mathbf{c}$
5. $\mathbf{a} \cdot (\mathbf{b} \times \mathbf{c}) = (\mathbf{a} \times \mathbf{b}) \cdot \mathbf{c}$
6. $\mathbf{a} \times (\mathbf{b} \times \mathbf{c}) = (\mathbf{a} \cdot \mathbf{c})\mathbf{b} - (\mathbf{a} \cdot \mathbf{b})\mathbf{c}$

These properties can be proved by writing the vectors in terms of their components and using the definition of a cross product. We give the proof of Property 5 and leave the remaining proofs as exercises.

PROOF OF PROPERTY 5 If $\mathbf{a} = \langle a_1, a_2, a_3 \rangle$, $\mathbf{b} = \langle b_1, b_2, b_3 \rangle$, and $\mathbf{c} = \langle c_1, c_2, c_3 \rangle$, then

$$\boxed{12} \quad \begin{aligned} \mathbf{a} \cdot (\mathbf{b} \times \mathbf{c}) &= a_1(b_2c_3 - b_3c_2) + a_2(b_3c_1 - b_1c_3) + a_3(b_1c_2 - b_2c_1) \\ &= a_1b_2c_3 - a_1b_3c_2 + a_2b_3c_1 - a_2b_1c_3 + a_3b_1c_2 - a_3b_2c_1 \\ &= (a_2b_3 - a_3b_2)c_1 + (a_3b_1 - a_1b_3)c_2 + (a_1b_2 - a_2b_1)c_3 \\ &= (\mathbf{a} \times \mathbf{b}) \cdot \mathbf{c} \end{aligned}$$

Triple Products

The product $\mathbf{a} \cdot (\mathbf{b} \times \mathbf{c})$ that occurs in Property 5 is called the **scalar triple product** of the vectors **a**, **b**, and **c**. Notice from Equation 12 that we can write the scalar triple product as a determinant:

$$\boxed{13} \qquad \mathbf{a} \cdot (\mathbf{b} \times \mathbf{c}) = \begin{vmatrix} a_1 & a_2 & a_3 \\ b_1 & b_2 & b_3 \\ c_1 & c_2 & c_3 \end{vmatrix}$$

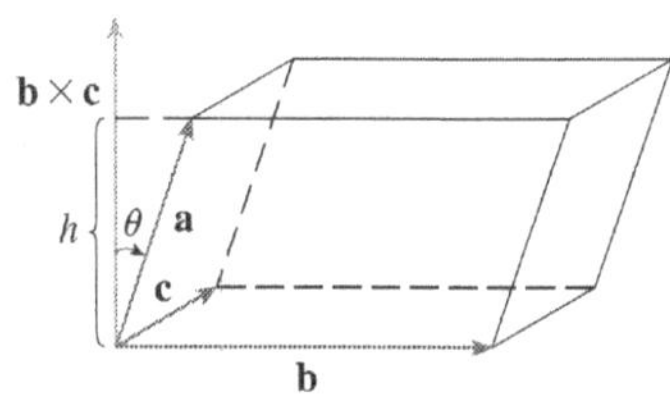

FIGURE 3

The geometric significance of the scalar triple product can be seen by considering the parallelepiped determined by the vectors **a**, **b**, and **c**. (See Figure 3.) The area of the base parallelogram is $A = |\mathbf{b} \times \mathbf{c}|$. If θ is the angle between **a** and $\mathbf{b} \times \mathbf{c}$, then the height h of the parallelepiped is $h = |\mathbf{a}||\cos\theta|$. (We must use $|\cos\theta|$ instead of $\cos\theta$ in case $\theta > \pi/2$.) Therefore the volume of the parallelepiped is

$$V = Ah = |\mathbf{b} \times \mathbf{c}||\mathbf{a}||\cos\theta| = |\mathbf{a} \cdot (\mathbf{b} \times \mathbf{c})|$$

Thus we have proved the following formula.

> **14** The volume of the parallelepiped determined by the vectors **a**, **b**, and **c** is the magnitude of their scalar triple product:
>
> $$V = |\mathbf{a} \cdot (\mathbf{b} \times \mathbf{c})|$$

If we use the formula in **14** and discover that the volume of the parallelepiped determined by **a**, **b**, and **c** is 0, then the vectors must lie in the same plane; that is, they are **coplanar**.

V EXAMPLE 5 Use the scalar triple product to show that the vectors $\mathbf{a} = \langle 1, 4, -7 \rangle$, $\mathbf{b} = \langle 2, -1, 4 \rangle$, and $\mathbf{c} = \langle 0, -9, 18 \rangle$ are coplanar.

SOLUTION We use Equation 13 to compute their scalar triple product:

$$\mathbf{a} \cdot (\mathbf{b} \times \mathbf{c}) = \begin{vmatrix} 1 & 4 & -7 \\ 2 & -1 & 4 \\ 0 & -9 & 18 \end{vmatrix}$$

$$= 1\begin{vmatrix} -1 & 4 \\ -9 & 18 \end{vmatrix} - 4\begin{vmatrix} 2 & 4 \\ 0 & 18 \end{vmatrix} - 7\begin{vmatrix} 2 & -1 \\ 0 & -9 \end{vmatrix}$$

$$= 1(18) - 4(36) - 7(-18) = 0$$

Therefore, by **14**, the volume of the parallelepiped determined by **a**, **b**, and **c** is 0. This means that **a**, **b**, and **c** are coplanar.

The product $\mathbf{a} \times (\mathbf{b} \times \mathbf{c})$ that occurs in Property 6 is called the **vector triple product** of **a**, **b**, and **c**. Property 6 will be used to derive Kepler's First Law of planetary motion in Chapter 13. Its proof is left as Exercise 50.

Torque

The idea of a cross product occurs often in physics. In particular, we consider a force **F** acting on a rigid body at a point given by a position vector **r**. (For instance, if we tighten a bolt by applying a force to a wrench as in Figure 4, we produce a turning effect.) The **torque** $\boldsymbol{\tau}$ (relative to the origin) is defined to be the cross product of the position and force vectors

$$\boldsymbol{\tau} = \mathbf{r} \times \mathbf{F}$$

and measures the tendency of the body to rotate about the origin. The direction of the torque vector indicates the axis of rotation. According to Theorem 9, the magnitude of the torque vector is

$$|\boldsymbol{\tau}| = |\mathbf{r} \times \mathbf{F}| = |\mathbf{r}||\mathbf{F}|\sin\theta$$

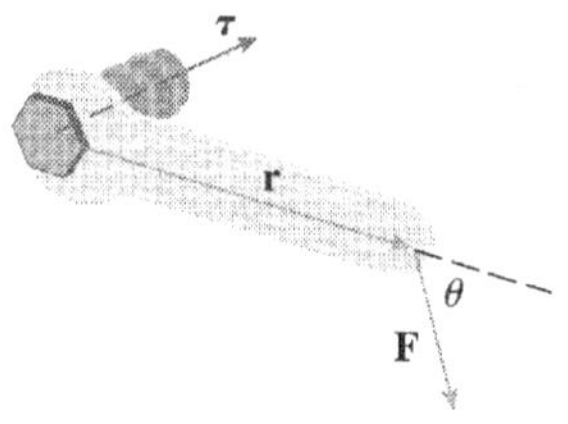

FIGURE 4

where θ is the angle between the position and force vectors. Observe that the only component of $\mathbf{F}$ that can cause a rotation is the one perpendicular to $\mathbf{r}$, that is, $|\mathbf{F}|\sin\theta$. The magnitude of the torque is equal to the area of the parallelogram determined by $\mathbf{r}$ and $\mathbf{F}$.

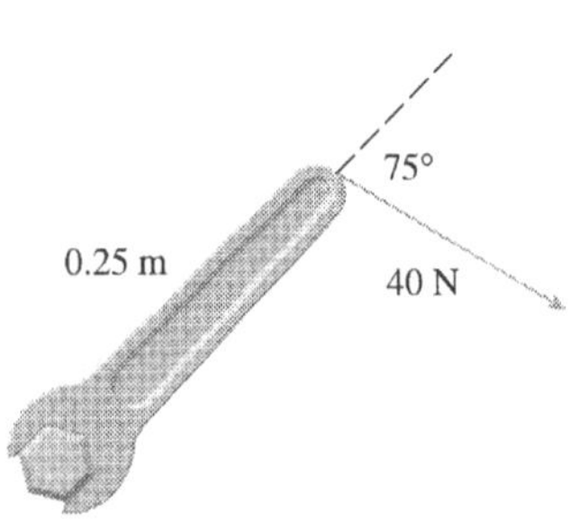

FIGURE 5

EXAMPLE 6 A bolt is tightened by applying a 40-N force to a 0.25-m wrench as shown in Figure 5. Find the magnitude of the torque about the center of the bolt.

SOLUTION The magnitude of the torque vector is

$$|\boldsymbol{\tau}| = |\mathbf{r}\times\mathbf{F}| = |\mathbf{r}||\mathbf{F}|\sin 75^\circ = (0.25)(40)\sin 75^\circ$$
$$= 10\sin 75^\circ \approx 9.66\ \text{N}\cdot\text{m}$$

If the bolt is right-threaded, then the torque vector itself is

$$\boldsymbol{\tau} = |\boldsymbol{\tau}|\,\mathbf{n} \approx 9.66\,\mathbf{n}$$

where $\mathbf{n}$ is a unit vector directed down into the page.

12.4 Exercises

1–7 Find the cross product $\mathbf{a}\times\mathbf{b}$ and verify that it is orthogonal to both $\mathbf{a}$ and $\mathbf{b}$.

1. $\mathbf{a} = \langle 6, 0, -2\rangle,\quad \mathbf{b} = \langle 0, 8, 0\rangle$
2. $\mathbf{a} = \langle 1, 1, -1\rangle,\quad \mathbf{b} = \langle 2, 4, 6\rangle$
3. $\mathbf{a} = \mathbf{i} + 3\mathbf{j} - 2\mathbf{k},\quad \mathbf{b} = -\mathbf{i} + 5\mathbf{k}$
4. $\mathbf{a} = \mathbf{j} + 7\mathbf{k},\quad \mathbf{b} = 2\mathbf{i} - \mathbf{j} + 4\mathbf{k}$
5. $\mathbf{a} = \mathbf{i} - \mathbf{j} - \mathbf{k},\quad \mathbf{b} = \frac{1}{2}\mathbf{i} + \mathbf{j} + \frac{1}{2}\mathbf{k}$
6. $\mathbf{a} = t\,\mathbf{i} + \cos t\,\mathbf{j} + \sin t\,\mathbf{k},\quad \mathbf{b} = \mathbf{i} - \sin t\,\mathbf{j} + \cos t\,\mathbf{k}$
7. $\mathbf{a} = \langle t, 1, 1/t\rangle,\quad \mathbf{b} = \langle t^2, t^2, 1\rangle$

8. If $\mathbf{a} = \mathbf{i} - 2\mathbf{k}$ and $\mathbf{b} = \mathbf{j} + \mathbf{k}$, find $\mathbf{a}\times\mathbf{b}$. Sketch $\mathbf{a}$, $\mathbf{b}$, and $\mathbf{a}\times\mathbf{b}$ as vectors starting at the origin.

9–12 Find the vector, not with determinants, but by using properties of cross products.

9. $(\mathbf{i}\times\mathbf{j})\times\mathbf{k}$
10. $\mathbf{k}\times(\mathbf{i} - 2\mathbf{j})$
11. $(\mathbf{j} - \mathbf{k})\times(\mathbf{k} - \mathbf{i})$
12. $(\mathbf{i} + \mathbf{j})\times(\mathbf{i} - \mathbf{j})$

13. State whether each expression is meaningful. If not, explain why. If so, state whether it is a vector or a scalar.
(a) $\mathbf{a}\cdot(\mathbf{b}\times\mathbf{c})$ (b) $\mathbf{a}\times(\mathbf{b}\cdot\mathbf{c})$
(c) $\mathbf{a}\times(\mathbf{b}\times\mathbf{c})$ (d) $\mathbf{a}\cdot(\mathbf{b}\cdot\mathbf{c})$
(e) $(\mathbf{a}\cdot\mathbf{b})\times(\mathbf{c}\cdot\mathbf{d})$ (f) $(\mathbf{a}\times\mathbf{b})\cdot(\mathbf{c}\times\mathbf{d})$

14–15 Find $|\mathbf{u}\times\mathbf{v}|$ and determine whether $\mathbf{u}\times\mathbf{v}$ is directed into the page or out of the page.

14.
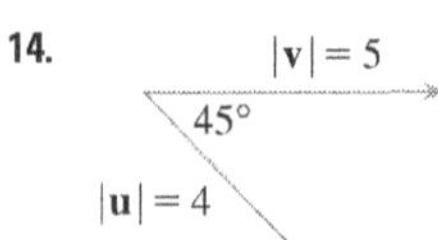

15.
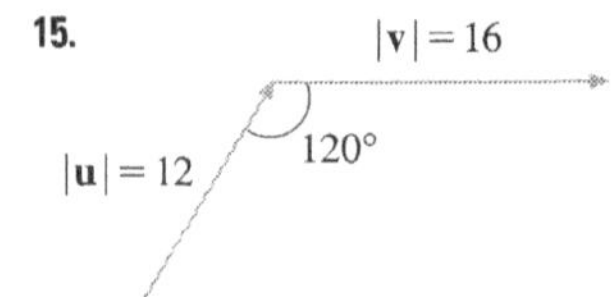

16. The figure shows a vector $\mathbf{a}$ in the xy-plane and a vector $\mathbf{b}$ in the direction of $\mathbf{k}$. Their lengths are $|\mathbf{a}| = 3$ and $|\mathbf{b}| = 2$.
(a) Find $|\mathbf{a}\times\mathbf{b}|$.
(b) Use the right-hand rule to decide whether the components of $\mathbf{a}\times\mathbf{b}$ are positive, negative, or 0.

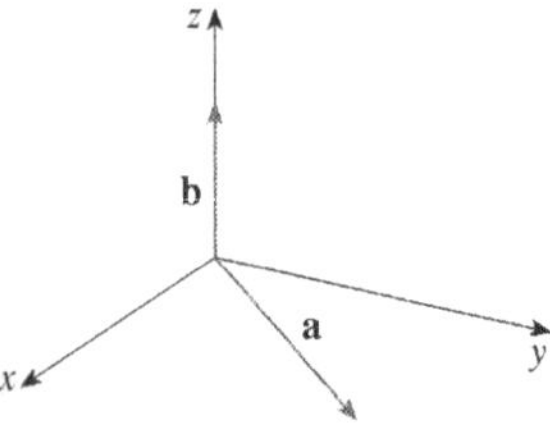

17. If $\mathbf{a} = \langle 2, -1, 3\rangle$ and $\mathbf{b} = \langle 4, 2, 1\rangle$, find $\mathbf{a}\times\mathbf{b}$ and $\mathbf{b}\times\mathbf{a}$.

18. If $\mathbf{a} = \langle 1, 0, 1\rangle$, $\mathbf{b} = \langle 2, 1, -1\rangle$, and $\mathbf{c} = \langle 0, 1, 3\rangle$, show that $\mathbf{a}\times(\mathbf{b}\times\mathbf{c}) \neq (\mathbf{a}\times\mathbf{b})\times\mathbf{c}$.

19. Find two unit vectors orthogonal to both $\langle 3, 2, 1\rangle$ and $\langle -1, 1, 0\rangle$.

1. Homework Hints available at stewartcalculus.com

20. Find two unit vectors orthogonal to both $\mathbf{j} - \mathbf{k}$ and $\mathbf{i} + \mathbf{j}$.

21. Show that $\mathbf{0} \times \mathbf{a} = \mathbf{0} = \mathbf{a} \times \mathbf{0}$ for any vector $\mathbf{a}$ in V_3.

22. Show that $(\mathbf{a} \times \mathbf{b}) \cdot \mathbf{b} = 0$ for all vectors $\mathbf{a}$ and $\mathbf{b}$ in V_3.

23. Prove Property 1 of Theorem 11.

24. Prove Property 2 of Theorem 11.

25. Prove Property 3 of Theorem 11.

26. Prove Property 4 of Theorem 11.

27. Find the area of the parallelogram with vertices $A(-2, 1)$, $B(0, 4)$, $C(4, 2)$, and $D(2, -1)$.

28. Find the area of the parallelogram with vertices $K(1, 2, 3)$, $L(1, 3, 6)$, $M(3, 8, 6)$, and $N(3, 7, 3)$.

29–32 (a) Find a nonzero vector orthogonal to the plane through the points P, Q, and R, and (b) find the area of triangle PQR.

29. $P(1, 0, 1)$, $Q(-2, 1, 3)$, $R(4, 2, 5)$

30. $P(0, 0, -3)$, $Q(4, 2, 0)$, $R(3, 3, 1)$

31. $P(0, -2, 0)$, $Q(4, 1, -2)$, $R(5, 3, 1)$

32. $P(-1, 3, 1)$, $Q(0, 5, 2)$, $R(4, 3, -1)$

33–34 Find the volume of the parallelepiped determined by the vectors $\mathbf{a}$, $\mathbf{b}$, and $\mathbf{c}$.

33. $\mathbf{a} = \langle 1, 2, 3 \rangle$, $\mathbf{b} = \langle -1, 1, 2 \rangle$, $\mathbf{c} = \langle 2, 1, 4 \rangle$

34. $\mathbf{a} = \mathbf{i} + \mathbf{j}$, $\mathbf{b} = \mathbf{j} + \mathbf{k}$, $\mathbf{c} = \mathbf{i} + \mathbf{j} + \mathbf{k}$

35–36 Find the volume of the parallelepiped with adjacent edges PQ, PR, and PS.

35. $P(-2, 1, 0)$, $Q(2, 3, 2)$, $R(1, 4, -1)$, $S(3, 6, 1)$

36. $P(3, 0, 1)$, $Q(-1, 2, 5)$, $R(5, 1, -1)$, $S(0, 4, 2)$

37. Use the scalar triple product to verify that the vectors $\mathbf{u} = \mathbf{i} + 5\mathbf{j} - 2\mathbf{k}$, $\mathbf{v} = 3\mathbf{i} - \mathbf{j}$, and $\mathbf{w} = 5\mathbf{i} + 9\mathbf{j} - 4\mathbf{k}$ are coplanar.

38. Use the scalar triple product to determine whether the points $A(1, 3, 2)$, $B(3, -1, 6)$, $C(5, 2, 0)$, and $D(3, 6, -4)$ lie in the same plane.

39. A bicycle pedal is pushed by a foot with a 60-N force as shown. The shaft of the pedal is 18 cm long. Find the magnitude of the torque about P.

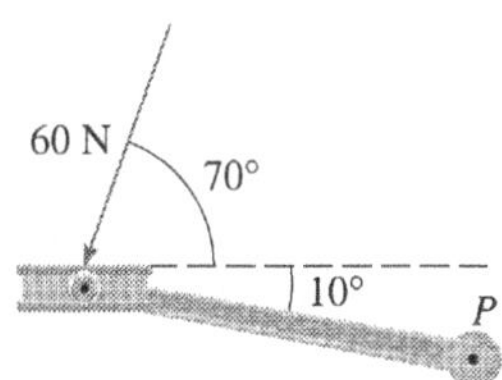

40. Find the magnitude of the torque about P if a 36-lb force is applied as shown.

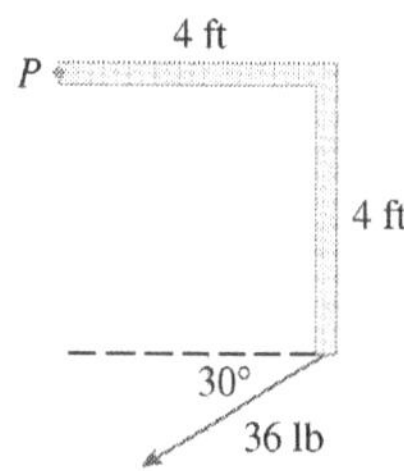

41. A wrench 30 cm long lies along the positive y-axis and grips a bolt at the origin. A force is applied in the direction $\langle 0, 3, -4 \rangle$ at the end of the wrench. Find the magnitude of the force needed to supply 100 N·m of torque to the bolt.

42. Let $\mathbf{v} = 5\mathbf{j}$ and let $\mathbf{u}$ be a vector with length 3 that starts at the origin and rotates in the xy-plane. Find the maximum and minimum values of the length of the vector $\mathbf{u} \times \mathbf{v}$. In what direction does $\mathbf{u} \times \mathbf{v}$ point?

43. If $\mathbf{a} \cdot \mathbf{b} = \sqrt{3}$ and $\mathbf{a} \times \mathbf{b} = \langle 1, 2, 2 \rangle$, find the angle between $\mathbf{a}$ and $\mathbf{b}$.

44. (a) Find all vectors $\mathbf{v}$ such that

$$\langle 1, 2, 1 \rangle \times \mathbf{v} = \langle 3, 1, -5 \rangle$$

(b) Explain why there is no vector $\mathbf{v}$ such that

$$\langle 1, 2, 1 \rangle \times \mathbf{v} = \langle 3, 1, 5 \rangle$$

45. (a) Let P be a point not on the line L that passes through the points Q and R. Show that the distance d from the point P to the line L is

$$d = \frac{|\mathbf{a} \times \mathbf{b}|}{|\mathbf{a}|}$$

where $\mathbf{a} = \overrightarrow{QR}$ and $\mathbf{b} = \overrightarrow{QP}$.

(b) Use the formula in part (a) to find the distance from the point $P(1, 1, 1)$ to the line through $Q(0, 6, 8)$ and $R(-1, 4, 7)$.

46. (a) Let P be a point not on the plane that passes through the points Q, R, and S. Show that the distance d from P to the plane is

$$d = \frac{|\mathbf{a} \cdot (\mathbf{b} \times \mathbf{c})|}{|\mathbf{a} \times \mathbf{b}|}$$

where $\mathbf{a} = \overrightarrow{QR}$, $\mathbf{b} = \overrightarrow{QS}$, and $\mathbf{c} = \overrightarrow{QP}$.

(b) Use the formula in part (a) to find the distance from the point $P(2, 1, 4)$ to the plane through the points $Q(1, 0, 0)$, $R(0, 2, 0)$, and $S(0, 0, 3)$.

47. Show that $|\mathbf{a} \times \mathbf{b}|^2 = |\mathbf{a}|^2 |\mathbf{b}|^2 - (\mathbf{a} \cdot \mathbf{b})^2$.

48. If $\mathbf{a} + \mathbf{b} + \mathbf{c} = \mathbf{0}$, show that

$$\mathbf{a} \times \mathbf{b} = \mathbf{b} \times \mathbf{c} = \mathbf{c} \times \mathbf{a}$$

49. Prove that $(\mathbf{a} - \mathbf{b}) \times (\mathbf{a} + \mathbf{b}) = 2(\mathbf{a} \times \mathbf{b})$.

50. Prove Property 6 of Theorem 11, that is,

$$\mathbf{a} \times (\mathbf{b} \times \mathbf{c}) = (\mathbf{a} \cdot \mathbf{c})\mathbf{b} - (\mathbf{a} \cdot \mathbf{b})\mathbf{c}$$

51. Use Exercise 50 to prove that

$$\mathbf{a} \times (\mathbf{b} \times \mathbf{c}) + \mathbf{b} \times (\mathbf{c} \times \mathbf{a}) + \mathbf{c} \times (\mathbf{a} \times \mathbf{b}) = \mathbf{0}$$

52. Prove that

$$(\mathbf{a} \times \mathbf{b}) \cdot (\mathbf{c} \times \mathbf{d}) = \begin{vmatrix} \mathbf{a} \cdot \mathbf{c} & \mathbf{b} \cdot \mathbf{c} \\ \mathbf{a} \cdot \mathbf{d} & \mathbf{b} \cdot \mathbf{d} \end{vmatrix}$$

53. Suppose that $\mathbf{a} \neq \mathbf{0}$.
(a) If $\mathbf{a} \cdot \mathbf{b} = \mathbf{a} \cdot \mathbf{c}$, does it follow that $\mathbf{b} = \mathbf{c}$?
(b) If $\mathbf{a} \times \mathbf{b} = \mathbf{a} \times \mathbf{c}$, does it follow that $\mathbf{b} = \mathbf{c}$?
(c) If $\mathbf{a} \cdot \mathbf{b} = \mathbf{a} \cdot \mathbf{c}$ and $\mathbf{a} \times \mathbf{b} = \mathbf{a} \times \mathbf{c}$, does it follow that $\mathbf{b} = \mathbf{c}$?

54. If $\mathbf{v}_1$, $\mathbf{v}_2$, and $\mathbf{v}_3$ are noncoplanar vectors, let

$$\mathbf{k}_1 = \frac{\mathbf{v}_2 \times \mathbf{v}_3}{\mathbf{v}_1 \cdot (\mathbf{v}_2 \times \mathbf{v}_3)} \qquad \mathbf{k}_2 = \frac{\mathbf{v}_3 \times \mathbf{v}_1}{\mathbf{v}_1 \cdot (\mathbf{v}_2 \times \mathbf{v}_3)}$$

$$\mathbf{k}_3 = \frac{\mathbf{v}_1 \times \mathbf{v}_2}{\mathbf{v}_1 \cdot (\mathbf{v}_2 \times \mathbf{v}_3)}$$

(These vectors occur in the study of crystallography. Vectors of the form $n_1\mathbf{v}_1 + n_2\mathbf{v}_2 + n_3\mathbf{v}_3$, where each n_i is an integer, form a *lattice* for a crystal. Vectors written similarly in terms of $\mathbf{k}_1$, $\mathbf{k}_2$, and $\mathbf{k}_3$ form the *reciprocal lattice*.)
(a) Show that $\mathbf{k}_i$ is perpendicular to $\mathbf{v}_j$ if $i \neq j$.
(b) Show that $\mathbf{k}_i \cdot \mathbf{v}_i = 1$ for $i = 1, 2, 3$.
(c) Show that $\mathbf{k}_1 \cdot (\mathbf{k}_2 \times \mathbf{k}_3) = \dfrac{1}{\mathbf{v}_1 \cdot (\mathbf{v}_2 \times \mathbf{v}_3)}$.

DISCOVERY PROJECT THE GEOMETRY OF A TETRAHEDRON

A tetrahedron is a solid with four vertices, P, Q, R, and S, and four triangular faces, as shown in the figure.

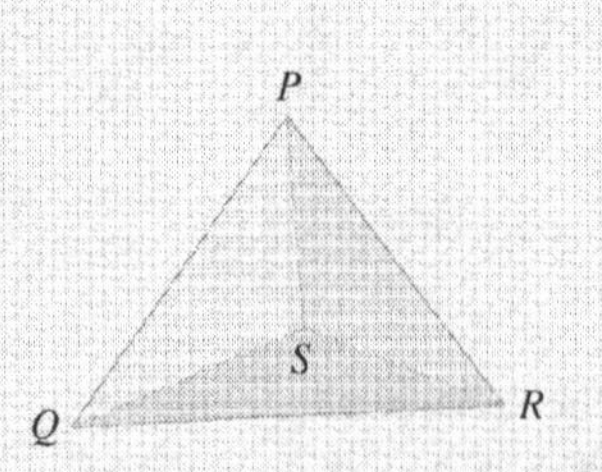

1. Let $\mathbf{v}_1$, $\mathbf{v}_2$, $\mathbf{v}_3$, and $\mathbf{v}_4$ be vectors with lengths equal to the areas of the faces opposite the vertices P, Q, R, and S, respectively, and directions perpendicular to the respective faces and pointing outward. Show that

$$\mathbf{v}_1 + \mathbf{v}_2 + \mathbf{v}_3 + \mathbf{v}_4 = \mathbf{0}$$

2. The volume V of a tetrahedron is one-third the distance from a vertex to the opposite face, times the area of that face.
(a) Find a formula for the volume of a tetrahedron in terms of the coordinates of its vertices P, Q, R, and S.
(b) Find the volume of the tetrahedron whose vertices are $P(1, 1, 1)$, $Q(1, 2, 3)$, $R(1, 1, 2)$, and $S(3, -1, 2)$.

3. Suppose the tetrahedron in the figure has a trirectangular vertex S. (This means that the three angles at S are all right angles.) Let A, B, and C be the areas of the three faces that meet at S, and let D be the area of the opposite face PQR. Using the result of Problem 1, or otherwise, show that

$$D^2 = A^2 + B^2 + C^2$$

(This is a three-dimensional version of the Pythagorean Theorem.)

12.5 Equations of Lines and Planes

A line in the xy-plane is determined when a point on the line and the direction of the line (its slope or angle of inclination) are given. The equation of the line can then be written using the point-slope form.

Likewise, a line L in three-dimensional space is determined when we know a point $P_0(x_0, y_0, z_0)$ on L and the direction of L. In three dimensions the direction of a line is conveniently described by a vector, so we let $\mathbf{v}$ be a vector parallel to L. Let $P(x, y, z)$ be an arbitrary point on L and let $\mathbf{r}_0$ and $\mathbf{r}$ be the position vectors of P_0 and P (that is, they have

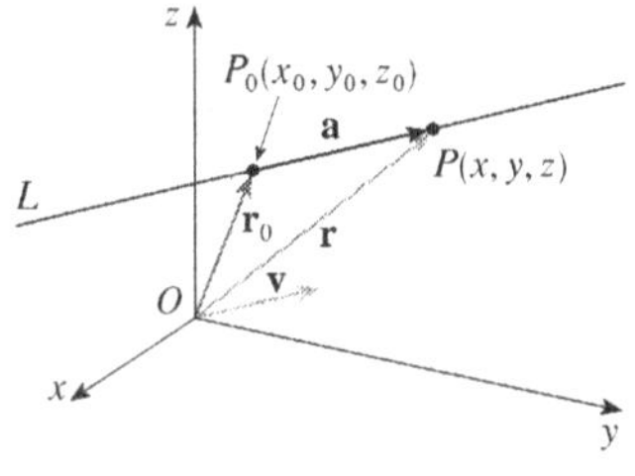

FIGURE 1

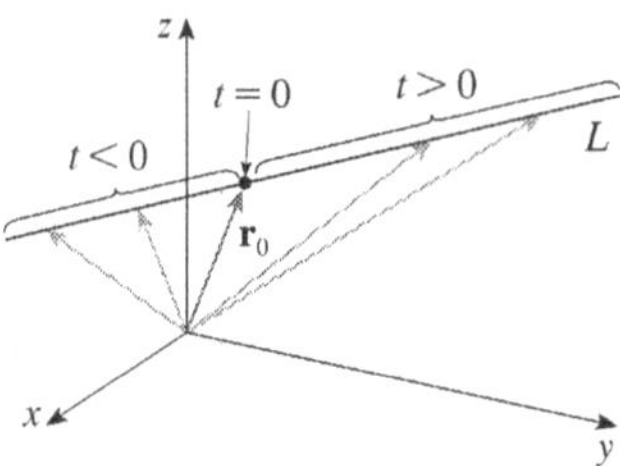

FIGURE 2

representations $\overrightarrow{OP_0}$ and $\overrightarrow{OP}$). If $\mathbf{a}$ is the vector with representation $\overrightarrow{P_0P}$, as in Figure 1, then the Triangle Law for vector addition gives $\mathbf{r} = \mathbf{r}_0 + \mathbf{a}$. But, since $\mathbf{a}$ and $\mathbf{v}$ are parallel vectors, there is a scalar t such that $\mathbf{a} = t\mathbf{v}$. Thus

1 $$\mathbf{r} = \mathbf{r}_0 + t\mathbf{v}$$

which is a **vector equation** of L. Each value of the **parameter** t gives the position vector $\mathbf{r}$ of a point on L. In other words, as t varies, the line is traced out by the tip of the vector $\mathbf{r}$. As Figure 2 indicates, positive values of t correspond to points on L that lie on one side of P_0, whereas negative values of t correspond to points that lie on the other side of P_0.

If the vector $\mathbf{v}$ that gives the direction of the line L is written in component form as $\mathbf{v} = \langle a, b, c \rangle$, then we have $t\mathbf{v} = \langle ta, tb, tc \rangle$. We can also write $\mathbf{r} = \langle x, y, z \rangle$ and $\mathbf{r}_0 = \langle x_0, y_0, z_0 \rangle$, so the vector equation 1 becomes

$$\langle x, y, z \rangle = \langle x_0 + ta, y_0 + tb, z_0 + tc \rangle$$

Two vectors are equal if and only if corresponding components are equal. Therefore we have the three scalar equations:

2 $$x = x_0 + at \qquad y = y_0 + bt \qquad z = z_0 + ct$$

where $t \in \mathbb{R}$. These equations are called **parametric equations** of the line L through the point $P_0(x_0, y_0, z_0)$ and parallel to the vector $\mathbf{v} = \langle a, b, c \rangle$. Each value of the parameter t gives a point (x, y, z) on L.

Figure 3 shows the line L in Example 1 and its relation to the given point and to the vector that gives its direction.

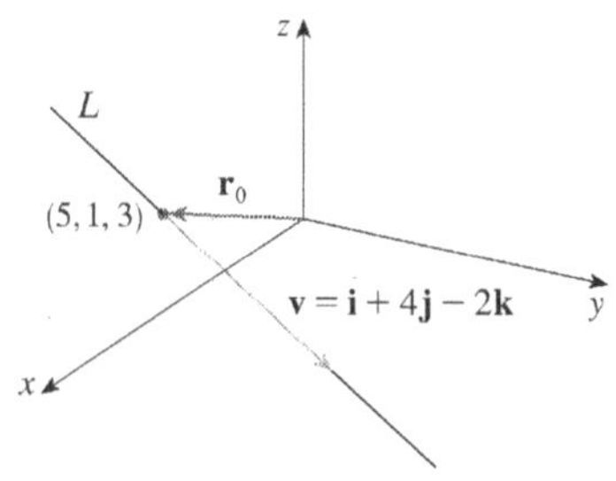

FIGURE 3

EXAMPLE 1

(a) Find a vector equation and parametric equations for the line that passes through the point (5, 1, 3) and is parallel to the vector $\mathbf{i} + 4\mathbf{j} - 2\mathbf{k}$.

(b) Find two other points on the line.

SOLUTION

(a) Here $\mathbf{r}_0 = \langle 5, 1, 3 \rangle = 5\mathbf{i} + \mathbf{j} + 3\mathbf{k}$ and $\mathbf{v} = \mathbf{i} + 4\mathbf{j} - 2\mathbf{k}$, so the vector equation 1 becomes

$$\mathbf{r} = (5\mathbf{i} + \mathbf{j} + 3\mathbf{k}) + t(\mathbf{i} + 4\mathbf{j} - 2\mathbf{k})$$

or

$$\mathbf{r} = (5 + t)\,\mathbf{i} + (1 + 4t)\,\mathbf{j} + (3 - 2t)\,\mathbf{k}$$

Parametric equations are

$$x = 5 + t \qquad y = 1 + 4t \qquad z = 3 - 2t$$

(b) Choosing the parameter value $t = 1$ gives $x = 6$, $y = 5$, and $z = 1$, so (6, 5, 1) is a point on the line. Similarly, $t = -1$ gives the point (4, −3, 5). ■

The vector equation and parametric equations of a line are not unique. If we change the point or the parameter or choose a different parallel vector, then the equations change. For instance, if, instead of (5, 1, 3), we choose the point (6, 5, 1) in Example 1, then the parametric equations of the line become

$$x = 6 + t \qquad y = 5 + 4t \qquad z = 1 - 2t$$

Or, if we stay with the point (5, 1, 3) but choose the parallel vector $2\mathbf{i} + 8\mathbf{j} - 4\mathbf{k}$, we arrive at the equations

$$x = 5 + 2t \qquad y = 1 + 8t \qquad z = 3 - 4t$$

In general, if a vector $\mathbf{v} = \langle a, b, c \rangle$ is used to describe the direction of a line L, then the numbers a, b, and c are called **direction numbers** of L. Since any vector parallel to $\mathbf{v}$ could also be used, we see that any three numbers proportional to a, b, and c could also be used as a set of direction numbers for L.

Another way of describing a line L is to eliminate the parameter t from Equations 2. If none of a, b, or c is 0, we can solve each of these equations for t, equate the results, and obtain

3

$$\frac{x - x_0}{a} = \frac{y - y_0}{b} = \frac{z - z_0}{c}$$

These equations are called **symmetric equations** of L. Notice that the numbers a, b, and c that appear in the denominators of Equations 3 are direction numbers of L, that is, components of a vector parallel to L. If one of a, b, or c is 0, we can still eliminate t. For instance, if $a = 0$, we could write the equations of L as

$$x = x_0 \qquad \frac{y - y_0}{b} = \frac{z - z_0}{c}$$

This means that L lies in the vertical plane $x = x_0$.

Figure 4 shows the line L in Example 2 and the point P where it intersects the xy-plane.

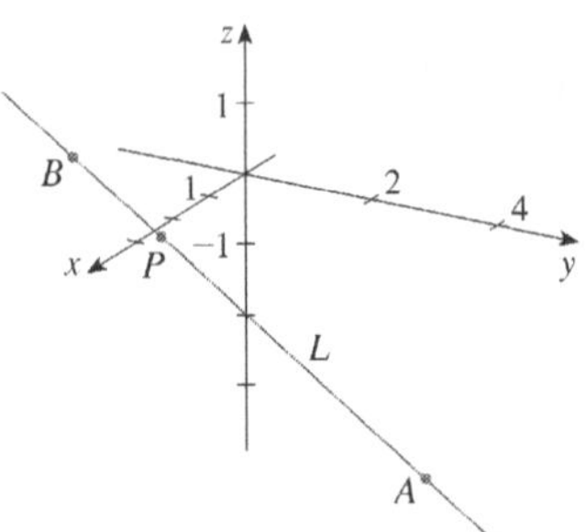

FIGURE 4

EXAMPLE 2

(a) Find parametric equations and symmetric equations of the line that passes through the points $A(2, 4, -3)$ and $B(3, -1, 1)$.

(b) At what point does this line intersect the xy-plane?

SOLUTION

(a) We are not explicitly given a vector parallel to the line, but observe that the vector $\mathbf{v}$ with representation $\overrightarrow{AB}$ is parallel to the line and

$$\mathbf{v} = \langle 3 - 2, -1 - 4, 1 - (-3) \rangle = \langle 1, -5, 4 \rangle$$

Thus direction numbers are $a = 1$, $b = -5$, and $c = 4$. Taking the point $(2, 4, -3)$ as P_0, we see that parametric equations 2 are

$$x = 2 + t \qquad y = 4 - 5t \qquad z = -3 + 4t$$

and symmetric equations 3 are

$$\frac{x - 2}{1} = \frac{y - 4}{-5} = \frac{z + 3}{4}$$

(b) The line intersects the xy-plane when $z = 0$, so we put $z = 0$ in the symmetric equations and obtain

$$\frac{x - 2}{1} = \frac{y - 4}{-5} = \frac{3}{4}$$

This gives $x = \frac{11}{4}$ and $y = \frac{1}{4}$, so the line intersects the xy-plane at the point $\left(\frac{11}{4}, \frac{1}{4}, 0\right)$.

In general, the procedure of Example 2 shows that direction numbers of the line L through the points $P_0(x_0, y_0, z_0)$ and $P_1(x_1, y_1, z_1)$ are $x_1 - x_0$, $y_1 - y_0$, and $z_1 - z_0$ and so symmetric equations of L are

$$\frac{x - x_0}{x_1 - x_0} = \frac{y - y_0}{y_1 - y_0} = \frac{z - z_0}{z_1 - z_0}$$

Often, we need a description, not of an entire line, but of just a line segment. How, for instance, could we describe the line segment AB in Example 2? If we put $t = 0$ in the parametric equations in Example 2(a), we get the point $(2, 4, -3)$ and if we put $t = 1$ we get $(3, -1, 1)$. So the line segment AB is described by the parametric equations

$$x = 2 + t \qquad y = 4 - 5t \qquad z = -3 + 4t \qquad 0 \leqslant t \leqslant 1$$

or by the corresponding vector equation

$$\mathbf{r}(t) = \langle 2 + t, 4 - 5t, -3 + 4t \rangle \qquad 0 \leqslant t \leqslant 1$$

In general, we know from Equation 1 that the vector equation of a line through the (tip of the) vector $\mathbf{r}_0$ in the direction of a vector $\mathbf{v}$ is $\mathbf{r} = \mathbf{r}_0 + t\mathbf{v}$. If the line also passes through (the tip of) $\mathbf{r}_1$, then we can take $\mathbf{v} = \mathbf{r}_1 - \mathbf{r}_0$ and so its vector equation is

$$\mathbf{r} = \mathbf{r}_0 + t(\mathbf{r}_1 - \mathbf{r}_0) = (1 - t)\mathbf{r}_0 + t\mathbf{r}_1$$

The line segment from $\mathbf{r}_0$ to $\mathbf{r}_1$ is given by the parameter interval $0 \leqslant t \leqslant 1$.

> **4** The line segment from $\mathbf{r}_0$ to $\mathbf{r}_1$ is given by the vector equation
>
> $$\mathbf{r}(t) = (1 - t)\mathbf{r}_0 + t\mathbf{r}_1 \qquad 0 \leqslant t \leqslant 1$$

The lines L_1 and L_2 in Example 3, shown in Figure 5, are skew lines.

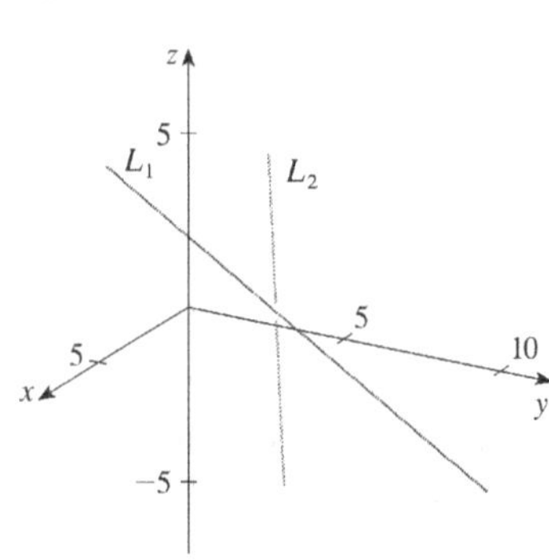

FIGURE 5

V EXAMPLE 3 Show that the lines L_1 and L_2 with parametric equations

$$x = 1 + t \qquad y = -2 + 3t \qquad z = 4 - t$$

$$x = 2s \qquad y = 3 + s \qquad z = -3 + 4s$$

are **skew lines**; that is, they do not intersect and are not parallel (and therefore do not lie in the same plane).

SOLUTION The lines are not parallel because the corresponding vectors $\langle 1, 3, -1 \rangle$ and $\langle 2, 1, 4 \rangle$ are not parallel. (Their components are not proportional.) If L_1 and L_2 had a point of intersection, there would be values of t and s such that

$$1 + \ t = 2s$$

$$-2 + 3t = 3 + s$$

$$4 - \ t = -3 + 4s$$

But if we solve the first two equations, we get $t = \frac{11}{5}$ and $s = \frac{8}{5}$, and these values don't satisfy the third equation. Therefore there are no values of t and s that satisfy the three equations, so L_1 and L_2 do not intersect. Thus L_1 and L_2 are skew lines.

Planes

Although a line in space is determined by a point and a direction, a plane in space is more difficult to describe. A single vector parallel to a plane is not enough to convey the "direction" of the plane, but a vector perpendicular to the plane does completely specify

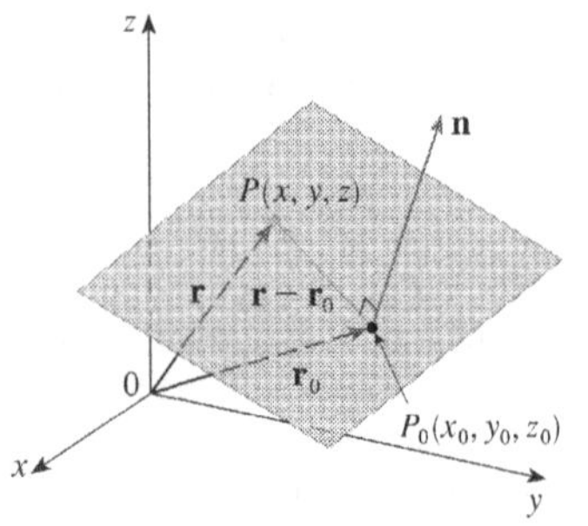

FIGURE 6

its direction. Thus a plane in space is determined by a point $P_0(x_0, y_0, z_0)$ in the plane and a vector $\mathbf{n}$ that is orthogonal to the plane. This orthogonal vector $\mathbf{n}$ is called a **normal vector**. Let $P(x, y, z)$ be an arbitrary point in the plane, and let $\mathbf{r}_0$ and $\mathbf{r}$ be the position vectors of P_0 and P. Then the vector $\mathbf{r} - \mathbf{r}_0$ is represented by $\overrightarrow{P_0P}$. (See Figure 6.) The normal vector $\mathbf{n}$ is orthogonal to every vector in the given plane. In particular, $\mathbf{n}$ is orthogonal to $\mathbf{r} - \mathbf{r}_0$ and so we have

5 $$\mathbf{n} \cdot (\mathbf{r} - \mathbf{r}_0) = 0$$

which can be rewritten as

6 $$\mathbf{n} \cdot \mathbf{r} = \mathbf{n} \cdot \mathbf{r}_0$$

Either Equation 5 or Equation 6 is called a **vector equation of the plane**.

To obtain a scalar equation for the plane, we write $\mathbf{n} = \langle a, b, c\rangle$, $\mathbf{r} = \langle x, y, z\rangle$, and $\mathbf{r}_0 = \langle x_0, y_0, z_0\rangle$. Then the vector equation 5 becomes

$$\langle a, b, c\rangle \cdot \langle x - x_0, y - y_0, z - z_0\rangle = 0$$

or

7 $$a(x - x_0) + b(y - y_0) + c(z - z_0) = 0$$

Equation 7 is the **scalar equation of the plane through $P_0(x_0, y_0, z_0)$ with normal vector $\mathbf{n} = \langle a, b, c\rangle$**.

V EXAMPLE 4 Find an equation of the plane through the point $(2, 4, -1)$ with normal vector $\mathbf{n} = \langle 2, 3, 4\rangle$. Find the intercepts and sketch the plane.

SOLUTION Putting $a = 2$, $b = 3$, $c = 4$, $x_0 = 2$, $y_0 = 4$, and $z_0 = -1$ in Equation 7, we see that an equation of the plane is

$$2(x - 2) + 3(y - 4) + 4(z + 1) = 0$$

or

$$2x + 3y + 4z = 12$$

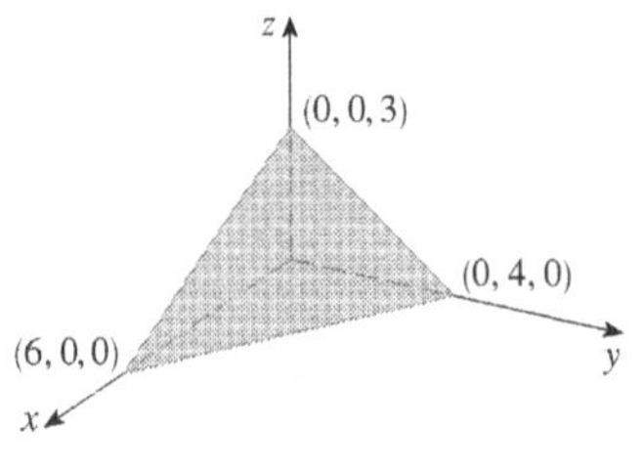

FIGURE 7

To find the x-intercept we set $y = z = 0$ in this equation and obtain $x = 6$. Similarly, the y-intercept is 4 and the z-intercept is 3. This enables us to sketch the portion of the plane that lies in the first octant (see Figure 7).

By collecting terms in Equation 7 as we did in Example 4, we can rewrite the equation of a plane as

8 $$ax + by + cz + d = 0$$

where $d = -(ax_0 + by_0 + cz_0)$. Equation 8 is called a **linear equation** in x, y, and z. Conversely, it can be shown that if a, b, and c are not all 0, then the linear equation 8 represents a plane with normal vector $\langle a, b, c\rangle$. (See Exercise 81.)

Figure 8 shows the portion of the plane in Example 5 that is enclosed by triangle PQR.

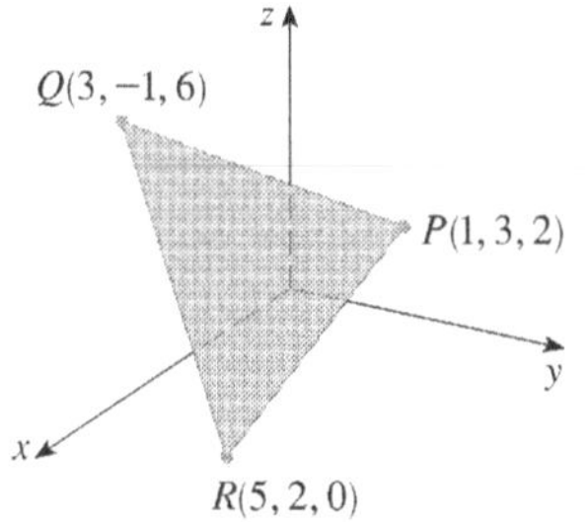

FIGURE 8

EXAMPLE 5 Find an equation of the plane that passes through the points $P(1, 3, 2)$, $Q(3, -1, 6)$, and $R(5, 2, 0)$.

SOLUTION The vectors $\mathbf{a}$ and $\mathbf{b}$ corresponding to $\overrightarrow{PQ}$ and $\overrightarrow{PR}$ are

$$\mathbf{a} = \langle 2, -4, 4 \rangle \qquad \mathbf{b} = \langle 4, -1, -2 \rangle$$

Since both $\mathbf{a}$ and $\mathbf{b}$ lie in the plane, their cross product $\mathbf{a} \times \mathbf{b}$ is orthogonal to the plane and can be taken as the normal vector. Thus

$$\mathbf{n} = \mathbf{a} \times \mathbf{b} = \begin{vmatrix} \mathbf{i} & \mathbf{j} & \mathbf{k} \\ 2 & -4 & 4 \\ 4 & -1 & -2 \end{vmatrix} = 12\,\mathbf{i} + 20\,\mathbf{j} + 14\,\mathbf{k}$$

With the point $P(1, 3, 2)$ and the normal vector $\mathbf{n}$, an equation of the plane is

$$12(x - 1) + 20(y - 3) + 14(z - 2) = 0$$

or

$$6x + 10y + 7z = 50$$

EXAMPLE 6 Find the point at which the line with parametric equations $x = 2 + 3t$, $y = -4t$, $z = 5 + t$ intersects the plane $4x + 5y - 2z = 18$.

SOLUTION We substitute the expressions for x, y, and z from the parametric equations into the equation of the plane:

$$4(2 + 3t) + 5(-4t) - 2(5 + t) = 18$$

This simplifies to $-10t = 20$, so $t = -2$. Therefore the point of intersection occurs when the parameter value is $t = -2$. Then $x = 2 + 3(-2) = -4$, $y = -4(-2) = 8$, $z = 5 - 2 = 3$ and so the point of intersection is $(-4, 8, 3)$.

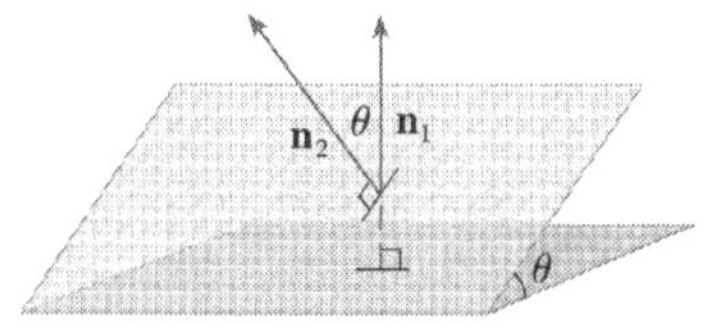

FIGURE 9

Two planes are **parallel** if their normal vectors are parallel. For instance, the planes $x + 2y - 3z = 4$ and $2x + 4y - 6z = 3$ are parallel because their normal vectors are $\mathbf{n}_1 = \langle 1, 2, -3 \rangle$ and $\mathbf{n}_2 = \langle 2, 4, -6 \rangle$ and $\mathbf{n}_2 = 2\mathbf{n}_1$. If two planes are not parallel, then they intersect in a straight line and the angle between the two planes is defined as the acute angle between their normal vectors (see angle θ in Figure 9).

Figure 10 shows the planes in Example 7 and their line of intersection L.

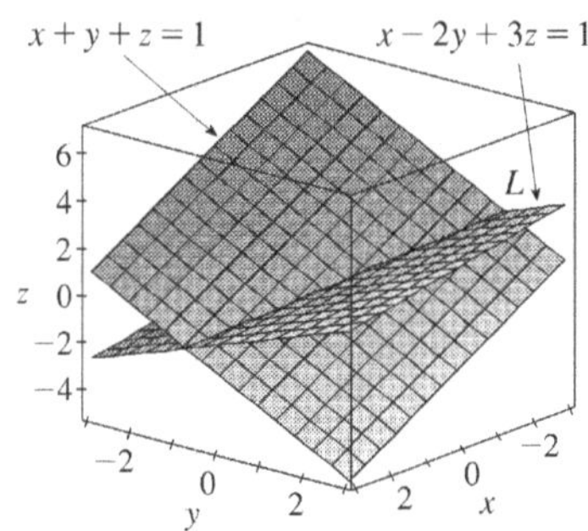

FIGURE 10

V EXAMPLE 7

(a) Find the angle between the planes $x + y + z = 1$ and $x - 2y + 3z = 1$.
(b) Find symmetric equations for the line of intersection L of these two planes.

SOLUTION

(a) The normal vectors of these planes are

$$\mathbf{n}_1 = \langle 1, 1, 1 \rangle \qquad \mathbf{n}_2 = \langle 1, -2, 3 \rangle$$

and so, if θ is the angle between the planes, Corollary 12.3.6 gives

$$\cos\theta = \frac{\mathbf{n}_1 \cdot \mathbf{n}_2}{|\mathbf{n}_1|\,|\mathbf{n}_2|} = \frac{1(1) + 1(-2) + 1(3)}{\sqrt{1 + 1 + 1}\,\sqrt{1 + 4 + 9}} = \frac{2}{\sqrt{42}}$$

$$\theta = \cos^{-1}\left(\frac{2}{\sqrt{42}}\right) \approx 72^\circ$$

(b) We first need to find a point on L. For instance, we can find the point where the line intersects the xy-plane by setting $z = 0$ in the equations of both planes. This gives the equations $x + y = 1$ and $x - 2y = 1$, whose solution is $x = 1$, $y = 0$. So the point $(1, 0, 0)$ lies on L.

Now we observe that, since L lies in both planes, it is perpendicular to both of the normal vectors. Thus a vector $\mathbf{v}$ parallel to L is given by the cross product

$$\mathbf{v} = \mathbf{n}_1 \times \mathbf{n}_2 = \begin{vmatrix} \mathbf{i} & \mathbf{j} & \mathbf{k} \\ 1 & 1 & 1 \\ 1 & -2 & 3 \end{vmatrix} = 5\mathbf{i} - 2\mathbf{j} - 3\mathbf{k}$$

Another way to find the line of intersection is to solve the equations of the planes for two of the variables in terms of the third, which can be taken as the parameter.

and so the symmetric equations of L can be written as

$$\frac{x - 1}{5} = \frac{y}{-2} = \frac{z}{-3}$$

NOTE Since a linear equation in x, y, and z represents a plane and two nonparallel planes intersect in a line, it follows that two linear equations can represent a line. The points (x, y, z) that satisfy both $a_1x + b_1y + c_1z + d_1 = 0$ and $a_2x + b_2y + c_2z + d_2 = 0$ lie on both of these planes, and so the pair of linear equations represents the line of intersection of the planes (if they are not parallel). For instance, in Example 7 the line L was given as the line of intersection of the planes $x + y + z = 1$ and $x - 2y + 3z = 1$. The symmetric equations that we found for L could be written as

$$\frac{x - 1}{5} = \frac{y}{-2} \qquad \text{and} \qquad \frac{y}{-2} = \frac{z}{-3}$$

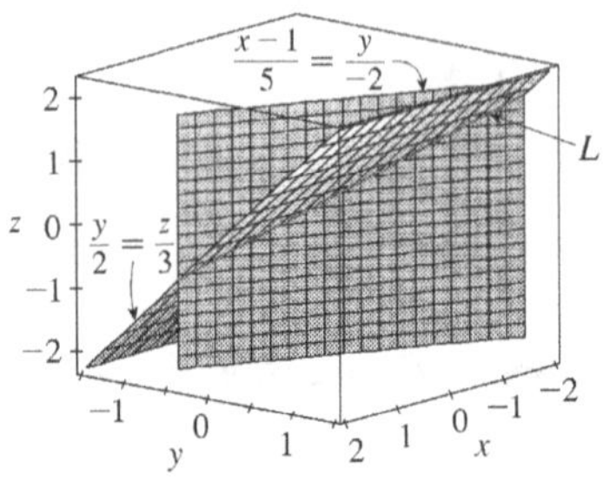

FIGURE 11

Figure 11 shows how the line L in Example 7 can also be regarded as the line of intersection of planes derived from its symmetric equations.

which is again a pair of linear equations. They exhibit L as the line of intersection of the planes $(x - 1)/5 = y/(-2)$ and $y/(-2) = z/(-3)$. (See Figure 11.)

In general, when we write the equations of a line in the symmetric form

$$\frac{x - x_0}{a} = \frac{y - y_0}{b} = \frac{z - z_0}{c}$$

we can regard the line as the line of intersection of the two planes

$$\frac{x - x_0}{a} = \frac{y - y_0}{b} \qquad \text{and} \qquad \frac{y - y_0}{b} = \frac{z - z_0}{c}$$

EXAMPLE 8 Find a formula for the distance D from a point $P_1(x_1, y_1, z_1)$ to the plane $ax + by + cz + d = 0$.

SOLUTION Let $P_0(x_0, y_0, z_0)$ be any point in the given plane and let $\mathbf{b}$ be the vector corresponding to $\overrightarrow{P_0P_1}$. Then

$$\mathbf{b} = \langle x_1 - x_0, y_1 - y_0, z_1 - z_0 \rangle$$

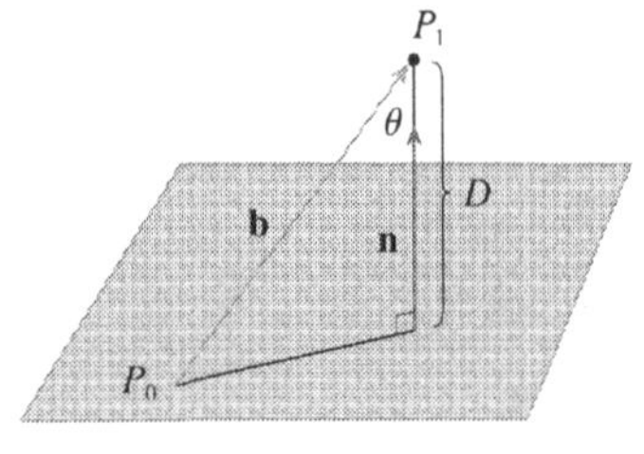

FIGURE 12

From Figure 12 you can see that the distance D from P_1 to the plane is equal to the absolute value of the scalar projection of $\mathbf{b}$ onto the normal vector $\mathbf{n} = \langle a, b, c \rangle$. (See Section 12.3.) Thus

$$\begin{aligned} D = |\text{comp}_{\mathbf{n}}\mathbf{b}| &= \frac{|\mathbf{n} \cdot \mathbf{b}|}{|\mathbf{n}|} \\ &= \frac{|a(x_1 - x_0) + b(y_1 - y_0) + c(z_1 - z_0)|}{\sqrt{a^2 + b^2 + c^2}} \\ &= \frac{|(ax_1 + by_1 + cz_1) - (ax_0 + by_0 + cz_0)|}{\sqrt{a^2 + b^2 + c^2}} \end{aligned}$$

Since P_0 lies in the plane, its coordinates satisfy the equation of the plane and so we have $ax_0 + by_0 + cz_0 + d = 0$. Thus the formula for D can be written as

9
$$D = \frac{|ax_1 + by_1 + cz_1 + d|}{\sqrt{a^2 + b^2 + c^2}}$$

EXAMPLE 9 Find the distance between the parallel planes $10x + 2y - 2z = 5$ and $5x + y - z = 1$.

SOLUTION First we note that the planes are parallel because their normal vectors $\langle 10, 2, -2 \rangle$ and $\langle 5, 1, -1 \rangle$ are parallel. To find the distance D between the planes, we choose any point on one plane and calculate its distance to the other plane. In particular, if we put $y = z = 0$ in the equation of the first plane, we get $10x = 5$ and so $\left(\frac{1}{2}, 0, 0\right)$ is a point in this plane. By Formula 9, the distance between $\left(\frac{1}{2}, 0, 0\right)$ and the plane $5x + y - z - 1 = 0$ is

$$D = \frac{\left|5\left(\frac{1}{2}\right) + 1(0) - 1(0) - 1\right|}{\sqrt{5^2 + 1^2 + (-1)^2}} = \frac{\frac{3}{2}}{3\sqrt{3}} = \frac{\sqrt{3}}{6}$$

So the distance between the planes is $\sqrt{3}/6$.

EXAMPLE 10 In Example 3 we showed that the lines

$$L_1: \quad x = 1 + t \qquad y = -2 + 3t \qquad z = 4 - t$$
$$L_2: \quad x = 2s \qquad y = 3 + s \qquad z = -3 + 4s$$

are skew. Find the distance between them.

SOLUTION Since the two lines L_1 and L_2 are skew, they can be viewed as lying on two parallel planes P_1 and P_2. The distance between L_1 and L_2 is the same as the distance between P_1 and P_2, which can be computed as in Example 9. The common normal vector to both planes must be orthogonal to both $\mathbf{v}_1 = \langle 1, 3, -1 \rangle$ (the direction of L_1) and $\mathbf{v}_2 = \langle 2, 1, 4 \rangle$ (the direction of L_2). So a normal vector is

$$\mathbf{n} = \mathbf{v}_1 \times \mathbf{v}_2 = \begin{vmatrix} \mathbf{i} & \mathbf{j} & \mathbf{k} \\ 1 & 3 & -1 \\ 2 & 1 & 4 \end{vmatrix} = 13\mathbf{i} - 6\mathbf{j} - 5\mathbf{k}$$

If we put $s = 0$ in the equations of L_2, we get the point $(0, 3, -3)$ on L_2 and so an equation for P_2 is

$$13(x - 0) - 6(y - 3) - 5(z + 3) = 0 \qquad \text{or} \qquad 13x - 6y - 5z + 3 = 0$$

If we now set $t = 0$ in the equations for L_1, we get the point $(1, -2, 4)$ on P_1. So the distance between L_1 and L_2 is the same as the distance from $(1, -2, 4)$ to $13x - 6y - 5z + 3 = 0$. By Formula 9, this distance is

$$D = \frac{|13(1) - 6(-2) - 5(4) + 3|}{\sqrt{13^2 + (-6)^2 + (-5)^2}} = \frac{8}{\sqrt{230}} \approx 0.53$$

12.5 Exercises

1. Determine whether each statement is true or false.
   (a) Two lines parallel to a third line are parallel.
   (b) Two lines perpendicular to a third line are parallel.
   (c) Two planes parallel to a third plane are parallel.
   (d) Two planes perpendicular to a third plane are parallel.
   (e) Two lines parallel to a plane are parallel.
   (f) Two lines perpendicular to a plane are parallel.
   (g) Two planes parallel to a line are parallel.
   (h) Two planes perpendicular to a line are parallel.
   (i) Two planes either intersect or are parallel.
   (j) Two lines either intersect or are parallel.
   (k) A plane and a line either intersect or are parallel.

2–5 Find a vector equation and parametric equations for the line.

2. The line through the point $(6, -5, 2)$ and parallel to the vector $\langle 1, 3, -\frac{2}{3} \rangle$

3. The line through the point $(2, 2.4, 3.5)$ and parallel to the vector $3\mathbf{i} + 2\mathbf{j} - \mathbf{k}$

4. The line through the point $(0, 14, -10)$ and parallel to the line $x = -1 + 2t$, $y = 6 - 3t$, $z = 3 + 9t$

5. The line through the point $(1, 0, 6)$ and perpendicular to the plane $x + 3y + z = 5$

6–12 Find parametric equations and symmetric equations for the line.

6. The line through the origin and the point $(4, 3, -1)$

7. The line through the points $(0, \frac{1}{2}, 1)$ and $(2, 1, -3)$

8. The line through the points $(1.0, 2.4, 4.6)$ and $(2.6, 1.2, 0.3)$

9. The line through the points $(-8, 1, 4)$ and $(3, -2, 4)$

10. The line through $(2, 1, 0)$ and perpendicular to both $\mathbf{i} + \mathbf{j}$ and $\mathbf{j} + \mathbf{k}$

11. The line through $(1, -1, 1)$ and parallel to the line $x + 2 = \frac{1}{2}y = z - 3$

12. The line of intersection of the planes $x + 2y + 3z = 1$ and $x - y + z = 1$

13. Is the line through $(-4, -6, 1)$ and $(-2, 0, -3)$ parallel to the line through $(10, 18, 4)$ and $(5, 3, 14)$?

14. Is the line through $(-2, 4, 0)$ and $(1, 1, 1)$ perpendicular to the line through $(2, 3, 4)$ and $(3, -1, -8)$?

15. (a) Find symmetric equations for the line that passes through the point $(1, -5, 6)$ and is parallel to the vector $\langle -1, 2, -3 \rangle$.
    (b) Find the points in which the required line in part (a) intersects the coordinate planes.

16. (a) Find parametric equations for the line through $(2, 4, 6)$ that is perpendicular to the plane $x - y + 3z = 7$.
    (b) In what points does this line intersect the coordinate planes?

17. Find a vector equation for the line segment from $(2, -1, 4)$ to $(4, 6, 1)$.

18. Find parametric equations for the line segment from $(10, 3, 1)$ to $(5, 6, -3)$.

19–22 Determine whether the lines L_1 and L_2 are parallel, skew, or intersecting. If they intersect, find the point of intersection.

19. L_1: $x = 3 + 2t$, $y = 4 - t$, $z = 1 + 3t$
    L_2: $x = 1 + 4s$, $y = 3 - 2s$, $z = 4 + 5s$

20. L_1: $x = 5 - 12t$, $y = 3 + 9t$, $z = 1 - 3t$
    L_2: $x = 3 + 8s$, $y = -6s$, $z = 7 + 2s$

21. L_1: $\dfrac{x - 2}{1} = \dfrac{y - 3}{-2} = \dfrac{z - 1}{-3}$
    L_2: $\dfrac{x - 3}{1} = \dfrac{y + 4}{3} = \dfrac{z - 2}{-7}$

22. L_1: $\dfrac{x}{1} = \dfrac{y - 1}{-1} = \dfrac{z - 2}{3}$
    L_2: $\dfrac{x - 2}{2} = \dfrac{y - 3}{-2} = \dfrac{z}{7}$

23–40 Find an equation of the plane.

23. The plane through the origin and perpendicular to the vector $\langle 1, -2, 5 \rangle$

24. The plane through the point $(5, 3, 5)$ and with normal vector $2\mathbf{i} + \mathbf{j} - \mathbf{k}$

25. The plane through the point $(-1, \frac{1}{2}, 3)$ and with normal vector $\mathbf{i} + 4\mathbf{j} + \mathbf{k}$

26. The plane through the point $(2, 0, 1)$ and perpendicular to the line $x = 3t$, $y = 2 - t$, $z = 3 + 4t$

27. The plane through the point $(1, -1, -1)$ and parallel to the plane $5x - y - z = 6$

28. The plane through the point $(2, 4, 6)$ and parallel to the plane $z = x + y$

29. The plane through the point $(1, \frac{1}{2}, \frac{1}{3})$ and parallel to the plane $x + y + z = 0$

30. The plane that contains the line $x = 1 + t$, $y = 2 - t$, $z = 4 - 3t$ and is parallel to the plane $5x + 2y + z = 1$

31. The plane through the points $(0, 1, 1)$, $(1, 0, 1)$, and $(1, 1, 0)$

32. The plane through the origin and the points $(2, -4, 6)$ and $(5, 1, 3)$

1. Homework Hints available at stewartcalculus.com

33. The plane through the points $(3, -1, 2)$, $(8, 2, 4)$, and $(-1, -2, -3)$

34. The plane that passes through the point $(1, 2, 3)$ and contains the line $x = 3t$, $y = 1 + t$, $z = 2 - t$

35. The plane that passes through the point $(6, 0, -2)$ and contains the line $x = 4 - 2t$, $y = 3 + 5t$, $z = 7 + 4t$

36. The plane that passes through the point $(1, -1, 1)$ and contains the line with symmetric equations $x = 2y = 3z$

37. The plane that passes through the point $(-1, 2, 1)$ and contains the line of intersection of the planes $x + y - z = 2$ and $2x - y + 3z = 1$

38. The plane that passes through the points $(0, -2, 5)$ and $(-1, 3, 1)$ and is perpendicular to the plane $2z = 5x + 4y$

39. The plane that passes through the point $(1, 5, 1)$ and is perpendicular to the planes $2x + y - 2z = 2$ and $x + 3z = 4$

40. The plane that passes through the line of intersection of the planes $x - z = 1$ and $y + 2z = 3$ and is perpendicular to the plane $x + y - 2z = 1$

41–44 Use intercepts to help sketch the plane.

41. $2x + 5y + z = 10$

42. $3x + y + 2z = 6$

43. $6x - 3y + 4z = 6$

44. $6x + 5y - 3z = 15$

45–47 Find the point at which the line intersects the given plane.

45. $x = 3 - t$, $y = 2 + t$, $z = 5t$; $x - y + 2z = 9$

46. $x = 1 + 2t$, $y = 4t$, $z = 2 - 3t$; $x + 2y - z + 1 = 0$

47. $x = y - 1 = 2z$; $4x - y + 3z = 8$

48. Where does the line through $(1, 0, 1)$ and $(4, -2, 2)$ intersect the plane $x + y + z = 6$?

49. Find direction numbers for the line of intersection of the planes $x + y + z = 1$ and $x + z = 0$.

50. Find the cosine of the angle between the planes $x + y + z = 0$ and $x + 2y + 3z = 1$.

51–56 Determine whether the planes are parallel, perpendicular, or neither. If neither, find the angle between them.

51. $x + 4y - 3z = 1$, $-3x + 6y + 7z = 0$

52. $2z = 4y - x$, $3x - 12y + 6z = 1$

53. $x + y + z = 1$, $x - y + z = 1$

54. $2x - 3y + 4z = 5$, $x + 6y + 4z = 3$

55. $x = 4y - 2z$, $8y = 1 + 2x + 4z$

56. $x + 2y + 2z = 1$, $2x - y + 2z = 1$

57–58 (a) Find parametric equations for the line of intersection of the planes and (b) find the angle between the planes.

57. $x + y + z = 1$, $x + 2y + 2z = 1$

58. $3x - 2y + z = 1$, $2x + y - 3z = 3$

59–60 Find symmetric equations for the line of intersection of the planes.

59. $5x - 2y - 2z = 1$, $4x + y + z = 6$

60. $z = 2x - y - 5$, $z = 4x + 3y - 5$

61. Find an equation for the plane consisting of all points that are equidistant from the points $(1, 0, -2)$ and $(3, 4, 0)$.

62. Find an equation for the plane consisting of all points that are equidistant from the points $(2, 5, 5)$ and $(-6, 3, 1)$.

63. Find an equation of the plane with x-intercept a, y-intercept b, and z-intercept c.

64. (a) Find the point at which the given lines intersect:

$$\mathbf{r} = \langle 1, 1, 0 \rangle + t\langle 1, -1, 2 \rangle$$

$$\mathbf{r} = \langle 2, 0, 2 \rangle + s\langle -1, 1, 0 \rangle$$

(b) Find an equation of the plane that contains these lines.

65. Find parametric equations for the line through the point $(0, 1, 2)$ that is parallel to the plane $x + y + z = 2$ and perpendicular to the line $x = 1 + t$, $y = 1 - t$, $z = 2t$.

66. Find parametric equations for the line through the point $(0, 1, 2)$ that is perpendicular to the line $x = 1 + t$, $y = 1 - t$, $z = 2t$ and intersects this line.

67. Which of the following four planes are parallel? Are any of them identical?

P_1: $3x + 6y - 3z = 6$ P_2: $4x - 12y + 8z = 5$

P_3: $9y = 1 + 3x + 6z$ P_4: $z = x + 2y - 2$

68. Which of the following four lines are parallel? Are any of them identical?

L_1: $x = 1 + 6t$, $y = 1 - 3t$, $z = 12t + 5$

L_2: $x = 1 + 2t$, $y = t$, $z = 1 + 4t$

L_3: $2x - 2 = 4 - 4y = z + 1$

L_4: $\mathbf{r} = \langle 3, 1, 5 \rangle + t\langle 4, 2, 8 \rangle$

69–70 Use the formula in Exercise 45 in Section 12.4 to find the distance from the point to the given line.

69. $(4, 1, -2)$; $x = 1 + t$, $y = 3 - 2t$, $z = 4 - 3t$

70. $(0, 1, 3)$; $x = 2t$, $y = 6 - 2t$, $z = 3 + t$

71–72 Find the distance from the point to the given plane.

71. $(1, -2, 4)$, $3x + 2y + 6z = 5$

72. $(-6, 3, 5)$, $x - 2y - 4z = 8$

73–74 Find the distance between the given parallel planes.

73. $2x - 3y + z = 4$, $4x - 6y + 2z = 3$

74. $6z = 4y - 2x$, $9z = 1 - 3x + 6y$

75. Show that the distance between the parallel planes $ax + by + cz + d_1 = 0$ and $ax + by + cz + d_2 = 0$ is

$$D = \frac{|d_1 - d_2|}{\sqrt{a^2 + b^2 + c^2}}$$

76. Find equations of the planes that are parallel to the plane $x + 2y - 2z = 1$ and two units away from it.

77. Show that the lines with symmetric equations $x = y = z$ and $x + 1 = y/2 = z/3$ are skew, and find the distance between these lines.

78. Find the distance between the skew lines with parametric equations $x = 1 + t$, $y = 1 + 6t$, $z = 2t$, and $x = 1 + 2s$, $y = 5 + 15s$, $z = -2 + 6s$.

79. Let L_1 be the line through the origin and the point $(2, 0, -1)$. Let L_2 be the line through the points $(1, -1, 1)$ and $(4, 1, 3)$. Find the distance between L_1 and L_2.

80. Let L_1 be the line through the points $(1, 2, 6)$ and $(2, 4, 8)$. Let L_2 be the line of intersection of the planes π_1 and π_2, where π_1 is the plane $x - y + 2z + 1 = 0$ and π_2 is the plane through the points $(3, 2, -1)$, $(0, 0, 1)$, and $(1, 2, 1)$. Calculate the distance between L_1 and L_2.

81. If a, b, and c are not all 0, show that the equation $ax + by + cz + d = 0$ represents a plane and $\langle a, b, c \rangle$ is a normal vector to the plane.
Hint: Suppose $a \neq 0$ and rewrite the equation in the form

$$a\left(x + \frac{d}{a}\right) + b(y - 0) + c(z - 0) = 0$$

82. Give a geometric description of each family of planes.
(a) $x + y + z = c$ (b) $x + y + cz = 1$
(c) $y \cos \theta + z \sin \theta = 1$

LABORATORY PROJECT PUTTING 3D IN PERSPECTIVE

Computer graphics programmers face the same challenge as the great painters of the past: how to represent a three-dimensional scene as a flat image on a two-dimensional plane (a screen or a canvas). To create the illusion of perspective, in which closer objects appear larger than those farther away, three-dimensional objects in the computer's memory are projected onto a rectangular screen window from a viewpoint where the eye, or camera, is located. The viewing volume—the portion of space that will be visible—is the region contained by the four planes that pass through the viewpoint and an edge of the screen window. If objects in the scene extend beyond these four planes, they must be truncated before pixel data are sent to the screen. These planes are therefore called *clipping planes*.

1. Suppose the screen is represented by a rectangle in the yz-plane with vertices $(0, \pm 400, 0)$ and $(0, \pm 400, 600)$, and the camera is placed at $(1000, 0, 0)$. A line L in the scene passes through the points $(230, -285, 102)$ and $(860, 105, 264)$. At what points should L be clipped by the clipping planes?

2. If the clipped line segment is projected on the screen window, identify the resulting line segment.

3. Use parametric equations to plot the edges of the screen window, the clipped line segment, and its projection on the screen window. Then add sight lines connecting the viewpoint to each end of the clipped segments to verify that the projection is correct.

4. A rectangle with vertices $(621, -147, 206)$, $(563, 31, 242)$, $(657, -111, 86)$, and $(599, 67, 122)$ is added to the scene. The line L intersects this rectangle. To make the rectangle appear opaque, a programmer can use *hidden line rendering*, which removes portions of objects that are behind other objects. Identify the portion of L that should be removed.

12.6 Cylinders and Quadric Surfaces

We have already looked at two special types of surfaces: planes (in Section 12.5) and spheres (in Section 12.1). Here we investigate two other types of surfaces: cylinders and quadric surfaces.

In order to sketch the graph of a surface, it is useful to determine the curves of intersection of the surface with planes parallel to the coordinate planes. These curves are called **traces** (or cross-sections) of the surface.

Cylinders

A **cylinder** is a surface that consists of all lines (called **rulings**) that are parallel to a given line and pass through a given plane curve.

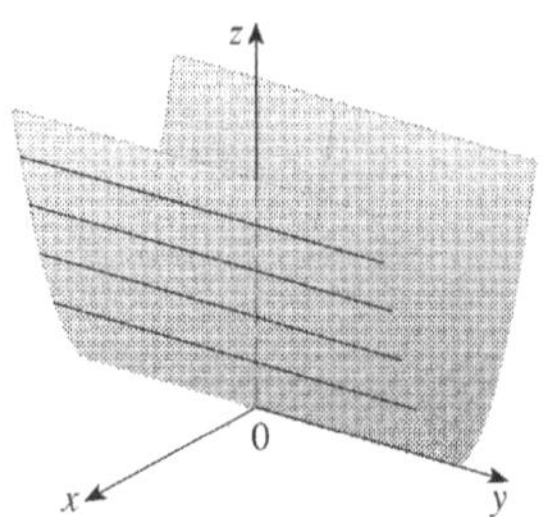

FIGURE 1
The surface $z = x^2$ is a parabolic cylinder.

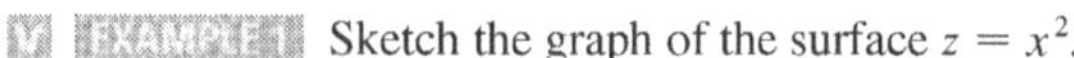

EXAMPLE 1 Sketch the graph of the surface $z = x^2$.

SOLUTION Notice that the equation of the graph, $z = x^2$, doesn't involve y. This means that any vertical plane with equation $y = k$ (parallel to the xz-plane) intersects the graph in a curve with equation $z = x^2$. So these vertical traces are parabolas. Figure 1 shows how the graph is formed by taking the parabola $z = x^2$ in the xz-plane and moving it in the direction of the y-axis. The graph is a surface, called a **parabolic cylinder**, made up of infinitely many shifted copies of the same parabola. Here the rulings of the cylinder are parallel to the y-axis.

We noticed that the variable y is missing from the equation of the cylinder in Example 1. This is typical of a surface whose rulings are parallel to one of the coordinate axes. If one of the variables x, y, or z is missing from the equation of a surface, then the surface is a cylinder.

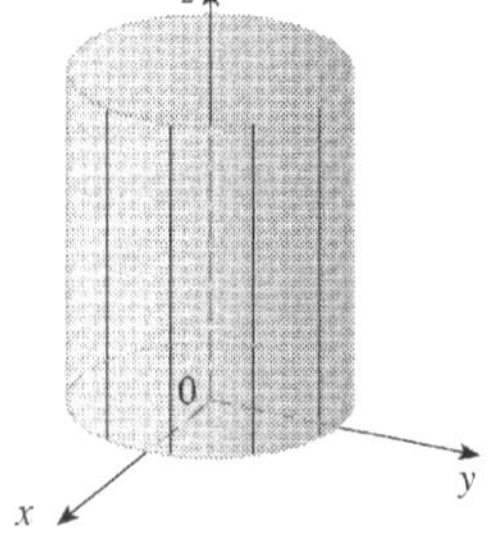

FIGURE 2 $x^2 + y^2 = 1$

EXAMPLE 2 Identify and sketch the surfaces.

(a) $x^2 + y^2 = 1$ (b) $y^2 + z^2 = 1$

SOLUTION

(a) Since z is missing and the equations $x^2 + y^2 = 1$, $z = k$ represent a circle with radius 1 in the plane $z = k$, the surface $x^2 + y^2 = 1$ is a circular cylinder whose axis is the z-axis. (See Figure 2.) Here the rulings are vertical lines.

(b) In this case x is missing and the surface is a circular cylinder whose axis is the x-axis. (See Figure 3.) It is obtained by taking the circle $y^2 + z^2 = 1$, $x = 0$ in the yz-plane and moving it parallel to the x-axis.

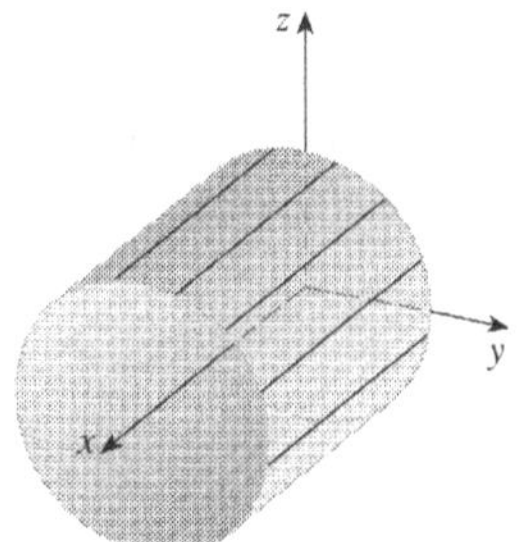

FIGURE 3 $y^2 + z^2 = 1$

NOTE When you are dealing with surfaces, it is important to recognize that an equation like $x^2 + y^2 = 1$ represents a cylinder and not a circle. The trace of the cylinder $x^2 + y^2 = 1$ in the xy-plane is the circle with equations $x^2 + y^2 = 1$, $z = 0$.

Quadric Surfaces

A **quadric surface** is the graph of a second-degree equation in three variables x, y, and z. The most general such equation is

$$Ax^2 + By^2 + Cz^2 + Dxy + Eyz + Fxz + Gx + Hy + Iz + J = 0$$

where $A, B, C, \dots, J$ are constants, but by translation and rotation it can be brought into one of the two standard forms

$$Ax^2 + By^2 + Cz^2 + J = 0 \qquad \text{or} \qquad Ax^2 + By^2 + Iz = 0$$

Quadric surfaces are the counterparts in three dimensions of the conic sections in the plane. (See Section 10.5 for a review of conic sections.)

EXAMPLE 3 Use traces to sketch the quadric surface with equation

$$x^2 + \frac{y^2}{9} + \frac{z^2}{4} = 1$$

SOLUTION By substituting $z = 0$, we find that the trace in the xy-plane is $x^2 + y^2/9 = 1$, which we recognize as an equation of an ellipse. In general, the horizontal trace in the plane $z = k$ is

$$x^2 + \frac{y^2}{9} = 1 - \frac{k^2}{4} \qquad z = k$$

which is an ellipse, provided that $k^2 < 4$, that is, $-2 < k < 2$.

Similarly, the vertical traces are also ellipses:

$$\frac{y^2}{9} + \frac{z^2}{4} = 1 - k^2 \qquad x = k \qquad (\text{if } -1 < k < 1)$$

$$x^2 + \frac{z^2}{4} = 1 - \frac{k^2}{9} \qquad y = k \qquad (\text{if } -3 < k < 3)$$

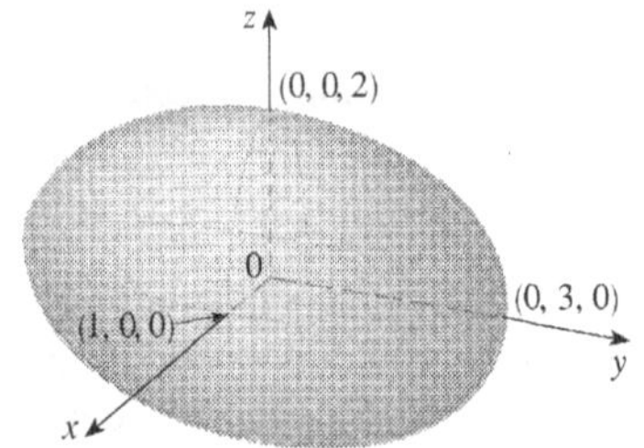

FIGURE 4

The ellipsoid $x^2 + \frac{y^2}{9} + \frac{z^2}{4} = 1$

Figure 4 shows how drawing some traces indicates the shape of the surface. It's called an **ellipsoid** because all of its traces are ellipses. Notice that it is symmetric with respect to each coordinate plane; this is a reflection of the fact that its equation involves only even powers of x, y, and z.

EXAMPLE 4 Use traces to sketch the surface $z = 4x^2 + y^2$.

SOLUTION If we put $x = 0$, we get $z = y^2$, so the yz-plane intersects the surface in a parabola. If we put $x = k$ (a constant), we get $z = y^2 + 4k^2$. This means that if we slice the graph with any plane parallel to the yz-plane, we obtain a parabola that opens upward. Similarly, if $y = k$, the trace is $z = 4x^2 + k^2$, which is again a parabola that opens upward. If we put $z = k$, we get the horizontal traces $4x^2 + y^2 = k$, which we recognize as a family of ellipses. Knowing the shapes of the traces, we can sketch the graph in Figure 5. Because of the elliptical and parabolic traces, the quadric surface $z = 4x^2 + y^2$ is called an **elliptic paraboloid**.

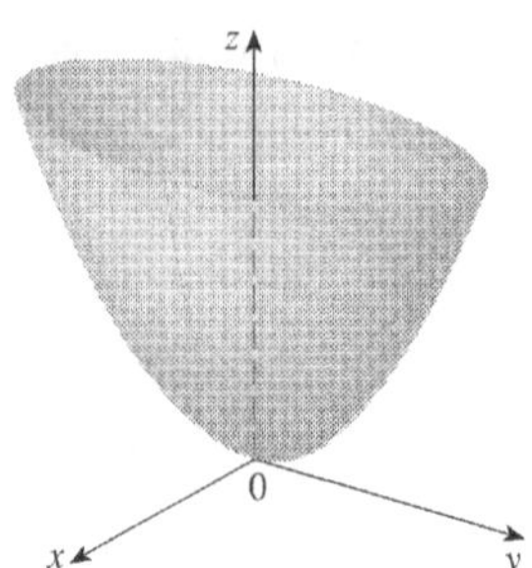

FIGURE 5

The surface $z = 4x^2 + y^2$ is an elliptic paraboloid. Horizontal traces are ellipses; vertical traces are parabolas.

V EXAMPLE 5 Sketch the surface $z = y^2 - x^2$.

SOLUTION The traces in the vertical planes $x = k$ are the parabolas $z = y^2 - k^2$, which open upward. The traces in $y = k$ are the parabolas $z = -x^2 + k^2$, which open downward. The horizontal traces are $y^2 - x^2 = k$, a family of hyperbolas. We draw the families of traces in Figure 6, and we show how the traces appear when placed in their correct planes in Figure 7.

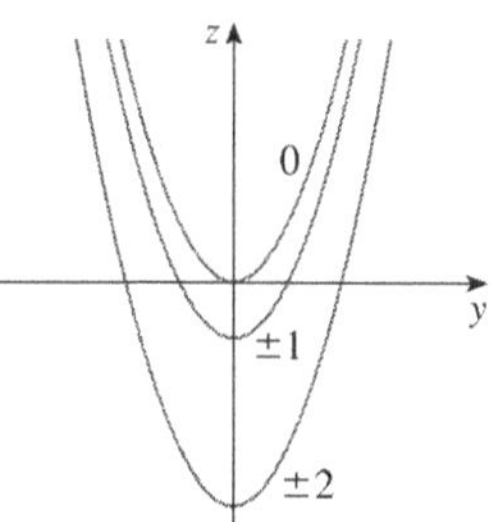

Traces in $x = k$ are $z = y^2 - k^2$

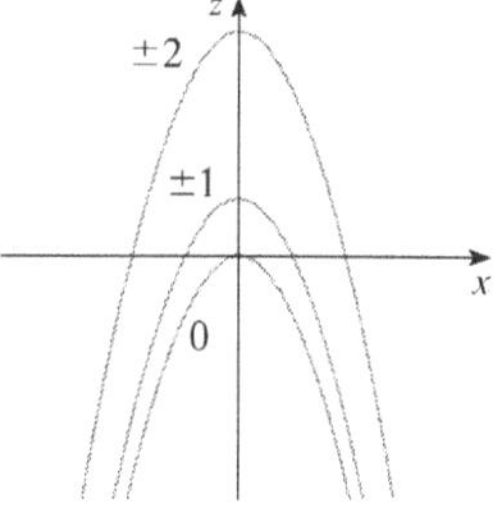

Traces in $y = k$ are $z = -x^2 + k^2$

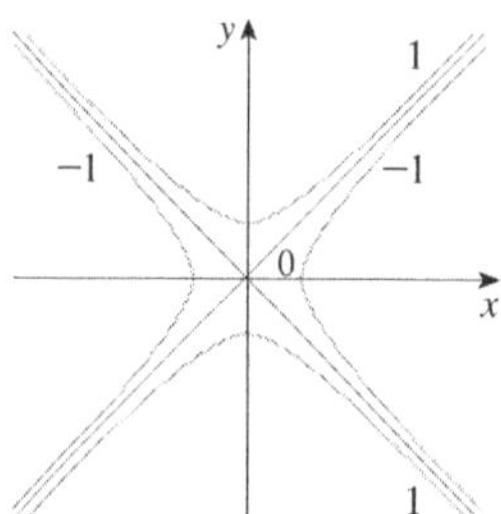

Traces in $z = k$ are $y^2 - x^2 = k$

FIGURE 6
Vertical traces are parabolas; horizontal traces are hyperbolas. All traces are labeled with the value of k.

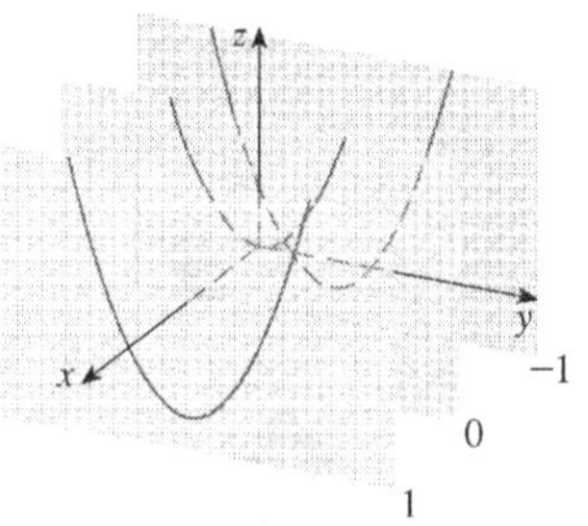

Traces in $x = k$

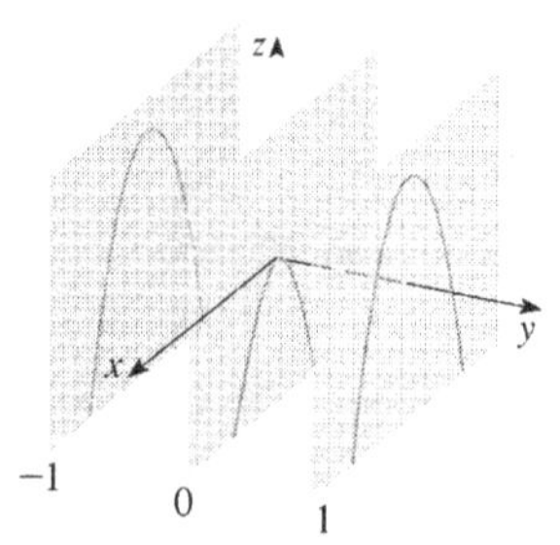

Traces in $y = k$

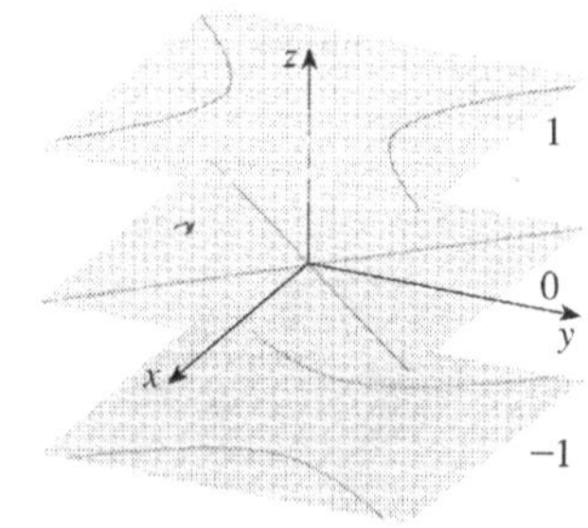

Traces in $z = k$

FIGURE 7
Traces moved to their correct planes

TEC In Module 12.6A you can investigate how traces determine the shape of a surface.

In Figure 8 we fit together the traces from Figure 7 to form the surface $z = y^2 - x^2$, a **hyperbolic paraboloid**. Notice that the shape of the surface near the origin resembles that of a saddle. This surface will be investigated further in Section 14.7 when we discuss saddle points.

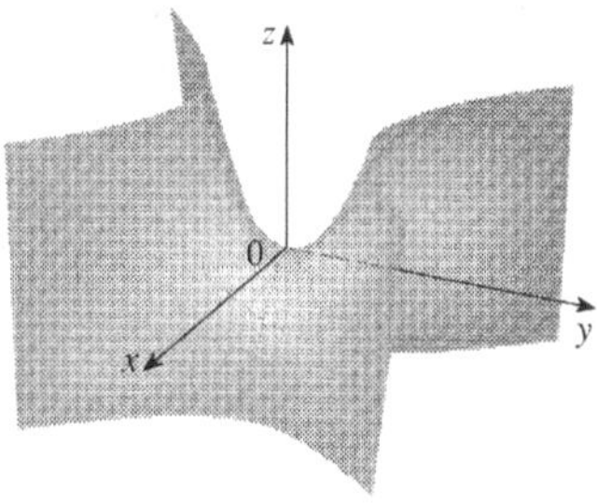

FIGURE 8
The surface $z = y^2 - x^2$ is a hyperbolic paraboloid.

EXAMPLE 6 Sketch the surface $\dfrac{x^2}{4} + y^2 - \dfrac{z^2}{4} = 1$.

SOLUTION The trace in any horizontal plane $z = k$ is the ellipse

$$\frac{x^2}{4} + y^2 = 1 + \frac{k^2}{4} \qquad z = k$$

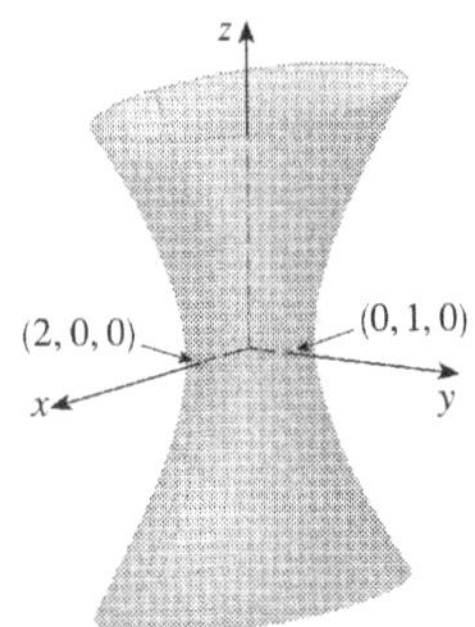

FIGURE 9

but the traces in the xz- and yz-planes are the hyperbolas

$$\frac{x^2}{4} - \frac{z^2}{4} = 1 \quad y = 0 \qquad \text{and} \qquad y^2 - \frac{z^2}{4} = 1 \quad x = 0$$

This surface is called a **hyperboloid of one sheet** and is sketched in Figure 9.

The idea of using traces to draw a surface is employed in three-dimensional graphing software for computers. In most such software, traces in the vertical planes $x = k$ and $y = k$ are drawn for equally spaced values of k, and parts of the graph are eliminated using hidden line removal. Table 1 shows computer-drawn graphs of the six basic types of quadric surfaces in standard form. All surfaces are symmetric with respect to the z-axis. If a quadric surface is symmetric about a different axis, its equation changes accordingly.

TABLE 1 Graphs of quadric surfaces

Surface	Equation	Surface	Equation
Ellipsoid	$\frac{x^2}{a^2} + \frac{y^2}{b^2} + \frac{z^2}{c^2} = 1$ All traces are ellipses. If $a = b = c$, the ellipsoid is a sphere.	Cone	$\frac{z^2}{c^2} = \frac{x^2}{a^2} + \frac{y^2}{b^2}$ Horizontal traces are ellipses. Vertical traces in the planes $x = k$ and $y = k$ are hyperbolas if $k \neq 0$ but are pairs of lines if $k = 0$.
Elliptic Paraboloid	$\frac{z}{c} = \frac{x^2}{a^2} + \frac{y^2}{b^2}$ Horizontal traces are ellipses. Vertical traces are parabolas. The variable raised to the first power indicates the axis of the paraboloid.	Hyperboloid of One Sheet	$\frac{x^2}{a^2} + \frac{y^2}{b^2} - \frac{z^2}{c^2} = 1$ Horizontal traces are ellipses. Vertical traces are hyperbolas. The axis of symmetry corresponds to the variable whose coefficient is negative.
Hyperbolic Paraboloid	$\frac{z}{c} = \frac{x^2}{a^2} - \frac{y^2}{b^2}$ Horizontal traces are hyperbolas. Vertical traces are parabolas. The case where $c < 0$ is illustrated.	Hyperboloid of Two Sheets	$-\frac{x^2}{a^2} - \frac{y^2}{b^2} + \frac{z^2}{c^2} = 1$ Horizontal traces in $z = k$ are ellipses if $k > c$ or $k < -c$. Vertical traces are hyperbolas. The two minus signs indicate two sheets.

TEC In Module 12.6B you can see how changing a, b, and c in Table 1 affects the shape of the quadric surface.

V EXAMPLE 7 Identify and sketch the surface $4x^2 - y^2 + 2z^2 + 4 = 0$.

SOLUTION Dividing by -4, we first put the equation in standard form:

$$-x^2 + \frac{y^2}{4} - \frac{z^2}{2} = 1$$

Comparing this equation with Table 1, we see that it represents a hyperboloid of two sheets, the only difference being that in this case the axis of the hyperboloid is the y-axis. The traces in the xy- and yz-planes are the hyperbolas

$$-x^2 + \frac{y^2}{4} = 1 \qquad z = 0 \qquad \text{and} \qquad \frac{y^2}{4} - \frac{z^2}{2} = 1 \qquad x = 0$$

The surface has no trace in the xz-plane, but traces in the vertical planes $y = k$ for $|k| > 2$ are the ellipses

$$x^2 + \frac{z^2}{2} = \frac{k^2}{4} - 1 \qquad y = k$$

which can be written as

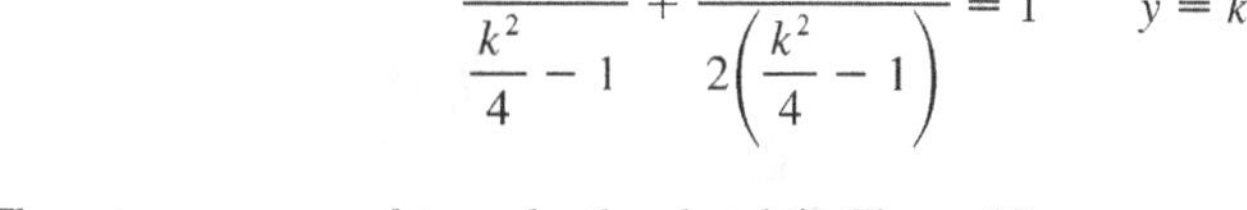

$$\frac{x^2}{\dfrac{k^2}{4} - 1} + \frac{z^2}{2\left(\dfrac{k^2}{4} - 1\right)} = 1 \qquad y = k$$

These traces are used to make the sketch in Figure 10.

FIGURE 10
$4x^2 - y^2 + 2z^2 + 4 = 0$

EXAMPLE 8 Classify the quadric surface $x^2 + 2z^2 - 6x - y + 10 = 0$.

SOLUTION By completing the square we rewrite the equation as

$$y - 1 = (x - 3)^2 + 2z^2$$

Comparing this equation with Table 1, we see that it represents an elliptic paraboloid. Here, however, the axis of the paraboloid is parallel to the y-axis, and it has been shifted so that its vertex is the point $(3, 1, 0)$. The traces in the plane $y = k$ $(k > 1)$ are the ellipses

$$(x - 3)^2 + 2z^2 = k - 1 \qquad y = k$$

The trace in the xy-plane is the parabola with equation $y = 1 + (x - 3)^2$, $z = 0$. The paraboloid is sketched in Figure 11.

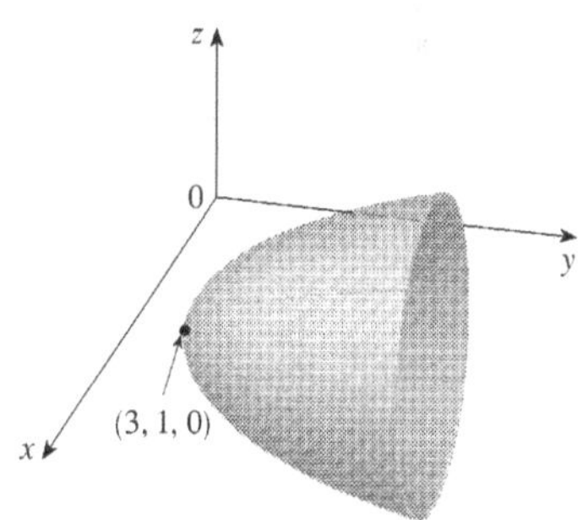

FIGURE 11
$x^2 + 2z^2 - 6x - y + 10 = 0$

Applications of Quadric Surfaces

Examples of quadric surfaces can be found in the world around us. In fact, the world itself is a good example. Although the earth is commonly modeled as a sphere, a more accurate model is an ellipsoid because the earth's rotation has caused a flattening at the poles. (See Exercise 47.)

Circular paraboloids, obtained by rotating a parabola about its axis, are used to collect and reflect light, sound, and radio and television signals. In a radio telescope, for instance, signals from distant stars that strike the bowl are all reflected to the receiver at the focus and are therefore amplified. (The idea is explained in Problem 20 on page 271.) The same principle applies to microphones and satellite dishes in the shape of paraboloids.

Cooling towers for nuclear reactors are usually designed in the shape of hyperboloids of one sheet for reasons of structural stability. Pairs of hyperboloids are used to transmit rotational motion between skew axes. (The cogs of the gears are the generating lines of the hyperboloids. See Exercise 49.)

A satellite dish reflects signals to the focus of a paraboloid.

Nuclear reactors have cooling towers in the shape of hyperboloids.

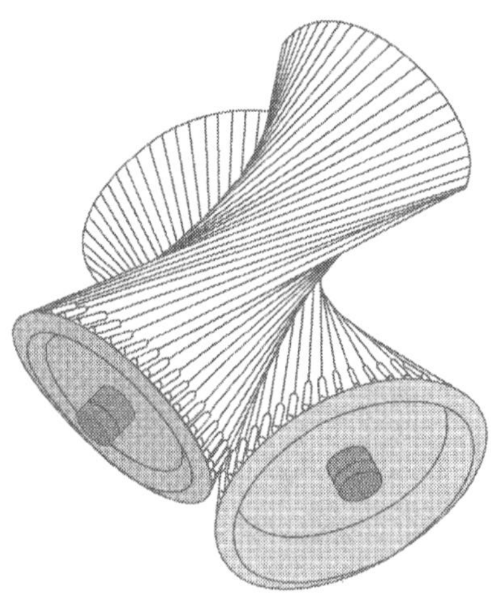

Hyperboloids produce gear transmission.

12.6 Exercises

1. (a) What does the equation $y = x^2$ represent as a curve in $\mathbb{R}^2$?
   (b) What does it represent as a surface in $\mathbb{R}^3$?
   (c) What does the equation $z = y^2$ represent?

2. (a) Sketch the graph of $y = e^x$ as a curve in $\mathbb{R}^2$.
   (b) Sketch the graph of $y = e^x$ as a surface in $\mathbb{R}^3$.
   (c) Describe and sketch the surface $z = e^y$.

3–8 Describe and sketch the surface.

3. $x^2 + z^2 = 1$

4. $4x^2 + y^2 = 4$

5. $z = 1 - y^2$

6. $y = z^2$

7. $xy = 1$

8. $z = \sin y$

9. (a) Find and identify the traces of the quadric surface $x^2 + y^2 - z^2 = 1$ and explain why the graph looks like the graph of the hyperboloid of one sheet in Table 1.
   (b) If we change the equation in part (a) to $x^2 - y^2 + z^2 = 1$, how is the graph affected?
   (c) What if we change the equation in part (a) to $x^2 + y^2 + 2y - z^2 = 0$?

Graphing calculator or computer required   1. Homework Hints available at stewartcalculus.com

11. For any vectors **u**, **v**, and **w** in V_3, $\mathbf{u} \cdot (\mathbf{v} \times \mathbf{w}) = (\mathbf{u} \times \mathbf{v}) \cdot \mathbf{w}$.

12. For any vectors **u**, **v**, and **w** in V_3, $\mathbf{u} \times (\mathbf{v} \times \mathbf{w}) = (\mathbf{u} \times \mathbf{v}) \times \mathbf{w}$.

13. For any vectors **u** and **v** in V_3, $(\mathbf{u} \times \mathbf{v}) \cdot \mathbf{u} = 0$.

14. For any vectors **u** and **v** in V_3, $(\mathbf{u} + \mathbf{v}) \times \mathbf{v} = \mathbf{u} \times \mathbf{v}$.

15. The vector $\langle 3, -1, 2 \rangle$ is parallel to the plane $6x - 2y + 4z = 1$.

16. A linear equation $Ax + By + Cz + D = 0$ represents a line in space.

17. The set of points $\{(x, y, z) \mid x^2 + y^2 = 1\}$ is a circle.

18. In $\mathbb{R}^3$ the graph of $y = x^2$ is a paraboloid.

19. If $\mathbf{u} \cdot \mathbf{v} = 0$, then $\mathbf{u} = \mathbf{0}$ or $\mathbf{v} = \mathbf{0}$.

20. If $\mathbf{u} \times \mathbf{v} = \mathbf{0}$, then $\mathbf{u} = \mathbf{0}$ or $\mathbf{v} = \mathbf{0}$.

21. If $\mathbf{u} \cdot \mathbf{v} = 0$ and $\mathbf{u} \times \mathbf{v} = \mathbf{0}$, then $\mathbf{u} = \mathbf{0}$ or $\mathbf{v} = \mathbf{0}$.

22. If **u** and **v** are in V_3, then $|\mathbf{u} \cdot \mathbf{v}| \leqslant |\mathbf{u}||\mathbf{v}|$.

Exercises

1. (a) Find an equation of the sphere that passes through the point $(6, -2, 3)$ and has center $(-1, 2, 1)$.
   (b) Find the curve in which this sphere intersects the yz-plane.
   (c) Find the center and radius of the sphere

$$x^2 + y^2 + z^2 - 8x + 2y + 6z + 1 = 0$$

2. Copy the vectors in the figure and use them to draw each of the following vectors.
   (a) $\mathbf{a} + \mathbf{b}$ (b) $\mathbf{a} - \mathbf{b}$ (c) $-\frac{1}{2}\mathbf{a}$ (d) $2\mathbf{a} + \mathbf{b}$

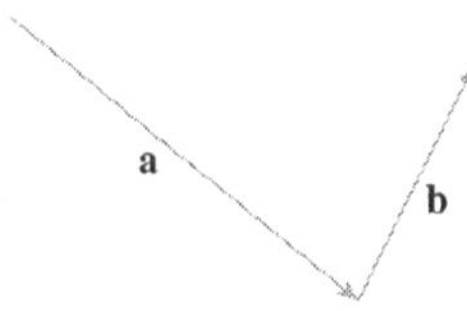

3. If **u** and **v** are the vectors shown in the figure, find $\mathbf{u} \cdot \mathbf{v}$ and $|\mathbf{u} \times \mathbf{v}|$. Is $\mathbf{u} \times \mathbf{v}$ directed into the page or out of it?

4. Calculate the given quantity if

$$\mathbf{a} = \mathbf{i} + \mathbf{j} - 2\mathbf{k}$$
$$\mathbf{b} = 3\mathbf{i} - 2\mathbf{j} + \mathbf{k}$$
$$\mathbf{c} = \mathbf{j} - 5\mathbf{k}$$

   (a) $2\mathbf{a} + 3\mathbf{b}$ (b) $|\mathbf{b}|$
   (c) $\mathbf{a} \cdot \mathbf{b}$ (d) $\mathbf{a} \times \mathbf{b}$
   (e) $|\mathbf{b} \times \mathbf{c}|$ (f) $\mathbf{a} \cdot (\mathbf{b} \times \mathbf{c})$
   (g) $\mathbf{c} \times \mathbf{c}$ (h) $\mathbf{a} \times (\mathbf{b} \times \mathbf{c})$
   (i) $\text{comp}_{\mathbf{a}}\, \mathbf{b}$ (j) $\text{proj}_{\mathbf{a}}\, \mathbf{b}$
   (k) The angle between **a** and **b** (correct to the nearest degree)

5. Find the values of x such that the vectors $\langle 3, 2, x \rangle$ and $\langle 2x, 4, x \rangle$ are orthogonal.

6. Find two unit vectors that are orthogonal to both $\mathbf{j} + 2\mathbf{k}$ and $\mathbf{i} - 2\mathbf{j} + 3\mathbf{k}$.

7. Suppose that $\mathbf{u} \cdot (\mathbf{v} \times \mathbf{w}) = 2$. Find
   (a) $(\mathbf{u} \times \mathbf{v}) \cdot \mathbf{w}$ (b) $\mathbf{u} \cdot (\mathbf{w} \times \mathbf{v})$
   (c) $\mathbf{v} \cdot (\mathbf{u} \times \mathbf{w})$ (d) $(\mathbf{u} \times \mathbf{v}) \cdot \mathbf{v}$

8. Show that if **a**, **b**, and **c** are in V_3, then

$$(\mathbf{a} \times \mathbf{b}) \cdot [(\mathbf{b} \times \mathbf{c}) \times (\mathbf{c} \times \mathbf{a})] = [\mathbf{a} \cdot (\mathbf{b} \times \mathbf{c})]^2$$

9. Find the acute angle between two diagonals of a cube.

10. Given the points $A(1, 0, 1)$, $B(2, 3, 0)$, $C(-1, 1, 4)$, and $D(0, 3, 2)$, find the volume of the parallelepiped with adjacent edges AB, AC, and AD.

11. (a) Find a vector perpendicular to the plane through the points $A(1, 0, 0)$, $B(2, 0, -1)$, and $C(1, 4, 3)$.
    (b) Find the area of triangle ABC.

12. A constant force $\mathbf{F} = 3\mathbf{i} + 5\mathbf{j} + 10\mathbf{k}$ moves an object along the line segment from $(1, 0, 2)$ to $(5, 3, 8)$. Find the work done if the distance is measured in meters and the force in newtons.

13. A boat is pulled onto shore using two ropes, as shown in the diagram. If a force of 255 N is needed, find the magnitude of the force in each rope.

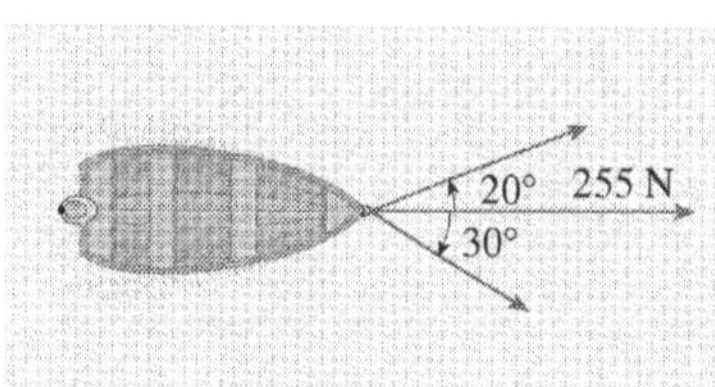

14. Find the magnitude of the torque about P if a 50-N force is applied as shown.

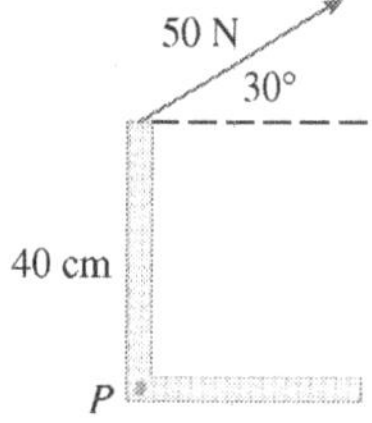

15–17 Find parametric equations for the line.

15. The line through $(4, -1, 2)$ and $(1, 1, 5)$

16. The line through $(1, 0, -1)$ and parallel to the line $\frac{1}{3}(x - 4) = \frac{1}{2}y = z + 2$

17. The line through $(-2, 2, 4)$ and perpendicular to the plane $2x - y + 5z = 12$

18–20 Find an equation of the plane.

18. The plane through $(2, 1, 0)$ and parallel to $x + 4y - 3z = 1$

19. The plane through $(3, -1, 1)$, $(4, 0, 2)$, and $(6, 3, 1)$

20. The plane through $(1, 2, -2)$ that contains the line $x = 2t$, $y = 3 - t$, $z = 1 + 3t$

21. Find the point in which the line with parametric equations $x = 2 - t$, $y = 1 + 3t$, $z = 4t$ intersects the plane $2x - y + z = 2$.

22. Find the distance from the origin to the line $x = 1 + t$, $y = 2 - t$, $z = -1 + 2t$.

23. Determine whether the lines given by the symmetric equations

$$\frac{x - 1}{2} = \frac{y - 2}{3} = \frac{z - 3}{4}$$

and

$$\frac{x + 1}{6} = \frac{y - 3}{-1} = \frac{z + 5}{2}$$

are parallel, skew, or intersecting.

24. (a) Show that the planes $x + y - z = 1$ and $2x - 3y + 4z = 5$ are neither parallel nor perpendicular.

(b) Find, correct to the nearest degree, the angle between these planes.

25. Find an equation of the plane through the line of intersection of the planes $x - z = 1$ and $y + 2z = 3$ and perpendicular to the plane $x + y - 2z = 1$.

26. (a) Find an equation of the plane that passes through the points $A(2, 1, 1)$, $B(-1, -1, 10)$, and $C(1, 3, -4)$.

(b) Find symmetric equations for the line through B that is perpendicular to the plane in part (a).

(c) A second plane passes through $(2, 0, 4)$ and has normal vector $\langle 2, -4, -3 \rangle$. Show that the acute angle between the planes is approximately 43°.

(d) Find parametric equations for the line of intersection of the two planes.

27. Find the distance between the planes $3x + y - 4z = 2$ and $3x + y - 4z = 24$.

28–36 Identify and sketch the graph of each surface.

28. $x = 3$

29. $x = z$

30. $y = z^2$

31. $x^2 = y^2 + 4z^2$

32. $4x - y + 2z = 4$

33. $-4x^2 + y^2 - 4z^2 = 4$

34. $y^2 + z^2 = 1 + x^2$

35. $4x^2 + 4y^2 - 8y + z^2 = 0$

36. $x = y^2 + z^2 - 2y - 4z + 5$

37. An ellipsoid is created by rotating the ellipse $4x^2 + y^2 = 16$ about the x-axis. Find an equation of the ellipsoid.

38. A surface consists of all points P such that the distance from P to the plane $y = 1$ is twice the distance from P to the point $(0, -1, 0)$. Find an equation for this surface and identify it.

Problems Plus

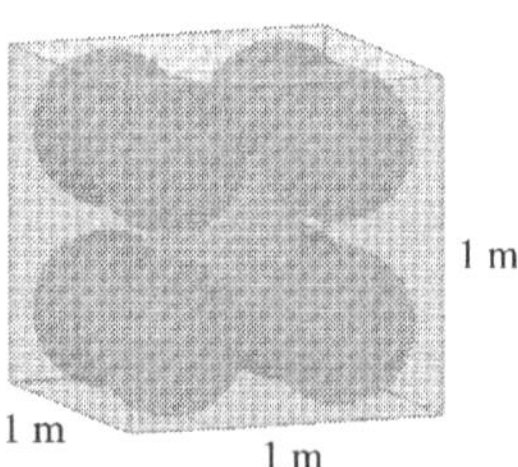

FIGURE FOR PROBLEM 1

1. Each edge of a cubical box has length 1 m. The box contains nine spherical balls with the same radius r. The center of one ball is at the center of the cube and it touches the other eight balls. Each of the other eight balls touches three sides of the box. Thus the balls are tightly packed in the box. (See the figure.) Find r. (If you have trouble with this problem, read about the problem-solving strategy entitled *Use Analogy* on page 75.)

2. Let B be a solid box with length L, width W, and height H. Let S be the set of all points that are a distance at most 1 from some point of B. Express the volume of S in terms of L, W, and H.

3. Let L be the line of intersection of the planes $cx + y + z = c$ and $x - cy + cz = -1$, where c is a real number.
   (a) Find symmetric equations for L.
   (b) As the number c varies, the line L sweeps out a surface S. Find an equation for the curve of intersection of S with the horizontal plane $z = t$ (the trace of S in the plane $z = t$).
   (c) Find the volume of the solid bounded by S and the planes $z = 0$ and $z = 1$.

4. A plane is capable of flying at a speed of 180 km/h in still air. The pilot takes off from an airfield and heads due north according to the plane's compass. After 30 minutes of flight time, the pilot notices that, due to the wind, the plane has actually traveled 80 km at an angle 5° east of north.
   (a) What is the wind velocity?
   (b) In what direction should the pilot have headed to reach the intended destination?

5. Suppose $\mathbf{v}_1$ and $\mathbf{v}_2$ are vectors with $|\mathbf{v}_1| = 2$, $|\mathbf{v}_2| = 3$, and $\mathbf{v}_1 \cdot \mathbf{v}_2 = 5$. Let $\mathbf{v}_3 = \text{proj}_{\mathbf{v}_1}\mathbf{v}_2$, $\mathbf{v}_4 = \text{proj}_{\mathbf{v}_2}\mathbf{v}_3$, $\mathbf{v}_5 = \text{proj}_{\mathbf{v}_3}\mathbf{v}_4$, and so on. Compute $\sum_{n=1}^{\infty} |\mathbf{v}_n|$.

6. Find an equation of the largest sphere that passes through the point $(-1, 1, 4)$ and is such that each of the points (x, y, z) inside the sphere satisfies the condition

$$x^2 + y^2 + z^2 < 136 + 2(x + 2y + 3z)$$

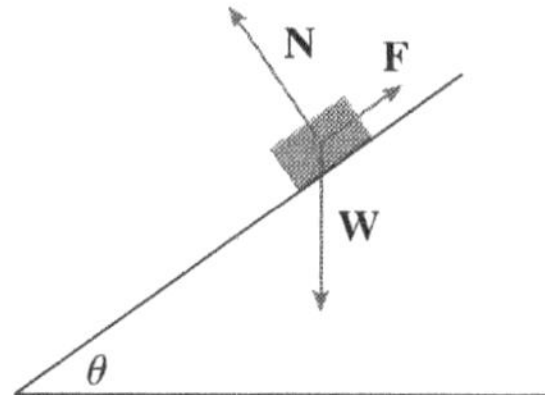

FIGURE FOR PROBLEM 7

7. Suppose a block of mass m is placed on an inclined plane, as shown in the figure. The block's descent down the plane is slowed by friction; if θ is not too large, friction will prevent the block from moving at all. The forces acting on the block are the weight $\mathbf{W}$, where $|\mathbf{W}| = mg$ (g is the acceleration due to gravity); the normal force $\mathbf{N}$ (the normal component of the reactionary force of the plane on the block), where $|\mathbf{N}| = n$; and the force $\mathbf{F}$ due to friction, which acts parallel to the inclined plane, opposing the direction of motion. If the block is at rest and θ is increased, $|\mathbf{F}|$ must also increase until ultimately $|\mathbf{F}|$ reaches its maximum, beyond which the block begins to slide. At this angle θ_s, it has been observed that $|\mathbf{F}|$ is proportional to n. Thus, when $|\mathbf{F}|$ is maximal, we can say that $|\mathbf{F}| = \mu_s n$, where μ_s is called the *coefficient of static friction* and depends on the materials that are in contact.
   (a) Observe that $\mathbf{N} + \mathbf{F} + \mathbf{W} = \mathbf{0}$ and deduce that $\mu_s = \tan(\theta_s)$.
   (b) Suppose that, for $\theta > \theta_s$, an additional outside force $\mathbf{H}$ is applied to the block, horizontally from the left, and let $|\mathbf{H}| = h$. If h is small, the block may still slide down the plane; if h is large enough, the block will move up the plane. Let $h_{\min}$ be the smallest value of h that allows the block to remain motionless (so that $|\mathbf{F}|$ is maximal).
   By choosing the coordinate axes so that $\mathbf{F}$ lies along the x-axis, resolve each force into components parallel and perpendicular to the inclined plane and show that

$$h_{\min} \sin\theta + mg\cos\theta = n \qquad \text{and} \qquad h_{\min}\cos\theta + \mu_s n = mg\sin\theta$$

   (c) Show that

$$h_{\min} = mg\tan(\theta - \theta_s)$$

   Does this equation seem reasonable? Does it make sense for $\theta = \theta_s$? As $\theta \to 90°$? Explain.

(d) Let h_{max} be the largest value of h that allows the block to remain motionless. (In which direction is **F** heading?) Show that

$$h_{max} = mg \tan(\theta + \theta_s)$$

Does this equation seem reasonable? Explain.

8. A solid has the following properties. When illuminated by rays parallel to the z-axis, its shadow is a circular disk. If the rays are parallel to the y-axis, its shadow is a square. If the rays are parallel to the x-axis, its shadow is an isosceles triangle. (In Exercise 44 in Section 12.1 you were asked to describe and sketch an example of such a solid, but there are many such solids.) Assume that the projection onto the xz-plane is a square whose sides have length 1.

(a) What is the volume of the largest such solid?

(b) Is there a smallest volume?

13 Vector Functions

Kepler's First Law says that the planets revolve around the sun in elliptical orbits. In Section 13.4 you will see how the material of this chapter is used in one of the great achievements of calculus: proving Kepler's Laws.

The functions that we have been using so far have been real-valued functions. We now study functions whose values are vectors because such functions are needed to describe curves and surfaces in space. We will also use vector-valued functions to describe the motion of objects through space. In particular, we will use them to derive Kepler's laws of planetary motion.

13.1 Vector Functions and Space Curves

In general, a function is a rule that assigns to each element in the domain an element in the range. A **vector-valued function**, or **vector function**, is simply a function whose domain is a set of real numbers and whose range is a set of vectors. We are most interested in vector functions $\mathbf{r}$ whose values are three-dimensional vectors. This means that for every number t in the domain of $\mathbf{r}$ there is a unique vector in V_3 denoted by $\mathbf{r}(t)$. If $f(t)$, $g(t)$, and $h(t)$ are the components of the vector $\mathbf{r}(t)$, then f, g, and h are real-valued functions called the **component functions** of $\mathbf{r}$ and we can write

$$\mathbf{r}(t) = \langle f(t), g(t), h(t) \rangle = f(t)\,\mathbf{i} + g(t)\,\mathbf{j} + h(t)\,\mathbf{k}$$

We use the letter t to denote the independent variable because it represents time in most applications of vector functions.

EXAMPLE 1 If

$$\mathbf{r}(t) = \langle t^3, \ln(3 - t), \sqrt{t} \rangle$$

then the component functions are

$$f(t) = t^3 \qquad g(t) = \ln(3 - t) \qquad h(t) = \sqrt{t}$$

By our usual convention, the domain of $\mathbf{r}$ consists of all values of t for which the expression for $\mathbf{r}(t)$ is defined. The expressions t^3, $\ln(3 - t)$, and $\sqrt{t}$ are all defined when $3 - t > 0$ and $t \geqslant 0$. Therefore the domain of $\mathbf{r}$ is the interval $[0, 3)$.

The **limit** of a vector function $\mathbf{r}$ is defined by taking the limits of its component functions as follows.

If $\lim_{t \to a} \mathbf{r}(t) = \mathbf{L}$, this definition is equivalent to saying that the length and direction of the vector $\mathbf{r}(t)$ approach the length and direction of the vector $\mathbf{L}$.

1 If $\mathbf{r}(t) = \langle f(t), g(t), h(t) \rangle$, then

$$\lim_{t \to a} \mathbf{r}(t) = \left\langle \lim_{t \to a} f(t), \lim_{t \to a} g(t), \lim_{t \to a} h(t) \right\rangle$$

provided the limits of the component functions exist.

Equivalently, we could have used an ε-δ definition (see Exercise 51). Limits of vector functions obey the same rules as limits of real-valued functions (see Exercise 49).

EXAMPLE 2 Find $\lim_{t \to 0} \mathbf{r}(t)$, where $\mathbf{r}(t) = (1 + t^3)\,\mathbf{i} + te^{-t}\,\mathbf{j} + \dfrac{\sin t}{t}\,\mathbf{k}$.

SOLUTION According to Definition 1, the limit of $\mathbf{r}$ is the vector whose components are the limits of the component functions of $\mathbf{r}$:

$$\lim_{t \to 0} \mathbf{r}(t) = \left[\lim_{t \to 0} (1 + t^3)\right]\mathbf{i} + \left[\lim_{t \to 0} te^{-t}\right]\mathbf{j} + \left[\lim_{t \to 0} \frac{\sin t}{t}\right]\mathbf{k}$$

$$= \mathbf{i} + \mathbf{k} \qquad \text{(by Equation 3.3.2)}$$

A vector function $\mathbf{r}$ is **continuous at a** if

$$\lim_{t \to a} \mathbf{r}(t) = \mathbf{r}(a)$$

In view of Definition 1, we see that $\mathbf{r}$ is continuous at a if and only if its component functions f, g, and h are continuous at a.

There is a close connection between continuous vector functions and space curves. Suppose that f, g, and h are continuous real-valued functions on an interval I. Then the set C of all points (x, y, z) in space, where

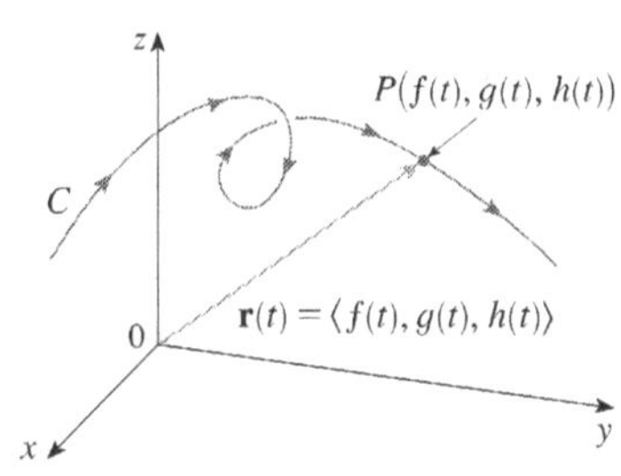

FIGURE 1
C is traced out by the tip of a moving position vector $\mathbf{r}(t)$.

2 $$x = f(t) \qquad y = g(t) \qquad z = h(t)$$

and t varies throughout the interval I, is called a **space curve**. The equations in **2** are called **parametric equations of C** and t is called a **parameter**. We can think of C as being traced out by a moving particle whose position at time t is $(f(t), g(t), h(t))$. If we now consider the vector function $\mathbf{r}(t) = \langle f(t), g(t), h(t) \rangle$, then $\mathbf{r}(t)$ is the position vector of the point $P(f(t), g(t), h(t))$ on C. Thus any continuous vector function $\mathbf{r}$ defines a space curve C that is traced out by the tip of the moving vector $\mathbf{r}(t)$, as shown in Figure 1.

V EXAMPLE 3 Describe the curve defined by the vector function

$$\mathbf{r}(t) = \langle 1 + t, 2 + 5t, -1 + 6t \rangle$$

TEC Visual 13.1A shows several curves being traced out by position vectors, including those in Figures 1 and 2.

SOLUTION The corresponding parametric equations are

$$x = 1 + t \qquad y = 2 + 5t \qquad z = -1 + 6t$$

which we recognize from Equations 12.5.2 as parametric equations of a line passing through the point $(1, 2, -1)$ and parallel to the vector $\langle 1, 5, 6 \rangle$. Alternatively, we could observe that the function can be written as $\mathbf{r} = \mathbf{r}_0 + t\mathbf{v}$, where $\mathbf{r}_0 = \langle 1, 2, -1 \rangle$ and $\mathbf{v} = \langle 1, 5, 6 \rangle$, and this is the vector equation of a line as given by Equation 12.5.1. ■

Plane curves can also be represented in vector notation. For instance, the curve given by the parametric equations $x = t^2 - 2t$ and $y = t + 1$ (see Example 1 in Section 10.1) could also be described by the vector equation

$$\mathbf{r}(t) = \langle t^2 - 2t, t + 1 \rangle = (t^2 - 2t)\,\mathbf{i} + (t + 1)\,\mathbf{j}$$

where $\mathbf{i} = \langle 1, 0 \rangle$ and $\mathbf{j} = \langle 0, 1 \rangle$.

V EXAMPLE 4 Sketch the curve whose vector equation is

$$\mathbf{r}(t) = \cos t\,\mathbf{i} + \sin t\,\mathbf{j} + t\,\mathbf{k}$$

SOLUTION The parametric equations for this curve are

$$x = \cos t \qquad y = \sin t \qquad z = t$$

Since $x^2 + y^2 = \cos^2 t + \sin^2 t = 1$, the curve must lie on the circular cylinder $x^2 + y^2 = 1$. The point (x, y, z) lies directly above the point $(x, y, 0)$, which moves counterclockwise around the circle $x^2 + y^2 = 1$ in the xy-plane. (The projection of the curve onto the xy-plane has vector equation $\mathbf{r}(t) = \langle \cos t, \sin t, 0 \rangle$. See Example 2 in Section 10.1.) Since $z = t$, the curve spirals upward around the cylinder as t increases. The curve, shown in Figure 2, is called a **helix**. ■

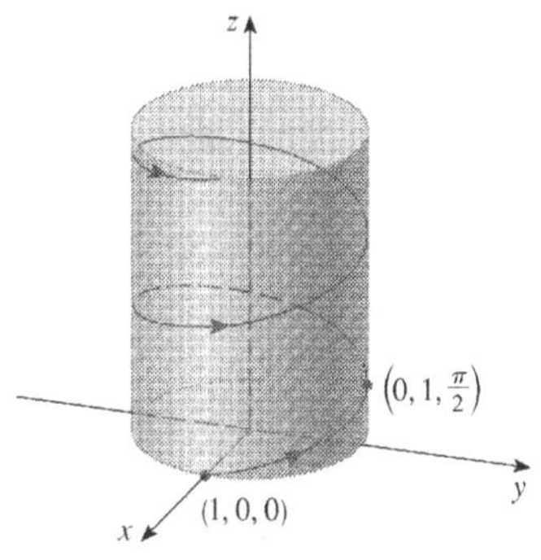

FIGURE 2

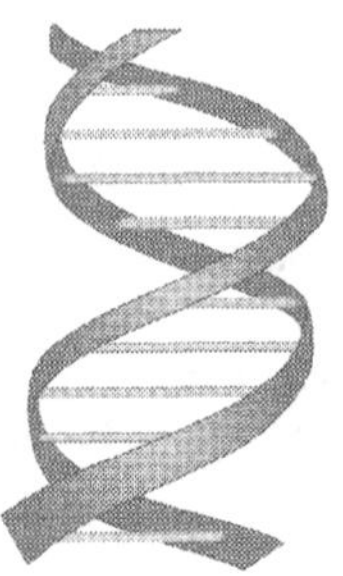

FIGURE 3
A double helix

The corkscrew shape of the helix in Example 4 is familiar from its occurrence in coiled springs. It also occurs in the model of DNA (deoxyribonucleic acid, the genetic material of living cells). In 1953 James Watson and Francis Crick showed that the structure of the DNA molecule is that of two linked, parallel helixes that are intertwined as in Figure 3.

In Examples 3 and 4 we were given vector equations of curves and asked for a geometric description or sketch. In the next two examples we are given a geometric description of a curve and are asked to find parametric equations for the curve.

EXAMPLE 5 Find a vector equation and parametric equations for the line segment that joins the point $P(1, 3, -2)$ to the point $Q(2, -1, 3)$.

Figure 4 shows the line segment PQ in Example 5.

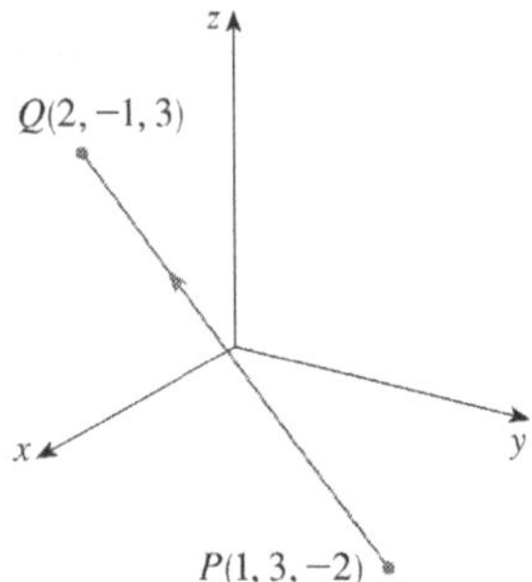

FIGURE 4

SOLUTION In Section 12.5 we found a vector equation for the line segment that joins the tip of the vector $\mathbf{r}_0$ to the tip of the vector $\mathbf{r}_1$:

$$\mathbf{r}(t) = (1 - t)\mathbf{r}_0 + t\mathbf{r}_1 \qquad 0 \leqslant t \leqslant 1$$

(See Equation 12.5.4.) Here we take $\mathbf{r}_0 = \langle 1, 3, -2 \rangle$ and $\mathbf{r}_1 = \langle 2, -1, 3 \rangle$ to obtain a vector equation of the line segment from P to Q:

$$\mathbf{r}(t) = (1 - t)\,\langle 1, 3, -2 \rangle + t\langle 2, -1, 3 \rangle \qquad 0 \leqslant t \leqslant 1$$

or

$$\mathbf{r}(t) = \langle 1 + t, 3 - 4t, -2 + 5t \rangle \qquad 0 \leqslant t \leqslant 1$$

The corresponding parametric equations are

$$x = 1 + t \qquad y = 3 - 4t \qquad z = -2 + 5t \qquad 0 \leqslant t \leqslant 1$$

V **EXAMPLE 6** Find a vector function that represents the curve of intersection of the cylinder $x^2 + y^2 = 1$ and the plane $y + z = 2$.

SOLUTION Figure 5 shows how the plane and the cylinder intersect, and Figure 6 shows the curve of intersection C, which is an ellipse.

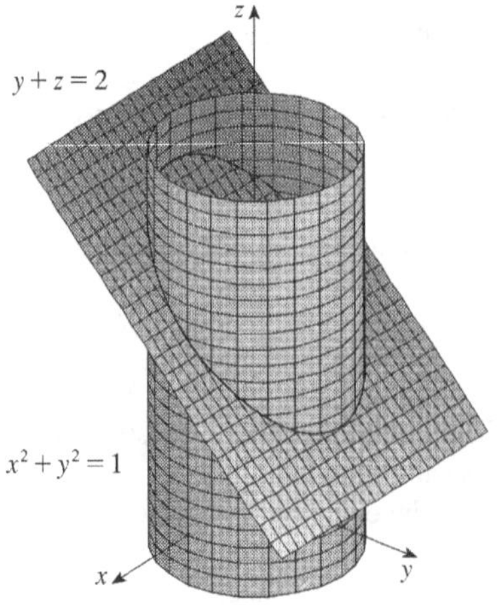

FIGURE 5

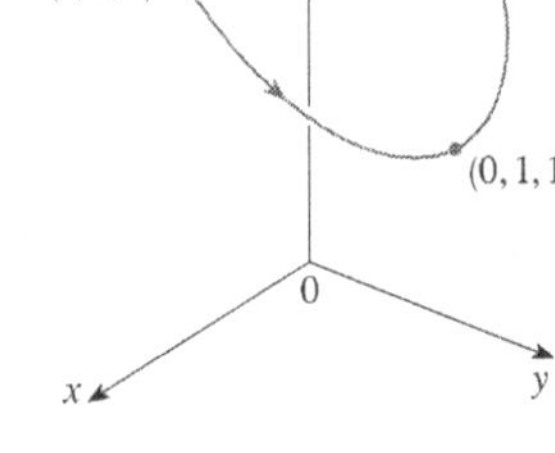

FIGURE 6

The projection of C onto the xy-plane is the circle $x^2 + y^2 = 1$, $z = 0$. So we know from Example 2 in Section 10.1 that we can write

$$x = \cos t \qquad y = \sin t \qquad 0 \leqslant t \leqslant 2\pi$$

From the equation of the plane, we have

$$z = 2 - y = 2 - \sin t$$

So we can write parametric equations for C as

$$x = \cos t \qquad y = \sin t \qquad z = 2 - \sin t \qquad 0 \leqslant t \leqslant 2\pi$$

The corresponding vector equation is

$$\mathbf{r}(t) = \cos t\,\mathbf{i} + \sin t\,\mathbf{j} + (2 - \sin t)\,\mathbf{k} \qquad 0 \leqslant t \leqslant 2\pi$$

This equation is called a *parametrization* of the curve C. The arrows in Figure 6 indicate the direction in which C is traced as the parameter t increases.

Using Computers to Draw Space Curves

Space curves are inherently more difficult to draw by hand than plane curves; for an accurate representation we need to use technology. For instance, Figure 7 shows a computer-generated graph of the curve with parametric equations

$$x = (4 + \sin 20t)\cos t \qquad y = (4 + \sin 20t)\sin t \qquad z = \cos 20t$$

It's called a **toroidal spiral** because it lies on a torus. Another interesting curve, the **trefoil knot**, with equations

$$x = (2 + \cos 1.5t)\cos t \qquad y = (2 + \cos 1.5t)\sin t \qquad z = \sin 1.5t$$

is graphed in Figure 8. It wouldn't be easy to plot either of these curves by hand.

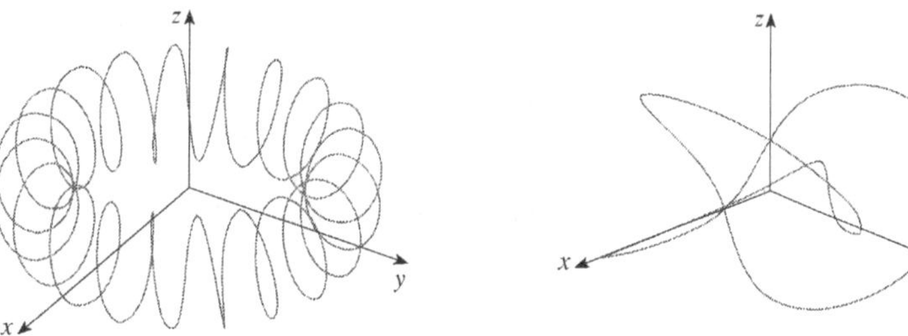

FIGURE 7 A toroidal spiral

FIGURE 8 A trefoil knot

Even when a computer is used to draw a space curve, optical illusions make it difficult to get a good impression of what the curve really looks like. (This is especially true in Figure 8. See Exercise 50.) The next example shows how to cope with this problem.

EXAMPLE 7 Use a computer to draw the curve with vector equation $\mathbf{r}(t) = \langle t, t^2, t^3 \rangle$. This curve is called a **twisted cubic**.

SOLUTION We start by using the computer to plot the curve with parametric equations $x = t$, $y = t^2$, $z = t^3$ for $-2 \leqslant t \leqslant 2$. The result is shown in Figure 9(a), but it's hard to see the true nature of the curve from that graph alone. Most three-dimensional computer graphing programs allow the user to enclose a curve or surface in a box instead of displaying the coordinate axes. When we look at the same curve in a box in Figure 9(b), we have a much clearer picture of the curve. We can see that it climbs from a lower corner of the box to the upper corner nearest us, and it twists as it climbs.

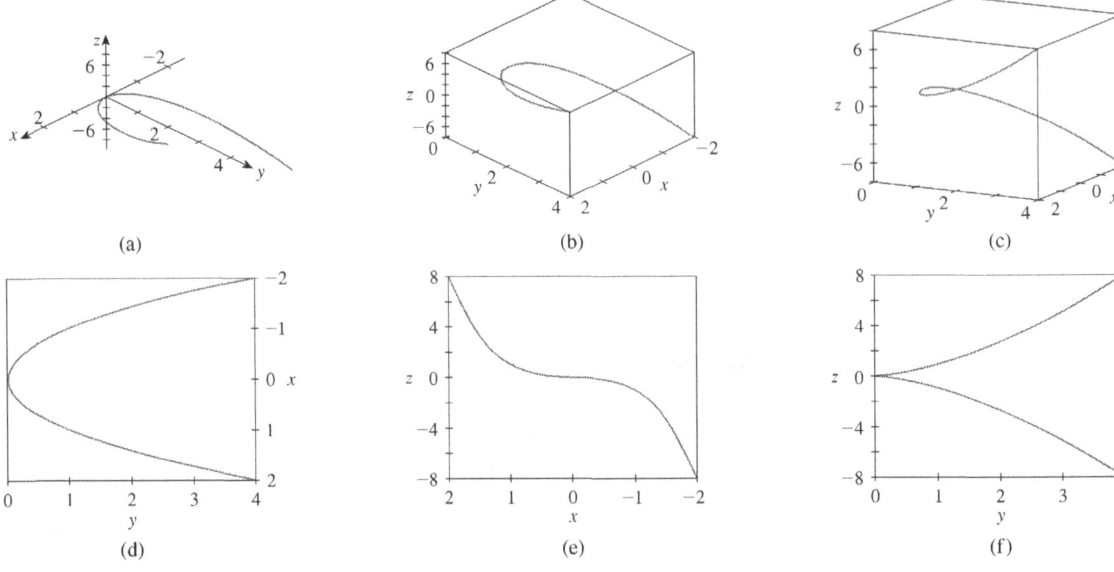

FIGURE 9 Views of the twisted cubic

TEC In Visual 13.1B you can rotate the box in Figure 9 to see the curve from any viewpoint.

We get an even better idea of the curve when we view it from different vantage points. Part (c) shows the result of rotating the box to give another viewpoint. Parts (d), (e), and (f) show the views we get when we look directly at a face of the box. In particular, part (d) shows the view from directly above the box. It is the projection of the curve on the xy-plane, namely, the parabola $y = x^2$. Part (e) shows the projection on the xz-plane, the cubic curve $z = x^3$. It's now obvious why the given curve is called a twisted cubic.

Another method of visualizing a space curve is to draw it on a surface. For instance, the twisted cubic in Example 7 lies on the parabolic cylinder $y = x^2$. (Eliminate the parameter from the first two parametric equations, $x = t$ and $y = t^2$.) Figure 10 shows both the cylinder and the twisted cubic, and we see that the curve moves upward from the origin along the surface of the cylinder. We also used this method in Example 4 to visualize the helix lying on the circular cylinder (see Figure 2).

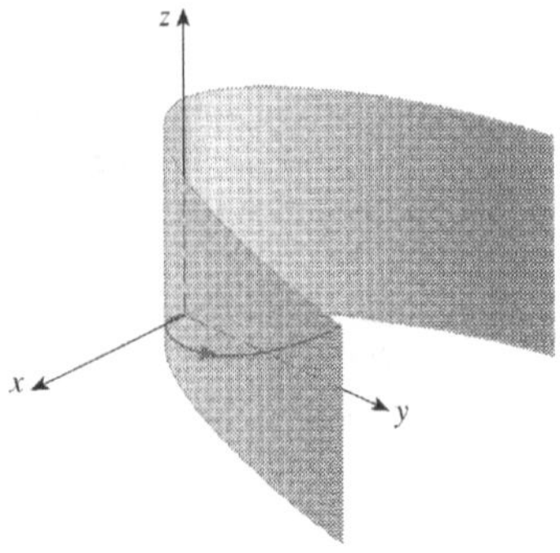

FIGURE 10

A third method for visualizing the twisted cubic is to realize that it also lies on the cylinder $z = x^3$. So it can be viewed as the curve of intersection of the cylinders $y = x^2$ and $z = x^3$. (See Figure 11.)

TEC Visual 13.1C shows how curves arise as intersections of surfaces.

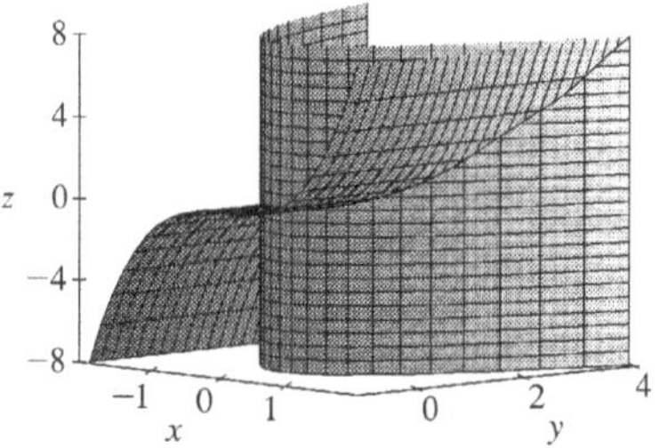

FIGURE 11

Some computer algebra systems provide us with a clearer picture of a space curve by enclosing it in a tube. Such a plot enables us to see whether one part of a curve passes in front of or behind another part of the curve. For example, Figure 13 shows the curve of Figure 12(b) as rendered by the `tubeplot` command in Maple.

We have seen that an interesting space curve, the helix, occurs in the model of DNA. Another notable example of a space curve in science is the trajectory of a positively charged particle in orthogonally oriented electric and magnetic fields **E** and **B**. Depending on the initial velocity given the particle at the origin, the path of the particle is either a space curve whose projection on the horizontal plane is the cycloid we studied in Section 10.1 [Figure 12(a)] or a curve whose projection is the trochoid investigated in Exercise 40 in Section 10.1 [Figure 12(b)].

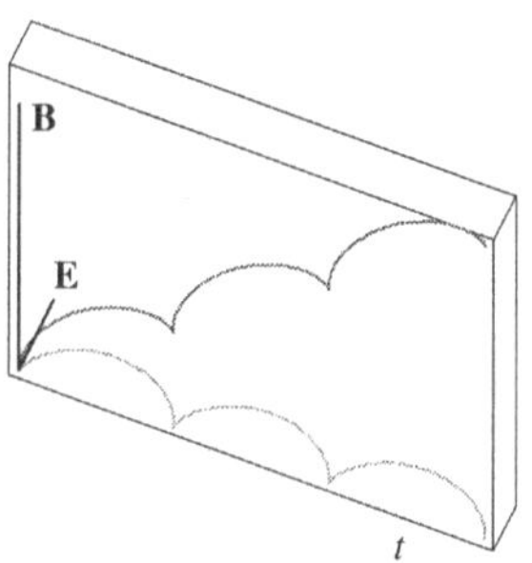

(a) $\mathbf{r}(t) = \langle t - \sin t, 1 - \cos t, t \rangle$

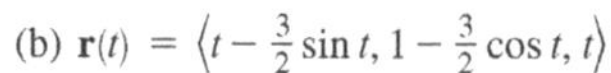

(b) $\mathbf{r}(t) = \left\langle t - \frac{3}{2}\sin t, 1 - \frac{3}{2}\cos t, t \right\rangle$

FIGURE 12
Motion of a charged particle in orthogonally oriented electric and magnetic fields

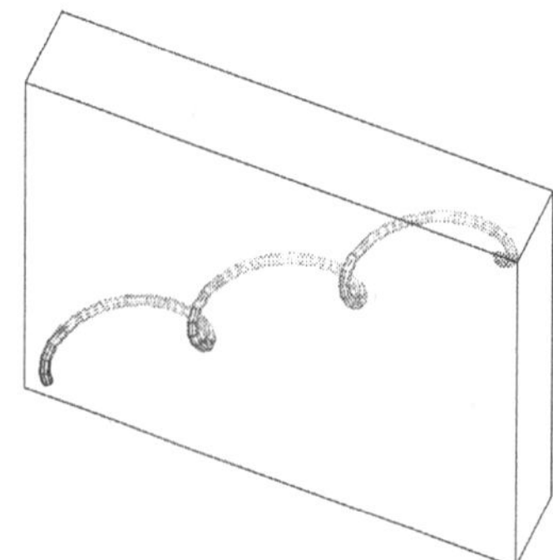

FIGURE 13

For further details concerning the physics involved and animations of the trajectories of the particles, see the following web sites:

- www.phy.ntnu.edu.tw/java/emField/emField.html
- www.physics.ucla.edu/plasma-exp/Beam/

13.1 Exercises

1–2 Find the domain of the vector function.

1. $\mathbf{r}(t) = \left\langle \sqrt{4 - t^2}, e^{-3t}, \ln(t + 1) \right\rangle$

2. $\mathbf{r}(t) = \dfrac{t - 2}{t + 2}\,\mathbf{i} + \sin t\,\mathbf{j} + \ln(9 - t^2)\,\mathbf{k}$

3–6 Find the limit.

3. $\displaystyle \lim_{t \to 0} \left(e^{-3t}\,\mathbf{i} + \frac{t^2}{\sin^2 t}\,\mathbf{j} + \cos 2t\,\mathbf{k} \right)$

4. $\displaystyle \lim_{t \to 1} \left(\frac{t^2 - t}{t - 1}\,\mathbf{i} + \sqrt{t + 8}\,\mathbf{j} + \frac{\sin \pi t}{\ln t}\,\mathbf{k} \right)$

5. $\lim_{t\to\infty} \left\langle \frac{1+t^2}{1-t^2}, \tan^{-1}t, \frac{1-e^{-2t}}{t} \right\rangle$

6. $\lim_{t\to\infty} \left\langle te^{-t}, \frac{t^3+t}{2t^3-1}, t\sin\frac{1}{t} \right\rangle$

7–14 Sketch the curve with the given vector equation. Indicate with an arrow the direction in which t increases.

7. $\mathbf{r}(t) = \langle \sin t, t \rangle$

8. $\mathbf{r}(t) = \langle t^3, t^2 \rangle$

9. $\mathbf{r}(t) = \langle t, 2-t, 2t \rangle$

10. $\mathbf{r}(t) = \langle \sin \pi t, t, \cos \pi t \rangle$

11. $\mathbf{r}(t) = \langle 1, \cos t, 2\sin t \rangle$

12. $\mathbf{r}(t) = t^2\mathbf{i} + t\mathbf{j} + 2\mathbf{k}$

13. $\mathbf{r}(t) = t^2\mathbf{i} + t^4\mathbf{j} + t^6\mathbf{k}$

14. $\mathbf{r}(t) = \cos t\,\mathbf{i} - \cos t\,\mathbf{j} + \sin t\,\mathbf{k}$

15–16 Draw the projections of the curve on the three coordinate planes. Use these projections to help sketch the curve.

15. $\mathbf{r}(t) = \langle t, \sin t, 2\cos t \rangle$

16. $\mathbf{r}(t) = \langle t, t, t^2 \rangle$

17–20 Find a vector equation and parametric equations for the line segment that joins P to Q.

17. $P(2, 0, 0)$, $Q(6, 2, -2)$

18. $P(-1, 2, -2)$, $Q(-3, 5, 1)$

19. $P(0, -1, 1)$, $Q(\frac{1}{2}, \frac{1}{3}, \frac{1}{4})$

20. $P(a, b, c)$, $Q(u, v, w)$

21–26 Match the parametric equations with the graphs (labeled I–VI). Give reasons for your choices.

I

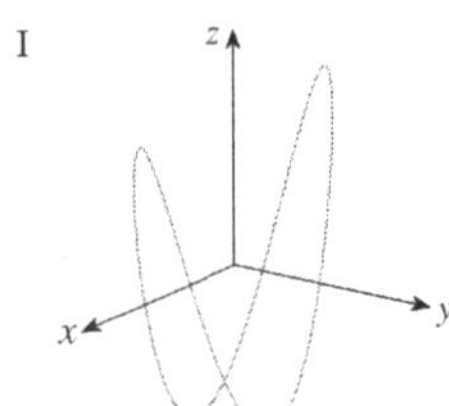

II

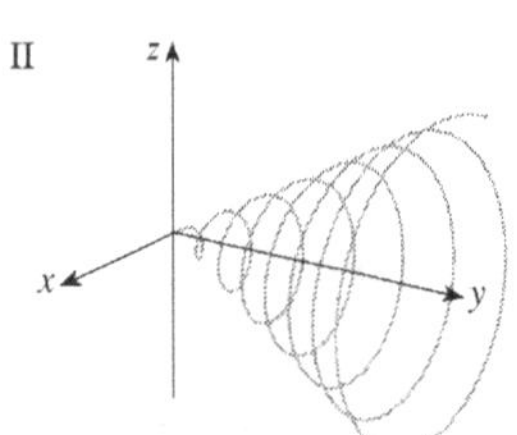

III

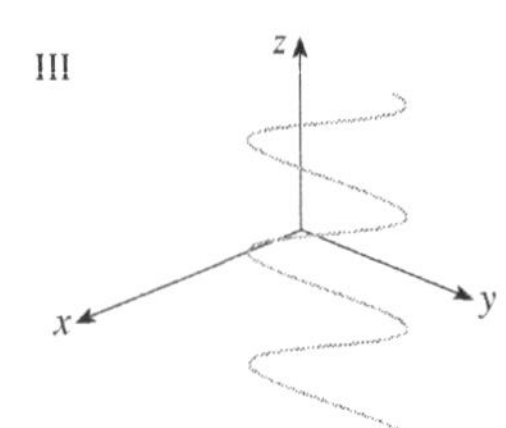

IV

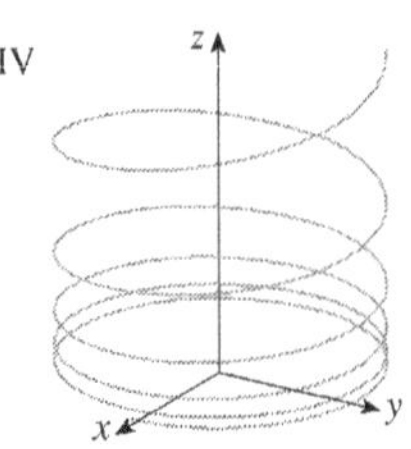

V

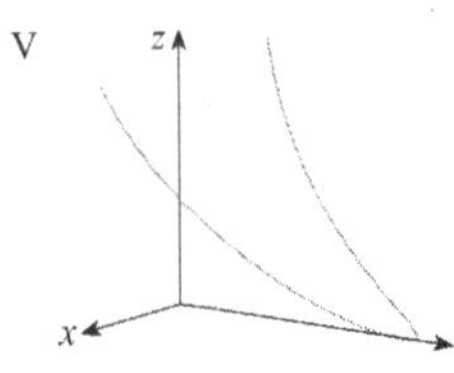

VI

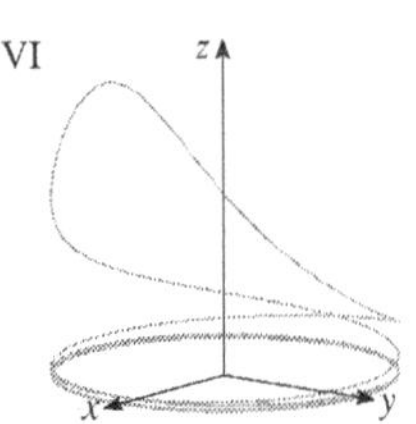

21. $x = t\cos t$, $y = t$, $z = t\sin t$, $t \geqslant 0$

22. $x = \cos t$, $y = \sin t$, $z = 1/(1+t^2)$

23. $x = t$, $y = 1/(1+t^2)$, $z = t^2$

24. $x = \cos t$, $y = \sin t$, $z = \cos 2t$

25. $x = \cos 8t$, $y = \sin 8t$, $z = e^{0.8t}$, $t \geqslant 0$

26. $x = \cos^2 t$, $y = \sin^2 t$, $z = t$

27. Show that the curve with parametric equations $x = t\cos t$, $y = t\sin t$, $z = t$ lies on the cone $z^2 = x^2 + y^2$, and use this fact to help sketch the curve.

28. Show that the curve with parametric equations $x = \sin t$, $y = \cos t$, $z = \sin^2 t$ is the curve of intersection of the surfaces $z = x^2$ and $x^2 + y^2 = 1$. Use this fact to help sketch the curve.

29. At what points does the curve $\mathbf{r}(t) = t\,\mathbf{i} + (2t - t^2)\,\mathbf{k}$ intersect the paraboloid $z = x^2 + y^2$?

30. At what points does the helix $\mathbf{r}(t) = \langle \sin t, \cos t, t \rangle$ intersect the sphere $x^2 + y^2 + z^2 = 5$?

31–35 Use a computer to graph the curve with the given vector equation. Make sure you choose a parameter domain and viewpoints that reveal the true nature of the curve.

31. $\mathbf{r}(t) = \langle \cos t \sin 2t, \sin t \sin 2t, \cos 2t \rangle$

32. $\mathbf{r}(t) = \langle t^2, \ln t, t \rangle$

33. $\mathbf{r}(t) = \langle t, t\sin t, t\cos t \rangle$

34. $\mathbf{r}(t) = \langle t, e^t, \cos t \rangle$

35. $\mathbf{r}(t) = \langle \cos 2t, \cos 3t, \cos 4t \rangle$

36. Graph the curve with parametric equations $x = \sin t$, $y = \sin 2t$, $z = \cos 4t$. Explain its shape by graphing its projections onto the three coordinate planes.

37. Graph the curve with parametric equations

$$x = (1 + \cos 16t)\cos t$$
$$y = (1 + \cos 16t)\sin t$$
$$z = 1 + \cos 16t$$

Explain the appearance of the graph by showing that it lies on a cone.

38. Graph the curve with parametric equations

$$x = \sqrt{1 - 0.25\cos^2 10t}\,\cos t$$
$$y = \sqrt{1 - 0.25\cos^2 10t}\,\sin t$$
$$z = 0.5\cos 10t$$

Explain the appearance of the graph by showing that it lies on a sphere.

39. Show that the curve with parametric equations $x = t^2$, $y = 1 - 3t$, $z = 1 + t^3$ passes through the points $(1, 4, 0)$ and $(9, -8, 28)$ but not through the point $(4, 7, -6)$.

40–44 Find a vector function that represents the curve of intersection of the two surfaces.

40. The cylinder $x^2 + y^2 = 4$ and the surface $z = xy$

41. The cone $z = \sqrt{x^2 + y^2}$ and the plane $z = 1 + y$

42. The paraboloid $z = 4x^2 + y^2$ and the parabolic cylinder $y = x^2$

43. The hyperboloid $z = x^2 - y^2$ and the cylinder $x^2 + y^2 = 1$

44. The semiellipsoid $x^2 + y^2 + 4z^2 = 4$, $y \geq 0$, and the cylinder $x^2 + z^2 = 1$

45. Try to sketch by hand the curve of intersection of the circular cylinder $x^2 + y^2 = 4$ and the parabolic cylinder $z = x^2$. Then find parametric equations for this curve and use these equations and a computer to graph the curve.

46. Try to sketch by hand the curve of intersection of the parabolic cylinder $y = x^2$ and the top half of the ellipsoid $x^2 + 4y^2 + 4z^2 = 16$. Then find parametric equations for this curve and use these equations and a computer to graph the curve.

47. If two objects travel through space along two different curves, it's often important to know whether they will collide. (Will a missile hit its moving target? Will two aircraft collide?) The curves might intersect, but we need to know whether the objects are in the same position *at the same time.* Suppose the trajectories of two particles are given by the vector functions

$$\mathbf{r}_1(t) = \langle t^2, 7t - 12, t^2 \rangle \qquad \mathbf{r}_2(t) = \langle 4t - 3, t^2, 5t - 6 \rangle$$

for $t \geq 0$. Do the particles collide?

48. Two particles travel along the space curves

$$\mathbf{r}_1(t) = \langle t, t^2, t^3 \rangle \qquad \mathbf{r}_2(t) = \langle 1 + 2t, 1 + 6t, 1 + 14t \rangle$$

Do the particles collide? Do their paths intersect?

49. Suppose **u** and **v** are vector functions that possess limits as $t \to a$ and let c be a constant. Prove the following properties of limits.

(a) $\lim_{t\to a} [\mathbf{u}(t) + \mathbf{v}(t)] = \lim_{t\to a} \mathbf{u}(t) + \lim_{t\to a} \mathbf{v}(t)$

(b) $\lim_{t\to a} c\mathbf{u}(t) = c \lim_{t\to a} \mathbf{u}(t)$

(c) $\lim_{t\to a} [\mathbf{u}(t) \cdot \mathbf{v}(t)] = \lim_{t\to a} \mathbf{u}(t) \cdot \lim_{t\to a} \mathbf{v}(t)$

(d) $\lim_{t\to a} [\mathbf{u}(t) \times \mathbf{v}(t)] = \lim_{t\to a} \mathbf{u}(t) \times \lim_{t\to a} \mathbf{v}(t)$

50. The view of the trefoil knot shown in Figure 8 is accurate, but it doesn't reveal the whole story. Use the parametric equations

$$x = (2 + \cos 1.5t) \cos t$$
$$y = (2 + \cos 1.5t) \sin t$$
$$z = \sin 1.5t$$

to sketch the curve by hand as viewed from above, with gaps indicating where the curve passes over itself. Start by showing that the projection of the curve onto the xy-plane has polar coordinates $r = 2 + \cos 1.5t$ and $\theta = t$, so r varies between 1 and 3. Then show that z has maximum and minimum values when the projection is halfway between $r = 1$ and $r = 3$.

When you have finished your sketch, use a computer to draw the curve with viewpoint directly above and compare with your sketch. Then use the computer to draw the curve from several other viewpoints. You can get a better impression of the curve if you plot a tube with radius 0.2 around the curve. (Use the `tubeplot` command in Maple or the `tubecurve` or `Tube` command in Mathematica.)

51. Show that $\lim_{t\to a} \mathbf{r}(t) = \mathbf{b}$ if and only if for every $\varepsilon > 0$ there is a number $\delta > 0$ such that

$$\text{if } 0 < |t - a| < \delta \quad \text{then} \quad |\mathbf{r}(t) - \mathbf{b}| < \varepsilon$$

13.2 Derivatives and Integrals of Vector Functions

Later in this chapter we are going to use vector functions to describe the motion of planets and other objects through space. Here we prepare the way by developing the calculus of vector functions.

Derivatives

The **derivative r′** of a vector function **r** is defined in much the same way as for real-valued functions:

$$\boxed{1} \qquad \frac{d\mathbf{r}}{dt} = \mathbf{r}'(t) = \lim_{h\to 0} \frac{\mathbf{r}(t + h) - \mathbf{r}(t)}{h}$$

if this limit exists. The geometric significance of this definition is shown in Figure 1. If the points P and Q have position vectors $\mathbf{r}(t)$ and $\mathbf{r}(t+h)$, then $\overrightarrow{PQ}$ represents the vector $\mathbf{r}(t+h) - \mathbf{r}(t)$, which can therefore be regarded as a secant vector. If $h > 0$, the scalar multiple $(1/h)(\mathbf{r}(t+h) - \mathbf{r}(t))$ has the same direction as $\mathbf{r}(t+h) - \mathbf{r}(t)$. As $h \to 0$, it appears that this vector approaches a vector that lies on the tangent line. For this reason, the vector $\mathbf{r}'(t)$ is called the **tangent vector** to the curve defined by $\mathbf{r}$ at the point P, provided that $\mathbf{r}'(t)$ exists and $\mathbf{r}'(t) \neq \mathbf{0}$. The **tangent line** to C at P is defined to be the line through P parallel to the tangent vector $\mathbf{r}'(t)$. We will also have occasion to consider the **unit tangent vector**, which is

$$\mathbf{T}(t) = \frac{\mathbf{r}'(t)}{|\mathbf{r}'(t)|}$$

TEC Visual 13.2 shows an animation of Figure 1.

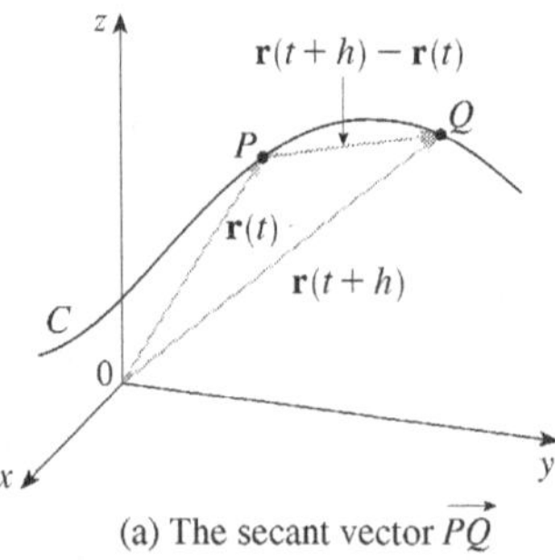

(a) The secant vector $\overrightarrow{PQ}$

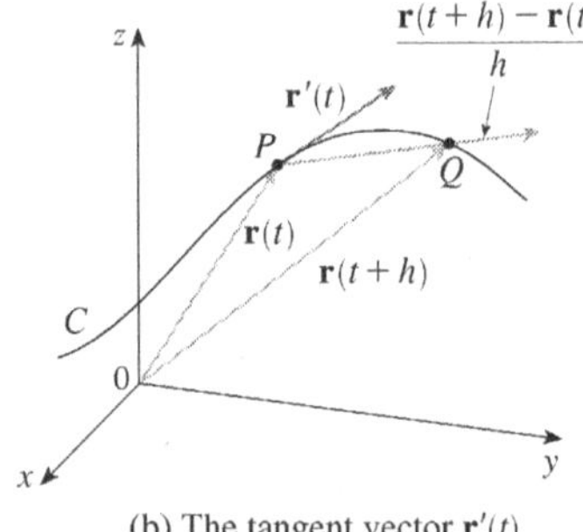

(b) The tangent vector $\mathbf{r}'(t)$

FIGURE 1

The following theorem gives us a convenient method for computing the derivative of a vector function $\mathbf{r}$: just differentiate each component of $\mathbf{r}$.

2 Theorem If $\mathbf{r}(t) = \langle f(t), g(t), h(t) \rangle = f(t)\,\mathbf{i} + g(t)\,\mathbf{j} + h(t)\,\mathbf{k}$, where f, g, and h are differentiable functions, then

$$\mathbf{r}'(t) = \langle f'(t), g'(t), h'(t) \rangle = f'(t)\,\mathbf{i} + g'(t)\,\mathbf{j} + h'(t)\,\mathbf{k}$$

PROOF

$$\begin{aligned}
\mathbf{r}'(t) &= \lim_{\Delta t \to 0} \frac{1}{\Delta t}[\mathbf{r}(t+\Delta t) - \mathbf{r}(t)] \\
&= \lim_{\Delta t \to 0} \frac{1}{\Delta t}[\langle f(t+\Delta t), g(t+\Delta t), h(t+\Delta t) \rangle - \langle f(t), g(t), h(t) \rangle] \\
&= \lim_{\Delta t \to 0} \left\langle \frac{f(t+\Delta t) - f(t)}{\Delta t}, \frac{g(t+\Delta t) - g(t)}{\Delta t}, \frac{h(t+\Delta t) - h(t)}{\Delta t} \right\rangle \\
&= \left\langle \lim_{\Delta t \to 0} \frac{f(t+\Delta t) - f(t)}{\Delta t}, \lim_{\Delta t \to 0} \frac{g(t+\Delta t) - g(t)}{\Delta t}, \lim_{\Delta t \to 0} \frac{h(t+\Delta t) - h(t)}{\Delta t} \right\rangle \\
&= \langle f'(t), g'(t), h'(t) \rangle
\end{aligned}$$

V EXAMPLE 1
(a) Find the derivative of $\mathbf{r}(t) = (1 + t^3)\mathbf{i} + te^{-t}\mathbf{j} + \sin 2t\,\mathbf{k}$.
(b) Find the unit tangent vector at the point where $t = 0$.

SOLUTION
(a) According to Theorem 2, we differentiate each component of $\mathbf{r}$:

$$\mathbf{r}'(t) = 3t^2\mathbf{i} + (1 - t)e^{-t}\mathbf{j} + 2\cos 2t\,\mathbf{k}$$

(b) Since $\mathbf{r}(0) = \mathbf{i}$ and $\mathbf{r}'(0) = \mathbf{j} + 2\mathbf{k}$, the unit tangent vector at the point $(1, 0, 0)$ is

$$\mathbf{T}(0) = \frac{\mathbf{r}'(0)}{|\mathbf{r}'(0)|} = \frac{\mathbf{j} + 2\mathbf{k}}{\sqrt{1 + 4}} = \frac{1}{\sqrt{5}}\mathbf{j} + \frac{2}{\sqrt{5}}\mathbf{k}$$

EXAMPLE 2 For the curve $\mathbf{r}(t) = \sqrt{t}\,\mathbf{i} + (2 - t)\,\mathbf{j}$, find $\mathbf{r}'(t)$ and sketch the position vector $\mathbf{r}(1)$ and the tangent vector $\mathbf{r}'(1)$.

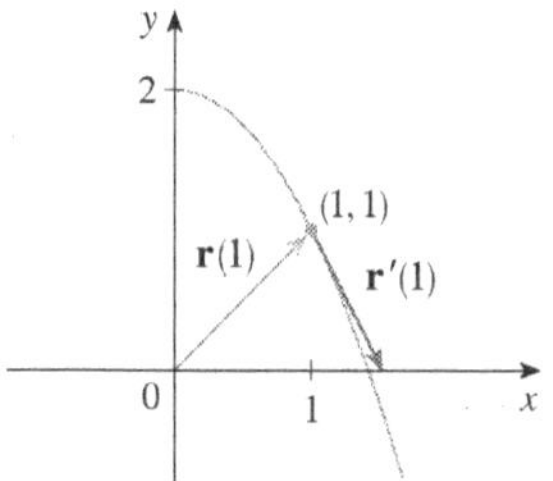

FIGURE 2

Notice from Figure 2 that the tangent vector points in the direction of increasing t. (See Exercise 56.)

SOLUTION We have

$$\mathbf{r}'(t) = \frac{1}{2\sqrt{t}}\mathbf{i} - \mathbf{j} \qquad \text{and} \qquad \mathbf{r}'(1) = \frac{1}{2}\mathbf{i} - \mathbf{j}$$

The curve is a plane curve and elimination of the parameter from the equations $x = \sqrt{t}$, $y = 2 - t$ gives $y = 2 - x^2$, $x \geqslant 0$. In Figure 2 we draw the position vector $\mathbf{r}(1) = \mathbf{i} + \mathbf{j}$ starting at the origin and the tangent vector $\mathbf{r}'(1)$ starting at the corresponding point $(1, 1)$.

V EXAMPLE 3 Find parametric equations for the tangent line to the helix with parametric equations

$$x = 2\cos t \qquad y = \sin t \qquad z = t$$

at the point $(0, 1, \pi/2)$.

SOLUTION The vector equation of the helix is $\mathbf{r}(t) = \langle 2\cos t, \sin t, t\rangle$, so

$$\mathbf{r}'(t) = \langle -2\sin t, \cos t, 1\rangle$$

The parameter value corresponding to the point $(0, 1, \pi/2)$ is $t = \pi/2$, so the tangent vector there is $\mathbf{r}'(\pi/2) = \langle -2, 0, 1\rangle$. The tangent line is the line through $(0, 1, \pi/2)$ parallel to the vector $\langle -2, 0, 1\rangle$, so by Equations 12.5.2 its parametric equations are

$$x = -2t \qquad y = 1 \qquad z = \frac{\pi}{2} + t$$

The helix and the tangent line in Example 3 are shown in Figure 3.

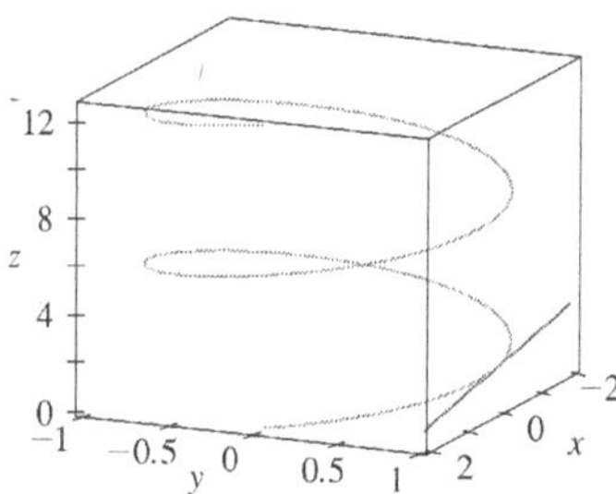

FIGURE 3

In Section 13.4 we will see how $\mathbf{r}'(t)$ and $\mathbf{r}''(t)$ can be interpreted as the velocity and acceleration vectors of a particle moving through space with position vector $\mathbf{r}(t)$ at time t.

Just as for real-valued functions, the **second derivative** of a vector function $\mathbf{r}$ is the derivative of $\mathbf{r}'$, that is, $\mathbf{r}'' = (\mathbf{r}')'$. For instance, the second derivative of the function in Example 3 is

$$\mathbf{r}''(t) = \langle -2\cos t, -\sin t, 0\rangle$$

Differentiation Rules

The next theorem shows that the differentiation formulas for real-valued functions have their counterparts for vector-valued functions.

3 Theorem Suppose $\mathbf{u}$ and $\mathbf{v}$ are differentiable vector functions, c is a scalar, and f is a real-valued function. Then

1. $\dfrac{d}{dt}[\mathbf{u}(t) + \mathbf{v}(t)] = \mathbf{u}'(t) + \mathbf{v}'(t)$

2. $\dfrac{d}{dt}[c\mathbf{u}(t)] = c\mathbf{u}'(t)$

3. $\dfrac{d}{dt}[f(t)\,\mathbf{u}(t)] = f'(t)\,\mathbf{u}(t) + f(t)\,\mathbf{u}'(t)$

4. $\dfrac{d}{dt}[\mathbf{u}(t)\cdot\mathbf{v}(t)] = \mathbf{u}'(t)\cdot\mathbf{v}(t) + \mathbf{u}(t)\cdot\mathbf{v}'(t)$

5. $\dfrac{d}{dt}[\mathbf{u}(t)\times\mathbf{v}(t)] = \mathbf{u}'(t)\times\mathbf{v}(t) + \mathbf{u}(t)\times\mathbf{v}'(t)$

6. $\dfrac{d}{dt}[\mathbf{u}(f(t))] = f'(t)\,\mathbf{u}'(f(t))$ (Chain Rule)

This theorem can be proved either directly from Definition 1 or by using Theorem 2 and the corresponding differentiation formulas for real-valued functions. The proof of Formula 4 follows; the remaining formulas are left as exercises.

PROOF OF FORMULA 4 Let

$$\mathbf{u}(t) = \langle f_1(t), f_2(t), f_3(t)\rangle \qquad \mathbf{v}(t) = \langle g_1(t), g_2(t), g_3(t)\rangle$$

Then

$$\mathbf{u}(t)\cdot\mathbf{v}(t) = f_1(t)\,g_1(t) + f_2(t)\,g_2(t) + f_3(t)\,g_3(t) = \sum_{i=1}^{3} f_i(t)\,g_i(t)$$

so the ordinary Product Rule gives

$$\begin{aligned}\frac{d}{dt}[\mathbf{u}(t)\cdot\mathbf{v}(t)] &= \frac{d}{dt}\sum_{i=1}^{3} f_i(t)\,g_i(t) = \sum_{i=1}^{3}\frac{d}{dt}[f_i(t)\,g_i(t)]\\ &= \sum_{i=1}^{3}[f_i'(t)\,g_i(t) + f_i(t)\,g_i'(t)]\\ &= \sum_{i=1}^{3} f_i'(t)\,g_i(t) + \sum_{i=1}^{3} f_i(t)\,g_i'(t)\\ &= \mathbf{u}'(t)\cdot\mathbf{v}(t) + \mathbf{u}(t)\cdot\mathbf{v}'(t)\end{aligned}$$

V EXAMPLE 4 Show that if $|\mathbf{r}(t)| = c$ (a constant), then $\mathbf{r}'(t)$ is orthogonal to $\mathbf{r}(t)$ for all t.

SOLUTION Since

$$\mathbf{r}(t) \cdot \mathbf{r}(t) = |\mathbf{r}(t)|^2 = c^2$$

and c^2 is a constant, Formula 4 of Theorem 3 gives

$$0 = \frac{d}{dt}[\mathbf{r}(t) \cdot \mathbf{r}(t)] = \mathbf{r}'(t) \cdot \mathbf{r}(t) + \mathbf{r}(t) \cdot \mathbf{r}'(t) = 2\mathbf{r}'(t) \cdot \mathbf{r}(t)$$

Thus $\mathbf{r}'(t) \cdot \mathbf{r}(t) = 0$, which says that $\mathbf{r}'(t)$ is orthogonal to $\mathbf{r}(t)$.

Geometrically, this result says that if a curve lies on a sphere with center the origin, then the tangent vector $\mathbf{r}'(t)$ is always perpendicular to the position vector $\mathbf{r}(t)$.

Integrals

The **definite integral** of a continuous vector function $\mathbf{r}(t)$ can be defined in much the same way as for real-valued functions except that the integral is a vector. But then we can express the integral of $\mathbf{r}$ in terms of the integrals of its component functions f, g, and h as follows. (We use the notation of Chapter 5.)

$$\int_a^b \mathbf{r}(t)\,dt = \lim_{n\to\infty} \sum_{i=1}^{n} \mathbf{r}(t_i^*)\,\Delta t$$

$$= \lim_{n\to\infty} \left[\left(\sum_{i=1}^{n} f(t_i^*)\,\Delta t \right) \mathbf{i} + \left(\sum_{i=1}^{n} g(t_i^*)\,\Delta t \right) \mathbf{j} + \left(\sum_{i=1}^{n} h(t_i^*)\,\Delta t \right) \mathbf{k} \right]$$

and so

$$\int_a^b \mathbf{r}(t)\,dt = \left(\int_a^b f(t)\,dt \right) \mathbf{i} + \left(\int_a^b g(t)\,dt \right) \mathbf{j} + \left(\int_a^b h(t)\,dt \right) \mathbf{k}$$

This means that we can evaluate an integral of a vector function by integrating each component function.

We can extend the Fundamental Theorem of Calculus to continuous vector functions as follows:

$$\int_a^b \mathbf{r}(t)\,dt = \mathbf{R}(t)\Big]_a^b = \mathbf{R}(b) - \mathbf{R}(a)$$

where $\mathbf{R}$ is an antiderivative of $\mathbf{r}$, that is, $\mathbf{R}'(t) = \mathbf{r}(t)$. We use the notation $\int \mathbf{r}(t)\,dt$ for indefinite integrals (antiderivatives).

EXAMPLE 5 If $\mathbf{r}(t) = 2\cos t\,\mathbf{i} + \sin t\,\mathbf{j} + 2t\,\mathbf{k}$, then

$$\int \mathbf{r}(t)\,dt = \left(\int 2\cos t\,dt \right) \mathbf{i} + \left(\int \sin t\,dt \right) \mathbf{j} + \left(\int 2t\,dt \right) \mathbf{k}$$

$$= 2\sin t\,\mathbf{i} - \cos t\,\mathbf{j} + t^2\mathbf{k} + \mathbf{C}$$

where $\mathbf{C}$ is a vector constant of integration, and

$$\int_0^{\pi/2} \mathbf{r}(t)\,dt = \left[2\sin t\,\mathbf{i} - \cos t\,\mathbf{j} + t^2\mathbf{k}\right]_0^{\pi/2} = 2\mathbf{i} + \mathbf{j} + \frac{\pi^2}{4}\mathbf{k}$$

13.2 Exercises

1. The figure shows a curve C given by a vector function $\mathbf{r}(t)$.
   (a) Draw the vectors $\mathbf{r}(4.5) - \mathbf{r}(4)$ and $\mathbf{r}(4.2) - \mathbf{r}(4)$.
   (b) Draw the vectors
   $$\frac{\mathbf{r}(4.5) - \mathbf{r}(4)}{0.5} \quad \text{and} \quad \frac{\mathbf{r}(4.2) - \mathbf{r}(4)}{0.2}$$
   (c) Write expressions for $\mathbf{r}'(4)$ and the unit tangent vector $\mathbf{T}(4)$.
   (d) Draw the vector $\mathbf{T}(4)$.

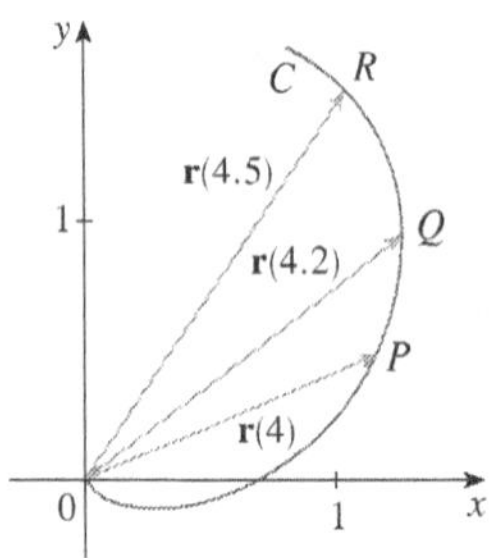

2. (a) Make a large sketch of the curve described by the vector function $\mathbf{r}(t) = \langle t^2, t \rangle$, $0 \leqslant t \leqslant 2$, and draw the vectors $\mathbf{r}(1)$, $\mathbf{r}(1.1)$, and $\mathbf{r}(1.1) - \mathbf{r}(1)$.
   (b) Draw the vector $\mathbf{r}'(1)$ starting at (1, 1), and compare it with the vector
   $$\frac{\mathbf{r}(1.1) - \mathbf{r}(1)}{0.1}$$
   Explain why these vectors are so close to each other in length and direction.

3–8
(a) Sketch the plane curve with the given vector equation.
(b) Find $\mathbf{r}'(t)$.
(c) Sketch the position vector $\mathbf{r}(t)$ and the tangent vector $\mathbf{r}'(t)$ for the given value of t.

3. $\mathbf{r}(t) = \langle t - 2, t^2 + 1 \rangle, \quad t = -1$

4. $\mathbf{r}(t) = \langle t^2, t^3 \rangle, \quad t = 1$

5. $\mathbf{r}(t) = \sin t\, \mathbf{i} + 2 \cos t\, \mathbf{j}, \quad t = \pi/4$

6. $\mathbf{r}(t) = e^t\, \mathbf{i} + e^{-t}\, \mathbf{j}, \quad t = 0$

7. $\mathbf{r}(t) = e^{2t}\, \mathbf{i} + e^t\, \mathbf{j}, \quad t = 0$

8. $\mathbf{r}(t) = (1 + \cos t)\, \mathbf{i} + (2 + \sin t)\, \mathbf{j}, \quad t = \pi/6$

9–16 Find the derivative of the vector function.

9. $\mathbf{r}(t) = \langle t \sin t, t^2, t \cos 2t \rangle$

10. $\mathbf{r}(t) = \langle \tan t, \sec t, 1/t^2 \rangle$

11. $\mathbf{r}(t) = t\, \mathbf{i} + \mathbf{j} + 2\sqrt{t}\, \mathbf{k}$

12. $\mathbf{r}(t) = \dfrac{1}{1 + t}\mathbf{i} + \dfrac{t}{1 + t}\mathbf{j} + \dfrac{t^2}{1 + t}\mathbf{k}$

13. $\mathbf{r}(t) = e^{t^2}\mathbf{i} - \mathbf{j} + \ln(1 + 3t)\, \mathbf{k}$

14. $\mathbf{r}(t) = at \cos 3t\, \mathbf{i} + b \sin^3 t\, \mathbf{j} + c \cos^3 t\, \mathbf{k}$

15. $\mathbf{r}(t) = \mathbf{a} + t\, \mathbf{b} + t^2 \mathbf{c}$

16. $\mathbf{r}(t) = t\, \mathbf{a} \times (\mathbf{b} + t\, \mathbf{c})$

17–20 Find the unit tangent vector $\mathbf{T}(t)$ at the point with the given value of the parameter t.

17. $\mathbf{r}(t) = \langle te^{-t}, 2 \arctan t, 2e^t \rangle, \quad t = 0$

18. $\mathbf{r}(t) = \langle t^3 + 3t, t^2 + 1, 3t + 4 \rangle, \quad t = 1$

19. $\mathbf{r}(t) = \cos t\, \mathbf{i} + 3t\, \mathbf{j} + 2 \sin 2t\, \mathbf{k}, \quad t = 0$

20. $\mathbf{r}(t) = \sin^2 t\, \mathbf{i} + \cos^2 t\, \mathbf{j} + \tan^2 t\, \mathbf{k}, \quad t = \pi/4$

21. If $\mathbf{r}(t) = \langle t, t^2, t^3 \rangle$, find $\mathbf{r}'(t)$, $\mathbf{T}(1)$, $\mathbf{r}''(t)$, and $\mathbf{r}'(t) \times \mathbf{r}''(t)$.

22. If $\mathbf{r}(t) = \langle e^{2t}, e^{-2t}, te^{2t} \rangle$, find $\mathbf{T}(0)$, $\mathbf{r}''(0)$, and $\mathbf{r}'(t) \cdot \mathbf{r}''(t)$.

23–26 Find parametric equations for the tangent line to the curve with the given parametric equations at the specified point.

23. $x = 1 + 2\sqrt{t}, \quad y = t^3 - t, \quad z = t^3 + t; \quad (3, 0, 2)$

24. $x = e^t, \quad y = te^t, \quad z = te^{t^2}; \quad (1, 0, 0)$

25. $x = e^{-t} \cos t, \quad y = e^{-t} \sin t, \quad z = e^{-t}; \quad (1, 0, 1)$

26. $x = \sqrt{t^2 + 3}, \quad y = \ln(t^2 + 3), \quad z = t; \quad (2, \ln 4, 1)$

27. Find a vector equation for the tangent line to the curve of intersection of the cylinders $x^2 + y^2 = 25$ and $y^2 + z^2 = 20$ at the point (3, 4, 2).

28. Find the point on the curve $\mathbf{r}(t) = \langle 2 \cos t, 2 \sin t, e^t \rangle$, $0 \leqslant t \leqslant \pi$, where the tangent line is parallel to the plane $\sqrt{3}x + y = 1$.

CAS 29–31 Find parametric equations for the tangent line to the curve with the given parametric equations at the specified point. Illustrate by graphing both the curve and the tangent line on a common screen.

29. $x = t, \ y = e^{-t}, \ z = 2t - t^2; \quad (0, 1, 0)$

30. $x = 2 \cos t, \ y = 2 \sin t, \ z = 4 \cos 2t; \quad (\sqrt{3}, 1, 2)$

31. $x = t \cos t, \ y = t, \ z = t \sin t; \quad (-\pi, \pi, 0)$

32. (a) Find the point of intersection of the tangent lines to the curve $\mathbf{r}(t) = \langle \sin \pi t, 2 \sin \pi t, \cos \pi t \rangle$ at the points where $t = 0$ and $t = 0.5$.
    (b) Illustrate by graphing the curve and both tangent lines.

33. The curves $\mathbf{r}_1(t) = \langle t, t^2, t^3 \rangle$ and $\mathbf{r}_2(t) = \langle \sin t, \sin 2t, t \rangle$ intersect at the origin. Find their angle of intersection correct to the nearest degree.

34. At what point do the curves $\mathbf{r}_1(t) = \langle t, 1 - t, 3 + t^2 \rangle$ and $\mathbf{r}_2(s) = \langle 3 - s, s - 2, s^2 \rangle$ intersect? Find their angle of intersection correct to the nearest degree.

35–40 Evaluate the integral.

35. $\int_0^2 (t\,\mathbf{i} - t^3\,\mathbf{j} + 3t^5\,\mathbf{k})\,dt$

36. $\int_0^1 \left(\frac{4}{1 + t^2}\,\mathbf{j} + \frac{2t}{1 + t^2}\,\mathbf{k} \right) dt$

37. $\int_0^{\pi/2} (3\sin^2 t \cos t\,\mathbf{i} + 3\sin t \cos^2 t\,\mathbf{j} + 2\sin t \cos t\,\mathbf{k})\,dt$

38. $\int_1^2 (t^2\,\mathbf{i} + t\sqrt{t - 1}\,\mathbf{j} + t\sin \pi t\,\mathbf{k})\,dt$

39. $\int (\sec^2 t\,\mathbf{i} + t(t^2 + 1)^3\,\mathbf{j} + t^2 \ln t\,\mathbf{k})\,dt$

40. $\int \left(te^{2t}\,\mathbf{i} + \frac{t}{1 - t}\,\mathbf{j} + \frac{1}{\sqrt{1 - t^2}}\,\mathbf{k} \right) dt$

41. Find $\mathbf{r}(t)$ if $\mathbf{r}'(t) = 2t\,\mathbf{i} + 3t^2\,\mathbf{j} + \sqrt{t}\,\mathbf{k}$ and $\mathbf{r}(1) = \mathbf{i} + \mathbf{j}$.

42. Find $\mathbf{r}(t)$ if $\mathbf{r}'(t) = t\,\mathbf{i} + e^t\,\mathbf{j} + te^t\,\mathbf{k}$ and $\mathbf{r}(0) = \mathbf{i} + \mathbf{j} + \mathbf{k}$.

43. Prove Formula 1 of Theorem 3.

44. Prove Formula 3 of Theorem 3.

45. Prove Formula 5 of Theorem 3.

46. Prove Formula 6 of Theorem 3.

47. If $\mathbf{u}(t) = \langle \sin t, \cos t, t \rangle$ and $\mathbf{v}(t) = \langle t, \cos t, \sin t \rangle$, use Formula 4 of Theorem 3 to find

$$\frac{d}{dt}[\mathbf{u}(t) \cdot \mathbf{v}(t)]$$

48. If $\mathbf{u}$ and $\mathbf{v}$ are the vector functions in Exercise 47, use Formula 5 of Theorem 3 to find

$$\frac{d}{dt}[\mathbf{u}(t) \times \mathbf{v}(t)]$$

49. Find $f'(2)$, where $f(t) = \mathbf{u}(t) \cdot \mathbf{v}(t)$, $\mathbf{u}(2) = \langle 1, 2, -1 \rangle$, $\mathbf{u}'(2) = \langle 3, 0, 4 \rangle$, and $\mathbf{v}(t) = \langle t, t^2, t^3 \rangle$.

50. If $\mathbf{r}(t) = \mathbf{u}(t) \times \mathbf{v}(t)$, where $\mathbf{u}$ and $\mathbf{v}$ are the vector functions in Exercise 49, find $\mathbf{r}'(2)$.

51. Show that if $\mathbf{r}$ is a vector function such that $\mathbf{r}''$ exists, then

$$\frac{d}{dt}[\mathbf{r}(t) \times \mathbf{r}'(t)] = \mathbf{r}(t) \times \mathbf{r}''(t)$$

52. Find an expression for $\frac{d}{dt}[\mathbf{u}(t) \cdot (\mathbf{v}(t) \times \mathbf{w}(t))]$.

53. If $\mathbf{r}(t) \neq \mathbf{0}$, show that $\frac{d}{dt}|\mathbf{r}(t)| = \frac{1}{|\mathbf{r}(t)|}\mathbf{r}(t) \cdot \mathbf{r}'(t)$.

[*Hint:* $|\mathbf{r}(t)|^2 = \mathbf{r}(t) \cdot \mathbf{r}(t)$]

54. If a curve has the property that the position vector $\mathbf{r}(t)$ is always perpendicular to the tangent vector $\mathbf{r}'(t)$, show that the curve lies on a sphere with center the origin.

55. If $\mathbf{u}(t) = \mathbf{r}(t) \cdot [\mathbf{r}'(t) \times \mathbf{r}''(t)]$, show that

$$\mathbf{u}'(t) = \mathbf{r}(t) \cdot [\mathbf{r}'(t) \times \mathbf{r}'''(t)]$$

56. Show that the tangent vector to a curve defined by a vector function $\mathbf{r}(t)$ points in the direction of increasing t. [*Hint:* Refer to Figure 1 and consider the cases $h > 0$ and $h < 0$ separately.]

13.3 Arc Length and Curvature

In Section 10.2 we defined the length of a plane curve with parametric equations $x = f(t)$, $y = g(t)$, $a \leqslant t \leqslant b$, as the limit of lengths of inscribed polygons and, for the case where f' and g' are continuous, we arrived at the formula

$$\boxed{1} \qquad L = \int_a^b \sqrt{[f'(t)]^2 + [g'(t)]^2}\,dt = \int_a^b \sqrt{\left(\frac{dx}{dt}\right)^2 + \left(\frac{dy}{dt}\right)^2}\,dt$$

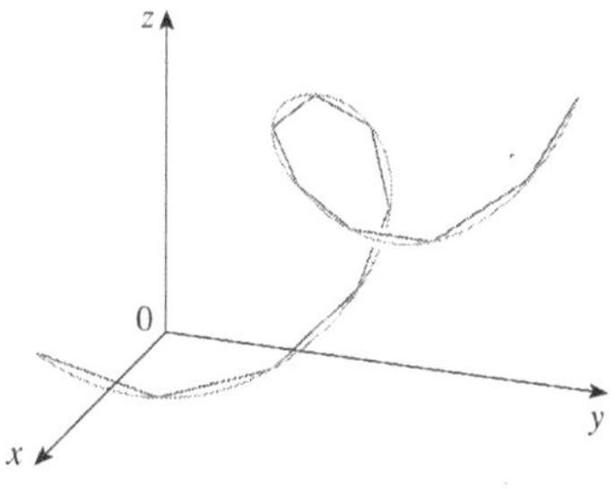

The length of a space curve is defined in exactly the same way (see Figure 1). Suppose that the curve has the vector equation $\mathbf{r}(t) = \langle f(t), g(t), h(t) \rangle$, $a \leqslant t \leqslant b$, or, equivalently, the parametric equations $x = f(t)$, $y = g(t)$, $z = h(t)$, where f', g', and h' are continuous. If the curve is traversed exactly once as t increases from a to b, then it can be shown that its length is

$$\boxed{2} \qquad L = \int_a^b \sqrt{[f'(t)]^2 + [g'(t)]^2 + [h'(t)]^2}\,dt = \int_a^b \sqrt{\left(\frac{dx}{dt}\right)^2 + \left(\frac{dy}{dt}\right)^2 + \left(\frac{dz}{dt}\right)^2}\,dt$$

FIGURE 1
The length of a space curve is the limit of lengths of inscribed polygons.

Notice that both of the arc length formulas [1] and [2] can be put into the more compact form

$$\boxed{3} \qquad L = \int_a^b |\mathbf{r}'(t)|\, dt$$

because, for plane curves $\mathbf{r}(t) = f(t)\,\mathbf{i} + g(t)\,\mathbf{j}$,

$$|\mathbf{r}'(t)| = |f'(t)\,\mathbf{i} + g'(t)\,\mathbf{j}| = \sqrt{[f'(t)]^2 + [g'(t)]^2}$$

and for space curves $\mathbf{r}(t) = f(t)\,\mathbf{i} + g(t)\,\mathbf{j} + h(t)\,\mathbf{k}$,

$$|\mathbf{r}'(t)| = |f'(t)\,\mathbf{i} + g'(t)\,\mathbf{j} + h'(t)\,\mathbf{k}| = \sqrt{[f'(t)]^2 + [g'(t)]^2 + [h'(t)]^2}$$

Figure 2 shows the arc of the helix whose length is computed in Example 1.

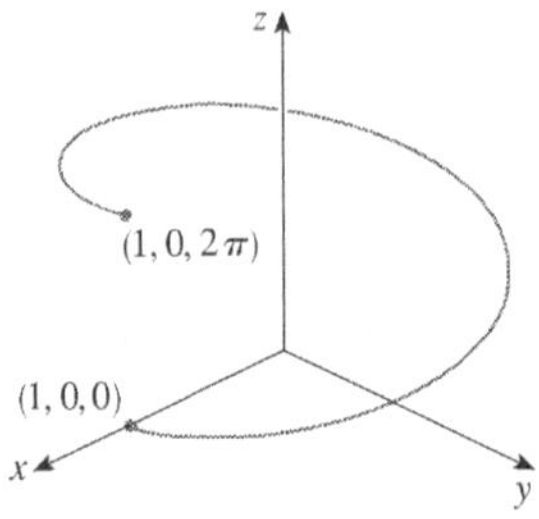

FIGURE 2

V EXAMPLE 1 Find the length of the arc of the circular helix with vector equation $\mathbf{r}(t) = \cos t\,\mathbf{i} + \sin t\,\mathbf{j} + t\,\mathbf{k}$ from the point $(1, 0, 0)$ to the point $(1, 0, 2\pi)$.

SOLUTION Since $\mathbf{r}'(t) = -\sin t\,\mathbf{i} + \cos t\,\mathbf{j} + \mathbf{k}$, we have

$$|\mathbf{r}'(t)| = \sqrt{(-\sin t)^2 + \cos^2 t + 1} = \sqrt{2}$$

The arc from $(1, 0, 0)$ to $(1, 0, 2\pi)$ is described by the parameter interval $0 \leqslant t \leqslant 2\pi$ and so, from Formula 3, we have

$$L = \int_0^{2\pi} |\mathbf{r}'(t)|\, dt = \int_0^{2\pi} \sqrt{2}\, dt = 2\sqrt{2}\,\pi$$

A single curve C can be represented by more than one vector function. For instance, the twisted cubic

$$\boxed{4} \qquad \mathbf{r}_1(t) = \langle t, t^2, t^3 \rangle \qquad 1 \leqslant t \leqslant 2$$

could also be represented by the function

$$\boxed{5} \qquad \mathbf{r}_2(u) = \langle e^u, e^{2u}, e^{3u} \rangle \qquad 0 \leqslant u \leqslant \ln 2$$

where the connection between the parameters t and u is given by $t = e^u$. We say that Equations 4 and 5 are **parametrizations** of the curve C. If we were to use Equation 3 to compute the length of C using Equations 4 and 5, we would get the same answer. In general, it can be shown that when Equation 3 is used to compute arc length, the answer is independent of the parametrization that is used.

Now we suppose that C is a curve given by a vector function

$$\mathbf{r}(t) = f(t)\mathbf{i} + g(t)\mathbf{j} + h(t)\mathbf{k} \qquad a \leqslant t \leqslant b$$

where $\mathbf{r}'$ is continuous and C is traversed exactly once as t increases from a to b. We define its **arc length function** s by

$$\boxed{6} \qquad s(t) = \int_a^t |\mathbf{r}'(u)|\, du = \int_a^t \sqrt{\left(\frac{dx}{du}\right)^2 + \left(\frac{dy}{du}\right)^2 + \left(\frac{dz}{du}\right)^2}\, du$$

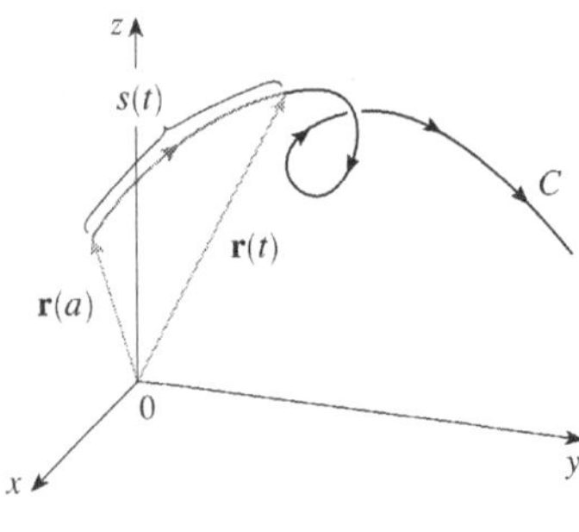

FIGURE 3

Thus $s(t)$ is the length of the part of C between $\mathbf{r}(a)$ and $\mathbf{r}(t)$. (See Figure 3.) If we differentiate both sides of Equation 6 using Part 1 of the Fundamental Theorem of Calculus, we obtain

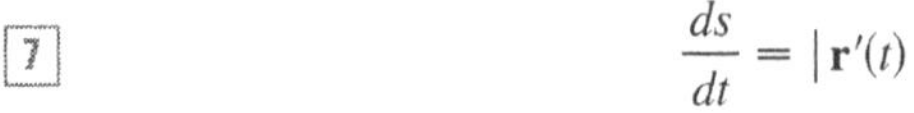

$$\boxed{7} \qquad \frac{ds}{dt} = |\mathbf{r}'(t)|$$

It is often useful to **parametrize a curve with respect to arc length** because arc length arises naturally from the shape of the curve and does not depend on a particular coordinate system. If a curve $\mathbf{r}(t)$ is already given in terms of a parameter t and $s(t)$ is the arc length function given by Equation 6, then we may be able to solve for t as a function of s: $t = t(s)$. Then the curve can be reparametrized in terms of s by substituting for t: $\mathbf{r} = \mathbf{r}(t(s))$. Thus, if $s = 3$ for instance, $\mathbf{r}(t(3))$ is the position vector of the point 3 units of length along the curve from its starting point.

EXAMPLE 2 Reparametrize the helix $\mathbf{r}(t) = \cos t\,\mathbf{i} + \sin t\,\mathbf{j} + t\,\mathbf{k}$ with respect to arc length measured from $(1, 0, 0)$ in the direction of increasing t.

SOLUTION The initial point $(1, 0, 0)$ corresponds to the parameter value $t = 0$. From Example 1 we have

$$\frac{ds}{dt} = |\mathbf{r}'(t)| = \sqrt{2}$$

and so

$$s = s(t) = \int_0^t |\mathbf{r}'(u)|\,du = \int_0^t \sqrt{2}\,du = \sqrt{2}\,t$$

Therefore $t = s/\sqrt{2}$ and the required reparametrization is obtained by substituting for t:

$$\mathbf{r}(t(s)) = \cos\left(s/\sqrt{2}\right)\mathbf{i} + \sin\left(s/\sqrt{2}\right)\mathbf{j} + \left(s/\sqrt{2}\right)\mathbf{k}$$

Curvature

A parametrization $\mathbf{r}(t)$ is called **smooth** on an interval I if $\mathbf{r}'$ is continuous and $\mathbf{r}'(t) \neq \mathbf{0}$ on I. A curve is called **smooth** if it has a smooth parametrization. A smooth curve has no sharp corners or cusps; when the tangent vector turns, it does so continuously.

TEC Visual 13.3A shows animated unit tangent vectors, like those in Figure 4, for a variety of plane curves and space curves.

If C is a smooth curve defined by the vector function $\mathbf{r}$, recall that the unit tangent vector $\mathbf{T}(t)$ is given by

$$\mathbf{T}(t) = \frac{\mathbf{r}'(t)}{|\mathbf{r}'(t)|}$$

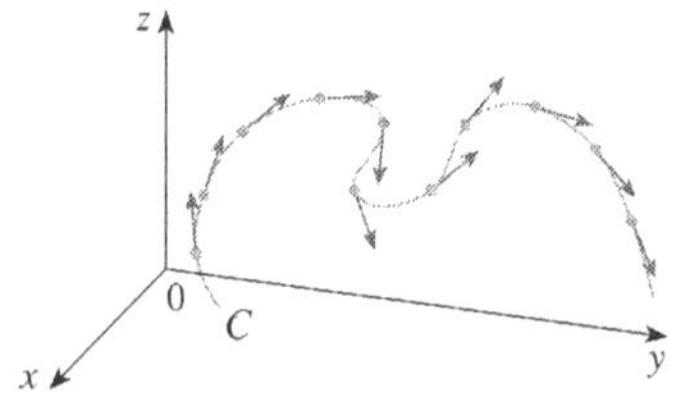

FIGURE 4
Unit tangent vectors at equally spaced points on C

and indicates the direction of the curve. From Figure 4 you can see that $\mathbf{T}(t)$ changes direction very slowly when C is fairly straight, but it changes direction more quickly when C bends or twists more sharply.

The curvature of C at a given point is a measure of how quickly the curve changes direction at that point. Specifically, we define it to be the magnitude of the rate of change of the unit tangent vector with respect to arc length. (We use arc length so that the curvature will be independent of the parametrization.)

8 Definition The **curvature** of a curve is

$$\kappa = \left|\frac{d\mathbf{T}}{ds}\right|$$

where $\mathbf{T}$ is the unit tangent vector.

The curvature is easier to compute if it is expressed in terms of the parameter t instead of s, so we use the Chain Rule (Theorem 13.2.3, Formula 6) to write

$$\frac{d\mathbf{T}}{dt} = \frac{d\mathbf{T}}{ds}\frac{ds}{dt} \qquad \text{and} \qquad \kappa = \left|\frac{d\mathbf{T}}{ds}\right| = \left|\frac{d\mathbf{T}/dt}{ds/dt}\right|$$

But $ds/dt = |\mathbf{r}'(t)|$ from Equation 7, so

9
$$\kappa(t) = \frac{|\mathbf{T}'(t)|}{|\mathbf{r}'(t)|}$$

V EXAMPLE 3 Show that the curvature of a circle of radius a is $1/a$.

SOLUTION We can take the circle to have center the origin, and then a parametrization is

$$\mathbf{r}(t) = a \cos t\,\mathbf{i} + a \sin t\,\mathbf{j}$$

Therefore $\qquad \mathbf{r}'(t) = -a \sin t\,\mathbf{i} + a \cos t\,\mathbf{j} \qquad$ and $\qquad |\mathbf{r}'(t)| = a$

so
$$\mathbf{T}(t) = \frac{\mathbf{r}'(t)}{|\mathbf{r}'(t)|} = -\sin t\,\mathbf{i} + \cos t\,\mathbf{j}$$

and
$$\mathbf{T}'(t) = -\cos t\,\mathbf{i} - \sin t\,\mathbf{j}$$

This gives $|\mathbf{T}'(t)| = 1$, so using Equation 9, we have

$$\kappa(t) = \frac{|\mathbf{T}'(t)|}{|\mathbf{r}'(t)|} = \frac{1}{a}$$

The result of Example 3 shows that small circles have large curvature and large circles have small curvature, in accordance with our intuition. We can see directly from the definition of curvature that the curvature of a straight line is always 0 because the tangent vector is constant.

Although Formula 9 can be used in all cases to compute the curvature, the formula given by the following theorem is often more convenient to apply.

10 Theorem The curvature of the curve given by the vector function $\mathbf{r}$ is

$$\kappa(t) = \frac{|\mathbf{r}'(t) \times \mathbf{r}''(t)|}{|\mathbf{r}'(t)|^3}$$

PROOF Since $\mathbf{T} = \mathbf{r}'/|\mathbf{r}'|$ and $|\mathbf{r}'| = ds/dt$, we have

$$\mathbf{r}' = |\mathbf{r}'|\mathbf{T} = \frac{ds}{dt}\mathbf{T}$$

so the Product Rule (Theorem 13.2.3, Formula 3) gives

$$\mathbf{r}'' = \frac{d^2s}{dt^2}\mathbf{T} + \frac{ds}{dt}\mathbf{T}'$$

Using the fact that $\mathbf{T} \times \mathbf{T} = \mathbf{0}$ (see Example 2 in Section 12.4), we have

$$\mathbf{r}' \times \mathbf{r}'' = \left(\frac{ds}{dt}\right)^2 (\mathbf{T} \times \mathbf{T}')$$

Now $|\mathbf{T}(t)| = 1$ for all t, so $\mathbf{T}$ and $\mathbf{T}'$ are orthogonal by Example 4 in Section 13.2. Therefore, by Theorem 12.4.9,

$$|\mathbf{r}' \times \mathbf{r}''| = \left(\frac{ds}{dt}\right)^2 |\mathbf{T} \times \mathbf{T}'| = \left(\frac{ds}{dt}\right)^2 |\mathbf{T}||\mathbf{T}'| = \left(\frac{ds}{dt}\right)^2 |\mathbf{T}'|$$

Thus

$$|\mathbf{T}'| = \frac{|\mathbf{r}' \times \mathbf{r}''|}{(ds/dt)^2} = \frac{|\mathbf{r}' \times \mathbf{r}''|}{|\mathbf{r}'|^2}$$

and

$$\kappa = \frac{|\mathbf{T}'|}{|\mathbf{r}'|} = \frac{|\mathbf{r}' \times \mathbf{r}''|}{|\mathbf{r}'|^3}$$

EXAMPLE 4 Find the curvature of the twisted cubic $\mathbf{r}(t) = \langle t, t^2, t^3 \rangle$ at a general point and at (0, 0, 0).

SOLUTION We first compute the required ingredients:

$$\mathbf{r}'(t) = \langle 1, 2t, 3t^2 \rangle \qquad \mathbf{r}''(t) = \langle 0, 2, 6t \rangle$$

$$|\mathbf{r}'(t)| = \sqrt{1 + 4t^2 + 9t^4}$$

$$\mathbf{r}'(t) \times \mathbf{r}''(t) = \begin{vmatrix} \mathbf{i} & \mathbf{j} & \mathbf{k} \\ 1 & 2t & 3t^2 \\ 0 & 2 & 6t \end{vmatrix} = 6t^2\,\mathbf{i} - 6t\,\mathbf{j} + 2\,\mathbf{k}$$

$$|\mathbf{r}'(t) \times \mathbf{r}''(t)| = \sqrt{36t^4 + 36t^2 + 4} = 2\sqrt{9t^4 + 9t^2 + 1}$$

Theorem 10 then gives

$$\kappa(t) = \frac{|\mathbf{r}'(t) \times \mathbf{r}''(t)|}{|\mathbf{r}'(t)|^3} = \frac{2\sqrt{1 + 9t^2 + 9t^4}}{(1 + 4t^2 + 9t^4)^{3/2}}$$

At the origin, where $t = 0$, the curvature is $\kappa(0) = 2$.

For the special case of a plane curve with equation $y = f(x)$, we choose x as the parameter and write $\mathbf{r}(x) = x\,\mathbf{i} + f(x)\,\mathbf{j}$. Then $\mathbf{r}'(x) = \mathbf{i} + f'(x)\,\mathbf{j}$ and $\mathbf{r}''(x) = f''(x)\,\mathbf{j}$. Since $\mathbf{i} \times \mathbf{j} = \mathbf{k}$ and $\mathbf{j} \times \mathbf{j} = \mathbf{0}$, it follows that $\mathbf{r}'(x) \times \mathbf{r}''(x) = f''(x)\,\mathbf{k}$. We also have $|\mathbf{r}'(x)| = \sqrt{1 + [f'(x)]^2}$ and so, by Theorem 10,

11
$$\kappa(x) = \frac{|f''(x)|}{[1 + (f'(x))^2]^{3/2}}$$

EXAMPLE 5 Find the curvature of the parabola $y = x^2$ at the points (0, 0), (1, 1), and (2, 4).

SOLUTION Since $y' = 2x$ and $y'' = 2$, Formula 11 gives

$$\kappa(x) = \frac{|y''|}{[1 + (y')^2]^{3/2}} = \frac{2}{(1 + 4x^2)^{3/2}}$$

The curvature at $(0, 0)$ is $\kappa(0) = 2$. At $(1, 1)$ it is $\kappa(1) = 2/5^{3/2} \approx 0.18$. At $(2, 4)$ it is $\kappa(2) = 2/17^{3/2} \approx 0.03$. Observe from the expression for $\kappa(x)$ or the graph of κ in Figure 5 that $\kappa(x) \to 0$ as $x \to \pm\infty$. This corresponds to the fact that the parabola appears to become flatter as $x \to \pm\infty$.

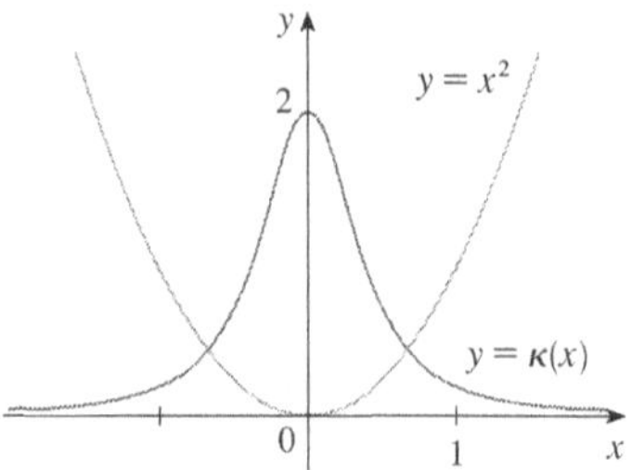

FIGURE 5
The parabola $y = x^2$ and its curvature function

The Normal and Binormal Vectors

We can think of the normal vector as indicating the direction in which the curve is turning at each point.

At a given point on a smooth space curve $\mathbf{r}(t)$, there are many vectors that are orthogonal to the unit tangent vector $\mathbf{T}(t)$. We single out one by observing that, because $|\mathbf{T}(t)| = 1$ for all t, we have $\mathbf{T}(t) \cdot \mathbf{T}'(t) = 0$ by Example 4 in Section 13.2, so $\mathbf{T}'(t)$ is orthogonal to $\mathbf{T}(t)$. Note that $\mathbf{T}'(t)$ is itself not a unit vector. But at any point where $\kappa \neq 0$ we can define the **principal unit normal vector** $\mathbf{N}(t)$ (or simply **unit normal**) as

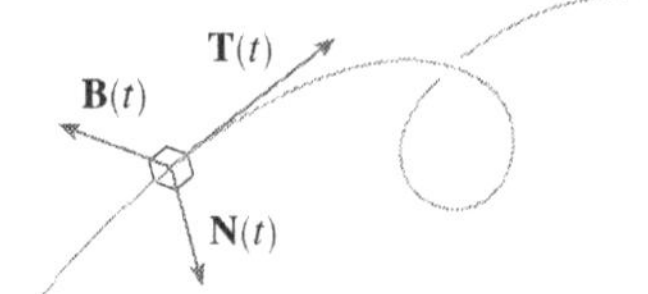

FIGURE 6

$$\mathbf{N}(t) = \frac{\mathbf{T}'(t)}{|\mathbf{T}'(t)|}$$

The vector $\mathbf{B}(t) = \mathbf{T}(t) \times \mathbf{N}(t)$ is called the **binormal vector**. It is perpendicular to both $\mathbf{T}$ and $\mathbf{N}$ and is also a unit vector. (See Figure 6.)

Figure 7 illustrates Example 6 by showing the vectors **T**, **N**, and **B** at two locations on the helix. In general, the vectors **T**, **N**, and **B**, starting at the various points on a curve, form a set of orthogonal vectors, called the **TNB** frame, that moves along the curve as t varies. This **TNB** frame plays an important role in the branch of mathematics known as differential geometry and in its applications to the motion of spacecraft.

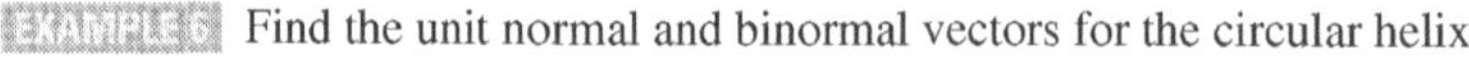

EXAMPLE 6 Find the unit normal and binormal vectors for the circular helix

$$\mathbf{r}(t) = \cos t\,\mathbf{i} + \sin t\,\mathbf{j} + t\,\mathbf{k}$$

SOLUTION We first compute the ingredients needed for the unit normal vector:

$$\mathbf{r}'(t) = -\sin t\,\mathbf{i} + \cos t\,\mathbf{j} + \mathbf{k} \qquad |\mathbf{r}'(t)| = \sqrt{2}$$

$$\mathbf{T}(t) = \frac{\mathbf{r}'(t)}{|\mathbf{r}'(t)|} = \frac{1}{\sqrt{2}}(-\sin t\,\mathbf{i} + \cos t\,\mathbf{j} + \mathbf{k})$$

$$\mathbf{T}'(t) = \frac{1}{\sqrt{2}}(-\cos t\,\mathbf{i} - \sin t\,\mathbf{j}) \qquad |\mathbf{T}'(t)| = \frac{1}{\sqrt{2}}$$

$$\mathbf{N}(t) = \frac{\mathbf{T}'(t)}{|\mathbf{T}'(t)|} = -\cos t\,\mathbf{i} - \sin t\,\mathbf{j} = \langle -\cos t, -\sin t, 0\rangle$$

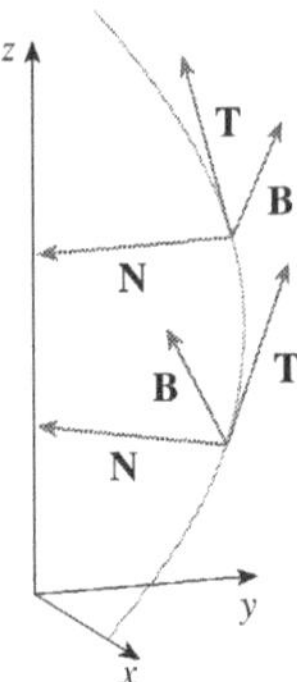

FIGURE 7

This shows that the normal vector at any point on the helix is horizontal and points toward the z-axis. The binormal vector is

$$\mathbf{B}(t) = \mathbf{T}(t) \times \mathbf{N}(t) = \frac{1}{\sqrt{2}}\begin{bmatrix} \mathbf{i} & \mathbf{j} & \mathbf{k} \\ -\sin t & \cos t & 1 \\ -\cos t & -\sin t & 0 \end{bmatrix} = \frac{1}{\sqrt{2}}\langle \sin t, -\cos t, 1\rangle$$

TEC Visual 13.3B shows how the TNB frame moves along several curves.

The plane determined by the normal and binormal vectors **N** and **B** at a point P on a curve C is called the **normal plane** of C at P. It consists of all lines that are orthogonal to the tangent vector **T**. The plane determined by the vectors **T** and **N** is called the **osculating plane** of C at P. The name comes from the Latin *osculum*, meaning "kiss." It is the plane that comes closest to containing the part of the curve near P. (For a plane curve, the osculating plane is simply the plane that contains the curve.)

The circle that lies in the osculating plane of C at P, has the same tangent as C at P, lies on the concave side of C (toward which **N** points), and has radius $\rho = 1/\kappa$ (the reciprocal of the curvature) is called the **osculating circle** (or the **circle of curvature**) of C at P. It is the circle that best describes how C behaves near P; it shares the same tangent, normal, and curvature at P.

V EXAMPLE 7 Find the equations of the normal plane and osculating plane of the helix in Example 6 at the point $P(0, 1, \pi/2)$.

Figure 8 shows the helix and the osculating plane in Example 7.

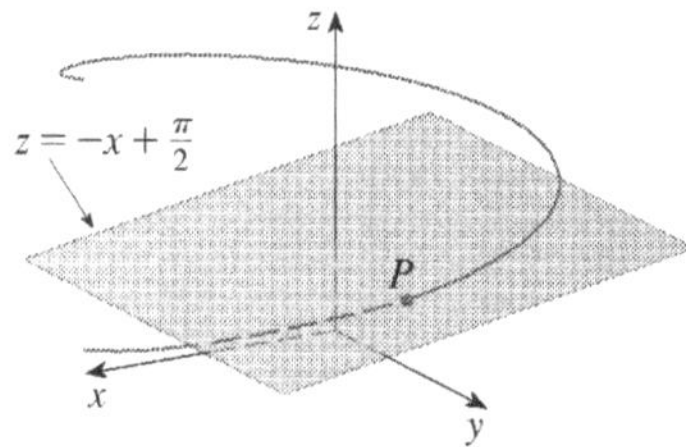

FIGURE 8

SOLUTION The normal plane at P has normal vector $\mathbf{r}'(\pi/2) = \langle -1, 0, 1 \rangle$, so an equation is

$$-1(x - 0) + 0(y - 1) + 1\left(z - \frac{\pi}{2}\right) = 0 \qquad \text{or} \qquad z = x + \frac{\pi}{2}$$

The osculating plane at P contains the vectors **T** and **N**, so its normal vector is $\mathbf{T} \times \mathbf{N} = \mathbf{B}$. From Example 6 we have

$$\mathbf{B}(t) = \frac{1}{\sqrt{2}} \langle \sin t, -\cos t, 1 \rangle \qquad \mathbf{B}\left(\frac{\pi}{2}\right) = \left\langle \frac{1}{\sqrt{2}}, 0, \frac{1}{\sqrt{2}} \right\rangle$$

A simpler normal vector is $\langle 1, 0, 1 \rangle$, so an equation of the osculating plane is

$$1(x - 0) + 0(y - 1) + 1\left(z - \frac{\pi}{2}\right) = 0 \qquad \text{or} \qquad z = -x + \frac{\pi}{2}$$

EXAMPLE 8 Find and graph the osculating circle of the parabola $y = x^2$ at the origin.

SOLUTION From Example 5, the curvature of the parabola at the origin is $\kappa(0) = 2$. So the radius of the osculating circle at the origin is $1/\kappa = \frac{1}{2}$ and its center is $\left(0, \frac{1}{2}\right)$. Its equation is therefore

$$x^2 + \left(y - \tfrac{1}{2}\right)^2 = \tfrac{1}{4}$$

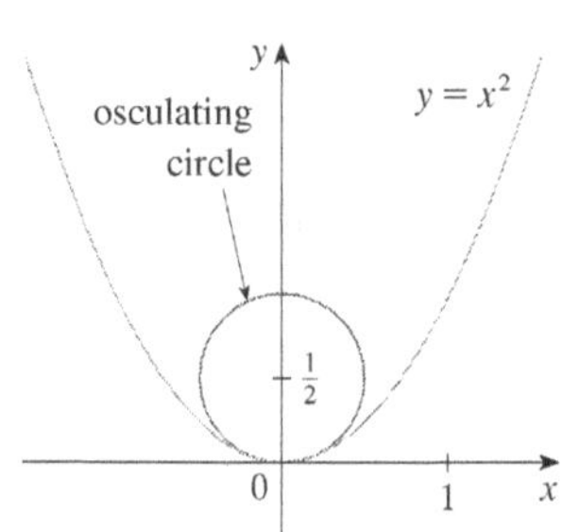

FIGURE 9

For the graph in Figure 9 we use parametric equations of this circle:

$$x = \tfrac{1}{2} \cos t \qquad y = \tfrac{1}{2} + \tfrac{1}{2} \sin t$$

We summarize here the formulas for unit tangent, unit normal and binormal vectors, and curvature.

$$\mathbf{T}(t) = \frac{\mathbf{r}'(t)}{|\mathbf{r}'(t)|} \qquad \mathbf{N}(t) = \frac{\mathbf{T}'(t)}{|\mathbf{T}'(t)|} \qquad \mathbf{B}(t) = \mathbf{T}(t) \times \mathbf{N}(t)$$

$$\kappa = \left| \frac{d\mathbf{T}}{ds} \right| = \frac{|\mathbf{T}'(t)|}{|\mathbf{r}'(t)|} = \frac{|\mathbf{r}'(t) \times \mathbf{r}''(t)|}{|\mathbf{r}'(t)|^3}$$

TEC Visual 13.3C shows how the osculating circle changes as a point moves along a curve.

13.3 Exercises

1–6 Find the length of the curve.

1. $\mathbf{r}(t) = \langle t, 3\cos t, 3\sin t\rangle, \quad -5 \leqslant t \leqslant 5$

2. $\mathbf{r}(t) = \langle 2t, t^2, \frac{1}{3}t^3\rangle, \quad 0 \leqslant t \leqslant 1$

3. $\mathbf{r}(t) = \sqrt{2}\,t\,\mathbf{i} + e^t\,\mathbf{j} + e^{-t}\,\mathbf{k}, \quad 0 \leqslant t \leqslant 1$

4. $\mathbf{r}(t) = \cos t\,\mathbf{i} + \sin t\,\mathbf{j} + \ln\cos t\,\mathbf{k}, \quad 0 \leqslant t \leqslant \pi/4$

5. $\mathbf{r}(t) = \mathbf{i} + t^2\,\mathbf{j} + t^3\,\mathbf{k}, \quad 0 \leqslant t \leqslant 1$

6. $\mathbf{r}(t) = 12t\,\mathbf{i} + 8t^{3/2}\,\mathbf{j} + 3t^2\,\mathbf{k}, \quad 0 \leqslant t \leqslant 1$

7–9 Find the length of the curve correct to four decimal places. (Use your calculator to approximate the integral.)

7. $\mathbf{r}(t) = \langle t^2, t^3, t^4\rangle, \quad 0 \leqslant t \leqslant 2$

8. $\mathbf{r}(t) = \langle t, e^{-t}, te^{-t}\rangle, \quad 1 \leqslant t \leqslant 3$

9. $\mathbf{r}(t) = \langle \sin t, \cos t, \tan t\rangle, \quad 0 \leqslant t \leqslant \pi/4$

10. Graph the curve with parametric equations $x = \sin t$, $y = \sin 2t$, $z = \sin 3t$. Find the total length of this curve correct to four decimal places.

11. Let C be the curve of intersection of the parabolic cylinder $x^2 = 2y$ and the surface $3z = xy$. Find the exact length of C from the origin to the point $(6, 18, 36)$.

12. Find, correct to four decimal places, the length of the curve of intersection of the cylinder $4x^2 + y^2 = 4$ and the plane $x + y + z = 2$.

13–14 Reparametrize the curve with respect to arc length measured from the point where $t = 0$ in the direction of increasing t.

13. $\mathbf{r}(t) = 2t\,\mathbf{i} + (1 - 3t)\,\mathbf{j} + (5 + 4t)\,\mathbf{k}$

14. $\mathbf{r}(t) = e^{2t}\cos 2t\,\mathbf{i} + 2\,\mathbf{j} + e^{2t}\sin 2t\,\mathbf{k}$

15. Suppose you start at the point $(0, 0, 3)$ and move 5 units along the curve $x = 3\sin t$, $y = 4t$, $z = 3\cos t$ in the positive direction. Where are you now?

16. Reparametrize the curve

$$\mathbf{r}(t) = \left(\frac{2}{t^2+1} - 1\right)\mathbf{i} + \frac{2t}{t^2+1}\mathbf{j}$$

with respect to arc length measured from the point $(1, 0)$ in the direction of increasing t. Express the reparametrization in its simplest form. What can you conclude about the curve?

17–20
(a) Find the unit tangent and unit normal vectors $\mathbf{T}(t)$ and $\mathbf{N}(t)$.
(b) Use Formula 9 to find the curvature.

17. $\mathbf{r}(t) = \langle t, 3\cos t, 3\sin t\rangle$

18. $\mathbf{r}(t) = \langle t^2, \sin t - t\cos t, \cos t + t\sin t\rangle, \quad t > 0$

19. $\mathbf{r}(t) = \langle \sqrt{2}\,t, e^t, e^{-t}\rangle$

20. $\mathbf{r}(t) = \langle t, \frac{1}{2}t^2, t^2\rangle$

21–23 Use Theorem 10 to find the curvature.

21. $\mathbf{r}(t) = t^3\,\mathbf{j} + t^2\,\mathbf{k}$

22. $\mathbf{r}(t) = t\,\mathbf{i} + t^2\,\mathbf{j} + e^t\,\mathbf{k}$

23. $\mathbf{r}(t) = 3t\,\mathbf{i} + 4\sin t\,\mathbf{j} + 4\cos t\,\mathbf{k}$

24. Find the curvature of $\mathbf{r}(t) = \langle t^2, \ln t, t\ln t\rangle$ at the point $(1, 0, 0)$.

25. Find the curvature of $\mathbf{r}(t) = \langle t, t^2, t^3\rangle$ at the point $(1, 1, 1)$.

26. Graph the curve with parametric equations $x = \cos t$, $y = \sin t$, $z = \sin 5t$ and find the curvature at the point $(1, 0, 0)$.

27–29 Use Formula 11 to find the curvature.

27. $y = x^4$ 28. $y = \tan x$ 29. $y = xe^x$

30–31 At what point does the curve have maximum curvature? What happens to the curvature as $x \to \infty$?

30. $y = \ln x$ 31. $y = e^x$

32. Find an equation of a parabola that has curvature 4 at the origin.

33. (a) Is the curvature of the curve C shown in the figure greater at P or at Q? Explain.
(b) Estimate the curvature at P and at Q by sketching the osculating circles at those points.

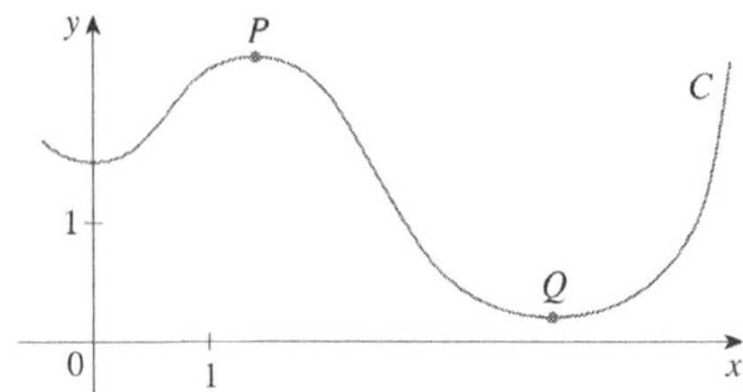

34–35 Use a graphing calculator or computer to graph both the curve and its curvature function $\kappa(x)$ on the same screen. Is the graph of κ what you would expect?

34. $y = x^4 - 2x^2$ **35.** $y = x^{-2}$

36–37 Plot the space curve and its curvature function $\kappa(t)$. Comment on how the curvature reflects the shape of the curve.

36. $\mathbf{r}(t) = \langle t - \sin t,\ 1 - \cos t,\ 4\cos(t/2)\rangle$, $0 \leqslant t \leqslant 8\pi$

37. $\mathbf{r}(t) = \langle te^t, e^{-t}, \sqrt{2}\,t\rangle$, $-5 \leqslant t \leqslant 5$

38–39 Two graphs, a and b, are shown. One is a curve $y = f(x)$ and the other is the graph of its curvature function $y = \kappa(x)$. Identify each curve and explain your choices.

38.

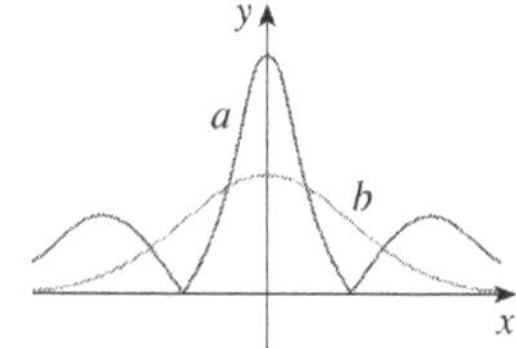

39.

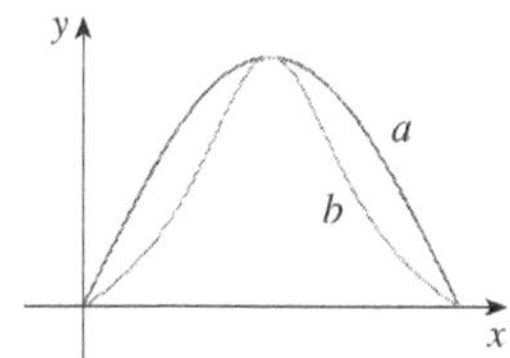

40. (a) Graph the curve $\mathbf{r}(t) = \langle \sin 3t, \sin 2t, \sin 3t\rangle$. At how many points on the curve does it appear that the curvature has a local or absolute maximum?
(b) Use a CAS to find and graph the curvature function. Does this graph confirm your conclusion from part (a)?

41. The graph of $\mathbf{r}(t) = \langle t - \frac{3}{2}\sin t,\ 1 - \frac{3}{2}\cos t,\ t\rangle$ is shown in Figure 12(b) in Section 13.1. Where do you think the curvature is largest? Use a CAS to find and graph the curvature function. For which values of t is the curvature largest?

42. Use Theorem 10 to show that the curvature of a plane parametric curve $x = f(t)$, $y = g(t)$ is

$$\kappa = \frac{|\dot{x}\ddot{y} - \dot{y}\ddot{x}|}{[\dot{x}^2 + \dot{y}^2]^{3/2}}$$

where the dots indicate derivatives with respect to t.

43–45 Use the formula in Exercise 42 to find the curvature.

43. $x = t^2$, $y = t^3$

44. $x = a\cos\omega t$, $y = b\sin\omega t$

45. $x = e^t\cos t$, $y = e^t\sin t$

46. Consider the curvature at $x = 0$ for each member of the family of functions $f(x) = e^{cx}$. For which members is $\kappa(0)$ largest?

47–48 Find the vectors $\mathbf{T}$, $\mathbf{N}$, and $\mathbf{B}$ at the given point.

47. $\mathbf{r}(t) = \langle t^2, \frac{2}{3}t^3, t\rangle$, $(1, \frac{2}{3}, 1)$

48. $\mathbf{r}(t) = \langle \cos t, \sin t, \ln\cos t\rangle$, $(1, 0, 0)$

49–50 Find equations of the normal plane and osculating plane of the curve at the given point.

49. $x = 2\sin 3t$, $y = t$, $z = 2\cos 3t$; $(0, \pi, -2)$

50. $x = t$, $y = t^2$, $z = t^3$; $(1, 1, 1)$

51. Find equations of the osculating circles of the ellipse $9x^2 + 4y^2 = 36$ at the points $(2, 0)$ and $(0, 3)$. Use a graphing calculator or computer to graph the ellipse and both osculating circles on the same screen.

52. Find equations of the osculating circles of the parabola $y = \frac{1}{2}x^2$ at the points $(0, 0)$ and $(1, \frac{1}{2})$. Graph both osculating circles and the parabola on the same screen.

53. At what point on the curve $x = t^3$, $y = 3t$, $z = t^4$ is the normal plane parallel to the plane $6x + 6y - 8z = 1$?

54. Is there a point on the curve in Exercise 53 where the osculating plane is parallel to the plane $x + y + z = 1$? [*Note:* You will need a CAS for differentiating, for simplifying, and for computing a cross product.]

55. Find equations of the normal and osculating planes of the curve of intersection of the parabolic cylinders $x = y^2$ and $z = x^2$ at the point $(1, 1, 1)$.

56. Show that the osculating plane at every point on the curve $\mathbf{r}(t) = \langle t + 2, 1 - t, \frac{1}{2}t^2\rangle$ is the same plane. What can you conclude about the curve?

57. Show that the curvature κ is related to the tangent and normal vectors by the equation

$$\frac{d\mathbf{T}}{ds} = \kappa\mathbf{N}$$

58. Show that the curvature of a plane curve is $\kappa = |d\phi/ds|$, where ϕ is the angle between $\mathbf{T}$ and $\mathbf{i}$; that is, ϕ is the angle of inclination of the tangent line. (This shows that the definition of curvature is consistent with the definition for plane curves given in Exercise 69 in Section 10.2.)

59. (a) Show that $d\mathbf{B}/ds$ is perpendicular to $\mathbf{B}$.
(b) Show that $d\mathbf{B}/ds$ is perpendicular to $\mathbf{T}$.
(c) Deduce from parts (a) and (b) that $d\mathbf{B}/ds = -\tau(s)\mathbf{N}$ for some number $\tau(s)$ called the **torsion** of the curve. (The torsion measures the degree of twisting of a curve.)
(d) Show that for a plane curve the torsion is $\tau(s) = 0$.

60. The following formulas, called the **Frenet-Serret formulas**, are of fundamental importance in differential geometry:

1. $d\mathbf{T}/ds = \kappa\mathbf{N}$
2. $d\mathbf{N}/ds = -\kappa\mathbf{T} + \tau\mathbf{B}$
3. $d\mathbf{B}/ds = -\tau\mathbf{N}$

(Formula 1 comes from Exercise 57 and Formula 3 comes from Exercise 59.) Use the fact that $\mathbf{N} = \mathbf{B} \times \mathbf{T}$ to deduce Formula 2 from Formulas 1 and 3.

61. Use the Frenet-Serret formulas to prove each of the following. (Primes denote derivatives with respect to t. Start as in the proof of Theorem 10.)

(a) $\mathbf{r}'' = s''\mathbf{T} + \kappa(s')^2\mathbf{N}$

(b) $\mathbf{r}' \times \mathbf{r}'' = \kappa(s')^3\mathbf{B}$

(c) $\mathbf{r}''' = [s''' - \kappa^2(s')^3]\mathbf{T} + [3\kappa s's'' + \kappa'(s')^2]\mathbf{N} + \kappa\tau(s')^3\mathbf{B}$

(d) $\tau = \dfrac{(\mathbf{r}' \times \mathbf{r}'') \cdot \mathbf{r}'''}{|\mathbf{r}' \times \mathbf{r}''|^2}$

62. Show that the circular helix $\mathbf{r}(t) = \langle a\cos t, a\sin t, bt\rangle$, where a and b are positive constants, has constant curvature and constant torsion. [Use the result of Exercise 61(d).]

63. Use the formula in Exercise 61(d) to find the torsion of the curve $\mathbf{r}(t) = \langle t, \frac{1}{2}t^2, \frac{1}{3}t^3\rangle$.

64. Find the curvature and torsion of the curve $x = \sinh t$, $y = \cosh t$, $z = t$ at the point $(0, 1, 0)$.

65. The DNA molecule has the shape of a double helix (see Figure 3 on page 842). The radius of each helix is about 10 angstroms ($1\ \text{Å} = 10^{-8}$ cm). Each helix rises about 34 Å during each complete turn, and there are about 2.9×10^8 complete turns. Estimate the length of each helix.

66. Let's consider the problem of designing a railroad track to make a smooth transition between sections of straight track. Existing track along the negative x-axis is to be joined smoothly to a track along the line $y = 1$ for $x \geqslant 1$.

(a) Find a polynomial $P = P(x)$ of degree 5 such that the function F defined by

$$F(x) = \begin{cases} 0 & \text{if } x \leqslant 0 \\ P(x) & \text{if } 0 < x < 1 \\ 1 & \text{if } x \geqslant 1 \end{cases}$$

is continuous and has continuous slope and continuous curvature.

(b) Use a graphing calculator or computer to draw the graph of F.

13.4 Motion in Space: Velocity and Acceleration

In this section we show how the ideas of tangent and normal vectors and curvature can be used in physics to study the motion of an object, including its velocity and acceleration, along a space curve. In particular, we follow in the footsteps of Newton by using these methods to derive Kepler's First Law of planetary motion.

Suppose a particle moves through space so that its position vector at time t is $\mathbf{r}(t)$. Notice from Figure 1 that, for small values of h, the vector

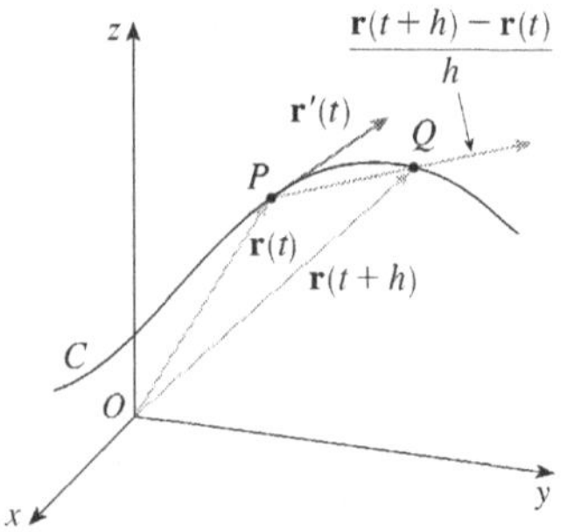

$$\boxed{1} \qquad \frac{\mathbf{r}(t + h) - \mathbf{r}(t)}{h}$$

approximates the direction of the particle moving along the curve $\mathbf{r}(t)$. Its magnitude measures the size of the displacement vector per unit time. The vector $\boxed{1}$ gives the average velocity over a time interval of length h and its limit is the **velocity vector** $\mathbf{v}(t)$ at time t:

$$\boxed{2} \qquad \mathbf{v}(t) = \lim_{h \to 0} \frac{\mathbf{r}(t + h) - \mathbf{r}(t)}{h} = \mathbf{r}'(t)$$

FIGURE 1

Thus the velocity vector is also the tangent vector and points in the direction of the tangent line.

The **speed** of the particle at time t is the magnitude of the velocity vector, that is, $|\mathbf{v}(t)|$. This is appropriate because, from $\boxed{2}$ and from Equation 13.3.7, we have

$$|\mathbf{v}(t)| = |\mathbf{r}'(t)| = \frac{ds}{dt} = \text{rate of change of distance with respect to time}$$

As in the case of one-dimensional motion, the **acceleration** of the particle is defined as the derivative of the velocity:

$$\mathbf{a}(t) = \mathbf{v}'(t) = \mathbf{r}''(t)$$

EXAMPLE 1 The position vector of an object moving in a plane is given by $\mathbf{r}(t) = t^3\,\mathbf{i} + t^2\,\mathbf{j}$. Find its velocity, speed, and acceleration when $t = 1$ and illustrate geometrically.

SOLUTION The velocity and acceleration at time t are

$$\mathbf{v}(t) = \mathbf{r}'(t) = 3t^2\,\mathbf{i} + 2t\,\mathbf{j}$$

$$\mathbf{a}(t) = \mathbf{r}''(t) = 6t\,\mathbf{i} + 2\,\mathbf{j}$$

and the speed is

$$|\mathbf{v}(t)| = \sqrt{(3t^2)^2 + (2t)^2} = \sqrt{9t^4 + 4t^2}$$

When $t = 1$, we have

$$\mathbf{v}(1) = 3\,\mathbf{i} + 2\,\mathbf{j} \qquad \mathbf{a}(1) = 6\,\mathbf{i} + 2\,\mathbf{j} \qquad |\mathbf{v}(1)| = \sqrt{13}$$

These velocity and acceleration vectors are shown in Figure 2.

FIGURE 2

TEC Visual 13.4 shows animated velocity and acceleration vectors for objects moving along various curves.

EXAMPLE 2 Find the velocity, acceleration, and speed of a particle with position vector $\mathbf{r}(t) = \langle t^2, e^t, te^t \rangle$.

SOLUTION

$$\mathbf{v}(t) = \mathbf{r}'(t) = \langle 2t, e^t, (1 + t)e^t \rangle$$

$$\mathbf{a}(t) = \mathbf{v}'(t) = \langle 2, e^t, (2 + t)e^t \rangle$$

$$|\mathbf{v}(t)| = \sqrt{4t^2 + e^{2t} + (1 + t)^2 e^{2t}}$$

Figure 3 shows the path of the particle in Example 2 with the velocity and acceleration vectors when $t = 1$.

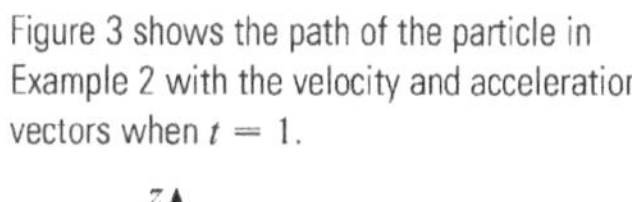

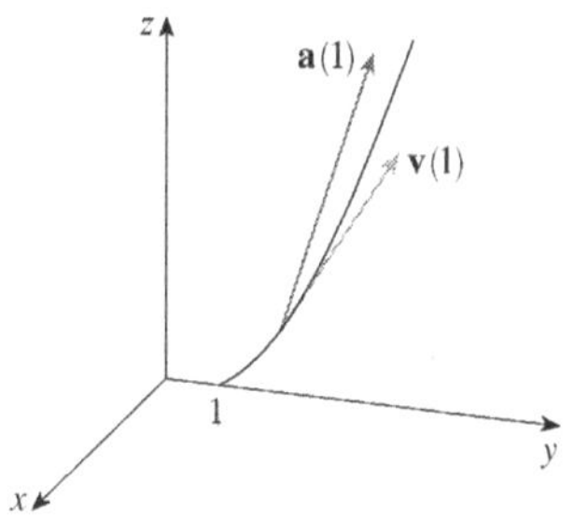

FIGURE 3

The vector integrals that were introduced in Section 13.2 can be used to find position vectors when velocity or acceleration vectors are known, as in the next example.

V EXAMPLE 3 A moving particle starts at an initial position $\mathbf{r}(0) = \langle 1, 0, 0 \rangle$ with initial velocity $\mathbf{v}(0) = \mathbf{i} - \mathbf{j} + \mathbf{k}$. Its acceleration is $\mathbf{a}(t) = 4t\,\mathbf{i} + 6t\,\mathbf{j} + \mathbf{k}$. Find its velocity and position at time t.

SOLUTION Since $\mathbf{a}(t) = \mathbf{v}'(t)$, we have

$$\mathbf{v}(t) = \int \mathbf{a}(t)\,dt = \int (4t\,\mathbf{i} + 6t\,\mathbf{j} + \mathbf{k})\,dt$$

$$= 2t^2\,\mathbf{i} + 3t^2\,\mathbf{j} + t\,\mathbf{k} + \mathbf{C}$$

To determine the value of the constant vector $\mathbf{C}$, we use the fact that $\mathbf{v}(0) = \mathbf{i} - \mathbf{j} + \mathbf{k}$. The preceding equation gives $\mathbf{v}(0) = \mathbf{C}$, so $\mathbf{C} = \mathbf{i} - \mathbf{j} + \mathbf{k}$ and

$$\mathbf{v}(t) = 2t^2\,\mathbf{i} + 3t^2\,\mathbf{j} + t\,\mathbf{k} + \mathbf{i} - \mathbf{j} + \mathbf{k}$$

$$= (2t^2 + 1)\,\mathbf{i} + (3t^2 - 1)\,\mathbf{j} + (t + 1)\,\mathbf{k}$$

The expression for $\mathbf{r}(t)$ that we obtained in Example 3 was used to plot the path of the particle in Figure 4 for $0 \leqslant t \leqslant 3$.

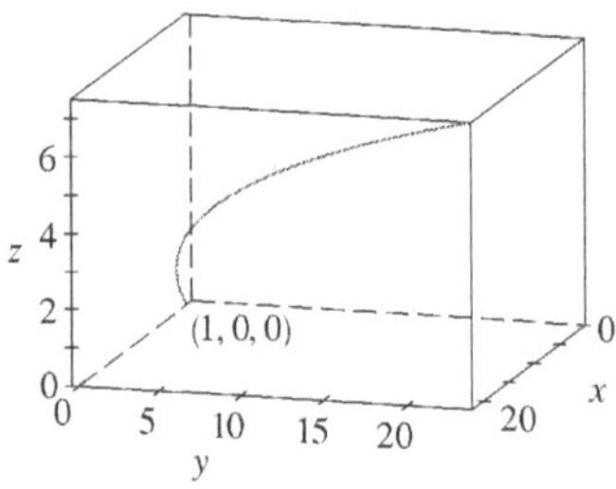

FIGURE 4

Since $\mathbf{v}(t) = \mathbf{r}'(t)$, we have

$$\mathbf{r}(t) = \int \mathbf{v}(t)\,dt$$

$$= \int [(2t^2 + 1)\,\mathbf{i} + (3t^2 - 1)\,\mathbf{j} + (t + 1)\,\mathbf{k}]\,dt$$

$$= \left(\tfrac{2}{3}t^3 + t\right)\mathbf{i} + (t^3 - t)\,\mathbf{j} + \left(\tfrac{1}{2}t^2 + t\right)\mathbf{k} + \mathbf{D}$$

Putting $t = 0$, we find that $\mathbf{D} = \mathbf{r}(0) = \mathbf{i}$, so the position at time t is given by

$$\mathbf{r}(t) = \left(\tfrac{2}{3}t^3 + t + 1\right)\mathbf{i} + (t^3 - t)\,\mathbf{j} + \left(\tfrac{1}{2}t^2 + t\right)\mathbf{k}$$

In general, vector integrals allow us to recover velocity when acceleration is known and position when velocity is known:

$$\mathbf{v}(t) = \mathbf{v}(t_0) + \int_{t_0}^{t} \mathbf{a}(u)\,du \qquad \mathbf{r}(t) = \mathbf{r}(t_0) + \int_{t_0}^{t} \mathbf{v}(u)\,du$$

If the force that acts on a particle is known, then the acceleration can be found from **Newton's Second Law of Motion**. The vector version of this law states that if, at any time t, a force $\mathbf{F}(t)$ acts on an object of mass m producing an acceleration $\mathbf{a}(t)$, then

$$\mathbf{F}(t) = m\mathbf{a}(t)$$

EXAMPLE 4 An object with mass m that moves in a circular path with constant angular speed ω has position vector $\mathbf{r}(t) = a\cos\omega t\,\mathbf{i} + a\sin\omega t\,\mathbf{j}$. Find the force acting on the object and show that it is directed toward the origin.

The angular speed of the object moving with position P is $\omega = d\theta/dt$, where θ is the angle shown in Figure 5.

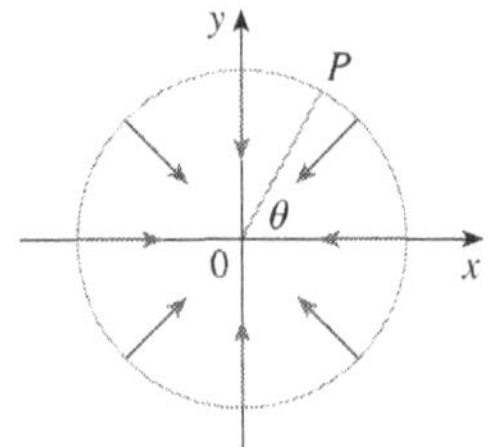

FIGURE 5

SOLUTION To find the force, we first need to know the acceleration:

$$\mathbf{v}(t) = \mathbf{r}'(t) = -a\omega\sin\omega t\,\mathbf{i} + a\omega\cos\omega t\,\mathbf{j}$$

$$\mathbf{a}(t) = \mathbf{v}'(t) = -a\omega^2\cos\omega t\,\mathbf{i} - a\omega^2\sin\omega t\,\mathbf{j}$$

Therefore Newton's Second Law gives the force as

$$\mathbf{F}(t) = m\mathbf{a}(t) = -m\omega^2(a\cos\omega t\,\mathbf{i} + a\sin\omega t\,\mathbf{j})$$

Notice that $\mathbf{F}(t) = -m\omega^2\mathbf{r}(t)$. This shows that the force acts in the direction opposite to the radius vector $\mathbf{r}(t)$ and therefore points toward the origin (see Figure 5). Such a force is called a *centripetal* (center-seeking) force.

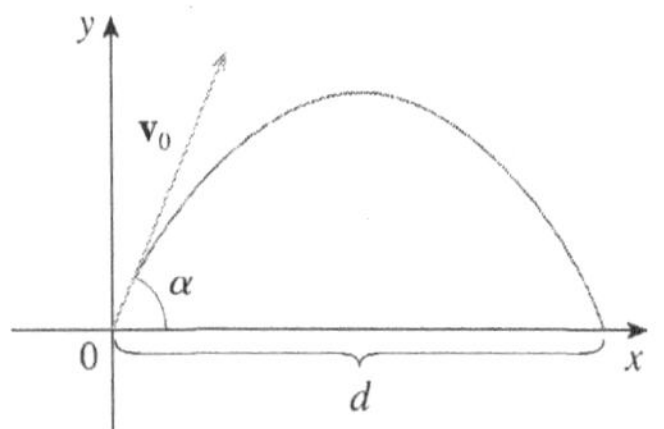

FIGURE 6

V **EXAMPLE 5** A projectile is fired with angle of elevation α and initial velocity $\mathbf{v}_0$. (See Figure 6.) Assuming that air resistance is negligible and the only external force is due to gravity, find the position function $\mathbf{r}(t)$ of the projectile. What value of α maximizes the range (the horizontal distance traveled)?

SOLUTION We set up the axes so that the projectile starts at the origin. Since the force due to gravity acts downward, we have

$$\mathbf{F} = m\mathbf{a} = -mg\,\mathbf{j}$$

where $g = |\mathbf{a}| \approx 9.8\ \text{m/s}^2$. Thus

$$\mathbf{a} = -g\,\mathbf{j}$$

Since $\mathbf{v}'(t) = \mathbf{a}$, we have

$$\mathbf{v}(t) = -gt\,\mathbf{j} + \mathbf{C}$$

where $\mathbf{C} = \mathbf{v}(0) = \mathbf{v}_0$. Therefore

$$\mathbf{r}'(t) = \mathbf{v}(t) = -gt\,\mathbf{j} + \mathbf{v}_0$$

Integrating again, we obtain

$$\mathbf{r}(t) = -\tfrac{1}{2}gt^2\,\mathbf{j} + t\,\mathbf{v}_0 + \mathbf{D}$$

But $\mathbf{D} = \mathbf{r}(0) = \mathbf{0}$, so the position vector of the projectile is given by

3 $$\mathbf{r}(t) = -\tfrac{1}{2}gt^2\,\mathbf{j} + t\,\mathbf{v}_0$$

If we write $|\mathbf{v}_0| = v_0$ (the initial speed of the projectile), then

$$\mathbf{v}_0 = v_0 \cos\alpha\,\mathbf{i} + v_0 \sin\alpha\,\mathbf{j}$$

and Equation 3 becomes

$$\mathbf{r}(t) = (v_0 \cos\alpha)t\,\mathbf{i} + \left[(v_0 \sin\alpha)t - \tfrac{1}{2}gt^2\right]\mathbf{j}$$

The parametric equations of the trajectory are therefore

If you eliminate t from Equations 4, you will see that y is a quadratic function of x. So the path of the projectile is part of a parabola.

4 $$x = (v_0 \cos\alpha)t \qquad y = (v_0 \sin\alpha)t - \tfrac{1}{2}gt^2$$

The horizontal distance d is the value of x when $y = 0$. Setting $y = 0$, we obtain $t = 0$ or $t = (2v_0 \sin\alpha)/g$. This second value of t then gives

$$d = x = (v_0 \cos\alpha)\frac{2v_0 \sin\alpha}{g} = \frac{v_0^2(2 \sin\alpha \cos\alpha)}{g} = \frac{v_0^2 \sin 2\alpha}{g}$$

Clearly, d has its maximum value when $\sin 2\alpha = 1$, that is, $\alpha = \pi/4$.

V EXAMPLE 6 A projectile is fired with muzzle speed 150 m/s and angle of elevation 45° from a position 10 m above ground level. Where does the projectile hit the ground, and with what speed?

SOLUTION If we place the origin at ground level, then the initial position of the projectile is (0, 10) and so we need to adjust Equations 4 by adding 10 to the expression for y. With $v_0 = 150$ m/s, $\alpha = 45°$, and $g = 9.8\ \text{m/s}^2$, we have

$$x = 150 \cos(\pi/4)t = 75\sqrt{2}\,t$$

$$y = 10 + 150 \sin(\pi/4)t - \tfrac{1}{2}(9.8)t^2 = 10 + 75\sqrt{2}\,t - 4.9t^2$$

Impact occurs when $y = 0$, that is, $4.9t^2 - 75\sqrt{2}\,t - 10 = 0$. Solving this quadratic equation (and using only the positive value of t), we get

$$t = \frac{75\sqrt{2} + \sqrt{11{,}250 + 196}}{9.8} \approx 21.74$$

Then $x \approx 75\sqrt{2}\,(21.74) \approx 2306$, so the projectile hits the ground about 2306 m away.

The velocity of the projectile is

$$\mathbf{v}(t) = \mathbf{r}'(t) = 75\sqrt{2}\,\mathbf{i} + \left(75\sqrt{2} - 9.8t\right)\mathbf{j}$$

So its speed at impact is

$$|\mathbf{v}(21.74)| = \sqrt{(75\sqrt{2})^2 + (75\sqrt{2} - 9.8 \cdot 21.74)^2} \approx 151 \text{ m/s}$$

Tangential and Normal Components of Acceleration

When we study the motion of a particle, it is often useful to resolve the acceleration into two components, one in the direction of the tangent and the other in the direction of the normal. If we write $v = |\mathbf{v}|$ for the speed of the particle, then

$$\mathbf{T}(t) = \frac{\mathbf{r}'(t)}{|\mathbf{r}'(t)|} = \frac{\mathbf{v}(t)}{|\mathbf{v}(t)|} = \frac{\mathbf{v}}{v}$$

and so

$$\mathbf{v} = v\mathbf{T}$$

If we differentiate both sides of this equation with respect to t, we get

5 $$\mathbf{a} = \mathbf{v}' = v'\mathbf{T} + v\mathbf{T}'$$

If we use the expression for the curvature given by Equation 13.3.9, then we have

6 $$\kappa = \frac{|\mathbf{T}'|}{|\mathbf{r}'|} = \frac{|\mathbf{T}'|}{v} \qquad \text{so} \qquad |\mathbf{T}'| = \kappa v$$

The unit normal vector was defined in the preceding section as $\mathbf{N} = \mathbf{T}'/|\mathbf{T}'|$, so 6 gives

$$\mathbf{T}' = |\mathbf{T}'|\mathbf{N} = \kappa v\mathbf{N}$$

and Equation 5 becomes

7 $$\mathbf{a} = v'\mathbf{T} + \kappa v^2\mathbf{N}$$

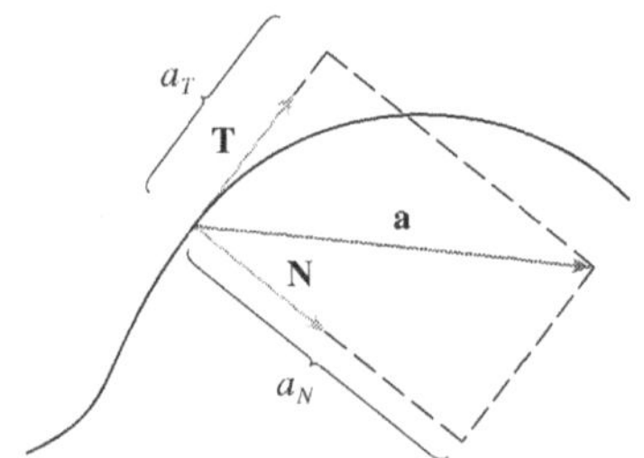

FIGURE 7

Writing a_T and a_N for the tangential and normal components of acceleration, we have

$$\mathbf{a} = a_T\mathbf{T} + a_N\mathbf{N}$$

where

8 $$a_T = v' \qquad \text{and} \qquad a_N = \kappa v^2$$

This resolution is illustrated in Figure 7.

Let's look at what Formula 7 says. The first thing to notice is that the binormal vector **B** is absent. No matter how an object moves through space, its acceleration always lies in the plane of **T** and **N** (the osculating plane). (Recall that **T** gives the direction of motion and **N** points in the direction the curve is turning.) Next we notice that the tangential component of acceleration is v', the rate of change of speed, and the normal component of acceleration is κv^2, the curvature times the square of the speed. This makes sense if we think of a passenger in a car—a sharp turn in a road means a large value of the curvature κ, so the component of the acceleration perpendicular to the motion is large and the passenger is thrown against a car door. High speed around the turn has the same effect; in fact, if you double your speed, a_N is increased by a factor of 4.

Although we have expressions for the tangential and normal components of acceleration in Equations 8, it's desirable to have expressions that depend only on $\mathbf{r}$, $\mathbf{r}'$, and $\mathbf{r}''$. To this end we take the dot product of $\mathbf{v} = v\mathbf{T}$ with $\mathbf{a}$ as given by Equation 7:

$$\begin{aligned}\mathbf{v} \cdot \mathbf{a} &= v\mathbf{T} \cdot (v'\mathbf{T} + \kappa v^2 \mathbf{N}) \\ &= vv'\mathbf{T} \cdot \mathbf{T} + \kappa v^3 \mathbf{T} \cdot \mathbf{N} \\ &= vv' \qquad \text{(since } \mathbf{T} \cdot \mathbf{T} = 1 \text{ and } \mathbf{T} \cdot \mathbf{N} = 0\text{)}\end{aligned}$$

Therefore

9 $$a_T = v' = \frac{\mathbf{v} \cdot \mathbf{a}}{v} = \frac{\mathbf{r}'(t) \cdot \mathbf{r}''(t)}{|\mathbf{r}'(t)|}$$

Using the formula for curvature given by Theorem 13.3.10, we have

10 $$a_N = \kappa v^2 = \frac{|\mathbf{r}'(t) \times \mathbf{r}''(t)|}{|\mathbf{r}'(t)|^3} |\mathbf{r}'(t)|^2 = \frac{|\mathbf{r}'(t) \times \mathbf{r}''(t)|}{|\mathbf{r}'(t)|}$$

EXAMPLE 7 A particle moves with position function $\mathbf{r}(t) = \langle t^2, t^2, t^3 \rangle$. Find the tangential and normal components of acceleration.

SOLUTION

$$\begin{aligned}\mathbf{r}(t) &= t^2\,\mathbf{i} + t^2\,\mathbf{j} + t^3\,\mathbf{k} \\ \mathbf{r}'(t) &= 2t\,\mathbf{i} + 2t\,\mathbf{j} + 3t^2\,\mathbf{k} \\ \mathbf{r}''(t) &= 2\,\mathbf{i} + 2\,\mathbf{j} + 6t\,\mathbf{k} \\ |\mathbf{r}'(t)| &= \sqrt{8t^2 + 9t^4}\end{aligned}$$

Therefore Equation 9 gives the tangential component as

$$a_T = \frac{\mathbf{r}'(t) \cdot \mathbf{r}''(t)}{|\mathbf{r}'(t)|} = \frac{8t + 18t^3}{\sqrt{8t^2 + 9t^4}}$$

Since $$\mathbf{r}'(t) \times \mathbf{r}''(t) = \begin{vmatrix} \mathbf{i} & \mathbf{j} & \mathbf{k} \\ 2t & 2t & 3t^2 \\ 2 & 2 & 6t \end{vmatrix} = 6t^2\,\mathbf{i} - 6t^2\,\mathbf{j}$$

Equation 10 gives the normal component as

$$a_N = \frac{|\mathbf{r}'(t) \times \mathbf{r}''(t)|}{|\mathbf{r}'(t)|} = \frac{6\sqrt{2}\,t^2}{\sqrt{8t^2 + 9t^4}}$$

Kepler's Laws of Planetary Motion

We now describe one of the great accomplishments of calculus by showing how the material of this chapter can be used to prove Kepler's laws of planetary motion. After 20 years of studying the astronomical observations of the Danish astronomer Tycho Brahe, the German mathematician and astronomer Johannes Kepler (1571–1630) formulated the following three laws.

Kepler's Laws

1. A planet revolves around the sun in an elliptical orbit with the sun at one focus.
2. The line joining the sun to a planet sweeps out equal areas in equal times.
3. The square of the period of revolution of a planet is proportional to the cube of the length of the major axis of its orbit.

In his book *Principia Mathematica* of 1687, Sir Isaac Newton was able to show that these three laws are consequences of two of his own laws, the Second Law of Motion and the Law of Universal Gravitation. In what follows we prove Kepler's First Law. The remaining laws are left as exercises (with hints).

Since the gravitational force of the sun on a planet is so much larger than the forces exerted by other celestial bodies, we can safely ignore all bodies in the universe except the sun and one planet revolving about it. We use a coordinate system with the sun at the origin and we let $\mathbf{r} = \mathbf{r}(t)$ be the position vector of the planet. (Equally well, $\mathbf{r}$ could be the position vector of the moon or a satellite moving around the earth or a comet moving around a star.) The velocity vector is $\mathbf{v} = \mathbf{r}'$ and the acceleration vector is $\mathbf{a} = \mathbf{r}''$. We use the following laws of Newton:

$$\text{Second Law of Motion:} \quad \mathbf{F} = m\mathbf{a}$$

$$\text{Law of Gravitation:} \quad \mathbf{F} = -\frac{GMm}{r^3}\mathbf{r} = -\frac{GMm}{r^2}\mathbf{u}$$

where $\mathbf{F}$ is the gravitational force on the planet, m and M are the masses of the planet and the sun, G is the gravitational constant, $r = |\mathbf{r}|$, and $\mathbf{u} = (1/r)\mathbf{r}$ is the unit vector in the direction of $\mathbf{r}$.

We first show that the planet moves in one plane. By equating the expressions for $\mathbf{F}$ in Newton's two laws, we find that

$$\mathbf{a} = -\frac{GM}{r^3}\mathbf{r}$$

and so $\mathbf{a}$ is parallel to $\mathbf{r}$. It follows that $\mathbf{r} \times \mathbf{a} = \mathbf{0}$. We use Formula 5 in Theorem 13.2.3 to write

$$\begin{aligned} \frac{d}{dt}(\mathbf{r} \times \mathbf{v}) &= \mathbf{r}' \times \mathbf{v} + \mathbf{r} \times \mathbf{v}' \\ &= \mathbf{v} \times \mathbf{v} + \mathbf{r} \times \mathbf{a} = \mathbf{0} + \mathbf{0} = \mathbf{0} \end{aligned}$$

Therefore

$$\mathbf{r} \times \mathbf{v} = \mathbf{h}$$

where $\mathbf{h}$ is a constant vector. (We may assume that $\mathbf{h} \neq \mathbf{0}$; that is, $\mathbf{r}$ and $\mathbf{v}$ are not parallel.) This means that the vector $\mathbf{r} = \mathbf{r}(t)$ is perpendicular to $\mathbf{h}$ for all values of t, so the planet always lies in the plane through the origin perpendicular to $\mathbf{h}$. Thus the orbit of the planet is a plane curve.

To prove Kepler's First Law we rewrite the vector $\mathbf{h}$ as follows:

$$\begin{aligned} \mathbf{h} &= \mathbf{r} \times \mathbf{v} = \mathbf{r} \times \mathbf{r}' = r\mathbf{u} \times (r\mathbf{u})' \\ &= r\mathbf{u} \times (r\mathbf{u}' + r'\mathbf{u}) = r^2(\mathbf{u} \times \mathbf{u}') + rr'(\mathbf{u} \times \mathbf{u}) \\ &= r^2(\mathbf{u} \times \mathbf{u}') \end{aligned}$$

Then

$$\mathbf{a} \times \mathbf{h} = \frac{-GM}{r^2} \mathbf{u} \times (r^2 \mathbf{u} \times \mathbf{u}') = -GM\, \mathbf{u} \times (\mathbf{u} \times \mathbf{u}')$$

$$= -GM[(\mathbf{u} \cdot \mathbf{u}')\mathbf{u} - (\mathbf{u} \cdot \mathbf{u})\mathbf{u}'] \qquad \text{(by Theorem 12.4.11, Property 6)}$$

But $\mathbf{u} \cdot \mathbf{u} = |\mathbf{u}|^2 = 1$ and, since $|\mathbf{u}(t)| = 1$, it follows from Example 4 in Section 13.2 that $\mathbf{u} \cdot \mathbf{u}' = 0$. Therefore

$$\mathbf{a} \times \mathbf{h} = GM\, \mathbf{u}'$$

and so

$$(\mathbf{v} \times \mathbf{h})' = \mathbf{v}' \times \mathbf{h} = \mathbf{a} \times \mathbf{h} = GM\, \mathbf{u}'$$

Integrating both sides of this equation, we get

11
$$\mathbf{v} \times \mathbf{h} = GM\, \mathbf{u} + \mathbf{c}$$

where $\mathbf{c}$ is a constant vector.

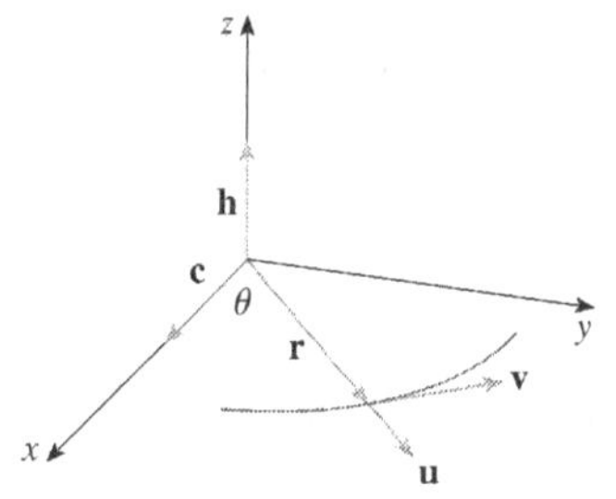

FIGURE 8

At this point it is convenient to choose the coordinate axes so that the standard basis vector $\mathbf{k}$ points in the direction of the vector $\mathbf{h}$. Then the planet moves in the xy-plane. Since both $\mathbf{v} \times \mathbf{h}$ and $\mathbf{u}$ are perpendicular to $\mathbf{h}$, Equation 11 shows that $\mathbf{c}$ lies in the xy-plane. This means that we can choose the x- and y-axes so that the vector $\mathbf{i}$ lies in the direction of $\mathbf{c}$, as shown in Figure 8.

If θ is the angle between $\mathbf{c}$ and $\mathbf{r}$, then (r, θ) are polar coordinates of the planet. From Equation 11 we have

$$\mathbf{r} \cdot (\mathbf{v} \times \mathbf{h}) = \mathbf{r} \cdot (GM\, \mathbf{u} + \mathbf{c}) = GM\, \mathbf{r} \cdot \mathbf{u} + \mathbf{r} \cdot \mathbf{c}$$

$$= GMr\, \mathbf{u} \cdot \mathbf{u} + |\mathbf{r}||\mathbf{c}|\cos\theta = GMr + rc\cos\theta$$

where $c = |\mathbf{c}|$. Then

$$r = \frac{\mathbf{r} \cdot (\mathbf{v} \times \mathbf{h})}{GM + c\cos\theta} = \frac{1}{GM} \frac{\mathbf{r} \cdot (\mathbf{v} \times \mathbf{h})}{1 + e\cos\theta}$$

where $e = c/(GM)$. But

$$\mathbf{r} \cdot (\mathbf{v} \times \mathbf{h}) = (\mathbf{r} \times \mathbf{v}) \cdot \mathbf{h} = \mathbf{h} \cdot \mathbf{h} = |\mathbf{h}|^2 = h^2$$

where $h = |\mathbf{h}|$. So

$$r = \frac{h^2/(GM)}{1 + e\cos\theta} = \frac{eh^2/c}{1 + e\cos\theta}$$

Writing $d = h^2/c$, we obtain the equation

12
$$r = \frac{ed}{1 + e\cos\theta}$$

Comparing with Theorem 10.6.6, we see that Equation 12 is the polar equation of a conic section with focus at the origin and eccentricity e. We know that the orbit of a planet is a closed curve and so the conic must be an ellipse.

This completes the derivation of Kepler's First Law. We will guide you through the derivation of the Second and Third Laws in the Applied Project on page 872. The proofs of these three laws show that the methods of this chapter provide a powerful tool for describing some of the laws of nature.

13.4 Exercises

1. The table gives coordinates of a particle moving through space along a smooth curve.
   (a) Find the average velocities over the time intervals $[0, 1]$, $[0.5, 1]$, $[1, 2]$, and $[1, 1.5]$.
   (b) Estimate the velocity and speed of the particle at $t = 1$.

t	x	y	z
0	2.7	9.8	3.7
0.5	3.5	7.2	3.3
1.0	4.5	6.0	3.0
1.5	5.9	6.4	2.8
2.0	7.3	7.8	2.7

2. The figure shows the path of a particle that moves with position vector $\mathbf{r}(t)$ at time t.
   (a) Draw a vector that represents the average velocity of the particle over the time interval $2 \leqslant t \leqslant 2.4$.
   (b) Draw a vector that represents the average velocity over the time interval $1.5 \leqslant t \leqslant 2$.
   (c) Write an expression for the velocity vector $\mathbf{v}(2)$.
   (d) Draw an approximation to the vector $\mathbf{v}(2)$ and estimate the speed of the particle at $t = 2$.

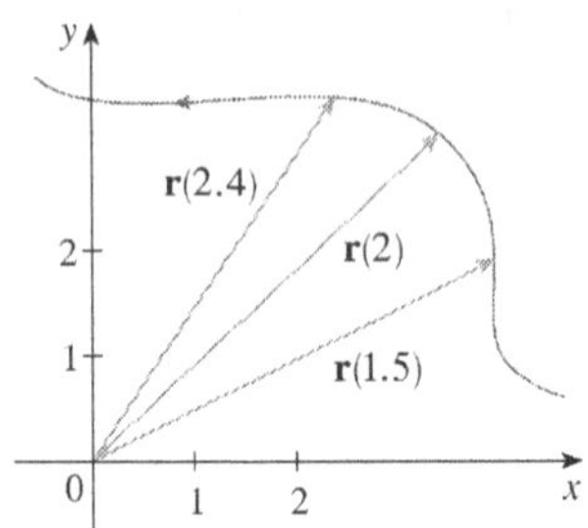

3–8 Find the velocity, acceleration, and speed of a particle with the given position function. Sketch the path of the particle and draw the velocity and acceleration vectors for the specified value of t.

3. $\mathbf{r}(t) = \left\langle -\frac{1}{2}t^2, t \right\rangle$, $t = 2$

4. $\mathbf{r}(t) = \left\langle 2 - t, 4\sqrt{t} \right\rangle$, $t = 1$

5. $\mathbf{r}(t) = 3\cos t\,\mathbf{i} + 2\sin t\,\mathbf{j}$, $t = \pi/3$

6. $\mathbf{r}(t) = e^t\,\mathbf{i} + e^{2t}\,\mathbf{j}$, $t = 0$

7. $\mathbf{r}(t) = t\,\mathbf{i} + t^2\mathbf{j} + 2\,\mathbf{k}$, $t = 1$

8. $\mathbf{r}(t) = t\,\mathbf{i} + 2\cos t\,\mathbf{j} + \sin t\,\mathbf{k}$, $t = 0$

9–14 Find the velocity, acceleration, and speed of a particle with the given position function.

9. $\mathbf{r}(t) = \langle t^2 + t, t^2 - t, t^3 \rangle$

10. $\mathbf{r}(t) = \langle 2\cos t, 3t, 2\sin t \rangle$

11. $\mathbf{r}(t) = \sqrt{2}\,t\,\mathbf{i} + e^t\,\mathbf{j} + e^{-t}\,\mathbf{k}$

12. $\mathbf{r}(t) = t^2\,\mathbf{i} + 2t\,\mathbf{j} + \ln t\,\mathbf{k}$

13. $\mathbf{r}(t) = e^t(\cos t\,\mathbf{i} + \sin t\,\mathbf{j} + t\,\mathbf{k})$

14. $\mathbf{r}(t) = \langle t^2, \sin t - t\cos t, \cos t + t\sin t \rangle$, $t \geqslant 0$

15–16 Find the velocity and position vectors of a particle that has the given acceleration and the given initial velocity and position.

15. $\mathbf{a}(t) = \mathbf{i} + 2\,\mathbf{j}$, $\mathbf{v}(0) = \mathbf{k}$, $\mathbf{r}(0) = \mathbf{i}$

16. $\mathbf{a}(t) = 2\,\mathbf{i} + 6t\,\mathbf{j} + 12t^2\,\mathbf{k}$, $\mathbf{v}(0) = \mathbf{i}$, $\mathbf{r}(0) = \mathbf{j} - \mathbf{k}$

17–18
(a) Find the position vector of a particle that has the given acceleration and the specified initial velocity and position.
(b) Use a computer to graph the path of the particle.

17. $\mathbf{a}(t) = 2t\,\mathbf{i} + \sin t\,\mathbf{j} + \cos 2t\,\mathbf{k}$, $\mathbf{v}(0) = \mathbf{i}$, $\mathbf{r}(0) = \mathbf{j}$

18. $\mathbf{a}(t) = t\,\mathbf{i} + e^t\,\mathbf{j} + e^{-t}\,\mathbf{k}$, $\mathbf{v}(0) = \mathbf{k}$, $\mathbf{r}(0) = \mathbf{j} + \mathbf{k}$

19. The position function of a particle is given by $\mathbf{r}(t) = \langle t^2, 5t, t^2 - 16t \rangle$. When is the speed a minimum?

20. What force is required so that a particle of mass m has the position function $\mathbf{r}(t) = t^3\,\mathbf{i} + t^2\,\mathbf{j} + t^3\,\mathbf{k}$?

21. A force with magnitude 20 N acts directly upward from the xy-plane on an object with mass 4 kg. The object starts at the origin with initial velocity $\mathbf{v}(0) = \mathbf{i} - \mathbf{j}$. Find its position function and its speed at time t.

22. Show that if a particle moves with constant speed, then the velocity and acceleration vectors are orthogonal.

23. A projectile is fired with an initial speed of 200 m/s and angle of elevation 60°. Find (a) the range of the projectile, (b) the maximum height reached, and (c) the speed at impact.

24. Rework Exercise 23 if the projectile is fired from a position 100 m above the ground.

25. A ball is thrown at an angle of 45° to the ground. If the ball lands 90 m away, what was the initial speed of the ball?

26. A gun is fired with angle of elevation 30°. What is the muzzle speed if the maximum height of the shell is 500 m?

27. A gun has muzzle speed 150 m/s. Find two angles of elevation that can be used to hit a target 800 m away.

28. A batter hits a baseball 3 ft above the ground toward the center field fence, which is 10 ft high and 400 ft from home plate. The ball leaves the bat with speed 115 ft/s at an angle 50° above the horizontal. Is it a home run? (In other words, does the ball clear the fence?)

29. A medieval city has the shape of a square and is protected by walls with length 500 m and height 15 m. You are the commander of an attacking army and the closest you can get to the wall is 100 m. Your plan is to set fire to the city by catapulting heated rocks over the wall (with an initial speed of 80 m/s). At what range of angles should you tell your men to set the catapult? (Assume the path of the rocks is perpendicular to the wall.)

30. Show that a projectile reaches three-quarters of its maximum height in half the time needed to reach its maximum height.

31. A ball is thrown eastward into the air from the origin (in the direction of the positive x-axis). The initial velocity is $50\,\mathbf{i} + 80\,\mathbf{k}$, with speed measured in feet per second. The spin of the ball results in a southward acceleration of 4 ft/s^2, so the acceleration vector is $\mathbf{a} = -4\mathbf{j} - 32\mathbf{k}$. Where does the ball land and with what speed?

32. A ball with mass 0.8 kg is thrown southward into the air with a speed of 30 m/s at an angle of 30° to the ground. A west wind applies a steady force of 4 N to the ball in an easterly direction. Where does the ball land and with what speed?

33. Water traveling along a straight portion of a river normally flows fastest in the middle, and the speed slows to almost zero at the banks. Consider a long straight stretch of river flowing north, with parallel banks 40 m apart. If the maximum water speed is 3 m/s, we can use a quadratic function as a basic model for the rate of water flow x units from the west bank: $f(x) = \frac{3}{400}x(40 - x)$.

(a) A boat proceeds at a constant speed of 5 m/s from a point A on the west bank while maintaining a heading perpendicular to the bank. How far down the river on the opposite bank will the boat touch shore? Graph the path of the boat.

(b) Suppose we would like to pilot the boat to land at the point B on the east bank directly opposite A. If we maintain a constant speed of 5 m/s and a constant heading, find the angle at which the boat should head. Then graph the actual path the boat follows. Does the path seem realistic?

34. Another reasonable model for the water speed of the river in Exercise 33 is a sine function: $f(x) = 3\sin(\pi x/40)$. If a boater would like to cross the river from A to B with constant heading and a constant speed of 5 m/s, determine the angle at which the boat should head.

35. A particle has position function $\mathbf{r}(t)$. If $\mathbf{r}'(t) = \mathbf{c} \times \mathbf{r}(t)$, where $\mathbf{c}$ is a constant vector, describe the path of the particle.

36. (a) If a particle moves along a straight line, what can you say about its acceleration vector?

(b) If a particle moves with constant speed along a curve, what can you say about its acceleration vector?

37–42 Find the tangential and normal components of the acceleration vector.

37. $\mathbf{r}(t) = (3t - t^3)\,\mathbf{i} + 3t^2\,\mathbf{j}$

38. $\mathbf{r}(t) = (1 + t)\,\mathbf{i} + (t^2 - 2t)\,\mathbf{j}$

39. $\mathbf{r}(t) = \cos t\,\mathbf{i} + \sin t\,\mathbf{j} + t\,\mathbf{k}$

40. $\mathbf{r}(t) = t\,\mathbf{i} + t^2\,\mathbf{j} + 3t\,\mathbf{k}$

41. $\mathbf{r}(t) = e^t\,\mathbf{i} + \sqrt{2}\,t\,\mathbf{j} + e^{-t}\,\mathbf{k}$

42. $\mathbf{r}(t) = t\,\mathbf{i} + \cos^2 t\,\mathbf{j} + \sin^2 t\,\mathbf{k}$

43. The magnitude of the acceleration vector $\mathbf{a}$ is 10 cm/s^2. Use the figure to estimate the tangential and normal components of $\mathbf{a}$.

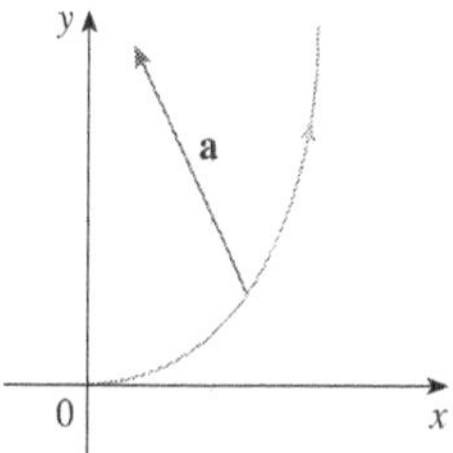

44. If a particle with mass m moves with position vector $\mathbf{r}(t)$, then its **angular momentum** is defined as $\mathbf{L}(t) = m\mathbf{r}(t) \times \mathbf{v}(t)$ and its **torque** as $\boldsymbol{\tau}(t) = m\mathbf{r}(t) \times \mathbf{a}(t)$. Show that $\mathbf{L}'(t) = \boldsymbol{\tau}(t)$. Deduce that if $\boldsymbol{\tau}(t) = \mathbf{0}$ for all t, then $\mathbf{L}(t)$ is constant. (This is the *law of conservation of angular momentum.*)

45. The position function of a spaceship is

$$\mathbf{r}(t) = (3 + t)\,\mathbf{i} + (2 + \ln t)\,\mathbf{j} + \left(7 - \frac{4}{t^2 + 1}\right)\mathbf{k}$$

and the coordinates of a space station are $(6, 4, 9)$. The captain wants the spaceship to coast into the space station. When should the engines be turned off?

46. A rocket burning its onboard fuel while moving through space has velocity $\mathbf{v}(t)$ and mass $m(t)$ at time t. If the exhaust gases escape with velocity $\mathbf{v}_e$ relative to the rocket, it can be deduced from Newton's Second Law of Motion that

$$m\frac{d\mathbf{v}}{dt} = \frac{dm}{dt}\mathbf{v}_e$$

(a) Show that $\mathbf{v}(t) = \mathbf{v}(0) - \ln\dfrac{m(0)}{m(t)}\mathbf{v}_e$.

(b) For the rocket to accelerate in a straight line from rest to twice the speed of its own exhaust gases, what fraction of its initial mass would the rocket have to burn as fuel?

APPLIED PROJECT

KEPLER'S LAWS

Johannes Kepler stated the following three laws of planetary motion on the basis of massive amounts of data on the positions of the planets at various times.

Kepler's Laws

1. A planet revolves around the sun in an elliptical orbit with the sun at one focus.
2. The line joining the sun to a planet sweeps out equal areas in equal times.
3. The square of the period of revolution of a planet is proportional to the cube of the length of the major axis of its orbit.

Kepler formulated these laws because they fitted the astronomical data. He wasn't able to see why they were true or how they related to each other. But Sir Isaac Newton, in his *Principia Mathematica* of 1687, showed how to deduce Kepler's three laws from two of Newton's own laws, the Second Law of Motion and the Law of Universal Gravitation. In Section 13.4 we proved Kepler's First Law using the calculus of vector functions. In this project we guide you through the proofs of Kepler's Second and Third Laws and explore some of their consequences.

1. Use the following steps to prove Kepler's Second Law. The notation is the same as in the proof of the First Law in Section 13.4. In particular, use polar coordinates so that $\mathbf{r} = (r\cos\theta)\,\mathbf{i} + (r\sin\theta)\,\mathbf{j}$.

   (a) Show that $\mathbf{h} = r^2 \dfrac{d\theta}{dt}\,\mathbf{k}$.

   (b) Deduce that $r^2 \dfrac{d\theta}{dt} = h$.

   (c) If $A = A(t)$ is the area swept out by the radius vector $\mathbf{r} = \mathbf{r}(t)$ in the time interval $[t_0, t]$ as in the figure, show that

$$\frac{dA}{dt} = \tfrac{1}{2}r^2\frac{d\theta}{dt}$$

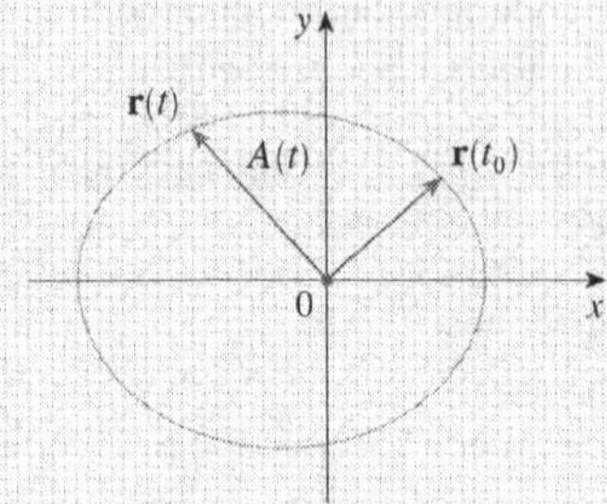

   (d) Deduce that

$$\frac{dA}{dt} = \tfrac{1}{2}h = \text{constant}$$

   This says that the rate at which A is swept out is constant and proves Kepler's Second Law.

2. Let T be the period of a planet about the sun; that is, T is the time required for it to travel once around its elliptical orbit. Suppose that the lengths of the major and minor axes of the ellipse are $2a$ and $2b$.

   (a) Use part (d) of Problem 1 to show that $T = 2\pi ab/h$.

   (b) Show that $\dfrac{h^2}{GM} = ed = \dfrac{b^2}{a}$.

   (c) Use parts (a) and (b) to show that $T^2 = \dfrac{4\pi^2}{GM}a^3$.

   This proves Kepler's Third Law. [Notice that the proportionality constant $4\pi^2/(GM)$ is independent of the planet.]

3. The period of the earth's orbit is approximately 365.25 days. Use this fact and Kepler's Third Law to find the length of the major axis of the earth's orbit. You will need the mass of the sun, $M = 1.99 \times 10^{30}$ kg, and the gravitational constant, $G = 6.67 \times 10^{-11}\ \text{N}\cdot\text{m}^2/\text{kg}^2$.

4. It's possible to place a satellite into orbit about the earth so that it remains fixed above a given location on the equator. Compute the altitude that is needed for such a satellite. The earth's mass is 5.98×10^{24} kg; its radius is 6.37×10^6 m. (This orbit is called the Clarke Geosynchronous Orbit after Arthur C. Clarke, who first proposed the idea in 1945. The first such satellite, *Syncom II*, was launched in July 1963.)

13 Review

Concept Check

1. What is a vector function? How do you find its derivative and its integral?

2. What is the connection between vector functions and space curves?

3. How do you find the tangent vector to a smooth curve at a point? How do you find the tangent line? The unit tangent vector?

4. If $\mathbf{u}$ and $\mathbf{v}$ are differentiable vector functions, c is a scalar, and f is a real-valued function, write the rules for differentiating the following vector functions.
   (a) $\mathbf{u}(t) + \mathbf{v}(t)$ (b) $c\mathbf{u}(t)$ (c) $f(t)\,\mathbf{u}(t)$
   (d) $\mathbf{u}(t) \cdot \mathbf{v}(t)$ (e) $\mathbf{u}(t) \times \mathbf{v}(t)$ (f) $\mathbf{u}(f(t))$

5. How do you find the length of a space curve given by a vector function $\mathbf{r}(t)$?

6. (a) What is the definition of curvature?
   (b) Write a formula for curvature in terms of $\mathbf{r}'(t)$ and $\mathbf{T}'(t)$.
   (c) Write a formula for curvature in terms of $\mathbf{r}'(t)$ and $\mathbf{r}''(t)$.
   (d) Write a formula for the curvature of a plane curve with equation $y = f(x)$.

7. (a) Write formulas for the unit normal and binormal vectors of a smooth space curve $\mathbf{r}(t)$.
   (b) What is the normal plane of a curve at a point? What is the osculating plane? What is the osculating circle?

8. (a) How do you find the velocity, speed, and acceleration of a particle that moves along a space curve?
   (b) Write the acceleration in terms of its tangential and normal components.

9. State Kepler's Laws.

True-False Quiz

Determine whether the statement is true or false. If it is true, explain why. If it is false, explain why or give an example that disproves the statement.

1. The curve with vector equation $\mathbf{r}(t) = t^3\mathbf{i} + 2t^3\mathbf{j} + 3t^3\mathbf{k}$ is a line.

2. The curve $\mathbf{r}(t) = \langle 0, t^2, 4t \rangle$ is a parabola.

3. The curve $\mathbf{r}(t) = \langle 2t, 3 - t, 0 \rangle$ is a line that passes through the origin.

4. The derivative of a vector function is obtained by differentiating each component function.

5. If $\mathbf{u}(t)$ and $\mathbf{v}(t)$ are differentiable vector functions, then
$$\frac{d}{dt}[\mathbf{u}(t) \times \mathbf{v}(t)] = \mathbf{u}'(t) \times \mathbf{v}'(t)$$

6. If $\mathbf{r}(t)$ is a differentiable vector function, then
$$\frac{d}{dt}|\mathbf{r}(t)| = |\mathbf{r}'(t)|$$

7. If $\mathbf{T}(t)$ is the unit tangent vector of a smooth curve, then the curvature is $\kappa = |d\mathbf{T}/dt|$.

8. The binormal vector is $\mathbf{B}(t) = \mathbf{N}(t) \times \mathbf{T}(t)$.

9. Suppose f is twice continuously differentiable. At an inflection point of the curve $y = f(x)$, the curvature is 0.

10. If $\kappa(t) = 0$ for all t, the curve is a straight line.

11. If $|\mathbf{r}(t)| = 1$ for all t, then $|\mathbf{r}'(t)|$ is a constant.

12. If $|\mathbf{r}(t)| = 1$ for all t, then $\mathbf{r}'(t)$ is orthogonal to $\mathbf{r}(t)$ for all t.

13. The osculating circle of a curve C at a point has the same tangent vector, normal vector, and curvature as C at that point.

14. Different parametrizations of the same curve result in identical tangent vectors at a given point on the curve.

Exercises

1. (a) Sketch the curve with vector function

$$\mathbf{r}(t) = t\,\mathbf{i} + \cos \pi t\,\mathbf{j} + \sin \pi t\,\mathbf{k} \qquad t \geqslant 0$$

(b) Find $\mathbf{r}'(t)$ and $\mathbf{r}''(t)$.

2. Let $\mathbf{r}(t) = \langle \sqrt{2 - t}, (e^t - 1)/t, \ln(t + 1) \rangle$.
(a) Find the domain of $\mathbf{r}$.
(b) Find $\lim_{t \to 0} \mathbf{r}(t)$.
(c) Find $\mathbf{r}'(t)$.

3. Find a vector function that represents the curve of intersection of the cylinder $x^2 + y^2 = 16$ and the plane $x + z = 5$.

 **4.** Find parametric equations for the tangent line to the curve $x = 2 \sin t$, $y = 2 \sin 2t$, $z = 2 \sin 3t$ at the point $(1, \sqrt{3}, 2)$. Graph the curve and the tangent line on a common screen.

5. If $\mathbf{r}(t) = t^2\,\mathbf{i} + t \cos \pi t\,\mathbf{j} + \sin \pi t\,\mathbf{k}$, evaluate $\int_0^1 \mathbf{r}(t)\,dt$.

6. Let C be the curve with equations $x = 2 - t^3$, $y = 2t - 1$, $z = \ln t$. Find (a) the point where C intersects the xz-plane, (b) parametric equations of the tangent line at $(1, 1, 0)$, and (c) an equation of the normal plane to C at $(1, 1, 0)$.

7. Use Simpson's Rule with $n = 6$ to estimate the length of the arc of the curve with equations $x = t^2$, $y = t^3$, $z = t^4$, $0 \leqslant t \leqslant 3$.

8. Find the length of the curve $\mathbf{r}(t) = \langle 2t^{3/2}, \cos 2t, \sin 2t \rangle$, $0 \leqslant t \leqslant 1$.

9. The helix $\mathbf{r}_1(t) = \cos t\,\mathbf{i} + \sin t\,\mathbf{j} + t\,\mathbf{k}$ intersects the curve $\mathbf{r}_2(t) = (1 + t)\mathbf{i} + t^2\mathbf{j} + t^3\mathbf{k}$ at the point $(1, 0, 0)$. Find the angle of intersection of these curves.

10. Reparametrize the curve $\mathbf{r}(t) = e^t\,\mathbf{i} + e^t \sin t\,\mathbf{j} + e^t \cos t\,\mathbf{k}$ with respect to arc length measured from the point $(1, 0, 1)$ in the direction of increasing t.

11. For the curve given by $\mathbf{r}(t) = \left\langle \frac{1}{3}t^3, \frac{1}{2}t^2, t \right\rangle$, find
(a) the unit tangent vector,
(b) the unit normal vector, and
(c) the curvature.

12. Find the curvature of the ellipse $x = 3 \cos t$, $y = 4 \sin t$ at the points $(3, 0)$ and $(0, 4)$.

13. Find the curvature of the curve $y = x^4$ at the point $(1, 1)$.

 **14.** Find an equation of the osculating circle of the curve $y = x^4 - x^2$ at the origin. Graph both the curve and its osculating circle.

15. Find an equation of the osculating plane of the curve $x = \sin 2t$, $y = t$, $z = \cos 2t$ at the point $(0, \pi, 1)$.

16. The figure shows the curve C traced by a particle with position vector $\mathbf{r}(t)$ at time t.
(a) Draw a vector that represents the average velocity of the particle over the time interval $3 \leqslant t \leqslant 3.2$.
(b) Write an expression for the velocity $\mathbf{v}(3)$.
(c) Write an expression for the unit tangent vector $\mathbf{T}(3)$ and draw it.

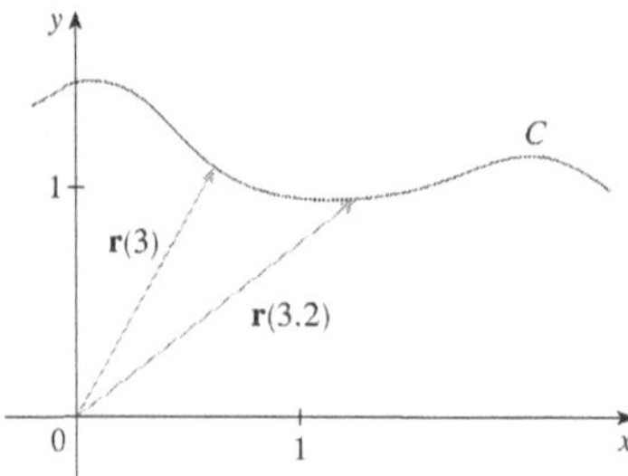

17. A particle moves with position function $\mathbf{r}(t) = t \ln t\,\mathbf{i} + t\,\mathbf{j} + e^{-t}\,\mathbf{k}$. Find the velocity, speed, and acceleration of the particle.

18. A particle starts at the origin with initial velocity $\mathbf{i} - \mathbf{j} + 3\mathbf{k}$. Its acceleration is $\mathbf{a}(t) = 6t\,\mathbf{i} + 12t^2\,\mathbf{j} - 6t\,\mathbf{k}$. Find its position function.

19. An athlete throws a shot at an angle of 45° to the horizontal at an initial speed of 43 ft/s. It leaves his hand 7 ft above the ground.
(a) Where is the shot 2 seconds later?
(b) How high does the shot go?
(c) Where does the shot land?

20. Find the tangential and normal components of the acceleration vector of a particle with position function

$$\mathbf{r}(t) = t\,\mathbf{i} + 2t\,\mathbf{j} + t^2\mathbf{k}$$

21. A disk of radius 1 is rotating in the counterclockwise direction at a constant angular speed ω. A particle starts at the center of the disk and moves toward the edge along a fixed radius so that its position at time t, $t \geqslant 0$, is given by $\mathbf{r}(t) = t\mathbf{R}(t)$, where

$$\mathbf{R}(t) = \cos \omega t\,\mathbf{i} + \sin \omega t\,\mathbf{j}$$

(a) Show that the velocity $\mathbf{v}$ of the particle is

$$\mathbf{v} = \cos \omega t\,\mathbf{i} + \sin \omega t\,\mathbf{j} + t\,\mathbf{v}_d$$

where $\mathbf{v}_d = \mathbf{R}'(t)$ is the velocity of a point on the edge of the disk.
(b) Show that the acceleration $\mathbf{a}$ of the particle is

$$\mathbf{a} = 2\mathbf{v}_d + t\,\mathbf{a}_d$$

where $\mathbf{a}_d = \mathbf{R}''(t)$ is the acceleration of a point on the rim of the disk. The extra term $2\mathbf{v}_d$ is called the *Coriolis acceleration*; it is the result of the interaction of the rotation of the disk and the motion of the particle. One can obtain a physical demonstration of this acceleration by walking toward the edge of a moving merry-go-round.

Graphing calculator or computer required

(c) Determine the Coriolis acceleration of a particle that moves on a rotating disk according to the equation

$$\mathbf{r}(t) = e^{-t}\cos \omega t\,\mathbf{i} + e^{-t}\sin \omega t\,\mathbf{j}$$

22. In designing *transfer curves* to connect sections of straight railroad tracks, it's important to realize that the acceleration of the train should be continuous so that the reactive force exerted by the train on the track is also continuous. Because of the formulas for the components of acceleration in Section 13.4, this will be the case if the curvature varies continuously.

(a) A logical candidate for a transfer curve to join existing tracks given by $y = 1$ for $x \leqslant 0$ and $y = \sqrt{2} - x$ for $x \geqslant 1/\sqrt{2}$ might be the function $f(x) = \sqrt{1 - x^2}$, $0 < x < 1/\sqrt{2}$, whose graph is the arc of the circle shown in the figure. It looks reasonable at first glance. Show that the function

$$F(x) = \begin{cases} 1 & \text{if } x \leqslant 0 \\ \sqrt{1 - x^2} & \text{if } 0 < x < 1/\sqrt{2} \\ \sqrt{2} - x & \text{if } x \geqslant 1/\sqrt{2} \end{cases}$$

is continuous and has continuous slope, but does not have continuous curvature. Therefore f is not an appropriate transfer curve.

(b) Find a fifth-degree polynomial to serve as a transfer curve between the following straight line segments: $y = 0$ for $x \leqslant 0$ and $y = x$ for $x \geqslant 1$. Could this be done with a fourth-degree polynomial? Use a graphing calculator or computer to sketch the graph of the "connected" function and check to see that it looks like the one in the figure.

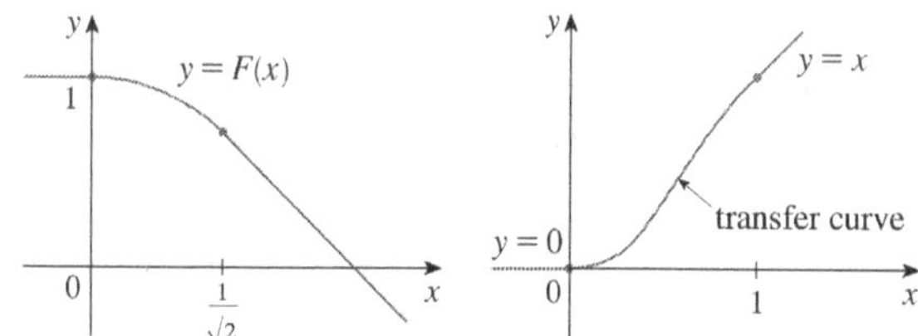

23. A particle P moves with constant angular speed ω around a circle whose center is at the origin and whose radius is R. The particle is said to be in *uniform circular motion*. Assume that the motion is counterclockwise and that the particle is at the point $(R, 0)$ when $t = 0$. The position vector at time $t \geqslant 0$ is $\mathbf{r}(t) = R\cos \omega t\,\mathbf{i} + R\sin \omega t\,\mathbf{j}$.

(a) Find the velocity vector $\mathbf{v}$ and show that $\mathbf{v} \cdot \mathbf{r} = 0$. Conclude that $\mathbf{v}$ is tangent to the circle and points in the direction of the motion.

(b) Show that the speed $|\mathbf{v}|$ of the particle is the constant ωR. The *period* T of the particle is the time required for one complete revolution. Conclude that

$$T = \frac{2\pi R}{|\mathbf{v}|} = \frac{2\pi}{\omega}$$

(c) Find the acceleration vector $\mathbf{a}$. Show that it is proportional to $\mathbf{r}$ and that it points toward the origin. An acceleration with this property is called a *centripetal acceleration*. Show that the magnitude of the acceleration vector is $|\mathbf{a}| = R\omega^2$.

(d) Suppose that the particle has mass m. Show that the magnitude of the force $\mathbf{F}$ that is required to produce this motion, called a *centripetal force*, is

$$|\mathbf{F}| = \frac{m|\mathbf{v}|^2}{R}$$

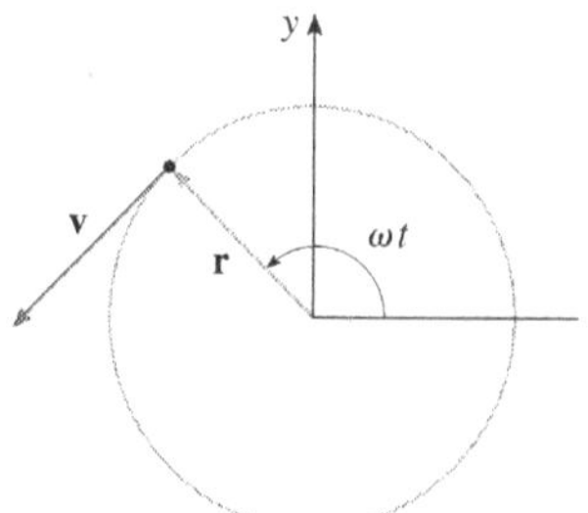

24. A circular curve of radius R on a highway is banked at an angle θ so that a car can safely traverse the curve without skidding when there is no friction between the road and the tires. The loss of friction could occur, for example, if the road is covered with a film of water or ice. The rated speed v_R of the curve is the maximum speed that a car can attain without skidding. Suppose a car of mass m is traversing the curve at the rated speed v_R. Two forces are acting on the car: the vertical force, mg, due to the weight of the car, and a force $\mathbf{F}$ exerted by, and normal to, the road (see the figure).

The vertical component of $\mathbf{F}$ balances the weight of the car, so that $|\mathbf{F}|\cos \theta = mg$. The horizontal component of $\mathbf{F}$ produces a centripetal force on the car so that, by Newton's Second Law and part (d) of Problem 23,

$$|\mathbf{F}|\sin \theta = \frac{mv_R^2}{R}$$

(a) Show that $v_R^2 = Rg\tan \theta$.

(b) Find the rated speed of a circular curve with radius 400 ft that is banked at an angle of 12°.

(c) Suppose the design engineers want to keep the banking at 12°, but wish to increase the rated speed by 50%. What should the radius of the curve be?

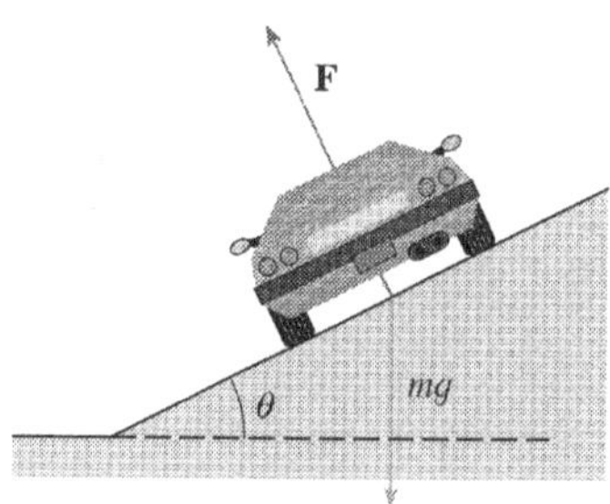

Problems Plus

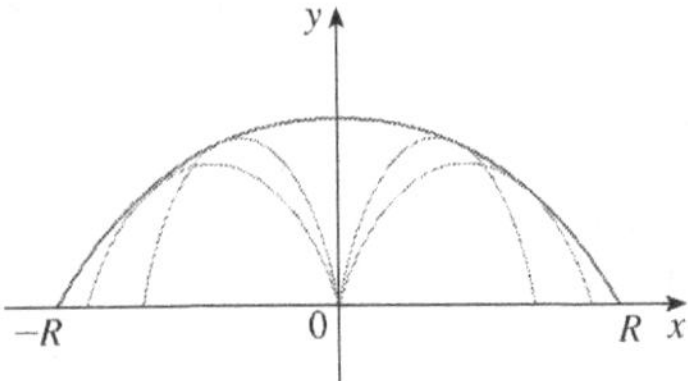

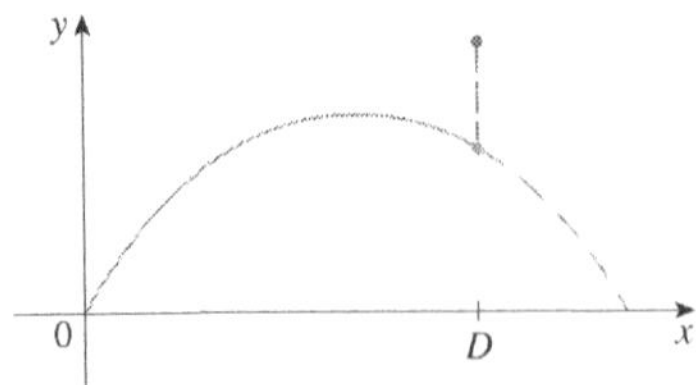

FIGURE FOR PROBLEM 1

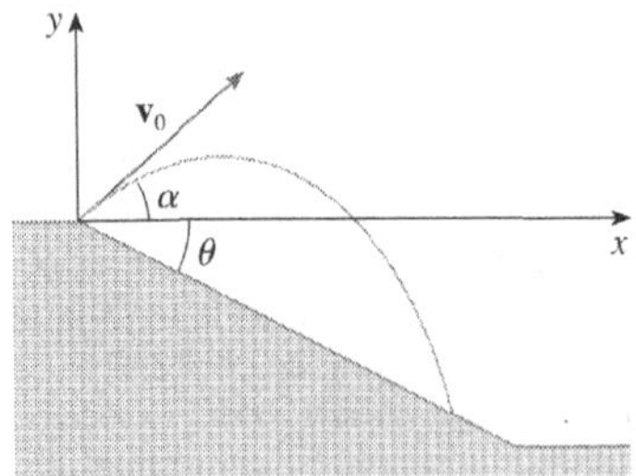

FIGURE FOR PROBLEM 2

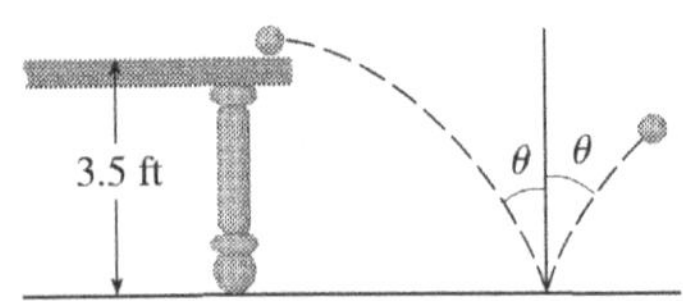

FIGURE FOR PROBLEM 3

1. A projectile is fired from the origin with angle of elevation α and initial speed v_0. Assuming that air resistance is negligible and that the only force acting on the projectile is gravity, g, we showed in Example 5 in Section 13.4 that the position vector of the projectile is $\mathbf{r}(t) = (v_0 \cos\alpha)t\,\mathbf{i} + \left[(v_0 \sin\alpha)t - \frac{1}{2}gt^2\right]\mathbf{j}$. We also showed that the maximum horizontal distance of the projectile is achieved when $\alpha = 45°$ and in this case the range is $R = v_0^2/g$.
   (a) At what angle should the projectile be fired to achieve maximum height and what is the maximum height?
   (b) Fix the initial speed v_0 and consider the parabola $x^2 + 2Ry - R^2 = 0$, whose graph is shown in the figure. Show that the projectile can hit any target inside or on the boundary of the region bounded by the parabola and the x-axis, and that it can't hit any target outside this region.
   (c) Suppose that the gun is elevated to an angle of inclination α in order to aim at a target that is suspended at a height h directly over a point D units downrange. The target is released at the instant the gun is fired. Show that the projectile always hits the target, regardless of the value v_0, provided the projectile does not hit the ground "before" D.

2. (a) A projectile is fired from the origin down an inclined plane that makes an angle θ with the horizontal. The angle of elevation of the gun and the initial speed of the projectile are α and v_0, respectively. Find the position vector of the projectile and the parametric equations of the path of the projectile as functions of the time t. (Ignore air resistance.)
   (b) Show that the angle of elevation α that will maximize the downhill range is the angle halfway between the plane and the vertical.
   (c) Suppose the projectile is fired up an inclined plane whose angle of inclination is θ. Show that, in order to maximize the (uphill) range, the projectile should be fired in the direction halfway between the plane and the vertical.
   (d) In a paper presented in 1686, Edmond Halley summarized the laws of gravity and projectile motion and applied them to gunnery. One problem he posed involved firing a projectile to hit a target a distance R up an inclined plane. Show that the angle at which the projectile should be fired to hit the target but use the least amount of energy is the same as the angle in part (c). (Use the fact that the energy needed to fire the projectile is proportional to the square of the initial speed, so minimizing the energy is equivalent to minimizing the initial speed.)

3. A ball rolls off a table with a speed of 2 ft/s. The table is 3.5 ft high.
   (a) Determine the point at which the ball hits the floor and find its speed at the instant of impact.
   (b) Find the angle θ between the path of the ball and the vertical line drawn through the point of impact (see the figure).
   (c) Suppose the ball rebounds from the floor at the same angle with which it hits the floor, but loses 20% of its speed due to energy absorbed by the ball on impact. Where does the ball strike the floor on the second bounce?

4. Find the curvature of the curve with parametric equations

$$x = \int_0^t \sin\left(\tfrac{1}{2}\pi\theta^2\right) d\theta \qquad y = \int_0^t \cos\left(\tfrac{1}{2}\pi\theta^2\right) d\theta$$

5. If a projectile is fired with angle of elevation α and initial speed v, then parametric equations for its trajectory are $x = (v\cos\alpha)t$, $y = (v \sin\alpha)t - \frac{1}{2}gt^2$. (See Example 5 in Section 13.4.) We know that the range (horizontal distance traveled) is maximized when $\alpha = 45°$. What value of α maximizes the total distance traveled by the projectile? (State your answer correct to the nearest degree.)

6. A cable has radius r and length L and is wound around a spool with radius R without overlapping. What is the shortest length along the spool that is covered by the cable?

7. Show that the curve with vector equation

$$\mathbf{r}(t) = \langle a_1t^2 + b_1t + c_1, a_2t^2 + b_2t + c_2, a_3t^2 + b_3t + c_3\rangle$$

lies in a plane and find an equation of the plane.

14 Partial Derivatives

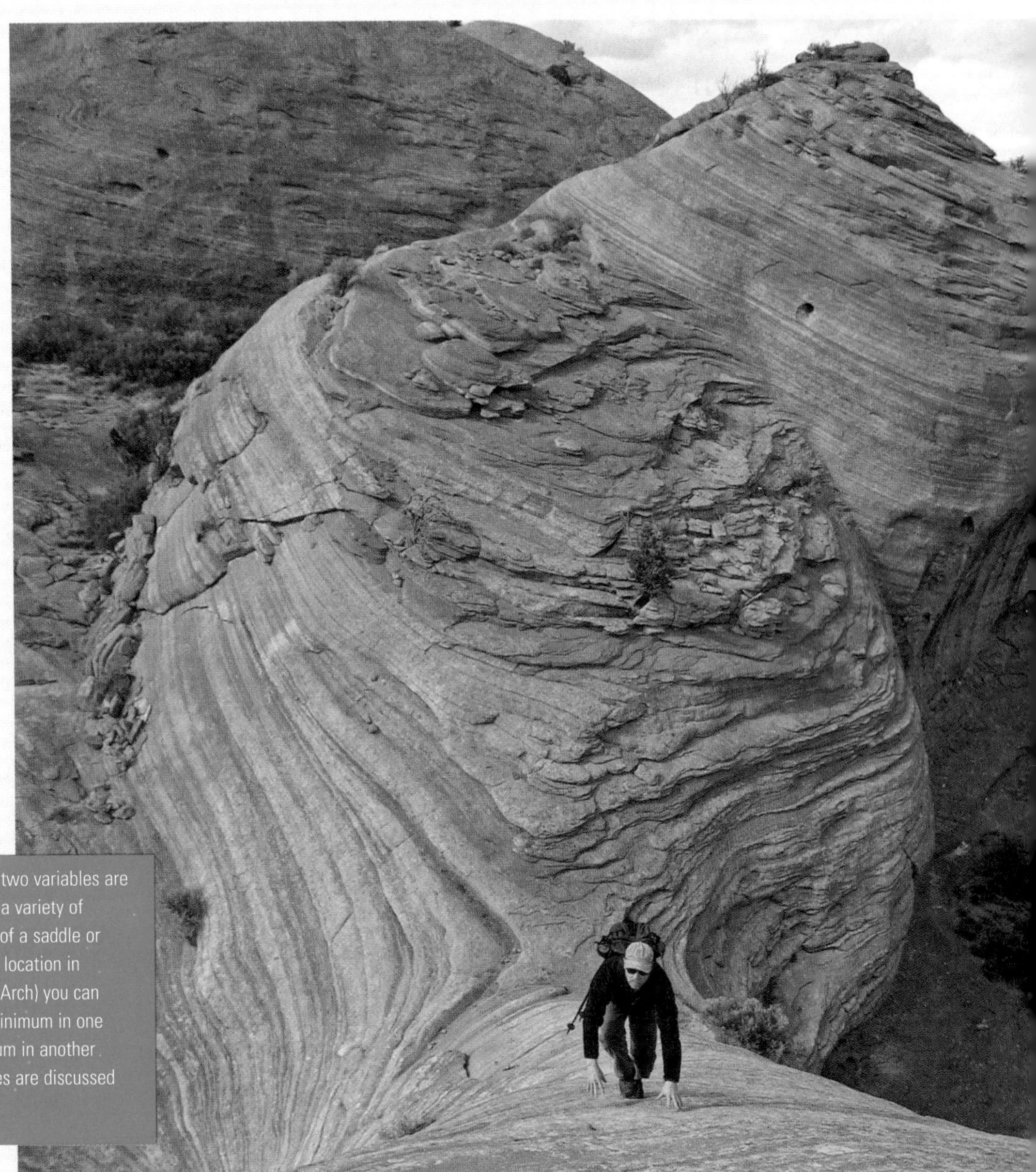

Graphs of functions of two variables are surfaces that can take a variety of shapes, including that of a saddle or mountain pass. At this location in southern Utah (Phipps Arch) you can see a point that is a minimum in one direction but a maximum in another direction. Such surfaces are discussed in Section 14.7.

Photo by Stan Wagon, Macalester College

So far we have dealt with the calculus of functions of a single variable. But, in the real world, physical quantities often depend on two or more variables, so in this chapter we turn our attention to functions of several variables and extend the basic ideas of differential calculus to such functions.

14.1 Functions of Several Variables

In this section we study functions of two or more variables from four points of view:

- verbally (by a description in words)
- numerically (by a table of values)
- algebraically (by an explicit formula)
- visually (by a graph or level curves)

Functions of Two Variables

The temperature T at a point on the surface of the earth at any given time depends on the longitude x and latitude y of the point. We can think of T as being a function of the two variables x and y, or as a function of the pair (x, y). We indicate this functional dependence by writing $T = f(x, y)$.

The volume V of a circular cylinder depends on its radius r and its height h. In fact, we know that $V = \pi r^2 h$. We say that V is a function of r and h, and we write $V(r, h) = \pi r^2 h$.

Definition A **function f of two variables** is a rule that assigns to each ordered pair of real numbers (x, y) in a set D a unique real number denoted by $f(x, y)$. The set D is the **domain** of f and its **range** is the set of values that f takes on, that is, $\{f(x, y) \mid (x, y) \in D\}$.

We often write $z = f(x, y)$ to make explicit the value taken on by f at the general point (x, y). The variables x and y are **independent variables** and z is the **dependent variable**. [Compare this with the notation $y = f(x)$ for functions of a single variable.]

A function of two variables is just a function whose domain is a subset of $\mathbb{R}^2$ and whose range is a subset of $\mathbb{R}$. One way of visualizing such a function is by means of an arrow diagram (see Figure 1), where the domain D is represented as a subset of the xy-plane and the range is a set of numbers on a real line, shown as a z-axis. For instance, if $f(x, y)$ represents the temperature at a point (x, y) in a flat metal plate with the shape of D, we can think of the z-axis as a thermometer displaying the recorded temperatures.

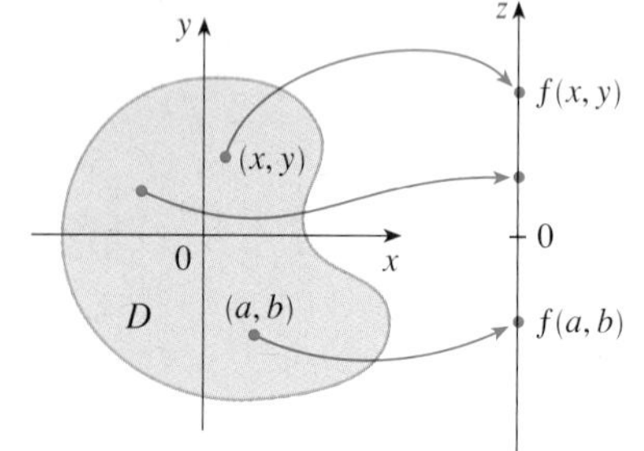

FIGURE 1

If a function f is given by a formula and no domain is specified, then the domain of f is understood to be the set of all pairs (x, y) for which the given expression is a well-defined real number.

EXAMPLE 1 For each of the following functions, evaluate $f(3, 2)$ and find and sketch the domain.

(a) $f(x, y) = \dfrac{\sqrt{x + y + 1}}{x - 1}$ (b) $f(x, y) = x \ln(y^2 - x)$

SOLUTION

(a)
$$f(3, 2) = \frac{\sqrt{3 + 2 + 1}}{3 - 1} = \frac{\sqrt{6}}{2}$$

The expression for f makes sense if the denominator is not 0 and the quantity under the square root sign is nonnegative. So the domain of f is

$$D = \{(x, y) \mid x + y + 1 \geqslant 0,\ x \neq 1\}$$

The inequality $x + y + 1 \geqslant 0$, or $y \geqslant -x - 1$, describes the points that lie on or above

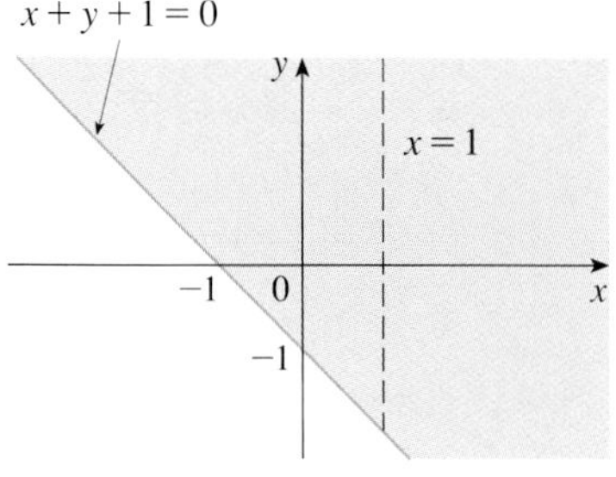

FIGURE 2
Domain of $f(x, y) = \dfrac{\sqrt{x + y + 1}}{x - 1}$

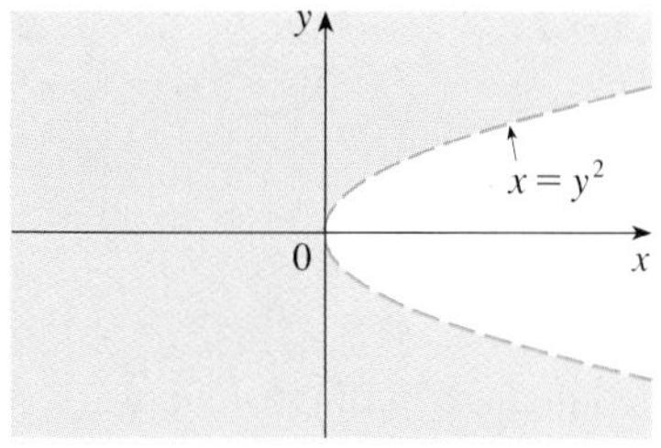

FIGURE 3
Domain of $f(x, y) = x \ln(y^2 - x)$

The New Wind-Chill Index

A new wind-chill index was introduced in November of 2001 and is more accurate than the old index for measuring how cold it feels when it's windy. The new index is based on a model of how fast a human face loses heat. It was developed through clinical trials in which volunteers were exposed to a variety of temperatures and wind speeds in a refrigerated wind tunnel.

the line $y = -x - 1$, while $x \neq 1$ means that the points on the line $x = 1$ must be excluded from the domain. (See Figure 2.)

(b)
$$f(3, 2) = 3 \ln(2^2 - 3) = 3 \ln 1 = 0$$

Since $\ln(y^2 - x)$ is defined only when $y^2 - x > 0$, that is, $x < y^2$, the domain of f is $D = \{(x, y) \mid x < y^2\}$. This is the set of points to the left of the parabola $x = y^2$. (See Figure 3.) ■

Not all functions can be represented by explicit formulas. The function in the next example is described verbally and by numerical estimates of its values.

EXAMPLE 2 In regions with severe winter weather, the *wind-chill index* is often used to describe the apparent severity of the cold. This index W is a subjective temperature that depends on the actual temperature T and the wind speed v. So W is a function of T and v, and we can write $W = f(T, v)$. Table 1 records values of W compiled by the National Weather Service of the US and the Meteorological Service of Canada.

TABLE 1 Wind-chill index as a function of air temperature and wind speed

Wind speed (km/h); rows: Actual temperature (°C)

T \ v	5	10	15	20	25	30	40	50	60	70	80
5	4	3	2	1	1	0	−1	−1	−2	−2	−3
0	−2	−3	−4	−5	−6	−6	−7	−8	−9	−9	−10
−5	−7	−9	−11	−12	−12	−13	−14	−15	−16	−16	−17
−10	−13	−15	−17	−18	−19	−20	−21	−22	−23	−23	−24
−15	−19	−21	−23	−24	−25	−26	−27	−29	−30	−30	−31
−20	−24	−27	−29	−30	−32	−33	−34	−35	−36	−37	−38
−25	−30	−33	−35	−37	−38	−39	−41	−42	−43	−44	−45
−30	−36	−39	−41	−43	−44	−46	−48	−49	−50	−51	−52
−35	−41	−45	−48	−49	−51	−52	−54	−56	−57	−58	−60
−40	−47	−51	−54	−56	−57	−59	−61	−63	−64	−65	−67

For instance, the table shows that if the temperature is −5°C and the wind speed is 50 km/h, then subjectively it would feel as cold as a temperature of about −15°C with no wind. So

$$f(-5, 50) = -15$$ ■

EXAMPLE 3 In 1928 Charles Cobb and Paul Douglas published a study in which they modeled the growth of the American economy during the period 1899–1922. They considered a simplified view of the economy in which production output is determined by the amount of labor involved and the amount of capital invested. While there are many other factors affecting economic performance, their model proved to be remarkably accurate. The function they used to model production was of the form

1 $$P(L, K) = bL^{\alpha}K^{1-\alpha}$$

where P is the total production (the monetary value of all goods produced in a year), L is the amount of labor (the total number of person-hours worked in a year), and K is

TABLE 2

Year	P	L	K
1899	100	100	100
1900	101	105	107
1901	112	110	114
1902	122	117	122
1903	124	122	131
1904	122	121	138
1905	143	125	149
1906	152	134	163
1907	151	140	176
1908	126	123	185
1909	155	143	198
1910	159	147	208
1911	153	148	216
1912	177	155	226
1913	184	156	236
1914	169	152	244
1915	189	156	266
1916	225	183	298
1917	227	198	335
1918	223	201	366
1919	218	196	387
1920	231	194	407
1921	179	146	417
1922	240	161	431

the amount of capital invested (the monetary worth of all machinery, equipment, and buildings). In Section 14.3 we will show how the form of Equation 1 follows from certain economic assumptions.

Cobb and Douglas used economic data published by the government to obtain Table 2. They took the year 1899 as a baseline and P, L, and K for 1899 were each assigned the value 100. The values for other years were expressed as percentages of the 1899 figures.

Cobb and Douglas used the method of least squares to fit the data of Table 2 to the function

$$\boxed{2} \qquad P(L, K) = 1.01L^{0.75}K^{0.25}$$

(See Exercise 79 for the details.)

If we use the model given by the function in Equation 2 to compute the production in the years 1910 and 1920, we get the values

$$P(147, 208) = 1.01(147)^{0.75}(208)^{0.25} \approx 161.9$$

$$P(194, 407) = 1.01(194)^{0.75}(407)^{0.25} \approx 235.8$$

which are quite close to the actual values, 159 and 231.

The production function $\boxed{1}$ has subsequently been used in many settings, ranging from individual firms to global economics. It has become known as the **Cobb-Douglas production function**. Its domain is $\{(L, K) \mid L \geqslant 0, K \geqslant 0\}$ because L and K represent labor and capital and are therefore never negative.

EXAMPLE 4 Find the domain and range of $g(x, y) = \sqrt{9 - x^2 - y^2}$.

SOLUTION The domain of g is

$$D = \{(x, y) \mid 9 - x^2 - y^2 \geqslant 0\} = \{(x, y) \mid x^2 + y^2 \leqslant 9\}$$

which is the disk with center (0, 0) and radius 3. (See Figure 4.) The range of g is

$$\left\{z \mid z = \sqrt{9 - x^2 - y^2}, (x, y) \in D\right\}$$

Since z is a positive square root, $z \geqslant 0$. Also, because $9 - x^2 - y^2 \leqslant 9$, we have

$$\sqrt{9 - x^2 - y^2} \leqslant 3$$

So the range is

$$\{z \mid 0 \leqslant z \leqslant 3\} = [0, 3]$$

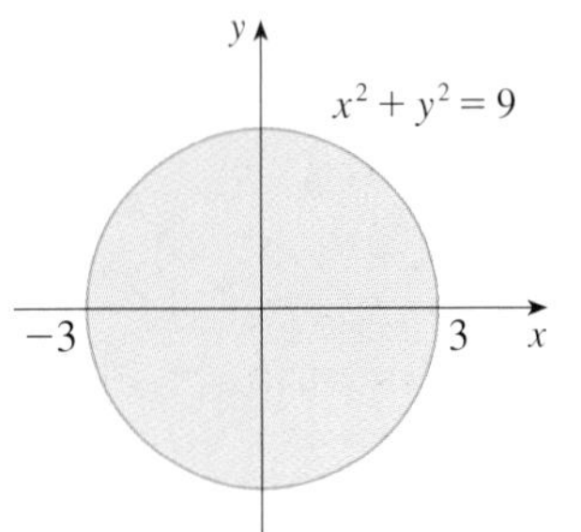

FIGURE 4
Domain of $g(x, y) = \sqrt{9 - x^2 - y^2}$

Graphs

Another way of visualizing the behavior of a function of two variables is to consider its graph.

Definition If f is a function of two variables with domain D, then the **graph** of f is the set of all points (x, y, z) in $\mathbb{R}^3$ such that $z = f(x, y)$ and (x, y) is in D.

Just as the graph of a function f of one variable is a curve C with equation $y = f(x)$, so the graph of a function f of two variables is a surface S with equation $z = f(x, y)$. We can visualize the graph S of f as lying directly above or below its domain D in the xy-plane (see Figure 5).

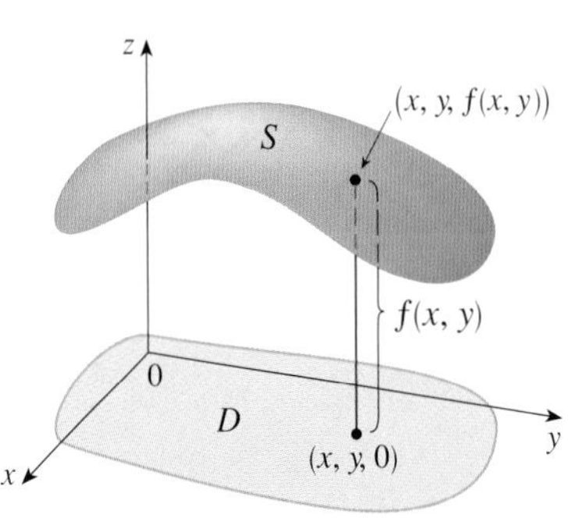

FIGURE 5

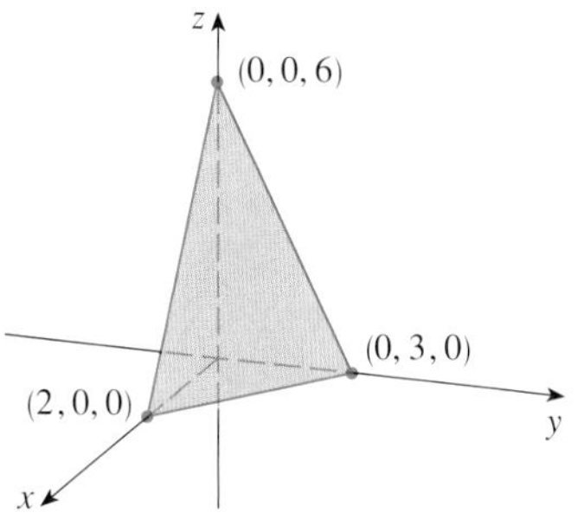

FIGURE 6

EXAMPLE 5 Sketch the graph of the function $f(x, y) = 6 - 3x - 2y$.

SOLUTION The graph of f has the equation $z = 6 - 3x - 2y$, or $3x + 2y + z = 6$, which represents a plane. To graph the plane we first find the intercepts. Putting $y = z = 0$ in the equation, we get $x = 2$ as the x-intercept. Similarly, the y-intercept is 3 and the z-intercept is 6. This helps us sketch the portion of the graph that lies in the first octant in Figure 6.

The function in Example 5 is a special case of the function

$$f(x, y) = ax + by + c$$

which is called a **linear function**. The graph of such a function has the equation

$$z = ax + by + c \qquad \text{or} \qquad ax + by - z + c = 0$$

so it is a plane. In much the same way that linear functions of one variable are important in single-variable calculus, we will see that linear functions of two variables play a central role in multivariable calculus.

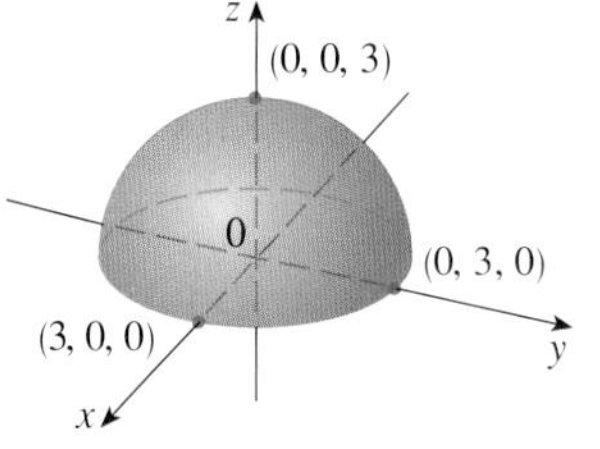

FIGURE 7
Graph of $g(x, y) = \sqrt{9 - x^2 - y^2}$

V EXAMPLE 6 Sketch the graph of $g(x, y) = \sqrt{9 - x^2 - y^2}$.

SOLUTION The graph has equation $z = \sqrt{9 - x^2 - y^2}$. We square both sides of this equation to obtain $z^2 = 9 - x^2 - y^2$, or $x^2 + y^2 + z^2 = 9$, which we recognize as an equation of the sphere with center the origin and radius 3. But, since $z \geqslant 0$, the graph of g is just the top half of this sphere (see Figure 7).

NOTE An entire sphere can't be represented by a single function of x and y. As we saw in Example 6, the upper hemisphere of the sphere $x^2 + y^2 + z^2 = 9$ is represented by the function $g(x, y) = \sqrt{9 - x^2 - y^2}$. The lower hemisphere is represented by the function $h(x, y) = -\sqrt{9 - x^2 - y^2}$.

EXAMPLE 7 Use a computer to draw the graph of the Cobb-Douglas production function $P(L, K) = 1.01L^{0.75}K^{0.25}$.

SOLUTION Figure 8 shows the graph of P for values of the labor L and capital K that lie between 0 and 300. The computer has drawn the surface by plotting vertical traces. We see from these traces that the value of the production P increases as either L or K increases, as is to be expected.

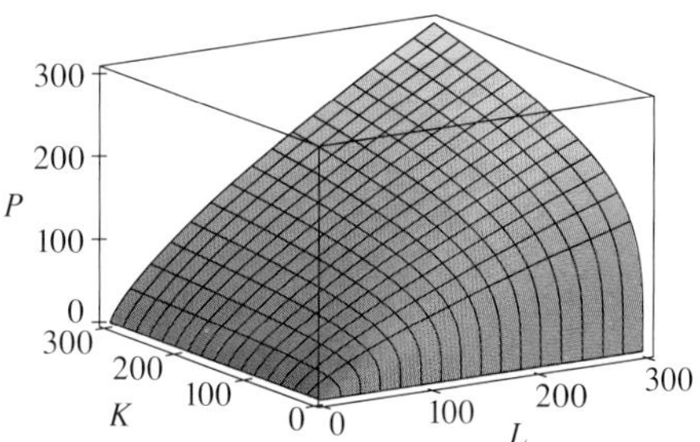

FIGURE 8

V EXAMPLE 8 Find the domain and range and sketch the graph of $h(x, y) = 4x^2 + y^2$.

SOLUTION Notice that $h(x, y)$ is defined for all possible ordered pairs of real numbers (x, y), so the domain is $\mathbb{R}^2$, the entire xy-plane. The range of h is the set $[0, \infty)$ of all nonnegative real numbers. [Notice that $x^2 \geqslant 0$ and $y^2 \geqslant 0$, so $h(x, y) \geqslant 0$ for all x and y.]

The graph of h has the equation $z = 4x^2 + y^2$, which is the elliptic paraboloid that we sketched in Example 4 in Section 12.6. Horizontal traces are ellipses and vertical traces are parabolas (see Figure 9).

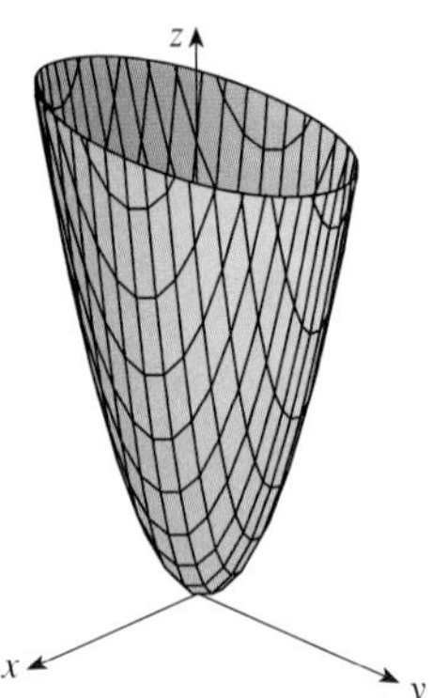

FIGURE 9
Graph of $h(x, y) = 4x^2 + y^2$

Computer programs are readily available for graphing functions of two variables. In most such programs, traces in the vertical planes $x = k$ and $y = k$ are drawn for equally spaced values of k and parts of the graph are eliminated using hidden line removal.

Figure 10 shows computer-generated graphs of several functions. Notice that we get an especially good picture of a function when rotation is used to give views from different

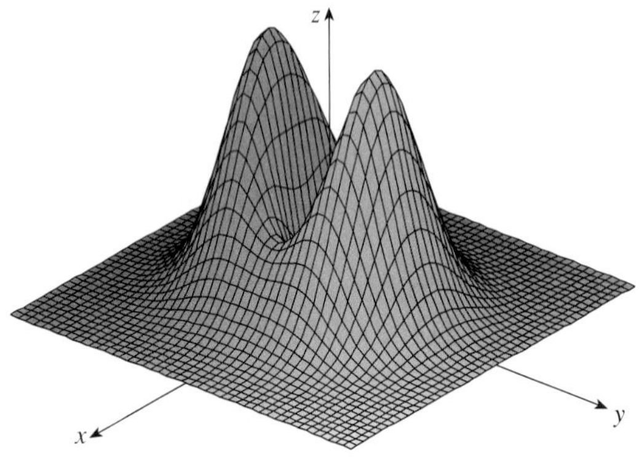

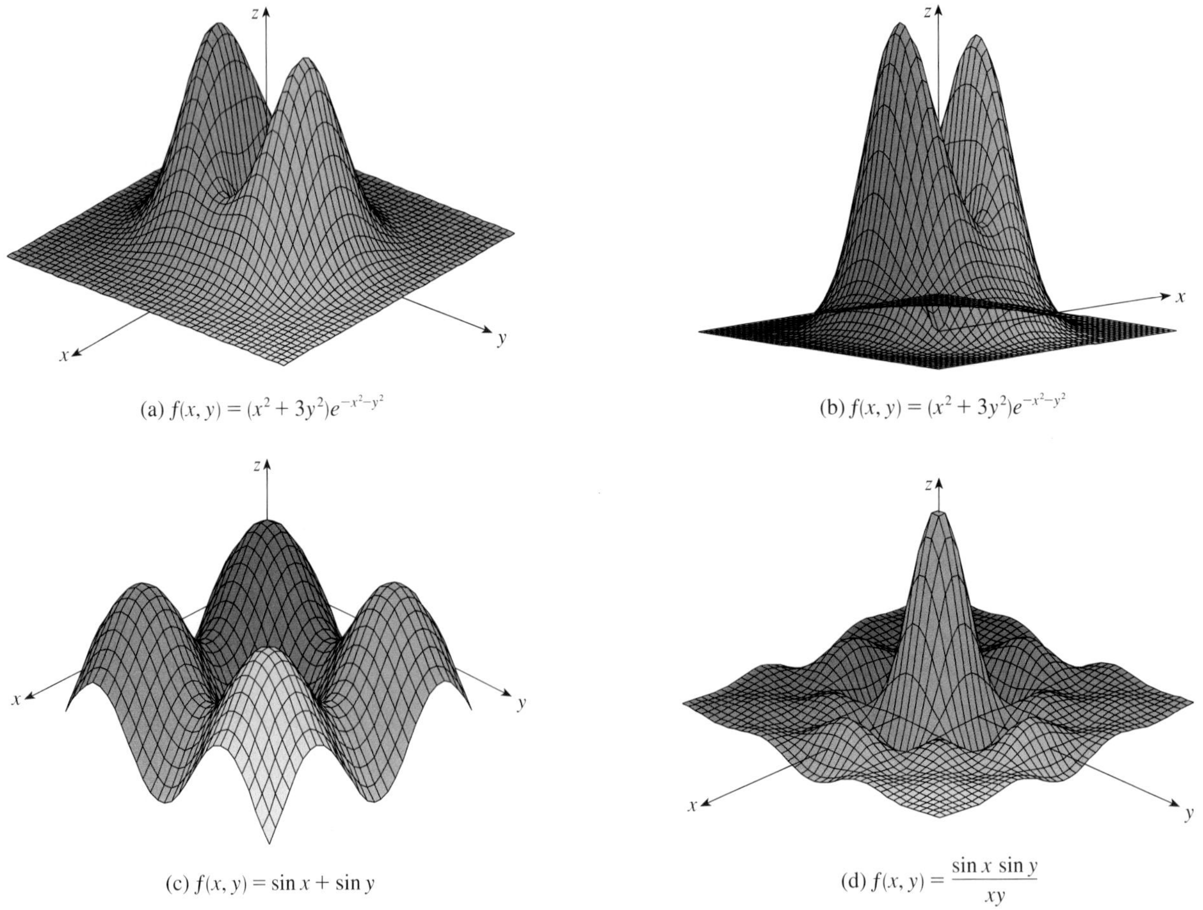

(a) $f(x, y) = (x^2 + 3y^2)e^{-x^2-y^2}$

(b) $f(x, y) = (x^2 + 3y^2)e^{-x^2-y^2}$

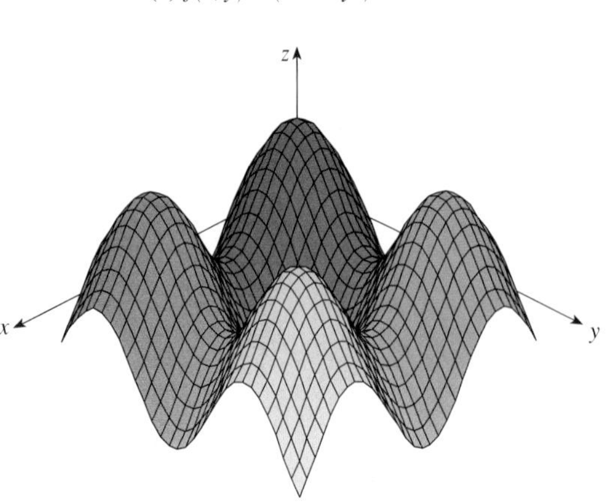

(c) $f(x, y) = \sin x + \sin y$

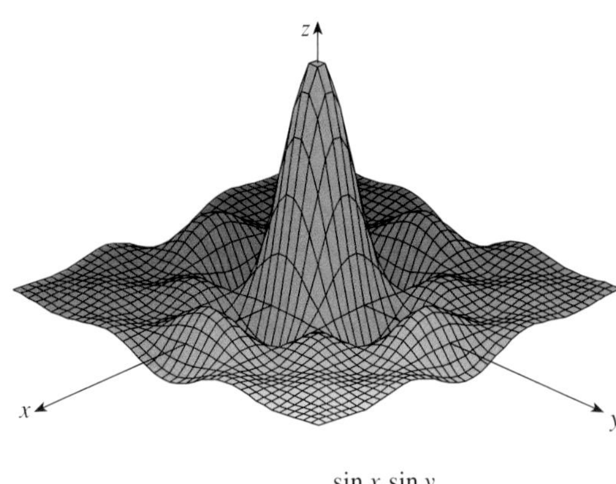

(d) $f(x, y) = \dfrac{\sin x \sin y}{xy}$

FIGURE 10

vantage points. In parts (a) and (b) the graph of f is very flat and close to the xy-plane except near the origin; this is because $e^{-x^2-y^2}$ is very small when x or y is large.

Level Curves

So far we have two methods for visualizing functions: arrow diagrams and graphs. A third method, borrowed from mapmakers, is a contour map on which points of constant elevation are joined to form *contour lines*, or *level curves*.

Definition The **level curves** of a function f of two variables are the curves with equations $f(x, y) = k$, where k is a constant (in the range of f).

A level curve $f(x, y) = k$ is the set of all points in the domain of f at which f takes on a given value k. In other words, it shows where the graph of f has height k.

You can see from Figure 11 the relation between level curves and horizontal traces. The level curves $f(x, y) = k$ are just the traces of the graph of f in the horizontal plane $z = k$ projected down to the xy-plane. So if you draw the level curves of a function and visualize them being lifted up to the surface at the indicated height, then you can mentally piece together a picture of the graph. The surface is steep where the level curves are close together. It is somewhat flatter where they are farther apart.

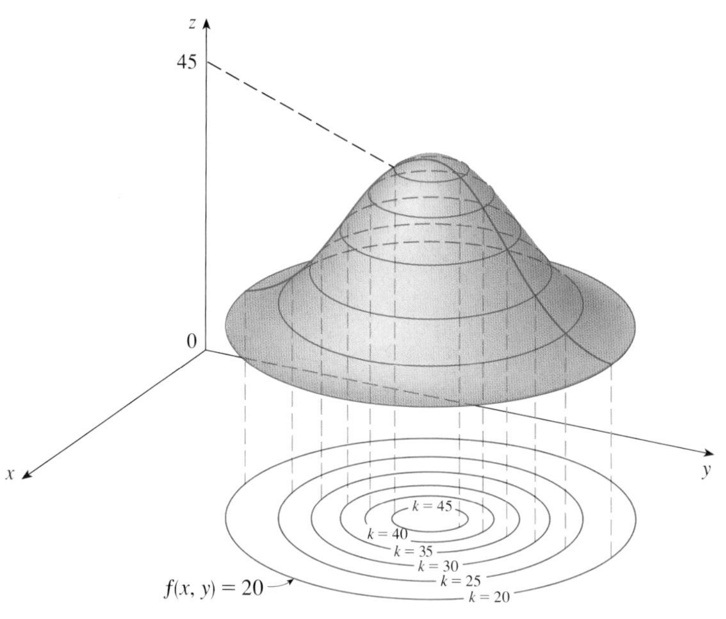

FIGURE 11

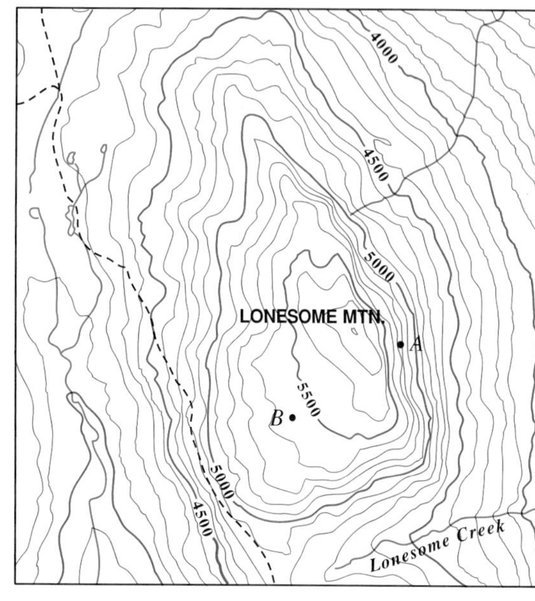

FIGURE 12

TEC Visual 14.1A animates Figure 11 by showing level curves being lifted up to graphs of functions.

One common example of level curves occurs in topographic maps of mountainous regions, such as the map in Figure 12. The level curves are curves of constant elevation above sea level. If you walk along one of these contour lines, you neither ascend nor descend. Another common example is the temperature function introduced in the opening paragraph of this section. Here the level curves are called **isothermals** and join locations with the same temperature. Figure 13 shows a weather map of the world indicating the average January temperatures. The isothermals are the curves that separate the colored bands.

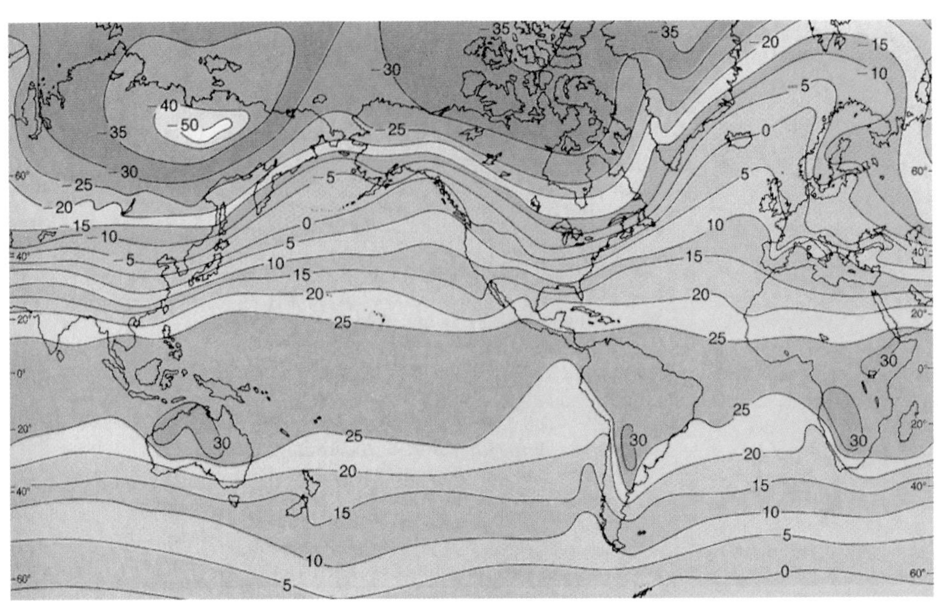

FIGURE 13
World mean sea-level temperatures in January in degrees Celsius

From *Atmosphere: Introduction to Meteorology*, 4th Edition, 1989.

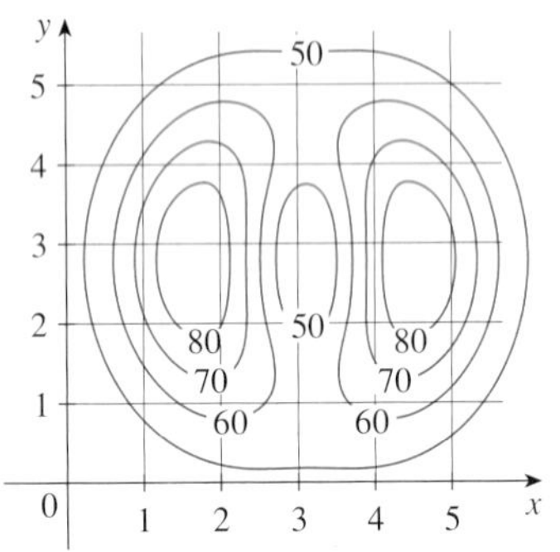

FIGURE 14

EXAMPLE 9 A contour map for a function f is shown in Figure 14. Use it to estimate the values of $f(1, 3)$ and $f(4, 5)$.

SOLUTION The point $(1, 3)$ lies partway between the level curves with z-values 70 and 80. We estimate that

$$f(1, 3) \approx 73$$

Similarly, we estimate that

$$f(4, 5) \approx 56$$

EXAMPLE 10 Sketch the level curves of the function $f(x, y) = 6 - 3x - 2y$ for the values $k = -6, 0, 6, 12$.

SOLUTION The level curves are

$$6 - 3x - 2y = k \qquad \text{or} \qquad 3x + 2y + (k - 6) = 0$$

This is a family of lines with slope $-\frac{3}{2}$. The four particular level curves with $k = -6, 0, 6$, and 12 are $3x + 2y - 12 = 0$, $3x + 2y - 6 = 0$, $3x + 2y = 0$, and $3x + 2y + 6 = 0$. They are sketched in Figure 15. The level curves are equally spaced parallel lines because the graph of f is a plane (see Figure 6).

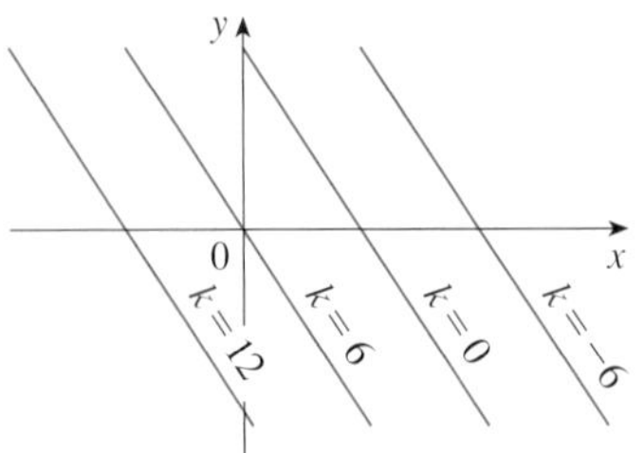

FIGURE 15
Contour map of $f(x, y) = 6 - 3x - 2y$

V EXAMPLE 11 Sketch the level curves of the function

$$g(x, y) = \sqrt{9 - x^2 - y^2} \qquad \text{for} \quad k = 0, 1, 2, 3$$

SOLUTION The level curves are

$$\sqrt{9 - x^2 - y^2} = k \qquad \text{or} \qquad x^2 + y^2 = 9 - k^2$$

This is a family of concentric circles with center $(0, 0)$ and radius $\sqrt{9 - k^2}$. The cases $k = 0, 1, 2, 3$ are shown in Figure 16. Try to visualize these level curves lifted up to form a surface and compare with the graph of g (a hemisphere) in Figure 7. (See TEC Visual 14.1A.)

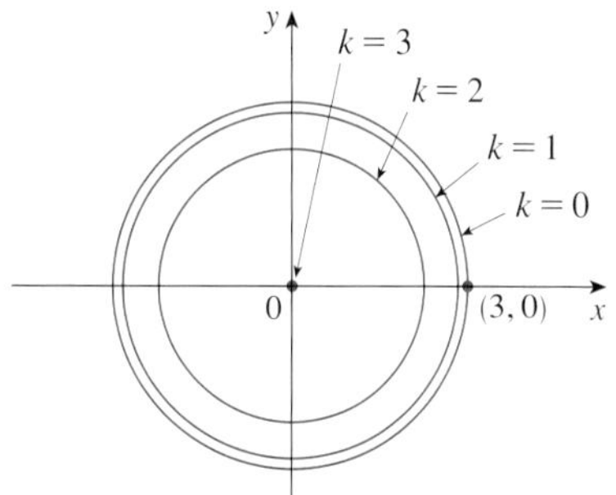

FIGURE 16
Contour map of $g(x, y) = \sqrt{9 - x^2 - y^2}$

EXAMPLE 12 Sketch some level curves of the function $h(x, y) = 4x^2 + y^2 + 1$.

SOLUTION The level curves are

$$4x^2 + y^2 + 1 = k \qquad \text{or} \qquad \frac{x^2}{\frac{1}{4}(k - 1)} + \frac{y^2}{k - 1} = 1$$

which, for $k > 1$, describes a family of ellipses with semiaxes $\frac{1}{2}\sqrt{k - 1}$ and $\sqrt{k - 1}$. Figure 17(a) shows a contour map of h drawn by a computer. Figure 17(b) shows these level curves lifted up to the graph of h (an elliptic paraboloid) where they become horizontal traces. We see from Figure 17 how the graph of h is put together from the level curves.

Visual 14.1B demonstrates the connection between surfaces and their contour maps.

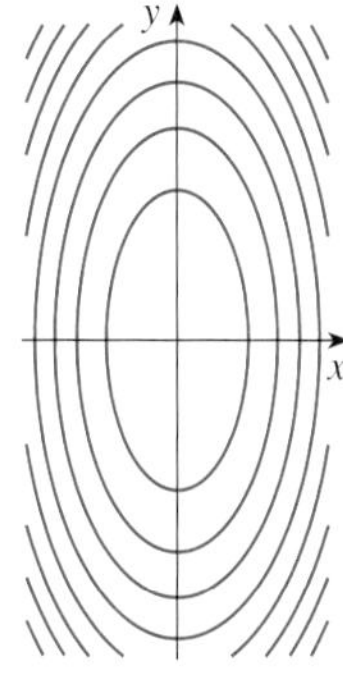

(a) Contour map

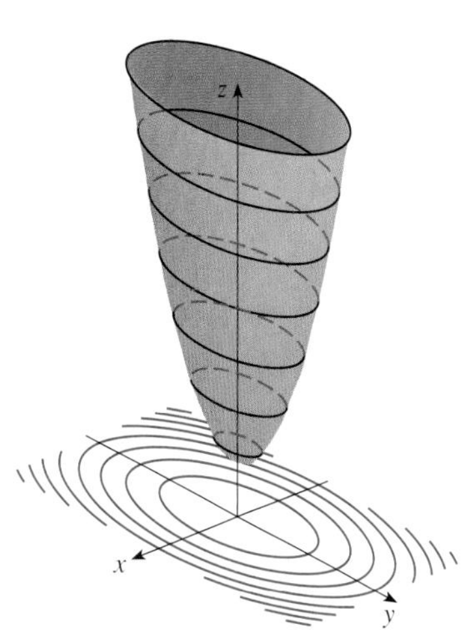

(b) Horizontal traces are raised level curves

FIGURE 17
The graph of $h(x, y) = 4x^2 + y^2 + 1$ is formed by lifting the level curves.

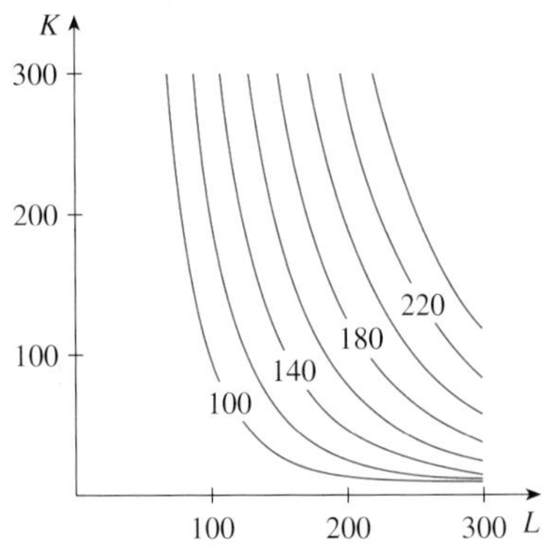

FIGURE 18

EXAMPLE 13 Plot level curves for the Cobb-Douglas production function of Example 3.

SOLUTION In Figure 18 we use a computer to draw a contour plot for the Cobb-Douglas production function

$$P(L, K) = 1.01L^{0.75}K^{0.25}$$

Level curves are labeled with the value of the production P. For instance, the level curve labeled 140 shows all values of the labor L and capital investment K that result in a production of $P = 140$. We see that, for a fixed value of P, as L increases K decreases, and vice versa.

For some purposes, a contour map is more useful than a graph. That is certainly true in Example 13. (Compare Figure 18 with Figure 8.) It is also true in estimating function values, as in Example 9.

Figure 19 shows some computer-generated level curves together with the corresponding computer-generated graphs. Notice that the level curves in part (c) crowd together near the origin. That corresponds to the fact that the graph in part (d) is very steep near the origin.

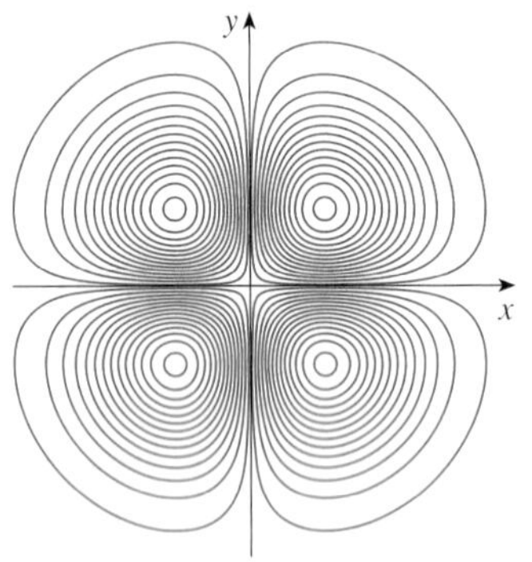

(a) Level curves of $f(x, y) = -xye^{-x^2-y^2}$

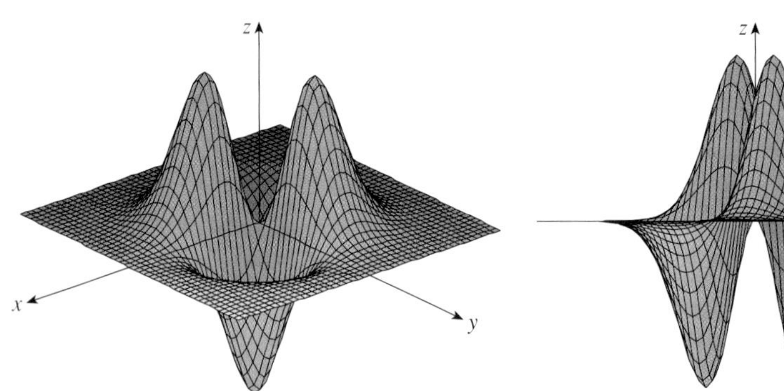

(b) Two views of $f(x, y) = -xye^{-x^2-y^2}$

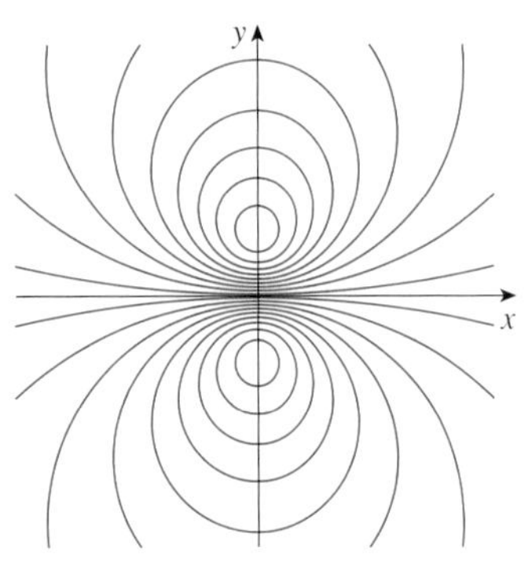

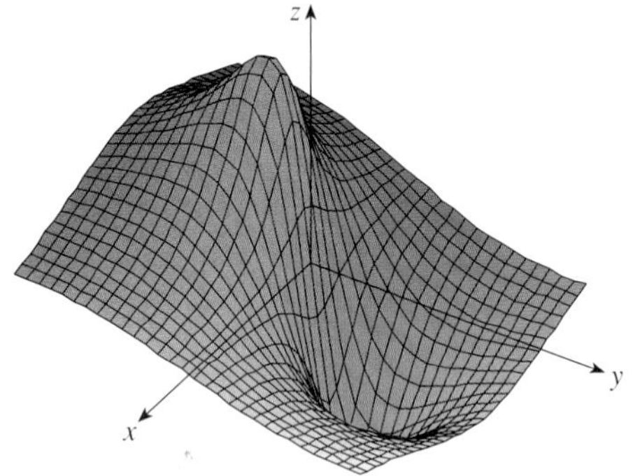

FIGURE 19

(c) Level curves of $f(x, y) = \dfrac{-3y}{x^2 + y^2 + 1}$

(d) $f(x, y) = \dfrac{-3y}{x^2 + y^2 + 1}$

Functions of Three or More Variables

A **function of three variables**, f, is a rule that assigns to each ordered triple (x, y, z) in a domain $D \subset \mathbb{R}^3$ a unique real number denoted by $f(x, y, z)$. For instance, the temperature T at a point on the surface of the earth depends on the longitude x and latitude y of the point and on the time t, so we could write $T = f(x, y, t)$.

EXAMPLE 14 Find the domain of f if

$$f(x, y, z) = \ln(z - y) + xy \sin z$$

SOLUTION The expression for $f(x, y, z)$ is defined as long as $z - y > 0$, so the domain of f is

$$D = \{(x, y, z) \in \mathbb{R}^3 \mid z > y\}$$

This is a **half-space** consisting of all points that lie above the plane $z = y$.

It's very difficult to visualize a function f of three variables by its graph, since that would lie in a four-dimensional space. However, we do gain some insight into f by examining its **level surfaces**, which are the surfaces with equations $f(x, y, z) = k$, where k is a constant. If the point (x, y, z) moves along a level surface, the value of $f(x, y, z)$ remains fixed.

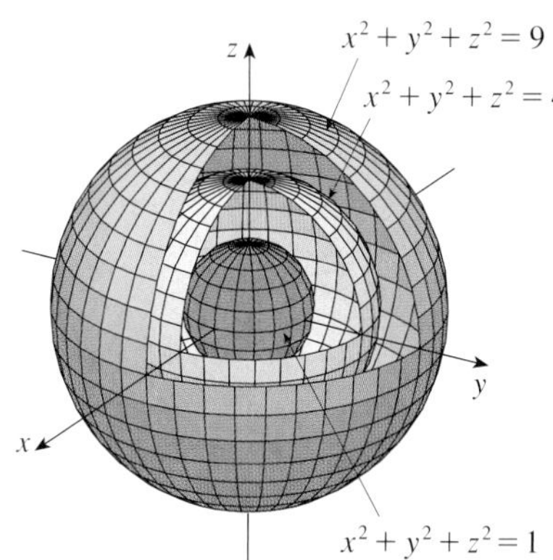

FIGURE 20

EXAMPLE 15 Find the level surfaces of the function

$$f(x, y, z) = x^2 + y^2 + z^2$$

SOLUTION The level surfaces are $x^2 + y^2 + z^2 = k$, where $k \geqslant 0$. These form a family of concentric spheres with radius $\sqrt{k}$. (See Figure 20.) Thus, as (x, y, z) varies over any sphere with center O, the value of $f(x, y, z)$ remains fixed.

Functions of any number of variables can be considered. A **function of *n* variables** is a rule that assigns a number $z = f(x_1, x_2, \ldots, x_n)$ to an n-tuple $(x_1, x_2, \ldots, x_n)$ of real numbers. We denote by $\mathbb{R}^n$ the set of all such n-tuples. For example, if a company uses n different ingredients in making a food product, c_i is the cost per unit of the ith ingredient, and x_i units of the ith ingredient are used, then the total cost C of the ingredients is a function of the n variables $x_1, x_2, \ldots, x_n$:

$$C = f(x_1, x_2, \ldots, x_n) = c_1x_1 + c_2x_2 + \cdots + c_nx_n \tag{3}$$

The function f is a real-valued function whose domain is a subset of $\mathbb{R}^n$. Sometimes we will use vector notation to write such functions more compactly: If $\mathbf{x} = \langle x_1, x_2, \ldots, x_n \rangle$, we often write $f(\mathbf{x})$ in place of $f(x_1, x_2, \ldots, x_n)$. With this notation we can rewrite the function defined in Equation 3 as

$$f(\mathbf{x}) = \mathbf{c} \cdot \mathbf{x}$$

where $\mathbf{c} = \langle c_1, c_2, \ldots, c_n \rangle$ and $\mathbf{c} \cdot \mathbf{x}$ denotes the dot product of the vectors $\mathbf{c}$ and $\mathbf{x}$ in V_n.

In view of the one-to-one correspondence between points $(x_1, x_2, \ldots, x_n)$ in $\mathbb{R}^n$ and their position vectors $\mathbf{x} = \langle x_1, x_2, \ldots, x_n \rangle$ in V_n, we have three ways of looking at a function f defined on a subset of $\mathbb{R}^n$:

1. As a function of n real variables $x_1, x_2, \ldots, x_n$
2. As a function of a single point variable $(x_1, x_2, \ldots, x_n)$
3. As a function of a single vector variable $\mathbf{x} = \langle x_1, x_2, \ldots, x_n \rangle$

We will see that all three points of view are useful.

14.1 Exercises

1. In Example 2 we considered the function $W = f(T, v)$, where W is the wind-chill index, T is the actual temperature, and v is the wind speed. A numerical representation is given in Table 1.
   (a) What is the value of $f(-15, 40)$? What is its meaning?
   (b) Describe in words the meaning of the question "For what value of v is $f(-20, v) = -30$?" Then answer the question.
   (c) Describe in words the meaning of the question "For what value of T is $f(T, 20) = -49$?" Then answer the question.
   (d) What is the meaning of the function $W = f(-5, v)$? Describe the behavior of this function.
   (e) What is the meaning of the function $W = f(T, 50)$? Describe the behavior of this function.

2. The *temperature-humidity index* I (or humidex, for short) is the perceived air temperature when the actual temperature is T and the relative humidity is h, so we can write $I = f(T, h)$. The following table of values of I is an excerpt from a table compiled by the National Oceanic & Atmospheric Administration.

TABLE 3 Apparent temperature as a function of temperature and humidity

Relative humidity (%) / Actual temperature (°F)

T \ h	20	30	40	50	60	70
80	77	78	79	81	82	83
85	82	84	86	88	90	93
90	87	90	93	96	100	106
95	93	96	101	107	114	124
100	99	104	110	120	132	144

   (a) What is the value of $f(95, 70)$? What is its meaning?
   (b) For what value of h is $f(90, h) = 100$?
   (c) For what value of T is $f(T, 50) = 88$?
   (d) What are the meanings of the functions $I = f(80, h)$ and $I = f(100, h)$? Compare the behavior of these two functions of h.

3. A manufacturer has modeled its yearly production function P (the monetary value of its entire production in millions of dollars) as a Cobb-Douglas function

$$P(L, K) = 1.47L^{0.65}K^{0.35}$$

where L is the number of labor hours (in thousands) and K is the invested capital (in millions of dollars). Find $P(120, 20)$ and interpret it.

4. Verify for the Cobb-Douglas production function

$$P(L, K) = 1.01L^{0.75}K^{0.25}$$

discussed in Example 3 that the production will be doubled if both the amount of labor and the amount of capital are doubled. Determine whether this is also true for the general production function

$$P(L, K) = bL^{\alpha}K^{1-\alpha}$$

5. A model for the surface area of a human body is given by the function

$$S = f(w, h) = 0.1091w^{0.425}h^{0.725}$$

where w is the weight (in pounds), h is the height (in inches), and S is measured in square feet.
   (a) Find $f(160, 70)$ and interpret it.
   (b) What is your own surface area?

6. The wind-chill index W discussed in Example 2 has been modeled by the following function:

$$W(T, v) = 13.12 + 0.6215T - 11.37v^{0.16} + 0.3965Tv^{0.16}$$

Check to see how closely this model agrees with the values in Table 1 for a few values of T and v.

7. The wave heights h in the open sea depend on the speed v of the wind and the length of time t that the wind has been blowing at that speed. Values of the function $h = f(v, t)$ are recorded in feet in Table 4.
   (a) What is the value of $f(40, 15)$? What is its meaning?
   (b) What is the meaning of the function $h = f(30, t)$? Describe the behavior of this function.
   (c) What is the meaning of the function $h = f(v, 30)$? Describe the behavior of this function.

TABLE 4

Duration (hours) / Wind speed (knots)

v \ t	5	10	15	20	30	40	50
10	2	2	2	2	2	2	2
15	4	4	5	5	5	5	5
20	5	7	8	8	9	9	9
30	9	13	16	17	18	19	19
40	14	21	25	28	31	33	33
50	19	29	36	40	45	48	50
60	24	37	47	54	62	67	69

8. A company makes three sizes of cardboard boxes: small, medium, and large. It costs \$2.50 to make a small box, \$4.00

for a medium box, and \$4.50 for a large box. Fixed costs are \$8000.

(a) Express the cost of making x small boxes, y medium boxes, and z large boxes as a function of three variables: $C = f(x, y, z)$.

(b) Find $f(3000, 5000, 4000)$ and interpret it.

(c) What is the domain of f?

9. Let $g(x, y) = \cos(x + 2y)$.

(a) Evaluate $g(2, -1)$.

(b) Find the domain of g.

(c) Find the range of g.

10. Let $F(x, y) = 1 + \sqrt{4 - y^2}$.

(a) Evaluate $F(3, 1)$.

(b) Find and sketch the domain of F.

(c) Find the range of F.

11. Let $f(x, y, z) = \sqrt{x} + \sqrt{y} + \sqrt{z} + \ln(4 - x^2 - y^2 - z^2)$.

(a) Evaluate $f(1, 1, 1)$.

(b) Find and describe the domain of f.

12. Let $g(x, y, z) = x^3y^2z\sqrt{10 - x - y - z}$.

(a) Evaluate $g(1, 2, 3)$.

(b) Find and describe the domain of g.

13–22 Find and sketch the domain of the function.

13. $f(x, y) = \sqrt{2x - y}$

14. $f(x, y) = \sqrt{xy}$

15. $f(x, y) = \ln(9 - x^2 - 9y^2)$

16. $f(x, y) = \sqrt{x^2 - y^2}$

17. $f(x, y) = \sqrt{1 - x^2} - \sqrt{1 - y^2}$

18. $f(x, y) = \sqrt{y} + \sqrt{25 - x^2 - y^2}$

19. $f(x, y) = \dfrac{\sqrt{y - x^2}}{1 - x^2}$

20. $f(x, y) = \arcsin(x^2 + y^2 - 2)$

21. $f(x, y, z) = \sqrt{1 - x^2 - y^2 - z^2}$

22. $f(x, y, z) = \ln(16 - 4x^2 - 4y^2 - z^2)$

23–31 Sketch the graph of the function.

23. $f(x, y) = 1 + y$

24. $f(x, y) = 2 - x$

25. $f(x, y) = 10 - 4x - 5y$

26. $f(x, y) = e^{-y}$

27. $f(x, y) = y^2 + 1$

28. $f(x, y) = 1 + 2x^2 + 2y^2$

29. $f(x, y) = 9 - x^2 - 9y^2$

30. $f(x, y) = \sqrt{4x^2 + y^2}$

31. $f(x, y) = \sqrt{4 - 4x^2 - y^2}$

32. Match the function with its graph (labeled I–VI). Give reasons for your choices.

(a) $f(x, y) = |x| + |y|$

(b) $f(x, y) = |xy|$

(c) $f(x, y) = \dfrac{1}{1 + x^2 + y^2}$

(d) $f(x, y) = (x^2 - y^2)^2$

(e) $f(x, y) = (x - y)^2$

(f) $f(x, y) = \sin(|x| + |y|)$

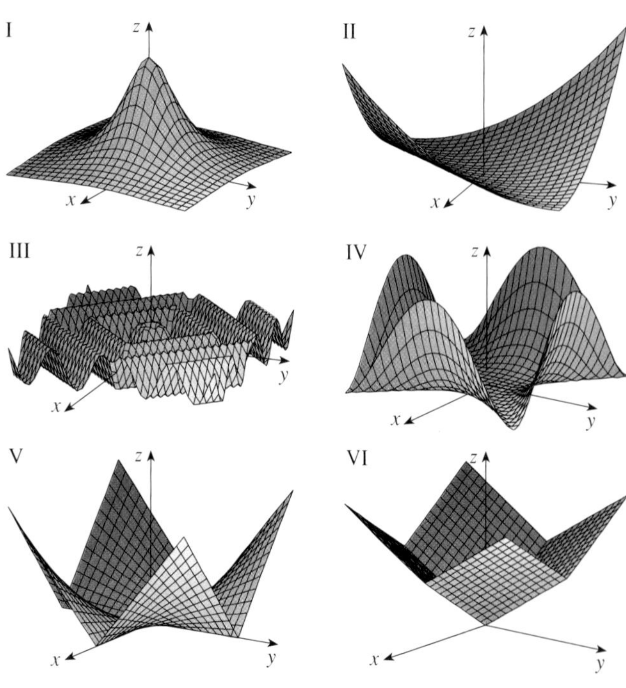

33. A contour map for a function f is shown. Use it to estimate the values of $f(-3, 3)$ and $f(3, -2)$. What can you say about the shape of the graph?

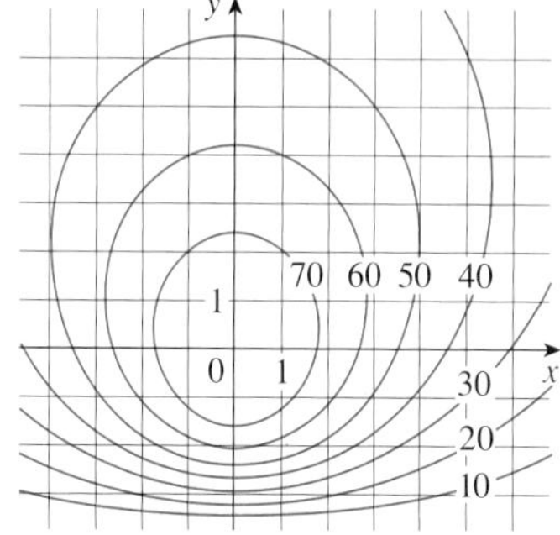

34. Shown is a contour map of atmospheric pressure in North America on August 12, 2008. On the level curves (called isobars) the pressure is indicated in millibars (mb).

(a) Estimate the pressure at C (Chicago), N (Nashville), S (San Francisco), and V (Vancouver).

(b) At which of these locations were the winds strongest?

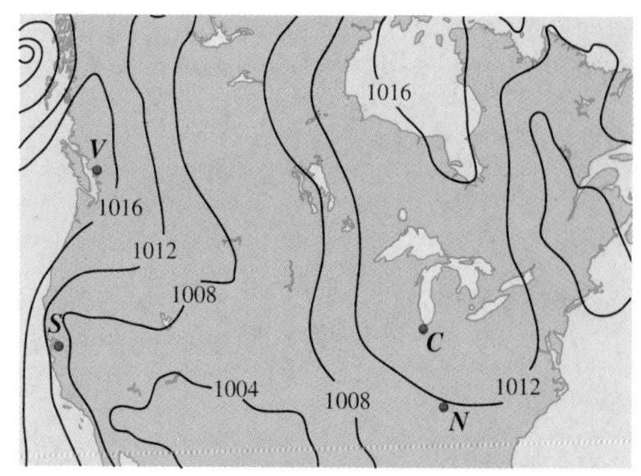

35. Level curves (isothermals) are shown for the water temperature (in °C) in Long Lake (Minnesota) in 1998 as a function of depth and time of year. Estimate the temperature in the lake on June 9 (day 160) at a depth of 10 m and on June 29 (day 180) at a depth of 5 m.

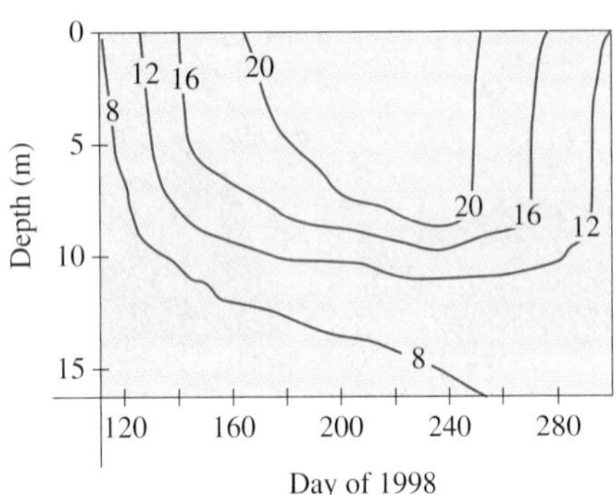

36. Two contour maps are shown. One is for a function f whose graph is a cone. The other is for a function g whose graph is a paraboloid. Which is which, and why?

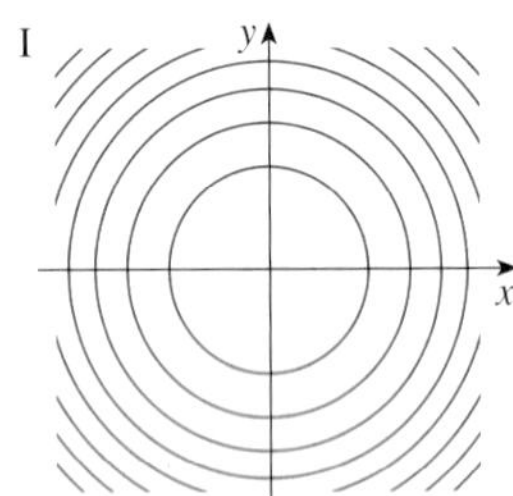

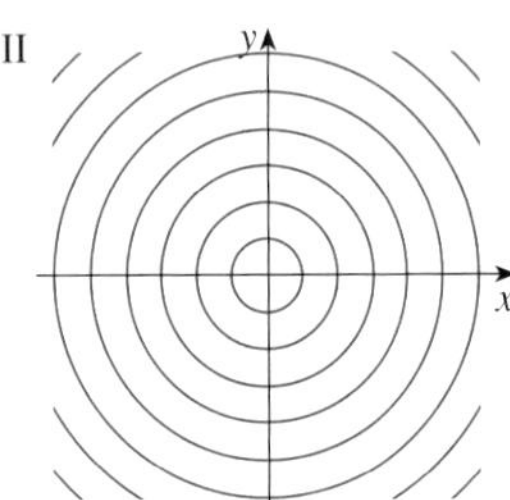

37. Locate the points A and B on the map of Lonesome Mountain (Figure 12). How would you describe the terrain near A? Near B?

38. Make a rough sketch of a contour map for the function whose graph is shown.

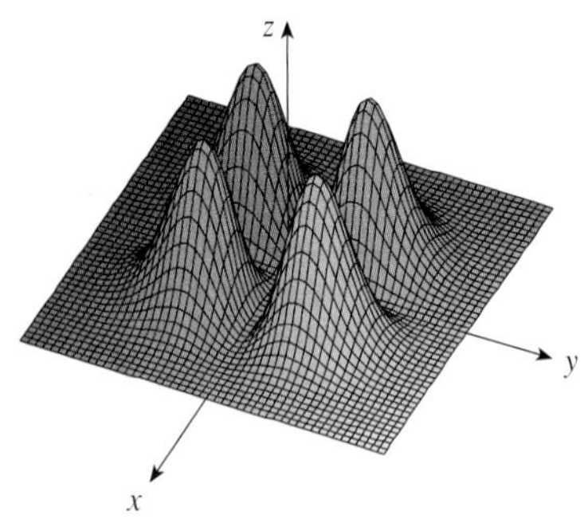

39–42 A contour map of a function is shown. Use it to make a rough sketch of the graph of f.

39.

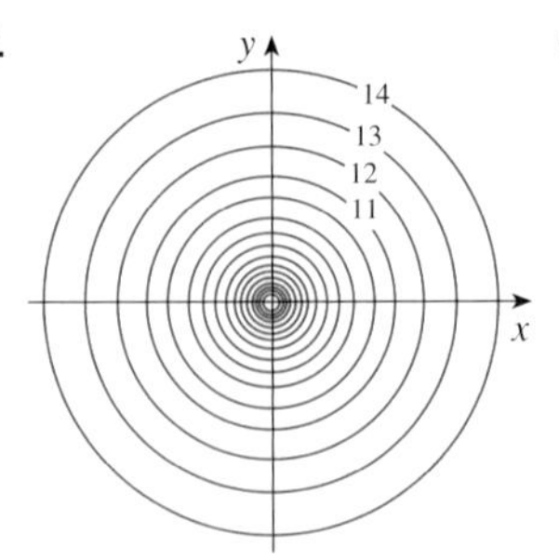

40.

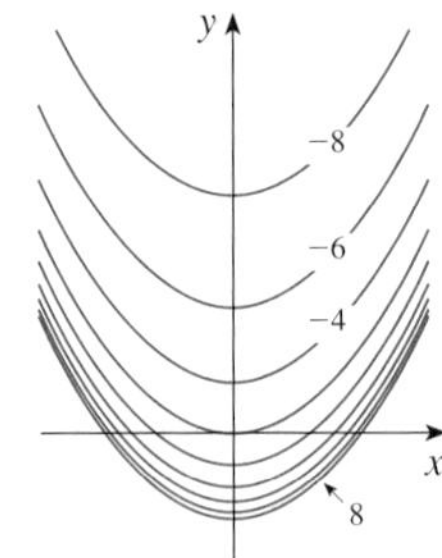

41.

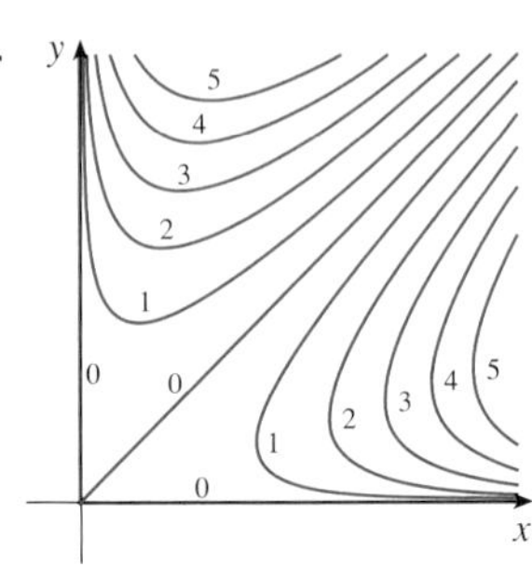

42.

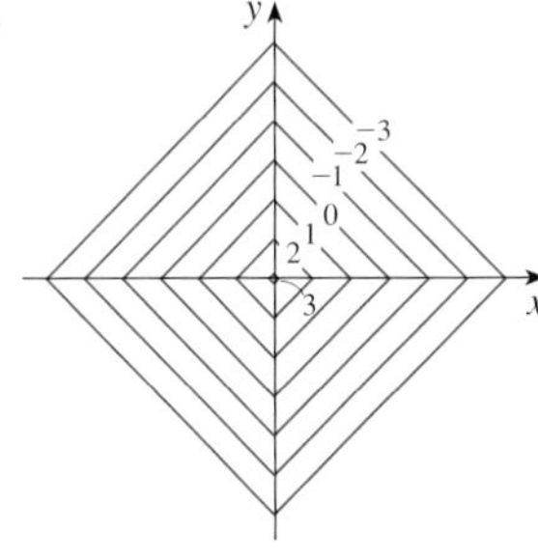

43–50 Draw a contour map of the function showing several level curves.

43. $f(x, y) = (y - 2x)^2$

44. $f(x, y) = x^3 - y$

45. $f(x, y) = \sqrt{x} + y$

46. $f(x, y) = \ln(x^2 + 4y^2)$

47. $f(x, y) = ye^x$

48. $f(x, y) = y \sec x$

49. $f(x, y) = \sqrt{y^2 - x^2}$

50. $f(x, y) = y/(x^2 + y^2)$

51–52 Sketch both a contour map and a graph of the function and compare them.

51. $f(x, y) = x^2 + 9y^2$

52. $f(x, y) = \sqrt{36 - 9x^2 - 4y^2}$

53. A thin metal plate, located in the xy-plane, has temperature $T(x, y)$ at the point (x, y). The level curves of T are called *isothermals* because at all points on such a curve the temperature is the same. Sketch some isothermals if the temperature function is given by

$$T(x, y) = \frac{100}{1 + x^2 + 2y^2}$$

54. If $V(x, y)$ is the electric potential at a point (x, y) in the xy-plane, then the level curves of V are called *equipotential curves* because at all points on such a curve the electric potential is the same. Sketch some equipotential curves if $V(x, y) = c/\sqrt{r^2 - x^2 - y^2}$, where c is a positive constant.

55–58 Use a computer to graph the function using various domains and viewpoints. Get a printout of one that, in your opinion, gives a good view. If your software also produces level curves, then plot some contour lines of the same function and compare with the graph.

55. $f(x, y) = xy^2 - x^3$ (monkey saddle)

56. $f(x, y) = xy^3 - yx^3$ (dog saddle)

57. $f(x, y) = e^{-(x^2+y^2)/3}(\sin(x^2) + \cos(y^2))$

58. $f(x, y) = \cos x \cos y$

59–64 Match the function (a) with its graph (labeled A–F below) and (b) with its contour map (labeled I–VI). Give reasons for your choices.

59. $z = \sin(xy)$

60. $z = e^x \cos y$

61. $z = \sin(x - y)$

62. $z = \sin x - \sin y$

63. $z = (1 - x^2)(1 - y^2)$

64. $z = \dfrac{x - y}{1 + x^2 + y^2}$

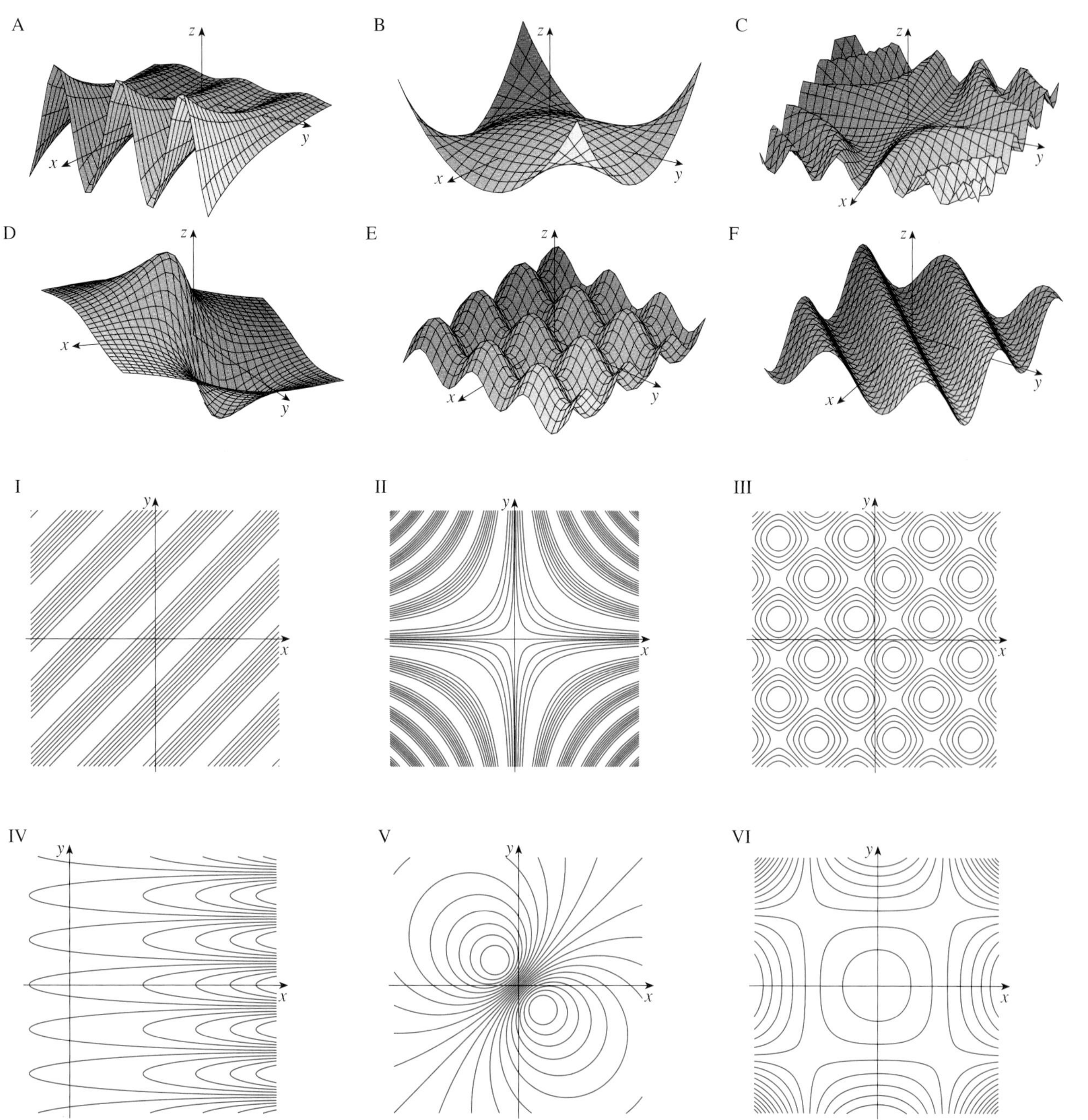

65–68 Describe the level surfaces of the function.

65. $f(x, y, z) = x + 3y + 5z$

66. $f(x, y, z) = x^2 + 3y^2 + 5z^2$

67. $f(x, y, z) = y^2 + z^2$

68. $f(x, y, z) = x^2 - y^2 - z^2$

69–70 Describe how the graph of g is obtained from the graph of f.

69. (a) $g(x, y) = f(x, y) + 2$
(b) $g(x, y) = 2f(x, y)$
(c) $g(x, y) = -f(x, y)$
(d) $g(x, y) = 2 - f(x, y)$

70. (a) $g(x, y) = f(x - 2, y)$
(b) $g(x, y) = f(x, y + 2)$
(c) $g(x, y) = f(x + 3, y - 4)$

71–72 Use a computer to graph the function using various domains and viewpoints. Get a printout that gives a good view of the "peaks and valleys." Would you say the function has a maximum value? Can you identify any points on the graph that you might consider to be "local maximum points"? What about "local minimum points"?

71. $f(x, y) = 3x - x^4 - 4y^2 - 10xy$

72. $f(x, y) = xye^{-x^2-y^2}$

73–74 Use a computer to graph the function using various domains and viewpoints. Comment on the limiting behavior of the function. What happens as both x and y become large? What happens as (x, y) approaches the origin?

73. $f(x, y) = \dfrac{x + y}{x^2 + y^2}$ **74.** $f(x, y) = \dfrac{xy}{x^2 + y^2}$

75. Use a computer to investigate the family of functions $f(x, y) = e^{cx^2+y^2}$. How does the shape of the graph depend on c?

76. Use a computer to investigate the family of surfaces

$$z = (ax^2 + by^2)e^{-x^2-y^2}$$

How does the shape of the graph depend on the numbers a and b?

77. Use a computer to investigate the family of surfaces $z = x^2 + y^2 + cxy$. In particular, you should determine the transitional values of c for which the surface changes from one type of quadric surface to another.

78. Graph the functions

$$f(x, y) = \sqrt{x^2 + y^2}$$

$$f(x, y) = e^{\sqrt{x^2+y^2}}$$

$$f(x, y) = \ln\sqrt{x^2 + y^2}$$

$$f(x, y) = \sin\left(\sqrt{x^2 + y^2}\right)$$

and $$f(x, y) = \frac{1}{\sqrt{x^2 + y^2}}$$

In general, if g is a function of one variable, how is the graph of

$$f(x, y) = g\left(\sqrt{x^2 + y^2}\right)$$

obtained from the graph of g?

79. (a) Show that, by taking logarithms, the general Cobb-Douglas function $P = bL^{\alpha}K^{1-\alpha}$ can be expressed as

$$\ln\frac{P}{K} = \ln b + \alpha \ln\frac{L}{K}$$

(b) If we let $x = \ln(L/K)$ and $y = \ln(P/K)$, the equation in part (a) becomes the linear equation $y = \alpha x + \ln b$. Use Table 2 (in Example 3) to make a table of values of $\ln(L/K)$ and $\ln(P/K)$ for the years 1899–1922. Then use a graphing calculator or computer to find the least squares regression line through the points $(\ln(L/K), \ln(P/K))$.
(c) Deduce that the Cobb-Douglas production function is $P = 1.01L^{0.75}K^{0.25}$.

14.2 Limits and Continuity

Let's compare the behavior of the functions

$$f(x, y) = \frac{\sin(x^2 + y^2)}{x^2 + y^2} \quad \text{and} \quad g(x, y) = \frac{x^2 - y^2}{x^2 + y^2}$$

as x and y both approach 0 [and therefore the point (x, y) approaches the origin].

Tables 1 and 2 show values of $f(x, y)$ and $g(x, y)$, correct to three decimal places, for points (x, y) near the origin. (Notice that neither function is defined at the origin.)

TABLE 1 Values of $f(x, y)$

x \ y	−1.0	−0.5	−0.2	0	0.2	0.5	1.0
−1.0	0.455	0.759	0.829	0.841	0.829	0.759	0.455
−0.5	0.759	0.959	0.986	0.990	0.986	0.959	0.759
−0.2	0.829	0.986	0.999	1.000	0.999	0.986	0.829
0	0.841	0.990	1.000		1.000	0.990	0.841
0.2	0.829	0.986	0.999	1.000	0.999	0.986	0.829
0.5	0.759	0.959	0.986	0.990	0.986	0.959	0.759
1.0	0.455	0.759	0.829	0.841	0.829	0.759	0.455

TABLE 2 Values of $g(x, y)$

x \ y	−1.0	−0.5	−0.2	0	0.2	0.5	1.0
−1.0	0.000	0.600	0.923	1.000	0.923	0.600	0.000
−0.5	−0.600	0.000	0.724	1.000	0.724	0.000	−0.600
−0.2	−0.923	−0.724	0.000	1.000	0.000	−0.724	−0.923
0	−1.000	−1.000	−1.000		−1.000	−1.000	−1.000
0.2	−0.923	−0.724	0.000	1.000	0.000	−0.724	−0.923
0.5	−0.600	0.000	0.724	1.000	0.724	0.000	−0.600
1.0	0.000	0.600	0.923	1.000	0.923	0.600	0.000

It appears that as (x, y) approaches $(0, 0)$, the values of $f(x, y)$ are approaching 1 whereas the values of $g(x, y)$ aren't approaching any number. It turns out that these guesses based on numerical evidence are correct, and we write

$$\lim_{(x, y) \to (0, 0)} \frac{\sin(x^2 + y^2)}{x^2 + y^2} = 1 \qquad \text{and} \qquad \lim_{(x, y) \to (0, 0)} \frac{x^2 - y^2}{x^2 + y^2} \quad \text{does not exist}$$

In general, we use the notation

$$\lim_{(x, y) \to (a, b)} f(x, y) = L$$

to indicate that the values of $f(x, y)$ approach the number L as the point (x, y) approaches the point (a, b) along any path that stays within the domain of f. In other words, we can make the values of $f(x, y)$ as close to L as we like by taking the point (x, y) sufficiently close to the point (a, b), but not equal to (a, b). A more precise definition follows.

1 Definition Let f be a function of two variables whose domain D includes points arbitrarily close to (a, b). Then we say that the **limit of $f(x, y)$ as (x, y) approaches (a, b)** is L and we write

$$\lim_{(x, y) \to (a, b)} f(x, y) = L$$

if for every number $\varepsilon > 0$ there is a corresponding number $\delta > 0$ such that

$$\text{if} \quad (x, y) \in D \quad \text{and} \quad 0 < \sqrt{(x - a)^2 + (y - b)^2} < \delta \quad \text{then} \quad |f(x, y) - L| < \varepsilon$$

Other notations for the limit in Definition 1 are

$$\lim_{\substack{x \to a \\ y \to b}} f(x, y) = L \qquad \text{and} \qquad f(x, y) \to L \text{ as } (x, y) \to (a, b)$$

Notice that $|f(x, y) - L|$ is the distance between the numbers $f(x, y)$ and L, and $\sqrt{(x - a)^2 + (y - b)^2}$ is the distance between the point (x, y) and the point (a, b). Thus Definition 1 says that the distance between $f(x, y)$ and L can be made arbitrarily small by making the distance from (x, y) to (a, b) sufficiently small (but not 0). Figure 1 illustrates Definition 1 by means of an arrow diagram. If any small interval $(L - \varepsilon, L + \varepsilon)$ is given around L, then we can find a disk D_δ with center (a, b) and radius $\delta > 0$ such that f maps all the points in D_δ [except possibly (a, b)] into the interval $(L - \varepsilon, L + \varepsilon)$.

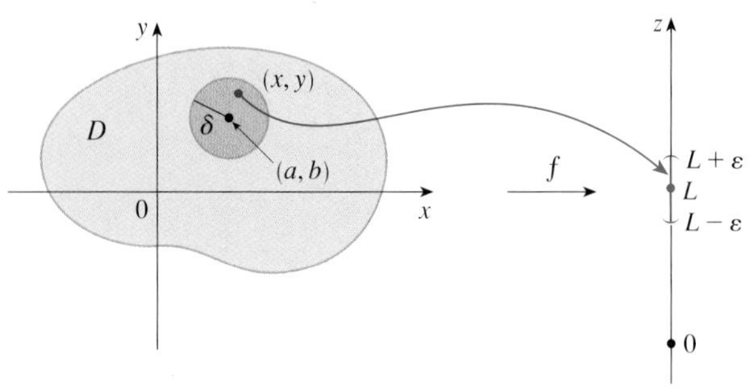

FIGURE 1

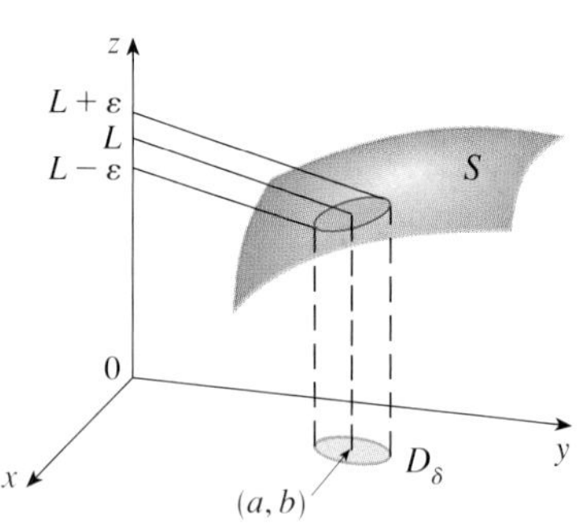

FIGURE 2

Another illustration of Definition 1 is given in Figure 2 where the surface S is the graph of f. If $\varepsilon > 0$ is given, we can find $\delta > 0$ such that if (x, y) is restricted to lie in the disk D_δ and $(x, y) \neq (a, b)$, then the corresponding part of S lies between the horizontal planes $z = L - \varepsilon$ and $z = L + \varepsilon$.

For functions of a single variable, when we let x approach a, there are only two possible directions of approach, from the left or from the right. We recall from Chapter 2 that if $\lim_{x \to a^-} f(x) \neq \lim_{x \to a^+} f(x)$, then $\lim_{x \to a} f(x)$ does not exist.

For functions of two variables the situation is not as simple because we can let (x, y) approach (a, b) from an infinite number of directions in any manner whatsoever (see Figure 3) as long as (x, y) stays within the domain of f.

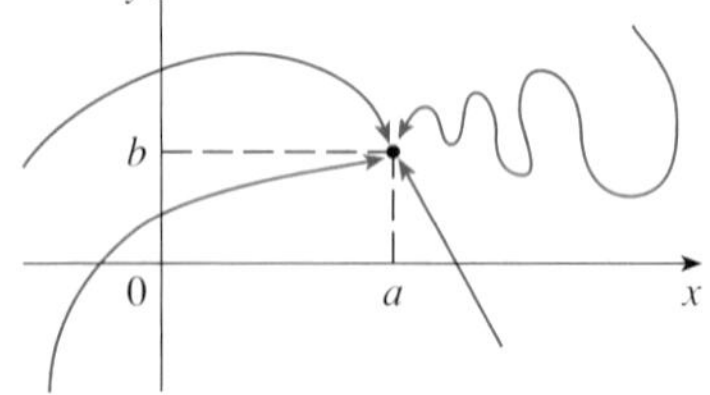

FIGURE 3

Definition 1 says that the distance between $f(x, y)$ and L can be made arbitrarily small by making the distance from (x, y) to (a, b) sufficiently small (but not 0). The definition refers only to the *distance* between (x, y) and (a, b). It does not refer to the direction of approach. Therefore, if the limit exists, then $f(x, y)$ must approach the same limit no matter how (x, y) approaches (a, b). Thus, if we can find two different paths of approach along which the function $f(x, y)$ has different limits, then it follows that $\lim_{(x, y) \to (a, b)} f(x, y)$ does not exist.

> If $f(x, y) \to L_1$ as $(x, y) \to (a, b)$ along a path C_1 and $f(x, y) \to L_2$ as $(x, y) \to (a, b)$ along a path C_2, where $L_1 \neq L_2$, then $\lim_{(x, y) \to (a, b)} f(x, y)$ does not exist.

V EXAMPLE 1 Show that $\displaystyle\lim_{(x, y) \to (0, 0)} \frac{x^2 - y^2}{x^2 + y^2}$ does not exist.

SOLUTION Let $f(x, y) = (x^2 - y^2)/(x^2 + y^2)$. First let's approach $(0, 0)$ along the x-axis. Then $y = 0$ gives $f(x, 0) = x^2/x^2 = 1$ for all $x \neq 0$, so

$$f(x, y) \to 1 \qquad \text{as} \qquad (x, y) \to (0, 0) \text{ along the } x\text{-axis}$$

We now approach along the y-axis by putting $x = 0$. Then $f(0, y) = \dfrac{-y^2}{y^2} = -1$ for all $y \neq 0$, so

$$f(x, y) \to -1 \qquad \text{as} \qquad (x, y) \to (0, 0) \text{ along the } y\text{-axis}$$

FIGURE 4

(See Figure 4.) Since f has two different limits along two different lines, the given limit does not exist. (This confirms the conjecture we made on the basis of numerical evidence at the beginning of this section.) ■

EXAMPLE 2 If $f(x, y) = xy/(x^2 + y^2)$, does $\lim_{(x, y) \to (0, 0)} f(x, y)$ exist?

SOLUTION If $y = 0$, then $f(x, 0) = 0/x^2 = 0$. Therefore

$$f(x, y) \to 0 \quad \text{as} \quad (x, y) \to (0, 0) \text{ along the } x\text{-axis}$$

If $x = 0$, then $f(0, y) = 0/y^2 = 0$, so

$$f(x, y) \to 0 \quad \text{as} \quad (x, y) \to (0, 0) \text{ along the } y\text{-axis}$$

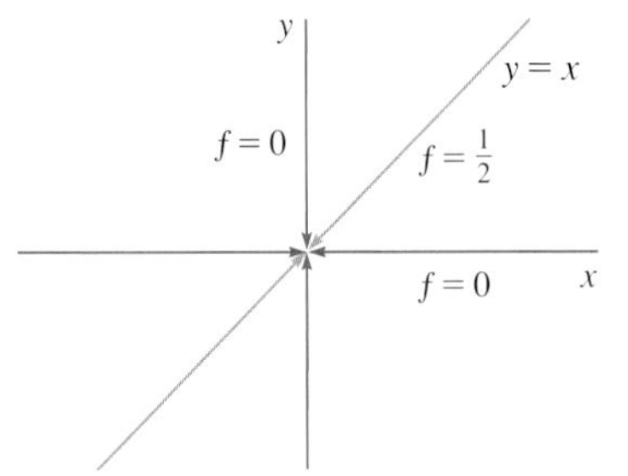

FIGURE 5

Although we have obtained identical limits along the axes, that does not show that the given limit is 0. Let's now approach $(0, 0)$ along another line, say $y = x$. For all $x \neq 0$,

$$f(x, x) = \frac{x^2}{x^2 + x^2} = \frac{1}{2}$$

Therefore $\qquad f(x, y) \to \frac{1}{2} \quad \text{as} \quad (x, y) \to (0, 0) \text{ along } y = x$

(See Figure 5.) Since we have obtained different limits along different paths, the given limit does not exist. ■

Figure 6 sheds some light on Example 2. The ridge that occurs above the line $y = x$ corresponds to the fact that $f(x, y) = \frac{1}{2}$ for all points (x, y) on that line except the origin.

TEC In Visual 14.2 a rotating line on the surface in Figure 6 shows different limits at the origin from different directions.

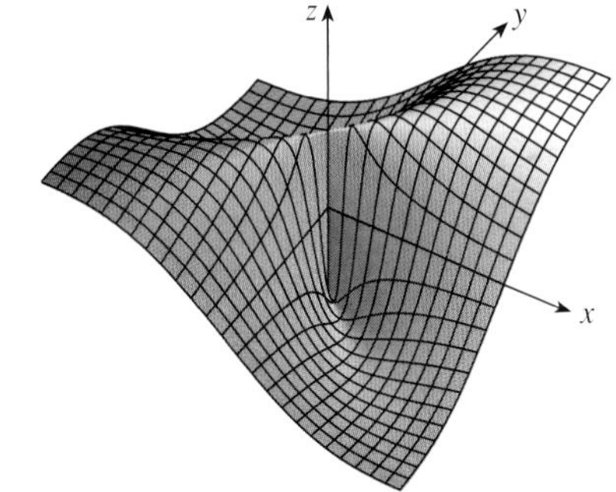

FIGURE 6

$f(x, y) = \dfrac{xy}{x^2 + y^2}$

V EXAMPLE 3 If $f(x, y) = \dfrac{xy^2}{x^2 + y^4}$, does $\lim_{(x, y) \to (0, 0)} f(x, y)$ exist?

SOLUTION With the solution of Example 2 in mind, let's try to save time by letting $(x, y) \to (0, 0)$ along any nonvertical line through the origin. Then $y = mx$, where m is the slope, and

$$f(x, y) = f(x, mx) = \frac{x(mx)^2}{x^2 + (mx)^4} = \frac{m^2x^3}{x^2 + m^4x^4} = \frac{m^2x}{1 + m^4x^2}$$

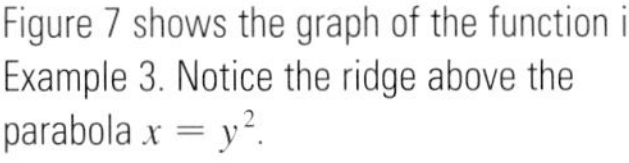
Figure 7 shows the graph of the function in Example 3. Notice the ridge above the parabola $x = y^2$.

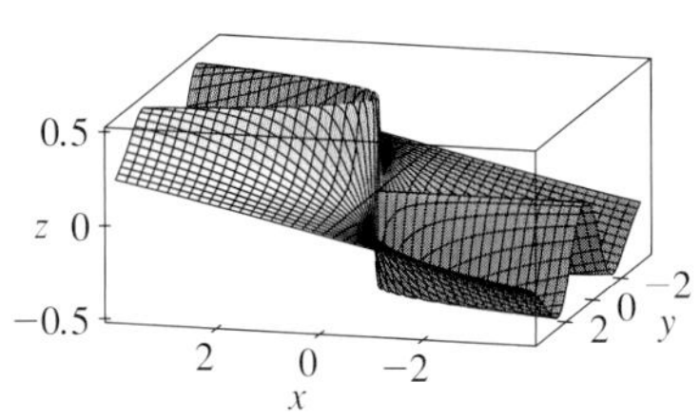

FIGURE 7

So $\qquad f(x, y) \to 0 \quad \text{as} \quad (x, y) \to (0, 0) \text{ along } y = mx$

Thus f has the same limiting value along every nonvertical line through the origin. But that does not show that the given limit is 0, for if we now let $(x, y) \to (0, 0)$ along the parabola $x = y^2$, we have

$$f(x, y) = f(y^2, y) = \frac{y^2 \cdot y^2}{(y^2)^2 + y^4} = \frac{y^4}{2y^4} = \frac{1}{2}$$

so $\qquad f(x, y) \to \frac{1}{2} \qquad$ as $\qquad (x, y) \to (0, 0)$ along $x = y^2$

Since different paths lead to different limiting values, the given limit does not exist.

Now let's look at limits that *do* exist. Just as for functions of one variable, the calculation of limits for functions of two variables can be greatly simplified by the use of properties of limits. The Limit Laws listed in Section 2.3 can be extended to functions of two variables: The limit of a sum is the sum of the limits, the limit of a product is the product of the limits, and so on. In particular, the following equations are true.

2
$$\lim_{(x, y) \to (a, b)} x = a \qquad \lim_{(x, y) \to (a, b)} y = b \qquad \lim_{(x, y) \to (a, b)} c = c$$

The Squeeze Theorem also holds.

EXAMPLE 4 Find $\displaystyle\lim_{(x, y) \to (0, 0)} \frac{3x^2y}{x^2 + y^2}$ if it exists.

SOLUTION As in Example 3, we could show that the limit along any line through the origin is 0. This doesn't prove that the given limit is 0, but the limits along the parabolas $y = x^2$ and $x = y^2$ also turn out to be 0, so we begin to suspect that the limit does exist and is equal to 0.

Let $\varepsilon > 0$. We want to find $\delta > 0$ such that

$$\text{if} \quad 0 < \sqrt{x^2 + y^2} < \delta \quad \text{then} \quad \left| \frac{3x^2y}{x^2 + y^2} - 0 \right| < \varepsilon$$

that is,
$$\text{if} \quad 0 < \sqrt{x^2 + y^2} < \delta \quad \text{then} \quad \frac{3x^2|y|}{x^2 + y^2} < \varepsilon$$

But $x^2 \leqslant x^2 + y^2$ since $y^2 \geqslant 0$, so $x^2/(x^2 + y^2) \leqslant 1$ and therefore

3
$$\frac{3x^2|y|}{x^2 + y^2} \leqslant 3|y| = 3\sqrt{y^2} \leqslant 3\sqrt{x^2 + y^2}$$

Thus if we choose $\delta = \varepsilon/3$ and let $0 < \sqrt{x^2 + y^2} < \delta$, then

$$\left| \frac{3x^2y}{x^2 + y^2} - 0 \right| \leqslant 3\sqrt{x^2 + y^2} < 3\delta = 3\left(\frac{\varepsilon}{3}\right) = \varepsilon$$

Hence, by Definition 1,

$$\lim_{(x, y) \to (0, 0)} \frac{3x^2y}{x^2 + y^2} = 0$$

Another way to do Example 4 is to use the Squeeze Theorem instead of Definition 1. From **2** it follows that

$$\lim_{(x, y) \to (0, 0)} 3|y| = 0$$

and so the first inequality in **3** shows that the given limit is 0.

Continuity

Recall that evaluating limits of *continuous* functions of a single variable is easy. It can be accomplished by direct substitution because the defining property of a continuous function is $\lim_{x \to a} f(x) = f(a)$. Continuous functions of two variables are also defined by the direct substitution property.

4 Definition A function f of two variables is called **continuous at** (a, b) if

$$\lim_{(x, y) \to (a, b)} f(x, y) = f(a, b)$$

We say f is **continuous on** D if f is continuous at every point (a, b) in D.

The intuitive meaning of continuity is that if the point (x, y) changes by a small amount, then the value of $f(x, y)$ changes by a small amount. This means that a surface that is the graph of a continuous function has no hole or break.

Using the properties of limits, you can see that sums, differences, products, and quotients of continuous functions are continuous on their domains. Let's use this fact to give examples of continuous functions.

A **polynomial function of two variables** (or polynomial, for short) is a sum of terms of the form $cx^m y^n$, where c is a constant and m and n are nonnegative integers. A **rational function** is a ratio of polynomials. For instance,

$$f(x, y) = x^4 + 5x^3y^2 + 6xy^4 - 7y + 6$$

is a polynomial, whereas

$$g(x, y) = \frac{2xy + 1}{x^2 + y^2}$$

is a rational function.

The limits in [2] show that the functions $f(x, y) = x$, $g(x, y) = y$, and $h(x, y) = c$ are continuous. Since any polynomial can be built up out of the simple functions f, g, and h by multiplication and addition, it follows that *all polynomials are continuous on* $\mathbb{R}^2$. Likewise, any rational function is continuous on its domain because it is a quotient of continuous functions.

V EXAMPLE 5 Evaluate $\lim_{(x, y) \to (1, 2)} (x^2y^3 - x^3y^2 + 3x + 2y)$.

SOLUTION Since $f(x, y) = x^2y^3 - x^3y^2 + 3x + 2y$ is a polynomial, it is continuous everywhere, so we can find the limit by direct substitution:

$$\lim_{(x, y) \to (1, 2)} (x^2y^3 - x^3y^2 + 3x + 2y) = 1^2 \cdot 2^3 - 1^3 \cdot 2^2 + 3 \cdot 1 + 2 \cdot 2 = 11$$

EXAMPLE 6 Where is the function $f(x, y) = \dfrac{x^2 - y^2}{x^2 + y^2}$ continuous?

SOLUTION The function f is discontinuous at $(0, 0)$ because it is not defined there. Since f is a rational function, it is continuous on its domain, which is the set $D = \{(x, y) \mid (x, y) \neq (0, 0)\}$.

EXAMPLE 7 Let

$$g(x, y) = \begin{cases} \dfrac{x^2 - y^2}{x^2 + y^2} & \text{if } (x, y) \neq (0, 0) \\ 0 & \text{if } (x, y) = (0, 0) \end{cases}$$

Here g is defined at $(0, 0)$ but g is still discontinuous there because $\lim_{(x, y) \to (0, 0)} g(x, y)$ does not exist (see Example 1).

Figure 8 shows the graph of the continuous function in Example 8.

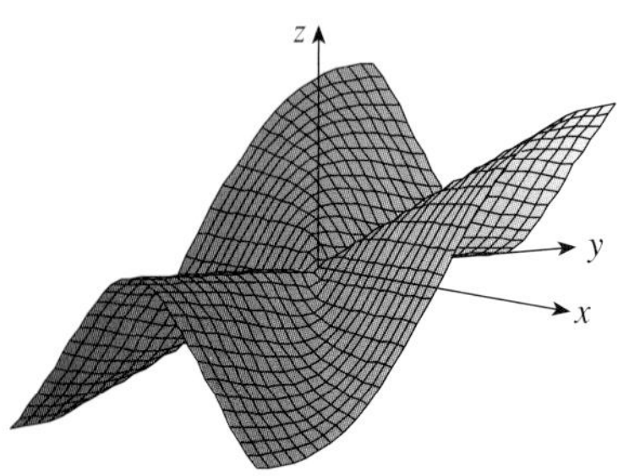

FIGURE 8

EXAMPLE 8 Let

$$f(x, y) = \begin{cases} \dfrac{3x^2y}{x^2 + y^2} & \text{if } (x, y) \neq (0, 0) \\ 0 & \text{if } (x, y) = (0, 0) \end{cases}$$

We know f is continuous for $(x, y) \neq (0, 0)$ since it is equal to a rational function there. Also, from Example 4, we have

$$\lim_{(x, y) \to (0, 0)} f(x, y) = \lim_{(x, y) \to (0, 0)} \frac{3x^2y}{x^2 + y^2} = 0 = f(0, 0)$$

Therefore f is continuous at $(0, 0)$, and so it is continuous on $\mathbb{R}^2$.

Just as for functions of one variable, composition is another way of combining two continuous functions to get a third. In fact, it can be shown that if f is a continuous function of two variables and g is a continuous function of a single variable that is defined on the range of f, then the composite function $h = g \circ f$ defined by $h(x, y) = g(f(x, y))$ is also a continuous function.

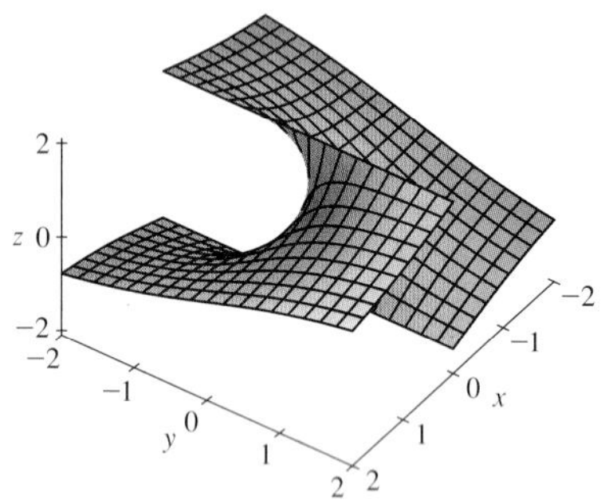

FIGURE 9
The function $h(x, y) = \arctan(y/x)$ is discontinuous where $x = 0$.

EXAMPLE 9 Where is the function $h(x, y) = \arctan(y/x)$ continuous?

SOLUTION The function $f(x, y) = y/x$ is a rational function and therefore continuous except on the line $x = 0$. The function $g(t) = \arctan t$ is continuous everywhere. So the composite function

$$g(f(x, y)) = \arctan(y/x) = h(x, y)$$

is continuous except where $x = 0$. The graph in Figure 9 shows the break in the graph of h above the y-axis.

Functions of Three or More Variables

Everything that we have done in this section can be extended to functions of three or more variables. The notation

$$\lim_{(x, y, z) \to (a, b, c)} f(x, y, z) = L$$

means that the values of $f(x, y, z)$ approach the number L as the point (x, y, z) approaches the point (a, b, c) along any path in the domain of f. Because the distance between two points (x, y, z) and (a, b, c) in $\mathbb{R}^3$ is given by $\sqrt{(x - a)^2 + (y - b)^2 + (z - c)^2}$, we can write the precise definition as follows: For every number $\varepsilon > 0$ there is a corresponding number $\delta > 0$ such that

$$\text{if } (x, y, z) \text{ is in the domain of } f \quad \text{and} \quad 0 < \sqrt{(x - a)^2 + (y - b)^2 + (z - c)^2} < \delta$$

$$\text{then} \quad |f(x, y, z) - L| < \varepsilon$$

The function f is **continuous** at (a, b, c) if

$$\lim_{(x, y, z) \to (a, b, c)} f(x, y, z) = f(a, b, c)$$

For instance, the function

$$f(x, y, z) = \frac{1}{x^2 + y^2 + z^2 - 1}$$

is a rational function of three variables and so is continuous at every point in $\mathbb{R}^3$ except where $x^2 + y^2 + z^2 = 1$. In other words, it is discontinuous on the sphere with center the origin and radius 1.

If we use the vector notation introduced at the end of Section 14.1, then we can write the definitions of a limit for functions of two or three variables in a single compact form as follows.

> **5** If f is defined on a subset D of $\mathbb{R}^n$, then $\lim_{\mathbf{x}\to\mathbf{a}} f(\mathbf{x}) = L$ means that for every number $\varepsilon > 0$ there is a corresponding number $\delta > 0$ such that
>
> $$\text{if} \quad \mathbf{x} \in D \quad \text{and} \quad 0 < |\mathbf{x} - \mathbf{a}| < \delta \quad \text{then} \quad |f(\mathbf{x}) - L| < \varepsilon$$

Notice that if $n = 1$, then $\mathbf{x} = x$ and $\mathbf{a} = a$, and **5** is just the definition of a limit for functions of a single variable. For the case $n = 2$, we have $\mathbf{x} = \langle x, y\rangle$, $\mathbf{a} = \langle a, b\rangle$, and $|\mathbf{x} - \mathbf{a}| = \sqrt{(x-a)^2 + (y-b)^2}$, so **5** becomes Definition 1. If $n = 3$, then $\mathbf{x} = \langle x, y, z\rangle$, $\mathbf{a} = \langle a, b, c\rangle$, and **5** becomes the definition of a limit of a function of three variables. In each case the definition of continuity can be written as

$$\lim_{\mathbf{x}\to\mathbf{a}} f(\mathbf{x}) = f(\mathbf{a})$$

14.2 Exercises

1. Suppose that $\lim_{(x,y)\to(3,1)} f(x, y) = 6$. What can you say about the value of $f(3, 1)$? What if f is continuous?

2. Explain why each function is continuous or discontinuous.
(a) The outdoor temperature as a function of longitude, latitude, and time
(b) Elevation (height above sea level) as a function of longitude, latitude, and time
(c) The cost of a taxi ride as a function of distance traveled and time

3–4 Use a table of numerical values of $f(x, y)$ for (x, y) near the origin to make a conjecture about the value of the limit of $f(x, y)$ as $(x, y) \to (0, 0)$. Then explain why your guess is correct.

3. $f(x, y) = \dfrac{x^2y^3 + x^3y^2 - 5}{2 - xy}$ **4.** $f(x, y) = \dfrac{2xy}{x^2 + 2y^2}$

5–22 Find the limit, if it exists, or show that the limit does not exist.

5. $\lim_{(x,y)\to(1,2)} (5x^3 - x^2y^2)$ **6.** $\lim_{(x,y)\to(1,-1)} e^{-xy}\cos(x + y)$

7. $\lim_{(x,y)\to(2,1)} \dfrac{4 - xy}{x^2 + 3y^2}$ **8.** $\lim_{(x,y)\to(1,0)} \ln\left(\dfrac{1 + y^2}{x^2 + xy}\right)$

9. $\lim_{(x,y)\to(0,0)} \dfrac{x^4 - 4y^2}{x^2 + 2y^2}$ **10.** $\lim_{(x,y)\to(0,0)} \dfrac{5y^4\cos^2 x}{x^4 + y^4}$

11. $\lim_{(x,y)\to(0,0)} \dfrac{y^2\sin^2 x}{x^4 + y^4}$ **12.** $\lim_{(x,y)\to(1,0)} \dfrac{xy - y}{(x-1)^2 + y^2}$

13. $\lim_{(x,y)\to(0,0)} \dfrac{xy}{\sqrt{x^2 + y^2}}$ **14.** $\lim_{(x,y)\to(0,0)} \dfrac{x^4 - y^4}{x^2 + y^2}$

15. $\lim_{(x,y)\to(0,0)} \dfrac{x^2ye^y}{x^4 + 4y^2}$ **16.** $\lim_{(x,y)\to(0,0)} \dfrac{x^2\sin^2 y}{x^2 + 2y^2}$

17. $\lim_{(x,y)\to(0,0)} \dfrac{x^2 + y^2}{\sqrt{x^2 + y^2 + 1} - 1}$ **18.** $\lim_{(x,y)\to(0,0)} \dfrac{xy^4}{x^2 + y^8}$

19. $\lim_{(x,y,z)\to(\pi,0,1/3)} e^{y^2}\tan(xz)$

20. $\lim_{(x,y,z)\to(0,0,0)} \dfrac{xy + yz}{x^2 + y^2 + z^2}$

21. $\lim_{(x,y,z)\to(0,0,0)} \dfrac{xy + yz^2 + xz^2}{x^2 + y^2 + z^4}$

22. $\lim_{(x,y,z)\to(0,0,0)} \dfrac{yz}{x^2 + 4y^2 + 9z^2}$

23–24 Use a computer graph of the function to explain why the limit does not exist.

23. $\lim_{(x,y)\to(0,0)} \dfrac{2x^2 + 3xy + 4y^2}{3x^2 + 5y^2}$ **24.** $\lim_{(x,y)\to(0,0)} \dfrac{xy^3}{x^2 + y^6}$

25–26 Find $h(x, y) = g(f(x, y))$ and the set on which h is continuous.

25. $g(t) = t^2 + \sqrt{t}, \quad f(x, y) = 2x + 3y - 6$

26. $g(t) = t + \ln t, \quad f(x, y) = \dfrac{1 - xy}{1 + x^2y^2}$

27–28 Graph the function and observe where it is discontinuous. Then use the formula to explain what you have observed.

27. $f(x, y) = e^{1/(x-y)}$

28. $f(x, y) = \dfrac{1}{1 - x^2 - y^2}$

29–38 Determine the set of points at which the function is continuous.

29. $F(x, y) = \dfrac{xy}{1 + e^{x-y}}$

30. $F(x, y) = \cos\sqrt{1 + x - y}$

31. $F(x, y) = \dfrac{1 + x^2 + y^2}{1 - x^2 - y^2}$

32. $H(x, y) = \dfrac{e^x + e^y}{e^{xy} - 1}$

33. $G(x, y) = \ln(x^2 + y^2 - 4)$

34. $G(x, y) = \tan^{-1}((x + y)^{-2})$

35. $f(x, y, z) = \arcsin(x^2 + y^2 + z^2)$

36. $f(x, y, z) = \sqrt{y - x^2}\,\ln z$

37. $f(x, y) = \begin{cases} \dfrac{x^2y^3}{2x^2 + y^2} & \text{if } (x, y) \neq (0, 0) \\ 1 & \text{if } (x, y) = (0, 0) \end{cases}$

38. $f(x, y) = \begin{cases} \dfrac{xy}{x^2 + xy + y^2} & \text{if } (x, y) \neq (0, 0) \\ 0 & \text{if } (x, y) = (0, 0) \end{cases}$

39–41 Use polar coordinates to find the limit. [If (r, θ) are polar coordinates of the point (x, y) with $r \geqslant 0$, note that $r \to 0^+$ as $(x, y) \to (0, 0)$.]

39. $\displaystyle\lim_{(x, y)\to(0, 0)} \frac{x^3 + y^3}{x^2 + y^2}$

40. $\displaystyle\lim_{(x, y)\to(0, 0)} (x^2 + y^2)\ln(x^2 + y^2)$

41. $\displaystyle\lim_{(x, y)\to(0, 0)} \frac{e^{-x^2-y^2} - 1}{x^2 + y^2}$

---

42. At the beginning of this section we considered the function

$$f(x, y) = \frac{\sin(x^2 + y^2)}{x^2 + y^2}$$

and guessed that $f(x, y) \to 1$ as $(x, y) \to (0, 0)$ on the basis of numerical evidence. Use polar coordinates to confirm the value of the limit. Then graph the function.

43. Graph and discuss the continuity of the function

$$f(x, y) = \begin{cases} \dfrac{\sin xy}{xy} & \text{if } xy \neq 0 \\ 1 & \text{if } xy = 0 \end{cases}$$

44. Let

$$f(x, y) = \begin{cases} 0 & \text{if } y \leqslant 0 \quad \text{or} \quad y \geqslant x^4 \\ 1 & \text{if } 0 < y < x^4 \end{cases}$$

(a) Show that $f(x, y) \to 0$ as $(x, y) \to (0, 0)$ along any path through $(0, 0)$ of the form $y = mx^a$ with $a < 4$.
(b) Despite part (a), show that f is discontinuous at $(0, 0)$.
(c) Show that f is discontinuous on two entire curves.

45. Show that the function f given by $f(\mathbf{x}) = |\mathbf{x}|$ is continuous on $\mathbb{R}^n$. [*Hint*: Consider $|\mathbf{x} - \mathbf{a}|^2 = (\mathbf{x} - \mathbf{a}) \cdot (\mathbf{x} - \mathbf{a})$.]

46. If $\mathbf{c} \in V_n$, show that the function f given by $f(\mathbf{x}) = \mathbf{c} \cdot \mathbf{x}$ is continuous on $\mathbb{R}^n$.

14.3 Partial Derivatives

On a hot day, extreme humidity makes us think the temperature is higher than it really is, whereas in very dry air we perceive the temperature to be lower than the thermometer indicates. The National Weather Service has devised the *heat index* (also called the temperature-humidity index, or humidex, in some countries) to describe the combined effects of temperature and humidity. The heat index I is the perceived air temperature when the actual temperature is T and the relative humidity is H. So I is a function of T and H and we can write $I = f(T, H)$. The following table of values of I is an excerpt from a table compiled by the National Weather Service.

TABLE 1
Heat index I as a function of temperature and humidity

Relative humidity (%)

Actual temperature (°F) T \ H	50	55	60	65	70	75	80	85	90
90	96	98	100	103	106	109	112	115	119
92	100	103	105	108	112	115	119	123	128
94	104	107	111	114	118	122	127	132	137
96	109	113	116	121	125	130	135	141	146
98	114	118	123	127	133	138	144	150	157
100	119	124	129	135	141	147	154	161	168

If we concentrate on the highlighted column of the table, which corresponds to a relative humidity of $H = 70\%$, we are considering the heat index as a function of the single variable T for a fixed value of H. Let's write $g(T) = f(T, 70)$. Then $g(T)$ describes how the heat index I increases as the actual temperature T increases when the relative humidity is 70%. The derivative of g when $T = 96°\text{F}$ is the rate of change of I with respect to T when $T = 96°\text{F}$:

$$g'(96) = \lim_{h \to 0} \frac{g(96 + h) - g(96)}{h} = \lim_{h \to 0} \frac{f(96 + h, 70) - f(96, 70)}{h}$$

We can approximate $g'(96)$ using the values in Table 1 by taking $h = 2$ and -2:

$$g'(96) \approx \frac{g(98) - g(96)}{2} = \frac{f(98, 70) - f(96, 70)}{2} = \frac{133 - 125}{2} = 4$$

$$g'(96) \approx \frac{g(94) - g(96)}{-2} = \frac{f(94, 70) - f(96, 70)}{-2} = \frac{118 - 125}{-2} = 3.5$$

Averaging these values, we can say that the derivative $g'(96)$ is approximately 3.75. This means that, when the actual temperature is 96°F and the relative humidity is 70%, the apparent temperature (heat index) rises by about 3.75°F for every degree that the actual temperature rises!

Now let's look at the highlighted row in Table 1, which corresponds to a fixed temperature of $T = 96°\text{F}$. The numbers in this row are values of the function $G(H) = f(96, H)$, which describes how the heat index increases as the relative humidity H increases when the actual temperature is $T = 96°\text{F}$. The derivative of this function when $H = 70\%$ is the rate of change of I with respect to H when $H = 70\%$:

$$G'(70) = \lim_{h \to 0} \frac{G(70 + h) - G(70)}{h} = \lim_{h \to 0} \frac{f(96, 70 + h) - f(96, 70)}{h}$$

By taking $h = 5$ and -5, we approximate $G'(70)$ using the tabular values:

$$G'(70) \approx \frac{G(75) - G(70)}{5} = \frac{f(96, 75) - f(96, 70)}{5} = \frac{130 - 125}{5} = 1$$

$$G'(70) \approx \frac{G(65) - G(70)}{-5} = \frac{f(96, 65) - f(96, 70)}{-5} = \frac{121 - 125}{-5} = 0.8$$

By averaging these values we get the estimate $G'(70) \approx 0.9$. This says that, when the temperature is 96°F and the relative humidity is 70%, the heat index rises about 0.9°F for every percent that the relative humidity rises.

In general, if f is a function of two variables x and y, suppose we let only x vary while keeping y fixed, say $y = b$, where b is a constant. Then we are really considering a function of a single variable x, namely, $g(x) = f(x, b)$. If g has a derivative at a, then we call it the **partial derivative of f with respect to x at (a, b)** and denote it by $f_x(a, b)$. Thus

1 $$f_x(a, b) = g'(a) \qquad \text{where} \qquad g(x) = f(x, b)$$

By the definition of a derivative, we have

$$g'(a) = \lim_{h \to 0} \frac{g(a + h) - g(a)}{h}$$

and so Equation 1 becomes

2 $$f_x(a, b) = \lim_{h \to 0} \frac{f(a + h, b) - f(a, b)}{h}$$

Similarly, the **partial derivative of f with respect to y at (a, b)**, denoted by $f_y(a, b)$, is obtained by keeping x fixed ($x = a$) and finding the ordinary derivative at b of the function $G(y) = f(a, y)$:

3 $$f_y(a, b) = \lim_{h \to 0} \frac{f(a, b + h) - f(a, b)}{h}$$

With this notation for partial derivatives, we can write the rates of change of the heat index I with respect to the actual temperature T and relative humidity H when $T = 96°\text{F}$ and $H = 70\%$ as follows:

$$f_T(96, 70) \approx 3.75 \qquad f_H(96, 70) \approx 0.9$$

If we now let the point (a, b) vary in Equations 2 and 3, f_x and f_y become functions of two variables.

4 If f is a function of two variables, its **partial derivatives** are the functions f_x and f_y defined by

$$f_x(x, y) = \lim_{h \to 0} \frac{f(x + h, y) - f(x, y)}{h}$$

$$f_y(x, y) = \lim_{h \to 0} \frac{f(x, y + h) - f(x, y)}{h}$$

There are many alternative notations for partial derivatives. For instance, instead of f_x we can write f_1 or $D_1 f$ (to indicate differentiation with respect to the *first* variable) or $\partial f/\partial x$. But here $\partial f/\partial x$ can't be interpreted as a ratio of differentials.

Notations for Partial Derivatives If $z = f(x, y)$, we write

$$f_x(x, y) = f_x = \frac{\partial f}{\partial x} = \frac{\partial}{\partial x} f(x, y) = \frac{\partial z}{\partial x} = f_1 = D_1 f = D_x f$$

$$f_y(x, y) = f_y = \frac{\partial f}{\partial y} = \frac{\partial}{\partial y} f(x, y) = \frac{\partial z}{\partial y} = f_2 = D_2 f = D_y f$$

To compute partial derivatives, all we have to do is remember from Equation 1 that the partial derivative with respect to x is just the *ordinary* derivative of the function g of a single variable that we get by keeping y fixed. Thus we have the following rule.

Rule for Finding Partial Derivatives of $z = f(x, y)$

1. To find f_x, regard y as a constant and differentiate $f(x, y)$ with respect to x.
2. To find f_y, regard x as a constant and differentiate $f(x, y)$ with respect to y.

EXAMPLE 1 If $f(x, y) = x^3 + x^2y^3 - 2y^2$, find $f_x(2, 1)$ and $f_y(2, 1)$.

SOLUTION Holding y constant and differentiating with respect to x, we get

$$f_x(x, y) = 3x^2 + 2xy^3$$

and so

$$f_x(2, 1) = 3 \cdot 2^2 + 2 \cdot 2 \cdot 1^3 = 16$$

Holding x constant and differentiating with respect to y, we get

$$f_y(x, y) = 3x^2y^2 - 4y$$

$$f_y(2, 1) = 3 \cdot 2^2 \cdot 1^2 - 4 \cdot 1 = 8$$

Interpretations of Partial Derivatives

To give a geometric interpretation of partial derivatives, we recall that the equation $z = f(x, y)$ represents a surface S (the graph of f). If $f(a, b) = c$, then the point $P(a, b, c)$ lies on S. By fixing $y = b$, we are restricting our attention to the curve C_1 in which the vertical plane $y = b$ intersects S. (In other words, C_1 is the trace of S in the plane $y = b$.) Likewise, the vertical plane $x = a$ intersects S in a curve C_2. Both of the curves C_1 and C_2 pass through the point P. (See Figure 1.)

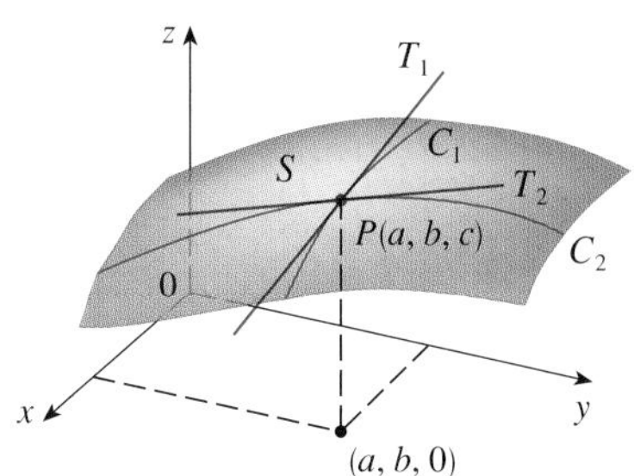

FIGURE 1
The partial derivatives of f at (a, b) are the slopes of the tangents to C_1 and C_2.

Notice that the curve C_1 is the graph of the function $g(x) = f(x, b)$, so the slope of its tangent T_1 at P is $g'(a) = f_x(a, b)$. The curve C_2 is the graph of the function $G(y) = f(a, y)$, so the slope of its tangent T_2 at P is $G'(b) = f_y(a, b)$.

Thus the partial derivatives $f_x(a, b)$ and $f_y(a, b)$ can be interpreted geometrically as the slopes of the tangent lines at $P(a, b, c)$ to the traces C_1 and C_2 of S in the planes $y = b$ and $x = a$.

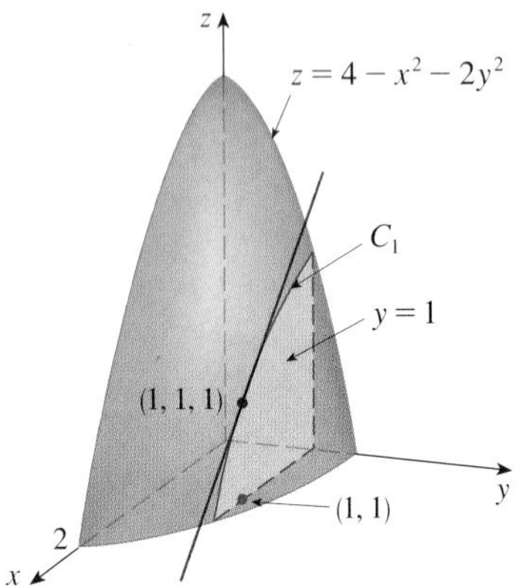

FIGURE 2

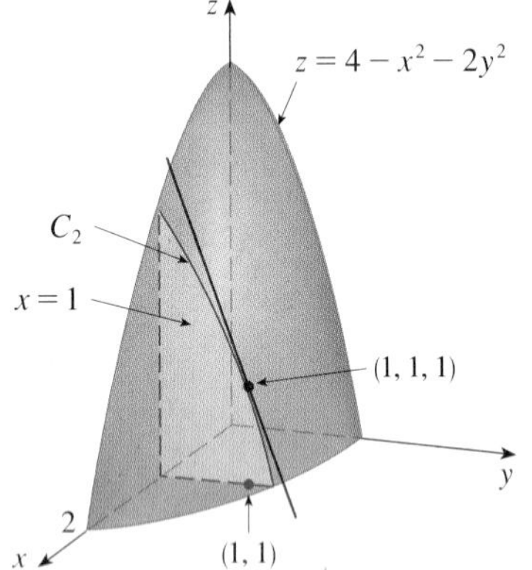

FIGURE 3

As we have seen in the case of the heat index function, partial derivatives can also be interpreted as *rates of change*. If $z = f(x, y)$, then $\partial z/\partial x$ represents the rate of change of z with respect to x when y is fixed. Similarly, $\partial z/\partial y$ represents the rate of change of z with respect to y when x is fixed.

EXAMPLE 2 If $f(x, y) = 4 - x^2 - 2y^2$, find $f_x(1, 1)$ and $f_y(1, 1)$ and interpret these numbers as slopes.

SOLUTION We have

$$f_x(x, y) = -2x \qquad f_y(x, y) = -4y$$

$$f_x(1, 1) = -2 \qquad f_y(1, 1) = -4$$

The graph of f is the paraboloid $z = 4 - x^2 - 2y^2$ and the vertical plane $y = 1$ intersects it in the parabola $z = 2 - x^2$, $y = 1$. (As in the preceding discussion, we label it C_1 in Figure 2.) The slope of the tangent line to this parabola at the point $(1, 1, 1)$ is $f_x(1, 1) = -2$. Similarly, the curve C_2 in which the plane $x = 1$ intersects the paraboloid is the parabola $z = 3 - 2y^2$, $x = 1$, and the slope of the tangent line at $(1, 1, 1)$ is $f_y(1, 1) = -4$. (See Figure 3.)

Figure 4 is a computer-drawn counterpart to Figure 2. Part (a) shows the plane $y = 1$ intersecting the surface to form the curve C_1 and part (b) shows C_1 and T_1. [We have used the vector equations $\mathbf{r}(t) = \langle t, 1, 2 - t^2 \rangle$ for C_1 and $\mathbf{r}(t) = \langle 1 + t, 1, 1 - 2t \rangle$ for T_1.] Similarly, Figure 5 corresponds to Figure 3.

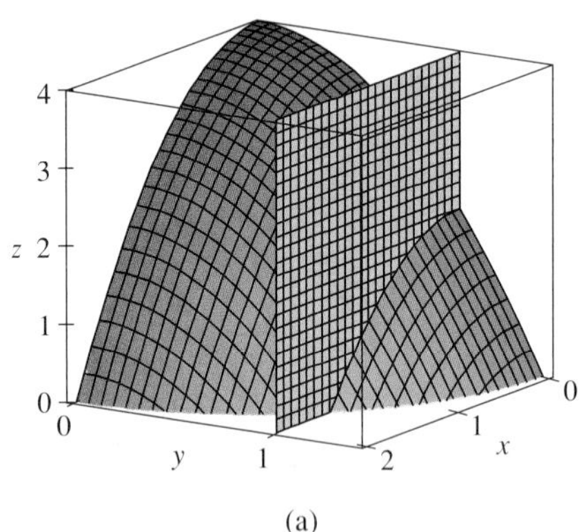

(a)

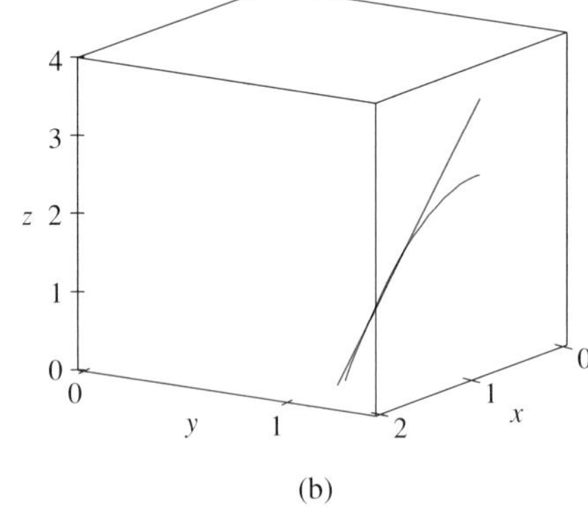

(b)

FIGURE 4

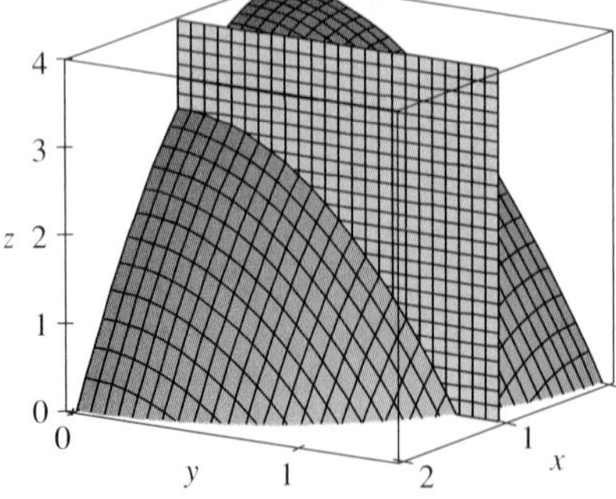

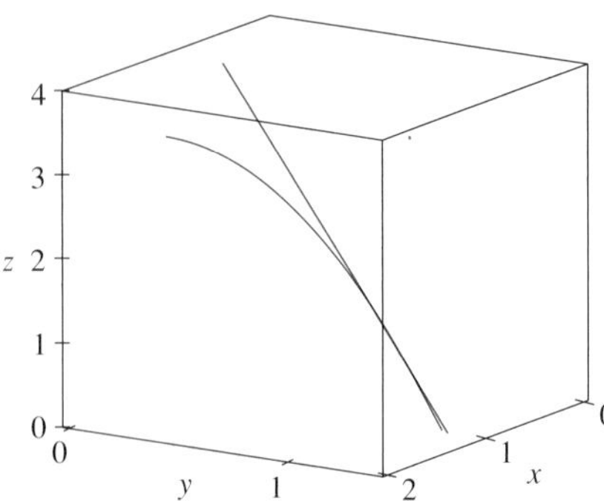

FIGURE 5

V EXAMPLE 3 If $f(x, y) = \sin\left(\dfrac{x}{1 + y}\right)$, calculate $\dfrac{\partial f}{\partial x}$ and $\dfrac{\partial f}{\partial y}$.

SOLUTION Using the Chain Rule for functions of one variable, we have

$$\frac{\partial f}{\partial x} = \cos\left(\frac{x}{1+y}\right) \cdot \frac{\partial}{\partial x}\left(\frac{x}{1+y}\right) = \cos\left(\frac{x}{1+y}\right) \cdot \frac{1}{1+y}$$

$$\frac{\partial f}{\partial y} = \cos\left(\frac{x}{1+y}\right) \cdot \frac{\partial}{\partial y}\left(\frac{x}{1+y}\right) = -\cos\left(\frac{x}{1+y}\right) \cdot \frac{x}{(1+y)^2}$$

Some computer algebra systems can plot surfaces defined by implicit equations in three variables. Figure 6 shows such a plot of the surface defined by the equation in Example 4.

FIGURE 6

V EXAMPLE 4 Find $\partial z/\partial x$ and $\partial z/\partial y$ if z is defined implicitly as a function of x and y by the equation

$$x^3 + y^3 + z^3 + 6xyz = 1$$

SOLUTION To find $\partial z/\partial x$, we differentiate implicitly with respect to x, being careful to treat y as a constant:

$$3x^2 + 3z^2\frac{\partial z}{\partial x} + 6yz + 6xy\frac{\partial z}{\partial x} = 0$$

Solving this equation for $\partial z/\partial x$, we obtain

$$\frac{\partial z}{\partial x} = -\frac{x^2 + 2yz}{z^2 + 2xy}$$

Similarly, implicit differentiation with respect to y gives

$$\frac{\partial z}{\partial y} = -\frac{y^2 + 2xz}{z^2 + 2xy}$$

Functions of More Than Two Variables

Partial derivatives can also be defined for functions of three or more variables. For example, if f is a function of three variables x, y, and z, then its partial derivative with respect to x is defined as

$$f_x(x, y, z) = \lim_{h \to 0} \frac{f(x + h, y, z) - f(x, y, z)}{h}$$

and it is found by regarding y and z as constants and differentiating $f(x, y, z)$ with respect to x. If $w = f(x, y, z)$, then $f_x = \partial w/\partial x$ can be interpreted as the rate of change of w with respect to x when y and z are held fixed. But we can't interpret it geometrically because the graph of f lies in four-dimensional space.

In general, if u is a function of n variables, $u = f(x_1, x_2, \ldots, x_n)$, its partial derivative with respect to the ith variable x_i is

$$\frac{\partial u}{\partial x_i} = \lim_{h \to 0} \frac{f(x_1, \ldots, x_{i-1}, x_i + h, x_{i+1}, \ldots, x_n) - f(x_1, \ldots, x_i, \ldots, x_n)}{h}$$

and we also write

$$\frac{\partial u}{\partial x_i} = \frac{\partial f}{\partial x_i} = f_{x_i} = f_i = D_i f$$

EXAMPLE 5 Find f_x, f_y, and f_z if $f(x, y, z) = e^{xy} \ln z$.

SOLUTION Holding y and z constant and differentiating with respect to x, we have

$$f_x = ye^{xy} \ln z$$

Similarly,

$$f_y = xe^{xy} \ln z \qquad \text{and} \qquad f_z = \frac{e^{xy}}{z}$$

Higher Derivatives

If f is a function of two variables, then its partial derivatives f_x and f_y are also functions of two variables, so we can consider their partial derivatives $(f_x)_x$, $(f_x)_y$, $(f_y)_x$, and $(f_y)_y$, which are called the **second partial derivatives** of f. If $z = f(x, y)$, we use the following notation:

$$(f_x)_x = f_{xx} = f_{11} = \frac{\partial}{\partial x}\left(\frac{\partial f}{\partial x}\right) = \frac{\partial^2 f}{\partial x^2} = \frac{\partial^2 z}{\partial x^2}$$

$$(f_x)_y = f_{xy} = f_{12} = \frac{\partial}{\partial y}\left(\frac{\partial f}{\partial x}\right) = \frac{\partial^2 f}{\partial y\, \partial x} = \frac{\partial^2 z}{\partial y\, \partial x}$$

$$(f_y)_x = f_{yx} = f_{21} = \frac{\partial}{\partial x}\left(\frac{\partial f}{\partial y}\right) = \frac{\partial^2 f}{\partial x\, \partial y} = \frac{\partial^2 z}{\partial x\, \partial y}$$

$$(f_y)_y = f_{yy} = f_{22} = \frac{\partial}{\partial y}\left(\frac{\partial f}{\partial y}\right) = \frac{\partial^2 f}{\partial y^2} = \frac{\partial^2 z}{\partial y^2}$$

Thus the notation f_{xy} (or $\partial^2 f/\partial y\, \partial x$) means that we first differentiate with respect to x and then with respect to y, whereas in computing f_{yx} the order is reversed.

EXAMPLE 6 Find the second partial derivatives of

$$f(x, y) = x^3 + x^2y^3 - 2y^2$$

SOLUTION In Example 1 we found that

$$f_x(x, y) = 3x^2 + 2xy^3 \qquad f_y(x, y) = 3x^2y^2 - 4y$$

Therefore

$$f_{xx} = \frac{\partial}{\partial x}(3x^2 + 2xy^3) = 6x + 2y^3 \qquad f_{xy} = \frac{\partial}{\partial y}(3x^2 + 2xy^3) = 6xy^2$$

$$f_{yx} = \frac{\partial}{\partial x}(3x^2y^2 - 4y) = 6xy^2 \qquad f_{yy} = \frac{\partial}{\partial y}(3x^2y^2 - 4y) = 6x^2y - 4$$

Figure 7 shows the graph of the function f in Example 6 and the graphs of its first- and second-order partial derivatives for $-2 \leqslant x \leqslant 2$, $-2 \leqslant y \leqslant 2$. Notice that these graphs are consistent with our interpretations of f_x and f_y as slopes of tangent lines to traces of the graph of f. For instance, the graph of f decreases if we start at $(0, -2)$ and move in the positive x-direction. This is reflected in the negative values of f_x. You should compare the graphs of f_{yx} and f_{yy} with the graph of f_y to see the relationships.

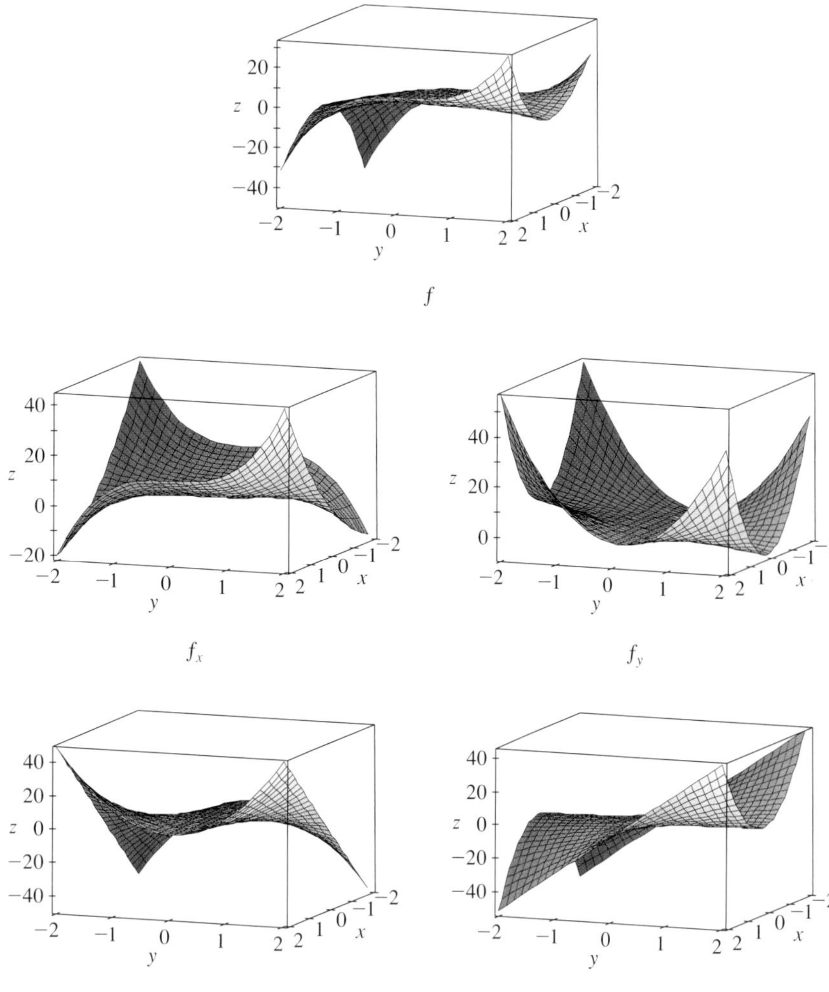

FIGURE 7

Notice that $f_{xy} = f_{yx}$ in Example 6. This is not just a coincidence. It turns out that the mixed partial derivatives f_{xy} and f_{yx} are equal for most functions that one meets in practice. The following theorem, which was discovered by the French mathematician Alexis Clairaut (1713–1765), gives conditions under which we can assert that $f_{xy} = f_{yx}$. The proof is given in Appendix F.

Clairaut's Theorem Suppose f is defined on a disk D that contains the point (a, b). If the functions f_{xy} and f_{yx} are both continuous on D, then

$$f_{xy}(a, b) = f_{yx}(a, b)$$

Clairaut

Alexis Clairaut was a child prodigy in mathematics: he read l'Hospital's textbook on calculus when he was ten and presented a paper on geometry to the French Academy of Sciences when he was 13. At the age of 18, Clairaut published *Recherches sur les courbes à double courbure,* which was the first systematic treatise on three-dimensional analytic geometry and included the calculus of space curves.

Partial derivatives of order 3 or higher can also be defined. For instance,

$$f_{xyy} = (f_{xy})_y = \frac{\partial}{\partial y}\left(\frac{\partial^2 f}{\partial y\, \partial x}\right) = \frac{\partial^3 f}{\partial y^2\, \partial x}$$

and using Clairaut's Theorem it can be shown that $f_{xyy} = f_{yxy} = f_{yyx}$ if these functions are continuous.

V EXAMPLE 7 Calculate f_{xxyz} if $f(x, y, z) = \sin(3x + yz)$.

SOLUTION

$$f_x = 3\cos(3x + yz)$$

$$f_{xx} = -9\sin(3x + yz)$$

$$f_{xxy} = -9z\cos(3x + yz)$$

$$f_{xxyz} = -9\cos(3x + yz) + 9yz\sin(3x + yz)$$

Partial Differential Equations

Partial derivatives occur in *partial differential equations* that express certain physical laws. For instance, the partial differential equation

$$\frac{\partial^2 u}{\partial x^2} + \frac{\partial^2 u}{\partial y^2} = 0$$

is called **Laplace's equation** after Pierre Laplace (1749–1827). Solutions of this equation are called **harmonic functions**; they play a role in problems of heat conduction, fluid flow, and electric potential.

EXAMPLE 8 Show that the function $u(x, y) = e^x \sin y$ is a solution of Laplace's equation.

SOLUTION We first compute the needed second-order partial derivatives:

$$u_x = e^x \sin y \qquad u_y = e^x \cos y$$

$$u_{xx} = e^x \sin y \qquad u_{yy} = -e^x \sin y$$

So

$$u_{xx} + u_{yy} = e^x \sin y - e^x \sin y = 0$$

Therefore u satisfies Laplace's equation.

The **wave equation**

$$\frac{\partial^2 u}{\partial t^2} = a^2 \frac{\partial^2 u}{\partial x^2}$$

describes the motion of a waveform, which could be an ocean wave, a sound wave, a light wave, or a wave traveling along a vibrating string. For instance, if $u(x, t)$ represents the displacement of a vibrating violin string at time t and at a distance x from one end of the string (as in Figure 8), then $u(x, t)$ satisfies the wave equation. Here the constant a depends on the density of the string and on the tension in the string.

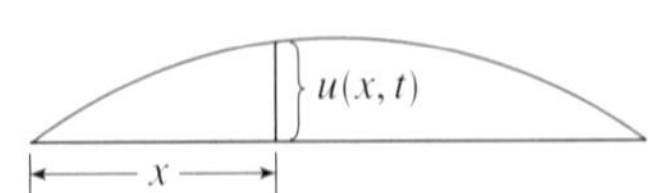

FIGURE 8

EXAMPLE 9 Verify that the function $u(x, t) = \sin(x - at)$ satisfies the wave equation.

SOLUTION

$$u_x = \cos(x - at) \qquad u_t = -a\cos(x - at)$$

$$u_{xx} = -\sin(x - at) \qquad u_{tt} = -a^2\sin(x - at) = a^2 u_{xx}$$

So u satisfies the wave equation.

Partial differential equations involving functions of three variables are also very important in science and engineering. The three-dimensional Laplace equation is

$$\boxed{5} \qquad \frac{\partial^2 u}{\partial x^2} + \frac{\partial^2 u}{\partial y^2} + \frac{\partial^2 u}{\partial z^2} = 0$$

and one place it occurs is in geophysics. If $u(x, y, z)$ represents magnetic field strength at position (x, y, z), then it satisfies Equation 5. The strength of the magnetic field indicates the distribution of iron-rich minerals and reflects different rock types and the location of faults. Figure 9 shows a contour map of the earth's magnetic field as recorded from an aircraft carrying a magnetometer and flying 200 m above the surface of the ground. The contour map is enhanced by color-coding of the regions between the level curves.

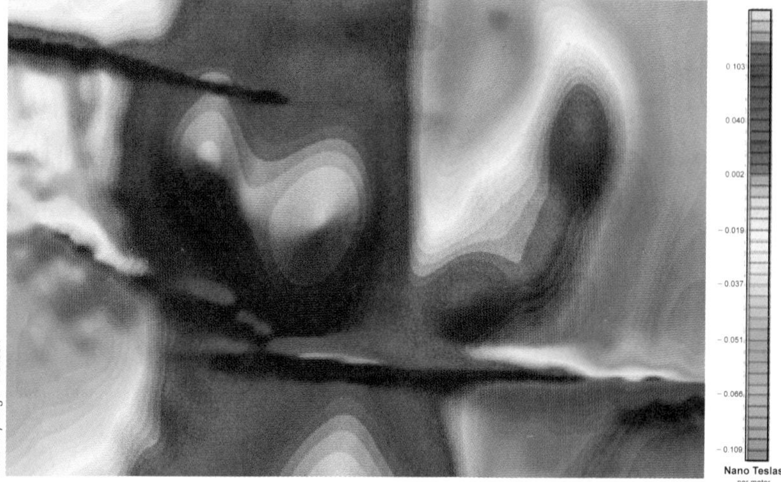

FIGURE 9
Magnetic field strength of the earth

Figure 10 shows a contour map for the second-order partial derivative of u in the vertical direction, that is, u_{zz}. It turns out that the values of the partial derivatives u_{xx} and u_{yy} are relatively easily measured from a map of the magnetic field. Then values of u_{zz} can be calculated from Laplace's equation $\boxed{5}$.

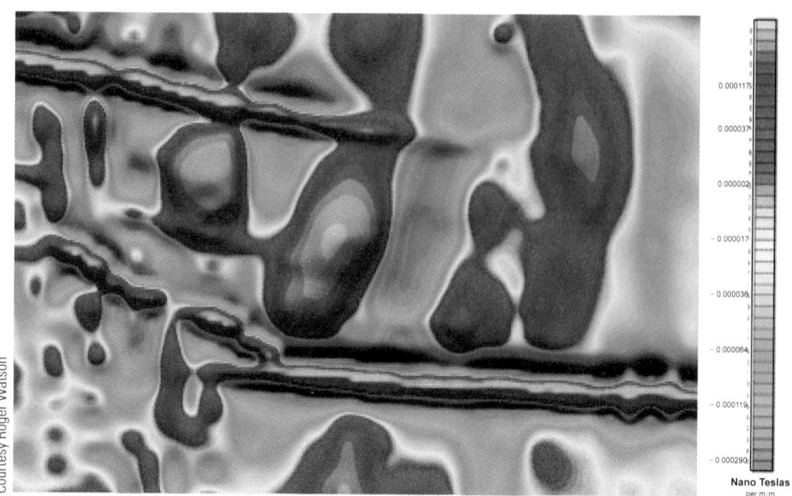

FIGURE 10
Second vertical derivative of the magnetic field

The Cobb-Douglas Production Function

In Example 3 in Section 14.1 we described the work of Cobb and Douglas in modeling the total production P of an economic system as a function of the amount of labor L and the capital investment K. Here we use partial derivatives to show how the particular form of their model follows from certain assumptions they made about the economy.

If the production function is denoted by $P = P(L, K)$, then the partial derivative $\partial P/\partial L$ is the rate at which production changes with respect to the amount of labor. Economists call it the marginal production with respect to labor or the **marginal productivity of labor**. Likewise, the partial derivative $\partial P/\partial K$ is the rate of change of production with respect to capital and is called the **marginal productivity of capital**. In these terms, the assumptions made by Cobb and Douglas can be stated as follows.

(i) If either labor or capital vanishes, then so will production.

(ii) The marginal productivity of labor is proportional to the amount of production per unit of labor.

(iii) The marginal productivity of capital is proportional to the amount of production per unit of capital.

Because the production per unit of labor is P/L, assumption (ii) says that

$$\frac{\partial P}{\partial L} = \alpha \frac{P}{L}$$

for some constant α. If we keep K constant ($K = K_0$), then this partial differential equation becomes an ordinary differential equation:

6 $$\frac{dP}{dL} = \alpha \frac{P}{L}$$

If we solve this separable differential equation by the methods of Section 9.3 (see also Exercise 85), we get

7 $$P(L, K_0) = C_1(K_0)L^{\alpha}$$

Notice that we have written the constant C_1 as a function of K_0 because it could depend on the value of K_0.

Similarly, assumption (iii) says that

$$\frac{\partial P}{\partial K} = \beta \frac{P}{K}$$

and we can solve this differential equation to get

8 $$P(L_0, K) = C_2(L_0)K^{\beta}$$

Comparing Equations 7 and 8, we have

9 $$P(L, K) = bL^{\alpha}K^{\beta}$$

where b is a constant that is independent of both L and K. Assumption (i) shows that $\alpha > 0$ and $\beta > 0$.

Notice from Equation 9 that if labor and capital are both increased by a factor m, then

$$P(mL, mK) = b(mL)^{\alpha}(mK)^{\beta} = m^{\alpha+\beta}bL^{\alpha}K^{\beta} = m^{\alpha+\beta}P(L, K)$$

If $\alpha + \beta = 1$, then $P(mL, mK) = mP(L, K)$, which means that production is also increased by a factor of m. That is why Cobb and Douglas assumed that $\alpha + \beta = 1$ and therefore

$$P(L, K) = bL^{\alpha}K^{1-\alpha}$$

This is the Cobb-Douglas production function that we discussed in Section 14.1.

14.3 Exercises

1. The temperature T (in °C) at a location in the Northern Hemisphere depends on the longitude x, latitude y, and time t, so we can write $T = f(x, y, t)$. Let's measure time in hours from the beginning of January.
   (a) What are the meanings of the partial derivatives $\partial T/\partial x$, $\partial T/\partial y$, and $\partial T/\partial t$?
   (b) Honolulu has longitude 158° W and latitude 21° N. Suppose that at 9:00 AM on January 1 the wind is blowing hot air to the northeast, so the air to the west and south is warm and the air to the north and east is cooler. Would you expect $f_x(158, 21, 9)$, $f_y(158, 21, 9)$, and $f_t(158, 21, 9)$ to be positive or negative? Explain.

2. At the beginning of this section we discussed the function $I = f(T, H)$, where I is the heat index, T is the temperature, and H is the relative humidity. Use Table 1 to estimate $f_T(92, 60)$ and $f_H(92, 60)$. What are the practical interpretations of these values?

3. The wind-chill index W is the perceived temperature when the actual temperature is T and the wind speed is v, so we can write $W = f(T, v)$. The following table of values is an excerpt from Table 1 in Section 14.1.

Wind speed (km/h); Actual temperature (°C)

T \ v	20	30	40	50	60	70
−10	−18	−20	−21	−22	−23	−23
−15	−24	−26	−27	−29	−30	−30
−20	−30	−33	−34	−35	−36	−37
−25	−37	−39	−41	−42	−43	−44

   (a) Estimate the values of $f_T(-15, 30)$ and $f_v(-15, 30)$. What are the practical interpretations of these values?
   (b) In general, what can you say about the signs of $\partial W/\partial T$ and $\partial W/\partial v$?
   (c) What appears to be the value of the following limit?

$$\lim_{v\to\infty} \frac{\partial W}{\partial v}$$

4. The wave heights h in the open sea depend on the speed v of the wind and the length of time t that the wind has been blowing at that speed. Values of the function $h = f(v, t)$ are recorded in feet in the following table.

Duration (hours); Wind speed (knots)

v \ t	5	10	15	20	30	40	50
10	2	2	2	2	2	2	2
15	4	4	5	5	5	5	5
20	5	7	8	8	9	9	9
30	9	13	16	17	18	19	19
40	14	21	25	28	31	33	33
50	19	29	36	40	45	48	50
60	24	37	47	54	62	67	69

   (a) What are the meanings of the partial derivatives $\partial h/\partial v$ and $\partial h/\partial t$?
   (b) Estimate the values of $f_v(40, 15)$ and $f_t(40, 15)$. What are the practical interpretations of these values?
   (c) What appears to be the value of the following limit?

$$\lim_{t\to\infty} \frac{\partial h}{\partial t}$$

5–8 Determine the signs of the partial derivatives for the function f whose graph is shown.

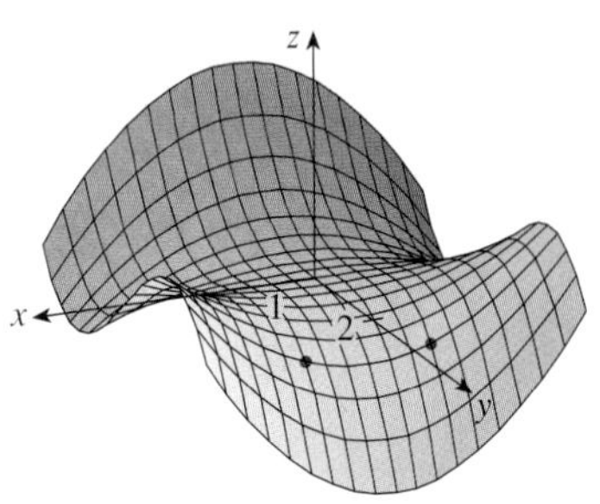

5. (a) $f_x(1, 2)$ (b) $f_y(1, 2)$

6. (a) $f_x(-1, 2)$ (b) $f_y(-1, 2)$

7. (a) $f_{xx}(-1, 2)$ (b) $f_{yy}(-1, 2)$

8. (a) $f_{xy}(1, 2)$ (b) $f_{xy}(-1, 2)$

9. The following surfaces, labeled a, b, and c, are graphs of a function f and its partial derivatives f_x and f_y. Identify each surface and give reasons for your choices.

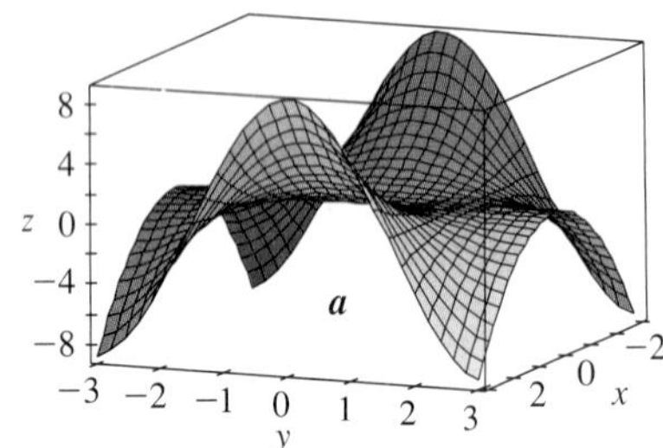

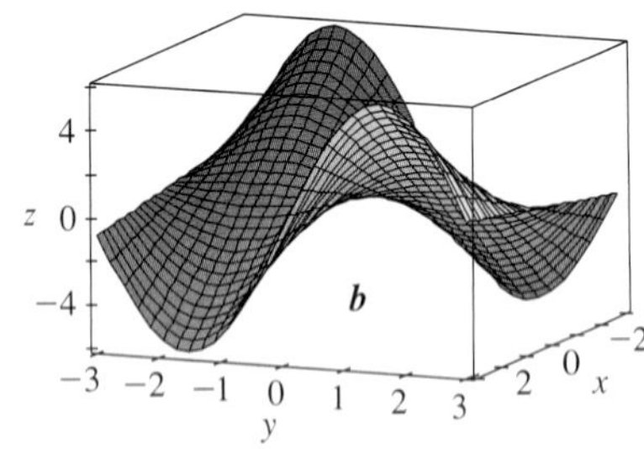

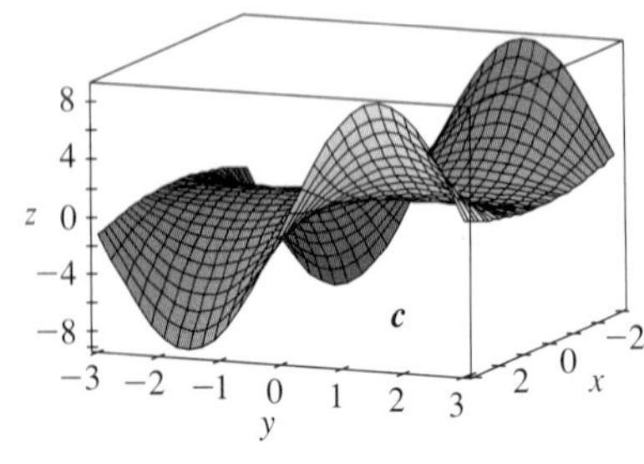

10. A contour map is given for a function f. Use it to estimate $f_x(2, 1)$ and $f_y(2, 1)$.

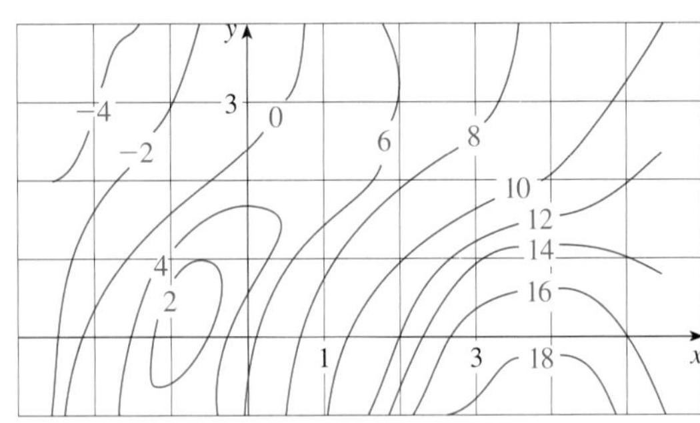

11. If $f(x, y) = 16 - 4x^2 - y^2$, find $f_x(1, 2)$ and $f_y(1, 2)$ and interpret these numbers as slopes. Illustrate with either hand-drawn sketches or computer plots.

12. If $f(x, y) = \sqrt{4 - x^2 - 4y^2}$, find $f_x(1, 0)$ and $f_y(1, 0)$ and interpret these numbers as slopes. Illustrate with either hand-drawn sketches or computer plots.

13–14 Find f_x and f_y and graph f, f_x, and f_y with domains and viewpoints that enable you to see the relationships between them.

13. $f(x, y) = x^2y^3$ **14.** $f(x, y) = \dfrac{y}{1 + x^2y^2}$

15–40 Find the first partial derivatives of the function.

15. $f(x, y) = y^5 - 3xy$ **16.** $f(x, y) = x^4y^3 + 8x^2y$

17. $f(x, t) = e^{-t}\cos \pi x$ **18.** $f(x, t) = \sqrt{x}\, \ln t$

19. $z = (2x + 3y)^{10}$ **20.** $z = \tan xy$

21. $f(x, y) = \dfrac{x}{y}$ **22.** $f(x, y) = \dfrac{x}{(x + y)^2}$

23. $f(x, y) = \dfrac{ax + by}{cx + dy}$ **24.** $w = \dfrac{e^v}{u + v^2}$

25. $g(u, v) = (u^2v - v^3)^5$ **26.** $u(r, \theta) = \sin(r\cos\theta)$

27. $R(p, q) = \tan^{-1}(pq^2)$ **28.** $f(x, y) = x^y$

29. $F(x, y) = \int_y^x \cos(e^t)\, dt$ **30.** $F(\alpha, \beta) = \int_\alpha^\beta \sqrt{t^3 + 1}\, dt$

31. $f(x, y, z) = xz - 5x^2y^3z^4$ **32.** $f(x, y, z) = x\sin(y - z)$

33. $w = \ln(x + 2y + 3z)$ **34.** $w = ze^{xyz}$

35. $u = xy\sin^{-1}(yz)$ **36.** $u = x^{y/z}$

37. $h(x, y, z, t) = x^2y\cos(z/t)$ **38.** $\phi(x, y, z, t) = \dfrac{\alpha x + \beta y^2}{\gamma z + \delta t^2}$

39. $u = \sqrt{x_1^2 + x_2^2 + \cdots + x_n^2}$

40. $u = \sin(x_1 + 2x_2 + \cdots + nx_n)$

41–44 Find the indicated partial derivative.

41. $f(x, y) = \ln\left(x + \sqrt{x^2 + y^2}\right)$; $f_x(3, 4)$

42. $f(x, y) = \arctan(y/x)$; $f_x(2, 3)$

43. $f(x, y, z) = \dfrac{y}{x + y + z}$; $f_y(2, 1, -1)$

44. $f(x, y, z) = \sqrt{\sin^2 x + \sin^2 y + \sin^2 z}$; $f_z(0, 0, \pi/4)$

45–46 Use the definition of partial derivatives as limits [4] to find $f_x(x, y)$ and $f_y(x, y)$.

45. $f(x, y) = xy^2 - x^3y$

46. $f(x, y) = \dfrac{x}{x + y^2}$

47–50 Use implicit differentiation to find $\partial z/\partial x$ and $\partial z/\partial y$.

47. $x^2 + 2y^2 + 3z^2 = 1$

48. $x^2 - y^2 + z^2 - 2z = 4$

49. $e^z = xyz$

50. $yz + x \ln y = z^2$

51–52 Find $\partial z/\partial x$ and $\partial z/\partial y$.

51. (a) $z = f(x) + g(y)$ (b) $z = f(x + y)$

52. (a) $z = f(x)g(y)$ (b) $z = f(xy)$
(c) $z = f(x/y)$

53–58 Find all the second partial derivatives.

53. $f(x, y) = x^3y^5 + 2x^4y$

54. $f(x, y) = \sin^2(mx + ny)$

55. $w = \sqrt{u^2 + v^2}$

56. $v = \dfrac{xy}{x - y}$

57. $z = \arctan \dfrac{x + y}{1 - xy}$

58. $v = e^{xe^y}$

59–62 Verify that the conclusion of Clairaut's Theorem holds, that is, $u_{xy} = u_{yx}$.

59. $u = x^4y^3 - y^4$

60. $u = e^{xy} \sin y$

61. $u = \cos(x^2y)$

62. $u = \ln(x + 2y)$

63–70 Find the indicated partial derivative(s).

63. $f(x, y) = x^4y^2 - x^3y$; f_{xxx}, f_{xyx}

64. $f(x, y) = \sin(2x + 5y)$; f_{yxy}

65. $f(x, y, z) = e^{xyz^2}$; f_{xyz}

66. $g(r, s, t) = e^r \sin(st)$; g_{rst}

67. $u = e^{r\theta} \sin \theta$; $\dfrac{\partial^3 u}{\partial r^2\, \partial\theta}$

68. $z = u\sqrt{v - w}$; $\dfrac{\partial^3 z}{\partial u\, \partial v\, \partial w}$

69. $w = \dfrac{x}{y + 2z}$; $\dfrac{\partial^3 w}{\partial z\, \partial y\, \partial x}$, $\dfrac{\partial^3 w}{\partial x^2\, \partial y}$

70. $u = x^a y^b z^c$; $\dfrac{\partial^6 u}{\partial x\, \partial y^2\, \partial z^3}$

71. If $f(x, y, z) = xy^2z^3 + \arcsin(x\sqrt{z})$, find f_{xzy}. [*Hint:* Which order of differentiation is easiest?]

72. If $g(x, y, z) = \sqrt{1 + xz} + \sqrt{1 - xy}$, find g_{xyz}. [*Hint:* Use a different order of differentiation for each term.]

73. Use the table of values of $f(x, y)$ to estimate the values of $f_x(3, 2)$, $f_x(3, 2.2)$, and $f_{xy}(3, 2)$.

x \ y	1.8	2.0	2.2
2.5	12.5	10.2	9.3
3.0	18.1	17.5	15.9
3.5	20.0	22.4	26.1

74. Level curves are shown for a function f. Determine whether the following partial derivatives are positive or negative at the point P.
(a) f_x (b) f_y (c) f_{xx}
(d) f_{xy} (e) f_{yy}

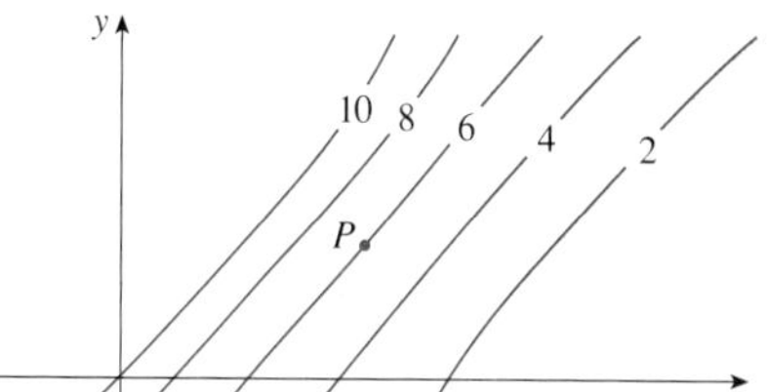

75. Verify that the function $u = e^{-\alpha^2k^2t} \sin kx$ is a solution of the *heat conduction equation* $u_t = \alpha^2 u_{xx}$.

76. Determine whether each of the following functions is a solution of Laplace's equation $u_{xx} + u_{yy} = 0$.
(a) $u = x^2 + y^2$ (b) $u = x^2 - y^2$
(c) $u = x^3 + 3xy^2$ (d) $u = \ln \sqrt{x^2 + y^2}$
(e) $u = \sin x \cosh y + \cos x \sinh y$
(f) $u = e^{-x} \cos y - e^{-y} \cos x$

77. Verify that the function $u = 1/\sqrt{x^2 + y^2 + z^2}$ is a solution of the three-dimensional Laplace equation $u_{xx} + u_{yy} + u_{zz} = 0$.

78. Show that each of the following functions is a solution of the wave equation $u_{tt} = a^2 u_{xx}$.
(a) $u = \sin(kx) \sin(akt)$ (b) $u = t/(a^2t^2 - x^2)$
(c) $u = (x - at)^6 + (x + at)^6$
(d) $u = \sin(x - at) + \ln(x + at)$

79. If f and g are twice differentiable functions of a single variable, show that the function

$$u(x, t) = f(x + at) + g(x - at)$$

is a solution of the wave equation given in Exercise 78.

80. If $u = e^{a_1x_1+a_2x_2+\cdots+a_nx_n}$, where $a_1^2 + a_2^2 + \cdots + a_n^2 = 1$, show that

$$\frac{\partial^2 u}{\partial x_1^2} + \frac{\partial^2 u}{\partial x_2^2} + \cdots + \frac{\partial^2 u}{\partial x_n^2} = u$$

81. Verify that the function $z = \ln(e^x + e^y)$ is a solution of the differential equations

$$\frac{\partial z}{\partial x} + \frac{\partial z}{\partial y} = 1$$

and

$$\frac{\partial^2 z}{\partial x^2}\frac{\partial^2 z}{\partial y^2} - \left(\frac{\partial^2 z}{\partial x\,\partial y}\right)^2 = 0$$

82. The temperature at a point (x, y) on a flat metal plate is given by $T(x, y) = 60/(1 + x^2 + y^2)$, where T is measured in °C and x, y in meters. Find the rate of change of temperature with respect to distance at the point $(2, 1)$ in (a) the x-direction and (b) the y-direction.

83. The total resistance R produced by three conductors with resistances R_1, R_2, R_3 connected in a parallel electrical circuit is given by the formula

$$\frac{1}{R} = \frac{1}{R_1} + \frac{1}{R_2} + \frac{1}{R_3}$$

Find $\partial R/\partial R_1$.

84. Show that the Cobb-Douglas production function $P = bL^\alpha K^\beta$ satisfies the equation

$$L\frac{\partial P}{\partial L} + K\frac{\partial P}{\partial K} = (\alpha + \beta)P$$

85. Show that the Cobb-Douglas production function satisfies $P(L, K_0) = C_1(K_0)L^\alpha$ by solving the differential equation

$$\frac{dP}{dL} = \alpha\frac{P}{L}$$

(See Equation 6.)

86. Cobb and Douglas used the equation $P(L, K) = 1.01L^{0.75}K^{0.25}$ to model the American economy from 1899 to 1922, where L is the amount of labor and K is the amount of capital. (See Example 3 in Section 14.1.)

(a) Calculate P_L and P_K.

(b) Find the marginal productivity of labor and the marginal productivity of capital in the year 1920, when $L = 194$ and $K = 407$ (compared with the assigned values $L = 100$ and $K = 100$ in 1899). Interpret the results.

(c) In the year 1920 which would have benefited production more, an increase in capital investment or an increase in spending on labor?

87. The *van der Waals equation* for n moles of a gas is

$$\left(P + \frac{n^2a}{V^2}\right)(V - nb) = nRT$$

where P is the pressure, V is the volume, and T is the temperature of the gas. The constant R is the universal gas constant and a and b are positive constants that are characteristic of a particular gas. Calculate $\partial T/\partial P$ and $\partial P/\partial V$.

88. The gas law for a fixed mass m of an ideal gas at absolute temperature T, pressure P, and volume V is $PV = mRT$, where R is the gas constant. Show that

$$\frac{\partial P}{\partial V}\frac{\partial V}{\partial T}\frac{\partial T}{\partial P} = -1$$

89. For the ideal gas of Exercise 88, show that

$$T\frac{\partial P}{\partial T}\frac{\partial V}{\partial T} = mR$$

90. The wind-chill index is modeled by the function

$$W = 13.12 + 0.6215T - 11.37v^{0.16} + 0.3965Tv^{0.16}$$

where T is the temperature (°C) and v is the wind speed (km/h). When $T = -15$°C and $v = 30$ km/h, by how much would you expect the apparent temperature W to drop if the actual temperature decreases by 1°C? What if the wind speed increases by 1 km/h?

91. The kinetic energy of a body with mass m and velocity v is $K = \frac{1}{2}mv^2$. Show that

$$\frac{\partial K}{\partial m}\frac{\partial^2 K}{\partial v^2} = K$$

92. If a, b, c are the sides of a triangle and A, B, C are the opposite angles, find $\partial A/\partial a$, $\partial A/\partial b$, $\partial A/\partial c$ by implicit differentiation of the Law of Cosines.

93. You are told that there is a function f whose partial derivatives are $f_x(x, y) = x + 4y$ and $f_y(x, y) = 3x - y$. Should you believe it?

94. The paraboloid $z = 6 - x - x^2 - 2y^2$ intersects the plane $x = 1$ in a parabola. Find parametric equations for the tangent line to this parabola at the point $(1, 2, -4)$. Use a computer to graph the paraboloid, the parabola, and the tangent line on the same screen.

95. The ellipsoid $4x^2 + 2y^2 + z^2 = 16$ intersects the plane $y = 2$ in an ellipse. Find parametric equations for the tangent line to this ellipse at the point $(1, 2, 2)$.

96. In a study of frost penetration it was found that the temperature T at time t (measured in days) at a depth x (measured in feet) can be modeled by the function

$$T(x, t) = T_0 + T_1e^{-\lambda x}\sin(\omega t - \lambda x)$$

where $\omega = 2\pi/365$ and λ is a positive constant.

(a) Find $\partial T/\partial x$. What is its physical significance?

(b) Find $\partial T/\partial t$. What is its physical significance?

(c) Show that T satisfies the heat equation $T_t = kT_{xx}$ for a certain constant k.

(d) If $\lambda = 0.2$, $T_0 = 0$, and $T_1 = 10$, use a computer to graph $T(x, t)$.

(e) What is the physical significance of the term $-\lambda x$ in the expression $\sin(\omega t - \lambda x)$?

97. Use Clairaut's Theorem to show that if the third-order partial derivatives of f are continuous, then

$$f_{xyy} = f_{yxy} = f_{yyx}$$

98. (a) How many nth-order partial derivatives does a function of two variables have?

(b) If these partial derivatives are all continuous, how many of them can be distinct?

(c) Answer the question in part (a) for a function of three variables.

99. If $f(x, y) = x(x^2 + y^2)^{-3/2}e^{\sin(x^2y)}$, find $f_x(1, 0)$. [*Hint:* Instead of finding $f_x(x, y)$ first, note that it's easier to use Equation 1 or Equation 2.]

100. If $f(x, y) = \sqrt[3]{x^3 + y^3}$, find $f_x(0, 0)$.

101. Let

$$f(x, y) = \begin{cases} \dfrac{x^3y - xy^3}{x^2 + y^2} & \text{if } (x, y) \neq (0, 0) \\ 0 & \text{if } (x, y) = (0, 0) \end{cases}$$

(a) Use a computer to graph f.

(b) Find $f_x(x, y)$ and $f_y(x, y)$ when $(x, y) \neq (0, 0)$.

(c) Find $f_x(0, 0)$ and $f_y(0, 0)$ using Equations 2 and 3.

(d) Show that $f_{xy}(0, 0) = -1$ and $f_{yx}(0, 0) = 1$.

CAS (e) Does the result of part (d) contradict Clairaut's Theorem? Use graphs of f_{xy} and f_{yx} to illustrate your answer.

14.4 Tangent Planes and Linear Approximations

One of the most important ideas in single-variable calculus is that as we zoom in toward a point on the graph of a differentiable function, the graph becomes indistinguishable from its tangent line and we can approximate the function by a linear function. (See Section 3.10.) Here we develop similar ideas in three dimensions. As we zoom in toward a point on a surface that is the graph of a differentiable function of two variables, the surface looks more and more like a plane (its tangent plane) and we can approximate the function by a linear function of two variables. We also extend the idea of a differential to functions of two or more variables.

Tangent Planes

Suppose a surface S has equation $z = f(x, y)$, where f has continuous first partial derivatives, and let $P(x_0, y_0, z_0)$ be a point on S. As in the preceding section, let C_1 and C_2 be the curves obtained by intersecting the vertical planes $y = y_0$ and $x = x_0$ with the surface S. Then the point P lies on both C_1 and C_2. Let T_1 and T_2 be the tangent lines to the curves C_1 and C_2 at the point P. Then the **tangent plane** to the surface S at the point P is defined to be the plane that contains both tangent lines T_1 and T_2. (See Figure 1.)

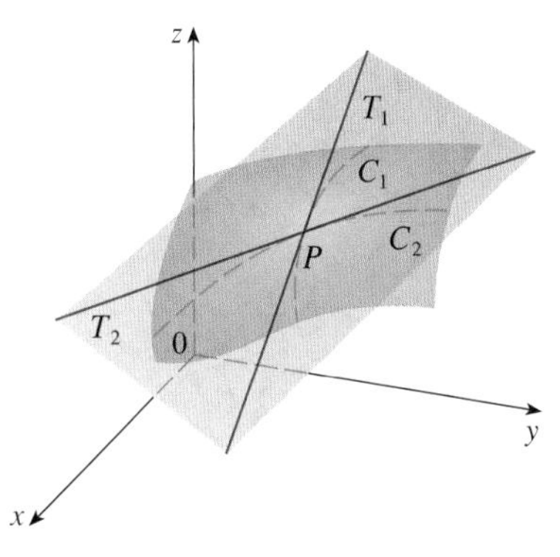

FIGURE 1
The tangent plane contains the tangent lines T_1 and T_2.

We will see in Section 14.6 that if C is any other curve that lies on the surface S and passes through P, then its tangent line at P also lies in the tangent plane. Therefore you can think of the tangent plane to S at P as consisting of all possible tangent lines at P to curves that lie on S and pass through P. The tangent plane at P is the plane that most closely approximates the surface S near the point P.

We know from Equation 12.5.7 that any plane passing through the point $P(x_0, y_0, z_0)$ has an equation of the form

$$A(x - x_0) + B(y - y_0) + C(z - z_0) = 0$$

By dividing this equation by C and letting $a = -A/C$ and $b = -B/C$, we can write it in the form

$$\boxed{1} \qquad z - z_0 = a(x - x_0) + b(y - y_0)$$

If Equation 1 represents the tangent plane at P, then its intersection with the plane $y = y_0$ must be the tangent line T_1. Setting $y = y_0$ in Equation 1 gives

$$z - z_0 = a(x - x_0) \qquad \text{where } y = y_0$$

and we recognize this as the equation (in point-slope form) of a line with slope a. But from Section 14.3 we know that the slope of the tangent T_1 is $f_x(x_0, y_0)$. Therefore $a = f_x(x_0, y_0)$.

Similarly, putting $x = x_0$ in Equation 1, we get $z - z_0 = b(y - y_0)$, which must represent the tangent line T_2, so $b = f_y(x_0, y_0)$.

Note the similarity between the equation of a tangent plane and the equation of a tangent line:

$$y - y_0 = f'(x_0)(x - x_0)$$

> **2** Suppose f has continuous partial derivatives. An equation of the tangent plane to the surface $z = f(x, y)$ at the point $P(x_0, y_0, z_0)$ is
>
> $$z - z_0 = f_x(x_0, y_0)(x - x_0) + f_y(x_0, y_0)(y - y_0)$$

V EXAMPLE 1 Find the tangent plane to the elliptic paraboloid $z = 2x^2 + y^2$ at the point $(1, 1, 3)$.

SOLUTION Let $f(x, y) = 2x^2 + y^2$. Then

$$f_x(x, y) = 4x \qquad f_y(x, y) = 2y$$

$$f_x(1, 1) = 4 \qquad f_y(1, 1) = 2$$

Then **2** gives the equation of the tangent plane at $(1, 1, 3)$ as

$$z - 3 = 4(x - 1) + 2(y - 1)$$

or

$$z = 4x + 2y - 3$$

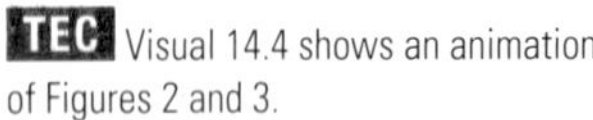
TEC Visual 14.4 shows an animation of Figures 2 and 3.

Figure 2(a) shows the elliptic paraboloid and its tangent plane at $(1, 1, 3)$ that we found in Example 1. In parts (b) and (c) we zoom in toward the point $(1, 1, 3)$ by restricting the domain of the function $f(x, y) = 2x^2 + y^2$. Notice that the more we zoom in, the flatter the graph appears and the more it resembles its tangent plane.

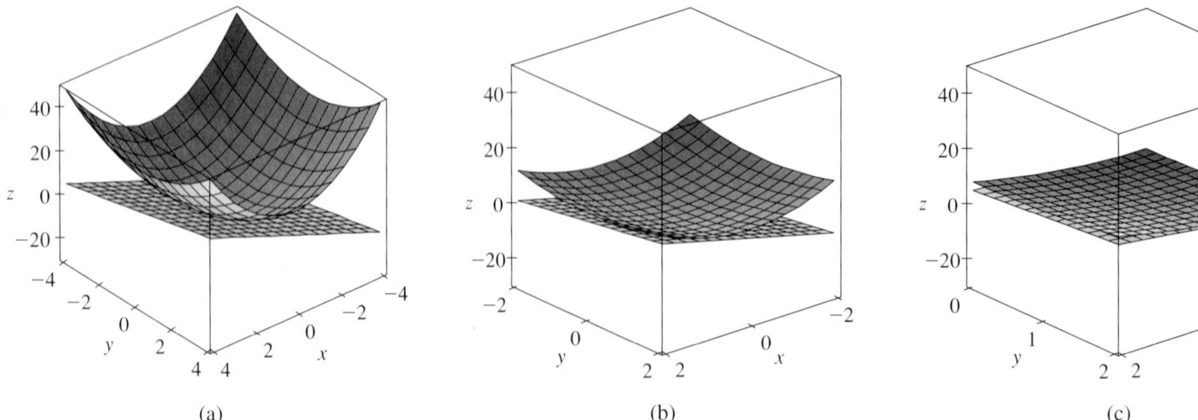

FIGURE 2 The elliptic paraboloid $z = 2x^2 + y^2$ appears to coincide with its tangent plane as we zoom in toward $(1, 1, 3)$.

In Figure 3 we corroborate this impression by zooming in toward the point (1, 1) on a contour map of the function $f(x, y) = 2x^2 + y^2$. Notice that the more we zoom in, the more the level curves look like equally spaced parallel lines, which is characteristic of a plane.

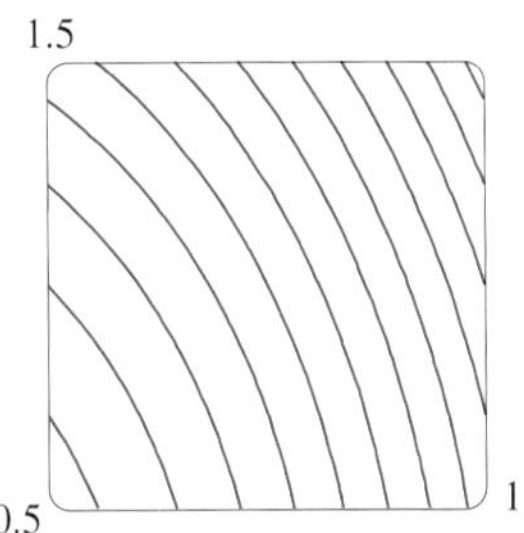

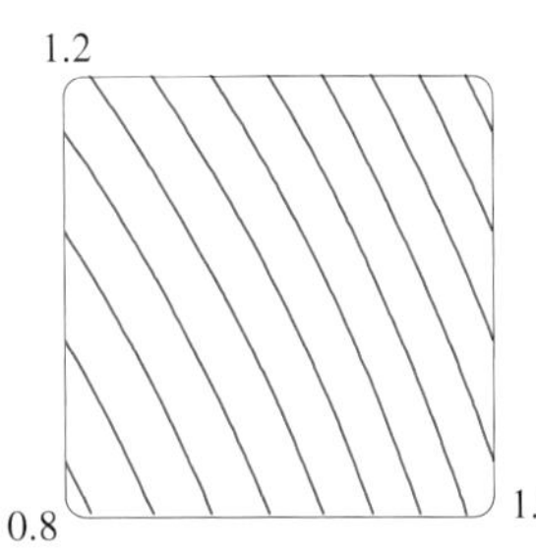

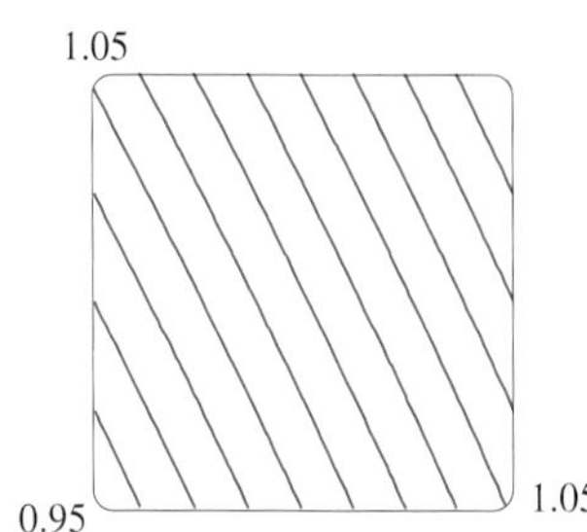

FIGURE 3
Zooming in toward (1, 1) on a contour map of $f(x, y) = 2x^2 + y^2$

Linear Approximations

In Example 1 we found that an equation of the tangent plane to the graph of the function $f(x, y) = 2x^2 + y^2$ at the point (1, 1, 3) is $z = 4x + 2y - 3$. Therefore, in view of the visual evidence in Figures 2 and 3, the linear function of two variables

$$L(x, y) = 4x + 2y - 3$$

is a good approximation to $f(x, y)$ when (x, y) is near (1, 1). The function L is called the *linearization* of f at (1, 1) and the approximation

$$f(x, y) \approx 4x + 2y - 3$$

is called the *linear approximation* or *tangent plane approximation* of f at (1, 1).

For instance, at the point (1.1, 0.95) the linear approximation gives

$$f(1.1, 0.95) \approx 4(1.1) + 2(0.95) - 3 = 3.3$$

which is quite close to the true value of $f(1.1, 0.95) = 2(1.1)^2 + (0.95)^2 = 3.3225$. But if we take a point farther away from (1, 1), such as (2, 3), we no longer get a good approximation. In fact, $L(2, 3) = 11$ whereas $f(2, 3) = 17$.

In general, we know from [2] that an equation of the tangent plane to the graph of a function f of two variables at the point $(a, b, f(a, b))$ is

$$z = f(a, b) + f_x(a, b)(x - a) + f_y(a, b)(y - b)$$

The linear function whose graph is this tangent plane, namely

3 $$L(x, y) = f(a, b) + f_x(a, b)(x - a) + f_y(a, b)(y - b)$$

is called the **linearization** of f at (a, b) and the approximation

4 $$f(x, y) \approx f(a, b) + f_x(a, b)(x - a) + f_y(a, b)(y - b)$$

is called the **linear approximation** or the **tangent plane approximation** of f at (a, b).

We have defined tangent planes for surfaces $z = f(x, y)$, where f has continuous first partial derivatives. What happens if f_x and f_y are not continuous? Figure 4 pictures such a function; its equation is

$$f(x, y) = \begin{cases} \dfrac{xy}{x^2 + y^2} & \text{if } (x, y) \neq (0, 0) \\ 0 & \text{if } (x, y) = (0, 0) \end{cases}$$

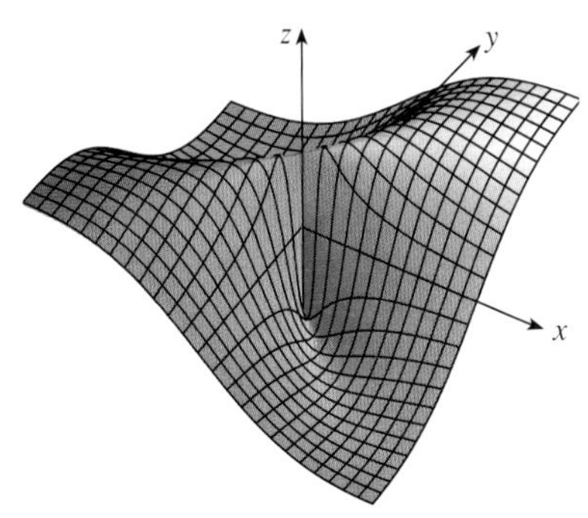

FIGURE 4
$f(x, y) = \dfrac{xy}{x^2 + y^2}$ if $(x, y) \neq (0, 0)$,
$f(0, 0) = 0$

You can verify (see Exercise 46) that its partial derivatives exist at the origin and, in fact, $f_x(0, 0) = 0$ and $f_y(0, 0) = 0$, but f_x and f_y are not continuous. The linear approximation would be $f(x, y) \approx 0$, but $f(x, y) = \frac{1}{2}$ at all points on the line $y = x$. So a function of two variables can behave badly even though both of its partial derivatives exist. To rule out such behavior, we formulate the idea of a differentiable function of two variables.

Recall that for a function of one variable, $y = f(x)$, if x changes from a to $a + \Delta x$, we defined the increment of y as

$$\Delta y = f(a + \Delta x) - f(a)$$

In Chapter 3 we showed that if f is differentiable at a, then

This is Equation 3.4.7.

$$\boxed{5} \qquad \Delta y = f'(a)\,\Delta x + \varepsilon\,\Delta x \qquad \text{where } \varepsilon \to 0 \text{ as } \Delta x \to 0$$

Now consider a function of two variables, $z = f(x, y)$, and suppose x changes from a to $a + \Delta x$ and y changes from b to $b + \Delta y$. Then the corresponding **increment** of z is

$$\boxed{6} \qquad \Delta z = f(a + \Delta x, b + \Delta y) - f(a, b)$$

Thus the increment Δz represents the change in the value of f when (x, y) changes from (a, b) to $(a + \Delta x, b + \Delta y)$. By analogy with $\boxed{5}$ we define the differentiability of a function of two variables as follows.

$\boxed{7}$ **Definition** If $z = f(x, y)$, then f is **differentiable** at (a, b) if Δz can be expressed in the form

$$\Delta z = f_x(a, b)\,\Delta x + f_y(a, b)\,\Delta y + \varepsilon_1\,\Delta x + \varepsilon_2\,\Delta y$$

where ε_1 and $\varepsilon_2 \to 0$ as $(\Delta x, \Delta y) \to (0, 0)$.

Definition 7 says that a differentiable function is one for which the linear approximation $\boxed{4}$ is a good approximation when (x, y) is near (a, b). In other words, the tangent plane approximates the graph of f well near the point of tangency.

It's sometimes hard to use Definition 7 directly to check the differentiability of a function, but the next theorem provides a convenient sufficient condition for differentiability.

Theorem 8 is proved in Appendix F.

$\boxed{8}$ **Theorem** If the partial derivatives f_x and f_y exist near (a, b) and are continuous at (a, b), then f is differentiable at (a, b).

Figure 5 shows the graphs of the function f and its linearization L in Example 2.

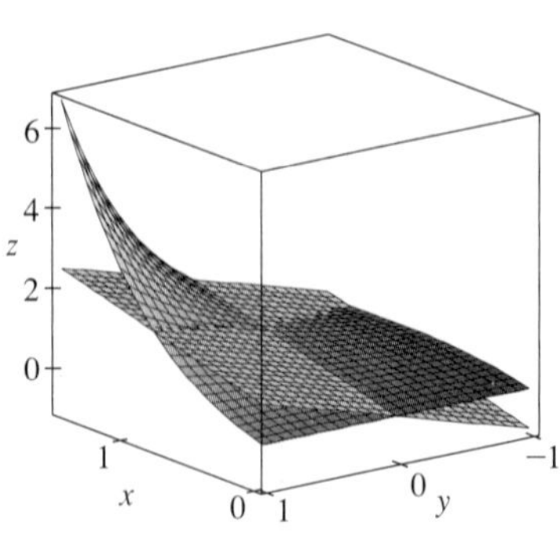

FIGURE 5

V EXAMPLE 2 Show that $f(x, y) = xe^{xy}$ is differentiable at $(1, 0)$ and find its linearization there. Then use it to approximate $f(1.1, -0.1)$.

SOLUTION The partial derivatives are

$$f_x(x, y) = e^{xy} + xye^{xy} \qquad f_y(x, y) = x^2e^{xy}$$

$$f_x(1, 0) = 1 \qquad f_y(1, 0) = 1$$

Both f_x and f_y are continuous functions, so f is differentiable by Theorem 8. The linearization is

$$L(x, y) = f(1, 0) + f_x(1, 0)(x - 1) + f_y(1, 0)(y - 0)$$

$$= 1 + 1(x - 1) + 1 \cdot y = x + y$$

The corresponding linear approximation is

$$xe^{xy} \approx x + y$$

so

$$f(1.1, -0.1) \approx 1.1 - 0.1 = 1$$

Compare this with the actual value of $f(1.1, -0.1) = 1.1e^{-0.11} \approx 0.98542$.

EXAMPLE 3 At the beginning of Section 14.3 we discussed the heat index (perceived temperature) I as a function of the actual temperature T and the relative humidity H and gave the following table of values from the National Weather Service.

Relative humidity (%)

Actual temperature (°F)

T \ H	50	55	60	65	70	75	80	85	90
90	96	98	100	103	106	109	112	115	119
92	100	103	105	108	112	115	119	123	128
94	104	107	111	114	118	122	127	132	137
96	109	113	116	121	125	130	135	141	146
98	114	118	123	127	133	138	144	150	157
100	119	124	129	135	141	147	154	161	168

Find a linear approximation for the heat index $I = f(T, H)$ when T is near 96°F and H is near 70%. Use it to estimate the heat index when the temperature is 97°F and the relative humidity is 72%.

SOLUTION We read from the table that $f(96, 70) = 125$. In Section 14.3 we used the tabular values to estimate that $f_T(96, 70) \approx 3.75$ and $f_H(96, 70) \approx 0.9$. (See pages 901–02.) So the linear approximation is

$$\begin{aligned} f(T, H) &\approx f(96, 70) + f_T(96, 70)(T - 96) + f_H(96, 70)(H - 70) \\ &\approx 125 + 3.75(T - 96) + 0.9(H - 70) \end{aligned}$$

In particular,

$$f(97, 72) \approx 125 + 3.75(1) + 0.9(2) = 130.55$$

Therefore, when $T = 97$°F and $H = 72$%, the heat index is

$$I \approx 131°\text{F}$$

Differentials

For a differentiable function of one variable, $y = f(x)$, we define the differential dx to be an independent variable; that is, dx can be given the value of any real number. The differential of y is then defined as

$$\boxed{9} \qquad dy = f'(x)\,dx$$

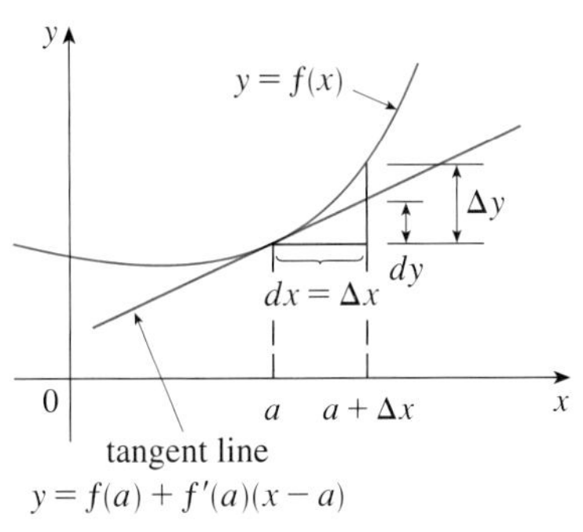

FIGURE 6

(See Section 3.10.) Figure 6 shows the relationship between the increment Δy and the differential dy: Δy represents the change in height of the curve $y = f(x)$ and dy represents the change in height of the tangent line when x changes by an amount $dx = \Delta x$.

For a differentiable function of two variables, $z = f(x, y)$, we define the **differentials** dx and dy to be independent variables; that is, they can be given any values. Then the

differential dz, also called the **total differential**, is defined by

10
$$dz = f_x(x, y)\,dx + f_y(x, y)\,dy = \frac{\partial z}{\partial x}\,dx + \frac{\partial z}{\partial y}\,dy$$

(Compare with Equation 9.) Sometimes the notation df is used in place of dz.

If we take $dx = \Delta x = x - a$ and $dy = \Delta y = y - b$ in Equation 10, then the differential of z is

$$dz = f_x(a, b)(x - a) + f_y(a, b)(y - b)$$

So, in the notation of differentials, the linear approximation 4 can be written as

$$f(x, y) \approx f(a, b) + dz$$

Figure 7 is the three-dimensional counterpart of Figure 6 and shows the geometric interpretation of the differential dz and the increment Δz: dz represents the change in height of the tangent plane, whereas Δz represents the change in height of the surface $z = f(x, y)$ when (x, y) changes from (a, b) to $(a + \Delta x, b + \Delta y)$.

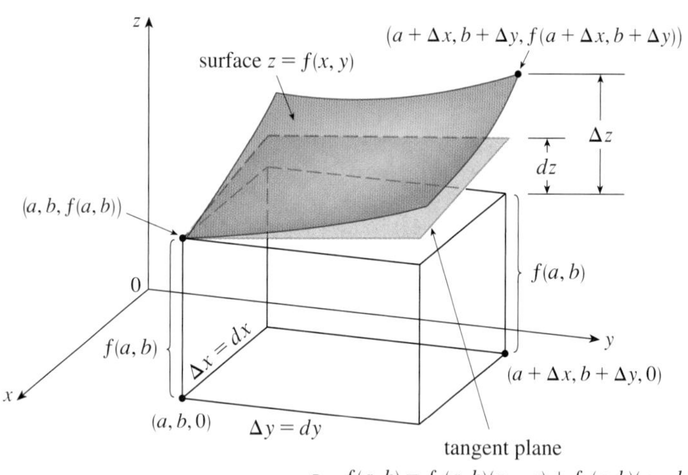

FIGURE 7

In Example 4, dz is close to Δz because the tangent plane is a good approximation to the surface $z = x^2 + 3xy - y^2$ near $(2, 3, 13)$. (See Figure 8.)

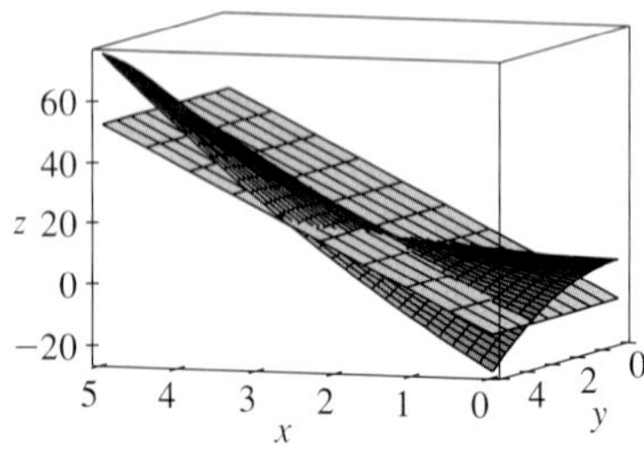

FIGURE 8

V EXAMPLE 4

(a) If $z = f(x, y) = x^2 + 3xy - y^2$, find the differential dz.

(b) If x changes from 2 to 2.05 and y changes from 3 to 2.96, compare the values of Δz and dz.

SOLUTION

(a) Definition 10 gives

$$dz = \frac{\partial z}{\partial x}\,dx + \frac{\partial z}{\partial y}\,dy = (2x + 3y)\,dx + (3x - 2y)\,dy$$

(b) Putting $x = 2$, $dx = \Delta x = 0.05$, $y = 3$, and $dy = \Delta y = -0.04$, we get

$$dz = [2(2) + 3(3)]0.05 + [3(2) - 2(3)](-0.04) = 0.65$$

The increment of z is

$$\begin{aligned}\Delta z &= f(2.05, 2.96) - f(2, 3) \\ &= [(2.05)^2 + 3(2.05)(2.96) - (2.96)^2] - [2^2 + 3(2)(3) - 3^2] \\ &= 0.6449\end{aligned}$$

Notice that $\Delta z \approx dz$ but dz is easier to compute. ■

EXAMPLE 5 The base radius and height of a right circular cone are measured as 10 cm and 25 cm, respectively, with a possible error in measurement of as much as 0.1 cm in each. Use differentials to estimate the maximum error in the calculated volume of the cone.

SOLUTION The volume V of a cone with base radius r and height h is $V = \pi r^2 h/3$. So the differential of V is

$$dV = \frac{\partial V}{\partial r}\,dr + \frac{\partial V}{\partial h}\,dh = \frac{2\pi rh}{3}\,dr + \frac{\pi r^2}{3}\,dh$$

Since each error is at most 0.1 cm, we have $|\Delta r| \leqslant 0.1$, $|\Delta h| \leqslant 0.1$. To estimate the largest error in the volume we take the largest error in the measurement of r and of h. Therefore we take $dr = 0.1$ and $dh = 0.1$ along with $r = 10$, $h = 25$. This gives

$$dV = \frac{500\pi}{3}(0.1) + \frac{100\pi}{3}(0.1) = 20\pi$$

Thus the maximum error in the calculated volume is about $20\pi\ \text{cm}^3 \approx 63\ \text{cm}^3$. ■

Functions of Three or More Variables

Linear approximations, differentiability, and differentials can be defined in a similar manner for functions of more than two variables. A differentiable function is defined by an expression similar to the one in Definition 7. For such functions the **linear approximation** is

$$f(x, y, z) \approx f(a, b, c) + f_x(a, b, c)(x - a) + f_y(a, b, c)(y - b) + f_z(a, b, c)(z - c)$$

and the linearization $L(x, y, z)$ is the right side of this expression.

If $w = f(x, y, z)$, then the **increment** of w is

$$\Delta w = f(x + \Delta x, y + \Delta y, z + \Delta z) - f(x, y, z)$$

The **differential** dw is defined in terms of the differentials dx, dy, and dz of the independent variables by

$$dw = \frac{\partial w}{\partial x}\,dx + \frac{\partial w}{\partial y}\,dy + \frac{\partial w}{\partial z}\,dz$$

EXAMPLE 6 The dimensions of a rectangular box are measured to be 75 cm, 60 cm, and 40 cm, and each measurement is correct to within 0.2 cm. Use differentials to estimate the largest possible error when the volume of the box is calculated from these measurements.

SOLUTION If the dimensions of the box are x, y, and z, its volume is $V = xyz$ and so

$$dV = \frac{\partial V}{\partial x}\,dx + \frac{\partial V}{\partial y}\,dy + \frac{\partial V}{\partial z}\,dz = yz\,dx + xz\,dy + xy\,dz$$

We are given that $|\Delta x| \leqslant 0.2$, $|\Delta y| \leqslant 0.2$, and $|\Delta z| \leqslant 0.2$. To estimate the largest error in the volume, we therefore use $dx = 0.2$, $dy = 0.2$, and $dz = 0.2$ together with $x = 75$, $y = 60$, and $z = 40$:

$$\Delta V \approx dV = (60)(40)(0.2) + (75)(40)(0.2) + (75)(60)(0.2) = 1980$$

Thus an error of only 0.2 cm in measuring each dimension could lead to an error of approximately 1980 cm^3 in the calculated volume! This may seem like a large error, but it's only about 1% of the volume of the box.

14.4 Exercises

1–6 Find an equation of the tangent plane to the given surface at the specified point.

1. $z = 3y^2 - 2x^2 + x, \quad (2, -1, -3)$

2. $z = 3(x - 1)^2 + 2(y + 3)^2 + 7, \quad (2, -2, 12)$

3. $z = \sqrt{xy}, \quad (1, 1, 1)$

4. $z = xe^{xy}, \quad (2, 0, 2)$

5. $z = x \sin(x + y), \quad (-1, 1, 0)$

6. $z = \ln(x - 2y), \quad (3, 1, 0)$

7–8 Graph the surface and the tangent plane at the given point. (Choose the domain and viewpoint so that you get a good view of both the surface and the tangent plane.) Then zoom in until the surface and the tangent plane become indistinguishable.

7. $z = x^2 + xy + 3y^2, \quad (1, 1, 5)$

8. $z = \arctan(xy^2), \quad (1, 1, \pi/4)$

CAS **9–10** Draw the graph of f and its tangent plane at the given point. (Use your computer algebra system both to compute the partial derivatives and to graph the surface and its tangent plane.) Then zoom in until the surface and the tangent plane become indistinguishable.

9. $f(x, y) = \dfrac{xy \sin(x - y)}{1 + x^2 + y^2}, \quad (1, 1, 0)$

10. $f(x, y) = e^{-xy/10}(\sqrt{x} + \sqrt{y} + \sqrt{xy}), \quad (1, 1, 3e^{-0.1})$

11–16 Explain why the function is differentiable at the given point. Then find the linearization $L(x, y)$ of the function at that point.

11. $f(x, y) = 1 + x \ln(xy - 5), \quad (2, 3)$

12. $f(x, y) = x^3y^4, \quad (1, 1)$

13. $f(x, y) = \dfrac{x}{x + y}, \quad (2, 1)$

14. $f(x, y) = \sqrt{x + e^{4y}}, \quad (3, 0)$

15. $f(x, y) = e^{-xy}\cos y, \quad (\pi, 0)$

16. $f(x, y) = y + \sin(x/y), \quad (0, 3)$

17–18 Verify the linear approximation at (0, 0).

17. $\dfrac{2x + 3}{4y + 1} \approx 3 + 2x - 12y$ **18.** $\sqrt{y + \cos^2 x} \approx 1 + \frac{1}{2}y$

19. Given that f is a differentiable function with $f(2, 5) = 6$, $f_x(2, 5) = 1$, and $f_y(2, 5) = -1$, use a linear approximation to estimate $f(2.2, 4.9)$.

20. Find the linear approximation of the function $f(x, y) = 1 - xy \cos \pi y$ at (1, 1) and use it to approximate $f(1.02, 0.97)$. Illustrate by graphing f and the tangent plane.

21. Find the linear approximation of the function $f(x, y, z) = \sqrt{x^2 + y^2 + z^2}$ at (3, 2, 6) and use it to approximate the number $\sqrt{(3.02)^2 + (1.97)^2 + (5.99)^2}$.

22. The wave heights h in the open sea depend on the speed v of the wind and the length of time t that the wind has been blowing at that speed. Values of the function $h = f(v, t)$ are recorded in feet in the following table. Use the table to find a linear approximation to the wave height function when v is near 40 knots and t is near 20 hours. Then estimate the wave heights when the wind has been blowing for 24 hours at 43 knots.

	Duration (hours)						
Wind speed (knots) v \ t	5	10	15	20	30	40	50
20	5	7	8	8	9	9	9
30	9	13	16	17	18	19	19
40	14	21	25	28	31	33	33
50	19	29	36	40	45	48	50
60	24	37	47	54	62	67	69

23. Use the table in Example 3 to find a linear approximation to the heat index function when the temperature is near 94°F and the relative humidity is near 80%. Then estimate the heat index when the temperature is 95°F and the relative humidity is 78%.

24. The wind-chill index W is the perceived temperature when the actual temperature is T and the wind speed is v, so we can write $W = f(T, v)$. The following table of values is an excerpt from Table 1 in Section 14.1. Use the table to find a linear approximation to the wind-chill index function when T is near $-15°C$ and v is near 50 km/h. Then estimate the wind-chill index when the temperature is $-17°C$ and the wind speed is 55 km/h.

Wind speed (km/h); Actual temperature (°C)

T \ v	20	30	40	50	60	70
−10	−18	−20	−21	−22	−23	−23
−15	−24	−26	−27	−29	−30	−30
−20	−30	−33	−34	−35	−36	−37
−25	−37	−39	−41	−42	−43	−44

25–30 Find the differential of the function.

25. $z = e^{-2x} \cos 2\pi t$

26. $u = \sqrt{x^2 + 3y^2}$

27. $m = p^5 q^3$

28. $T = \dfrac{v}{1 + uvw}$

29. $R = \alpha\beta^2 \cos \gamma$

30. $L = xze^{-y^2-z^2}$

31. If $z = 5x^2 + y^2$ and (x, y) changes from $(1, 2)$ to $(1.05, 2.1)$, compare the values of Δz and dz.

32. If $z = x^2 - xy + 3y^2$ and (x, y) changes from $(3, -1)$ to $(2.96, -0.95)$, compare the values of Δz and dz.

33. The length and width of a rectangle are measured as 30 cm and 24 cm, respectively, with an error in measurement of at most 0.1 cm in each. Use differentials to estimate the maximum error in the calculated area of the rectangle.

34. Use differentials to estimate the amount of metal in a closed cylindrical can that is 10 cm high and 4 cm in diameter if the metal in the top and bottom is 0.1 cm thick and the metal in the sides is 0.05 cm thick.

35. Use differentials to estimate the amount of tin in a closed tin can with diameter 8 cm and height 12 cm if the tin is 0.04 cm thick.

36. The wind-chill index is modeled by the function

$$W = 13.12 + 0.6215T - 11.37v^{0.16} + 0.3965Tv^{0.16}$$

where T is the temperature (in °C) and v is the wind speed (in km/h). The wind speed is measured as 26 km/h, with a possible error of ± 2 km/h, and the temperature is measured as $-11°C$, with a possible error of $\pm 1°C$. Use differentials to estimate the maximum error in the calculated value of W due to the measurement errors in T and v.

37. The tension T in the string of the yo-yo in the figure is

$$T = \frac{mgR}{2r^2 + R^2}$$

where m is the mass of the yo-yo and g is acceleration due to gravity. Use differentials to estimate the change in the tension if R is increased from 3 cm to 3.1 cm and r is increased from 0.7 cm to 0.8 cm. Does the tension increase or decrease?

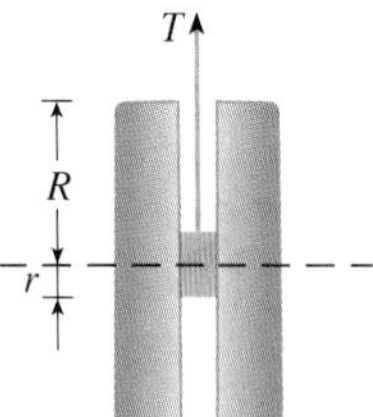

38. The pressure, volume, and temperature of a mole of an ideal gas are related by the equation $PV = 8.31T$, where P is measured in kilopascals, V in liters, and T in kelvins. Use differentials to find the approximate change in the pressure if the volume increases from 12 L to 12.3 L and the temperature decreases from 310 K to 305 K.

39. If R is the total resistance of three resistors, connected in parallel, with resistances R_1, R_2, R_3, then

$$\frac{1}{R} = \frac{1}{R_1} + \frac{1}{R_2} + \frac{1}{R_3}$$

If the resistances are measured in ohms as $R_1 = 25\ \Omega$, $R_2 = 40\ \Omega$, and $R_3 = 50\ \Omega$, with a possible error of 0.5% in each case, estimate the maximum error in the calculated value of R.

40. Four positive numbers, each less than 50, are rounded to the first decimal place and then multiplied together. Use differentials to estimate the maximum possible error in the computed product that might result from the rounding.

41. A model for the surface area of a human body is given by $S = 0.1091w^{0.425}h^{0.725}$, where w is the weight (in pounds), h is the height (in inches), and S is measured in square feet. If the errors in measurement of w and h are at most 2%, use differentials to estimate the maximum percentage error in the calculated surface area.

42. Suppose you need to know an equation of the tangent plane to a surface S at the point $P(2, 1, 3)$. You don't have an equation for S but you know that the curves

$$\mathbf{r}_1(t) = \langle 2 + 3t, 1 - t^2, 3 - 4t + t^2 \rangle$$

$$\mathbf{r}_2(u) = \langle 1 + u^2, 2u^3 - 1, 2u + 1 \rangle$$

both lie on S. Find an equation of the tangent plane at P.

43–44 Show that the function is differentiable by finding values of ε_1 and ε_2 that satisfy Definition 7.

43. $f(x, y) = x^2 + y^2$

44. $f(x, y) = xy - 5y^2$

45. Prove that if f is a function of two variables that is differentiable at (a, b), then f is continuous at (a, b).
Hint: Show that

$$\lim_{(\Delta x,\, \Delta y) \to (0,\, 0)} f(a + \Delta x, b + \Delta y) = f(a, b)$$

46. (a) The function

$$f(x, y) = \begin{cases} \dfrac{xy}{x^2 + y^2} & \text{if } (x, y) \neq (0, 0) \\ 0 & \text{if } (x, y) = (0, 0) \end{cases}$$

was graphed in Figure 4. Show that $f_x(0, 0)$ and $f_y(0, 0)$ both exist but f is not differentiable at $(0, 0)$. [*Hint:* Use the result of Exercise 45.]

(b) Explain why f_x and f_y are not continuous at $(0, 0)$.

14.5 The Chain Rule

Recall that the Chain Rule for functions of a single variable gives the rule for differentiating a composite function: If $y = f(x)$ and $x = g(t)$, where f and g are differentiable functions, then y is indirectly a differentiable function of t and

$$\boxed{1} \qquad \frac{dy}{dt} = \frac{dy}{dx}\frac{dx}{dt}$$

For functions of more than one variable, the Chain Rule has several versions, each of them giving a rule for differentiating a composite function. The first version (Theorem 2) deals with the case where $z = f(x, y)$ and each of the variables x and y is, in turn, a function of a variable t. This means that z is indirectly a function of t, $z = f(g(t), h(t))$, and the Chain Rule gives a formula for differentiating z as a function of t. We assume that f is differentiable (Definition 14.4.7). Recall that this is the case when f_x and f_y are continuous (Theorem 14.4.8).

$\boxed{2}$ **The Chain Rule (Case 1)** Suppose that $z = f(x, y)$ is a differentiable function of x and y, where $x = g(t)$ and $y = h(t)$ are both differentiable functions of t. Then z is a differentiable function of t and

$$\frac{dz}{dt} = \frac{\partial f}{\partial x}\frac{dx}{dt} + \frac{\partial f}{\partial y}\frac{dy}{dt}$$

PROOF A change of Δt in t produces changes of Δx in x and Δy in y. These, in turn, produce a change of Δz in z, and from Definition 14.4.7 we have

$$\Delta z = \frac{\partial f}{\partial x}\Delta x + \frac{\partial f}{\partial y}\Delta y + \varepsilon_1\,\Delta x + \varepsilon_2\,\Delta y$$

where $\varepsilon_1 \to 0$ and $\varepsilon_2 \to 0$ as $(\Delta x,\ \Delta y) \to (0, 0)$. [If the functions ε_1 and ε_2 are not defined at $(0, 0)$, we can define them to be 0 there.] Dividing both sides of this equation by Δt, we have

$$\frac{\Delta z}{\Delta t} = \frac{\partial f}{\partial x}\frac{\Delta x}{\Delta t} + \frac{\partial f}{\partial y}\frac{\Delta y}{\Delta t} + \varepsilon_1\frac{\Delta x}{\Delta t} + \varepsilon_2\frac{\Delta y}{\Delta t}$$

If we now let $\Delta t \to 0$, then $\Delta x = g(t + \Delta t) - g(t) \to 0$ because g is differentiable and therefore continuous. Similarly, $\Delta y \to 0$. This, in turn, means that $\varepsilon_1 \to 0$ and $\varepsilon_2 \to 0$, so

$$\begin{aligned}\frac{dz}{dt} &= \lim_{\Delta t \to 0} \frac{\Delta z}{\Delta t} \\ &= \frac{\partial f}{\partial x} \lim_{\Delta t \to 0} \frac{\Delta x}{\Delta t} + \frac{\partial f}{\partial y} \lim_{\Delta t \to 0} \frac{\Delta y}{\Delta t} + \left(\lim_{\Delta t \to 0} \varepsilon_1\right) \lim_{\Delta t \to 0} \frac{\Delta x}{\Delta t} + \left(\lim_{\Delta t \to 0} \varepsilon_2\right) \lim_{\Delta t \to 0} \frac{\Delta y}{\Delta t} \\ &= \frac{\partial f}{\partial x}\frac{dx}{dt} + \frac{\partial f}{\partial y}\frac{dy}{dt} + 0 \cdot \frac{dx}{dt} + 0 \cdot \frac{dy}{dt} \\ &= \frac{\partial f}{\partial x}\frac{dx}{dt} + \frac{\partial f}{\partial y}\frac{dy}{dt}\end{aligned}$$

Since we often write $\partial z/\partial x$ in place of $\partial f/\partial x$, we can rewrite the Chain Rule in the form

Notice the similarity to the definition of the differential:

$$dz = \frac{\partial z}{\partial x}dx + \frac{\partial z}{\partial y}dy$$

$$\boxed{\frac{dz}{dt} = \frac{\partial z}{\partial x}\frac{dx}{dt} + \frac{\partial z}{\partial y}\frac{dy}{dt}}$$

EXAMPLE 1 If $z = x^2y + 3xy^4$, where $x = \sin 2t$ and $y = \cos t$, find dz/dt when $t = 0$.

SOLUTION The Chain Rule gives

$$\begin{aligned}\frac{dz}{dt} &= \frac{\partial z}{\partial x}\frac{dx}{dt} + \frac{\partial z}{\partial y}\frac{dy}{dt} \\ &= (2xy + 3y^4)(2\cos 2t) + (x^2 + 12xy^3)(-\sin t)\end{aligned}$$

It's not necessary to substitute the expressions for x and y in terms of t. We simply observe that when $t = 0$, we have $x = \sin 0 = 0$ and $y = \cos 0 = 1$. Therefore

$$\left.\frac{dz}{dt}\right|_{t=0} = (0 + 3)(2\cos 0) + (0 + 0)(-\sin 0) = 6$$

The derivative in Example 1 can be interpreted as the rate of change of z with respect to t as the point (x, y) moves along the curve C with parametric equations $x = \sin 2t$, $y = \cos t$. (See Figure 1.) In particular, when $t = 0$, the point (x, y) is $(0, 1)$ and $dz/dt = 6$ is the rate of increase as we move along the curve C through $(0, 1)$. If, for instance, $z = T(x, y) = x^2y + 3xy^4$ represents the temperature at the point (x, y), then the composite function $z = T(\sin 2t, \cos t)$ represents the temperature at points on C and the derivative dz/dt represents the rate at which the temperature changes along C.

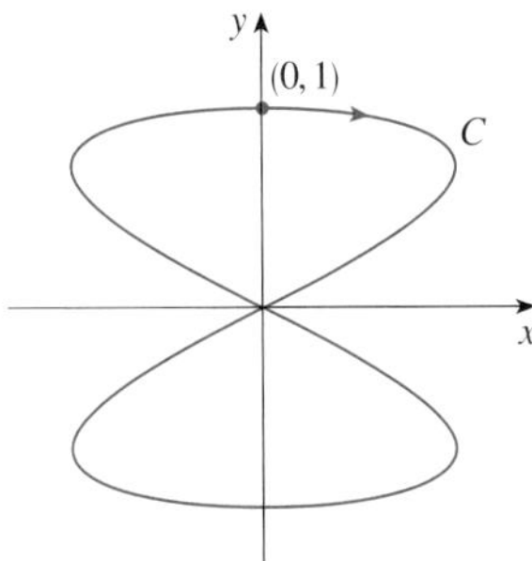

FIGURE 1
The curve $x = \sin 2t$, $y = \cos t$

V EXAMPLE 2 The pressure P (in kilopascals), volume V (in liters), and temperature T (in kelvins) of a mole of an ideal gas are related by the equation $PV = 8.31T$. Find the rate at which the pressure is changing when the temperature is 300 K and increasing at a rate of 0.1 K/s and the volume is 100 L and increasing at a rate of 0.2 L/s.

SOLUTION If t represents the time elapsed in seconds, then at the given instant we have $T = 300$, $dT/dt = 0.1$, $V = 100$, $dV/dt = 0.2$. Since

$$P = 8.31\frac{T}{V}$$

the Chain Rule gives

$$\frac{dP}{dt} = \frac{\partial P}{\partial T}\frac{dT}{dt} + \frac{\partial P}{\partial V}\frac{dV}{dt} = \frac{8.31}{V}\frac{dT}{dt} - \frac{8.31T}{V^2}\frac{dV}{dt}$$

$$= \frac{8.31}{100}(0.1) - \frac{8.31(300)}{100^2}(0.2) = -0.04155$$

The pressure is decreasing at a rate of about 0.042 kPa/s. ■

We now consider the situation where $z = f(x, y)$ but each of x and y is a function of two variables s and t: $x = g(s, t)$, $y = h(s, t)$. Then z is indirectly a function of s and t and we wish to find $\partial z/\partial s$ and $\partial z/\partial t$. Recall that in computing $\partial z/\partial t$ we hold s fixed and compute the ordinary derivative of z with respect to t. Therefore we can apply Theorem 2 to obtain

$$\frac{\partial z}{\partial t} = \frac{\partial z}{\partial x}\frac{\partial x}{\partial t} + \frac{\partial z}{\partial y}\frac{\partial y}{\partial t}$$

A similar argument holds for $\partial z/\partial s$ and so we have proved the following version of the Chain Rule.

3 **The Chain Rule (Case 2)** Suppose that $z = f(x, y)$ is a differentiable function of x and y, where $x = g(s, t)$ and $y = h(s, t)$ are differentiable functions of s and t. Then

$$\frac{\partial z}{\partial s} = \frac{\partial z}{\partial x}\frac{\partial x}{\partial s} + \frac{\partial z}{\partial y}\frac{\partial y}{\partial s} \qquad \frac{\partial z}{\partial t} = \frac{\partial z}{\partial x}\frac{\partial x}{\partial t} + \frac{\partial z}{\partial y}\frac{\partial y}{\partial t}$$

EXAMPLE 3 If $z = e^x \sin y$, where $x = st^2$ and $y = s^2t$, find $\partial z/\partial s$ and $\partial z/\partial t$.

SOLUTION Applying Case 2 of the Chain Rule, we get

$$\frac{\partial z}{\partial s} = \frac{\partial z}{\partial x}\frac{\partial x}{\partial s} + \frac{\partial z}{\partial y}\frac{\partial y}{\partial s} = (e^x \sin y)(t^2) + (e^x \cos y)(2st)$$

$$= t^2 e^{st^2} \sin(s^2t) + 2ste^{st^2} \cos(s^2t)$$

$$\frac{\partial z}{\partial t} = \frac{\partial z}{\partial x}\frac{\partial x}{\partial t} + \frac{\partial z}{\partial y}\frac{\partial y}{\partial t} = (e^x \sin y)(2st) + (e^x \cos y)(s^2)$$

$$= 2ste^{st^2} \sin(s^2t) + s^2 e^{st^2} \cos(s^2t)$$ ■

Case 2 of the Chain Rule contains three types of variables: s and t are **independent** variables, x and y are called **intermediate** variables, and z is the **dependent** variable. Notice that Theorem 3 has one term for each intermediate variable and each of these terms resembles the one-dimensional Chain Rule in Equation 1.

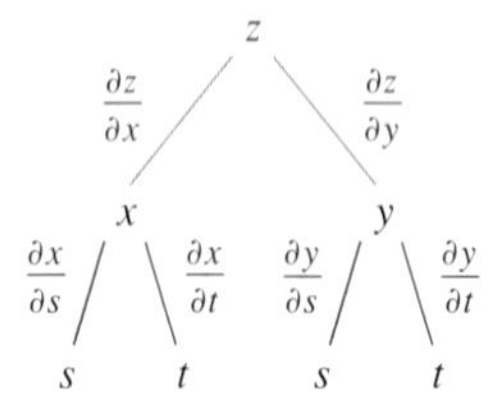

FIGURE 2

To remember the Chain Rule, it's helpful to draw the **tree diagram** in Figure 2. We draw branches from the dependent variable z to the intermediate variables x and y to indicate that z is a function of x and y. Then we draw branches from x and y to the independent variables s and t. On each branch we write the corresponding partial derivative. To find $\partial z/\partial s$, we find the product of the partial derivatives along each path from z to s and then add these products:

$$\frac{\partial z}{\partial s} = \frac{\partial z}{\partial x}\frac{\partial x}{\partial s} + \frac{\partial z}{\partial y}\frac{\partial y}{\partial s}$$

Similarly, we find $\partial z/\partial t$ by using the paths from z to t.

Now we consider the general situation in which a dependent variable u is a function of n intermediate variables $x_1, \ldots, x_n$, each of which is, in turn, a function of m independent variables $t_1, \ldots, t_m$. Notice that there are n terms, one for each intermediate variable. The proof is similar to that of Case 1.

> **4 The Chain Rule (General Version)** Suppose that u is a differentiable function of the n variables $x_1, x_2, \ldots, x_n$ and each x_j is a differentiable function of the m variables $t_1, t_2, \ldots, t_m$. Then u is a function of $t_1, t_2, \ldots, t_m$ and
>
> $$\frac{\partial u}{\partial t_i} = \frac{\partial u}{\partial x_1}\frac{\partial x_1}{\partial t_i} + \frac{\partial u}{\partial x_2}\frac{\partial x_2}{\partial t_i} + \cdots + \frac{\partial u}{\partial x_n}\frac{\partial x_n}{\partial t_i}$$
>
> for each $i = 1, 2, \ldots, m$.

V EXAMPLE 4 Write out the Chain Rule for the case where $w = f(x, y, z, t)$ and $x = x(u, v)$, $y = y(u, v)$, $z = z(u, v)$, and $t = t(u, v)$.

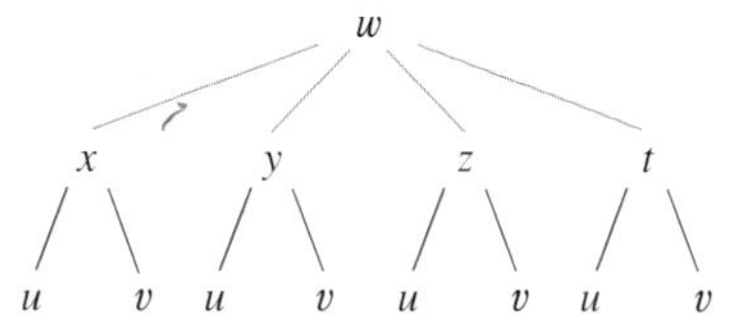

FIGURE 3

SOLUTION We apply Theorem 4 with $n = 4$ and $m = 2$. Figure 3 shows the tree diagram. Although we haven't written the derivatives on the branches, it's understood that if a branch leads from y to u, then the partial derivative for that branch is $\partial y/\partial u$. With the aid of the tree diagram, we can now write the required expressions:

$$\frac{\partial w}{\partial u} = \frac{\partial w}{\partial x}\frac{\partial x}{\partial u} + \frac{\partial w}{\partial y}\frac{\partial y}{\partial u} + \frac{\partial w}{\partial z}\frac{\partial z}{\partial u} + \frac{\partial w}{\partial t}\frac{\partial t}{\partial u}$$

$$\frac{\partial w}{\partial v} = \frac{\partial w}{\partial x}\frac{\partial x}{\partial v} + \frac{\partial w}{\partial y}\frac{\partial y}{\partial v} + \frac{\partial w}{\partial z}\frac{\partial z}{\partial v} + \frac{\partial w}{\partial t}\frac{\partial t}{\partial v}$$

V EXAMPLE 5 If $u = x^4y + y^2z^3$, where $x = rse^t$, $y = rs^2e^{-t}$, and $z = r^2s\sin t$, find the value of $\partial u/\partial s$ when $r = 2$, $s = 1$, $t = 0$.

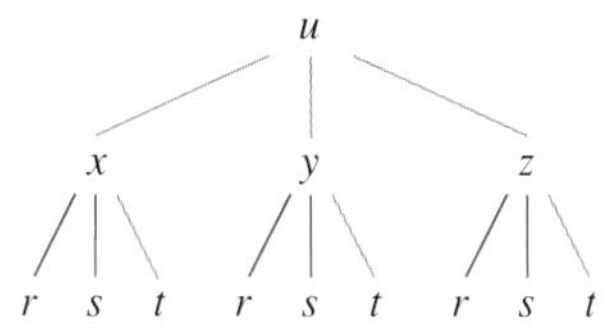

FIGURE 4

SOLUTION With the help of the tree diagram in Figure 4, we have

$$\frac{\partial u}{\partial s} = \frac{\partial u}{\partial x}\frac{\partial x}{\partial s} + \frac{\partial u}{\partial y}\frac{\partial y}{\partial s} + \frac{\partial u}{\partial z}\frac{\partial z}{\partial s}$$

$$= (4x^3y)(re^t) + (x^4 + 2yz^3)(2rse^{-t}) + (3y^2z^2)(r^2\sin t)$$

When $r = 2$, $s = 1$, and $t = 0$, we have $x = 2$, $y = 2$, and $z = 0$, so

$$\frac{\partial u}{\partial s} = (64)(2) + (16)(4) + (0)(0) = 192$$

EXAMPLE 6 If $g(s, t) = f(s^2 - t^2, t^2 - s^2)$ and f is differentiable, show that g satisfies the equation

$$t\frac{\partial g}{\partial s} + s\frac{\partial g}{\partial t} = 0$$

SOLUTION Let $x = s^2 - t^2$ and $y = t^2 - s^2$. Then $g(s, t) = f(x, y)$ and the Chain Rule gives

$$\frac{\partial g}{\partial s} = \frac{\partial f}{\partial x}\frac{\partial x}{\partial s} + \frac{\partial f}{\partial y}\frac{\partial y}{\partial s} = \frac{\partial f}{\partial x}(2s) + \frac{\partial f}{\partial y}(-2s)$$

$$\frac{\partial g}{\partial t} = \frac{\partial f}{\partial x}\frac{\partial x}{\partial t} + \frac{\partial f}{\partial y}\frac{\partial y}{\partial t} = \frac{\partial f}{\partial x}(-2t) + \frac{\partial f}{\partial y}(2t)$$

Therefore

$$t\frac{\partial g}{\partial s} + s\frac{\partial g}{\partial t} = \left(2st\frac{\partial f}{\partial x} - 2st\frac{\partial f}{\partial y}\right) + \left(-2st\frac{\partial f}{\partial x} + 2st\frac{\partial f}{\partial y}\right) = 0$$

EXAMPLE 7 If $z = f(x, y)$ has continuous second-order partial derivatives and $x = r^2 + s^2$ and $y = 2rs$, find (a) $\partial z/\partial r$ and (b) $\partial^2 z/\partial r^2$.

SOLUTION

(a) The Chain Rule gives

$$\frac{\partial z}{\partial r} = \frac{\partial z}{\partial x}\frac{\partial x}{\partial r} + \frac{\partial z}{\partial y}\frac{\partial y}{\partial r} = \frac{\partial z}{\partial x}(2r) + \frac{\partial z}{\partial y}(2s)$$

(b) Applying the Product Rule to the expression in part (a), we get

5
$$\begin{aligned}\frac{\partial^2 z}{\partial r^2} &= \frac{\partial}{\partial r}\left(2r\frac{\partial z}{\partial x} + 2s\frac{\partial z}{\partial y}\right)\\ &= 2\frac{\partial z}{\partial x} + 2r\frac{\partial}{\partial r}\left(\frac{\partial z}{\partial x}\right) + 2s\frac{\partial}{\partial r}\left(\frac{\partial z}{\partial y}\right)\end{aligned}$$

But, using the Chain Rule again (see Figure 5), we have

$$\frac{\partial}{\partial r}\left(\frac{\partial z}{\partial x}\right) = \frac{\partial}{\partial x}\left(\frac{\partial z}{\partial x}\right)\frac{\partial x}{\partial r} + \frac{\partial}{\partial y}\left(\frac{\partial z}{\partial x}\right)\frac{\partial y}{\partial r} = \frac{\partial^2 z}{\partial x^2}(2r) + \frac{\partial^2 z}{\partial y\,\partial x}(2s)$$

$$\frac{\partial}{\partial r}\left(\frac{\partial z}{\partial y}\right) = \frac{\partial}{\partial x}\left(\frac{\partial z}{\partial y}\right)\frac{\partial x}{\partial r} + \frac{\partial}{\partial y}\left(\frac{\partial z}{\partial y}\right)\frac{\partial y}{\partial r} = \frac{\partial^2 z}{\partial x\,\partial y}(2r) + \frac{\partial^2 z}{\partial y^2}(2s)$$

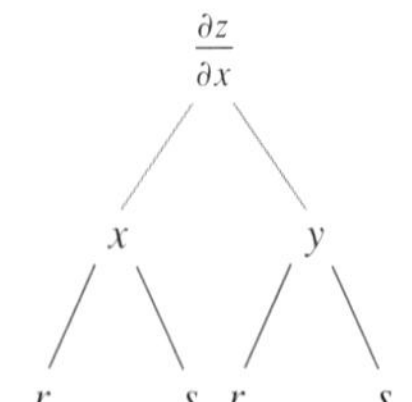

FIGURE 5

Putting these expressions into Equation 5 and using the equality of the mixed second-order derivatives, we obtain

$$\begin{aligned}\frac{\partial^2 z}{\partial r^2} &= 2\frac{\partial z}{\partial x} + 2r\left(2r\frac{\partial^2 z}{\partial x^2} + 2s\frac{\partial^2 z}{\partial y\,\partial x}\right) + 2s\left(2r\frac{\partial^2 z}{\partial x\,\partial y} + 2s\frac{\partial^2 z}{\partial y^2}\right)\\ &= 2\frac{\partial z}{\partial x} + 4r^2\frac{\partial^2 z}{\partial x^2} + 8rs\frac{\partial^2 z}{\partial x\,\partial y} + 4s^2\frac{\partial^2 z}{\partial y^2}\end{aligned}$$

Implicit Differentiation

The Chain Rule can be used to give a more complete description of the process of implicit differentiation that was introduced in Sections 3.5 and 14.3. We suppose that an equation of the form $F(x, y) = 0$ defines y implicitly as a differentiable function of x, that is,

$y = f(x)$, where $F(x, f(x)) = 0$ for all x in the domain of f. If F is differentiable, we can apply Case 1 of the Chain Rule to differentiate both sides of the equation $F(x, y) = 0$ with respect to x. Since both x and y are functions of x, we obtain

$$\frac{\partial F}{\partial x}\frac{dx}{dx} + \frac{\partial F}{\partial y}\frac{dy}{dx} = 0$$

But $dx/dx = 1$, so if $\partial F/\partial y \neq 0$ we solve for dy/dx and obtain

6 $$\frac{dy}{dx} = -\frac{\dfrac{\partial F}{\partial x}}{\dfrac{\partial F}{\partial y}} = -\frac{F_x}{F_y}$$

To derive this equation we assumed that $F(x, y) = 0$ defines y implicitly as a function of x. The **Implicit Function Theorem**, proved in advanced calculus, gives conditions under which this assumption is valid: It states that if F is defined on a disk containing (a, b), where $F(a, b) = 0$, $F_y(a, b) \neq 0$, and F_x and F_y are continuous on the disk, then the equation $F(x, y) = 0$ defines y as a function of x near the point (a, b) and the derivative of this function is given by Equation 6.

EXAMPLE 8 Find y' if $x^3 + y^3 = 6xy$.

SOLUTION The given equation can be written as

$$F(x, y) = x^3 + y^3 - 6xy = 0$$

so Equation 6 gives

The solution to Example 8 should be compared to the one in Example 2 in Section 3.5.

$$\frac{dy}{dx} = -\frac{F_x}{F_y} = -\frac{3x^2 - 6y}{3y^2 - 6x} = -\frac{x^2 - 2y}{y^2 - 2x}$$

Now we suppose that z is given implicitly as a function $z = f(x, y)$ by an equation of the form $F(x, y, z) = 0$. This means that $F(x, y, f(x, y)) = 0$ for all (x, y) in the domain of f. If F and f are differentiable, then we can use the Chain Rule to differentiate the equation $F(x, y, z) = 0$ as follows:

$$\frac{\partial F}{\partial x}\frac{\partial x}{\partial x} + \frac{\partial F}{\partial y}\frac{\partial y}{\partial x} + \frac{\partial F}{\partial z}\frac{\partial z}{\partial x} = 0$$

But $$\frac{\partial}{\partial x}(x) = 1 \qquad \text{and} \qquad \frac{\partial}{\partial x}(y) = 0$$

so this equation becomes

$$\frac{\partial F}{\partial x} + \frac{\partial F}{\partial z}\frac{\partial z}{\partial x} = 0$$

If $\partial F/\partial z \neq 0$, we solve for $\partial z/\partial x$ and obtain the first formula in Equations 7 on page 930. The formula for $\partial z/\partial y$ is obtained in a similar manner.

$$\boxed{7} \qquad \frac{\partial z}{\partial x} = -\frac{\dfrac{\partial F}{\partial x}}{\dfrac{\partial F}{\partial z}} \qquad \frac{\partial z}{\partial y} = -\frac{\dfrac{\partial F}{\partial y}}{\dfrac{\partial F}{\partial z}}$$

Again, a version of the **Implicit Function Theorem** stipulates conditions under which our assumption is valid: If F is defined within a sphere containing (a, b, c), where $F(a, b, c) = 0$, $F_z(a, b, c) \neq 0$, and F_x, F_y, and F_z are continuous inside the sphere, then the equation $F(x, y, z) = 0$ defines z as a function of x and y near the point (a, b, c) and this function is differentiable, with partial derivatives given by 7.

EXAMPLE 9 Find $\dfrac{\partial z}{\partial x}$ and $\dfrac{\partial z}{\partial y}$ if $x^3 + y^3 + z^3 + 6xyz = 1$.

The solution to Example 9 should be compared to the one in Example 4 in Section 14.3.

SOLUTION Let $F(x, y, z) = x^3 + y^3 + z^3 + 6xyz - 1$. Then, from Equations 7, we have

$$\frac{\partial z}{\partial x} = -\frac{F_x}{F_z} = -\frac{3x^2 + 6yz}{3z^2 + 6xy} = -\frac{x^2 + 2yz}{z^2 + 2xy}$$

$$\frac{\partial z}{\partial y} = -\frac{F_y}{F_z} = -\frac{3y^2 + 6xz}{3z^2 + 6xy} = -\frac{y^2 + 2xz}{z^2 + 2xy}$$

14.5 Exercises

1–6 Use the Chain Rule to find dz/dt or dw/dt.

1. $z = x^2 + y^2 + xy$, $x = \sin t$, $y = e^t$

2. $z = \cos(x + 4y)$, $x = 5t^4$, $y = 1/t$

3. $z = \sqrt{1 + x^2 + y^2}$, $x = \ln t$, $y = \cos t$

4. $z = \tan^{-1}(y/x)$, $x = e^t$, $y = 1 - e^{-t}$

5. $w = xe^{y/z}$, $x = t^2$, $y = 1 - t$, $z = 1 + 2t$

6. $w = \ln\sqrt{x^2 + y^2 + z^2}$, $x = \sin t$, $y = \cos t$, $z = \tan t$

7–12 Use the Chain Rule to find $\partial z/\partial s$ and $\partial z/\partial t$.

7. $z = x^2y^3$, $x = s\cos t$, $y = s\sin t$

8. $z = \arcsin(x - y)$, $x = s^2 + t^2$, $y = 1 - 2st$

9. $z = \sin\theta\cos\phi$, $\theta = st^2$, $\phi = s^2t$

10. $z = e^{x+2y}$, $x = s/t$, $y = t/s$

11. $z = e^r\cos\theta$, $r = st$, $\theta = \sqrt{s^2 + t^2}$

12. $z = \tan(u/v)$, $u = 2s + 3t$, $v = 3s - 2t$

13. If $z = f(x, y)$, where f is differentiable, and

$$\begin{array}{ll} x = g(t) & y = h(t) \\ g(3) = 2 & h(3) = 7 \\ g'(3) = 5 & h'(3) = -4 \\ f_x(2, 7) = 6 & f_y(2, 7) = -8 \end{array}$$

find dz/dt when $t = 3$.

14. Let $W(s, t) = F(u(s, t), v(s, t))$, where F, u, and v are differentiable, and

$$\begin{array}{ll} u(1, 0) = 2 & v(1, 0) = 3 \\ u_s(1, 0) = -2 & v_s(1, 0) = 5 \\ u_t(1, 0) = 6 & v_t(1, 0) = 4 \\ F_u(2, 3) = -1 & F_v(2, 3) = 10 \end{array}$$

Find $W_s(1, 0)$ and $W_t(1, 0)$.

15. Suppose f is a differentiable function of x and y, and $g(u, v) = f(e^u + \sin v, e^u + \cos v)$. Use the table of values to calculate $g_u(0, 0)$ and $g_v(0, 0)$.

	f	g	f_x	f_y
(0, 0)	3	6	4	8
(1, 2)	6	3	2	5

16. Suppose f is a differentiable function of x and y, and $g(r, s) = f(2r - s, s^2 - 4r)$. Use the table of values in Exercise 15 to calculate $g_r(1, 2)$ and $g_s(1, 2)$.

1. Homework Hints available at stewartcalculus.com

17–20 Use a tree diagram to write out the Chain Rule for the given case. Assume all functions are differentiable.

17. $u = f(x, y)$, where $x = x(r, s, t)$, $y = y(r, s, t)$

18. $R = f(x, y, z, t)$, where $x = x(u, v, w)$, $y = y(u, v, w)$, $z = z(u, v, w)$, $t = t(u, v, w)$

19. $w = f(r, s, t)$, where $r = r(x, y)$, $s = s(x, y)$, $t = t(x, y)$

20. $t = f(u, v, w)$, where $u = u(p, q, r, s)$, $v = v(p, q, r, s)$, $w = w(p, q, r, s)$

21–26 Use the Chain Rule to find the indicated partial derivatives.

21. $z = x^4 + x^2y$, $x = s + 2t - u$, $y = stu^2$;

$\dfrac{\partial z}{\partial s}, \dfrac{\partial z}{\partial t}, \dfrac{\partial z}{\partial u}$ when $s = 4, t = 2, u = 1$

22. $T = \dfrac{v}{2u + v}$, $u = pq\sqrt{r}$, $v = p\sqrt{q}r$;

$\dfrac{\partial T}{\partial p}, \dfrac{\partial T}{\partial q}, \dfrac{\partial T}{\partial r}$ when $p = 2, q = 1, r = 4$

23. $w = xy + yz + zx$, $x = r\cos\theta$, $y = r\sin\theta$, $z = r\theta$;

$\dfrac{\partial w}{\partial r}, \dfrac{\partial w}{\partial \theta}$ when $r = 2, \theta = \pi/2$

24. $P = \sqrt{u^2 + v^2 + w^2}$, $u = xe^y$, $v = ye^x$, $w = e^{xy}$;

$\dfrac{\partial P}{\partial x}, \dfrac{\partial P}{\partial y}$ when $x = 0, y = 2$

25. $N = \dfrac{p + q}{p + r}$, $p = u + vw$, $q = v + uw$, $r = w + uv$;

$\dfrac{\partial N}{\partial u}, \dfrac{\partial N}{\partial v}, \dfrac{\partial N}{\partial w}$ when $u = 2, v = 3, w = 4$

26. $u = xe^{ty}$, $x = \alpha^2\beta$, $y = \beta^2\gamma$, $t = \gamma^2\alpha$;

$\dfrac{\partial u}{\partial \alpha}, \dfrac{\partial u}{\partial \beta}, \dfrac{\partial u}{\partial \gamma}$ when $\alpha = -1, \beta = 2, \gamma = 1$

27–30 Use Equation 6 to find dy/dx.

27. $y\cos x = x^2 + y^2$

28. $\cos(xy) = 1 + \sin y$

29. $\tan^{-1}(x^2y) = x + xy^2$

30. $e^y \sin x = x + xy$

31–34 Use Equations 7 to find $\partial z/\partial x$ and $\partial z/\partial y$.

31. $x^2 + 2y^2 + 3z^2 = 1$

32. $x^2 - y^2 + z^2 - 2z = 4$

33. $e^z = xyz$

34. $yz + x\ln y = z^2$

---

35. The temperature at a point (x, y) is $T(x, y)$, measured in degrees Celsius. A bug crawls so that its position after t seconds is given by $x = \sqrt{1 + t}$, $y = 2 + \frac{1}{3}t$, where x and y are measured in centimeters. The temperature function satisfies $T_x(2, 3) = 4$ and $T_y(2, 3) = 3$. How fast is the temperature rising on the bug's path after 3 seconds?

36. Wheat production W in a given year depends on the average temperature T and the annual rainfall R. Scientists estimate that the average temperature is rising at a rate of 0.15°C/year and rainfall is decreasing at a rate of 0.1 cm/year. They also estimate that, at current production levels, $\partial W/\partial T = -2$ and $\partial W/\partial R = 8$.

(a) What is the significance of the signs of these partial derivatives?

(b) Estimate the current rate of change of wheat production, dW/dt.

37. The speed of sound traveling through ocean water with salinity 35 parts per thousand has been modeled by the equation

$$C = 1449.2 + 4.6T - 0.055T^2 + 0.00029T^3 + 0.016D$$

where C is the speed of sound (in meters per second), T is the temperature (in degrees Celsius), and D is the depth below the ocean surface (in meters). A scuba diver began a leisurely dive into the ocean water; the diver's depth and the surrounding water temperature over time are recorded in the following graphs. Estimate the rate of change (with respect to time) of the speed of sound through the ocean water experienced by the diver 20 minutes into the dive. What are the units?

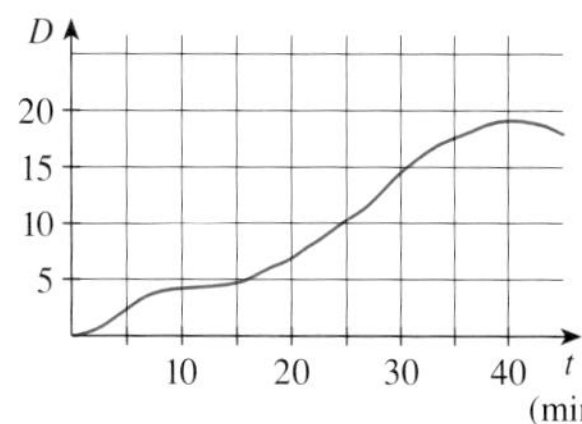

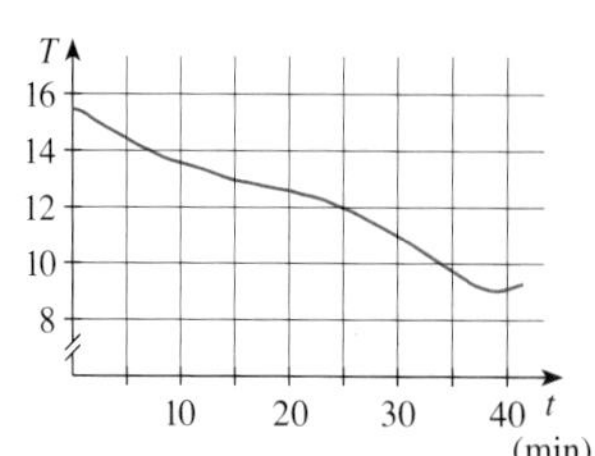

38. The radius of a right circular cone is increasing at a rate of 1.8 in/s while its height is decreasing at a rate of 2.5 in/s. At what rate is the volume of the cone changing when the radius is 120 in. and the height is 140 in.?

39. The length ℓ, width w, and height h of a box change with time. At a certain instant the dimensions are $\ell = 1$ m and $w = h = 2$ m, and ℓ and w are increasing at a rate of 2 m/s while h is decreasing at a rate of 3 m/s. At that instant find the rates at which the following quantities are changing.

(a) The volume

(b) The surface area

(c) The length of a diagonal

40. The voltage V in a simple electrical circuit is slowly decreasing as the battery wears out. The resistance R is slowly increasing as the resistor heats up. Use Ohm's Law, $V = IR$, to find how the current I is changing at the moment when $R = 400\ \Omega$, $I = 0.08$ A, $dV/dt = -0.01$ V/s, and $dR/dt = 0.03\ \Omega$/s.

41. The pressure of 1 mole of an ideal gas is increasing at a rate of 0.05 kPa/s and the temperature is increasing at a rate of 0.15 K/s. Use the equation in Example 2 to find the rate of change of the volume when the pressure is 20 kPa and the temperature is 320 K.

42. A manufacturer has modeled its yearly production function P (the value of its entire production in millions of dollars) as a Cobb-Douglas function

$$P(L, K) = 1.47L^{0.65}K^{0.35}$$

where L is the number of labor hours (in thousands) and K is the invested capital (in millions of dollars). Suppose that when $L = 30$ and $K = 8$, the labor force is decreasing at a rate of 2000 labor hours per year and capital is increasing at a rate of \$500,000 per year. Find the rate of change of production.

43. One side of a triangle is increasing at a rate of 3 cm/s and a second side is decreasing at a rate of 2 cm/s. If the area of the triangle remains constant, at what rate does the angle between the sides change when the first side is 20 cm long, the second side is 30 cm, and the angle is $\pi/6$?

44. If a sound with frequency f_s is produced by a source traveling along a line with speed v_s and an observer is traveling with speed v_o along the same line from the opposite direction toward the source, then the frequency of the sound heard by the observer is

$$f_o = \left(\frac{c + v_o}{c - v_s}\right) f_s$$

where c is the speed of sound, about 332 m/s. (This is the **Doppler effect**.) Suppose that, at a particular moment, you are in a train traveling at 34 m/s and accelerating at 1.2 m/s^2. A train is approaching you from the opposite direction on the other track at 40 m/s, accelerating at 1.4 m/s^2, and sounds its whistle, which has a frequency of 460 Hz. At that instant, what is the perceived frequency that you hear and how fast is it changing?

45–48 Assume that all the given functions are differentiable.

45. If $z = f(x, y)$, where $x = r\cos\theta$ and $y = r\sin\theta$, (a) find $\partial z/\partial r$ and $\partial z/\partial\theta$ and (b) show that

$$\left(\frac{\partial z}{\partial x}\right)^2 + \left(\frac{\partial z}{\partial y}\right)^2 = \left(\frac{\partial z}{\partial r}\right)^2 + \frac{1}{r^2}\left(\frac{\partial z}{\partial \theta}\right)^2$$

46. If $u = f(x, y)$, where $x = e^s\cos t$ and $y = e^s\sin t$, show that

$$\left(\frac{\partial u}{\partial x}\right)^2 + \left(\frac{\partial u}{\partial y}\right)^2 = e^{-2s}\left[\left(\frac{\partial u}{\partial s}\right)^2 + \left(\frac{\partial u}{\partial t}\right)^2\right]$$

47. If $z = f(x - y)$, show that $\dfrac{\partial z}{\partial x} + \dfrac{\partial z}{\partial y} = 0$.

48. If $z = f(x, y)$, where $x = s + t$ and $y = s - t$, show that

$$\left(\frac{\partial z}{\partial x}\right)^2 - \left(\frac{\partial z}{\partial y}\right)^2 = \frac{\partial z}{\partial s}\frac{\partial z}{\partial t}$$

49–54 Assume that all the given functions have continuous second-order partial derivatives.

49. Show that any function of the form

$$z = f(x + at) + g(x - at)$$

is a solution of the wave equation

$$\frac{\partial^2 z}{\partial t^2} = a^2\frac{\partial^2 z}{\partial x^2}$$

[*Hint:* Let $u = x + at$, $v = x - at$.]

50. If $u = f(x, y)$, where $x = e^s\cos t$ and $y = e^s\sin t$, show that

$$\frac{\partial^2 u}{\partial x^2} + \frac{\partial^2 u}{\partial y^2} = e^{-2s}\left[\frac{\partial^2 u}{\partial s^2} + \frac{\partial^2 u}{\partial t^2}\right]$$

51. If $z = f(x, y)$, where $x = r^2 + s^2$ and $y = 2rs$, find $\partial^2 z/\partial r\,\partial s$. (Compare with Example 7.)

52. If $z = f(x, y)$, where $x = r\cos\theta$ and $y = r\sin\theta$, find (a) $\partial z/\partial r$, (b) $\partial z/\partial\theta$, and (c) $\partial^2 z/\partial r\,\partial\theta$.

53. If $z = f(x, y)$, where $x = r\cos\theta$ and $y = r\sin\theta$, show that

$$\frac{\partial^2 z}{\partial x^2} + \frac{\partial^2 z}{\partial y^2} = \frac{\partial^2 z}{\partial r^2} + \frac{1}{r^2}\frac{\partial^2 z}{\partial \theta^2} + \frac{1}{r}\frac{\partial z}{\partial r}$$

54. Suppose $z = f(x, y)$, where $x = g(s, t)$ and $y = h(s, t)$.
(a) Show that

$$\frac{\partial^2 z}{\partial t^2} = \frac{\partial^2 z}{\partial x^2}\left(\frac{\partial x}{\partial t}\right)^2 + 2\frac{\partial^2 z}{\partial x\,\partial y}\frac{\partial x}{\partial t}\frac{\partial y}{\partial t} + \frac{\partial^2 z}{\partial y^2}\left(\frac{\partial y}{\partial t}\right)^2 + \frac{\partial z}{\partial x}\frac{\partial^2 x}{\partial t^2} + \frac{\partial z}{\partial y}\frac{\partial^2 y}{\partial t^2}$$

(b) Find a similar formula for $\partial^2 z/\partial s\,\partial t$.

55. A function f is called **homogeneous of degree n** if it satisfies the equation $f(tx, ty) = t^n f(x, y)$ for all t, where n is a positive integer and f has continuous second-order partial derivatives.
(a) Verify that $f(x, y) = x^2y + 2xy^2 + 5y^3$ is homogeneous of degree 3.
(b) Show that if f is homogeneous of degree n, then

$$x\frac{\partial f}{\partial x} + y\frac{\partial f}{\partial y} = nf(x, y)$$

[*Hint:* Use the Chain Rule to differentiate $f(tx, ty)$ with respect to t.]

56. If f is homogeneous of degree n, show that

$$x^2\frac{\partial^2 f}{\partial x^2} + 2xy\frac{\partial^2 f}{\partial x\,\partial y} + y^2\frac{\partial^2 f}{\partial y^2} = n(n - 1)f(x, y)$$

57. If f is homogeneous of degree n, show that

$$f_x(tx, ty) = t^{n-1}f_x(x, y)$$

58. Suppose that the equation $F(x, y, z) = 0$ implicitly defines each of the three variables x, y, and z as functions of the other two: $z = f(x, y)$, $y = g(x, z)$, $x = h(y, z)$. If F is differentiable and F_x, F_y, and F_z are all nonzero, show that

$$\frac{\partial z}{\partial x}\frac{\partial x}{\partial y}\frac{\partial y}{\partial z} = -1$$

59. Equation 6 is a formula for the derivative dy/dx of a function defined implicitly by an equation $F(x, y) = 0$, provided that F is differentiable and $F_y \neq 0$. Prove that if F has continuous second derivatives, then a formula for the second derivative of y is

$$\frac{d^2y}{dx^2} = -\frac{F_{xx}F_y^2 - 2F_{xy}F_xF_y + F_{yy}F_x^2}{F_y^3}$$

14.6 Directional Derivatives and the Gradient Vector

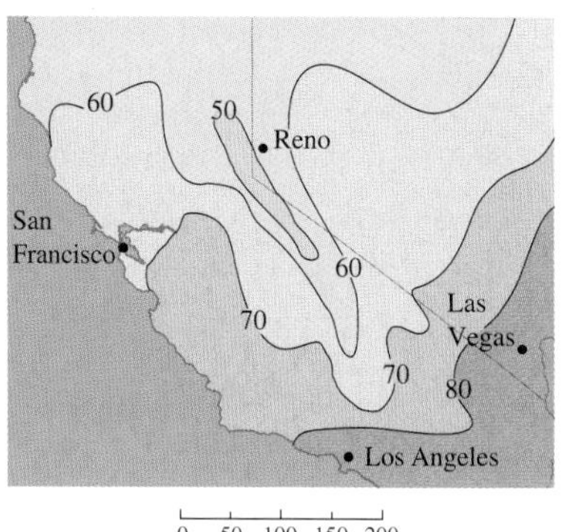

FIGURE 1

The weather map in Figure 1 shows a contour map of the temperature function $T(x, y)$ for the states of California and Nevada at 3:00 PM on a day in October. The level curves, or isothermals, join locations with the same temperature. The partial derivative T_x at a location such as Reno is the rate of change of temperature with respect to distance if we travel east from Reno; T_y is the rate of change of temperature if we travel north. But what if we want to know the rate of change of temperature when we travel southeast (toward Las Vegas), or in some other direction? In this section we introduce a type of derivative, called a *directional derivative,* that enables us to find the rate of change of a function of two or more variables in any direction.

Directional Derivatives

Recall that if $z = f(x, y)$, then the partial derivatives f_x and f_y are defined as

1
$$f_x(x_0, y_0) = \lim_{h \to 0} \frac{f(x_0 + h, y_0) - f(x_0, y_0)}{h}$$
$$f_y(x_0, y_0) = \lim_{h \to 0} \frac{f(x_0, y_0 + h) - f(x_0, y_0)}{h}$$

and represent the rates of change of z in the x- and y-directions, that is, in the directions of the unit vectors **i** and **j**.

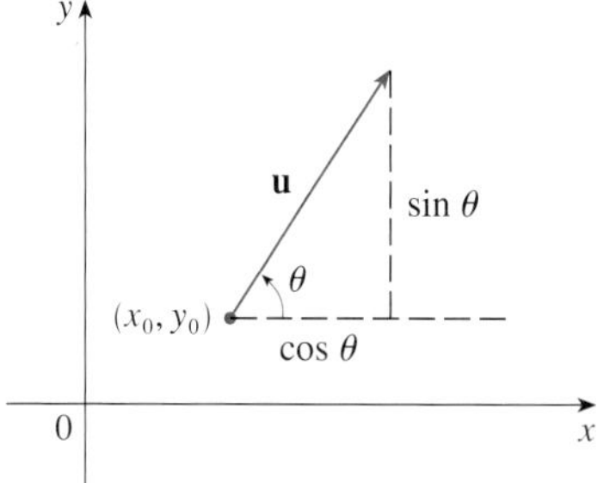

FIGURE 2
A unit vector $\mathbf{u} = \langle a, b \rangle = \langle \cos\theta, \sin\theta \rangle$

Suppose that we now wish to find the rate of change of z at (x_0, y_0) in the direction of an arbitrary unit vector $\mathbf{u} = \langle a, b \rangle$. (See Figure 2.) To do this we consider the surface S with the equation $z = f(x, y)$ (the graph of f) and we let $z_0 = f(x_0, y_0)$. Then the point $P(x_0, y_0, z_0)$ lies on S. The vertical plane that passes through P in the direction of **u** intersects S in a curve C. (See Figure 3.) The slope of the tangent line T to C at the point P is the rate of change of z in the direction of **u**.

TEC Visual 14.6A animates Figure 3 by rotating **u** and therefore T.

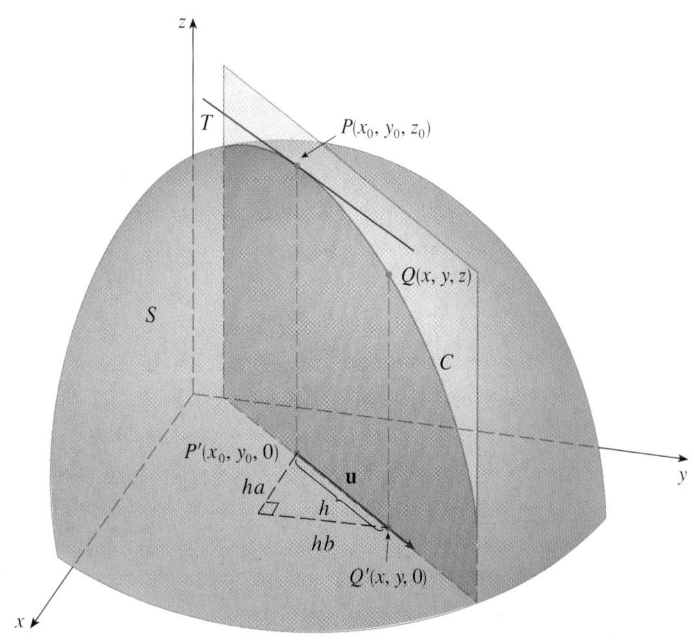

FIGURE 3

If $Q(x, y, z)$ is another point on C and P', Q' are the projections of P, Q onto the xy-plane, then the vector $\overrightarrow{P'Q'}$ is parallel to $\mathbf{u}$ and so

$$\overrightarrow{P'Q'} = h\mathbf{u} = \langle ha, hb \rangle$$

for some scalar h. Therefore $x - x_0 = ha$, $y - y_0 = hb$, so $x = x_0 + ha$, $y = y_0 + hb$, and

$$\frac{\Delta z}{h} = \frac{z - z_0}{h} = \frac{f(x_0 + ha, y_0 + hb) - f(x_0, y_0)}{h}$$

If we take the limit as $h \to 0$, we obtain the rate of change of z (with respect to distance) in the direction of $\mathbf{u}$, which is called the directional derivative of f in the direction of $\mathbf{u}$.

2 Definition The **directional derivative** of f at (x_0, y_0) in the direction of a unit vector $\mathbf{u} = \langle a, b \rangle$ is

$$D_{\mathbf{u}} f(x_0, y_0) = \lim_{h \to 0} \frac{f(x_0 + ha, y_0 + hb) - f(x_0, y_0)}{h}$$

if this limit exists.

By comparing Definition 2 with Equations 1, we see that if $\mathbf{u} = \mathbf{i} = \langle 1, 0 \rangle$, then $D_{\mathbf{i}} f = f_x$ and if $\mathbf{u} = \mathbf{j} = \langle 0, 1 \rangle$, then $D_{\mathbf{j}} f = f_y$. In other words, the partial derivatives of f with respect to x and y are just special cases of the directional derivative.

EXAMPLE 1 Use the weather map in Figure 1 to estimate the value of the directional derivative of the temperature function at Reno in the southeasterly direction.

SOLUTION The unit vector directed toward the southeast is $\mathbf{u} = (\mathbf{i} - \mathbf{j})/\sqrt{2}$, but we won't need to use this expression. We start by drawing a line through Reno toward the southeast (see Figure 4).

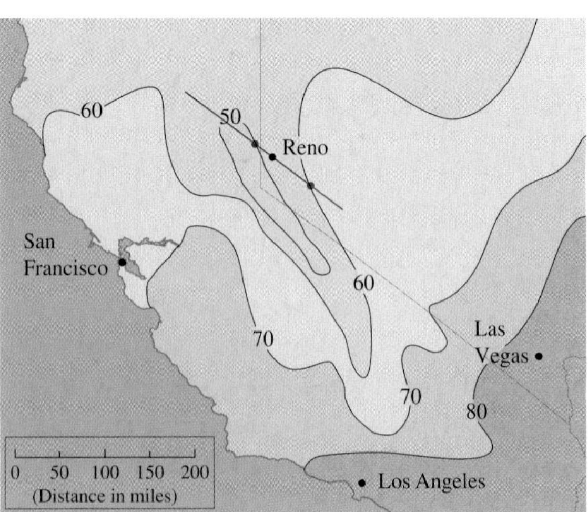

FIGURE 4

We approximate the directional derivative $D_{\mathbf{u}} T$ by the average rate of change of the temperature between the points where this line intersects the isothermals $T = 50$ and

$T = 60$. The temperature at the point southeast of Reno is $T = 60°\text{F}$ and the temperature at the point northwest of Reno is $T = 50°\text{F}$. The distance between these points looks to be about 75 miles. So the rate of change of the temperature in the southeasterly direction is

$$D_{\mathbf{u}}T \approx \frac{60 - 50}{75} = \frac{10}{75} \approx 0.13°\text{F/mi}$$

When we compute the directional derivative of a function defined by a formula, we generally use the following theorem.

3 Theorem If f is a differentiable function of x and y, then f has a directional derivative in the direction of any unit vector $\mathbf{u} = \langle a, b \rangle$ and

$$D_{\mathbf{u}}f(x, y) = f_x(x, y)\,a + f_y(x, y)\,b$$

PROOF If we define a function g of the single variable h by

$$g(h) = f(x_0 + ha, y_0 + hb)$$

then, by the definition of a derivative, we have

4
$$g'(0) = \lim_{h \to 0} \frac{g(h) - g(0)}{h} = \lim_{h \to 0} \frac{f(x_0 + ha, y_0 + hb) - f(x_0, y_0)}{h}$$
$$= D_{\mathbf{u}}f(x_0, y_0)$$

On the other hand, we can write $g(h) = f(x, y)$, where $x = x_0 + ha$, $y = y_0 + hb$, so the Chain Rule (Theorem 14.5.2) gives

$$g'(h) = \frac{\partial f}{\partial x}\frac{dx}{dh} + \frac{\partial f}{\partial y}\frac{dy}{dh} = f_x(x, y)\,a + f_y(x, y)\,b$$

If we now put $h = 0$, then $x = x_0$, $y = y_0$, and

5
$$g'(0) = f_x(x_0, y_0)\,a + f_y(x_0, y_0)\,b$$

Comparing Equations 4 and 5, we see that

$$D_{\mathbf{u}}f(x_0, y_0) = f_x(x_0, y_0)\,a + f_y(x_0, y_0)\,b$$

If the unit vector $\mathbf{u}$ makes an angle θ with the positive x-axis (as in Figure 2), then we can write $\mathbf{u} = \langle \cos\theta, \sin\theta \rangle$ and the formula in Theorem 3 becomes

6
$$D_{\mathbf{u}}f(x, y) = f_x(x, y)\cos\theta + f_y(x, y)\sin\theta$$

EXAMPLE 2 Find the directional derivative $D_{\mathbf{u}}f(x, y)$ if

$$f(x, y) = x^3 - 3xy + 4y^2$$

and $\mathbf{u}$ is the unit vector given by angle $\theta = \pi/6$. What is $D_{\mathbf{u}}f(1, 2)$?

The directional derivative $D_{\mathbf{u}} f(1, 2)$ in Example 2 represents the rate of change of z in the direction of $\mathbf{u}$. This is the slope of the tangent line to the curve of intersection of the surface $z = x^3 - 3xy + 4y^2$ and the vertical plane through $(1, 2, 0)$ in the direction of $\mathbf{u}$ shown in Figure 5.

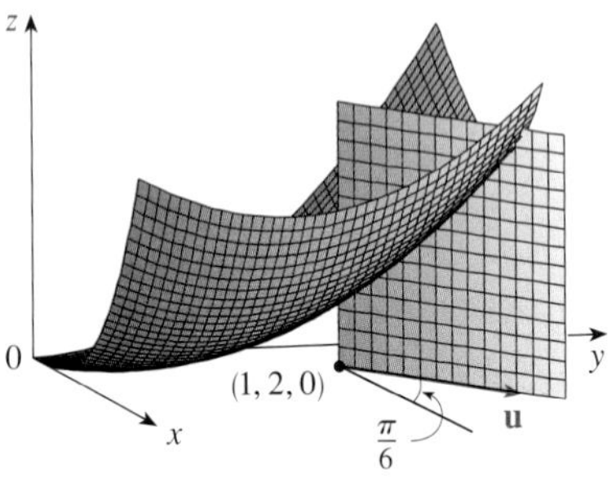

FIGURE 5

SOLUTION Formula 6 gives

$$\begin{aligned} D_{\mathbf{u}} f(x, y) &= f_x(x, y) \cos \frac{\pi}{6} + f_y(x, y) \sin \frac{\pi}{6} \\ &= (3x^2 - 3y) \frac{\sqrt{3}}{2} + (-3x + 8y) \tfrac{1}{2} \\ &= \tfrac{1}{2} \left[3\sqrt{3}\, x^2 - 3x + \left(8 - 3\sqrt{3}\right) y \right] \end{aligned}$$

Therefore

$$D_{\mathbf{u}} f(1, 2) = \tfrac{1}{2} \left[3\sqrt{3}(1)^2 - 3(1) + \left(8 - 3\sqrt{3}\right)(2) \right] = \frac{13 - 3\sqrt{3}}{2}$$

The Gradient Vector

Notice from Theorem 3 that the directional derivative of a differentiable function can be written as the dot product of two vectors:

7
$$\begin{aligned} D_{\mathbf{u}} f(x, y) &= f_x(x, y)\, a + f_y(x, y)\, b \\ &= \langle f_x(x, y), f_y(x, y) \rangle \cdot \langle a, b \rangle \\ &= \langle f_x(x, y), f_y(x, y) \rangle \cdot \mathbf{u} \end{aligned}$$

The first vector in this dot product occurs not only in computing directional derivatives but in many other contexts as well. So we give it a special name (the *gradient* of f) and a special notation (**grad** f or ∇f, which is read "del f").

8 Definition If f is a function of two variables x and y, then the **gradient** of f is the vector function ∇f defined by

$$\nabla f(x, y) = \langle f_x(x, y), f_y(x, y) \rangle = \frac{\partial f}{\partial x} \mathbf{i} + \frac{\partial f}{\partial y} \mathbf{j}$$

EXAMPLE 3 If $f(x, y) = \sin x + e^{xy}$, then

$$\nabla f(x, y) = \langle f_x, f_y \rangle = \langle \cos x + y e^{xy}, x e^{xy} \rangle$$

and

$$\nabla f(0, 1) = \langle 2, 0 \rangle$$

With this notation for the gradient vector, we can rewrite Equation 7 for the directional derivative of a differentiable function as

9
$$D_{\mathbf{u}} f(x, y) = \nabla f(x, y) \cdot \mathbf{u}$$

This expresses the directional derivative in the direction of a unit vector $\mathbf{u}$ as the scalar projection of the gradient vector onto $\mathbf{u}$.

The gradient vector $\nabla f(2, -1)$ in Example 4 is shown in Figure 6 with initial point $(2, -1)$. Also shown is the vector **v** that gives the direction of the directional derivative. Both of these vectors are superimposed on a contour plot of the graph of f.

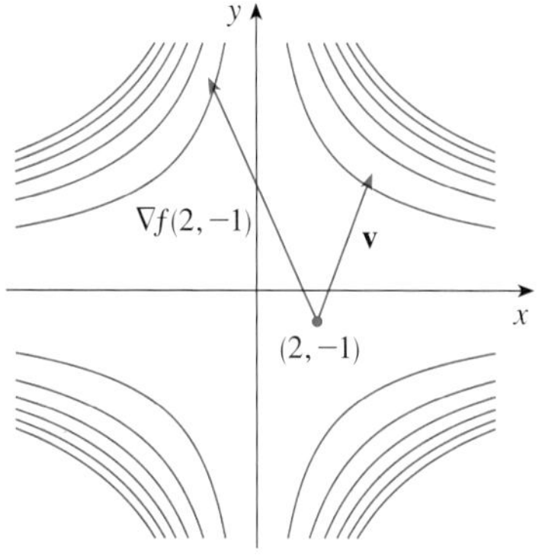

FIGURE 6

V EXAMPLE 4 Find the directional derivative of the function $f(x, y) = x^2y^3 - 4y$ at the point $(2, -1)$ in the direction of the vector $\mathbf{v} = 2\mathbf{i} + 5\mathbf{j}$.

SOLUTION We first compute the gradient vector at $(2, -1)$:

$$\nabla f(x, y) = 2xy^3\,\mathbf{i} + (3x^2y^2 - 4)\,\mathbf{j}$$

$$\nabla f(2, -1) = -4\,\mathbf{i} + 8\,\mathbf{j}$$

Note that **v** is not a unit vector, but since $|\mathbf{v}| = \sqrt{29}$, the unit vector in the direction of **v** is

$$\mathbf{u} = \frac{\mathbf{v}}{|\mathbf{v}|} = \frac{2}{\sqrt{29}}\,\mathbf{i} + \frac{5}{\sqrt{29}}\,\mathbf{j}$$

Therefore, by Equation 9, we have

$$D_{\mathbf{u}} f(2, -1) = \nabla f(2, -1) \cdot \mathbf{u} = (-4\,\mathbf{i} + 8\,\mathbf{j}) \cdot \left(\frac{2}{\sqrt{29}}\,\mathbf{i} + \frac{5}{\sqrt{29}}\,\mathbf{j} \right)$$

$$= \frac{-4 \cdot 2 + 8 \cdot 5}{\sqrt{29}} = \frac{32}{\sqrt{29}}$$

Functions of Three Variables

For functions of three variables we can define directional derivatives in a similar manner. Again $D_{\mathbf{u}} f(x, y, z)$ can be interpreted as the rate of change of the function in the direction of a unit vector **u**.

10 Definition The **directional derivative** of f at (x_0, y_0, z_0) in the direction of a unit vector $\mathbf{u} = \langle a, b, c \rangle$ is

$$D_{\mathbf{u}} f(x_0, y_0, z_0) = \lim_{h \to 0} \frac{f(x_0 + ha, y_0 + hb, z_0 + hc) - f(x_0, y_0, z_0)}{h}$$

if this limit exists.

If we use vector notation, then we can write both definitions (2 and 10) of the directional derivative in the compact form

11
$$D_{\mathbf{u}} f(\mathbf{x}_0) = \lim_{h \to 0} \frac{f(\mathbf{x}_0 + h\mathbf{u}) - f(\mathbf{x}_0)}{h}$$

where $\mathbf{x}_0 = \langle x_0, y_0 \rangle$ if $n = 2$ and $\mathbf{x}_0 = \langle x_0, y_0, z_0 \rangle$ if $n = 3$. This is reasonable because the vector equation of the line through $\mathbf{x}_0$ in the direction of the vector **u** is given by $\mathbf{x} = \mathbf{x}_0 + t\mathbf{u}$ (Equation 12.5.1) and so $f(\mathbf{x}_0 + h\mathbf{u})$ represents the value of f at a point on this line.

If $f(x, y, z)$ is differentiable and $\mathbf{u} = \langle a, b, c \rangle$, then the same method that was used to prove Theorem 3 can be used to show that

12 $$D_{\mathbf{u}} f(x, y, z) = f_x(x, y, z)\, a + f_y(x, y, z)\, b + f_z(x, y, z)\, c$$

For a function f of three variables, the **gradient vector**, denoted by ∇f or **grad** f, is

$$\nabla f(x, y, z) = \langle f_x(x, y, z), f_y(x, y, z), f_z(x, y, z) \rangle$$

or, for short,

13 $$\nabla f = \langle f_x, f_y, f_z \rangle = \frac{\partial f}{\partial x}\mathbf{i} + \frac{\partial f}{\partial y}\mathbf{j} + \frac{\partial f}{\partial z}\mathbf{k}$$

Then, just as with functions of two variables, Formula 12 for the directional derivative can be rewritten as

14 $$D_{\mathbf{u}} f(x, y, z) = \nabla f(x, y, z) \cdot \mathbf{u}$$

V **EXAMPLE 5** If $f(x, y, z) = x \sin yz$, (a) find the gradient of f and (b) find the directional derivative of f at $(1, 3, 0)$ in the direction of $\mathbf{v} = \mathbf{i} + 2\mathbf{j} - \mathbf{k}$.

SOLUTION

(a) The gradient of f is

$$\begin{aligned} \nabla f(x, y, z) &= \langle f_x(x, y, z), f_y(x, y, z), f_z(x, y, z) \rangle \\ &= \langle \sin yz, xz \cos yz, xy \cos yz \rangle \end{aligned}$$

(b) At $(1, 3, 0)$ we have $\nabla f(1, 3, 0) = \langle 0, 0, 3 \rangle$. The unit vector in the direction of $\mathbf{v} = \mathbf{i} + 2\mathbf{j} - \mathbf{k}$ is

$$\mathbf{u} = \frac{1}{\sqrt{6}}\mathbf{i} + \frac{2}{\sqrt{6}}\mathbf{j} - \frac{1}{\sqrt{6}}\mathbf{k}$$

Therefore Equation 14 gives

$$\begin{aligned} D_{\mathbf{u}} f(1, 3, 0) &= \nabla f(1, 3, 0) \cdot \mathbf{u} \\ &= 3\mathbf{k} \cdot \left(\frac{1}{\sqrt{6}}\mathbf{i} + \frac{2}{\sqrt{6}}\mathbf{j} - \frac{1}{\sqrt{6}}\mathbf{k} \right) \\ &= 3\left(-\frac{1}{\sqrt{6}} \right) = -\sqrt{\frac{3}{2}} \end{aligned}$$

Maximizing the Directional Derivative

Suppose we have a function f of two or three variables and we consider all possible directional derivatives of f at a given point. These give the rates of change of f in all possible directions. We can then ask the questions: In which of these directions does f change fastest and what is the maximum rate of change? The answers are provided by the following theorem.

TEC Visual 14.6B provides visual confirmation of Theorem 15.

> **15 Theorem** Suppose f is a differentiable function of two or three variables. The maximum value of the directional derivative $D_{\mathbf{u}}f(\mathbf{x})$ is $|\nabla f(\mathbf{x})|$ and it occurs when $\mathbf{u}$ has the same direction as the gradient vector $\nabla f(\mathbf{x})$.

PROOF From Equation 9 or 14 we have

$$D_{\mathbf{u}}f = \nabla f \cdot \mathbf{u} = |\nabla f||\mathbf{u}|\cos\theta = |\nabla f|\cos\theta$$

where θ is the angle between ∇f and $\mathbf{u}$. The maximum value of $\cos\theta$ is 1 and this occurs when $\theta = 0$. Therefore the maximum value of $D_{\mathbf{u}}f$ is $|\nabla f|$ and it occurs when $\theta = 0$, that is, when $\mathbf{u}$ has the same direction as ∇f. ∎

EXAMPLE 6

(a) If $f(x, y) = xe^y$, find the rate of change of f at the point $P(2, 0)$ in the direction from P to $Q(\frac{1}{2}, 2)$.

(b) In what direction does f have the maximum rate of change? What is this maximum rate of change?

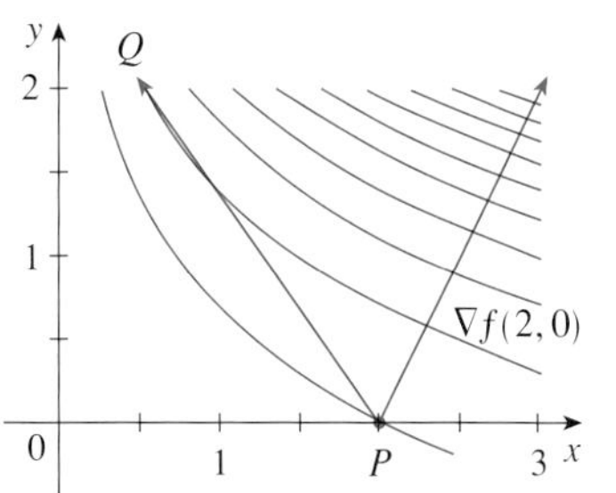

FIGURE 7

At (2, 0) the function in Example 6 increases fastest in the direction of the gradient vector $\nabla f(2, 0) = \langle 1, 2\rangle$. Notice from Figure 7 that this vector appears to be perpendicular to the level curve through (2, 0). Figure 8 shows the graph of f and the gradient vector.

SOLUTION

(a) We first compute the gradient vector:

$$\nabla f(x, y) = \langle f_x, f_y\rangle = \langle e^y, xe^y\rangle$$

$$\nabla f(2, 0) = \langle 1, 2\rangle$$

The unit vector in the direction of $\overrightarrow{PQ} = \langle -1.5, 2\rangle$ is $\mathbf{u} = \langle -\frac{3}{5}, \frac{4}{5}\rangle$, so the rate of change of f in the direction from P to Q is

$$D_{\mathbf{u}}f(2, 0) = \nabla f(2, 0) \cdot \mathbf{u} = \langle 1, 2\rangle \cdot \langle -\tfrac{3}{5}, \tfrac{4}{5}\rangle$$

$$= 1\left(-\tfrac{3}{5}\right) + 2\left(\tfrac{4}{5}\right) = 1$$

(b) According to Theorem 15, f increases fastest in the direction of the gradient vector $\nabla f(2, 0) = \langle 1, 2\rangle$. The maximum rate of change is

$$|\nabla f(2, 0)| = |\langle 1, 2\rangle| = \sqrt{5}$$ ∎

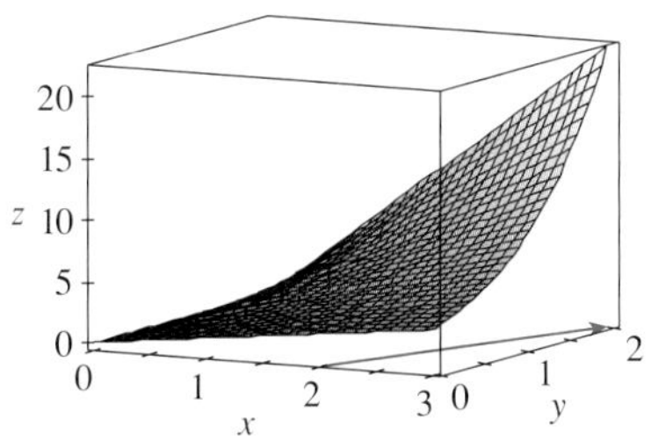

FIGURE 8

EXAMPLE 7 Suppose that the temperature at a point (x, y, z) in space is given by $T(x, y, z) = 80/(1 + x^2 + 2y^2 + 3z^2)$, where T is measured in degrees Celsius and x, y, z in meters. In which direction does the temperature increase fastest at the point $(1, 1, -2)$? What is the maximum rate of increase?

SOLUTION The gradient of T is

$$\nabla T = \frac{\partial T}{\partial x}\mathbf{i} + \frac{\partial T}{\partial y}\mathbf{j} + \frac{\partial T}{\partial z}\mathbf{k}$$

$$= -\frac{160x}{(1 + x^2 + 2y^2 + 3z^2)^2}\mathbf{i} - \frac{320y}{(1 + x^2 + 2y^2 + 3z^2)^2}\mathbf{j} - \frac{480z}{(1 + x^2 + 2y^2 + 3z^2)^2}\mathbf{k}$$

$$= \frac{160}{(1 + x^2 + 2y^2 + 3z^2)^2}(-x\,\mathbf{i} - 2y\,\mathbf{j} - 3z\,\mathbf{k})$$

At the point $(1, 1, -2)$ the gradient vector is

$$\nabla T(1, 1, -2) = \tfrac{160}{256}(-\mathbf{i} - 2\mathbf{j} + 6\mathbf{k}) = \tfrac{5}{8}(-\mathbf{i} - 2\mathbf{j} + 6\mathbf{k})$$

By Theorem 15 the temperature increases fastest in the direction of the gradient vector $\nabla T(1, 1, -2) = \frac{5}{8}(-\mathbf{i} - 2\mathbf{j} + 6\mathbf{k})$ or, equivalently, in the direction of $-\mathbf{i} - 2\mathbf{j} + 6\mathbf{k}$ or the unit vector $(-\mathbf{i} - 2\mathbf{j} + 6\mathbf{k})/\sqrt{41}$. The maximum rate of increase is the length of the gradient vector:

$$|\nabla T(1, 1, -2)| = \tfrac{5}{8}|-\mathbf{i} - 2\mathbf{j} + 6\mathbf{k}| = \tfrac{5}{8}\sqrt{41}$$

Therefore the maximum rate of increase of temperature is $\frac{5}{8}\sqrt{41} \approx 4°\text{C/m}$.

Tangent Planes to Level Surfaces

Suppose S is a surface with equation $F(x, y, z) = k$, that is, it is a level surface of a function F of three variables, and let $P(x_0, y_0, z_0)$ be a point on S. Let C be any curve that lies on the surface S and passes through the point P. Recall from Section 13.1 that the curve C is described by a continuous vector function $\mathbf{r}(t) = \langle x(t), y(t), z(t)\rangle$. Let t_0 be the parameter value corresponding to P; that is, $\mathbf{r}(t_0) = \langle x_0, y_0, z_0\rangle$. Since C lies on S, any point $(x(t), y(t), z(t))$ must satisfy the equation of S, that is,

16 $$F(x(t), y(t), z(t)) = k$$

If x, y, and z are differentiable functions of t and F is also differentiable, then we can use the Chain Rule to differentiate both sides of Equation 16 as follows:

17 $$\frac{\partial F}{\partial x}\frac{dx}{dt} + \frac{\partial F}{\partial y}\frac{dy}{dt} + \frac{\partial F}{\partial z}\frac{dz}{dt} = 0$$

But, since $\nabla F = \langle F_x, F_y, F_z\rangle$ and $\mathbf{r}'(t) = \langle x'(t), y'(t), z'(t)\rangle$, Equation 17 can be written in terms of a dot product as

$$\nabla F \cdot \mathbf{r}'(t) = 0$$

In particular, when $t = t_0$ we have $\mathbf{r}(t_0) = \langle x_0, y_0, z_0\rangle$, so

18 $$\nabla F(x_0, y_0, z_0) \cdot \mathbf{r}'(t_0) = 0$$

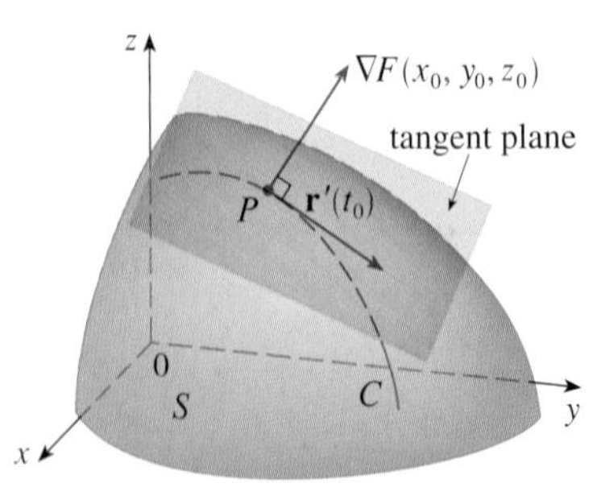

FIGURE 9

Equation 18 says that *the gradient vector at P, $\nabla F(x_0, y_0, z_0)$, is perpendicular to the tangent vector $\mathbf{r}'(t_0)$ to any curve C on S that passes through P.* (See Figure 9.) If $\nabla F(x_0, y_0, z_0) \neq \mathbf{0}$, it is therefore natural to define the **tangent plane to the level surface $F(x, y, z) = k$ at $P(x_0, y_0, z_0)$** as the plane that passes through P and has normal vector $\nabla F(x_0, y_0, z_0)$. Using the standard equation of a plane (Equation 12.5.7), we can write the equation of this tangent plane as

19 $$F_x(x_0, y_0, z_0)(x - x_0) + F_y(x_0, y_0, z_0)(y - y_0) + F_z(x_0, y_0, z_0)(z - z_0) = 0$$

The **normal line** to S at P is the line passing through P and perpendicular to the tangent plane. The direction of the normal line is therefore given by the gradient vector $\nabla F(x_0, y_0, z_0)$ and so, by Equation 12.5.3, its symmetric equations are

20
$$\frac{x - x_0}{F_x(x_0, y_0, z_0)} = \frac{y - y_0}{F_y(x_0, y_0, z_0)} = \frac{z - z_0}{F_z(x_0, y_0, z_0)}$$

In the special case in which the equation of a surface S is of the form $z = f(x, y)$ (that is, S is the graph of a function f of two variables), we can rewrite the equation as

$$F(x, y, z) = f(x, y) - z = 0$$

and regard S as a level surface (with $k = 0$) of F. Then

$$F_x(x_0, y_0, z_0) = f_x(x_0, y_0)$$
$$F_y(x_0, y_0, z_0) = f_y(x_0, y_0)$$
$$F_z(x_0, y_0, z_0) = -1$$

so Equation 19 becomes

$$f_x(x_0, y_0)(x - x_0) + f_y(x_0, y_0)(y - y_0) - (z - z_0) = 0$$

which is equivalent to Equation 14.4.2. Thus our new, more general, definition of a tangent plane is consistent with the definition that was given for the special case of Section 14.4.

V EXAMPLE 8 Find the equations of the tangent plane and normal line at the point $(-2, 1, -3)$ to the ellipsoid

$$\frac{x^2}{4} + y^2 + \frac{z^2}{9} = 3$$

SOLUTION The ellipsoid is the level surface (with $k = 3$) of the function

$$F(x, y, z) = \frac{x^2}{4} + y^2 + \frac{z^2}{9}$$

Figure 10 shows the ellipsoid, tangent plane, and normal line in Example 8.

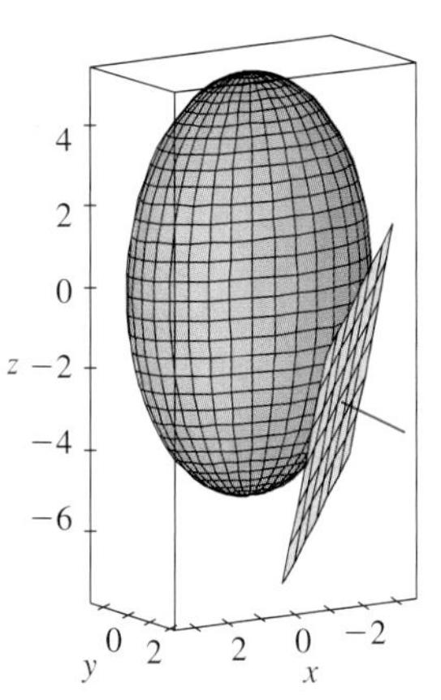

FIGURE 10

Therefore we have

$$F_x(x, y, z) = \frac{x}{2} \qquad F_y(x, y, z) = 2y \qquad F_z(x, y, z) = \frac{2z}{9}$$

$$F_x(-2, 1, -3) = -1 \qquad F_y(-2, 1, -3) = 2 \qquad F_z(-2, 1, -3) = -\tfrac{2}{3}$$

Then Equation 19 gives the equation of the tangent plane at $(-2, 1, -3)$ as

$$-1(x + 2) + 2(y - 1) - \tfrac{2}{3}(z + 3) = 0$$

which simplifies to $3x - 6y + 2z + 18 = 0$.

By Equation 20, symmetric equations of the normal line are

$$\frac{x + 2}{-1} = \frac{y - 1}{2} = \frac{z + 3}{-\frac{2}{3}}$$

Significance of the Gradient Vector

We now summarize the ways in which the gradient vector is significant. We first consider a function f of three variables and a point $P(x_0, y_0, z_0)$ in its domain. On the one hand, we know from Theorem 15 that the gradient vector $\nabla f(x_0, y_0, z_0)$ gives the direction of fastest increase of f. On the other hand, we know that $\nabla f(x_0, y_0, z_0)$ is orthogonal to the level surface S of f through P. (Refer to Figure 9.) These two properties are quite compatible intuitively because as we move away from P on the level surface S, the value of f does not change at all. So it seems reasonable that if we move in the perpendicular direction, we get the maximum increase.

In like manner we consider a function f of two variables and a point $P(x_0, y_0)$ in its domain. Again the gradient vector $\nabla f(x_0, y_0)$ gives the direction of fastest increase of f. Also, by considerations similar to our discussion of tangent planes, it can be shown that $\nabla f(x_0, y_0)$ is perpendicular to the level curve $f(x, y) = k$ that passes through P. Again this is intuitively plausible because the values of f remain constant as we move along the curve. (See Figure 11.)

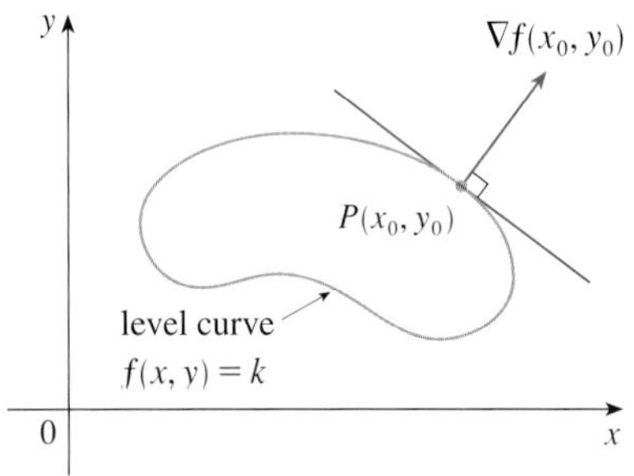

FIGURE 11

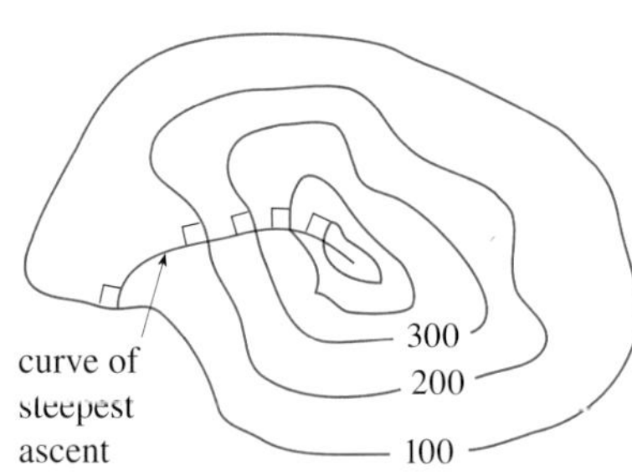

FIGURE 12

If we consider a topographical map of a hill and let $f(x, y)$ represent the height above sea level at a point with coordinates (x, y), then a curve of steepest ascent can be drawn as in Figure 12 by making it perpendicular to all of the contour lines. This phenomenon can also be noticed in Figure 12 in Section 14.1, where Lonesome Creek follows a curve of steepest descent.

Computer algebra systems have commands that plot sample gradient vectors. Each gradient vector $\nabla f(a, b)$ is plotted starting at the point (a, b). Figure 13 shows such a plot (called a *gradient vector field*) for the function $f(x, y) = x^2 - y^2$ superimposed on a contour map of f. As expected, the gradient vectors point "uphill" and are perpendicular to the level curves.

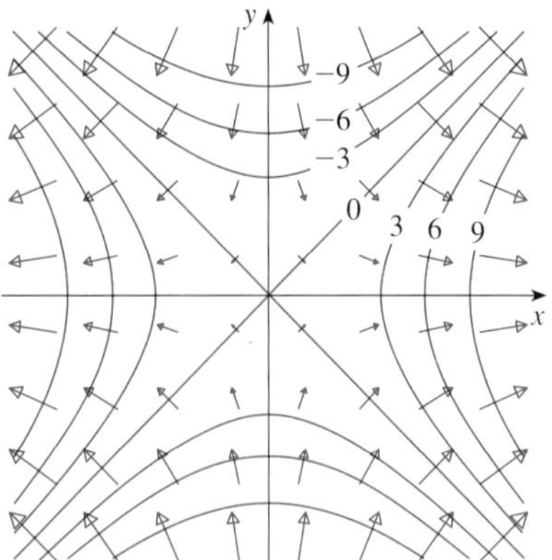

FIGURE 13

14.6 Exercises

1. Level curves for barometric pressure (in millibars) are shown for 6:00 AM on November 10, 1998. A deep low with pressure 972 mb is moving over northeast Iowa. The distance along the red line from K (Kearney, Nebraska) to S (Sioux City, Iowa) is 300 km. Estimate the value of the directional derivative of the pressure function at Kearney in the direction of Sioux City. What are the units of the directional derivative?

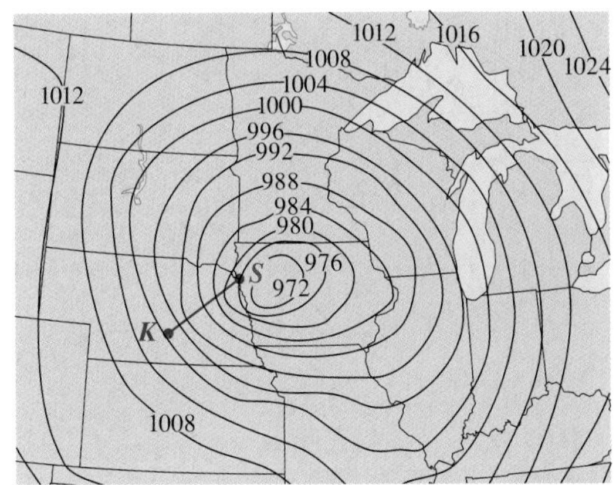

2. The contour map shows the average maximum temperature for November 2004 (in °C). Estimate the value of the directional derivative of this temperature function at Dubbo, New South Wales, in the direction of Sydney. What are the units?

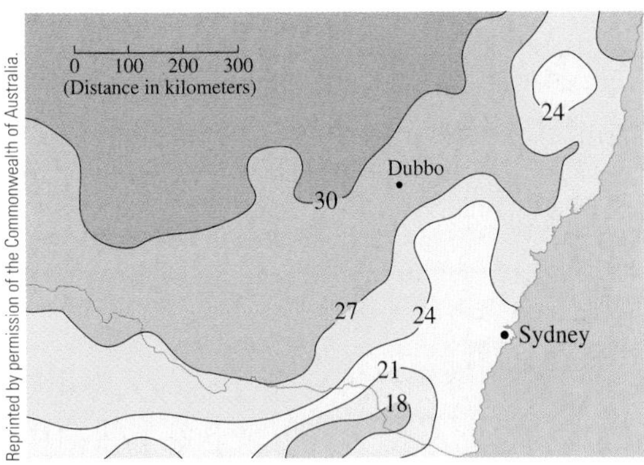

3. A table of values for the wind-chill index $W = f(T, v)$ is given in Exercise 3 on page 911. Use the table to estimate the value of $D_{\mathbf{u}} f(-20, 30)$, where $\mathbf{u} = (\mathbf{i} + \mathbf{j})/\sqrt{2}$.

4–6 Find the directional derivative of f at the given point in the direction indicated by the angle θ.

4. $f(x, y) = x^3y^4 + x^4y^3$, $(1, 1)$, $\theta = \pi/6$

5. $f(x, y) = ye^{-x}$, $(0, 4)$, $\theta = 2\pi/3$

6. $f(x, y) = e^x \cos y$, $(0, 0)$, $\theta = \pi/4$

7–10
(a) Find the gradient of f.
(b) Evaluate the gradient at the point P.
(c) Find the rate of change of f at P in the direction of the vector $\mathbf{u}$.

7. $f(x, y) = \sin(2x + 3y)$, $P(-6, 4)$, $\mathbf{u} = \frac{1}{2}(\sqrt{3}\,\mathbf{i} - \mathbf{j})$

8. $f(x, y) = y^2/x$, $P(1, 2)$, $\mathbf{u} = \frac{1}{3}(2\mathbf{i} + \sqrt{5}\,\mathbf{j})$

9. $f(x, y, z) = x^2yz - xyz^3$, $P(2, -1, 1)$, $\mathbf{u} = \langle 0, \frac{4}{5}, -\frac{3}{5} \rangle$

10. $f(x, y, z) = y^2e^{xyz}$, $P(0, 1, -1)$, $\mathbf{u} = \langle \frac{3}{13}, \frac{4}{13}, \frac{12}{13} \rangle$

11–17 Find the directional derivative of the function at the given point in the direction of the vector $\mathbf{v}$.

11. $f(x, y) = e^x \sin y$, $(0, \pi/3)$, $\mathbf{v} = \langle -6, 8 \rangle$

12. $f(x, y) = \dfrac{x}{x^2 + y^2}$, $(1, 2)$, $\mathbf{v} = \langle 3, 5 \rangle$

13. $g(p, q) = p^4 - p^2q^3$, $(2, 1)$, $\mathbf{v} = \mathbf{i} + 3\mathbf{j}$

14. $g(r, s) = \tan^{-1}(rs)$, $(1, 2)$, $\mathbf{v} = 5\mathbf{i} + 10\mathbf{j}$

15. $f(x, y, z) = xe^y + ye^z + ze^x$, $(0, 0, 0)$, $\mathbf{v} = \langle 5, 1, -2 \rangle$

16. $f(x, y, z) = \sqrt{xyz}$, $(3, 2, 6)$, $\mathbf{v} = \langle -1, -2, 2 \rangle$

17. $h(r, s, t) = \ln(3r + 6s + 9t)$, $(1, 1, 1)$, $\mathbf{v} = 4\mathbf{i} + 12\mathbf{j} + 6\mathbf{k}$

18. Use the figure to estimate $D_{\mathbf{u}} f(2, 2)$.

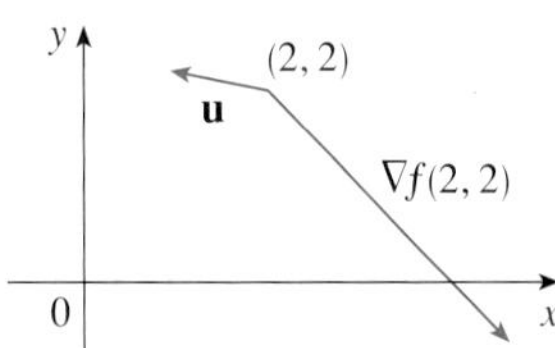

19. Find the directional derivative of $f(x, y) = \sqrt{xy}$ at $P(2, 8)$ in the direction of $Q(5, 4)$.

20. Find the directional derivative of $f(x, y, z) = xy + yz + zx$ at $P(1, -1, 3)$ in the direction of $Q(2, 4, 5)$.

21–26 Find the maximum rate of change of f at the given point and the direction in which it occurs.

21. $f(x, y) = 4y\sqrt{x}$, $(4, 1)$

22. $f(s, t) = te^{st}$, $(0, 2)$

23. $f(x, y) = \sin(xy)$, $(1, 0)$

24. $f(x, y, z) = (x + y)/z$, $(1, 1, -1)$

25. $f(x, y, z) = \sqrt{x^2 + y^2 + z^2}$, $(3, 6, -2)$

26. $f(p, q, r) = \arctan(pqr)$, $(1, 2, 1)$

27. (a) Show that a differentiable function f decreases most rapidly at $\mathbf{x}$ in the direction opposite to the gradient vector, that is, in the direction of $-\nabla f(\mathbf{x})$.
(b) Use the result of part (a) to find the direction in which the function $f(x, y) = x^4y - x^2y^3$ decreases fastest at the point $(2, -3)$.

28. Find the directions in which the directional derivative of $f(x, y) = ye^{-xy}$ at the point $(0, 2)$ has the value 1.

29. Find all points at which the direction of fastest change of the function $f(x, y) = x^2 + y^2 - 2x - 4y$ is $\mathbf{i} + \mathbf{j}$.

30. Near a buoy, the depth of a lake at the point with coordinates (x, y) is $z = 200 + 0.02x^2 - 0.001y^3$, where x, y, and z are measured in meters. A fisherman in a small boat starts at the point $(80, 60)$ and moves toward the buoy, which is located at $(0, 0)$. Is the water under the boat getting deeper or shallower when he departs? Explain.

31. The temperature T in a metal ball is inversely proportional to the distance from the center of the ball, which we take to be the origin. The temperature at the point $(1, 2, 2)$ is $120°$.
(a) Find the rate of change of T at $(1, 2, 2)$ in the direction toward the point $(2, 1, 3)$.
(b) Show that at any point in the ball the direction of greatest increase in temperature is given by a vector that points toward the origin.

32. The temperature at a point (x, y, z) is given by

$$T(x, y, z) = 200e^{-x^2-3y^2-9z^2}$$

where T is measured in °C and x, y, z in meters.
(a) Find the rate of change of temperature at the point $P(2, -1, 2)$ in the direction toward the point $(3, -3, 3)$.
(b) In which direction does the temperature increase fastest at P?
(c) Find the maximum rate of increase at P.

33. Suppose that over a certain region of space the electrical potential V is given by $V(x, y, z) = 5x^2 - 3xy + xyz$.
(a) Find the rate of change of the potential at $P(3, 4, 5)$ in the direction of the vector $\mathbf{v} = \mathbf{i} + \mathbf{j} - \mathbf{k}$.
(b) In which direction does V change most rapidly at P?
(c) What is the maximum rate of change at P?

34. Suppose you are climbing a hill whose shape is given by the equation $z = 1000 - 0.005x^2 - 0.01y^2$, where x, y, and z are measured in meters, and you are standing at a point with coordinates $(60, 40, 966)$. The positive x-axis points east and the positive y-axis points north.
(a) If you walk due south, will you start to ascend or descend? At what rate?
(b) If you walk northwest, will you start to ascend or descend? At what rate?
(c) In which direction is the slope largest? What is the rate of ascent in that direction? At what angle above the horizontal does the path in that direction begin?

35. Let f be a function of two variables that has continuous partial derivatives and consider the points $A(1, 3)$, $B(3, 3)$, $C(1, 7)$, and $D(6, 15)$. The directional derivative of f at A in the direction of the vector $\overrightarrow{AB}$ is 3 and the directional derivative at A in the direction of $\overrightarrow{AC}$ is 26. Find the directional derivative of f at A in the direction of the vector $\overrightarrow{AD}$.

36. Shown is a topographic map of Blue River Pine Provincial Park in British Columbia. Draw curves of steepest descent from point A (descending to Mud Lake) and from point B.

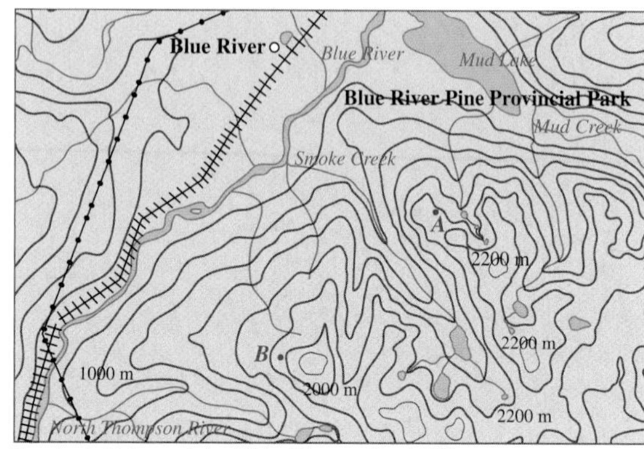

Reproduced with the permission of Natural Resources Canada 2009, courtesy of the Centre of Topographic Information.

37. Show that the operation of taking the gradient of a function has the given property. Assume that u and v are differentiable functions of x and y and that a, b are constants.
(a) $\nabla(au + bv) = a\,\nabla u + b\,\nabla v$ (b) $\nabla(uv) = u\,\nabla v + v\,\nabla u$
(c) $\nabla\left(\dfrac{u}{v}\right) = \dfrac{v\,\nabla u - u\,\nabla v}{v^2}$ (d) $\nabla u^n = nu^{n-1}\,\nabla u$

38. Sketch the gradient vector $\nabla f(4, 6)$ for the function f whose level curves are shown. Explain how you chose the direction and length of this vector.

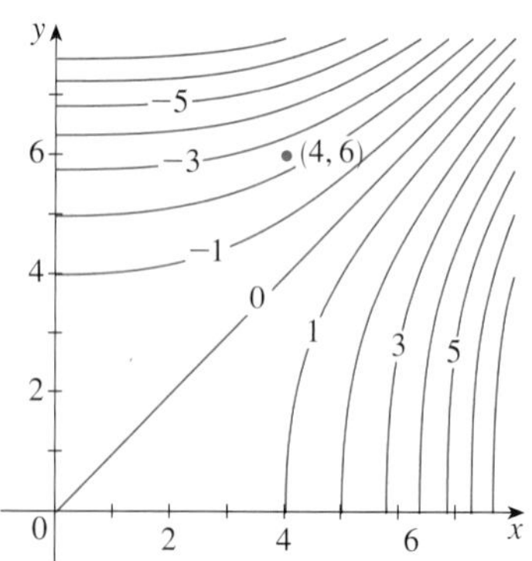

39. The **second directional derivative** of $f(x, y)$ is

$$D_{\mathbf{u}}^2 f(x, y) = D_{\mathbf{u}}[D_{\mathbf{u}} f(x, y)]$$

If $f(x, y) = x^3 + 5x^2y + y^3$ and $\mathbf{u} = \left\langle \frac{3}{5}, \frac{4}{5} \right\rangle$, calculate $D_{\mathbf{u}}^2 f(2, 1)$.

40. (a) If $\mathbf{u} = \langle a, b \rangle$ is a unit vector and f has continuous second partial derivatives, show that

$$D_{\mathbf{u}}^2 f = f_{xx}a^2 + 2f_{xy}ab + f_{yy}b^2$$

(b) Find the second directional derivative of $f(x, y) = xe^{2y}$ in the direction of $\mathbf{v} = \langle 4, 6 \rangle$.

41–46 Find equations of (a) the tangent plane and (b) the normal line to the given surface at the specified point.

41. $2(x - 2)^2 + (y - 1)^2 + (z - 3)^2 = 10$, $(3, 3, 5)$

42. $y = x^2 - z^2$, $(4, 7, 3)$

43. $xyz^2 = 6$, $(3, 2, 1)$

44. $xy + yz + zx = 5$, $(1, 2, 1)$

45. $x + y + z = e^{xyz}$, $(0, 0, 1)$

46. $x^4 + y^4 + z^4 = 3x^2y^2z^2$, $(1, 1, 1)$

47–48 Use a computer to graph the surface, the tangent plane, and the normal line on the same screen. Choose the domain carefully so that you avoid extraneous vertical planes. Choose the viewpoint so that you get a good view of all three objects.

47. $xy + yz + zx = 3$, $(1, 1, 1)$ **48.** $xyz = 6$, $(1, 2, 3)$

---

49. If $f(x, y) = xy$, find the gradient vector $\nabla f(3, 2)$ and use it to find the tangent line to the level curve $f(x, y) = 6$ at the point $(3, 2)$. Sketch the level curve, the tangent line, and the gradient vector.

50. If $g(x, y) = x^2 + y^2 - 4x$, find the gradient vector $\nabla g(1, 2)$ and use it to find the tangent line to the level curve $g(x, y) = 1$ at the point $(1, 2)$. Sketch the level curve, the tangent line, and the gradient vector.

51. Show that the equation of the tangent plane to the ellipsoid $x^2/a^2 + y^2/b^2 + z^2/c^2 = 1$ at the point (x_0, y_0, z_0) can be written as

$$\frac{xx_0}{a^2} + \frac{yy_0}{b^2} + \frac{zz_0}{c^2} = 1$$

52. Find the equation of the tangent plane to the hyperboloid $x^2/a^2 + y^2/b^2 - z^2/c^2 = 1$ at (x_0, y_0, z_0) and express it in a form similar to the one in Exercise 51.

53. Show that the equation of the tangent plane to the elliptic paraboloid $z/c = x^2/a^2 + y^2/b^2$ at the point (x_0, y_0, z_0) can be written as

$$\frac{2xx_0}{a^2} + \frac{2yy_0}{b^2} = \frac{z + z_0}{c}$$

54. At what point on the paraboloid $y = x^2 + z^2$ is the tangent plane parallel to the plane $x + 2y + 3z = 1$?

55. Are there any points on the hyperboloid $x^2 - y^2 - z^2 = 1$ where the tangent plane is parallel to the plane $z = x + y$?

56. Show that the ellipsoid $3x^2 + 2y^2 + z^2 = 9$ and the sphere $x^2 + y^2 + z^2 - 8x - 6y - 8z + 24 = 0$ are tangent to each other at the point $(1, 1, 2)$. (This means that they have a common tangent plane at the point.)

57. Show that every plane that is tangent to the cone $x^2 + y^2 = z^2$ passes through the origin.

58. Show that every normal line to the sphere $x^2 + y^2 + z^2 = r^2$ passes through the center of the sphere.

59. Where does the normal line to the paraboloid $z = x^2 + y^2$ at the point $(1, 1, 2)$ intersect the paraboloid a second time?

60. At what points does the normal line through the point $(1, 2, 1)$ on the ellipsoid $4x^2 + y^2 + 4z^2 = 12$ intersect the sphere $x^2 + y^2 + z^2 = 102$?

61. Show that the sum of the x-, y-, and z-intercepts of any tangent plane to the surface $\sqrt{x} + \sqrt{y} + \sqrt{z} = \sqrt{c}$ is a constant.

62. Show that the pyramids cut off from the first octant by any tangent planes to the surface $xyz = 1$ at points in the first octant must all have the same volume.

63. Find parametric equations for the tangent line to the curve of intersection of the paraboloid $z = x^2 + y^2$ and the ellipsoid $4x^2 + y^2 + z^2 = 9$ at the point $(-1, 1, 2)$.

64. (a) The plane $y + z = 3$ intersects the cylinder $x^2 + y^2 = 5$ in an ellipse. Find parametric equations for the tangent line to this ellipse at the point $(1, 2, 1)$.

(b) Graph the cylinder, the plane, and the tangent line on the same screen.

65. (a) Two surfaces are called **orthogonal** at a point of intersection if their normal lines are perpendicular at that point. Show that surfaces with equations $F(x, y, z) = 0$ and $G(x, y, z) = 0$ are orthogonal at a point P where $\nabla F \neq \mathbf{0}$ and $\nabla G \neq \mathbf{0}$ if and only if

$$F_xG_x + F_yG_y + F_zG_z = 0 \quad \text{at } P$$

(b) Use part (a) to show that the surfaces $z^2 = x^2 + y^2$ and $x^2 + y^2 + z^2 = r^2$ are orthogonal at every point of intersection. Can you see why this is true without using calculus?

66. (a) Show that the function $f(x, y) = \sqrt[3]{xy}$ is continuous and the partial derivatives f_x and f_y exist at the origin but the directional derivatives in all other directions do not exist.

(b) Graph f near the origin and comment on how the graph confirms part (a).

67. Suppose that the directional derivatives of $f(x, y)$ are known at a given point in two nonparallel directions given by unit vectors $\mathbf{u}$ and $\mathbf{v}$. Is it possible to find ∇f at this point? If so, how would you do it?

68. Show that if $z = f(x, y)$ is differentiable at $\mathbf{x}_0 = \langle x_0, y_0 \rangle$, then

$$\lim_{\mathbf{x} \to \mathbf{x}_0} \frac{f(\mathbf{x}) - f(\mathbf{x}_0) - \nabla f(\mathbf{x}_0) \cdot (\mathbf{x} - \mathbf{x}_0)}{|\mathbf{x} - \mathbf{x}_0|} = 0$$

[*Hint:* Use Definition 14.4.7 directly.]

14.7 Maximum and Minimum Values

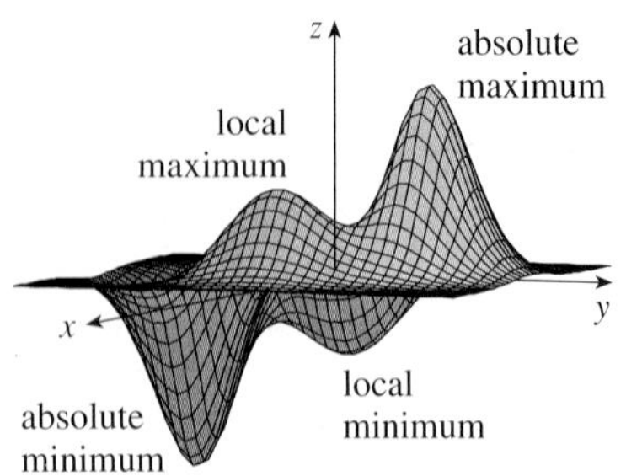

FIGURE 1

As we saw in Chapter 4, one of the main uses of ordinary derivatives is in finding maximum and minimum values (extreme values). In this section we see how to use partial derivatives to locate maxima and minima of functions of two variables. In particular, in Example 6 we will see how to maximize the volume of a box without a lid if we have a fixed amount of cardboard to work with.

Look at the hills and valleys in the graph of f shown in Figure 1. There are two points (a, b) where f has a *local maximum*, that is, where $f(a, b)$ is larger than nearby values of $f(x, y)$. The larger of these two values is the *absolute maximum*. Likewise, f has two *local minima*, where $f(a, b)$ is smaller than nearby values. The smaller of these two values is the *absolute minimum*.

1 Definition A function of two variables has a **local maximum** at (a, b) if $f(x, y) \leqslant f(a, b)$ when (x, y) is near (a, b). [This means that $f(x, y) \leqslant f(a, b)$ for all points (x, y) in some disk with center (a, b).] The number $f(a, b)$ is called a **local maximum value**. If $f(x, y) \geqslant f(a, b)$ when (x, y) is near (a, b), then f has a **local minimum** at (a, b) and $f(a, b)$ is a **local minimum value**.

If the inequalities in Definition 1 hold for *all* points (x, y) in the domain of f, then f has an **absolute maximum** (or **absolute minimum**) at (a, b).

Notice that the conclusion of Theorem 2 can be stated in the notation of gradient vectors as $\nabla f(a, b) = \mathbf{0}$.

2 Theorem If f has a local maximum or minimum at (a, b) and the first-order partial derivatives of f exist there, then $f_x(a, b) = 0$ and $f_y(a, b) = 0$.

PROOF Let $g(x) = f(x, b)$. If f has a local maximum (or minimum) at (a, b), then g has a local maximum (or minimum) at a, so $g'(a) = 0$ by Fermat's Theorem (see Theorem 4.1.4). But $g'(a) = f_x(a, b)$ (see Equation 14.3.1) and so $f_x(a, b) = 0$. Similarly, by applying Fermat's Theorem to the function $G(y) = f(a, y)$, we obtain $f_y(a, b) = 0$. ■

If we put $f_x(a, b) = 0$ and $f_y(a, b) = 0$ in the equation of a tangent plane (Equation 14.4.2), we get $z = z_0$. Thus the geometric interpretation of Theorem 2 is that if the graph of f has a tangent plane at a local maximum or minimum, then the tangent plane must be horizontal.

A point (a, b) is called a **critical point** (or *stationary point*) of f if $f_x(a, b) = 0$ and $f_y(a, b) = 0$, or if one of these partial derivatives does not exist. Theorem 2 says that if f has a local maximum or minimum at (a, b), then (a, b) is a critical point of f. However, as in single-variable calculus, not all critical points give rise to maxima or minima. At a critical point, a function could have a local maximum or a local minimum or neither.

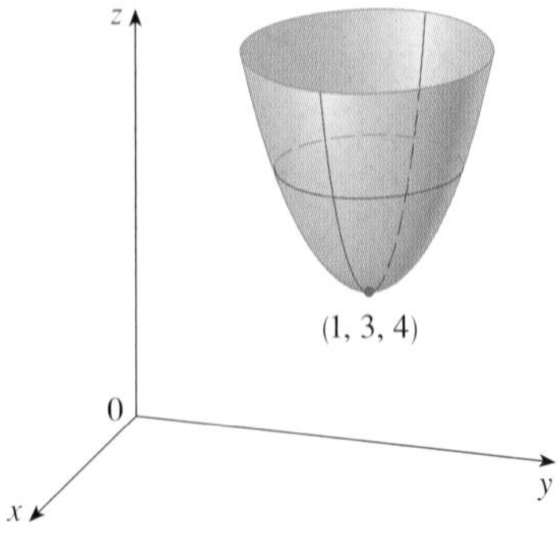

EXAMPLE 1 Let $f(x, y) = x^2 + y^2 - 2x - 6y + 14$. Then

$$f_x(x, y) = 2x - 2 \qquad f_y(x, y) = 2y - 6$$

These partial derivatives are equal to 0 when $x = 1$ and $y = 3$, so the only critical point is $(1, 3)$. By completing the square, we find that

$$f(x, y) = 4 + (x - 1)^2 + (y - 3)^2$$

Since $(x - 1)^2 \geqslant 0$ and $(y - 3)^2 \geqslant 0$, we have $f(x, y) \geqslant 4$ for all values of x and y. Therefore $f(1, 3) = 4$ is a local minimum, and in fact it is the absolute minimum of f.

z
(1, 3, 4)
0
x
y

FIGURE 2
$z = x^2 + y^2 - 2x - 6y + 14$

This can be confirmed geometrically from the graph of f, which is the elliptic paraboloid with vertex $(1, 3, 4)$ shown in Figure 2.

EXAMPLE 2 Find the extreme values of $f(x, y) = y^2 - x^2$.

SOLUTION Since $f_x = -2x$ and $f_y = 2y$, the only critical point is $(0, 0)$. Notice that for points on the x-axis we have $y = 0$, so $f(x, y) = -x^2 < 0$ (if $x \neq 0$). However, for points on the y-axis we have $x = 0$, so $f(x, y) = y^2 > 0$ (if $y \neq 0$). Thus every disk with center $(0, 0)$ contains points where f takes positive values as well as points where f takes negative values. Therefore $f(0, 0) = 0$ can't be an extreme value for f, so f has no extreme value.

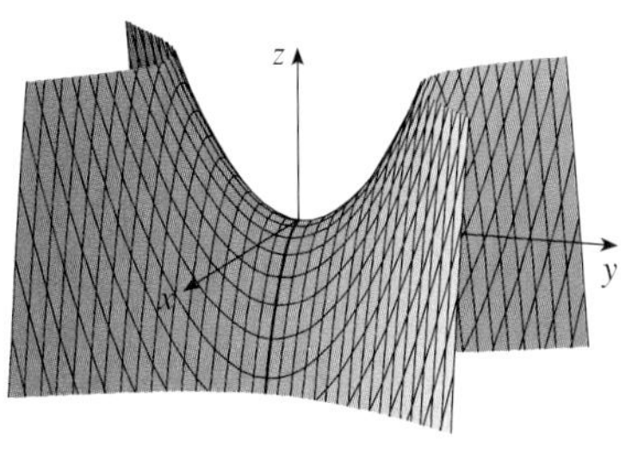

FIGURE 3
$z = y^2 - x^2$

Example 2 illustrates the fact that a function need not have a maximum or minimum value at a critical point. Figure 3 shows how this is possible. The graph of f is the hyperbolic paraboloid $z = y^2 - x^2$, which has a horizontal tangent plane ($z = 0$) at the origin. You can see that $f(0, 0) = 0$ is a maximum in the direction of the x-axis but a minimum in the direction of the y-axis. Near the origin the graph has the shape of a saddle and so $(0, 0)$ is called a *saddle point* of f.

A mountain pass also has the shape of a saddle. As the photograph of the geological formation illustrates, for people hiking in one direction the saddle point is the lowest point on their route, while for those traveling in a different direction the saddle point is the highest point.

We need to be able to determine whether or not a function has an extreme value at a critical point. The following test, which is proved at the end of this section, is analogous to the Second Derivative Test for functions of one variable.

3 Second Derivatives Test Suppose the second partial derivatives of f are continuous on a disk with center (a, b), and suppose that $f_x(a, b) = 0$ and $f_y(a, b) = 0$ [that is, (a, b) is a critical point of f]. Let

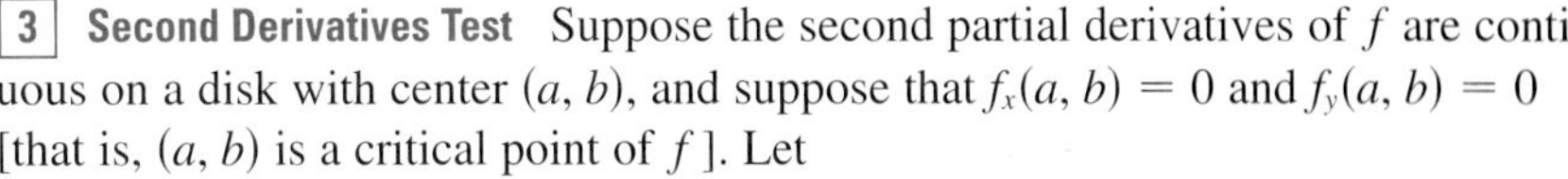

$$D = D(a, b) = f_{xx}(a, b)\,f_{yy}(a, b) - [f_{xy}(a, b)]^2$$

(a) If $D > 0$ and $f_{xx}(a, b) > 0$, then $f(a, b)$ is a local minimum.

(b) If $D > 0$ and $f_{xx}(a, b) < 0$, then $f(a, b)$ is a local maximum.

(c) If $D < 0$, then $f(a, b)$ is not a local maximum or minimum.

NOTE 1 In case (c) the point (a, b) is called a **saddle point** of f and the graph of f crosses its tangent plane at (a, b).

NOTE 2 If $D = 0$, the test gives no information: f could have a local maximum or local minimum at (a, b), or (a, b) could be a saddle point of f.

NOTE 3 To remember the formula for D, it's helpful to write it as a determinant:

$$D = \begin{vmatrix} f_{xx} & f_{xy} \\ f_{yx} & f_{yy} \end{vmatrix} = f_{xx} f_{yy} - (f_{xy})^2$$

V EXAMPLE 3 Find the local maximum and minimum values and saddle points of $f(x, y) = x^4 + y^4 - 4xy + 1$.

SOLUTION We first locate the critical points:

$$f_x = 4x^3 - 4y \qquad f_y = 4y^3 - 4x$$

Setting these partial derivatives equal to 0, we obtain the equations

$$x^3 - y = 0 \qquad \text{and} \qquad y^3 - x = 0$$

To solve these equations we substitute $y = x^3$ from the first equation into the second one. This gives

$$0 = x^9 - x = x(x^8 - 1) = x(x^4 - 1)(x^4 + 1) = x(x^2 - 1)(x^2 + 1)(x^4 + 1)$$

so there are three real roots: $x = 0, 1, -1$. The three critical points are $(0, 0)$, $(1, 1)$, and $(-1, -1)$.

Next we calculate the second partial derivatives and $D(x, y)$:

$$f_{xx} = 12x^2 \qquad f_{xy} = -4 \qquad f_{yy} = 12y^2$$

$$D(x, y) = f_{xx}f_{yy} - (f_{xy})^2 = 144x^2y^2 - 16$$

Since $D(0, 0) = -16 < 0$, it follows from case (c) of the Second Derivatives Test that the origin is a saddle point; that is, f has no local maximum or minimum at $(0, 0)$. Since $D(1, 1) = 128 > 0$ and $f_{xx}(1, 1) = 12 > 0$, we see from case (a) of the test that $f(1, 1) = -1$ is a local minimum. Similarly, we have $D(-1, -1) = 128 > 0$ and $f_{xx}(-1, -1) = 12 > 0$, so $f(-1, -1) = -1$ is also a local minimum.

The graph of f is shown in Figure 4.

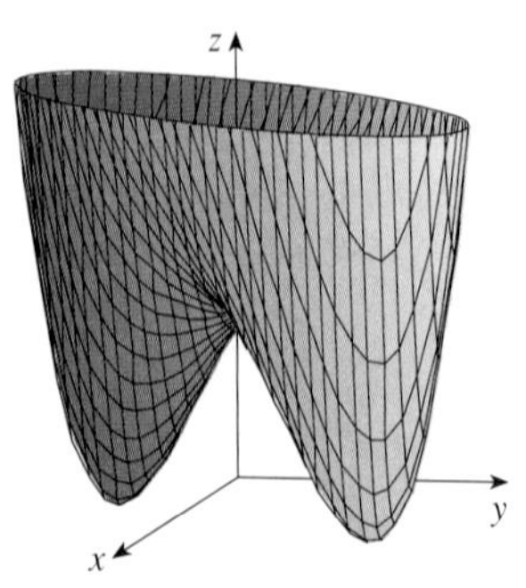

FIGURE 4
$z = x^4 + y^4 - 4xy + 1$

A contour map of the function f in Example 3 is shown in Figure 5. The level curves near $(1, 1)$ and $(-1, -1)$ are oval in shape and indicate that as we move away from $(1, 1)$ or $(-1, -1)$ in any direction the values of f are increasing. The level curves near $(0, 0)$, on the other hand, resemble hyperbolas. They reveal that as we move away from the origin (where the value of f is 1), the values of f decrease in some directions but increase in other directions. Thus the contour map suggests the presence of the minima and saddle point that we found in Example 3.

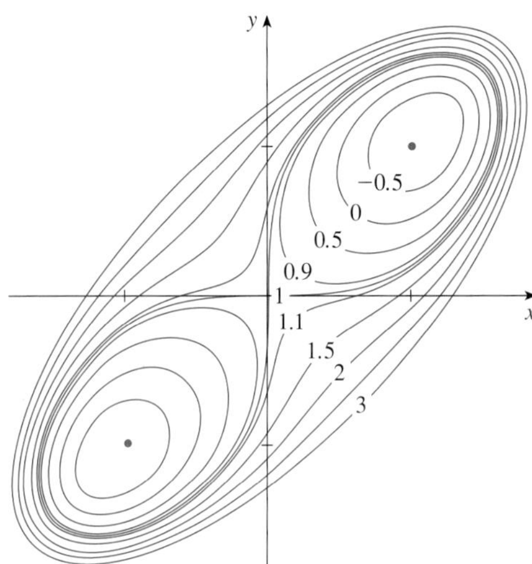

FIGURE 5

TEC In Module 14.7 you can use contour maps to estimate the locations of critical points.

EXAMPLE 4 Find and classify the critical points of the function

$$f(x, y) = 10x^2y - 5x^2 - 4y^2 - x^4 - 2y^4$$

Also find the highest point on the graph of f.

SOLUTION The first-order partial derivatives are

$$f_x = 20xy - 10x - 4x^3 \qquad f_y = 10x^2 - 8y - 8y^3$$

So to find the critical points we need to solve the equations

$$\boxed{4} \qquad 2x(10y - 5 - 2x^2) = 0$$

$$\boxed{5} \qquad 5x^2 - 4y - 4y^3 = 0$$

From Equation 4 we see that either

$$x = 0 \qquad \text{or} \qquad 10y - 5 - 2x^2 = 0$$

In the first case ($x = 0$), Equation 5 becomes $-4y(1 + y^2) = 0$, so $y = 0$ and we have the critical point $(0, 0)$.

In the second case ($10y - 5 - 2x^2 = 0$), we get

$$\boxed{6} \qquad x^2 = 5y - 2.5$$

and, putting this in Equation 5, we have $25y - 12.5 - 4y - 4y^3 = 0$. So we have to solve the cubic equation

$$\boxed{7} \qquad 4y^3 - 21y + 12.5 = 0$$

Using a graphing calculator or computer to graph the function

$$g(y) = 4y^3 - 21y + 12.5$$

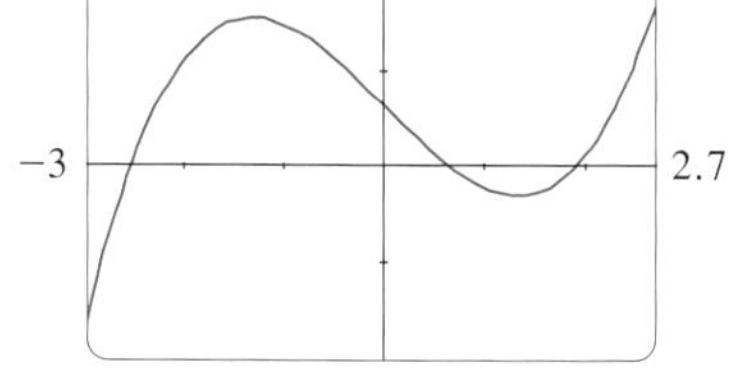

FIGURE 6

as in Figure 6, we see that Equation 7 has three real roots. By zooming in, we can find the roots to four decimal places:

$$y \approx -2.5452 \qquad y \approx 0.6468 \qquad y \approx 1.8984$$

(Alternatively, we could have used Newton's method or a rootfinder to locate these roots.) From Equation 6, the corresponding x-values are given by

$$x = \pm\sqrt{5y - 2.5}$$

If $y \approx -2.5452$, then x has no corresponding real values. If $y \approx 0.6468$, then $x \approx \pm 0.8567$. If $y \approx 1.8984$, then $x \approx \pm 2.6442$. So we have a total of five critical points, which are analyzed in the following chart. All quantities are rounded to two decimal places.

Critical point	Value of f	f_{xx}	D	Conclusion
$(0, 0)$	0.00	−10.00	80.00	local maximum
$(\pm 2.64, 1.90)$	8.50	−55.93	2488.72	local maximum
$(\pm 0.86, 0.65)$	−1.48	−5.87	−187.64	saddle point

Figures 7 and 8 give two views of the graph of f and we see that the surface opens downward. [This can also be seen from the expression for $f(x, y)$: The dominant terms are $-x^4 - 2y^4$ when $|x|$ and $|y|$ are large.] Comparing the values of f at its local maximum points, we see that the absolute maximum value of f is $f(\pm 2.64, 1.90) \approx 8.50$. In other words, the highest points on the graph of f are $(\pm 2.64, 1.90, 8.50)$.

TEC Visual 14.7 shows several families of surfaces. The surface in Figures 7 and 8 is a member of one of these families.

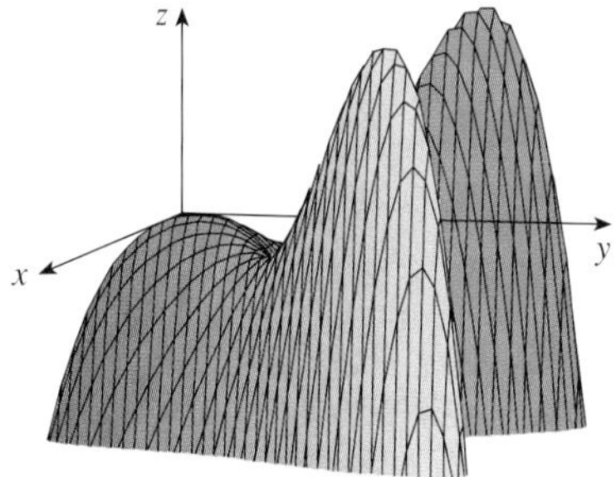

FIGURE 7

FIGURE 8

The five critical points of the function f in Example 4 are shown in red in the contour map of f in Figure 9.

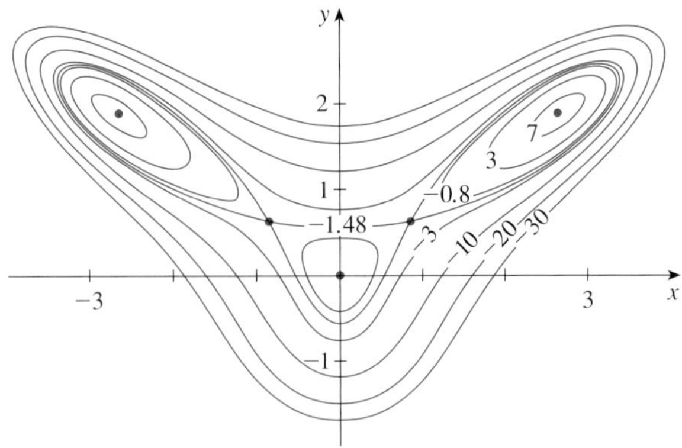

FIGURE 9

V EXAMPLE 5 Find the shortest distance from the point $(1, 0, -2)$ to the plane $x + 2y + z = 4$.

SOLUTION The distance from any point (x, y, z) to the point $(1, 0, -2)$ is

$$d = \sqrt{(x - 1)^2 + y^2 + (z + 2)^2}$$

but if (x, y, z) lies on the plane $x + 2y + z = 4$, then $z = 4 - x - 2y$ and so we have $d = \sqrt{(x - 1)^2 + y^2 + (6 - x - 2y)^2}$. We can minimize d by minimizing the simpler expression

$$d^2 = f(x, y) = (x - 1)^2 + y^2 + (6 - x - 2y)^2$$

By solving the equations

$$f_x = 2(x - 1) - 2(6 - x - 2y) = 4x + 4y - 14 = 0$$

$$f_y = 2y - 4(6 - x - 2y) = 4x + 10y - 24 = 0$$

we find that the only critical point is $\left(\frac{11}{6}, \frac{5}{3}\right)$. Since $f_{xx} = 4$, $f_{xy} = 4$, and $f_{yy} = 10$, we have $D(x, y) = f_{xx}f_{yy} - (f_{xy})^2 = 24 > 0$ and $f_{xx} > 0$, so by the Second Derivatives Test f has a local minimum at $\left(\frac{11}{6}, \frac{5}{3}\right)$. Intuitively, we can see that this local minimum is actually an absolute minimum because there must be a point on the given plane that is closest to $(1, 0, -2)$. If $x = \frac{11}{6}$ and $y = \frac{5}{3}$, then

Example 5 could also be solved using vectors. Compare with the methods of Section 12.5.

$$d = \sqrt{(x - 1)^2 + y^2 + (6 - x - 2y)^2} = \sqrt{\left(\tfrac{5}{6}\right)^2 + \left(\tfrac{5}{3}\right)^2 + \left(\tfrac{5}{6}\right)^2} = \tfrac{5}{6}\sqrt{6}$$

The shortest distance from $(1, 0, -2)$ to the plane $x + 2y + z = 4$ is $\frac{5}{6}\sqrt{6}$.

V EXAMPLE 6 A rectangular box without a lid is to be made from 12 m^2 of cardboard. Find the maximum volume of such a box.

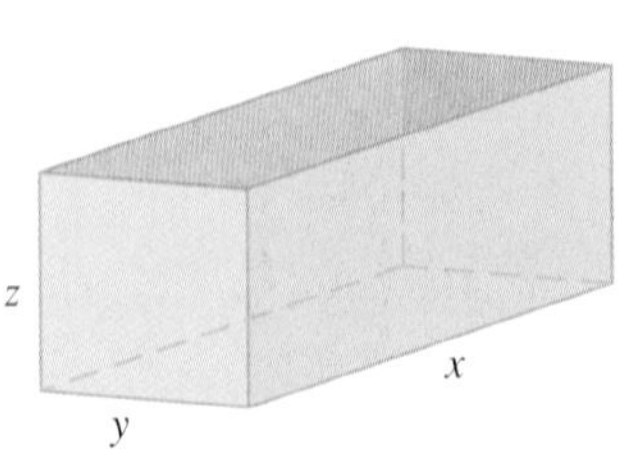

FIGURE 10

SOLUTION Let the length, width, and height of the box (in meters) be x, y, and z, as shown in Figure 10. Then the volume of the box is

$$V = xyz$$

We can express V as a function of just two variables x and y by using the fact that the area of the four sides and the bottom of the box is

$$2xz + 2yz + xy = 12$$

Solving this equation for z, we get $z = (12 - xy)/[2(x + y)]$, so the expression for V becomes

$$V = xy \frac{12 - xy}{2(x + y)} = \frac{12xy - x^2y^2}{2(x + y)}$$

We compute the partial derivatives:

$$\frac{\partial V}{\partial x} = \frac{y^2(12 - 2xy - x^2)}{2(x + y)^2} \qquad \frac{\partial V}{\partial y} = \frac{x^2(12 - 2xy - y^2)}{2(x + y)^2}$$

If V is a maximum, then $\partial V/\partial x = \partial V/\partial y = 0$, but $x = 0$ or $y = 0$ gives $V = 0$, so we must solve the equations

$$12 - 2xy - x^2 = 0 \qquad 12 - 2xy - y^2 = 0$$

These imply that $x^2 = y^2$ and so $x = y$. (Note that x and y must both be positive in this problem.) If we put $x = y$ in either equation we get $12 - 3x^2 = 0$, which gives $x = 2$, $y = 2$, and $z = (12 - 2 \cdot 2)/[2(2 + 2)] = 1$.

We could use the Second Derivatives Test to show that this gives a local maximum of V, or we could simply argue from the physical nature of this problem that there must be an absolute maximum volume, which has to occur at a critical point of V, so it must occur when $x = 2$, $y = 2$, $z = 1$. Then $V = 2 \cdot 2 \cdot 1 = 4$, so the maximum volume of the box is 4 m^3.

Absolute Maximum and Minimum Values

For a function f of one variable, the Extreme Value Theorem says that if f is continuous on a closed interval $[a, b]$, then f has an absolute minimum value and an absolute maximum value. According to the Closed Interval Method in Section 4.1, we found these by evaluating f not only at the critical numbers but also at the endpoints a and b.

There is a similar situation for functions of two variables. Just as a closed interval contains its endpoints, a **closed set** in $\mathbb{R}^2$ is one that contains all its boundary points. [A boundary point of D is a point (a, b) such that every disk with center (a, b) contains points in D and also points not in D.] For instance, the disk

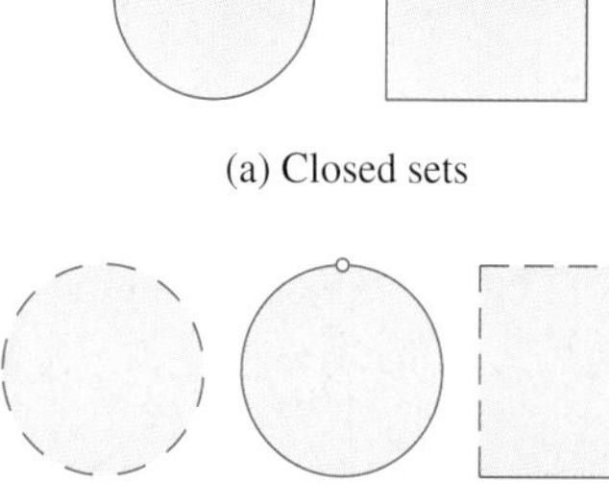

(a) Closed sets

(b) Sets that are not closed

FIGURE 11

$$D = \{(x, y) \mid x^2 + y^2 \leqslant 1\}$$

which consists of all points on and inside the circle $x^2 + y^2 = 1$, is a closed set because it contains all of its boundary points (which are the points on the circle $x^2 + y^2 = 1$). But if even one point on the boundary curve were omitted, the set would not be closed. (See Figure 11.)

A **bounded set** in $\mathbb{R}^2$ is one that is contained within some disk. In other words, it is finite in extent. Then, in terms of closed and bounded sets, we can state the following counterpart of the Extreme Value Theorem in two dimensions.

> **8 Extreme Value Theorem for Functions of Two Variables** If f is continuous on a closed, bounded set D in $\mathbb{R}^2$, then f attains an absolute maximum value $f(x_1, y_1)$ and an absolute minimum value $f(x_2, y_2)$ at some points (x_1, y_1) and (x_2, y_2) in D.

To find the extreme values guaranteed by Theorem 8, we note that, by Theorem 2, if f has an extreme value at (x_1, y_1), then (x_1, y_1) is either a critical point of f or a boundary point of D. Thus we have the following extension of the Closed Interval Method.

9 To find the absolute maximum and minimum values of a continuous function f on a closed, bounded set D:

1. Find the values of f at the critical points of f in D.
2. Find the extreme values of f on the boundary of D.
3. The largest of the values from steps 1 and 2 is the absolute maximum value; the smallest of these values is the absolute minimum value.

EXAMPLE 7 Find the absolute maximum and minimum values of the function $f(x, y) = x^2 - 2xy + 2y$ on the rectangle $D = \{(x, y) \mid 0 \leqslant x \leqslant 3, 0 \leqslant y \leqslant 2\}$.

SOLUTION Since f is a polynomial, it is continuous on the closed, bounded rectangle D, so Theorem 8 tells us there is both an absolute maximum and an absolute minimum. According to step 1 in **9**, we first find the critical points. These occur when

$$f_x = 2x - 2y = 0 \qquad f_y = -2x + 2 = 0$$

so the only critical point is $(1, 1)$, and the value of f there is $f(1, 1) = 1$.

In step 2 we look at the values of f on the boundary of D, which consists of the four line segments L_1, L_2, L_3, L_4 shown in Figure 12. On L_1 we have $y = 0$ and

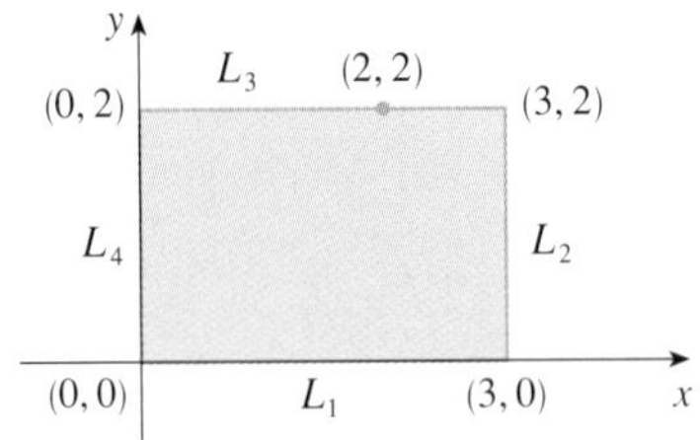

FIGURE 12

$$f(x, 0) = x^2 \qquad 0 \leqslant x \leqslant 3$$

This is an increasing function of x, so its minimum value is $f(0, 0) = 0$ and its maximum value is $f(3, 0) = 9$. On L_2 we have $x = 3$ and

$$f(3, y) = 9 - 4y \qquad 0 \leqslant y \leqslant 2$$

This is a decreasing function of y, so its maximum value is $f(3, 0) = 9$ and its minimum value is $f(3, 2) = 1$. On L_3 we have $y = 2$ and

$$f(x, 2) = x^2 - 4x + 4 \qquad 0 \leqslant x \leqslant 3$$

By the methods of Chapter 4, or simply by observing that $f(x, 2) = (x - 2)^2$, we see that the minimum value of this function is $f(2, 2) = 0$ and the maximum value is $f(0, 2) = 4$. Finally, on L_4 we have $x = 0$ and

$$f(0, y) = 2y \qquad 0 \leqslant y \leqslant 2$$

with maximum value $f(0, 2) = 4$ and minimum value $f(0, 0) = 0$. Thus, on the boundary, the minimum value of f is 0 and the maximum is 9.

In step 3 we compare these values with the value $f(1, 1) = 1$ at the critical point and conclude that the absolute maximum value of f on D is $f(3, 0) = 9$ and the absolute minimum value is $f(0, 0) = f(2, 2) = 0$. Figure 13 shows the graph of f.

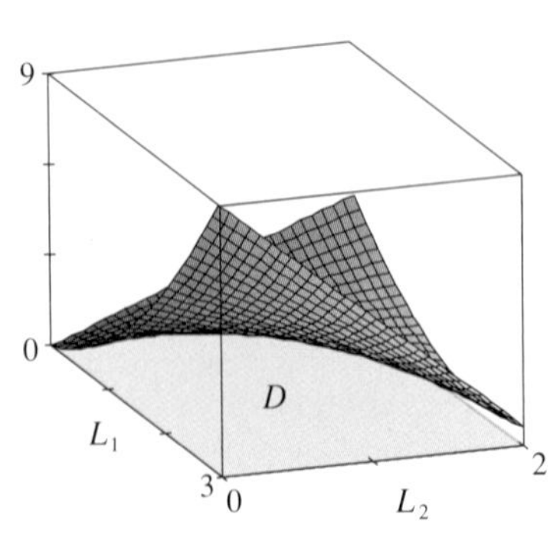

FIGURE 13
$f(x, y) = x^2 - 2xy + 2y$

We close this section by giving a proof of the first part of the Second Derivatives Test. Part (b) has a similar proof.

PROOF OF THEOREM 3, PART (a) We compute the second-order directional derivative of f in the direction of $\mathbf{u} = \langle h, k \rangle$. The first-order derivative is given by Theorem 14.6.3:

$$D_{\mathbf{u}} f = f_x h + f_y k$$

Applying this theorem a second time, we have

$$\begin{aligned} D_{\mathbf{u}}^2 f = D_{\mathbf{u}}(D_{\mathbf{u}} f) &= \frac{\partial}{\partial x}(D_{\mathbf{u}} f)h + \frac{\partial}{\partial y}(D_{\mathbf{u}} f)k \\ &= (f_{xx}h + f_{yx}k)h + (f_{xy}h + f_{yy}k)k \\ &= f_{xx}h^2 + 2f_{xy}hk + f_{yy}k^2 \qquad \text{(by Clairaut's Theorem)} \end{aligned}$$

If we complete the square in this expression, we obtain

$$\boxed{10} \qquad D_{\mathbf{u}}^2 f = f_{xx}\left(h + \frac{f_{xy}}{f_{xx}}k\right)^2 + \frac{k^2}{f_{xx}}(f_{xx}f_{yy} - f_{xy}^2)$$

We are given that $f_{xx}(a, b) > 0$ and $D(a, b) > 0$. But f_{xx} and $D = f_{xx}f_{yy} - f_{xy}^2$ are continuous functions, so there is a disk B with center (a, b) and radius $\delta > 0$ such that $f_{xx}(x, y) > 0$ and $D(x, y) > 0$ whenever (x, y) is in B. Therefore, by looking at Equation 10, we see that $D_{\mathbf{u}}^2 f(x, y) > 0$ whenever (x, y) is in B. This means that if C is the curve obtained by intersecting the graph of f with the vertical plane through $P(a, b, f(a, b))$ in the direction of $\mathbf{u}$, then C is concave upward on an interval of length 2δ. This is true in the direction of every vector $\mathbf{u}$, so if we restrict (x, y) to lie in B, the graph of f lies above its horizontal tangent plane at P. Thus $f(x, y) \geqslant f(a, b)$ whenever (x, y) is in B. This shows that $f(a, b)$ is a local minimum. ∎

14.7 Exercises

1. Suppose $(1, 1)$ is a critical point of a function f with continuous second derivatives. In each case, what can you say about f?

(a) $f_{xx}(1, 1) = 4, \quad f_{xy}(1, 1) = 1, \quad f_{yy}(1, 1) = 2$

(b) $f_{xx}(1, 1) = 4, \quad f_{xy}(1, 1) = 3, \quad f_{yy}(1, 1) = 2$

2. Suppose $(0, 2)$ is a critical point of a function g with continuous second derivatives. In each case, what can you say about g?

(a) $g_{xx}(0, 2) = -1, \quad g_{xy}(0, 2) = 6, \quad g_{yy}(0, 2) = 1$

(b) $g_{xx}(0, 2) = -1, \quad g_{xy}(0, 2) = 2, \quad g_{yy}(0, 2) = -8$

(c) $g_{xx}(0, 2) = 4, \quad g_{xy}(0, 2) = 6, \quad g_{yy}(0, 2) = 9$

3–4 Use the level curves in the figure to predict the location of the critical points of f and whether f has a saddle point or a local maximum or minimum at each critical point. Explain your reasoning. Then use the Second Derivatives Test to confirm your predictions.

3. $f(x, y) = 4 + x^3 + y^3 - 3xy$

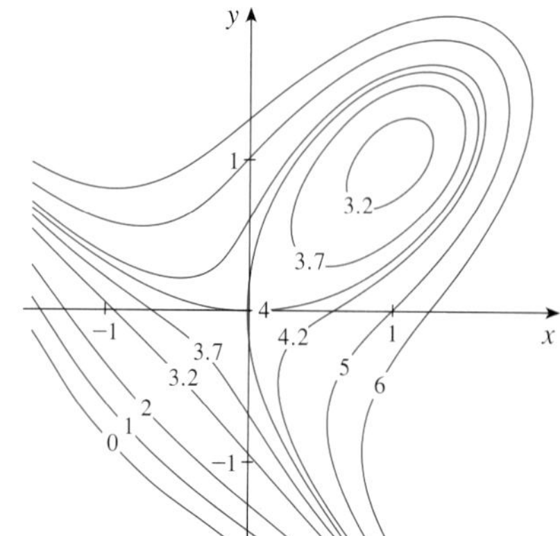

4. $f(x, y) = 3x - x^3 - 2y^2 + y^4$

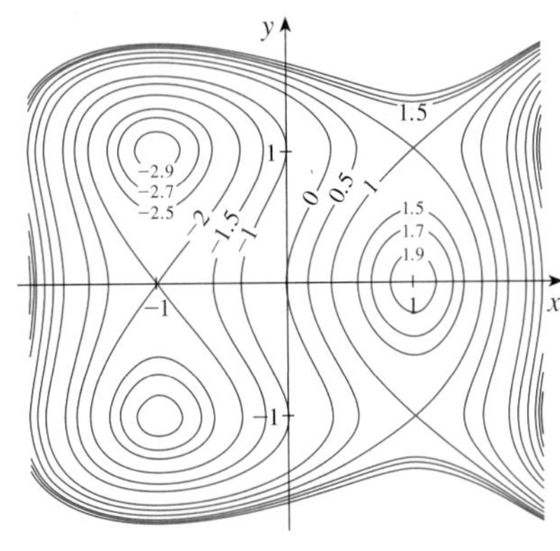

5–18 Find the local maximum and minimum values and saddle point(s) of the function. If you have three-dimensional graphing software, graph the function with a domain and viewpoint that reveal all the important aspects of the function.

5. $f(x, y) = x^2 + xy + y^2 + y$

6. $f(x, y) = xy - 2x - 2y - x^2 - y^2$

7. $f(x, y) = (x - y)(1 - xy)$

8. $f(x, y) = xe^{-2x^2-2y^2}$

9. $f(x, y) = y^3 + 3x^2y - 6x^2 - 6y^2 + 2$

10. $f(x, y) = xy(1 - x - y)$

11. $f(x, y) = x^3 - 12xy + 8y^3$

12. $f(x, y) = xy + \dfrac{1}{x} + \dfrac{1}{y}$

13. $f(x, y) = e^x \cos y$

14. $f(x, y) = y \cos x$

15. $f(x, y) = (x^2 + y^2)e^{y^2-x^2}$

16. $f(x, y) = e^y(y^2 - x^2)$

17. $f(x, y) = y^2 - 2y \cos x, \quad -1 \leqslant x \leqslant 7$

18. $f(x, y) = \sin x \sin y, \quad -\pi < x < \pi, \quad -\pi < y < \pi$

19. Show that $f(x, y) = x^2 + 4y^2 - 4xy + 2$ has an infinite number of critical points and that $D = 0$ at each one. Then show that f has a local (and absolute) minimum at each critical point.

20. Show that $f(x, y) = x^2ye^{-x^2-y^2}$ has maximum values at $(\pm 1, 1/\sqrt{2})$ and minimum values at $(\pm 1, -1/\sqrt{2})$. Show also that f has infinitely many other critical points and $D = 0$ at each of them. Which of them give rise to maximum values? Minimum values? Saddle points?

21–24 Use a graph or level curves or both to estimate the local maximum and minimum values and saddle point(s) of the function. Then use calculus to find these values precisely.

21. $f(x, y) = x^2 + y^2 + x^{-2}y^{-2}$

22. $f(x, y) = xye^{-x^2-y^2}$

23. $f(x, y) = \sin x + \sin y + \sin(x + y)$,
$0 \leqslant x \leqslant 2\pi, \ 0 \leqslant y \leqslant 2\pi$

24. $f(x, y) = \sin x + \sin y + \cos(x + y)$,
$0 \leqslant x \leqslant \pi/4, \ 0 \leqslant y \leqslant \pi/4$

25–28 Use a graphing device as in Example 4 (or Newton's method or a rootfinder) to find the critical points of f correct to three decimal places. Then classify the critical points and find the highest or lowest points on the graph, if any.

25. $f(x, y) = x^4 + y^4 - 4x^2y + 2y$

26. $f(x, y) = y^6 - 2y^4 + x^2 - y^2 + y$

27. $f(x, y) = x^4 + y^3 - 3x^2 + y^2 + x - 2y + 1$

28. $f(x, y) = 20e^{-x^2-y^2} \sin 3x \cos 3y, \quad |x| \leqslant 1, \quad |y| \leqslant 1$

29–36 Find the absolute maximum and minimum values of f on the set D.

29. $f(x, y) = x^2 + y^2 - 2x$, D is the closed triangular region with vertices $(2, 0)$, $(0, 2)$, and $(0, -2)$

30. $f(x, y) = x + y - xy$, D is the closed triangular region with vertices $(0, 0)$, $(0, 2)$, and $(4, 0)$

31. $f(x, y) = x^2 + y^2 + x^2y + 4$,
$D = \{(x, y) \mid |x| \leqslant 1, |y| \leqslant 1\}$

32. $f(x, y) = 4x + 6y - x^2 - y^2$,
$D = \{(x, y) \mid 0 \leqslant x \leqslant 4, 0 \leqslant y \leqslant 5\}$

33. $f(x, y) = x^4 + y^4 - 4xy + 2$,
$D = \{(x, y) \mid 0 \leqslant x \leqslant 3, 0 \leqslant y \leqslant 2\}$

34. $f(x, y) = xy^2, \quad D = \{(x, y) \mid x \geqslant 0, \ y \geqslant 0, x^2 + y^2 \leqslant 3\}$

35. $f(x, y) = 2x^3 + y^4, \quad D = \{(x, y) \mid x^2 + y^2 \leqslant 1\}$

36. $f(x, y) = x^3 - 3x - y^3 + 12y$, D is the quadrilateral whose vertices are $(-2, 3)$, $(2, 3)$, $(2, 2)$, and $(-2, -2)$.

37. For functions of one variable it is impossible for a continuous function to have two local maxima and no local minimum. But for functions of two variables such functions exist. Show that the function

$$f(x, y) = -(x^2 - 1)^2 - (x^2y - x - 1)^2$$

has only two critical points, but has local maxima at both of them. Then use a computer to produce a graph with a carefully chosen domain and viewpoint to see how this is possible.

38. If a function of one variable is continuous on an interval and has only one critical number, then a local maximum has to be

an absolute maximum. But this is not true for functions of two variables. Show that the function

$$f(x, y) = 3xe^y - x^3 - e^{3y}$$

has exactly one critical point, and that f has a local maximum there that is not an absolute maximum. Then use a computer to produce a graph with a carefully chosen domain and viewpoint to see how this is possible.

39. Find the shortest distance from the point $(2, 0, -3)$ to the plane $x + y + z = 1$.

40. Find the point on the plane $x - 2y + 3z = 6$ that is closest to the point $(0, 1, 1)$.

41. Find the points on the cone $z^2 = x^2 + y^2$ that are closest to the point $(4, 2, 0)$.

42. Find the points on the surface $y^2 = 9 + xz$ that are closest to the origin.

43. Find three positive numbers whose sum is 100 and whose product is a maximum.

44. Find three positive numbers whose sum is 12 and the sum of whose squares is as small as possible.

45. Find the maximum volume of a rectangular box that is inscribed in a sphere of radius r.

46. Find the dimensions of the box with volume 1000 cm^3 that has minimal surface area.

47. Find the volume of the largest rectangular box in the first octant with three faces in the coordinate planes and one vertex in the plane $x + 2y + 3z = 6$.

48. Find the dimensions of the rectangular box with largest volume if the total surface area is given as 64 cm^2.

49. Find the dimensions of a rectangular box of maximum volume such that the sum of the lengths of its 12 edges is a constant c.

50. The base of an aquarium with given volume V is made of slate and the sides are made of glass. If slate costs five times as much (per unit area) as glass, find the dimensions of the aquarium that minimize the cost of the materials.

51. A cardboard box without a lid is to have a volume of 32,000 cm^3. Find the dimensions that minimize the amount of cardboard used.

52. A rectangular building is being designed to minimize heat loss. The east and west walls lose heat at a rate of 10 units/m^2 per day, the north and south walls at a rate of 8 units/m^2 per day, the floor at a rate of 1 unit/m^2 per day, and the roof at a rate of 5 units/m^2 per day. Each wall must be at least 30 m long, the height must be at least 4 m, and the volume must be exactly 4000 m^3.

(a) Find and sketch the domain of the heat loss as a function of the lengths of the sides.

(b) Find the dimensions that minimize heat loss. (Check both the critical points and the points on the boundary of the domain.)

(c) Could you design a building with even less heat loss if the restrictions on the lengths of the walls were removed?

53. If the length of the diagonal of a rectangular box must be L, what is the largest possible volume?

54. Three alleles (alternative versions of a gene) A, B, and O determine the four blood types A (AA or AO), B (BB or BO), O (OO), and AB. The Hardy-Weinberg Law states that the proportion of individuals in a population who carry two different alleles is

$$P = 2pq + 2pr + 2rq$$

where p, q, and r are the proportions of A, B, and O in the population. Use the fact that $p + q + r = 1$ to show that P is at most $\frac{2}{3}$.

55. Suppose that a scientist has reason to believe that two quantities x and y are related linearly, that is, $y = mx + b$, at least approximately, for some values of m and b. The scientist performs an experiment and collects data in the form of points (x_1, y_1), (x_2, y_2), . . . , (x_n, y_n), and then plots these points. The points don't lie exactly on a straight line, so the scientist wants to find constants m and b so that the line $y = mx + b$ "fits" the points as well as possible (see the figure).

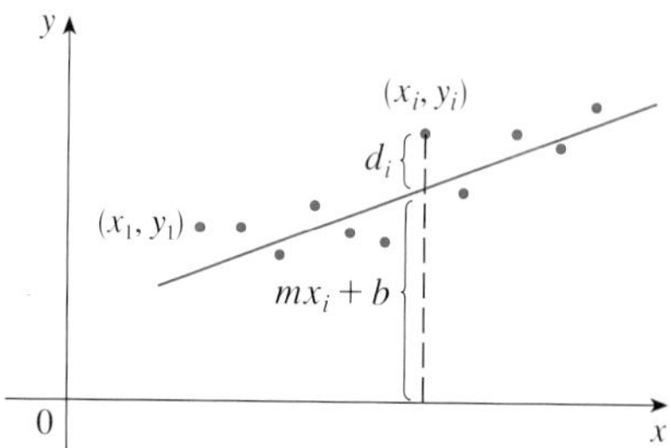

Let $d_i = y_i - (mx_i + b)$ be the vertical deviation of the point (x_i, y_i) from the line. The **method of least squares** determines m and b so as to minimize $\sum_{i=1}^{n} d_i^2$, the sum of the squares of these deviations. Show that, according to this method, the line of best fit is obtained when

$$m \sum_{i=1}^{n} x_i + bn = \sum_{i=1}^{n} y_i$$

$$m \sum_{i=1}^{n} x_i^2 + b \sum_{i=1}^{n} x_i = \sum_{i=1}^{n} x_i y_i$$

Thus the line is found by solving these two equations in the two unknowns m and b. (See Section 1.2 for a further discussion and applications of the method of least squares.)

56. Find an equation of the plane that passes through the point $(1, 2, 3)$ and cuts off the smallest volume in the first octant.

APPLIED PROJECT

DESIGNING A DUMPSTER

For this project we locate a rectangular trash Dumpster in order to study its shape and construction. We then attempt to determine the dimensions of a container of similar design that minimize construction cost.

1. First locate a trash Dumpster in your area. Carefully study and describe all details of its construction, and determine its volume. Include a sketch of the container.
2. While maintaining the general shape and method of construction, determine the dimensions such a container of the same volume should have in order to minimize the cost of construction. Use the following assumptions in your analysis:
   - The sides, back, and front are to be made from 12-gauge (0.1046 inch thick) steel sheets, which cost \$0.70 per square foot (including any required cuts or bends).
   - The base is to be made from a 10-gauge (0.1345 inch thick) steel sheet, which costs \$0.90 per square foot.
   - Lids cost approximately \$50.00 each, regardless of dimensions.
   - Welding costs approximately \$0.18 per foot for material and labor combined.

   Give justification of any further assumptions or simplifications made of the details of construction.
3. Describe how any of your assumptions or simplifications may affect the final result.
4. If you were hired as a consultant on this investigation, what would your conclusions be? Would you recommend altering the design of the Dumpster? If so, describe the savings that would result.

DISCOVERY PROJECT

QUADRATIC APPROXIMATIONS AND CRITICAL POINTS

The Taylor polynomial approximation to functions of one variable that we discussed in Chapter 11 can be extended to functions of two or more variables. Here we investigate quadratic approximations to functions of two variables and use them to give insight into the Second Derivatives Test for classifying critical points.

In Section 14.4 we discussed the linearization of a function f of two variables at a point (a, b):

$$L(x, y) = f(a, b) + f_x(a, b)(x - a) + f_y(a, b)(y - b)$$

Recall that the graph of L is the tangent plane to the surface $z = f(x, y)$ at $(a, b, f(a, b))$ and the corresponding linear approximation is $f(x, y) \approx L(x, y)$. The linearization L is also called the **first-degree Taylor polynomial** of f at (a, b).

1. If f has continuous second-order partial derivatives at (a, b), then the **second-degree Taylor polynomial** of f at (a, b) is

$$\begin{aligned} Q(x, y) &= f(a, b) + f_x(a, b)(x - a) + f_y(a, b)(y - b) \\ &\quad + \tfrac{1}{2} f_{xx}(a, b)(x - a)^2 + f_{xy}(a, b)(x - a)(y - b) + \tfrac{1}{2} f_{yy}(a, b)(y - b)^2 \end{aligned}$$

   and the approximation $f(x, y) \approx Q(x, y)$ is called the **quadratic approximation** to f at (a, b). Verify that Q has the same first- and second-order partial derivatives as f at (a, b).

2. (a) Find the first- and second-degree Taylor polynomials L and Q of $f(x, y) = e^{-x^2-y^2}$ at (0, 0).
(b) Graph f, L, and Q. Comment on how well L and Q approximate f.

3. (a) Find the first- and second-degree Taylor polynomials L and Q for $f(x, y) = xe^y$ at (1, 0).
(b) Compare the values of L, Q, and f at (0.9, 0.1).
(c) Graph f, L, and Q. Comment on how well L and Q approximate f.

4. In this problem we analyze the behavior of the polynomial $f(x, y) = ax^2 + bxy + cy^2$ (without using the Second Derivatives Test) by identifying the graph as a paraboloid.
(a) By completing the square, show that if $a \neq 0$, then

$$f(x, y) = ax^2 + bxy + cy^2 = a\left[\left(x + \frac{b}{2a}y\right)^2 + \left(\frac{4ac - b^2}{4a^2}\right)y^2\right]$$

(b) Let $D = 4ac - b^2$. Show that if $D > 0$ and $a > 0$, then f has a local minimum at (0, 0).
(c) Show that if $D > 0$ and $a < 0$, then f has a local maximum at (0, 0).
(d) Show that if $D < 0$, then (0, 0) is a saddle point.

5. (a) Suppose f is any function with continuous second-order partial derivatives such that $f(0, 0) = 0$ and (0, 0) is a critical point of f. Write an expression for the second-degree Taylor polynomial, Q, of f at (0, 0).
(b) What can you conclude about Q from Problem 4?
(c) In view of the quadratic approximation $f(x, y) \approx Q(x, y)$, what does part (b) suggest about f?

Graphing calculator or computer required

14.8 Lagrange Multipliers

In Example 6 in Section 14.7 we maximized a volume function $V = xyz$ subject to the constraint $2xz + 2yz + xy = 12$, which expressed the side condition that the surface area was 12 m^2. In this section we present Lagrange's method for maximizing or minimizing a general function $f(x, y, z)$ subject to a constraint (or side condition) of the form $g(x, y, z) = k$.

It's easier to explain the geometric basis of Lagrange's method for functions of two variables. So we start by trying to find the extreme values of $f(x, y)$ subject to a constraint of the form $g(x, y) = k$. In other words, we seek the extreme values of $f(x, y)$ when the point (x, y) is restricted to lie on the level curve $g(x, y) = k$. Figure 1 shows this curve together with several level curves of f. These have the equations $f(x, y) = c$, where $c = 7, 8, 9, 10, 11$. To maximize $f(x, y)$ subject to $g(x, y) = k$ is to find the largest value of c such that the level curve $f(x, y) = c$ intersects $g(x, y) = k$. It appears from Figure 1 that this happens when these curves just touch each other, that is, when they have a common tangent line. (Otherwise, the value of c could be increased further.) This means that the normal lines at the point (x_0, y_0) where they touch are identical. So the gradient vectors are parallel; that is, $\nabla f(x_0, y_0) = \lambda \nabla g(x_0, y_0)$ for some scalar λ.

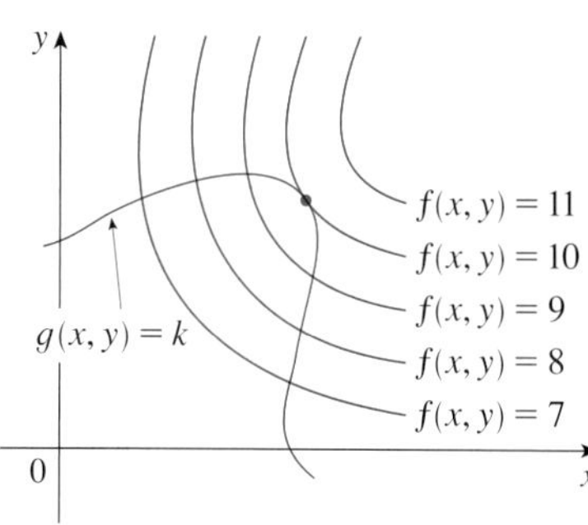

FIGURE 1

TEC Visual 14.8 animates Figure 1 for both level curves and level surfaces.

This kind of argument also applies to the problem of finding the extreme values of $f(x, y, z)$ subject to the constraint $g(x, y, z) = k$. Thus the point (x, y, z) is restricted to lie on the level surface S with equation $g(x, y, z) = k$. Instead of the level curves in Figure 1, we consider the level surfaces $f(x, y, z) = c$ and argue that if the maximum value of f is $f(x_0, y_0, z_0) = c$, then the level surface $f(x, y, z) = c$ is tangent to the level surface $g(x, y, z) = k$ and so the corresponding gradient vectors are parallel.

This intuitive argument can be made precise as follows. Suppose that a function f has an extreme value at a point $P(x_0, y_0, z_0)$ on the surface S and let C be a curve with vector equation $\mathbf{r}(t) = \langle x(t), y(t), z(t) \rangle$ that lies on S and passes through P. If t_0 is the parameter value corresponding to the point P, then $\mathbf{r}(t_0) = \langle x_0, y_0, z_0 \rangle$. The composite function $h(t) = f(x(t), y(t), z(t))$ represents the values that f takes on the curve C. Since f has an extreme value at (x_0, y_0, z_0), it follows that h has an extreme value at t_0, so $h'(t_0) = 0$. But if f is differentiable, we can use the Chain Rule to write

$$\begin{aligned} 0 &= h'(t_0) \\ &= f_x(x_0, y_0, z_0)x'(t_0) + f_y(x_0, y_0, z_0)y'(t_0) + f_z(x_0, y_0, z_0)z'(t_0) \\ &= \nabla f(x_0, y_0, z_0) \cdot \mathbf{r}'(t_0) \end{aligned}$$

This shows that the gradient vector $\nabla f(x_0, y_0, z_0)$ is orthogonal to the tangent vector $\mathbf{r}'(t_0)$ to every such curve C. But we already know from Section 14.6 that the gradient vector of g, $\nabla g(x_0, y_0, z_0)$, is also orthogonal to $\mathbf{r}'(t_0)$ for every such curve. (See Equation 14.6.18.) This means that the gradient vectors $\nabla f(x_0, y_0, z_0)$ and $\nabla g(x_0, y_0, z_0)$ must be parallel. Therefore, if $\nabla g(x_0, y_0, z_0) \neq \mathbf{0}$, there is a number λ such that

1 $$\nabla f(x_0, y_0, z_0) = \lambda \nabla g(x_0, y_0, z_0)$$

Lagrange multipliers are named after the French-Italian mathematician Joseph-Louis Lagrange (1736–1813). See page 286 for a biographical sketch of Lagrange.

The number λ in Equation 1 is called a **Lagrange multiplier**. The procedure based on Equation 1 is as follows.

In deriving Lagrange's method we assumed that $\nabla g \neq \mathbf{0}$. In each of our examples you can check that $\nabla g \neq \mathbf{0}$ at all points where $g(x, y, z) = k$. See Exercise 23 for what can go wrong if $\nabla g = \mathbf{0}$.

Method of Lagrange Multipliers To find the maximum and minimum values of $f(x, y, z)$ subject to the constraint $g(x, y, z) = k$ [assuming that these extreme values exist and $\nabla g \neq \mathbf{0}$ on the surface $g(x, y, z) = k$]:

(a) Find all values of x, y, z, and λ such that

$$\nabla f(x, y, z) = \lambda \nabla g(x, y, z)$$

and

$$g(x, y, z) = k$$

(b) Evaluate f at all the points (x, y, z) that result from step (a). The largest of these values is the maximum value of f; the smallest is the minimum value of f.

If we write the vector equation $\nabla f = \lambda \nabla g$ in terms of components, then the equations in step (a) become

$$f_x = \lambda g_x \qquad f_y = \lambda g_y \qquad f_z = \lambda g_z \qquad g(x, y, z) = k$$

This is a system of four equations in the four unknowns x, y, z, and λ, but it is not necessary to find explicit values for λ.

For functions of two variables the method of Lagrange multipliers is similar to the method just described. To find the extreme values of $f(x, y)$ subject to the constraint $g(x, y) = k$, we look for values of x, y, and λ such that

$$\nabla f(x, y) = \lambda \nabla g(x, y) \qquad \text{and} \qquad g(x, y) = k$$

This amounts to solving three equations in three unknowns:

$$f_x = \lambda g_x \qquad f_y = \lambda g_y \qquad g(x, y) = k$$

Our first illustration of Lagrange's method is to reconsider the problem given in Example 6 in Section 14.7.

V EXAMPLE 1 A rectangular box without a lid is to be made from 12 m^2 of cardboard. Find the maximum volume of such a box.

SOLUTION As in Example 6 in Section 14.7, we let x, y, and z be the length, width, and height, respectively, of the box in meters. Then we wish to maximize

$$V = xyz$$

subject to the constraint

$$g(x, y, z) = 2xz + 2yz + xy = 12$$

Using the method of Lagrange multipliers, we look for values of x, y, z, and λ such that $\nabla V = \lambda \nabla g$ and $g(x, y, z) = 12$. This gives the equations

$$V_x = \lambda g_x$$
$$V_y = \lambda g_y$$
$$V_z = \lambda g_z$$
$$2xz + 2yz + xy = 12$$

which become

$$\boxed{2} \qquad yz = \lambda(2z + y)$$

$$\boxed{3} \qquad xz = \lambda(2z + x)$$

$$\boxed{4} \qquad xy = \lambda(2x + 2y)$$

$$\boxed{5} \qquad 2xz + 2yz + xy = 12$$

There are no general rules for solving systems of equations. Sometimes some ingenuity is required. In the present example you might notice that if we multiply $\boxed{2}$ by x, $\boxed{3}$ by y, and $\boxed{4}$ by z, then the left sides of these equations will be identical. Doing this, we have

Another method for solving the system of equations (2–5) is to solve each of Equations 2, 3, and 4 for λ and then to equate the resulting expressions.

$$\boxed{6} \qquad xyz = \lambda(2xz + xy)$$

$$\boxed{7} \qquad xyz = \lambda(2yz + xy)$$

$$\boxed{8} \qquad xyz = \lambda(2xz + 2yz)$$

We observe that $\lambda \neq 0$ because $\lambda = 0$ would imply $yz = xz = xy = 0$ from $\boxed{2}$, $\boxed{3}$, and $\boxed{4}$ and this would contradict $\boxed{5}$. Therefore, from $\boxed{6}$ and $\boxed{7}$, we have

$$2xz + xy = 2yz + xy$$

which gives $xz = yz$. But $z \neq 0$ (since $z = 0$ would give $V = 0$), so $x = y$. From [7] and [8] we have

$$2yz + xy = 2xz + 2yz$$

which gives $2xz = xy$ and so (since $x \neq 0$) $y = 2z$. If we now put $x = y = 2z$ in [5], we get

$$4z^2 + 4z^2 + 4z^2 = 12$$

Since x, y, and z are all positive, we therefore have $z = 1$ and so $x = 2$ and $y = 2$. This agrees with our answer in Section 14.7.

In geometric terms, Example 2 asks for the highest and lowest points on the curve C in Figure 2 that lie on the paraboloid $z = x^2 + 2y^2$ and directly above the constraint circle $x^2 + y^2 = 1$.

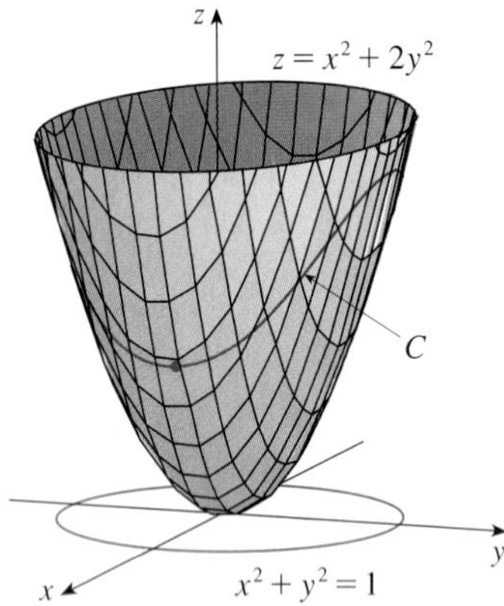

FIGURE 2

V EXAMPLE 2 Find the extreme values of the function $f(x, y) = x^2 + 2y^2$ on the circle $x^2 + y^2 = 1$.

SOLUTION We are asked for the extreme values of f subject to the constraint $g(x, y) = x^2 + y^2 = 1$. Using Lagrange multipliers, we solve the equations $\nabla f = \lambda \nabla g$ and $g(x, y) = 1$, which can be written as

$$f_x = \lambda g_x \qquad f_y = \lambda g_y \qquad g(x, y) = 1$$

or as

9 $$2x = 2x\lambda$$

10 $$4y = 2y\lambda$$

11 $$x^2 + y^2 = 1$$

From [9] we have $x = 0$ or $\lambda = 1$. If $x = 0$, then [11] gives $y = \pm 1$. If $\lambda = 1$, then $y = 0$ from [10], so then [11] gives $x = \pm 1$. Therefore f has possible extreme values at the points $(0, 1)$, $(0, -1)$, $(1, 0)$, and $(-1, 0)$. Evaluating f at these four points, we find that

$$f(0, 1) = 2 \qquad f(0, -1) = 2 \qquad f(1, 0) = 1 \qquad f(-1, 0) = 1$$

Therefore the maximum value of f on the circle $x^2 + y^2 = 1$ is $f(0, \pm 1) = 2$ and the minimum value is $f(\pm 1, 0) = 1$. Checking with Figure 2, we see that these values look reasonable.

The geometry behind the use of Lagrange multipliers in Example 2 is shown in Figure 3. The extreme values of $f(x, y) = x^2 + 2y^2$ correspond to the level curves that touch the circle $x^2 + y^2 = 1$.

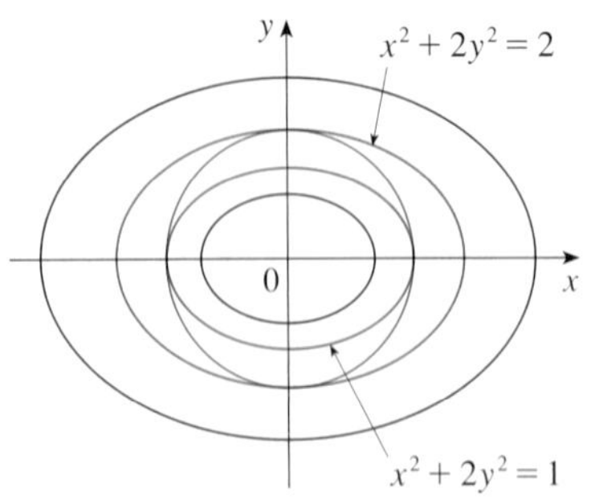

FIGURE 3

EXAMPLE 3 Find the extreme values of $f(x, y) = x^2 + 2y^2$ on the disk $x^2 + y^2 \leqslant 1$.

SOLUTION According to the procedure in (14.7.9), we compare the values of f at the critical points with values at the points on the boundary. Since $f_x = 2x$ and $f_y = 4y$, the only critical point is $(0, 0)$. We compare the value of f at that point with the extreme values on the boundary from Example 2:

$$f(0, 0) = 0 \qquad f(\pm 1, 0) = 1 \qquad f(0, \pm 1) = 2$$

Therefore the maximum value of f on the disk $x^2 + y^2 \leqslant 1$ is $f(0, \pm 1) = 2$ and the minimum value is $f(0, 0) = 0$.

EXAMPLE 4 Find the points on the sphere $x^2 + y^2 + z^2 = 4$ that are closest to and farthest from the point $(3, 1, -1)$.

SOLUTION The distance from a point (x, y, z) to the point $(3, 1, -1)$ is

$$d = \sqrt{(x - 3)^2 + (y - 1)^2 + (z + 1)^2}$$

but the algebra is simpler if we instead maximize and minimize the square of the distance:

$$d^2 = f(x, y, z) = (x - 3)^2 + (y - 1)^2 + (z + 1)^2$$

The constraint is that the point (x, y, z) lies on the sphere, that is,

$$g(x, y, z) = x^2 + y^2 + z^2 = 4$$

According to the method of Lagrange multipliers, we solve $\nabla f = \lambda \nabla g$, $g = 4$. This gives

$$\boxed{12} \qquad 2(x - 3) = 2x\lambda$$

$$\boxed{13} \qquad 2(y - 1) = 2y\lambda$$

$$\boxed{14} \qquad 2(z + 1) = 2z\lambda$$

$$\boxed{15} \qquad x^2 + y^2 + z^2 = 4$$

The simplest way to solve these equations is to solve for x, y, and z in terms of λ from $\boxed{12}$, $\boxed{13}$, and $\boxed{14}$, and then substitute these values into $\boxed{15}$. From $\boxed{12}$ we have

$$x - 3 = x\lambda \qquad \text{or} \qquad x(1 - \lambda) = 3 \qquad \text{or} \qquad x = \frac{3}{1 - \lambda}$$

Figure 4 shows the sphere and the nearest point P in Example 4. Can you see how to find the coordinates of P without using calculus?

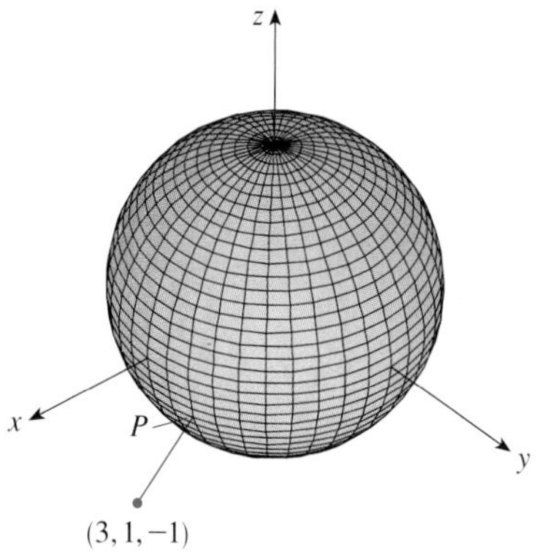

FIGURE 4

[Note that $1 - \lambda \neq 0$ because $\lambda = 1$ is impossible from $\boxed{12}$.] Similarly, $\boxed{13}$ and $\boxed{14}$ give

$$y = \frac{1}{1 - \lambda} \qquad z = -\frac{1}{1 - \lambda}$$

Therefore, from $\boxed{15}$, we have

$$\frac{3^2}{(1 - \lambda)^2} + \frac{1^2}{(1 - \lambda)^2} + \frac{(-1)^2}{(1 - \lambda)^2} = 4$$

which gives $(1 - \lambda)^2 = \frac{11}{4}$, $1 - \lambda = \pm\sqrt{11}/2$, so

$$\lambda = 1 \pm \frac{\sqrt{11}}{2}$$

These values of λ then give the corresponding points (x, y, z):

$$\left(\frac{6}{\sqrt{11}}, \frac{2}{\sqrt{11}}, -\frac{2}{\sqrt{11}}\right) \qquad \text{and} \qquad \left(-\frac{6}{\sqrt{11}}, -\frac{2}{\sqrt{11}}, \frac{2}{\sqrt{11}}\right)$$

It's easy to see that f has a smaller value at the first of these points, so the closest point is $\left(6/\sqrt{11}, 2/\sqrt{11}, -2/\sqrt{11}\right)$ and the farthest is $\left(-6/\sqrt{11}, -2/\sqrt{11}, 2/\sqrt{11}\right)$.

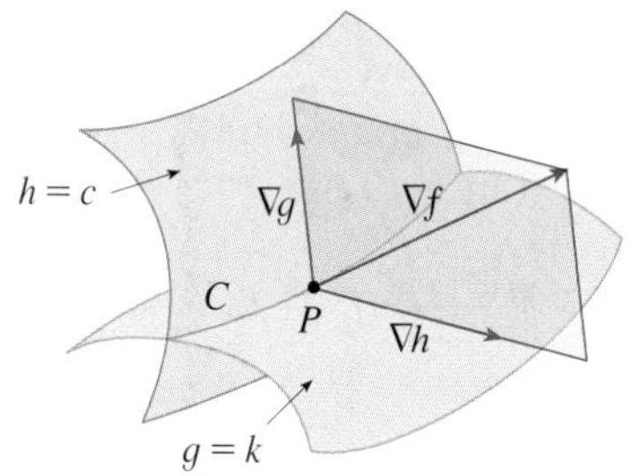

FIGURE 5

Two Constraints

Suppose now that we want to find the maximum and minimum values of a function $f(x, y, z)$ subject to two constraints (side conditions) of the form $g(x, y, z) = k$ and $h(x, y, z) = c$. Geometrically, this means that we are looking for the extreme values of f when (x, y, z) is restricted to lie on the curve of intersection C of the level surfaces $g(x, y, z) = k$ and $h(x, y, z) = c$. (See Figure 5.) Suppose f has such an extreme value at a point $P(x_0, y_0, z_0)$.

We know from the beginning of this section that ∇f is orthogonal to C at P. But we also know that ∇g is orthogonal to $g(x, y, z) = k$ and ∇h is orthogonal to $h(x, y, z) = c$, so ∇g and ∇h are both orthogonal to C. This means that the gradient vector $\nabla f(x_0, y_0, z_0)$ is in the plane determined by $\nabla g(x_0, y_0, z_0)$ and $\nabla h(x_0, y_0, z_0)$. (We assume that these gradient vectors are not zero and not parallel.) So there are numbers λ and μ (called Lagrange multipliers) such that

$$\boxed{16} \qquad \boxed{\nabla f(x_0, y_0, z_0) = \lambda \nabla g(x_0, y_0, z_0) + \mu \nabla h(x_0, y_0, z_0)}$$

In this case Lagrange's method is to look for extreme values by solving five equations in the five unknowns x, y, z, λ, and μ. These equations are obtained by writing Equation 16 in terms of its components and using the constraint equations:

$$f_x = \lambda g_x + \mu h_x$$

$$f_y = \lambda g_y + \mu h_y$$

$$f_z = \lambda g_z + \mu h_z$$

$$g(x, y, z) = k$$

$$h(x, y, z) = c$$

The cylinder $x^2 + y^2 = 1$ intersects the plane $x - y + z = 1$ in an ellipse (Figure 6). Example 5 asks for the maximum value of f when (x, y, z) is restricted to lie on the ellipse.

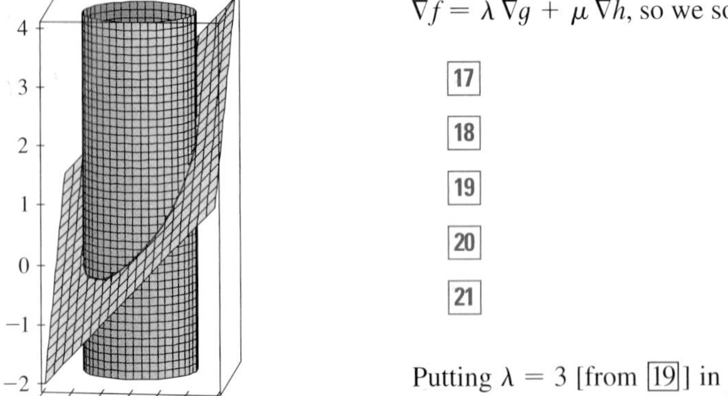

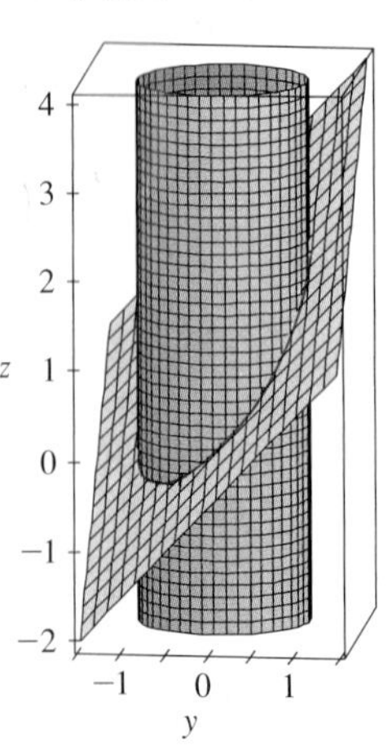

FIGURE 6

V EXAMPLE 5 Find the maximum value of the function $f(x, y, z) = x + 2y + 3z$ on the curve of intersection of the plane $x - y + z = 1$ and the cylinder $x^2 + y^2 = 1$.

SOLUTION We maximize the function $f(x, y, z) = x + 2y + 3z$ subject to the constraints $g(x, y, z) = x - y + z = 1$ and $h(x, y, z) = x^2 + y^2 = 1$. The Lagrange condition is $\nabla f = \lambda \nabla g + \mu \nabla h$, so we solve the equations

$$\boxed{17} \qquad 1 = \lambda + 2x\mu$$

$$\boxed{18} \qquad 2 = -\lambda + 2y\mu$$

$$\boxed{19} \qquad 3 = \lambda$$

$$\boxed{20} \qquad x - y + z = 1$$

$$\boxed{21} \qquad x^2 + y^2 = 1$$

Putting $\lambda = 3$ [from $\boxed{19}$] in $\boxed{17}$, we get $2x\mu = -2$, so $x = -1/\mu$. Similarly, $\boxed{18}$ gives $y = 5/(2\mu)$. Substitution in $\boxed{21}$ then gives

$$\frac{1}{\mu^2} + \frac{25}{4\mu^2} = 1$$

and so $\mu^2 = \frac{29}{4}$, $\mu = \pm\sqrt{29}/2$. Then $x = \mp 2/\sqrt{29}$, $y = \pm 5/\sqrt{29}$, and, from $\boxed{20}$, $z = 1 - x + y = 1 \pm 7/\sqrt{29}$. The corresponding values of f are

$$\mp\frac{2}{\sqrt{29}} + 2\left(\pm\frac{5}{\sqrt{29}}\right) + 3\left(1 \pm \frac{7}{\sqrt{29}}\right) = 3 \pm \sqrt{29}$$

Therefore the maximum value of f on the given curve is $3 + \sqrt{29}$. ∎

14.8 Exercises

1. Pictured are a contour map of f and a curve with equation $g(x, y) = 8$. Estimate the maximum and minimum values of f subject to the constraint that $g(x, y) = 8$. Explain your reasoning.

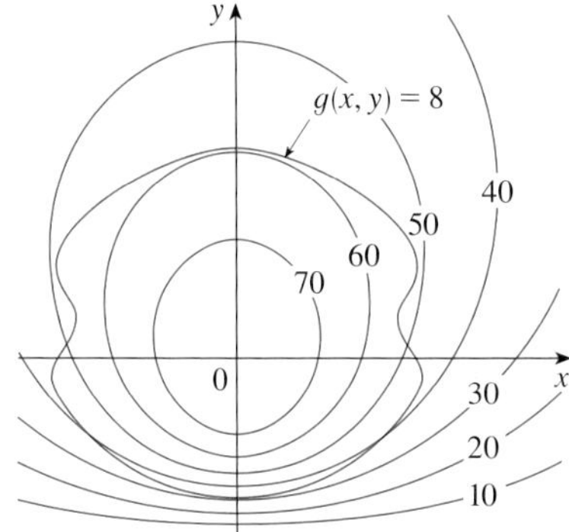

2. (a) Use a graphing calculator or computer to graph the circle $x^2 + y^2 = 1$. On the same screen, graph several curves of the form $x^2 + y = c$ until you find two that just touch the circle. What is the significance of the values of c for these two curves?
   (b) Use Lagrange multipliers to find the extreme values of $f(x, y) = x^2 + y$ subject to the constraint $x^2 + y^2 = 1$. Compare your answers with those in part (a).

3–14 Use Lagrange multipliers to find the maximum and minimum values of the function subject to the given constraint.

3. $f(x, y) = x^2 + y^2; \quad xy = 1$

4. $f(x, y) = 3x + y; \quad x^2 + y^2 = 10$

5. $f(x, y) = y^2 - x^2; \quad \frac{1}{4}x^2 + y^2 = 1$

6. $f(x, y) = e^{xy}; \quad x^3 + y^3 = 16$

7. $f(x, y, z) = 2x + 2y + z; \quad x^2 + y^2 + z^2 = 9$

8. $f(x, y, z) = x^2 + y^2 + z^2; \quad x + y + z = 12$

9. $f(x, y, z) = xyz; \quad x^2 + 2y^2 + 3z^2 = 6$

10. $f(x, y, z) = x^2y^2z^2; \quad x^2 + y^2 + z^2 = 1$

11. $f(x, y, z) = x^2 + y^2 + z^2; \quad x^4 + y^4 + z^4 = 1$

12. $f(x, y, z) = x^4 + y^4 + z^4; \quad x^2 + y^2 + z^2 = 1$

13. $f(x, y, z, t) = x + y + z + t; \quad x^2 + y^2 + z^2 + t^2 = 1$

14. $f(x_1, x_2, \ldots, x_n) = x_1 + x_2 + \cdots + x_n;$
    $x_1^2 + x_2^2 + \cdots + x_n^2 = 1$

15–18 Find the extreme values of f subject to both constraints.

15. $f(x, y, z) = x + 2y; \quad x + y + z = 1, \quad y^2 + z^2 = 4$

16. $f(x, y, z) = 3x - y - 3z;$
    $x + y - z = 0, \quad x^2 + 2z^2 = 1$

17. $f(x, y, z) = yz + xy; \quad xy = 1, \quad y^2 + z^2 = 1$

18. $f(x, y, z) = x^2 + y^2 + z^2; \quad x - y = 1, \quad y^2 - z^2 = 1$

19–21 Find the extreme values of f on the region described by the inequality.

19. $f(x, y) = x^2 + y^2 + 4x - 4y, \quad x^2 + y^2 \leqslant 9$

20. $f(x, y) = 2x^2 + 3y^2 - 4x - 5, \quad x^2 + y^2 \leqslant 16$

21. $f(x, y) = e^{-xy}, \quad x^2 + 4y^2 \leqslant 1$

22. Consider the problem of maximizing the function $f(x, y) = 2x + 3y$ subject to the constraint $\sqrt{x} + \sqrt{y} = 5$.
    (a) Try using Lagrange multipliers to solve the problem.
    (b) Does $f(25, 0)$ give a larger value than the one in part (a)?
    (c) Solve the problem by graphing the constraint equation and several level curves of f.
    (d) Explain why the method of Lagrange multipliers fails to solve the problem.
    (e) What is the significance of $f(9, 4)$?

23. Consider the problem of minimizing the function $f(x, y) = x$ on the curve $y^2 + x^4 - x^3 = 0$ (a piriform).
    (a) Try using Lagrange multipliers to solve the problem.
    (b) Show that the minimum value is $f(0, 0) = 0$ but the Lagrange condition $\nabla f(0, 0) = \lambda \nabla g(0, 0)$ is not satisfied for any value of λ.
    (c) Explain why Lagrange multipliers fail to find the minimum value in this case.

CAS 24. (a) If your computer algebra system plots implicitly defined curves, use it to estimate the minimum and maximum values of $f(x, y) = x^3 + y^3 + 3xy$ subject to the constraint $(x - 3)^2 + (y - 3)^2 = 9$ by graphical methods.
    (b) Solve the problem in part (a) with the aid of Lagrange multipliers. Use your CAS to solve the equations numerically. Compare your answers with those in part (a).

25. The total production P of a certain product depends on the amount L of labor used and the amount K of capital investment. In Sections 14.1 and 14.3 we discussed how the Cobb-Douglas model $P = bL^{\alpha}K^{1-\alpha}$ follows from certain economic assumptions, where b and α are positive constants and $\alpha < 1$. If the cost of a unit of labor is m and the cost of a unit of capital is n, and the company can spend only p dollars as its total budget, then maximizing the production P is subject to the constraint $mL + nK = p$. Show that the maximum production occurs when

$$L = \frac{\alpha p}{m} \qquad \text{and} \qquad K = \frac{(1 - \alpha)p}{n}$$

26. Referring to Exercise 25, we now suppose that the production is fixed at $bL^{\alpha}K^{1-\alpha} = Q$, where Q is a constant. What values of L and K minimize the cost function $C(L, K) = mL + nK$?

27. Use Lagrange multipliers to prove that the rectangle with maximum area that has a given perimeter p is a square.

28. Use Lagrange multipliers to prove that the triangle with maximum area that has a given perimeter p is equilateral.

Hint: Use Heron's formula for the area:

$$A = \sqrt{s(s - x)(s - y)(s - z)}$$

where $s = p/2$ and x, y, z are the lengths of the sides.

29–41 Use Lagrange multipliers to give an alternate solution to the indicated exercise in Section 14.7.

29. Exercise 39 **30.** Exercise 40

31. Exercise 41 **32.** Exercise 42

33. Exercise 43 **34.** Exercise 44

35. Exercise 45 **36.** Exercise 46

37. Exercise 47 **38.** Exercise 48

39. Exercise 49 **40.** Exercise 50

41. Exercise 53

---

42. Find the maximum and minimum volumes of a rectangular box whose surface area is 1500 cm^2 and whose total edge length is 200 cm.

43. The plane $x + y + 2z = 2$ intersects the paraboloid $z = x^2 + y^2$ in an ellipse. Find the points on this ellipse that are nearest to and farthest from the origin.

44. The plane $4x - 3y + 8z = 5$ intersects the cone $z^2 = x^2 + y^2$ in an ellipse.

(a) Graph the cone, the plane, and the ellipse.

(b) Use Lagrange multipliers to find the highest and lowest points on the ellipse.

CAS **45–46** Find the maximum and minimum values of f subject to the given constraints. Use a computer algebra system to solve the system of equations that arises in using Lagrange multipliers. (If your CAS finds only one solution, you may need to use additional commands.)

45. $f(x, y, z) = ye^{x-z}$; $9x^2 + 4y^2 + 36z^2 = 36$, $xy + yz = 1$

46. $f(x, y, z) = x + y + z$; $x^2 - y^2 = z$, $x^2 + z^2 = 4$

---

47. (a) Find the maximum value of

$$f(x_1, x_2, \ldots, x_n) = \sqrt[n]{x_1 x_2 \cdots x_n}$$

given that $x_1, x_2, \ldots, x_n$ are positive numbers and $x_1 + x_2 + \cdots + x_n = c$, where c is a constant.

(b) Deduce from part (a) that if $x_1, x_2, \ldots, x_n$ are positive numbers, then

$$\sqrt[n]{x_1 x_2 \cdots x_n} \leqslant \frac{x_1 + x_2 + \cdots + x_n}{n}$$

This inequality says that the geometric mean of n numbers is no larger than the arithmetic mean of the numbers. Under what circumstances are these two means equal?

48. (a) Maximize $\sum_{i=1}^{n} x_i y_i$ subject to the constraints $\sum_{i=1}^{n} x_i^2 = 1$ and $\sum_{i=1}^{n} y_i^2 = 1$.

(b) Put

$$x_i = \frac{a_i}{\sqrt{\sum a_j^2}} \quad \text{and} \quad y_i = \frac{b_i}{\sqrt{\sum b_j^2}}$$

to show that

$$\sum a_i b_i \leqslant \sqrt{\sum a_j^2}\,\sqrt{\sum b_j^2}$$

for any numbers $a_1, \ldots, a_n, b_1, \ldots, b_n$. This inequality is known as the Cauchy-Schwarz Inequality.

APPLIED PROJECT ROCKET SCIENCE

Many rockets, such as the *Pegasus XL* currently used to launch satellites and the *Saturn V* that first put men on the moon, are designed to use three stages in their ascent into space. A large first stage initially propels the rocket until its fuel is consumed, at which point the stage is jettisoned to reduce the mass of the rocket. The smaller second and third stages function similarly in order to place the rocket's payload into orbit about the earth. (With this design, at least two stages are required in order to reach the necessary velocities, and using three stages has proven to be a good compromise between cost and performance.) Our goal here is to determine the individual masses of the three stages, which are to be designed in such a way as to minimize the total mass of the rocket while enabling it to reach a desired velocity.

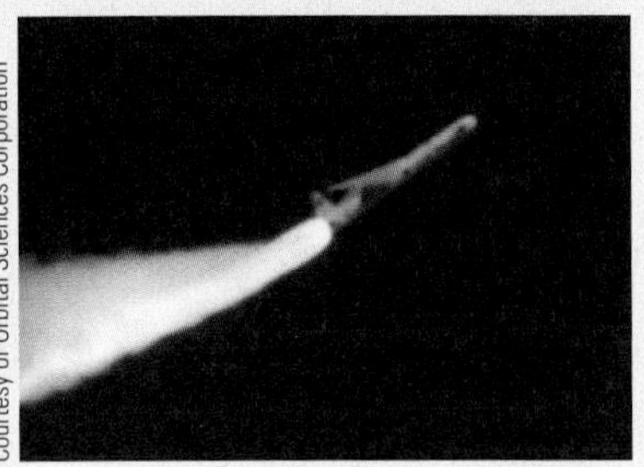

For a single-stage rocket consuming fuel at a constant rate, the change in velocity resulting from the acceleration of the rocket vehicle has been modeled by

$$\Delta V = -c \ln\left(1 - \frac{(1 - S)M_r}{P + M_r}\right)$$

where M_r is the mass of the rocket engine including initial fuel, P is the mass of the payload, S is a *structural factor* determined by the design of the rocket (specifically, it is the ratio of the mass of the rocket vehicle without fuel to the total mass of the rocket with payload), and c is the (constant) speed of exhaust relative to the rocket.

Now consider a rocket with three stages and a payload of mass A. Assume that outside forces are negligible and that c and S remain constant for each stage. If M_i is the mass of the ith stage, we can initially consider the rocket engine to have mass M_1 and its payload to have mass $M_2 + M_3 + A$; the second and third stages can be handled similarly.

1. Show that the velocity attained after all three stages have been jettisoned is given by

$$v_f = c\left[\ln\left(\frac{M_1 + M_2 + M_3 + A}{SM_1 + M_2 + M_3 + A}\right) + \ln\left(\frac{M_2 + M_3 + A}{SM_2 + M_3 + A}\right) + \ln\left(\frac{M_3 + A}{SM_3 + A}\right)\right]$$

2. We wish to minimize the total mass $M = M_1 + M_2 + M_3$ of the rocket engine subject to the constraint that the desired velocity v_f from Problem 1 is attained. The method of Lagrange multipliers is appropriate here, but difficult to implement using the current expressions. To simplify, we define variables N_i so that the constraint equation may be expressed as $v_f = c(\ln N_1 + \ln N_2 + \ln N_3)$. Since M is now difficult to express in terms of the N_i's, we wish to use a simpler function that will be minimized at the same place as M. Show that

$$\frac{M_1 + M_2 + M_3 + A}{M_2 + M_3 + A} = \frac{(1 - S)N_1}{1 - SN_1}$$

$$\frac{M_2 + M_3 + A}{M_3 + A} = \frac{(1 - S)N_2}{1 - SN_2}$$

$$\frac{M_3 + A}{A} = \frac{(1 - S)N_3}{1 - SN_3}$$

and conclude that

$$\frac{M + A}{A} = \frac{(1 - S)^3 N_1 N_2 N_3}{(1 - SN_1)(1 - SN_2)(1 - SN_3)}$$

3. Verify that $\ln((M + A)/A)$ is minimized at the same location as M; use Lagrange multipliers and the results of Problem 2 to find expressions for the values of N_i where the minimum occurs subject to the constraint $v_f = c(\ln N_1 + \ln N_2 + \ln N_3)$. [*Hint:* Use properties of logarithms to help simplify the expressions.]

4. Find an expression for the minimum value of M as a function of v_f.

5. If we want to put a three-stage rocket into orbit 100 miles above the earth's surface, a final velocity of approximately 17,500 mi/h is required. Suppose that each stage is built with a structural factor $S = 0.2$ and an exhaust speed of $c = 6000$ mi/h.
   (a) Find the minimum total mass M of the rocket engines as a function of A.
   (b) Find the mass of each individual stage as a function of A. (They are not equally sized!)

6. The same rocket would require a final velocity of approximately 24,700 mi/h in order to escape earth's gravity. Find the mass of each individual stage that would minimize the total mass of the rocket engines and allow the rocket to propel a 500-pound probe into deep space.

APPLIED PROJECT HYDRO-TURBINE OPTIMIZATION

The Katahdin Paper Company in Millinocket, Maine, operates a hydroelectric generating station on the Penobscot River. Water is piped from a dam to the power station. The rate at which the water flows through the pipe varies, depending on external conditions.

The power station has three different hydroelectric turbines, each with a known (and unique) power function that gives the amount of electric power generated as a function of the water flow arriving at the turbine. The incoming water can be apportioned in different volumes to each turbine, so the goal is to determine how to distribute water among the turbines to give the maximum total energy production for any rate of flow.

Using experimental evidence and *Bernoulli's equation*, the following quadratic models were determined for the power output of each turbine, along with the allowable flows of operation:

$$KW_1 = (-18.89 + 0.1277Q_1 - 4.08 \cdot 10^{-5}Q_1^2)(170 - 1.6 \cdot 10^{-6}Q_T^2)$$

$$KW_2 = (-24.51 + 0.1358Q_2 - 4.69 \cdot 10^{-5}Q_2^2)(170 - 1.6 \cdot 10^{-6}Q_T^2)$$

$$KW_3 = (-27.02 + 0.1380Q_3 - 3.84 \cdot 10^{-5}Q_3^2)(170 - 1.6 \cdot 10^{-6}Q_T^2)$$

$$250 \leqslant Q_1 \leqslant 1110, \quad 250 \leqslant Q_2 \leqslant 1110, \quad 250 \leqslant Q_3 \leqslant 1225$$

where

$$Q_i = \text{flow through turbine } i \text{ in cubic feet per second}$$

$$KW_i = \text{power generated by turbine } i \text{ in kilowatts}$$

$$Q_T = \text{total flow through the station in cubic feet per second}$$

1. If all three turbines are being used, we wish to determine the flow Q_i to each turbine that will give the maximum total energy production. Our limitations are that the flows must sum to the total incoming flow and the given domain restrictions must be observed. Consequently, use Lagrange multipliers to find the values for the individual flows (as functions of Q_T) that maximize the total energy production $KW_1 + KW_2 + KW_3$ subject to the constraints $Q_1 + Q_2 + Q_3 = Q_T$ and the domain restrictions on each Q_i.

2. For which values of Q_T is your result valid?

3. For an incoming flow of 2500 ft^3/s, determine the distribution to the turbines and verify (by trying some nearby distributions) that your result is indeed a maximum.

4. Until now we have assumed that all three turbines are operating; is it possible in some situations that more power could be produced by using only one turbine? Make a graph of the three power functions and use it to help decide if an incoming flow of 1000 ft^3/s should be distributed to all three turbines or routed to just one. (If you determine that only one turbine should be used, which one would it be?) What if the flow is only 600 ft^3/s?

5. Perhaps for some flow levels it would be advantageous to use two turbines. If the incoming flow is 1500 ft^3/s, which two turbines would you recommend using? Use Lagrange multipliers to determine how the flow should be distributed between the two turbines to maximize the energy produced. For this flow, is using two turbines more efficient than using all three?

6. If the incoming flow is 3400 ft^3/s, what would you recommend to the company?

14 Review

Concept Check

1. (a) What is a function of two variables?
   (b) Describe three methods for visualizing a function of two variables.

2. What is a function of three variables? How can you visualize such a function?

3. What does

$$\lim_{(x, y) \to (a, b)} f(x, y) = L$$

mean? How can you show that such a limit does not exist?

4. (a) What does it mean to say that f is continuous at (a, b)?
   (b) If f is continuous on $\mathbb{R}^2$, what can you say about its graph?

5. (a) Write expressions for the partial derivatives $f_x(a, b)$ and $f_y(a, b)$ as limits.
   (b) How do you interpret $f_x(a, b)$ and $f_y(a, b)$ geometrically? How do you interpret them as rates of change?
   (c) If $f(x, y)$ is given by a formula, how do you calculate f_x and f_y?

6. What does Clairaut's Theorem say?

7. How do you find a tangent plane to each of the following types of surfaces?
   (a) A graph of a function of two variables, $z = f(x, y)$
   (b) A level surface of a function of three variables, $F(x, y, z) = k$

8. Define the linearization of f at (a, b). What is the corresponding linear approximation? What is the geometric interpretation of the linear approximation?

9. (a) What does it mean to say that f is differentiable at (a, b)?
   (b) How do you usually verify that f is differentiable?

10. If $z = f(x, y)$, what are the differentials dx, dy, and dz?

11. State the Chain Rule for the case where $z = f(x, y)$ and x and y are functions of one variable. What if x and y are functions of two variables?

12. If z is defined implicitly as a function of x and y by an equation of the form $F(x, y, z) = 0$, how do you find $\partial z/\partial x$ and $\partial z/\partial y$?

13. (a) Write an expression as a limit for the directional derivative of f at (x_0, y_0) in the direction of a unit vector $\mathbf{u} = \langle a, b \rangle$. How do you interpret it as a rate? How do you interpret it geometrically?
    (b) If f is differentiable, write an expression for $D_{\mathbf{u}} f(x_0, y_0)$ in terms of f_x and f_y.

14. (a) Define the gradient vector ∇f for a function f of two or three variables.
    (b) Express $D_{\mathbf{u}} f$ in terms of ∇f.
    (c) Explain the geometric significance of the gradient.

15. What do the following statements mean?
    (a) f has a local maximum at (a, b).
    (b) f has an absolute maximum at (a, b).
    (c) f has a local minimum at (a, b).
    (d) f has an absolute minimum at (a, b).
    (e) f has a saddle point at (a, b).

16. (a) If f has a local maximum at (a, b), what can you say about its partial derivatives at (a, b)?
    (b) What is a critical point of f?

17. State the Second Derivatives Test.

18. (a) What is a closed set in $\mathbb{R}^2$? What is a bounded set?
    (b) State the Extreme Value Theorem for functions of two variables.
    (c) How do you find the values that the Extreme Value Theorem guarantees?

19. Explain how the method of Lagrange multipliers works in finding the extreme values of $f(x, y, z)$ subject to the constraint $g(x, y, z) = k$. What if there is a second constraint $h(x, y, z) = c$?

True-False Quiz

Determine whether the statement is true or false. If it is true, explain why. If it is false, explain why or give an example that disproves the statement.

1. $f_y(a, b) = \lim_{y \to b} \dfrac{f(a, y) - f(a, b)}{y - b}$

2. There exists a function f with continuous second-order partial derivatives such that $f_x(x, y) = x + y^2$ and $f_y(x, y) = x - y^2$.

3. $f_{xy} = \dfrac{\partial^2 f}{\partial x \, \partial y}$

4. $D_{\mathbf{k}} f(x, y, z) = f_z(x, y, z)$

5. If $f(x, y) \to L$ as $(x, y) \to (a, b)$ along every straight line through (a, b), then $\lim_{(x, y) \to (a, b)} f(x, y) = L$.

6. If $f_x(a, b)$ and $f_y(a, b)$ both exist, then f is differentiable at (a, b).

7. If f has a local minimum at (a, b) and f is differentiable at (a, b), then $\nabla f(a, b) = \mathbf{0}$.

8. If f is a function, then

$$\lim_{(x, y) \to (2, 5)} f(x, y) = f(2, 5)$$

9. If $f(x, y) = \ln y$, then $\nabla f(x, y) = 1/y$.

10. If (2, 1) is a critical point of f and

$$f_{xx}(2, 1)\, f_{yy}(2, 1) < [f_{xy}(2, 1)]^2$$

then f has a saddle point at (2, 1).

11. If $f(x, y) = \sin x + \sin y$, then $-\sqrt{2} \leqslant D_{\mathbf{u}} f(x, y) \leqslant \sqrt{2}$.

12. If $f(x, y)$ has two local maxima, then f must have a local minimum.

Exercises

1–2 Find and sketch the domain of the function.

1. $f(x, y) = \ln(x + y + 1)$

2. $f(x, y) = \sqrt{4 - x^2 - y^2} + \sqrt{1 - x^2}$

3–4 Sketch the graph of the function.

3. $f(x, y) = 1 - y^2$

4. $f(x, y) = x^2 + (y - 2)^2$

5–6 Sketch several level curves of the function.

5. $f(x, y) = \sqrt{4x^2 + y^2}$ **6.** $f(x, y) = e^x + y$

7. Make a rough sketch of a contour map for the function whose graph is shown.

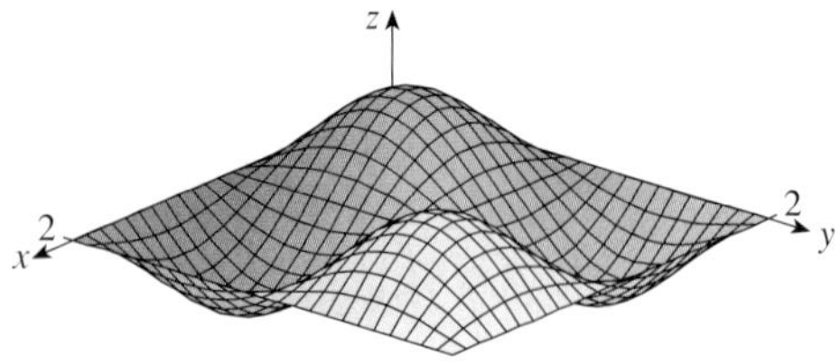

8. A contour map of a function f is shown. Use it to make a rough sketch of the graph of f.

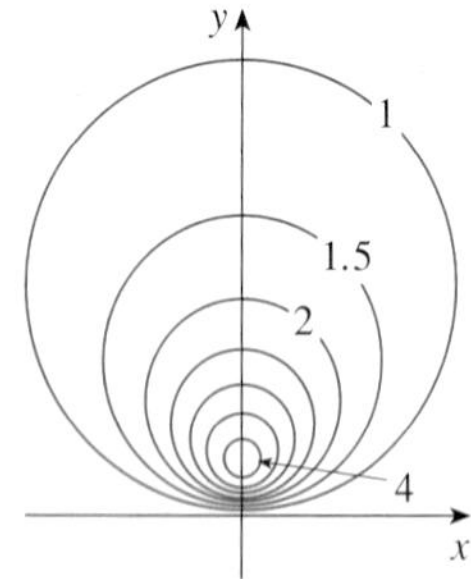

9–10 Evaluate the limit or show that it does not exist.

9. $\displaystyle\lim_{(x, y)\to(1, 1)} \frac{2xy}{x^2 + 2y^2}$ **10.** $\displaystyle\lim_{(x, y)\to(0, 0)} \frac{2xy}{x^2 + 2y^2}$

11. A metal plate is situated in the xy-plane and occupies the rectangle $0 \leqslant x \leqslant 10$, $0 \leqslant y \leqslant 8$, where x and y are measured in meters. The temperature at the point (x, y) in the plate is $T(x, y)$, where T is measured in degrees Celsius. Temperatures at equally spaced points were measured and recorded in the table.

(a) Estimate the values of the partial derivatives $T_x(6, 4)$ and $T_y(6, 4)$. What are the units?

(b) Estimate the value of $D_{\mathbf{u}} T(6, 4)$, where $\mathbf{u} = (\mathbf{i} + \mathbf{j})/\sqrt{2}$. Interpret your result.

(c) Estimate the value of $T_{xy}(6, 4)$.

x \ y	0	2	4	6	8
0	30	38	45	51	55
2	52	56	60	62	61
4	78	74	72	68	66
6	98	87	80	75	71
8	96	90	86	80	75
10	92	92	91	87	78

12. Find a linear approximation to the temperature function $T(x, y)$ in Exercise 11 near the point (6, 4). Then use it to estimate the temperature at the point (5, 3.8).

13–17 Find the first partial derivatives.

13. $f(x, y) = (5y^3 + 2x^2y)^8$ **14.** $g(u, v) = \dfrac{u + 2v}{u^2 + v^2}$

15. $F(\alpha, \beta) = \alpha^2 \ln(\alpha^2 + \beta^2)$ **16.** $G(x, y, z) = e^{xz} \sin(y/z)$

17. $S(u, v, w) = u \arctan\left(v\sqrt{w}\right)$

18. The speed of sound traveling through ocean water is a function of temperature, salinity, and pressure. It has been modeled by the function

$$C = 1449.2 + 4.6T - 0.055T^2 + 0.00029T^3 + (1.34 - 0.01T)(S - 35) + 0.016D$$

where C is the speed of sound (in meters per second), T is the temperature (in degrees Celsius), S is the salinity (the concentration of salts in parts per thousand, which means the number of grams of dissolved solids per 1000 g of water), and D is the depth below the ocean surface (in meters). Compute $\partial C/\partial T$, $\partial C/\partial S$, and $\partial C/\partial D$ when $T = 10°$C, $S = 35$ parts per thousand, and $D = 100$ m. Explain the physical significance of these partial derivatives.

Graphing calculator or computer required

19–22 Find all second partial derivatives of f.

19. $f(x, y) = 4x^3 - xy^2$

20. $z = xe^{-2y}$

21. $f(x, y, z) = x^k y^l z^m$

22. $v = r\cos(s + 2t)$

23. If $z = xy + xe^{y/x}$, show that $x\dfrac{\partial z}{\partial x} + y\dfrac{\partial z}{\partial y} = xy + z$.

24. If $z = \sin(x + \sin t)$, show that

$$\frac{\partial z}{\partial x}\frac{\partial^2 z}{\partial x\,\partial t} = \frac{\partial z}{\partial t}\frac{\partial^2 z}{\partial x^2}$$

25–29 Find equations of (a) the tangent plane and (b) the normal line to the given surface at the specified point.

25. $z = 3x^2 - y^2 + 2x$, $(1, -2, 1)$

26. $z = e^x \cos y$, $(0, 0, 1)$

27. $x^2 + 2y^2 - 3z^2 = 3$, $(2, -1, 1)$

28. $xy + yz + zx = 3$, $(1, 1, 1)$

29. $\sin(xyz) = x + 2y + 3z$, $(2, -1, 0)$

30. Use a computer to graph the surface $z = x^2 + y^4$ and its tangent plane and normal line at $(1, 1, 2)$ on the same screen. Choose the domain and viewpoint so that you get a good view of all three objects.

31. Find the points on the hyperboloid $x^2 + 4y^2 - z^2 = 4$ where the tangent plane is parallel to the plane $2x + 2y + z = 5$.

32. Find du if $u = \ln(1 + se^{2t})$.

33. Find the linear approximation of the function $f(x, y, z) = x^3\sqrt{y^2 + z^2}$ at the point $(2, 3, 4)$ and use it to estimate the number $(1.98)^3\sqrt{(3.01)^2 + (3.97)^2}$.

34. The two legs of a right triangle are measured as 5 m and 12 m with a possible error in measurement of at most 0.2 cm in each. Use differentials to estimate the maximum error in the calculated value of (a) the area of the triangle and (b) the length of the hypotenuse.

35. If $u = x^2y^3 + z^4$, where $x = p + 3p^2$, $y = pe^p$, and $z = p\sin p$, use the Chain Rule to find du/dp.

36. If $v = x^2\sin y + ye^{xy}$, where $x = s + 2t$ and $y = st$, use the Chain Rule to find $\partial v/\partial s$ and $\partial v/\partial t$ when $s = 0$ and $t = 1$.

37. Suppose $z = f(x, y)$, where $x = g(s, t)$, $y = h(s, t)$, $g(1, 2) = 3$, $g_s(1, 2) = -1$, $g_t(1, 2) = 4$, $h(1, 2) = 6$, $h_s(1, 2) = -5$, $h_t(1, 2) = 10$, $f_x(3, 6) = 7$, and $f_y(3, 6) = 8$. Find $\partial z/\partial s$ and $\partial z/\partial t$ when $s = 1$ and $t = 2$.

38. Use a tree diagram to write out the Chain Rule for the case where $w = f(t, u, v)$, $t = t(p, q, r, s)$, $u = u(p, q, r, s)$, and $v = v(p, q, r, s)$ are all differentiable functions.

39. If $z = y + f(x^2 - y^2)$, where f is differentiable, show that

$$y\frac{\partial z}{\partial x} + x\frac{\partial z}{\partial y} = x$$

40. The length x of a side of a triangle is increasing at a rate of 3 in/s, the length y of another side is decreasing at a rate of 2 in/s, and the contained angle θ is increasing at a rate of 0.05 radian/s. How fast is the area of the triangle changing when $x = 40$ in, $y = 50$ in, and $\theta = \pi/6$?

41. If $z = f(u, v)$, where $u = xy$, $v = y/x$, and f has continuous second partial derivatives, show that

$$x^2\frac{\partial^2 z}{\partial x^2} - y^2\frac{\partial^2 z}{\partial y^2} = -4uv\frac{\partial^2 z}{\partial u\,\partial v} + 2v\frac{\partial z}{\partial v}$$

42. If $\cos(xyz) = 1 + x^2y^2 + z^2$, find $\dfrac{\partial z}{\partial x}$ and $\dfrac{\partial z}{\partial y}$.

43. Find the gradient of the function $f(x, y, z) = x^2e^{yz^2}$.

44. (a) When is the directional derivative of f a maximum?
(b) When is it a minimum?
(c) When is it 0?
(d) When is it half of its maximum value?

45–46 Find the directional derivative of f at the given point in the indicated direction.

45. $f(x, y) = x^2e^{-y}$, $(-2, 0)$, in the direction toward the point $(2, -3)$

46. $f(x, y, z) = x^2y + x\sqrt{1 + z}$, $(1, 2, 3)$, in the direction of $\mathbf{v} = 2\mathbf{i} + \mathbf{j} - 2\mathbf{k}$

47. Find the maximum rate of change of $f(x, y) = x^2y + \sqrt{y}$ at the point $(2, 1)$. In which direction does it occur?

48. Find the direction in which $f(x, y, z) = ze^{xy}$ increases most rapidly at the point $(0, 1, 2)$. What is the maximum rate of increase?

49. The contour map shows wind speed in knots during Hurricane Andrew on August 24, 1992. Use it to estimate the value of the directional derivative of the wind speed at Homestead, Florida, in the direction of the eye of the hurricane.

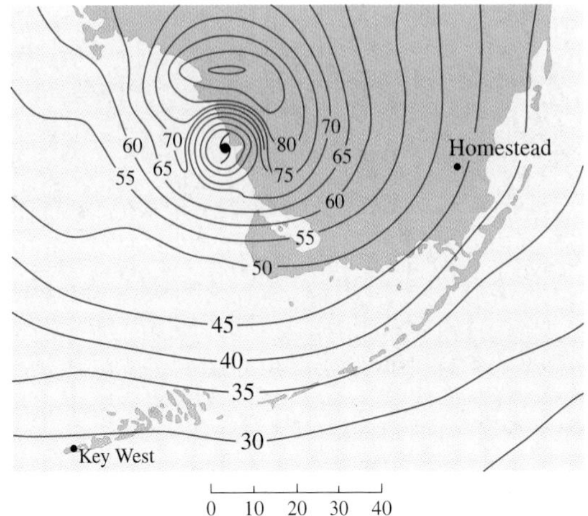

50. Find parametric equations of the tangent line at the point $(-2, 2, 4)$ to the curve of intersection of the surface $z = 2x^2 - y^2$ and the plane $z = 4$.

51–54 Find the local maximum and minimum values and saddle points of the function. If you have three-dimensional graphing software, graph the function with a domain and viewpoint that reveal all the important aspects of the function.

51. $f(x, y) = x^2 - xy + y^2 + 9x - 6y + 10$

52. $f(x, y) = x^3 - 6xy + 8y^3$

53. $f(x, y) = 3xy - x^2y - xy^2$

54. $f(x, y) = (x^2 + y)e^{y/2}$

55–56 Find the absolute maximum and minimum values of f on the set D.

55. $f(x, y) = 4xy^2 - x^2y^2 - xy^3$; D is the closed triangular region in the xy-plane with vertices $(0, 0)$, $(0, 6)$, and $(6, 0)$

56. $f(x, y) = e^{-x^2-y^2}(x^2 + 2y^2)$; D is the disk $x^2 + y^2 \leqslant 4$

57. Use a graph or level curves or both to estimate the local maximum and minimum values and saddle points of $f(x, y) = x^3 - 3x + y^4 - 2y^2$. Then use calculus to find these values precisely.

58. Use a graphing calculator or computer (or Newton's method or a computer algebra system) to find the critical points of $f(x, y) = 12 + 10y - 2x^2 - 8xy - y^4$ correct to three decimal places. Then classify the critical points and find the highest point on the graph.

59–62 Use Lagrange multipliers to find the maximum and minimum values of f subject to the given constraint(s).

59. $f(x, y) = x^2y$; $x^2 + y^2 = 1$

60. $f(x, y) = \dfrac{1}{x} + \dfrac{1}{y}$; $\dfrac{1}{x^2} + \dfrac{1}{y^2} = 1$

61. $f(x, y, z) = xyz$; $x^2 + y^2 + z^2 = 3$

62. $f(x, y, z) = x^2 + 2y^2 + 3z^2$; $x + y + z = 1$, $x - y + 2z = 2$

63. Find the points on the surface $xy^2z^3 = 2$ that are closest to the origin.

64. A package in the shape of a rectangular box can be mailed by the US Postal Service if the sum of its length and girth (the perimeter of a cross-section perpendicular to the length) is at most 108 in. Find the dimensions of the package with largest volume that can be mailed.

65. A pentagon is formed by placing an isosceles triangle on a rectangle, as shown in the figure. If the pentagon has fixed perimeter P, find the lengths of the sides of the pentagon that maximize the area of the pentagon.

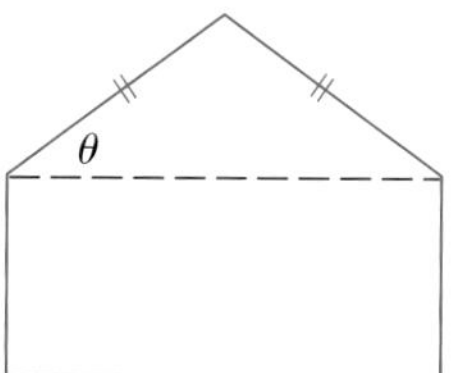

66. A particle of mass m moves on the surface $z = f(x, y)$. Let $x = x(t)$ and $y = y(t)$ be the x- and y-coordinates of the particle at time t.

(a) Find the velocity vector $\mathbf{v}$ and the kinetic energy $K = \frac{1}{2}m|\mathbf{v}|^2$ of the particle.

(b) Determine the acceleration vector $\mathbf{a}$.

(c) Let $z = x^2 + y^2$ and $x(t) = t\cos t$, $y(t) = t\sin t$. Find the velocity vector, the kinetic energy, and the acceleration vector.

Problems Plus

1. A rectangle with length L and width W is cut into four smaller rectangles by two lines parallel to the sides. Find the maximum and minimum values of the sum of the squares of the areas of the smaller rectangles.

2. Marine biologists have determined that when a shark detects the presence of blood in the water, it will swim in the direction in which the concentration of the blood increases most rapidly. Based on certain tests, the concentration of blood (in parts per million) at a point $P(x, y)$ on the surface of seawater is approximated by

$$C(x, y) = e^{-(x^2+2y^2)/10^4}$$

where x and y are measured in meters in a rectangular coordinate system with the blood source at the origin.

(a) Identify the level curves of the concentration function and sketch several members of this family together with a path that a shark will follow to the source.

(b) Suppose a shark is at the point (x_0, y_0) when it first detects the presence of blood in the water. Find an equation of the shark's path by setting up and solving a differential equation.

3. A long piece of galvanized sheet metal with width w is to be bent into a symmetric form with three straight sides to make a rain gutter. A cross-section is shown in the figure.

(a) Determine the dimensions that allow the maximum possible flow; that is, find the dimensions that give the maximum possible cross-sectional area.

(b) Would it be better to bend the metal into a gutter with a semicircular cross-section?

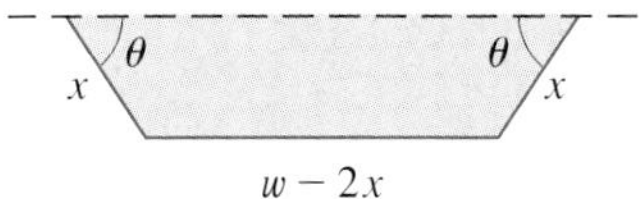

4. For what values of the number r is the function

$$f(x, y, z) = \begin{cases} \dfrac{(x + y + z)^r}{x^2 + y^2 + z^2} & \text{if } (x, y, z) \neq (0, 0, 0) \\ 0 & \text{if } (x, y, z) = (0, 0, 0) \end{cases}$$

continuous on $\mathbb{R}^3$?

5. Suppose f is a differentiable function of one variable. Show that all tangent planes to the surface $z = xf(y/x)$ intersect in a common point.

6. (a) Newton's method for approximating a root of an equation $f(x) = 0$ (see Section 4.8) can be adapted to approximating a solution of a system of equations $f(x, y) = 0$ and $g(x, y) = 0$. The surfaces $z = f(x, y)$ and $z = g(x, y)$ intersect in a curve that intersects the xy-plane at the point (r, s), which is the solution of the system. If an initial approximation (x_1, y_1) is close to this point, then the tangent planes to the surfaces at (x_1, y_1) intersect in a straight line that intersects the xy-plane in a point (x_2, y_2), which should be closer to (r, s). (Compare with Figure 2 in Section 4.8.) Show that

$$x_2 = x_1 - \frac{fg_y - f_y g}{f_x g_y - f_y g_x} \qquad \text{and} \qquad y_2 = y_1 - \frac{f_x g - fg_x}{f_x g_y - f_y g_x}$$

where f, g, and their partial derivatives are evaluated at (x_1, y_1). If we continue this procedure, we obtain successive approximations (x_n, y_n).

(b) It was Thomas Simpson (1710–1761) who formulated Newton's method as we know it today and who extended it to functions of two variables as in part (a). (See the biography of Simpson on page 513.) The example that he gave to illustrate the method was to solve the system of equations

$$x^x + y^y = 1000 \qquad x^y + y^x = 100$$

In other words, he found the points of intersection of the curves in the figure. Use the method of part (a) to find the coordinates of the points of intersection correct to six decimal places.

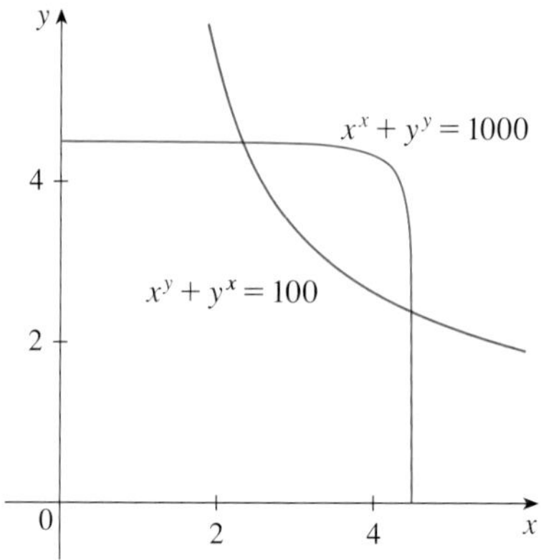

7. If the ellipse $x^2/a^2 + y^2/b^2 = 1$ is to enclose the circle $x^2 + y^2 = 2y$, what values of a and b minimize the area of the ellipse?

8. Among all planes that are tangent to the surface $xy^2z^2 = 1$, find the ones that are farthest from the origin.

15 Multiple Integrals

Geologists study how mountain ranges were formed and estimate the work required to lift them from sea level. In Section 15.8 you are asked to use a triple integral to compute the work done in the formation of Mount Fuji in Japan.

In this chapter we extend the idea of a definite integral to double and triple integrals of functions of two or three variables. These ideas are then used to compute volumes, masses, and centroids of more general regions than we were able to consider in Chapters 6 and 8. We also use double integrals to calculate probabilities when two random variables are involved.

We will see that polar coordinates are useful in computing double integrals over some types of regions. In a similar way, we will introduce two new coordinate systems in three-dimensional space—cylindrical coordinates and spherical coordinates—that greatly simplify the computation of triple integrals over certain commonly occurring solid regions.

15.1 Double Integrals over Rectangles

In much the same way that our attempt to solve the area problem led to the definition of a definite integral, we now seek to find the volume of a solid and in the process we arrive at the definition of a double integral.

Review of the Definite Integral

First let's recall the basic facts concerning definite integrals of functions of a single variable. If $f(x)$ is defined for $a \leqslant x \leqslant b$, we start by dividing the interval $[a, b]$ into n subintervals $[x_{i-1}, x_i]$ of equal width $\Delta x = (b - a)/n$ and we choose sample points x_i^* in these subintervals. Then we form the Riemann sum

$$\boxed{1} \qquad \sum_{i=1}^{n} f(x_i^*)\,\Delta x$$

and take the limit of such sums as $n \to \infty$ to obtain the definite integral of f from a to b:

$$\boxed{2} \qquad \int_a^b f(x)\,dx = \lim_{n\to\infty} \sum_{i=1}^{n} f(x_i^*)\,\Delta x$$

In the special case where $f(x) \geqslant 0$, the Riemann sum can be interpreted as the sum of the areas of the approximating rectangles in Figure 1, and $\int_a^b f(x)\,dx$ represents the area under the curve $y = f(x)$ from a to b.

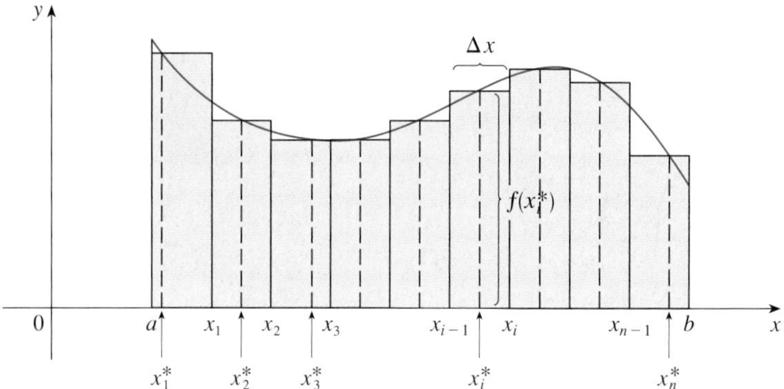

FIGURE 1

Volumes and Double Integrals

In a similar manner we consider a function f of two variables defined on a closed rectangle

$$R = [a, b] \times [c, d] = \{(x, y) \in \mathbb{R}^2 \mid a \leqslant x \leqslant b,\ c \leqslant y \leqslant d\}$$

and we first suppose that $f(x, y) \geqslant 0$. The graph of f is a surface with equation $z = f(x, y)$. Let S be the solid that lies above R and under the graph of f, that is,

$$S = \{(x, y, z) \in \mathbb{R}^3 \mid 0 \leqslant z \leqslant f(x, y),\ (x, y) \in R\}$$

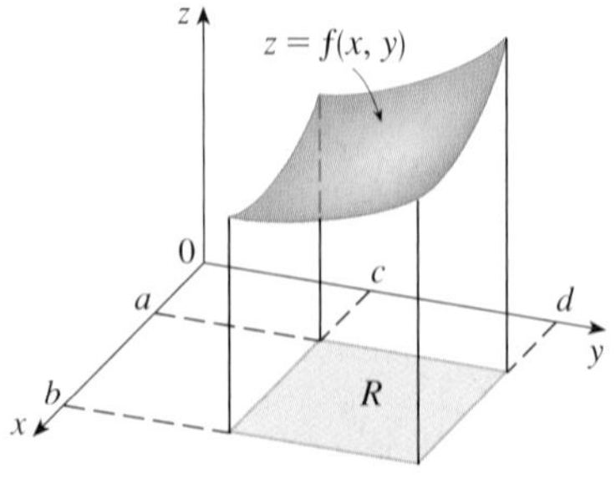

FIGURE 2

(See Figure 2.) Our goal is to find the volume of S.

The first step is to divide the rectangle R into subrectangles. We accomplish this by dividing the interval $[a, b]$ into m subintervals $[x_{i-1}, x_i]$ of equal width $\Delta x = (b - a)/m$ and dividing $[c, d]$ into n subintervals $[y_{j-1}, y_j]$ of equal width $\Delta y = (d - c)/n$. By drawing lines parallel to the coordinate axes through the endpoints of these subintervals, as in

Figure 3, we form the subrectangles

$$R_{ij} = [x_{i-1}, x_i] \times [y_{j-1}, y_j] = \{(x, y) \mid x_{i-1} \leqslant x \leqslant x_i,\ y_{j-1} \leqslant y \leqslant y_j\}$$

each with area $\Delta A = \Delta x \, \Delta y$.

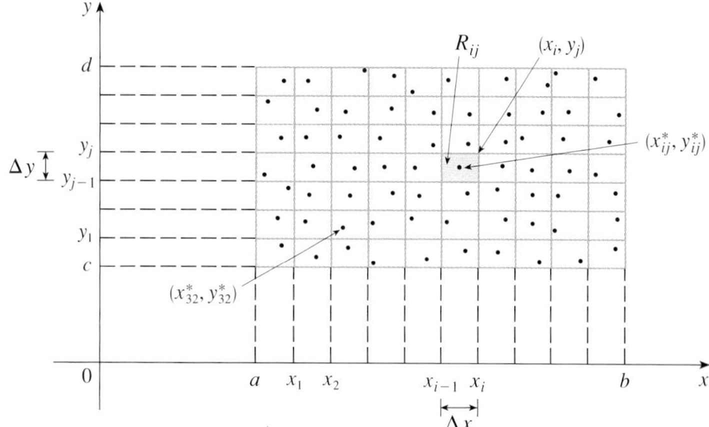

FIGURE 3
Dividing R into subrectangles

If we choose a **sample point** (x_{ij}^*, y_{ij}^*) in each R_{ij}, then we can approximate the part of S that lies above each R_{ij} by a thin rectangular box (or "column") with base R_{ij} and height $f(x_{ij}^*, y_{ij}^*)$ as shown in Figure 4. (Compare with Figure 1.) The volume of this box is the height of the box times the area of the base rectangle:

$$f(x_{ij}^*, y_{ij}^*) \, \Delta A$$

If we follow this procedure for all the rectangles and add the volumes of the corresponding boxes, we get an approximation to the total volume of S:

$$\boxed{3} \qquad V \approx \sum_{i=1}^{m} \sum_{j=1}^{n} f(x_{ij}^*, y_{ij}^*) \, \Delta A$$

(See Figure 5.) This double sum means that for each subrectangle we evaluate f at the chosen point and multiply by the area of the subrectangle, and then we add the results.

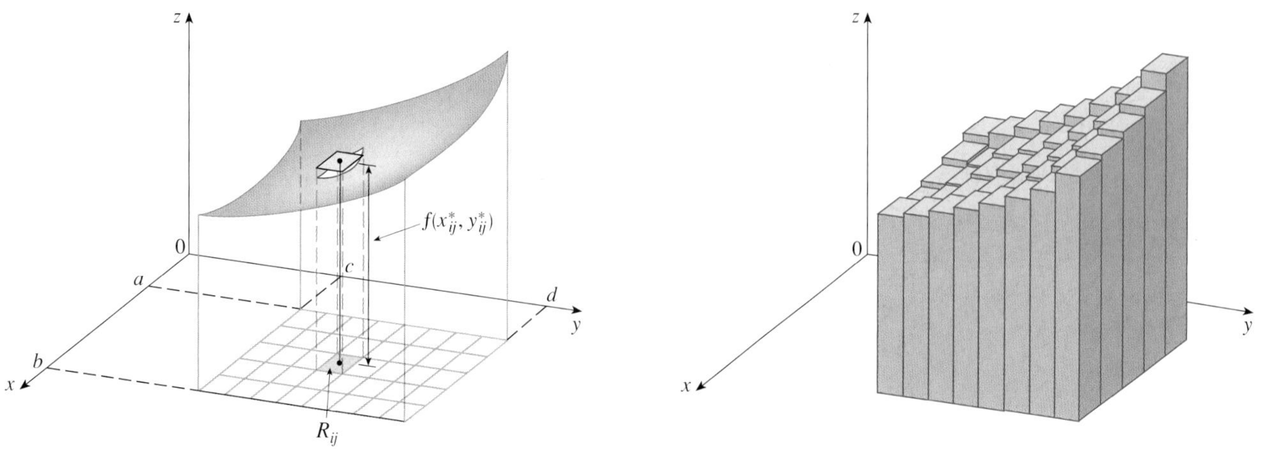

FIGURE 4

FIGURE 5

Our intuition tells us that the approximation given in 3 becomes better as m and n become larger and so we would expect that

The meaning of the double limit in Equation 4 is that we can make the double sum as close as we like to the number V [for any choice of (x_{ij}^*, y_{ij}^*) in R_{ij}] by taking m and n sufficiently large.

4
$$V = \lim_{m,\, n \to \infty} \sum_{i=1}^{m} \sum_{j=1}^{n} f(x_{ij}^*, y_{ij}^*)\, \Delta A$$

We use the expression in Equation 4 to define the **volume** of the solid S that lies under the graph of f and above the rectangle R. (It can be shown that this definition is consistent with our formula for volume in Section 6.2.)

Limits of the type that appear in Equation 4 occur frequently, not just in finding volumes but in a variety of other situations as well—as we will see in Section 15.5—even when f is not a positive function. So we make the following definition.

5 Definition The **double integral** of f over the rectangle R is

$$\iint_R f(x, y)\, dA = \lim_{m,\, n \to \infty} \sum_{i=1}^{m} \sum_{j=1}^{n} f(x_{ij}^*, y_{ij}^*)\, \Delta A$$

if this limit exists.

Notice the similarity between Definition 5 and the definition of a single integral in Equation 2.

The precise meaning of the limit in Definition 5 is that for every number $\varepsilon > 0$ there is an integer N such that

$$\left| \iint_R f(x, y)\, dA - \sum_{i=1}^{m} \sum_{j=1}^{n} f(x_{ij}^*, y_{ij}^*)\, \Delta A \right| < \varepsilon$$

Although we have defined the double integral by dividing R into equal-sized subrectangles, we could have used subrectangles R_{ij} of unequal size. But then we would have to ensure that all of their dimensions approach 0 in the limiting process.

for all integers m and n greater than N and for any choice of sample points (x_{ij}^*, y_{ij}^*) in R_{ij}.

A function f is called **integrable** if the limit in Definition 5 exists. It is shown in courses on advanced calculus that all continuous functions are integrable. In fact, the double integral of f exists provided that f is "not too discontinuous." In particular, if f is bounded [that is, there is a constant M such that $|f(x, y)| \leqslant M$ for all (x, y) in R], and f is continuous there, except on a finite number of smooth curves, then f is integrable over R.

The sample point (x_{ij}^*, y_{ij}^*) can be chosen to be any point in the subrectangle R_{ij}, but if we choose it to be the upper right-hand corner of R_{ij} [namely (x_i, y_j), see Figure 3], then the expression for the double integral looks simpler:

6
$$\iint_R f(x, y)\, dA = \lim_{m,\, n \to \infty} \sum_{i=1}^{m} \sum_{j=1}^{n} f(x_i, y_j)\, \Delta A$$

By comparing Definitions 4 and 5, we see that a volume can be written as a double integral:

If $f(x, y) \geqslant 0$, then the volume V of the solid that lies above the rectangle R and below the surface $z = f(x, y)$ is

$$V = \iint_R f(x, y)\, dA$$

The sum in Definition 5,

$$\sum_{i=1}^{m} \sum_{j=1}^{n} f(x_{ij}^*, y_{ij}^*) \, \Delta A$$

is called a **double Riemann sum** and is used as an approximation to the value of the double integral. [Notice how similar it is to the Riemann sum in [1] for a function of a single variable.] If f happens to be a *positive* function, then the double Riemann sum represents the sum of volumes of columns, as in Figure 5, and is an approximation to the volume under the graph of f.

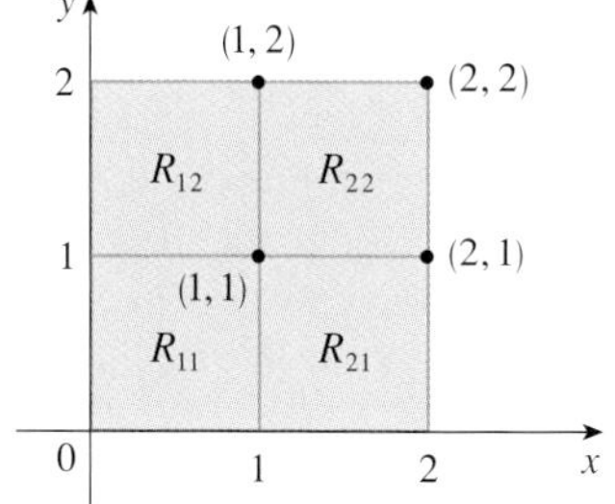

FIGURE 6

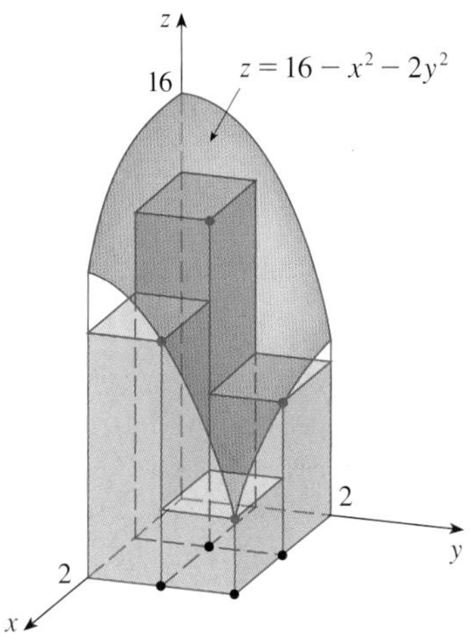

FIGURE 7

V EXAMPLE 1 Estimate the volume of the solid that lies above the square $R = [0, 2] \times [0, 2]$ and below the elliptic paraboloid $z = 16 - x^2 - 2y^2$. Divide R into four equal squares and choose the sample point to be the upper right corner of each square R_{ij}. Sketch the solid and the approximating rectangular boxes.

SOLUTION The squares are shown in Figure 6. The paraboloid is the graph of $f(x, y) = 16 - x^2 - 2y^2$ and the area of each square is $\Delta A = 1$. Approximating the volume by the Riemann sum with $m = n = 2$, we have

$$V \approx \sum_{i=1}^{2} \sum_{j=1}^{2} f(x_i, y_j) \, \Delta A$$

$$= f(1, 1) \, \Delta A + f(1, 2) \, \Delta A + f(2, 1) \, \Delta A + f(2, 2) \, \Delta A$$

$$= 13(1) + 7(1) + 10(1) + 4(1) = 34$$

This is the volume of the approximating rectangular boxes shown in Figure 7. ■

We get better approximations to the volume in Example 1 if we increase the number of squares. Figure 8 shows how the columns start to look more like the actual solid and the corresponding approximations become more accurate when we use 16, 64, and 256 squares. In the next section we will be able to show that the exact volume is 48.

(a) $m = n = 4$, $V \approx 41.5$

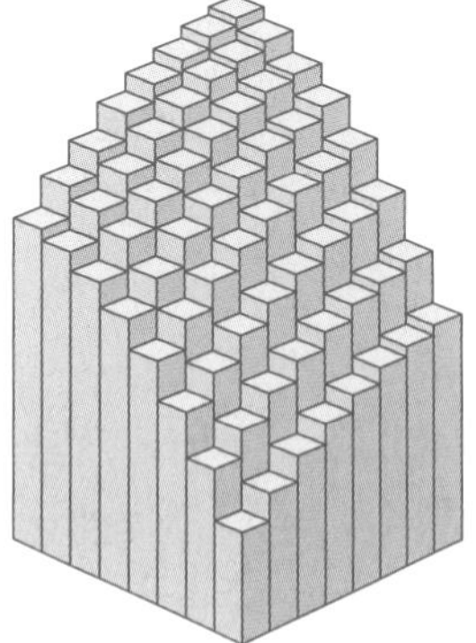

(b) $m = n = 8$, $V \approx 44.875$

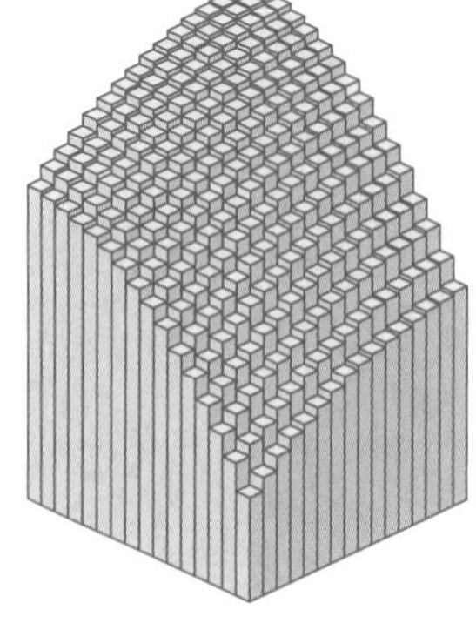

(c) $m = n = 16$, $V \approx 46.46875$

FIGURE 8
The Riemann sum approximations to the volume under $z = 16 - x^2 - 2y^2$ become more accurate as m and n increase.

V EXAMPLE 2 If $R = \{(x, y) \mid -1 \leqslant x \leqslant 1, -2 \leqslant y \leqslant 2\}$, evaluate the integral

$$\iint_R \sqrt{1 - x^2} \, dA$$

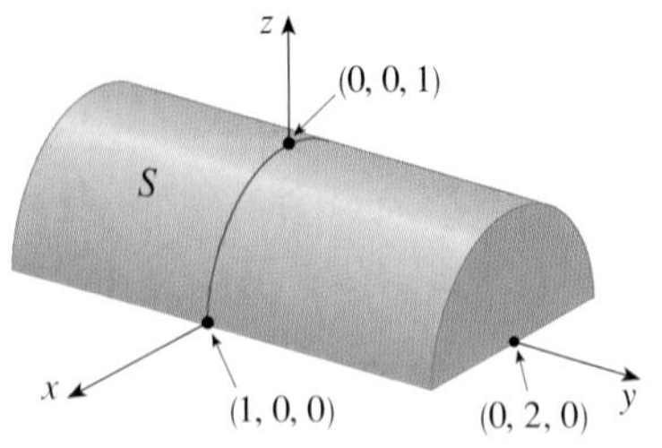

FIGURE 9

SOLUTION It would be very difficult to evaluate this integral directly from Definition 5 but, because $\sqrt{1-x^2} \geqslant 0$, we can compute the integral by interpreting it as a volume. If $z = \sqrt{1-x^2}$, then $x^2 + z^2 = 1$ and $z \geqslant 0$, so the given double integral represents the volume of the solid S that lies below the circular cylinder $x^2 + z^2 = 1$ and above the rectangle R. (See Figure 9.) The volume of S is the area of a semicircle with radius 1 times the length of the cylinder. Thus

$$\iint_R \sqrt{1-x^2}\, dA = \tfrac{1}{2}\pi(1)^2 \times 4 = 2\pi$$

The Midpoint Rule

The methods that we used for approximating single integrals (the Midpoint Rule, the Trapezoidal Rule, Simpson's Rule) all have counterparts for double integrals. Here we consider only the Midpoint Rule for double integrals. This means that we use a double Riemann sum to approximate the double integral, where the sample point (x_{ij}^*, y_{ij}^*) in R_{ij} is chosen to be the center $(\bar{x}_i, \bar{y}_j)$ of R_{ij}. In other words, $\bar{x}_i$ is the midpoint of $[x_{i-1}, x_i]$ and $\bar{y}_j$ is the midpoint of $[y_{j-1}, y_j]$.

Midpoint Rule for Double Integrals

$$\iint_R f(x, y)\, dA \approx \sum_{i=1}^{m} \sum_{j=1}^{n} f(\bar{x}_i, \bar{y}_j)\, \Delta A$$

where $\bar{x}_i$ is the midpoint of $[x_{i-1}, x_i]$ and $\bar{y}_j$ is the midpoint of $[y_{j-1}, y_j]$.

V EXAMPLE 3 Use the Midpoint Rule with $m = n = 2$ to estimate the value of the integral $\iint_R (x - 3y^2)\, dA$, where $R = \{(x, y) \mid 0 \leqslant x \leqslant 2,\ 1 \leqslant y \leqslant 2\}$.

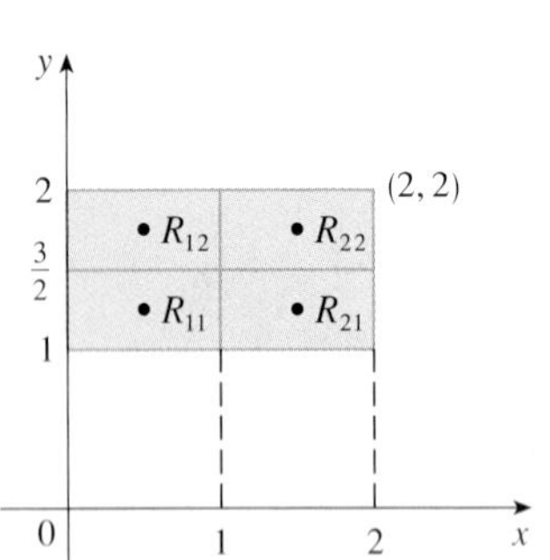

FIGURE 10

SOLUTION In using the Midpoint Rule with $m = n = 2$, we evaluate $f(x, y) = x - 3y^2$ at the centers of the four subrectangles shown in Figure 10. So $\bar{x}_1 = \frac{1}{2}$, $\bar{x}_2 = \frac{3}{2}$, $\bar{y}_1 = \frac{5}{4}$, and $\bar{y}_2 = \frac{7}{4}$. The area of each subrectangle is $\Delta A = \frac{1}{2}$. Thus

$$\begin{aligned}
\iint_R (x - 3y^2)\, dA &\approx \sum_{i=1}^{2} \sum_{j=1}^{2} f(\bar{x}_i, \bar{y}_j)\, \Delta A \\
&= f(\bar{x}_1, \bar{y}_1)\, \Delta A + f(\bar{x}_1, \bar{y}_2)\, \Delta A + f(\bar{x}_2, \bar{y}_1)\, \Delta A + f(\bar{x}_2, \bar{y}_2)\, \Delta A \\
&= f\left(\tfrac{1}{2}, \tfrac{5}{4}\right) \Delta A + f\left(\tfrac{1}{2}, \tfrac{7}{4}\right) \Delta A + f\left(\tfrac{3}{2}, \tfrac{5}{4}\right) \Delta A + f\left(\tfrac{3}{2}, \tfrac{7}{4}\right) \Delta A \\
&= \left(-\tfrac{67}{16}\right)\tfrac{1}{2} + \left(-\tfrac{139}{16}\right)\tfrac{1}{2} + \left(-\tfrac{51}{16}\right)\tfrac{1}{2} + \left(-\tfrac{123}{16}\right)\tfrac{1}{2} \\
&= -\tfrac{95}{8} = -11.875
\end{aligned}$$

Thus we have

$$\iint_R (x - 3y^2)\, dA \approx -11.875$$

NOTE In the next section we will develop an efficient method for computing double integrals and then we will see that the exact value of the double integral in Example 3 is -12. (Remember that the interpretation of a double integral as a volume is valid only when the integrand f is a *positive* function. The integrand in Example 3 is not a positive function, so its integral is not a volume. In Examples 2 and 3 in Section 15.2 we will discuss how to interpret integrals of functions that are not always positive in terms of volumes.) If we keep dividing each subrectangle in Figure 10 into four smaller ones with similar shape,

Number of subrectangles	Midpoint Rule approximation
1	-11.5000
4	-11.8750
16	-11.9687
64	-11.9922
256	-11.9980
1024	-11.9995

we get the Midpoint Rule approximations displayed in the chart in the margin. Notice how these approximations approach the exact value of the double integral, -12.

Average Value

Recall from Section 6.5 that the average value of a function f of one variable defined on an interval $[a, b]$ is

$$f_{\text{ave}} = \frac{1}{b-a}\int_a^b f(x)\,dx$$

In a similar fashion we define the **average value** of a function f of two variables defined on a rectangle R to be

$$f_{\text{ave}} = \frac{1}{A(R)}\iint_R f(x, y)\,dA$$

where $A(R)$ is the area of R.

If $f(x, y) \geqslant 0$, the equation

$$A(R) \times f_{\text{ave}} = \iint_R f(x, y)\,dA$$

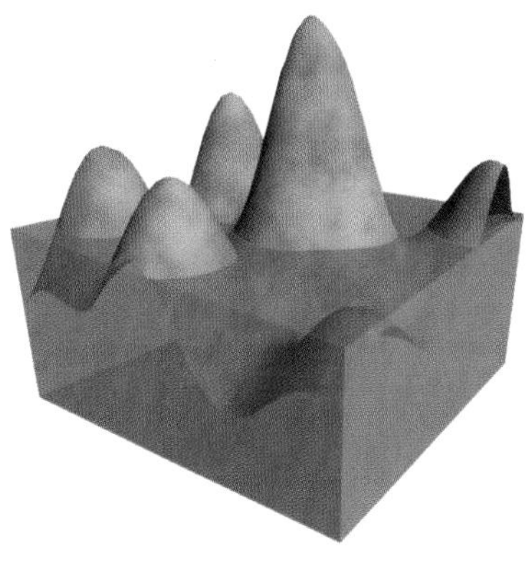

FIGURE 11

says that the box with base R and height f_{ave} has the same volume as the solid that lies under the graph of f. [If $z = f(x, y)$ describes a mountainous region and you chop off the tops of the mountains at height f_{ave}, then you can use them to fill in the valleys so that the region becomes completely flat. See Figure 11.]

EXAMPLE 4 The contour map in Figure 12 shows the snowfall, in inches, that fell on the state of Colorado on December 20 and 21, 2006. (The state is in the shape of a rectangle that measures 388 mi west to east and 276 mi south to north.) Use the contour map to estimate the average snowfall for the entire state of Colorado on those days.

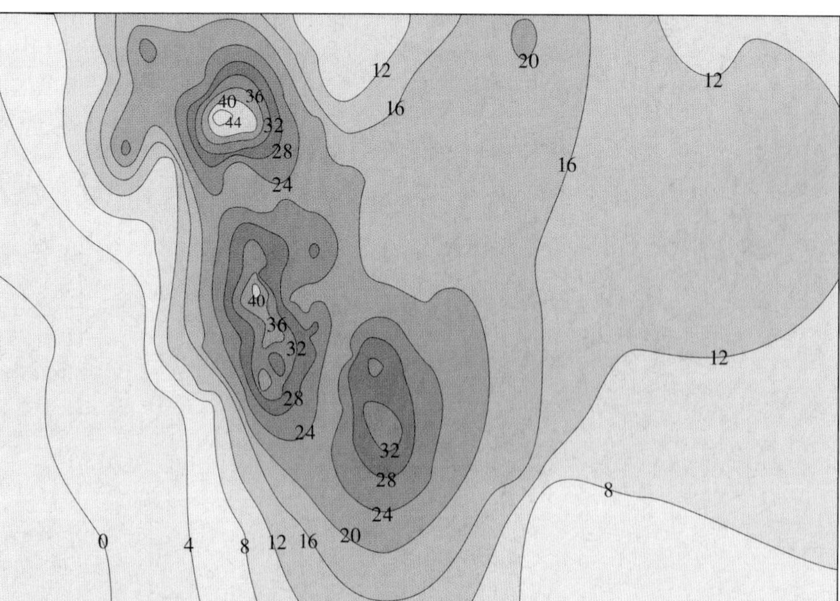

FIGURE 12

SOLUTION Let's place the origin at the southwest corner of the state. Then $0 \leqslant x \leqslant 388$, $0 \leqslant y \leqslant 276$, and $f(x, y)$ is the snowfall, in inches, at a location x miles to the east and

y miles to the north of the origin. If R is the rectangle that represents Colorado, then the average snowfall for the state on December 20–21 was

$$f_{\text{ave}} = \frac{1}{A(R)} \iint_R f(x, y)\, dA$$

where $A(R) = 388 \cdot 276$. To estimate the value of this double integral, let's use the Midpoint Rule with $m = n = 4$. In other words, we divide R into 16 subrectangles of equal size, as in Figure 13. The area of each subrectangle is

$$\Delta A = \tfrac{1}{16}(388)(276) = 6693 \text{ mi}^2$$

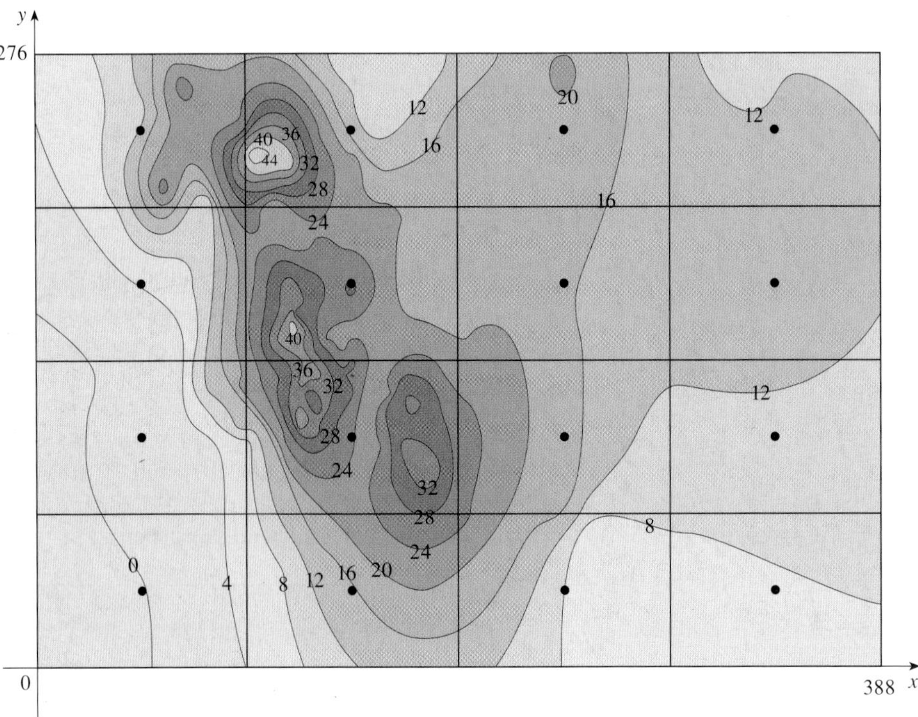

FIGURE 13

Using the contour map to estimate the value of f at the center of each subrectangle, we get

$$\begin{aligned}\iint_R f(x, y)\, dA &\approx \sum_{i=1}^{4} \sum_{j=1}^{4} f(\bar{x}_i, \bar{y}_j)\, \Delta A \\ &\approx \Delta A[0 + 15 + 8 + 7 + 2 + 25 + 18.5 + 11 \\ &\qquad + 4.5 + 28 + 17 + 13.5 + 12 + 15 + 17.5 + 13] \\ &= (6693)(207)\end{aligned}$$

Therefore $$f_{\text{ave}} \approx \frac{(6693)(207)}{(388)(276)} \approx 12.9$$

On December 20–21, 2006, Colorado received an average of approximately 13 inches of snow.

Properties of Double Integrals

We list here three properties of double integrals that can be proved in the same manner as in Section 5.2. We assume that all of the integrals exist. Properties 7 and 8 are referred to as the *linearity* of the integral.

Double integrals behave this way because the double sums that define them behave this way.

$$\boxed{7} \qquad \iint_R [f(x, y) + g(x, y)]\, dA = \iint_R f(x, y)\, dA + \iint_R g(x, y)\, dA$$

$$\boxed{8} \qquad \iint_R cf(x, y)\, dA = c \iint_R f(x, y)\, dA \qquad \text{where } c \text{ is a constant}$$

If $f(x, y) \geqslant g(x, y)$ for all (x, y) in R, then

$$\boxed{9} \qquad \iint_R f(x, y)\, dA \geqslant \iint_R g(x, y)\, dA$$

15.1 Exercises

1. (a) Estimate the volume of the solid that lies below the surface $z = xy$ and above the rectangle

$$R = \{(x, y) \mid 0 \leqslant x \leqslant 6,\ 0 \leqslant y \leqslant 4\}$$

Use a Riemann sum with $m = 3$, $n = 2$, and take the sample point to be the upper right corner of each square.
(b) Use the Midpoint Rule to estimate the volume of the solid in part (a).

2. If $R = [0, 4] \times [-1, 2]$, use a Riemann sum with $m = 2$, $n = 3$ to estimate the value of $\iint_R (1 - xy^2)\, dA$. Take the sample points to be (a) the lower right corners and (b) the upper left corners of the rectangles.

3. (a) Use a Riemann sum with $m = n = 2$ to estimate the value of $\iint_R xe^{-xy}\, dA$, where $R = [0, 2] \times [0, 1]$. Take the sample points to be upper right corners.
(b) Use the Midpoint Rule to estimate the integral in part (a).

4. (a) Estimate the volume of the solid that lies below the surface $z = 1 + x^2 + 3y$ and above the rectangle $R = [1, 2] \times [0, 3]$. Use a Riemann sum with $m = n = 2$ and choose the sample points to be lower left corners.
(b) Use the Midpoint Rule to estimate the volume in part (a).

5. A table of values is given for a function $f(x, y)$ defined on $R = [0, 4] \times [2, 4]$.
(a) Estimate $\iint_R f(x, y)\, dA$ using the Midpoint Rule with $m = n = 2$.
(b) Estimate the double integral with $m = n = 4$ by choosing the sample points to be the points closest to the origin.

x \ y	2.0	2.5	3.0	3.5	4.0
0	−3	−5	−6	−4	−1
1	−1	−2	−3	−1	1
2	1	0	−1	1	4
3	2	2	1	3	7
4	3	4	2	5	9

6. A 20-ft-by-30-ft swimming pool is filled with water. The depth is measured at 5-ft intervals, starting at one corner of the pool, and the values are recorded in the table. Estimate the volume of water in the pool.

	0	5	10	15	20	25	30
0	2	3	4	6	7	8	8
5	2	3	4	7	8	10	8
10	2	4	6	8	10	12	10
15	2	3	4	5	6	8	7
20	2	2	2	2	3	4	4

7. Let V be the volume of the solid that lies under the graph of $f(x, y) = \sqrt{52 - x^2 - y^2}$ and above the rectangle given by $2 \leqslant x \leqslant 4$, $2 \leqslant y \leqslant 6$. We use the lines $x = 3$ and $y = 4$ to

1. Homework Hints available at stewartcalculus.com

divide R into subrectangles. Let L and U be the Riemann sums computed using lower left corners and upper right corners, respectively. Without calculating the numbers V, L, and U, arrange them in increasing order and explain your reasoning.

8. The figure shows level curves of a function f in the square $R = [0, 2] \times [0, 2]$. Use the Midpoint Rule with $m = n = 2$ to estimate $\iint_R f(x, y)\, dA$. How could you improve your estimate?

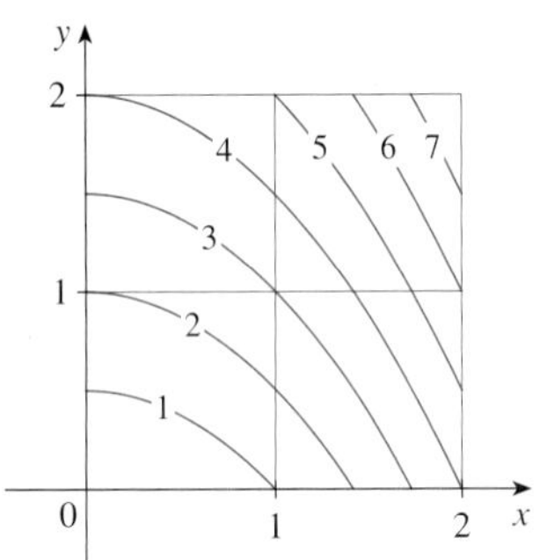

9. A contour map is shown for a function f on the square $R = [0, 4] \times [0, 4]$.
(a) Use the Midpoint Rule with $m = n = 2$ to estimate the value of $\iint_R f(x, y)\, dA$.
(b) Estimate the average value of f.

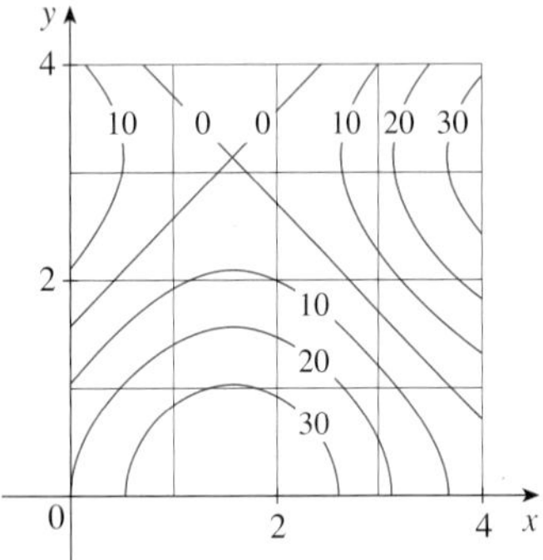

10. The contour map shows the temperature, in degrees Fahrenheit, at 4:00 PM on February 26, 2007, in Colorado. (The state measures 388 mi west to east and 276 mi south to north.) Use the Midpoint Rule with $m = n = 4$ to estimate the average temperature in Colorado at that time.

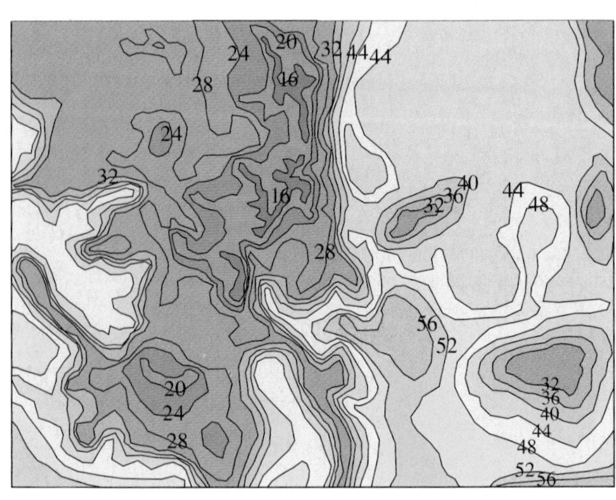

11–13 Evaluate the double integral by first identifying it as the volume of a solid.

11. $\iint_R 3\, dA, \quad R = \{(x, y) \mid -2 \leqslant x \leqslant 2, 1 \leqslant y \leqslant 6\}$

12. $\iint_R (5 - x)\, dA, \quad R = \{(x, y) \mid 0 \leqslant x \leqslant 5, 0 \leqslant y \leqslant 3\}$

13. $\iint_R (4 - 2y)\, dA, \quad R = [0, 1] \times [0, 1]$

---

14. The integral $\iint_R \sqrt{9 - y^2}\, dA$, where $R = [0, 4] \times [0, 2]$, represents the volume of a solid. Sketch the solid.

15. Use a programmable calculator or computer (or the sum command on a CAS) to estimate

$$\iint_R \sqrt{1 + xe^{-y}}\, dA$$

where $R = [0, 1] \times [0, 1]$. Use the Midpoint Rule with the following numbers of squares of equal size: 1, 4, 16, 64, 256, and 1024.

16. Repeat Exercise 15 for the integral $\iint_R \sin\left(x + \sqrt{y}\right) dA$.

17. If f is a constant function, $f(x, y) = k$, and $R = [a, b] \times [c, d]$, show that

$$\iint_R k\, dA = k(b - a)(d - c)$$

18. Use the result of Exercise 17 to show that

$$0 \leqslant \iint_R \sin \pi x \cos \pi y\, dA \leqslant \frac{1}{32}$$

where $R = \left[0, \frac{1}{4}\right] \times \left[\frac{1}{4}, \frac{1}{2}\right]$.

15.2 Iterated Integrals

Recall that it is usually difficult to evaluate single integrals directly from the definition of an integral, but the Fundamental Theorem of Calculus provides a much easier method. The evaluation of double integrals from first principles is even more difficult, but in this section we see how to express a double integral as an iterated integral, which can then be evaluated by calculating two single integrals.

Suppose that f is a function of two variables that is integrable on the rectangle $R = [a, b] \times [c, d]$. We use the notation $\int_c^d f(x, y)\, dy$ to mean that x is held fixed and $f(x, y)$ is integrated with respect to y from $y = c$ to $y = d$. This procedure is called *partial integration with respect to y*. (Notice its similarity to partial differentiation.) Now $\int_c^d f(x, y)\, dy$ is a number that depends on the value of x, so it defines a function of x:

$$A(x) = \int_c^d f(x, y)\, dy$$

If we now integrate the function A with respect to x from $x = a$ to $x = b$, we get

1 $$\int_a^b A(x)\, dx = \int_a^b \left[\int_c^d f(x, y)\, dy \right] dx$$

The integral on the right side of Equation 1 is called an **iterated integral**. Usually the brackets are omitted. Thus

2 $$\int_a^b \int_c^d f(x, y)\, dy\, dx = \int_a^b \left[\int_c^d f(x, y)\, dy \right] dx$$

means that we first integrate with respect to y from c to d and then with respect to x from a to b.

Similarly, the iterated integral

3 $$\int_c^d \int_a^b f(x, y)\, dx\, dy = \int_c^d \left[\int_a^b f(x, y)\, dx \right] dy$$

means that we first integrate with respect to x (holding y fixed) from $x = a$ to $x = b$ and then we integrate the resulting function of y with respect to y from $y = c$ to $y = d$. Notice that in both Equations 2 and 3 we work *from the inside out*.

EXAMPLE 1 Evaluate the iterated integrals.

(a) $\int_0^3 \int_1^2 x^2 y\, dy\, dx$ (b) $\int_1^2 \int_0^3 x^2 y\, dx\, dy$

SOLUTION

(a) Regarding x as a constant, we obtain

$$\int_1^2 x^2 y\, dy = \left[x^2 \frac{y^2}{2} \right]_{y=1}^{y=2} = x^2 \left(\frac{2^2}{2} \right) - x^2 \left(\frac{1^2}{2} \right) = \tfrac{3}{2} x^2$$

Thus the function A in the preceding discussion is given by $A(x) = \frac{3}{2}x^2$ in this example. We now integrate this function of x from 0 to 3:

$$\int_0^3 \int_1^2 x^2 y\, dy\, dx = \int_0^3 \left[\int_1^2 x^2 y\, dy \right] dx$$

$$= \int_0^3 \tfrac{3}{2} x^2\, dx = \frac{x^3}{2} \Bigg]_0^3 = \frac{27}{2}$$

(b) Here we first integrate with respect to x:

$$\int_1^2 \int_0^3 x^2 y \, dx \, dy = \int_1^2 \left[\int_0^3 x^2 y \, dx \right] dy = \int_1^2 \left[\frac{x^3}{3} y \right]_{x=0}^{x=3} dy$$

$$= \int_1^2 9y \, dy = 9 \frac{y^2}{2} \bigg]_1^2 = \frac{27}{2}$$

Notice that in Example 1 we obtained the same answer whether we integrated with respect to y or x first. In general, it turns out (see Theorem 4) that the two iterated integrals in Equations 2 and 3 are always equal; that is, the order of integration does not matter. (This is similar to Clairaut's Theorem on the equality of the mixed partial derivatives.)

The following theorem gives a practical method for evaluating a double integral by expressing it as an iterated integral (in either order).

Theorem 4 is named after the Italian mathematician Guido Fubini (1879–1943), who proved a very general version of this theorem in 1907. But the version for continuous functions was known to the French mathematician Augustin-Louis Cauchy almost a century earlier.

4 Fubini's Theorem If f is continuous on the rectangle $R = \{(x, y) \mid a \leqslant x \leqslant b, c \leqslant y \leqslant d\}$, then

$$\iint_R f(x, y) \, dA = \int_a^b \int_c^d f(x, y) \, dy \, dx = \int_c^d \int_a^b f(x, y) \, dx \, dy$$

More generally, this is true if we assume that f is bounded on R, f is discontinuous only on a finite number of smooth curves, and the iterated integrals exist.

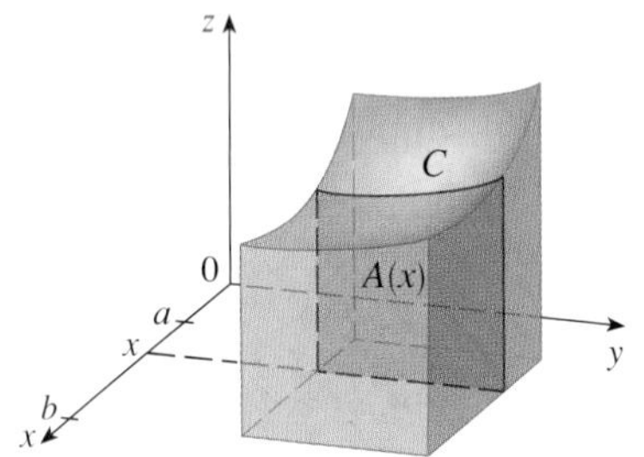

FIGURE 1

The proof of Fubini's Theorem is too difficult to include in this book, but we can at least give an intuitive indication of why it is true for the case where $f(x, y) \geqslant 0$. Recall that if f is positive, then we can interpret the double integral $\iint_R f(x, y) \, dA$ as the volume V of the solid S that lies above R and under the surface $z = f(x, y)$. But we have another formula that we used for volume in Chapter 6, namely,

$$V = \int_a^b A(x) \, dx$$

where $A(x)$ is the area of a cross-section of S in the plane through x perpendicular to the x-axis. From Figure 1 you can see that $A(x)$ is the area under the curve C whose equation is $z = f(x, y)$, where x is held constant and $c \leqslant y \leqslant d$. Therefore

TEC Visual 15.2 illustrates Fubini's Theorem by showing an animation of Figures 1 and 2.

$$A(x) = \int_c^d f(x, y) \, dy$$

and we have

$$\iint_R f(x, y) \, dA = V = \int_a^b A(x) \, dx = \int_a^b \int_c^d f(x, y) \, dy \, dx$$

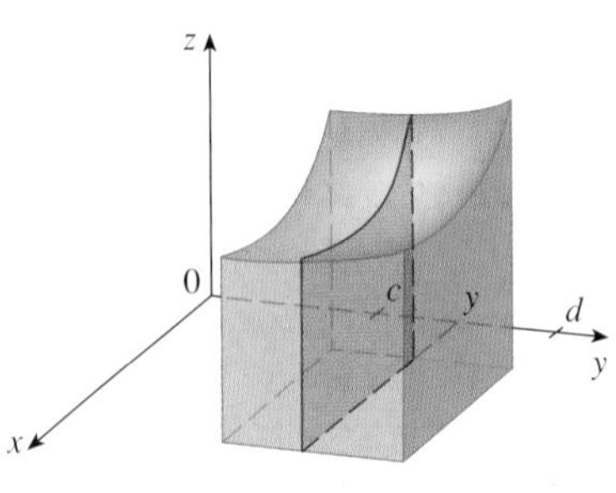

FIGURE 2

A similar argument, using cross-sections perpendicular to the y-axis as in Figure 2, shows that

$$\iint_R f(x, y) \, dA = \int_c^d \int_a^b f(x, y) \, dx \, dy$$

V EXAMPLE 2 Evaluate the double integral $\iint_R (x - 3y^2)\,dA$, where $R = \{(x, y) \mid 0 \leqslant x \leqslant 2, 1 \leqslant y \leqslant 2\}$. (Compare with Example 3 in Section 15.1.)

Notice the negative answer in Example 2; nothing is wrong with that. The function f is not a positive function, so its integral doesn't represent a volume. From Figure 3 we see that f is always negative on R, so the value of the integral is the *negative* of the volume that lies *above* the graph of f and *below* R.

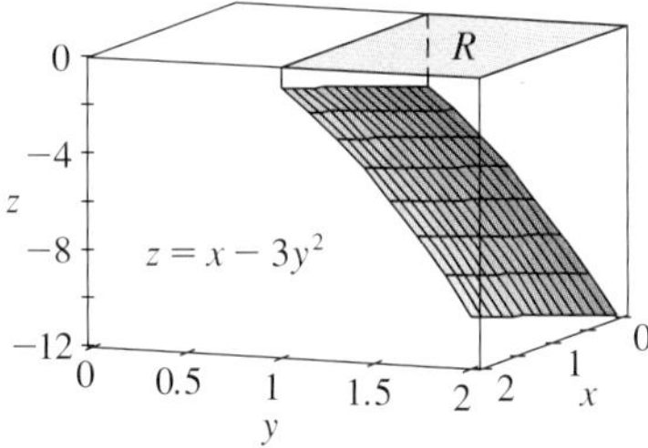

FIGURE 3

SOLUTION 1 Fubini's Theorem gives

$$\iint_R (x - 3y^2)\,dA = \int_0^2 \int_1^2 (x - 3y^2)\,dy\,dx = \int_0^2 \Big[xy - y^3\Big]_{y=1}^{y=2} dx$$

$$= \int_0^2 (x - 7)\,dx = \frac{x^2}{2} - 7x \Bigg]_0^2 = -12$$

SOLUTION 2 Again applying Fubini's Theorem, but this time integrating with respect to x first, we have

$$\iint_R (x - 3y^2)\,dA = \int_1^2 \int_0^2 (x - 3y^2)\,dx\,dy$$

$$= \int_1^2 \left[\frac{x^2}{2} - 3xy^2\right]_{x=0}^{x=2} dy$$

$$= \int_1^2 (2 - 6y^2)\,dy = 2y - 2y^3\Big]_1^2 = -12$$

V EXAMPLE 3 Evaluate $\iint_R y \sin(xy)\,dA$, where $R = [1, 2] \times [0, \pi]$.

SOLUTION 1 If we first integrate with respect to x, we get

$$\iint_R y \sin(xy)\,dA = \int_0^\pi \int_1^2 y \sin(xy)\,dx\,dy = \int_0^\pi \Big[-\cos(xy)\Big]_{x=1}^{x=2} dy$$

$$= \int_0^\pi (-\cos 2y + \cos y)\,dy$$

$$= -\tfrac{1}{2} \sin 2y + \sin y\Big]_0^\pi = 0$$

SOLUTION 2 If we reverse the order of integration, we get

$$\iint_R y \sin(xy)\,dA = \int_1^2 \int_0^\pi y \sin(xy)\,dy\,dx$$

For a function f that takes on both positive and negative values, $\iint_R f(x, y)\,dA$ is a difference of volumes: $V_1 - V_2$, where V_1 is the volume above R and below the graph of f, and V_2 is the volume below R and above the graph. The fact that the integral in Example 3 is 0 means that these two volumes V_1 and V_2 are equal. (See Figure 4.)

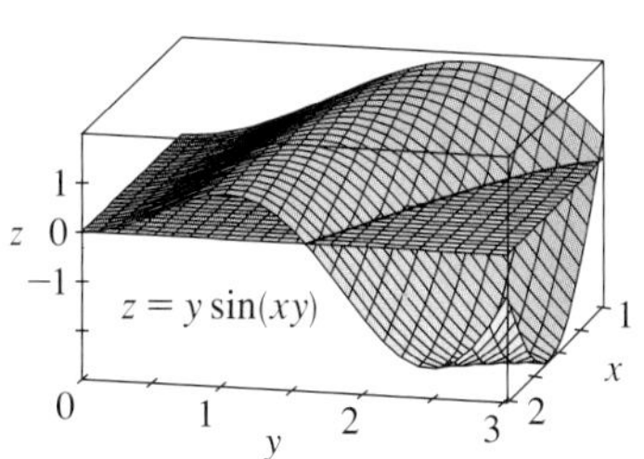

FIGURE 4

To evaluate the inner integral, we use integration by parts with

$$u = y \qquad dv = \sin(xy)\,dy$$

$$du = dy \qquad v = -\frac{\cos(xy)}{x}$$

and so

$$\int_0^\pi y \sin(xy)\,dy = -\frac{y \cos(xy)}{x}\Bigg]_{y=0}^{y=\pi} + \frac{1}{x}\int_0^\pi \cos(xy)\,dy$$

$$= -\frac{\pi \cos \pi x}{x} + \frac{1}{x^2}\Big[\sin(xy)\Big]_{y=0}^{y=\pi}$$

$$= -\frac{\pi \cos \pi x}{x} + \frac{\sin \pi x}{x^2}$$

If we now integrate the first term by parts with $u = -1/x$ and $dv = \pi \cos \pi x \, dx$, we get $du = dx/x^2$, $v = \sin \pi x$, and

$$\int \left(-\frac{\pi \cos \pi x}{x} \right) dx = -\frac{\sin \pi x}{x} - \int \frac{\sin \pi x}{x^2} \, dx$$

Therefore

$$\int \left(-\frac{\pi \cos \pi x}{x} + \frac{\sin \pi x}{x^2} \right) dx = -\frac{\sin \pi x}{x}$$

and so

$$\int_1^2 \int_0^\pi y \sin(xy) \, dy \, dx = \left[-\frac{\sin \pi x}{x} \right]_1^2$$

$$= -\frac{\sin 2\pi}{2} + \sin \pi = 0$$

In Example 2, Solutions 1 and 2 are equally straightforward, but in Example 3 the first solution is much easier than the second one. Therefore, when we evaluate double integrals, it's wise to choose the order of integration that gives simpler integrals.

V EXAMPLE 4 Find the volume of the solid S that is bounded by the elliptic paraboloid $x^2 + 2y^2 + z = 16$, the planes $x = 2$ and $y = 2$, and the three coordinate planes.

SOLUTION We first observe that S is the solid that lies under the surface $z = 16 - x^2 - 2y^2$ and above the square $R = [0, 2] \times [0, 2]$. (See Figure 5.) This solid was considered in Example 1 in Section 15.1, but we are now in a position to evaluate the double integral using Fubini's Theorem. Therefore

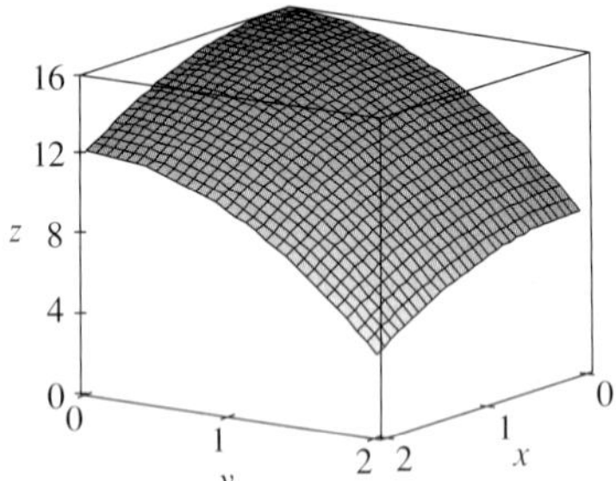

FIGURE 5

$$V = \iint_R (16 - x^2 - 2y^2) \, dA = \int_0^2 \int_0^2 (16 - x^2 - 2y^2) \, dx \, dy$$

$$= \int_0^2 \left[16x - \tfrac{1}{3}x^3 - 2y^2 x \right]_{x=0}^{x=2} dy$$

$$= \int_0^2 \left(\tfrac{88}{3} - 4y^2 \right) dy = \left[\tfrac{88}{3}y - \tfrac{4}{3}y^3 \right]_0^2 = 48$$

In the special case where $f(x, y)$ can be factored as the product of a function of x only and a function of y only, the double integral of f can be written in a particularly simple form. To be specific, suppose that $f(x, y) = g(x)h(y)$ and $R = [a, b] \times [c, d]$. Then Fubini's Theorem gives

$$\iint_R f(x, y) \, dA = \int_c^d \int_a^b g(x)h(y) \, dx \, dy = \int_c^d \left[\int_a^b g(x)h(y) \, dx \right] dy$$

In the inner integral, y is a constant, so $h(y)$ is a constant and we can write

$$\int_c^d \left[\int_a^b g(x)h(y) \, dx \right] dy = \int_c^d \left[h(y) \left(\int_a^b g(x) \, dx \right) \right] dy = \int_a^b g(x) \, dx \int_c^d h(y) \, dy$$

since $\int_a^b g(x) \, dx$ is a constant. Therefore, in this case, the double integral of f can be written as the product of two single integrals:

5 $$\iint_R g(x)h(y) \, dA = \int_a^b g(x) \, dx \int_c^d h(y) \, dy \qquad \text{where } R = [a, b] \times [c, d]$$

EXAMPLE 5 If $R = [0, \pi/2] \times [0, \pi/2]$, then, by Equation 5,

$$\iint_R \sin x \cos y \, dA = \int_0^{\pi/2} \sin x \, dx \int_0^{\pi/2} \cos y \, dy$$

$$= \Big[-\cos x\Big]_0^{\pi/2} \Big[\sin y\Big]_0^{\pi/2} = 1 \cdot 1 = 1$$

The function $f(x, y) = \sin x \cos y$ in Example 5 is positive on R, so the integral represents the volume of the solid that lies above R and below the graph of f shown in Figure 6.

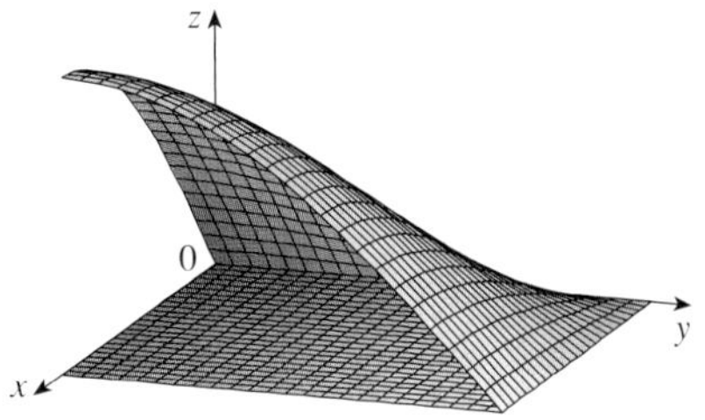

FIGURE 6

15.2 Exercises

1–2 Find $\int_0^5 f(x, y)\, dx$ and $\int_0^1 f(x, y)\, dy$.

1. $f(x, y) = 12x^2y^3$

2. $f(x, y) = y + xe^y$

3–14 Calculate the iterated integral.

3. $\int_1^4 \int_0^2 (6x^2y - 2x)\, dy\, dx$

4. $\int_0^1 \int_1^2 (4x^3 - 9x^2y^2)\, dy\, dx$

5. $\int_0^2 \int_0^4 y^3 e^{2x}\, dy\, dx$

6. $\int_{\pi/6}^{\pi/2} \int_{-1}^5 \cos y\, dx\, dy$

7. $\int_{-3}^3 \int_0^{\pi/2} (y + y^2 \cos x)\, dx\, dy$

8. $\int_1^3 \int_1^5 \frac{\ln y}{xy}\, dy\, dx$

9. $\int_1^4 \int_1^2 \left(\frac{x}{y} + \frac{y}{x}\right) dy\, dx$

10. $\int_0^1 \int_0^3 e^{x+3y}\, dx\, dy$

11. $\int_0^1 \int_0^1 v(u + v^2)^4\, du\, dv$

12. $\int_0^1 \int_0^1 xy\sqrt{x^2 + y^2}\, dy\, dx$

13. $\int_0^2 \int_0^\pi r \sin^2\theta\, d\theta\, dr$

14. $\int_0^1 \int_0^1 \sqrt{s + t}\, ds\, dt$

15–22 Calculate the double integral.

15. $\iint_R \sin(x - y)\, dA, \quad R = \{(x, y) \mid 0 \leqslant x \leqslant \pi/2, 0 \leqslant y \leqslant \pi/2\}$

16. $\iint_R (y + xy^{-2})\, dA, \quad R = \{(x, y) \mid 0 \leqslant x \leqslant 2, 1 \leqslant y \leqslant 2\}$

17. $\iint_R \frac{xy^2}{x^2 + 1}\, dA, \quad R = \{(x, y) \mid 0 \leqslant x \leqslant 1, -3 \leqslant y \leqslant 3\}$

18. $\iint_R \frac{1 + x^2}{1 + y^2}\, dA, \quad R = \{(x, y) \mid 0 \leqslant x \leqslant 1, 0 \leqslant y \leqslant 1\}$

19. $\iint_R x \sin(x + y)\, dA, \quad R = [0, \pi/6] \times [0, \pi/3]$

20. $\iint_R \frac{x}{1 + xy}\, dA, \quad R = [0, 1] \times [0, 1]$

21. $\iint_R ye^{-xy}\, dA, \quad R = [0, 2] \times [0, 3]$

22. $\iint_R \frac{1}{1 + x + y}\, dA, \quad R = [1, 3] \times [1, 2]$

23–24 Sketch the solid whose volume is given by the iterated integral.

23. $\int_0^1 \int_0^1 (4 - x - 2y)\, dx\, dy$

24. $\int_0^1 \int_0^1 (2 - x^2 - y^2)\, dy\, dx$

25. Find the volume of the solid that lies under the plane $4x + 6y - 2z + 15 = 0$ and above the rectangle $R = \{(x, y) \mid -1 \leqslant x \leqslant 2, -1 \leqslant y \leqslant 1\}$.

26. Find the volume of the solid that lies under the hyperbolic paraboloid $z = 3y^2 - x^2 + 2$ and above the rectangle $R = [-1, 1] \times [1, 2]$.

27. Find the volume of the solid lying under the elliptic paraboloid $x^2/4 + y^2/9 + z = 1$ and above the rectangle $R = [-1, 1] \times [-2, 2]$.

28. Find the volume of the solid enclosed by the surface $z = 1 + e^x \sin y$ and the planes $x = \pm 1$, $y = 0$, $y = \pi$, and $z = 0$.

29. Find the volume of the solid enclosed by the surface $z = x \sec^2 y$ and the planes $z = 0$, $x = 0$, $x = 2$, $y = 0$, and $y = \pi/4$.

30. Find the volume of the solid in the first octant bounded by the cylinder $z = 16 - x^2$ and the plane $y = 5$.

31. Find the volume of the solid enclosed by the paraboloid $z = 2 + x^2 + (y - 2)^2$ and the planes $z = 1$, $x = 1$, $x = -1$, $y = 0$, and $y = 4$.

32. Graph the solid that lies between the surface $z = 2xy/(x^2 + 1)$ and the plane $z = x + 2y$ and is bounded by the planes $x = 0$, $x = 2$, $y = 0$, and $y = 4$. Then find its volume.

CAS **33.** Use a computer algebra system to find the exact value of the integral $\iint_R x^5 y^3 e^{xy}\, dA$, where $R = [0, 1] \times [0, 1]$. Then use the CAS to draw the solid whose volume is given by the integral.

CAS **34.** Graph the solid that lies between the surfaces $z = e^{-x^2} \cos(x^2 + y^2)$ and $z = 2 - x^2 - y^2$ for $|x| \leqslant 1$, $|y| \leqslant 1$. Use a computer algebra system to approximate the volume of this solid correct to four decimal places.

35–36 Find the average value of f over the given rectangle.

35. $f(x, y) = x^2 y$, R has vertices $(-1, 0)$, $(-1, 5)$, $(1, 5)$, $(1, 0)$

36. $f(x, y) = e^y \sqrt{x + e^y}$, $R = [0, 4] \times [0, 1]$

37–38 Use symmetry to evaluate the double integral.

37. $\displaystyle\iint_R \frac{xy}{1 + x^4}\, dA$, $R = \{(x, y) \mid -1 \leqslant x \leqslant 1, 0 \leqslant y \leqslant 1\}$

38. $\displaystyle\iint_R (1 + x^2 \sin y + y^2 \sin x)\, dA$, $R = [-\pi, \pi] \times [-\pi, \pi]$

CAS **39.** Use your CAS to compute the iterated integrals

$$\int_0^1 \int_0^1 \frac{x - y}{(x + y)^3}\, dy\, dx \quad \text{and} \quad \int_0^1 \int_0^1 \frac{x - y}{(x + y)^3}\, dx\, dy$$

Do the answers contradict Fubini's Theorem? Explain what is happening.

40. (a) In what way are the theorems of Fubini and Clairaut similar?

(b) If $f(x, y)$ is continuous on $[a, b] \times [c, d]$ and

$$g(x, y) = \int_a^x \int_c^y f(s, t)\, dt\, ds$$

for $a < x < b$, $c < y < d$, show that $g_{xy} = g_{yx} = f(x, y)$.

15.3 Double Integrals over General Regions

For single integrals, the region over which we integrate is always an interval. But for double integrals, we want to be able to integrate a function f not just over rectangles but also over regions D of more general shape, such as the one illustrated in Figure 1. We suppose that D is a bounded region, which means that D can be enclosed in a rectangular region R as in Figure 2. Then we define a new function F with domain R by

$$\boxed{1} \qquad F(x, y) = \begin{cases} f(x, y) & \text{if } (x, y) \text{ is in } D \\ 0 & \text{if } (x, y) \text{ is in } R \text{ but not in } D \end{cases}$$

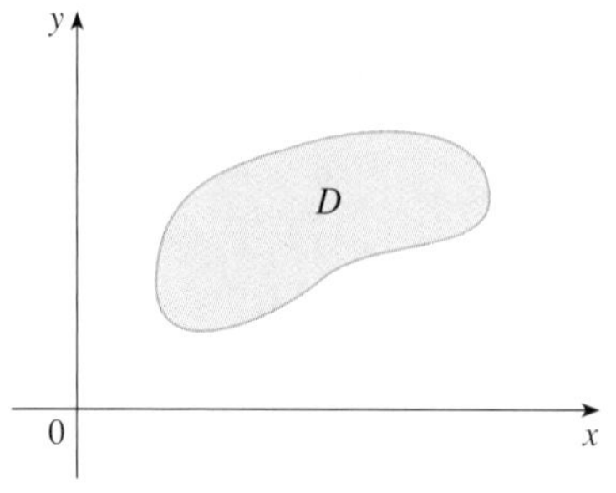

FIGURE 1

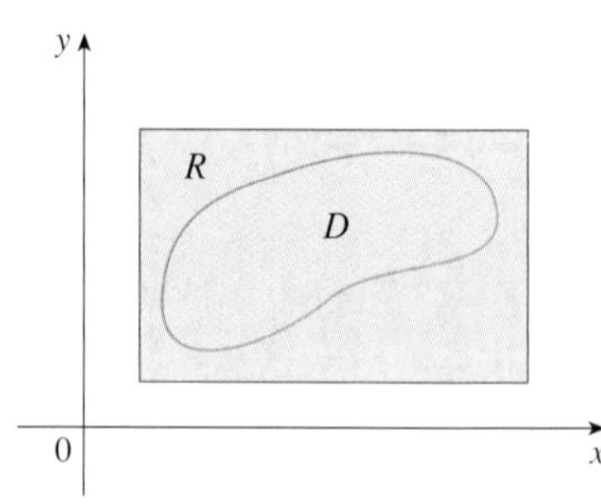

FIGURE 2

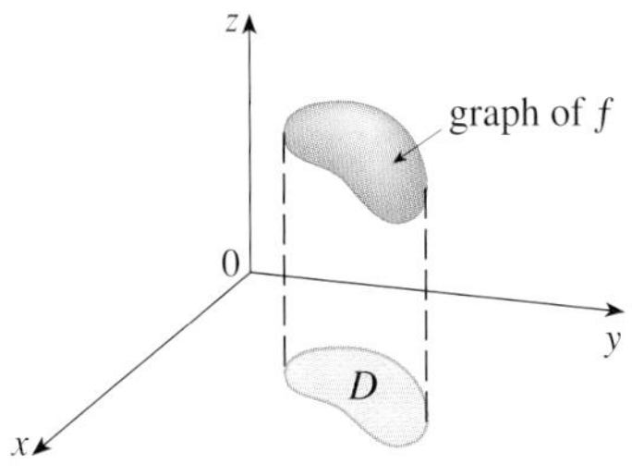

FIGURE 3

If F is integrable over R, then we define the **double integral of f over D** by

$$\boxed{2} \qquad \iint_D f(x, y)\, dA = \iint_R F(x, y)\, dA \qquad \text{where } F \text{ is given by Equation 1}$$

Definition 2 makes sense because R is a rectangle and so $\iint_R F(x, y)\, dA$ has been previously defined in Section 15.1. The procedure that we have used is reasonable because the values of $F(x, y)$ are 0 when (x, y) lies outside D and so they contribute nothing to the integral. This means that it doesn't matter what rectangle R we use as long as it contains D.

In the case where $f(x, y) \geqslant 0$, we can still interpret $\iint_D f(x, y)\, dA$ as the volume of the solid that lies above D and under the surface $z = f(x, y)$ (the graph of f). You can see that this is reasonable by comparing the graphs of f and F in Figures 3 and 4 and remembering that $\iint_R F(x, y)\, dA$ is the volume under the graph of F.

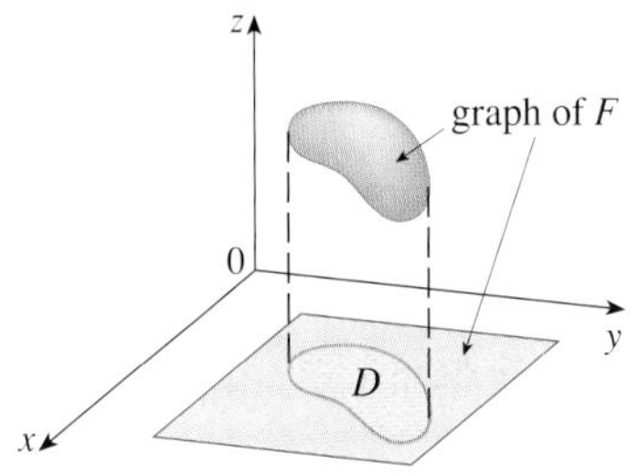

FIGURE 4

Figure 4 also shows that F is likely to have discontinuities at the boundary points of D. Nonetheless, if f is continuous on D and the boundary curve of D is "well behaved" (in a sense outside the scope of this book), then it can be shown that $\iint_R F(x, y)\, dA$ exists and therefore $\iint_D f(x, y)\, dA$ exists. In particular, this is the case for the following two types of regions.

A plane region D is said to be of **type I** if it lies between the graphs of two continuous functions of x, that is,

$$D = \{(x, y) \mid a \leqslant x \leqslant b,\ g_1(x) \leqslant y \leqslant g_2(x)\}$$

where g_1 and g_2 are continuous on $[a, b]$. Some examples of type I regions are shown in Figure 5.

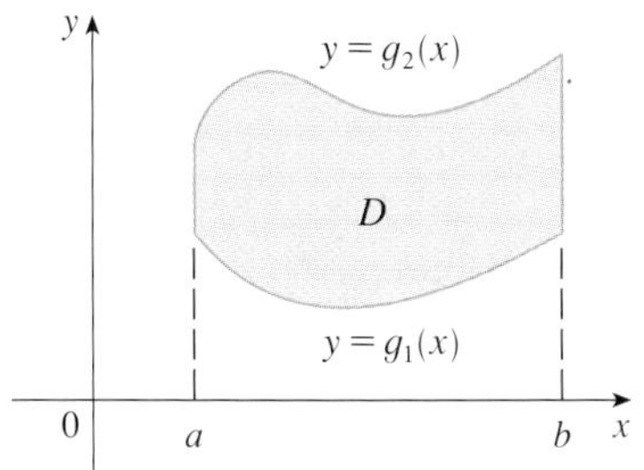

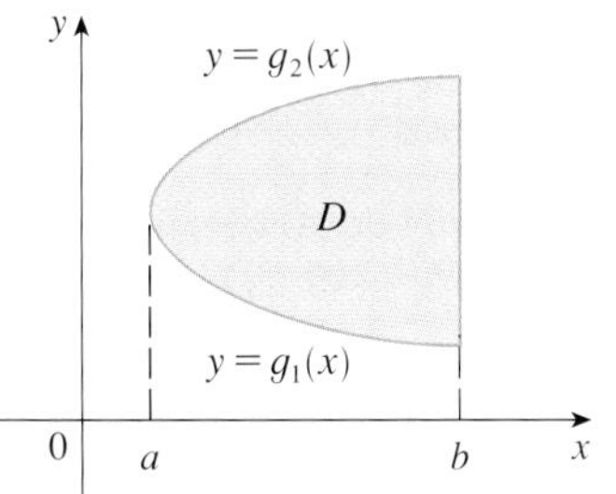

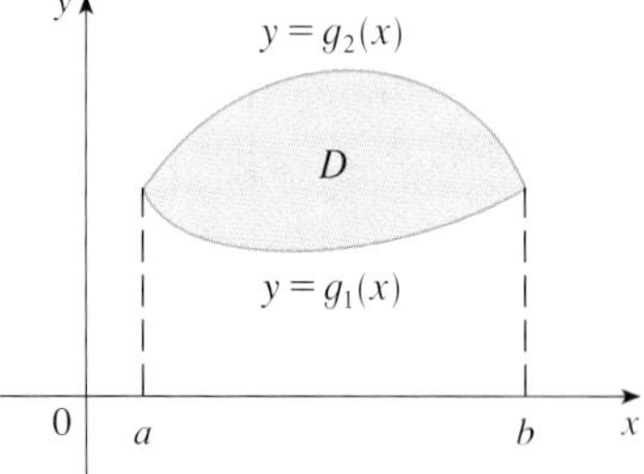

FIGURE 5 Some type I regions

In order to evaluate $\iint_D f(x, y)\, dA$ when D is a region of type I, we choose a rectangle $R = [a, b] \times [c, d]$ that contains D, as in Figure 6, and we let F be the function given by Equation 1; that is, F agrees with f on D and F is 0 outside D. Then, by Fubini's Theorem,

$$\iint_D f(x, y)\, dA = \iint_R F(x, y)\, dA = \int_a^b \int_c^d F(x, y)\, dy\, dx$$

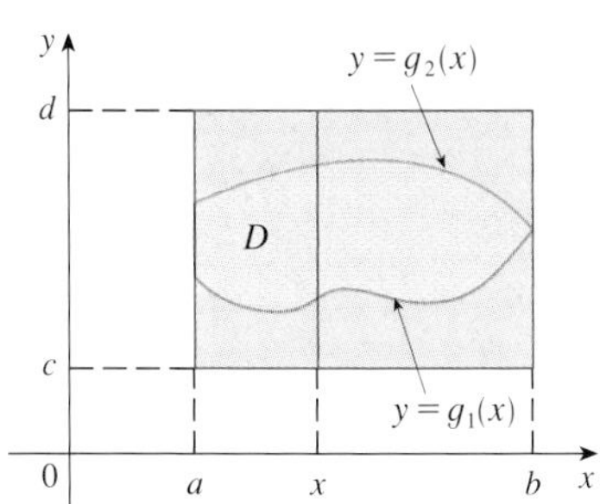

FIGURE 6

Observe that $F(x, y) = 0$ if $y < g_1(x)$ or $y > g_2(x)$ because (x, y) then lies outside D. Therefore

$$\int_c^d F(x, y)\, dy = \int_{g_1(x)}^{g_2(x)} F(x, y)\, dy = \int_{g_1(x)}^{g_2(x)} f(x, y)\, dy$$

because $F(x, y) = f(x, y)$ when $g_1(x) \leqslant y \leqslant g_2(x)$. Thus we have the following formula that enables us to evaluate the double integral as an iterated integral.

> **3** If f is continuous on a type I region D such that
>
> $$D = \{(x, y) \mid a \leqslant x \leqslant b,\ g_1(x) \leqslant y \leqslant g_2(x)\}$$
>
> then $$\iint_D f(x, y)\, dA = \int_a^b \int_{g_1(x)}^{g_2(x)} f(x, y)\, dy\, dx$$

The integral on the right side of [3] is an iterated integral that is similar to the ones we considered in the preceding section, except that in the inner integral we regard x as being constant not only in $f(x, y)$ but also in the limits of integration, $g_1(x)$ and $g_2(x)$.

We also consider plane regions of **type II**, which can be expressed as

$$\boxed{4} \qquad D = \{(x, y) \mid c \leqslant y \leqslant d,\ h_1(y) \leqslant x \leqslant h_2(y)\}$$

where h_1 and h_2 are continuous. Two such regions are illustrated in Figure 7.

Using the same methods that were used in establishing [3], we can show that

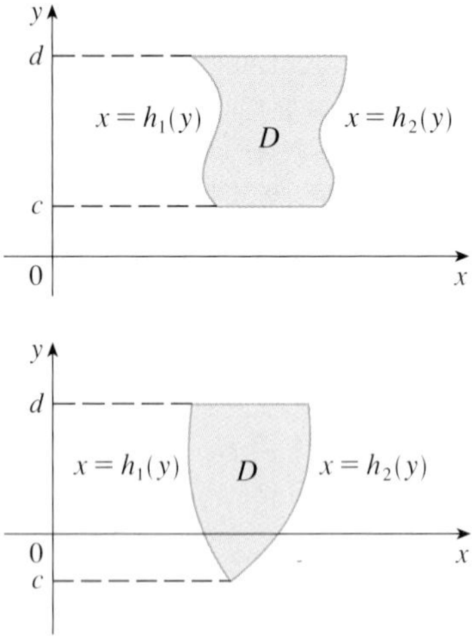

FIGURE 7
Some type II regions

> **5** $$\iint_D f(x, y)\, dA = \int_c^d \int_{h_1(y)}^{h_2(y)} f(x, y)\, dx\, dy$$
>
> where D is a type II region given by Equation 4.

V EXAMPLE 1 Evaluate $\iint_D (x + 2y)\, dA$, where D is the region bounded by the parabolas $y = 2x^2$ and $y = 1 + x^2$.

SOLUTION The parabolas intersect when $2x^2 = 1 + x^2$, that is, $x^2 = 1$, so $x = \pm 1$. We note that the region D, sketched in Figure 8, is a type I region but not a type II region and we can write

$$D = \{(x, y) \mid -1 \leqslant x \leqslant 1,\ 2x^2 \leqslant y \leqslant 1 + x^2\}$$

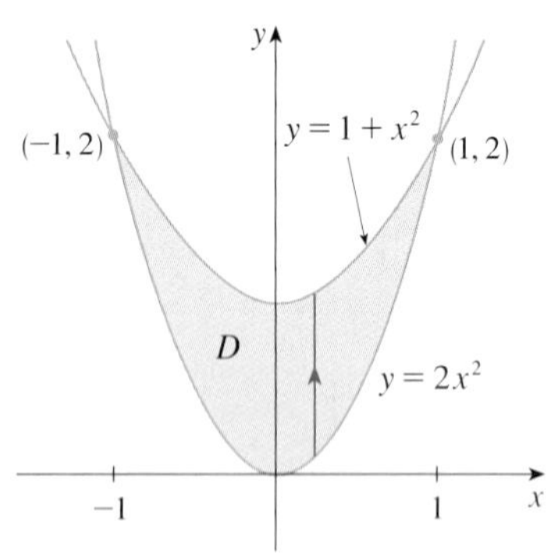

FIGURE 8

Since the lower boundary is $y = 2x^2$ and the upper boundary is $y = 1 + x^2$, Equation 3 gives

$$\begin{aligned}
\iint_D (x + 2y)\, dA &= \int_{-1}^{1} \int_{2x^2}^{1+x^2} (x + 2y)\, dy\, dx \\
&= \int_{-1}^{1} \Big[xy + y^2\Big]_{y=2x^2}^{y=1+x^2} dx \\
&= \int_{-1}^{1} [x(1 + x^2) + (1 + x^2)^2 - x(2x^2) - (2x^2)^2]\, dx \\
&= \int_{-1}^{1} (-3x^4 - x^3 + 2x^2 + x + 1)\, dx \\
&= -3\frac{x^5}{5} - \frac{x^4}{4} + 2\frac{x^3}{3} + \frac{x^2}{2} + x \Big]_{-1}^{1} = \frac{32}{15}
\end{aligned}$$

NOTE When we set up a double integral as in Example 1, it is essential to draw a diagram. Often it is helpful to draw a vertical arrow as in Figure 8. Then the limits of integration for the *inner* integral can be read from the diagram as follows: The arrow starts at the lower boundary $y = g_1(x)$, which gives the lower limit in the integral, and the arrow ends at the upper boundary $y = g_2(x)$, which gives the upper limit of integration. For a type II region the arrow is drawn horizontally from the left boundary to the right boundary.

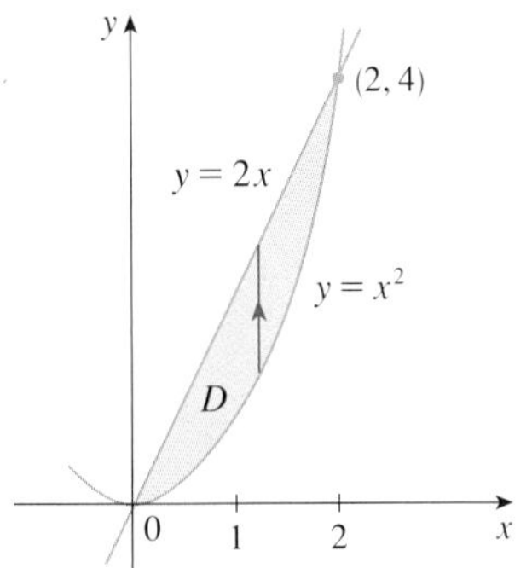

FIGURE 9
D as a type I region

EXAMPLE 2 Find the volume of the solid that lies under the paraboloid $z = x^2 + y^2$ and above the region D in the xy-plane bounded by the line $y = 2x$ and the parabola $y = x^2$.

SOLUTION 1 From Figure 9 we see that D is a type I region and

$$D = \{(x, y) \mid 0 \leqslant x \leqslant 2,\ x^2 \leqslant y \leqslant 2x\}$$

Therefore the volume under $z = x^2 + y^2$ and above D is

$$V = \iint_D (x^2 + y^2)\, dA = \int_0^2 \int_{x^2}^{2x} (x^2 + y^2)\, dy\, dx$$

$$= \int_0^2 \left[x^2 y + \frac{y^3}{3} \right]_{y=x^2}^{y=2x} dx$$

$$= \int_0^2 \left[x^2(2x) + \frac{(2x)^3}{3} - x^2x^2 - \frac{(x^2)^3}{3} \right] dx$$

$$= \int_0^2 \left(-\frac{x^6}{3} - x^4 + \frac{14x^3}{3} \right) dx$$

$$= -\frac{x^7}{21} - \frac{x^5}{5} + \frac{7x^4}{6} \Bigg]_0^2 = \frac{216}{35}$$

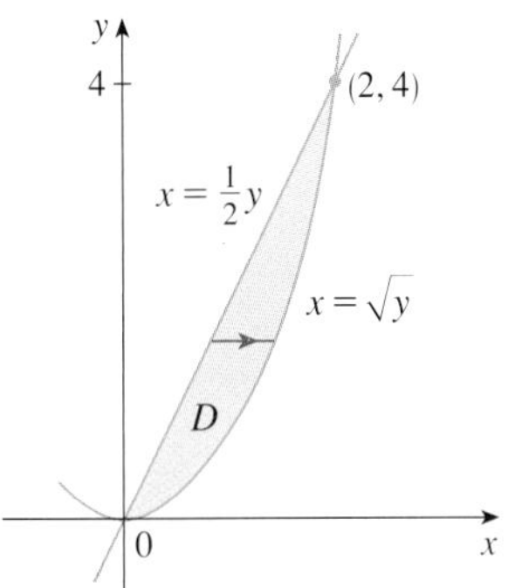

FIGURE 10
D as a type II region

SOLUTION 2 From Figure 10 we see that D can also be written as a type II region:

$$D = \{(x, y) \mid 0 \leqslant y \leqslant 4,\ \tfrac{1}{2}y \leqslant x \leqslant \sqrt{y}\}$$

Figure 11 shows the solid whose volume is calculated in Example 2. It lies above the xy-plane, below the paraboloid $z = x^2 + y^2$, and between the plane $y = 2x$ and the parabolic cylinder $y = x^2$.

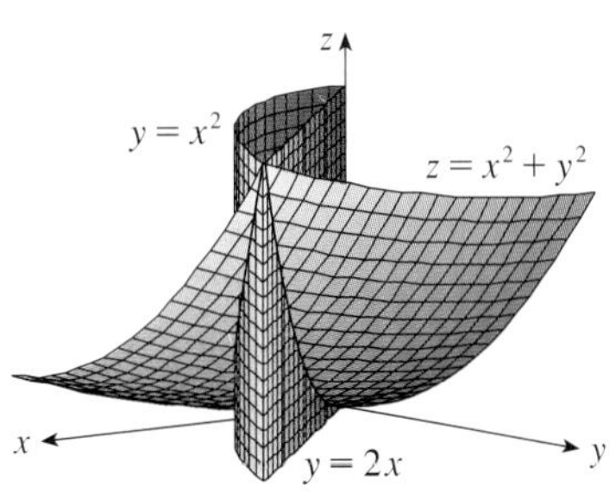

FIGURE 11

Therefore another expression for V is

$$V = \iint_D (x^2 + y^2)\, dA = \int_0^4 \int_{\frac{1}{2}y}^{\sqrt{y}} (x^2 + y^2)\, dx\, dy$$

$$= \int_0^4 \left[\frac{x^3}{3} + y^2 x \right]_{x=\frac{1}{2}y}^{x=\sqrt{y}} dy = \int_0^4 \left(\frac{y^{3/2}}{3} + y^{5/2} - \frac{y^3}{24} - \frac{y^3}{2} \right) dy$$

$$= \tfrac{2}{15}y^{5/2} + \tfrac{2}{7}y^{7/2} - \tfrac{13}{96}y^4 \Big]_0^4 = \tfrac{216}{35}$$

V EXAMPLE 3 Evaluate $\iint_D xy\,dA$, where D is the region bounded by the line $y = x - 1$ and the parabola $y^2 = 2x + 6$.

SOLUTION The region D is shown in Figure 12. Again D is both type I and type II, but the description of D as a type I region is more complicated because the lower boundary consists of two parts. Therefore we prefer to express D as a type II region:

$$D = \left\{(x, y) \mid -2 \leq y \leq 4,\ \tfrac{1}{2}y^2 - 3 \leq x \leq y + 1\right\}$$

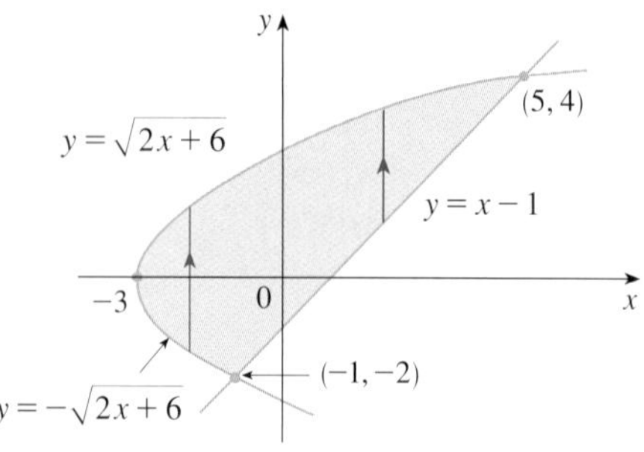

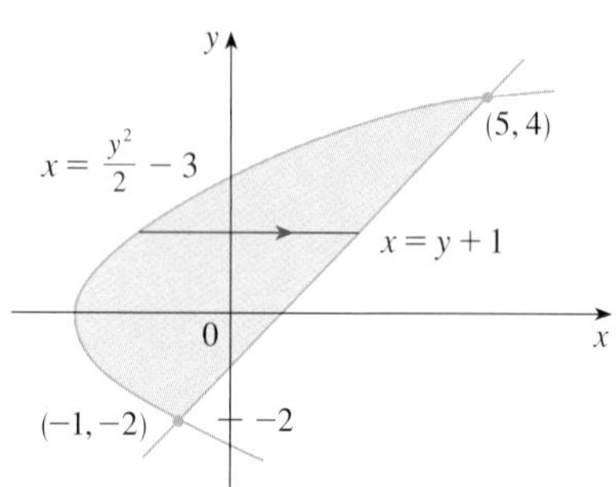

FIGURE 12 (a) D as a type I region (b) D as a type II region

Then [5] gives

$$\begin{aligned}
\iint_D xy\,dA &= \int_{-2}^{4}\int_{\frac{1}{2}y^2-3}^{y+1} xy\,dx\,dy = \int_{-2}^{4}\left[\frac{x^2}{2}y\right]_{x=\frac{1}{2}y^2-3}^{x=y+1} dy \\
&= \tfrac{1}{2}\int_{-2}^{4} y\left[(y+1)^2 - \left(\tfrac{1}{2}y^2 - 3\right)^2\right] dy \\
&= \tfrac{1}{2}\int_{-2}^{4}\left(-\frac{y^5}{4} + 4y^3 + 2y^2 - 8y\right) dy \\
&= \frac{1}{2}\left[-\frac{y^6}{24} + y^4 + 2\frac{y^3}{3} - 4y^2\right]_{-2}^{4} = 36
\end{aligned}$$

If we had expressed D as a type I region using Figure 12(a), then we would have obtained

$$\iint_D xy\,dA = \int_{-3}^{-1}\int_{-\sqrt{2x+6}}^{\sqrt{2x+6}} xy\,dy\,dx + \int_{-1}^{5}\int_{x-1}^{\sqrt{2x+6}} xy\,dy\,dx$$

but this would have involved more work than the other method.

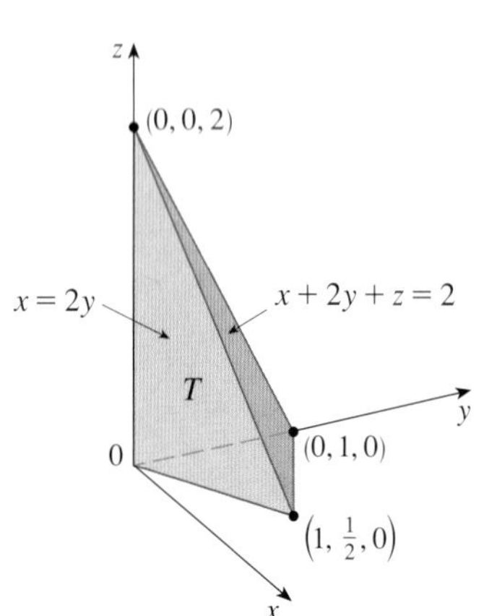

FIGURE 13

EXAMPLE 4 Find the volume of the tetrahedron bounded by the planes $x + 2y + z = 2$, $x = 2y$, $x = 0$, and $z = 0$.

SOLUTION In a question such as this, it's wise to draw two diagrams: one of the three-dimensional solid and another of the plane region D over which it lies. Figure 13 shows the tetrahedron T bounded by the coordinate planes $x = 0$, $z = 0$, the vertical plane $x = 2y$, and the plane $x + 2y + z = 2$. Since the plane $x + 2y + z = 2$ intersects the xy-plane (whose equation is $z = 0$) in the line $x + 2y = 2$, we see that T lies above the triangular region D in the xy-plane bounded by the lines $x = 2y$, $x + 2y = 2$, and $x = 0$. (See Figure 14.)

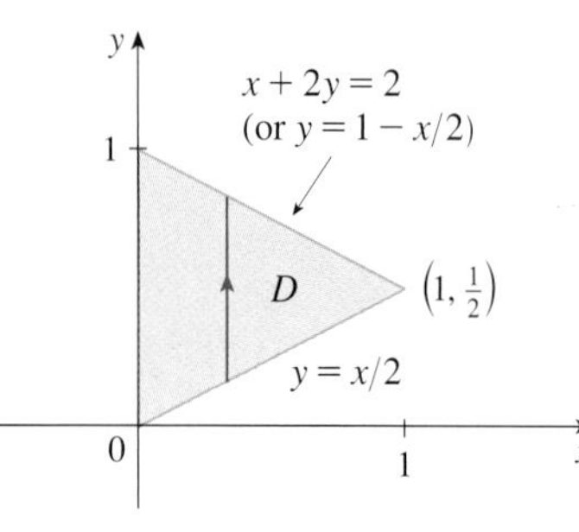

FIGURE 14

The plane $x + 2y + z = 2$ can be written as $z = 2 - x - 2y$, so the required volume lies under the graph of the function $z = 2 - x - 2y$ and above

$$D = \left\{(x, y) \mid 0 \leq x \leq 1,\ x/2 \leq y \leq 1 - x/2\right\}$$

Therefore

$$V = \iint_D (2 - x - 2y)\, dA$$

$$= \int_0^1 \int_{x/2}^{1-x/2} (2 - x - 2y)\, dy\, dx$$

$$= \int_0^1 \Big[2y - xy - y^2\Big]_{y=x/2}^{y=1-x/2} dx$$

$$= \int_0^1 \left[2 - x - x\left(1 - \frac{x}{2}\right) - \left(1 - \frac{x}{2}\right)^2 - x + \frac{x^2}{2} + \frac{x^2}{4}\right] dx$$

$$= \int_0^1 (x^2 - 2x + 1)\, dx = \frac{x^3}{3} - x^2 + x\Big]_0^1 = \frac{1}{3}$$

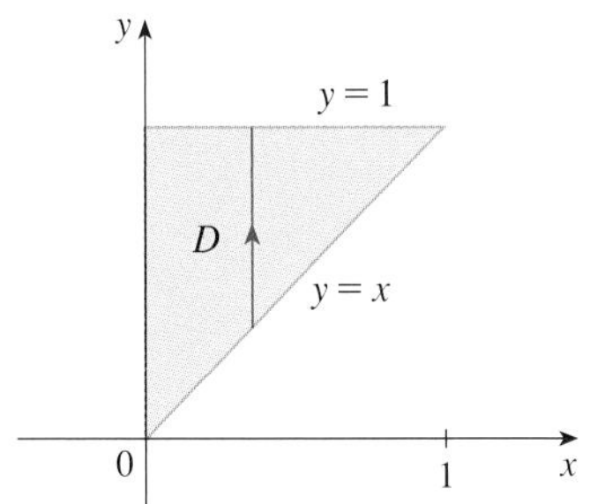

FIGURE 15
D as a type I region

V EXAMPLE 5 Evaluate the iterated integral $\int_0^1 \int_x^1 \sin(y^2)\, dy\, dx$.

SOLUTION If we try to evaluate the integral as it stands, we are faced with the task of first evaluating $\int \sin(y^2)\, dy$. But it's impossible to do so in finite terms since $\int \sin(y^2)\, dy$ is not an elementary function. (See the end of Section 7.5.) So we must change the order of integration. This is accomplished by first expressing the given iterated integral as a double integral. Using [3] backward, we have

$$\int_0^1 \int_x^1 \sin(y^2)\, dy\, dx = \iint_D \sin(y^2)\, dA$$

where $$D = \{(x, y) \mid 0 \leqslant x \leqslant 1,\ x \leqslant y \leqslant 1\}$$

We sketch this region D in Figure 15. Then from Figure 16 we see that an alternative description of D is

$$D = \{(x, y) \mid 0 \leqslant y \leqslant 1,\ 0 \leqslant x \leqslant y\}$$

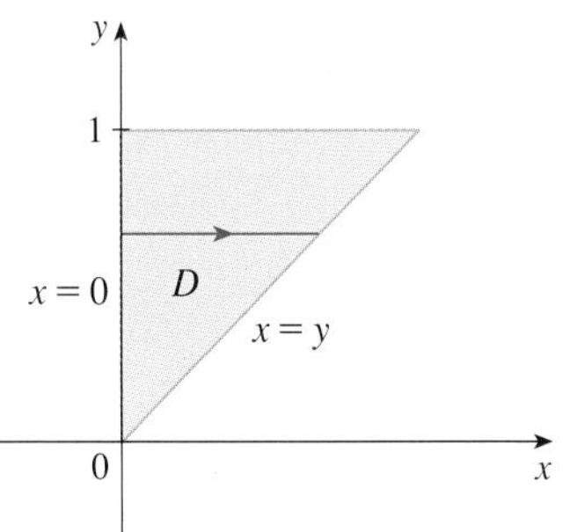

FIGURE 16
D as a type II region

This enables us to use [5] to express the double integral as an iterated integral in the reverse order:

$$\int_0^1 \int_x^1 \sin(y^2)\, dy\, dx = \iint_D \sin(y^2)\, dA$$

$$= \int_0^1 \int_0^y \sin(y^2)\, dx\, dy = \int_0^1 \Big[x \sin(y^2)\Big]_{x=0}^{x=y} dy$$

$$= \int_0^1 y \sin(y^2)\, dy = -\tfrac{1}{2} \cos(y^2)\Big]_0^1 = \tfrac{1}{2}(1 - \cos 1)$$

Properties of Double Integrals

We assume that all of the following integrals exist. The first three properties of double integrals over a region D follow immediately from Definition 2 in this section and Properties 7, 8, and 9 in Section 15.1.

6 $$\iint_D [f(x, y) + g(x, y)]\, dA = \iint_D f(x, y)\, dA + \iint_D g(x, y)\, dA$$

7 $$\iint_D c f(x, y)\, dA = c \iint_D f(x, y)\, dA$$

If $f(x, y) \geqslant g(x, y)$ for all (x, y) in D, then

8 $$\iint_D f(x, y)\, dA \geqslant \iint_D g(x, y)\, dA$$

The next property of double integrals is similar to the property of single integrals given by the equation $\int_a^b f(x)\, dx = \int_a^c f(x)\, dx + \int_c^b f(x)\, dx$.

If $D = D_1 \cup D_2$, where D_1 and D_2 don't overlap except perhaps on their boundaries (see Figure 17), then

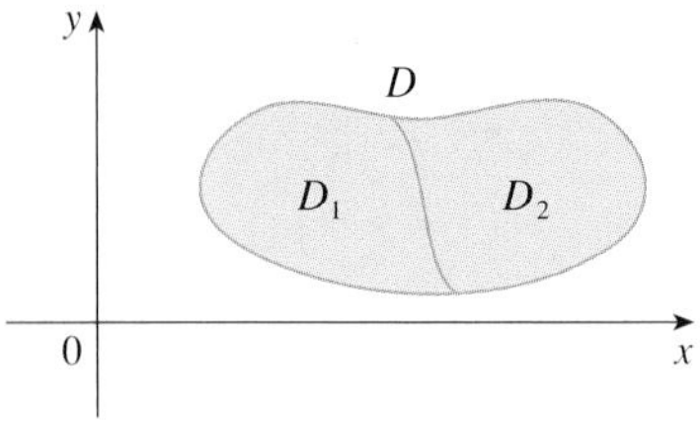

FIGURE 17

9 $$\iint_D f(x, y)\, dA = \iint_{D_1} f(x, y)\, dA + \iint_{D_2} f(x, y)\, dA$$

Property 9 can be used to evaluate double integrals over regions D that are neither type I nor type II but can be expressed as a union of regions of type I or type II. Figure 18 illustrates this procedure. (See Exercises 55 and 56.)

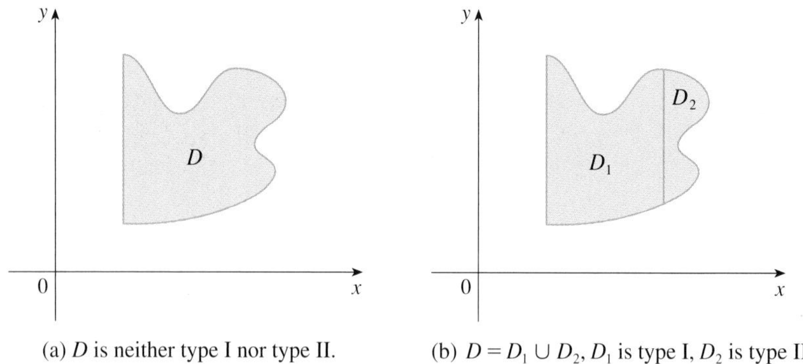

FIGURE 18

(a) D is neither type I nor type II.

(b) $D = D_1 \cup D_2$, D_1 is type I, D_2 is type II.

The next property of integrals says that if we integrate the constant function $f(x, y) = 1$ over a region D, we get the area of D:

10 $$\iint_D 1\, dA = A(D)$$

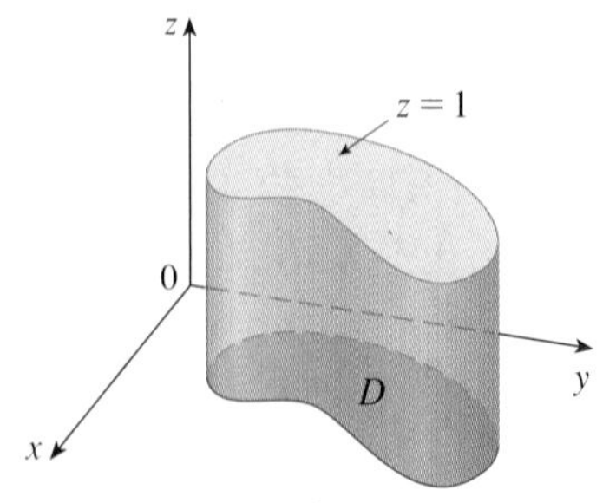

FIGURE 19
Cylinder with base D and height 1

Figure 19 illustrates why Equation 10 is true: A solid cylinder whose base is D and whose height is 1 has volume $A(D) \cdot 1 = A(D)$, but we know that we can also write its volume as $\iint_D 1\, dA$.

Finally, we can combine Properties 7, 8, and 10 to prove the following property. (See Exercise 61.)

11 If $m \leqslant f(x, y) \leqslant M$ for all (x, y) in D, then

$$mA(D) \leqslant \iint_D f(x, y)\, dA \leqslant MA(D)$$

EXAMPLE 6 Use Property 11 to estimate the integral $\iint_D e^{\sin x \cos y}\, dA$, where D is the disk with center the origin and radius 2.

SOLUTION Since $-1 \leqslant \sin x \leqslant 1$ and $-1 \leqslant \cos y \leqslant 1$, we have $-1 \leqslant \sin x \cos y \leqslant 1$ and therefore

$$e^{-1} \leqslant e^{\sin x \cos y} \leqslant e^1 = e$$

Thus, using $m = e^{-1} = 1/e$, $M = e$, and $A(D) = \pi(2)^2$ in Property 11, we obtain

$$\frac{4\pi}{e} \leqslant \iint_D e^{\sin x \cos y}\, dA \leqslant 4\pi e$$

15.3 Exercises

1–6 Evaluate the iterated integral.

1. $\int_0^4 \int_0^{\sqrt{y}} xy^2\, dx\, dy$
2. $\int_0^1 \int_{2x}^2 (x - y)\, dy\, dx$
3. $\int_0^1 \int_{x^2}^x (1 + 2y)\, dy\, dx$
4. $\int_0^2 \int_y^{2y} xy\, dx\, dy$
5. $\int_0^1 \int_0^{s^2} \cos(s^3)\, dt\, ds$
6. $\int_0^1 \int_0^{e^v} \sqrt{1 + e^v}\, dw\, dv$

7–10 Evaluate the double integral.

7. $\iint_D y^2\, dA, \quad D = \{(x, y) \mid -1 \leqslant y \leqslant 1,\ -y - 2 \leqslant x \leqslant y\}$
8. $\iint_D \frac{y}{x^5 + 1}\, dA, \quad D = \{(x, y) \mid 0 \leqslant x \leqslant 1,\ 0 \leqslant y \leqslant x^2\}$
9. $\iint_D x\, dA, \quad D = \{(x, y) \mid 0 \leqslant x \leqslant \pi, 0 \leqslant y \leqslant \sin x\}$
10. $\iint_D x^3\, dA, \quad D = \{(x, y) \mid 1 \leqslant x \leqslant e, 0 \leqslant y \leqslant \ln x\}$

11. Draw an example of a region that is
(a) type I but not type II
(b) type II but not type I

12. Draw an example of a region that is
(a) both type I and type II
(b) neither type I nor type II

13–14 Express D as a region of type I and also as a region of type II. Then evaluate the double integral in two ways.

13. $\iint_D x\, dA$, D is enclosed by the lines $y = x$, $y = 0$, $x = 1$
14. $\iint_D xy\, dA$, D is enclosed by the curves $y = x^2$, $y = 3x$

15–16 Set up iterated integrals for both orders of integration. Then evaluate the double integral using the easier order and explain why it's easier.

15. $\iint_D y\, dA$, D is bounded by $y = x - 2$, $x = y^2$
16. $\iint_D y^2 e^{xy}\, dA$, D is bounded by $y = x$, $y = 4$, $x = 0$

17–22 Evaluate the double integral.

17. $\iint_D x \cos y\, dA$, D is bounded by $y = 0$, $y = x^2$, $x = 1$
18. $\iint_D (x^2 + 2y)\, dA$, D is bounded by $y = x$, $y = x^3$, $x \geqslant 0$
19. $\iint_D y^2\, dA$,
D is the triangular region with vertices (0, 1), (1, 2), (4, 1)
20. $\iint_D xy^2\, dA$, D is enclosed by $x = 0$ and $x = \sqrt{1 - y^2}$
21. $\iint_D (2x - y)\, dA$,
D is bounded by the circle with center the origin and radius 2
22. $\iint_D 2xy\, dA$, D is the triangular region with vertices (0, 0), (1, 2), and (0, 3)

23–32 Find the volume of the given solid.

23. Under the plane $x - 2y + z = 1$ and above the region bounded by $x + y = 1$ and $x^2 + y = 1$

24. Under the surface $z = 1 + x^2y^2$ and above the region enclosed by $x = y^2$ and $x = 4$

25. Under the surface $z = xy$ and above the triangle with vertices $(1, 1)$, $(4, 1)$, and $(1, 2)$

26. Enclosed by the paraboloid $z = x^2 + 3y^2$ and the planes $x = 0$, $y = 1$, $y = x$, $z = 0$

27. Bounded by the coordinate planes and the plane $3x + 2y + z = 6$

28. Bounded by the planes $z = x$, $y = x$, $x + y = 2$, and $z = 0$

29. Enclosed by the cylinders $z = x^2$, $y = x^2$ and the planes $z = 0$, $y = 4$

30. Bounded by the cylinder $y^2 + z^2 = 4$ and the planes $x = 2y$, $x = 0$, $z = 0$ in the first octant

31. Bounded by the cylinder $x^2 + y^2 = 1$ and the planes $y = z$, $x = 0$, $z = 0$ in the first octant

32. Bounded by the cylinders $x^2 + y^2 = r^2$ and $y^2 + z^2 = r^2$

33. Use a graphing calculator or computer to estimate the x-coordinates of the points of intersection of the curves $y = x^4$ and $y = 3x - x^2$. If D is the region bounded by these curves, estimate $\iint_D x\, dA$.

34. Find the approximate volume of the solid in the first octant that is bounded by the planes $y = x$, $z = 0$, and $z = x$ and the cylinder $y = \cos x$. (Use a graphing device to estimate the points of intersection.)

35–36 Find the volume of the solid by subtracting two volumes.

35. The solid enclosed by the parabolic cylinders $y = 1 - x^2$, $y = x^2 - 1$ and the planes $x + y + z = 2$, $2x + 2y - z + 10 = 0$

36. The solid enclosed by the parabolic cylinder $y = x^2$ and the planes $z = 3y$, $z = 2 + y$

37–38 Sketch the solid whose volume is given by the iterated integral.

37. $\int_0^1 \int_0^{1-x} (1 - x - y)\, dy\, dx$

38. $\int_0^1 \int_0^{1-x^2} (1 - x)\, dy\, dx$

CAS **39–42** Use a computer algebra system to find the exact volume of the solid.

39. Under the surface $z = x^3y^4 + xy^2$ and above the region bounded by the curves $y = x^3 - x$ and $y = x^2 + x$ for $x \geqslant 0$

40. Between the paraboloids $z = 2x^2 + y^2$ and $z = 8 - x^2 - 2y^2$ and inside the cylinder $x^2 + y^2 = 1$

41. Enclosed by $z = 1 - x^2 - y^2$ and $z = 0$

42. Enclosed by $z = x^2 + y^2$ and $z = 2y$

43–48 Sketch the region of integration and change the order of integration.

43. $\int_0^1 \int_0^y f(x, y)\, dx\, dy$

44. $\int_0^2 \int_{x^2}^4 f(x, y)\, dy\, dx$

45. $\int_0^{\pi/2} \int_0^{\cos x} f(x, y)\, dy\, dx$

46. $\int_{-2}^2 \int_0^{\sqrt{4-y^2}} f(x, y)\, dx\, dy$

47. $\int_1^2 \int_0^{\ln x} f(x, y)\, dy\, dx$

48. $\int_0^1 \int_{\arctan x}^{\pi/4} f(x, y)\, dy\, dx$

49–54 Evaluate the integral by reversing the order of integration.

49. $\int_0^1 \int_{3y}^3 e^{x^2}\, dx\, dy$

50. $\int_0^{\sqrt{\pi}} \int_y^{\sqrt{\pi}} \cos(x^2)\, dx\, dy$

51. $\int_0^4 \int_{\sqrt{x}}^2 \frac{1}{y^3 + 1}\, dy\, dx$

52. $\int_0^1 \int_x^1 e^{x/y}\, dy\, dx$

53. $\int_0^1 \int_{\arcsin y}^{\pi/2} \cos x \sqrt{1 + \cos^2 x}\, dx\, dy$

54. $\int_0^8 \int_{\sqrt[3]{y}}^2 e^{x^4}\, dx\, dy$

55–56 Express D as a union of regions of type I or type II and evaluate the integral.

55. $\iint_D x^2\, dA$

56. $\iint_D y\, dA$

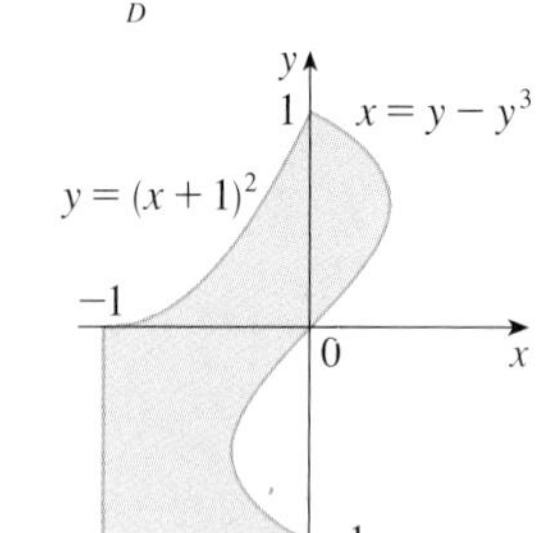

57–58 Use Property 11 to estimate the value of the integral.

57. $\iint_Q e^{-(x^2+y^2)^2}\, dA$, Q is the quarter-circle with center the origin and radius $\frac{1}{2}$ in the first quadrant

58. $\iint_T \sin^4(x + y)\, dA$, T is the triangle enclosed by the lines $y = 0$, $y = 2x$, and $x = 1$

59–60 Find the average value of f over the region D.

59. $f(x, y) = xy$, D is the triangle with vertices $(0, 0)$, $(1, 0)$, and $(1, 3)$

60. $f(x, y) = x \sin y$, D is enclosed by the curves $y = 0$, $y = x^2$, and $x = 1$

---

61. Prove Property 11.

62. In evaluating a double integral over a region D, a sum of iterated integrals was obtained as follows:

$$\iint_D f(x, y)\, dA = \int_0^1 \int_0^{2y} f(x, y)\, dx\, dy + \int_1^3 \int_0^{3-y} f(x, y)\, dx\, dy$$

Sketch the region D and express the double integral as an iterated integral with reversed order of integration.

63–67 Use geometry or symmetry, or both, to evaluate the double integral.

63. $\displaystyle\iint_D (x + 2)\, dA$, $D = \{(x, y) \mid 0 \leqslant y \leqslant \sqrt{9 - x^2}\}$

64. $\displaystyle\iint_D \sqrt{R^2 - x^2 - y^2}\, dA$,
D is the disk with center the origin and radius R

65. $\displaystyle\iint_D (2x + 3y)\, dA$,
D is the rectangle $0 \leqslant x \leqslant a$, $0 \leqslant y \leqslant b$

66. $\displaystyle\iint_D (2 + x^2y^3 - y^2 \sin x)\, dA$,
$D = \{(x, y) \mid |x| + |y| \leqslant 1\}$

67. $\displaystyle\iint_D \left(ax^3 + by^3 + \sqrt{a^2 - x^2}\right) dA$,
$D = [-a, a] \times [-b, b]$

CAS **68.** Graph the solid bounded by the plane $x + y + z = 1$ and the paraboloid $z = 4 - x^2 - y^2$ and find its exact volume. (Use your CAS to do the graphing, to find the equations of the boundary curves of the region of integration, and to evaluate the double integral.)

15.4 Double Integrals in Polar Coordinates

Suppose that we want to evaluate a double integral $\iint_R f(x, y)\, dA$, where R is one of the regions shown in Figure 1. In either case the description of R in terms of rectangular coordinates is rather complicated, but R is easily described using polar coordinates.

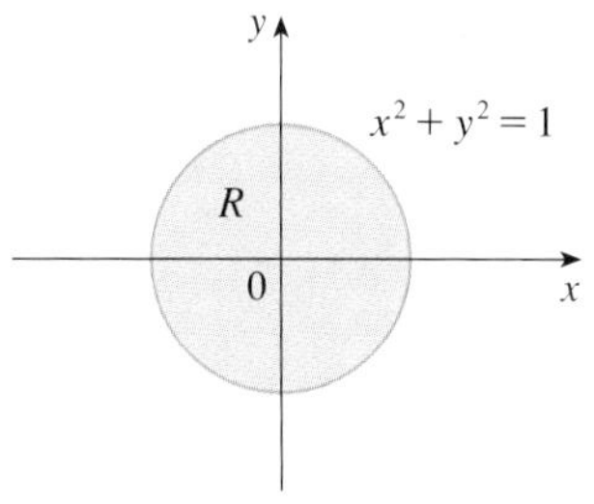

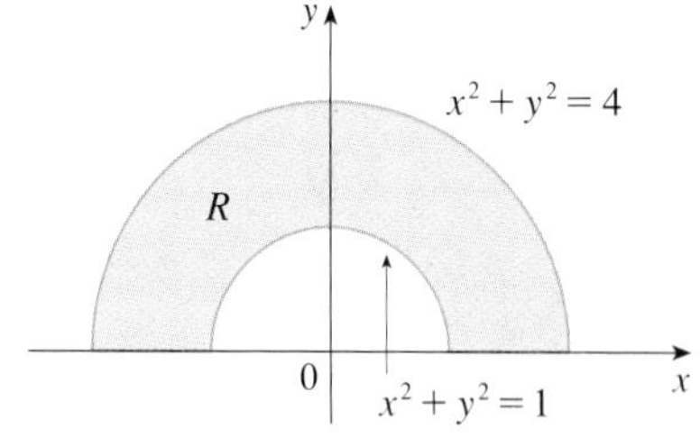

FIGURE 1 (a) $R = \{(r, \theta) \mid 0 \leqslant r \leqslant 1, 0 \leqslant \theta \leqslant 2\pi\}$ (b) $R = \{(r, \theta) \mid 1 \leqslant r \leqslant 2, 0 \leqslant \theta \leqslant \pi\}$

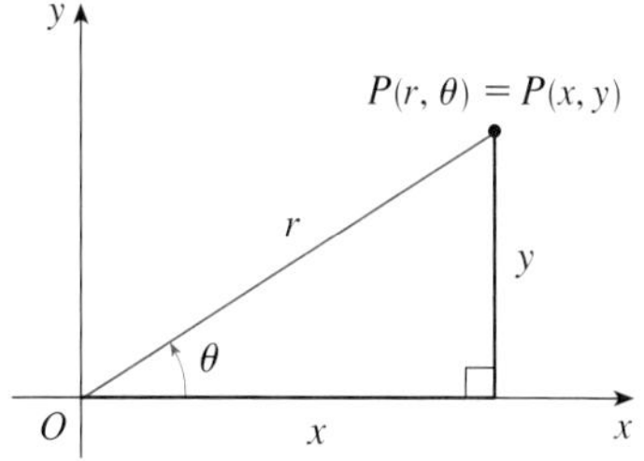

FIGURE 2

Recall from Figure 2 that the polar coordinates (r, θ) of a point are related to the rectangular coordinates (x, y) by the equations

$$r^2 = x^2 + y^2 \qquad x = r \cos \theta \qquad y = r \sin \theta$$

(See Section 10.3.)

The regions in Figure 1 are special cases of a **polar rectangle**

$$R = \{(r, \theta) \mid a \leqslant r \leqslant b,\ \alpha \leqslant \theta \leqslant \beta\}$$

which is shown in Figure 3. In order to compute the double integral $\iint_R f(x, y)\,dA$, where R is a polar rectangle, we divide the interval $[a, b]$ into m subintervals $[r_{i-1}, r_i]$ of equal width $\Delta r = (b - a)/m$ and we divide the interval $[\alpha, \beta]$ into n subintervals $[\theta_{j-1}, \theta_j]$ of equal width $\Delta\theta = (\beta - \alpha)/n$. Then the circles $r = r_i$ and the rays $\theta = \theta_j$ divide the polar rectangle R into the small polar rectangles R_{ij} shown in Figure 4.

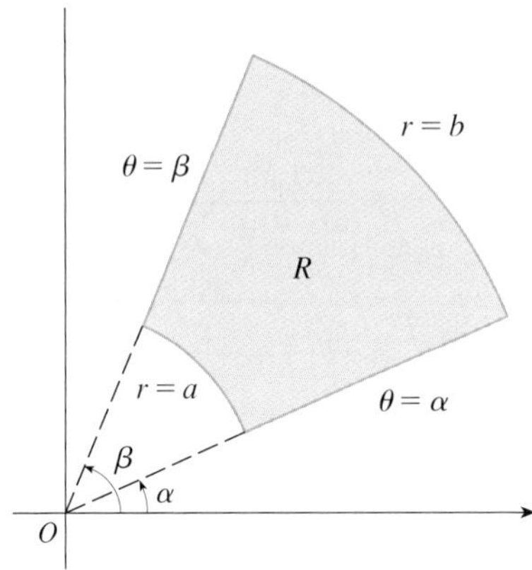

FIGURE 3 Polar rectangle

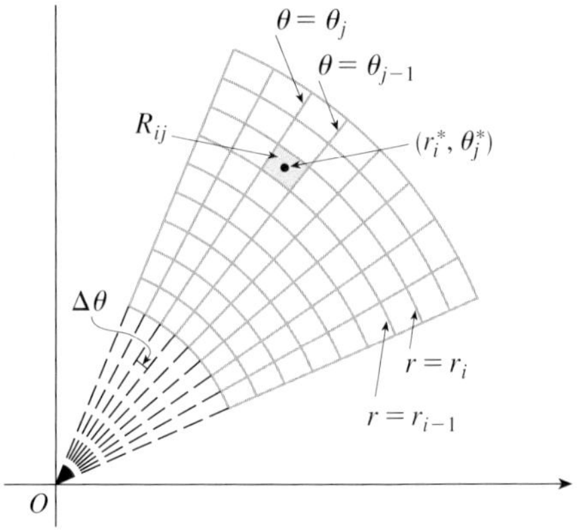

FIGURE 4 Dividing R into polar subrectangles

The "center" of the polar subrectangle

$$R_{ij} = \{(r, \theta) \mid r_{i-1} \leqslant r \leqslant r_i,\ \theta_{j-1} \leqslant \theta \leqslant \theta_j\}$$

has polar coordinates

$$r_i^* = \tfrac{1}{2}(r_{i-1} + r_i) \qquad \theta_j^* = \tfrac{1}{2}(\theta_{j-1} + \theta_j)$$

We compute the area of R_{ij} using the fact that the area of a sector of a circle with radius r and central angle θ is $\frac{1}{2}r^2\theta$. Subtracting the areas of two such sectors, each of which has central angle $\Delta\theta = \theta_j - \theta_{j-1}$, we find that the area of R_{ij} is

$$\begin{aligned}\Delta A_i &= \tfrac{1}{2}r_i^2\,\Delta\theta - \tfrac{1}{2}r_{i-1}^2\,\Delta\theta = \tfrac{1}{2}(r_i^2 - r_{i-1}^2)\,\Delta\theta \\ &= \tfrac{1}{2}(r_i + r_{i-1})(r_i - r_{i-1})\,\Delta\theta = r_i^*\,\Delta r\,\Delta\theta\end{aligned}$$

Although we have defined the double integral $\iint_R f(x, y)\,dA$ in terms of ordinary rectangles, it can be shown that, for continuous functions f, we always obtain the same answer using polar rectangles. The rectangular coordinates of the center of R_{ij} are $(r_i^* \cos\theta_j^*, r_i^* \sin\theta_j^*)$, so a typical Riemann sum is

$$\boxed{1} \qquad \sum_{i=1}^{m}\sum_{j=1}^{n} f(r_i^* \cos\theta_j^*, r_i^* \sin\theta_j^*)\,\Delta A_i = \sum_{i=1}^{m}\sum_{j=1}^{n} f(r_i^* \cos\theta_j^*, r_i^* \sin\theta_j^*)\,r_i^*\,\Delta r\,\Delta\theta$$

If we write $g(r, \theta) = rf(r\cos\theta, r\sin\theta)$, then the Riemann sum in Equation 1 can be written as

$$\sum_{i=1}^{m}\sum_{j=1}^{n} g(r_i^*, \theta_j^*)\,\Delta r\,\Delta\theta$$

which is a Riemann sum for the double integral

$$\int_{\alpha}^{\beta} \int_{a}^{b} g(r, \theta)\, dr\, d\theta$$

Therefore we have

$$\iint_R f(x, y)\, dA = \lim_{m,n \to \infty} \sum_{i=1}^{m} \sum_{j=1}^{n} f(r_i^* \cos \theta_j^*, r_i^* \sin \theta_j^*)\, \Delta A_i$$

$$= \lim_{m,n \to \infty} \sum_{i=1}^{m} \sum_{j=1}^{n} g(r_i^*, \theta_j^*)\, \Delta r\, \Delta \theta = \int_{\alpha}^{\beta} \int_{a}^{b} g(r, \theta)\, dr\, d\theta$$

$$= \int_{\alpha}^{\beta} \int_{a}^{b} f(r \cos \theta, r \sin \theta)\, r\, dr\, d\theta$$

> **2** **Change to Polar Coordinates in a Double Integral** If f is continuous on a polar rectangle R given by $0 \leqslant a \leqslant r \leqslant b$, $\alpha \leqslant \theta \leqslant \beta$, where $0 \leqslant \beta - \alpha \leqslant 2\pi$, then
>
> $$\iint_R f(x, y)\, dA = \int_{\alpha}^{\beta} \int_{a}^{b} f(r \cos \theta, r \sin \theta)\, r\, dr\, d\theta$$

The formula in 2 says that we convert from rectangular to polar coordinates in a double integral by writing $x = r \cos \theta$ and $y = r \sin \theta$, using the appropriate limits of integration for r and θ, and replacing dA by $r\, dr\, d\theta$. Be careful not to forget the additional factor r on the right side of Formula 2. A classical method for remembering this is shown in Figure 5, where the "infinitesimal" polar rectangle can be thought of as an ordinary rectangle with dimensions $r\, d\theta$ and dr and therefore has "area" $dA = r\, dr\, d\theta$.

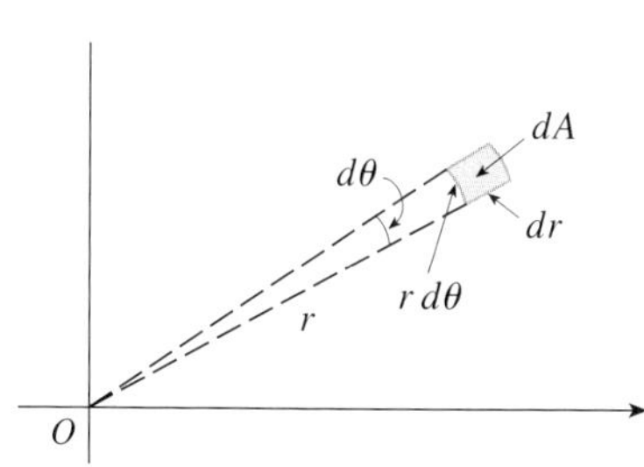

FIGURE 5

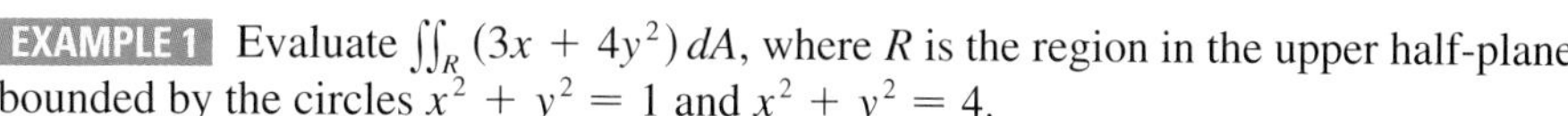

EXAMPLE 1 Evaluate $\iint_R (3x + 4y^2)\, dA$, where R is the region in the upper half-plane bounded by the circles $x^2 + y^2 = 1$ and $x^2 + y^2 = 4$.

SOLUTION The region R can be described as

$$R = \{(x, y) \mid y \geqslant 0,\ 1 \leqslant x^2 + y^2 \leqslant 4\}$$

It is the half-ring shown in Figure 1(b), and in polar coordinates it is given by $1 \leqslant r \leqslant 2$, $0 \leqslant \theta \leqslant \pi$. Therefore, by Formula 2,

$$\iint_R (3x + 4y^2)\, dA = \int_0^{\pi} \int_1^2 (3r \cos \theta + 4r^2 \sin^2\theta)\, r\, dr\, d\theta$$

$$= \int_0^{\pi} \int_1^2 (3r^2 \cos \theta + 4r^3 \sin^2\theta)\, dr\, d\theta$$

$$= \int_0^{\pi} \left[r^3 \cos \theta + r^4 \sin^2\theta\right]_{r=1}^{r=2} d\theta = \int_0^{\pi} (7 \cos \theta + 15 \sin^2\theta)\, d\theta$$

Here we use the trigonometric identity

$$\sin^2\theta = \tfrac{1}{2}(1 - \cos 2\theta)$$

See Section 7.2 for advice on integrating trigonometric functions.

$$= \int_0^{\pi} \left[7 \cos \theta + \tfrac{15}{2}(1 - \cos 2\theta)\right] d\theta$$

$$= 7 \sin \theta + \frac{15\theta}{2} - \frac{15}{4} \sin 2\theta \Big]_0^{\pi} = \frac{15\pi}{2}$$

V EXAMPLE 2 Find the volume of the solid bounded by the plane $z = 0$ and the paraboloid $z = 1 - x^2 - y^2$.

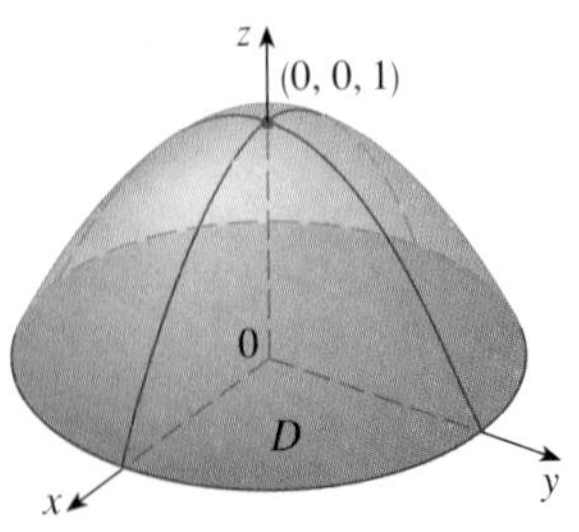

FIGURE 6

SOLUTION If we put $z = 0$ in the equation of the paraboloid, we get $x^2 + y^2 = 1$. This means that the plane intersects the paraboloid in the circle $x^2 + y^2 = 1$, so the solid lies under the paraboloid and above the circular disk D given by $x^2 + y^2 \leqslant 1$ [see Figures 6 and 1(a)]. In polar coordinates D is given by $0 \leqslant r \leqslant 1$, $0 \leqslant \theta \leqslant 2\pi$. Since $1 - x^2 - y^2 = 1 - r^2$, the volume is

$$V = \iint_D (1 - x^2 - y^2)\, dA = \int_0^{2\pi} \int_0^1 (1 - r^2)\, r\, dr\, d\theta$$

$$= \int_0^{2\pi} d\theta \int_0^1 (r - r^3)\, dr = 2\pi \left[\frac{r^2}{2} - \frac{r^4}{4} \right]_0^1 = \frac{\pi}{2}$$

If we had used rectangular coordinates instead of polar coordinates, then we would have obtained

$$V = \iint_D (1 - x^2 - y^2)\, dA = \int_{-1}^{1} \int_{-\sqrt{1-x^2}}^{\sqrt{1-x^2}} (1 - x^2 - y^2)\, dy\, dx$$

which is not easy to evaluate because it involves finding $\int (1 - x^2)^{3/2}\, dx$.

What we have done so far can be extended to the more complicated type of region shown in Figure 7. It's similar to the type II rectangular regions considered in Section 15.3. In fact, by combining Formula 2 in this section with Formula 15.3.5, we obtain the following formula.

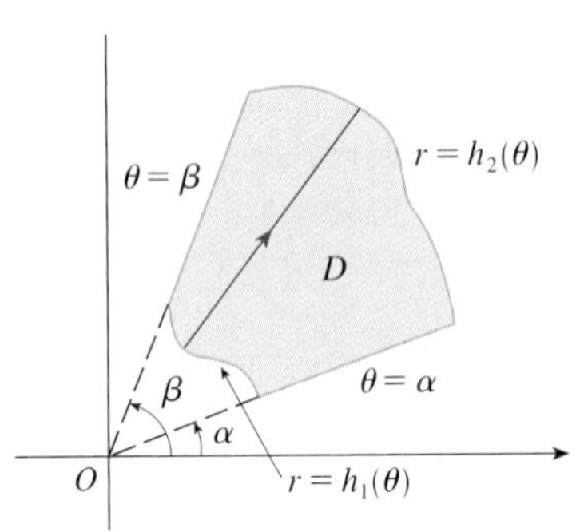

FIGURE 7
$D = \{(r, \theta) \mid \alpha \leqslant \theta \leqslant \beta, h_1(\theta) \leqslant r \leqslant h_2(\theta)\}$

3 If f is continuous on a polar region of the form

$$D = \{(r, \theta) \mid \alpha \leqslant \theta \leqslant \beta,\ h_1(\theta) \leqslant r \leqslant h_2(\theta)\}$$

then
$$\iint_D f(x, y)\, dA = \int_\alpha^\beta \int_{h_1(\theta)}^{h_2(\theta)} f(r\cos\theta, r\sin\theta)\, r\, dr\, d\theta$$

In particular, taking $f(x, y) = 1$, $h_1(\theta) = 0$, and $h_2(\theta) = h(\theta)$ in this formula, we see that the area of the region D bounded by $\theta = \alpha$, $\theta = \beta$, and $r = h(\theta)$ is

$$A(D) = \iint_D 1\, dA = \int_\alpha^\beta \int_0^{h(\theta)} r\, dr\, d\theta$$

$$= \int_\alpha^\beta \left[\frac{r^2}{2} \right]_0^{h(\theta)} d\theta = \int_\alpha^\beta \tfrac{1}{2}[h(\theta)]^2\, d\theta$$

and this agrees with Formula 10.4.3.

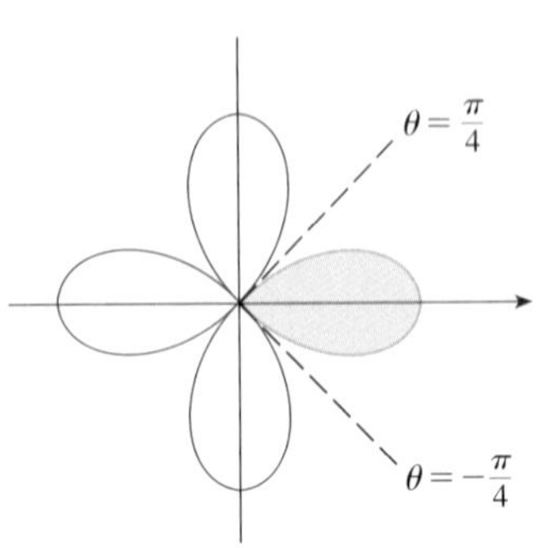

FIGURE 8

V EXAMPLE 3 Use a double integral to find the area enclosed by one loop of the four-leaved rose $r = \cos 2\theta$.

SOLUTION From the sketch of the curve in Figure 8, we see that a loop is given by the region

$$D = \{(r, \theta) \mid -\pi/4 \leqslant \theta \leqslant \pi/4,\ 0 \leqslant r \leqslant \cos 2\theta\}$$

So the area is

$$A(D) = \iint_D dA = \int_{-\pi/4}^{\pi/4} \int_0^{\cos 2\theta} r \, dr \, d\theta$$

$$= \int_{-\pi/4}^{\pi/4} \left[\tfrac{1}{2} r^2\right]_0^{\cos 2\theta} d\theta = \tfrac{1}{2} \int_{-\pi/4}^{\pi/4} \cos^2 2\theta \, d\theta$$

$$= \tfrac{1}{4} \int_{-\pi/4}^{\pi/4} (1 + \cos 4\theta) \, d\theta = \tfrac{1}{4}\left[\theta + \tfrac{1}{4} \sin 4\theta\right]_{-\pi/4}^{\pi/4} = \frac{\pi}{8}$$

V EXAMPLE 4 Find the volume of the solid that lies under the paraboloid $z = x^2 + y^2$, above the xy-plane, and inside the cylinder $x^2 + y^2 = 2x$.

SOLUTION The solid lies above the disk D whose boundary circle has equation $x^2 + y^2 = 2x$ or, after completing the square,

$$(x - 1)^2 + y^2 = 1$$

(See Figures 9 and 10.)

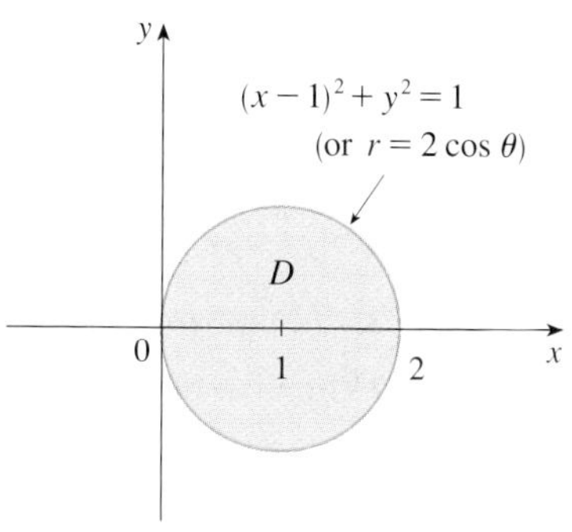

FIGURE 9

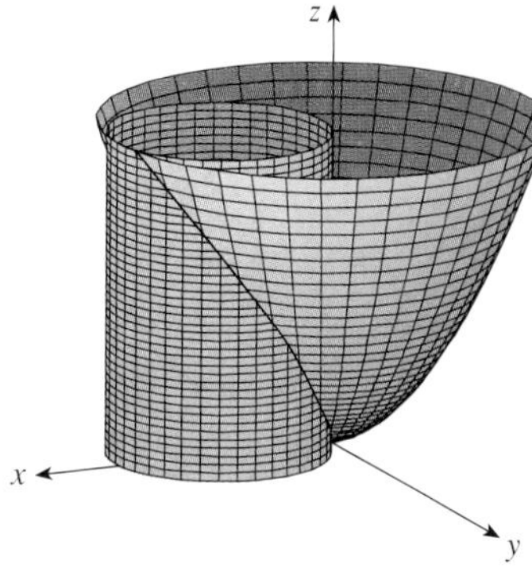

FIGURE 10

In polar coordinates we have $x^2 + y^2 = r^2$ and $x = r \cos \theta$, so the boundary circle becomes $r^2 = 2r \cos \theta$, or $r = 2 \cos \theta$. Thus the disk D is given by

$$D = \left\{(r, \theta) \mid -\pi/2 \leqslant \theta \leqslant \pi/2, \ 0 \leqslant r \leqslant 2 \cos \theta\right\}$$

and, by Formula 3, we have

$$V = \iint_D (x^2 + y^2) \, dA = \int_{-\pi/2}^{\pi/2} \int_0^{2\cos\theta} r^2 \, r \, dr \, d\theta = \int_{-\pi/2}^{\pi/2} \left[\frac{r^4}{4}\right]_0^{2\cos\theta} d\theta$$

$$= 4 \int_{-\pi/2}^{\pi/2} \cos^4\theta \, d\theta = 8 \int_0^{\pi/2} \cos^4\theta \, d\theta = 8 \int_0^{\pi/2} \left(\frac{1 + \cos 2\theta}{2}\right)^2 d\theta$$

$$= 2 \int_0^{\pi/2} \left[1 + 2\cos 2\theta + \tfrac{1}{2}(1 + \cos 4\theta)\right] d\theta$$

$$= 2\left[\tfrac{3}{2}\theta + \sin 2\theta + \tfrac{1}{8} \sin 4\theta\right]_0^{\pi/2} = 2\left(\frac{3}{2}\right)\left(\frac{\pi}{2}\right) = \frac{3\pi}{2}$$

15.4 Exercises

1–4 A region R is shown. Decide whether to use polar coordinates or rectangular coordinates and write $\iint_R f(x, y)\, dA$ as an iterated integral, where f is an arbitrary continuous function on R.

1.

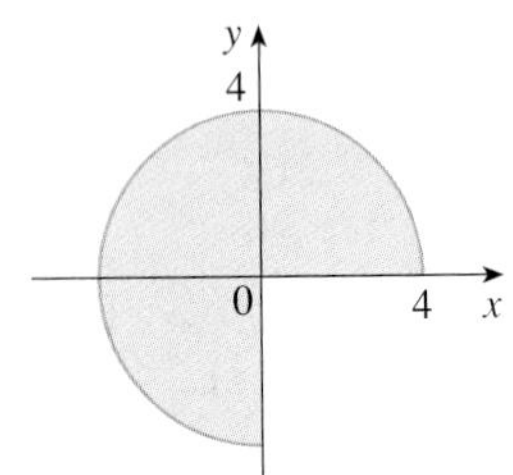

2.

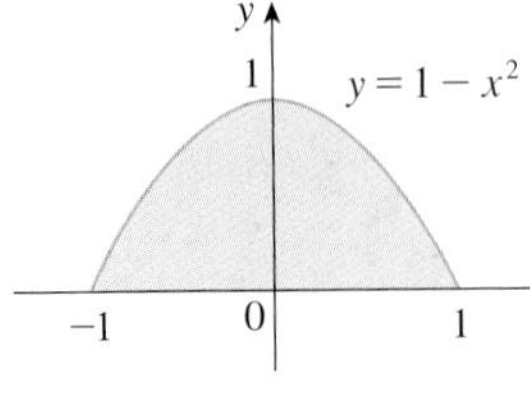

3.

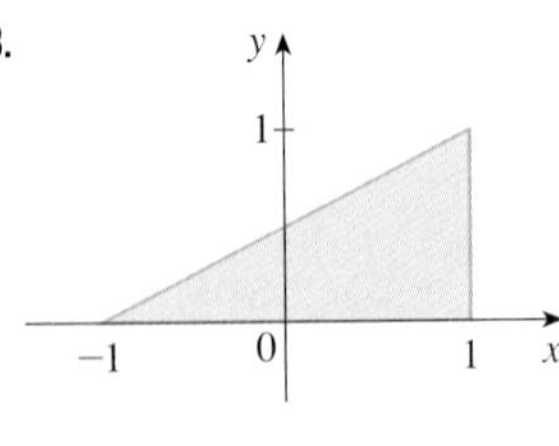

4.

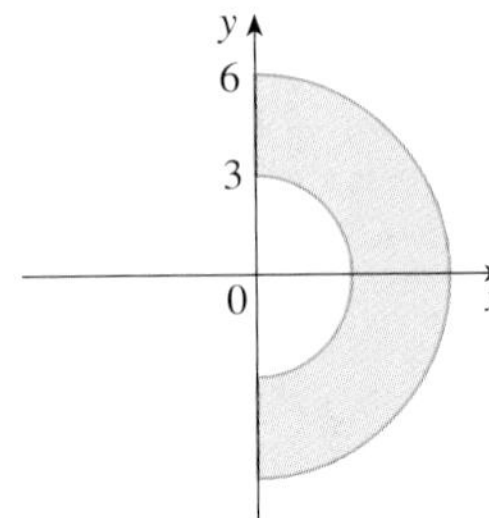

5–6 Sketch the region whose area is given by the integral and evaluate the integral.

5. $\int_{\pi/4}^{3\pi/4} \int_1^2 r\, dr\, d\theta$

6. $\int_{\pi/2}^{\pi} \int_0^{2\sin\theta} r\, dr\, d\theta$

7–14 Evaluate the given integral by changing to polar coordinates.

7. $\iint_D x^2 y\, dA$, where D is the top half of the disk with center the origin and radius 5

8. $\iint_R (2x - y)\, dA$, where R is the region in the first quadrant enclosed by the circle $x^2 + y^2 = 4$ and the lines $x = 0$ and $y = x$

9. $\iint_R \sin(x^2 + y^2)\, dA$, where R is the region in the first quadrant between the circles with center the origin and radii 1 and 3

10. $\iint_R \dfrac{y^2}{x^2 + y^2}\, dA$, where R is the region that lies between the circles $x^2 + y^2 = a^2$ and $x^2 + y^2 = b^2$ with $0 < a < b$

11. $\iint_D e^{-x^2-y^2}\, dA$, where D is the region bounded by the semicircle $x = \sqrt{4 - y^2}$ and the y-axis

12. $\iint_D \cos\sqrt{x^2 + y^2}\, dA$, where D is the disk with center the origin and radius 2

13. $\iint_R \arctan(y/x)\, dA$, where $R = \{(x, y) \mid 1 \leqslant x^2 + y^2 \leqslant 4,\ 0 \leqslant y \leqslant x\}$

14. $\iint_D x\, dA$, where D is the region in the first quadrant that lies between the circles $x^2 + y^2 = 4$ and $x^2 + y^2 = 2x$

15–18 Use a double integral to find the area of the region.

15. One loop of the rose $r = \cos 3\theta$

16. The region enclosed by both of the cardioids $r = 1 + \cos\theta$ and $r = 1 - \cos\theta$

17. The region inside the circle $(x - 1)^2 + y^2 = 1$ and outside the circle $x^2 + y^2 = 1$

18. The region inside the cardioid $r = 1 + \cos\theta$ and outside the circle $r = 3\cos\theta$

19–27 Use polar coordinates to find the volume of the given solid.

19. Under the cone $z = \sqrt{x^2 + y^2}$ and above the disk $x^2 + y^2 \leqslant 4$

20. Below the paraboloid $z = 18 - 2x^2 - 2y^2$ and above the xy-plane

21. Enclosed by the hyperboloid $-x^2 - y^2 + z^2 = 1$ and the plane $z = 2$

22. Inside the sphere $x^2 + y^2 + z^2 = 16$ and outside the cylinder $x^2 + y^2 = 4$

23. A sphere of radius a

24. Bounded by the paraboloid $z = 1 + 2x^2 + 2y^2$ and the plane $z = 7$ in the first octant

25. Above the cone $z = \sqrt{x^2 + y^2}$ and below the sphere $x^2 + y^2 + z^2 = 1$

26. Bounded by the paraboloids $z = 3x^2 + 3y^2$ and $z = 4 - x^2 - y^2$

27. Inside both the cylinder $x^2 + y^2 = 4$ and the ellipsoid $4x^2 + 4y^2 + z^2 = 64$

28. (a) A cylindrical drill with radius r_1 is used to bore a hole through the center of a sphere of radius r_2. Find the volume of the ring-shaped solid that remains.
(b) Express the volume in part (a) in terms of the height h of the ring. Notice that the volume depends only on h, not on r_1 or r_2.

29–32 Evaluate the iterated integral by converting to polar coordinates.

29. $\int_{-3}^{3} \int_0^{\sqrt{9-x^2}} \sin(x^2 + y^2)\, dy\, dx$

30. $\int_0^a \int_{-\sqrt{a^2-y^2}}^{0} x^2 y\, dx\, dy$

31. $\int_0^1 \int_y^{\sqrt{2-y^2}} (x + y)\, dx\, dy$

32. $\int_0^2 \int_0^{\sqrt{2x-x^2}} \sqrt{x^2 + y^2}\, dy\, dx$

1. Homework Hints available at stewartcalculus.com

33–34 Express the double integral in terms of a single integral with respect to r. Then use your calculator to evaluate the integral correct to four decimal places.

33. $\iint_D e^{(x^2+y^2)^2}\,dA$, where D is the disk with center the origin and radius 1

34. $\iint_D xy\sqrt{1+x^2+y^2}\,dA$, where D is the portion of the disk $x^2+y^2 \leqslant 1$ that lies in the first quadrant

---

35. A swimming pool is circular with a 40-ft diameter. The depth is constant along east-west lines and increases linearly from 2 ft at the south end to 7 ft at the north end. Find the volume of water in the pool.

36. An agricultural sprinkler distributes water in a circular pattern of radius 100 ft. It supplies water to a depth of e^{-r} feet per hour at a distance of r feet from the sprinkler.

(a) If $0 < R \leqslant 100$, what is the total amount of water supplied per hour to the region inside the circle of radius R centered at the sprinkler?

(b) Determine an expression for the average amount of water per hour per square foot supplied to the region inside the circle of radius R.

37. Find the average value of the function $f(x, y) = 1/\sqrt{x^2+y^2}$ on the annular region $a^2 \leqslant x^2+y^2 \leqslant b^2$, where $0 < a < b$.

38. Let D be the disk with center the origin and radius a. What is the average distance from points in D to the origin?

39. Use polar coordinates to combine the sum

$$\int_{1/\sqrt{2}}^{1}\int_{\sqrt{1-x^2}}^{x} xy\,dy\,dx + \int_{1}^{\sqrt{2}}\int_{0}^{x} xy\,dy\,dx + \int_{\sqrt{2}}^{2}\int_{0}^{\sqrt{4-x^2}} xy\,dy\,dx$$

into one double integral. Then evaluate the double integral.

40. (a) We define the improper integral (over the entire plane $\mathbb{R}^2$)

$$\begin{aligned} I &= \iint_{\mathbb{R}^2} e^{-(x^2+y^2)}\,dA = \int_{-\infty}^{\infty}\int_{-\infty}^{\infty} e^{-(x^2+y^2)}\,dy\,dx \\ &= \lim_{a\to\infty}\iint_{D_a} e^{-(x^2+y^2)}\,dA \end{aligned}$$

where D_a is the disk with radius a and center the origin. Show that

$$\int_{-\infty}^{\infty}\int_{-\infty}^{\infty} e^{-(x^2+y^2)}\,dA = \pi$$

(b) An equivalent definition of the improper integral in part (a) is

$$\iint_{\mathbb{R}^2} e^{-(x^2+y^2)}\,dA = \lim_{a\to\infty}\iint_{S_a} e^{-(x^2+y^2)}\,dA$$

where S_a is the square with vertices $(\pm a, \pm a)$. Use this to show that

$$\int_{-\infty}^{\infty} e^{-x^2}\,dx \int_{-\infty}^{\infty} e^{-y^2}\,dy = \pi$$

(c) Deduce that

$$\int_{-\infty}^{\infty} e^{-x^2}\,dx = \sqrt{\pi}$$

(d) By making the change of variable $t = \sqrt{2}\,x$, show that

$$\int_{-\infty}^{\infty} e^{-x^2/2}\,dx = \sqrt{2\pi}$$

(This is a fundamental result for probability and statistics.)

41. Use the result of Exercise 40 part (c) to evaluate the following integrals.

(a) $\int_0^{\infty} x^2 e^{-x^2}\,dx$ (b) $\int_0^{\infty} \sqrt{x}\,e^{-x}\,dx$

15.5 Applications of Double Integrals

We have already seen one application of double integrals: computing volumes. Another geometric application is finding areas of surfaces and this will be done in the next section. In this section we explore physical applications such as computing mass, electric charge, center of mass, and moment of inertia. We will see that these physical ideas are also important when applied to probability density functions of two random variables.

Density and Mass

In Section 8.3 we were able to use single integrals to compute moments and the center of mass of a thin plate or lamina with constant density. But now, equipped with the double integral, we can consider a lamina with variable density. Suppose the lamina occupies a region D of the xy-plane and its **density** (in units of mass per unit area) at a point (x, y) in D is given by $\rho(x, y)$, where ρ is a continuous function on D. This means that

$$\rho(x, y) = \lim \frac{\Delta m}{\Delta A}$$

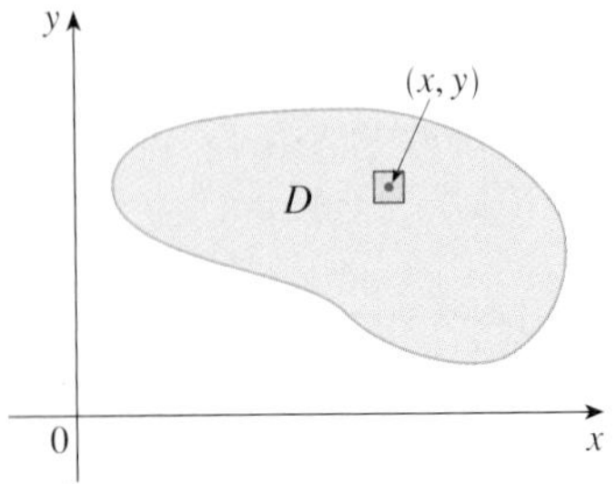

FIGURE 1

where Δm and ΔA are the mass and area of a small rectangle that contains (x, y) and the limit is taken as the dimensions of the rectangle approach 0. (See Figure 1.)

To find the total mass m of the lamina we divide a rectangle R containing D into subrectangles R_{ij} of the same size (as in Figure 2) and consider $\rho(x, y)$ to be 0 outside D. If we choose a point (x_{ij}^*, y_{ij}^*) in R_{ij}, then the mass of the part of the lamina that occupies R_{ij} is approximately $\rho(x_{ij}^*, y_{ij}^*)\,\Delta A$, where ΔA is the area of R_{ij}. If we add all such masses, we get an approximation to the total mass:

$$m \approx \sum_{i=1}^{k} \sum_{j=1}^{l} \rho(x_{ij}^*, y_{ij}^*)\,\Delta A$$

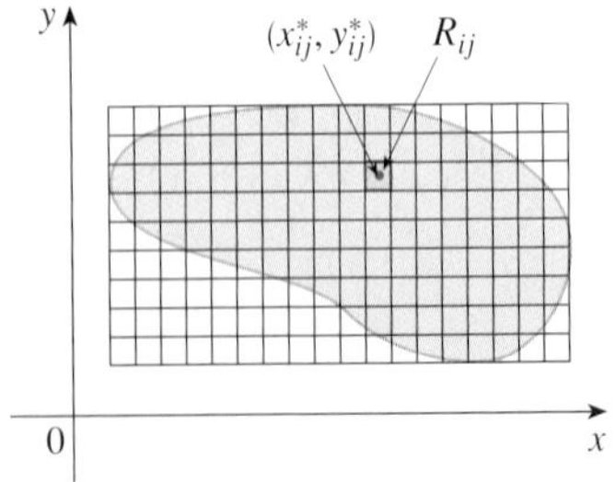

FIGURE 2

If we now increase the number of subrectangles, we obtain the total mass m of the lamina as the limiting value of the approximations:

1
$$m = \lim_{k,\,l \to \infty} \sum_{i=1}^{k} \sum_{j=1}^{l} \rho(x_{ij}^*, y_{ij}^*)\,\Delta A = \iint_D \rho(x, y)\,dA$$

Physicists also consider other types of density that can be treated in the same manner. For example, if an electric charge is distributed over a region D and the charge density (in units of charge per unit area) is given by $\sigma(x, y)$ at a point (x, y) in D, then the total charge Q is given by

2
$$Q = \iint_D \sigma(x, y)\,dA$$

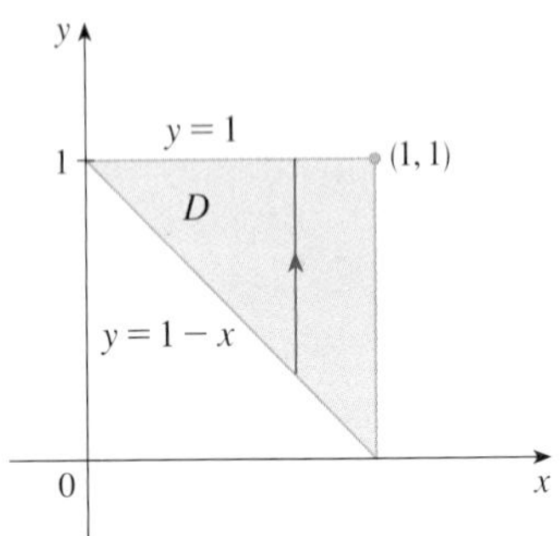

FIGURE 3

EXAMPLE 1 Charge is distributed over the triangular region D in Figure 3 so that the charge density at (x, y) is $\sigma(x, y) = xy$, measured in coulombs per square meter (C/m^2). Find the total charge.

SOLUTION From Equation 2 and Figure 3 we have

$$Q = \iint_D \sigma(x, y)\,dA = \int_0^1 \int_{1-x}^1 xy\,dy\,dx$$

$$= \int_0^1 \left[x\frac{y^2}{2} \right]_{y=1-x}^{y=1} dx = \int_0^1 \frac{x}{2}\left[1^2 - (1 - x)^2\right]dx$$

$$= \tfrac{1}{2}\int_0^1 (2x^2 - x^3)\,dx = \frac{1}{2}\left[\frac{2x^3}{3} - \frac{x^4}{4} \right]_0^1 = \frac{5}{24}$$

Thus the total charge is $\frac{5}{24}$ C.

Moments and Centers of Mass

In Section 8.3 we found the center of mass of a lamina with constant density; here we consider a lamina with variable density. Suppose the lamina occupies a region D and has density function $\rho(x, y)$. Recall from Chapter 8 that we defined the moment of a particle about an axis as the product of its mass and its directed distance from the axis. We divide D into small rectangles as in Figure 2. Then the mass of R_{ij} is approximately $\rho(x_{ij}^*, y_{ij}^*)\,\Delta A$, so we can approximate the moment of R_{ij} with respect to the x-axis by

$$[\rho(x_{ij}^*, y_{ij}^*)\,\Delta A]\,y_{ij}^*$$

If we now add these quantities and take the limit as the number of subrectangles becomes large, we obtain the **moment** of the entire lamina **about the x-axis**:

3
$$M_x = \lim_{m,\,n\to\infty} \sum_{i=1}^{m} \sum_{j=1}^{n} y_{ij}^* \, \rho(x_{ij}^*, y_{ij}^*) \, \Delta A = \iint_D y\, \rho(x, y)\, dA$$

Similarly, the **moment about the y-axis** is

4
$$M_y = \lim_{m,\,n\to\infty} \sum_{i=1}^{m} \sum_{j=1}^{n} x_{ij}^* \, \rho(x_{ij}^*, y_{ij}^*) \, \Delta A = \iint_D x\, \rho(x, y)\, dA$$

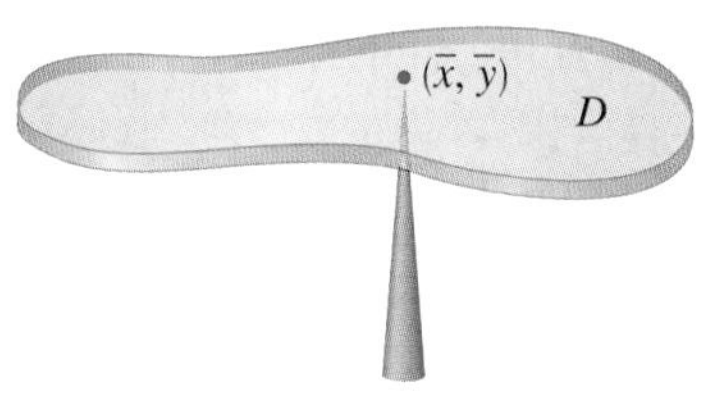

FIGURE 4

As before, we define the center of mass $(\bar{x}, \bar{y})$ so that $m\bar{x} = M_y$ and $m\bar{y} = M_x$. The physical significance is that the lamina behaves as if its entire mass is concentrated at its center of mass. Thus the lamina balances horizontally when supported at its center of mass (see Figure 4).

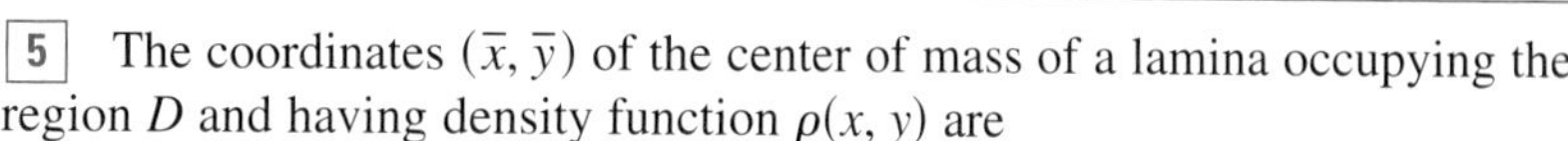

5 The coordinates $(\bar{x}, \bar{y})$ of the center of mass of a lamina occupying the region D and having density function $\rho(x, y)$ are

$$\bar{x} = \frac{M_y}{m} = \frac{1}{m} \iint_D x\rho(x, y)\, dA \qquad \bar{y} = \frac{M_x}{m} = \frac{1}{m} \iint_D y\rho(x, y)\, dA$$

where the mass m is given by

$$m = \iint_D \rho(x, y)\, dA$$

V EXAMPLE 2 Find the mass and center of mass of a triangular lamina with vertices $(0, 0)$, $(1, 0)$, and $(0, 2)$ if the density function is $\rho(x, y) = 1 + 3x + y$.

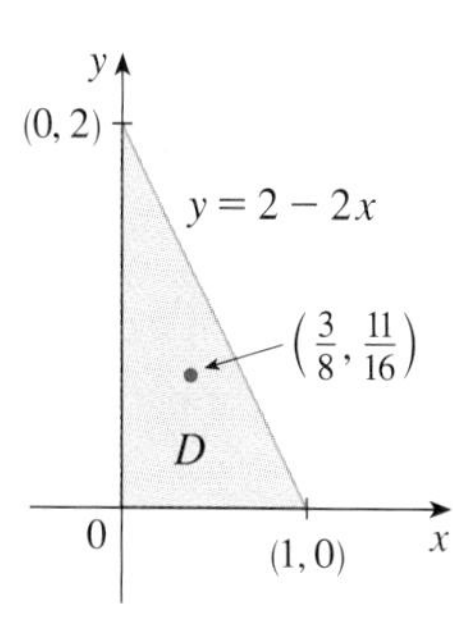

FIGURE 5

SOLUTION The triangle is shown in Figure 5. (Note that the equation of the upper boundary is $y = 2 - 2x$.) The mass of the lamina is

$$\begin{aligned} m &= \iint_D \rho(x, y)\, dA = \int_0^1 \int_0^{2-2x} (1 + 3x + y)\, dy\, dx \\ &= \int_0^1 \left[y + 3xy + \frac{y^2}{2} \right]_{y=0}^{y=2-2x} dx \\ &= 4 \int_0^1 (1 - x^2)\, dx = 4\left[x - \frac{x^3}{3} \right]_0^1 = \frac{8}{3} \end{aligned}$$

Then the formulas in **5** give

$$\begin{aligned} \bar{x} &= \frac{1}{m} \iint_D x\rho(x, y)\, dA = \tfrac{3}{8} \int_0^1 \int_0^{2-2x} (x + 3x^2 + xy)\, dy\, dx \\ &= \frac{3}{8} \int_0^1 \left[xy + 3x^2 y + x\frac{y^2}{2} \right]_{y=0}^{y=2-2x} dx \\ &= \frac{3}{2} \int_0^1 (x - x^3)\, dx = \frac{3}{2}\left[\frac{x^2}{2} - \frac{x^4}{4} \right]_0^1 = \frac{3}{8} \end{aligned}$$

$$\bar{y} = \frac{1}{m} \iint_D y\rho(x, y)\, dA = \tfrac{3}{8} \int_0^1 \int_0^{2-2x} (y + 3xy + y^2)\, dy\, dx$$

$$= \frac{3}{8} \int_0^1 \left[\frac{y^2}{2} + 3x\frac{y^2}{2} + \frac{y^3}{3} \right]_{y=0}^{y=2-2x} dx = \tfrac{1}{4} \int_0^1 (7 - 9x - 3x^2 + 5x^3)\, dx$$

$$= \frac{1}{4} \left[7x - 9\frac{x^2}{2} - x^3 + 5\frac{x^4}{4} \right]_0^1 = \frac{11}{16}$$

The center of mass is at the point $\left(\frac{3}{8}, \frac{11}{16}\right)$.

V EXAMPLE 3 The density at any point on a semicircular lamina is proportional to the distance from the center of the circle. Find the center of mass of the lamina.

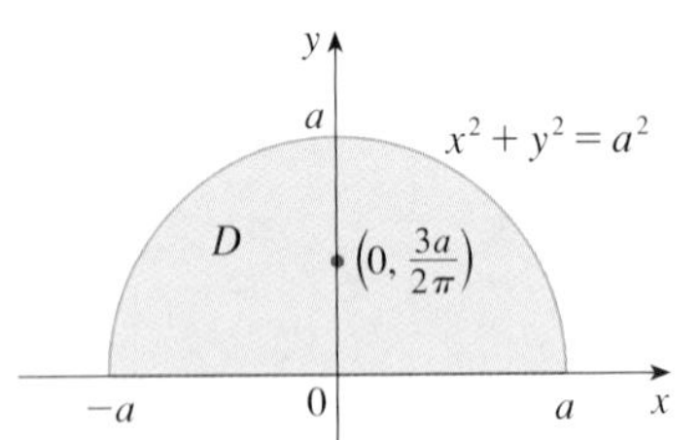

FIGURE 6

SOLUTION Let's place the lamina as the upper half of the circle $x^2 + y^2 = a^2$. (See Figure 6.) Then the distance from a point (x, y) to the center of the circle (the origin) is $\sqrt{x^2 + y^2}$. Therefore the density function is

$$\rho(x, y) = K\sqrt{x^2 + y^2}$$

where K is some constant. Both the density function and the shape of the lamina suggest that we convert to polar coordinates. Then $\sqrt{x^2 + y^2} = r$ and the region D is given by $0 \leqslant r \leqslant a$, $0 \leqslant \theta \leqslant \pi$. Thus the mass of the lamina is

$$m = \iint_D \rho(x, y)\, dA = \iint_D K\sqrt{x^2 + y^2}\, dA$$

$$= \int_0^{\pi} \int_0^a (Kr)\, r\, dr\, d\theta = K \int_0^{\pi} d\theta \int_0^a r^2\, dr$$

$$= K\pi \frac{r^3}{3}\bigg]_0^a = \frac{K\pi a^3}{3}$$

Both the lamina and the density function are symmetric with respect to the y-axis, so the center of mass must lie on the y-axis, that is, $\bar{x} = 0$. The y-coordinate is given by

$$\bar{y} = \frac{1}{m} \iint_D y\rho(x, y)\, dA = \frac{3}{K\pi a^3} \int_0^{\pi} \int_0^a r \sin\theta\, (Kr)\, r\, dr\, d\theta$$

$$= \frac{3}{\pi a^3} \int_0^{\pi} \sin\theta\, d\theta \int_0^a r^3\, dr = \frac{3}{\pi a^3} \Big[-\cos\theta\Big]_0^{\pi} \left[\frac{r^4}{4}\right]_0^a$$

$$= \frac{3}{\pi a^3} \frac{2a^4}{4} = \frac{3a}{2\pi}$$

Compare the location of the center of mass in Example 3 with Example 4 in Section 8.3, where we found that the center of mass of a lamina with the same shape but uniform density is located at the point $(0, 4a/(3\pi))$.

Therefore the center of mass is located at the point $(0, 3a/(2\pi))$.

Moment of Inertia

The **moment of inertia** (also called the **second moment**) of a particle of mass m about an axis is defined to be mr^2, where r is the distance from the particle to the axis. We extend this concept to a lamina with density function $\rho(x, y)$ and occupying a region D by proceeding as we did for ordinary moments. We divide D into small rectangles, approximate the moment of inertia of each subrectangle about the x-axis, and take the limit of the sum as the number of subrectangles becomes large. The result is the **moment of inertia** of the lamina **about the x-axis**:

6 $$I_x = \lim_{m,n\to\infty} \sum_{i=1}^{m}\sum_{j=1}^{n} (y_{ij}^*)^2 \rho(x_{ij}^*, y_{ij}^*)\,\Delta A = \iint_D y^2 \rho(x, y)\,dA$$

Similarly, the **moment of inertia about the y-axis** is

7 $$I_y = \lim_{m,n\to\infty} \sum_{i=1}^{m}\sum_{j=1}^{n} (x_{ij}^*)^2 \rho(x_{ij}^*, y_{ij}^*)\,\Delta A = \iint_D x^2 \rho(x, y)\,dA$$

It is also of interest to consider the **moment of inertia about the origin**, also called the **polar moment of inertia**:

8 $$I_0 = \lim_{m,n\to\infty} \sum_{i=1}^{m}\sum_{j=1}^{n} \left[(x_{ij}^*)^2 + (y_{ij}^*)^2\right] \rho(x_{ij}^*, y_{ij}^*)\,\Delta A = \iint_D (x^2 + y^2)\,\rho(x, y)\,dA$$

Note that $I_0 = I_x + I_y$.

V EXAMPLE 4 Find the moments of inertia I_x, I_y, and I_0 of a homogeneous disk D with density $\rho(x, y) = \rho$, center the origin, and radius a.

SOLUTION The boundary of D is the circle $x^2 + y^2 = a^2$ and in polar coordinates D is described by $0 \leqslant \theta \leqslant 2\pi$, $0 \leqslant r \leqslant a$. Let's compute I_0 first:

$$\begin{aligned} I_0 &= \iint_D (x^2 + y^2)\rho\,dA = \rho \int_0^{2\pi}\int_0^a r^2\,r\,dr\,d\theta \\ &= \rho \int_0^{2\pi} d\theta \int_0^a r^3\,dr = 2\pi\rho\left[\frac{r^4}{4}\right]_0^a = \frac{\pi\rho a^4}{2} \end{aligned}$$

Instead of computing I_x and I_y directly, we use the facts that $I_x + I_y = I_0$ and $I_x = I_y$ (from the symmetry of the problem). Thus

$$I_x = I_y = \frac{I_0}{2} = \frac{\pi\rho a^4}{4}$$ ■

In Example 4 notice that the mass of the disk is

$$m = \text{density} \times \text{area} = \rho(\pi a^2)$$

so the moment of inertia of the disk about the origin (like a wheel about its axle) can be written as

$$I_0 = \frac{\pi\rho a^4}{2} = \tfrac{1}{2}(\rho\pi a^2)a^2 = \tfrac{1}{2}ma^2$$

Thus if we increase the mass or the radius of the disk, we thereby increase the moment of inertia. In general, the moment of inertia plays much the same role in rotational motion that mass plays in linear motion. The moment of inertia of a wheel is what makes it difficult to start or stop the rotation of the wheel, just as the mass of a car is what makes it difficult to start or stop the motion of the car.

The **radius of gyration of a lamina about an axis** is the number R such that

$$mR^2 = I \tag{9}$$

where m is the mass of the lamina and I is the moment of inertia about the given axis. Equation 9 says that if the mass of the lamina were concentrated at a distance R from the axis, then the moment of inertia of this "point mass" would be the same as the moment of inertia of the lamina.

In particular, the radius of gyration $\bar{\bar{y}}$ with respect to the x-axis and the radius of gyration $\bar{\bar{x}}$ with respect to the y-axis are given by the equations

$$m\bar{\bar{y}}^2 = I_x \qquad m\bar{\bar{x}}^2 = I_y \tag{10}$$

Thus $(\bar{\bar{x}}, \bar{\bar{y}})$ is the point at which the mass of the lamina can be concentrated without changing the moments of inertia with respect to the coordinate axes. (Note the analogy with the center of mass.)

V EXAMPLE 5 Find the radius of gyration about the x-axis of the disk in Example 4.

SOLUTION As noted, the mass of the disk is $m = \rho\pi a^2$, so from Equations 10 we have

$$\bar{\bar{y}}^2 = \frac{I_x}{m} = \frac{\frac{1}{4}\pi\rho a^4}{\rho\pi a^2} = \frac{a^2}{4}$$

Therefore the radius of gyration about the x-axis is $\bar{\bar{y}} = \frac{1}{2}a$, which is half the radius of the disk.

Probability

In Section 8.5 we considered the *probability density function* f of a continuous random variable X. This means that $f(x) \geqslant 0$ for all x, $\int_{-\infty}^{\infty} f(x)\,dx = 1$, and the probability that X lies between a and b is found by integrating f from a to b:

$$P(a \leqslant X \leqslant b) = \int_a^b f(x)\,dx$$

Now we consider a pair of continuous random variables X and Y, such as the lifetimes of two components of a machine or the height and weight of an adult female chosen at random. The **joint density function** of X and Y is a function f of two variables such that the probability that (X, Y) lies in a region D is

$$P\big((X, Y) \in D\big) = \iint_D f(x, y)\,dA$$

In particular, if the region is a rectangle, the probability that X lies between a and b and Y lies between c and d is

$$P(a \leqslant X \leqslant b,\ c \leqslant Y \leqslant d) = \int_a^b \int_c^d f(x, y)\,dy\,dx$$

(See Figure 7.)

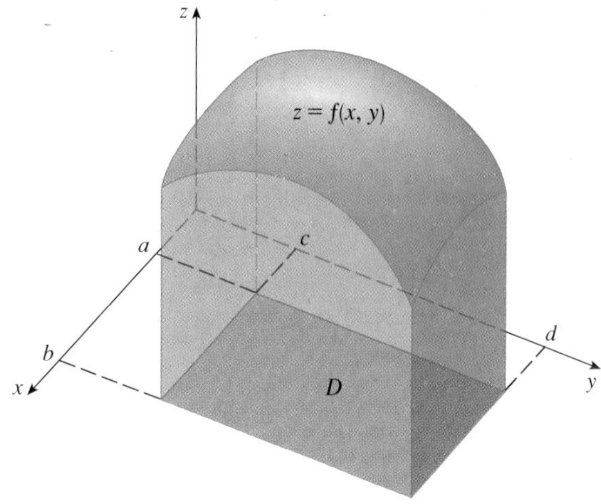

FIGURE 7
The probability that X lies between a and b and Y lies between c and d is the volume that lies above the rectangle $D = [a, b] \times [c, d]$ and below the graph of the joint density function.

Because probabilities aren't negative and are measured on a scale from 0 to 1, the joint density function has the following properties:

$$f(x, y) \geqslant 0 \qquad \iint_{\mathbb{R}^2} f(x, y)\, dA = 1$$

As in Exercise 40 in Section 15.4, the double integral over $\mathbb{R}^2$ is an improper integral defined as the limit of double integrals over expanding circles or squares, and we can write

$$\iint_{\mathbb{R}^2} f(x, y)\, dA = \int_{-\infty}^{\infty} \int_{-\infty}^{\infty} f(x, y)\, dx\, dy = 1$$

EXAMPLE 6 If the joint density function for X and Y is given by

$$f(x, y) = \begin{cases} C(x + 2y) & \text{if } 0 \leqslant x \leqslant 10,\ 0 \leqslant y \leqslant 10 \\ 0 & \text{otherwise} \end{cases}$$

find the value of the constant C. Then find $P(X \leqslant 7, Y \geqslant 2)$.

SOLUTION We find the value of C by ensuring that the double integral of f is equal to 1. Because $f(x, y) = 0$ outside the rectangle $[0, 10] \times [0, 10]$, we have

$$\int_{-\infty}^{\infty} \int_{-\infty}^{\infty} f(x, y)\, dy\, dx = \int_{0}^{10} \int_{0}^{10} C(x + 2y)\, dy\, dx = C \int_{0}^{10} \left[xy + y^2\right]_{y=0}^{y=10} dx$$

$$= C \int_{0}^{10} (10x + 100)\, dx = 1500C$$

Therefore $1500C = 1$ and so $C = \frac{1}{1500}$.

Now we can compute the probability that X is at most 7 and Y is at least 2:

$$P(X \leqslant 7, Y \geqslant 2) = \int_{-\infty}^{7} \int_{2}^{\infty} f(x, y)\, dy\, dx = \int_{0}^{7} \int_{2}^{10} \tfrac{1}{1500}(x + 2y)\, dy\, dx$$

$$= \tfrac{1}{1500} \int_{0}^{7} \left[xy + y^2\right]_{y=2}^{y=10} dx = \tfrac{1}{1500} \int_{0}^{7} (8x + 96)\, dx$$

$$= \tfrac{868}{1500} \approx 0.5787$$

Suppose X is a random variable with probability density function $f_1(x)$ and Y is a random variable with density function $f_2(y)$. Then X and Y are called **independent random variables** if their joint density function is the product of their individual density functions:

$$f(x, y) = f_1(x)f_2(y)$$

In Section 8.5 we modeled waiting times by using exponential density functions

$$f(t) = \begin{cases} 0 & \text{if } t < 0 \\ \mu^{-1}e^{-t/\mu} & \text{if } t \geqslant 0 \end{cases}$$

where μ is the mean waiting time. In the next example we consider a situation with two independent waiting times.

EXAMPLE 7 The manager of a movie theater determines that the average time moviegoers wait in line to buy a ticket for this week's film is 10 minutes and the average time they wait to buy popcorn is 5 minutes. Assuming that the waiting times are independent, find the probability that a moviegoer waits a total of less than 20 minutes before taking his or her seat.

SOLUTION Assuming that both the waiting time X for the ticket purchase and the waiting time Y in the refreshment line are modeled by exponential probability density functions, we can write the individual density functions as

$$f_1(x) = \begin{cases} 0 & \text{if } x < 0 \\ \frac{1}{10}e^{-x/10} & \text{if } x \geqslant 0 \end{cases} \qquad f_2(y) = \begin{cases} 0 & \text{if } y < 0 \\ \frac{1}{5}e^{-y/5} & \text{if } y \geqslant 0 \end{cases}$$

Since X and Y are independent, the joint density function is the product:

$$f(x, y) = f_1(x)f_2(y) = \begin{cases} \frac{1}{50}e^{-x/10}e^{-y/5} & \text{if } x \geqslant 0,\ y \geqslant 0 \\ 0 & \text{otherwise} \end{cases}$$

We are asked for the probability that $X + Y < 20$:

$$P(X + Y < 20) = P\big((X, Y) \in D\big)$$

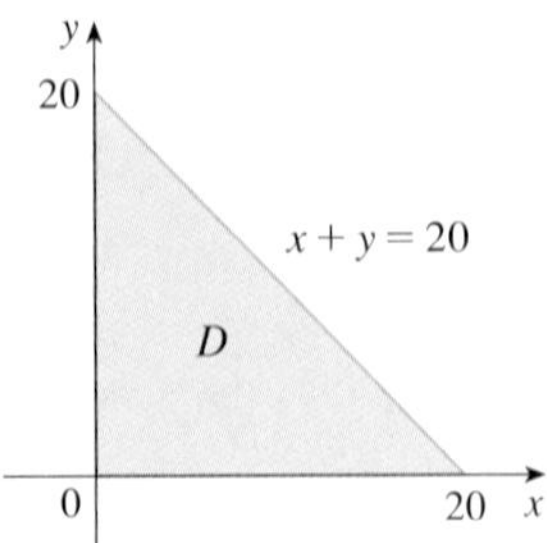

FIGURE 8

where D is the triangular region shown in Figure 8. Thus

$$\begin{aligned} P(X + Y < 20) &= \iint_D f(x, y)\, dA = \int_0^{20}\int_0^{20-x} \tfrac{1}{50}e^{-x/10}e^{-y/5}\, dy\, dx \\ &= \tfrac{1}{50}\int_0^{20} \Big[e^{-x/10}(-5)e^{-y/5}\Big]_{y=0}^{y=20-x} dx \\ &= \tfrac{1}{10}\int_0^{20} e^{-x/10}\big(1 - e^{(x-20)/5}\big)\, dx \\ &= \tfrac{1}{10}\int_0^{20} \big(e^{-x/10} - e^{-4}e^{x/10}\big)\, dx \\ &= 1 + e^{-4} - 2e^{-2} \approx 0.7476 \end{aligned}$$

This means that about 75% of the moviegoers wait less than 20 minutes before taking their seats.

Expected Values

Recall from Section 8.5 that if X is a random variable with probability density function f, then its *mean* is

$$\mu = \int_{-\infty}^{\infty} xf(x)\,dx$$

Now if X and Y are random variables with joint density function f, we define the **X-mean** and **Y-mean**, also called the **expected values** of X and Y, to be

$$\boxed{11} \qquad \mu_1 = \iint_{\mathbb{R}^2} xf(x, y)\,dA \qquad \mu_2 = \iint_{\mathbb{R}^2} yf(x, y)\,dA$$

Notice how closely the expressions for μ_1 and μ_2 in $\boxed{11}$ resemble the moments M_x and M_y of a lamina with density function ρ in Equations 3 and 4. In fact, we can think of probability as being like continuously distributed mass. We calculate probability the way we calculate mass—by integrating a density function. And because the total "probability mass" is 1, the expressions for $\bar{x}$ and $\bar{y}$ in $\boxed{5}$ show that we can think of the expected values of X and Y, μ_1 and μ_2, as the coordinates of the "center of mass" of the probability distribution.

In the next example we deal with normal distributions. As in Section 8.5, a single random variable is *normally distributed* if its probability density function is of the form

$$f(x) = \frac{1}{\sigma\sqrt{2\pi}}\, e^{-(x-\mu)^2/(2\sigma^2)}$$

where μ is the mean and σ is the standard deviation.

EXAMPLE 8 A factory produces (cylindrically shaped) roller bearings that are sold as having diameter 4.0 cm and length 6.0 cm. In fact, the diameters X are normally distributed with mean 4.0 cm and standard deviation 0.01 cm while the lengths Y are normally distributed with mean 6.0 cm and standard deviation 0.01 cm. Assuming that X and Y are independent, write the joint density function and graph it. Find the probability that a bearing randomly chosen from the production line has either length or diameter that differs from the mean by more than 0.02 cm.

SOLUTION We are given that X and Y are normally distributed with $\mu_1 = 4.0$, $\mu_2 = 6.0$, and $\sigma_1 = \sigma_2 = 0.01$. So the individual density functions for X and Y are

$$f_1(x) = \frac{1}{0.01\sqrt{2\pi}}\, e^{-(x-4)^2/0.0002} \qquad f_2(y) = \frac{1}{0.01\sqrt{2\pi}}\, e^{-(y-6)^2/0.0002}$$

Since X and Y are independent, the joint density function is the product:

$$\begin{aligned} f(x, y) &= f_1(x)f_2(y) \\ &= \frac{1}{0.0002\pi}\, e^{-(x-4)^2/0.0002} e^{-(y-6)^2/0.0002} \\ &= \frac{5000}{\pi}\, e^{-5000[(x-4)^2+(y-6)^2]} \end{aligned}$$

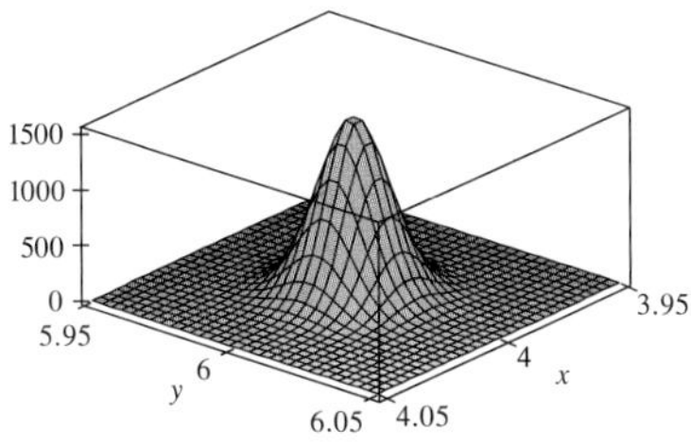

FIGURE 9
Graph of the bivariate normal joint density function in Example 8

A graph of this function is shown in Figure 9.

Let's first calculate the probability that both X and Y differ from their means by less than 0.02 cm. Using a calculator or computer to estimate the integral, we have

$$P(3.98 < X < 4.02,\ 5.98 < Y < 6.02) = \int_{3.98}^{4.02} \int_{5.98}^{6.02} f(x, y)\, dy\, dx$$

$$= \frac{5000}{\pi} \int_{3.98}^{4.02} \int_{5.98}^{6.02} e^{-5000[(x-4)^2+(y-6)^2]}\, dy\, dx$$

$$\approx 0.91$$

Then the probability that either X or Y differs from its mean by more than 0.02 cm is approximately

$$1 - 0.91 = 0.09$$

15.5 Exercises

1. Electric charge is distributed over the rectangle $0 \leqslant x \leqslant 5$, $2 \leqslant y \leqslant 5$ so that the charge density at (x, y) is $\sigma(x, y) = 2x + 4y$ (measured in coulombs per square meter). Find the total charge on the rectangle.

2. Electric charge is distributed over the disk $x^2 + y^2 \leqslant 1$ so that the charge density at (x, y) is $\sigma(x, y) = \sqrt{x^2 + y^2}$ (measured in coulombs per square meter). Find the total charge on the disk.

3–10 Find the mass and center of mass of the lamina that occupies the region D and has the given density function ρ.

3. $D = \{(x, y) \mid 1 \leqslant x \leqslant 3,\ 1 \leqslant y \leqslant 4\}$; $\rho(x, y) = ky^2$

4. $D = \{(x, y) \mid 0 \leqslant x \leqslant a,\ 0 \leqslant y \leqslant b\}$; $\rho(x, y) = 1 + x^2 + y^2$

5. D is the triangular region with vertices $(0, 0)$, $(2, 1)$, $(0, 3)$; $\rho(x, y) = x + y$

6. D is the triangular region enclosed by the lines $x = 0$, $y = x$, and $2x + y = 6$; $\rho(x, y) = x^2$

7. D is bounded by $y = 1 - x^2$ and $y = 0$; $\rho(x, y) = ky$

8. D is bounded by $y = x^2$ and $y = x + 2$; $\rho(x, y) = kx$

9. $D = \{(x, y) \mid 0 \leqslant y \leqslant \sin(\pi x/L),\ 0 \leqslant x \leqslant L\}$; $\rho(x, y) = y$

10. D is bounded by the parabolas $y = x^2$ and $x = y^2$; $\rho(x, y) = \sqrt{x}$

---

11. A lamina occupies the part of the disk $x^2 + y^2 \leqslant 1$ in the first quadrant. Find its center of mass if the density at any point is proportional to its distance from the x-axis.

12. Find the center of mass of the lamina in Exercise 11 if the density at any point is proportional to the square of its distance from the origin.

13. The boundary of a lamina consists of the semicircles $y = \sqrt{1 - x^2}$ and $y = \sqrt{4 - x^2}$ together with the portions of the x-axis that join them. Find the center of mass of the lamina if the density at any point is proportional to its distance from the origin.

14. Find the center of mass of the lamina in Exercise 13 if the density at any point is inversely proportional to its distance from the origin.

15. Find the center of mass of a lamina in the shape of an isosceles right triangle with equal sides of length a if the density at any point is proportional to the square of the distance from the vertex opposite the hypotenuse.

16. A lamina occupies the region inside the circle $x^2 + y^2 = 2y$ but outside the circle $x^2 + y^2 = 1$. Find the center of mass if the density at any point is inversely proportional to its distance from the origin.

17. Find the moments of inertia I_x, I_y, I_0 for the lamina of Exercise 7.

18. Find the moments of inertia I_x, I_y, I_0 for the lamina of Exercise 12.

19. Find the moments of inertia I_x, I_y, I_0 for the lamina of Exercise 15.

20. Consider a square fan blade with sides of length 2 and the lower left corner placed at the origin. If the density of the blade is $\rho(x, y) = 1 + 0.1x$, is it more difficult to rotate the blade about the x-axis or the y-axis?

21–24 A lamina with constant density $\rho(x, y) = \rho$ occupies the given region. Find the moments of inertia I_x and I_y and the radii of gyration $\bar{\bar{x}}$ and $\bar{\bar{y}}$.

21. The rectangle $0 \leqslant x \leqslant b$, $0 \leqslant y \leqslant h$

22. The triangle with vertices $(0, 0)$, $(b, 0)$, and $(0, h)$

23. The part of the disk $x^2 + y^2 \leq a^2$ in the first quadrant

24. The region under the curve $y = \sin x$ from $x = 0$ to $x = \pi$

CAS **25–26** Use a computer algebra system to find the mass, center of mass, and moments of inertia of the lamina that occupies the region D and has the given density function.

25. D is enclosed by the right loop of the four-leaved rose $r = \cos 2\theta$; $\rho(x, y) = x^2 + y^2$

26. $D = \{(x, y) \mid 0 \leq y \leq xe^{-x},\ 0 \leq x \leq 2\}$; $\rho(x, y) = x^2y^2$

---

27. The joint density function for a pair of random variables X and Y is

$$f(x, y) = \begin{cases} Cx(1 + y) & \text{if } 0 \leq x \leq 1,\ 0 \leq y \leq 2 \\ 0 & \text{otherwise} \end{cases}$$

(a) Find the value of the constant C.
(b) Find $P(X \leq 1, Y \leq 1)$.
(c) Find $P(X + Y \leq 1)$.

28. (a) Verify that

$$f(x, y) = \begin{cases} 4xy & \text{if } 0 \leq x \leq 1,\ 0 \leq y \leq 1 \\ 0 & \text{otherwise} \end{cases}$$

is a joint density function.
(b) If X and Y are random variables whose joint density function is the function f in part (a), find
(i) $P\left(X \geq \frac{1}{2}\right)$ (ii) $P\left(X \geq \frac{1}{2}, Y \leq \frac{1}{2}\right)$
(c) Find the expected values of X and Y.

29. Suppose X and Y are random variables with joint density function

$$f(x, y) = \begin{cases} 0.1e^{-(0.5x+0.2y)} & \text{if } x \geq 0,\ y \geq 0 \\ 0 & \text{otherwise} \end{cases}$$

(a) Verify that f is indeed a joint density function.
(b) Find the following probabilities.
(i) $P(Y \geq 1)$ (ii) $P(X \leq 2, Y \leq 4)$
(c) Find the expected values of X and Y.

30. (a) A lamp has two bulbs of a type with an average lifetime of 1000 hours. Assuming that we can model the probability of failure of these bulbs by an exponential density function with mean $\mu = 1000$, find the probability that both of the lamp's bulbs fail within 1000 hours.
(b) Another lamp has just one bulb of the same type as in part (a). If one bulb burns out and is replaced by a bulb of the same type, find the probability that the two bulbs fail within a total of 1000 hours.

CAS **31.** Suppose that X and Y are independent random variables, where X is normally distributed with mean 45 and standard deviation 0.5 and Y is normally distributed with mean 20 and standard deviation 0.1.
(a) Find $P(40 \leq X \leq 50, 20 \leq Y \leq 25)$.
(b) Find $P\left(4(X - 45)^2 + 100(Y - 20)^2 \leq 2\right)$.

32. Xavier and Yolanda both have classes that end at noon and they agree to meet every day after class. They arrive at the coffee shop independently. Xavier's arrival time is X and Yolanda's arrival time is Y, where X and Y are measured in minutes after noon. The individual density functions are

$$f_1(x) = \begin{cases} e^{-x} & \text{if } x \geq 0 \\ 0 & \text{if } x < 0 \end{cases} \qquad f_2(y) = \begin{cases} \frac{1}{50}y & \text{if } 0 \leq y \leq 10 \\ 0 & \text{otherwise} \end{cases}$$

(Xavier arrives sometime after noon and is more likely to arrive promptly than late. Yolanda always arrives by 12:10 PM and is more likely to arrive late than promptly.) After Yolanda arrives, she'll wait for up to half an hour for Xavier, but he won't wait for her. Find the probability that they meet.

33. When studying the spread of an epidemic, we assume that the probability that an infected individual will spread the disease to an uninfected individual is a function of the distance between them. Consider a circular city of radius 10 miles in which the population is uniformly distributed. For an uninfected individual at a fixed point $A(x_0, y_0)$, assume that the probability function is given by

$$f(P) = \tfrac{1}{20}[20 - d(P, A)]$$

where $d(P, A)$ denotes the distance between points P and A.
(a) Suppose the exposure of a person to the disease is the sum of the probabilities of catching the disease from all members of the population. Assume that the infected people are uniformly distributed throughout the city, with k infected individuals per square mile. Find a double integral that represents the exposure of a person residing at A.
(b) Evaluate the integral for the case in which A is the center of the city and for the case in which A is located on the edge of the city. Where would you prefer to live?

15.6 Surface Area

In Section 16.6 we will deal with areas of more general surfaces, called parametric surfaces, and so this section need not be covered if that later section will be covered.

In this section we apply double integrals to the problem of computing the area of a surface. In Section 8.2 we found the area of a very special type of surface—a surface of revolution—by the methods of single-variable calculus. Here we compute the area of a surface with equation $z = f(x, y)$, the graph of a function of two variables.

Let S be a surface with equation $z = f(x, y)$, where f has continuous partial derivatives. For simplicity in deriving the surface area formula, we assume that $f(x, y) \geq 0$ and the

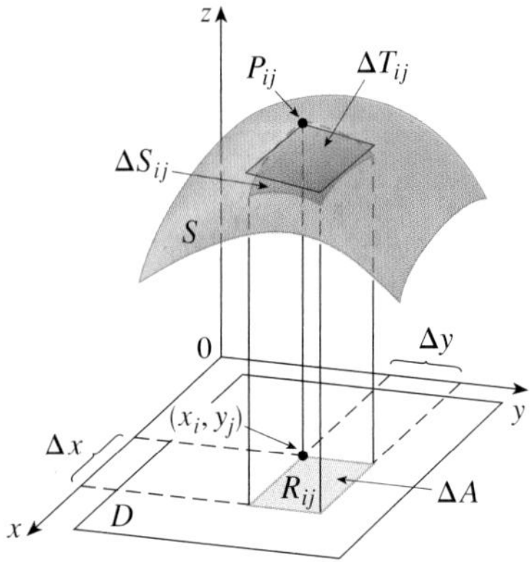

FIGURE 1

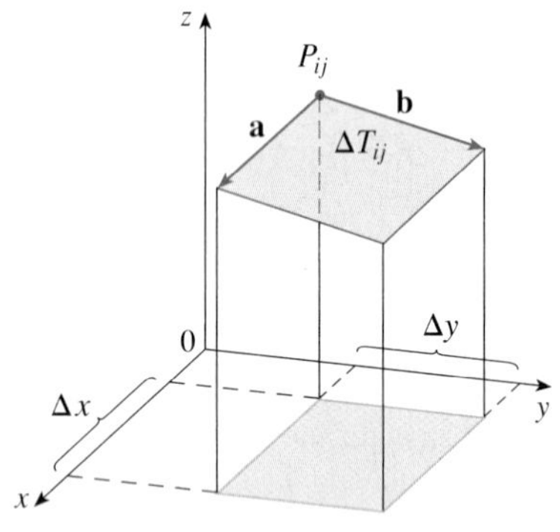

FIGURE 2

domain D of f is a rectangle. We divide D into small rectangles R_{ij} with area $\Delta A = \Delta x\, \Delta y$. If (x_i, y_j) is the corner of R_{ij} closest to the origin, let $P_{ij}(x_i, y_j, f(x_i, y_j))$ be the point on S directly above it (see Figure 1). The tangent plane to S at P_{ij} is an approximation to S near P_{ij}. So the area ΔT_{ij} of the part of this tangent plane (a parallelogram) that lies directly above R_{ij} is an approximation to the area ΔS_{ij} of the part of S that lies directly above R_{ij}. Thus the sum $\Sigma\Sigma\, \Delta T_{ij}$ is an approximation to the total area of S, and this approximation appears to improve as the number of rectangles increases. Therefore we define the **surface area** of S to be

$$\boxed{1} \qquad A(S) = \lim_{m,\, n \to \infty} \sum_{i=1}^{m} \sum_{j=1}^{n} \Delta T_{ij}$$

To find a formula that is more convenient than Equation 1 for computational purposes, we let **a** and **b** be the vectors that start at P_{ij} and lie along the sides of the parallelogram with area ΔT_{ij}. (See Figure 2.) Then $\Delta T_{ij} = |\mathbf{a} \times \mathbf{b}|$. Recall from Section 14.3 that $f_x(x_i, y_j)$ and $f_y(x_i, y_j)$ are the slopes of the tangent lines through P_{ij} in the directions of **a** and **b**. Therefore

$$\mathbf{a} = \Delta x\, \mathbf{i} + f_x(x_i, y_j)\, \Delta x\, \mathbf{k}$$

$$\mathbf{b} = \Delta y\, \mathbf{j} + f_y(x_i, y_j)\, \Delta y\, \mathbf{k}$$

and

$$\begin{aligned} \mathbf{a} \times \mathbf{b} &= \begin{vmatrix} \mathbf{i} & \mathbf{j} & \mathbf{k} \\ \Delta x & 0 & f_x(x_i, y_j)\, \Delta x \\ 0 & \Delta y & f_y(x_i, y_j)\, \Delta y \end{vmatrix} \\ &= -f_x(x_i, y_j)\, \Delta x\, \Delta y\, \mathbf{i} - f_y(x_i, y_j)\, \Delta x\, \Delta y\, \mathbf{j} + \Delta x\, \Delta y\, \mathbf{k} \\ &= [-f_x(x_i, y_j)\mathbf{i} - f_y(x_i, y_j)\mathbf{j} + \mathbf{k}]\, \Delta A \end{aligned}$$

Thus

$$\Delta T_{ij} = |\mathbf{a} \times \mathbf{b}| = \sqrt{[f_x(x_i, y_j)]^2 + [f_y(x_i, y_j)]^2 + 1}\, \Delta A$$

From Definition 1 we then have

$$\begin{aligned} A(S) &= \lim_{m,n \to \infty} \sum_{i=1}^{m} \sum_{j=1}^{n} \Delta T_{ij} \\ &= \lim_{m,n \to \infty} \sum_{i=1}^{m} \sum_{j=1}^{n} \sqrt{[f_x(x_i, y_j)]^2 + [f_y(x_i, y_j)]^2 + 1}\, \Delta A \end{aligned}$$

and by the definition of a double integral we get the following formula.

> **2** The area of the surface with equation $z = f(x, y)$, $(x, y) \in D$, where f_x and f_y are continuous, is
>
> $$A(S) = \iint_D \sqrt{[f_x(x, y)]^2 + [f_y(x, y)]^2 + 1}\, dA$$

We will verify in Section 16.6 that this formula is consistent with our previous formula for the area of a surface of revolution. If we use the alternative notation for partial derivatives, we can rewrite Formula 2 as follows:

3
$$A(s) = \iint_D \sqrt{1 + \left(\frac{\partial z}{\partial x}\right)^2 + \left(\frac{\partial z}{\partial y}\right)^2}\, dA$$

Notice the similarity between the surface area formula in Equation 3 and the arc length formula from Section 8.1:

$$L = \int_a^b \sqrt{1 + \left(\frac{dy}{dx}\right)^2}\, dx$$

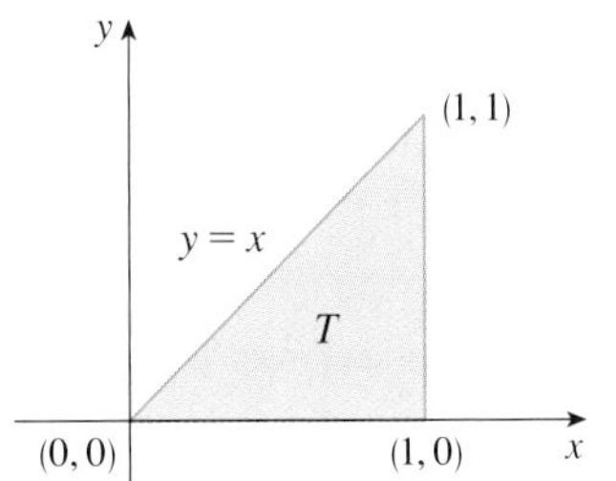

FIGURE 3

EXAMPLE 1 Find the surface area of the part of the surface $z = x^2 + 2y$ that lies above the triangular region T in the xy-plane with vertices (0, 0), (1, 0), and (1, 1).

SOLUTION The region T is shown in Figure 3 and is described by

$$T = \{(x, y) \mid 0 \leqslant x \leqslant 1,\ 0 \leqslant y \leqslant x\}$$

Using Formula 2 with $f(x, y) = x^2 + 2y$, we get

$$A = \iint_T \sqrt{(2x)^2 + (2)^2 + 1}\, dA = \int_0^1 \int_0^x \sqrt{4x^2 + 5}\, dy\, dx$$

$$= \int_0^1 x\sqrt{4x^2 + 5}\, dx = \tfrac{1}{8} \cdot \tfrac{2}{3}(4x^2 + 5)^{3/2}\Big]_0^1 = \tfrac{1}{12}\left(27 - 5\sqrt{5}\right)$$

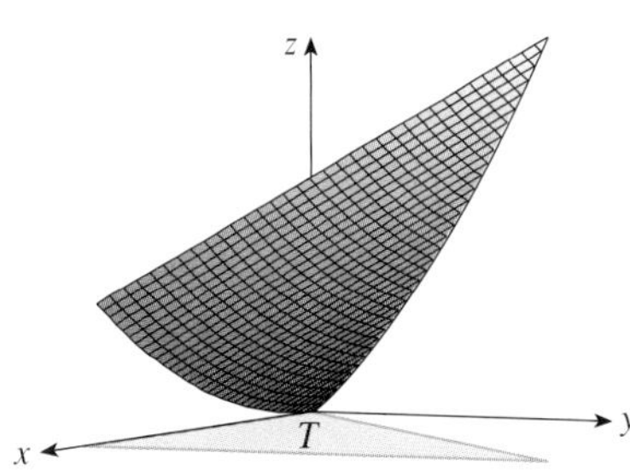

FIGURE 4

Figure 4 shows the portion of the surface whose area we have just computed.

EXAMPLE 2 Find the area of the part of the paraboloid $z = x^2 + y^2$ that lies under the plane $z = 9$.

SOLUTION The plane intersects the paraboloid in the circle $x^2 + y^2 = 9$, $z = 9$. Therefore the given surface lies above the disk D with center the origin and radius 3. (See Figure 5.) Using Formula 3, we have

$$A = \iint_D \sqrt{1 + \left(\frac{\partial z}{\partial x}\right)^2 + \left(\frac{\partial z}{\partial y}\right)^2}\, dA = \iint_D \sqrt{1 + (2x)^2 + (2y)^2}\, dA$$

$$= \iint_D \sqrt{1 + 4(x^2 + y^2)}\, dA$$

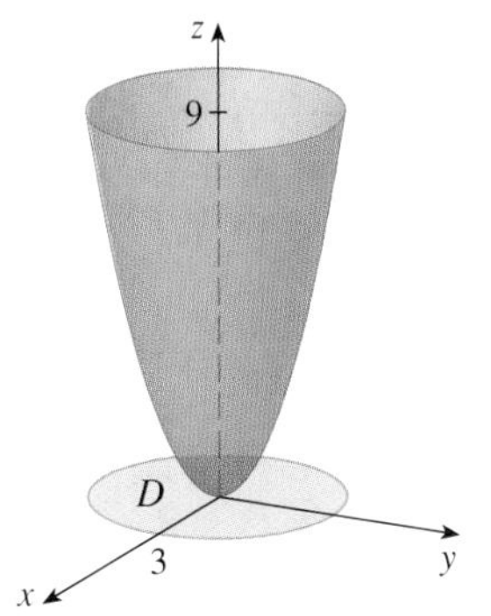

FIGURE 5

Converting to polar coordinates, we obtain

$$A = \int_0^{2\pi} \int_0^3 \sqrt{1 + 4r^2}\, r\, dr\, d\theta = \int_0^{2\pi} d\theta \int_0^3 \tfrac{1}{8}\sqrt{1 + 4r^2}\,(8r)\, dr$$

$$= 2\pi\left(\tfrac{1}{8}\right)\tfrac{2}{3}(1 + 4r^2)^{3/2}\Big]_0^3 = \frac{\pi}{6}\left(37\sqrt{37} - 1\right)$$

15.6 Exercises

1–12 Find the area of the surface.

1. The part of the plane $z = 2 + 3x + 4y$ that lies above the rectangle $[0, 5] \times [1, 4]$

2. The part of the plane $2x + 5y + z = 10$ that lies inside the cylinder $x^2 + y^2 = 9$

3. The part of the plane $3x + 2y + z = 6$ that lies in the first octant

4. The part of the surface $z = 1 + 3x + 2y^2$ that lies above the triangle with vertices $(0, 0)$, $(0, 1)$, and $(2, 1)$

5. The part of the cylinder $y^2 + z^2 = 9$ that lies above the rectangle with vertices $(0, 0)$, $(4, 0)$, $(0, 2)$, and $(4, 2)$

6. The part of the paraboloid $z = 4 - x^2 - y^2$ that lies above the xy-plane

7. The part of the hyperbolic paraboloid $z = y^2 - x^2$ that lies between the cylinders $x^2 + y^2 = 1$ and $x^2 + y^2 = 4$

8. The surface $z = \frac{2}{3}(x^{3/2} + y^{3/2})$, $0 \leqslant x \leqslant 1$, $0 \leqslant y \leqslant 1$

9. The part of the surface $z = xy$ that lies within the cylinder $x^2 + y^2 = 1$

10. The part of the sphere $x^2 + y^2 + z^2 = 4$ that lies above the plane $z = 1$

11. The part of the sphere $x^2 + y^2 + z^2 = a^2$ that lies within the cylinder $x^2 + y^2 = ax$ and above the xy-plane

12. The part of the sphere $x^2 + y^2 + z^2 = 4z$ that lies inside the paraboloid $z = x^2 + y^2$

13–14 Find the area of the surface correct to four decimal places by expressing the area in terms of a single integral and using your calculator to estimate the integral.

13. The part of the surface $z = e^{-x^2-y^2}$ that lies above the disk $x^2 + y^2 \leqslant 4$

14. The part of the surface $z = \cos(x^2 + y^2)$ that lies inside the cylinder $x^2 + y^2 = 1$

15. (a) Use the Midpoint Rule for double integrals (see Section 15.1) with four squares to estimate the surface area of the portion of the paraboloid $z = x^2 + y^2$ that lies above the square $[0, 1] \times [0, 1]$.

CAS (b) Use a computer algebra system to approximate the surface area in part (a) to four decimal places. Compare with the answer to part (a).

16. (a) Use the Midpoint Rule for double integrals with $m = n = 2$ to estimate the area of the surface $z = xy + x^2 + y^2$, $0 \leqslant x \leqslant 2$, $0 \leqslant y \leqslant 2$.

CAS (b) Use a computer algebra system to approximate the surface area in part (a) to four decimal places. Compare with the answer to part (a).

CAS **17.** Find the exact area of the surface $z = 1 + 2x + 3y + 4y^2$, $1 \leqslant x \leqslant 4$, $0 \leqslant y \leqslant 1$.

CAS **18.** Find the exact area of the surface

$$z = 1 + x + y + x^2 \qquad -2 \leqslant x \leqslant 1 \quad -1 \leqslant y \leqslant 1$$

Illustrate by graphing the surface.

CAS **19.** Find, to four decimal places, the area of the part of the surface $z = 1 + x^2y^2$ that lies above the disk $x^2 + y^2 \leqslant 1$.

CAS **20.** Find, to four decimal places, the area of the part of the surface $z = (1 + x^2)/(1 + y^2)$ that lies above the square $|x| + |y| \leqslant 1$. Illustrate by graphing this part of the surface.

21. Show that the area of the part of the plane $z = ax + by + c$ that projects onto a region D in the xy-plane with area $A(D)$ is $\sqrt{a^2 + b^2 + 1}\, A(D)$.

22. If you attempt to use Formula 2 to find the area of the top half of the sphere $x^2 + y^2 + z^2 = a^2$, you have a slight problem because the double integral is improper. In fact, the integrand has an infinite discontinuity at every point of the boundary circle $x^2 + y^2 = a^2$. However, the integral can be computed as the limit of the integral over the disk $x^2 + y^2 \leqslant t^2$ as $t \to a^-$. Use this method to show that the area of a sphere of radius a is $4\pi a^2$.

23. Find the area of the finite part of the paraboloid $y = x^2 + z^2$ cut off by the plane $y = 25$. [*Hint:* Project the surface onto the xz-plane.]

24. The figure shows the surface created when the cylinder $y^2 + z^2 = 1$ intersects the cylinder $x^2 + z^2 = 1$. Find the area of this surface.

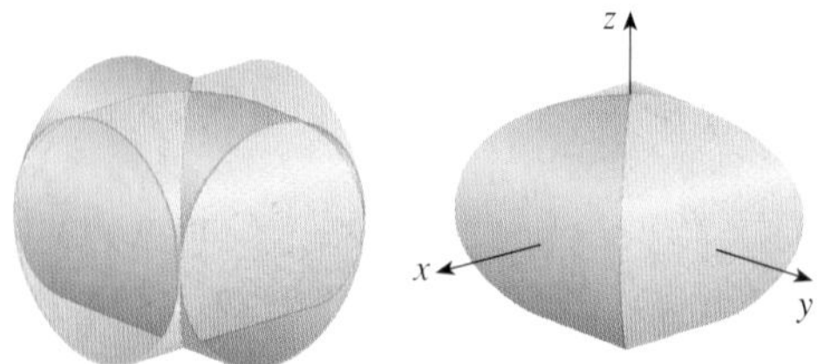

15.7 Triple Integrals

Just as we defined single integrals for functions of one variable and double integrals for functions of two variables, so we can define triple integrals for functions of three variables. Let's first deal with the simplest case where f is defined on a rectangular box:

1 $$B = \{(x, y, z) \mid a \leqslant x \leqslant b,\ c \leqslant y \leqslant d,\ r \leqslant z \leqslant s\}$$

The first step is to divide B into sub-boxes. We do this by dividing the interval $[a, b]$ into l subintervals $[x_{i-1}, x_i]$ of equal width Δx, dividing $[c, d]$ into m subintervals of width Δy, and dividing $[r, s]$ into n subintervals of width Δz. The planes through the endpoints of these subintervals parallel to the coordinate planes divide the box B into lmn sub-boxes

$$B_{ijk} = [x_{i-1}, x_i] \times [y_{j-1}, y_j] \times [z_{k-1}, z_k]$$

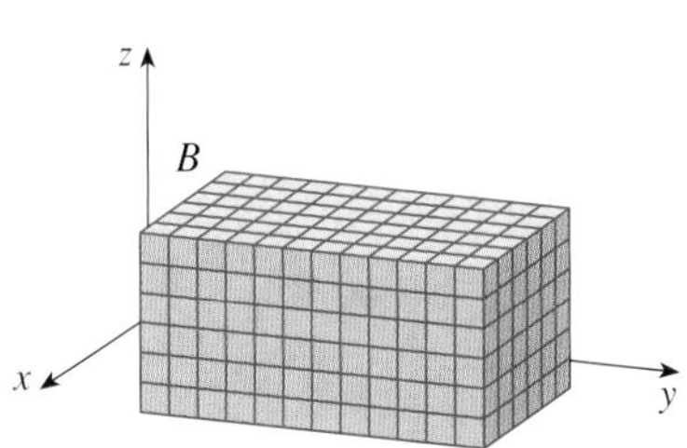

which are shown in Figure 1. Each sub-box has volume $\Delta V = \Delta x\, \Delta y\, \Delta z$.

Then we form the **triple Riemann sum**

2 $$\sum_{i=1}^{l} \sum_{j=1}^{m} \sum_{k=1}^{n} f(x_{ijk}^*, y_{ijk}^*, z_{ijk}^*)\, \Delta V$$

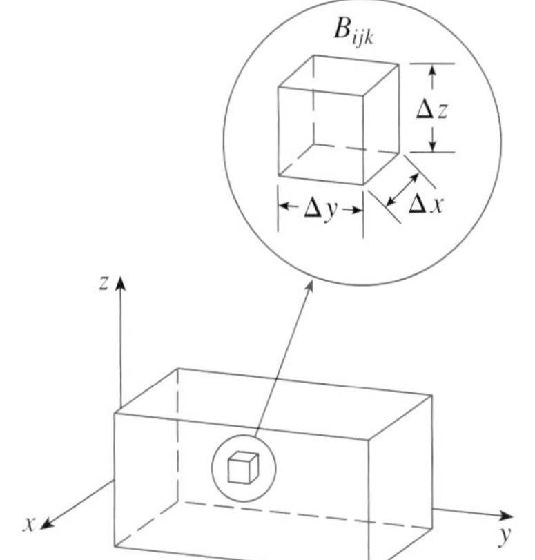

FIGURE 1

where the sample point $(x_{ijk}^*, y_{ijk}^*, z_{ijk}^*)$ is in B_{ijk}. By analogy with the definition of a double integral (15.1.5), we define the triple integral as the limit of the triple Riemann sums in **2**.

3 Definition The **triple integral** of f over the box B is

$$\iiint_B f(x, y, z)\, dV = \lim_{l, m, n \to \infty} \sum_{i=1}^{l} \sum_{j=1}^{m} \sum_{k=1}^{n} f(x_{ijk}^*, y_{ijk}^*, z_{ijk}^*)\, \Delta V$$

if this limit exists.

Again, the triple integral always exists if f is continuous. We can choose the sample point to be any point in the sub-box, but if we choose it to be the point (x_i, y_j, z_k) we get a simpler-looking expression for the triple integral:

$$\iiint_B f(x, y, z)\, dV = \lim_{l, m, n \to \infty} \sum_{i=1}^{l} \sum_{j=1}^{m} \sum_{k=1}^{n} f(x_i, y_j, z_k)\, \Delta V$$

Just as for double integrals, the practical method for evaluating triple integrals is to express them as iterated integrals as follows.

4 Fubini's Theorem for Triple Integrals If f is continuous on the rectangular box $B = [a, b] \times [c, d] \times [r, s]$, then

$$\iiint_B f(x, y, z)\, dV = \int_r^s \int_c^d \int_a^b f(x, y, z)\, dx\, dy\, dz$$

The iterated integral on the right side of Fubini's Theorem means that we integrate first with respect to x (keeping y and z fixed), then we integrate with respect to y (keeping z fixed), and finally we integrate with respect to z. There are five other possible orders in which we can integrate, all of which give the same value. For instance, if we integrate with respect to y, then z, and then x, we have

$$\iiint_B f(x, y, z)\, dV = \int_a^b \int_r^s \int_c^d f(x, y, z)\, dy\, dz\, dx$$

V EXAMPLE 1 Evaluate the triple integral $\iiint_B xyz^2\, dV$, where B is the rectangular box given by

$$B = \{(x, y, z) \mid 0 \leqslant x \leqslant 1,\ -1 \leqslant y \leqslant 2,\ 0 \leqslant z \leqslant 3\}$$

SOLUTION We could use any of the six possible orders of integration. If we choose to integrate with respect to x, then y, and then z, we obtain

$$\iiint_B xyz^2\, dV = \int_0^3 \int_{-1}^2 \int_0^1 xyz^2\, dx\, dy\, dz = \int_0^3 \int_{-1}^2 \left[\frac{x^2yz^2}{2}\right]_{x=0}^{x=1} dy\, dz$$

$$= \int_0^3 \int_{-1}^2 \frac{yz^2}{2}\, dy\, dz = \int_0^3 \left[\frac{y^2z^2}{4}\right]_{y=-1}^{y=2} dz$$

$$= \int_0^3 \frac{3z^2}{4}\, dz = \frac{z^3}{4}\bigg]_0^3 = \frac{27}{4}$$

Now we define the **triple integral over a general bounded region E** in three-dimensional space (a solid) by much the same procedure that we used for double integrals (15.3.2). We enclose E in a box B of the type given by Equation 1. Then we define F so that it agrees with f on E but is 0 for points in B that are outside E. By definition,

$$\iiint_E f(x, y, z)\, dV = \iiint_B F(x, y, z)\, dV$$

This integral exists if f is continuous and the boundary of E is "reasonably smooth." The triple integral has essentially the same properties as the double integral (Properties 6–9 in Section 15.3).

We restrict our attention to continuous functions f and to certain simple types of regions. A solid region E is said to be of **type 1** if it lies between the graphs of two continuous functions of x and y, that is,

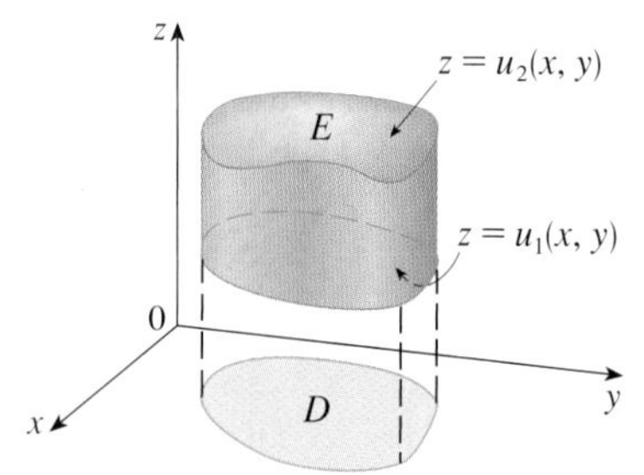

FIGURE 2
A type 1 solid region

5 $$E = \{(x, y, z) \mid (x, y) \in D,\ u_1(x, y) \leqslant z \leqslant u_2(x, y)\}$$

where D is the projection of E onto the xy-plane as shown in Figure 2. Notice that the upper boundary of the solid E is the surface with equation $z = u_2(x, y)$, while the lower boundary is the surface $z = u_1(x, y)$.

By the same sort of argument that led to (15.3.3), it can be shown that if E is a type 1 region given by Equation 5, then

6 $$\iiint_E f(x, y, z)\, dV = \iint_D \left[\int_{u_1(x, y)}^{u_2(x, y)} f(x, y, z)\, dz\right] dA$$

The meaning of the inner integral on the right side of Equation 6 is that x and y are held fixed, and therefore $u_1(x, y)$ and $u_2(x, y)$ are regarded as constants, while $f(x, y, z)$ is integrated with respect to z.

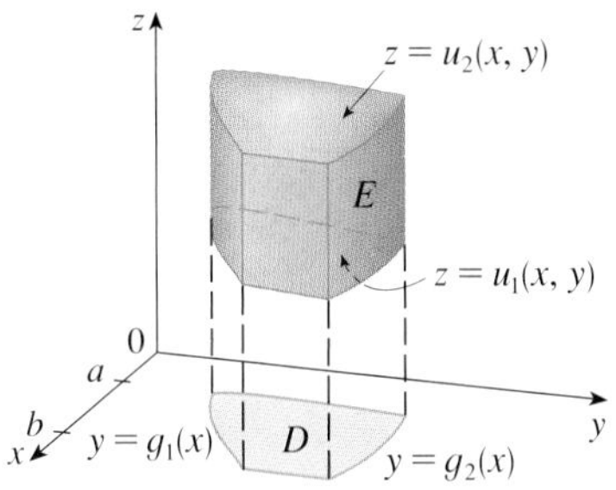

FIGURE 3
A type 1 solid region where the projection D is a type I plane region

In particular, if the projection D of E onto the xy-plane is a type I plane region (as in Figure 3), then

$$E = \{(x, y, z) \mid a \leqslant x \leqslant b,\ g_1(x) \leqslant y \leqslant g_2(x),\ u_1(x, y) \leqslant z \leqslant u_2(x, y)\}$$

and Equation 6 becomes

7
$$\iiint_E f(x, y, z)\, dV = \int_a^b \int_{g_1(x)}^{g_2(x)} \int_{u_1(x, y)}^{u_2(x, y)} f(x, y, z)\, dz\, dy\, dx$$

If, on the other hand, D is a type II plane region (as in Figure 4), then

$$E = \{(x, y, z) \mid c \leqslant y \leqslant d,\ h_1(y) \leqslant x \leqslant h_2(y),\ u_1(x, y) \leqslant z \leqslant u_2(x, y)\}$$

and Equation 6 becomes

8
$$\iiint_E f(x, y, z)\, dV = \int_c^d \int_{h_1(y)}^{h_2(y)} \int_{u_1(x, y)}^{u_2(x, y)} f(x, y, z)\, dz\, dx\, dy$$

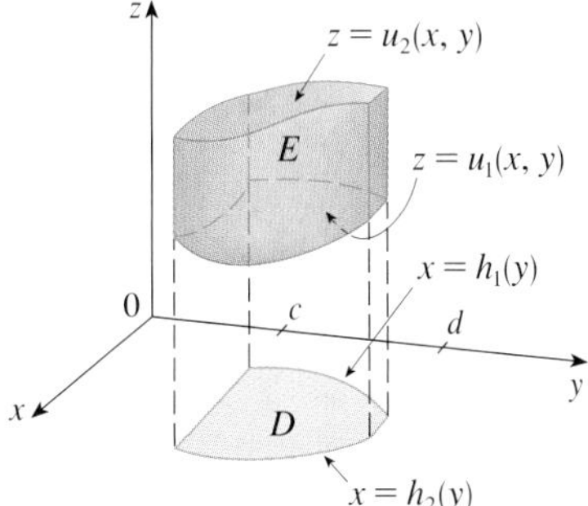

FIGURE 4
A type 1 solid region with a type II projection

EXAMPLE 2 Evaluate $\iiint_E z\, dV$, where E is the solid tetrahedron bounded by the four planes $x = 0$, $y = 0$, $z = 0$, and $x + y + z = 1$.

SOLUTION When we set up a triple integral it's wise to draw *two* diagrams: one of the solid region E (see Figure 5) and one of its projection D onto the xy-plane (see Figure 6). The lower boundary of the tetrahedron is the plane $z = 0$ and the upper boundary is the plane $x + y + z = 1$ (or $z = 1 - x - y$), so we use $u_1(x, y) = 0$ and $u_2(x, y) = 1 - x - y$ in Formula 7. Notice that the planes $x + y + z = 1$ and $z = 0$ intersect in the line $x + y = 1$ (or $y = 1 - x$) in the xy-plane. So the projection of E is the triangular region shown in Figure 6, and we have

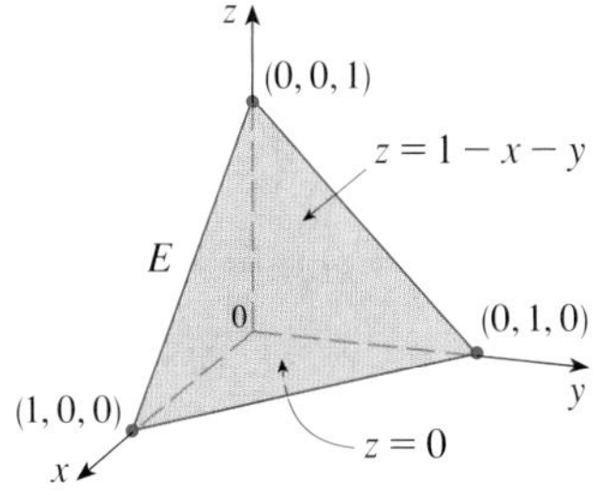

FIGURE 5

9
$$E = \{(x, y, z) \mid 0 \leqslant x \leqslant 1,\ 0 \leqslant y \leqslant 1 - x,\ 0 \leqslant z \leqslant 1 - x - y\}$$

This description of E as a type 1 region enables us to evaluate the integral as follows:

$$\iiint_E z\, dV = \int_0^1 \int_0^{1-x} \int_0^{1-x-y} z\, dz\, dy\, dx = \int_0^1 \int_0^{1-x} \left[\frac{z^2}{2}\right]_{z=0}^{z=1-x-y} dy\, dx$$

$$= \tfrac{1}{2} \int_0^1 \int_0^{1-x} (1 - x - y)^2\, dy\, dx = \tfrac{1}{2} \int_0^1 \left[-\frac{(1 - x - y)^3}{3}\right]_{y=0}^{y=1-x} dx$$

$$= \tfrac{1}{6} \int_0^1 (1 - x)^3\, dx = \frac{1}{6}\left[-\frac{(1 - x)^4}{4}\right]_0^1 = \frac{1}{24}$$

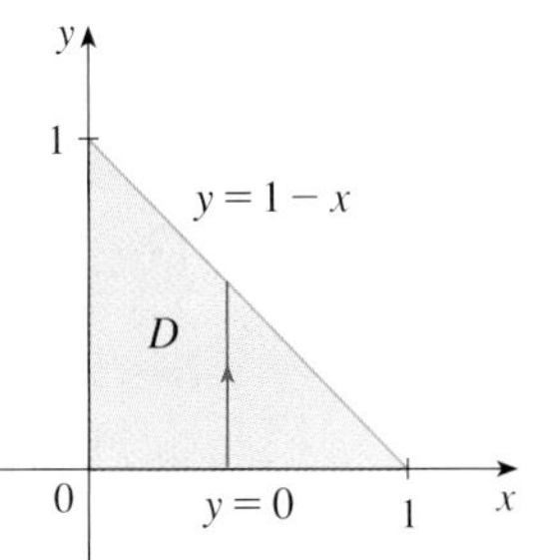

FIGURE 6

A solid region E is of **type 2** if it is of the form

$$E = \{(x, y, z) \mid (y, z) \in D,\ u_1(y, z) \leqslant x \leqslant u_2(y, z)\}$$

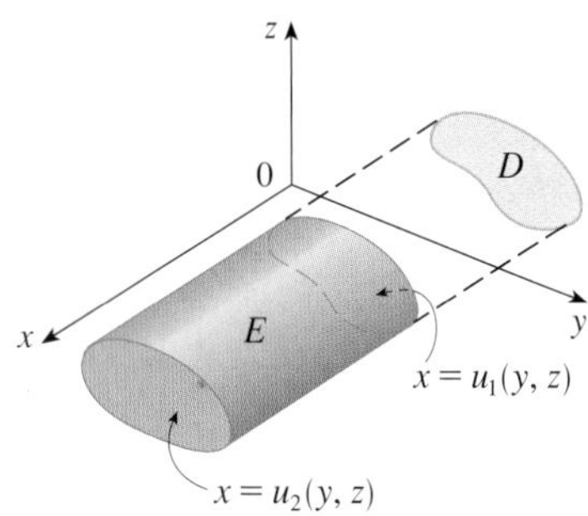

FIGURE 7
A type 2 region

where, this time, D is the projection of E onto the yz-plane (see Figure 7). The back surface is $x = u_1(y, z)$, the front surface is $x = u_2(y, z)$, and we have

$$\boxed{10} \qquad \iiint_E f(x, y, z)\, dV = \iint_D \left[\int_{u_1(y, z)}^{u_2(y, z)} f(x, y, z)\, dx \right] dA$$

Finally, a **type 3** region is of the form

$$E = \{(x, y, z) \mid (x, z) \in D,\ u_1(x, z) \leqslant y \leqslant u_2(x, z)\}$$

where D is the projection of E onto the xz-plane, $y = u_1(x, z)$ is the left surface, and $y = u_2(x, z)$ is the right surface (see Figure 8). For this type of region we have

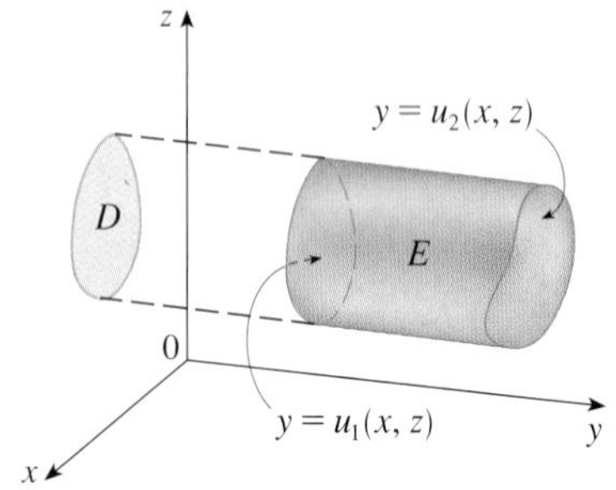

FIGURE 8
A type 3 region

$$\boxed{11} \qquad \iiint_E f(x, y, z)\, dV = \iint_D \left[\int_{u_1(x, z)}^{u_2(x, z)} f(x, y, z)\, dy \right] dA$$

In each of Equations 10 and 11 there may be two possible expressions for the integral depending on whether D is a type I or type II plane region (and corresponding to Equations 7 and 8).

V EXAMPLE 3 Evaluate $\iiint_E \sqrt{x^2 + z^2}\, dV$, where E is the region bounded by the paraboloid $y = x^2 + z^2$ and the plane $y = 4$.

SOLUTION The solid E is shown in Figure 9. If we regard it as a type 1 region, then we need to consider its projection D_1 onto the xy-plane, which is the parabolic region in Figure 10. (The trace of $y = x^2 + z^2$ in the plane $z = 0$ is the parabola $y = x^2$.)

TEC Visual 15.7 illustrates how solid regions (including the one in Figure 9) project onto coordinate planes.

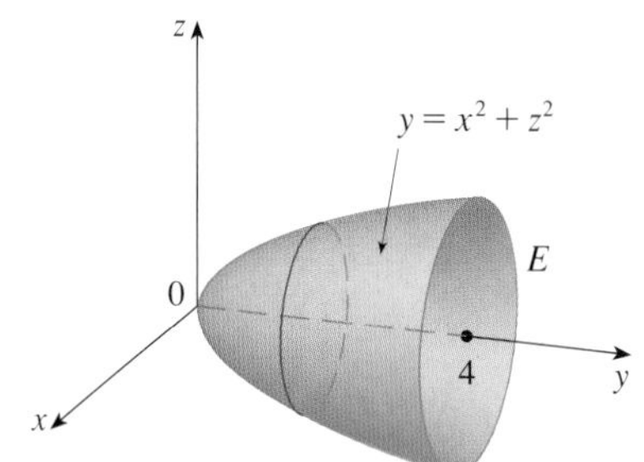

FIGURE 9
Region of integration

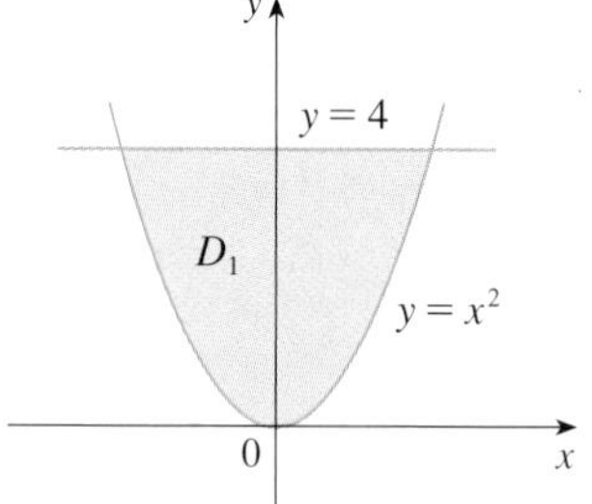

FIGURE 10
Projection onto xy-plane

From $y = x^2 + z^2$ we obtain $z = \pm\sqrt{y - x^2}$, so the lower boundary surface of E is $z = -\sqrt{y - x^2}$ and the upper surface is $z = \sqrt{y - x^2}$. Therefore the description of E as a type 1 region is

$$E = \left\{(x, y, z) \mid -2 \leqslant x \leqslant 2,\ x^2 \leqslant y \leqslant 4,\ -\sqrt{y - x^2} \leqslant z \leqslant \sqrt{y - x^2}\right\}$$

and so we obtain

$$\iiint_E \sqrt{x^2 + z^2}\, dV = \int_{-2}^{2} \int_{x^2}^{4} \int_{-\sqrt{y-x^2}}^{\sqrt{y-x^2}} \sqrt{x^2 + z^2}\, dz\, dy\, dx$$

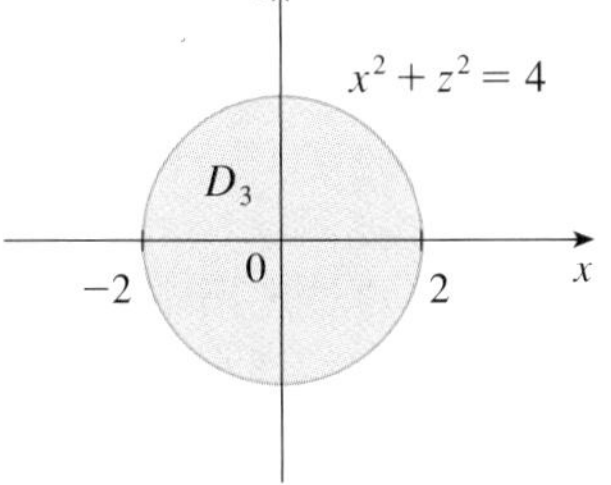

FIGURE 11
Projection onto xz-plane

Although this expression is correct, it is extremely difficult to evaluate. So let's instead consider E as a type 3 region. As such, its projection D_3 onto the xz-plane is the disk $x^2 + z^2 \leqslant 4$ shown in Figure 11.

Then the left boundary of E is the paraboloid $y = x^2 + z^2$ and the right boundary is the plane $y = 4$, so taking $u_1(x, z) = x^2 + z^2$ and $u_2(x, z) = 4$ in Equation 11, we have

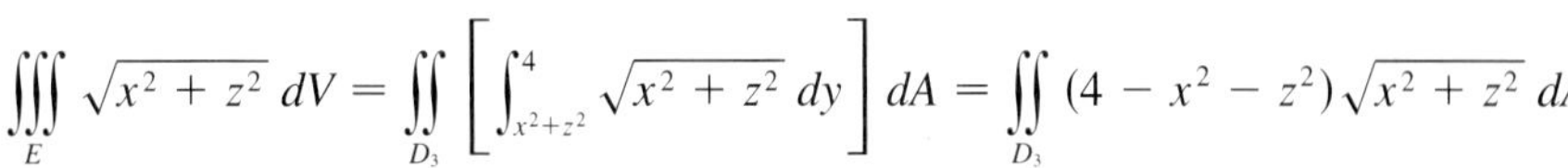

$$\iiint_E \sqrt{x^2 + z^2}\, dV = \iint_{D_3} \left[\int_{x^2+z^2}^{4} \sqrt{x^2 + z^2}\, dy \right] dA = \iint_{D_3} (4 - x^2 - z^2)\sqrt{x^2 + z^2}\, dA$$

Although this integral could be written as

$$\int_{-2}^{2} \int_{-\sqrt{4-x^2}}^{\sqrt{4-x^2}} (4 - x^2 - z^2)\sqrt{x^2 + z^2}\, dz\, dx$$

it's easier to convert to polar coordinates in the xz-plane: $x = r\cos\theta$, $z = r\sin\theta$. This gives

$$\begin{aligned}\iiint_E \sqrt{x^2 + z^2}\, dV &= \iint_{D_3} (4 - x^2 - z^2)\sqrt{x^2 + z^2}\, dA \\ &= \int_0^{2\pi} \int_0^2 (4 - r^2) r\, r\, dr\, d\theta = \int_0^{2\pi} d\theta \int_0^2 (4r^2 - r^4)\, dr \\ &= 2\pi \left[\frac{4r^3}{3} - \frac{r^5}{5} \right]_0^2 = \frac{128\pi}{15}\end{aligned}$$

⊘ The most difficult step in evaluating a triple integral is setting up an expression for the region of integration (such as Equation 9 in Example 2). Remember that the limits of integration in the inner integral contain at most two variables, the limits of integration in the middle integral contain at most one variable, and the limits of integration in the outer integral must be constants.

EXAMPLE 4 Express the iterated integral $\int_0^1 \int_0^{x^2} \int_0^y f(x, y, z)\, dz\, dy\, dx$ as a triple integral and then rewrite it as an iterated integral in a different order, integrating first with respect to x, then z, and then y.

SOLUTION We can write

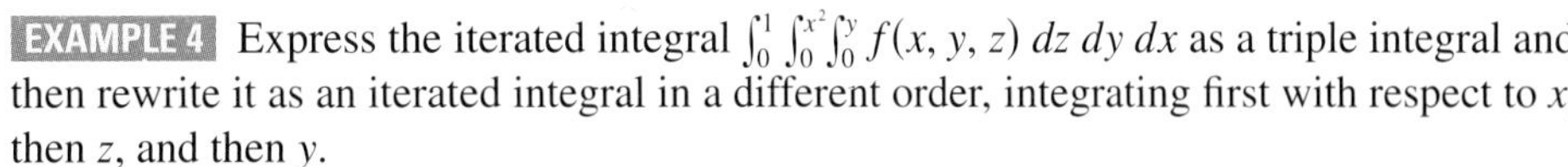

$$\int_0^1 \int_0^{x^2} \int_0^y f(x, y, z)\, dz\, dy\, dx = \iiint_E f(x, y, z)\, dV$$

where $E = \{(x, y, z) \mid 0 \leqslant x \leqslant 1, 0 \leqslant y \leqslant x^2, 0 \leqslant z \leqslant y\}$. This description of E enables us to write projections onto the three coordinate planes as follows:

on the xy-plane: $D_1 = \{(x, y) \mid 0 \leqslant x \leqslant 1, 0 \leqslant y \leqslant x^2\}$
$= \left\{(x, y) \mid 0 \leqslant y \leqslant 1, \sqrt{y} \leqslant x \leqslant 1\right\}$

on the yz-plane: $D_2 = \{(x, y) \mid 0 \leqslant y \leqslant 1, 0 \leqslant z \leqslant y\}$

on the xz-plane: $D_3 = \{(x, y) \mid 0 \leqslant x \leqslant 1, 0 \leqslant z \leqslant x^2\}$

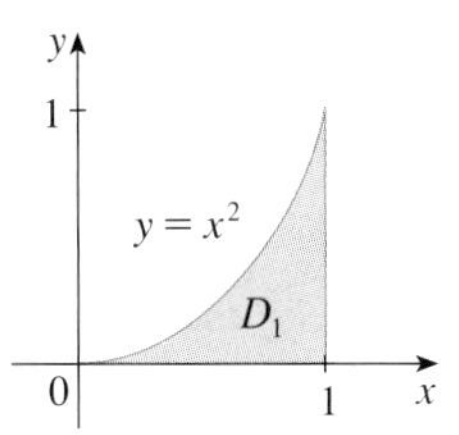

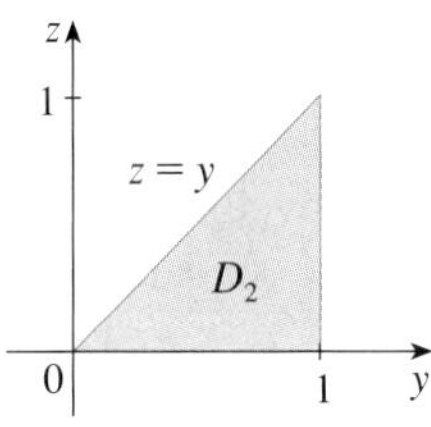

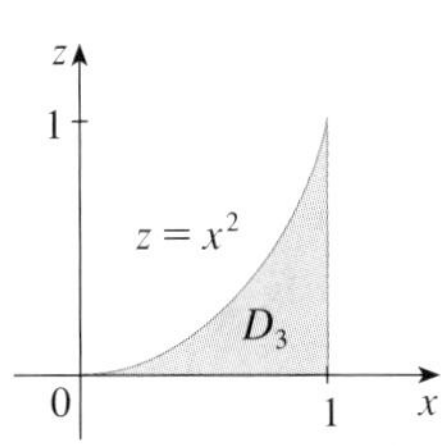

FIGURE 12
Projections of E

From the resulting sketches of the projections in Figure 12 we sketch the solid E in Figure 13. We see that it is the solid enclosed by the planes $z = 0$, $x = 1$, $y = z$ and the parabolic cylinder $y = x^2$ $\left(\text{or } x = \sqrt{y}\right)$.

If we integrate first with respect to x, then z, and then y, we use an alternate description of E:

$$E = \left\{(x, y, z) \mid 0 \leqslant x \leqslant 1, 0 \leqslant z \leqslant y, \sqrt{y} \leqslant x \leqslant 1\right\}$$

Thus

$$\iiint_E f(x, y, z)\, dV = \int_0^1 \int_0^y \int_{\sqrt{y}}^1 f(x, y, z)\, dx\, dz\, dy$$

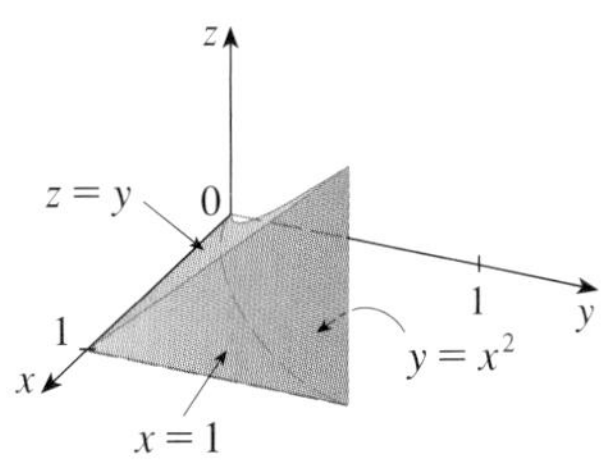

FIGURE 13
The solid E

Applications of Triple Integrals

Recall that if $f(x) \geqslant 0$, then the single integral $\int_a^b f(x)\,dx$ represents the area under the curve $y = f(x)$ from a to b, and if $f(x, y) \geqslant 0$, then the double integral $\iint_D f(x, y)\,dA$ represents the volume under the surface $z = f(x, y)$ and above D. The corresponding interpretation of a triple integral $\iiint_E f(x, y, z)\,dV$, where $f(x, y, z) \geqslant 0$, is not very useful because it would be the "hypervolume" of a four-dimensional object and, of course, that is very difficult to visualize. (Remember that E is just the *domain* of the function f; the graph of f lies in four-dimensional space.) Nonetheless, the triple integral $\iiint_E f(x, y, z)\,dV$ can be interpreted in different ways in different physical situations, depending on the physical interpretations of x, y, z, and $f(x, y, z)$.

Let's begin with the special case where $f(x, y, z) = 1$ for all points in E. Then the triple integral does represent the volume of E:

12
$$V(E) = \iiint_E dV$$

For example, you can see this in the case of a type 1 region by putting $f(x, y, z) = 1$ in Formula 6:

$$\iiint_E 1\,dV = \iint_D \left[\int_{u_1(x,y)}^{u_2(x,y)} dz \right] dA = \iint_D [u_2(x, y) - u_1(x, y)]\,dA$$

and from Section 15.3 we know this represents the volume that lies between the surfaces $z = u_1(x, y)$ and $z = u_2(x, y)$.

EXAMPLE 5 Use a triple integral to find the volume of the tetrahedron T bounded by the planes $x + 2y + z = 2$, $x = 2y$, $x = 0$, and $z = 0$.

SOLUTION The tetrahedron T and its projection D onto the xy-plane are shown in Figures 14 and 15. The lower boundary of T is the plane $z = 0$ and the upper boundary is the plane $x + 2y + z = 2$, that is, $z = 2 - x - 2y$.

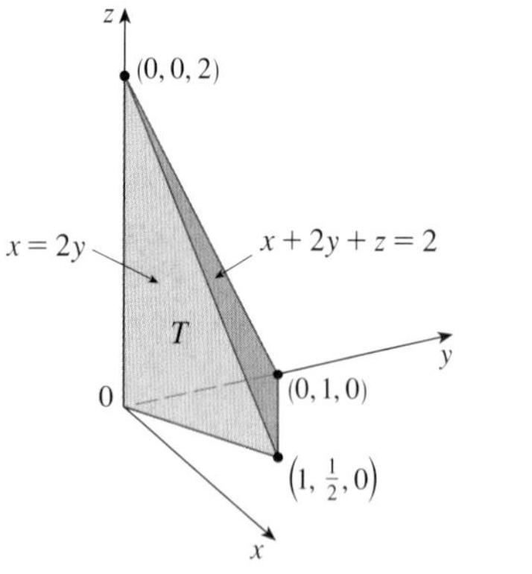

FIGURE 14

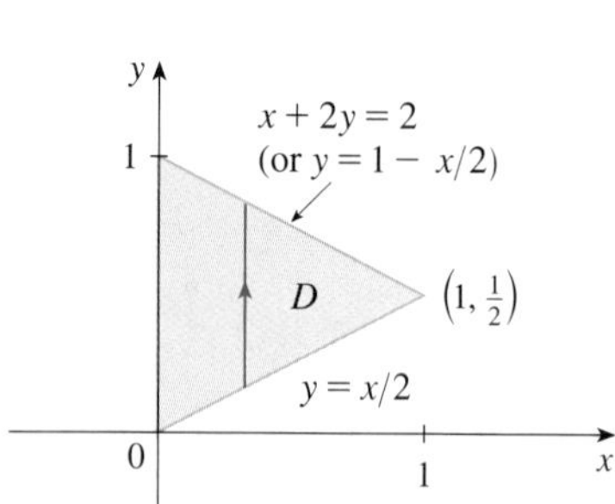

FIGURE 15

Therefore we have

$$V(T) = \iiint_T dV = \int_0^1 \int_{x/2}^{1-x/2} \int_0^{2-x-2y} dz\,dy\,dx$$

$$= \int_0^1 \int_{x/2}^{1-x/2} (2 - x - 2y)\,dy\,dx = \tfrac{1}{3}$$

by the same calculation as in Example 4 in Section 15.3.

(Notice that it is not necessary to use triple integrals to compute volumes. They simply give an alternative method for setting up the calculation.)

All the applications of double integrals in Section 15.5 can be immediately extended to triple integrals. For example, if the density function of a solid object that occupies the region E is $\rho(x, y, z)$, in units of mass per unit volume, at any given point (x, y, z), then its **mass** is

13 $$m = \iiint_E \rho(x, y, z)\, dV$$

and its **moments** about the three coordinate planes are

14 $$M_{yz} = \iiint_E x\,\rho(x, y, z)\, dV \qquad M_{xz} = \iiint_E y\,\rho(x, y, z)\, dV$$

$$M_{xy} = \iiint_E z\,\rho(x, y, z)\, dV$$

The **center of mass** is located at the point $(\bar{x}, \bar{y}, \bar{z})$, where

15 $$\bar{x} = \frac{M_{yz}}{m} \qquad \bar{y} = \frac{M_{xz}}{m} \qquad \bar{z} = \frac{M_{xy}}{m}$$

If the density is constant, the center of mass of the solid is called the **centroid** of E. The **moments of inertia** about the three coordinate axes are

16 $$I_x = \iiint_E (y^2 + z^2)\,\rho(x, y, z)\, dV \qquad I_y = \iiint_E (x^2 + z^2)\,\rho(x, y, z)\, dV$$

$$I_z = \iiint_E (x^2 + y^2)\,\rho(x, y, z)\, dV$$

As in Section 15.5, the total **electric charge** on a solid object occupying a region E and having charge density $\sigma(x, y, z)$ is

$$Q = \iiint_E \sigma(x, y, z)\, dV$$

If we have three continuous random variables X, Y, and Z, their **joint density function** is a function of three variables such that the probability that (X, Y, Z) lies in E is

$$P\big((X, Y, Z) \in E\big) = \iiint_E f(x, y, z)\, dV$$

In particular,

$$P(a \leqslant X \leqslant b,\ c \leqslant Y \leqslant d,\ r \leqslant Z \leqslant s) = \int_a^b \int_c^d \int_r^s f(x, y, z)\, dz\, dy\, dx$$

The joint density function satisfies

$$f(x, y, z) \geqslant 0 \qquad \int_{-\infty}^{\infty} \int_{-\infty}^{\infty} \int_{-\infty}^{\infty} f(x, y, z)\, dz\, dy\, dx = 1$$

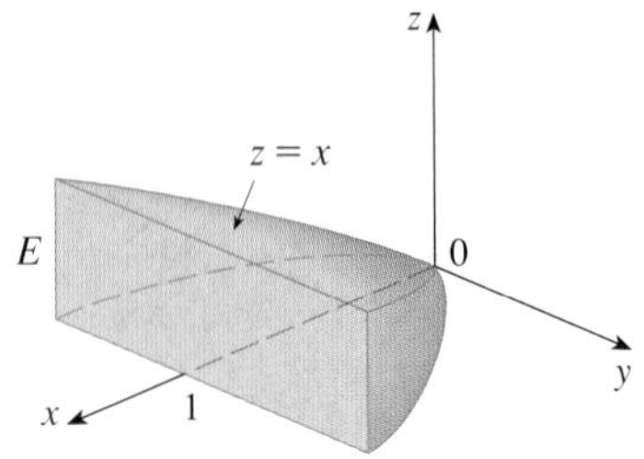

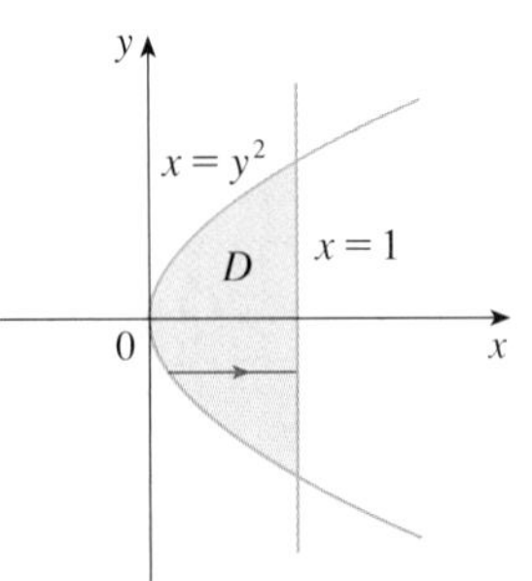

FIGURE 16

V EXAMPLE 6 Find the center of mass of a solid of constant density that is bounded by the parabolic cylinder $x = y^2$ and the planes $x = z$, $z = 0$, and $x = 1$.

SOLUTION The solid E and its projection onto the xy-plane are shown in Figure 16. The lower and upper surfaces of E are the planes $z = 0$ and $z = x$, so we describe E as a type 1 region:

$$E = \{(x, y, z) \mid -1 \leqslant y \leqslant 1,\ y^2 \leqslant x \leqslant 1,\ 0 \leqslant z \leqslant x\}$$

Then, if the density is $\rho(x, y, z) = \rho$, the mass is

$$m = \iiint_E \rho\, dV = \int_{-1}^{1} \int_{y^2}^{1} \int_{0}^{x} \rho\, dz\, dx\, dy$$

$$= \rho \int_{-1}^{1} \int_{y^2}^{1} x\, dx\, dy = \rho \int_{-1}^{1} \left[\frac{x^2}{2} \right]_{x=y^2}^{x=1} dy$$

$$= \frac{\rho}{2} \int_{-1}^{1} (1 - y^4)\, dy = \rho \int_{0}^{1} (1 - y^4)\, dy$$

$$= \rho \left[y - \frac{y^5}{5} \right]_0^1 = \frac{4\rho}{5}$$

Because of the symmetry of E and ρ about the xz-plane, we can immediately say that $M_{xz} = 0$ and therefore $\bar{y} = 0$. The other moments are

$$M_{yz} = \iiint_E x\rho\, dV = \int_{-1}^{1} \int_{y^2}^{1} \int_{0}^{x} x\rho\, dz\, dx\, dy$$

$$= \rho \int_{-1}^{1} \int_{y^2}^{1} x^2\, dx\, dy = \rho \int_{-1}^{1} \left[\frac{x^3}{3} \right]_{x=y^2}^{x=1} dy$$

$$= \frac{2\rho}{3} \int_{0}^{1} (1 - y^6)\, dy = \frac{2\rho}{3} \left[y - \frac{y^7}{7} \right]_0^1 = \frac{4\rho}{7}$$

$$M_{xy} = \iiint_E z\rho\, dV = \int_{-1}^{1} \int_{y^2}^{1} \int_{0}^{x} z\rho\, dz\, dx\, dy$$

$$= \rho \int_{-1}^{1} \int_{y^2}^{1} \left[\frac{z^2}{2} \right]_{z=0}^{z=x} dx\, dy = \frac{\rho}{2} \int_{-1}^{1} \int_{y^2}^{1} x^2\, dx\, dy$$

$$= \frac{\rho}{3} \int_{0}^{1} (1 - y^6)\, dy = \frac{2\rho}{7}$$

Therefore the center of mass is

$$(\bar{x}, \bar{y}, \bar{z}) = \left(\frac{M_{yz}}{m}, \frac{M_{xz}}{m}, \frac{M_{xy}}{m} \right) = \left(\tfrac{5}{7}, 0, \tfrac{5}{14} \right)$$

15.7 Exercises

1. Evaluate the integral in Example 1, integrating first with respect to y, then z, and then x.

2. Evaluate the integral $\iiint_E (xy + z^2)\, dV$, where

$$E = \{(x, y, z) \mid 0 \leqslant x \leqslant 2, 0 \leqslant y \leqslant 1, 0 \leqslant z \leqslant 3\}$$

using three different orders of integration.

3–8 Evaluate the iterated integral.

3. $\int_0^2 \int_0^{z^2} \int_0^{y-z} (2x - y)\, dx\, dy\, dz$

4. $\int_0^1 \int_x^{2x} \int_0^y 2xyz\, dz\, dy\, dx$

5. $\int_1^2 \int_0^{2z} \int_0^{\ln x} xe^{-y}\, dy\, dx\, dz$

6. $\int_0^1 \int_0^1 \int_0^{\sqrt{1-z^2}} \dfrac{z}{y+1}\, dx\, dz\, dy$

7. $\int_0^{\pi/2} \int_0^y \int_0^x \cos(x + y + z)\, dz\, dx\, dy$

8. $\int_0^{\sqrt{\pi}} \int_0^x \int_0^{xz} x^2 \sin y\, dy\, dz\, dx$

9–18 Evaluate the triple integral.

9. $\iiint_E y\, dV$, where
$$E = \{(x, y, z) \mid 0 \leqslant x \leqslant 3,\ 0 \leqslant y \leqslant x, x - y \leqslant z \leqslant x + y\}$$

10. $\iiint_E e^{z/y}\, dV$, where
$$E = \{(x, y, z) \mid 0 \leqslant y \leqslant 1, y \leqslant x \leqslant 1, 0 \leqslant z \leqslant xy\}$$

11. $\iiint_E \dfrac{z}{x^2 + z^2}\, dV$, where
$$E = \{(x, y, z) \mid 1 \leqslant y \leqslant 4, y \leqslant z \leqslant 4, 0 \leqslant x \leqslant z\}$$

12. $\iiint_E \sin y\, dV$, where E lies below the plane $z = x$ and above the triangular region with vertices $(0, 0, 0)$, $(\pi, 0, 0)$, and $(0, \pi, 0)$

13. $\iiint_E 6xy\, dV$, where E lies under the plane $z = 1 + x + y$ and above the region in the xy-plane bounded by the curves $y = \sqrt{x}$, $y = 0$, and $x = 1$

14. $\iiint_E xy\, dV$, where E is bounded by the parabolic cylinders $y = x^2$ and $x = y^2$ and the planes $z = 0$ and $z = x + y$

15. $\iiint_T x^2\, dV$, where T is the solid tetrahedron with vertices $(0, 0, 0)$, $(1, 0, 0)$, $(0, 1, 0)$, and $(0, 0, 1)$

16. $\iiint_T xyz\, dV$, where T is the solid tetrahedron with vertices $(0, 0, 0)$, $(1, 0, 0)$, $(1, 1, 0)$, and $(1, 0, 1)$

17. $\iiint_E x\, dV$, where E is bounded by the paraboloid $x = 4y^2 + 4z^2$ and the plane $x = 4$

18. $\iiint_E z\, dV$, where E is bounded by the cylinder $y^2 + z^2 = 9$ and the planes $x = 0$, $y = 3x$, and $z = 0$ in the first octant

19–22 Use a triple integral to find the volume of the given solid.

19. The tetrahedron enclosed by the coordinate planes and the plane $2x + y + z = 4$

20. The solid enclosed by the paraboloids $y = x^2 + z^2$ and $y = 8 - x^2 - z^2$

21. The solid enclosed by the cylinder $y = x^2$ and the planes $z = 0$ and $y + z = 1$

22. The solid enclosed by the cylinder $x^2 + z^2 = 4$ and the planes $y = -1$ and $y + z = 4$

23. (a) Express the volume of the wedge in the first octant that is cut from the cylinder $y^2 + z^2 = 1$ by the planes $y = x$ and $x = 1$ as a triple integral.
CAS (b) Use either the Table of Integrals (on Reference Pages 6–10) or a computer algebra system to find the exact value of the triple integral in part (a).

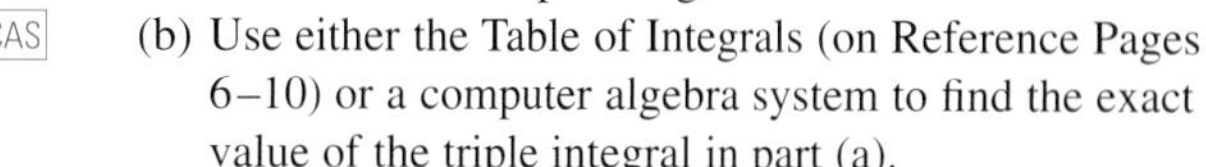

24. (a) In the **Midpoint Rule for triple integrals** we use a triple Riemann sum to approximate a triple integral over a box B, where $f(x, y, z)$ is evaluated at the center $(\bar{x}_i, \bar{y}_j, \bar{z}_k)$ of the box B_{ijk}. Use the Midpoint Rule to estimate $\iiint_B \sqrt{x^2 + y^2 + z^2}\, dV$, where B is the cube defined by $0 \leqslant x \leqslant 4,\ 0 \leqslant y \leqslant 4,\ 0 \leqslant z \leqslant 4$. Divide B into eight cubes of equal size.
CAS (b) Use a computer algebra system to approximate the integral in part (a) correct to the nearest integer. Compare with the answer to part (a).

25–26 Use the Midpoint Rule for triple integrals (Exercise 24) to estimate the value of the integral. Divide B into eight sub-boxes of equal size.

25. $\iiint_B \cos(xyz)\, dV$, where
$$B = \{(x, y, z) \mid 0 \leqslant x \leqslant 1,\ 0 \leqslant y \leqslant 1, 0 \leqslant z \leqslant 1\}$$

26. $\iiint_B \sqrt{x}\, e^{xyz}\, dV$, where
$$B = \{(x, y, z) \mid 0 \leqslant x \leqslant 4,\ 0 \leqslant y \leqslant 1, 0 \leqslant z \leqslant 2\}$$

27–28 Sketch the solid whose volume is given by the iterated integral.

27. $\int_0^1 \int_0^{1-x} \int_0^{2-2z} dy\, dz\, dx$

28. $\int_0^2 \int_0^{2-y} \int_0^{4-y^2} dx\, dz\, dy$

29–32 Express the integral $\iiint_E f(x, y, z)\, dV$ as an iterated integral in six different ways, where E is the solid bounded by the given surfaces.

29. $y = 4 - x^2 - 4z^2, \quad y = 0$

30. $y^2 + z^2 = 9, \quad x = -2, \quad x = 2$

31. $y = x^2, \quad z = 0, \quad y + 2z = 4$

32. $x = 2, \quad y = 2, \quad z = 0, \quad x + y - 2z = 2$

33. The figure shows the region of integration for the integral

$$\int_0^1 \int_{\sqrt{x}}^1 \int_0^{1-y} f(x, y, z)\, dz\, dy\, dx$$

Rewrite this integral as an equivalent iterated integral in the five other orders.

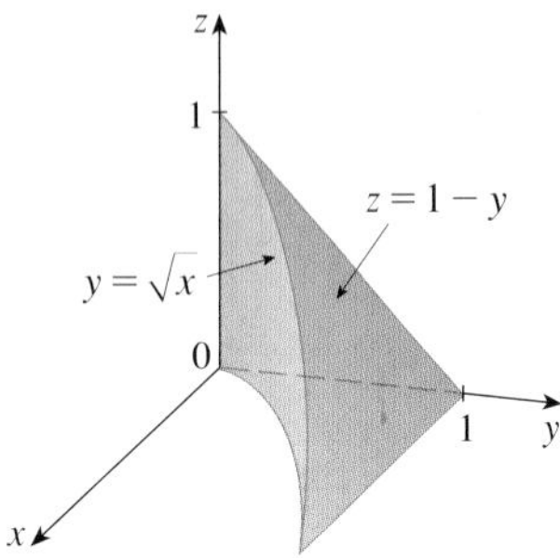

34. The figure shows the region of integration for the integral

$$\int_0^1 \int_0^{1-x^2} \int_0^{1-x} f(x, y, z)\, dy\, dz\, dx$$

Rewrite this integral as an equivalent iterated integral in the five other orders.

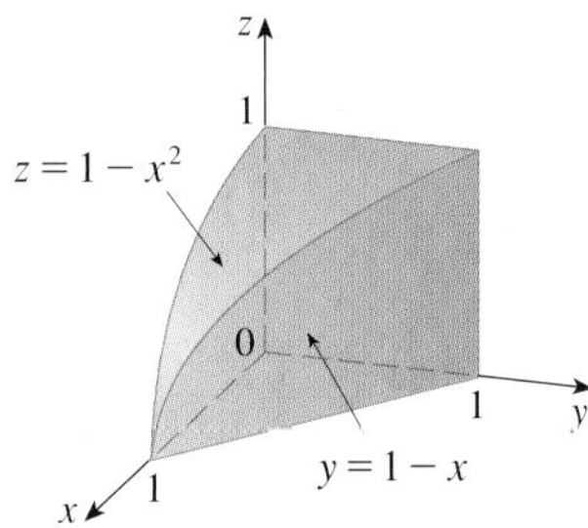

35–36 Write five other iterated integrals that are equal to the given iterated integral.

35. $\int_0^1 \int_y^1 \int_0^y f(x, y, z)\, dz\, dx\, dy$

36. $\int_0^1 \int_y^1 \int_0^z f(x, y, z)\, dx\, dz\, dy$

37–38 Evaluate the triple integral using only geometric interpretation and symmetry.

37. $\iiint_C (4 + 5x^2yz^2)\, dV$, where C is the cylindrical region $x^2 + y^2 \leqslant 4, -2 \leqslant z \leqslant 2$

38. $\iiint_B (z^3 + \sin y + 3)\, dV$, where B is the unit ball $x^2 + y^2 + z^2 \leqslant 1$

39–42 Find the mass and center of mass of the solid E with the given density function ρ.

39. E is the solid of Exercise 13; $\rho(x, y, z) = 2$

40. E is bounded by the parabolic cylinder $z = 1 - y^2$ and the planes $x + z = 1$, $x = 0$, and $z = 0$; $\rho(x, y, z) = 4$

41. E is the cube given by $0 \leqslant x \leqslant a,\ 0 \leqslant y \leqslant a,\ 0 \leqslant z \leqslant a$; $\rho(x, y, z) = x^2 + y^2 + z^2$

42. E is the tetrahedron bounded by the planes $x = 0$, $y = 0$, $z = 0$, $x + y + z = 1$; $\rho(x, y, z) = y$

43–46 Assume that the solid has constant density k.

43. Find the moments of inertia for a cube with side length L if one vertex is located at the origin and three edges lie along the coordinate axes.

44. Find the moments of inertia for a rectangular brick with dimensions a, b, and c and mass M if the center of the brick is situated at the origin and the edges are parallel to the coordinate axes.

45. Find the moment of inertia about the z-axis of the solid cylinder $x^2 + y^2 \leqslant a^2,\ 0 \leqslant z \leqslant h$.

46. Find the moment of inertia about the z-axis of the solid cone $\sqrt{x^2 + y^2} \leqslant z \leqslant h$.

47–48 Set up, but do not evaluate, integral expressions for (a) the mass, (b) the center of mass, and (c) the moment of inertia about the z-axis.

47. The solid of Exercise 21; $\rho(x, y, z) = \sqrt{x^2 + y^2}$

48. The hemisphere $x^2 + y^2 + z^2 \leqslant 1,\ z \geqslant 0$; $\rho(x, y, z) = \sqrt{x^2 + y^2 + z^2}$

CAS **49.** Let E be the solid in the first octant bounded by the cylinder $x^2 + y^2 = 1$ and the planes $y = z$, $x = 0$, and $z = 0$ with the density function $\rho(x, y, z) = 1 + x + y + z$. Use a computer algebra system to find the exact values of the following quantities for E.
(a) The mass
(b) The center of mass
(c) The moment of inertia about the z-axis

CAS **50.** If E is the solid of Exercise 18 with density function $\rho(x, y, z) = x^2 + y^2$, find the following quantities, correct to three decimal places.
(a) The mass
(b) The center of mass
(c) The moment of inertia about the z-axis

51. The joint density function for random variables X, Y, and Z is $f(x, y, z) = Cxyz$ if $0 \leqslant x \leqslant 2$, $0 \leqslant y \leqslant 2$, $0 \leqslant z \leqslant 2$, and $f(x, y, z) = 0$ otherwise.
(a) Find the value of the constant C.
(b) Find $P(X \leqslant 1, Y \leqslant 1, Z \leqslant 1)$.
(c) Find $P(X + Y + Z \leqslant 1)$.

52. Suppose X, Y, and Z are random variables with joint density function $f(x, y, z) = Ce^{-(0.5x+0.2y+0.1z)}$ if $x \geqslant 0$, $y \geqslant 0$, $z \geqslant 0$, and $f(x, y, z) = 0$ otherwise.
(a) Find the value of the constant C.
(b) Find $P(X \leqslant 1, Y \leqslant 1)$.
(c) Find $P(X \leqslant 1, Y \leqslant 1, Z \leqslant 1)$.

53–54 The **average value** of a function $f(x, y, z)$ over a solid region E is defined to be

$$f_{\text{ave}} = \frac{1}{V(E)} \iiint_E f(x, y, z)\, dV$$

where $V(E)$ is the volume of E. For instance, if ρ is a density function, then ρ_{ave} is the average density of E.

53. Find the average value of the function $f(x, y, z) = xyz$ over the cube with side length L that lies in the first octant with one vertex at the origin and edges parallel to the coordinate axes.

54. Find the average value of the function $f(x, y, z) = x^2z + y^2z$ over the region enclosed by the paraboloid $z = 1 - x^2 - y^2$ and the plane $z = 0$.

55. (a) Find the region E for which the triple integral

$$\iiint_E (1 - x^2 - 2y^2 - 3z^2)\, dV$$

is a maximum.

CAS (b) Use a computer algebra system to calculate the exact maximum value of the triple integral in part (a).

DISCOVERY PROJECT VOLUMES OF HYPERSPHERES

In this project we find formulas for the volume enclosed by a hypersphere in n-dimensional space.

1. Use a double integral and trigonometric substitution, together with Formula 64 in the Table of Integrals, to find the area of a circle with radius r.
2. Use a triple integral and trigonometric substitution to find the volume of a sphere with radius r.
3. Use a quadruple integral to find the hypervolume enclosed by the hypersphere $x^2 + y^2 + z^2 + w^2 = r^2$ in $\mathbb{R}^4$. (Use only trigonometric substitution and the reduction formulas for $\int \sin^n x\, dx$ or $\int \cos^n x\, dx$.)
4. Use an n-tuple integral to find the volume enclosed by a hypersphere of radius r in n-dimensional space $\mathbb{R}^n$. [*Hint:* The formulas are different for n even and n odd.]

15.8 Triple Integrals in Cylindrical Coordinates

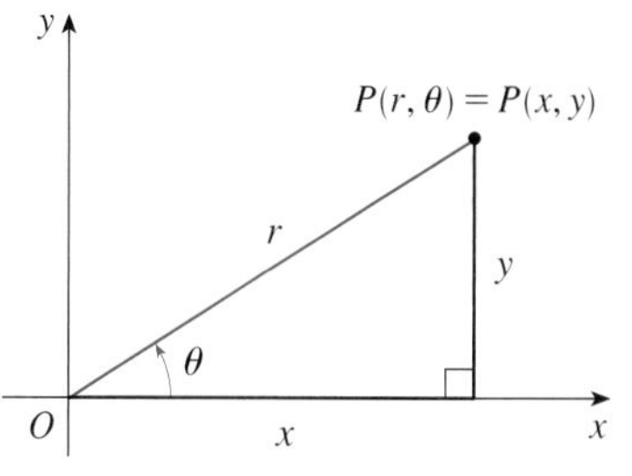

FIGURE 1

In plane geometry the polar coordinate system is used to give a convenient description of certain curves and regions. (See Section 10.3.) Figure 1 enables us to recall the connection between polar and Cartesian coordinates. If the point P has Cartesian coordinates (x, y) and polar coordinates (r, θ), then, from the figure,

$$x = r\cos\theta \qquad y = r\sin\theta$$

$$r^2 = x^2 + y^2 \qquad \tan\theta = \frac{y}{x}$$

In three dimensions there is a coordinate system, called *cylindrical coordinates*, that is similar to polar coordinates and gives convenient descriptions of some commonly occurring surfaces and solids. As we will see, some triple integrals are much easier to evaluate in cylindrical coordinates.

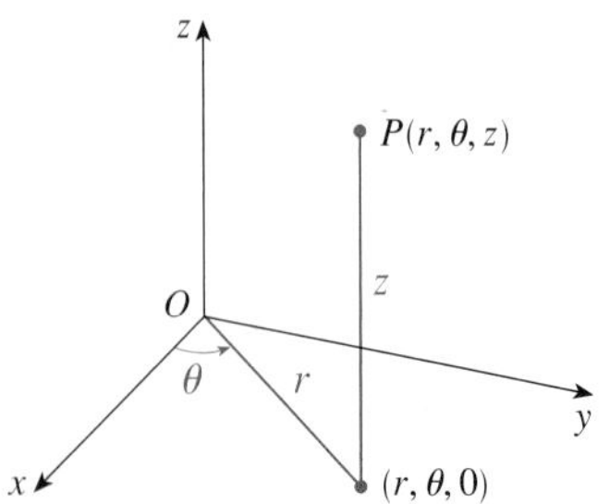

FIGURE 2
The cylindrical coordinates of a point

Cylindrical Coordinates

In the **cylindrical coordinate system**, a point P in three-dimensional space is represented by the ordered triple (r, θ, z), where r and θ are polar coordinates of the projection of P onto the xy-plane and z is the directed distance from the xy-plane to P. (See Figure 2.)

To convert from cylindrical to rectangular coordinates, we use the equations

1 $$x = r\cos\theta \qquad y = r\sin\theta \qquad z = z$$

whereas to convert from rectangular to cylindrical coordinates, we use

2 $$r^2 = x^2 + y^2 \qquad \tan\theta = \frac{y}{x} \qquad z = z$$

EXAMPLE 1

(a) Plot the point with cylindrical coordinates $(2, 2\pi/3, 1)$ and find its rectangular coordinates.

(b) Find cylindrical coordinates of the point with rectangular coordinates $(3, -3, -7)$.

SOLUTION

(a) The point with cylindrical coordinates $(2, 2\pi/3, 1)$ is plotted in Figure 3. From Equations 1, its rectangular coordinates are

$$x = 2\cos\frac{2\pi}{3} = 2\left(-\frac{1}{2}\right) = -1$$

$$y = 2\sin\frac{2\pi}{3} = 2\left(\frac{\sqrt{3}}{2}\right) = \sqrt{3}$$

$$z = 1$$

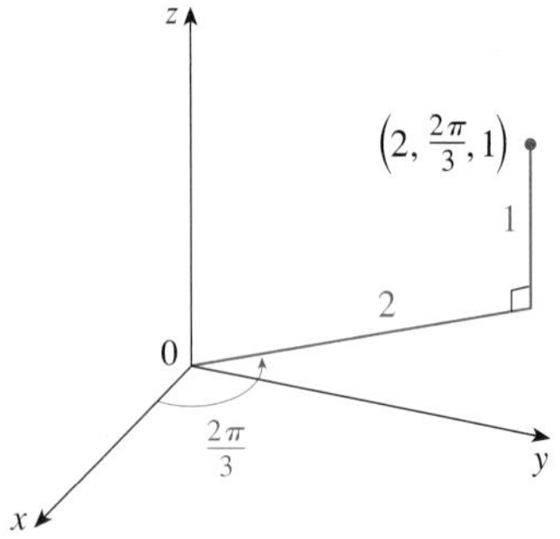

FIGURE 3

Thus the point is $(-1, \sqrt{3}, 1)$ in rectangular coordinates.

(b) From Equations 2 we have

$$r = \sqrt{3^2 + (-3)^2} = 3\sqrt{2}$$

$$\tan\theta = \frac{-3}{3} = -1 \qquad \text{so} \qquad \theta = \frac{7\pi}{4} + 2n\pi$$

$$z = -7$$

Therefore one set of cylindrical coordinates is $(3\sqrt{2}, 7\pi/4, -7)$. Another is $(3\sqrt{2}, -\pi/4, -7)$. As with polar coordinates, there are infinitely many choices.

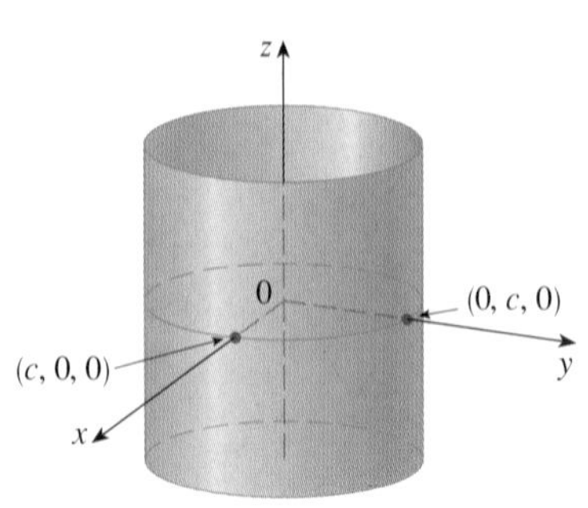

FIGURE 4
$r = c$, a cylinder

Cylindrical coordinates are useful in problems that involve symmetry about an axis, and the z-axis is chosen to coincide with this axis of symmetry. For instance, the axis of the circular cylinder with Cartesian equation $x^2 + y^2 = c^2$ is the z-axis. In cylindrical coordinates this cylinder has the very simple equation $r = c$. (See Figure 4.) This is the reason for the name "cylindrical" coordinates.

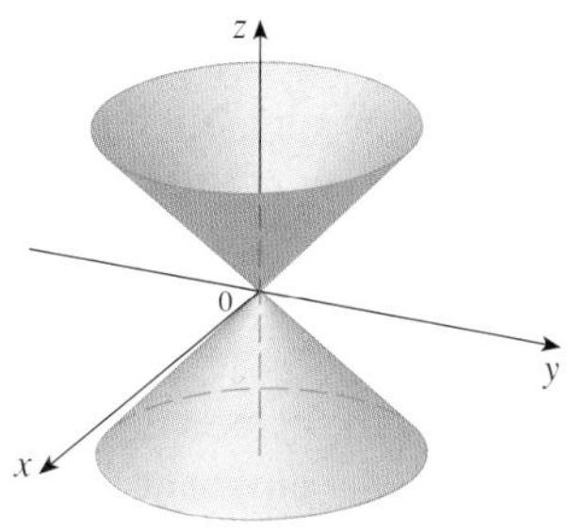

FIGURE 5
$z = r$, a cone

V EXAMPLE 2 Describe the surface whose equation in cylindrical coordinates is $z = r$.

SOLUTION The equation says that the z-value, or height, of each point on the surface is the same as r, the distance from the point to the z-axis. Because θ doesn't appear, it can vary. So any horizontal trace in the plane $z = k$ $(k > 0)$ is a circle of radius k. These traces suggest that the surface is a cone. This prediction can be confirmed by converting the equation into rectangular coordinates. From the first equation in [2] we have

$$z^2 = r^2 = x^2 + y^2$$

We recognize the equation $z^2 = x^2 + y^2$ (by comparison with Table 1 in Section 12.6) as being a circular cone whose axis is the z-axis (see Figure 5).

Evaluating Triple Integrals with Cylindrical Coordinates

Suppose that E is a type 1 region whose projection D onto the xy-plane is conveniently described in polar coordinates (see Figure 6). In particular, suppose that f is continuous and

$$E = \{(x, y, z) \mid (x, y) \in D,\ u_1(x, y) \leqslant z \leqslant u_2(x, y)\}$$

where D is given in polar coordinates by

$$D = \{(r, \theta) \mid \alpha \leqslant \theta \leqslant \beta,\ h_1(\theta) \leqslant r \leqslant h_2(\theta)\}$$

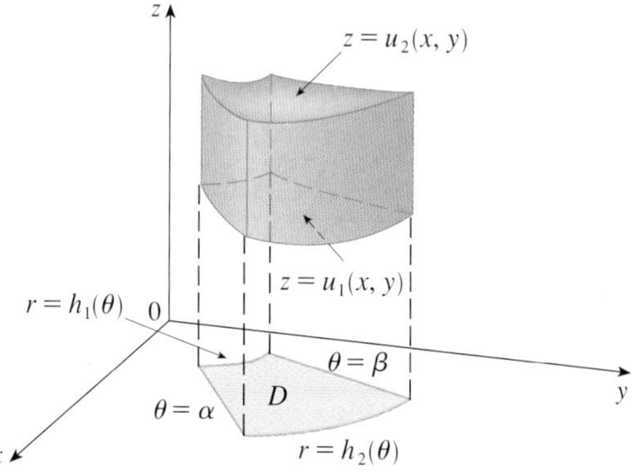

FIGURE 6

We know from Equation 15.7.6 that

3
$$\iiint_E f(x, y, z)\, dV = \iint_D \left[\int_{u_1(x, y)}^{u_2(x, y)} f(x, y, z)\, dz \right] dA$$

But we also know how to evaluate double integrals in polar coordinates. In fact, combining Equation 3 with Equation 15.4.3, we obtain

4
$$\iiint_E f(x, y, z)\, dV = \int_\alpha^\beta \int_{h_1(\theta)}^{h_2(\theta)} \int_{u_1(r\cos\theta,\, r\sin\theta)}^{u_2(r\cos\theta,\, r\sin\theta)} f(r\cos\theta, r\sin\theta, z)\, r\, dz\, dr\, d\theta$$

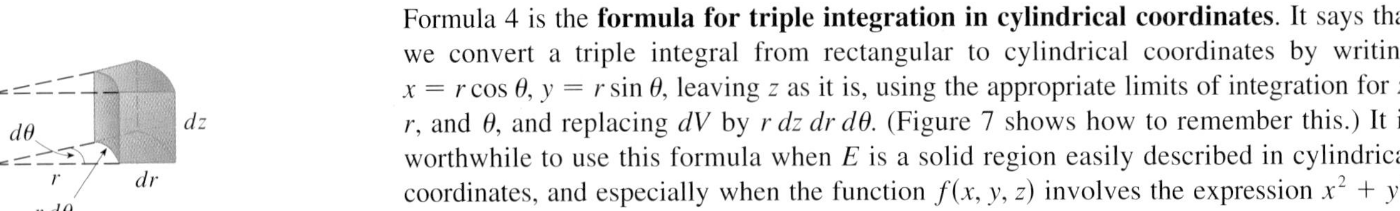

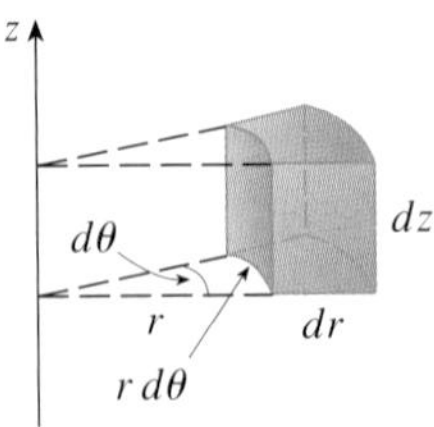

FIGURE 7
Volume element in cylindrical coordinates: $dV = r\,dz\,dr\,d\theta$

Formula 4 is the **formula for triple integration in cylindrical coordinates**. It says that we convert a triple integral from rectangular to cylindrical coordinates by writing $x = r\cos\theta$, $y = r\sin\theta$, leaving z as it is, using the appropriate limits of integration for z, r, and θ, and replacing dV by $r\,dz\,dr\,d\theta$. (Figure 7 shows how to remember this.) It is worthwhile to use this formula when E is a solid region easily described in cylindrical coordinates, and especially when the function $f(x, y, z)$ involves the expression $x^2 + y^2$.

V EXAMPLE 3 A solid E lies within the cylinder $x^2 + y^2 = 1$, below the plane $z = 4$, and above the paraboloid $z = 1 - x^2 - y^2$. (See Figure 8.) The density at any point is proportional to its distance from the axis of the cylinder. Find the mass of E.

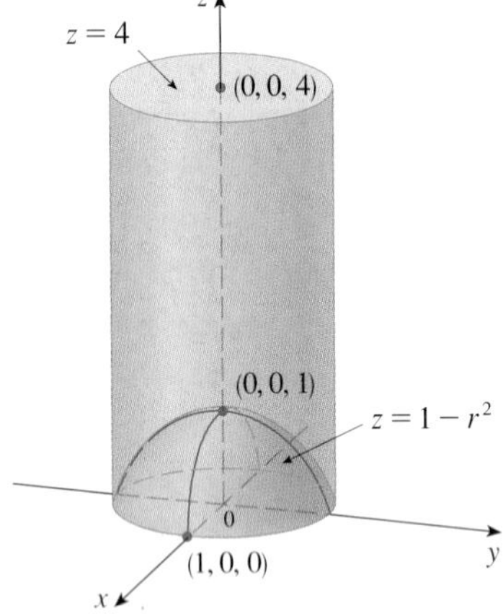

FIGURE 8

SOLUTION In cylindrical coordinates the cylinder is $r = 1$ and the paraboloid is $z = 1 - r^2$, so we can write

$$E = \{(r, \theta, z) \mid 0 \leqslant \theta \leqslant 2\pi,\ 0 \leqslant r \leqslant 1,\ 1 - r^2 \leqslant z \leqslant 4\}$$

Since the density at (x, y, z) is proportional to the distance from the z-axis, the density function is

$$f(x, y, z) = K\sqrt{x^2 + y^2} = Kr$$

where K is the proportionality constant. Therefore, from Formula 15.7.13, the mass of E is

$$m = \iiint_E K\sqrt{x^2 + y^2}\,dV = \int_0^{2\pi}\int_0^1\int_{1-r^2}^4 (Kr)\,r\,dz\,dr\,d\theta$$

$$= \int_0^{2\pi}\int_0^1 Kr^2[4 - (1 - r^2)]\,dr\,d\theta = K\int_0^{2\pi} d\theta \int_0^1 (3r^2 + r^4)\,dr$$

$$= 2\pi K\left[r^3 + \frac{r^5}{5}\right]_0^1 = \frac{12\pi K}{5}$$

EXAMPLE 4 Evaluate $\displaystyle\int_{-2}^{2}\int_{-\sqrt{4-x^2}}^{\sqrt{4-x^2}}\int_{\sqrt{x^2+y^2}}^{2} (x^2 + y^2)\,dz\,dy\,dx$.

SOLUTION This iterated integral is a triple integral over the solid region

$$E = \{(x, y, z) \mid -2 \leqslant x \leqslant 2,\ -\sqrt{4 - x^2} \leqslant y \leqslant \sqrt{4 - x^2},\ \sqrt{x^2 + y^2} \leqslant z \leqslant 2\}$$

and the projection of E onto the xy-plane is the disk $x^2 + y^2 \leqslant 4$. The lower surface of E is the cone $z = \sqrt{x^2 + y^2}$ and its upper surface is the plane $z = 2$. (See Figure 9.) This region has a much simpler description in cylindrical coordinates:

$$E = \{(r, \theta, z) \mid 0 \leqslant \theta \leqslant 2\pi,\ 0 \leqslant r \leqslant 2,\ r \leqslant z \leqslant 2\}$$

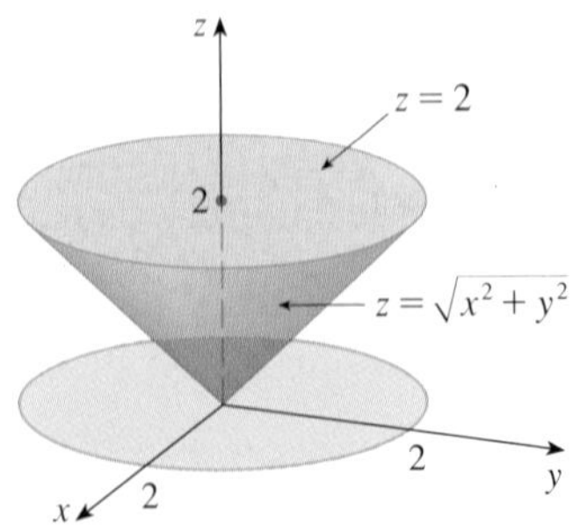

FIGURE 9

Therefore we have

$$\int_{-2}^{2}\int_{-\sqrt{4-x^2}}^{\sqrt{4-x^2}}\int_{\sqrt{x^2+y^2}}^{2} (x^2 + y^2)\,dz\,dy\,dx = \iiint_E (x^2 + y^2)\,dV$$

$$= \int_0^{2\pi}\int_0^2\int_r^2 r^2\,r\,dz\,dr\,d\theta$$

$$= \int_0^{2\pi} d\theta \int_0^2 r^3(2 - r)\,dr$$

$$= 2\pi\left[\tfrac{1}{2}r^4 - \tfrac{1}{5}r^5\right]_0^2 = \tfrac{16}{5}\pi$$

15.8 Exercises

1–2 Plot the point whose cylindrical coordinates are given. Then find the rectangular coordinates of the point.

1. (a) $(4, \pi/3, -2)$ (b) $(2, -\pi/2, 1)$

2. (a) $(\sqrt{2}, 3\pi/4, 2)$ (b) $(1, 1, 1)$

3–4 Change from rectangular to cylindrical coordinates.

3. (a) $(-1, 1, 1)$ (b) $(-2, 2\sqrt{3}, 3)$

4. (a) $(2\sqrt{3}, 2, -1)$ (b) $(4, -3, 2)$

5–6 Describe in words the surface whose equation is given.

5. $\theta = \pi/4$ **6.** $r = 5$

7–8 Identify the surface whose equation is given.

7. $z = 4 - r^2$ **8.** $2r^2 + z^2 = 1$

9–10 Write the equations in cylindrical coordinates.

9. (a) $x^2 - x + y^2 + z^2 = 1$ (b) $z = x^2 - y^2$

10. (a) $3x + 2y + z = 6$ (b) $-x^2 - y^2 + z^2 = 1$

11–12 Sketch the solid described by the given inequalities.

11. $0 \leqslant r \leqslant 2, \quad -\pi/2 \leqslant \theta \leqslant \pi/2, \quad 0 \leqslant z \leqslant 1$

12. $0 \leqslant \theta \leqslant \pi/2, \quad r \leqslant z \leqslant 2$

13. A cylindrical shell is 20 cm long, with inner radius 6 cm and outer radius 7 cm. Write inequalities that describe the shell in an appropriate coordinate system. Explain how you have positioned the coordinate system with respect to the shell.

14. Use a graphing device to draw the solid enclosed by the paraboloids $z = x^2 + y^2$ and $z = 5 - x^2 - y^2$.

15–16 Sketch the solid whose volume is given by the integral and evaluate the integral.

15. $\displaystyle\int_{-\pi/2}^{\pi/2} \int_0^2 \int_0^{r^2} r\, dz\, dr\, d\theta$ **16.** $\displaystyle\int_0^2 \int_0^{2\pi} \int_0^r r\, dz\, d\theta\, dr$

17–28 Use cylindrical coordinates.

17. Evaluate $\iiint_E \sqrt{x^2 + y^2}\, dV$, where E is the region that lies inside the cylinder $x^2 + y^2 = 16$ and between the planes $z = -5$ and $z = 4$.

18. Evaluate $\iiint_E z\, dV$, where E is enclosed by the paraboloid $z = x^2 + y^2$ and the plane $z = 4$.

19. Evaluate $\iiint_E (x + y + z)\, dV$, where E is the solid in the first octant that lies under the paraboloid $z = 4 - x^2 - y^2$.

20. Evaluate $\iiint_E x\, dV$, where E is enclosed by the planes $z = 0$ and $z = x + y + 5$ and by the cylinders $x^2 + y^2 = 4$ and $x^2 + y^2 = 9$.

21. Evaluate $\iiint_E x^2\, dV$, where E is the solid that lies within the cylinder $x^2 + y^2 = 1$, above the plane $z = 0$, and below the cone $z^2 = 4x^2 + 4y^2$.

22. Find the volume of the solid that lies within both the cylinder $x^2 + y^2 = 1$ and the sphere $x^2 + y^2 + z^2 = 4$.

23. Find the volume of the solid that is enclosed by the cone $z = \sqrt{x^2 + y^2}$ and the sphere $x^2 + y^2 + z^2 = 2$.

24. Find the volume of the solid that lies between the paraboloid $z = x^2 + y^2$ and the sphere $x^2 + y^2 + z^2 = 2$.

25. (a) Find the volume of the region E bounded by the paraboloids $z = x^2 + y^2$ and $z = 36 - 3x^2 - 3y^2$.
(b) Find the centroid of E (the center of mass in the case where the density is constant).

26. (a) Find the volume of the solid that the cylinder $r = a \cos \theta$ cuts out of the sphere of radius a centered at the origin.
(b) Illustrate the solid of part (a) by graphing the sphere and the cylinder on the same screen.

27. Find the mass and center of mass of the solid S bounded by the paraboloid $z = 4x^2 + 4y^2$ and the plane $z = a$ $(a > 0)$ if S has constant density K.

28. Find the mass of a ball B given by $x^2 + y^2 + z^2 \leqslant a^2$ if the density at any point is proportional to its distance from the z-axis.

29–30 Evaluate the integral by changing to cylindrical coordinates.

29. $\displaystyle\int_{-2}^{2} \int_{-\sqrt{4-y^2}}^{\sqrt{4-y^2}} \int_{\sqrt{x^2+y^2}}^{2} xz\, dz\, dx\, dy$

30. $\displaystyle\int_{-3}^{3} \int_0^{\sqrt{9-x^2}} \int_0^{9-x^2-y^2} \sqrt{x^2 + y^2}\, dz\, dy\, dx$

31. When studying the formation of mountain ranges, geologists estimate the amount of work required to lift a mountain from sea level. Consider a mountain that is essentially in the shape of a right circular cone. Suppose that the weight density of the material in the vicinity of a point P is $g(P)$ and the height is $h(P)$.
    (a) Find a definite integral that represents the total work done in forming the mountain.
    (b) Assume that Mount Fuji in Japan is in the shape of a right circular cone with radius 62,000 ft, height 12,400 ft, and density a constant 200 lb/ft^3. How much work was done in forming Mount Fuji if the land was initially at sea level?

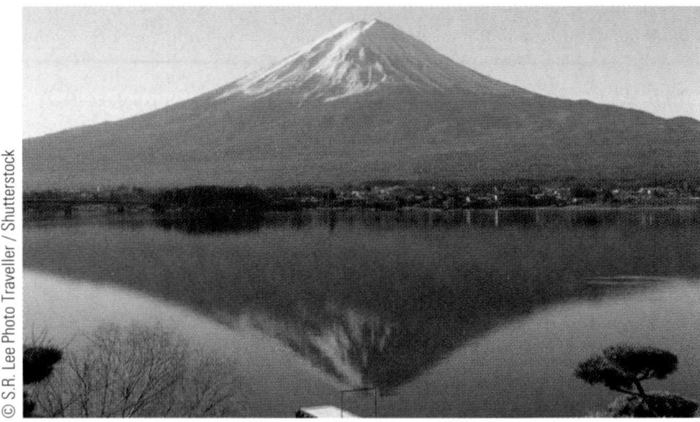

LABORATORY PROJECT THE INTERSECTION OF THREE CYLINDERS

The figure shows the solid enclosed by three circular cylinders with the same diameter that intersect at right angles. In this project we compute its volume and determine how its shape changes if the cylinders have different diameters.

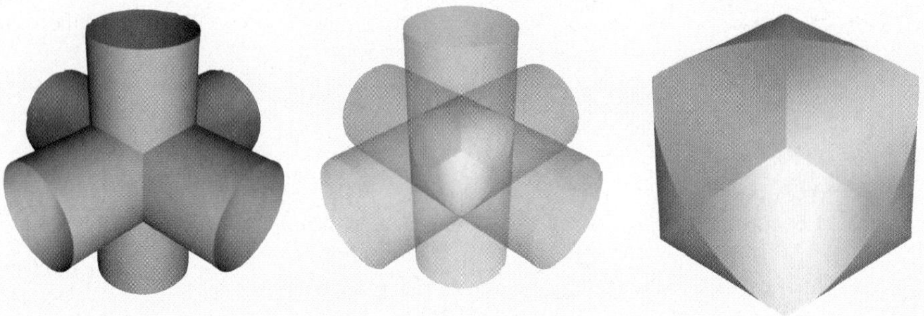

1. Sketch carefully the solid enclosed by the three cylinders $x^2 + y^2 = 1$, $x^2 + z^2 = 1$, and $y^2 + z^2 = 1$. Indicate the positions of the coordinate axes and label the faces with the equations of the corresponding cylinders.

2. Find the volume of the solid in Problem 1.

CAS 3. Use a computer algebra system to draw the edges of the solid.

4. What happens to the solid in Problem 1 if the radius of the first cylinder is different from 1? Illustrate with a hand-drawn sketch or a computer graph.

5. If the first cylinder is $x^2 + y^2 = a^2$, where $a < 1$, set up, but do not evaluate, a double integral for the volume of the solid. What if $a > 1$?

CAS Computer algebra system required

15.9 Triple Integrals in Spherical Coordinates

Another useful coordinate system in three dimensions is the *spherical coordinate system*. It simplifies the evaluation of triple integrals over regions bounded by spheres or cones.

Spherical Coordinates

The **spherical coordinates** (ρ, θ, ϕ) of a point P in space are shown in Figure 1, where $\rho = |OP|$ is the distance from the origin to P, θ is the same angle as in cylindrical coordinates, and ϕ is the angle between the positive z-axis and the line segment OP. Note that

$$\rho \geqslant 0 \qquad 0 \leqslant \phi \leqslant \pi$$

FIGURE 1
The spherical coordinates of a point

The spherical coordinate system is especially useful in problems where there is symmetry about a point, and the origin is placed at this point. For example, the sphere with center the origin and radius c has the simple equation $\rho = c$ (see Figure 2); this is the reason for the name "spherical" coordinates. The graph of the equation $\theta = c$ is a vertical half-plane (see Figure 3), and the equation $\phi = c$ represents a half-cone with the z-axis as its axis (see Figure 4).

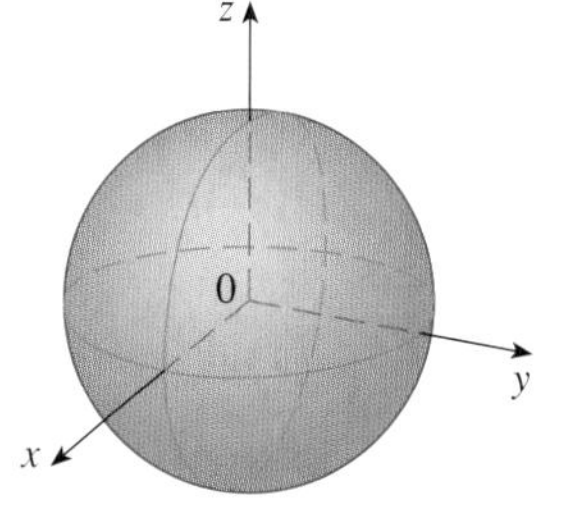

FIGURE 2 $\rho = c$, a sphere

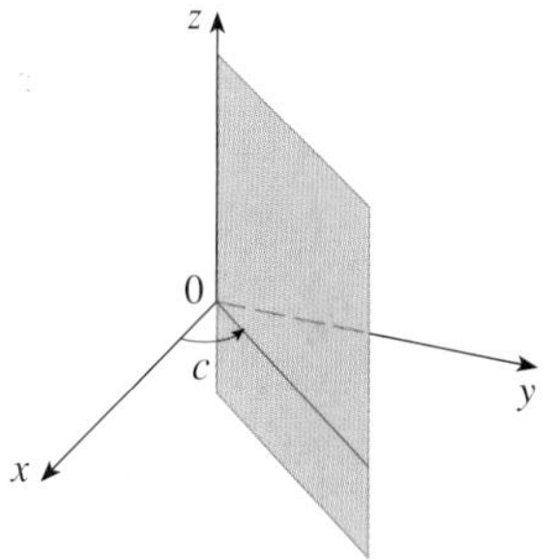

FIGURE 3 $\theta = c$, a half-plane

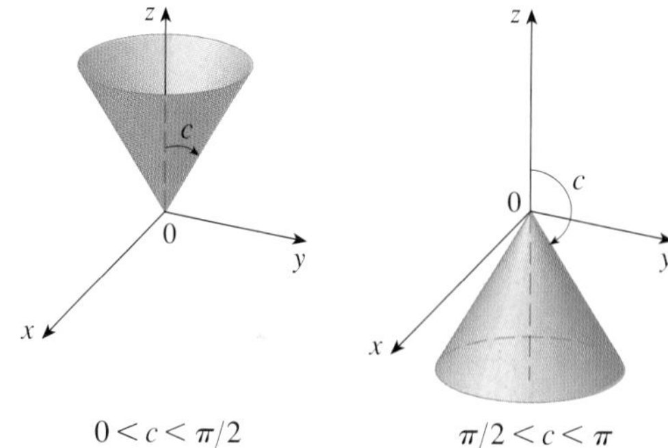

FIGURE 4 $\phi = c$, a half-cone

The relationship between rectangular and spherical coordinates can be seen from Figure 5. From triangles OPQ and OPP' we have

$$z = \rho \cos \phi \qquad r = \rho \sin \phi$$

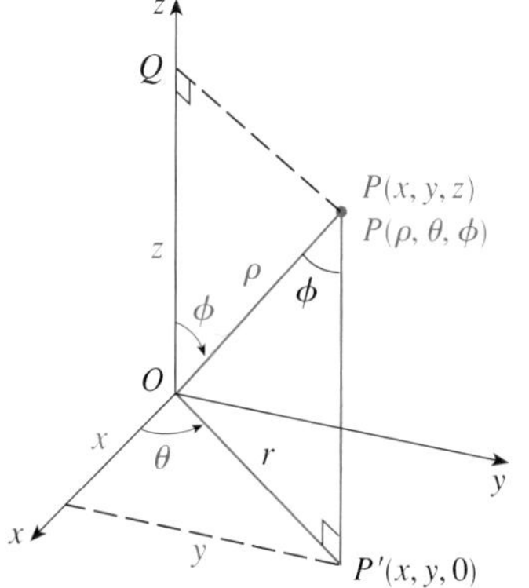

FIGURE 5

But $x = r \cos \theta$ and $y = r \sin \theta$, so to convert from spherical to rectangular coordinates, we use the equations

1
$$x = \rho \sin \phi \cos \theta \qquad y = \rho \sin \phi \sin \theta \qquad z = \rho \cos \phi$$

Also, the distance formula shows that

2
$$\rho^2 = x^2 + y^2 + z^2$$

We use this equation in converting from rectangular to spherical coordinates.

V EXAMPLE 1 The point $(2, \pi/4, \pi/3)$ is given in spherical coordinates. Plot the point and find its rectangular coordinates.

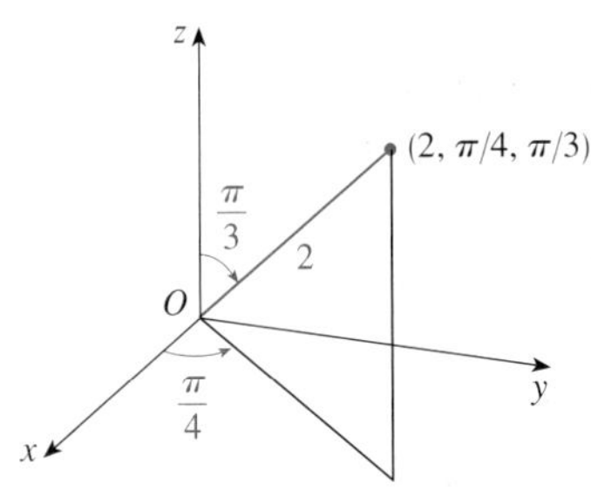

FIGURE 6

SOLUTION We plot the point in Figure 6. From Equations 1 we have

$$x = \rho \sin\phi \cos\theta = 2 \sin\frac{\pi}{3} \cos\frac{\pi}{4} = 2\left(\frac{\sqrt{3}}{2}\right)\left(\frac{1}{\sqrt{2}}\right) = \sqrt{\frac{3}{2}}$$

$$y = \rho \sin\phi \sin\theta = 2 \sin\frac{\pi}{3} \sin\frac{\pi}{4} = 2\left(\frac{\sqrt{3}}{2}\right)\left(\frac{1}{\sqrt{2}}\right) = \sqrt{\frac{3}{2}}$$

$$z = \rho \cos\phi = 2 \cos\frac{\pi}{3} = 2\left(\tfrac{1}{2}\right) = 1$$

Thus the point $(2, \pi/4, \pi/3)$ is $\left(\sqrt{3/2}, \sqrt{3/2}, 1\right)$ in rectangular coordinates.

V EXAMPLE 2 The point $\left(0, 2\sqrt{3}, -2\right)$ is given in rectangular coordinates. Find spherical coordinates for this point.

SOLUTION From Equation 2 we have

$$\rho = \sqrt{x^2 + y^2 + z^2} = \sqrt{0 + 12 + 4} = 4$$

WARNING There is not universal agreement on the notation for spherical coordinates. Most books on physics reverse the meanings of θ and ϕ and use r in place of ρ.

and so Equations 1 give

$$\cos\phi = \frac{z}{\rho} = \frac{-2}{4} = -\frac{1}{2} \qquad \phi = \frac{2\pi}{3}$$

$$\cos\theta = \frac{x}{\rho \sin\phi} = 0 \qquad \theta = \frac{\pi}{2}$$

TEC In Module 15.9 you can investigate families of surfaces in cylindrical and spherical coordinates.

(Note that $\theta \neq 3\pi/2$ because $y = 2\sqrt{3} > 0$.) Therefore spherical coordinates of the given point are $(4, \pi/2, 2\pi/3)$.

Evaluating Triple Integrals with Spherical Coordinates

In the spherical coordinate system the counterpart of a rectangular box is a **spherical wedge**

$$E = \{(\rho, \theta, \phi) \mid a \leqslant \rho \leqslant b,\ \alpha \leqslant \theta \leqslant \beta,\ c \leqslant \phi \leqslant d\}$$

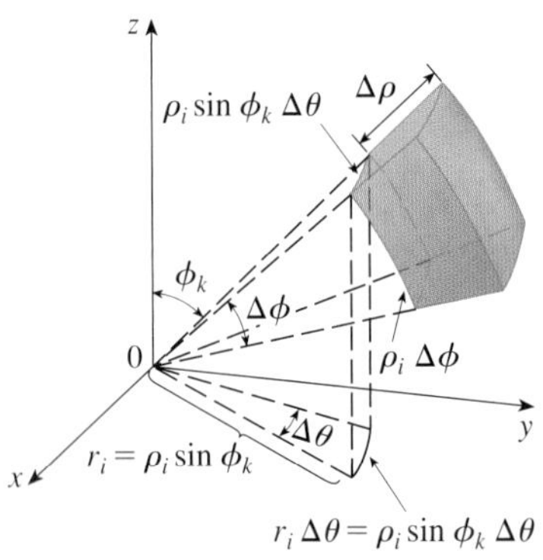

FIGURE 7

where $a \geqslant 0$ and $\beta - \alpha \leqslant 2\pi$, and $d - c \leqslant \pi$. Although we defined triple integrals by dividing solids into small boxes, it can be shown that dividing a solid into small spherical wedges always gives the same result. So we divide E into smaller spherical wedges E_{ijk} by means of equally spaced spheres $\rho = \rho_i$, half-planes $\theta = \theta_j$, and half-cones $\phi = \phi_k$. Figure 7 shows that E_{ijk} is approximately a rectangular box with dimensions $\Delta\rho$, $\rho_i\, \Delta\phi$ (arc of a circle with radius ρ_i, angle $\Delta\phi$), and $\rho_i \sin\phi_k\, \Delta\theta$ (arc of a circle with radius $\rho_i \sin\phi_k$, angle $\Delta\theta$). So an approximation to the volume of E_{ijk} is given by

$$\Delta V_{ijk} \approx (\Delta\rho)(\rho_i\, \Delta\phi)(\rho_i \sin\phi_k\, \Delta\theta) = \rho_i^2 \sin\phi_k\, \Delta\rho\, \Delta\theta\, \Delta\phi$$

In fact, it can be shown, with the aid of the Mean Value Theorem (Exercise 47), that the volume of E_{ijk} is given exactly by

$$\Delta V_{ijk} = \tilde{\rho}_i^2 \sin\tilde{\phi}_k\, \Delta\rho\, \Delta\theta\, \Delta\phi$$

where $(\tilde{\rho}_i, \tilde{\theta}_j, \tilde{\phi}_k)$ is some point in E_{ijk}. Let $(x^*_{ijk}, y^*_{ijk}, z^*_{ijk})$ be the rectangular coordinates of this point. Then

$$\iiint_E f(x, y, z)\, dV = \lim_{l, m, n \to \infty} \sum_{i=1}^{l} \sum_{j=1}^{m} \sum_{k=1}^{n} f(x^*_{ijk}, y^*_{ijk}, z^*_{ijk})\, \Delta V_{ijk}$$

$$= \lim_{l, m, n \to \infty} \sum_{i=1}^{l} \sum_{j=1}^{m} \sum_{k=1}^{n} f(\tilde{\rho}_i \sin \tilde{\phi}_k \cos \tilde{\theta}_j, \tilde{\rho}_i \sin \tilde{\phi}_k \sin \tilde{\theta}_j, \tilde{\rho}_i \cos \tilde{\phi}_k)\, \tilde{\rho}_i^2 \sin \tilde{\phi}_k\, \Delta\rho\, \Delta\theta\, \Delta\phi$$

But this sum is a Riemann sum for the function

$$F(\rho, \theta, \phi) = f(\rho \sin \phi \cos \theta, \rho \sin \phi \sin \theta, \rho \cos \phi)\, \rho^2 \sin \phi$$

Consequently, we have arrived at the following **formula for triple integration in spherical coordinates**.

3 $$\iiint_E f(x, y, z)\, dV$$

$$= \int_c^d \int_\alpha^\beta \int_a^b f(\rho \sin \phi \cos \theta, \rho \sin \phi \sin \theta, \rho \cos \phi)\, \rho^2 \sin \phi\, d\rho\, d\theta\, d\phi$$

where E is a spherical wedge given by

$$E = \{(\rho, \theta, \phi) \mid a \leqslant \rho \leqslant b,\ \alpha \leqslant \theta \leqslant \beta,\ c \leqslant \phi \leqslant d\}$$

Formula 3 says that we convert a triple integral from rectangular coordinates to spherical coordinates by writing

$$x = \rho \sin \phi \cos \theta \qquad y = \rho \sin \phi \sin \theta \qquad z = \rho \cos \phi$$

using the appropriate limits of integration, and replacing dV by $\rho^2 \sin \phi\, d\rho\, d\theta\, d\phi$. This is illustrated in Figure 8.

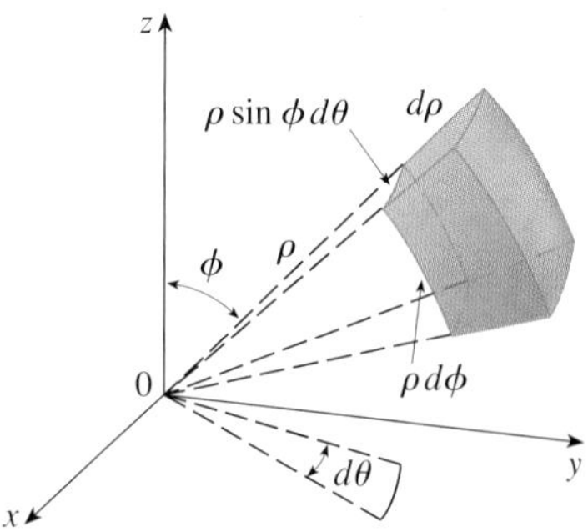

FIGURE 8
Volume element in spherical coordinates: $dV = \rho^2 \sin \phi\, d\rho\, d\theta\, d\phi$

This formula can be extended to include more general spherical regions such as

$$E = \{(\rho, \theta, \phi) \mid \alpha \leqslant \theta \leqslant \beta,\ c \leqslant \phi \leqslant d,\ g_1(\theta, \phi) \leqslant \rho \leqslant g_2(\theta, \phi)\}$$

In this case the formula is the same as in [3] except that the limits of integration for ρ are $g_1(\theta, \phi)$ and $g_2(\theta, \phi)$.

Usually, spherical coordinates are used in triple integrals when surfaces such as cones and spheres form the boundary of the region of integration.

V EXAMPLE 3 Evaluate $\iiint_B e^{(x^2+y^2+z^2)^{3/2}}\, dV$, where B is the unit ball:

$$B = \{(x, y, z) \mid x^2 + y^2 + z^2 \leqslant 1\}$$

SOLUTION Since the boundary of B is a sphere, we use spherical coordinates:

$$B = \{(\rho, \theta, \phi) \mid 0 \leqslant \rho \leqslant 1,\ 0 \leqslant \theta \leqslant 2\pi,\ 0 \leqslant \phi \leqslant \pi\}$$

In addition, spherical coordinates are appropriate because

$$x^2 + y^2 + z^2 = \rho^2$$

Thus [3] gives

$$\iiint_B e^{(x^2+y^2+z^2)^{3/2}}\, dV = \int_0^\pi \int_0^{2\pi} \int_0^1 e^{(\rho^2)^{3/2}} \rho^2 \sin\phi \, d\rho \, d\theta \, d\phi$$

$$= \int_0^\pi \sin\phi \, d\phi \int_0^{2\pi} d\theta \int_0^1 \rho^2 e^{\rho^3}\, d\rho$$

$$= \big[-\cos\phi\big]_0^\pi (2\pi) \big[\tfrac{1}{3} e^{\rho^3}\big]_0^1 = \tfrac{4}{3}\pi(e - 1)$$

NOTE It would have been extremely awkward to evaluate the integral in Example 3 without spherical coordinates. In rectangular coordinates the iterated integral would have been

$$\int_{-1}^1 \int_{-\sqrt{1-x^2}}^{\sqrt{1-x^2}} \int_{-\sqrt{1-x^2-y^2}}^{\sqrt{1-x^2-y^2}} e^{(x^2+y^2+z^2)^{3/2}}\, dz\, dy\, dx$$

V EXAMPLE 4 Use spherical coordinates to find the volume of the solid that lies above the cone $z = \sqrt{x^2 + y^2}$ and below the sphere $x^2 + y^2 + z^2 = z$. (See Figure 9.)

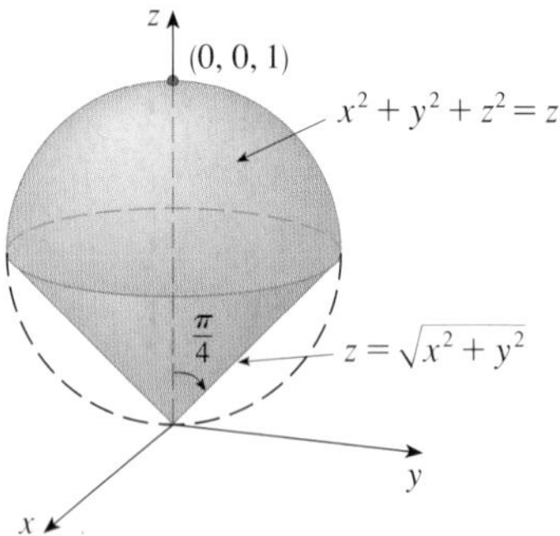

FIGURE 9

SOLUTION Notice that the sphere passes through the origin and has center $(0, 0, \frac{1}{2})$. We write the equation of the sphere in spherical coordinates as

$$\rho^2 = \rho\cos\phi \qquad \text{or} \qquad \rho = \cos\phi$$

The equation of the cone can be written as

$$\rho\cos\phi = \sqrt{\rho^2\sin^2\phi\,\cos^2\theta + \rho^2\sin^2\phi\,\sin^2\theta} = \rho\sin\phi$$

This gives $\sin\phi = \cos\phi$, or $\phi = \pi/4$. Therefore the description of the solid E in spherical coordinates is

$$E = \{(\rho, \theta, \phi) \mid 0 \leqslant \theta \leqslant 2\pi,\ 0 \leqslant \phi \leqslant \pi/4,\ 0 \leqslant \rho \leqslant \cos\phi\}$$

Figure 10 gives another look (this time drawn by Maple) at the solid of Example 4.

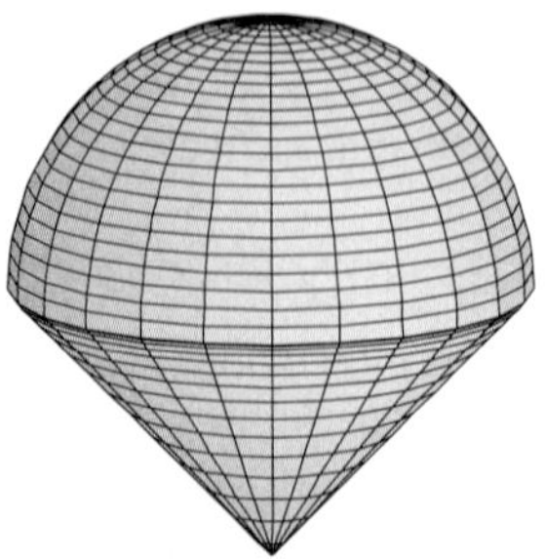

FIGURE 10

Figure 11 shows how E is swept out if we integrate first with respect to ρ, then ϕ, and then θ. The volume of E is

$$V(E) = \iiint_E dV = \int_0^{2\pi} \int_0^{\pi/4} \int_0^{\cos\phi} \rho^2 \sin\phi \, d\rho \, d\phi \, d\theta$$

$$= \int_0^{2\pi} d\theta \int_0^{\pi/4} \sin\phi \left[\frac{\rho^3}{3} \right]_{\rho=0}^{\rho=\cos\phi} d\phi$$

$$= \frac{2\pi}{3} \int_0^{\pi/4} \sin\phi \cos^3\phi \, d\phi = \frac{2\pi}{3} \left[-\frac{\cos^4\phi}{4} \right]_0^{\pi/4} = \frac{\pi}{8}$$

TEC Visual 15.9 shows an animation of Figure 11.

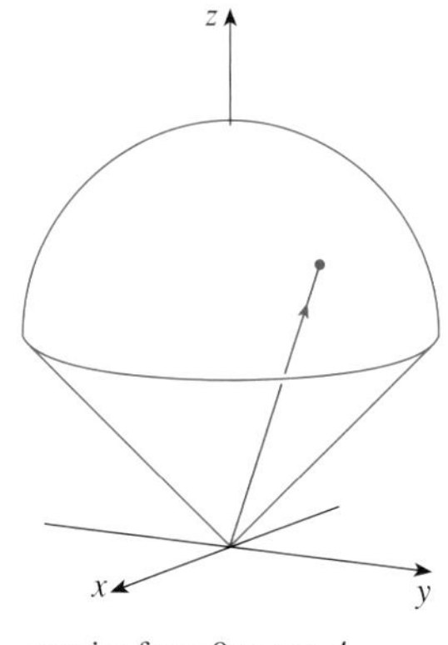

ρ varies from 0 to $\cos\phi$ while ϕ and θ are constant.

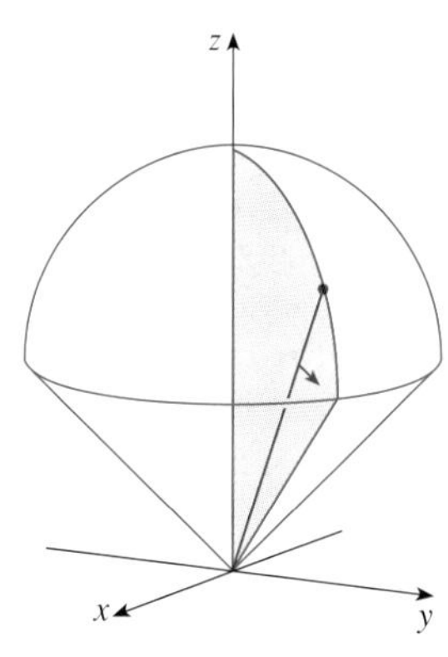

ϕ varies from 0 to $\pi/4$ while θ is constant.

θ varies from 0 to 2π.

FIGURE 11

15.9 Exercises

1–2 Plot the point whose spherical coordinates are given. Then find the rectangular coordinates of the point.

1. (a) $(6, \pi/3, \pi/6)$ (b) $(3, \pi/2, 3\pi/4)$

2. (a) $(2, \pi/2, \pi/2)$ (b) $(4, -\pi/4, \pi/3)$

3–4 Change from rectangular to spherical coordinates.

3. (a) $(0, -2, 0)$ (b) $\left(-1, 1, -\sqrt{2}\right)$

4. (a) $\left(1, 0, \sqrt{3}\right)$ (b) $\left(\sqrt{3}, -1, 2\sqrt{3}\right)$

5–6 Describe in words the surface whose equation is given.

5. $\phi = \pi/3$ **6.** $\rho = 3$

7–8 Identify the surface whose equation is given.

7. $\rho = \sin\theta \sin\phi$ **8.** $\rho^2(\sin^2\phi \sin^2\theta + \cos^2\phi) = 9$

9–10 Write the equation in spherical coordinates.

9. (a) $z^2 = x^2 + y^2$ (b) $x^2 + z^2 = 9$

10. (a) $x^2 - 2x + y^2 + z^2 = 0$ (b) $x + 2y + 3z = 1$

11–14 Sketch the solid described by the given inequalities.

11. $2 \leqslant \rho \leqslant 4, \quad 0 \leqslant \phi \leqslant \pi/3, \quad 0 \leqslant \theta \leqslant \pi$

12. $1 \leqslant \rho \leqslant 2, \quad 0 \leqslant \phi \leqslant \pi/2, \quad \pi/2 \leqslant \theta \leqslant 3\pi/2$

13. $\rho \leqslant 1, \quad 3\pi/4 \leqslant \phi \leqslant \pi$

14. $\rho \leqslant 2, \quad \rho \leqslant \csc\phi$

15. A solid lies above the cone $z = \sqrt{x^2 + y^2}$ and below the sphere $x^2 + y^2 + z^2 = z$. Write a description of the solid in terms of inequalities involving spherical coordinates.

16. (a) Find inequalities that describe a hollow ball with diameter 30 cm and thickness 0.5 cm. Explain how you have positioned the coordinate system that you have chosen.

(b) Suppose the ball is cut in half. Write inequalities that describe one of the halves.

17–18 Sketch the solid whose volume is given by the integral and evaluate the integral.

17. $\int_0^{\pi/6} \int_0^{\pi/2} \int_0^3 \rho^2 \sin\phi \, d\rho \, d\theta \, d\phi$

18. $\int_0^{2\pi} \int_{\pi/2}^{\pi} \int_1^2 \rho^2 \sin\phi \, d\rho \, d\phi \, d\theta$

19–20 Set up the triple integral of an arbitrary continuous function $f(x, y, z)$ in cylindrical or spherical coordinates over the solid shown.

19.

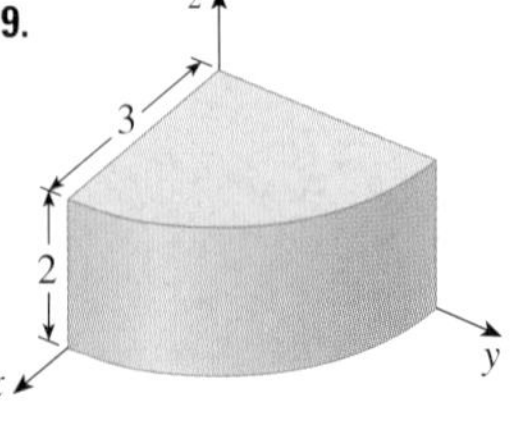

20.

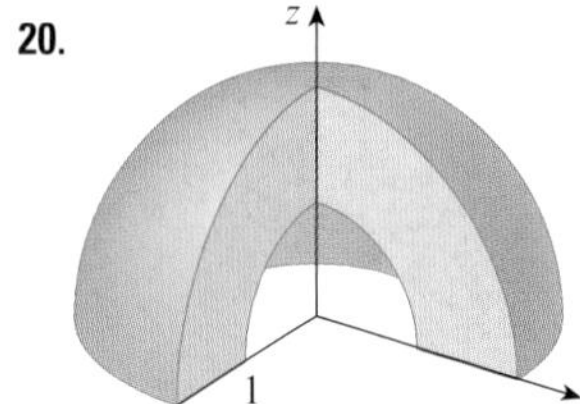

21–34 Use spherical coordinates.

21. Evaluate $\iiint_B (x^2 + y^2 + z^2)^2 \, dV$, where B is the ball with center the origin and radius 5.

22. Evaluate $\iiint_H (9 - x^2 - y^2) \, dV$, where H is the solid hemisphere $x^2 + y^2 + z^2 \leqslant 9$, $z \geqslant 0$.

23. Evaluate $\iiint_E (x^2 + y^2) \, dV$, where E lies between the spheres $x^2 + y^2 + z^2 = 4$ and $x^2 + y^2 + z^2 = 9$.

24. Evaluate $\iiint_E y^2 \, dV$, where E is the solid hemisphere $x^2 + y^2 + z^2 \leqslant 9$, $y \geqslant 0$.

25. Evaluate $\iiint_E xe^{x^2+y^2+z^2} \, dV$, where E is the portion of the unit ball $x^2 + y^2 + z^2 \leqslant 1$ that lies in the first octant.

26. Evaluate $\iiint_E xyz \, dV$, where E lies between the spheres $\rho = 2$ and $\rho = 4$ and above the cone $\phi = \pi/3$.

27. Find the volume of the part of the ball $\rho \leqslant a$ that lies between the cones $\phi = \pi/6$ and $\phi = \pi/3$.

28. Find the average distance from a point in a ball of radius a to its center.

29. (a) Find the volume of the solid that lies above the cone $\phi = \pi/3$ and below the sphere $\rho = 4\cos\phi$.
(b) Find the centroid of the solid in part (a).

30. Find the volume of the solid that lies within the sphere $x^2 + y^2 + z^2 = 4$, above the xy-plane, and below the cone $z = \sqrt{x^2 + y^2}$.

31. (a) Find the centroid of the solid in Example 4.
(b) Find the moment of inertia about the z-axis for this solid.

32. Let H be a solid hemisphere of radius a whose density at any point is proportional to its distance from the center of the base.
(a) Find the mass of H.
(b) Find the center of mass of H.
(c) Find the moment of inertia of H about its axis.

33. (a) Find the centroid of a solid homogeneous hemisphere of radius a.
(b) Find the moment of inertia of the solid in part (a) about a diameter of its base.

34. Find the mass and center of mass of a solid hemisphere of radius a if the density at any point is proportional to its distance from the base.

35–38 Use cylindrical or spherical coordinates, whichever seems more appropriate.

35. Find the volume and centroid of the solid E that lies above the cone $z = \sqrt{x^2 + y^2}$ and below the sphere $x^2 + y^2 + z^2 = 1$.

36. Find the volume of the smaller wedge cut from a sphere of radius a by two planes that intersect along a diameter at an angle of $\pi/6$.

CAS **37.** Evaluate $\iiint_E z \, dV$, where E lies above the paraboloid $z = x^2 + y^2$ and below the plane $z = 2y$. Use either the Table of Integrals (on Reference Pages 6–10) or a computer algebra system to evaluate the integral.

CAS **38.** (a) Find the volume enclosed by the torus $\rho = \sin\phi$.
(b) Use a computer to draw the torus.

39–41 Evaluate the integral by changing to spherical coordinates.

39. $\int_0^1 \int_0^{\sqrt{1-x^2}} \int_{\sqrt{x^2+y^2}}^{\sqrt{2-x^2-y^2}} xy \, dz \, dy \, dx$

40. $\int_{-a}^{a} \int_{-\sqrt{a^2-y^2}}^{\sqrt{a^2-y^2}} \int_{-\sqrt{a^2-x^2-y^2}}^{\sqrt{a^2-x^2-y^2}} (x^2z + y^2z + z^3) \, dz \, dx \, dy$

41. $\int_{-2}^{2} \int_{-\sqrt{4-x^2}}^{\sqrt{4-x^2}} \int_{2-\sqrt{4-x^2-y^2}}^{2+\sqrt{4-x^2-y^2}} (x^2 + y^2 + z^2)^{3/2} \, dz \, dy \, dx$

---

42. A model for the density δ of the earth's atmosphere near its surface is

$$\delta = 619.09 - 0.000097\rho$$

where ρ (the distance from the center of the earth) is measured in meters and δ is measured in kilograms per cubic meter. If we take the surface of the earth to be a sphere with radius 6370 km, then this model is a reasonable one for $6.370 \times 10^6 \leqslant \rho \leqslant 6.375 \times 10^6$. Use this model to estimate the mass of the atmosphere between the ground and an altitude of 5 km.

43. Use a graphing device to draw a silo consisting of a cylinder with radius 3 and height 10 surmounted by a hemisphere.

44. The latitude and longitude of a point P in the Northern Hemisphere are related to spherical coordinates ρ, θ, ϕ as follows. We take the origin to be the center of the earth and the positive z-axis to pass through the North Pole. The positive x-axis passes through the point where the prime meridian (the meridian through Greenwich, England) intersects the equator. Then the latitude of P is $\alpha = 90° - \phi°$ and the longitude is $\beta = 360° - \theta°$. Find the great-circle distance from Los Angeles (lat. 34.06° N, long. 118.25° W) to Montréal (lat. 45.50° N, long. 73.60° W). Take the radius of the earth to be 3960 mi. (A *great circle* is the circle of intersection of a sphere and a plane through the center of the sphere.)

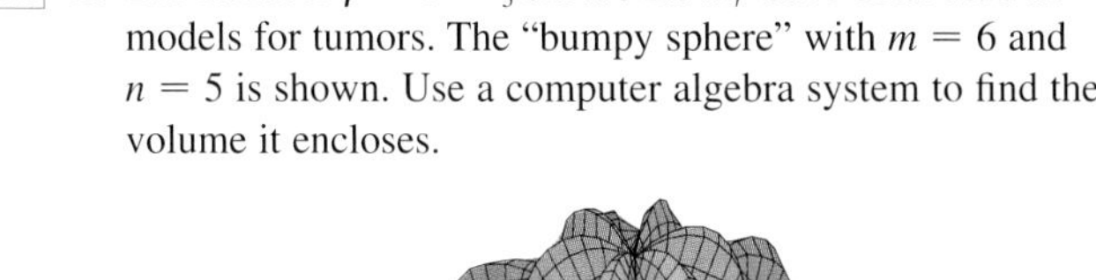

CAS **45.** The surfaces $\rho = 1 + \frac{1}{5}\sin m\theta \sin n\phi$ have been used as models for tumors. The "bumpy sphere" with $m = 6$ and $n = 5$ is shown. Use a computer algebra system to find the volume it encloses.

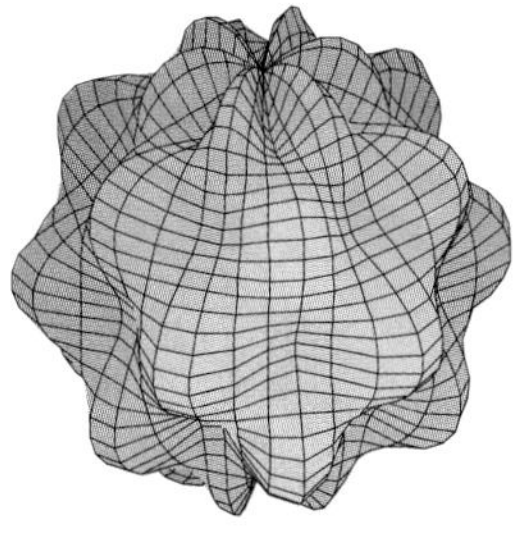

46. Show that

$$\int_{-\infty}^{\infty}\int_{-\infty}^{\infty}\int_{-\infty}^{\infty} \sqrt{x^2 + y^2 + z^2}\, e^{-(x^2+y^2+z^2)}\, dx\, dy\, dz = 2\pi$$

(The improper triple integral is defined as the limit of a triple integral over a solid sphere as the radius of the sphere increases indefinitely.)

47. (a) Use cylindrical coordinates to show that the volume of the solid bounded above by the sphere $r^2 + z^2 = a^2$ and below by the cone $z = r\cot\phi_0$ (or $\phi = \phi_0$), where $0 < \phi_0 < \pi/2$, is

$$V = \frac{2\pi a^3}{3}(1 - \cos\phi_0)$$

(b) Deduce that the volume of the spherical wedge given by $\rho_1 \leqslant \rho \leqslant \rho_2$, $\theta_1 \leqslant \theta \leqslant \theta_2$, $\phi_1 \leqslant \phi \leqslant \phi_2$ is

$$\Delta V = \frac{\rho_2^3 - \rho_1^3}{3}(\cos\phi_1 - \cos\phi_2)(\theta_2 - \theta_1)$$

(c) Use the Mean Value Theorem to show that the volume in part (b) can be written as

$$\Delta V = \tilde{\rho}^2 \sin\tilde{\phi}\, \Delta\rho\, \Delta\theta\, \Delta\phi$$

where $\tilde{\rho}$ lies between ρ_1 and ρ_2, $\tilde{\phi}$ lies between ϕ_1 and ϕ_2, $\Delta\rho = \rho_2 - \rho_1$, $\Delta\theta = \theta_2 - \theta_1$, and $\Delta\phi = \phi_2 - \phi_1$.

APPLIED PROJECT ROLLER DERBY

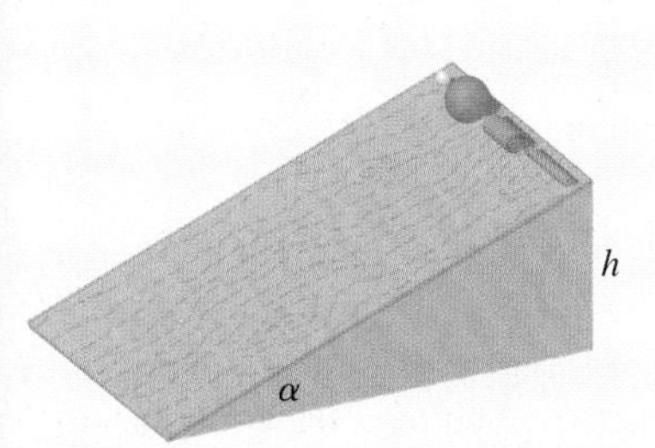

Suppose that a solid ball (a marble), a hollow ball (a squash ball), a solid cylinder (a steel bar), and a hollow cylinder (a lead pipe) roll down a slope. Which of these objects reaches the bottom first? (Make a guess before proceeding.)

To answer this question, we consider a ball or cylinder with mass m, radius r, and moment of inertia I (about the axis of rotation). If the vertical drop is h, then the potential energy at the top is mgh. Suppose the object reaches the bottom with velocity v and angular velocity ω, so $v = \omega r$. The kinetic energy at the bottom consists of two parts: $\frac{1}{2}mv^2$ from translation (moving down the slope) and $\frac{1}{2}I\omega^2$ from rotation. If we assume that energy loss from rolling friction is negligible, then conservation of energy gives

$$mgh = \tfrac{1}{2}mv^2 + \tfrac{1}{2}I\omega^2$$

1. Show that

$$v^2 = \frac{2gh}{1 + I^*} \qquad \text{where } I^* = \frac{I}{mr^2}$$

2. If $y(t)$ is the vertical distance traveled at time t, then the same reasoning as used in Problem 1 shows that $v^2 = 2gy/(1 + I^*)$ at any time t. Use this result to show that y satisfies the differential equation

$$\frac{dy}{dt} = \sqrt{\frac{2g}{1 + I^*}}\,(\sin\alpha)\sqrt{y}$$

where α is the angle of inclination of the plane.

3. By solving the differential equation in Problem 2, show that the total travel time is

$$T = \sqrt{\frac{2h(1 + I^*)}{g \sin^2\alpha}}$$

This shows that the object with the smallest value of I^* wins the race.

4. Show that $I^* = \frac{1}{2}$ for a solid cylinder and $I^* = 1$ for a hollow cylinder.

5. Calculate I^* for a partly hollow ball with inner radius a and outer radius r. Express your answer in terms of $b = a/r$. What happens as $a \to 0$ and as $a \to r$?

6. Show that $I^* = \frac{2}{5}$ for a solid ball and $I^* = \frac{2}{3}$ for a hollow ball. Thus the objects finish in the following order: solid ball, solid cylinder, hollow ball, hollow cylinder.

15.10 Change of Variables in Multiple Integrals

In one-dimensional calculus we often use a change of variable (a substitution) to simplify an integral. By reversing the roles of x and u, we can write the Substitution Rule (5.5.6) as

$$\boxed{1} \qquad \int_a^b f(x)\,dx = \int_c^d f(g(u))\,g'(u)\,du$$

where $x = g(u)$ and $a = g(c)$, $b = g(d)$. Another way of writing Formula 1 is as follows:

$$\boxed{2} \qquad \int_a^b f(x)\,dx = \int_c^d f(x(u))\,\frac{dx}{du}\,du$$

A change of variables can also be useful in double integrals. We have already seen one example of this: conversion to polar coordinates. The new variables r and θ are related to the old variables x and y by the equations

$$x = r\cos\theta \qquad y = r\sin\theta$$

and the change of variables formula (15.4.2) can be written as

$$\iint_R f(x, y)\,dA = \iint_S f(r\cos\theta, r\sin\theta)\,r\,dr\,d\theta$$

where S is the region in the $r\theta$-plane that corresponds to the region R in the xy-plane.

More generally, we consider a change of variables that is given by a **transformation**> T from the uv-plane to the xy-plane:

$$T(u, v) = (x, y)$$

where x and y are related to u and v by the equations

$$\boxed{3} \qquad x = g(u, v) \qquad y = h(u, v)$$

or, as we sometimes write,

$$x = x(u, v) \qquad y = y(u, v)$$

We usually assume that T is a C^1 **transformation**, which means that g and h have continuous first-order partial derivatives.

A transformation T is really just a function whose domain and range are both subsets of $\mathbb{R}^2$. If $T(u_1, v_1) = (x_1, y_1)$, then the point (x_1, y_1) is called the **image** of the point (u_1, v_1). If no two points have the same image, T is called **one-to-one**. Figure 1 shows the effect of a transformation T on a region S in the uv-plane. T transforms S into a region R in the xy-plane called the **image of S**, consisting of the images of all points in S.

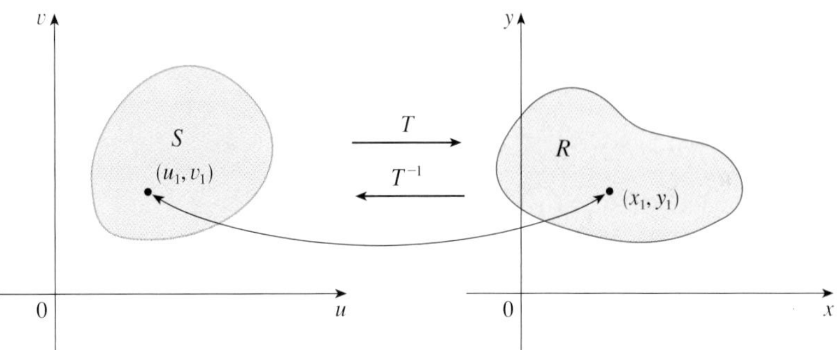

FIGURE 1

If T is a one-to-one transformation, then it has an **inverse transformation** T^{-1} from the xy-plane to the uv-plane and it may be possible to solve Equations 3 for u and v in terms of x and y:

$$u = G(x, y) \qquad v = H(x, y)$$

V EXAMPLE 1 A transformation is defined by the equations

$$x = u^2 - v^2 \qquad y = 2uv$$

Find the image of the square $S = \{(u, v) \mid 0 \leqslant u \leqslant 1,\ 0 \leqslant v \leqslant 1\}$.

SOLUTION The transformation maps the boundary of S into the boundary of the image. So we begin by finding the images of the sides of S. The first side, S_1, is given by $v = 0$ $(0 \leqslant u \leqslant 1)$. (See Figure 2.) From the given equations we have $x = u^2$, $y = 0$, and so $0 \leqslant x \leqslant 1$. Thus S_1 is mapped into the line segment from $(0, 0)$ to $(1, 0)$ in the xy-plane. The second side, S_2, is $u = 1$ $(0 \leqslant v \leqslant 1)$ and, putting $u = 1$ in the given equations, we get

$$x = 1 - v^2 \qquad y = 2v$$

Eliminating v, we obtain

$$\boxed{4} \qquad x = 1 - \frac{y^2}{4} \qquad 0 \leqslant x \leqslant 1$$

which is part of a parabola. Similarly, S_3 is given by $v = 1$ $(0 \leqslant u \leqslant 1)$, whose image is the parabolic arc

$$\boxed{5} \qquad x = \frac{y^2}{4} - 1 \qquad -1 \leqslant x \leqslant 0$$

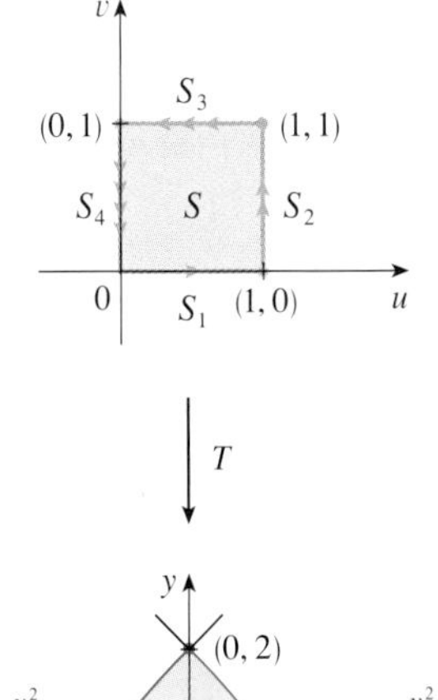

FIGURE 2

Finally, S_4 is given by $u = 0$ $(0 \leqslant v \leqslant 1)$ whose image is $x = -v^2$, $y = 0$, that is, $-1 \leqslant x \leqslant 0$. (Notice that as we move around the square in the counterclockwise direction, we also move around the parabolic region in the counterclockwise direction.) The image of S is the region R (shown in Figure 2) bounded by the x-axis and the parabolas given by Equations 4 and 5.

Now let's see how a change of variables affects a double integral. We start with a small rectangle S in the uv-plane whose lower left corner is the point (u_0, v_0) and whose dimensions are Δu and Δv. (See Figure 3.)

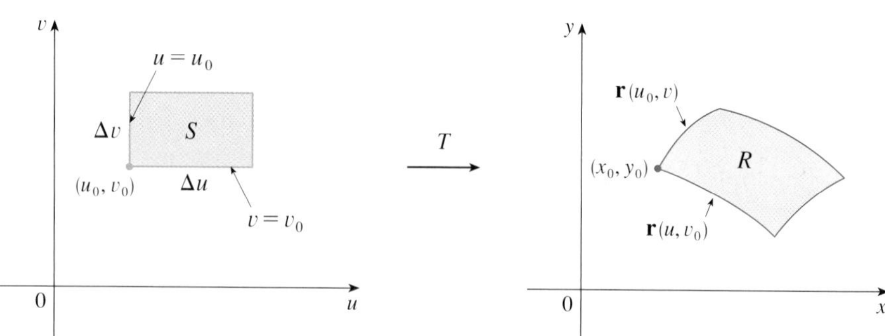

FIGURE 3

The image of S is a region R in the xy-plane, one of whose boundary points is $(x_0, y_0) = T(u_0, v_0)$. The vector

$$\mathbf{r}(u, v) = g(u, v)\,\mathbf{i} + h(u, v)\,\mathbf{j}$$

is the position vector of the image of the point (u, v). The equation of the lower side of S is $v = v_0$, whose image curve is given by the vector function $\mathbf{r}(u, v_0)$. The tangent vector at (x_0, y_0) to this image curve is

$$\mathbf{r}_u = g_u(u_0, v_0)\,\mathbf{i} + h_u(u_0, v_0)\,\mathbf{j} = \frac{\partial x}{\partial u}\,\mathbf{i} + \frac{\partial y}{\partial u}\,\mathbf{j}$$

Similarly, the tangent vector at (x_0, y_0) to the image curve of the left side of S (namely, $u = u_0$) is

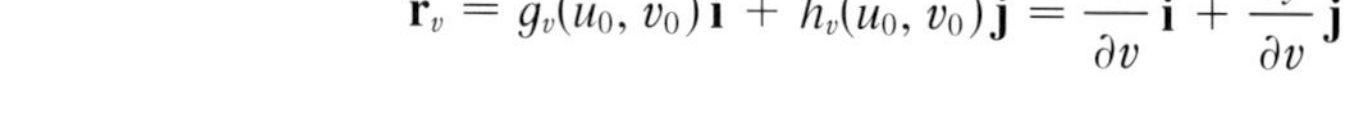

$$\mathbf{r}_v = g_v(u_0, v_0)\,\mathbf{i} + h_v(u_0, v_0)\,\mathbf{j} = \frac{\partial x}{\partial v}\,\mathbf{i} + \frac{\partial y}{\partial v}\,\mathbf{j}$$

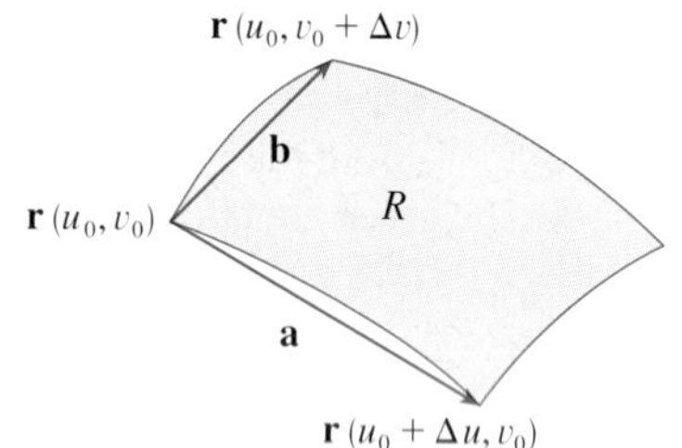

FIGURE 4

We can approximate the image region $R = T(S)$ by a parallelogram determined by the secant vectors

$$\mathbf{a} = \mathbf{r}(u_0 + \Delta u, v_0) - \mathbf{r}(u_0, v_0) \qquad \mathbf{b} = \mathbf{r}(u_0, v_0 + \Delta v) - \mathbf{r}(u_0, v_0)$$

shown in Figure 4. But

$$\mathbf{r}_u = \lim_{\Delta u \to 0} \frac{\mathbf{r}(u_0 + \Delta u, v_0) - \mathbf{r}(u_0, v_0)}{\Delta u}$$

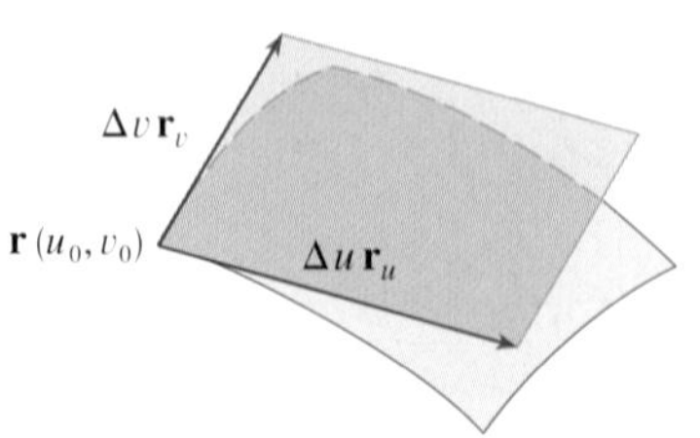

FIGURE 5

and so

$$\mathbf{r}(u_0 + \Delta u, v_0) - \mathbf{r}(u_0, v_0) \approx \Delta u\, \mathbf{r}_u$$

Similarly

$$\mathbf{r}(u_0, v_0 + \Delta v) - \mathbf{r}(u_0, v_0) \approx \Delta v\, \mathbf{r}_v$$

This means that we can approximate R by a parallelogram determined by the vectors $\Delta u\, \mathbf{r}_u$ and $\Delta v\, \mathbf{r}_v$. (See Figure 5.) Therefore we can approximate the area of R by the area of this parallelogram, which, from Section 12.4, is

6 $$|(\Delta u\, \mathbf{r}_u) \times (\Delta v\, \mathbf{r}_v)| = |\mathbf{r}_u \times \mathbf{r}_v|\, \Delta u\, \Delta v$$

Computing the cross product, we obtain

$$\mathbf{r}_u \times \mathbf{r}_v = \begin{vmatrix} \mathbf{i} & \mathbf{j} & \mathbf{k} \\ \dfrac{\partial x}{\partial u} & \dfrac{\partial y}{\partial u} & 0 \\ \dfrac{\partial x}{\partial v} & \dfrac{\partial y}{\partial v} & 0 \end{vmatrix} = \begin{vmatrix} \dfrac{\partial x}{\partial u} & \dfrac{\partial y}{\partial u} \\ \dfrac{\partial x}{\partial v} & \dfrac{\partial y}{\partial v} \end{vmatrix} \mathbf{k} = \begin{vmatrix} \dfrac{\partial x}{\partial u} & \dfrac{\partial x}{\partial v} \\ \dfrac{\partial y}{\partial u} & \dfrac{\partial y}{\partial v} \end{vmatrix} \mathbf{k}$$

The determinant that arises in this calculation is called the *Jacobian* of the transformation and is given a special notation.

The Jacobian is named after the German mathematician Carl Gustav Jacob Jacobi (1804–1851). Although the French mathematician Cauchy first used these special determinants involving partial derivatives, Jacobi developed them into a method for evaluating multiple integrals.

7 Definition The **Jacobian** of the transformation T given by $x = g(u, v)$ and $y = h(u, v)$ is

$$\frac{\partial(x, y)}{\partial(u, v)} = \begin{vmatrix} \dfrac{\partial x}{\partial u} & \dfrac{\partial x}{\partial v} \\ \dfrac{\partial y}{\partial u} & \dfrac{\partial y}{\partial v} \end{vmatrix} = \frac{\partial x}{\partial u}\frac{\partial y}{\partial v} - \frac{\partial x}{\partial v}\frac{\partial y}{\partial u}$$

With this notation we can use Equation 6 to give an approximation to the area ΔA of R:

8 $$\Delta A \approx \left| \frac{\partial(x, y)}{\partial(u, v)} \right| \Delta u \, \Delta v$$

where the Jacobian is evaluated at (u_0, v_0).

Next we divide a region S in the uv-plane into rectangles S_{ij} and call their images in the xy-plane R_{ij}. (See Figure 6.)

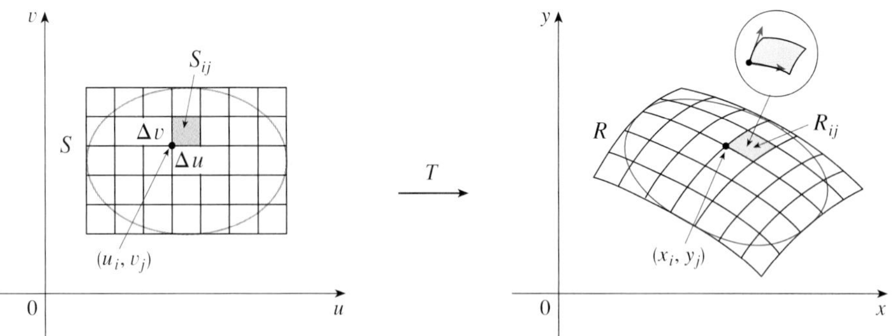

FIGURE 6

Applying the approximation **8** to each R_{ij}, we approximate the double integral of f over R as follows:

$$\iint_R f(x, y)\, dA \approx \sum_{i=1}^{m} \sum_{j=1}^{n} f(x_i, y_j)\, \Delta A$$

$$\approx \sum_{i=1}^{m} \sum_{j=1}^{n} f\big(g(u_i, v_j), h(u_i, v_j)\big) \left| \frac{\partial(x, y)}{\partial(u, v)} \right| \Delta u \, \Delta v$$

where the Jacobian is evaluated at (u_i, v_j). Notice that this double sum is a Riemann sum for the integral

$$\iint_S f(g(u, v), h(u, v)) \left| \frac{\partial(x, y)}{\partial(u, v)} \right| du\, dv$$

The foregoing argument suggests that the following theorem is true. (A full proof is given in books on advanced calculus.)

9 Change of Variables in a Double Integral Suppose that T is a C^1 transformation whose Jacobian is nonzero and that maps a region S in the uv-plane onto a region R in the xy-plane. Suppose that f is continuous on R and that R and S are type I or type II plane regions. Suppose also that T is one-to-one, except perhaps on the boundary of S. Then

$$\iint_R f(x, y)\, dA = \iint_S f(x(u, v), y(u, v)) \left| \frac{\partial(x, y)}{\partial(u, v)} \right| du\, dv$$

Theorem 9 says that we change from an integral in x and y to an integral in u and v by expressing x and y in terms of u and v and writing

$$dA = \left| \frac{\partial(x, y)}{\partial(u, v)} \right| du\, dv$$

Notice the similarity between Theorem 9 and the one-dimensional formula in Equation 2. Instead of the derivative dx/du, we have the absolute value of the Jacobian, that is, $|\partial(x, y)/\partial(u, v)|$.

As a first illustration of Theorem 9, we show that the formula for integration in polar coordinates is just a special case. Here the transformation T from the $r\theta$-plane to the xy-plane is given by

$$x = g(r, \theta) = r\cos\theta \qquad y = h(r, \theta) = r\sin\theta$$

and the geometry of the transformation is shown in Figure 7. T maps an ordinary rectangle in the $r\theta$-plane to a polar rectangle in the xy-plane. The Jacobian of T is

$$\frac{\partial(x, y)}{\partial(r, \theta)} = \begin{vmatrix} \dfrac{\partial x}{\partial r} & \dfrac{\partial x}{\partial \theta} \\ \dfrac{\partial y}{\partial r} & \dfrac{\partial y}{\partial \theta} \end{vmatrix} = \begin{vmatrix} \cos\theta & -r\sin\theta \\ \sin\theta & r\cos\theta \end{vmatrix} = r\cos^2\theta + r\sin^2\theta = r > 0$$

Thus Theorem 9 gives

$$\iint_R f(x, y)\, dx\, dy = \iint_S f(r\cos\theta, r\sin\theta) \left| \frac{\partial(x, y)}{\partial(r, \theta)} \right| dr\, d\theta$$
$$= \int_\alpha^\beta \int_a^b f(r\cos\theta, r\sin\theta)\, r\, dr\, d\theta$$

which is the same as Formula 15.4.2.

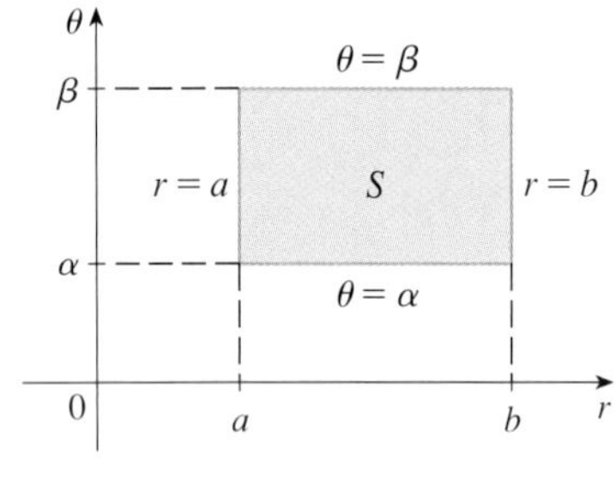

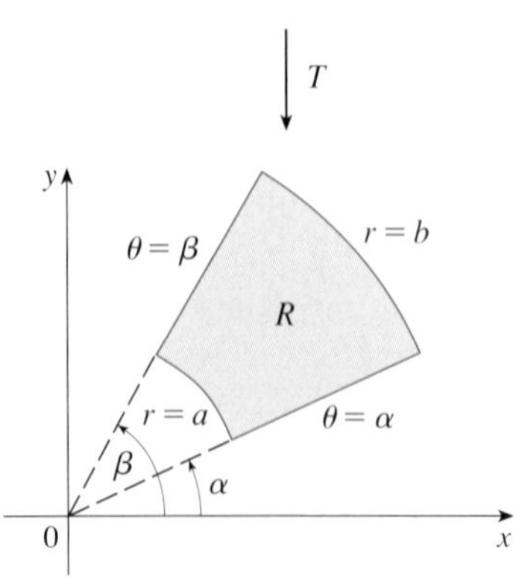

FIGURE 7
The polar coordinate transformation

EXAMPLE 2 Use the change of variables $x = u^2 - v^2$, $y = 2uv$ to evaluate the integral $\iint_R y\, dA$, where R is the region bounded by the x-axis and the parabolas $y^2 = 4 - 4x$ and $y^2 = 4 + 4x$, $y \geqslant 0$.

SOLUTION The region R is pictured in Figure 2 (on page 1041). In Example 1 we discovered that $T(S) = R$, where S is the square $[0, 1] \times [0, 1]$. Indeed, the reason for making the change of variables to evaluate the integral is that S is a much simpler region than R. First we need to compute the Jacobian:

$$\frac{\partial(x, y)}{\partial(u, v)} = \begin{vmatrix} \dfrac{\partial x}{\partial u} & \dfrac{\partial x}{\partial v} \\ \dfrac{\partial y}{\partial u} & \dfrac{\partial y}{\partial v} \end{vmatrix} = \begin{vmatrix} 2u & -2v \\ 2v & 2u \end{vmatrix} = 4u^2 + 4v^2 > 0$$

Therefore, by Theorem 9,

$$\iint_R y\, dA = \iint_S 2uv \left| \frac{\partial(x, y)}{\partial(u, v)} \right| dA = \int_0^1 \int_0^1 (2uv)4(u^2 + v^2)\, du\, dv$$

$$= 8 \int_0^1 \int_0^1 (u^3 v + uv^3)\, du\, dv = 8 \int_0^1 \left[\tfrac{1}{4}u^4 v + \tfrac{1}{2}u^2 v^3 \right]_{u=0}^{u=1} dv$$

$$= \int_0^1 (2v + 4v^3)\, dv = \left[v^2 + v^4 \right]_0^1 = 2$$

NOTE Example 2 was not a very difficult problem to solve because we were given a suitable change of variables. If we are not supplied with a transformation, then the first step is to think of an appropriate change of variables. If $f(x, y)$ is difficult to integrate, then the form of $f(x, y)$ may suggest a transformation. If the region of integration R is awkward, then the transformation should be chosen so that the corresponding region S in the uv-plane has a convenient description.

EXAMPLE 3 Evaluate the integral $\iint_R e^{(x+y)/(x-y)}\, dA$, where R is the trapezoidal region with vertices $(1, 0)$, $(2, 0)$, $(0, -2)$, and $(0, -1)$.

SOLUTION Since it isn't easy to integrate $e^{(x+y)/(x-y)}$, we make a change of variables suggested by the form of this function:

10 $$u = x + y \qquad v = x - y$$

These equations define a transformation T^{-1} from the xy-plane to the uv-plane. Theorem 9 talks about a transformation T from the uv-plane to the xy-plane. It is obtained by solving Equations 10 for x and y:

11 $$x = \tfrac{1}{2}(u + v) \qquad y = \tfrac{1}{2}(u - v)$$

The Jacobian of T is

$$\frac{\partial(x, y)}{\partial(u, v)} = \begin{vmatrix} \dfrac{\partial x}{\partial u} & \dfrac{\partial x}{\partial v} \\ \dfrac{\partial y}{\partial u} & \dfrac{\partial y}{\partial v} \end{vmatrix} = \begin{vmatrix} \frac{1}{2} & \frac{1}{2} \\ \frac{1}{2} & -\frac{1}{2} \end{vmatrix} = -\frac{1}{2}$$

To find the region S in the uv-plane corresponding to R, we note that the sides of R lie on the lines

$$y = 0 \qquad x - y = 2 \qquad x = 0 \qquad x - y = 1$$

and, from either Equations 10 or Equations 11, the image lines in the uv-plane are

$$u = v \qquad v = 2 \qquad u = -v \qquad v = 1$$

Thus the region S is the trapezoidal region with vertices (1, 1), (2, 2), (−2, 2), and (−1, 1) shown in Figure 8. Since

$$S = \{(u, v) \mid 1 \leqslant v \leqslant 2,\ -v \leqslant u \leqslant v\}$$

Theorem 9 gives

$$\iint_R e^{(x+y)/(x-y)}\, dA = \iint_S e^{u/v} \left| \frac{\partial(x, y)}{\partial(u, v)} \right| du\, dv$$

$$= \int_1^2 \int_{-v}^{v} e^{u/v} \left(\tfrac{1}{2}\right) du\, dv = \tfrac{1}{2} \int_1^2 \left[v e^{u/v} \right]_{u=-v}^{u=v} dv$$

$$= \tfrac{1}{2} \int_1^2 (e - e^{-1}) v\, dv = \tfrac{3}{4}(e - e^{-1})$$

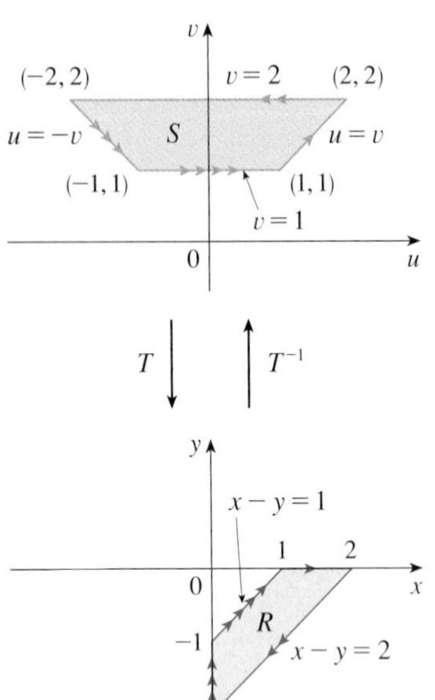

FIGURE 8

Triple Integrals

There is a similar change of variables formula for triple integrals. Let T be a transformation that maps a region S in uvw-space onto a region R in xyz-space by means of the equations

$$x = g(u, v, w) \qquad y = h(u, v, w) \qquad z = k(u, v, w)$$

The **Jacobian** of T is the following 3×3 determinant:

$$\boxed{12} \qquad \frac{\partial(x, y, z)}{\partial(u, v, w)} = \begin{vmatrix} \dfrac{\partial x}{\partial u} & \dfrac{\partial x}{\partial v} & \dfrac{\partial x}{\partial w} \\ \dfrac{\partial y}{\partial u} & \dfrac{\partial y}{\partial v} & \dfrac{\partial y}{\partial w} \\ \dfrac{\partial z}{\partial u} & \dfrac{\partial z}{\partial v} & \dfrac{\partial z}{\partial w} \end{vmatrix}$$

Under hypotheses similar to those in Theorem 9, we have the following formula for triple integrals:

$$\boxed{13} \qquad \iiint_R f(x, y, z)\, dV = \iiint_S f(x(u, v, w), y(u, v, w), z(u, v, w)) \left| \frac{\partial(x, y, z)}{\partial(u, v, w)} \right| du\, dv\, dw$$

V EXAMPLE 4 Use Formula 13 to derive the formula for triple integration in spherical coordinates.

SOLUTION Here the change of variables is given by

$$x = \rho \sin\phi \cos\theta \qquad y = \rho \sin\phi \sin\theta \qquad z = \rho \cos\phi$$

We compute the Jacobian as follows:

$$\frac{\partial(x, y, z)}{\partial(\rho, \theta, \phi)} = \begin{vmatrix} \sin\phi\cos\theta & -\rho\sin\phi\sin\theta & \rho\cos\phi\cos\theta \\ \sin\phi\sin\theta & \rho\sin\phi\cos\theta & \rho\cos\phi\sin\theta \\ \cos\phi & 0 & -\rho\sin\phi \end{vmatrix}$$

$$= \cos\phi \begin{vmatrix} -\rho\sin\phi\sin\theta & \rho\cos\phi\cos\theta \\ \rho\sin\phi\cos\theta & \rho\cos\phi\sin\theta \end{vmatrix} - \rho\sin\phi \begin{vmatrix} \sin\phi\cos\theta & -\rho\sin\phi\sin\theta \\ \sin\phi\sin\theta & \rho\sin\phi\cos\theta \end{vmatrix}$$

$$= \cos\phi\,(-\rho^2\sin\phi\cos\phi\sin^2\theta - \rho^2\sin\phi\cos\phi\cos^2\theta)$$

$$- \rho\sin\phi\,(\rho\sin^2\phi\cos^2\theta + \rho\sin^2\phi\sin^2\theta)$$

$$= -\rho^2\sin\phi\cos^2\phi - \rho^2\sin\phi\sin^2\phi = -\rho^2\sin\phi$$

Since $0 \leqslant \phi \leqslant \pi$, we have $\sin\phi \geqslant 0$. Therefore

$$\left|\frac{\partial(x, y, z)}{\partial(\rho, \theta, \phi)}\right| = |-\rho^2\sin\phi| = \rho^2\sin\phi$$

and Formula 13 gives

$$\iiint_R f(x, y, z)\,dV = \iiint_S f(\rho\sin\phi\cos\theta, \rho\sin\phi\sin\theta, \rho\cos\phi)\,\rho^2\sin\phi\,d\rho\,d\theta\,d\phi$$

which is equivalent to Formula 15.9.3. ■

15.10 Exercises

1–6 Find the Jacobian of the transformation.

1. $x = 5u - v, \quad y = u + 3v$

2. $x = uv, \quad y = u/v$

3. $x = e^{-r}\sin\theta, \quad y = e^{r}\cos\theta$

4. $x = e^{s+t}, \quad y = e^{s-t}$

5. $x = u/v, \quad y = v/w, \quad z = w/u$

6. $x = v + w^2, \quad y = w + u^2, \quad z = u + v^2$

7–10 Find the image of the set S under the given transformation.

7. $S = \{(u, v) \mid 0 \leqslant u \leqslant 3,\ 0 \leqslant v \leqslant 2\}$;
$x = 2u + 3v, \ y = u - v$

8. S is the square bounded by the lines $u = 0$, $u = 1$, $v = 0$, $v = 1$; $\ x = v, \ y = u(1 + v^2)$

9. S is the triangular region with vertices $(0, 0)$, $(1, 1)$, $(0, 1)$; $x = u^2, \ y = v$

10. S is the disk given by $u^2 + v^2 \leqslant 1$; $\ x = au, \ y = bv$

11–14 A region R in the xy-plane is given. Find equations for a transformation T that maps a rectangular region S in the uv-plane onto R, where the sides of S are parallel to the u- and v- axes.

11. R is bounded by $y = 2x - 1$, $y = 2x + 1$, $y = 1 - x$, $y = 3 - x$

12. R is the parallelogram with vertices $(0, 0)$, $(4, 3)$, $(2, 4)$, $(-2, 1)$

13. R lies between the circles $x^2 + y^2 = 1$ and $x^2 + y^2 = 2$ in the first quadrant

14. R is bounded by the hyperbolas $y = 1/x$, $y = 4/x$ and the lines $y = x$, $y = 4x$ in the first quadrant

15–20 Use the given transformation to evaluate the integral.

15. $\iint_R (x - 3y)\,dA$, where R is the triangular region with vertices $(0, 0)$, $(2, 1)$, and $(1, 2)$; $\ x = 2u + v, \ y = u + 2v$

16. $\iint_R (4x + 8y)\,dA$, where R is the parallelogram with vertices $(-1, 3)$, $(1, -3)$, $(3, -1)$, and $(1, 5)$; $x = \frac{1}{4}(u + v), \ y = \frac{1}{4}(v - 3u)$

17. $\iint_R x^2\,dA$, where R is the region bounded by the ellipse $9x^2 + 4y^2 = 36$; $\ x = 2u, \ y = 3v$

18. $\iint_R (x^2 - xy + y^2)\,dA$, where R is the region bounded by the ellipse $x^2 - xy + y^2 = 2$;
$x = \sqrt{2}\,u - \sqrt{2/3}\,v,\ y = \sqrt{2}\,u + \sqrt{2/3}\,v$

19. $\iint_R xy\,dA$, where R is the region in the first quadrant bounded by the lines $y = x$ and $y = 3x$ and the hyperbolas $xy = 1$, $xy = 3$; $x = u/v,\ y = v$

20. $\iint_R y^2\,dA$, where R is the region bounded by the curves $xy = 1$, $xy = 2$, $xy^2 = 1$, $xy^2 = 2$; $u = xy,\ v = xy^2$. Illustrate by using a graphing calculator or computer to draw R.

---

21. (a) Evaluate $\iiint_E dV$, where E is the solid enclosed by the ellipsoid $x^2/a^2 + y^2/b^2 + z^2/c^2 = 1$. Use the transformation $x = au,\ y = bv,\ z = cw$.

(b) The earth is not a perfect sphere; rotation has resulted in flattening at the poles. So the shape can be approximated by an ellipsoid with $a = b = 6378$ km and $c = 6356$ km. Use part (a) to estimate the volume of the earth.

(c) If the solid of part (a) has constant density k, find its moment of inertia about the z-axis.

22. An important problem in thermodynamics is to find the work done by an ideal Carnot engine. A cycle consists of alternating expansion and compression of gas in a piston. The work done by the engine is equal to the area of the region R enclosed by two isothermal curves $xy = a$, $xy = b$ and two adiabatic curves $xy^{1.4} = c$, $xy^{1.4} = d$, where $0 < a < b$ and $0 < c < d$. Compute the work done by determining the area of R.

23–27 Evaluate the integral by making an appropriate change of variables.

23. $\displaystyle\iint_R \frac{x - 2y}{3x - y}\,dA$, where R is the parallelogram enclosed by the lines $x - 2y = 0$, $x - 2y = 4$, $3x - y = 1$, and $3x - y = 8$

24. $\iint_R (x + y)e^{x^2 - y^2}\,dA$, where R is the rectangle enclosed by the lines $x - y = 0$, $x - y = 2$, $x + y = 0$, and $x + y = 3$

25. $\displaystyle\iint_R \cos\left(\frac{y - x}{y + x}\right)dA$, where R is the trapezoidal region with vertices $(1, 0)$, $(2, 0)$, $(0, 2)$, and $(0, 1)$

26. $\iint_R \sin(9x^2 + 4y^2)\,dA$, where R is the region in the first quadrant bounded by the ellipse $9x^2 + 4y^2 = 1$

27. $\iint_R e^{x+y}\,dA$, where R is given by the inequality $|x| + |y| \leqslant 1$

---

28. Let f be continuous on $[0, 1]$ and let R be the triangular region with vertices $(0, 0)$, $(1, 0)$, and $(0, 1)$. Show that

$$\iint_R f(x + y)\,dA = \int_0^1 u f(u)\,du$$

15 Review

Concept Check

1. Suppose f is a continuous function defined on a rectangle $R = [a, b] \times [c, d]$.
   (a) Write an expression for a double Riemann sum of f. If $f(x, y) \geqslant 0$, what does the sum represent?
   (b) Write the definition of $\iint_R f(x, y)\, dA$ as a limit.
   (c) What is the geometric interpretation of $\iint_R f(x, y)\, dA$ if $f(x, y) \geqslant 0$? What if f takes on both positive and negative values?
   (d) How do you evaluate $\iint_R f(x, y)\, dA$?
   (e) What does the Midpoint Rule for double integrals say?
   (f) Write an expression for the average value of f.

2. (a) How do you define $\iint_D f(x, y)\, dA$ if D is a bounded region that is not a rectangle?
   (b) What is a type I region? How do you evaluate $\iint_D f(x, y)\, dA$ if D is a type I region?
   (c) What is a type II region? How do you evaluate $\iint_D f(x, y)\, dA$ if D is a type II region?
   (d) What properties do double integrals have?

3. How do you change from rectangular coordinates to polar coordinates in a double integral? Why would you want to make the change?

4. If a lamina occupies a plane region D and has density function $\rho(x, y)$, write expressions for each of the following in terms of double integrals.
   (a) The mass
   (b) The moments about the axes
   (c) The center of mass
   (d) The moments of inertia about the axes and the origin

5. Let f be a joint density function of a pair of continuous random variables X and Y.
   (a) Write a double integral for the probability that X lies between a and b and Y lies between c and d.
   (b) What properties does f possess?
   (c) What are the expected values of X and Y?

6. Write an expression for the area of a surface with equation $z = f(x, y)$, $(x, y) \in D$.

7. (a) Write the definition of the triple integral of f over a rectangular box B.
   (b) How do you evaluate $\iiint_B f(x, y, z)\, dV$?
   (c) How do you define $\iiint_E f(x, y, z)\, dV$ if E is a bounded solid region that is not a box?
   (d) What is a type 1 solid region? How do you evaluate $\iiint_E f(x, y, z)\, dV$ if E is such a region?
   (e) What is a type 2 solid region? How do you evaluate $\iiint_E f(x, y, z)\, dV$ if E is such a region?
   (f) What is a type 3 solid region? How do you evaluate $\iiint_E f(x, y, z)\, dV$ if E is such a region?

8. Suppose a solid object occupies the region E and has density function $\rho(x, y, z)$. Write expressions for each of the following.
   (a) The mass
   (b) The moments about the coordinate planes
   (c) The coordinates of the center of mass
   (d) The moments of inertia about the axes

9. (a) How do you change from rectangular coordinates to cylindrical coordinates in a triple integral?
   (b) How do you change from rectangular coordinates to spherical coordinates in a triple integral?
   (c) In what situations would you change to cylindrical or spherical coordinates?

10. (a) If a transformation T is given by $x = g(u, v)$, $y = h(u, v)$, what is the Jacobian of T?
    (b) How do you change variables in a double integral?
    (c) How do you change variables in a triple integral?

True-False Quiz

Determine whether the statement is true or false. If it is true, explain why. If it is false, explain why or give an example that disproves the statement.

1. $\int_{-1}^{2} \int_{0}^{6} x^2 \sin(x - y)\, dx\, dy = \int_{0}^{6} \int_{-1}^{2} x^2 \sin(x - y)\, dy\, dx$

2. $\int_{0}^{1} \int_{0}^{x} \sqrt{x + y^2}\, dy\, dx = \int_{0}^{x} \int_{0}^{1} \sqrt{x + y^2}\, dx\, dy$

3. $\int_{1}^{2} \int_{3}^{4} x^2 e^y\, dy\, dx = \int_{1}^{2} x^2\, dx \int_{3}^{4} e^y\, dy$

4. $\int_{-1}^{1} \int_{0}^{1} e^{x^2+y^2} \sin y\, dx\, dy = 0$

5. If f is continuous on $[0, 1]$, then

$$\int_0^1 \int_0^1 f(x) f(y)\, dy\, dx = \left[\int_0^1 f(x)\, dx \right]^2$$

6. $\int_{1}^{4} \int_{0}^{1} \left(x^2 + \sqrt{y}\right) \sin(x^2 y^2)\, dx\, dy \leqslant 9$

7. If D is the disk given by $x^2 + y^2 \leqslant 4$, then

$$\iint_D \sqrt{4 - x^2 - y^2}\, dA = \tfrac{16}{3}\pi$$

8. The integral $\iiint_E kr^3\, dz\, dr\, d\theta$ represents the moment of inertia about the z-axis of a solid E with constant density k.

9. The integral

$$\int_0^{2\pi} \int_0^2 \int_r^2 dz\, dr\, d\theta$$

represents the volume enclosed by the cone $z = \sqrt{x^2 + y^2}$ and the plane $z = 2$.

Exercises

1. A contour map is shown for a function f on the square $R = [0, 3] \times [0, 3]$. Use a Riemann sum with nine terms to estimate the value of $\iint_R f(x, y)\, dA$. Take the sample points to be the upper right corners of the squares.

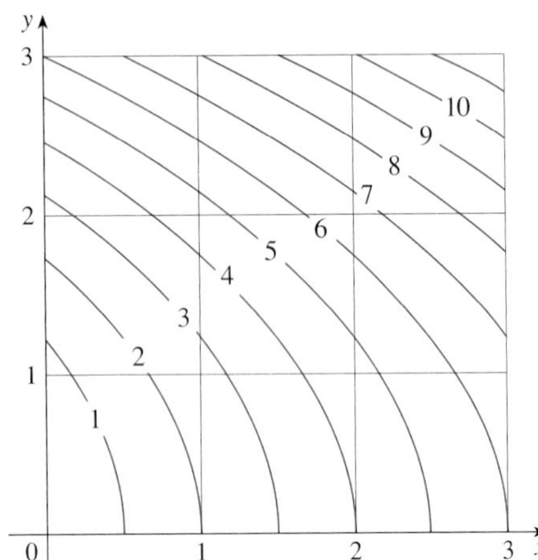

2. Use the Midpoint Rule to estimate the integral in Exercise 1.

3–8 Calculate the iterated integral.

3. $\int_1^2 \int_0^2 (y + 2xe^y)\, dx\, dy$

4. $\int_0^1 \int_0^1 ye^{xy}\, dx\, dy$

5. $\int_0^1 \int_0^x \cos(x^2)\, dy\, dx$

6. $\int_0^1 \int_x^{e^x} 3xy^2\, dy\, dx$

7. $\int_0^\pi \int_0^1 \int_0^{\sqrt{1-y^2}} y \sin x\, dz\, dy\, dx$

8. $\int_0^1 \int_0^y \int_x^1 6xyz\, dz\, dx\, dy$

9–10 Write $\iint_R f(x, y)\, dA$ as an iterated integral, where R is the region shown and f is an arbitrary continuous function on R.

9.

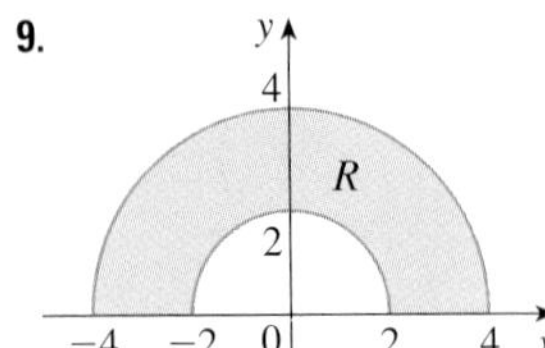

10.

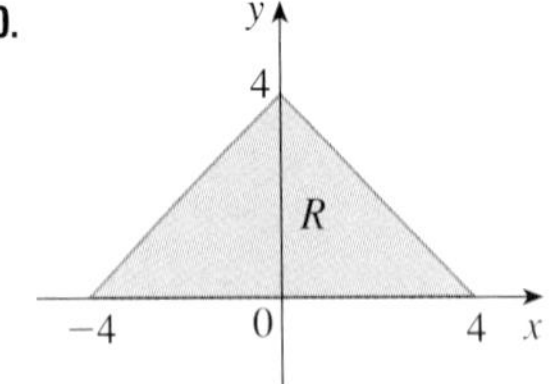

11. Describe the region whose area is given by the integral

$$\int_0^{\pi/2} \int_0^{\sin 2\theta} r\, dr\, d\theta$$

12. Describe the solid whose volume is given by the integral

$$\int_0^{\pi/2} \int_0^{\pi/2} \int_1^2 \rho^2 \sin \phi\, d\rho\, d\phi\, d\theta$$

and evaluate the integral.

13–14 Calculate the iterated integral by first reversing the order of integration.

13. $\int_0^1 \int_x^1 \cos(y^2)\, dy\, dx$

14. $\int_0^1 \int_{\sqrt{y}}^1 \dfrac{ye^{x^2}}{x^3}\, dx\, dy$

15–28 Calculate the value of the multiple integral.

15. $\iint_R ye^{xy}\, dA$, where $R = \{(x, y) \mid 0 \leqslant x \leqslant 2,\ 0 \leqslant y \leqslant 3\}$

16. $\iint_D xy\, dA$, where $D = \{(x, y) \mid 0 \leqslant y \leqslant 1,\ y^2 \leqslant x \leqslant y + 2\}$

17. $\displaystyle\iint_D \frac{y}{1 + x^2}\, dA$,
where D is bounded by $y = \sqrt{x}$, $y = 0$, $x = 1$

18. $\displaystyle\iint_D \frac{1}{1 + x^2}\, dA$, where D is the triangular region with vertices $(0, 0)$, $(1, 1)$, and $(0, 1)$

19. $\iint_D y\, dA$, where D is the region in the first quadrant bounded by the parabolas $x = y^2$ and $x = 8 - y^2$

20. $\iint_D y\, dA$, where D is the region in the first quadrant that lies above the hyperbola $xy = 1$ and the line $y = x$ and below the line $y = 2$

21. $\iint_D (x^2 + y^2)^{3/2}\, dA$, where D is the region in the first quadrant bounded by the lines $y = 0$ and $y = \sqrt{3}\, x$ and the circle $x^2 + y^2 = 9$

22. $\iint_D x\, dA$, where D is the region in the first quadrant that lies between the circles $x^2 + y^2 = 1$ and $x^2 + y^2 = 2$

23. $\iiint_E xy\, dV$, where
$E = \{(x, y, z) \mid 0 \leqslant x \leqslant 3,\ 0 \leqslant y \leqslant x,\ 0 \leqslant z \leqslant x + y\}$

24. $\iiint_T xy\, dV$, where T is the solid tetrahedron with vertices $(0, 0, 0)$, $(\frac{1}{3}, 0, 0)$, $(0, 1, 0)$, and $(0, 0, 1)$

25. $\iiint_E y^2 z^2\, dV$, where E is bounded by the paraboloid $x = 1 - y^2 - z^2$ and the plane $x = 0$

26. $\iiint_E z\,dV$, where E is bounded by the planes $y = 0$, $z = 0$, $x + y = 2$ and the cylinder $y^2 + z^2 = 1$ in the first octant

27. $\iiint_E yz\,dV$, where E lies above the plane $z = 0$, below the plane $z = y$, and inside the cylinder $x^2 + y^2 = 4$

28. $\iiint_H z^3\sqrt{x^2 + y^2 + z^2}\,dV$, where H is the solid hemisphere that lies above the xy-plane and has center the origin and radius 1

29–34 Find the volume of the given solid.

29. Under the paraboloid $z = x^2 + 4y^2$ and above the rectangle $R = [0, 2] \times [1, 4]$

30. Under the surface $z = x^2y$ and above the triangle in the xy-plane with vertices $(1, 0)$, $(2, 1)$, and $(4, 0)$

31. The solid tetrahedron with vertices $(0, 0, 0)$, $(0, 0, 1)$, $(0, 2, 0)$, and $(2, 2, 0)$

32. Bounded by the cylinder $x^2 + y^2 = 4$ and the planes $z = 0$ and $y + z = 3$

33. One of the wedges cut from the cylinder $x^2 + 9y^2 = a^2$ by the planes $z = 0$ and $z = mx$

34. Above the paraboloid $z = x^2 + y^2$ and below the half-cone $z = \sqrt{x^2 + y^2}$

35. Consider a lamina that occupies the region D bounded by the parabola $x = 1 - y^2$ and the coordinate axes in the first quadrant with density function $\rho(x, y) = y$.
(a) Find the mass of the lamina.
(b) Find the center of mass.
(c) Find the moments of inertia and radii of gyration about the x- and y-axes.

36. A lamina occupies the part of the disk $x^2 + y^2 \leqslant a^2$ that lies in the first quadrant.
(a) Find the centroid of the lamina.
(b) Find the center of mass of the lamina if the density function is $\rho(x, y) = xy^2$.

37. (a) Find the centroid of a right circular cone with height h and base radius a. (Place the cone so that its base is in the xy-plane with center the origin and its axis along the positive z-axis.)
(b) Find the moment of inertia of the cone about its axis (the z-axis).

38. Find the area of the part of the cone $z^2 = a^2(x^2 + y^2)$ between the planes $z = 1$ and $z = 2$.

39. Find the area of the part of the surface $z = x^2 + y$ that lies above the triangle with vertices $(0, 0)$, $(1, 0)$, and $(0, 2)$.

CAS 40. Graph the surface $z = x \sin y$, $-3 \leqslant x \leqslant 3$, $-\pi \leqslant y \leqslant \pi$, and find its surface area correct to four decimal places.

41. Use polar coordinates to evaluate

$$\int_0^3 \int_{-\sqrt{9-x^2}}^{\sqrt{9-x^2}} (x^3 + xy^2)\,dy\,dx$$

42. Use spherical coordinates to evaluate

$$\int_{-2}^{2} \int_0^{\sqrt{4-y^2}} \int_{-\sqrt{4-x^2-y^2}}^{\sqrt{4-x^2-y^2}} y^2\sqrt{x^2 + y^2 + z^2}\,dz\,dx\,dy$$

43. If D is the region bounded by the curves $y = 1 - x^2$ and $y = e^x$, find the approximate value of the integral $\iint_D y^2\,dA$. (Use a graphing device to estimate the points of intersection of the curves.)

CAS 44. Find the center of mass of the solid tetrahedron with vertices $(0, 0, 0)$, $(1, 0, 0)$, $(0, 2, 0)$, $(0, 0, 3)$ and density function $\rho(x, y, z) = x^2 + y^2 + z^2$.

45. The joint density function for random variables X and Y is

$$f(x, y) = \begin{cases} C(x + y) & \text{if } 0 \leqslant x \leqslant 3,\ 0 \leqslant y \leqslant 2 \\ 0 & \text{otherwise} \end{cases}$$

(a) Find the value of the constant C.
(b) Find $P(X \leqslant 2, Y \geqslant 1)$.
(c) Find $P(X + Y \leqslant 1)$.

46. A lamp has three bulbs, each of a type with average lifetime 800 hours. If we model the probability of failure of the bulbs by an exponential density function with mean 800, find the probability that all three bulbs fail within a total of 1000 hours.

47. Rewrite the integral

$$\int_{-1}^{1} \int_{x^2}^{1} \int_0^{1-y} f(x, y, z)\,dz\,dy\,dx$$

as an iterated integral in the order $dx\,dy\,dz$.

48. Give five other iterated integrals that are equal to

$$\int_0^2 \int_0^{y^3} \int_0^{y^2} f(x, y, z)\,dz\,dx\,dy$$

49. Use the transformation $u = x - y$, $v = x + y$ to evaluate

$$\iint_R \frac{x - y}{x + y}\,dA$$

where R is the square with vertices $(0, 2)$, $(1, 1)$, $(2, 2)$, and $(1, 3)$.

50. Use the transformation $x = u^2$, $y = v^2$, $z = w^2$ to find the volume of the region bounded by the surface $\sqrt{x} + \sqrt{y} + \sqrt{z} = 1$ and the coordinate planes.

51. Use the change of variables formula and an appropriate transformation to evaluate $\iint_R xy\,dA$, where R is the square with vertices $(0, 0)$, $(1, 1)$, $(2, 0)$, and $(1, -1)$.

52. The **Mean Value Theorem for double integrals** says that if f is a continuous function on a plane region D that is of type I or II, then there exists a point (x_0, y_0) in D such that

$$\iint_D f(x, y)\,dA = f(x_0, y_0)A(D)$$

Use the Extreme Value Theorem (14.7.8) and Property 15.3.11 of integrals to prove this theorem. (Use the proof of the single-variable version in Section 6.5 as a guide.)

53. Suppose that f is continuous on a disk that contains the point (a, b). Let D_r be the closed disk with center (a, b) and radius r. Use the Mean Value Theorem for double integrals (see Exercise 52) to show that

$$\lim_{r \to 0} \frac{1}{\pi r^2} \iint_{D_r} f(x, y)\,dA = f(a, b)$$

54. (a) Evaluate $\displaystyle\iint_D \frac{1}{(x^2 + y^2)^{n/2}}\,dA$, where n is an integer and D is the region bounded by the circles with center the origin and radii r and R, $0 < r < R$.

(b) For what values of n does the integral in part (a) have a limit as $r \to 0^+$?

(c) Find $\displaystyle\iiint_E \frac{1}{(x^2 + y^2 + z^2)^{n/2}}\,dV$, where E is the region bounded by the spheres with center the origin and radii r and R, $0 < r < R$.

(d) For what values of n does the integral in part (c) have a limit as $r \to 0^+$?

1. If $[\![x]\!]$ denotes the greatest integer in x, evaluate the integral

$$\iint_R [\![x + y]\!]\, dA$$

where $R = \{(x, y) \mid 1 \leqslant x \leqslant 3,\ 2 \leqslant y \leqslant 5\}$.

2. Evaluate the integral

$$\int_0^1 \int_0^1 e^{\max\{x^2, y^2\}}\, dy\, dx$$

where $\max\{x^2, y^2\}$ means the larger of the numbers x^2 and y^2.

3. Find the average value of the function $f(x) = \int_x^1 \cos(t^2)\, dt$ on the interval $[0, 1]$.

4. If $\mathbf{a}$, $\mathbf{b}$, and $\mathbf{c}$ are constant vectors, $\mathbf{r}$ is the position vector $x\,\mathbf{i} + y\,\mathbf{j} + z\,\mathbf{k}$, and E is given by the inequalities $0 \leqslant \mathbf{a} \cdot \mathbf{r} \leqslant \alpha$, $0 \leqslant \mathbf{b} \cdot \mathbf{r} \leqslant \beta$, $0 \leqslant \mathbf{c} \cdot \mathbf{r} \leqslant \gamma$, show that

$$\iiint_E (\mathbf{a} \cdot \mathbf{r})(\mathbf{b} \cdot \mathbf{r})(\mathbf{c} \cdot \mathbf{r})\, dV = \frac{(\alpha\beta\gamma)^2}{8\,|\mathbf{a} \cdot (\mathbf{b} \times \mathbf{c})|}$$

5. The double integral $\displaystyle\int_0^1 \int_0^1 \frac{1}{1 - xy}\, dx\, dy$ is an improper integral and could be defined as the limit of double integrals over the rectangle $[0, t] \times [0, t]$ as $t \to 1^-$. But if we expand the integrand as a geometric series, we can express the integral as the sum of an infinite series. Show that

$$\int_0^1 \int_0^1 \frac{1}{1 - xy}\, dx\, dy = \sum_{n=1}^{\infty} \frac{1}{n^2}$$

6. Leonhard Euler was able to find the exact sum of the series in Problem 5. In 1736 he proved that

$$\sum_{n=1}^{\infty} \frac{1}{n^2} = \frac{\pi^2}{6}$$

In this problem we ask you to prove this fact by evaluating the double integral in Problem 5. Start by making the change of variables

$$x = \frac{u - v}{\sqrt{2}} \qquad y = \frac{u + v}{\sqrt{2}}$$

This gives a rotation about the origin through the angle $\pi/4$. You will need to sketch the corresponding region in the uv-plane.

[*Hint:* If, in evaluating the integral, you encounter either of the expressions $(1 - \sin\theta)/\cos\theta$ or $(\cos\theta)/(1 + \sin\theta)$, you might like to use the identity $\cos\theta = \sin((\pi/2) - \theta)$ and the corresponding identity for $\sin\theta$.]

7. (a) Show that

$$\int_0^1 \int_0^1 \int_0^1 \frac{1}{1 - xyz}\, dx\, dy\, dz = \sum_{n=1}^{\infty} \frac{1}{n^3}$$

(Nobody has ever been able to find the exact value of the sum of this series.)

(b) Show that

$$\int_0^1 \int_0^1 \int_0^1 \frac{1}{1 + xyz}\, dx\, dy\, dz = \sum_{n=1}^{\infty} \frac{(-1)^{n-1}}{n^3}$$

Use this equation to evaluate the triple integral correct to two decimal places.

8. Show that

$$\int_0^\infty \frac{\arctan \pi x - \arctan x}{x}\,dx = \frac{\pi}{2}\ln \pi$$

by first expressing the integral as an iterated integral.

9. (a) Show that when Laplace's equation

$$\frac{\partial^2 u}{\partial x^2} + \frac{\partial^2 u}{\partial y^2} + \frac{\partial^2 u}{\partial z^2} = 0$$

is written in cylindrical coordinates, it becomes

$$\frac{\partial^2 u}{\partial r^2} + \frac{1}{r}\frac{\partial u}{\partial r} + \frac{1}{r^2}\frac{\partial^2 u}{\partial \theta^2} + \frac{\partial^2 u}{\partial z^2} = 0$$

(b) Show that when Laplace's equation is written in spherical coordinates, it becomes

$$\frac{\partial^2 u}{\partial \rho^2} + \frac{2}{\rho}\frac{\partial u}{\partial \rho} + \frac{\cot \phi}{\rho^2}\frac{\partial u}{\partial \phi} + \frac{1}{\rho^2}\frac{\partial^2 u}{\partial \phi^2} + \frac{1}{\rho^2 \sin^2\phi}\frac{\partial^2 u}{\partial \theta^2} = 0$$

10. (a) A lamina has constant density ρ and takes the shape of a disk with center the origin and radius R. Use Newton's Law of Gravitation (see Section 13.4) to show that the magnitude of the force of attraction that the lamina exerts on a body with mass m located at the point $(0, 0, d)$ on the positive z-axis is

$$F = 2\pi G m \rho d\left(\frac{1}{d} - \frac{1}{\sqrt{R^2 + d^2}}\right)$$

[*Hint:* Divide the disk as in Figure 4 in Section 15.4 and first compute the vertical component of the force exerted by the polar subrectangle R_{ij}.]

(b) Show that the magnitude of the force of attraction of a lamina with density ρ that occupies an entire plane on an object with mass m located at a distance d from the plane is

$$F = 2\pi G m \rho$$

Notice that this expression does not depend on d.

11. If f is continuous, show that

$$\int_0^x \int_0^y \int_0^z f(t)\,dt\,dz\,dy = \tfrac{1}{2}\int_0^x (x - t)^2 f(t)\,dt$$

12. Evaluate $\displaystyle\lim_{n\to\infty} n^{-2}\sum_{i=1}^{n}\sum_{j=1}^{n^2}\frac{1}{\sqrt{n^2 + ni + j}}$.

13. The plane

$$\frac{x}{a} + \frac{y}{b} + \frac{z}{c} = 1 \qquad a > 0, \quad b > 0, \quad c > 0$$

cuts the solid ellipsoid

$$\frac{x^2}{a^2} + \frac{y^2}{b^2} + \frac{z^2}{c^2} \leqslant 1$$

into two pieces. Find the volume of the smaller piece.

16 Vector Calculus

Parametric surfaces, which are studied in Section 16.6, are frequently used by programmers creating animated films. In this scene from Antz, Princess Bala is about to try to rescue Z, who is trapped in a dewdrop. A parametric surface represents the dewdrop and a family of such surfaces depicts its motion. One of the programmers for this film was heard to say, "I wish I had paid more attention in calculus class when we were studying parametric surfaces. It would sure have helped me today."

In this chapter we study the calculus of vector fields. (These are functions that assign vectors to points in space.) In particular we define line integrals (which can be used to find the work done by a force field in moving an object along a curve). Then we define surface integrals (which can be used to find the rate of fluid flow across a surface). The connections between these new types of integrals and the single, double, and triple integrals that we have already met are given by the higher-dimensional versions of the Fundamental Theorem of Calculus: Green's Theorem, Stokes' Theorem, and the Divergence Theorem.

16.1 Vector Fields

The vectors in Figure 1 are air velocity vectors that indicate the wind speed and direction at points 10 m above the surface elevation in the San Francisco Bay area. We see at a glance from the largest arrows in part (a) that the greatest wind speeds at that time occurred as the winds entered the bay across the Golden Gate Bridge. Part (b) shows the very different wind pattern 12 hours earlier. Associated with every point in the air we can imagine a wind velocity vector. This is an example of a *velocity vector field.*

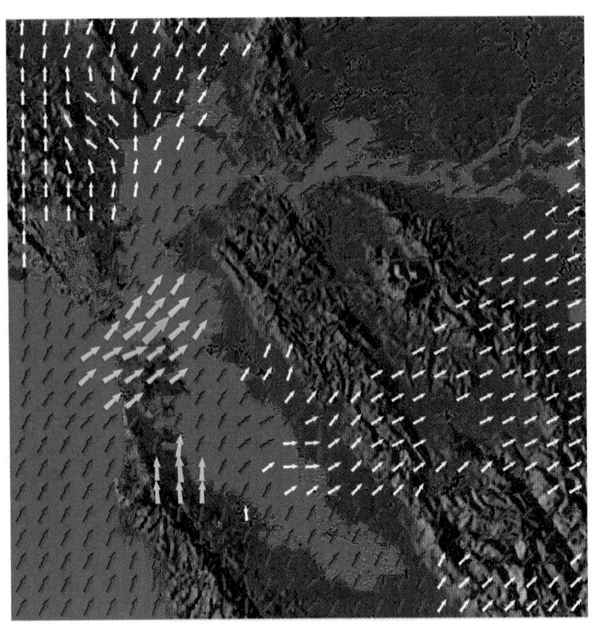

(a) 6:00 PM, March 1, 2010

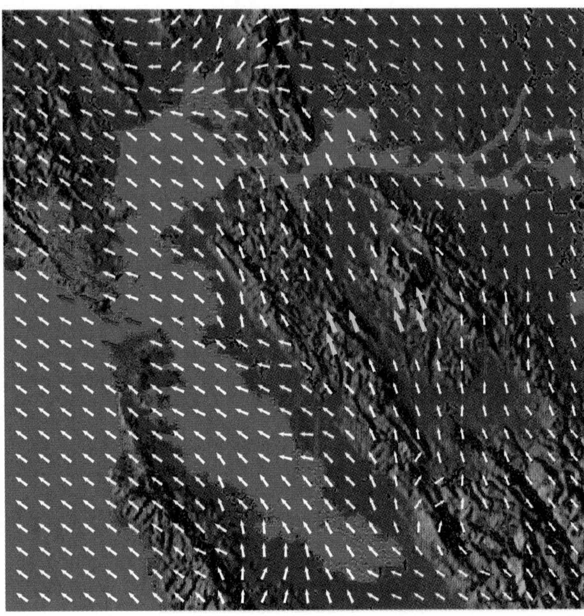

(b) 6:00 AM, March 1, 2010

FIGURE 1 Velocity vector fields showing San Francisco Bay wind patterns

Other examples of velocity vector fields are illustrated in Figure 2: ocean currents and flow past an airfoil.

(a) Ocean currents off the coast of Nova Scotia

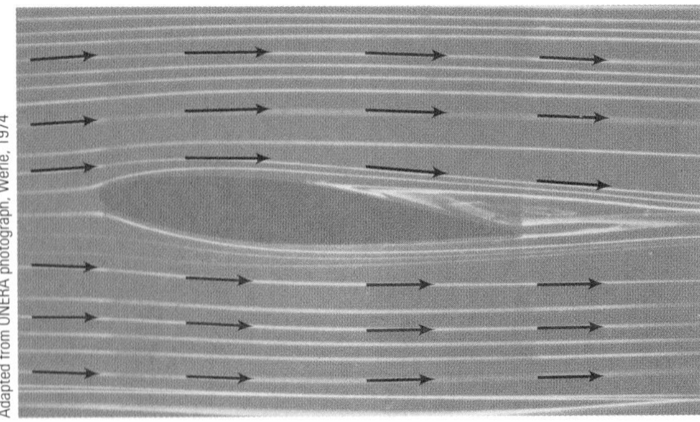

(b) Airflow past an inclined airfoil

FIGURE 2 Velocity vector fields

Another type of vector field, called a *force field*, associates a force vector with each point in a region. An example is the gravitational force field that we will look at in Example 4.

In general, a vector field is a function whose domain is a set of points in $\mathbb{R}^2$ (or $\mathbb{R}^3$) and whose range is a set of vectors in V_2 (or V_3).

> **1 Definition** Let D be a set in $\mathbb{R}^2$ (a plane region). A **vector field on** $\mathbb{R}^2$ is a function $\mathbf{F}$ that assigns to each point (x, y) in D a two-dimensional vector $\mathbf{F}(x, y)$.

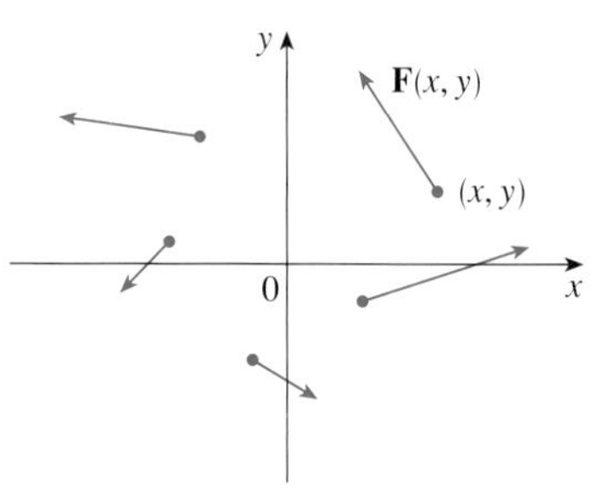

FIGURE 3
Vector field on $\mathbb{R}^2$

The best way to picture a vector field is to draw the arrow representing the vector $\mathbf{F}(x, y)$ starting at the point (x, y). Of course, it's impossible to do this for all points (x, y), but we can gain a reasonable impression of $\mathbf{F}$ by doing it for a few representative points in D as in Figure 3. Since $\mathbf{F}(x, y)$ is a two-dimensional vector, we can write it in terms of its **component functions** P and Q as follows:

$$\mathbf{F}(x, y) = P(x, y)\,\mathbf{i} + Q(x, y)\,\mathbf{j} = \langle P(x, y), Q(x, y)\rangle$$

or, for short,

$$\mathbf{F} = P\,\mathbf{i} + Q\,\mathbf{j}$$

Notice that P and Q are scalar functions of two variables and are sometimes called **scalar fields** to distinguish them from vector fields.

> **2 Definition** Let E be a subset of $\mathbb{R}^3$. A **vector field on** $\mathbb{R}^3$ is a function $\mathbf{F}$ that assigns to each point (x, y, z) in E a three-dimensional vector $\mathbf{F}(x, y, z)$.

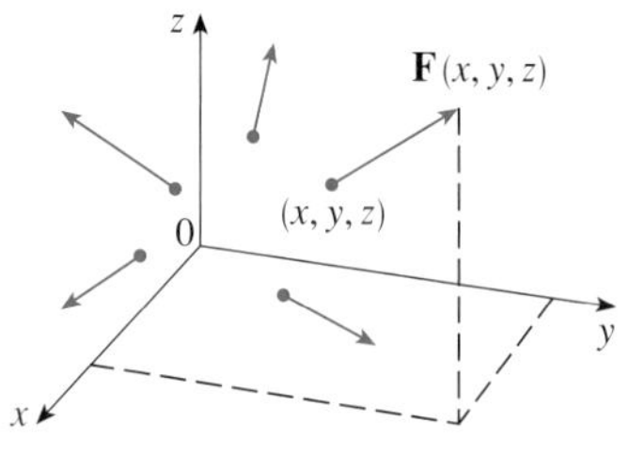

FIGURE 4
Vector field on $\mathbb{R}^3$

A vector field $\mathbf{F}$ on $\mathbb{R}^3$ is pictured in Figure 4. We can express it in terms of its component functions P, Q, and R as

$$\mathbf{F}(x, y, z) = P(x, y, z)\,\mathbf{i} + Q(x, y, z)\,\mathbf{j} + R(x, y, z)\,\mathbf{k}$$

As with the vector functions in Section 13.1, we can define continuity of vector fields and show that $\mathbf{F}$ is continuous if and only if its component functions P, Q, and R are continuous.

We sometimes identify a point (x, y, z) with its position vector $\mathbf{x} = \langle x, y, z\rangle$ and write $\mathbf{F}(\mathbf{x})$ instead of $\mathbf{F}(x, y, z)$. Then $\mathbf{F}$ becomes a function that assigns a vector $\mathbf{F}(\mathbf{x})$ to a vector $\mathbf{x}$.

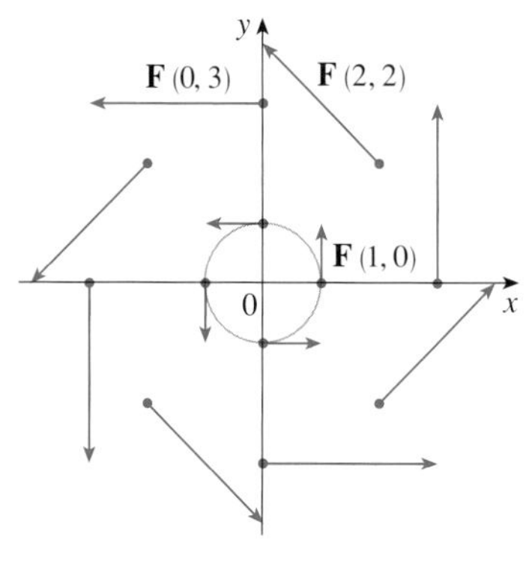

FIGURE 5
$\mathbf{F}(x, y) = -y\,\mathbf{i} + x\,\mathbf{j}$

V EXAMPLE 1 A vector field on $\mathbb{R}^2$ is defined by $\mathbf{F}(x, y) = -y\,\mathbf{i} + x\,\mathbf{j}$. Describe $\mathbf{F}$ by sketching some of the vectors $\mathbf{F}(x, y)$ as in Figure 3.

SOLUTION Since $\mathbf{F}(1, 0) = \mathbf{j}$, we draw the vector $\mathbf{j} = \langle 0, 1\rangle$ starting at the point $(1, 0)$ in Figure 5. Since $\mathbf{F}(0, 1) = -\mathbf{i}$, we draw the vector $\langle -1, 0\rangle$ with starting point $(0, 1)$. Continuing in this way, we calculate several other representative values of $\mathbf{F}(x, y)$ in the table and draw the corresponding vectors to represent the vector field in Figure 5.

(x, y)	$\mathbf{F}(x, y)$	(x, y)	$\mathbf{F}(x, y)$
$(1, 0)$	$\langle 0, 1\rangle$	$(-1, 0)$	$\langle 0, -1\rangle$
$(2, 2)$	$\langle -2, 2\rangle$	$(-2, -2)$	$\langle 2, -2\rangle$
$(3, 0)$	$\langle 0, 3\rangle$	$(-3, 0)$	$\langle 0, -3\rangle$
$(0, 1)$	$\langle -1, 0\rangle$	$(0, -1)$	$\langle 1, 0\rangle$
$(-2, 2)$	$\langle -2, -2\rangle$	$(2, -2)$	$\langle 2, 2\rangle$
$(0, 3)$	$\langle -3, 0\rangle$	$(0, -3)$	$\langle 3, 0\rangle$

It appears from Figure 5 that each arrow is tangent to a circle with center the origin. To confirm this, we take the dot product of the position vector $\mathbf{x} = x\,\mathbf{i} + y\,\mathbf{j}$ with the vector $\mathbf{F}(\mathbf{x}) = \mathbf{F}(x, y)$:

$$\mathbf{x} \cdot \mathbf{F}(\mathbf{x}) = (x\,\mathbf{i} + y\,\mathbf{j}) \cdot (-y\,\mathbf{i} + x\,\mathbf{j}) = -xy + yx = 0$$

This shows that $\mathbf{F}(x, y)$ is perpendicular to the position vector $\langle x, y \rangle$ and is therefore tangent to a circle with center the origin and radius $|\mathbf{x}| = \sqrt{x^2 + y^2}$. Notice also that

$$|\mathbf{F}(x, y)| = \sqrt{(-y)^2 + x^2} = \sqrt{x^2 + y^2} = |\mathbf{x}|$$

so the magnitude of the vector $\mathbf{F}(x, y)$ is equal to the radius of the circle. ■

Some computer algebra systems are capable of plotting vector fields in two or three dimensions. They give a better impression of the vector field than is possible by hand because the computer can plot a large number of representative vectors. Figure 6 shows a computer plot of the vector field in Example 1; Figures 7 and 8 show two other vector fields. Notice that the computer scales the lengths of the vectors so they are not too long and yet are proportional to their true lengths.

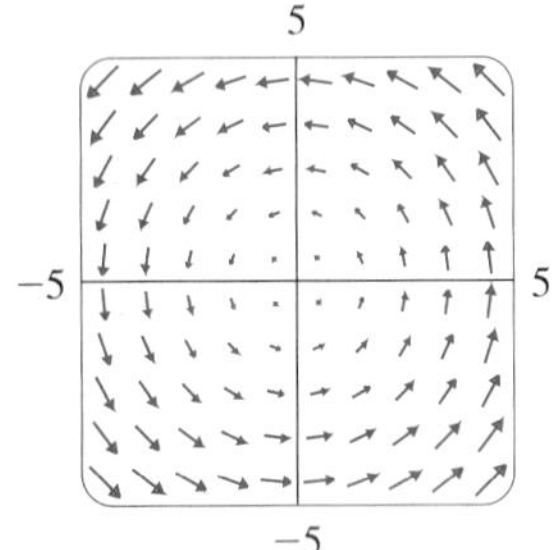

FIGURE 6
$\mathbf{F}(x, y) = \langle -y, x \rangle$

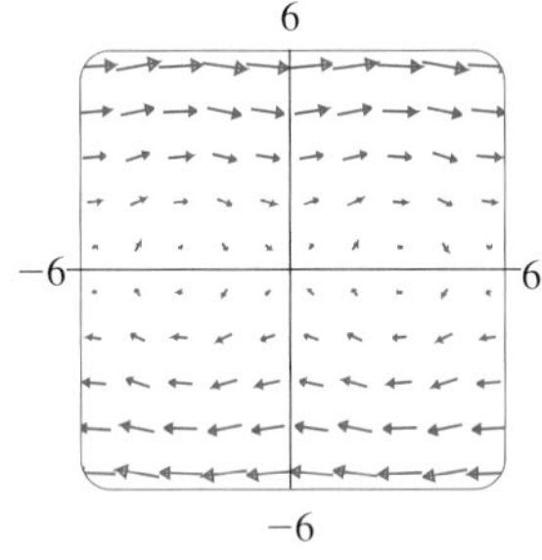

FIGURE 7
$\mathbf{F}(x, y) = \langle y, \sin x \rangle$

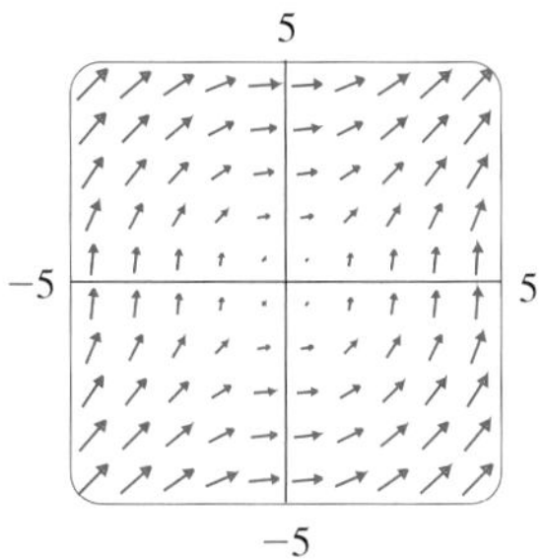

FIGURE 8
$\mathbf{F}(x, y) = \langle \ln(1 + y^2), \ln(1 + x^2) \rangle$

V EXAMPLE 2 Sketch the vector field on $\mathbb{R}^3$ given by $\mathbf{F}(x, y, z) = z\,\mathbf{k}$.

SOLUTION The sketch is shown in Figure 9. Notice that all vectors are vertical and point upward above the xy-plane or downward below it. The magnitude increases with the distance from the xy-plane.

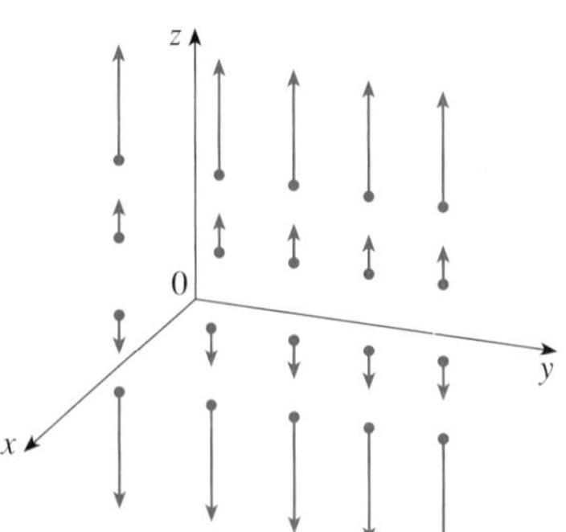

FIGURE 9
$\mathbf{F}(x, y, z) = z\,\mathbf{k}$

■

We were able to draw the vector field in Example 2 by hand because of its particularly simple formula. Most three-dimensional vector fields, however, are virtually impossible to sketch by hand and so we need to resort to a computer algebra system. Examples are shown in Figures 10, 11, and 12. Notice that the vector fields in Figures 10 and 11 have similar formulas, but all the vectors in Figure 11 point in the general direction of the negative y-axis because their y-components are all -2. If the vector field in Figure 12 represents a velocity field, then a particle would be swept upward and would spiral around the z-axis in the clockwise direction as viewed from above.

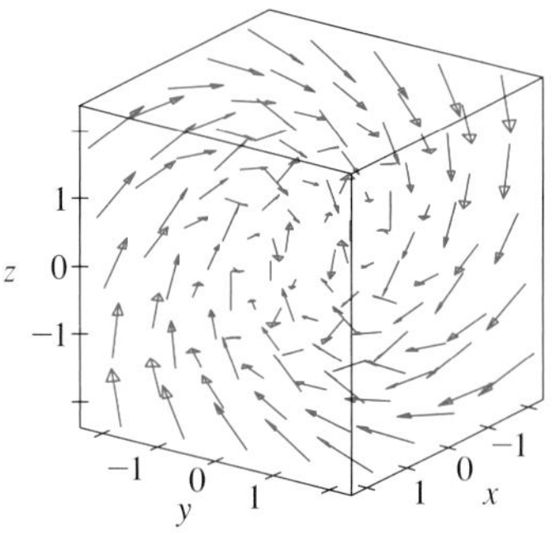

FIGURE 10
$\mathbf{F}(x, y, z) = y\,\mathbf{i} + z\,\mathbf{j} + x\,\mathbf{k}$

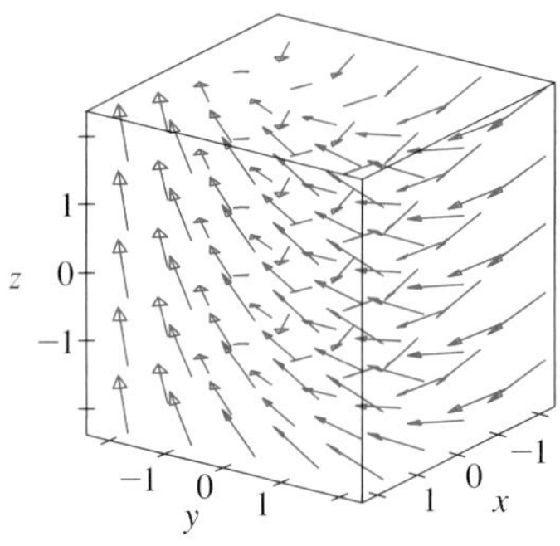

FIGURE 11
$\mathbf{F}(x, y, z) = y\,\mathbf{i} - 2\,\mathbf{j} + x\,\mathbf{k}$

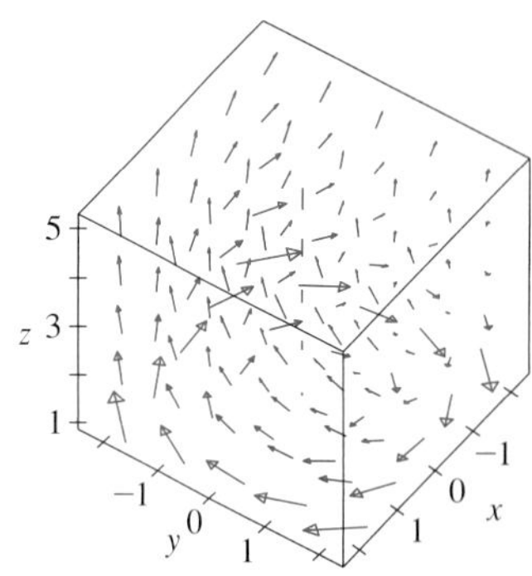

FIGURE 12
$\mathbf{F}(x, y, z) = \frac{y}{z}\,\mathbf{i} - \frac{x}{z}\,\mathbf{j} + \frac{z}{4}\,\mathbf{k}$

TEC In Visual 16.1 you can rotate the vector fields in Figures 10–12 as well as additional fields.

EXAMPLE 3 Imagine a fluid flowing steadily along a pipe and let $\mathbf{V}(x, y, z)$ be the velocity vector at a point (x, y, z). Then $\mathbf{V}$ assigns a vector to each point (x, y, z) in a certain domain E (the interior of the pipe) and so $\mathbf{V}$ is a vector field on $\mathbb{R}^3$ called a **velocity field**. A possible velocity field is illustrated in Figure 13. The speed at any given point is indicated by the length of the arrow.

Velocity fields also occur in other areas of physics. For instance, the vector field in Example 1 could be used as the velocity field describing the counterclockwise rotation of a wheel. We have seen other examples of velocity fields in Figures 1 and 2. ■

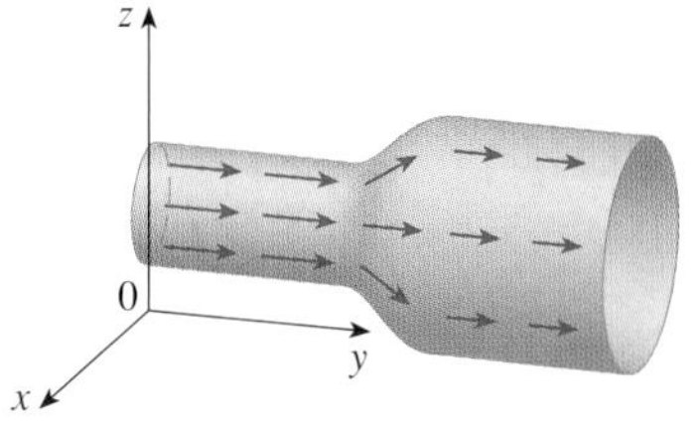

FIGURE 13
Velocity field in fluid flow

EXAMPLE 4 Newton's Law of Gravitation states that the magnitude of the gravitational force between two objects with masses m and M is

$$|\mathbf{F}| = \frac{mMG}{r^2}$$

where r is the distance between the objects and G is the gravitational constant. (This is an example of an inverse square law.) Let's assume that the object with mass M is located at the origin in $\mathbb{R}^3$. (For instance, M could be the mass of the earth and the origin would be at its center.) Let the position vector of the object with mass m be $\mathbf{x} = \langle x, y, z\rangle$. Then $r = |\mathbf{x}|$, so $r^2 = |\mathbf{x}|^2$. The gravitational force exerted on this second object acts toward the origin, and the unit vector in this direction is

$$-\frac{\mathbf{x}}{|\mathbf{x}|}$$

Therefore the gravitational force acting on the object at $\mathbf{x} = \langle x, y, z\rangle$ is

$$\boxed{3} \qquad \mathbf{F}(\mathbf{x}) = -\frac{mMG}{|\mathbf{x}|^3}\,\mathbf{x}$$

[Physicists often use the notation $\mathbf{r}$ instead of $\mathbf{x}$ for the position vector, so you may see

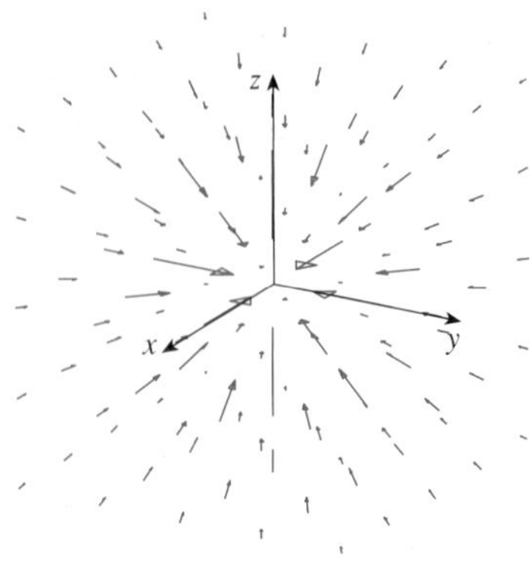

FIGURE 14
Gravitational force field

Formula 3 written in the form $\mathbf{F} = -(mMG/r^3)\mathbf{r}$.] The function given by Equation 3 is an example of a vector field, called the **gravitational field**, because it associates a vector [the force $\mathbf{F}(\mathbf{x})$] with every point $\mathbf{x}$ in space.

Formula 3 is a compact way of writing the gravitational field, but we can also write it in terms of its component functions by using the facts that $\mathbf{x} = x\,\mathbf{i} + y\,\mathbf{j} + z\,\mathbf{k}$ and $|\mathbf{x}| = \sqrt{x^2 + y^2 + z^2}$:

$$\mathbf{F}(x, y, z) = \frac{-mMGx}{(x^2 + y^2 + z^2)^{3/2}}\,\mathbf{i} + \frac{-mMGy}{(x^2 + y^2 + z^2)^{3/2}}\,\mathbf{j} + \frac{-mMGz}{(x^2 + y^2 + z^2)^{3/2}}\,\mathbf{k}$$

The gravitational field $\mathbf{F}$ is pictured in Figure 14.

EXAMPLE 5 Suppose an electric charge Q is located at the origin. According to Coulomb's Law, the electric force $\mathbf{F}(\mathbf{x})$ exerted by this charge on a charge q located at a point (x, y, z) with position vector $\mathbf{x} = \langle x, y, z \rangle$ is

$$\boxed{4} \qquad \mathbf{F}(\mathbf{x}) = \frac{\varepsilon q Q}{|\mathbf{x}|^3}\,\mathbf{x}$$

where ε is a constant (that depends on the units used). For like charges, we have $qQ > 0$ and the force is repulsive; for unlike charges, we have $qQ < 0$ and the force is attractive. Notice the similarity between Formulas 3 and 4. Both vector fields are examples of **force fields**.

Instead of considering the electric force $\mathbf{F}$, physicists often consider the force per unit charge:

$$\mathbf{E}(\mathbf{x}) = \frac{1}{q}\,\mathbf{F}(\mathbf{x}) = \frac{\varepsilon Q}{|\mathbf{x}|^3}\,\mathbf{x}$$

Then $\mathbf{E}$ is a vector field on $\mathbb{R}^3$ called the **electric field** of Q.

Gradient Fields

If f is a scalar function of two variables, recall from Section 14.6 that its gradient ∇f (or grad f) is defined by

$$\nabla f(x, y) = f_x(x, y)\,\mathbf{i} + f_y(x, y)\,\mathbf{j}$$

Therefore ∇f is really a vector field on $\mathbb{R}^2$ and is called a **gradient vector field**. Likewise, if f is a scalar function of three variables, its gradient is a vector field on $\mathbb{R}^3$ given by

$$\nabla f(x, y, z) = f_x(x, y, z)\,\mathbf{i} + f_y(x, y, z)\,\mathbf{j} + f_z(x, y, z)\,\mathbf{k}$$

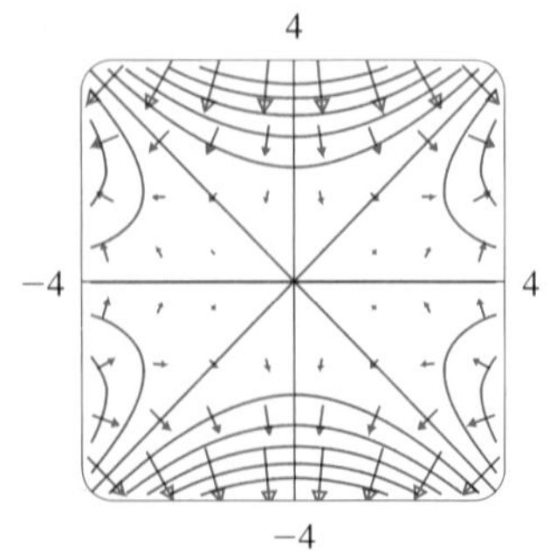

FIGURE 15

V EXAMPLE 6 Find the gradient vector field of $f(x, y) = x^2y - y^3$. Plot the gradient vector field together with a contour map of f. How are they related?

SOLUTION The gradient vector field is given by

$$\nabla f(x, y) = \frac{\partial f}{\partial x}\,\mathbf{i} + \frac{\partial f}{\partial y}\,\mathbf{j} = 2xy\,\mathbf{i} + (x^2 - 3y^2)\,\mathbf{j}$$

Figure 15 shows a contour map of f with the gradient vector field. Notice that the gradient vectors are perpendicular to the level curves, as we would expect from Section 14.6.

Notice also that the gradient vectors are long where the level curves are close to each other and short where the curves are farther apart. That's because the length of the gradient vector is the value of the directional derivative of f and closely spaced level curves indicate a steep graph. ■

A vector field $\mathbf{F}$ is called a **conservative vector field** if it is the gradient of some scalar function, that is, if there exists a function f such that $\mathbf{F} = \nabla f$. In this situation f is called a **potential function** for $\mathbf{F}$.

Not all vector fields are conservative, but such fields do arise frequently in physics. For example, the gravitational field $\mathbf{F}$ in Example 4 is conservative because if we define

$$f(x, y, z) = \frac{mMG}{\sqrt{x^2 + y^2 + z^2}}$$

then

$$\begin{aligned}\nabla f(x, y, z) &= \frac{\partial f}{\partial x}\,\mathbf{i} + \frac{\partial f}{\partial y}\,\mathbf{j} + \frac{\partial f}{\partial z}\,\mathbf{k} \\ &= \frac{-mMGx}{(x^2 + y^2 + z^2)^{3/2}}\,\mathbf{i} + \frac{-mMGy}{(x^2 + y^2 + z^2)^{3/2}}\,\mathbf{j} + \frac{-mMGz}{(x^2 + y^2 + z^2)^{3/2}}\,\mathbf{k} \\ &= \mathbf{F}(x, y, z)\end{aligned}$$

In Sections 16.3 and 16.5 we will learn how to tell whether or not a given vector field is conservative.

16.1 Exercises

1–10 Sketch the vector field $\mathbf{F}$ by drawing a diagram like Figure 5 or Figure 9.

1. $\mathbf{F}(x, y) = 0.3\,\mathbf{i} - 0.4\,\mathbf{j}$

2. $\mathbf{F}(x, y) = \frac{1}{2}x\,\mathbf{i} + y\,\mathbf{j}$

3. $\mathbf{F}(x, y) = -\frac{1}{2}\,\mathbf{i} + (y - x)\,\mathbf{j}$

4. $\mathbf{F}(x, y) = y\,\mathbf{i} + (x + y)\,\mathbf{j}$

5. $\mathbf{F}(x, y) = \dfrac{y\,\mathbf{i} + x\,\mathbf{j}}{\sqrt{x^2 + y^2}}$

6. $\mathbf{F}(x, y) = \dfrac{y\,\mathbf{i} - x\,\mathbf{j}}{\sqrt{x^2 + y^2}}$

7. $\mathbf{F}(x, y, z) = \mathbf{k}$

8. $\mathbf{F}(x, y, z) = -y\,\mathbf{k}$

9. $\mathbf{F}(x, y, z) = x\,\mathbf{k}$

10. $\mathbf{F}(x, y, z) = \mathbf{j} - \mathbf{i}$

11–14 Match the vector fields $\mathbf{F}$ with the plots labeled I–IV. Give reasons for your choices.

11. $\mathbf{F}(x, y) = \langle x, -y \rangle$

12. $\mathbf{F}(x, y) = \langle y, x - y \rangle$

13. $\mathbf{F}(x, y) = \langle y, y + 2 \rangle$

14. $\mathbf{F}(x, y) = \langle \cos(x + y), x \rangle$

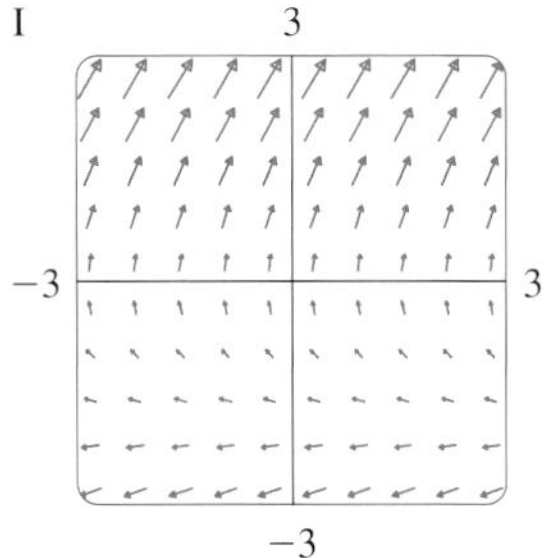

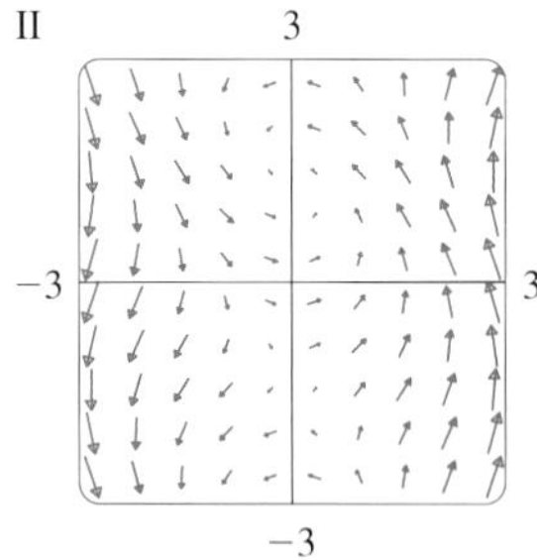

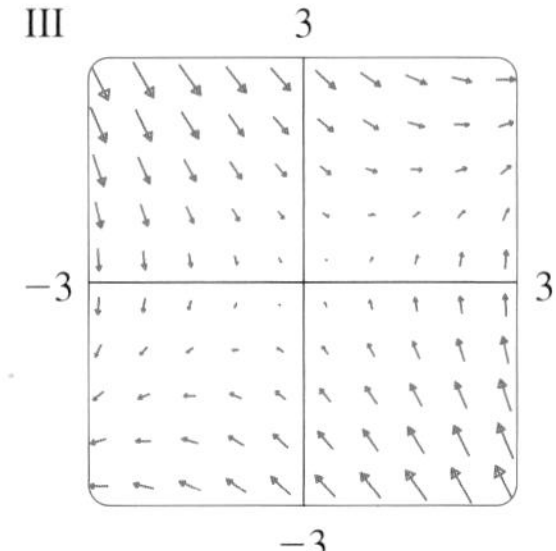

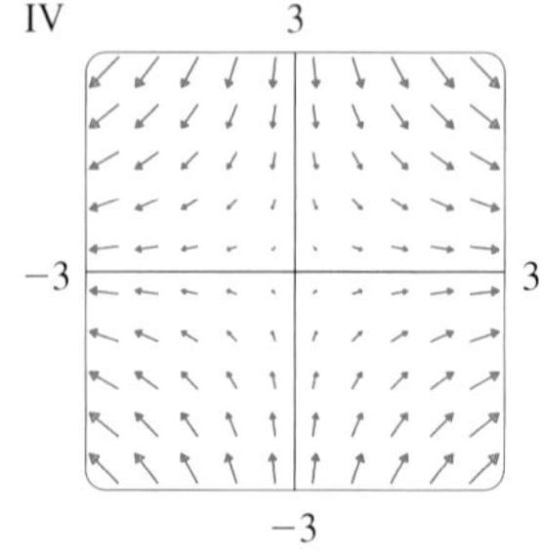

15–18 Match the vector fields $\mathbf{F}$ on $\mathbb{R}^3$ with the plots labeled I–IV. Give reasons for your choices.

15. $\mathbf{F}(x, y, z) = \mathbf{i} + 2\,\mathbf{j} + 3\,\mathbf{k}$ **16.** $\mathbf{F}(x, y, z) = \mathbf{i} + 2\,\mathbf{j} + z\,\mathbf{k}$

17. $\mathbf{F}(x, y, z) = x\,\mathbf{i} + y\,\mathbf{j} + 3\,\mathbf{k}$

18. $\mathbf{F}(x, y, z) = x\,\mathbf{i} + y\,\mathbf{j} + z\,\mathbf{k}$

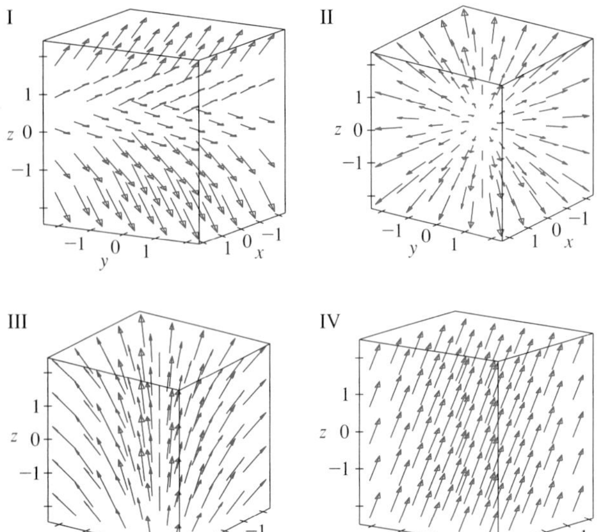

CAS 19. If you have a CAS that plots vector fields (the command is `fieldplot` in Maple and `PlotVectorField` or `VectorPlot` in Mathematica), use it to plot

$$\mathbf{F}(x, y) = (y^2 - 2xy)\,\mathbf{i} + (3xy - 6x^2)\,\mathbf{j}$$

Explain the appearance by finding the set of points (x, y) such that $\mathbf{F}(x, y) = \mathbf{0}$.

CAS 20. Let $\mathbf{F}(\mathbf{x}) = (r^2 - 2r)\mathbf{x}$, where $\mathbf{x} = \langle x, y \rangle$ and $r = |\mathbf{x}|$. Use a CAS to plot this vector field in various domains until you can see what is happening. Describe the appearance of the plot and explain it by finding the points where $\mathbf{F}(\mathbf{x}) = \mathbf{0}$.

21–24 Find the gradient vector field of f.

21. $f(x, y) = xe^{xy}$ **22.** $f(x, y) = \tan(3x - 4y)$

23. $f(x, y, z) = \sqrt{x^2 + y^2 + z^2}$

24. $f(x, y, z) = x \ln(y - 2z)$

25–26 Find the gradient vector field ∇f of f and sketch it.

25. $f(x, y) = x^2 - y$ **26.** $f(x, y) = \sqrt{x^2 + y^2}$

CAS 27–28 Plot the gradient vector field of f together with a contour map of f. Explain how they are related to each other.

27. $f(x, y) = \ln(1 + x^2 + 2y^2)$ **28.** $f(x, y) = \cos x - 2 \sin y$

29–32 Match the functions f with the plots of their gradient vector fields labeled I–IV. Give reasons for your choices.

29. $f(x, y) = x^2 + y^2$ **30.** $f(x, y) = x(x + y)$

31. $f(x, y) = (x + y)^2$ **32.** $f(x, y) = \sin\sqrt{x^2 + y^2}$

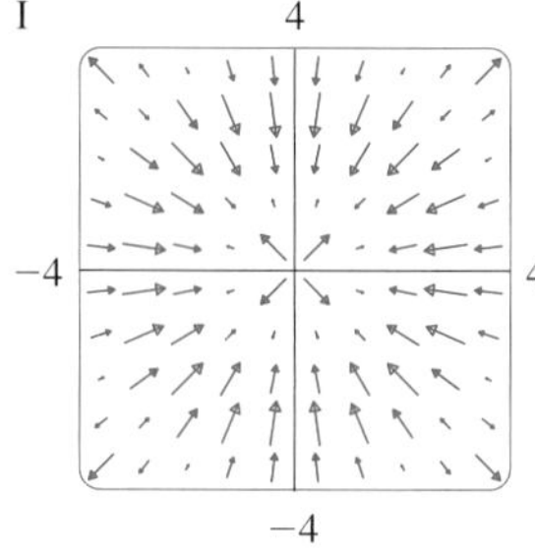

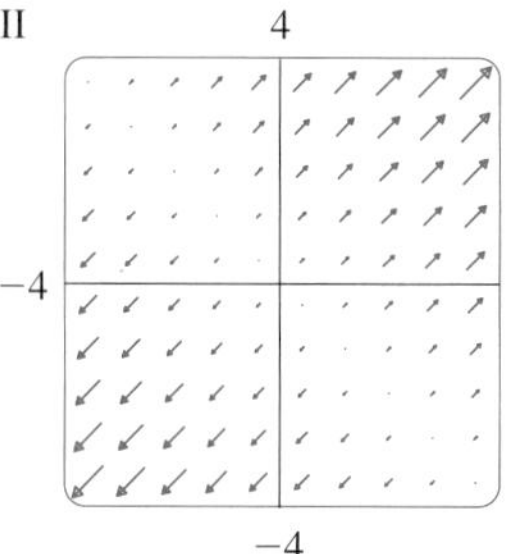

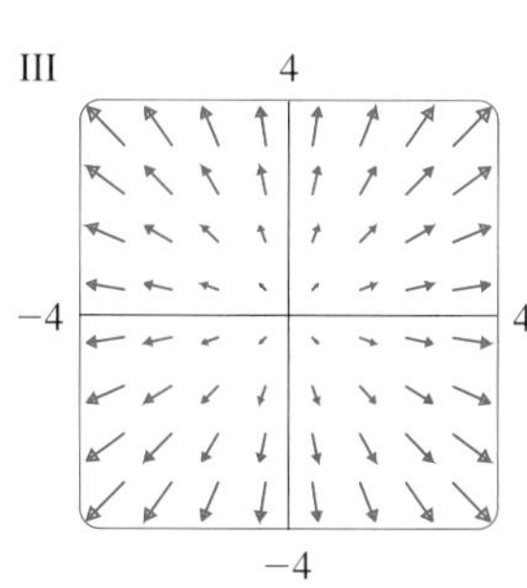

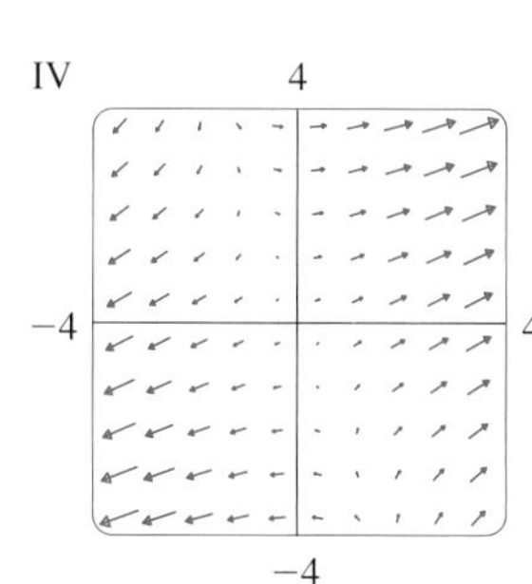

33. A particle moves in a velocity field $\mathbf{V}(x, y) = \langle x^2, x + y^2 \rangle$. If it is at position $(2, 1)$ at time $t = 3$, estimate its location at time $t = 3.01$.

34. At time $t = 1$, a particle is located at position $(1, 3)$. If it moves in a velocity field

$$\mathbf{F}(x, y) = \langle xy - 2, y^2 - 10 \rangle$$

find its approximate location at time $t = 1.05$.

35. The **flow lines** (or **streamlines**) of a vector field are the paths followed by a particle whose velocity field is the given vector field. Thus the vectors in a vector field are tangent to the flow lines.

(a) Use a sketch of the vector field $\mathbf{F}(x, y) = x\,\mathbf{i} - y\,\mathbf{j}$ to draw some flow lines. From your sketches, can you guess the equations of the flow lines?

(b) If parametric equations of a flow line are $x = x(t)$, $y = y(t)$, explain why these functions satisfy the differential equations $dx/dt = x$ and $dy/dt = -y$. Then solve the differential equations to find an equation of the flow line that passes through the point $(1, 1)$.

36. (a) Sketch the vector field $\mathbf{F}(x, y) = \mathbf{i} + x\,\mathbf{j}$ and then sketch some flow lines. What shape do these flow lines appear to have?

(b) If parametric equations of the flow lines are $x = x(t)$, $y = y(t)$, what differential equations do these functions satisfy? Deduce that $dy/dx = x$.

(c) If a particle starts at the origin in the velocity field given by $\mathbf{F}$, find an equation of the path it follows.

16.2 Line Integrals

In this section we define an integral that is similar to a single integral except that instead of integrating over an interval $[a, b]$, we integrate over a curve C. Such integrals are called *line integrals,* although "curve integrals" would be better terminology. They were invented in the early 19th century to solve problems involving fluid flow, forces, electricity, and magnetism.

We start with a plane curve C given by the parametric equations

$$\boxed{1} \qquad x = x(t) \qquad y = y(t) \qquad a \leqslant t \leqslant b$$

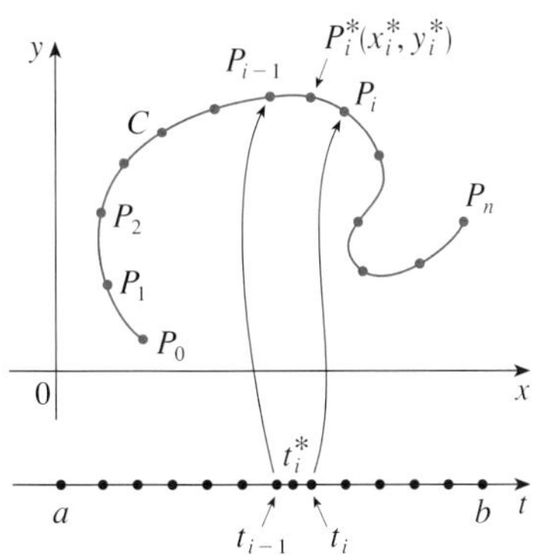

FIGURE 1

or, equivalently, by the vector equation $\mathbf{r}(t) = x(t)\,\mathbf{i} + y(t)\,\mathbf{j}$, and we assume that C is a smooth curve. [This means that $\mathbf{r}'$ is continuous and $\mathbf{r}'(t) \neq \mathbf{0}$. See Section 13.3.] If we divide the parameter interval $[a, b]$ into n subintervals $[t_{i-1}, t_i]$ of equal width and we let $x_i = x(t_i)$ and $y_i = y(t_i)$, then the corresponding points $P_i(x_i, y_i)$ divide C into n subarcs with lengths $\Delta s_1, \Delta s_2, \ldots, \Delta s_n$. (See Figure 1.) We choose any point $P_i^*(x_i^*, y_i^*)$ in the ith subarc. (This corresponds to a point t_i^* in $[t_{i-1}, t_i]$.) Now if f is any function of two variables whose domain includes the curve C, we evaluate f at the point (x_i^*, y_i^*), multiply by the length Δs_i of the subarc, and form the sum

$$\sum_{i=1}^{n} f(x_i^*, y_i^*)\,\Delta s_i$$

which is similar to a Riemann sum. Then we take the limit of these sums and make the following definition by analogy with a single integral.

> $\boxed{2}$ **Definition** If f is defined on a smooth curve C given by Equations 1, then the **line integral of *f* along *C*** is
>
> $$\int_C f(x, y)\,ds = \lim_{n \to \infty} \sum_{i=1}^{n} f(x_i^*, y_i^*)\,\Delta s_i$$
>
> if this limit exists.

In Section 10.2 we found that the length of C is

$$L = \int_a^b \sqrt{\left(\frac{dx}{dt}\right)^2 + \left(\frac{dy}{dt}\right)^2}\,dt$$

A similar type of argument can be used to show that if f is a continuous function, then the limit in Definition 2 always exists and the following formula can be used to evaluate the line integral:

$$\boxed{3} \qquad \int_C f(x, y)\,ds = \int_a^b f(x(t), y(t))\sqrt{\left(\frac{dx}{dt}\right)^2 + \left(\frac{dy}{dt}\right)^2}\,dt$$

The value of the line integral does not depend on the parametrization of the curve, provided that the curve is traversed exactly once as t increases from a to b.

The arc length function s is discussed in Section 13.3.

If $s(t)$ is the length of C between $\mathbf{r}(a)$ and $\mathbf{r}(t)$, then

$$\frac{ds}{dt} = \sqrt{\left(\frac{dx}{dt}\right)^2 + \left(\frac{dy}{dt}\right)^2}$$

So the way to remember Formula 3 is to express everything in terms of the parameter t: Use the parametric equations to express x and y in terms of t and write ds as

$$ds = \sqrt{\left(\frac{dx}{dt}\right)^2 + \left(\frac{dy}{dt}\right)^2}\,dt$$

In the special case where C is the line segment that joins $(a, 0)$ to $(b, 0)$, using x as the parameter, we can write the parametric equations of C as follows: $x = x$, $y = 0$, $a \leqslant x \leqslant b$. Formula 3 then becomes

$$\int_C f(x, y)\,ds = \int_a^b f(x, 0)\,dx$$

and so the line integral reduces to an ordinary single integral in this case.

Just as for an ordinary single integral, we can interpret the line integral of a *positive* function as an area. In fact, if $f(x, y) \geqslant 0$, $\int_C f(x, y)\,ds$ represents the area of one side of the "fence" or "curtain" in Figure 2, whose base is C and whose height above the point (x, y) is $f(x, y)$.

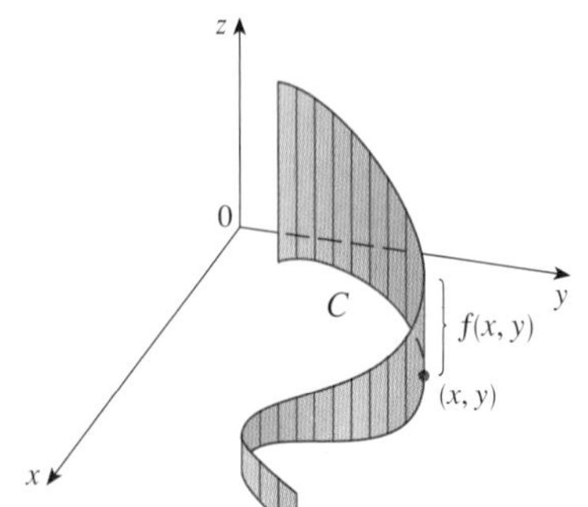

FIGURE 2

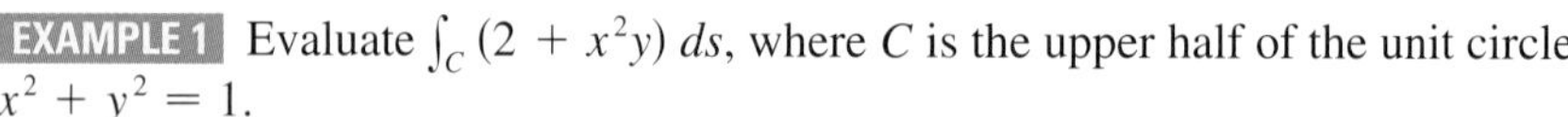
EXAMPLE 1 Evaluate $\int_C (2 + x^2 y)\,ds$, where C is the upper half of the unit circle $x^2 + y^2 = 1$.

SOLUTION In order to use Formula 3, we first need parametric equations to represent C. Recall that the unit circle can be parametrized by means of the equations

$$x = \cos t \qquad y = \sin t$$

and the upper half of the circle is described by the parameter interval $0 \leqslant t \leqslant \pi$. (See Figure 3.) Therefore Formula 3 gives

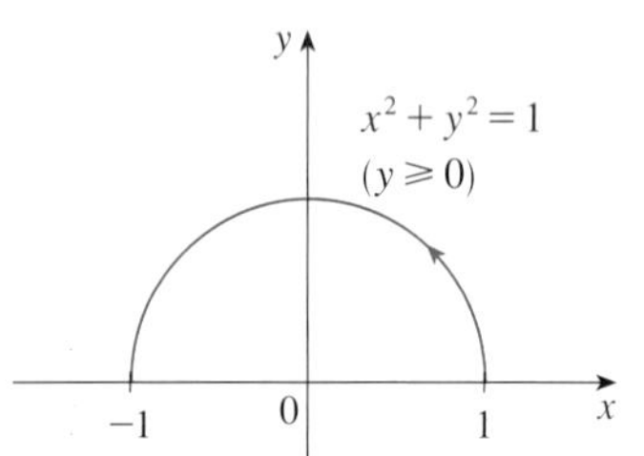

FIGURE 3

$$\begin{aligned}\int_C (2 + x^2 y)\,ds &= \int_0^\pi (2 + \cos^2 t \sin t)\sqrt{\left(\frac{dx}{dt}\right)^2 + \left(\frac{dy}{dt}\right)^2}\,dt \\ &= \int_0^\pi (2 + \cos^2 t \sin t)\sqrt{\sin^2 t + \cos^2 t}\,dt \\ &= \int_0^\pi (2 + \cos^2 t \sin t)\,dt = \left[2t - \frac{\cos^3 t}{3}\right]_0^\pi \\ &= 2\pi + \tfrac{2}{3}\end{aligned}$$

Suppose now that C is a **piecewise-smooth curve**; that is, C is a union of a finite number of smooth curves $C_1, C_2, \ldots, C_n$, where, as illustrated in Figure 4, the initial point of C_{i+1} is the terminal point of C_i. Then we define the integral of f along C as the sum of the integrals of f along each of the smooth pieces of C:

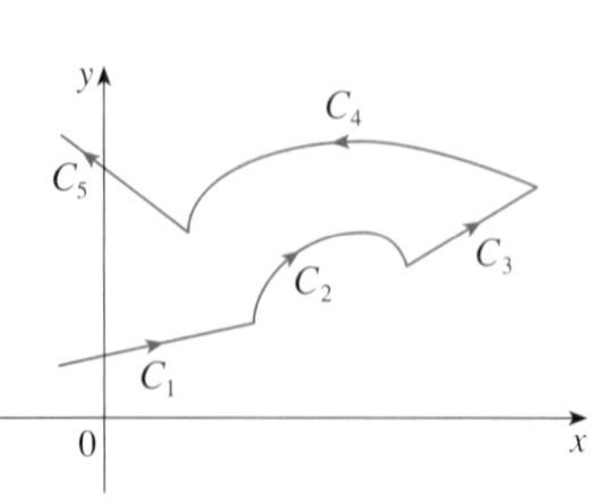

FIGURE 4
A piecewise-smooth curve

$$\int_C f(x, y)\,ds = \int_{C_1} f(x, y)\,ds + \int_{C_2} f(x, y)\,ds + \cdots + \int_{C_n} f(x, y)\,ds$$

EXAMPLE 2 Evaluate $\int_C 2x\,ds$, where C consists of the arc C_1 of the parabola $y = x^2$ from $(0, 0)$ to $(1, 1)$ followed by the vertical line segment C_2 from $(1, 1)$ to $(1, 2)$.

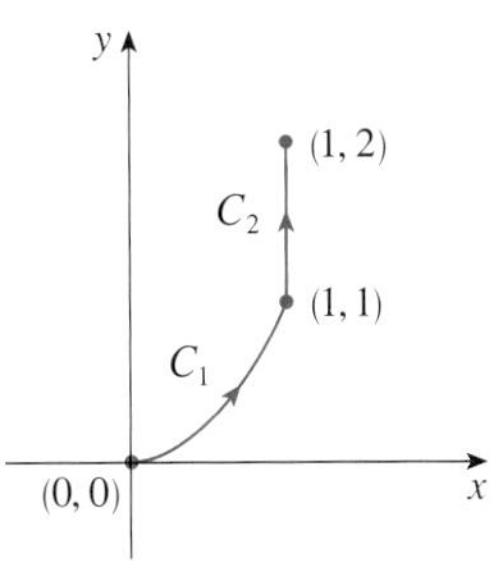

FIGURE 5
$C = C_1 \cup C_2$

SOLUTION The curve C is shown in Figure 5. C_1 is the graph of a function of x, so we can choose x as the parameter and the equations for C_1 become

$$x = x \qquad y = x^2 \qquad 0 \leqslant x \leqslant 1$$

Therefore

$$\int_{C_1} 2x\,ds = \int_0^1 2x\sqrt{\left(\frac{dx}{dx}\right)^2 + \left(\frac{dy}{dx}\right)^2}\,dx = \int_0^1 2x\sqrt{1 + 4x^2}\,dx$$

$$= \tfrac{1}{4} \cdot \tfrac{2}{3}(1 + 4x^2)^{3/2}\Big]_0^1 = \frac{5\sqrt{5} - 1}{6}$$

On C_2 we choose y as the parameter, so the equations of C_2 are

$$x = 1 \qquad y = y \qquad 1 \leqslant y \leqslant 2$$

and

$$\int_{C_2} 2x\,ds = \int_1^2 2(1)\sqrt{\left(\frac{dx}{dy}\right)^2 + \left(\frac{dy}{dy}\right)^2}\,dy = \int_1^2 2\,dy = 2$$

Thus

$$\int_C 2x\,ds = \int_{C_1} 2x\,ds + \int_{C_2} 2x\,ds = \frac{5\sqrt{5} - 1}{6} + 2$$

Any physical interpretation of a line integral $\int_C f(x, y)\,ds$ depends on the physical interpretation of the function f. Suppose that $\rho(x, y)$ represents the linear density at a point (x, y) of a thin wire shaped like a curve C. Then the mass of the part of the wire from P_{i-1} to P_i in Figure 1 is approximately $\rho(x_i^*, y_i^*)\,\Delta s_i$ and so the total mass of the wire is approximately $\Sigma\, \rho(x_i^*, y_i^*)\,\Delta s_i$. By taking more and more points on the curve, we obtain the **mass** m of the wire as the limiting value of these approximations:

$$m = \lim_{n \to \infty} \sum_{i=1}^{n} \rho(x_i^*, y_i^*)\,\Delta s_i = \int_C \rho(x, y)\,ds$$

[For example, if $f(x, y) = 2 + x^2y$ represents the density of a semicircular wire, then the integral in Example 1 would represent the mass of the wire.] The **center of mass** of the wire with density function ρ is located at the point $(\bar{x}, \bar{y})$, where

4
$$\bar{x} = \frac{1}{m}\int_C x\,\rho(x, y)\,ds \qquad \bar{y} = \frac{1}{m}\int_C y\,\rho(x, y)\,ds$$

Other physical interpretations of line integrals will be discussed later in this chapter.

V EXAMPLE 3 A wire takes the shape of the semicircle $x^2 + y^2 = 1$, $y \geqslant 0$, and is thicker near its base than near the top. Find the center of mass of the wire if the linear density at any point is proportional to its distance from the line $y = 1$.

SOLUTION As in Example 1 we use the parametrization $x = \cos t$, $y = \sin t$, $0 \leqslant t \leqslant \pi$, and find that $ds = dt$. The linear density is

$$\rho(x, y) = k(1 - y)$$

where k is a constant, and so the mass of the wire is

$$m = \int_C k(1 - y)\, ds = \int_0^{\pi} k(1 - \sin t)\, dt = k\Big[t + \cos t\Big]_0^{\pi} = k(\pi - 2)$$

From Equations 4 we have

$$\begin{aligned}\bar{y} &= \frac{1}{m}\int_C y\,\rho(x, y)\, ds = \frac{1}{k(\pi - 2)}\int_C y\,k(1 - y)\, ds \\ &= \frac{1}{\pi - 2}\int_0^{\pi} (\sin t - \sin^2 t)\, dt = \frac{1}{\pi - 2}\Big[-\cos t - \tfrac{1}{2}t + \tfrac{1}{4}\sin 2t\Big]_0^{\pi} \\ &= \frac{4 - \pi}{2(\pi - 2)}\end{aligned}$$

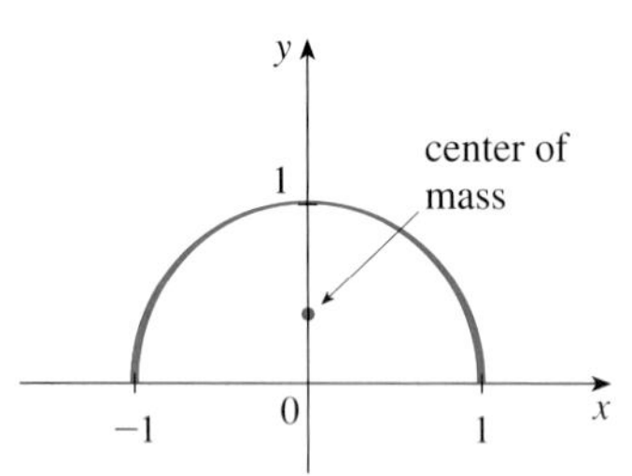

FIGURE 6

By symmetry we see that $\bar{x} = 0$, so the center of mass is

$$\left(0, \frac{4 - \pi}{2(\pi - 2)}\right) \approx (0, 0.38)$$

See Figure 6.

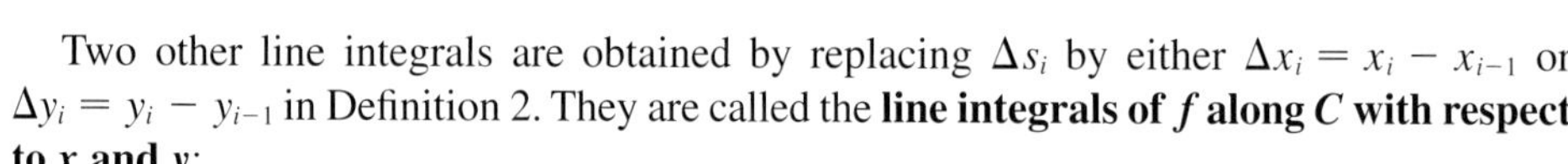

Two other line integrals are obtained by replacing Δs_i by either $\Delta x_i = x_i - x_{i-1}$ or $\Delta y_i = y_i - y_{i-1}$ in Definition 2. They are called the **line integrals of *f* along *C* with respect to *x* and *y***:

5 $$\int_C f(x, y)\, dx = \lim_{n \to \infty} \sum_{i=1}^{n} f(x_i^*, y_i^*)\, \Delta x_i$$

6 $$\int_C f(x, y)\, dy = \lim_{n \to \infty} \sum_{i=1}^{n} f(x_i^*, y_i^*)\, \Delta y_i$$

When we want to distinguish the original line integral $\int_C f(x, y)\, ds$ from those in Equations 5 and 6, we call it the **line integral with respect to arc length**.

The following formulas say that line integrals with respect to x and y can also be evaluated by expressing everything in terms of t: $x = x(t)$, $y = y(t)$, $dx = x'(t)\, dt$, $dy = y'(t)\, dt$.

7 $$\int_C f(x, y)\, dx = \int_a^b f\big(x(t), y(t)\big)\, x'(t)\, dt$$

$$\int_C f(x, y)\, dy = \int_a^b f\big(x(t), y(t)\big)\, y'(t)\, dt$$

It frequently happens that line integrals with respect to x and y occur together. When this happens, it's customary to abbreviate by writing

$$\int_C P(x, y)\, dx + \int_C Q(x, y)\, dy = \int_C P(x, y)\, dx + Q(x, y)\, dy$$

When we are setting up a line integral, sometimes the most difficult thing is to think of a parametric representation for a curve whose geometric description is given. In particular, we often need to parametrize a line segment, so it's useful to remember that a vector representation of the line segment that starts at $\mathbf{r}_0$ and ends at $\mathbf{r}_1$ is given by

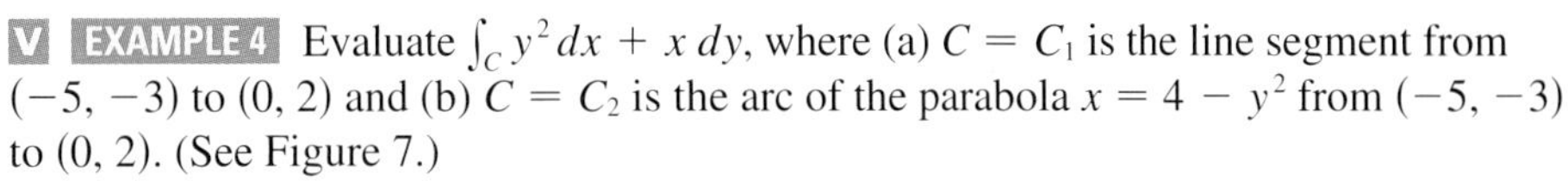

8 $$\mathbf{r}(t) = (1 - t)\mathbf{r}_0 + t\,\mathbf{r}_1 \qquad 0 \leqslant t \leqslant 1$$

(See Equation 12.5.4.)

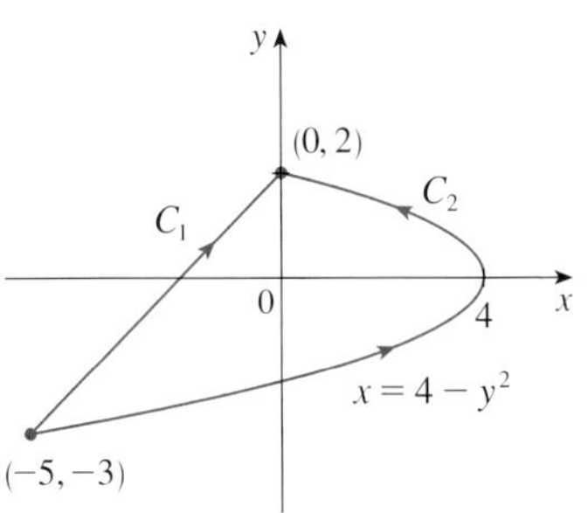

FIGURE 7

V EXAMPLE 4 Evaluate $\int_C y^2\,dx + x\,dy$, where (a) $C = C_1$ is the line segment from $(-5, -3)$ to $(0, 2)$ and (b) $C = C_2$ is the arc of the parabola $x = 4 - y^2$ from $(-5, -3)$ to $(0, 2)$. (See Figure 7.)

SOLUTION

(a) A parametric representation for the line segment is

$$x = 5t - 5 \qquad y = 5t - 3 \qquad 0 \leqslant t \leqslant 1$$

(Use Equation 8 with $\mathbf{r}_0 = \langle -5, -3 \rangle$ and $\mathbf{r}_1 = \langle 0, 2 \rangle$.) Then $dx = 5\,dt$, $dy = 5\,dt$, and Formulas 7 give

$$\begin{aligned} \int_{C_1} y^2\,dx + x\,dy &= \int_0^1 (5t - 3)^2(5\,dt) + (5t - 5)(5\,dt) \\ &= 5\int_0^1 (25t^2 - 25t + 4)\,dt \\ &= 5\left[\frac{25t^3}{3} - \frac{25t^2}{2} + 4t\right]_0^1 = -\frac{5}{6} \end{aligned}$$

(b) Since the parabola is given as a function of y, let's take y as the parameter and write C_2 as

$$x = 4 - y^2 \qquad y = y \qquad -3 \leqslant y \leqslant 2$$

Then $dx = -2y\,dy$ and by Formulas 7 we have

$$\begin{aligned} \int_{C_2} y^2\,dx + x\,dy &= \int_{-3}^2 y^2(-2y)\,dy + (4 - y^2)\,dy \\ &= \int_{-3}^2 (-2y^3 - y^2 + 4)\,dy \\ &= \left[-\frac{y^4}{2} - \frac{y^3}{3} + 4y\right]_{-3}^2 = 40\tfrac{5}{6} \end{aligned}$$

Notice that we got different answers in parts (a) and (b) of Example 4 even though the two curves had the same endpoints. Thus, in general, the value of a line integral depends not just on the endpoints of the curve but also on the path. (But see Section 16.3 for conditions under which the integral is independent of the path.)

Notice also that the answers in Example 4 depend on the direction, or orientation, of the curve. If $-C_1$ denotes the line segment from $(0, 2)$ to $(-5, -3)$, you can verify, using the parametrization

$$x = -5t \qquad y = 2 - 5t \qquad 0 \leqslant t \leqslant 1$$

that

$$\int_{-C_1} y^2\,dx + x\,dy = \tfrac{5}{6}$$

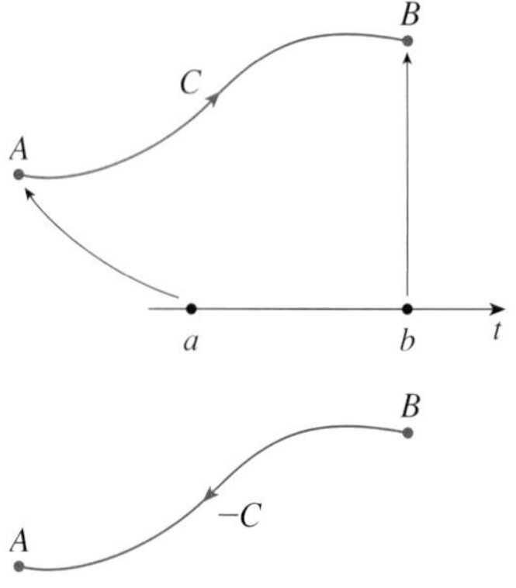

FIGURE 8

In general, a given parametrization $x = x(t)$, $y = y(t)$, $a \leqslant t \leqslant b$, determines an **orientation** of a curve C, with the positive direction corresponding to increasing values of the parameter t. (See Figure 8, where the initial point A corresponds to the parameter value a and the terminal point B corresponds to $t = b$.)

If $-C$ denotes the curve consisting of the same points as C but with the opposite orientation (from initial point B to terminal point A in Figure 8), then we have

$$\int_{-C} f(x, y)\, dx = -\int_C f(x, y)\, dx \qquad \int_{-C} f(x, y)\, dy = -\int_C f(x, y)\, dy$$

But if we integrate with respect to arc length, the value of the line integral does *not* change when we reverse the orientation of the curve:

$$\int_{-C} f(x, y)\, ds = \int_C f(x, y)\, ds$$

This is because Δs_i is always positive, whereas Δx_i and Δy_i change sign when we reverse the orientation of C.

Line Integrals in Space

We now suppose that C is a smooth space curve given by the parametric equations

$$x = x(t) \qquad y = y(t) \qquad z = z(t) \qquad a \leqslant t \leqslant b$$

or by a vector equation $\mathbf{r}(t) = x(t)\,\mathbf{i} + y(t)\,\mathbf{j} + z(t)\,\mathbf{k}$. If f is a function of three variables that is continuous on some region containing C, then we define the **line integral of f along C** (with respect to arc length) in a manner similar to that for plane curves:

$$\int_C f(x, y, z)\, ds = \lim_{n \to \infty} \sum_{i=1}^{n} f(x_i^*, y_i^*, z_i^*)\, \Delta s_i$$

We evaluate it using a formula similar to Formula 3:

9 $$\int_C f(x, y, z)\, ds = \int_a^b f\big(x(t), y(t), z(t)\big) \sqrt{\left(\frac{dx}{dt}\right)^2 + \left(\frac{dy}{dt}\right)^2 + \left(\frac{dz}{dt}\right)^2}\, dt$$

Observe that the integrals in both Formulas 3 and 9 can be written in the more compact vector notation

$$\int_a^b f(\mathbf{r}(t))\, |\mathbf{r}'(t)|\, dt$$

For the special case $f(x, y, z) = 1$, we get

$$\int_C ds = \int_a^b |\mathbf{r}'(t)|\, dt = L$$

where L is the length of the curve C (see Formula 13.3.3).

Line integrals along C with respect to x, y, and z can also be defined. For example,

$$\int_C f(x, y, z)\, dz = \lim_{n \to \infty} \sum_{i=1}^{n} f(x_i^*, y_i^*, z_i^*)\, \Delta z_i$$

$$= \int_a^b f\big(x(t), y(t), z(t)\big)\, z'(t)\, dt$$

Therefore, as with line integrals in the plane, we evaluate integrals of the form

10 $$\int_C P(x, y, z)\, dx + Q(x, y, z)\, dy + R(x, y, z)\, dz$$

by expressing everything (x, y, z, dx, dy, dz) in terms of the parameter t.

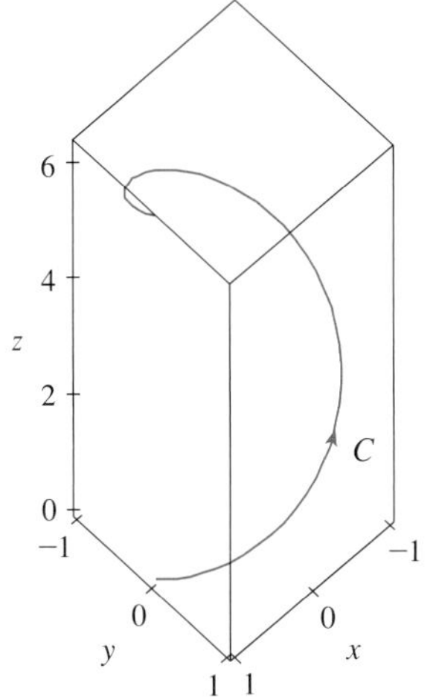

FIGURE 9

V EXAMPLE 5 Evaluate $\int_C y \sin z\, ds$, where C is the circular helix given by the equations $x = \cos t$, $y = \sin t$, $z = t$, $0 \leqslant t \leqslant 2\pi$. (See Figure 9.)

SOLUTION Formula 9 gives

$$\int_C y \sin z\, ds = \int_0^{2\pi} (\sin t) \sin t \sqrt{\left(\frac{dx}{dt}\right)^2 + \left(\frac{dy}{dt}\right)^2 + \left(\frac{dz}{dt}\right)^2}\, dt$$

$$= \int_0^{2\pi} \sin^2 t \sqrt{\sin^2 t + \cos^2 t + 1}\, dt = \sqrt{2} \int_0^{2\pi} \tfrac{1}{2}(1 - \cos 2t)\, dt$$

$$= \frac{\sqrt{2}}{2} \Big[t - \tfrac{1}{2} \sin 2t\Big]_0^{2\pi} = \sqrt{2}\, \pi$$

EXAMPLE 6 Evaluate $\int_C y\, dx + z\, dy + x\, dz$, where C consists of the line segment C_1 from $(2, 0, 0)$ to $(3, 4, 5)$, followed by the vertical line segment C_2 from $(3, 4, 5)$ to $(3, 4, 0)$.

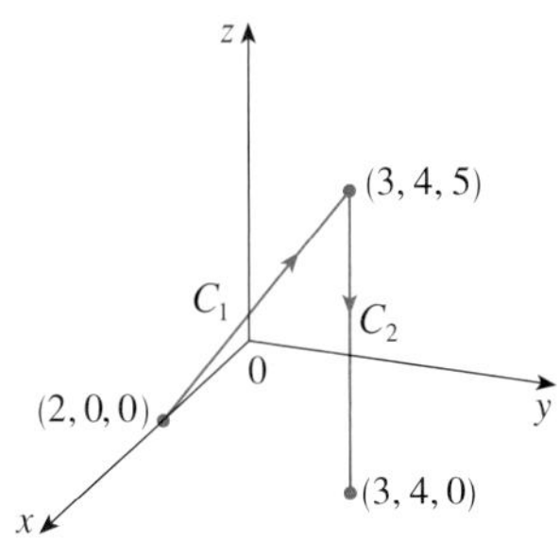

FIGURE 10

SOLUTION The curve C is shown in Figure 10. Using Equation 8, we write C_1 as

$$\mathbf{r}(t) = (1 - t)\langle 2, 0, 0 \rangle + t\langle 3, 4, 5 \rangle = \langle 2 + t, 4t, 5t \rangle$$

or, in parametric form, as

$$x = 2 + t \qquad y = 4t \qquad z = 5t \qquad 0 \leqslant t \leqslant 1$$

Thus

$$\int_{C_1} y\, dx + z\, dy + x\, dz = \int_0^1 (4t)\, dt + (5t)4\, dt + (2 + t)5\, dt$$

$$= \int_0^1 (10 + 29t)\, dt = 10t + 29 \frac{t^2}{2} \bigg]_0^1 = 24.5$$

Likewise, C_2 can be written in the form

$$\mathbf{r}(t) = (1 - t)\langle 3, 4, 5 \rangle + t\langle 3, 4, 0 \rangle = \langle 3, 4, 5 - 5t \rangle$$

or

$$x = 3 \qquad y = 4 \qquad z = 5 - 5t \qquad 0 \leqslant t \leqslant 1$$

Then $dx = 0 = dy$, so

$$\int_{C_2} y\,dx + z\,dy + x\,dz = \int_0^1 3(-5)\,dt = -15$$

Adding the values of these integrals, we obtain

$$\int_C y\,dx + z\,dy + x\,dz = 24.5 - 15 = 9.5$$

Line Integrals of Vector Fields

Recall from Section 6.4 that the work done by a variable force $f(x)$ in moving a particle from a to b along the x-axis is $W = \int_a^b f(x)\,dx$. Then in Section 12.3 we found that the work done by a constant force $\mathbf{F}$ in moving an object from a point P to another point Q in space is $W = \mathbf{F} \cdot \mathbf{D}$, where $\mathbf{D} = \overrightarrow{PQ}$ is the displacement vector.

Now suppose that $\mathbf{F} = P\,\mathbf{i} + Q\,\mathbf{j} + R\,\mathbf{k}$ is a continuous force field on $\mathbb{R}^3$, such as the gravitational field of Example 4 in Section 16.1 or the electric force field of Example 5 in Section 16.1. (A force field on $\mathbb{R}^2$ could be regarded as a special case where $R = 0$ and P and Q depend only on x and y.) We wish to compute the work done by this force in moving a particle along a smooth curve C.

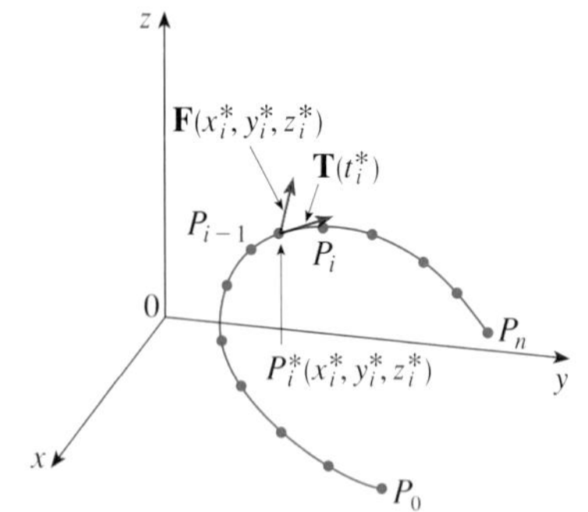

FIGURE 11

We divide C into subarcs $P_{i-1}P_i$ with lengths Δs_i by dividing the parameter interval $[a, b]$ into subintervals of equal width. (See Figure 1 for the two-dimensional case or Figure 11 for the three-dimensional case.) Choose a point $P_i^*(x_i^*, y_i^*, z_i^*)$ on the ith subarc corresponding to the parameter value t_i^*. If Δs_i is small, then as the particle moves from P_{i-1} to P_i along the curve, it proceeds approximately in the direction of $\mathbf{T}(t_i^*)$, the unit tangent vector at P_i^*. Thus the work done by the force $\mathbf{F}$ in moving the particle from P_{i-1} to P_i is approximately

$$\mathbf{F}(x_i^*, y_i^*, z_i^*) \cdot [\Delta s_i\,\mathbf{T}(t_i^*)] = [\mathbf{F}(x_i^*, y_i^*, z_i^*) \cdot \mathbf{T}(t_i^*)]\,\Delta s_i$$

and the total work done in moving the particle along C is approximately

11
$$\sum_{i=1}^{n} [\mathbf{F}(x_i^*, y_i^*, z_i^*) \cdot \mathbf{T}(x_i^*, y_i^*, z_i^*)]\,\Delta s_i$$

where $\mathbf{T}(x, y, z)$ is the unit tangent vector at the point (x, y, z) on C. Intuitively, we see that these approximations ought to become better as n becomes larger. Therefore we define the **work** W done by the force field $\mathbf{F}$ as the limit of the Riemann sums in 11, namely,

12
$$W = \int_C \mathbf{F}(x, y, z) \cdot \mathbf{T}(x, y, z)\,ds = \int_C \mathbf{F} \cdot \mathbf{T}\,ds$$

Equation 12 says that *work is the line integral with respect to arc length of the tangential component of the force.*

If the curve C is given by the vector equation $\mathbf{r}(t) = x(t)\,\mathbf{i} + y(t)\,\mathbf{j} + z(t)\,\mathbf{k}$, then $\mathbf{T}(t) = \mathbf{r}'(t)/|\mathbf{r}'(t)|$, so using Equation 9 we can rewrite Equation 12 in the form

$$W = \int_a^b \left[\mathbf{F}(\mathbf{r}(t)) \cdot \frac{\mathbf{r}'(t)}{|\mathbf{r}'(t)|}\right] |\mathbf{r}'(t)|\,dt = \int_a^b \mathbf{F}(\mathbf{r}(t)) \cdot \mathbf{r}'(t)\,dt$$

This integral is often abbreviated as $\int_C \mathbf{F} \cdot d\mathbf{r}$ and occurs in other areas of physics as well. Therefore we make the following definition for the line integral of *any* continuous vector field.

> **13 Definition** Let $\mathbf{F}$ be a continuous vector field defined on a smooth curve C given by a vector function $\mathbf{r}(t)$, $a \leqslant t \leqslant b$. Then the **line integral of F along *C*** is
>
> $$\int_C \mathbf{F} \cdot d\mathbf{r} = \int_a^b \mathbf{F}(\mathbf{r}(t)) \cdot \mathbf{r}'(t)\, dt = \int_C \mathbf{F} \cdot \mathbf{T}\, ds$$

When using Definition 13, bear in mind that $\mathbf{F}(\mathbf{r}(t))$ is just an abbreviation for $\mathbf{F}(x(t), y(t), z(t))$, so we evaluate $\mathbf{F}(\mathbf{r}(t))$ simply by putting $x = x(t)$, $y = y(t)$, and $z = z(t)$ in the expression for $\mathbf{F}(x, y, z)$. Notice also that we can formally write $d\mathbf{r} = \mathbf{r}'(t)\, dt$.

Figure 12 shows the force field and the curve in Example 7. The work done is negative because the field impedes movement along the curve.

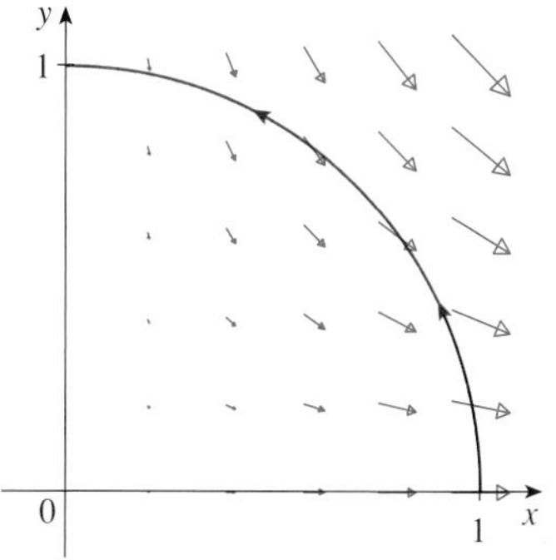

FIGURE 12

EXAMPLE 7 Find the work done by the force field $\mathbf{F}(x, y) = x^2\,\mathbf{i} - xy\,\mathbf{j}$ in moving a particle along the quarter-circle $\mathbf{r}(t) = \cos t\,\mathbf{i} + \sin t\,\mathbf{j}$, $0 \leqslant t \leqslant \pi/2$.

SOLUTION Since $x = \cos t$ and $y = \sin t$, we have

$$\mathbf{F}(\mathbf{r}(t)) = \cos^2 t\,\mathbf{i} - \cos t \sin t\,\mathbf{j}$$

and

$$\mathbf{r}'(t) = -\sin t\,\mathbf{i} + \cos t\,\mathbf{j}$$

Therefore the work done is

$$\int_C \mathbf{F} \cdot d\mathbf{r} = \int_0^{\pi/2} \mathbf{F}(\mathbf{r}(t)) \cdot \mathbf{r}'(t)\, dt = \int_0^{\pi/2} (-2\cos^2 t \sin t)\, dt$$

$$= 2\frac{\cos^3 t}{3}\Big]_0^{\pi/2} = -\frac{2}{3}$$

NOTE Even though $\int_C \mathbf{F} \cdot d\mathbf{r} = \int_C \mathbf{F} \cdot \mathbf{T}\, ds$ and integrals with respect to arc length are unchanged when orientation is reversed, it is still true that

$$\int_{-C} \mathbf{F} \cdot d\mathbf{r} = -\int_C \mathbf{F} \cdot d\mathbf{r}$$

because the unit tangent vector $\mathbf{T}$ is replaced by its negative when C is replaced by $-C$.

Figure 13 shows the twisted cubic C in Example 8 and some typical vectors acting at three points on C.

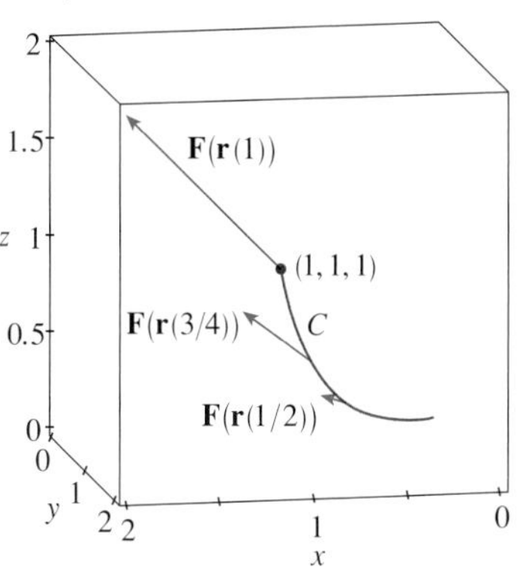

FIGURE 13

EXAMPLE 8 Evaluate $\int_C \mathbf{F} \cdot d\mathbf{r}$, where $\mathbf{F}(x, y, z) = xy\,\mathbf{i} + yz\,\mathbf{j} + zx\,\mathbf{k}$ and C is the twisted cubic given by

$$x = t \qquad y = t^2 \qquad z = t^3 \qquad 0 \leqslant t \leqslant 1$$

SOLUTION We have

$$\mathbf{r}(t) = t\,\mathbf{i} + t^2\,\mathbf{j} + t^3\,\mathbf{k}$$

$$\mathbf{r}'(t) = \mathbf{i} + 2t\,\mathbf{j} + 3t^2\,\mathbf{k}$$

$$\mathbf{F}(\mathbf{r}(t)) = t^3\,\mathbf{i} + t^5\,\mathbf{j} + t^4\,\mathbf{k}$$

Thus $$\int_C \mathbf{F} \cdot d\mathbf{r} = \int_0^1 \mathbf{F}(\mathbf{r}(t)) \cdot \mathbf{r}'(t)\, dt$$

$$= \int_0^1 (t^3 + 5t^6)\, dt = \frac{t^4}{4} + \frac{5t^7}{7}\bigg]_0^1 = \frac{27}{28}$$

Finally, we note the connection between line integrals of vector fields and line integrals of scalar fields. Suppose the vector field $\mathbf{F}$ on $\mathbb{R}^3$ is given in component form by the equation $\mathbf{F} = P\,\mathbf{i} + Q\,\mathbf{j} + R\,\mathbf{k}$. We use Definition 13 to compute its line integral along C:

$$\int_C \mathbf{F} \cdot d\mathbf{r} = \int_a^b \mathbf{F}(\mathbf{r}(t)) \cdot \mathbf{r}'(t)\, dt$$

$$= \int_a^b (P\,\mathbf{i} + Q\,\mathbf{j} + R\,\mathbf{k}) \cdot \big(x'(t)\,\mathbf{i} + y'(t)\,\mathbf{j} + z'(t)\,\mathbf{k}\big)\, dt$$

$$= \int_a^b \big[P\big(x(t), y(t), z(t)\big)\,x'(t) + Q\big(x(t), y(t), z(t)\big)\,y'(t) + R\big(x(t), y(t), z(t)\big)\,z'(t)\big]\, dt$$

But this last integral is precisely the line integral in [10]. Therefore we have

$$\int_C \mathbf{F} \cdot d\mathbf{r} = \int_C P\,dx + Q\,dy + R\,dz \qquad \text{where } \mathbf{F} = P\,\mathbf{i} + Q\,\mathbf{j} + R\,\mathbf{k}$$

For example, the integral $\int_C y\,dx + z\,dy + x\,dz$ in Example 6 could be expressed as $\int_C \mathbf{F} \cdot d\mathbf{r}$ where

$$\mathbf{F}(x, y, z) = y\,\mathbf{i} + z\,\mathbf{j} + x\,\mathbf{k}$$

16.2 Exercises

1–16 Evaluate the line integral, where C is the given curve.

1. $\int_C y^3\, ds$, $C: x = t^3,\ y = t,\ 0 \leqslant t \leqslant 2$

2. $\int_C xy\, ds$, $C: x = t^2,\ y = 2t,\ 0 \leqslant t \leqslant 1$

3. $\int_C xy^4\, ds$, C is the right half of the circle $x^2 + y^2 = 16$

4. $\int_C x \sin y\, ds$, C is the line segment from $(0, 3)$ to $(4, 6)$

5. $\int_C \big(x^2y^3 - \sqrt{x}\big)\, dy$,
C is the arc of the curve $y = \sqrt{x}$ from $(1, 1)$ to $(4, 2)$

6. $\int_C e^x\, dx$,
C is the arc of the curve $x = y^3$ from $(-1, -1)$ to $(1, 1)$

7. $\int_C (x + 2y)\, dx + x^2\, dy$, C consists of line segments from $(0, 0)$ to $(2, 1)$ and from $(2, 1)$ to $(3, 0)$

8. $\int_C x^2\, dx + y^2\, dy$, C consists of the arc of the circle $x^2 + y^2 = 4$ from $(2, 0)$ to $(0, 2)$ followed by the line segment from $(0, 2)$ to $(4, 3)$

9. $\int_C xyz\, ds$,
$C: x = 2\sin t,\ y = t,\ z = -2\cos t,\ 0 \leqslant t \leqslant \pi$

10. $\int_C xyz^2\, ds$,
C is the line segment from $(-1, 5, 0)$ to $(1, 6, 4)$

11. $\int_C xe^{yz}\, ds$,
C is the line segment from $(0, 0, 0)$ to $(1, 2, 3)$

12. $\int_C (x^2 + y^2 + z^2)\, ds$,
$C: x = t,\ y = \cos 2t,\ z = \sin 2t,\ 0 \leqslant t \leqslant 2\pi$

13. $\int_C xye^{yz}\, dy$, $C: x = t,\ y = t^2,\ z = t^3,\ 0 \leqslant t \leqslant 1$

14. $\int_C y\, dx + z\, dy + x\, dz$,
$C: x = \sqrt{t},\ y = t,\ z = t^2,\ 1 \leqslant t \leqslant 4$

15. $\int_C z^2\, dx + x^2\, dy + y^2\, dz$, C is the line segment from $(1, 0, 0)$ to $(4, 1, 2)$

16. $\int_C (y + z)\, dx + (x + z)\, dy + (x + y)\, dz$, C consists of line segments from $(0, 0, 0)$ to $(1, 0, 1)$ and from $(1, 0, 1)$ to $(0, 1, 2)$

17. Let $\mathbf{F}$ be the vector field shown in the figure.
(a) If C_1 is the vertical line segment from $(-3, -3)$ to $(-3, 3)$, determine whether $\int_{C_1} \mathbf{F} \cdot d\mathbf{r}$ is positive, negative, or zero.
(b) If C_2 is the counterclockwise-oriented circle with radius 3 and center the origin, determine whether $\int_{C_2} \mathbf{F} \cdot d\mathbf{r}$ is positive, negative, or zero.

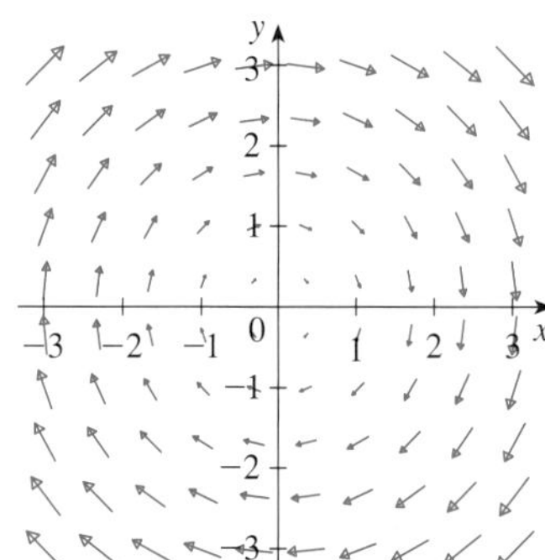

18. The figure shows a vector field $\mathbf{F}$ and two curves C_1 and C_2. Are the line integrals of $\mathbf{F}$ over C_1 and C_2 positive, negative, or zero? Explain.

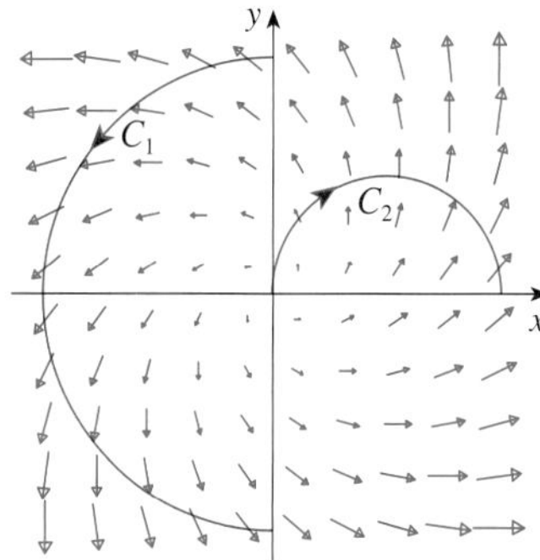

19–22 Evaluate the line integral $\int_C \mathbf{F} \cdot d\mathbf{r}$, where C is given by the vector function $\mathbf{r}(t)$.

19. $\mathbf{F}(x, y) = xy\,\mathbf{i} + 3y^2\,\mathbf{j}$,
$\mathbf{r}(t) = 11t^4\,\mathbf{i} + t^3\,\mathbf{j}, \quad 0 \leqslant t \leqslant 1$

20. $\mathbf{F}(x, y, z) = (x + y)\,\mathbf{i} + (y - z)\,\mathbf{j} + z^2\,\mathbf{k}$,
$\mathbf{r}(t) = t^2\,\mathbf{i} + t^3\,\mathbf{j} + t^2\,\mathbf{k}, \quad 0 \leqslant t \leqslant 1$

21. $\mathbf{F}(x, y, z) = \sin x\,\mathbf{i} + \cos y\,\mathbf{j} + xz\,\mathbf{k}$,
$\mathbf{r}(t) = t^3\,\mathbf{i} - t^2\,\mathbf{j} + t\,\mathbf{k}, \quad 0 \leqslant t \leqslant 1$

22. $\mathbf{F}(x, y, z) = x\,\mathbf{i} + y\,\mathbf{j} + xy\,\mathbf{k}$,
$\mathbf{r}(t) = \cos t\,\mathbf{i} + \sin t\,\mathbf{j} + t\,\mathbf{k}, \quad 0 \leqslant t \leqslant \pi$

23–26 Use a calculator or CAS to evaluate the line integral correct to four decimal places.

23. $\int_C \mathbf{F} \cdot d\mathbf{r}$, where $\mathbf{F}(x, y) = xy\,\mathbf{i} + \sin y\,\mathbf{j}$ and $\mathbf{r}(t) = e^t\,\mathbf{i} + e^{-t^2}\,\mathbf{j}, \ 1 \leqslant t \leqslant 2$

24. $\int_C \mathbf{F} \cdot d\mathbf{r}$, where $\mathbf{F}(x, y, z) = y \sin z\,\mathbf{i} + z \sin x\,\mathbf{j} + x \sin y\,\mathbf{k}$ and $\mathbf{r}(t) = \cos t\,\mathbf{i} + \sin t\,\mathbf{j} + \sin 5t\,\mathbf{k}, \ 0 \leqslant t \leqslant \pi$

25. $\int_C x \sin(y + z)\, ds$, where C has parametric equations $x = t^2$, $y = t^3$, $z = t^4$, $0 \leqslant t \leqslant 5$

26. $\int_C ze^{-xy}\, ds$, where C has parametric equations $x = t$, $y = t^2$, $z = e^{-t}$, $0 \leqslant t \leqslant 1$

CAS **27–28** Use a graph of the vector field $\mathbf{F}$ and the curve C to guess whether the line integral of $\mathbf{F}$ over C is positive, negative, or zero. Then evaluate the line integral.

27. $\mathbf{F}(x, y) = (x - y)\,\mathbf{i} + xy\,\mathbf{j}$,
C is the arc of the circle $x^2 + y^2 = 4$ traversed counterclockwise from $(2, 0)$ to $(0, -2)$

28. $\mathbf{F}(x, y) = \dfrac{x}{\sqrt{x^2 + y^2}}\,\mathbf{i} + \dfrac{y}{\sqrt{x^2 + y^2}}\,\mathbf{j}$,
C is the parabola $y = 1 + x^2$ from $(-1, 2)$ to $(1, 2)$

29. (a) Evaluate the line integral $\int_C \mathbf{F} \cdot d\mathbf{r}$, where $\mathbf{F}(x, y) = e^{x-1}\,\mathbf{i} + xy\,\mathbf{j}$ and C is given by $\mathbf{r}(t) = t^2\,\mathbf{i} + t^3\,\mathbf{j}, 0 \leqslant t \leqslant 1$.
(b) Illustrate part (a) by using a graphing calculator or computer to graph C and the vectors from the vector field corresponding to $t = 0$, $1/\sqrt{2}$, and 1 (as in Figure 13).

30. (a) Evaluate the line integral $\int_C \mathbf{F} \cdot d\mathbf{r}$, where $\mathbf{F}(x, y, z) = x\,\mathbf{i} - z\,\mathbf{j} + y\,\mathbf{k}$ and C is given by $\mathbf{r}(t) = 2t\,\mathbf{i} + 3t\,\mathbf{j} - t^2\,\mathbf{k}, -1 \leqslant t \leqslant 1$.
(b) Illustrate part (a) by using a computer to graph C and the vectors from the vector field corresponding to $t = \pm 1$ and $\pm\frac{1}{2}$ (as in Figure 13).

CAS 31. Find the exact value of $\int_C x^3 y^2 z\, ds$, where C is the curve with parametric equations $x = e^{-t} \cos 4t$, $y = e^{-t} \sin 4t$, $z = e^{-t}$, $0 \leqslant t \leqslant 2\pi$.

32. (a) Find the work done by the force field $\mathbf{F}(x, y) = x^2\,\mathbf{i} + xy\,\mathbf{j}$ on a particle that moves once around the circle $x^2 + y^2 = 4$ oriented in the counter-clockwise direction.
CAS (b) Use a computer algebra system to graph the force field and circle on the same screen. Use the graph to explain your answer to part (a).

33. A thin wire is bent into the shape of a semicircle $x^2 + y^2 = 4$, $x \geqslant 0$. If the linear density is a constant k, find the mass and center of mass of the wire.

34. A thin wire has the shape of the first-quadrant part of the circle with center the origin and radius a. If the density function is $\rho(x, y) = kxy$, find the mass and center of mass of the wire.

35. (a) Write the formulas similar to Equations 4 for the center of mass $(\bar{x}, \bar{y}, \bar{z})$ of a thin wire in the shape of a space curve C if the wire has density function $\rho(x, y, z)$.

(b) Find the center of mass of a wire in the shape of the helix $x = 2\sin t$, $y = 2\cos t$, $z = 3t$, $0 \leqslant t \leqslant 2\pi$, if the density is a constant k.

36. Find the mass and center of mass of a wire in the shape of the helix $x = t$, $y = \cos t$, $z = \sin t$, $0 \leqslant t \leqslant 2\pi$, if the density at any point is equal to the square of the distance from the origin.

37. If a wire with linear density $\rho(x, y)$ lies along a plane curve C, its **moments of inertia** about the x- and y-axes are defined as

$$I_x = \int_C y^2 \rho(x, y)\, ds \qquad I_y = \int_C x^2 \rho(x, y)\, ds$$

Find the moments of inertia for the wire in Example 3.

38. If a wire with linear density $\rho(x, y, z)$ lies along a space curve C, its **moments of inertia** about the x-, y-, and z-axes are defined as

$$I_x = \int_C (y^2 + z^2)\rho(x, y, z)\, ds$$

$$I_y = \int_C (x^2 + z^2)\rho(x, y, z)\, ds$$

$$I_z = \int_C (x^2 + y^2)\rho(x, y, z)\, ds$$

Find the moments of inertia for the wire in Exercise 35.

39. Find the work done by the force field $\mathbf{F}(x, y) = x\,\mathbf{i} + (y + 2)\,\mathbf{j}$ in moving an object along an arch of the cycloid $\mathbf{r}(t) = (t - \sin t)\,\mathbf{i} + (1 - \cos t)\,\mathbf{j}$, $0 \leqslant t \leqslant 2\pi$.

40. Find the work done by the force field $\mathbf{F}(x, y) = x^2\,\mathbf{i} + ye^x\,\mathbf{j}$ on a particle that moves along the parabola $x = y^2 + 1$ from $(1, 0)$ to $(2, 1)$.

41. Find the work done by the force field $\mathbf{F}(x, y, z) = \langle x - y^2, y - z^2, z - x^2 \rangle$ on a particle that moves along the line segment from $(0, 0, 1)$ to $(2, 1, 0)$.

42. The force exerted by an electric charge at the origin on a charged particle at a point (x, y, z) with position vector $\mathbf{r} = \langle x, y, z \rangle$ is $\mathbf{F}(\mathbf{r}) = K\mathbf{r}/|\mathbf{r}|^3$ where K is a constant. (See Example 5 in Section 16.1.) Find the work done as the particle moves along a straight line from $(2, 0, 0)$ to $(2, 1, 5)$.

43. The position of an object with mass m at time t is $\mathbf{r}(t) = at^2\,\mathbf{i} + bt^3\,\mathbf{j}$, $0 \leqslant t \leqslant 1$.

(a) What is the force acting on the object at time t?

(b) What is the work done by the force during the time interval $0 \leqslant t \leqslant 1$?

44. An object with mass m moves with position function $\mathbf{r}(t) = a\sin t\,\mathbf{i} + b\cos t\,\mathbf{j} + ct\,\mathbf{k}$, $0 \leqslant t \leqslant \pi/2$. Find the work done on the object during this time period.

45. A 160-lb man carries a 25-lb can of paint up a helical staircase that encircles a silo with a radius of 20 ft. If the silo is 90 ft high and the man makes exactly three complete revolutions climbing to the top, how much work is done by the man against gravity?

46. Suppose there is a hole in the can of paint in Exercise 45 and 9 lb of paint leaks steadily out of the can during the man's ascent. How much work is done?

47. (a) Show that a constant force field does zero work on a particle that moves once uniformly around the circle $x^2 + y^2 = 1$.

(b) Is this also true for a force field $\mathbf{F}(\mathbf{x}) = k\mathbf{x}$, where k is a constant and $\mathbf{x} = \langle x, y \rangle$?

48. The base of a circular fence with radius 10 m is given by $x = 10\cos t$, $y = 10\sin t$. The height of the fence at position (x, y) is given by the function $h(x, y) = 4 + 0.01(x^2 - y^2)$, so the height varies from 3 m to 5 m. Suppose that 1 L of paint covers 100 m^2. Sketch the fence and determine how much paint you will need if you paint both sides of the fence.

49. If C is a smooth curve given by a vector function $\mathbf{r}(t)$, $a \leqslant t \leqslant b$, and $\mathbf{v}$ is a constant vector, show that

$$\int_C \mathbf{v} \cdot d\mathbf{r} = \mathbf{v} \cdot [\mathbf{r}(b) - \mathbf{r}(a)]$$

50. If C is a smooth curve given by a vector function $\mathbf{r}(t)$, $a \leqslant t \leqslant b$, show that

$$\int_C \mathbf{r} \cdot d\mathbf{r} = \tfrac{1}{2}\left[|\mathbf{r}(b)|^2 - |\mathbf{r}(a)|^2\right]$$

51. An object moves along the curve C shown in the figure from $(1, 2)$ to $(9, 8)$. The lengths of the vectors in the force field $\mathbf{F}$ are measured in newtons by the scales on the axes. Estimate the work done by $\mathbf{F}$ on the object.

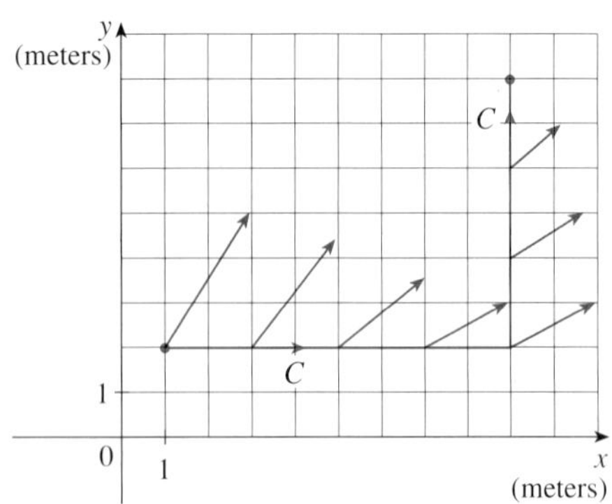

52. Experiments show that a steady current I in a long wire produces a magnetic field $\mathbf{B}$ that is tangent to any circle that lies in the plane perpendicular to the wire and whose center is the axis of the wire (as in the figure). *Ampère's Law* relates the electric current to its magnetic effects and states that

$$\int_C \mathbf{B} \cdot d\mathbf{r} = \mu_0 I$$

where I is the net current that passes through any surface bounded by a closed curve C, and μ_0 is a constant called the permeability of free space. By taking C to be a circle with radius r, show that the magnitude $B = |\mathbf{B}|$ of the magnetic field at a distance r from the center of the wire is

$$B = \frac{\mu_0 I}{2\pi r}$$

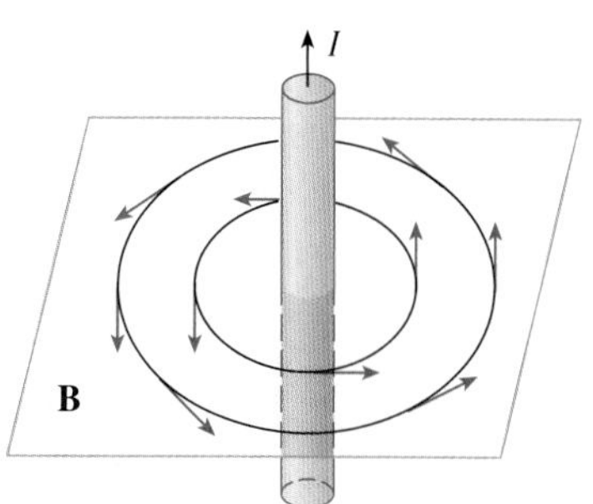

16.3 The Fundamental Theorem for Line Integrals

Recall from Section 5.3 that Part 2 of the Fundamental Theorem of Calculus can be written as

$$\boxed{1} \qquad \int_a^b F'(x)\,dx = F(b) - F(a)$$

where F' is continuous on $[a, b]$. We also called Equation 1 the Net Change Theorem: The integral of a rate of change is the net change.

If we think of the gradient vector ∇f of a function f of two or three variables as a sort of derivative of f, then the following theorem can be regarded as a version of the Fundamental Theorem for line integrals.

> **2 Theorem** Let C be a smooth curve given by the vector function $\mathbf{r}(t)$, $a \leqslant t \leqslant b$. Let f be a differentiable function of two or three variables whose gradient vector ∇f is continuous on C. Then
>
> $$\int_C \nabla f \cdot d\mathbf{r} = f(\mathbf{r}(b)) - f(\mathbf{r}(a))$$

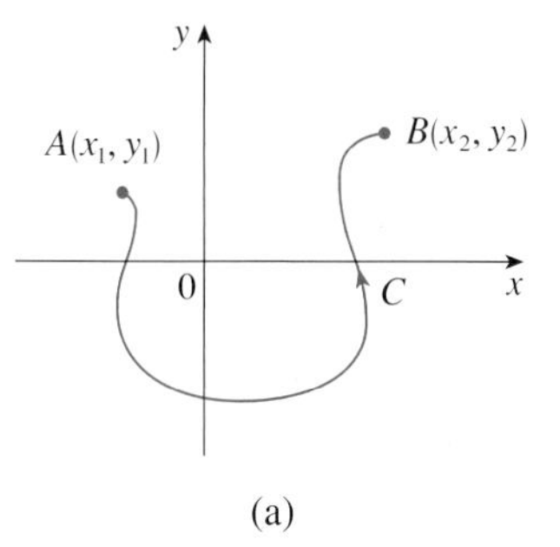

(a)

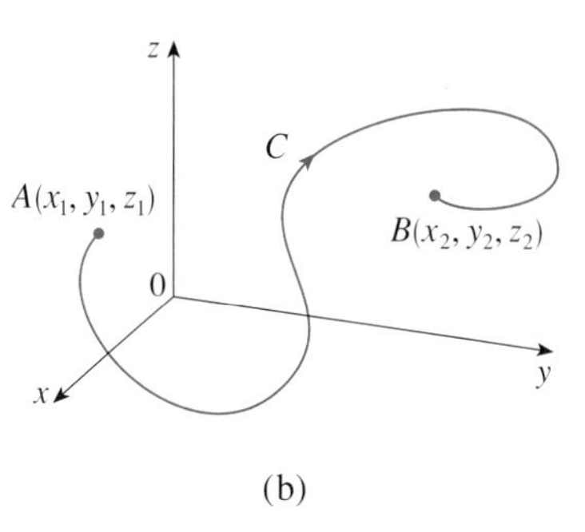

(b)

FIGURE 1

NOTE Theorem 2 says that we can evaluate the line integral of a conservative vector field (the gradient vector field of the potential function f) simply by knowing the value of f at the endpoints of C. In fact, Theorem 2 says that the line integral of ∇f is the net change in f. If f is a function of two variables and C is a plane curve with initial point $A(x_1, y_1)$ and terminal point $B(x_2, y_2)$, as in Figure 1, then Theorem 2 becomes

$$\int_C \nabla f \cdot d\mathbf{r} = f(x_2, y_2) - f(x_1, y_1)$$

If f is a function of three variables and C is a space curve joining the point $A(x_1, y_1, z_1)$ to the point $B(x_2, y_2, z_2)$, then we have

$$\int_C \nabla f \cdot d\mathbf{r} = f(x_2, y_2, z_2) - f(x_1, y_1, z_1)$$

Let's prove Theorem 2 for this case.

PROOF OF THEOREM 2 Using Definition 16.2.13, we have

$$\int_C \nabla f \cdot d\mathbf{r} = \int_a^b \nabla f(\mathbf{r}(t)) \cdot \mathbf{r}'(t)\, dt$$

$$= \int_a^b \left(\frac{\partial f}{\partial x}\frac{dx}{dt} + \frac{\partial f}{\partial y}\frac{dy}{dt} + \frac{\partial f}{\partial z}\frac{dz}{dt} \right) dt$$

$$= \int_a^b \frac{d}{dt} f(\mathbf{r}(t))\, dt \qquad \text{(by the Chain Rule)}$$

$$= f(\mathbf{r}(b)) - f(\mathbf{r}(a))$$

The last step follows from the Fundamental Theorem of Calculus (Equation 1).

Although we have proved Theorem 2 for smooth curves, it is also true for piecewise-smooth curves. This can be seen by subdividing C into a finite number of smooth curves and adding the resulting integrals.

EXAMPLE 1 Find the work done by the gravitational field

$$\mathbf{F}(\mathbf{x}) = -\frac{mMG}{|\mathbf{x}|^3}\mathbf{x}$$

in moving a particle with mass m from the point (3, 4, 12) to the point (2, 2, 0) along a piecewise-smooth curve C. (See Example 4 in Section 16.1.)

SOLUTION From Section 16.1 we know that $\mathbf{F}$ is a conservative vector field and, in fact, $\mathbf{F} = \nabla f$, where

$$f(x, y, z) = \frac{mMG}{\sqrt{x^2 + y^2 + z^2}}$$

Therefore, by Theorem 2, the work done is

$$W = \int_C \mathbf{F} \cdot d\mathbf{r} = \int_C \nabla f \cdot d\mathbf{r}$$

$$= f(2, 2, 0) - f(3, 4, 12)$$

$$= \frac{mMG}{\sqrt{2^2 + 2^2}} - \frac{mMG}{\sqrt{3^2 + 4^2 + 12^2}} = mMG\left(\frac{1}{2\sqrt{2}} - \frac{1}{13}\right)$$

Independence of Path

Suppose C_1 and C_2 are two piecewise-smooth curves (which are called **paths**) that have the same initial point A and terminal point B. We know from Example 4 in Section 16.2 that, in general, $\int_{C_1} \mathbf{F} \cdot d\mathbf{r} \neq \int_{C_2} \mathbf{F} \cdot d\mathbf{r}$. But one implication of Theorem 2 is that

$$\int_{C_1} \nabla f \cdot d\mathbf{r} = \int_{C_2} \nabla f \cdot d\mathbf{r}$$

whenever ∇f is continuous. In other words, the line integral of a *conservative* vector field depends only on the initial point and terminal point of a curve.

In general, if $\mathbf{F}$ is a continuous vector field with domain D, we say that the line integral $\int_C \mathbf{F} \cdot d\mathbf{r}$ is **independent of path** if $\int_{C_1} \mathbf{F} \cdot d\mathbf{r} = \int_{C_2} \mathbf{F} \cdot d\mathbf{r}$ for any two paths C_1 and C_2 in D that have the same initial and terminal points. With this terminology we can say that *line integrals of conservative vector fields are independent of path.*

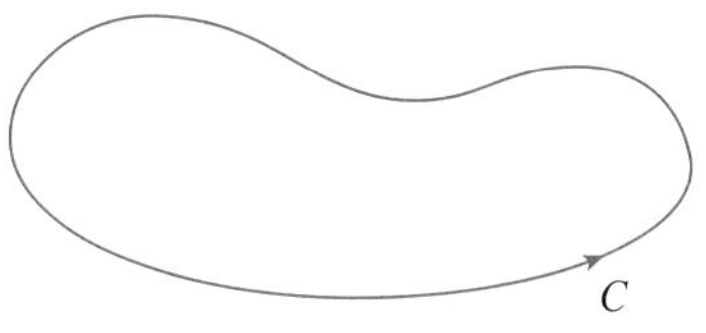

FIGURE 2
A closed curve

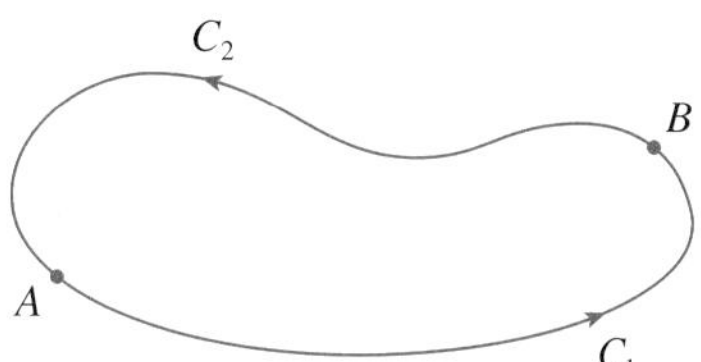

FIGURE 3

A curve is called **closed** if its terminal point coincides with its initial point, that is, $\mathbf{r}(b) = \mathbf{r}(a)$. (See Figure 2.) If $\int_C \mathbf{F} \cdot d\mathbf{r}$ is independent of path in D and C is any closed path in D, we can choose any two points A and B on C and regard C as being composed of the path C_1 from A to B followed by the path C_2 from B to A. (See Figure 3.) Then

$$\int_C \mathbf{F} \cdot d\mathbf{r} = \int_{C_1} \mathbf{F} \cdot d\mathbf{r} + \int_{C_2} \mathbf{F} \cdot d\mathbf{r} = \int_{C_1} \mathbf{F} \cdot d\mathbf{r} - \int_{-C_2} \mathbf{F} \cdot d\mathbf{r} = 0$$

since C_1 and $-C_2$ have the same initial and terminal points.

Conversely, if it is true that $\int_C \mathbf{F} \cdot d\mathbf{r} = 0$ whenever C is a closed path in D, then we demonstrate independence of path as follows. Take any two paths C_1 and C_2 from A to B in D and define C to be the curve consisting of C_1 followed by $-C_2$. Then

$$0 = \int_C \mathbf{F} \cdot d\mathbf{r} = \int_{C_1} \mathbf{F} \cdot d\mathbf{r} + \int_{-C_2} \mathbf{F} \cdot d\mathbf{r} = \int_{C_1} \mathbf{F} \cdot d\mathbf{r} - \int_{C_2} \mathbf{F} \cdot d\mathbf{r}$$

and so $\int_{C_1} \mathbf{F} \cdot d\mathbf{r} = \int_{C_2} \mathbf{F} \cdot d\mathbf{r}$. Thus we have proved the following theorem.

3 Theorem $\int_C \mathbf{F} \cdot d\mathbf{r}$ is independent of path in D if and only if $\int_C \mathbf{F} \cdot d\mathbf{r} = 0$ for every closed path C in D.

Since we know that the line integral of any conservative vector field $\mathbf{F}$ is independent of path, it follows that $\int_C \mathbf{F} \cdot d\mathbf{r} = 0$ for any closed path. The physical interpretation is that the work done by a conservative force field (such as the gravitational or electric field in Section 16.1) as it moves an object around a closed path is 0.

The following theorem says that the *only* vector fields that are independent of path are conservative. It is stated and proved for plane curves, but there is a similar version for space curves. We assume that D is **open**, which means that for every point P in D there is a disk with center P that lies entirely in D. (So D doesn't contain any of its boundary points.) In addition, we assume that D is **connected**: This means that any two points in D can be joined by a path that lies in D.

4 Theorem Suppose $\mathbf{F}$ is a vector field that is continuous on an open connected region D. If $\int_C \mathbf{F} \cdot d\mathbf{r}$ is independent of path in D, then $\mathbf{F}$ is a conservative vector field on D; that is, there exists a function f such that $\nabla f = \mathbf{F}$.

PROOF Let $A(a, b)$ be a fixed point in D. We construct the desired potential function f by defining

$$f(x, y) = \int_{(a, b)}^{(x, y)} \mathbf{F} \cdot d\mathbf{r}$$

for any point (x, y) in D. Since $\int_C \mathbf{F} \cdot d\mathbf{r}$ is independent of path, it does not matter which path C from (a, b) to (x, y) is used to evaluate $f(x, y)$. Since D is open, there exists a disk contained in D with center (x, y). Choose any point (x_1, y) in the disk with $x_1 < x$ and let C consist of any path C_1 from (a, b) to (x_1, y) followed by the horizontal line segment C_2 from (x_1, y) to (x, y). (See Figure 4.) Then

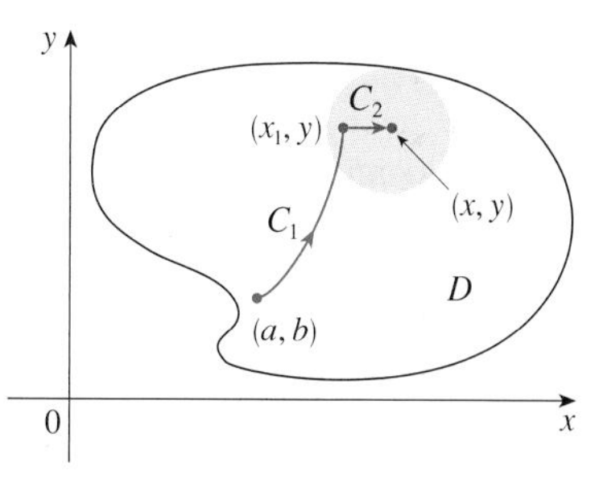

FIGURE 4

$$f(x, y) = \int_{C_1} \mathbf{F} \cdot d\mathbf{r} + \int_{C_2} \mathbf{F} \cdot d\mathbf{r} = \int_{(a, b)}^{(x_1, y)} \mathbf{F} \cdot d\mathbf{r} + \int_{C_2} \mathbf{F} \cdot d\mathbf{r}$$

Notice that the first of these integrals does not depend on x, so

$$\frac{\partial}{\partial x} f(x, y) = 0 + \frac{\partial}{\partial x} \int_{C_2} \mathbf{F} \cdot d\mathbf{r}$$

If we write $\mathbf{F} = P\,\mathbf{i} + Q\,\mathbf{j}$, then

$$\int_{C_2} \mathbf{F} \cdot d\mathbf{r} = \int_{C_2} P\,dx + Q\,dy$$

On C_2, y is constant, so $dy = 0$. Using t as the parameter, where $x_1 \leqslant t \leqslant x$, we have

$$\frac{\partial}{\partial x} f(x, y) = \frac{\partial}{\partial x} \int_{C_2} P\,dx + Q\,dy = \frac{\partial}{\partial x} \int_{x_1}^{x} P(t, y)\,dt = P(x, y)$$

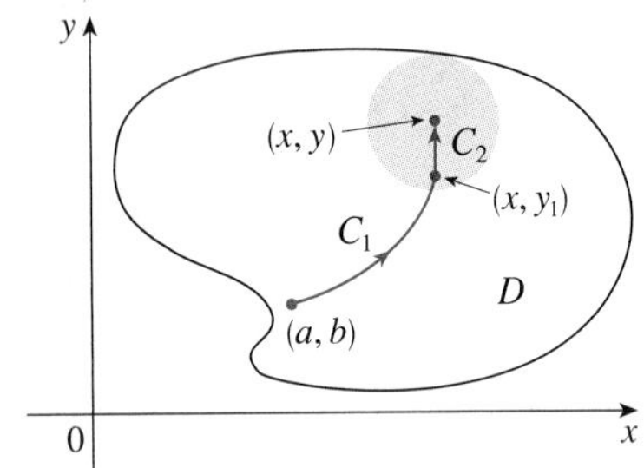

FIGURE 5

by Part 1 of the Fundamental Theorem of Calculus (see Section 5.3). A similar argument, using a vertical line segment (see Figure 5), shows that

$$\frac{\partial}{\partial y} f(x, y) = \frac{\partial}{\partial y} \int_{C_2} P\,dx + Q\,dy = \frac{\partial}{\partial y} \int_{y_1}^{y} Q(x, t)\,dt = Q(x, y)$$

Thus
$$\mathbf{F} = P\,\mathbf{i} + Q\,\mathbf{j} = \frac{\partial f}{\partial x}\,\mathbf{i} + \frac{\partial f}{\partial y}\,\mathbf{j} = \nabla f$$

which says that $\mathbf{F}$ is conservative. ■

The question remains: How is it possible to determine whether or not a vector field $\mathbf{F}$ is conservative? Suppose it is known that $\mathbf{F} = P\,\mathbf{i} + Q\,\mathbf{j}$ is conservative, where P and Q have continuous first-order partial derivatives. Then there is a function f such that $\mathbf{F} = \nabla f$, that is,

$$P = \frac{\partial f}{\partial x} \qquad \text{and} \qquad Q = \frac{\partial f}{\partial y}$$

Therefore, by Clairaut's Theorem,

$$\frac{\partial P}{\partial y} = \frac{\partial^2 f}{\partial y\,\partial x} = \frac{\partial^2 f}{\partial x\,\partial y} = \frac{\partial Q}{\partial x}$$

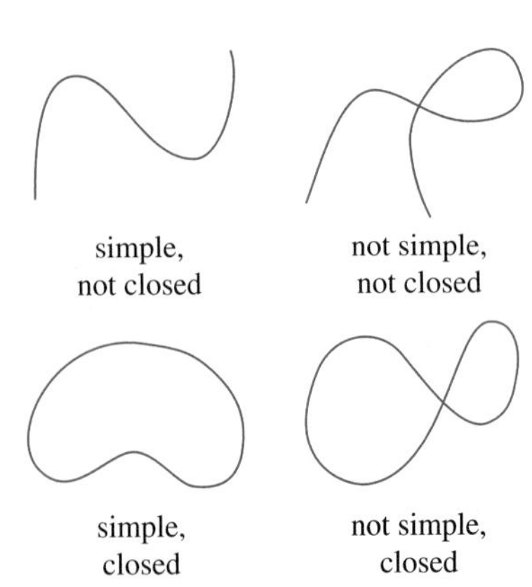

FIGURE 6
Types of curves

5 Theorem If $\mathbf{F}(x, y) = P(x, y)\,\mathbf{i} + Q(x, y)\,\mathbf{j}$ is a conservative vector field, where P and Q have continuous first-order partial derivatives on a domain D, then throughout D we have

$$\frac{\partial P}{\partial y} = \frac{\partial Q}{\partial x}$$

The converse of Theorem 5 is true only for a special type of region. To explain this, we first need the concept of a **simple curve**, which is a curve that doesn't intersect itself anywhere between its endpoints. [See Figure 6; $\mathbf{r}(a) = \mathbf{r}(b)$ for a simple closed curve, but $\mathbf{r}(t_1) \neq \mathbf{r}(t_2)$ when $a < t_1 < t_2 < b$.]

In Theorem 4 we needed an open connected region. For the next theorem we need a stronger condition. A **simply-connected region** in the plane is a connected region D such that every simple closed curve in D encloses only points that are in D. Notice from Figure 7 that, intuitively speaking, a simply-connected region contains no hole and can't consist of two separate pieces.

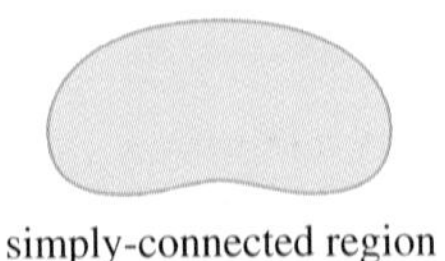
simply-connected region

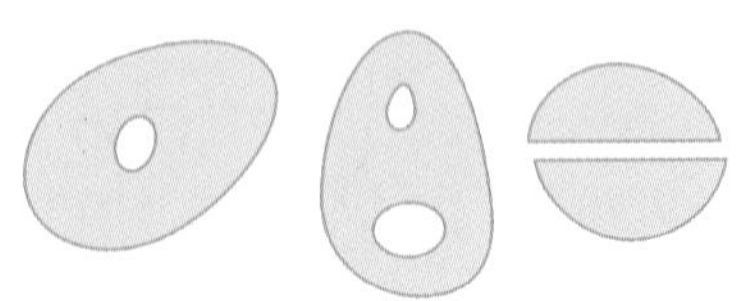
regions that are not simply-connected

FIGURE 7

In terms of simply-connected regions, we can now state a partial converse to Theorem 5 that gives a convenient method for verifying that a vector field on $\mathbb{R}^2$ is conservative. The proof will be sketched in the next section as a consequence of Green's Theorem.

6 Theorem Let $\mathbf{F} = P\,\mathbf{i} + Q\,\mathbf{j}$ be a vector field on an open simply-connected region D. Suppose that P and Q have continuous first-order derivatives and

$$\frac{\partial P}{\partial y} = \frac{\partial Q}{\partial x} \qquad \text{throughout } D$$

Then $\mathbf{F}$ is conservative.

V EXAMPLE 2 Determine whether or not the vector field

$$\mathbf{F}(x, y) = (x - y)\,\mathbf{i} + (x - 2)\,\mathbf{j}$$

is conservative.

SOLUTION Let $P(x, y) = x - y$ and $Q(x, y) = x - 2$. Then

$$\frac{\partial P}{\partial y} = -1 \qquad \frac{\partial Q}{\partial x} = 1$$

Since $\partial P/\partial y \neq \partial Q/\partial x$, $\mathbf{F}$ is not conservative by Theorem 5.

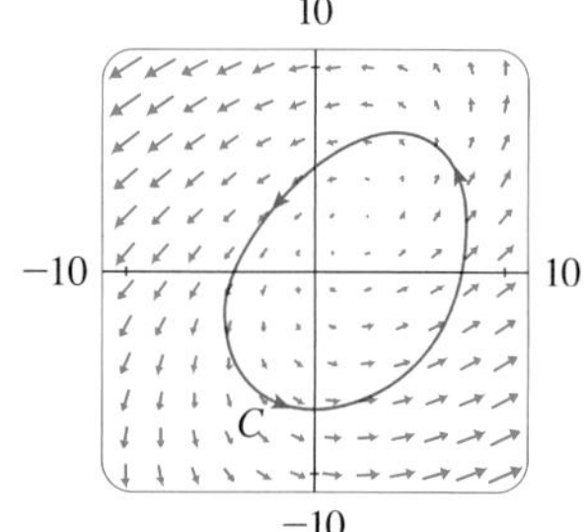

FIGURE 8

Figures 8 and 9 show the vector fields in Examples 2 and 3, respectively. The vectors in Figure 8 that start on the closed curve C all appear to point in roughly the same direction as C. So it looks as if $\int_C \mathbf{F} \cdot d\mathbf{r} > 0$ and therefore $\mathbf{F}$ is not conservative. The calculation in Example 2 confirms this impression. Some of the vectors near the curves C_1 and C_2 in Figure 9 point in approximately the same direction as the curves, whereas others point in the opposite direction. So it appears plausible that line integrals around all closed paths are 0. Example 3 shows that $\mathbf{F}$ is indeed conservative.

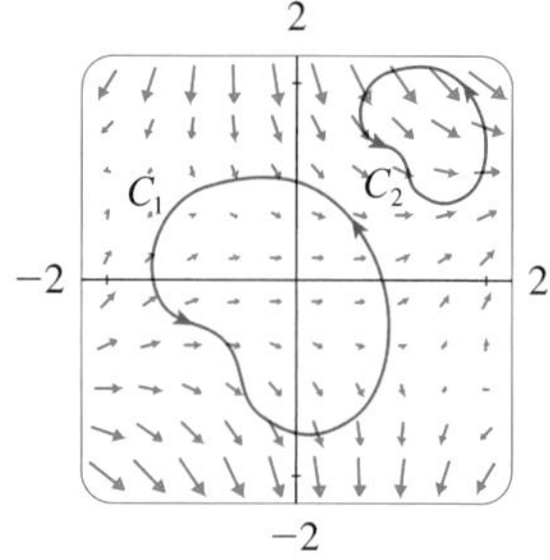

FIGURE 9

V EXAMPLE 3 Determine whether or not the vector field

$$\mathbf{F}(x, y) = (3 + 2xy)\,\mathbf{i} + (x^2 - 3y^2)\,\mathbf{j}$$

is conservative.

SOLUTION Let $P(x, y) = 3 + 2xy$ and $Q(x, y) = x^2 - 3y^2$. Then

$$\frac{\partial P}{\partial y} = 2x = \frac{\partial Q}{\partial x}$$

Also, the domain of $\mathbf{F}$ is the entire plane ($D = \mathbb{R}^2$), which is open and simply-connected. Therefore we can apply Theorem 6 and conclude that $\mathbf{F}$ is conservative.

In Example 3, Theorem 6 told us that $\mathbf{F}$ is conservative, but it did not tell us how to find the (potential) function f such that $\mathbf{F} = \nabla f$. The proof of Theorem 4 gives us a clue as to how to find f. We use "partial integration" as in the following example.

EXAMPLE 4

(a) If $\mathbf{F}(x, y) = (3 + 2xy)\,\mathbf{i} + (x^2 - 3y^2)\,\mathbf{j}$, find a function f such that $\mathbf{F} = \nabla f$.

(b) Evaluate the line integral $\int_C \mathbf{F} \cdot d\mathbf{r}$, where C is the curve given by

$$\mathbf{r}(t) = e^t \sin t\,\mathbf{i} + e^t \cos t\,\mathbf{j} \qquad 0 \leqslant t \leqslant \pi$$

SOLUTION

(a) From Example 3 we know that $\mathbf{F}$ is conservative and so there exists a function f with $\nabla f = \mathbf{F}$, that is,

$$\boxed{7} \qquad f_x(x, y) = 3 + 2xy$$

$$\boxed{8} \qquad f_y(x, y) = x^2 - 3y^2$$

Integrating [7] with respect to x, we obtain

9 $$f(x, y) = 3x + x^2y + g(y)$$

Notice that the constant of integration is a constant with respect to x, that is, a function of y, which we have called $g(y)$. Next we differentiate both sides of [9] with respect to y:

10 $$f_y(x, y) = x^2 + g'(y)$$

Comparing [8] and [10], we see that

$$g'(y) = -3y^2$$

Integrating with respect to y, we have

$$g(y) = -y^3 + K$$

where K is a constant. Putting this in [9], we have

$$f(x, y) = 3x + x^2y - y^3 + K$$

as the desired potential function.

(b) To use Theorem 2 all we have to know are the initial and terminal points of C, namely, $\mathbf{r}(0) = (0, 1)$ and $\mathbf{r}(\pi) = (0, -e^{\pi})$. In the expression for $f(x, y)$ in part (a), any value of the constant K will do, so let's choose $K = 0$. Then we have

$$\int_C \mathbf{F} \cdot d\mathbf{r} = \int_C \nabla f \cdot d\mathbf{r} = f(0, -e^{\pi}) - f(0, 1) = e^{3\pi} - (-1) = e^{3\pi} + 1$$

This method is much shorter than the straightforward method for evaluating line integrals that we learned in Section 16.2.

A criterion for determining whether or not a vector field $\mathbf{F}$ on $\mathbb{R}^3$ is conservative is given in Section 16.5. Meanwhile, the next example shows that the technique for finding the potential function is much the same as for vector fields on $\mathbb{R}^2$.

V EXAMPLE 5 If $\mathbf{F}(x, y, z) = y^2\,\mathbf{i} + (2xy + e^{3z})\,\mathbf{j} + 3ye^{3z}\,\mathbf{k}$, find a function f such that $\nabla f = \mathbf{F}$.

SOLUTION If there is such a function f, then

11 $$f_x(x, y, z) = y^2$$

12 $$f_y(x, y, z) = 2xy + e^{3z}$$

13 $$f_z(x, y, z) = 3ye^{3z}$$

Integrating [11] with respect to x, we get

14 $$f(x, y, z) = xy^2 + g(y, z)$$

where $g(y, z)$ is a constant with respect to x. Then differentiating [14] with respect to y, we have

$$f_y(x, y, z) = 2xy + g_y(y, z)$$

and comparison with [12] gives

$$g_y(y, z) = e^{3z}$$

Thus $g(y, z) = ye^{3z} + h(z)$ and we rewrite [14] as

$$f(x, y, z) = xy^2 + ye^{3z} + h(z)$$

Finally, differentiating with respect to z and comparing with [13], we obtain $h'(z) = 0$ and therefore $h(z) = K$, a constant. The desired function is

$$f(x, y, z) = xy^2 + ye^{3z} + K$$

It is easily verified that $\nabla f = \mathbf{F}$.

Conservation of Energy

Let's apply the ideas of this chapter to a continuous force field $\mathbf{F}$ that moves an object along a path C given by $\mathbf{r}(t)$, $a \leqslant t \leqslant b$, where $\mathbf{r}(a) = A$ is the initial point and $\mathbf{r}(b) = B$ is the terminal point of C. According to Newton's Second Law of Motion (see Section 13.4), the force $\mathbf{F}(\mathbf{r}(t))$ at a point on C is related to the acceleration $\mathbf{a}(t) = \mathbf{r}''(t)$ by the equation

$$\mathbf{F}(\mathbf{r}(t)) = m\mathbf{r}''(t)$$

So the work done by the force on the object is

$$\begin{aligned} W &= \int_C \mathbf{F} \cdot d\mathbf{r} = \int_a^b \mathbf{F}(\mathbf{r}(t)) \cdot \mathbf{r}'(t)\,dt = \int_a^b m\mathbf{r}''(t) \cdot \mathbf{r}'(t)\,dt \\ &= \frac{m}{2} \int_a^b \frac{d}{dt} [\mathbf{r}'(t) \cdot \mathbf{r}'(t)]\,dt && \text{(Theorem 13.2.3, Formula 4)} \\ &= \frac{m}{2} \int_a^b \frac{d}{dt} |\mathbf{r}'(t)|^2\,dt = \frac{m}{2} \left[|\mathbf{r}'(t)|^2 \right]_a^b && \text{(Fundamental Theorem of Calculus)} \\ &= \frac{m}{2} \left(|\mathbf{r}'(b)|^2 - |\mathbf{r}'(a)|^2 \right) \end{aligned}$$

Therefore

15 $$W = \tfrac{1}{2}m\,|\mathbf{v}(b)|^2 - \tfrac{1}{2}m\,|\mathbf{v}(a)|^2$$

where $\mathbf{v} = \mathbf{r}'$ is the velocity.

The quantity $\frac{1}{2}m\,|\mathbf{v}(t)|^2$, that is, half the mass times the square of the speed, is called the **kinetic energy** of the object. Therefore we can rewrite Equation 15 as

16 $$W = K(B) - K(A)$$

which says that the work done by the force field along C is equal to the change in kinetic energy at the endpoints of C.

Now let's further assume that $\mathbf{F}$ is a conservative force field; that is, we can write $\mathbf{F} = \nabla f$. In physics, the **potential energy** of an object at the point (x, y, z) is defined as $P(x, y, z) = -f(x, y, z)$, so we have $\mathbf{F} = -\nabla P$. Then by Theorem 2 we have

$$W = \int_C \mathbf{F} \cdot d\mathbf{r} = -\int_C \nabla P \cdot d\mathbf{r} = -[P(\mathbf{r}(b)) - P(\mathbf{r}(a))] = P(A) - P(B)$$

Comparing this equation with Equation 16, we see that

$$P(A) + K(A) = P(B) + K(B)$$

which says that if an object moves from one point A to another point B under the influence of a conservative force field, then the sum of its potential energy and its kinetic energy remains constant. This is called the **Law of Conservation of Energy** and it is the reason the vector field is called *conservative.*

16.3 Exercises

1. The figure shows a curve C and a contour map of a function f whose gradient is continuous. Find $\int_C \nabla f \cdot d\mathbf{r}$.

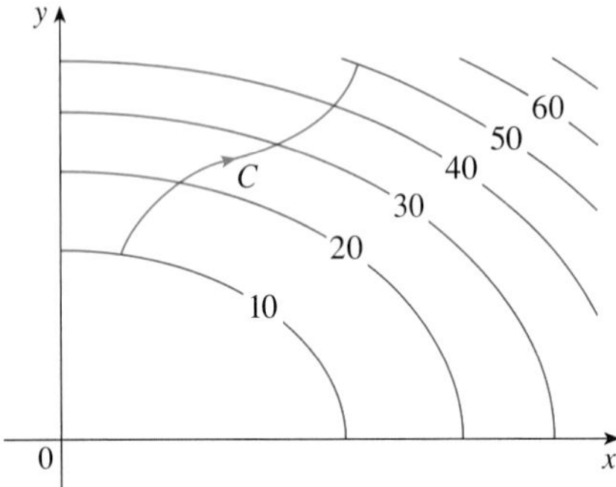

2. A table of values of a function f with continuous gradient is given. Find $\int_C \nabla f \cdot d\mathbf{r}$, where C has parametric equations

$$x = t^2 + 1 \qquad y = t^3 + t \qquad 0 \leqslant t \leqslant 1$$

x \ y	0	1	2
0	1	6	4
1	3	5	7
2	8	2	9

3–10 Determine whether or not $\mathbf{F}$ is a conservative vector field. If it is, find a function f such that $\mathbf{F} = \nabla f$.

3. $\mathbf{F}(x, y) = (2x - 3y)\,\mathbf{i} + (-3x + 4y - 8)\,\mathbf{j}$

4. $\mathbf{F}(x, y) = e^x \sin y\,\mathbf{i} + e^x \cos y\,\mathbf{j}$

5. $\mathbf{F}(x, y) = e^x \cos y\,\mathbf{i} + e^x \sin y\,\mathbf{j}$

6. $\mathbf{F}(x, y) = (3x^2 - 2y^2)\,\mathbf{i} + (4xy + 3)\,\mathbf{j}$

7. $\mathbf{F}(x, y) = (ye^x + \sin y)\,\mathbf{i} + (e^x + x\cos y)\,\mathbf{j}$

8. $\mathbf{F}(x, y) = (2xy + y^{-2})\,\mathbf{i} + (x^2 - 2xy^{-3})\,\mathbf{j}, \quad y > 0$

9. $\mathbf{F}(x, y) = (\ln y + 2xy^3)\,\mathbf{i} + (3x^2y^2 + x/y)\,\mathbf{j}$

10. $\mathbf{F}(x, y) = (xy\cosh xy + \sinh xy)\,\mathbf{i} + (x^2\cosh xy)\,\mathbf{j}$

11. The figure shows the vector field $\mathbf{F}(x, y) = \langle 2xy, x^2 \rangle$ and three curves that start at $(1, 2)$ and end at $(3, 2)$.

(a) Explain why $\int_C \mathbf{F} \cdot d\mathbf{r}$ has the same value for all three curves.

(b) What is this common value?

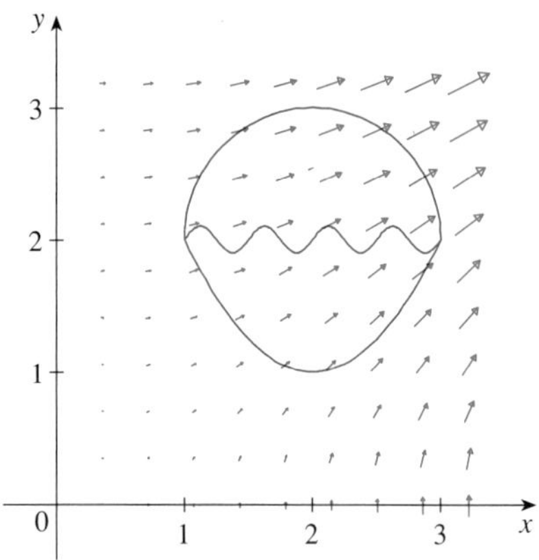

12–18 (a) Find a function f such that $\mathbf{F} = \nabla f$ and (b) use part (a) to evaluate $\int_C \mathbf{F} \cdot d\mathbf{r}$ along the given curve C.

12. $\mathbf{F}(x, y) = x^2\,\mathbf{i} + y^2\,\mathbf{j}$,
C is the arc of the parabola $y = 2x^2$ from $(-1, 2)$ to $(2, 8)$

13. $\mathbf{F}(x, y) = xy^2\,\mathbf{i} + x^2y\,\mathbf{j}$,
C: $\mathbf{r}(t) = \langle t + \sin \tfrac{1}{2}\pi t, t + \cos \tfrac{1}{2}\pi t \rangle, \quad 0 \leqslant t \leqslant 1$

14. $\mathbf{F}(x, y) = (1 + xy)e^{xy}\,\mathbf{i} + x^2e^{xy}\,\mathbf{j}$,
C: $\mathbf{r}(t) = \cos t\,\mathbf{i} + 2\sin t\,\mathbf{j}, \quad 0 \leqslant t \leqslant \pi/2$

15. $\mathbf{F}(x, y, z) = yz\,\mathbf{i} + xz\,\mathbf{j} + (xy + 2z)\,\mathbf{k}$,
C is the line segment from $(1, 0, -2)$ to $(4, 6, 3)$

16. $\mathbf{F}(x, y, z) = (y^2z + 2xz^2)\,\mathbf{i} + 2xyz\,\mathbf{j} + (xy^2 + 2x^2z)\,\mathbf{k}$,
C: $x = \sqrt{t}$, $y = t + 1$, $z = t^2$, $0 \leqslant t \leqslant 1$

17. $\mathbf{F}(x, y, z) = yze^{xz}\,\mathbf{i} + e^{xz}\,\mathbf{j} + xye^{xz}\,\mathbf{k}$,
C: $\mathbf{r}(t) = (t^2 + 1)\,\mathbf{i} + (t^2 - 1)\,\mathbf{j} + (t^2 - 2t)\,\mathbf{k}$, $0 \leqslant t \leqslant 2$

18. $\mathbf{F}(x, y, z) = \sin y\,\mathbf{i} + (x \cos y + \cos z)\,\mathbf{j} - y \sin z\,\mathbf{k}$,
C: $\mathbf{r}(t) = \sin t\,\mathbf{i} + t\,\mathbf{j} + 2t\,\mathbf{k}$, $0 \leqslant t \leqslant \pi/2$

19–20 Show that the line integral is independent of path and evaluate the integral.

19. $\int_C 2xe^{-y}\,dx + (2y - x^2e^{-y})\,dy$,
C is any path from $(1, 0)$ to $(2, 1)$

20. $\int_C \sin y\,dx + (x \cos y - \sin y)\,dy$,
C is any path from $(2, 0)$ to $(1, \pi)$

---

21. Suppose you're asked to determine the curve that requires the least work for a force field $\mathbf{F}$ to move a particle from one point to another point. You decide to check first whether $\mathbf{F}$ is conservative, and indeed it turns out that it is. How would you reply to the request?

22. Suppose an experiment determines that the amount of work required for a force field $\mathbf{F}$ to move a particle from the point $(1, 2)$ to the point $(5, -3)$ along a curve C_1 is 1.2 J and the work done by $\mathbf{F}$ in moving the particle along another curve C_2 between the same two points is 1.4 J. What can you say about $\mathbf{F}$? Why?

23–24 Find the work done by the force field $\mathbf{F}$ in moving an object from P to Q.

23. $\mathbf{F}(x, y) = 2y^{3/2}\,\mathbf{i} + 3x\sqrt{y}\,\mathbf{j}$; $P(1, 1)$, $Q(2, 4)$

24. $\mathbf{F}(x, y) = e^{-y}\,\mathbf{i} - xe^{-y}\,\mathbf{j}$; $P(0, 1)$, $Q(2, 0)$

25–26 Is the vector field shown in the figure conservative? Explain.

25. **26.**

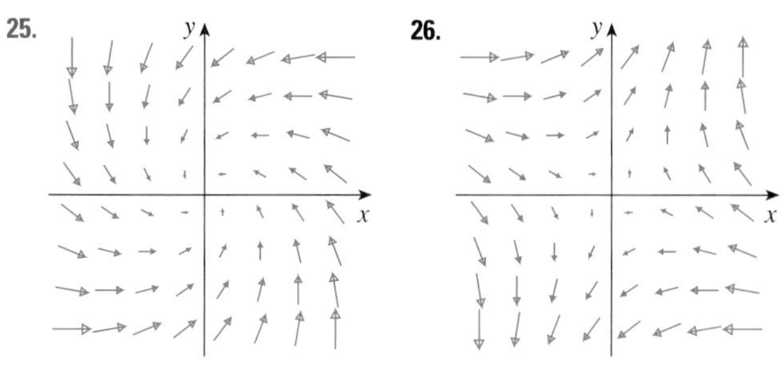

CAS **27.** If $\mathbf{F}(x, y) = \sin y\,\mathbf{i} + (1 + x \cos y)\,\mathbf{j}$, use a plot to guess whether $\mathbf{F}$ is conservative. Then determine whether your guess is correct.

28. Let $\mathbf{F} = \nabla f$, where $f(x, y) = \sin(x - 2y)$. Find curves C_1 and C_2 that are not closed and satisfy the equation.

(a) $\int_{C_1} \mathbf{F} \cdot d\mathbf{r} = 0$ (b) $\int_{C_2} \mathbf{F} \cdot d\mathbf{r} = 1$

29. Show that if the vector field $\mathbf{F} = P\,\mathbf{i} + Q\,\mathbf{j} + R\,\mathbf{k}$ is conservative and P, Q, R have continuous first-order partial derivatives, then

$$\frac{\partial P}{\partial y} = \frac{\partial Q}{\partial x} \qquad \frac{\partial P}{\partial z} = \frac{\partial R}{\partial x} \qquad \frac{\partial Q}{\partial z} = \frac{\partial R}{\partial y}$$

30. Use Exercise 29 to show that the line integral $\int_C y\,dx + x\,dy + xyz\,dz$ is not independent of path.

31–34 Determine whether or not the given set is (a) open, (b) connected, and (c) simply-connected.

31. $\{(x, y) \mid 0 < y < 3\}$

32. $\{(x, y) \mid 1 < |x| < 2\}$

33. $\{(x, y) \mid 1 \leqslant x^2 + y^2 \leqslant 4, y \geqslant 0\}$

34. $\{(x, y) \mid (x, y) \neq (2, 3)\}$

---

35. Let $\mathbf{F}(x, y) = \dfrac{-y\,\mathbf{i} + x\,\mathbf{j}}{x^2 + y^2}$.

(a) Show that $\partial P/\partial y = \partial Q/\partial x$.

(b) Show that $\int_C \mathbf{F} \cdot d\mathbf{r}$ is not independent of path. [*Hint:* Compute $\int_{C_1} \mathbf{F} \cdot d\mathbf{r}$ and $\int_{C_2} \mathbf{F} \cdot d\mathbf{r}$, where C_1 and C_2 are the upper and lower halves of the circle $x^2 + y^2 = 1$ from $(1, 0)$ to $(-1, 0)$.] Does this contradict Theorem 6?

36. (a) Suppose that $\mathbf{F}$ is an inverse square force field, that is,

$$\mathbf{F}(\mathbf{r}) = \frac{c\mathbf{r}}{|\mathbf{r}|^3}$$

for some constant c, where $\mathbf{r} = x\,\mathbf{i} + y\,\mathbf{j} + z\,\mathbf{k}$. Find the work done by $\mathbf{F}$ in moving an object from a point P_1 along a path to a point P_2 in terms of the distances d_1 and d_2 from these points to the origin.

(b) An example of an inverse square field is the gravitational field $\mathbf{F} = -(mMG)\mathbf{r}/|\mathbf{r}|^3$ discussed in Example 4 in Section 16.1. Use part (a) to find the work done by the gravitational field when the earth moves from aphelion (at a maximum distance of 1.52×10^8 km from the sun) to perihelion (at a minimum distance of 1.47×10^8 km). (Use the values $m = 5.97 \times 10^{24}$ kg, $M = 1.99 \times 10^{30}$ kg, and $G = 6.67 \times 10^{-11}$ N·m²/kg².)

(c) Another example of an inverse square field is the electric force field $\mathbf{F} = \varepsilon qQ\mathbf{r}/|\mathbf{r}|^3$ discussed in Example 5 in Section 16.1. Suppose that an electron with a charge of -1.6×10^{-19} C is located at the origin. A positive unit charge is positioned a distance 10^{-12} m from the electron and moves to a position half that distance from the electron. Use part (a) to find the work done by the electric force field. (Use the value $\varepsilon = 8.985 \times 10^9$.)

16.4 Green's Theorem

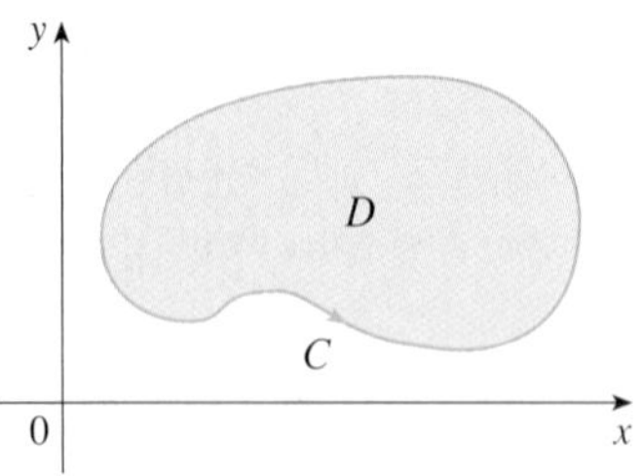

FIGURE 1

Green's Theorem gives the relationship between a line integral around a simple closed curve C and a double integral over the plane region D bounded by C. (See Figure 1. We assume that D consists of all points inside C as well as all points on C.) In stating Green's Theorem we use the convention that the **positive orientation** of a simple closed curve C refers to a single *counterclockwise* traversal of C. Thus if C is given by the vector function $\mathbf{r}(t)$, $a \leqslant t \leqslant b$, then the region D is always on the left as the point $\mathbf{r}(t)$ traverses C. (See Figure 2.)

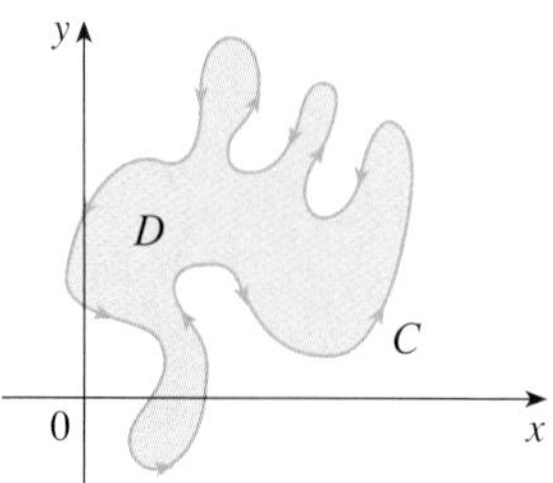

(a) Positive orientation

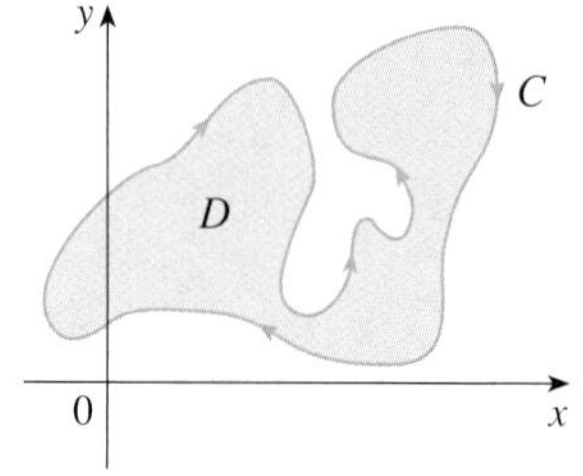

(b) Negative orientation

FIGURE 2

Green's Theorem Let C be a positively oriented, piecewise-smooth, simple closed curve in the plane and let D be the region bounded by C. If P and Q have continuous partial derivatives on an open region that contains D, then

$$\int_C P\,dx + Q\,dy = \iint_D \left(\frac{\partial Q}{\partial x} - \frac{\partial P}{\partial y} \right) dA$$

Recall that the left side of this equation is another way of writing $\int_C \mathbf{F} \cdot d\mathbf{r}$, where $\mathbf{F} = P\,\mathbf{i} + Q\,\mathbf{j}$.

NOTE The notation

$$\oint_C P\,dx + Q\,dy \qquad \text{or} \qquad \ointctrclockwise_C P\,dx + Q\,dy$$

is sometimes used to indicate that the line integral is calculated using the positive orientation of the closed curve C. Another notation for the positively oriented boundary curve of D is ∂D, so the equation in Green's Theorem can be written as

1
$$\iint_D \left(\frac{\partial Q}{\partial x} - \frac{\partial P}{\partial y} \right) dA = \int_{\partial D} P\,dx + Q\,dy$$

Green's Theorem should be regarded as the counterpart of the Fundamental Theorem of Calculus for double integrals. Compare Equation 1 with the statement of the Fundamental Theorem of Calculus, Part 2, in the following equation:

$$\int_a^b F'(x)\,dx = F(b) - F(a)$$

In both cases there is an integral involving derivatives (F', $\partial Q/\partial x$, and $\partial P/\partial y$) on the left side of the equation. And in both cases the right side involves the values of the original functions (F, Q, and P) only on the *boundary* of the domain. (In the one-dimensional case, the domain is an interval $[a, b]$ whose boundary consists of just two points, a and b.)

Green's Theorem is not easy to prove in general, but we can give a proof for the special case where the region is both type I and type II (see Section 15.3). Let's call such regions **simple regions**.

George Green

Green's Theorem is named after the self-taught English scientist George Green (1793–1841). He worked full-time in his father's bakery from the age of nine and taught himself mathematics from library books. In 1828 he published privately *An Essay on the Application of Mathematical Analysis to the Theories of Electricity and Magnetism*, but only 100 copies were printed and most of those went to his friends. This pamphlet contained a theorem that is equivalent to what we know as Green's Theorem, but it didn't become widely known at that time. Finally, at age 40, Green entered Cambridge University as an undergraduate but died four years after graduation. In 1846 William Thomson (Lord Kelvin) located a copy of Green's essay, realized its significance, and had it reprinted. Green was the first person to try to formulate a mathematical theory of electricity and magnetism. His work was the basis for the subsequent electromagnetic theories of Thomson, Stokes, Rayleigh, and Maxwell.

PROOF OF GREEN'S THEOREM FOR THE CASE IN WHICH D IS A SIMPLE REGION Notice that Green's Theorem will be proved if we can show that

2
$$\int_C P\,dx = -\iint_D \frac{\partial P}{\partial y}\,dA$$

and

3
$$\int_C Q\,dy = \iint_D \frac{\partial Q}{\partial x}\,dA$$

We prove Equation 2 by expressing D as a type I region:

$$D = \{(x, y) \mid a \leqslant x \leqslant b,\ g_1(x) \leqslant y \leqslant g_2(x)\}$$

where g_1 and g_2 are continuous functions. This enables us to compute the double integral on the right side of Equation 2 as follows:

4
$$\iint_D \frac{\partial P}{\partial y}\,dA = \int_a^b \int_{g_1(x)}^{g_2(x)} \frac{\partial P}{\partial y}(x, y)\,dy\,dx = \int_a^b [P(x, g_2(x)) - P(x, g_1(x))]\,dx$$

where the last step follows from the Fundamental Theorem of Calculus.

Now we compute the left side of Equation 2 by breaking up C as the union of the four curves C_1, C_2, C_3, and C_4 shown in Figure 3. On C_1 we take x as the parameter and write the parametric equations as $x = x$, $y = g_1(x)$, $a \leqslant x \leqslant b$. Thus

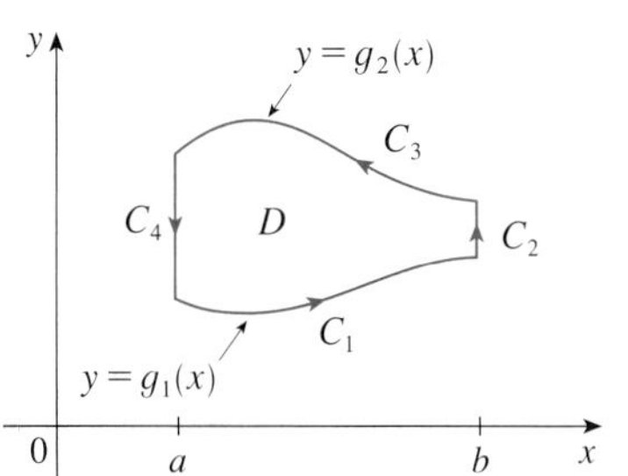

FIGURE 3

$$\int_{C_1} P(x, y)\,dx = \int_a^b P(x, g_1(x))\,dx$$

Observe that C_3 goes from right to left but $-C_3$ goes from left to right, so we can write the parametric equations of $-C_3$ as $x = x$, $y = g_2(x)$, $a \leqslant x \leqslant b$. Therefore

$$\int_{C_3} P(x, y)\,dx = -\int_{-C_3} P(x, y)\,dx = -\int_a^b P(x, g_2(x))\,dx$$

On C_2 or C_4 (either of which might reduce to just a single point), x is constant, so $dx = 0$ and

$$\int_{C_2} P(x, y)\,dx = 0 = \int_{C_4} P(x, y)\,dx$$

Hence

$$\begin{aligned}\int_C P(x, y)\,dx &= \int_{C_1} P(x, y)\,dx + \int_{C_2} P(x, y)\,dx + \int_{C_3} P(x, y)\,dx + \int_{C_4} P(x, y)\,dx \\ &= \int_a^b P(x, g_1(x))\,dx - \int_a^b P(x, g_2(x))\,dx\end{aligned}$$

Comparing this expression with the one in Equation 4, we see that

$$\int_C P(x, y)\, dx = -\iint_D \frac{\partial P}{\partial y}\, dA$$

Equation 3 can be proved in much the same way by expressing D as a type II region (see Exercise 30). Then, by adding Equations 2 and 3, we obtain Green's Theorem. ■

EXAMPLE 1 Evaluate $\int_C x^4\, dx + xy\, dy$, where C is the triangular curve consisting of the line segments from $(0, 0)$ to $(1, 0)$, from $(1, 0)$ to $(0, 1)$, and from $(0, 1)$ to $(0, 0)$.

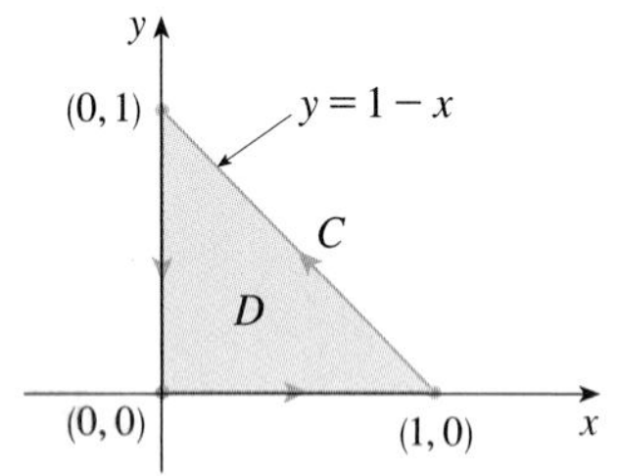

FIGURE 4

SOLUTION Although the given line integral could be evaluated as usual by the methods of Section 16.2, that would involve setting up three separate integrals along the three sides of the triangle, so let's use Green's Theorem instead. Notice that the region D enclosed by C is simple and C has positive orientation (see Figure 4). If we let $P(x, y) = x^4$ and $Q(x, y) = xy$, then we have

$$\int_C x^4\, dx + xy\, dy = \iint_D \left(\frac{\partial Q}{\partial x} - \frac{\partial P}{\partial y} \right) dA = \int_0^1 \int_0^{1-x} (y - 0)\, dy\, dx$$

$$= \int_0^1 \left[\tfrac{1}{2} y^2 \right]_{y=0}^{y=1-x} dx = \tfrac{1}{2} \int_0^1 (1 - x)^2\, dx$$

$$= -\tfrac{1}{6}(1 - x)^3 \Big]_0^1 = \tfrac{1}{6}$$ ■

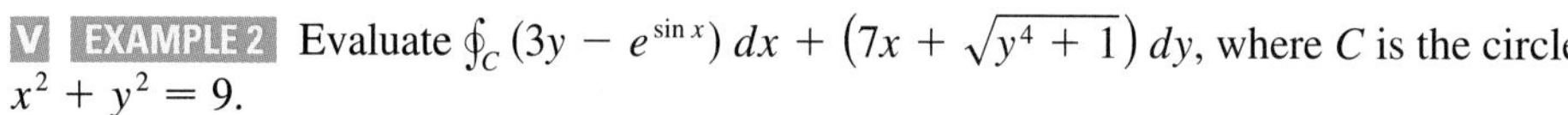

V EXAMPLE 2 Evaluate $\oint_C (3y - e^{\sin x})\, dx + \left(7x + \sqrt{y^4 + 1}\right) dy$, where C is the circle $x^2 + y^2 = 9$.

SOLUTION The region D bounded by C is the disk $x^2 + y^2 \le 9$, so let's change to polar coordinates after applying Green's Theorem:

$$\oint_C (3y - e^{\sin x})\, dx + \left(7x + \sqrt{y^4 + 1}\right) dy$$

$$= \iint_D \left[\frac{\partial}{\partial x}\left(7x + \sqrt{y^4 + 1}\right) - \frac{\partial}{\partial y}(3y - e^{\sin x}) \right] dA$$

$$= \int_0^{2\pi} \int_0^3 (7 - 3)\, r\, dr\, d\theta = 4 \int_0^{2\pi} d\theta \int_0^3 r\, dr = 36\pi$$ ■

Instead of using polar coordinates, we could simply use the fact that D is a disk of radius 3 and write

$$\iint_D 4\, dA = 4 \cdot \pi(3)^2 = 36\pi$$

In Examples 1 and 2 we found that the double integral was easier to evaluate than the line integral. (Try setting up the line integral in Example 2 and you'll soon be convinced!) But sometimes it's easier to evaluate the line integral, and Green's Theorem is used in the reverse direction. For instance, if it is known that $P(x, y) = Q(x, y) = 0$ on the curve C, then Green's Theorem gives

$$\iint_D \left(\frac{\partial Q}{\partial x} - \frac{\partial P}{\partial y} \right) dA = \int_C P\, dx + Q\, dy = 0$$

no matter what values P and Q assume in the region D.

Another application of the reverse direction of Green's Theorem is in computing areas. Since the area of D is $\iint_D 1\, dA$, we wish to choose P and Q so that

$$\frac{\partial Q}{\partial x} - \frac{\partial P}{\partial y} = 1$$

There are several possibilities:

$$P(x, y) = 0 \qquad P(x, y) = -y \qquad P(x, y) = -\tfrac{1}{2}y$$
$$Q(x, y) = x \qquad Q(x, y) = 0 \qquad Q(x, y) = \tfrac{1}{2}x$$

Then Green's Theorem gives the following formulas for the area of D:

5 $$A = \oint_C x\,dy = -\oint_C y\,dx = \tfrac{1}{2}\oint_C x\,dy - y\,dx$$

EXAMPLE 3 Find the area enclosed by the ellipse $\dfrac{x^2}{a^2} + \dfrac{y^2}{b^2} = 1$.

SOLUTION The ellipse has parametric equations $x = a\cos t$ and $y = b\sin t$, where $0 \leqslant t \leqslant 2\pi$. Using the third formula in Equation 5, we have

$$\begin{aligned} A &= \tfrac{1}{2}\int_C x\,dy - y\,dx \\ &= \tfrac{1}{2}\int_0^{2\pi} (a\cos t)(b\cos t)\,dt - (b\sin t)(-a\sin t)\,dt \\ &= \frac{ab}{2}\int_0^{2\pi} dt = \pi ab \end{aligned}$$

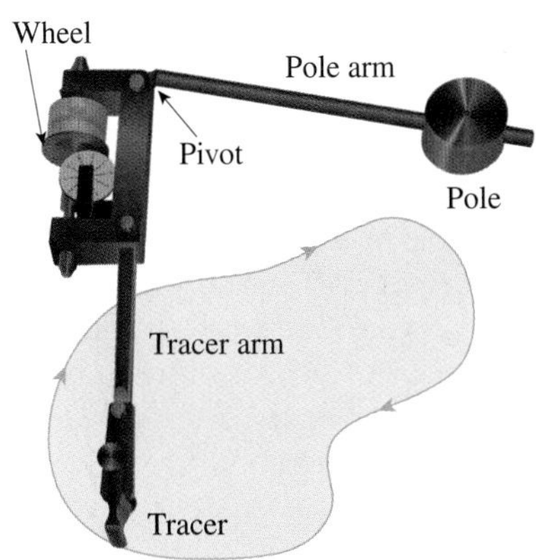

FIGURE 5
A Keuffel and Esser polar planimeter

Formula 5 can be used to explain how planimeters work. A **planimeter** is a mechanical instrument used for measuring the area of a region by tracing its boundary curve. These devices are useful in all the sciences: in biology for measuring the area of leaves or wings, in medicine for measuring the size of cross-sections of organs or tumors, in forestry for estimating the size of forested regions from photographs.

Figure 5 shows the operation of a polar planimeter: The pole is fixed and, as the tracer is moved along the boundary curve of the region, the wheel partly slides and partly rolls perpendicular to the tracer arm. The planimeter measures the distance that the wheel rolls and this is proportional to the area of the enclosed region. The explanation as a consequence of Formula 5 can be found in the following articles:

- R. W. Gatterman, "The planimeter as an example of Green's Theorem" *Amer. Math. Monthly,* Vol. 88 (1981), pp. 701–4.
- Tanya Leise, "As the planimeter wheel turns" *College Math. Journal,* Vol. 38 (2007), pp. 24–31.

Extended Versions of Green's Theorem

Although we have proved Green's Theorem only for the case where D is simple, we can now extend it to the case where D is a finite union of simple regions. For example, if D is the region shown in Figure 6, then we can write $D = D_1 \cup D_2$, where D_1 and D_2 are both simple. The boundary of D_1 is $C_1 \cup C_3$ and the boundary of D_2 is $C_2 \cup (-C_3)$ so, applying Green's Theorem to D_1 and D_2 separately, we get

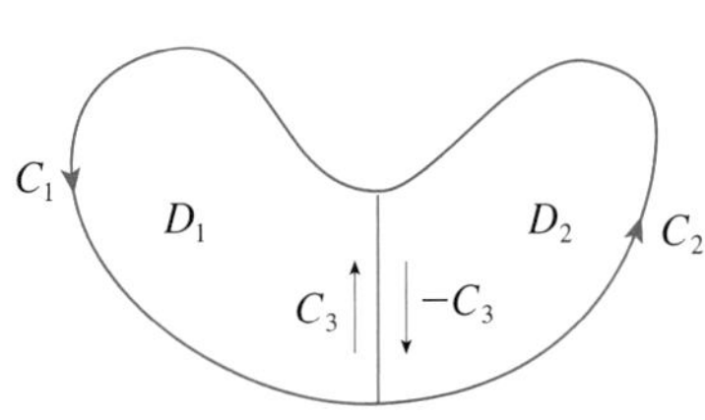

FIGURE 6

$$\int_{C_1 \cup C_3} P\,dx + Q\,dy = \iint_{D_1} \left(\frac{\partial Q}{\partial x} - \frac{\partial P}{\partial y}\right) dA$$

$$\int_{C_2 \cup (-C_3)} P\,dx + Q\,dy = \iint_{D_2} \left(\frac{\partial Q}{\partial x} - \frac{\partial P}{\partial y}\right) dA$$

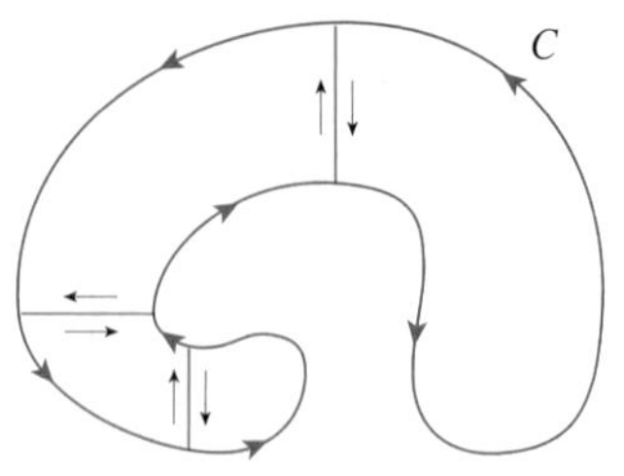

FIGURE 7

If we add these two equations, the line integrals along C_3 and $-C_3$ cancel, so we get

$$\int_{C_1 \cup C_2} P\,dx + Q\,dy = \iint_D \left(\frac{\partial Q}{\partial x} - \frac{\partial P}{\partial y} \right) dA$$

which is Green's Theorem for $D = D_1 \cup D_2$, since its boundary is $C = C_1 \cup C_2$.

The same sort of argument allows us to establish Green's Theorem for any finite union of nonoverlapping simple regions (see Figure 7).

V EXAMPLE 4 Evaluate $\oint_C y^2\,dx + 3xy\,dy$, where C is the boundary of the semiannular region D in the upper half-plane between the circles $x^2 + y^2 = 1$ and $x^2 + y^2 = 4$.

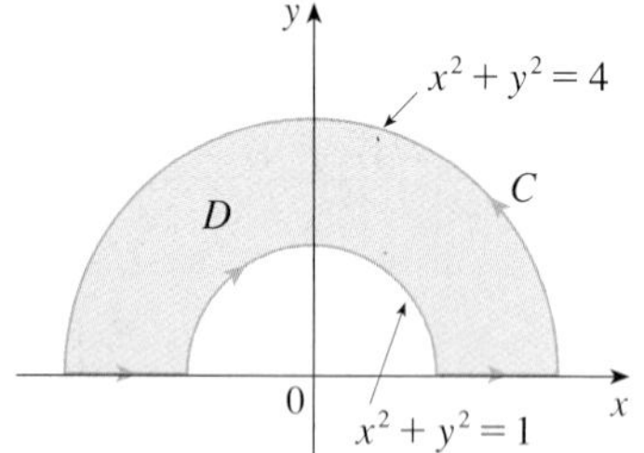

FIGURE 8

SOLUTION Notice that although D is not simple, the y-axis divides it into two simple regions (see Figure 8). In polar coordinates we can write

$$D = \{(r, \theta) \mid 1 \leqslant r \leqslant 2,\ 0 \leqslant \theta \leqslant \pi\}$$

Therefore Green's Theorem gives

$$\begin{aligned}\oint_C y^2\,dx + 3xy\,dy &= \iint_D \left[\frac{\partial}{\partial x}(3xy) - \frac{\partial}{\partial y}(y^2) \right] dA \\ &= \iint_D y\,dA = \int_0^\pi \int_1^2 (r \sin\theta)\, r\,dr\,d\theta \\ &= \int_0^\pi \sin\theta\,d\theta \int_1^2 r^2\,dr = \left[-\cos\theta\right]_0^\pi \left[\tfrac{1}{3}r^3\right]_1^2 = \frac{14}{3}\end{aligned}$$

Green's Theorem can be extended to apply to regions with holes, that is, regions that are not simply-connected. Observe that the boundary C of the region D in Figure 9 consists of two simple closed curves C_1 and C_2. We assume that these boundary curves are oriented so that the region D is always on the left as the curve C is traversed. Thus the positive direction is counterclockwise for the outer curve C_1 but clockwise for the inner curve C_2. If we divide D into two regions D' and D'' by means of the lines shown in Figure 10 and then apply Green's Theorem to each of D' and D'', we get

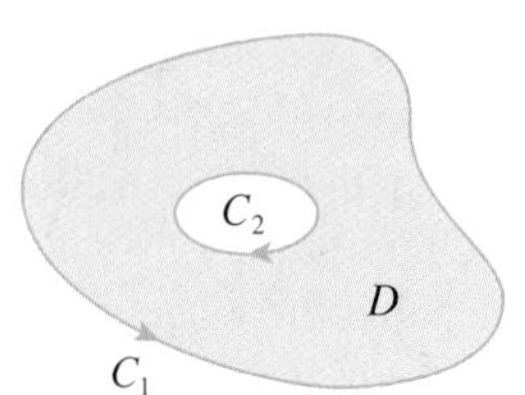

FIGURE 9

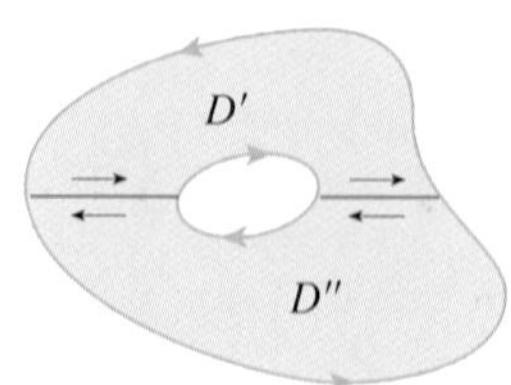

FIGURE 10

$$\begin{aligned}\iint_D \left(\frac{\partial Q}{\partial x} - \frac{\partial P}{\partial y} \right) dA &= \iint_{D'} \left(\frac{\partial Q}{\partial x} - \frac{\partial P}{\partial y} \right) dA + \iint_{D''} \left(\frac{\partial Q}{\partial x} - \frac{\partial P}{\partial y} \right) dA \\ &= \int_{\partial D'} P\,dx + Q\,dy + \int_{\partial D''} P\,dx + Q\,dy\end{aligned}$$

Since the line integrals along the common boundary lines are in opposite directions, they cancel and we get

$$\iint_D \left(\frac{\partial Q}{\partial x} - \frac{\partial P}{\partial y} \right) dA = \int_{C_1} P\,dx + Q\,dy + \int_{C_2} P\,dx + Q\,dy = \int_C P\,dx + Q\,dy$$

which is Green's Theorem for the region D.

V EXAMPLE 5 If $\mathbf{F}(x, y) = (-y\,\mathbf{i} + x\,\mathbf{j})/(x^2 + y^2)$, show that $\int_C \mathbf{F} \cdot d\mathbf{r} = 2\pi$ for every positively oriented simple closed path that encloses the origin.

SOLUTION Since C is an *arbitrary* closed path that encloses the origin, it's difficult to compute the given integral directly. So let's consider a counterclockwise-oriented circle C'

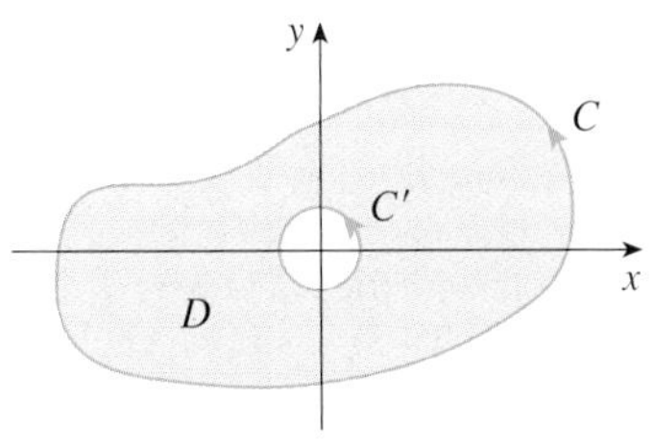

FIGURE 11

with center the origin and radius a, where a is chosen to be small enough that C' lies inside C. (See Figure 11.) Let D be the region bounded by C and C'. Then its positively oriented boundary is $C \cup (-C')$ and so the general version of Green's Theorem gives

$$\int_C P\,dx + Q\,dy + \int_{-C'} P\,dx + Q\,dy = \iint_D \left(\frac{\partial Q}{\partial x} - \frac{\partial P}{\partial y}\right) dA$$

$$= \iint_D \left[\frac{y^2 - x^2}{(x^2 + y^2)^2} - \frac{y^2 - x^2}{(x^2 + y^2)^2}\right] dA = 0$$

Therefore $$\int_C P\,dx + Q\,dy = \int_{C'} P\,dx + Q\,dy$$

that is, $$\int_C \mathbf{F} \cdot d\mathbf{r} = \int_{C'} \mathbf{F} \cdot d\mathbf{r}$$

We now easily compute this last integral using the parametrization given by $\mathbf{r}(t) = a\cos t\,\mathbf{i} + a\sin t\,\mathbf{j}$, $0 \leqslant t \leqslant 2\pi$. Thus

$$\int_C \mathbf{F} \cdot d\mathbf{r} = \int_{C'} \mathbf{F} \cdot d\mathbf{r} = \int_0^{2\pi} \mathbf{F}(\mathbf{r}(t)) \cdot \mathbf{r}'(t)\,dt$$

$$= \int_0^{2\pi} \frac{(-a\sin t)(-a\sin t) + (a\cos t)(a\cos t)}{a^2\cos^2 t + a^2\sin^2 t}\,dt = \int_0^{2\pi} dt = 2\pi$$

We end this section by using Green's Theorem to discuss a result that was stated in the preceding section.

SKETCH OF PROOF OF THEOREM 16.3.6 We're assuming that $\mathbf{F} = P\,\mathbf{i} + Q\,\mathbf{j}$ is a vector field on an open simply-connected region D, that P and Q have continuous first-order partial derivatives, and that

$$\frac{\partial P}{\partial y} = \frac{\partial Q}{\partial x} \qquad \text{throughout } D$$

If C is any simple closed path in D and R is the region that C encloses, then Green's Theorem gives

$$\oint_C \mathbf{F} \cdot d\mathbf{r} = \oint_C P\,dx + Q\,dy = \iint_R \left(\frac{\partial Q}{\partial x} - \frac{\partial P}{\partial y}\right) dA = \iint_R 0\,dA = 0$$

A curve that is not simple crosses itself at one or more points and can be broken up into a number of simple curves. We have shown that the line integrals of $\mathbf{F}$ around these simple curves are all 0 and, adding these integrals, we see that $\int_C \mathbf{F} \cdot d\mathbf{r} = 0$ for any closed curve C. Therefore $\int_C \mathbf{F} \cdot d\mathbf{r}$ is independent of path in D by Theorem 16.3.3. It follows that $\mathbf{F}$ is a conservative vector field.

16.4 Exercises

1–4 Evaluate the line integral by two methods: (a) directly and (b) using Green's Theorem.

1. $\oint_C (x - y)\,dx + (x + y)\,dy$,
C is the circle with center the origin and radius 2

2. $\oint_C xy\,dx + x^2\,dy$,
C is the rectangle with vertices $(0, 0)$, $(3, 0)$, $(3, 1)$, and $(0, 1)$

3. $\oint_C xy\,dx + x^2y^3\,dy$,
C is the triangle with vertices $(0, 0)$, $(1, 0)$, and $(1, 2)$

4. $\oint_C x^2y^2\,dx + xy\,dy$, C consists of the arc of the parabola $y = x^2$ from $(0, 0)$ to $(1, 1)$ and the line segments from $(1, 1)$ to $(0, 1)$ and from $(0, 1)$ to $(0, 0)$

5–10 Use Green's Theorem to evaluate the line integral along the given positively oriented curve.

5. $\int_C xy^2\,dx + 2x^2y\,dy$,
C is the triangle with vertices $(0, 0)$, $(2, 2)$, and $(2, 4)$

6. $\int_C \cos y\,dx + x^2\sin y\,dy$,
C is the rectangle with vertices $(0, 0)$, $(5, 0)$, $(5, 2)$, and $(0, 2)$

7. $\int_C (y + e^{\sqrt{x}})\,dx + (2x + \cos y^2)\,dy$,
C is the boundary of the region enclosed by the parabolas $y = x^2$ and $x = y^2$

8. $\int_C y^4\,dx + 2xy^3\,dy$, C is the ellipse $x^2 + 2y^2 = 2$

9. $\int_C y^3\,dx - x^3\,dy$, C is the circle $x^2 + y^2 = 4$

10. $\int_C (1 - y^3)\,dx + (x^3 + e^{y^2})\,dy$, C is the boundary of the region between the circles $x^2 + y^2 = 4$ and $x^2 + y^2 = 9$

11–14 Use Green's Theorem to evaluate $\int_C \mathbf{F} \cdot d\mathbf{r}$. (Check the orientation of the curve before applying the theorem.)

11. $\mathbf{F}(x, y) = \langle y\cos x - xy\sin x,\, xy + x\cos x\rangle$,
C is the triangle from $(0, 0)$ to $(0, 4)$ to $(2, 0)$ to $(0, 0)$

12. $\mathbf{F}(x, y) = \langle e^{-x} + y^2,\, e^{-y} + x^2\rangle$,
C consists of the arc of the curve $y = \cos x$ from $(-\pi/2, 0)$ to $(\pi/2, 0)$ and the line segment from $(\pi/2, 0)$ to $(-\pi/2, 0)$

13. $\mathbf{F}(x, y) = \langle y - \cos y,\, x\sin y\rangle$,
C is the circle $(x - 3)^2 + (y + 4)^2 = 4$ oriented clockwise

14. $\mathbf{F}(x, y) = \langle \sqrt{x^2 + 1},\, \tan^{-1}x\rangle$, C is the triangle from $(0, 0)$ to $(1, 1)$ to $(0, 1)$ to $(0, 0)$

CAS **15–16** Verify Green's Theorem by using a computer algebra system to evaluate both the line integral and the double integral.

15. $P(x, y) = y^2e^x$, $Q(x, y) = x^2e^y$,
C consists of the line segment from $(-1, 1)$ to $(1, 1)$ followed by the arc of the parabola $y = 2 - x^2$ from $(1, 1)$ to $(-1, 1)$

16. $P(x, y) = 2x - x^3y^5$, $Q(x, y) = x^3y^8$,
C is the ellipse $4x^2 + y^2 = 4$

17. Use Green's Theorem to find the work done by the force $\mathbf{F}(x, y) = x(x + y)\,\mathbf{i} + xy^2\,\mathbf{j}$ in moving a particle from the origin along the x-axis to $(1, 0)$, then along the line segment to $(0, 1)$, and then back to the origin along the y-axis.

18. A particle starts at the point $(-2, 0)$, moves along the x-axis to $(2, 0)$, and then along the semicircle $y = \sqrt{4 - x^2}$ to the starting point. Use Green's Theorem to find the work done on this particle by the force field $\mathbf{F}(x, y) = \langle x,\, x^3 + 3xy^2\rangle$.

19. Use one of the formulas in [5] to find the area under one arch of the cycloid $x = t - \sin t$, $y = 1 - \cos t$.

20. If a circle C with radius 1 rolls along the outside of the circle $x^2 + y^2 = 16$, a fixed point P on C traces out a curve called an *epicycloid*, with parametric equations $x = 5\cos t - \cos 5t$, $y = 5\sin t - \sin 5t$. Graph the epicycloid and use [5] to find the area it encloses.

21. (a) If C is the line segment connecting the point (x_1, y_1) to the point (x_2, y_2), show that

$$\int_C x\,dy - y\,dx = x_1y_2 - x_2y_1$$

(b) If the vertices of a polygon, in counterclockwise order, are $(x_1, y_1), (x_2, y_2), \ldots, (x_n, y_n)$, show that the area of the polygon is

$$A = \tfrac{1}{2}[(x_1y_2 - x_2y_1) + (x_2y_3 - x_3y_2) + \cdots + (x_{n-1}y_n - x_ny_{n-1}) + (x_ny_1 - x_1y_n)]$$

(c) Find the area of the pentagon with vertices $(0, 0)$, $(2, 1)$, $(1, 3)$, $(0, 2)$, and $(-1, 1)$.

22. Let D be a region bounded by a simple closed path C in the xy-plane. Use Green's Theorem to prove that the coordinates of the centroid $(\bar{x}, \bar{y})$ of D are

$$\bar{x} = \frac{1}{2A}\oint_C x^2\,dy \qquad \bar{y} = -\frac{1}{2A}\oint_C y^2\,dx$$

where A is the area of D.

23. Use Exercise 22 to find the centroid of a quarter-circular region of radius a.

24. Use Exercise 22 to find the centroid of the triangle with vertices $(0, 0)$, $(a, 0)$, and (a, b), where $a > 0$ and $b > 0$.

25. A plane lamina with constant density $\rho(x, y) = \rho$ occupies a region in the xy-plane bounded by a simple closed path C. Show that its moments of inertia about the axes are

$$I_x = -\frac{\rho}{3}\oint_C y^3\,dx \qquad I_y = \frac{\rho}{3}\oint_C x^3\,dy$$

26. Use Exercise 25 to find the moment of inertia of a circular disk of radius a with constant density ρ about a diameter. (Compare with Example 4 in Section 15.5.)

27. Use the method of Example 5 to calculate $\int_C \mathbf{F} \cdot d\mathbf{r}$, where

$$\mathbf{F}(x, y) = \frac{2xy\,\mathbf{i} + (y^2 - x^2)\,\mathbf{j}}{(x^2 + y^2)^2}$$

and C is any positively oriented simple closed curve that encloses the origin.

28. Calculate $\int_C \mathbf{F} \cdot d\mathbf{r}$, where $\mathbf{F}(x, y) = \langle x^2 + y,\, 3x - y^2\rangle$ and C is the positively oriented boundary curve of a region D that has area 6.

29. If $\mathbf{F}$ is the vector field of Example 5, show that $\int_C \mathbf{F} \cdot d\mathbf{r} = 0$ for every simple closed path that does not pass through or enclose the origin.

30. Complete the proof of the special case of Green's Theorem by proving Equation 3.

31. Use Green's Theorem to prove the change of variables formula for a double integral (Formula 15.10.9) for the case where $f(x, y) = 1$:

$$\iint_R dx\,dy = \iint_S \left| \frac{\partial(x, y)}{\partial(u, v)} \right| du\,dv$$

Here R is the region in the xy-plane that corresponds to the region S in the uv-plane under the transformation given by $x = g(u, v)$, $y = h(u, v)$.

[*Hint:* Note that the left side is $A(R)$ and apply the first part of Equation 5. Convert the line integral over ∂R to a line integral over ∂S and apply Green's Theorem in the uv-plane.]

16.5 Curl and Divergence

In this section we define two operations that can be performed on vector fields and that play a basic role in the applications of vector calculus to fluid flow and electricity and magnetism. Each operation resembles differentiation, but one produces a vector field whereas the other produces a scalar field.

Curl

If $\mathbf{F} = P\,\mathbf{i} + Q\,\mathbf{j} + R\,\mathbf{k}$ is a vector field on $\mathbb{R}^3$ and the partial derivatives of P, Q, and R all exist, then the **curl** of $\mathbf{F}$ is the vector field on $\mathbb{R}^3$ defined by

1
$$\text{curl}\,\mathbf{F} = \left(\frac{\partial R}{\partial y} - \frac{\partial Q}{\partial z}\right)\mathbf{i} + \left(\frac{\partial P}{\partial z} - \frac{\partial R}{\partial x}\right)\mathbf{j} + \left(\frac{\partial Q}{\partial x} - \frac{\partial P}{\partial y}\right)\mathbf{k}$$

As an aid to our memory, let's rewrite Equation 1 using operator notation. We introduce the vector differential operator ∇ ("del") as

$$\nabla = \mathbf{i}\frac{\partial}{\partial x} + \mathbf{j}\frac{\partial}{\partial y} + \mathbf{k}\frac{\partial}{\partial z}$$

It has meaning when it operates on a scalar function to produce the gradient of f:

$$\nabla f = \mathbf{i}\frac{\partial f}{\partial x} + \mathbf{j}\frac{\partial f}{\partial y} + \mathbf{k}\frac{\partial f}{\partial z} = \frac{\partial f}{\partial x}\mathbf{i} + \frac{\partial f}{\partial y}\mathbf{j} + \frac{\partial f}{\partial z}\mathbf{k}$$

If we think of ∇ as a vector with components $\partial/\partial x$, $\partial/\partial y$, and $\partial/\partial z$, we can also consider the formal cross product of ∇ with the vector field $\mathbf{F}$ as follows:

$$\begin{aligned}\nabla \times \mathbf{F} &= \begin{vmatrix} \mathbf{i} & \mathbf{j} & \mathbf{k} \\ \dfrac{\partial}{\partial x} & \dfrac{\partial}{\partial y} & \dfrac{\partial}{\partial z} \\ P & Q & R \end{vmatrix} \\ &= \left(\frac{\partial R}{\partial y} - \frac{\partial Q}{\partial z}\right)\mathbf{i} + \left(\frac{\partial P}{\partial z} - \frac{\partial R}{\partial x}\right)\mathbf{j} + \left(\frac{\partial Q}{\partial x} - \frac{\partial P}{\partial y}\right)\mathbf{k} \\ &= \text{curl}\,\mathbf{F}\end{aligned}$$

So the easiest way to remember Definition 1 is by means of the symbolic expression

2
$$\text{curl}\,\mathbf{F} = \nabla \times \mathbf{F}$$

EXAMPLE 1 If $\mathbf{F}(x, y, z) = xz\,\mathbf{i} + xyz\,\mathbf{j} - y^2\,\mathbf{k}$, find curl $\mathbf{F}$.

SOLUTION Using Equation 2, we have

$$\text{curl}\,\mathbf{F} = \nabla \times \mathbf{F} = \begin{vmatrix} \mathbf{i} & \mathbf{j} & \mathbf{k} \\ \dfrac{\partial}{\partial x} & \dfrac{\partial}{\partial y} & \dfrac{\partial}{\partial z} \\ xz & xyz & -y^2 \end{vmatrix}$$

$$= \left[\frac{\partial}{\partial y}(-y^2) - \frac{\partial}{\partial z}(xyz)\right]\mathbf{i} - \left[\frac{\partial}{\partial x}(-y^2) - \frac{\partial}{\partial z}(xz)\right]\mathbf{j}$$

$$+ \left[\frac{\partial}{\partial x}(xyz) - \frac{\partial}{\partial y}(xz)\right]\mathbf{k}$$

$$= (-2y - xy)\,\mathbf{i} - (0 - x)\,\mathbf{j} + (yz - 0)\,\mathbf{k}$$

$$= -y(2 + x)\,\mathbf{i} + x\,\mathbf{j} + yz\,\mathbf{k}$$

CAS Most computer algebra systems have commands that compute the curl and divergence of vector fields. If you have access to a CAS, use these commands to check the answers to the examples and exercises in this section.

Recall that the gradient of a function f of three variables is a vector field on $\mathbb{R}^3$ and so we can compute its curl. The following theorem says that the curl of a gradient vector field is $\mathbf{0}$.

3 Theorem If f is a function of three variables that has continuous second-order partial derivatives, then

$$\text{curl}(\nabla f) = \mathbf{0}$$

PROOF We have

$$\text{curl}(\nabla f) = \nabla \times (\nabla f) = \begin{vmatrix} \mathbf{i} & \mathbf{j} & \mathbf{k} \\ \dfrac{\partial}{\partial x} & \dfrac{\partial}{\partial y} & \dfrac{\partial}{\partial z} \\ \dfrac{\partial f}{\partial x} & \dfrac{\partial f}{\partial y} & \dfrac{\partial f}{\partial z} \end{vmatrix}$$

$$= \left(\frac{\partial^2 f}{\partial y\,\partial z} - \frac{\partial^2 f}{\partial z\,\partial y}\right)\mathbf{i} + \left(\frac{\partial^2 f}{\partial z\,\partial x} - \frac{\partial^2 f}{\partial x\,\partial z}\right)\mathbf{j} + \left(\frac{\partial^2 f}{\partial x\,\partial y} - \frac{\partial^2 f}{\partial y\,\partial x}\right)\mathbf{k}$$

$$= 0\,\mathbf{i} + 0\,\mathbf{j} + 0\,\mathbf{k} = \mathbf{0}$$

by Clairaut's Theorem.

Notice the similarity to what we know from Section 12.4: $\mathbf{a} \times \mathbf{a} = \mathbf{0}$ for every three-dimensional vector $\mathbf{a}$.

Since a conservative vector field is one for which $\mathbf{F} = \nabla f$, Theorem 3 can be rephrased as follows:

If $\mathbf{F}$ is conservative, then curl $\mathbf{F} = \mathbf{0}$.

Compare this with Exercise 29 in Section 16.3.

This gives us a way of verifying that a vector field is not conservative.

V EXAMPLE 2 Show that the vector field $\mathbf{F}(x, y, z) = xz\,\mathbf{i} + xyz\,\mathbf{j} - y^2\,\mathbf{k}$ is not conservative.

SOLUTION In Example 1 we showed that

$$\text{curl } \mathbf{F} = -y(2 + x)\,\mathbf{i} + x\,\mathbf{j} + yz\,\mathbf{k}$$

This shows that curl $\mathbf{F} \neq \mathbf{0}$ and so, by Theorem 3, $\mathbf{F}$ is not conservative. ■

The converse of Theorem 3 is not true in general, but the following theorem says the converse is true if $\mathbf{F}$ is defined everywhere. (More generally it is true if the domain is simply-connected, that is, "has no hole.") Theorem 4 is the three-dimensional version of Theorem 16.3.6. Its proof requires Stokes' Theorem and is sketched at the end of Section 16.8.

> **4 Theorem** If $\mathbf{F}$ is a vector field defined on all of $\mathbb{R}^3$ whose component functions have continuous partial derivatives and curl $\mathbf{F} = \mathbf{0}$, then $\mathbf{F}$ is a conservative vector field.

V EXAMPLE 3

(a) Show that

$$\mathbf{F}(x, y, z) = y^2z^3\,\mathbf{i} + 2xyz^3\,\mathbf{j} + 3xy^2z^2\,\mathbf{k}$$

is a conservative vector field.

(b) Find a function f such that $\mathbf{F} = \nabla f$.

SOLUTION

(a) We compute the curl of $\mathbf{F}$:

$$\begin{aligned} \text{curl } \mathbf{F} = \nabla \times \mathbf{F} &= \begin{vmatrix} \mathbf{i} & \mathbf{j} & \mathbf{k} \\ \dfrac{\partial}{\partial x} & \dfrac{\partial}{\partial y} & \dfrac{\partial}{\partial z} \\ y^2z^3 & 2xyz^3 & 3xy^2z^2 \end{vmatrix} \\ &= (6xyz^2 - 6xyz^2)\,\mathbf{i} - (3y^2z^2 - 3y^2z^2)\,\mathbf{j} + (2yz^3 - 2yz^3)\,\mathbf{k} \\ &= \mathbf{0} \end{aligned}$$

Since curl $\mathbf{F} = \mathbf{0}$ and the domain of $\mathbf{F}$ is $\mathbb{R}^3$, $\mathbf{F}$ is a conservative vector field by Theorem 4.

(b) The technique for finding f was given in Section 16.3. We have

5 $$f_x(x, y, z) = y^2z^3$$

6 $$f_y(x, y, z) = 2xyz^3$$

7 $$f_z(x, y, z) = 3xy^2z^2$$

Integrating 5 with respect to x, we obtain

8 $$f(x, y, z) = xy^2z^3 + g(y, z)$$

Differentiating [8] with respect to y, we get $f_y(x, y, z) = 2xyz^3 + g_y(y, z)$, so comparison with [6] gives $g_y(y, z) = 0$. Thus $g(y, z) = h(z)$ and

$$f_z(x, y, z) = 3xy^2z^2 + h'(z)$$

Then [7] gives $h'(z) = 0$. Therefore

$$f(x, y, z) = xy^2z^3 + K$$

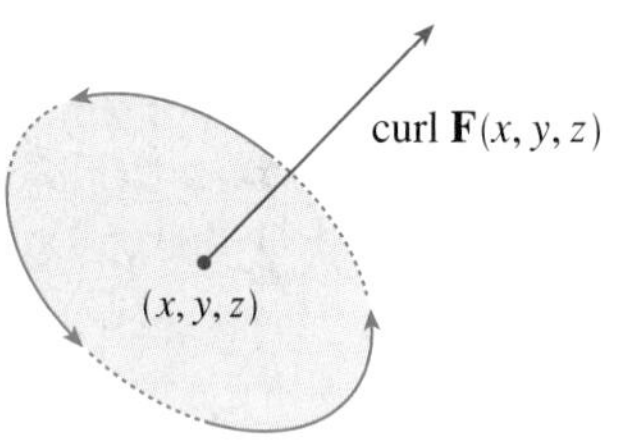

FIGURE 1

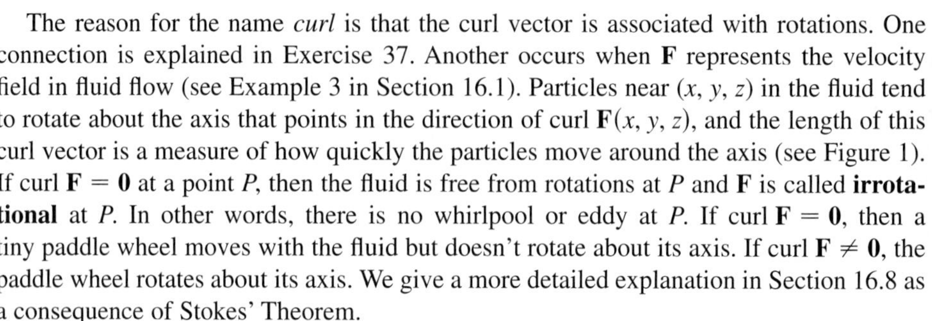

The reason for the name *curl* is that the curl vector is associated with rotations. One connection is explained in Exercise 37. Another occurs when $\mathbf{F}$ represents the velocity field in fluid flow (see Example 3 in Section 16.1). Particles near (x, y, z) in the fluid tend to rotate about the axis that points in the direction of curl $\mathbf{F}(x, y, z)$, and the length of this curl vector is a measure of how quickly the particles move around the axis (see Figure 1). If curl $\mathbf{F} = \mathbf{0}$ at a point P, then the fluid is free from rotations at P and $\mathbf{F}$ is called **irrotational** at P. In other words, there is no whirlpool or eddy at P. If curl $\mathbf{F} = \mathbf{0}$, then a tiny paddle wheel moves with the fluid but doesn't rotate about its axis. If curl $\mathbf{F} \neq \mathbf{0}$, the paddle wheel rotates about its axis. We give a more detailed explanation in Section 16.8 as a consequence of Stokes' Theorem.

Divergence

If $\mathbf{F} = P\,\mathbf{i} + Q\,\mathbf{j} + R\,\mathbf{k}$ is a vector field on $\mathbb{R}^3$ and $\partial P/\partial x$, $\partial Q/\partial y$, and $\partial R/\partial z$ exist, then the **divergence of F** is the function of three variables defined by

9 $$\operatorname{div} \mathbf{F} = \frac{\partial P}{\partial x} + \frac{\partial Q}{\partial y} + \frac{\partial R}{\partial z}$$

Observe that curl $\mathbf{F}$ is a vector field but div $\mathbf{F}$ is a scalar field. In terms of the gradient operator $\nabla = (\partial/\partial x)\,\mathbf{i} + (\partial/\partial y)\,\mathbf{j} + (\partial/\partial z)\,\mathbf{k}$, the divergence of $\mathbf{F}$ can be written symbolically as the dot product of ∇ and $\mathbf{F}$:

10 $$\operatorname{div} \mathbf{F} = \nabla \cdot \mathbf{F}$$

EXAMPLE 4 If $\mathbf{F}(x, y, z) = xz\,\mathbf{i} + xyz\,\mathbf{j} - y^2\,\mathbf{k}$, find div $\mathbf{F}$.

SOLUTION By the definition of divergence (Equation 9 or 10) we have

$$\operatorname{div} \mathbf{F} = \nabla \cdot \mathbf{F} = \frac{\partial}{\partial x}(xz) + \frac{\partial}{\partial y}(xyz) + \frac{\partial}{\partial z}(-y^2) = z + xz$$

If $\mathbf{F}$ is a vector field on $\mathbb{R}^3$, then curl $\mathbf{F}$ is also a vector field on $\mathbb{R}^3$. As such, we can compute its divergence. The next theorem shows that the result is 0.

11 Theorem If $\mathbf{F} = P\,\mathbf{i} + Q\,\mathbf{j} + R\,\mathbf{k}$ is a vector field on $\mathbb{R}^3$ and P, Q, and R have continuous second-order partial derivatives, then

$$\operatorname{div} \operatorname{curl} \mathbf{F} = 0$$

Note the analogy with the scalar triple product: $\mathbf{a} \cdot (\mathbf{a} \times \mathbf{b}) = 0$.

PROOF Using the definitions of divergence and curl, we have

$$\begin{aligned}
\text{div curl } \mathbf{F} &= \nabla \cdot (\nabla \times \mathbf{F}) \\
&= \frac{\partial}{\partial x}\left(\frac{\partial R}{\partial y} - \frac{\partial Q}{\partial z}\right) + \frac{\partial}{\partial y}\left(\frac{\partial P}{\partial z} - \frac{\partial R}{\partial x}\right) + \frac{\partial}{\partial z}\left(\frac{\partial Q}{\partial x} - \frac{\partial P}{\partial y}\right) \\
&= \frac{\partial^2 R}{\partial x\,\partial y} - \frac{\partial^2 Q}{\partial x\,\partial z} + \frac{\partial^2 P}{\partial y\,\partial z} - \frac{\partial^2 R}{\partial y\,\partial x} + \frac{\partial^2 Q}{\partial z\,\partial x} - \frac{\partial^2 P}{\partial z\,\partial y} \\
&= 0
\end{aligned}$$

because the terms cancel in pairs by Clairaut's Theorem. ■

V **EXAMPLE 5** Show that the vector field $\mathbf{F}(x, y, z) = xz\,\mathbf{i} + xyz\,\mathbf{j} - y^2\,\mathbf{k}$ can't be written as the curl of another vector field, that is, $\mathbf{F} \neq \text{curl } \mathbf{G}$.

SOLUTION In Example 4 we showed that

$$\text{div } \mathbf{F} = z + xz$$

and therefore $\text{div } \mathbf{F} \neq 0$. If it were true that $\mathbf{F} = \text{curl } \mathbf{G}$, then Theorem 11 would give

$$\text{div } \mathbf{F} = \text{div curl } \mathbf{G} = 0$$

which contradicts $\text{div } \mathbf{F} \neq 0$. Therefore $\mathbf{F}$ is not the curl of another vector field. ■

The reason for this interpretation of div $\mathbf{F}$ will be explained at the end of Section 16.9 as a consequence of the Divergence Theorem.

Again, the reason for the name *divergence* can be understood in the context of fluid flow. If $\mathbf{F}(x, y, z)$ is the velocity of a fluid (or gas), then $\text{div } \mathbf{F}(x, y, z)$ represents the net rate of change (with respect to time) of the mass of fluid (or gas) flowing from the point (x, y, z) per unit volume. In other words, $\text{div } \mathbf{F}(x, y, z)$ measures the tendency of the fluid to diverge from the point (x, y, z). If $\text{div } \mathbf{F} = 0$, then $\mathbf{F}$ is said to be **incompressible**.

Another differential operator occurs when we compute the divergence of a gradient vector field ∇f. If f is a function of three variables, we have

$$\text{div}(\nabla f) = \nabla \cdot (\nabla f) = \frac{\partial^2 f}{\partial x^2} + \frac{\partial^2 f}{\partial y^2} + \frac{\partial^2 f}{\partial z^2}$$

and this expression occurs so often that we abbreviate it as $\nabla^2 f$. The operator

$$\nabla^2 = \nabla \cdot \nabla$$

is called the **Laplace operator** because of its relation to **Laplace's equation**

$$\nabla^2 f = \frac{\partial^2 f}{\partial x^2} + \frac{\partial^2 f}{\partial y^2} + \frac{\partial^2 f}{\partial z^2} = 0$$

We can also apply the Laplace operator ∇^2 to a vector field

$$\mathbf{F} = P\,\mathbf{i} + Q\,\mathbf{j} + R\,\mathbf{k}$$

in terms of its components:

$$\nabla^2 \mathbf{F} = \nabla^2 P\,\mathbf{i} + \nabla^2 Q\,\mathbf{j} + \nabla^2 R\,\mathbf{k}$$

Vector Forms of Green's Theorem

The curl and divergence operators allow us to rewrite Green's Theorem in versions that will be useful in our later work. We suppose that the plane region D, its boundary curve C, and the functions P and Q satisfy the hypotheses of Green's Theorem. Then we consider the vector field $\mathbf{F} = P\,\mathbf{i} + Q\,\mathbf{j}$. Its line integral is

$$\oint_C \mathbf{F} \cdot d\mathbf{r} = \oint_C P\,dx + Q\,dy$$

and, regarding $\mathbf{F}$ as a vector field on $\mathbb{R}^3$ with third component 0, we have

$$\text{curl}\,\mathbf{F} = \begin{vmatrix} \mathbf{i} & \mathbf{j} & \mathbf{k} \\ \dfrac{\partial}{\partial x} & \dfrac{\partial}{\partial y} & \dfrac{\partial}{\partial z} \\ P(x, y) & Q(x, y) & 0 \end{vmatrix} = \left(\frac{\partial Q}{\partial x} - \frac{\partial P}{\partial y} \right) \mathbf{k}$$

Therefore

$$(\text{curl}\,\mathbf{F}) \cdot \mathbf{k} = \left(\frac{\partial Q}{\partial x} - \frac{\partial P}{\partial y} \right) \mathbf{k} \cdot \mathbf{k} = \frac{\partial Q}{\partial x} - \frac{\partial P}{\partial y}$$

and we can now rewrite the equation in Green's Theorem in the vector form

12 $$\oint_C \mathbf{F} \cdot d\mathbf{r} = \iint_D (\text{curl}\,\mathbf{F}) \cdot \mathbf{k}\,dA$$

Equation 12 expresses the line integral of the tangential component of $\mathbf{F}$ along C as the double integral of the vertical component of curl $\mathbf{F}$ over the region D enclosed by C. We now derive a similar formula involving the *normal* component of $\mathbf{F}$.

If C is given by the vector equation

$$\mathbf{r}(t) = x(t)\,\mathbf{i} + y(t)\,\mathbf{j} \qquad a \leqslant t \leqslant b$$

then the unit tangent vector (see Section 13.2) is

$$\mathbf{T}(t) = \frac{x'(t)}{|\mathbf{r}'(t)|}\,\mathbf{i} + \frac{y'(t)}{|\mathbf{r}'(t)|}\,\mathbf{j}$$

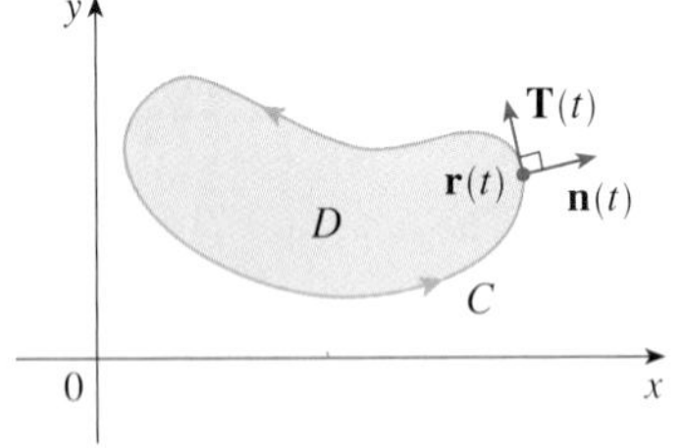

FIGURE 2

You can verify that the outward unit normal vector to C is given by

$$\mathbf{n}(t) = \frac{y'(t)}{|\mathbf{r}'(t)|}\,\mathbf{i} - \frac{x'(t)}{|\mathbf{r}'(t)|}\,\mathbf{j}$$

(See Figure 2.) Then, from Equation 16.2.3, we have

$$\begin{aligned} \oint_C \mathbf{F} \cdot \mathbf{n}\,ds &= \int_a^b (\mathbf{F} \cdot \mathbf{n})(t)\,|\mathbf{r}'(t)|\,dt \\ &= \int_a^b \left[\frac{P(x(t), y(t))\,y'(t)}{|\mathbf{r}'(t)|} - \frac{Q(x(t), y(t))\,x'(t)}{|\mathbf{r}'(t)|} \right] |\mathbf{r}'(t)|\,dt \\ &= \int_a^b P(x(t), y(t))\,y'(t)\,dt - Q(x(t), y(t))\,x'(t)\,dt \\ &= \int_C P\,dy - Q\,dx = \iint_D \left(\frac{\partial P}{\partial x} + \frac{\partial Q}{\partial y} \right) dA \end{aligned}$$

by Green's Theorem. But the integrand in this double integral is just the divergence of **F**. So we have a second vector form of Green's Theorem.

13
$$\oint_C \mathbf{F} \cdot \mathbf{n}\, ds = \iint_D \operatorname{div} \mathbf{F}(x, y)\, dA$$

This version says that the line integral of the normal component of **F** along C is equal to the double integral of the divergence of **F** over the region D enclosed by C.

16.5 Exercises

1–8 Find (a) the curl and (b) the divergence of the vector field.

1. $\mathbf{F}(x, y, z) = (x + yz)\,\mathbf{i} + (y + xz)\,\mathbf{j} + (z + xy)\,\mathbf{k}$

2. $\mathbf{F}(x, y, z) = xy^2z^3\,\mathbf{i} + x^3yz^2\,\mathbf{j} + x^2y^3z\,\mathbf{k}$

3. $\mathbf{F}(x, y, z) = xye^z\,\mathbf{i} + yze^x\,\mathbf{k}$

4. $\mathbf{F}(x, y, z) = \sin yz\,\mathbf{i} + \sin zx\,\mathbf{j} + \sin xy\,\mathbf{k}$

5. $\mathbf{F}(x, y, z) = \dfrac{1}{\sqrt{x^2 + y^2 + z^2}}(x\,\mathbf{i} + y\,\mathbf{j} + z\,\mathbf{k})$

6. $\mathbf{F}(x, y, z) = e^{xy}\sin z\,\mathbf{j} + y\tan^{-1}(x/z)\,\mathbf{k}$

7. $\mathbf{F}(x, y, z) = \langle e^x \sin y, e^y \sin z, e^z \sin x\rangle$

8. $\mathbf{F}(x, y, z) = \left\langle \dfrac{x}{y}, \dfrac{y}{z}, \dfrac{z}{x} \right\rangle$

9–11 The vector field **F** is shown in the xy-plane and looks the same in all other horizontal planes. (In other words, **F** is independent of z and its z-component is 0.)
(a) Is div **F** positive, negative, or zero? Explain.
(b) Determine whether curl **F** = **0**. If not, in which direction does curl **F** point?

9.
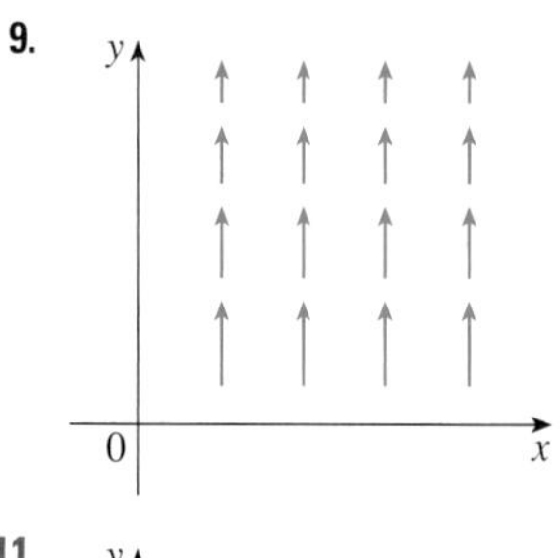

10.
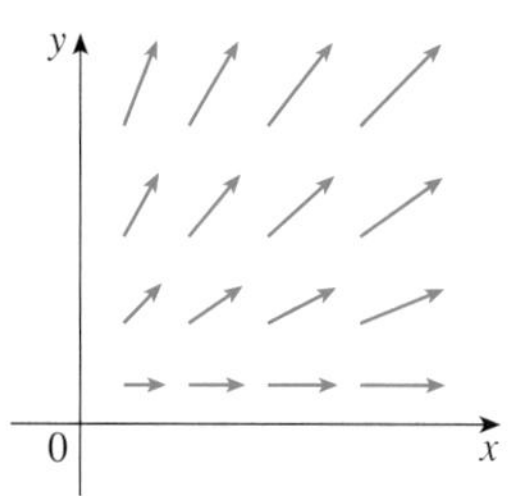

11.
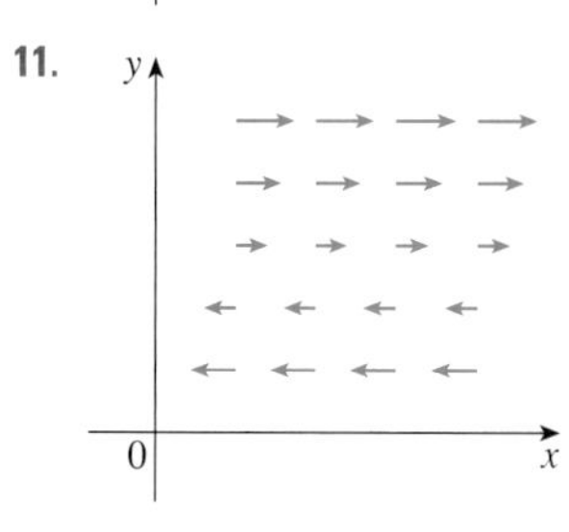

12. Let f be a scalar field and **F** a vector field. State whether each expression is meaningful. If not, explain why. If so, state whether it is a scalar field or a vector field.
(a) curl f (b) grad f
(c) div **F** (d) curl(grad f)
(e) grad **F** (f) grad(div **F**)
(g) div(grad f) (h) grad(div f)
(i) curl(curl **F**) (j) div(div **F**)
(k) (grad f) × (div **F**) (l) div(curl(grad f))

13–18 Determine whether or not the vector field is conservative. If it is conservative, find a function f such that $\mathbf{F} = \nabla f$.

13. $\mathbf{F}(x, y, z) = y^2z^3\,\mathbf{i} + 2xyz^3\,\mathbf{j} + 3xy^2z^2\,\mathbf{k}$

14. $\mathbf{F}(x, y, z) = xyz^2\,\mathbf{i} + x^2yz^2\,\mathbf{j} + x^2y^2z\,\mathbf{k}$

15. $\mathbf{F}(x, y, z) = 3xy^2z^2\,\mathbf{i} + 2x^2yz^3\,\mathbf{j} + 3x^2y^2z^2\,\mathbf{k}$

16. $\mathbf{F}(x, y, z) = \mathbf{i} + \sin z\,\mathbf{j} + y\cos z\,\mathbf{k}$

17. $\mathbf{F}(x, y, z) = e^{yz}\,\mathbf{i} + xze^{yz}\,\mathbf{j} + xye^{yz}\,\mathbf{k}$

18. $\mathbf{F}(x, y, z) = e^x\sin yz\,\mathbf{i} + ze^x\cos yz\,\mathbf{j} + ye^x\cos yz\,\mathbf{k}$

19. Is there a vector field **G** on $\mathbb{R}^3$ such that curl **G** = $\langle x\sin y, \cos y, z - xy\rangle$? Explain.

20. Is there a vector field **G** on $\mathbb{R}^3$ such that curl **G** = $\langle xyz, -y^2z, yz^2\rangle$? Explain.

21. Show that any vector field of the form

$$\mathbf{F}(x, y, z) = f(x)\,\mathbf{i} + g(y)\,\mathbf{j} + h(z)\,\mathbf{k}$$

where f, g, h are differentiable functions, is irrotational.

22. Show that any vector field of the form

$$\mathbf{F}(x, y, z) = f(y, z)\,\mathbf{i} + g(x, z)\,\mathbf{j} + h(x, y)\,\mathbf{k}$$

is incompressible.

1. Homework Hints available at stewartcalculus.com

23–29 Prove the identity, assuming that the appropriate partial derivatives exist and are continuous. If f is a scalar field and $\mathbf{F}$, $\mathbf{G}$ are vector fields, then $f\mathbf{F}$, $\mathbf{F} \cdot \mathbf{G}$, and $\mathbf{F} \times \mathbf{G}$ are defined by

$$(f\mathbf{F})(x, y, z) = f(x, y, z)\,\mathbf{F}(x, y, z)$$

$$(\mathbf{F} \cdot \mathbf{G})(x, y, z) = \mathbf{F}(x, y, z) \cdot \mathbf{G}(x, y, z)$$

$$(\mathbf{F} \times \mathbf{G})(x, y, z) = \mathbf{F}(x, y, z) \times \mathbf{G}(x, y, z)$$

23. $\operatorname{div}(\mathbf{F} + \mathbf{G}) = \operatorname{div} \mathbf{F} + \operatorname{div} \mathbf{G}$

24. $\operatorname{curl}(\mathbf{F} + \mathbf{G}) = \operatorname{curl} \mathbf{F} + \operatorname{curl} \mathbf{G}$

25. $\operatorname{div}(f\mathbf{F}) = f \operatorname{div} \mathbf{F} + \mathbf{F} \cdot \nabla f$

26. $\operatorname{curl}(f\mathbf{F}) = f \operatorname{curl} \mathbf{F} + (\nabla f) \times \mathbf{F}$

27. $\operatorname{div}(\mathbf{F} \times \mathbf{G}) = \mathbf{G} \cdot \operatorname{curl} \mathbf{F} - \mathbf{F} \cdot \operatorname{curl} \mathbf{G}$

28. $\operatorname{div}(\nabla f \times \nabla g) = 0$

29. $\operatorname{curl}(\operatorname{curl} \mathbf{F}) = \operatorname{grad}(\operatorname{div} \mathbf{F}) - \nabla^2 \mathbf{F}$

30–32 Let $\mathbf{r} = x\,\mathbf{i} + y\,\mathbf{j} + z\,\mathbf{k}$ and $r = |\mathbf{r}|$.

30. Verify each identity.
(a) $\nabla \cdot \mathbf{r} = 3$
(b) $\nabla \cdot (r\mathbf{r}) = 4r$
(c) $\nabla^2 r^3 = 12r$

31. Verify each identity.
(a) $\nabla r = \mathbf{r}/r$
(b) $\nabla \times \mathbf{r} = \mathbf{0}$
(c) $\nabla(1/r) = -\mathbf{r}/r^3$
(d) $\nabla \ln r = \mathbf{r}/r^2$

32. If $\mathbf{F} = \mathbf{r}/r^p$, find div $\mathbf{F}$. Is there a value of p for which div $\mathbf{F} = 0$?

33. Use Green's Theorem in the form of Equation 13 to prove **Green's first identity**:

$$\iint_D f\nabla^2 g\, dA = \oint_C f(\nabla g) \cdot \mathbf{n}\, ds - \iint_D \nabla f \cdot \nabla g\, dA$$

where D and C satisfy the hypotheses of Green's Theorem and the appropriate partial derivatives of f and g exist and are continuous. (The quantity $\nabla g \cdot \mathbf{n} = D_{\mathbf{n}}\, g$ occurs in the line integral. This is the directional derivative in the direction of the normal vector $\mathbf{n}$ and is called the **normal derivative** of g.)

34. Use Green's first identity (Exercise 33) to prove **Green's second identity**:

$$\iint_D (f\nabla^2 g - g\nabla^2 f)\, dA = \oint_C (f\nabla g - g\nabla f) \cdot \mathbf{n}\, ds$$

where D and C satisfy the hypotheses of Green's Theorem and the appropriate partial derivatives of f and g exist and are continuous.

35. Recall from Section 14.3 that a function g is called *harmonic* on D if it satisfies Laplace's equation, that is, $\nabla^2 g = 0$ on D. Use Green's first identity (with the same hypotheses as in Exercise 33) to show that if g is harmonic on D, then $\oint_C D_{\mathbf{n}}\, g\, ds = 0$. Here $D_{\mathbf{n}}\, g$ is the normal derivative of g defined in Exercise 33.

36. Use Green's first identity to show that if f is harmonic on D, and if $f(x, y) = 0$ on the boundary curve C, then $\iint_D |\nabla f|^2\, dA = 0$. (Assume the same hypotheses as in Exercise 33.)

37. This exercise demonstrates a connection between the curl vector and rotations. Let B be a rigid body rotating about the z-axis. The rotation can be described by the vector $\mathbf{w} = \omega\mathbf{k}$, where ω is the angular speed of B, that is, the tangential speed of any point P in B divided by the distance d from the axis of rotation. Let $\mathbf{r} = \langle x, y, z \rangle$ be the position vector of P.
(a) By considering the angle θ in the figure, show that the velocity field of B is given by $\mathbf{v} = \mathbf{w} \times \mathbf{r}$.
(b) Show that $\mathbf{v} = -\omega y\,\mathbf{i} + \omega x\,\mathbf{j}$.
(c) Show that curl $\mathbf{v} = 2\mathbf{w}$.

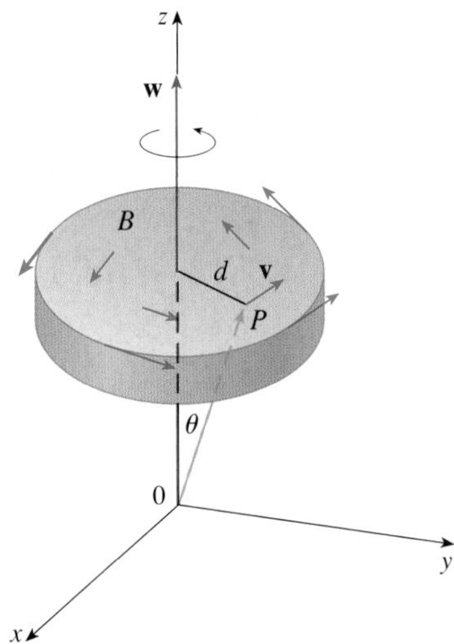

38. Maxwell's equations relating the electric field $\mathbf{E}$ and magnetic field $\mathbf{H}$ as they vary with time in a region containing no charge and no current can be stated as follows:

$$\operatorname{div} \mathbf{E} = 0 \qquad \operatorname{div} \mathbf{H} = 0$$

$$\operatorname{curl} \mathbf{E} = -\frac{1}{c}\frac{\partial \mathbf{H}}{\partial t} \qquad \operatorname{curl} \mathbf{H} = \frac{1}{c}\frac{\partial \mathbf{E}}{\partial t}$$

where c is the speed of light. Use these equations to prove the following:

(a) $\nabla \times (\nabla \times \mathbf{E}) = -\dfrac{1}{c^2}\dfrac{\partial^2 \mathbf{E}}{\partial t^2}$

(b) $\nabla \times (\nabla \times \mathbf{H}) = -\dfrac{1}{c^2}\dfrac{\partial^2 \mathbf{H}}{\partial t^2}$

(c) $\nabla^2 \mathbf{E} = \dfrac{1}{c^2}\dfrac{\partial^2 \mathbf{E}}{\partial t^2}$ [*Hint:* Use Exercise 29.]

(d) $\nabla^2 \mathbf{H} = \dfrac{1}{c^2}\dfrac{\partial^2 \mathbf{H}}{\partial t^2}$

39. We have seen that all vector fields of the form $\mathbf{F} = \nabla g$ satisfy the equation curl $\mathbf{F} = \mathbf{0}$ and that all vector fields of the form $\mathbf{F} = \text{curl } \mathbf{G}$ satisfy the equation div $\mathbf{F} = 0$ (assuming continuity of the appropriate partial derivatives). This suggests the question: Are there any equations that all functions of the form $f = \text{div } \mathbf{G}$ must satisfy? Show that the answer to this question is "No" by proving that *every* continuous function f on $\mathbb{R}^3$ is the divergence of some vector field.
[*Hint:* Let $\mathbf{G}(x, y, z) = \langle g(x, y, z), 0, 0 \rangle$, where $g(x, y, z) = \int_0^x f(t, y, z)\, dt$.]

16.6 Parametric Surfaces and Their Areas

So far we have considered special types of surfaces: cylinders, quadric surfaces, graphs of functions of two variables, and level surfaces of functions of three variables. Here we use vector functions to describe more general surfaces, called *parametric surfaces*, and compute their areas. Then we take the general surface area formula and see how it applies to special surfaces.

Parametric Surfaces

In much the same way that we describe a space curve by a vector function $\mathbf{r}(t)$ of a single parameter t, we can describe a surface by a vector function $\mathbf{r}(u, v)$ of two parameters u and v. We suppose that

$$\boxed{1} \qquad \mathbf{r}(u, v) = x(u, v)\,\mathbf{i} + y(u, v)\,\mathbf{j} + z(u, v)\,\mathbf{k}$$

is a vector-valued function defined on a region D in the uv-plane. So x, y, and z, the component functions of $\mathbf{r}$, are functions of the two variables u and v with domain D. The set of all points (x, y, z) in $\mathbb{R}^3$ such that

$$\boxed{2} \qquad x = x(u, v) \qquad y = y(u, v) \qquad z = z(u, v)$$

and (u, v) varies throughout D, is called a **parametric surface** S and Equations 2 are called **parametric equations** of S. Each choice of u and v gives a point on S; by making all choices, we get all of S. In other words, the surface S is traced out by the tip of the position vector $\mathbf{r}(u, v)$ as (u, v) moves throughout the region D. (See Figure 1.)

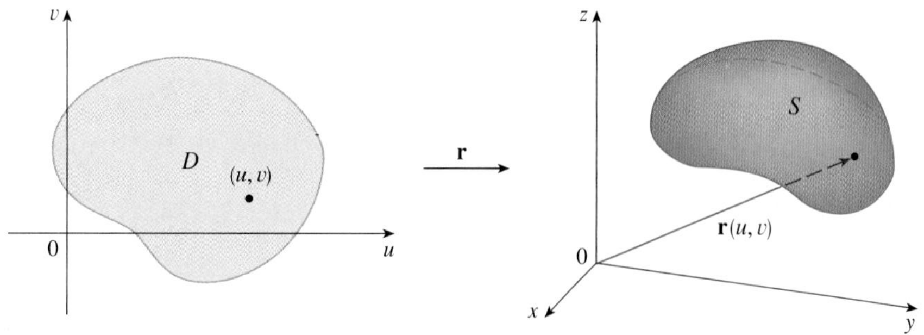

FIGURE 1
A parametric surface

EXAMPLE 1 Identify and sketch the surface with vector equation

$$\mathbf{r}(u, v) = 2 \cos u\,\mathbf{i} + v\,\mathbf{j} + 2 \sin u\,\mathbf{k}$$

SOLUTION The parametric equations for this surface are

$$x = 2 \cos u \qquad y = v \qquad z = 2 \sin u$$

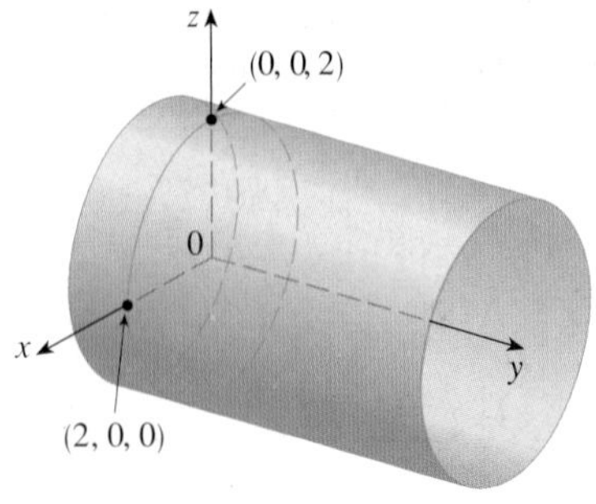

FIGURE 2

So for any point (x, y, z) on the surface, we have

$$x^2 + z^2 = 4\cos^2 u + 4\sin^2 u = 4$$

This means that vertical cross-sections parallel to the xz-plane (that is, with y constant) are all circles with radius 2. Since $y = v$ and no restriction is placed on v, the surface is a circular cylinder with radius 2 whose axis is the y-axis (see Figure 2).

In Example 1 we placed no restrictions on the parameters u and v and so we obtained the entire cylinder. If, for instance, we restrict u and v by writing the parameter domain as

$$0 \leqslant u \leqslant \pi/2 \qquad 0 \leqslant v \leqslant 3$$

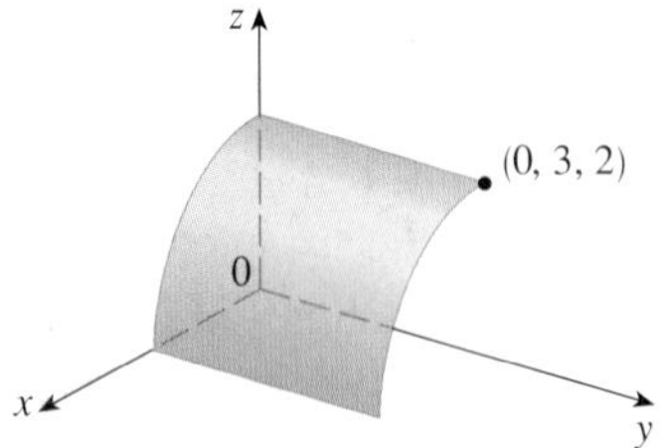

FIGURE 3

then $x \geqslant 0$, $z \geqslant 0$, $0 \leqslant y \leqslant 3$, and we get the quarter-cylinder with length 3 illustrated in Figure 3.

If a parametric surface S is given by a vector function $\mathbf{r}(u, v)$, then there are two useful families of curves that lie on S, one family with u constant and the other with v constant. These families correspond to vertical and horizontal lines in the uv-plane. If we keep u constant by putting $u = u_0$, then $\mathbf{r}(u_0, v)$ becomes a vector function of the single parameter v and defines a curve C_1 lying on S. (See Figure 4.)

TEC Visual 16.6 shows animated versions of Figures 4 and 5, with moving grid curves, for several parametric surfaces.

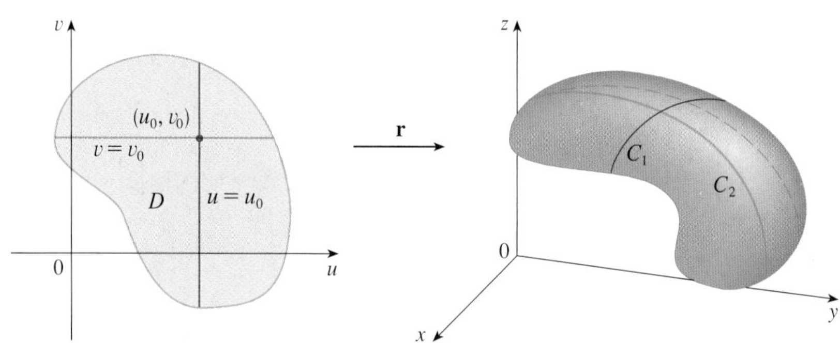

FIGURE 4

Similarly, if we keep v constant by putting $v = v_0$, we get a curve C_2 given by $\mathbf{r}(u, v_0)$ that lies on S. We call these curves **grid curves**. (In Example 1, for instance, the grid curves obtained by letting u be constant are horizontal lines whereas the grid curves with v constant are circles.) In fact, when a computer graphs a parametric surface, it usually depicts the surface by plotting these grid curves, as we see in the following example.

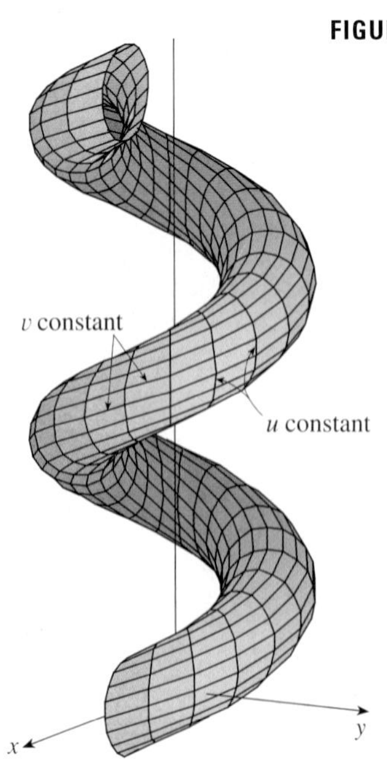

FIGURE 5

EXAMPLE 2 Use a computer algebra system to graph the surface

$$\mathbf{r}(u, v) = \langle (2 + \sin v)\cos u,\ (2 + \sin v)\sin u,\ u + \cos v \rangle$$

Which grid curves have u constant? Which have v constant?

SOLUTION We graph the portion of the surface with parameter domain $0 \leqslant u \leqslant 4\pi$, $0 \leqslant v \leqslant 2\pi$ in Figure 5. It has the appearance of a spiral tube. To identify the grid curves, we write the corresponding parametric equations:

$$x = (2 + \sin v)\cos u \qquad y = (2 + \sin v)\sin u \qquad z = u + \cos v$$

If v is constant, then $\sin v$ and $\cos v$ are constant, so the parametric equations resemble those of the helix in Example 4 in Section 13.1. Thus the grid curves with v constant are the spiral curves in Figure 5. We deduce that the grid curves with u constant must be curves that look like circles in the figure. Further evidence for this assertion is that if u is kept constant, $u = u_0$, then the equation $z = u_0 + \cos v$ shows that the z-values vary from $u_0 - 1$ to $u_0 + 1$.

In Examples 1 and 2 we were given a vector equation and asked to graph the corresponding parametric surface. In the following examples, however, we are given the more challenging problem of finding a vector function to represent a given surface. In the rest of this chapter we will often need to do exactly that.

EXAMPLE 3 Find a vector function that represents the plane that passes through the point P_0 with position vector $\mathbf{r}_0$ and that contains two nonparallel vectors $\mathbf{a}$ and $\mathbf{b}$.

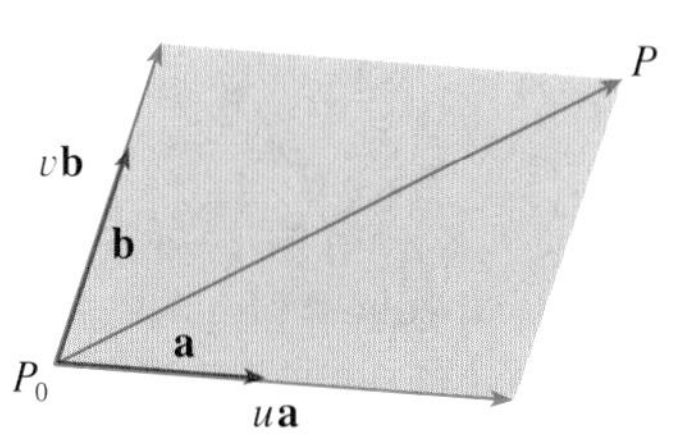

FIGURE 6

SOLUTION If P is any point in the plane, we can get from P_0 to P by moving a certain distance in the direction of $\mathbf{a}$ and another distance in the direction of $\mathbf{b}$. So there are scalars u and v such that $\overrightarrow{P_0P} = u\mathbf{a} + v\mathbf{b}$. (Figure 6 illustrates how this works, by means of the Parallelogram Law, for the case where u and v are positive. See also Exercise 46 in Section 12.2.) If $\mathbf{r}$ is the position vector of P, then

$$\mathbf{r} = \overrightarrow{OP_0} + \overrightarrow{P_0P} = \mathbf{r}_0 + u\mathbf{a} + v\mathbf{b}$$

So the vector equation of the plane can be written as

$$\mathbf{r}(u, v) = \mathbf{r}_0 + u\mathbf{a} + v\mathbf{b}$$

where u and v are real numbers.

If we write $\mathbf{r} = \langle x, y, z \rangle$, $\mathbf{r}_0 = \langle x_0, y_0, z_0 \rangle$, $\mathbf{a} = \langle a_1, a_2, a_3 \rangle$, and $\mathbf{b} = \langle b_1, b_2, b_3 \rangle$, then we can write the parametric equations of the plane through the point (x_0, y_0, z_0) as follows:

$$x = x_0 + ua_1 + vb_1 \qquad y = y_0 + ua_2 + vb_2 \qquad z = z_0 + ua_3 + vb_3$$

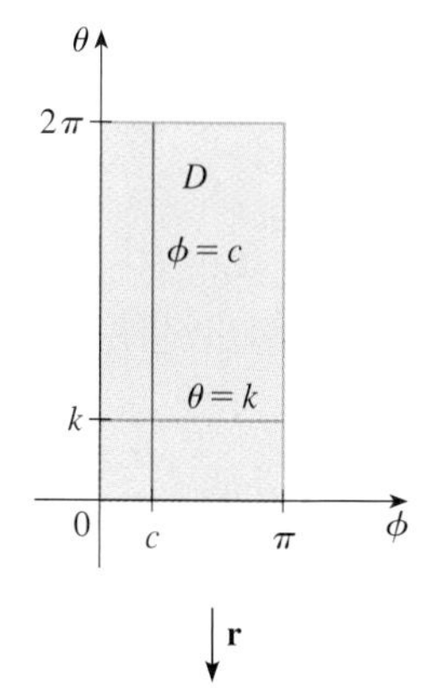

V EXAMPLE 4 Find a parametric representation of the sphere

$$x^2 + y^2 + z^2 = a^2$$

SOLUTION The sphere has a simple representation $\rho = a$ in spherical coordinates, so let's choose the angles ϕ and θ in spherical coordinates as the parameters (see Section 15.9). Then, putting $\rho = a$ in the equations for conversion from spherical to rectangular coordinates (Equations 15.9.1), we obtain

$$x = a \sin \phi \cos \theta \qquad y = a \sin \phi \sin \theta \qquad z = a \cos \phi$$

as the parametric equations of the sphere. The corresponding vector equation is

$$\mathbf{r}(\phi, \theta) = a \sin \phi \cos \theta\, \mathbf{i} + a \sin \phi \sin \theta\, \mathbf{j} + a \cos \phi\, \mathbf{k}$$

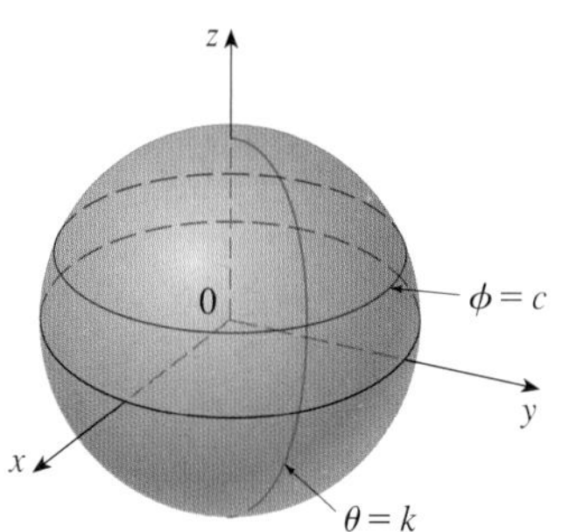

FIGURE 7

We have $0 \leq \phi \leq \pi$ and $0 \leq \theta \leq 2\pi$, so the parameter domain is the rectangle $D = [0, \pi] \times [0, 2\pi]$. The grid curves with ϕ constant are the circles of constant latitude (including the equator). The grid curves with θ constant are the meridians (semicircles), which connect the north and south poles (see Figure 7).

NOTE We saw in Example 4 that the grid curves for a sphere are curves of constant latitude and longitude. For a general parametric surface we are really making a map and the grid curves are similar to lines of latitude and longitude. Describing a point on a parametric surface (like the one in Figure 5) by giving specific values of u and v is like giving the latitude and longitude of a point.

One of the uses of parametric surfaces is in computer graphics. Figure 8 shows the result of trying to graph the sphere $x^2 + y^2 + z^2 = 1$ by solving the equation for z and graphing the top and bottom hemispheres separately. Part of the sphere appears to be missing because of the rectangular grid system used by the computer. The much better picture in Figure 9 was produced by a computer using the parametric equations found in Example 4.

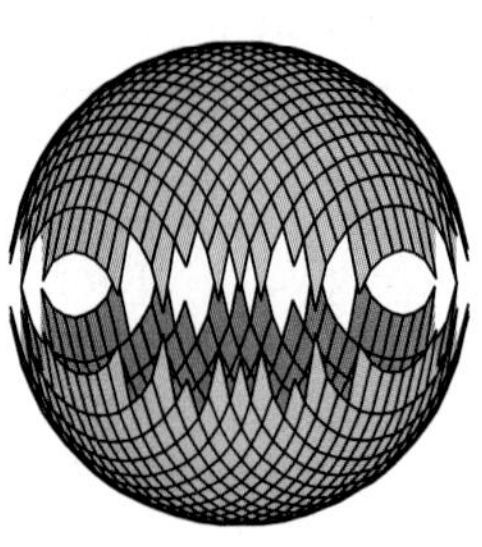

FIGURE 8

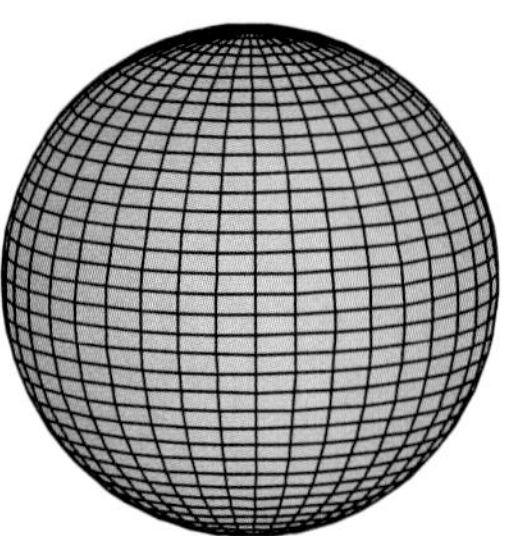

FIGURE 9

EXAMPLE 5 Find a parametric representation for the cylinder

$$x^2 + y^2 = 4 \qquad 0 \leqslant z \leqslant 1$$

SOLUTION The cylinder has a simple representation $r = 2$ in cylindrical coordinates, so we choose as parameters θ and z in cylindrical coordinates. Then the parametric equations of the cylinder are

$$x = 2\cos\theta \qquad y = 2\sin\theta \qquad z = z$$

where $0 \leqslant \theta \leqslant 2\pi$ and $0 \leqslant z \leqslant 1$.

V EXAMPLE 6 Find a vector function that represents the elliptic paraboloid $z = x^2 + 2y^2$.

SOLUTION If we regard x and y as parameters, then the parametric equations are simply

$$x = x \qquad y = y \qquad z = x^2 + 2y^2$$

and the vector equation is

$$\mathbf{r}(x, y) = x\,\mathbf{i} + y\,\mathbf{j} + (x^2 + 2y^2)\,\mathbf{k}$$

TEC In Module 16.6 you can investigate several families of parametric surfaces.

In general, a surface given as the graph of a function of x and y, that is, with an equation of the form $z = f(x, y)$, can always be regarded as a parametric surface by taking x and y as parameters and writing the parametric equations as

$$x = x \qquad y = y \qquad z = f(x, y)$$

Parametric representations (also called parametrizations) of surfaces are not unique. The next example shows two ways to parametrize a cone.

EXAMPLE 7 Find a parametric representation for the surface $z = 2\sqrt{x^2 + y^2}$, that is, the top half of the cone $z^2 = 4x^2 + 4y^2$.

SOLUTION 1 One possible representation is obtained by choosing x and y as parameters:

$$x = x \qquad y = y \qquad z = 2\sqrt{x^2 + y^2}$$

So the vector equation is

$$\mathbf{r}(x, y) = x\,\mathbf{i} + y\,\mathbf{j} + 2\sqrt{x^2 + y^2}\,\mathbf{k}$$

SOLUTION 2 Another representation results from choosing as parameters the polar coordinates r and θ. A point (x, y, z) on the cone satisfies $x = r\cos\theta$, $y = r\sin\theta$, and

For some purposes the parametric representations in Solutions 1 and 2 are equally good, but Solution 2 might be preferable in certain situations. If we are interested only in the part of the cone that lies below the plane $z = 1$, for instance, all we have to do in Solution 2 is change the parameter domain to

$$0 \leqslant r \leqslant \tfrac{1}{2} \qquad 0 \leqslant \theta \leqslant 2\pi$$

$z = 2\sqrt{x^2 + y^2} = 2r$. So a vector equation for the cone is

$$\mathbf{r}(r, \theta) = r\cos\theta\,\mathbf{i} + r\sin\theta\,\mathbf{j} + 2r\,\mathbf{k}$$

where $r \geqslant 0$ and $0 \leqslant \theta \leqslant 2\pi$.

Surfaces of Revolution

Surfaces of revolution can be represented parametrically and thus graphed using a computer. For instance, let's consider the surface S obtained by rotating the curve $y = f(x)$, $a \leqslant x \leqslant b$, about the x-axis, where $f(x) \geqslant 0$. Let θ be the angle of rotation as shown in Figure 10. If (x, y, z) is a point on S, then

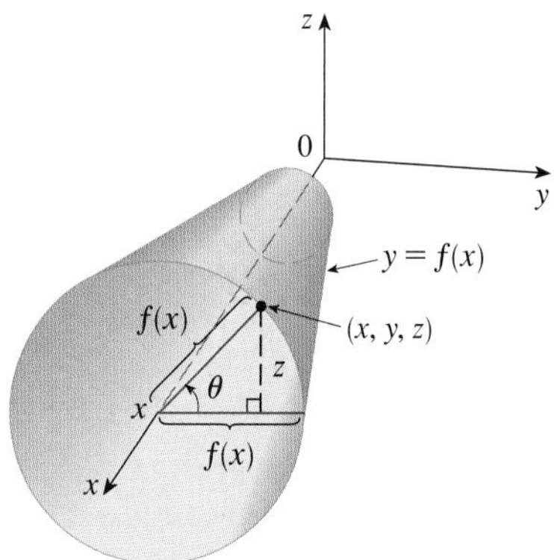

FIGURE 10

3 $$x = x \qquad y = f(x)\cos\theta \qquad z = f(x)\sin\theta$$

Therefore we take x and θ as parameters and regard Equations 3 as parametric equations of S. The parameter domain is given by $a \leqslant x \leqslant b$, $0 \leqslant \theta \leqslant 2\pi$.

EXAMPLE 8 Find parametric equations for the surface generated by rotating the curve $y = \sin x$, $0 \leqslant x \leqslant 2\pi$, about the x-axis. Use these equations to graph the surface of revolution.

SOLUTION From Equations 3, the parametric equations are

$$x = x \qquad y = \sin x\cos\theta \qquad z = \sin x\sin\theta$$

and the parameter domain is $0 \leqslant x \leqslant 2\pi$, $0 \leqslant \theta \leqslant 2\pi$. Using a computer to plot these equations and rotate the image, we obtain the graph in Figure 11.

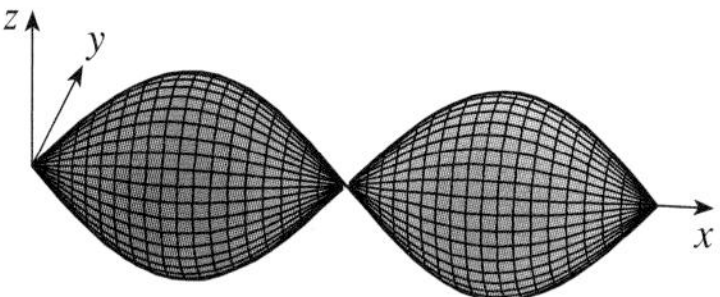

FIGURE 11

We can adapt Equations 3 to represent a surface obtained through revolution about the y- or z-axis (see Exercise 30).

Tangent Planes

We now find the tangent plane to a parametric surface S traced out by a vector function

$$\mathbf{r}(u, v) = x(u, v)\,\mathbf{i} + y(u, v)\,\mathbf{j} + z(u, v)\,\mathbf{k}$$

at a point P_0 with position vector $\mathbf{r}(u_0, v_0)$. If we keep u constant by putting $u = u_0$, then $\mathbf{r}(u_0, v)$ becomes a vector function of the single parameter v and defines a grid curve C_1 lying on S. (See Figure 12.) The tangent vector to C_1 at P_0 is obtained by taking the partial derivative of $\mathbf{r}$ with respect to v:

4 $$\mathbf{r}_v = \frac{\partial x}{\partial v}(u_0, v_0)\,\mathbf{i} + \frac{\partial y}{\partial v}(u_0, v_0)\,\mathbf{j} + \frac{\partial z}{\partial v}(u_0, v_0)\,\mathbf{k}$$

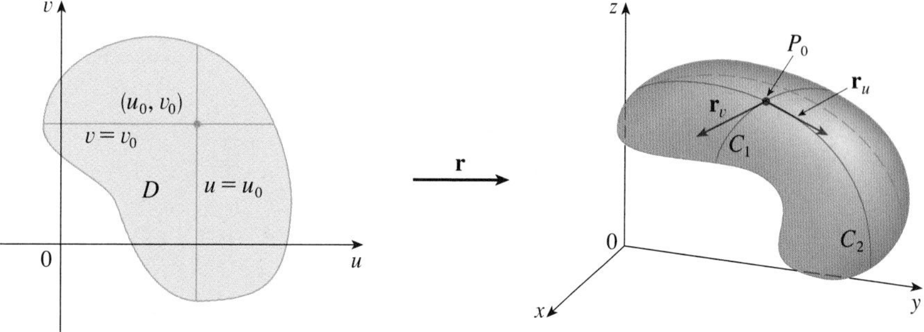

FIGURE 12

Similarly, if we keep v constant by putting $v = v_0$, we get a grid curve C_2 given by $\mathbf{r}(u, v_0)$ that lies on S, and its tangent vector at P_0 is

$$\boxed{5} \qquad \mathbf{r}_u = \frac{\partial x}{\partial u}(u_0, v_0)\,\mathbf{i} + \frac{\partial y}{\partial u}(u_0, v_0)\,\mathbf{j} + \frac{\partial z}{\partial u}(u_0, v_0)\,\mathbf{k}$$

If $\mathbf{r}_u \times \mathbf{r}_v$ is not $\mathbf{0}$, then the surface S is called **smooth** (it has no "corners"). For a smooth surface, the **tangent plane** is the plane that contains the tangent vectors $\mathbf{r}_u$ and $\mathbf{r}_v$, and the vector $\mathbf{r}_u \times \mathbf{r}_v$ is a normal vector to the tangent plane.

Figure 13 shows the self-intersecting surface in Example 9 and its tangent plane at $(1, 1, 3)$.

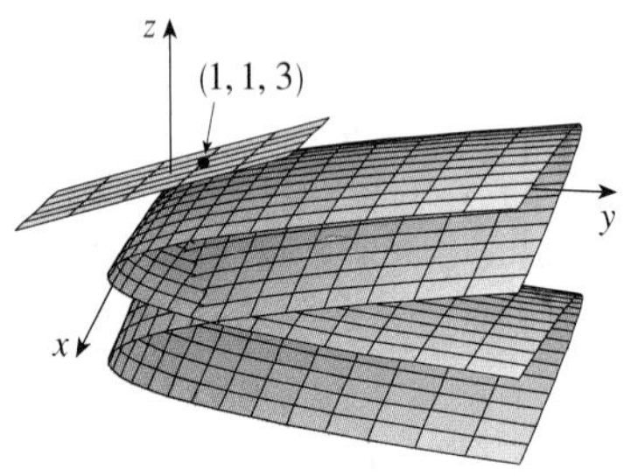

FIGURE 13

V EXAMPLE 9 Find the tangent plane to the surface with parametric equations $x = u^2$, $y = v^2$, $z = u + 2v$ at the point $(1, 1, 3)$.

SOLUTION We first compute the tangent vectors:

$$\mathbf{r}_u = \frac{\partial x}{\partial u}\mathbf{i} + \frac{\partial y}{\partial u}\mathbf{j} + \frac{\partial z}{\partial u}\mathbf{k} = 2u\,\mathbf{i} + \mathbf{k}$$

$$\mathbf{r}_v = \frac{\partial x}{\partial v}\mathbf{i} + \frac{\partial y}{\partial v}\mathbf{j} + \frac{\partial z}{\partial v}\mathbf{k} = 2v\,\mathbf{j} + 2\,\mathbf{k}$$

Thus a normal vector to the tangent plane is

$$\mathbf{r}_u \times \mathbf{r}_v = \begin{vmatrix} \mathbf{i} & \mathbf{j} & \mathbf{k} \\ 2u & 0 & 1 \\ 0 & 2v & 2 \end{vmatrix} = -2v\,\mathbf{i} - 4u\,\mathbf{j} + 4uv\,\mathbf{k}$$

Notice that the point $(1, 1, 3)$ corresponds to the parameter values $u = 1$ and $v = 1$, so the normal vector there is

$$-2\,\mathbf{i} - 4\,\mathbf{j} + 4\,\mathbf{k}$$

Therefore an equation of the tangent plane at $(1, 1, 3)$ is

$$-2(x - 1) - 4(y - 1) + 4(z - 3) = 0$$

or

$$x + 2y - 2z + 3 = 0$$

Surface Area

Now we define the surface area of a general parametric surface given by Equation 1. For simplicity we start by considering a surface whose parameter domain D is a rectangle, and we divide it into subrectangles R_{ij}. Let's choose (u_i^*, v_j^*) to be the lower left corner of R_{ij}. (See Figure 14.)

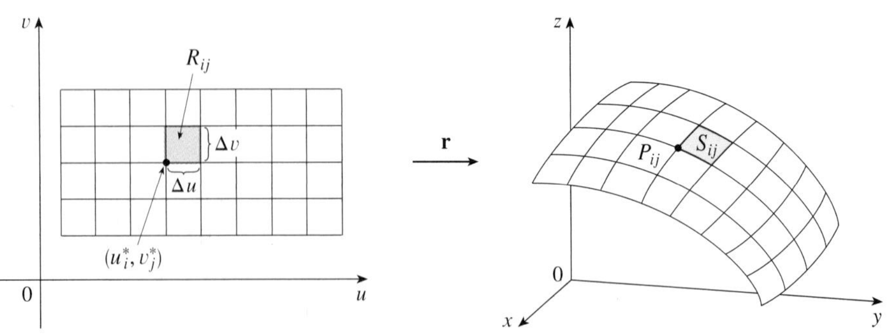

FIGURE 14
The image of the subrectangle R_{ij} is the patch S_{ij}.

The part S_{ij} of the surface S that corresponds to R_{ij} is called a *patch* and has the point P_{ij} with position vector $\mathbf{r}(u_i^*, v_j^*)$ as one of its corners. Let

$$\mathbf{r}_u^* = \mathbf{r}_u(u_i^*, v_j^*) \qquad \text{and} \qquad \mathbf{r}_v^* = \mathbf{r}_v(u_i^*, v_j^*)$$

be the tangent vectors at P_{ij} as given by Equations 5 and 4.

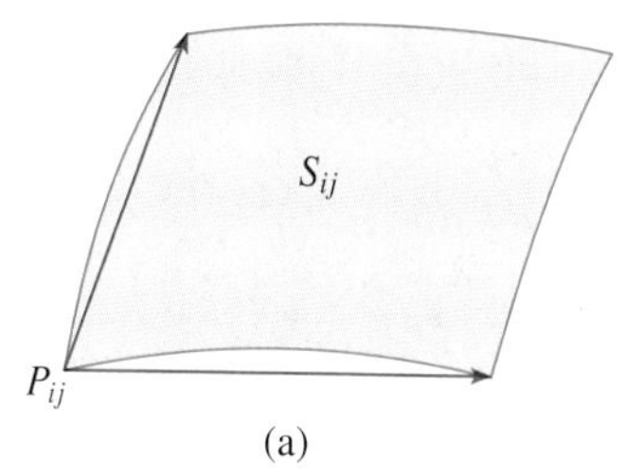

(a)

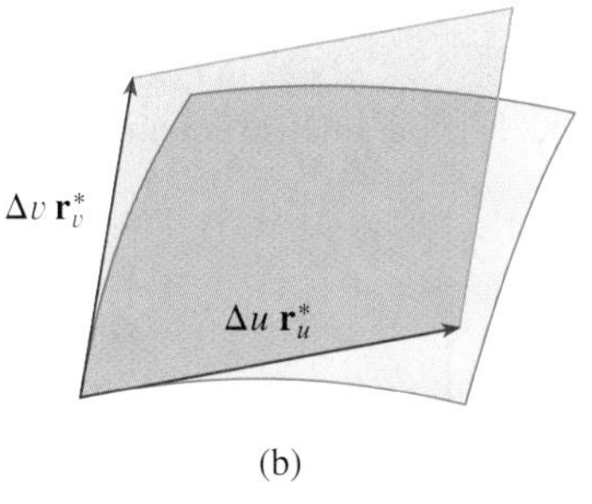

(b)

FIGURE 15
Approximating a patch by a parallelogram

Figure 15(a) shows how the two edges of the patch that meet at P_{ij} can be approximated by vectors. These vectors, in turn, can be approximated by the vectors $\Delta u\, \mathbf{r}_u^*$ and $\Delta v\, \mathbf{r}_v^*$ because partial derivatives can be approximated by difference quotients. So we approximate S_{ij} by the parallelogram determined by the vectors $\Delta u\, \mathbf{r}_u^*$ and $\Delta v\, \mathbf{r}_v^*$. This parallelogram is shown in Figure 15(b) and lies in the tangent plane to S at P_{ij}. The area of this parallelogram is

$$|(\Delta u\, \mathbf{r}_u^*) \times (\Delta v\, \mathbf{r}_v^*)| = |\mathbf{r}_u^* \times \mathbf{r}_v^*|\, \Delta u\, \Delta v$$

and so an approximation to the area of S is

$$\sum_{i=1}^{m} \sum_{j=1}^{n} |\mathbf{r}_u^* \times \mathbf{r}_v^*|\, \Delta u\, \Delta v$$

Our intuition tells us that this approximation gets better as we increase the number of subrectangles, and we recognize the double sum as a Riemann sum for the double integral $\iint_D |\mathbf{r}_u \times \mathbf{r}_v|\, du\, dv$. This motivates the following definition.

6 Definition If a smooth parametric surface S is given by the equation

$$\mathbf{r}(u, v) = x(u, v)\, \mathbf{i} + y(u, v)\, \mathbf{j} + z(u, v)\, \mathbf{k} \qquad (u, v) \in D$$

and S is covered just once as (u, v) ranges throughout the parameter domain D, then the **surface area** of S is

$$A(S) = \iint_D |\mathbf{r}_u \times \mathbf{r}_v|\, dA$$

where $\quad \mathbf{r}_u = \dfrac{\partial x}{\partial u}\mathbf{i} + \dfrac{\partial y}{\partial u}\mathbf{j} + \dfrac{\partial z}{\partial u}\mathbf{k} \qquad \mathbf{r}_v = \dfrac{\partial x}{\partial v}\mathbf{i} + \dfrac{\partial y}{\partial v}\mathbf{j} + \dfrac{\partial z}{\partial v}\mathbf{k}$

EXAMPLE 10 Find the surface area of a sphere of radius a.

SOLUTION In Example 4 we found the parametric representation

$$x = a \sin\phi \cos\theta \qquad y = a \sin\phi \sin\theta \qquad z = a \cos\phi$$

where the parameter domain is

$$D = \{(\phi, \theta) \mid 0 \leqslant \phi \leqslant \pi,\ 0 \leqslant \theta \leqslant 2\pi\}$$

We first compute the cross product of the tangent vectors:

$$\mathbf{r}_\phi \times \mathbf{r}_\theta = \begin{vmatrix} \mathbf{i} & \mathbf{j} & \mathbf{k} \\ \dfrac{\partial x}{\partial \phi} & \dfrac{\partial y}{\partial \phi} & \dfrac{\partial z}{\partial \phi} \\ \dfrac{\partial x}{\partial \theta} & \dfrac{\partial y}{\partial \theta} & \dfrac{\partial z}{\partial \theta} \end{vmatrix} = \begin{vmatrix} \mathbf{i} & \mathbf{j} & \mathbf{k} \\ a\cos\phi\cos\theta & a\cos\phi\sin\theta & -a\sin\phi \\ -a\sin\phi\sin\theta & a\sin\phi\cos\theta & 0 \end{vmatrix}$$

$$= a^2 \sin^2\phi \cos\theta\, \mathbf{i} + a^2 \sin^2\phi \sin\theta\, \mathbf{j} + a^2 \sin\phi \cos\phi\, \mathbf{k}$$

Thus

$$|\mathbf{r}_\phi \times \mathbf{r}_\theta| = \sqrt{a^4 \sin^4\phi \cos^2\theta + a^4 \sin^4\phi \sin^2\theta + a^4 \sin^2\phi \cos^2\phi}$$
$$= \sqrt{a^4 \sin^4\phi + a^4 \sin^2\phi \cos^2\phi} = a^2\sqrt{\sin^2\phi} = a^2 \sin\phi$$

since $\sin\phi \geqslant 0$ for $0 \leqslant \phi \leqslant \pi$. Therefore, by Definition 6, the area of the sphere is

$$A = \iint_D |\mathbf{r}_\phi \times \mathbf{r}_\theta|\, dA = \int_0^{2\pi}\int_0^{\pi} a^2 \sin\phi\, d\phi\, d\theta$$
$$= a^2 \int_0^{2\pi} d\theta \int_0^{\pi} \sin\phi\, d\phi = a^2(2\pi)2 = 4\pi a^2$$

Surface Area of the Graph of a Function

For the special case of a surface S with equation $z = f(x, y)$, where (x, y) lies in D and f has continuous partial derivatives, we take x and y as parameters. The parametric equations are

$$x = x \qquad y = y \qquad z = f(x, y)$$

so

$$\mathbf{r}_x = \mathbf{i} + \left(\frac{\partial f}{\partial x}\right)\mathbf{k} \qquad \mathbf{r}_y = \mathbf{j} + \left(\frac{\partial f}{\partial y}\right)\mathbf{k}$$

and

$$\boxed{7} \qquad \mathbf{r}_x \times \mathbf{r}_y = \begin{vmatrix} \mathbf{i} & \mathbf{j} & \mathbf{k} \\ 1 & 0 & \dfrac{\partial f}{\partial x} \\ 0 & 1 & \dfrac{\partial f}{\partial y} \end{vmatrix} = -\frac{\partial f}{\partial x}\mathbf{i} - \frac{\partial f}{\partial y}\mathbf{j} + \mathbf{k}$$

Thus we have

$$\boxed{8} \qquad |\mathbf{r}_x \times \mathbf{r}_y| = \sqrt{\left(\frac{\partial f}{\partial x}\right)^2 + \left(\frac{\partial f}{\partial y}\right)^2 + 1} = \sqrt{1 + \left(\frac{\partial z}{\partial x}\right)^2 + \left(\frac{\partial z}{\partial y}\right)^2}$$

and the surface area formula in Definition 6 becomes

Notice the similarity between the surface area formula in Equation 9 and the arc length formula

$$L = \int_a^b \sqrt{1 + \left(\frac{dy}{dx}\right)^2}\, dx$$

from Section 8.1.

$$\boxed{9} \qquad \boxed{A(S) = \iint_D \sqrt{1 + \left(\frac{\partial z}{\partial x}\right)^2 + \left(\frac{\partial z}{\partial y}\right)^2}\, dA}$$

V EXAMPLE 11 Find the area of the part of the paraboloid $z = x^2 + y^2$ that lies under the plane $z = 9$.

SOLUTION The plane intersects the paraboloid in the circle $x^2 + y^2 = 9$, $z = 9$. Therefore the given surface lies above the disk D with center the origin and radius 3. (See

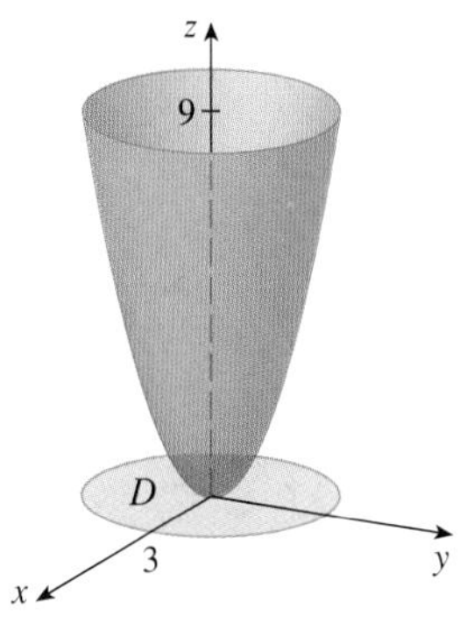

FIGURE 16

Figure 16.) Using Formula 9, we have

$$A = \iint_D \sqrt{1 + \left(\frac{\partial z}{\partial x}\right)^2 + \left(\frac{\partial z}{\partial y}\right)^2}\, dA$$

$$= \iint_D \sqrt{1 + (2x)^2 + (2y)^2}\, dA$$

$$= \iint_D \sqrt{1 + 4(x^2 + y^2)}\, dA$$

Converting to polar coordinates, we obtain

$$A = \int_0^{2\pi} \int_0^3 \sqrt{1 + 4r^2}\, r\, dr\, d\theta = \int_0^{2\pi} d\theta \int_0^3 r\sqrt{1 + 4r^2}\, dr$$

$$= 2\pi\left(\tfrac{1}{8}\right)\tfrac{2}{3}(1 + 4r^2)^{3/2}\Big]_0^3 = \frac{\pi}{6}\left(37\sqrt{37} - 1\right)$$

The question remains whether our definition of surface area [6] is consistent with the surface area formula from single-variable calculus (8.2.4).

We consider the surface S obtained by rotating the curve $y = f(x)$, $a \leqslant x \leqslant b$, about the x-axis, where $f(x) \geqslant 0$ and f' is continuous. From Equations 3 we know that parametric equations of S are

$$x = x \qquad y = f(x)\cos\theta \qquad z = f(x)\sin\theta \qquad a \leqslant x \leqslant b \qquad 0 \leqslant \theta \leqslant 2\pi$$

To compute the surface area of S we need the tangent vectors

$$\mathbf{r}_x = \mathbf{i} + f'(x)\cos\theta\,\mathbf{j} + f'(x)\sin\theta\,\mathbf{k}$$

$$\mathbf{r}_\theta = -f(x)\sin\theta\,\mathbf{j} + f(x)\cos\theta\,\mathbf{k}$$

Thus

$$\mathbf{r}_x \times \mathbf{r}_\theta = \begin{vmatrix} \mathbf{i} & \mathbf{j} & \mathbf{k} \\ 1 & f'(x)\cos\theta & f'(x)\sin\theta \\ 0 & -f(x)\sin\theta & f(x)\cos\theta \end{vmatrix}$$

$$= f(x)f'(x)\,\mathbf{i} - f(x)\cos\theta\,\mathbf{j} - f(x)\sin\theta\,\mathbf{k}$$

and so

$$|\mathbf{r}_x \times \mathbf{r}_\theta| = \sqrt{[f(x)]^2[f'(x)]^2 + [f(x)]^2\cos^2\theta + [f(x)]^2\sin^2\theta}$$

$$= \sqrt{[f(x)]^2[1 + [f'(x)]^2]} = f(x)\sqrt{1 + [f'(x)]^2}$$

because $f(x) \geqslant 0$. Therefore the area of S is

$$A = \iint_D |\mathbf{r}_x \times \mathbf{r}_\theta|\, dA$$

$$= \int_0^{2\pi} \int_a^b f(x)\sqrt{1 + [f'(x)]^2}\, dx\, d\theta$$

$$= 2\pi \int_a^b f(x)\sqrt{1 + [f'(x)]^2}\, dx$$

This is precisely the formula that was used to define the area of a surface of revolution in single-variable calculus (8.2.4).

16.6 Exercises

1–2 Determine whether the points P and Q lie on the given surface.

1. $\mathbf{r}(u, v) = \langle 2u + 3v, 1 + 5u - v, 2 + u + v \rangle$
$P(7, 10, 4)$, $Q(5, 22, 5)$

2. $\mathbf{r}(u, v) = \langle u + v, u^2 - v, u + v^2 \rangle$
$P(3, -1, 5)$, $Q(-1, 3, 4)$

3–6 Identify the surface with the given vector equation.

3. $\mathbf{r}(u, v) = (u + v)\,\mathbf{i} + (3 - v)\,\mathbf{j} + (1 + 4u + 5v)\,\mathbf{k}$

4. $\mathbf{r}(u, v) = 2 \sin u\,\mathbf{i} + 3 \cos u\,\mathbf{j} + v\,\mathbf{k}, \quad 0 \leqslant v \leqslant 2$

5. $\mathbf{r}(s, t) = \langle s, t, t^2 - s^2 \rangle$

6. $\mathbf{r}(s, t) = \langle s \sin 2t, s^2, s \cos 2t \rangle$

7–12 Use a computer to graph the parametric surface. Get a printout and indicate on it which grid curves have u constant and which have v constant.

7. $\mathbf{r}(u, v) = \langle u^2, v^2, u + v \rangle$,
$-1 \leqslant u \leqslant 1, \ -1 \leqslant v \leqslant 1$

8. $\mathbf{r}(u, v) = \langle u, v^3, -v \rangle$,
$-2 \leqslant u \leqslant 2, \ -2 \leqslant v \leqslant 2$

9. $\mathbf{r}(u, v) = \langle u \cos v, u \sin v, u^5 \rangle$,
$-1 \leqslant u \leqslant 1, \ 0 \leqslant v \leqslant 2\pi$

10. $\mathbf{r}(u, v) = \langle u, \sin(u + v), \sin v \rangle$,
$-\pi \leqslant u \leqslant \pi, \ -\pi \leqslant v \leqslant \pi$

11. $x = \sin v, \quad y = \cos u \sin 4v, \quad z = \sin 2u \sin 4v$,
$0 \leqslant u \leqslant 2\pi, \ -\pi/2 \leqslant v \leqslant \pi/2$

12. $x = \sin u, \quad y = \cos u \sin v, \quad z = \sin v$,
$0 \leqslant u \leqslant 2\pi, 0 \leqslant v \leqslant 2\pi$

13–18 Match the equations with the graphs labeled I–VI and give reasons for your answers. Determine which families of grid curves have u constant and which have v constant.

13. $\mathbf{r}(u, v) = u \cos v\,\mathbf{i} + u \sin v\,\mathbf{j} + v\,\mathbf{k}$

14. $\mathbf{r}(u, v) = u \cos v\,\mathbf{i} + u \sin v\,\mathbf{j} + \sin u\,\mathbf{k}, \quad -\pi \leqslant u \leqslant \pi$

15. $\mathbf{r}(u, v) = \sin v\,\mathbf{i} + \cos u \sin 2v\,\mathbf{j} + \sin u \sin 2v\,\mathbf{k}$

16. $x = (1 - u)(3 + \cos v) \cos 4\pi u$,
$y = (1 - u)(3 + \cos v) \sin 4\pi u$,
$z = 3u + (1 - u) \sin v$

17. $x = \cos^3 u \cos^3 v, \quad y = \sin^3 u \cos^3 v, \quad z = \sin^3 v$

18. $x = (1 - |u|)\cos v, \quad y = (1 - |u|)\sin v, \quad z = u$

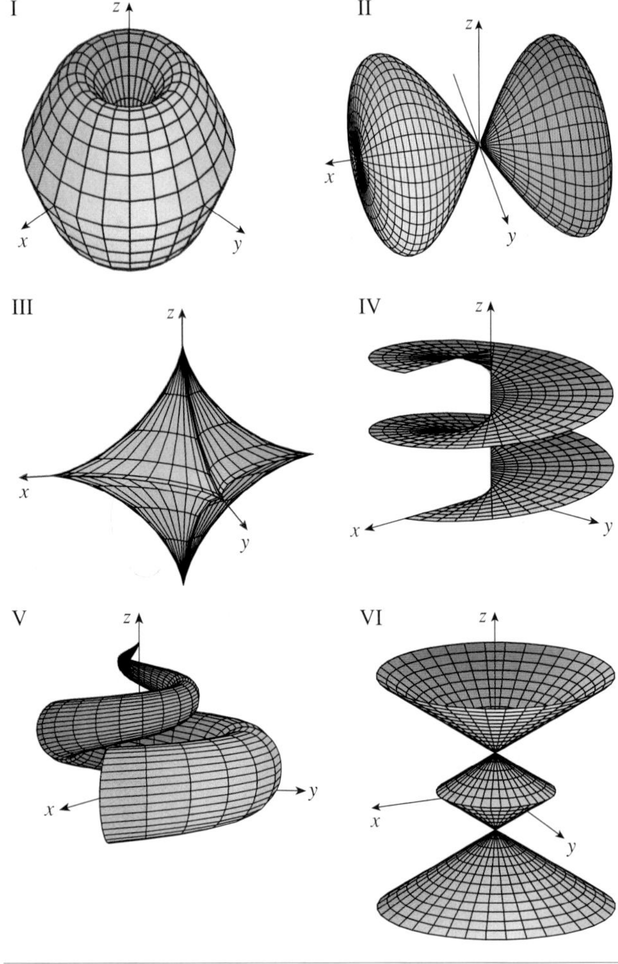

19–26 Find a parametric representation for the surface.

19. The plane through the origin that contains the vectors $\mathbf{i} - \mathbf{j}$ and $\mathbf{j} - \mathbf{k}$

20. The plane that passes through the point $(0, -1, 5)$ and contains the vectors $\langle 2, 1, 4 \rangle$ and $\langle -3, 2, 5 \rangle$

21. The part of the hyperboloid $4x^2 - 4y^2 - z^2 = 4$ that lies in front of the yz-plane

22. The part of the ellipsoid $x^2 + 2y^2 + 3z^2 = 1$ that lies to the left of the xz-plane

23. The part of the sphere $x^2 + y^2 + z^2 = 4$ that lies above the cone $z = \sqrt{x^2 + y^2}$

24. The part of the sphere $x^2 + y^2 + z^2 = 16$ that lies between the planes $z = -2$ and $z = 2$

25. The part of the cylinder $y^2 + z^2 = 16$ that lies between the planes $x = 0$ and $x = 5$

26. The part of the plane $z = x + 3$ that lies inside the cylinder $x^2 + y^2 = 1$

CAS **27–28** Use a computer algebra system to produce a graph that looks like the given one.

27.

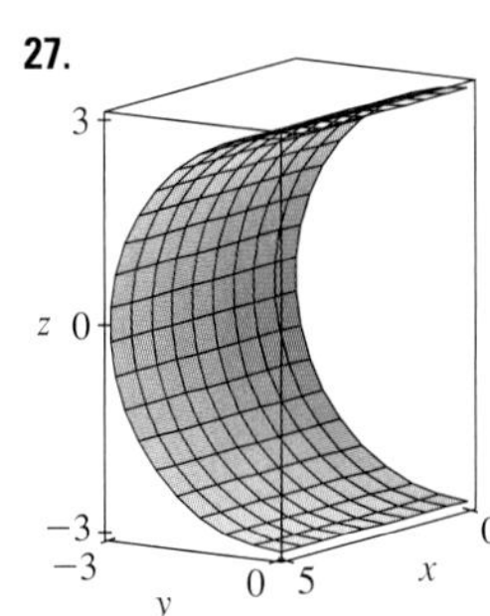

28.

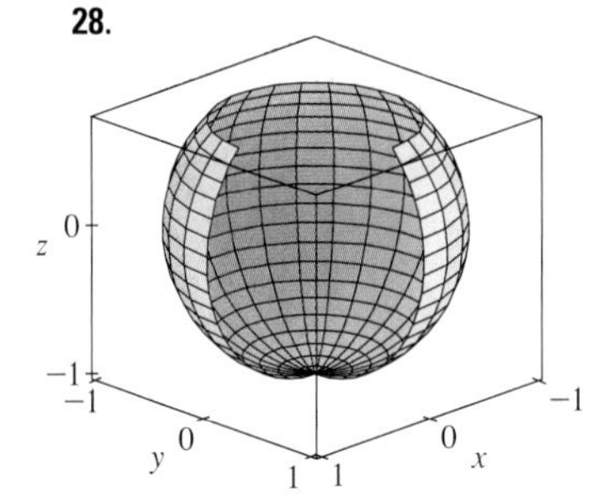

29. Find parametric equations for the surface obtained by rotating the curve $y = e^{-x}$, $0 \leqslant x \leqslant 3$, about the x-axis and use them to graph the surface.

30. Find parametric equations for the surface obtained by rotating the curve $x = 4y^2 - y^4$, $-2 \leqslant y \leqslant 2$, about the y-axis and use them to graph the surface.

31. (a) What happens to the spiral tube in Example 2 (see Figure 5) if we replace $\cos u$ by $\sin u$ and $\sin u$ by $\cos u$?
(b) What happens if we replace $\cos u$ by $\cos 2u$ and $\sin u$ by $\sin 2u$?

32. The surface with parametric equations

$$x = 2\cos\theta + r\cos(\theta/2)$$
$$y = 2\sin\theta + r\cos(\theta/2)$$
$$z = r\sin(\theta/2)$$

where $-\frac{1}{2} \leqslant r \leqslant \frac{1}{2}$ and $0 \leqslant \theta \leqslant 2\pi$, is called a **Möbius strip**. Graph this surface with several viewpoints. What is unusual about it?

33–36 Find an equation of the tangent plane to the given parametric surface at the specified point.

33. $x = u + v$, $y = 3u^2$, $z = u - v$; $(2, 3, 0)$

34. $x = u^2 + 1$, $y = v^3 + 1$, $z = u + v$; $(5, 2, 3)$

35. $\mathbf{r}(u, v) = u\cos v\,\mathbf{i} + u\sin v\,\mathbf{j} + v\,\mathbf{k}$; $u = 1$, $v = \pi/3$

36. $\mathbf{r}(u, v) = \sin u\,\mathbf{i} + \cos u\sin v\,\mathbf{j} + \sin v\,\mathbf{k}$; $u = \pi/6$, $v = \pi/6$

CAS **37–38** Find an equation of the tangent plane to the given parametric surface at the specified point. Graph the surface and the tangent plane.

37. $\mathbf{r}(u, v) = u^2\,\mathbf{i} + 2u\sin v\,\mathbf{j} + u\cos v\,\mathbf{k}$; $u = 1$, $v = 0$

38. $\mathbf{r}(u, v) = (1 - u^2 - v^2)\,\mathbf{i} - v\,\mathbf{j} - u\,\mathbf{k}$; $(-1, -1, -1)$

39–50 Find the area of the surface.

39. The part of the plane $3x + 2y + z = 6$ that lies in the first octant

40. The part of the plane with vector equation $\mathbf{r}(u, v) = \langle u + v, 2 - 3u, 1 + u - v\rangle$ that is given by $0 \leqslant u \leqslant 2$, $-1 \leqslant v \leqslant 1$

41. The part of the plane $x + 2y + 3z = 1$ that lies inside the cylinder $x^2 + y^2 = 3$

42. The part of the cone $z = \sqrt{x^2 + y^2}$ that lies between the plane $y = x$ and the cylinder $y = x^2$

43. The surface $z = \frac{2}{3}(x^{3/2} + y^{3/2})$, $0 \leqslant x \leqslant 1$, $0 \leqslant y \leqslant 1$

44. The part of the surface $z = 1 + 3x + 2y^2$ that lies above the triangle with vertices $(0, 0)$, $(0, 1)$, and $(2, 1)$

45. The part of the surface $z = xy$ that lies within the cylinder $x^2 + y^2 = 1$

46. The part of the paraboloid $x = y^2 + z^2$ that lies inside the cylinder $y^2 + z^2 = 9$

47. The part of the surface $y = 4x + z^2$ that lies between the planes $x = 0$, $x = 1$, $z = 0$, and $z = 1$

48. The helicoid (or spiral ramp) with vector equation $\mathbf{r}(u, v) = u\cos v\,\mathbf{i} + u\sin v\,\mathbf{j} + v\,\mathbf{k}$, $0 \leqslant u \leqslant 1$, $0 \leqslant v \leqslant \pi$

49. The surface with parametric equations $x = u^2$, $y = uv$, $z = \frac{1}{2}v^2$, $0 \leqslant u \leqslant 1$, $0 \leqslant v \leqslant 2$

50. The part of the sphere $x^2 + y^2 + z^2 = b^2$ that lies inside the cylinder $x^2 + y^2 = a^2$, where $0 < a < b$

51. If the equation of a surface S is $z = f(x, y)$, where $x^2 + y^2 \leqslant R^2$, and you know that $|f_x| \leqslant 1$ and $|f_y| \leqslant 1$, what can you say about $A(S)$?

52–53 Find the area of the surface correct to four decimal places by expressing the area in terms of a single integral and using your calculator to estimate the integral.

52. The part of the surface $z = \cos(x^2 + y^2)$ that lies inside the cylinder $x^2 + y^2 = 1$

53. The part of the surface $z = e^{-x^2-y^2}$ that lies above the disk $x^2 + y^2 \leqslant 4$

CAS **54.** Find, to four decimal places, the area of the part of the surface $z = (1 + x^2)/(1 + y^2)$ that lies above the square $|x| + |y| \leqslant 1$. Illustrate by graphing this part of the surface.

55. (a) Use the Midpoint Rule for double integrals (see Section 15.1) with six squares to estimate the area of the surface $z = 1/(1 + x^2 + y^2)$, $0 \leqslant x \leqslant 6$, $0 \leqslant y \leqslant 4$.

CAS (b) Use a computer algebra system to approximate the surface area in part (a) to four decimal places. Compare with the answer to part (a).

CAS **56.** Find the area of the surface with vector equation $\mathbf{r}(u, v) = \langle \cos^3 u \cos^3 v, \sin^3 u \cos^3 v, \sin^3 v \rangle$, $0 \leqslant u \leqslant \pi$, $0 \leqslant v \leqslant 2\pi$. State your answer correct to four decimal places.

CAS **57.** Find the exact area of the surface $z = 1 + 2x + 3y + 4y^2$, $1 \leqslant x \leqslant 4$, $0 \leqslant y \leqslant 1$.

58. (a) Set up, but do not evaluate, a double integral for the area of the surface with parametric equations $x = au \cos v$, $y = bu \sin v$, $z = u^2$, $0 \leqslant u \leqslant 2$, $0 \leqslant v \leqslant 2\pi$.

(b) Eliminate the parameters to show that the surface is an elliptic paraboloid and set up another double integral for the surface area.

(c) Use the parametric equations in part (a) with $a = 2$ and $b = 3$ to graph the surface.

(d) For the case $a = 2$, $b = 3$, use a computer algebra system to find the surface area correct to four decimal places.

59. (a) Show that the parametric equations $x = a \sin u \cos v$, $y = b \sin u \sin v$, $z = c \cos u$, $0 \leqslant u \leqslant \pi$, $0 \leqslant v \leqslant 2\pi$, represent an ellipsoid.

(b) Use the parametric equations in part (a) to graph the ellipsoid for the case $a = 1$, $b = 2$, $c = 3$.

(c) Set up, but do not evaluate, a double integral for the surface area of the ellipsoid in part (b).

60. (a) Show that the parametric equations $x = a \cosh u \cos v$, $y = b \cosh u \sin v$, $z = c \sinh u$, represent a hyperboloid of one sheet.

(b) Use the parametric equations in part (a) to graph the hyperboloid for the case $a = 1$, $b = 2$, $c = 3$.

(c) Set up, but do not evaluate, a double integral for the surface area of the part of the hyperboloid in part (b) that lies between the planes $z = -3$ and $z = 3$.

61. Find the area of the part of the sphere $x^2 + y^2 + z^2 = 4z$ that lies inside the paraboloid $z = x^2 + y^2$.

62. The figure shows the surface created when the cylinder $y^2 + z^2 = 1$ intersects the cylinder $x^2 + z^2 = 1$. Find the area of this surface.

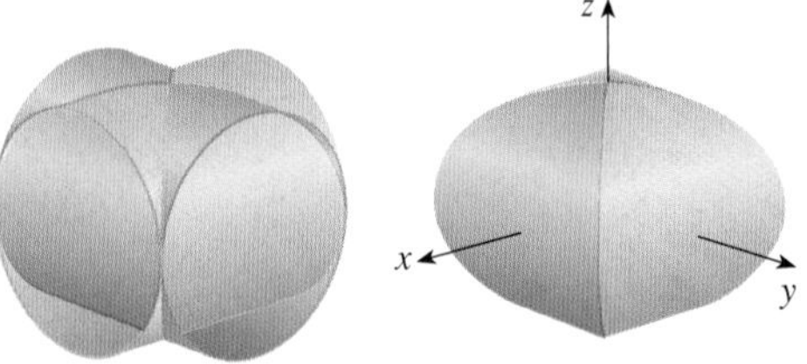

63. Find the area of the part of the sphere $x^2 + y^2 + z^2 = a^2$ that lies inside the cylinder $x^2 + y^2 = ax$.

64. (a) Find a parametric representation for the torus obtained by rotating about the z-axis the circle in the xz-plane with center $(b, 0, 0)$ and radius $a < b$. [*Hint:* Take as parameters the angles θ and α shown in the figure.]

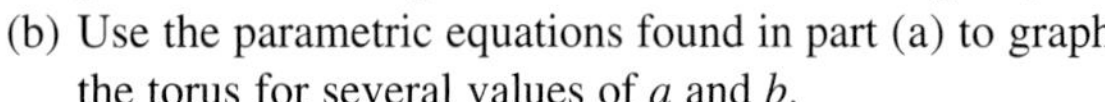
(b) Use the parametric equations found in part (a) to graph the torus for several values of a and b.

(c) Use the parametric representation from part (a) to find the surface area of the torus.

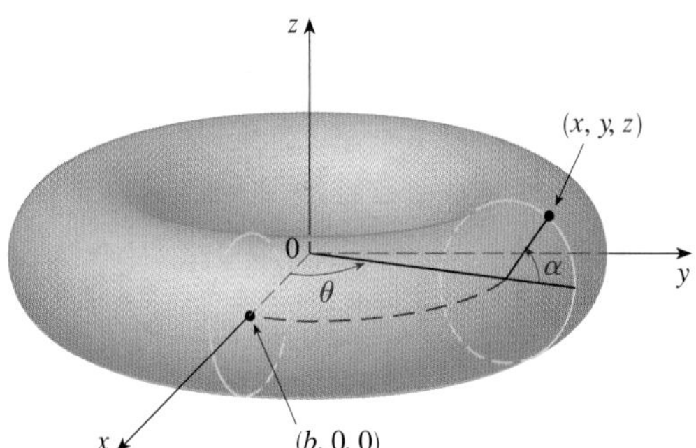

16.7 Surface Integrals

The relationship between surface integrals and surface area is much the same as the relationship between line integrals and arc length. Suppose f is a function of three variables whose domain includes a surface S. We will define the surface integral of f over S in such a way that, in the case where $f(x, y, z) = 1$, the value of the surface integral is equal to the surface area of S. We start with parametric surfaces and then deal with the special case where S is the graph of a function of two variables.

Parametric Surfaces

Suppose that a surface S has a vector equation

$$\mathbf{r}(u, v) = x(u, v)\,\mathbf{i} + y(u, v)\,\mathbf{j} + z(u, v)\,\mathbf{k} \qquad (u, v) \in D$$

We first assume that the parameter domain D is a rectangle and we divide it into subrect-

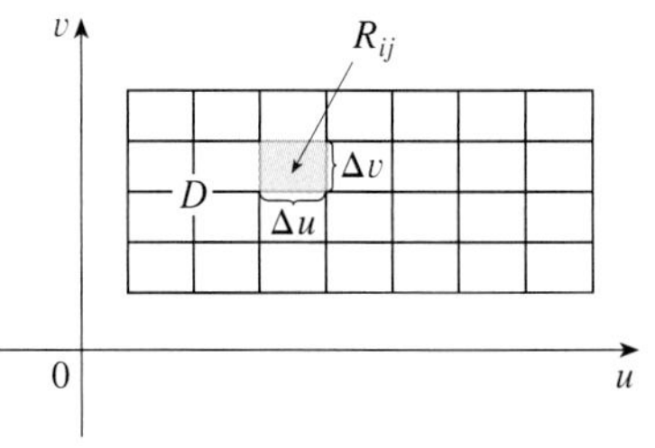

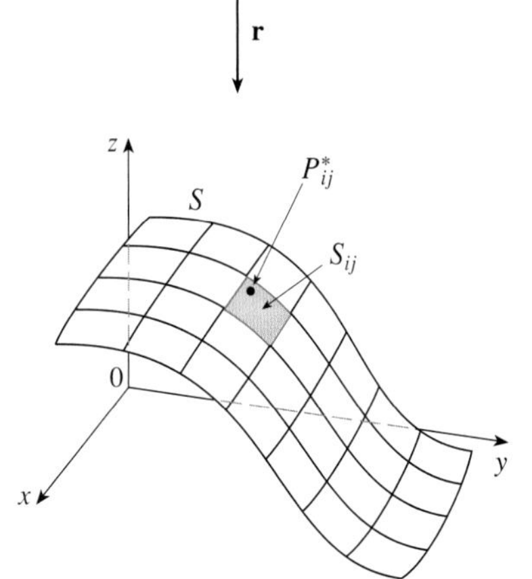

FIGURE 1

angles R_{ij} with dimensions Δu and Δv. Then the surface S is divided into corresponding patches S_{ij} as in Figure 1. We evaluate f at a point P_{ij}^* in each patch, multiply by the area ΔS_{ij} of the patch, and form the Riemann sum

$$\sum_{i=1}^{m} \sum_{j=1}^{n} f(P_{ij}^*) \, \Delta S_{ij}$$

Then we take the limit as the number of patches increases and define the **surface integral of f over the surface S** as

$$\boxed{1} \qquad \iint_S f(x, y, z) \, dS = \lim_{m, n \to \infty} \sum_{i=1}^{m} \sum_{j=1}^{n} f(P_{ij}^*) \, \Delta S_{ij}$$

Notice the analogy with the definition of a line integral (16.2.2) and also the analogy with the definition of a double integral (15.1.5).

To evaluate the surface integral in Equation 1 we approximate the patch area ΔS_{ij} by the area of an approximating parallelogram in the tangent plane. In our discussion of surface area in Section 16.6 we made the approximation

$$\Delta S_{ij} \approx |\mathbf{r}_u \times \mathbf{r}_v| \, \Delta u \, \Delta v$$

where $\qquad \mathbf{r}_u = \dfrac{\partial x}{\partial u}\mathbf{i} + \dfrac{\partial y}{\partial u}\mathbf{j} + \dfrac{\partial z}{\partial u}\mathbf{k} \qquad \mathbf{r}_v = \dfrac{\partial x}{\partial v}\mathbf{i} + \dfrac{\partial y}{\partial v}\mathbf{j} + \dfrac{\partial z}{\partial v}\mathbf{k}$

are the tangent vectors at a corner of S_{ij}. If the components are continuous and $\mathbf{r}_u$ and $\mathbf{r}_v$ are nonzero and nonparallel in the interior of D, it can be shown from Definition 1, even when D is not a rectangle, that

We assume that the surface is covered only once as (u, v) ranges throughout D. The value of the surface integral does not depend on the parametrization that is used.

$$\boxed{2} \qquad \iint_S f(x, y, z) \, dS = \iint_D f(\mathbf{r}(u, v)) \, |\mathbf{r}_u \times \mathbf{r}_v| \, dA$$

This should be compared with the formula for a line integral:

$$\int_C f(x, y, z) \, ds = \int_a^b f(\mathbf{r}(t)) \, |\mathbf{r}'(t)| \, dt$$

Observe also that

$$\iint_S 1 \, dS = \iint_D |\mathbf{r}_u \times \mathbf{r}_v| \, dA = A(S)$$

Formula 2 allows us to compute a surface integral by converting it into a double integral over the parameter domain D. When using this formula, remember that $f(\mathbf{r}(u, v))$ is evaluated by writing $x = x(u, v)$, $y = y(u, v)$, and $z = z(u, v)$ in the formula for $f(x, y, z)$.

EXAMPLE 1 Compute the surface integral $\iint_S x^2 \, dS$, where S is the unit sphere $x^2 + y^2 + z^2 = 1$.

SOLUTION As in Example 4 in Section 16.6, we use the parametric representation

$$x = \sin\phi \, \cos\theta \qquad y = \sin\phi \, \sin\theta \qquad z = \cos\phi \qquad 0 \leqslant \phi \leqslant \pi \qquad 0 \leqslant \theta \leqslant 2\pi$$

that is, $$\mathbf{r}(\phi, \theta) = \sin\phi\cos\theta\,\mathbf{i} + \sin\phi\sin\theta\,\mathbf{j} + \cos\phi\,\mathbf{k}$$

As in Example 10 in Section 16.6, we can compute that

$$|\mathbf{r}_\phi \times \mathbf{r}_\theta| = \sin\phi$$

Therefore, by Formula 2,

$$\begin{aligned}
\iint_S x^2\,dS &= \iint_D (\sin\phi\cos\theta)^2\,|\mathbf{r}_\phi \times \mathbf{r}_\theta|\,dA \\
&= \int_0^{2\pi}\int_0^{\pi} \sin^2\phi\cos^2\theta\sin\phi\,d\phi\,d\theta = \int_0^{2\pi}\cos^2\theta\,d\theta\int_0^{\pi}\sin^3\phi\,d\phi \\
&= \int_0^{2\pi}\tfrac{1}{2}(1+\cos 2\theta)\,d\theta\int_0^{\pi}(\sin\phi - \sin\phi\cos^2\phi)\,d\phi \\
&= \tfrac{1}{2}\left[\theta + \tfrac{1}{2}\sin 2\theta\right]_0^{2\pi}\left[-\cos\phi + \tfrac{1}{3}\cos^3\phi\right]_0^{\pi} = \frac{4\pi}{3}
\end{aligned}$$

Here we use the identities

$$\cos^2\theta = \tfrac{1}{2}(1 + \cos 2\theta)$$

$$\sin^2\phi = 1 - \cos^2\phi$$

Instead, we could use Formulas 64 and 67 in the Table of Integrals.

Surface integrals have applications similar to those for the integrals we have previously considered. For example, if a thin sheet (say, of aluminum foil) has the shape of a surface S and the density (mass per unit area) at the point (x, y, z) is $\rho(x, y, z)$, then the total **mass** of the sheet is

$$m = \iint_S \rho(x, y, z)\,dS$$

and the **center of mass** is $(\bar{x}, \bar{y}, \bar{z})$, where

$$\bar{x} = \frac{1}{m}\iint_S x\,\rho(x, y, z)\,dS \qquad \bar{y} = \frac{1}{m}\iint_S y\,\rho(x, y, z)\,dS \qquad \bar{z} = \frac{1}{m}\iint_S z\,\rho(x, y, z)\,dS$$

Moments of inertia can also be defined as before (see Exercise 41).

Graphs

Any surface S with equation $z = g(x, y)$ can be regarded as a parametric surface with parametric equations

$$x = x \qquad y = y \qquad z = g(x, y)$$

and so we have $$\mathbf{r}_x = \mathbf{i} + \left(\frac{\partial g}{\partial x}\right)\mathbf{k} \qquad \mathbf{r}_y = \mathbf{j} + \left(\frac{\partial g}{\partial y}\right)\mathbf{k}$$

Thus

3 $$\mathbf{r}_x \times \mathbf{r}_y = -\frac{\partial g}{\partial x}\mathbf{i} - \frac{\partial g}{\partial y}\mathbf{j} + \mathbf{k}$$

and $$|\mathbf{r}_x \times \mathbf{r}_y| = \sqrt{\left(\frac{\partial z}{\partial x}\right)^2 + \left(\frac{\partial z}{\partial y}\right)^2 + 1}$$

Therefore, in this case, Formula 2 becomes

$$\boxed{4} \qquad \iint_S f(x, y, z)\, dS = \iint_D f(x, y, g(x, y)) \sqrt{\left(\frac{\partial z}{\partial x}\right)^2 + \left(\frac{\partial z}{\partial y}\right)^2 + 1}\, dA$$

Similar formulas apply when it is more convenient to project S onto the yz-plane or xz-plane. For instance, if S is a surface with equation $y = h(x, z)$ and D is its projection onto the xz-plane, then

$$\iint_S f(x, y, z)\, dS = \iint_D f(x, h(x, z), z) \sqrt{\left(\frac{\partial y}{\partial x}\right)^2 + \left(\frac{\partial y}{\partial z}\right)^2 + 1}\, dA$$

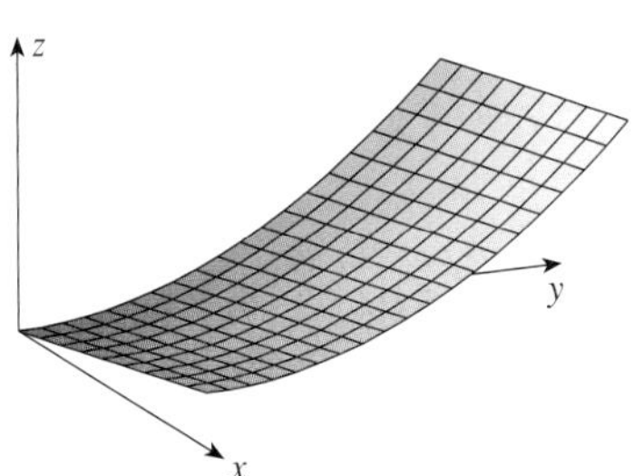

FIGURE 2

EXAMPLE 2 Evaluate $\iint_S y\, dS$, where S is the surface $z = x + y^2$, $0 \leqslant x \leqslant 1$, $0 \leqslant y \leqslant 2$. (See Figure 2.)

SOLUTION Since

$$\frac{\partial z}{\partial x} = 1 \qquad \text{and} \qquad \frac{\partial z}{\partial y} = 2y$$

Formula 4 gives

$$\begin{aligned} \iint_S y\, dS &= \iint_D y \sqrt{1 + \left(\frac{\partial z}{\partial x}\right)^2 + \left(\frac{\partial z}{\partial y}\right)^2}\, dA \\ &= \int_0^1 \int_0^2 y\sqrt{1 + 1 + 4y^2}\, dy\, dx \\ &= \int_0^1 dx\, \sqrt{2} \int_0^2 y\sqrt{1 + 2y^2}\, dy \\ &= \sqrt{2}\left(\tfrac{1}{4}\right)\tfrac{2}{3}(1 + 2y^2)^{3/2}\Big]_0^2 = \frac{13\sqrt{2}}{3} \end{aligned}$$

If S is a piecewise-smooth surface, that is, a finite union of smooth surfaces $S_1, S_2, \ldots, S_n$ that intersect only along their boundaries, then the surface integral of f over S is defined by

$$\iint_S f(x, y, z)\, dS = \iint_{S_1} f(x, y, z)\, dS + \cdots + \iint_{S_n} f(x, y, z)\, dS$$

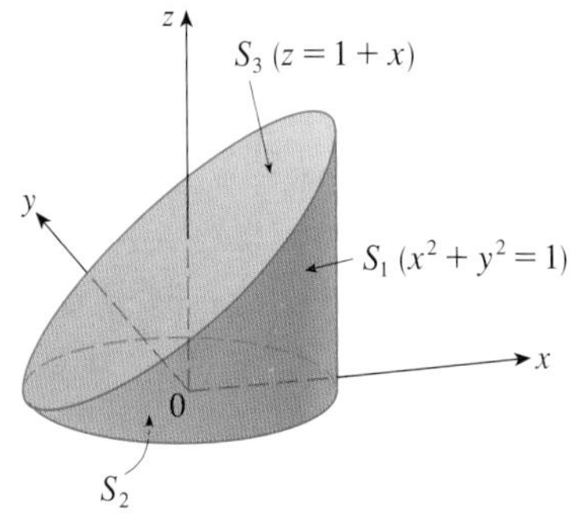

FIGURE 3

V EXAMPLE 3 Evaluate $\iint_S z\, dS$, where S is the surface whose sides S_1 are given by the cylinder $x^2 + y^2 = 1$, whose bottom S_2 is the disk $x^2 + y^2 \leqslant 1$ in the plane $z = 0$, and whose top S_3 is the part of the plane $z = 1 + x$ that lies above S_2.

SOLUTION The surface S is shown in Figure 3. (We have changed the usual position of the axes to get a better look at S.) For S_1 we use θ and z as parameters (see Example 5 in Section 16.6) and write its parametric equations as

$$x = \cos\theta \qquad y = \sin\theta \qquad z = z$$

where

$$0 \leqslant \theta \leqslant 2\pi \qquad \text{and} \qquad 0 \leqslant z \leqslant 1 + x = 1 + \cos\theta$$

Therefore

$$\mathbf{r}_\theta \times \mathbf{r}_z = \begin{vmatrix} \mathbf{i} & \mathbf{j} & \mathbf{k} \\ -\sin\theta & \cos\theta & 0 \\ 0 & 0 & 1 \end{vmatrix} = \cos\theta\,\mathbf{i} + \sin\theta\,\mathbf{j}$$

and

$$|\mathbf{r}_\theta \times \mathbf{r}_z| = \sqrt{\cos^2\theta + \sin^2\theta} = 1$$

Thus the surface integral over S_1 is

$$\begin{aligned}\iint_{S_1} z\,dS &= \iint_D z\,|\mathbf{r}_\theta \times \mathbf{r}_z|\,dA \\ &= \int_0^{2\pi}\int_0^{1+\cos\theta} z\,dz\,d\theta = \int_0^{2\pi} \tfrac{1}{2}(1+\cos\theta)^2\,d\theta \\ &= \tfrac{1}{2}\int_0^{2\pi}\left[1 + 2\cos\theta + \tfrac{1}{2}(1+\cos 2\theta)\right]d\theta \\ &= \tfrac{1}{2}\left[\tfrac{3}{2}\theta + 2\sin\theta + \tfrac{1}{4}\sin 2\theta\right]_0^{2\pi} = \frac{3\pi}{2}\end{aligned}$$

Since S_2 lies in the plane $z = 0$, we have

$$\iint_{S_2} z\,dS = \iint_{S_2} 0\,dS = 0$$

The top surface S_3 lies above the unit disk D and is part of the plane $z = 1 + x$. So, taking $g(x, y) = 1 + x$ in Formula 4 and converting to polar coordinates, we have

$$\begin{aligned}\iint_{S_3} z\,dS &= \iint_D (1+x)\sqrt{1 + \left(\frac{\partial z}{\partial x}\right)^2 + \left(\frac{\partial z}{\partial y}\right)^2}\,dA \\ &= \int_0^{2\pi}\int_0^1 (1 + r\cos\theta)\sqrt{1+1+0}\;r\,dr\,d\theta \\ &= \sqrt{2}\int_0^{2\pi}\int_0^1 (r + r^2\cos\theta)\,dr\,d\theta \\ &= \sqrt{2}\int_0^{2\pi}\left(\tfrac{1}{2} + \tfrac{1}{3}\cos\theta\right)d\theta \\ &= \sqrt{2}\left[\frac{\theta}{2} + \frac{\sin\theta}{3}\right]_0^{2\pi} = \sqrt{2}\,\pi\end{aligned}$$

Therefore

$$\begin{aligned}\iint_S z\,dS &= \iint_{S_1} z\,dS + \iint_{S_2} z\,dS + \iint_{S_3} z\,dS \\ &= \frac{3\pi}{2} + 0 + \sqrt{2}\,\pi = \left(\tfrac{3}{2} + \sqrt{2}\right)\pi\end{aligned}$$

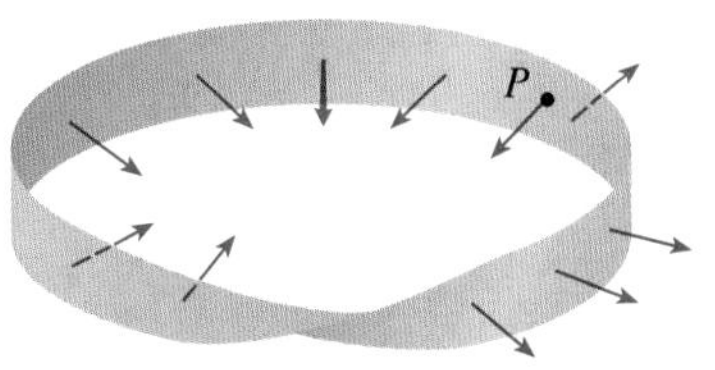

FIGURE 4
A Möbius strip

Oriented Surfaces

To define surface integrals of vector fields, we need to rule out nonorientable surfaces such as the Möbius strip shown in Figure 4. [It is named after the German geometer August Möbius (1790–1868).] You can construct one for yourself by taking a long rectangular strip of paper, giving it a half-twist, and taping the short edges together as in Figure 5. If an ant were to crawl along the Möbius strip starting at a point P, it would end up on the "other side" of the strip (that is, with its upper side pointing in the opposite direction). Then, if the ant continued to crawl in the same direction, it would end up back at the same point P without ever having crossed an edge. (If you have constructed a Möbius strip, try drawing a pencil line down the middle.) Therefore a Möbius strip really has only one side. You can graph the Möbius strip using the parametric equations in Exercise 32 in Section 16.6.

TEC Visual 16.7 shows a Möbius strip with a normal vector that can be moved along the surface.

FIGURE 5
Constructing a Möbius strip

From now on we consider only orientable (two-sided) surfaces. We start with a surface S that has a tangent plane at every point (x, y, z) on S (except at any boundary point). There are two unit normal vectors $\mathbf{n}_1$ and $\mathbf{n}_2 = -\mathbf{n}_1$ at (x, y, z). (See Figure 6.)

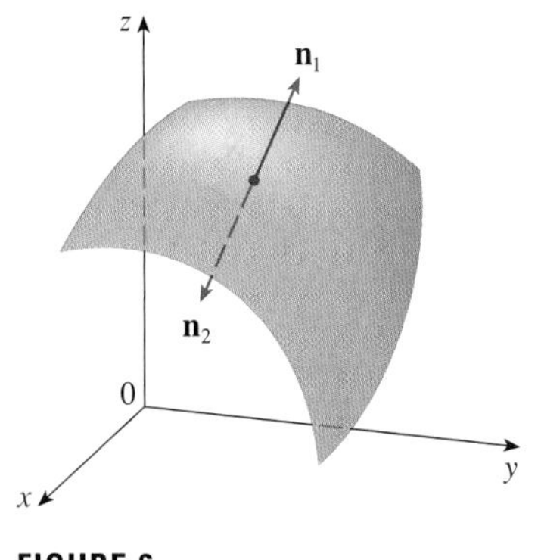

FIGURE 6

If it is possible to choose a unit normal vector $\mathbf{n}$ at every such point (x, y, z) so that $\mathbf{n}$ varies continuously over S, then S is called an **oriented surface** and the given choice of $\mathbf{n}$ provides S with an **orientation**. There are two possible orientations for any orientable surface (see Figure 7).

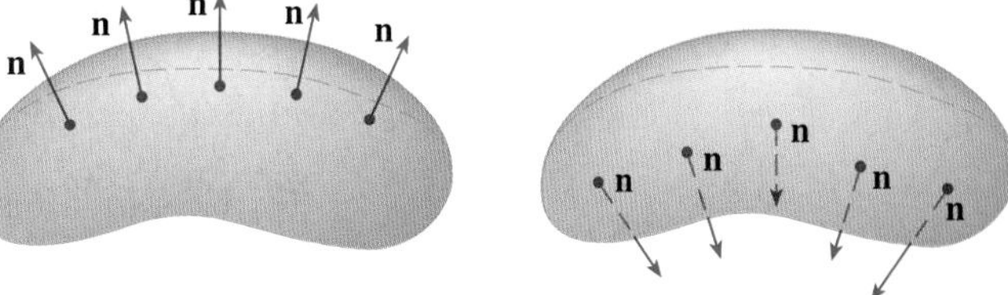

FIGURE 7
The two orientations of an orientable surface

For a surface $z = g(x, y)$ given as the graph of g, we use Equation 3 to associate with the surface a natural orientation given by the unit normal vector

$$\boxed{5} \qquad \mathbf{n} = \frac{-\dfrac{\partial g}{\partial x}\mathbf{i} - \dfrac{\partial g}{\partial y}\mathbf{j} + \mathbf{k}}{\sqrt{1 + \left(\dfrac{\partial g}{\partial x}\right)^2 + \left(\dfrac{\partial g}{\partial y}\right)^2}}$$

Since the $\mathbf{k}$-component is positive, this gives the *upward* orientation of the surface.

If S is a smooth orientable surface given in parametric form by a vector function $\mathbf{r}(u, v)$, then it is automatically supplied with the orientation of the unit normal vector

$$\boxed{6} \qquad \mathbf{n} = \frac{\mathbf{r}_u \times \mathbf{r}_v}{|\mathbf{r}_u \times \mathbf{r}_v|}$$

and the opposite orientation is given by $-\mathbf{n}$. For instance, in Example 4 in Section 16.6 we found the parametric representation

$$\mathbf{r}(\phi, \theta) = a \sin \phi \cos \theta \, \mathbf{i} + a \sin \phi \sin \theta \, \mathbf{j} + a \cos \phi \, \mathbf{k}$$

for the sphere $x^2 + y^2 + z^2 = a^2$. Then in Example 10 in Section 16.6 we found that

$$\mathbf{r}_\phi \times \mathbf{r}_\theta = a^2 \sin^2\phi \cos \theta \, \mathbf{i} + a^2 \sin^2\phi \sin \theta \, \mathbf{j} + a^2 \sin \phi \cos \phi \, \mathbf{k}$$

and

$$|\mathbf{r}_\phi \times \mathbf{r}_\theta| = a^2 \sin \phi$$

So the orientation induced by $\mathbf{r}(\phi, \theta)$ is defined by the unit normal vector

$$\mathbf{n} = \frac{\mathbf{r}_\phi \times \mathbf{r}_\theta}{|\mathbf{r}_\phi \times \mathbf{r}_\theta|} = \sin \phi \cos \theta \, \mathbf{i} + \sin \phi \sin \theta \, \mathbf{j} + \cos \phi \, \mathbf{k} = \frac{1}{a} \mathbf{r}(\phi, \theta)$$

Observe that $\mathbf{n}$ points in the same direction as the position vector, that is, outward from the sphere (see Figure 8). The opposite (inward) orientation would have been obtained (see Figure 9) if we had reversed the order of the parameters because $\mathbf{r}_\theta \times \mathbf{r}_\phi = -\mathbf{r}_\phi \times \mathbf{r}_\theta$.

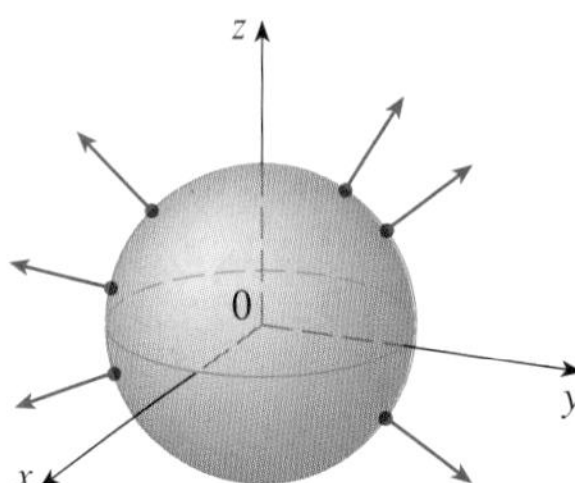

FIGURE 8
Positive orientation

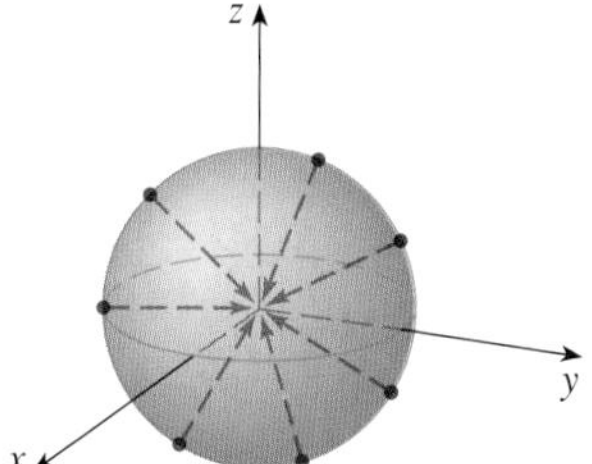

FIGURE 9
Negative orientation

For a **closed surface**, that is, a surface that is the boundary of a solid region E, the convention is that the **positive orientation** is the one for which the normal vectors point *outward* from E, and inward-pointing normals give the negative orientation (see Figures 8 and 9).

Surface Integrals of Vector Fields

Suppose that S is an oriented surface with unit normal vector $\mathbf{n}$, and imagine a fluid with density $\rho(x, y, z)$ and velocity field $\mathbf{v}(x, y, z)$ flowing through S. (Think of S as an imaginary surface that doesn't impede the fluid flow, like a fishing net across a stream.) Then the rate of flow (mass per unit time) per unit area is $\rho\mathbf{v}$. If we divide S into small patches S_{ij}, as in Figure 10 (compare with Figure 1), then S_{ij} is nearly planar and so we can approximate the mass of fluid per unit time crossing S_{ij} in the direction of the normal $\mathbf{n}$ by the quantity

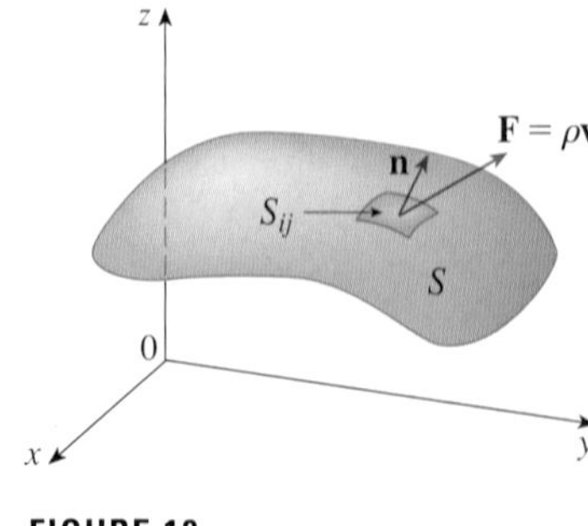

FIGURE 10

$$(\rho\mathbf{v} \cdot \mathbf{n})A(S_{ij})$$

where ρ, $\mathbf{v}$, and $\mathbf{n}$ are evaluated at some point on S_{ij}. (Recall that the component of the vector $\rho\mathbf{v}$ in the direction of the unit vector $\mathbf{n}$ is $\rho\mathbf{v} \cdot \mathbf{n}$.) By summing these quantities and taking the limit we get, according to Definition 1, the surface integral of the function $\rho\mathbf{v} \cdot \mathbf{n}$ over S:

$$\boxed{7} \qquad \iint_S \rho\mathbf{v} \cdot \mathbf{n} \, dS = \iint_S \rho(x, y, z)\mathbf{v}(x, y, z) \cdot \mathbf{n}(x, y, z) \, dS$$

and this is interpreted physically as the rate of flow through S.

If we write $\mathbf{F} = \rho\mathbf{v}$, then $\mathbf{F}$ is also a vector field on $\mathbb{R}^3$ and the integral in Equation 7 becomes

$$\iint_S \mathbf{F} \cdot \mathbf{n}\, dS$$

A surface integral of this form occurs frequently in physics, even when $\mathbf{F}$ is not $\rho\mathbf{v}$, and is called the *surface integral* (or *flux integral*) of $\mathbf{F}$ over S.

8 Definition If $\mathbf{F}$ is a continuous vector field defined on an oriented surface S with unit normal vector $\mathbf{n}$, then the **surface integral of F over *S*** is

$$\iint_S \mathbf{F} \cdot d\mathbf{S} = \iint_S \mathbf{F} \cdot \mathbf{n}\, dS$$

This integral is also called the **flux** of $\mathbf{F}$ across S.

In words, Definition 8 says that the surface integral of a vector field over S is equal to the surface integral of its normal component over S (as previously defined).

If S is given by a vector function $\mathbf{r}(u, v)$, then $\mathbf{n}$ is given by Equation 6, and from Definition 8 and Equation 2 we have

$$\iint_S \mathbf{F} \cdot d\mathbf{S} = \iint_S \mathbf{F} \cdot \frac{\mathbf{r}_u \times \mathbf{r}_v}{|\mathbf{r}_u \times \mathbf{r}_v|}\, dS$$

$$= \iint_D \left[\mathbf{F}(\mathbf{r}(u, v)) \cdot \frac{\mathbf{r}_u \times \mathbf{r}_v}{|\mathbf{r}_u \times \mathbf{r}_v|} \right] |\mathbf{r}_u \times \mathbf{r}_v|\, dA$$

where D is the parameter domain. Thus we have

Compare Equation 9 to the similar expression for evaluating line integrals of vector fields in Definition 16.2.13:

$$\int_C \mathbf{F} \cdot d\mathbf{r} = \int_a^b \mathbf{F}(\mathbf{r}(t)) \cdot \mathbf{r}'(t)\, dt$$

9 $$\iint_S \mathbf{F} \cdot d\mathbf{S} = \iint_D \mathbf{F} \cdot (\mathbf{r}_u \times \mathbf{r}_v)\, dA$$

EXAMPLE 4 Find the flux of the vector field $\mathbf{F}(x, y, z) = z\,\mathbf{i} + y\,\mathbf{j} + x\,\mathbf{k}$ across the unit sphere $x^2 + y^2 + z^2 = 1$.

Figure 11 shows the vector field $\mathbf{F}$ in Example 4 at points on the unit sphere.

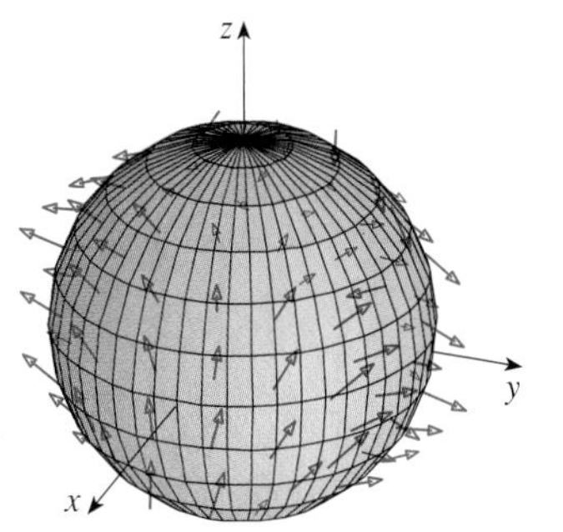

FIGURE 11

SOLUTION As in Example 1, we use the parametric representation

$$\mathbf{r}(\phi, \theta) = \sin\phi\cos\theta\,\mathbf{i} + \sin\phi\sin\theta\,\mathbf{j} + \cos\phi\,\mathbf{k} \qquad 0 \leqslant \phi \leqslant \pi \qquad 0 \leqslant \theta \leqslant 2\pi$$

Then $$\mathbf{F}(\mathbf{r}(\phi, \theta)) = \cos\phi\,\mathbf{i} + \sin\phi\sin\theta\,\mathbf{j} + \sin\phi\cos\theta\,\mathbf{k}$$

and, from Example 10 in Section 16.6,

$$\mathbf{r}_\phi \times \mathbf{r}_\theta = \sin^2\phi\cos\theta\,\mathbf{i} + \sin^2\phi\sin\theta\,\mathbf{j} + \sin\phi\cos\phi\,\mathbf{k}$$

Therefore

$$\mathbf{F}(\mathbf{r}(\phi, \theta)) \cdot (\mathbf{r}_\phi \times \mathbf{r}_\theta) = \cos\phi\sin^2\phi\cos\theta + \sin^3\phi\sin^2\theta + \sin^2\phi\cos\phi\cos\theta$$

and, by Formula 9, the flux is

$$\begin{aligned}\iint_S \mathbf{F} \cdot d\mathbf{S} &= \iint_D \mathbf{F} \cdot (\mathbf{r}_\phi \times \mathbf{r}_\theta)\, dA \\ &= \int_0^{2\pi} \int_0^{\pi} (2 \sin^2\phi \cos\phi \cos\theta + \sin^3\phi \sin^2\theta)\, d\phi\, d\theta \\ &= 2 \int_0^{\pi} \sin^2\phi \cos\phi\, d\phi \int_0^{2\pi} \cos\theta\, d\theta + \int_0^{\pi} \sin^3\phi\, d\phi \int_0^{2\pi} \sin^2\theta\, d\theta \\ &= 0 + \int_0^{\pi} \sin^3\phi\, d\phi \int_0^{2\pi} \sin^2\theta\, d\theta \qquad \left(\text{since } \int_0^{2\pi} \cos\theta\, d\theta = 0\right) \\ &= \frac{4\pi}{3}\end{aligned}$$

by the same calculation as in Example 1.

If, for instance, the vector field in Example 4 is a velocity field describing the flow of a fluid with density 1, then the answer, $4\pi/3$, represents the rate of flow through the unit sphere in units of mass per unit time.

In the case of a surface S given by a graph $z = g(x, y)$, we can think of x and y as parameters and use Equation 3 to write

$$\mathbf{F} \cdot (\mathbf{r}_x \times \mathbf{r}_y) = (P\,\mathbf{i} + Q\,\mathbf{j} + R\,\mathbf{k}) \cdot \left(-\frac{\partial g}{\partial x}\,\mathbf{i} - \frac{\partial g}{\partial y}\,\mathbf{j} + \mathbf{k}\right)$$

Thus Formula 9 becomes

10
$$\iint_S \mathbf{F} \cdot d\mathbf{S} = \iint_D \left(-P\frac{\partial g}{\partial x} - Q\frac{\partial g}{\partial y} + R\right) dA$$

This formula assumes the upward orientation of S; for a downward orientation we multiply by -1. Similar formulas can be worked out if S is given by $y = h(x, z)$ or $x = k(y, z)$. (See Exercises 37 and 38.)

V EXAMPLE 5 Evaluate $\iint_S \mathbf{F} \cdot d\mathbf{S}$, where $\mathbf{F}(x, y, z) = y\,\mathbf{i} + x\,\mathbf{j} + z\,\mathbf{k}$ and S is the boundary of the solid region E enclosed by the paraboloid $z = 1 - x^2 - y^2$ and the plane $z = 0$.

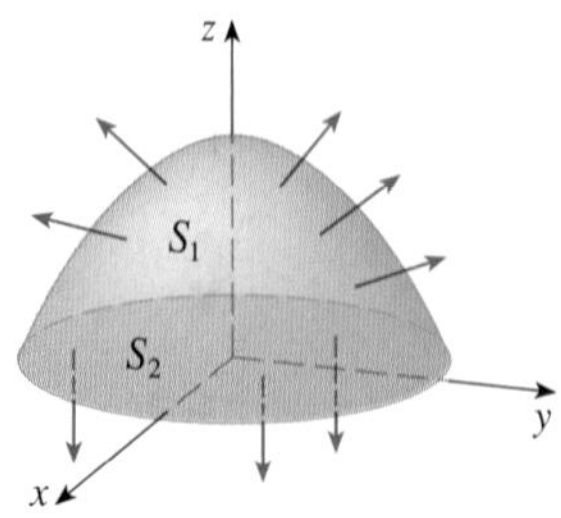

FIGURE 12

SOLUTION S consists of a parabolic top surface S_1 and a circular bottom surface S_2. (See Figure 12.) Since S is a closed surface, we use the convention of positive (outward) orientation. This means that S_1 is oriented upward and we can use Equation 10 with D being the projection of S_1 onto the xy-plane, namely, the disk $x^2 + y^2 \leqslant 1$. Since

$$P(x, y, z) = y \qquad Q(x, y, z) = x \qquad R(x, y, z) = z = 1 - x^2 - y^2$$

on S_1 and

$$\frac{\partial g}{\partial x} = -2x \qquad \frac{\partial g}{\partial y} = -2y$$

we have

$$\begin{aligned}
\iint_{S_1} \mathbf{F} \cdot d\mathbf{S} &= \iint_D \left(-P\frac{\partial g}{\partial x} - Q\frac{\partial g}{\partial y} + R\right) dA \\
&= \iint_D [-y(-2x) - x(-2y) + 1 - x^2 - y^2]\, dA \\
&= \iint_D (1 + 4xy - x^2 - y^2)\, dA \\
&= \int_0^{2\pi} \int_0^1 (1 + 4r^2\cos\theta \sin\theta - r^2)\, r\, dr\, d\theta \\
&= \int_0^{2\pi} \int_0^1 (r - r^3 + 4r^3 \cos\theta \sin\theta)\, dr\, d\theta \\
&= \int_0^{2\pi} \left(\tfrac{1}{4} + \cos\theta \sin\theta\right) d\theta = \tfrac{1}{4}(2\pi) + 0 = \frac{\pi}{2}
\end{aligned}$$

The disk S_2 is oriented downward, so its unit normal vector is $\mathbf{n} = -\mathbf{k}$ and we have

$$\iint_{S_2} \mathbf{F} \cdot d\mathbf{S} = \iint_{S_2} \mathbf{F} \cdot (-\mathbf{k})\, dS = \iint_D (-z)\, dA = \iint_D 0\, dA = 0$$

since $z = 0$ on S_2. Finally, we compute, by definition, $\iint_S \mathbf{F} \cdot d\mathbf{S}$ as the sum of the surface integrals of $\mathbf{F}$ over the pieces S_1 and S_2:

$$\iint_S \mathbf{F} \cdot d\mathbf{S} = \iint_{S_1} \mathbf{F} \cdot d\mathbf{S} + \iint_{S_2} \mathbf{F} \cdot d\mathbf{S} = \frac{\pi}{2} + 0 = \frac{\pi}{2}$$

Although we motivated the surface integral of a vector field using the example of fluid flow, this concept also arises in other physical situations. For instance, if $\mathbf{E}$ is an electric field (see Example 5 in Section 16.1), then the surface integral

$$\iint_S \mathbf{E} \cdot d\mathbf{S}$$

is called the **electric flux** of $\mathbf{E}$ through the surface S. One of the important laws of electrostatics is **Gauss's Law**, which says that the net charge enclosed by a closed surface S is

11 $$Q = \varepsilon_0 \iint_S \mathbf{E} \cdot d\mathbf{S}$$

where ε_0 is a constant (called the permittivity of free space) that depends on the units used. (In the SI system, $\varepsilon_0 \approx 8.8542 \times 10^{-12}\ \text{C}^2/\text{N}\cdot\text{m}^2$.) Therefore, if the vector field $\mathbf{F}$ in Example 4 represents an electric field, we can conclude that the charge enclosed by S is $Q = \frac{4}{3}\pi\varepsilon_0$.

Another application of surface integrals occurs in the study of heat flow. Suppose the temperature at a point (x, y, z) in a body is $u(x, y, z)$. Then the **heat flow** is defined as the vector field

$$\mathbf{F} = -K\,\nabla u$$

where K is an experimentally determined constant called the **conductivity** of the substance. The rate of heat flow across the surface S in the body is then given by the surface integral

$$\iint_S \mathbf{F} \cdot d\mathbf{S} = -K \iint_S \nabla u \cdot d\mathbf{S}$$

V EXAMPLE 6 The temperature u in a metal ball is proportional to the square of the distance from the center of the ball. Find the rate of heat flow across a sphere S of radius a with center at the center of the ball.

SOLUTION Taking the center of the ball to be at the origin, we have

$$u(x, y, z) = C(x^2 + y^2 + z^2)$$

where C is the proportionality constant. Then the heat flow is

$$\mathbf{F}(x, y, z) = -K \nabla u = -KC(2x\,\mathbf{i} + 2y\,\mathbf{j} + 2z\,\mathbf{k})$$

where K is the conductivity of the metal. Instead of using the usual parametrization of the sphere as in Example 4, we observe that the outward unit normal to the sphere $x^2 + y^2 + z^2 = a^2$ at the point (x, y, z) is

$$\mathbf{n} = \frac{1}{a}(x\,\mathbf{i} + y\,\mathbf{j} + z\,\mathbf{k})$$

and so
$$\mathbf{F} \cdot \mathbf{n} = -\frac{2KC}{a}(x^2 + y^2 + z^2)$$

But on S we have $x^2 + y^2 + z^2 = a^2$, so $\mathbf{F} \cdot \mathbf{n} = -2aKC$. Therefore the rate of heat flow across S is

$$\iint_S \mathbf{F} \cdot d\mathbf{S} = \iint_S \mathbf{F} \cdot \mathbf{n}\, dS = -2aKC \iint_S dS$$
$$= -2aKCA(S) = -2aKC(4\pi a^2) = -8KC\pi a^3$$

16.7 Exercises

1. Let S be the boundary surface of the box enclosed by the planes $x = 0$, $x = 2$, $y = 0$, $y = 4$, $z = 0$, and $z = 6$. Approximate $\iint_S e^{-0.1(x+y+z)}\, dS$ by using a Riemann sum as in Definition 1, taking the patches S_{ij} to be the rectangles that are the faces of the box S and the points P_{ij}^* to be the centers of the rectangles.

2. A surface S consists of the cylinder $x^2 + y^2 = 1$, $-1 \leqslant z \leqslant 1$, together with its top and bottom disks. Suppose you know that f is a continuous function with

   $f(\pm 1, 0, 0) = 2 \qquad f(0, \pm 1, 0) = 3 \qquad f(0, 0, \pm 1) = 4$

   Estimate the value of $\iint_S f(x, y, z)\, dS$ by using a Riemann sum, taking the patches S_{ij} to be four quarter-cylinders and the top and bottom disks.

3. Let H be the hemisphere $x^2 + y^2 + z^2 = 50$, $z \geqslant 0$, and suppose f is a continuous function with $f(3, 4, 5) = 7$, $f(3, -4, 5) = 8$, $f(-3, 4, 5) = 9$, and $f(-3, -4, 5) = 12$. By dividing H into four patches, estimate the value of $\iint_H f(x, y, z)\, dS$.

4. Suppose that $f(x, y, z) = g\left(\sqrt{x^2 + y^2 + z^2}\right)$, where g is a function of one variable such that $g(2) = -5$. Evaluate $\iint_S f(x, y, z)\, dS$, where S is the sphere $x^2 + y^2 + z^2 = 4$.

5–20 Evaluate the surface integral.

5. $\iint_S (x + y + z)\, dS$,
   S is the parallelogram with parametric equations $x = u + v$, $y = u - v$, $z = 1 + 2u + v$, $0 \leqslant u \leqslant 2$, $0 \leqslant v \leqslant 1$

6. $\iint_S xyz\,dS$,
S is the cone with parametric equations $x = u\cos v$, $y = u\sin v$, $z = u$, $0 \leqslant u \leqslant 1$, $0 \leqslant v \leqslant \pi/2$

7. $\iint_S y\,dS$, S is the helicoid with vector equation $\mathbf{r}(u, v) = \langle u\cos v, u\sin v, v\rangle$, $0 \leqslant u \leqslant 1$, $0 \leqslant v \leqslant \pi$

8. $\iint_S (x^2 + y^2)\,dS$,
S is the surface with vector equation $\mathbf{r}(u, v) = \langle 2uv, u^2 - v^2, u^2 + v^2\rangle$, $u^2 + v^2 \leqslant 1$

9. $\iint_S x^2yz\,dS$,
S is the part of the plane $z = 1 + 2x + 3y$ that lies above the rectangle $[0, 3] \times [0, 2]$

10. $\iint_S xz\,dS$,
S is the part of the plane $2x + 2y + z = 4$ that lies in the first octant

11. $\iint_S x\,dS$,
S is the triangular region with vertices $(1, 0, 0)$, $(0, -2, 0)$, and $(0, 0, 4)$

12. $\iint_S y\,dS$,
S is the surface $z = \frac{2}{3}(x^{3/2} + y^{3/2})$, $0 \leqslant x \leqslant 1$, $0 \leqslant y \leqslant 1$

13. $\iint_S x^2z^2\,dS$,
S is the part of the cone $z^2 = x^2 + y^2$ that lies between the planes $z = 1$ and $z = 3$

14. $\iint_S z\,dS$,
S is the surface $x = y + 2z^2$, $0 \leqslant y \leqslant 1$, $0 \leqslant z \leqslant 1$

15. $\iint_S y\,dS$,
S is the part of the paraboloid $y = x^2 + z^2$ that lies inside the cylinder $x^2 + z^2 = 4$

16. $\iint_S y^2\,dS$,
S is the part of the sphere $x^2 + y^2 + z^2 = 4$ that lies inside the cylinder $x^2 + y^2 = 1$ and above the xy-plane

17. $\iint_S (x^2z + y^2z)\,dS$,
S is the hemisphere $x^2 + y^2 + z^2 = 4$, $z \geqslant 0$

18. $\iint_S xz\,dS$,
S is the boundary of the region enclosed by the cylinder $y^2 + z^2 = 9$ and the planes $x = 0$ and $x + y = 5$

19. $\iint_S (z + x^2y)\,dS$,
S is the part of the cylinder $y^2 + z^2 = 1$ that lies between the planes $x = 0$ and $x = 3$ in the first octant

20. $\iint_S (x^2 + y^2 + z^2)\,dS$,
S is the part of the cylinder $x^2 + y^2 = 9$ between the planes $z = 0$ and $z = 2$, together with its top and bottom disks

21–32 Evaluate the surface integral $\iint_S \mathbf{F} \cdot d\mathbf{S}$ for the given vector field $\mathbf{F}$ and the oriented surface S. In other words, find the flux of $\mathbf{F}$ across S. For closed surfaces, use the positive (outward) orientation.

21. $\mathbf{F}(x, y, z) = ze^{xy}\,\mathbf{i} - 3ze^{xy}\,\mathbf{j} + xy\,\mathbf{k}$,
S is the parallelogram of Exercise 5 with upward orientation

22. $\mathbf{F}(x, y, z) = z\,\mathbf{i} + y\,\mathbf{j} + x\,\mathbf{k}$,
S is the helicoid of Exercise 7 with upward orientation

23. $\mathbf{F}(x, y, z) = xy\,\mathbf{i} + yz\,\mathbf{j} + zx\,\mathbf{k}$, S is the part of the paraboloid $z = 4 - x^2 - y^2$ that lies above the square $0 \leqslant x \leqslant 1$, $0 \leqslant y \leqslant 1$, and has upward orientation

24. $\mathbf{F}(x, y, z) = -x\,\mathbf{i} - y\,\mathbf{j} + z^3\,\mathbf{k}$,
S is the part of the cone $z = \sqrt{x^2 + y^2}$ between the planes $z = 1$ and $z = 3$ with downward orientation

25. $\mathbf{F}(x, y, z) = x\,\mathbf{i} - z\,\mathbf{j} + y\,\mathbf{k}$,
S is the part of the sphere $x^2 + y^2 + z^2 = 4$ in the first octant, with orientation toward the origin

26. $\mathbf{F}(x, y, z) = xz\,\mathbf{i} + x\,\mathbf{j} + y\,\mathbf{k}$,
S is the hemisphere $x^2 + y^2 + z^2 = 25$, $y \geqslant 0$, oriented in the direction of the positive y-axis

27. $\mathbf{F}(x, y, z) = y\,\mathbf{j} - z\,\mathbf{k}$,
S consists of the paraboloid $y = x^2 + z^2$, $0 \leqslant y \leqslant 1$, and the disk $x^2 + z^2 \leqslant 1$, $y = 1$

28. $\mathbf{F}(x, y, z) = xy\,\mathbf{i} + 4x^2\,\mathbf{j} + yz\,\mathbf{k}$, S is the surface $z = xe^y$, $0 \leqslant x \leqslant 1$, $0 \leqslant y \leqslant 1$, with upward orientation

29. $\mathbf{F}(x, y, z) = x\,\mathbf{i} + 2y\,\mathbf{j} + 3z\,\mathbf{k}$,
S is the cube with vertices $(\pm 1, \pm 1, \pm 1)$

30. $\mathbf{F}(x, y, z) = x\,\mathbf{i} + y\,\mathbf{j} + 5\,\mathbf{k}$, S is the boundary of the region enclosed by the cylinder $x^2 + z^2 = 1$ and the planes $y = 0$ and $x + y = 2$

31. $\mathbf{F}(x, y, z) = x^2\,\mathbf{i} + y^2\mathbf{j} + z^2\,\mathbf{k}$, S is the boundary of the solid half-cylinder $0 \leqslant z \leqslant \sqrt{1 - y^2}$, $0 \leqslant x \leqslant 2$

32. $\mathbf{F}(x, y, z) = y\,\mathbf{i} + (z - y)\,\mathbf{j} + x\,\mathbf{k}$,
S is the surface of the tetrahedron with vertices $(0, 0, 0)$, $(1, 0, 0)$, $(0, 1, 0)$, and $(0, 0, 1)$

CAS **33.** Evaluate $\iint_S (x^2 + y^2 + z^2)\,dS$ correct to four decimal places, where S is the surface $z = xe^y$, $0 \leqslant x \leqslant 1$, $0 \leqslant y \leqslant 1$.

CAS **34.** Find the exact value of $\iint_S x^2yz\,dS$, where S is the surface $z = xy$, $0 \leqslant x \leqslant 1$, $0 \leqslant y \leqslant 1$.

CAS **35.** Find the value of $\iint_S x^2y^2z^2\,dS$ correct to four decimal places, where S is the part of the paraboloid $z = 3 - 2x^2 - y^2$ that lies above the xy-plane.

CAS **36.** Find the flux of

$$\mathbf{F}(x, y, z) = \sin(xyz)\,\mathbf{i} + x^2y\,\mathbf{j} + z^2e^{x/5}\,\mathbf{k}$$

across the part of the cylinder $4y^2 + z^2 = 4$ that lies above the xy-plane and between the planes $x = -2$ and $x = 2$ with upward orientation. Illustrate by using a computer algebra system to draw the cylinder and the vector field on the same screen.

37. Find a formula for $\iint_S \mathbf{F} \cdot d\mathbf{S}$ similar to Formula 10 for the case where S is given by $y = h(x, z)$ and $\mathbf{n}$ is the unit normal that points toward the left.

38. Find a formula for $\iint_S \mathbf{F} \cdot d\mathbf{S}$ similar to Formula 10 for the case where S is given by $x = k(y, z)$ and $\mathbf{n}$ is the unit normal that points forward (that is, toward the viewer when the axes are drawn in the usual way).

39. Find the center of mass of the hemisphere $x^2 + y^2 + z^2 = a^2$, $z \geqslant 0$, if it has constant density.

40. Find the mass of a thin funnel in the shape of a cone $z = \sqrt{x^2 + y^2}$, $1 \leqslant z \leqslant 4$, if its density function is $\rho(x, y, z) = 10 - z$.

41. (a) Give an integral expression for the moment of inertia I_z about the z-axis of a thin sheet in the shape of a surface S if the density function is ρ.
(b) Find the moment of inertia about the z-axis of the funnel in Exercise 40.

42. Let S be the part of the sphere $x^2 + y^2 + z^2 = 25$ that lies above the plane $z = 4$. If S has constant density k, find (a) the center of mass and (b) the moment of inertia about the z-axis.

43. A fluid has density 870 kg/m^3 and flows with velocity $\mathbf{v} = z\,\mathbf{i} + y^2\,\mathbf{j} + x^2\,\mathbf{k}$, where x, y, and z are measured in meters and the components of $\mathbf{v}$ in meters per second. Find the rate of flow outward through the cylinder $x^2 + y^2 = 4$, $0 \leqslant z \leqslant 1$.

44. Seawater has density 1025 kg/m^3 and flows in a velocity field $\mathbf{v} = y\,\mathbf{i} + x\,\mathbf{j}$, where x, y, and z are measured in meters and the components of $\mathbf{v}$ in meters per second. Find the rate of flow outward through the hemisphere $x^2 + y^2 + z^2 = 9$, $z \geqslant 0$.

45. Use Gauss's Law to find the charge contained in the solid hemisphere $x^2 + y^2 + z^2 \leqslant a^2$, $z \geqslant 0$, if the electric field is

$$\mathbf{E}(x, y, z) = x\,\mathbf{i} + y\,\mathbf{j} + 2z\,\mathbf{k}$$

46. Use Gauss's Law to find the charge enclosed by the cube with vertices $(\pm 1, \pm 1, \pm 1)$ if the electric field is

$$\mathbf{E}(x, y, z) = x\,\mathbf{i} + y\,\mathbf{j} + z\,\mathbf{k}$$

47. The temperature at the point (x, y, z) in a substance with conductivity $K = 6.5$ is $u(x, y, z) = 2y^2 + 2z^2$. Find the rate of heat flow inward across the cylindrical surface $y^2 + z^2 = 6$, $0 \leqslant x \leqslant 4$.

48. The temperature at a point in a ball with conductivity K is inversely proportional to the distance from the center of the ball. Find the rate of heat flow across a sphere S of radius a with center at the center of the ball.

49. Let $\mathbf{F}$ be an inverse square field, that is, $\mathbf{F}(\mathbf{r}) = c\mathbf{r}/|\mathbf{r}|^3$ for some constant c, where $\mathbf{r} = x\,\mathbf{i} + y\,\mathbf{j} + z\,\mathbf{k}$. Show that the flux of $\mathbf{F}$ across a sphere S with center the origin is independent of the radius of S.

16.8 Stokes' Theorem

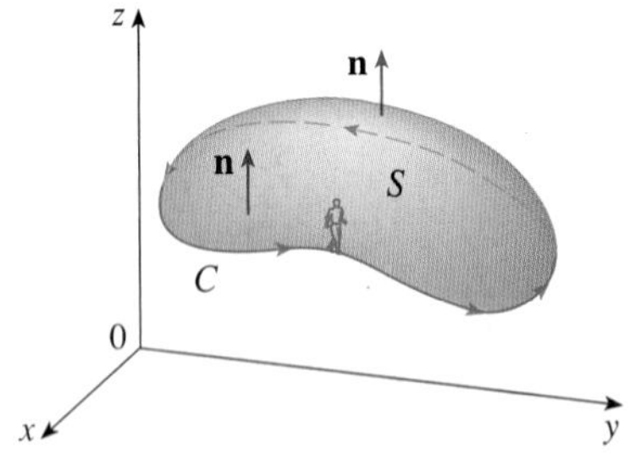

FIGURE 1

Stokes' Theorem can be regarded as a higher-dimensional version of Green's Theorem. Whereas Green's Theorem relates a double integral over a plane region D to a line integral around its plane boundary curve, Stokes' Theorem relates a surface integral over a surface S to a line integral around the boundary curve of S (which is a space curve). Figure 1 shows an oriented surface with unit normal vector $\mathbf{n}$. The orientation of S induces the **positive orientation of the boundary curve C** shown in the figure. This means that if you walk in the positive direction around C with your head pointing in the direction of $\mathbf{n}$, then the surface will always be on your left.

> **Stokes' Theorem** Let S be an oriented piecewise-smooth surface that is bounded by a simple, closed, piecewise-smooth boundary curve C with positive orientation. Let $\mathbf{F}$ be a vector field whose components have continuous partial derivatives on an open region in $\mathbb{R}^3$ that contains S. Then
>
> $$\int_C \mathbf{F} \cdot d\mathbf{r} = \iint_S \operatorname{curl} \mathbf{F} \cdot d\mathbf{S}$$

Since

$$\int_C \mathbf{F} \cdot d\mathbf{r} = \int_C \mathbf{F} \cdot \mathbf{T}\, ds \qquad \text{and} \qquad \iint_S \operatorname{curl} \mathbf{F} \cdot d\mathbf{S} = \iint_S \operatorname{curl} \mathbf{F} \cdot \mathbf{n}\, dS$$

George Stokes

Stokes' Theorem is named after the Irish mathematical physicist Sir George Stokes (1819–1903). Stokes was a professor at Cambridge University (in fact he held the same position as Newton, Lucasian Professor of Mathematics) and was especially noted for his studies of fluid flow and light. What we call Stokes' Theorem was actually discovered by the Scottish physicist Sir William Thomson (1824–1907, known as Lord Kelvin). Stokes learned of this theorem in a letter from Thomson in 1850 and asked students to prove it on an examination at Cambridge University in 1854. We don't know if any of those students was able to do so.

Stokes' Theorem says that the line integral around the boundary curve of S of the tangential component of $\mathbf{F}$ is equal to the surface integral over S of the normal component of the curl of $\mathbf{F}$.

The positively oriented boundary curve of the oriented surface S is often written as ∂S, so Stokes' Theorem can be expressed as

$$\boxed{1} \qquad \iint_S \operatorname{curl} \mathbf{F} \cdot d\mathbf{S} = \int_{\partial S} \mathbf{F} \cdot d\mathbf{r}$$

There is an analogy among Stokes' Theorem, Green's Theorem, and the Fundamental Theorem of Calculus. As before, there is an integral involving derivatives on the left side of Equation 1 (recall that curl $\mathbf{F}$ is a sort of derivative of $\mathbf{F}$) and the right side involves the values of $\mathbf{F}$ only on the *boundary* of S.

In fact, in the special case where the surface S is flat and lies in the xy-plane with upward orientation, the unit normal is $\mathbf{k}$, the surface integral becomes a double integral, and Stokes' Theorem becomes

$$\int_C \mathbf{F} \cdot d\mathbf{r} = \iint_S \operatorname{curl} \mathbf{F} \cdot d\mathbf{S} = \iint_S (\operatorname{curl} \mathbf{F}) \cdot \mathbf{k}\, dA$$

This is precisely the vector form of Green's Theorem given in Equation 16.5.12. Thus we see that Green's Theorem is really a special case of Stokes' Theorem.

Although Stokes' Theorem is too difficult for us to prove in its full generality, we can give a proof when S is a graph and $\mathbf{F}$, S, and C are well behaved.

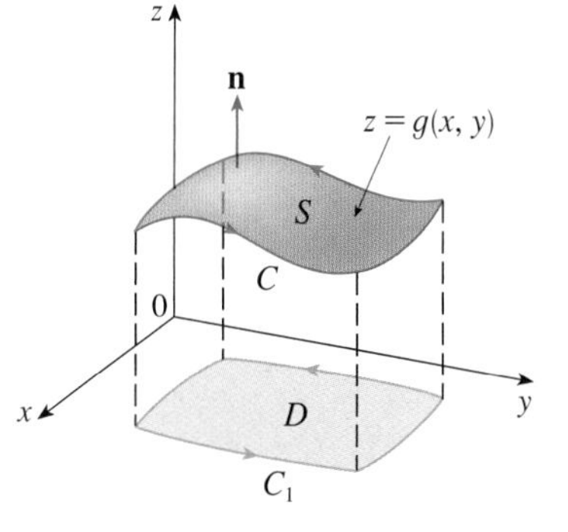

FIGURE 2

PROOF OF A SPECIAL CASE OF STOKES' THEOREM We assume that the equation of S is $z = g(x, y)$, $(x, y) \in D$, where g has continuous second-order partial derivatives and D is a simple plane region whose boundary curve C_1 corresponds to C. If the orientation of S is upward, then the positive orientation of C corresponds to the positive orientation of C_1. (See Figure 2.) We are also given that $\mathbf{F} = P\,\mathbf{i} + Q\,\mathbf{j} + R\,\mathbf{k}$, where the partial derivatives of P, Q, and R are continuous.

Since S is a graph of a function, we can apply Formula 16.7.10 with $\mathbf{F}$ replaced by curl $\mathbf{F}$. The result is

$$\boxed{2} \qquad \iint_S \operatorname{curl} \mathbf{F} \cdot d\mathbf{S}$$

$$= \iint_D \left[-\left(\frac{\partial R}{\partial y} - \frac{\partial Q}{\partial z} \right) \frac{\partial z}{\partial x} - \left(\frac{\partial P}{\partial z} - \frac{\partial R}{\partial x} \right) \frac{\partial z}{\partial y} + \left(\frac{\partial Q}{\partial x} - \frac{\partial P}{\partial y} \right) \right] dA$$

where the partial derivatives of P, Q, and R are evaluated at $(x, y, g(x, y))$. If

$$x = x(t) \qquad y = y(t) \qquad a \leqslant t \leqslant b$$

is a parametric representation of C_1, then a parametric representation of C is

$$x = x(t) \qquad y = y(t) \qquad z = g\big(x(t), y(t)\big) \qquad a \leqslant t \leqslant b$$

This allows us, with the aid of the Chain Rule, to evaluate the line integral as follows:

$$\begin{aligned}
\int_C \mathbf{F} \cdot d\mathbf{r} &= \int_a^b \left(P \frac{dx}{dt} + Q \frac{dy}{dt} + R \frac{dz}{dt} \right) dt \\
&= \int_a^b \left[P \frac{dx}{dt} + Q \frac{dy}{dt} + R \left(\frac{\partial z}{\partial x} \frac{dx}{dt} + \frac{\partial z}{\partial y} \frac{dy}{dt} \right) \right] dt \\
&= \int_a^b \left[\left(P + R \frac{\partial z}{\partial x} \right) \frac{dx}{dt} + \left(Q + R \frac{\partial z}{\partial y} \right) \frac{dy}{dt} \right] dt \\
&= \int_{C_1} \left(P + R \frac{\partial z}{\partial x} \right) dx + \left(Q + R \frac{\partial z}{\partial y} \right) dy \\
&= \iint_D \left[\frac{\partial}{\partial x} \left(Q + R \frac{\partial z}{\partial y} \right) - \frac{\partial}{\partial y} \left(P + R \frac{\partial z}{\partial x} \right) \right] dA
\end{aligned}$$

where we have used Green's Theorem in the last step. Then, using the Chain Rule again and remembering that P, Q, and R are functions of x, y, and z and that z is itself a function of x and y, we get

$$\begin{aligned}
\int_C \mathbf{F} \cdot d\mathbf{r} = \iint_D \Bigg[&\left(\frac{\partial Q}{\partial x} + \frac{\partial Q}{\partial z} \frac{\partial z}{\partial x} + \frac{\partial R}{\partial x} \frac{\partial z}{\partial y} + \frac{\partial R}{\partial z} \frac{\partial z}{\partial x} \frac{\partial z}{\partial y} + R \frac{\partial^2 z}{\partial x \, \partial y} \right) \\
&- \left(\frac{\partial P}{\partial y} + \frac{\partial P}{\partial z} \frac{\partial z}{\partial y} + \frac{\partial R}{\partial y} \frac{\partial z}{\partial x} + \frac{\partial R}{\partial z} \frac{\partial z}{\partial y} \frac{\partial z}{\partial x} + R \frac{\partial^2 z}{\partial y \, \partial x} \right) \Bigg] dA
\end{aligned}$$

Four of the terms in this double integral cancel and the remaining six terms can be arranged to coincide with the right side of Equation 2. Therefore

$$\int_C \mathbf{F} \cdot d\mathbf{r} = \iint_S \operatorname{curl} \mathbf{F} \cdot d\mathbf{S}$$

V **EXAMPLE 1** Evaluate $\int_C \mathbf{F} \cdot d\mathbf{r}$, where $\mathbf{F}(x, y, z) = -y^2 \mathbf{i} + x \mathbf{j} + z^2 \mathbf{k}$ and C is the curve of intersection of the plane $y + z = 2$ and the cylinder $x^2 + y^2 = 1$. (Orient C to be counterclockwise when viewed from above.)

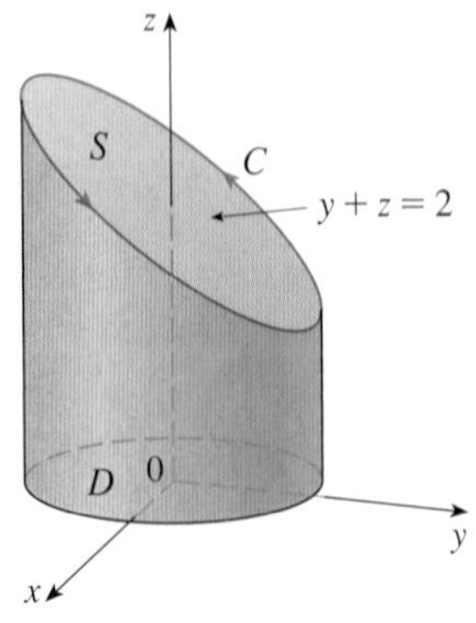

FIGURE 3

SOLUTION The curve C (an ellipse) is shown in Figure 3. Although $\int_C \mathbf{F} \cdot d\mathbf{r}$ could be evaluated directly, it's easier to use Stokes' Theorem. We first compute

$$\operatorname{curl} \mathbf{F} = \begin{vmatrix} \mathbf{i} & \mathbf{j} & \mathbf{k} \\ \dfrac{\partial}{\partial x} & \dfrac{\partial}{\partial y} & \dfrac{\partial}{\partial z} \\ -y^2 & x & z^2 \end{vmatrix} = (1 + 2y)\,\mathbf{k}$$

Although there are many surfaces with boundary C, the most convenient choice is the elliptical region S in the plane $y + z = 2$ that is bounded by C. If we orient S upward, then C has the induced positive orientation. The projection D of S onto the xy-plane is the disk $x^2 + y^2 \leqslant 1$ and so using Equation 16.7.10 with $z = g(x, y) = 2 - y$, we have

$$\begin{aligned}\int_C \mathbf{F} \cdot d\mathbf{r} &= \iint_S \operatorname{curl} \mathbf{F} \cdot d\mathbf{S} = \iint_D (1 + 2y)\, dA \\ &= \int_0^{2\pi} \int_0^1 (1 + 2r \sin \theta)\, r\, dr\, d\theta \\ &= \int_0^{2\pi} \left[\frac{r^2}{2} + 2\frac{r^3}{3} \sin \theta \right]_0^1 d\theta = \int_0^{2\pi} \left(\tfrac{1}{2} + \tfrac{2}{3} \sin \theta\right) d\theta \\ &= \tfrac{1}{2}(2\pi) + 0 = \pi\end{aligned}$$

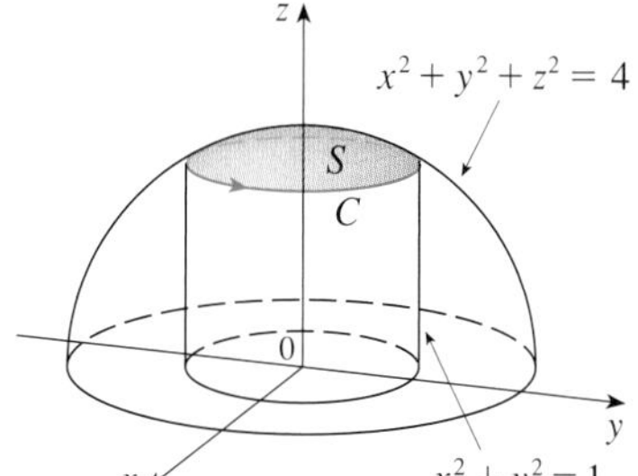

FIGURE 4

V EXAMPLE 2 Use Stokes' Theorem to compute the integral $\iint_S \operatorname{curl} \mathbf{F} \cdot d\mathbf{S}$, where $\mathbf{F}(x, y, z) = xz\,\mathbf{i} + yz\,\mathbf{j} + xy\,\mathbf{k}$ and S is the part of the sphere $x^2 + y^2 + z^2 = 4$ that lies inside the cylinder $x^2 + y^2 = 1$ and above the xy-plane. (See Figure 4.)

SOLUTION To find the boundary curve C we solve the equations $x^2 + y^2 + z^2 = 4$ and $x^2 + y^2 = 1$. Subtracting, we get $z^2 = 3$ and so $z = \sqrt{3}$ (since $z > 0$). Thus C is the circle given by the equations $x^2 + y^2 = 1$, $z = \sqrt{3}$. A vector equation of C is

$$\mathbf{r}(t) = \cos t\,\mathbf{i} + \sin t\,\mathbf{j} + \sqrt{3}\,\mathbf{k} \qquad 0 \leqslant t \leqslant 2\pi$$

so

$$\mathbf{r}'(t) = -\sin t\,\mathbf{i} + \cos t\,\mathbf{j}$$

Also, we have

$$\mathbf{F}(\mathbf{r}(t)) = \sqrt{3} \cos t\,\mathbf{i} + \sqrt{3} \sin t\,\mathbf{j} + \cos t \sin t\,\mathbf{k}$$

Therefore, by Stokes' Theorem,

$$\begin{aligned}\iint_S \operatorname{curl} \mathbf{F} \cdot d\mathbf{S} &= \int_C \mathbf{F} \cdot d\mathbf{r} = \int_0^{2\pi} \mathbf{F}(\mathbf{r}(t)) \cdot \mathbf{r}'(t)\, dt \\ &= \int_0^{2\pi} \left(-\sqrt{3} \cos t \sin t + \sqrt{3} \sin t \cos t\right) dt \\ &= \sqrt{3} \int_0^{2\pi} 0\, dt = 0\end{aligned}$$

Note that in Example 2 we computed a surface integral simply by knowing the values of $\mathbf{F}$ on the boundary curve C. This means that if we have another oriented surface with the same boundary curve C, then we get exactly the same value for the surface integral!

In general, if S_1 and S_2 are oriented surfaces with the same oriented boundary curve C and both satisfy the hypotheses of Stokes' Theorem, then

$$\boxed{3} \qquad \iint_{S_1} \operatorname{curl} \mathbf{F} \cdot d\mathbf{S} = \int_C \mathbf{F} \cdot d\mathbf{r} = \iint_{S_2} \operatorname{curl} \mathbf{F} \cdot d\mathbf{S}$$

This fact is useful when it is difficult to integrate over one surface but easy to integrate over the other.

We now use Stokes' Theorem to throw some light on the meaning of the curl vector. Suppose that C is an oriented closed curve and $\mathbf{v}$ represents the velocity field in fluid flow. Consider the line integral

$$\int_C \mathbf{v} \cdot d\mathbf{r} = \int_C \mathbf{v} \cdot \mathbf{T}\, ds$$

and recall that $\mathbf{v} \cdot \mathbf{T}$ is the component of $\mathbf{v}$ in the direction of the unit tangent vector $\mathbf{T}$. This means that the closer the direction of $\mathbf{v}$ is to the direction of $\mathbf{T}$, the larger the value of $\mathbf{v} \cdot \mathbf{T}$. Thus $\int_C \mathbf{v} \cdot d\mathbf{r}$ is a measure of the tendency of the fluid to move around C and is called the **circulation** of $\mathbf{v}$ around C. (See Figure 5.)

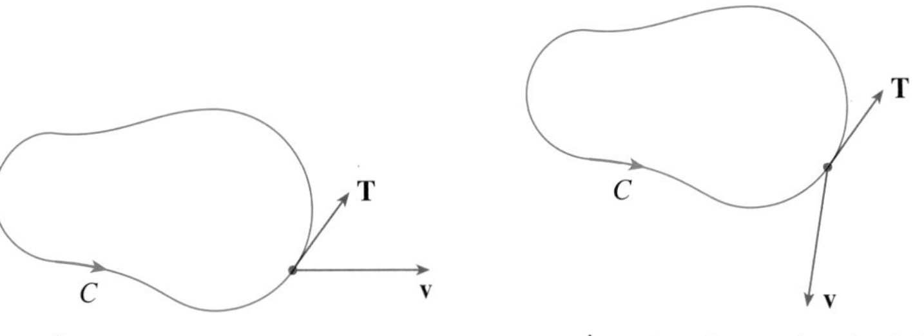

FIGURE 5 (a) $\int_C \mathbf{v} \cdot d\mathbf{r} > 0$, positive circulation (b) $\int_C \mathbf{v} \cdot d\mathbf{r} < 0$, negative circulation

Now let $P_0(x_0, y_0, z_0)$ be a point in the fluid and let S_a be a small disk with radius a and center P_0. Then $(\text{curl } \mathbf{F})(P) \approx (\text{curl } \mathbf{F})(P_0)$ for all points P on S_a because curl $\mathbf{F}$ is continuous. Thus, by Stokes' Theorem, we get the following approximation to the circulation around the boundary circle C_a:

$$\int_{C_a} \mathbf{v} \cdot d\mathbf{r} = \iint_{S_a} \text{curl } \mathbf{v} \cdot d\mathbf{S} = \iint_{S_a} \text{curl } \mathbf{v} \cdot \mathbf{n} \, dS$$

$$\approx \iint_{S_a} \text{curl } \mathbf{v}(P_0) \cdot \mathbf{n}(P_0) \, dS = \text{curl } \mathbf{v}(P_0) \cdot \mathbf{n}(P_0) \pi a^2$$

Imagine a tiny paddle wheel placed in the fluid at a point P, as in Figure 6; the paddle wheel rotates fastest when its axis is parallel to curl $\mathbf{v}$.

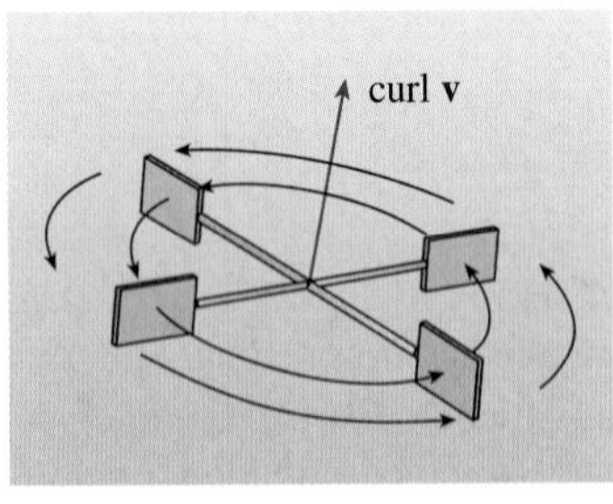

FIGURE 6

This approximation becomes better as $a \to 0$ and we have

$$\boxed{4} \qquad \text{curl } \mathbf{v}(P_0) \cdot \mathbf{n}(P_0) = \lim_{a \to 0} \frac{1}{\pi a^2} \int_{C_a} \mathbf{v} \cdot d\mathbf{r}$$

Equation 4 gives the relationship between the curl and the circulation. It shows that curl $\mathbf{v} \cdot \mathbf{n}$ is a measure of the rotating effect of the fluid about the axis $\mathbf{n}$. The curling effect is greatest about the axis parallel to curl $\mathbf{v}$.

Finally, we mention that Stokes' Theorem can be used to prove Theorem 16.5.4 (which states that if curl $\mathbf{F} = \mathbf{0}$ on all of $\mathbb{R}^3$, then $\mathbf{F}$ is conservative). From our previous work (Theorems 16.3.3 and 16.3.4), we know that $\mathbf{F}$ is conservative if $\int_C \mathbf{F} \cdot d\mathbf{r} = 0$ for every closed path C. Given C, suppose we can find an orientable surface S whose boundary is C. (This can be done, but the proof requires advanced techniques.) Then Stokes' Theorem gives

$$\int_C \mathbf{F} \cdot d\mathbf{r} = \iint_S \text{curl } \mathbf{F} \cdot d\mathbf{S} = \iint_S \mathbf{0} \cdot d\mathbf{S} = 0$$

A curve that is not simple can be broken into a number of simple curves, and the integrals around these simple curves are all 0. Adding these integrals, we obtain $\int_C \mathbf{F} \cdot d\mathbf{r} = 0$ for any closed curve C.

16.8 Exercises

1. A hemisphere H and a portion P of a paraboloid are shown. Suppose $\mathbf{F}$ is a vector field on $\mathbb{R}^3$ whose components have continuous partial derivatives. Explain why

$$\iint_H \operatorname{curl} \mathbf{F} \cdot d\mathbf{S} = \iint_P \operatorname{curl} \mathbf{F} \cdot d\mathbf{S}$$

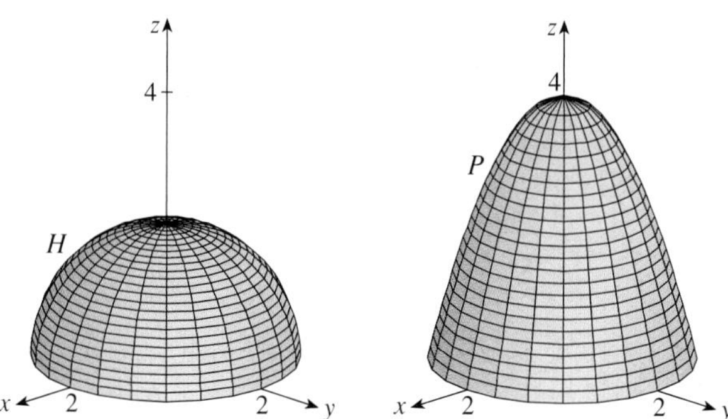

2–6 Use Stokes' Theorem to evaluate $\iint_S \operatorname{curl} \mathbf{F} \cdot d\mathbf{S}$.

2. $\mathbf{F}(x, y, z) = 2y \cos z\, \mathbf{i} + e^x \sin z\, \mathbf{j} + xe^y\, \mathbf{k}$, S is the hemisphere $x^2 + y^2 + z^2 = 9$, $z \geqslant 0$, oriented upward

3. $\mathbf{F}(x, y, z) = x^2z^2\, \mathbf{i} + y^2z^2\, \mathbf{j} + xyz\, \mathbf{k}$, S is the part of the paraboloid $z = x^2 + y^2$ that lies inside the cylinder $x^2 + y^2 = 4$, oriented upward

4. $\mathbf{F}(x, y, z) = \tan^{-1}(x^2yz^2)\, \mathbf{i} + x^2y\, \mathbf{j} + x^2z^2\, \mathbf{k}$, S is the cone $x = \sqrt{y^2 + z^2}$, $0 \leqslant x \leqslant 2$, oriented in the direction of the positive x-axis

5. $\mathbf{F}(x, y, z) = xyz\, \mathbf{i} + xy\, \mathbf{j} + x^2yz\, \mathbf{k}$, S consists of the top and the four sides (but not the bottom) of the cube with vertices $(\pm1, \pm1, \pm1)$, oriented outward

6. $\mathbf{F}(x, y, z) = e^{xy}\, \mathbf{i} + e^{xz}\, \mathbf{j} + x^2z\, \mathbf{k}$, S is the half of the ellipsoid $4x^2 + y^2 + 4z^2 = 4$ that lies to the right of the xz-plane, oriented in the direction of the positive y-axis

7–10 Use Stokes' Theorem to evaluate $\int_C \mathbf{F} \cdot d\mathbf{r}$. In each case C is oriented counterclockwise as viewed from above.

7. $\mathbf{F}(x, y, z) = (x + y^2)\, \mathbf{i} + (y + z^2)\, \mathbf{j} + (z + x^2)\, \mathbf{k}$, C is the triangle with vertices $(1, 0, 0)$, $(0, 1, 0)$, and $(0, 0, 1)$

8. $\mathbf{F}(x, y, z) = \mathbf{i} + (x + yz)\, \mathbf{j} + \left(xy - \sqrt{z}\right) \mathbf{k}$, C is the boundary of the part of the plane $3x + 2y + z = 1$ in the first octant

9. $\mathbf{F}(x, y, z) = yz\, \mathbf{i} + 2xz\, \mathbf{j} + e^{xy}\, \mathbf{k}$, C is the circle $x^2 + y^2 = 16$, $z = 5$

10. $\mathbf{F}(x, y, z) = xy\, \mathbf{i} + 2z\, \mathbf{j} + 3y\, \mathbf{k}$, C is the curve of intersection of the plane $x + z = 5$ and the cylinder $x^2 + y^2 = 9$

11. (a) Use Stokes' Theorem to evaluate $\int_C \mathbf{F} \cdot d\mathbf{r}$, where

$$\mathbf{F}(x, y, z) = x^2z\, \mathbf{i} + xy^2\, \mathbf{j} + z^2\, \mathbf{k}$$

and C is the curve of intersection of the plane $x + y + z = 1$ and the cylinder $x^2 + y^2 = 9$ oriented counterclockwise as viewed from above.

(b) Graph both the plane and the cylinder with domains chosen so that you can see the curve C and the surface that you used in part (a).

(c) Find parametric equations for C and use them to graph C.

12. (a) Use Stokes' Theorem to evaluate $\int_C \mathbf{F} \cdot d\mathbf{r}$, where $\mathbf{F}(x, y, z) = x^2y\, \mathbf{i} + \frac{1}{3}x^3\, \mathbf{j} + xy\, \mathbf{k}$ and C is the curve of intersection of the hyperbolic paraboloid $z = y^2 - x^2$ and the cylinder $x^2 + y^2 = 1$ oriented counterclockwise as viewed from above.

(b) Graph both the hyperbolic paraboloid and the cylinder with domains chosen so that you can see the curve C and the surface that you used in part (a).

(c) Find parametric equations for C and use them to graph C.

13–15 Verify that Stokes' Theorem is true for the given vector field $\mathbf{F}$ and surface S.

13. $\mathbf{F}(x, y, z) = -y\, \mathbf{i} + x\, \mathbf{j} - 2\, \mathbf{k}$, S is the cone $z^2 = x^2 + y^2$, $0 \leqslant z \leqslant 4$, oriented downward

14. $\mathbf{F}(x, y, z) = -2yz\, \mathbf{i} + y\, \mathbf{j} + 3x\, \mathbf{k}$, S is the part of the paraboloid $z = 5 - x^2 - y^2$ that lies above the plane $z = 1$, oriented upward

15. $\mathbf{F}(x, y, z) = y\, \mathbf{i} + z\, \mathbf{j} + x\, \mathbf{k}$, S is the hemisphere $x^2 + y^2 + z^2 = 1$, $y \geqslant 0$, oriented in the direction of the positive y-axis

16. Let C be a simple closed smooth curve that lies in the plane $x + y + z = 1$. Show that the line integral

$$\int_C z\, dx - 2x\, dy + 3y\, dz$$

depends only on the area of the region enclosed by C and not on the shape of C or its location in the plane.

17. A particle moves along line segments from the origin to the points $(1, 0, 0)$, $(1, 2, 1)$, $(0, 2, 1)$, and back to the origin under the influence of the force field

$$\mathbf{F}(x, y, z) = z^2\, \mathbf{i} + 2xy\, \mathbf{j} + 4y^2\, \mathbf{k}$$

Find the work done.

18. Evaluate

$$\int_C (y + \sin x)\,dx + (z^2 + \cos y)\,dy + x^3\,dz$$

where C is the curve $\mathbf{r}(t) = \langle \sin t, \cos t, \sin 2t \rangle$, $0 \leqslant t \leqslant 2\pi$. [*Hint:* Observe that C lies on the surface $z = 2xy$.]

19. If S is a sphere and $\mathbf{F}$ satisfies the hypotheses of Stokes' Theorem, show that $\iint_S \operatorname{curl} \mathbf{F} \cdot d\mathbf{S} = 0$.

20. Suppose S and C satisfy the hypotheses of Stokes' Theorem and f, g have continuous second-order partial derivatives. Use Exercises 24 and 26 in Section 16.5 to show the following.

(a) $\int_C (f\nabla g) \cdot d\mathbf{r} = \iint_S (\nabla f \times \nabla g) \cdot d\mathbf{S}$

(b) $\int_C (f\nabla f) \cdot d\mathbf{r} = 0$

(c) $\int_C (f\nabla g + g\nabla f) \cdot d\mathbf{r} = 0$

WRITING PROJECT THREE MEN AND TWO THEOREMS

The photograph shows a stained-glass window at Cambridge University in honor of George Green.

Courtesy of the Masters and Fellows of Gonville and Caius College, Cambridge University, England

Although two of the most important theorems in vector calculus are named after George Green and George Stokes, a third man, William Thomson (also known as Lord Kelvin), played a large role in the formulation, dissemination, and application of both of these results. All three men were interested in how the two theorems could help to explain and predict physical phenomena in electricity and magnetism and fluid flow. The basic facts of the story are given in the margin notes on pages 1085 and 1123.

Write a report on the historical origins of Green's Theorem and Stokes' Theorem. Explain the similarities and relationship between the theorems. Discuss the roles that Green, Thomson, and Stokes played in discovering these theorems and making them widely known. Show how both theorems arose from the investigation of electricity and magnetism and were later used to study a variety of physical problems.

The dictionary edited by Gillispie [2] is a good source for both biographical and scientific information. The book by Hutchinson [5] gives an account of Stokes' life and the book by Thompson [8] is a biography of Lord Kelvin. The articles by Grattan-Guinness [3] and Gray [4] and the book by Cannell [1] give background on the extraordinary life and works of Green. Additional historical and mathematical information is found in the books by Katz [6] and Kline [7].

1. D. M. Cannell, *George Green, Mathematician and Physicist 1793–1841: The Background to His Life and Work* (Philadelphia: Society for Industrial and Applied Mathematics, 2001).
2. C. C. Gillispie, ed., *Dictionary of Scientific Biography* (New York: Scribner's, 1974). See the article on Green by P. J. Wallis in Volume XV and the articles on Thomson by Jed Buchwald and on Stokes by E. M. Parkinson in Volume XIII.
3. I. Grattan-Guinness, "Why did George Green write his essay of 1828 on electricity and magnetism?" *Amer. Math. Monthly*, Vol. 102 (1995), pp. 387–96.
4. J. Gray, "There was a jolly miller." *The New Scientist*, Vol. 139 (1993), pp. 24–27.
5. G. E. Hutchinson, *The Enchanted Voyage and Other Studies* (Westport, CT: Greenwood Press, 1978).
6. Victor Katz, *A History of Mathematics: An Introduction* (New York: HarperCollins, 1993), pp. 678–80.
7. Morris Kline, *Mathematical Thought from Ancient to Modern Times* (New York: Oxford University Press, 1972), pp. 683–85.
8. Sylvanus P. Thompson, *The Life of Lord Kelvin* (New York: Chelsea, 1976).

16.9 The Divergence Theorem

In Section 16.5 we rewrote Green's Theorem in a vector version as

$$\int_C \mathbf{F} \cdot \mathbf{n}\,ds = \iint_D \operatorname{div} \mathbf{F}(x, y)\,dA$$

where C is the positively oriented boundary curve of the plane region D. If we were seeking to extend this theorem to vector fields on $\mathbb{R}^3$, we might make the guess that

$$\boxed{1} \qquad \iint_S \mathbf{F} \cdot \mathbf{n}\, dS = \iiint_E \operatorname{div} \mathbf{F}(x, y, z)\, dV$$

where S is the boundary surface of the solid region E. It turns out that Equation 1 is true, under appropriate hypotheses, and is called the Divergence Theorem. Notice its similarity to Green's Theorem and Stokes' Theorem in that it relates the integral of a derivative of a function (div $\mathbf{F}$ in this case) over a region to the integral of the original function $\mathbf{F}$ over the boundary of the region.

At this stage you may wish to review the various types of regions over which we were able to evaluate triple integrals in Section 15.7. We state and prove the Divergence Theorem for regions E that are simultaneously of types 1, 2, and 3 and we call such regions **simple solid regions**. (For instance, regions bounded by ellipsoids or rectangular boxes are simple solid regions.) The boundary of E is a closed surface, and we use the convention, introduced in Section 16.7, that the positive orientation is outward; that is, the unit normal vector $\mathbf{n}$ is directed outward from E.

The Divergence Theorem is sometimes called Gauss's Theorem after the great German mathematician Karl Friedrich Gauss (1777–1855), who discovered this theorem during his investigation of electrostatics. In Eastern Europe the Divergence Theorem is known as Ostrogradsky's Theorem after the Russian mathematician Mikhail Ostrogradsky (1801–1862), who published this result in 1826.

The Divergence Theorem Let E be a simple solid region and let S be the boundary surface of E, given with positive (outward) orientation. Let $\mathbf{F}$ be a vector field whose component functions have continuous partial derivatives on an open region that contains E. Then

$$\iint_S \mathbf{F} \cdot d\mathbf{S} = \iiint_E \operatorname{div} \mathbf{F}\, dV$$

Thus the Divergence Theorem states that, under the given conditions, the flux of $\mathbf{F}$ across the boundary surface of E is equal to the triple integral of the divergence of $\mathbf{F}$ over E.

PROOF Let $\mathbf{F} = P\,\mathbf{i} + Q\,\mathbf{j} + R\,\mathbf{k}$. Then

$$\operatorname{div} \mathbf{F} = \frac{\partial P}{\partial x} + \frac{\partial Q}{\partial y} + \frac{\partial R}{\partial z}$$

so

$$\iiint_E \operatorname{div} \mathbf{F}\, dV = \iiint_E \frac{\partial P}{\partial x}\, dV + \iiint_E \frac{\partial Q}{\partial y}\, dV + \iiint_E \frac{\partial R}{\partial z}\, dV$$

If $\mathbf{n}$ is the unit outward normal of S, then the surface integral on the left side of the Divergence Theorem is

$$\begin{aligned} \iint_S \mathbf{F} \cdot d\mathbf{S} &= \iint_S \mathbf{F} \cdot \mathbf{n}\, dS = \iint_S (P\,\mathbf{i} + Q\,\mathbf{j} + R\,\mathbf{k}) \cdot \mathbf{n}\, dS \\ &= \iint_S P\,\mathbf{i} \cdot \mathbf{n}\, dS + \iint_S Q\,\mathbf{j} \cdot \mathbf{n}\, dS + \iint_S R\,\mathbf{k} \cdot \mathbf{n}\, dS \end{aligned}$$

Therefore, to prove the Divergence Theorem, it suffices to prove the following three equations:

$$\boxed{2} \qquad \iint_S P\,\mathbf{i}\cdot\mathbf{n}\,dS = \iiint_E \frac{\partial P}{\partial x}\,dV$$

$$\boxed{3} \qquad \iint_S Q\,\mathbf{j}\cdot\mathbf{n}\,dS = \iiint_E \frac{\partial Q}{\partial y}\,dV$$

$$\boxed{4} \qquad \iint_S R\,\mathbf{k}\cdot\mathbf{n}\,dS = \iiint_E \frac{\partial R}{\partial z}\,dV$$

To prove Equation 4 we use the fact that E is a type 1 region:

$$E = \{(x, y, z) \mid (x, y) \in D,\, u_1(x, y) \leqslant z \leqslant u_2(x, y)\}$$

where D is the projection of E onto the xy-plane. By Equation 15.7.6, we have

$$\iiint_E \frac{\partial R}{\partial z}\,dV = \iint_D \left[\int_{u_1(x,y)}^{u_2(x,y)} \frac{\partial R}{\partial z}(x, y, z)\,dz\right] dA$$

and therefore, by the Fundamental Theorem of Calculus,

$$\boxed{5} \qquad \iiint_E \frac{\partial R}{\partial z}\,dV = \iint_D \left[R(x, y, u_2(x, y)) - R(x, y, u_1(x, y))\right] dA$$

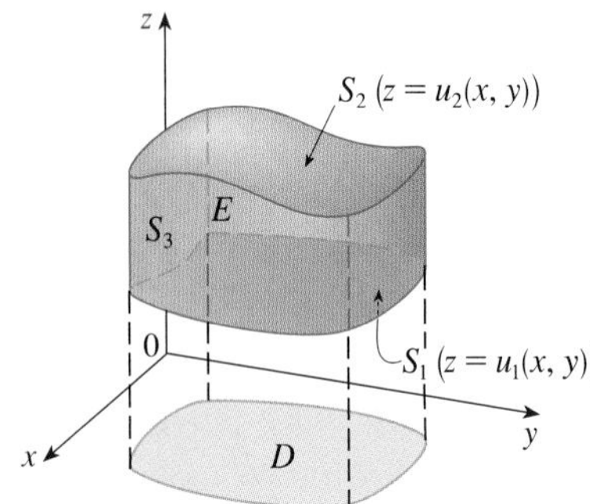
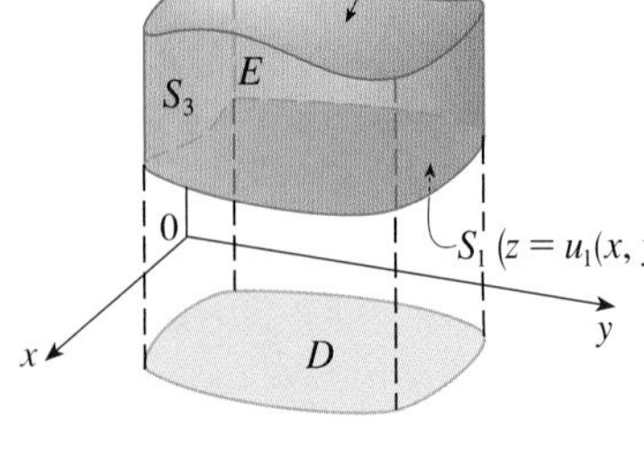

FIGURE 1

The boundary surface S consists of three pieces: the bottom surface S_1, the top surface S_2, and possibly a vertical surface S_3, which lies above the boundary curve of D. (See Figure 1. It might happen that S_3 doesn't appear, as in the case of a sphere.) Notice that on S_3 we have $\mathbf{k}\cdot\mathbf{n} = 0$, because $\mathbf{k}$ is vertical and $\mathbf{n}$ is horizontal, and so

$$\iint_{S_3} R\,\mathbf{k}\cdot\mathbf{n}\,dS = \iint_{S_3} 0\,dS = 0$$

Thus, regardless of whether there is a vertical surface, we can write

$$\boxed{6} \qquad \iint_S R\,\mathbf{k}\cdot\mathbf{n}\,dS = \iint_{S_1} R\,\mathbf{k}\cdot\mathbf{n}\,dS + \iint_{S_2} R\,\mathbf{k}\cdot\mathbf{n}\,dS$$

The equation of S_2 is $z = u_2(x, y)$, $(x, y) \in D$, and the outward normal $\mathbf{n}$ points upward, so from Equation 16.7.10 (with $\mathbf{F}$ replaced by $R\,\mathbf{k}$) we have

$$\iint_{S_2} R\,\mathbf{k}\cdot\mathbf{n}\,dS = \iint_D R(x, y, u_2(x, y))\,dA$$

On S_1 we have $z = u_1(x, y)$, but here the outward normal $\mathbf{n}$ points downward, so we multiply by -1:

$$\iint_{S_1} R\,\mathbf{k}\cdot\mathbf{n}\,dS = -\iint_D R(x, y, u_1(x, y))\,dA$$

Therefore Equation 6 gives

$$\iint_S R\,\mathbf{k}\cdot\mathbf{n}\,dS = \iint_D \left[R(x, y, u_2(x, y)) - R(x, y, u_1(x, y))\right] dA$$

Comparison with Equation 5 shows that

$$\iint_S R\,\mathbf{k}\cdot\mathbf{n}\,dS = \iiint_E \frac{\partial R}{\partial z}\,dV$$

Notice that the method of proof of the Divergence Theorem is very similar to that of Green's Theorem.

Equations 2 and 3 are proved in a similar manner using the expressions for E as a type 2 or type 3 region, respectively. ■

V EXAMPLE 1 Find the flux of the vector field $\mathbf{F}(x, y, z) = z\,\mathbf{i} + y\,\mathbf{j} + x\,\mathbf{k}$ over the unit sphere $x^2 + y^2 + z^2 = 1$.

SOLUTION First we compute the divergence of $\mathbf{F}$:

$$\text{div}\,\mathbf{F} = \frac{\partial}{\partial x}(z) + \frac{\partial}{\partial y}(y) + \frac{\partial}{\partial z}(x) = 1$$

The unit sphere S is the boundary of the unit ball B given by $x^2 + y^2 + z^2 \leqslant 1$. Thus the Divergence Theorem gives the flux as

The solution in Example 1 should be compared with the solution in Example 4 in Section 16.7.

$$\iint_S \mathbf{F}\cdot d\mathbf{S} = \iiint_B \text{div}\,\mathbf{F}\,dV = \iiint_B 1\,dV = V(B) = \tfrac{4}{3}\pi(1)^3 = \frac{4\pi}{3}$$ ■

V EXAMPLE 2 Evaluate $\iint_S \mathbf{F}\cdot d\mathbf{S}$, where

$$\mathbf{F}(x, y, z) = xy\,\mathbf{i} + \left(y^2 + e^{xz^2}\right)\mathbf{j} + \sin(xy)\,\mathbf{k}$$

and S is the surface of the region E bounded by the parabolic cylinder $z = 1 - x^2$ and the planes $z = 0$, $y = 0$, and $y + z = 2$. (See Figure 2.)

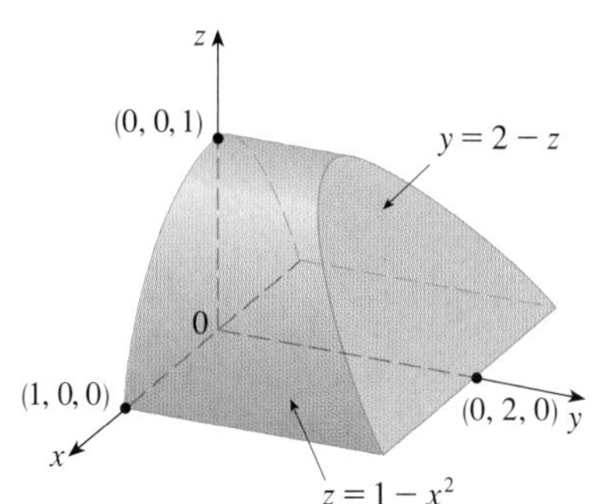

FIGURE 2

SOLUTION It would be extremely difficult to evaluate the given surface integral directly. (We would have to evaluate four surface integrals corresponding to the four pieces of S.) Furthermore, the divergence of $\mathbf{F}$ is much less complicated than $\mathbf{F}$ itself:

$$\text{div}\,\mathbf{F} = \frac{\partial}{\partial x}(xy) + \frac{\partial}{\partial y}\left(y^2 + e^{xz^2}\right) + \frac{\partial}{\partial z}(\sin xy) = y + 2y = 3y$$

Therefore we use the Divergence Theorem to transform the given surface integral into a triple integral. The easiest way to evaluate the triple integral is to express E as a type 3 region:

$$E = \left\{(x, y, z) \mid -1 \leqslant x \leqslant 1,\ 0 \leqslant z \leqslant 1 - x^2,\ 0 \leqslant y \leqslant 2 - z\right\}$$

Then we have

$$\iint_S \mathbf{F}\cdot d\mathbf{S} = \iiint_E \text{div}\,\mathbf{F}\,dV = \iiint_E 3y\,dV$$

$$= 3\int_{-1}^{1}\int_0^{1-x^2}\int_0^{2-z} y\,dy\,dz\,dx = 3\int_{-1}^{1}\int_0^{1-x^2} \frac{(2-z)^2}{2}\,dz\,dx$$

$$= \frac{3}{2}\int_{-1}^{1}\left[-\frac{(2-z)^3}{3}\right]_0^{1-x^2} dx = -\tfrac{1}{2}\int_{-1}^{1}\left[(x^2+1)^3 - 8\right]dx$$

$$= -\int_0^1 (x^6 + 3x^4 + 3x^2 - 7)\,dx = \frac{184}{35}$$ ■

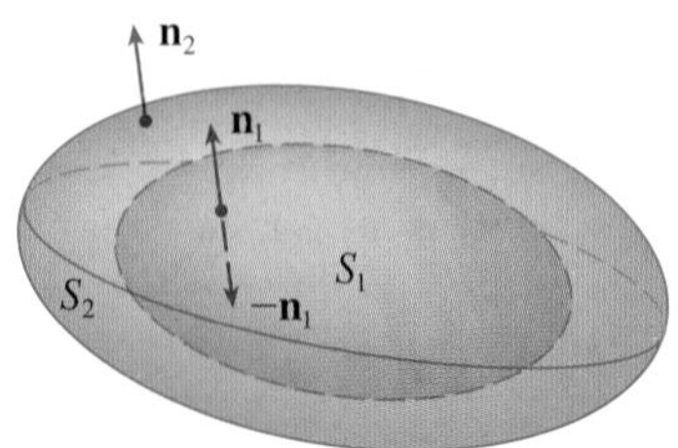

FIGURE 3

Although we have proved the Divergence Theorem only for simple solid regions, it can be proved for regions that are finite unions of simple solid regions. (The procedure is similar to the one we used in Section 16.4 to extend Green's Theorem.)

For example, let's consider the region E that lies between the closed surfaces S_1 and S_2, where S_1 lies inside S_2. Let $\mathbf{n}_1$ and $\mathbf{n}_2$ be outward normals of S_1 and S_2. Then the boundary surface of E is $S = S_1 \cup S_2$ and its normal $\mathbf{n}$ is given by $\mathbf{n} = -\mathbf{n}_1$ on S_1 and $\mathbf{n} = \mathbf{n}_2$ on S_2. (See Figure 3.) Applying the Divergence Theorem to S, we get

7
$$\begin{aligned}\iiint_E \operatorname{div} \mathbf{F}\, dV &= \iint_S \mathbf{F} \cdot d\mathbf{S} = \iint_S \mathbf{F} \cdot \mathbf{n}\, dS \\ &= \iint_{S_1} \mathbf{F} \cdot (-\mathbf{n}_1)\, dS + \iint_{S_2} \mathbf{F} \cdot \mathbf{n}_2\, dS \\ &= -\iint_{S_1} \mathbf{F} \cdot d\mathbf{S} + \iint_{S_2} \mathbf{F} \cdot d\mathbf{S}\end{aligned}$$

EXAMPLE 3 In Example 5 in Section 16.1 we considered the electric field

$$\mathbf{E}(\mathbf{x}) = \frac{\varepsilon Q}{|\mathbf{x}|^3}\mathbf{x}$$

where the electric charge Q is located at the origin and $\mathbf{x} = \langle x, y, z \rangle$ is a position vector. Use the Divergence Theorem to show that the electric flux of $\mathbf{E}$ through any closed surface S_2 that encloses the origin is

$$\iint_{S_2} \mathbf{E} \cdot d\mathbf{S} = 4\pi\varepsilon Q$$

SOLUTION The difficulty is that we don't have an explicit equation for S_2 because it is *any* closed surface enclosing the origin. The simplest such surface would be a sphere, so we let S_1 be a small sphere with radius a and center the origin. You can verify that $\operatorname{div} \mathbf{E} = 0$. (See Exercise 23.) Therefore Equation 7 gives

$$\iint_{S_2} \mathbf{E} \cdot d\mathbf{S} = \iint_{S_1} \mathbf{E} \cdot d\mathbf{S} + \iiint_E \operatorname{div} \mathbf{E}\, dV = \iint_{S_1} \mathbf{E} \cdot d\mathbf{S} = \iint_{S_1} \mathbf{E} \cdot \mathbf{n}\, dS$$

The point of this calculation is that we can compute the surface integral over S_1 because S_1 is a sphere. The normal vector at $\mathbf{x}$ is $\mathbf{x}/|\mathbf{x}|$. Therefore

$$\mathbf{E} \cdot \mathbf{n} = \frac{\varepsilon Q}{|\mathbf{x}|^3}\mathbf{x} \cdot \left(\frac{\mathbf{x}}{|\mathbf{x}|}\right) = \frac{\varepsilon Q}{|\mathbf{x}|^4}\mathbf{x} \cdot \mathbf{x} = \frac{\varepsilon Q}{|\mathbf{x}|^2} = \frac{\varepsilon Q}{a^2}$$

since the equation of S_1 is $|\mathbf{x}| = a$. Thus we have

$$\iint_{S_2} \mathbf{E} \cdot d\mathbf{S} = \iint_{S_1} \mathbf{E} \cdot \mathbf{n}\, dS = \frac{\varepsilon Q}{a^2}\iint_{S_1} dS = \frac{\varepsilon Q}{a^2} A(S_1) = \frac{\varepsilon Q}{a^2} 4\pi a^2 = 4\pi\varepsilon Q$$

This shows that the electric flux of $\mathbf{E}$ is $4\pi\varepsilon Q$ through *any* closed surface S_2 that contains the origin. [This is a special case of Gauss's Law (Equation 16.7.11) for a single charge. The relationship between ε and ε_0 is $\varepsilon = 1/(4\pi\varepsilon_0)$.]

Another application of the Divergence Theorem occurs in fluid flow. Let $\mathbf{v}(x, y, z)$ be the velocity field of a fluid with constant density ρ. Then $\mathbf{F} = \rho\mathbf{v}$ is the rate of flow per unit area. If $P_0(x_0, y_0, z_0)$ is a point in the fluid and B_a is a ball with center P_0 and very small radius a, then $\text{div}\,\mathbf{F}(P) \approx \text{div}\,\mathbf{F}(P_0)$ for all points in B_a since div $\mathbf{F}$ is continuous. We approximate the flux over the boundary sphere S_a as follows:

$$\iint_{S_a} \mathbf{F} \cdot d\mathbf{S} = \iiint_{B_a} \text{div}\,\mathbf{F}\, dV \approx \iiint_{B_a} \text{div}\,\mathbf{F}(P_0)\, dV = \text{div}\,\mathbf{F}(P_0)V(B_a)$$

This approximation becomes better as $a \to 0$ and suggests that

$$\boxed{8} \qquad \text{div}\,\mathbf{F}(P_0) = \lim_{a\to 0} \frac{1}{V(B_a)} \iint_{S_a} \mathbf{F} \cdot d\mathbf{S}$$

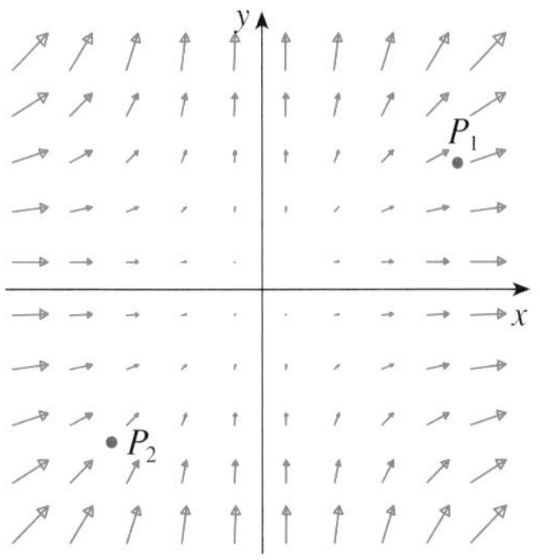

Equation 8 says that div $\mathbf{F}(P_0)$ is the net rate of outward flux per unit volume at P_0. (This is the reason for the name *divergence*.) If $\text{div}\,\mathbf{F}(P) > 0$, the net flow is outward near P and P is called a **source**. If $\text{div}\,\mathbf{F}(P) < 0$, the net flow is inward near P and P is called a **sink**.

For the vector field in Figure 4, it appears that the vectors that end near P_1 are shorter than the vectors that start near P_1. Thus the net flow is outward near P_1, so $\text{div}\,\mathbf{F}(P_1) > 0$ and P_1 is a source. Near P_2, on the other hand, the incoming arrows are longer than the outgoing arrows. Here the net flow is inward, so $\text{div}\,\mathbf{F}(P_2) < 0$ and P_2 is a sink. We can use the formula for $\mathbf{F}$ to confirm this impression. Since $\mathbf{F} = x^2\,\mathbf{i} + y^2\,\mathbf{j}$, we have $\text{div}\,\mathbf{F} = 2x + 2y$, which is positive when $y > -x$. So the points above the line $y = -x$ are sources and those below are sinks.

FIGURE 4
The vector field $\mathbf{F} = x^2\,\mathbf{i} + y^2\,\mathbf{j}$

16.9 Exercises

1–4 Verify that the Divergence Theorem is true for the vector field $\mathbf{F}$ on the region E.

1. $\mathbf{F}(x, y, z) = 3x\,\mathbf{i} + xy\,\mathbf{j} + 2xz\,\mathbf{k}$,
E is the cube bounded by the planes $x = 0$, $x = 1$, $y = 0$, $y = 1$, $z = 0$, and $z = 1$

2. $\mathbf{F}(x, y, z) = x^2\,\mathbf{i} + xy\,\mathbf{j} + z\,\mathbf{k}$,
E is the solid bounded by the paraboloid $z = 4 - x^2 - y^2$ and the xy-plane

3. $\mathbf{F}(x, y, z) = \langle z, y, x\rangle$,
E is the solid ball $x^2 + y^2 + z^2 \leqslant 16$

4. $\mathbf{F}(x, y, z) = \langle x^2, -y, z\rangle$,
E is the solid cylinder $y^2 + z^2 \leqslant 9$, $0 \leqslant x \leqslant 2$

5–15 Use the Divergence Theorem to calculate the surface integral $\iint_S \mathbf{F} \cdot d\mathbf{S}$; that is, calculate the flux of $\mathbf{F}$ across S.

5. $\mathbf{F}(x, y, z) = xye^z\,\mathbf{i} + xy^2z^3\,\mathbf{j} - ye^z\,\mathbf{k}$,
S is the surface of the box bounded by the coordinate planes and the planes $x = 3$, $y = 2$, and $z = 1$

6. $\mathbf{F}(x, y, z) = x^2yz\,\mathbf{i} + xy^2z\,\mathbf{j} + xyz^2\,\mathbf{k}$,
S is the surface of the box enclosed by the planes $x = 0$, $x = a$, $y = 0$, $y = b$, $z = 0$, and $z = c$, where a, b, and c are positive numbers

7. $\mathbf{F}(x, y, z) = 3xy^2\,\mathbf{i} + xe^z\,\mathbf{j} + z^3\,\mathbf{k}$,
S is the surface of the solid bounded by the cylinder $y^2 + z^2 = 1$ and the planes $x = -1$ and $x = 2$

8. $\mathbf{F}(x, y, z) = (x^3 + y^3)\,\mathbf{i} + (y^3 + z^3)\,\mathbf{j} + (z^3 + x^3)\,\mathbf{k}$,
S is the sphere with center the origin and radius 2

9. $\mathbf{F}(x, y, z) = x^2\sin y\,\mathbf{i} + x\cos y\,\mathbf{j} - xz\sin y\,\mathbf{k}$,
S is the "fat sphere" $x^8 + y^8 + z^8 = 8$

10. $\mathbf{F}(x, y, z) = z\,\mathbf{i} + y\,\mathbf{j} + zx\,\mathbf{k}$,
S is the surface of the tetrahedron enclosed by the coordinate planes and the plane

$$\frac{x}{a} + \frac{y}{b} + \frac{z}{c} = 1$$

where a, b, and c are positive numbers

11. $\mathbf{F}(x, y, z) = (\cos z + xy^2)\,\mathbf{i} + xe^{-z}\,\mathbf{j} + (\sin y + x^2z)\,\mathbf{k}$,
S is the surface of the solid bounded by the paraboloid $z = x^2 + y^2$ and the plane $z = 4$

12. $\mathbf{F}(x, y, z) = x^4\,\mathbf{i} - x^3z^2\,\mathbf{j} + 4xy^2z\,\mathbf{k}$,
S is the surface of the solid bounded by the cylinder $x^2 + y^2 = 1$ and the planes $z = x + 2$ and $z = 0$

13. $\mathbf{F} = |\mathbf{r}|\,\mathbf{r}$, where $\mathbf{r} = x\,\mathbf{i} + y\,\mathbf{j} + z\,\mathbf{k}$,
S consists of the hemisphere $z = \sqrt{1 - x^2 - y^2}$ and the disk $x^2 + y^2 \leqslant 1$ in the xy-plane

14. $\mathbf{F} = |\mathbf{r}|^2\mathbf{r}$, where $\mathbf{r} = x\,\mathbf{i} + y\,\mathbf{j} + z\,\mathbf{k}$,
S is the sphere with radius R and center the origin

CAS **15.** $\mathbf{F}(x, y, z) = e^y \tan z\,\mathbf{i} + y\sqrt{3 - x^2}\,\mathbf{j} + x \sin y\,\mathbf{k}$,
S is the surface of the solid that lies above the xy-plane and below the surface $z = 2 - x^4 - y^4$, $-1 \leqslant x \leqslant 1$, $-1 \leqslant y \leqslant 1$

CAS **16.** Use a computer algebra system to plot the vector field $\mathbf{F}(x, y, z) = \sin x \cos^2 y\,\mathbf{i} + \sin^3 y \cos^4 z\,\mathbf{j} + \sin^5 z \cos^6 x\,\mathbf{k}$ in the cube cut from the first octant by the planes $x = \pi/2$, $y = \pi/2$, and $z = \pi/2$. Then compute the flux across the surface of the cube.

17. Use the Divergence Theorem to evaluate $\iint_S \mathbf{F} \cdot d\mathbf{S}$, where
$\mathbf{F}(x, y, z) = z^2x\,\mathbf{i} + \left(\frac{1}{3}y^3 + \tan z\right)\mathbf{j} + (x^2z + y^2)\,\mathbf{k}$
and S is the top half of the sphere $x^2 + y^2 + z^2 = 1$. [*Hint:* Note that S is not a closed surface. First compute integrals over S_1 and S_2, where S_1 is the disk $x^2 + y^2 \leqslant 1$, oriented downward, and $S_2 = S \cup S_1$.]

18. Let $\mathbf{F}(x, y, z) = z \tan^{-1}(y^2)\,\mathbf{i} + z^3 \ln(x^2 + 1)\,\mathbf{j} + z\,\mathbf{k}$. Find the flux of $\mathbf{F}$ across the part of the paraboloid $x^2 + y^2 + z = 2$ that lies above the plane $z = 1$ and is oriented upward.

19. A vector field $\mathbf{F}$ is shown. Use the interpretation of divergence derived in this section to determine whether div $\mathbf{F}$ is positive or negative at P_1 and at P_2.

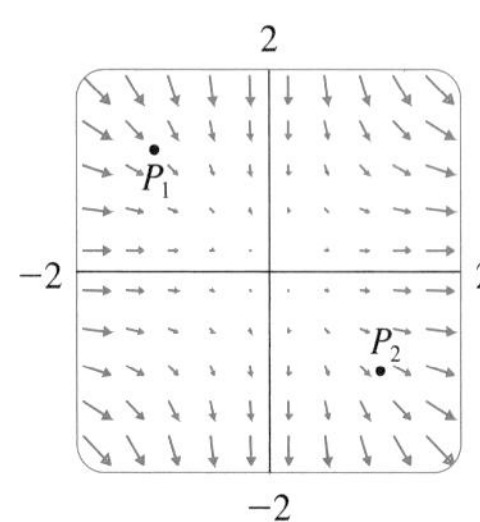

20. (a) Are the points P_1 and P_2 sources or sinks for the vector field $\mathbf{F}$ shown in the figure? Give an explanation based solely on the picture.
(b) Given that $\mathbf{F}(x, y) = \langle x, y^2\rangle$, use the definition of divergence to verify your answer to part (a).

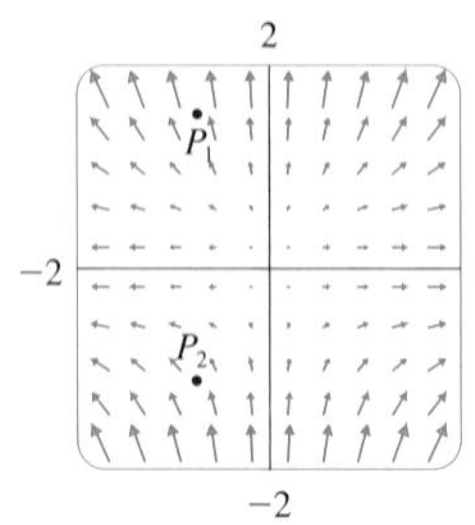

CAS **21–22** Plot the vector field and guess where div $\mathbf{F} > 0$ and where div $\mathbf{F} < 0$. Then calculate div $\mathbf{F}$ to check your guess.

21. $\mathbf{F}(x, y) = \langle xy, x + y^2\rangle$ **22.** $\mathbf{F}(x, y) = \langle x^2, y^2\rangle$

23. Verify that div $\mathbf{E} = 0$ for the electric field $\mathbf{E}(\mathbf{x}) = \dfrac{\varepsilon Q}{|\mathbf{x}|^3}\,\mathbf{x}$.

24. Use the Divergence Theorem to evaluate

$$\iint_S (2x + 2y + z^2)\, dS$$

where S is the sphere $x^2 + y^2 + z^2 = 1$.

25–30 Prove each identity, assuming that S and E satisfy the conditions of the Divergence Theorem and the scalar functions and components of the vector fields have continuous second-order partial derivatives.

25. $\displaystyle\iint_S \mathbf{a} \cdot \mathbf{n}\, dS = 0$, where $\mathbf{a}$ is a constant vector

26. $V(E) = \frac{1}{3}\displaystyle\iint_S \mathbf{F} \cdot d\mathbf{S}$, where $\mathbf{F}(x, y, z) = x\,\mathbf{i} + y\,\mathbf{j} + z\,\mathbf{k}$

27. $\displaystyle\iint_S \operatorname{curl} \mathbf{F} \cdot d\mathbf{S} = 0$ **28.** $\displaystyle\iint_S D_{\mathbf{n}} f\, dS = \iiint_E \nabla^2 f\, dV$

29. $\displaystyle\iint_S (f \nabla g) \cdot \mathbf{n}\, dS = \iiint_E (f \nabla^2 g + \nabla f \cdot \nabla g)\, dV$

30. $\displaystyle\iint_S (f \nabla g - g \nabla f) \cdot \mathbf{n}\, dS = \iiint_E (f \nabla^2 g - g \nabla^2 f)\, dV$

31. Suppose S and E satisfy the conditions of the Divergence Theorem and f is a scalar function with continuous partial derivatives. Prove that

$$\iint_S f\mathbf{n}\, dS = \iiint_E \nabla f\, dV$$

These surface and triple integrals of vector functions are vectors defined by integrating each component function. [*Hint:* Start by applying the Divergence Theorem to $\mathbf{F} = f\mathbf{c}$, where $\mathbf{c}$ is an arbitrary constant vector.]

32. A solid occupies a region E with surface S and is immersed in a liquid with constant density ρ. We set up a coordinate system so that the xy-plane coincides with the surface of the liquid, and positive values of z are measured downward into the liquid. Then the pressure at depth z is $p = \rho g z$, where g is the acceleration due to gravity (see Section 8.3). The total buoyant force on the solid due to the pressure distribution is given by the surface integral

$$\mathbf{F} = -\iint_S p\mathbf{n}\, dS$$

where $\mathbf{n}$ is the outer unit normal. Use the result of Exercise 31 to show that $\mathbf{F} = -W\mathbf{k}$, where W is the weight of the liquid displaced by the solid. (Note that $\mathbf{F}$ is directed upward because z is directed downward.) The result is *Archimedes' Principle:* The buoyant force on an object equals the weight of the displaced liquid.

16.10 Summary

The main results of this chapter are all higher-dimensional versions of the Fundamental Theorem of Calculus. To help you remember them, we collect them together here (without hypotheses) so that you can see more easily their essential similarity. Notice that in each case we have an integral of a "derivative" over a region on the left side, and the right side involves the values of the original function only on the *boundary* of the region.

Fundamental Theorem of Calculus	$\int_a^b F'(x)\,dx = F(b) - F(a)$	
Fundamental Theorem for Line Integrals	$\int_C \nabla f \cdot d\mathbf{r} = f(\mathbf{r}(b)) - f(\mathbf{r}(a))$	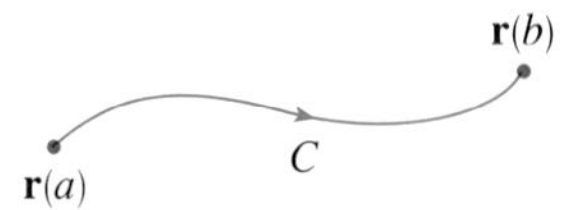
Green's Theorem	$\iint_D \left(\frac{\partial Q}{\partial x} - \frac{\partial P}{\partial y} \right) dA = \int_C P\,dx + Q\,dy$	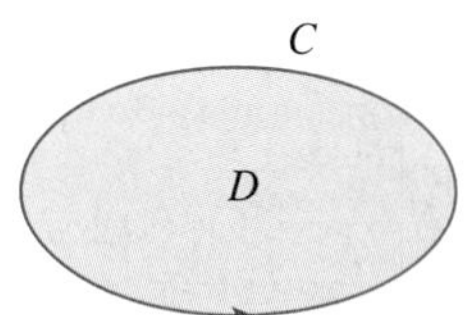
Stokes' Theorem	$\iint_S \operatorname{curl} \mathbf{F} \cdot d\mathbf{S} = \int_C \mathbf{F} \cdot d\mathbf{r}$	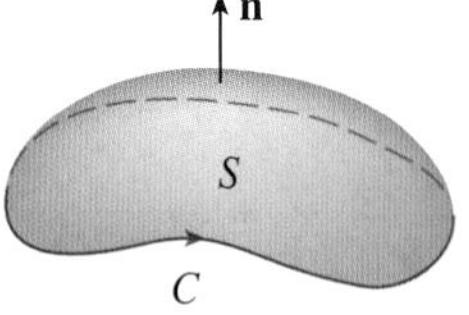
Divergence Theorem	$\iiint_E \operatorname{div} \mathbf{F}\,dV = \iint_S \mathbf{F} \cdot d\mathbf{S}$	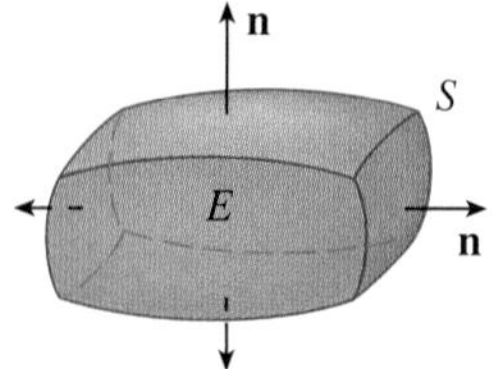

16 Review

Concept Check

1. What is a vector field? Give three examples that have physical meaning.

2. (a) What is a conservative vector field?
(b) What is a potential function?

3. (a) Write the definition of the line integral of a scalar function f along a smooth curve C with respect to arc length.
(b) How do you evaluate such a line integral?
(c) Write expressions for the mass and center of mass of a thin wire shaped like a curve C if the wire has linear density function $\rho(x, y)$.
(d) Write the definitions of the line integrals along C of a scalar function f with respect to x, y, and z.
(e) How do you evaluate these line integrals?

4. (a) Define the line integral of a vector field $\mathbf{F}$ along a smooth curve C given by a vector function $\mathbf{r}(t)$.
(b) If $\mathbf{F}$ is a force field, what does this line integral represent?
(c) If $\mathbf{F} = \langle P, Q, R \rangle$, what is the connection between the line integral of $\mathbf{F}$ and the line integrals of the component functions P, Q, and R?

5. State the Fundamental Theorem for Line Integrals.

6. (a) What does it mean to say that $\int_C \mathbf{F} \cdot d\mathbf{r}$ is independent of path?
(b) If you know that $\int_C \mathbf{F} \cdot d\mathbf{r}$ is independent of path, what can you say about $\mathbf{F}$?

7. State Green's Theorem.

8. Write expressions for the area enclosed by a curve C in terms of line integrals around C.

9. Suppose $\mathbf{F}$ is a vector field on $\mathbb{R}^3$.
(a) Define curl $\mathbf{F}$. (b) Define div $\mathbf{F}$.
(c) If $\mathbf{F}$ is a velocity field in fluid flow, what are the physical interpretations of curl $\mathbf{F}$ and div $\mathbf{F}$?

10. If $\mathbf{F} = P\,\mathbf{i} + Q\,\mathbf{j}$, how do you test to determine whether $\mathbf{F}$ is conservative? What if $\mathbf{F}$ is a vector field on $\mathbb{R}^3$?

11. (a) What is a parametric surface? What are its grid curves?
(b) Write an expression for the area of a parametric surface.
(c) What is the area of a surface given by an equation $z = g(x, y)$?

12. (a) Write the definition of the surface integral of a scalar function f over a surface S.
(b) How do you evaluate such an integral if S is a parametric surface given by a vector function $\mathbf{r}(u, v)$?
(c) What if S is given by an equation $z = g(x, y)$?
(d) If a thin sheet has the shape of a surface S, and the density at (x, y, z) is $\rho(x, y, z)$, write expressions for the mass and center of mass of the sheet.

13. (a) What is an oriented surface? Give an example of a non-orientable surface.
(b) Define the surface integral (or flux) of a vector field $\mathbf{F}$ over an oriented surface S with unit normal vector $\mathbf{n}$.
(c) How do you evaluate such an integral if S is a parametric surface given by a vector function $\mathbf{r}(u, v)$?
(d) What if S is given by an equation $z = g(x, y)$?

14. State Stokes' Theorem.

15. State the Divergence Theorem.

16. In what ways are the Fundamental Theorem for Line Integrals, Green's Theorem, Stokes' Theorem, and the Divergence Theorem similar?

True-False Quiz

Determine whether the statement is true or false. If it is true, explain why. If it is false, explain why or give an example that disproves the statement.

1. If $\mathbf{F}$ is a vector field, then div $\mathbf{F}$ is a vector field.

2. If $\mathbf{F}$ is a vector field, then curl $\mathbf{F}$ is a vector field.

3. If f has continuous partial derivatives of all orders on $\mathbb{R}^3$, then $\operatorname{div}(\operatorname{curl} \nabla f) = 0$.

4. If f has continuous partial derivatives on $\mathbb{R}^3$ and C is any circle, then $\int_C \nabla f \cdot d\mathbf{r} = 0$.

5. If $\mathbf{F} = P\,\mathbf{i} + Q\,\mathbf{j}$ and $P_y = Q_x$ in an open region D, then $\mathbf{F}$ is conservative.

6. $\int_{-C} f(x, y)\, ds = -\int_C f(x, y)\, ds$

7. If $\mathbf{F}$ and $\mathbf{G}$ are vector fields and $\operatorname{div}\mathbf{F} = \operatorname{div}\mathbf{G}$, then $\mathbf{F} = \mathbf{G}$.

8. The work done by a conservative force field in moving a particle around a closed path is zero.

9. If $\mathbf{F}$ and $\mathbf{G}$ are vector fields, then
$$\operatorname{curl}(\mathbf{F} + \mathbf{G}) = \operatorname{curl} \mathbf{F} + \operatorname{curl} \mathbf{G}$$

10. If $\mathbf{F}$ and $\mathbf{G}$ are vector fields, then
$$\operatorname{curl}(\mathbf{F} \cdot \mathbf{G}) = \operatorname{curl} \mathbf{F} \cdot \operatorname{curl} \mathbf{G}$$

11. If S is a sphere and $\mathbf{F}$ is a constant vector field, then $\iint_S \mathbf{F} \cdot d\mathbf{S} = 0$.

12. There is a vector field $\mathbf{F}$ such that
$$\operatorname{curl} \mathbf{F} = x\,\mathbf{i} + y\,\mathbf{j} + z\,\mathbf{k}$$

Exercises

1. A vector field **F**, a curve C, and a point P are shown.
(a) Is $\int_C \mathbf{F} \cdot d\mathbf{r}$ positive, negative, or zero? Explain.
(b) Is div $\mathbf{F}(P)$ positive, negative, or zero? Explain.

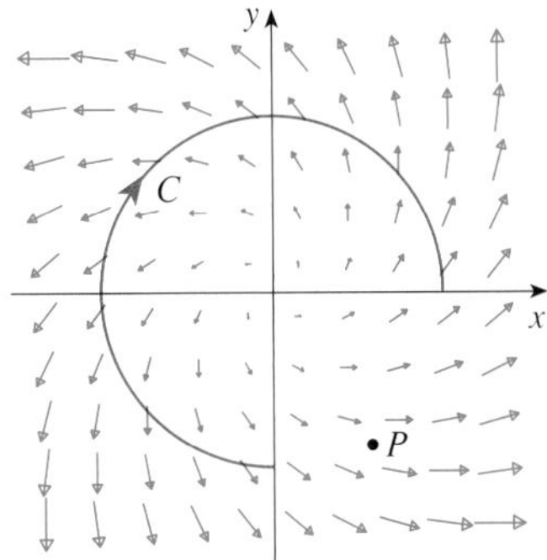

2–9 Evaluate the line integral.

2. $\int_C x\, ds$,
C is the arc of the parabola $y = x^2$ from $(0, 0)$ to $(1, 1)$

3. $\int_C yz \cos x\, ds$,
C: $x = t$, $y = 3 \cos t$, $z = 3 \sin t$, $0 \leqslant t \leqslant \pi$

4. $\int_C y\, dx + (x + y^2)\, dy$, C is the ellipse $4x^2 + 9y^2 = 36$ with counterclockwise orientation

5. $\int_C y^3\, dx + x^2\, dy$, C is the arc of the parabola $x = 1 - y^2$ from $(0, -1)$ to $(0, 1)$

6. $\int_C \sqrt{xy}\, dx + e^y\, dy + xz\, dz$,
C is given by $\mathbf{r}(t) = t^4\,\mathbf{i} + t^2\,\mathbf{j} + t^3\,\mathbf{k}$, $0 \leqslant t \leqslant 1$

7. $\int_C xy\, dx + y^2\, dy + yz\, dz$,
C is the line segment from $(1, 0, -1)$, to $(3, 4, 2)$

8. $\int_C \mathbf{F} \cdot d\mathbf{r}$, where $\mathbf{F}(x, y) = xy\,\mathbf{i} + x^2\,\mathbf{j}$ and C is given by $\mathbf{r}(t) = \sin t\,\mathbf{i} + (1 + t)\,\mathbf{j}$, $0 \leqslant t \leqslant \pi$

9. $\int_C \mathbf{F} \cdot d\mathbf{r}$, where $\mathbf{F}(x, y, z) = e^z\,\mathbf{i} + xz\,\mathbf{j} + (x + y)\,\mathbf{k}$ and C is given by $\mathbf{r}(t) = t^2\,\mathbf{i} + t^3\,\mathbf{j} - t\,\mathbf{k}$, $0 \leqslant t \leqslant 1$

10. Find the work done by the force field

$$\mathbf{F}(x, y, z) = z\,\mathbf{i} + x\,\mathbf{j} + y\,\mathbf{k}$$

in moving a particle from the point $(3, 0, 0)$ to the point $(0, \pi/2, 3)$ along
(a) a straight line
(b) the helix $x = 3 \cos t$, $y = t$, $z = 3 \sin t$

11–12 Show that **F** is a conservative vector field. Then find a function f such that $\mathbf{F} = \nabla f$.

11. $\mathbf{F}(x, y) = (1 + xy)e^{xy}\,\mathbf{i} + (e^y + x^2 e^{xy})\,\mathbf{j}$

12. $\mathbf{F}(x, y, z) = \sin y\,\mathbf{i} + x \cos y\,\mathbf{j} - \sin z\,\mathbf{k}$

13–14 Show that **F** is conservative and use this fact to evaluate $\int_C \mathbf{F} \cdot d\mathbf{r}$ along the given curve.

13. $\mathbf{F}(x, y) = (4x^3y^2 - 2xy^3)\,\mathbf{i} + (2x^4y - 3x^2y^2 + 4y^3)\,\mathbf{j}$,
C: $\mathbf{r}(t) = (t + \sin \pi t)\,\mathbf{i} + (2t + \cos \pi t)\,\mathbf{j}$, $0 \leqslant t \leqslant 1$

14. $\mathbf{F}(x, y, z) = e^y\,\mathbf{i} + (xe^y + e^z)\,\mathbf{j} + ye^z\,\mathbf{k}$,
C is the line segment from $(0, 2, 0)$ to $(4, 0, 3)$

15. Verify that Green's Theorem is true for the line integral $\int_C xy^2\, dx - x^2y\, dy$, where C consists of the parabola $y = x^2$ from $(-1, 1)$ to $(1, 1)$ and the line segment from $(1, 1)$ to $(-1, 1)$.

16. Use Green's Theorem to evaluate

$$\int_C \sqrt{1 + x^3}\, dx + 2xy\, dy$$

where C is the triangle with vertices $(0, 0)$, $(1, 0)$, and $(1, 3)$.

17. Use Green's Theorem to evaluate $\int_C x^2y\, dx - xy^2\, dy$, where C is the circle $x^2 + y^2 = 4$ with counterclockwise orientation.

18. Find curl **F** and div **F** if

$$\mathbf{F}(x, y, z) = e^{-x} \sin y\,\mathbf{i} + e^{-y} \sin z\,\mathbf{j} + e^{-z} \sin x\,\mathbf{k}$$

19. Show that there is no vector field **G** such that

$$\text{curl}\,\mathbf{G} = 2x\,\mathbf{i} + 3yz\,\mathbf{j} - xz^2\,\mathbf{k}$$

20. Show that, under conditions to be stated on the vector fields **F** and **G**,

$$\text{curl}(\mathbf{F} \times \mathbf{G}) = \mathbf{F}\,\text{div}\,\mathbf{G} - \mathbf{G}\,\text{div}\,\mathbf{F} + (\mathbf{G} \cdot \nabla)\mathbf{F} - (\mathbf{F} \cdot \nabla)\mathbf{G}$$

21. If C is any piecewise-smooth simple closed plane curve and f and g are differentiable functions, show that $\int_C f(x)\, dx + g(y)\, dy = 0$.

22. If f and g are twice differentiable functions, show that

$$\nabla^2(fg) = f\nabla^2 g + g\nabla^2 f + 2\nabla f \cdot \nabla g$$

23. If f is a harmonic function, that is, $\nabla^2 f = 0$, show that the line integral $\int f_y\, dx - f_x\, dy$ is independent of path in any simple region D.

24. (a) Sketch the curve C with parametric equations

$$x = \cos t \qquad y = \sin t \qquad z = \sin t \qquad 0 \leqslant t \leqslant 2\pi$$

(b) Find $\int_C 2xe^{2y}\, dx + (2x^2e^{2y} + 2y \cot z)\, dy - y^2 \csc^2 z\, dz$.

Graphing calculator or computer required CAS Computer algebra system required

25. Find the area of the part of the surface $z = x^2 + 2y$ that lies above the triangle with vertices $(0, 0)$, $(1, 0)$, and $(1, 2)$.

26. (a) Find an equation of the tangent plane at the point $(4, -2, 1)$ to the parametric surface S given by

$$\mathbf{r}(u, v) = v^2\,\mathbf{i} - uv\,\mathbf{j} + u^2\,\mathbf{k} \qquad 0 \leqslant u \leqslant 3,\ -3 \leqslant v \leqslant 3$$

(b) Use a computer to graph the surface S and the tangent plane found in part (a).

(c) Set up, but do not evaluate, an integral for the surface area of S.

CAS (d) If

$$\mathbf{F}(x, y, z) = \frac{z^2}{1 + x^2}\,\mathbf{i} + \frac{x^2}{1 + y^2}\,\mathbf{j} + \frac{y^2}{1 + z^2}\,\mathbf{k}$$

find $\iint_S \mathbf{F} \cdot d\mathbf{S}$ correct to four decimal places.

27–30 Evaluate the surface integral.

27. $\iint_S z\, dS$, where S is the part of the paraboloid $z = x^2 + y^2$ that lies under the plane $z = 4$

28. $\iint_S (x^2z + y^2z)\, dS$, where S is the part of the plane $z = 4 + x + y$ that lies inside the cylinder $x^2 + y^2 = 4$

29. $\iint_S \mathbf{F} \cdot d\mathbf{S}$, where $\mathbf{F}(x, y, z) = xz\,\mathbf{i} - 2y\,\mathbf{j} + 3x\,\mathbf{k}$ and S is the sphere $x^2 + y^2 + z^2 = 4$ with outward orientation

30. $\iint_S \mathbf{F} \cdot d\mathbf{S}$, where $\mathbf{F}(x, y, z) = x^2\,\mathbf{i} + xy\,\mathbf{j} + z\,\mathbf{k}$ and S is the part of the paraboloid $z = x^2 + y^2$ below the plane $z = 1$ with upward orientation

31. Verify that Stokes' Theorem is true for the vector field $\mathbf{F}(x, y, z) = x^2\,\mathbf{i} + y^2\,\mathbf{j} + z^2\,\mathbf{k}$, where S is the part of the paraboloid $z = 1 - x^2 - y^2$ that lies above the xy-plane and S has upward orientation.

32. Use Stokes' Theorem to evaluate $\iint_S \operatorname{curl} \mathbf{F} \cdot d\mathbf{S}$, where $\mathbf{F}(x, y, z) = x^2yz\,\mathbf{i} + yz^2\,\mathbf{j} + z^3e^{xy}\,\mathbf{k}$, S is the part of the sphere $x^2 + y^2 + z^2 = 5$ that lies above the plane $z = 1$, and S is oriented upward.

33. Use Stokes' Theorem to evaluate $\int_C \mathbf{F} \cdot d\mathbf{r}$, where $\mathbf{F}(x, y, z) = xy\,\mathbf{i} + yz\,\mathbf{j} + zx\,\mathbf{k}$, and C is the triangle with vertices $(1, 0, 0)$, $(0, 1, 0)$, and $(0, 0, 1)$, oriented counterclockwise as viewed from above.

34. Use the Divergence Theorem to calculate the surface integral $\iint_S \mathbf{F} \cdot d\mathbf{S}$, where $\mathbf{F}(x, y, z) = x^3\,\mathbf{i} + y^3\,\mathbf{j} + z^3\,\mathbf{k}$ and S is the surface of the solid bounded by the cylinder $x^2 + y^2 = 1$ and the planes $z = 0$ and $z = 2$.

35. Verify that the Divergence Theorem is true for the vector field $\mathbf{F}(x, y, z) = x\,\mathbf{i} + y\,\mathbf{j} + z\,\mathbf{k}$, where E is the unit ball $x^2 + y^2 + z^2 \leqslant 1$.

36. Compute the outward flux of

$$\mathbf{F}(x, y, z) = \frac{x\,\mathbf{i} + y\,\mathbf{j} + z\,\mathbf{k}}{(x^2 + y^2 + z^2)^{3/2}}$$

through the ellipsoid $4x^2 + 9y^2 + 6z^2 = 36$.

37. Let

$$\mathbf{F}(x, y, z) = (3x^2yz - 3y)\,\mathbf{i} + (x^3z - 3x)\,\mathbf{j} + (x^3y + 2z)\,\mathbf{k}$$

Evaluate $\int_C \mathbf{F} \cdot d\mathbf{r}$, where C is the curve with initial point $(0, 0, 2)$ and terminal point $(0, 3, 0)$ shown in the figure.

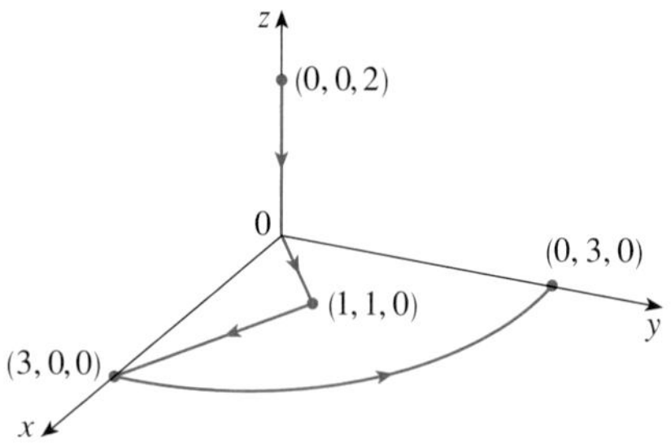

38. Let

$$\mathbf{F}(x, y) = \frac{(2x^3 + 2xy^2 - 2y)\,\mathbf{i} + (2y^3 + 2x^2y + 2x)\,\mathbf{j}}{x^2 + y^2}$$

Evaluate $\oint_C \mathbf{F} \cdot d\mathbf{r}$, where C is shown in the figure.

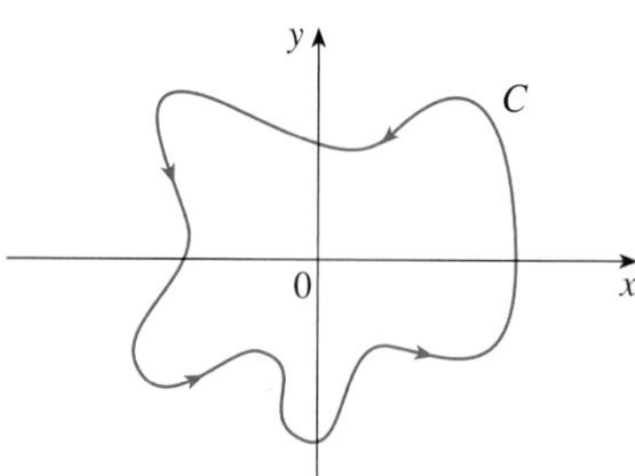

39. Find $\iint_S \mathbf{F} \cdot \mathbf{n}\, dS$, where $\mathbf{F}(x, y, z) = x\,\mathbf{i} + y\,\mathbf{j} + z\,\mathbf{k}$ and S is the outwardly oriented surface shown in the figure (the boundary surface of a cube with a unit corner cube removed).

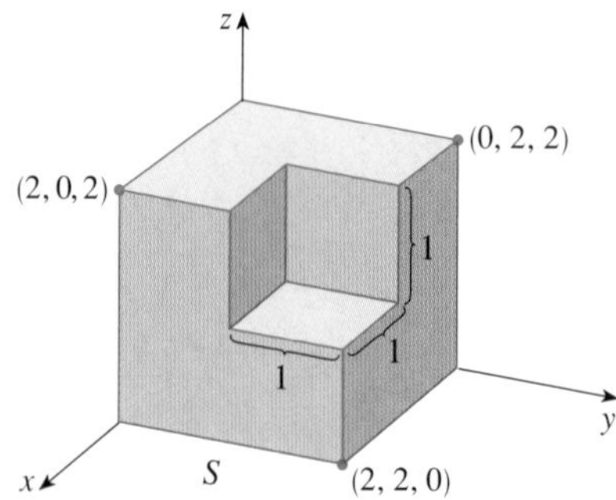

40. If the components of $\mathbf{F}$ have continuous second partial derivatives and S is the boundary surface of a simple solid region, show that $\iint_S \operatorname{curl} \mathbf{F} \cdot d\mathbf{S} = 0$.

41. If $\mathbf{a}$ is a constant vector, $\mathbf{r} = x\,\mathbf{i} + y\,\mathbf{j} + z\,\mathbf{k}$, and S is an oriented, smooth surface with a simple, closed, smooth, positively oriented boundary curve C, show that

$$\iint_S 2\mathbf{a} \cdot d\mathbf{S} = \int_C (\mathbf{a} \times \mathbf{r}) \cdot d\mathbf{r}$$

Problems Plus

1. Let S be a smooth parametric surface and let P be a point such that each line that starts at P intersects S at most once. The **solid angle** $\Omega(S)$ subtended by S at P is the set of lines starting at P and passing through S. Let $S(a)$ be the intersection of $\Omega(S)$ with the surface of the sphere with center P and radius a. Then the measure of the solid angle (in *steradians*) is defined to be

$$|\Omega(S)| = \frac{\text{area of } S(a)}{a^2}$$

Apply the Divergence Theorem to the part of $\Omega(S)$ between $S(a)$ and S to show that

$$|\Omega(S)| = \iint_S \frac{\mathbf{r} \cdot \mathbf{n}}{r^3}\, dS$$

where $\mathbf{r}$ is the radius vector from P to any point on S, $r = |\mathbf{r}|$, and the unit normal vector $\mathbf{n}$ is directed away from P.

This shows that the definition of the measure of a solid angle is independent of the radius a of the sphere. Thus the measure of the solid angle is equal to the area subtended on a *unit* sphere. (Note the analogy with the definition of radian measure.) The total solid angle subtended by a sphere at its center is thus 4π steradians.

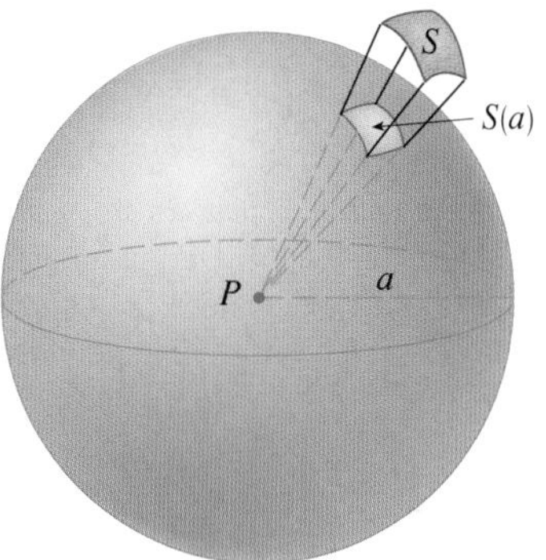

2. Find the positively oriented simple closed curve C for which the value of the line integral

$$\int_C (y^3 - y)\, dx - 2x^3\, dy$$

is a maximum.

3. Let C be a simple closed piecewise-smooth space curve that lies in a plane with unit normal vector $\mathbf{n} = \langle a, b, c \rangle$ and has positive orientation with respect to $\mathbf{n}$. Show that the plane area enclosed by C is

$$\tfrac{1}{2}\int_C (bz - cy)\, dx + (cx - az)\, dy + (ay - bx)\, dz$$

4. Investigate the shape of the surface with parametric equations $x = \sin u$, $y = \sin v$, $z = \sin(u + v)$. Start by graphing the surface from several points of view. Explain the appearance of the graphs by determining the traces in the horizontal planes $z = 0$, $z = \pm 1$, and $z = \pm\frac{1}{2}$.

5. Prove the following identity:

$$\nabla(\mathbf{F} \cdot \mathbf{G}) = (\mathbf{F} \cdot \nabla)\mathbf{G} + (\mathbf{G} \cdot \nabla)\mathbf{F} + \mathbf{F} \times \operatorname{curl} \mathbf{G} + \mathbf{G} \times \operatorname{curl} \mathbf{F}$$

Graphing calculator or computer required

6. The figure depicts the sequence of events in each cylinder of a four-cylinder internal combustion engine. Each piston moves up and down and is connected by a pivoted arm to a rotating crankshaft. Let $P(t)$ and $V(t)$ be the pressure and volume within a cylinder at time t, where $a \leqslant t \leqslant b$ gives the time required for a complete cycle. The graph shows how P and V vary through one cycle of a four-stroke engine.

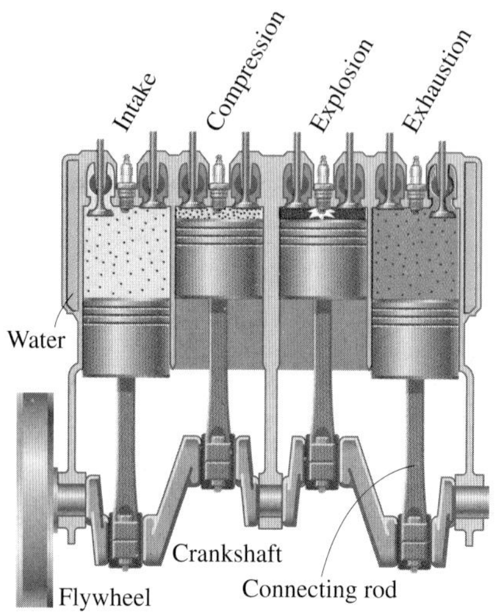

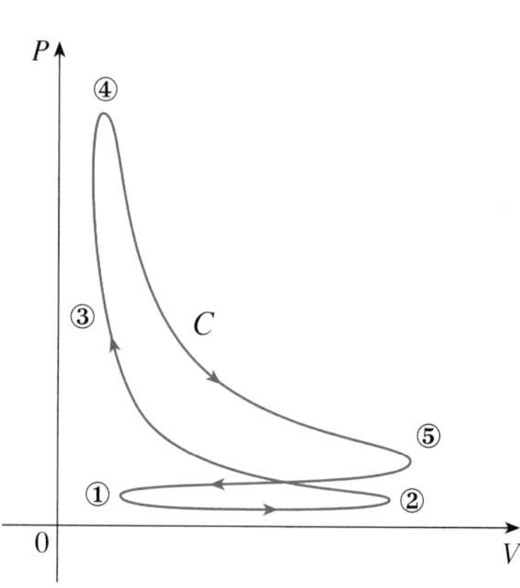

During the intake stroke (from ① to ②) a mixture of air and gasoline at atmospheric pressure is drawn into a cylinder through the intake valve as the piston moves downward. Then the piston rapidly compresses the mix with the valves closed in the compression stroke (from ② to ③) during which the pressure rises and the volume decreases. At ③ the sparkplug ignites the fuel, raising the temperature and pressure at almost constant volume to ④. Then, with valves closed, the rapid expansion forces the piston downward during the power stroke (from ④ to ⑤). The exhaust valve opens, temperature and pressure drop, and mechanical energy stored in a rotating flywheel pushes the piston upward, forcing the waste products out of the exhaust valve in the exhaust stroke. The exhaust valve closes and the intake valve opens. We're now back at ① and the cycle starts again.

(a) Show that the work done on the piston during one cycle of a four-stroke engine is $W = \int_C P\, dV$, where C is the curve in the PV-plane shown in the figure.

[*Hint:* Let $x(t)$ be the distance from the piston to the top of the cylinder and note that the force on the piston is $\mathbf{F} = AP(t)\,\mathbf{i}$, where A is the area of the top of the piston. Then $W = \int_{C_1} \mathbf{F} \cdot d\mathbf{r}$, where C_1 is given by $\mathbf{r}(t) = x(t)\,\mathbf{i}$, $a \leqslant t \leqslant b$. An alternative approach is to work directly with Riemann sums.]

(b) Use Formula 16.4.5 to show that the work is the difference of the areas enclosed by the two loops of C.

17 Second-Order Differential Equations

The motion of a shock absorber in a car is described by the differential equations that we solve in Section 17.3.

© Christoff / Shutterstock

The basic ideas of differential equations were explained in Chapter 9; there we concentrated on first-order equations. In this chapter we study second-order linear differential equations and learn how they can be applied to solve problems concerning the vibrations of springs and the analysis of electric circuits. We will also see how infinite series can be used to solve differential equations.

17.1 Second-Order Linear Equations

A **second-order linear differential equation** has the form

$$\boxed{1} \qquad P(x)\frac{d^2y}{dx^2} + Q(x)\frac{dy}{dx} + R(x)y = G(x)$$

where P, Q, R, and G are continuous functions. We saw in Section 9.1 that equations of this type arise in the study of the motion of a spring. In Section 17.3 we will further pursue this application as well as the application to electric circuits.

In this section we study the case where $G(x) = 0$, for all x, in Equation 1. Such equations are called **homogeneous** linear equations. Thus the form of a second-order linear homogeneous differential equation is

$$\boxed{2} \qquad P(x)\frac{d^2y}{dx^2} + Q(x)\frac{dy}{dx} + R(x)y = 0$$

If $G(x) \neq 0$ for some x, Equation 1 is **nonhomogeneous** and is discussed in Section 17.2.

Two basic facts enable us to solve homogeneous linear equations. The first of these says that if we know two solutions y_1 and y_2 of such an equation, then the **linear combination** $y = c_1y_1 + c_2y_2$ is also a solution.

> **3 Theorem** If $y_1(x)$ and $y_2(x)$ are both solutions of the linear homogeneous equation $\boxed{2}$ and c_1 and c_2 are any constants, then the function
>
> $$y(x) = c_1y_1(x) + c_2y_2(x)$$
>
> is also a solution of Equation 2.

PROOF Since y_1 and y_2 are solutions of Equation 2, we have

$$P(x)y_1'' + Q(x)y_1' + R(x)y_1 = 0$$

and

$$P(x)y_2'' + Q(x)y_2' + R(x)y_2 = 0$$

Therefore, using the basic rules for differentiation, we have

$$\begin{aligned} P(x)y'' &+ Q(x)y' + R(x)y \\ &= P(x)(c_1y_1 + c_2y_2)'' + Q(x)(c_1y_1 + c_2y_2)' + R(x)(c_1y_1 + c_2y_2) \\ &= P(x)(c_1y_1'' + c_2y_2'') + Q(x)(c_1y_1' + c_2y_2') + R(x)(c_1y_1 + c_2y_2) \\ &= c_1[P(x)y_1'' + Q(x)y_1' + R(x)y_1] + c_2[P(x)y_2'' + Q(x)y_2' + R(x)y_2] \\ &= c_1(0) + c_2(0) = 0 \end{aligned}$$

Thus $y = c_1y_1 + c_2y_2$ is a solution of Equation 2. ∎

The other fact we need is given by the following theorem, which is proved in more advanced courses. It says that the general solution is a linear combination of two **linearly independent** solutions y_1 and y_2. This means that neither y_1 nor y_2 is a constant multiple of the other. For instance, the functions $f(x) = x^2$ and $g(x) = 5x^2$ are linearly dependent, but $f(x) = e^x$ and $g(x) = xe^x$ are linearly independent.

4 Theorem If y_1 and y_2 are linearly independent solutions of Equation 2 on an interval, and $P(x)$ is never 0, then the general solution is given by

$$y(x) = c_1 y_1(x) + c_2 y_2(x)$$

where c_1 and c_2 are arbitrary constants.

Theorem 4 is very useful because it says that if we know *two* particular linearly independent solutions, then we know *every* solution.

In general, it's not easy to discover particular solutions to a second-order linear equation. But it is always possible to do so if the coefficient functions P, Q, and R are constant functions, that is, if the differential equation has the form

5
$$ay'' + by' + cy = 0$$

where a, b, and c are constants and $a \neq 0$.

It's not hard to think of some likely candidates for particular solutions of Equation 5 if we state the equation verbally. We are looking for a function y such that a constant times its second derivative y'' plus another constant times y' plus a third constant times y is equal to 0. We know that the exponential function $y = e^{rx}$ (where r is a constant) has the property that its derivative is a constant multiple of itself: $y' = re^{rx}$. Furthermore, $y'' = r^2e^{rx}$. If we substitute these expressions into Equation 5, we see that $y = e^{rx}$ is a solution if

$$ar^2e^{rx} + bre^{rx} + ce^{rx} = 0$$

or

$$(ar^2 + br + c)e^{rx} = 0$$

But e^{rx} is never 0. Thus $y = e^{rx}$ is a solution of Equation 5 if r is a root of the equation

6
$$ar^2 + br + c = 0$$

Equation 6 is called the **auxiliary equation** (or **characteristic equation**) of the differential equation $ay'' + by' + cy = 0$. Notice that it is an algebraic equation that is obtained from the differential equation by replacing y'' by r^2, y' by r, and y by 1.

Sometimes the roots r_1 and r_2 of the auxiliary equation can be found by factoring. In other cases they are found by using the quadratic formula:

7
$$r_1 = \frac{-b + \sqrt{b^2 - 4ac}}{2a} \qquad r_2 = \frac{-b - \sqrt{b^2 - 4ac}}{2a}$$

We distinguish three cases according to the sign of the discriminant $b^2 - 4ac$.

CASE I $b^2 - 4ac > 0$

In this case the roots r_1 and r_2 of the auxiliary equation are real and distinct, so $y_1 = e^{r_1 x}$ and $y_2 = e^{r_2 x}$ are two linearly independent solutions of Equation 5. (Note that $e^{r_2 x}$ is not a constant multiple of $e^{r_1 x}$.) Therefore, by Theorem 4, we have the following fact.

> **8** If the roots r_1 and r_2 of the auxiliary equation $ar^2 + br + c = 0$ are real and unequal, then the general solution of $ay'' + by' + cy = 0$ is
>
> $$y = c_1 e^{r_1 x} + c_2 e^{r_2 x}$$

In Figure 1 the graphs of the basic solutions $f(x) = e^{2x}$ and $g(x) = e^{-3x}$ of the differential equation in Example 1 are shown in blue and red, respectively. Some of the other solutions, linear combinations of f and g, are shown in black.

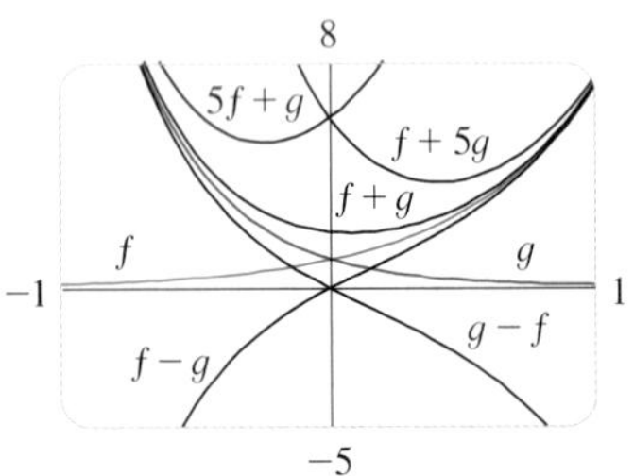

FIGURE 1

EXAMPLE 1 Solve the equation $y'' + y' - 6y = 0$.

SOLUTION The auxiliary equation is

$$r^2 + r - 6 = (r - 2)(r + 3) = 0$$

whose roots are $r = 2, -3$. Therefore, by **8**, the general solution of the given differential equation is

$$y = c_1 e^{2x} + c_2 e^{-3x}$$

We could verify that this is indeed a solution by differentiating and substituting into the differential equation.

EXAMPLE 2 Solve $3\dfrac{d^2y}{dx^2} + \dfrac{dy}{dx} - y = 0$.

SOLUTION To solve the auxiliary equation $3r^2 + r - 1 = 0$, we use the quadratic formula:

$$r = \frac{-1 \pm \sqrt{13}}{6}$$

Since the roots are real and distinct, the general solution is

$$y = c_1 e^{(-1+\sqrt{13})x/6} + c_2 e^{(-1-\sqrt{13})x/6}$$

CASE II $b^2 - 4ac = 0$

In this case $r_1 = r_2$; that is, the roots of the auxiliary equation are real and equal. Let's denote by r the common value of r_1 and r_2. Then, from Equations 7, we have

9 $$r = -\frac{b}{2a} \qquad \text{so} \qquad 2ar + b = 0$$

We know that $y_1 = e^{rx}$ is one solution of Equation 5. We now verify that $y_2 = xe^{rx}$ is also a solution:

$$\begin{aligned} ay_2'' + by_2' + cy_2 &= a(2re^{rx} + r^2xe^{rx}) + b(e^{rx} + rxe^{rx}) + cxe^{rx} \\ &= (2ar + b)e^{rx} + (ar^2 + br + c)xe^{rx} \\ &= 0(e^{rx}) + 0(xe^{rx}) = 0 \end{aligned}$$

The first term is 0 by Equations 9; the second term is 0 because r is a root of the auxiliary equation. Since $y_1 = e^{rx}$ and $y_2 = xe^{rx}$ are linearly independent solutions, Theorem 4 provides us with the general solution.

> **10** If the auxiliary equation $ar^2 + br + c = 0$ has only one real root r, then the general solution of $ay'' + by' + cy = 0$ is
>
> $$y = c_1e^{rx} + c_2xe^{rx}$$

Figure 2 shows the basic solutions $f(x) = e^{-3x/2}$ and $g(x) = xe^{-3x/2}$ in Example 3 and some other members of the family of solutions. Notice that all of them approach 0 as $x \to \infty$.

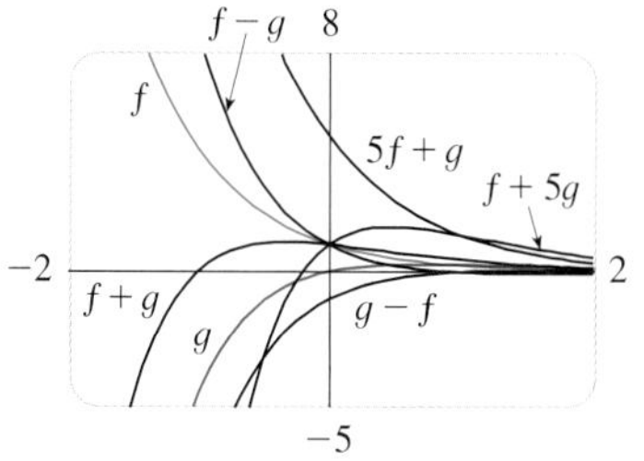

FIGURE 2

V EXAMPLE 3 Solve the equation $4y'' + 12y' + 9y = 0$.

SOLUTION The auxiliary equation $4r^2 + 12r + 9 = 0$ can be factored as

$$(2r + 3)^2 = 0$$

so the only root is $r = -\frac{3}{2}$. By $\boxed{10}$, the general solution is

$$y = c_1e^{-3x/2} + c_2xe^{-3x/2}$$

CASE III $b^2 - 4ac < 0$

In this case the roots r_1 and r_2 of the auxiliary equation are complex numbers. (See Appendix H for information about complex numbers.) We can write

$$r_1 = \alpha + i\beta \qquad r_2 = \alpha - i\beta$$

where α and β are real numbers. [In fact, $\alpha = -b/(2a)$, $\beta = \sqrt{4ac - b^2}/(2a)$.] Then, using Euler's equation

$$e^{i\theta} = \cos\theta + i\sin\theta$$

from Appendix H, we write the solution of the differential equation as

$$\begin{aligned} y &= C_1e^{r_1x} + C_2e^{r_2x} = C_1e^{(\alpha+i\beta)x} + C_2e^{(\alpha-i\beta)x} \\ &= C_1e^{\alpha x}(\cos\beta x + i\sin\beta x) + C_2e^{\alpha x}(\cos\beta x - i\sin\beta x) \\ &= e^{\alpha x}[(C_1 + C_2)\cos\beta x + i(C_1 - C_2)\sin\beta x] \\ &= e^{\alpha x}(c_1\cos\beta x + c_2\sin\beta x) \end{aligned}$$

where $c_1 = C_1 + C_2$, $c_2 = i(C_1 - C_2)$. This gives all solutions (real or complex) of the differential equation. The solutions are real when the constants c_1 and c_2 are real. We summarize the discussion as follows.

> **11** If the roots of the auxiliary equation $ar^2 + br + c = 0$ are the complex numbers $r_1 = \alpha + i\beta$, $r_2 = \alpha - i\beta$, then the general solution of $ay'' + by' + cy = 0$ is
>
> $$y = e^{\alpha x}(c_1\cos\beta x + c_2\sin\beta x)$$

Figure 3 shows the graphs of the solutions in Example 4, $f(x) = e^{3x} \cos 2x$ and $g(x) = e^{3x} \sin 2x$, together with some linear combinations. All solutions approach 0 as $x \to -\infty$.

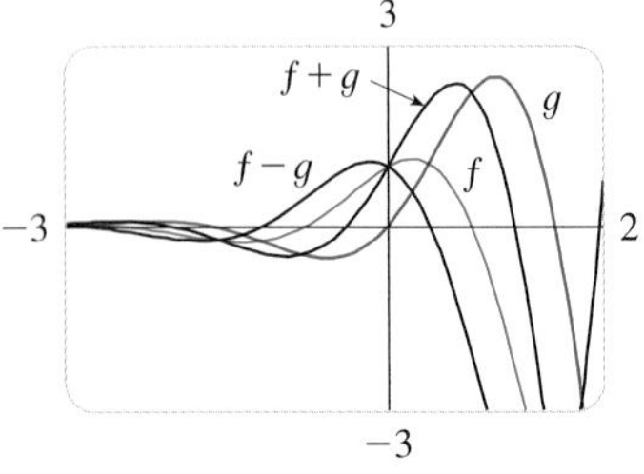

FIGURE 3

V EXAMPLE 4 Solve the equation $y'' - 6y' + 13y = 0$.

SOLUTION The auxiliary equation is $r^2 - 6r + 13 = 0$. By the quadratic formula, the roots are

$$r = \frac{6 \pm \sqrt{36 - 52}}{2} = \frac{6 \pm \sqrt{-16}}{2} = 3 \pm 2i$$

By [11], the general solution of the differential equation is

$$y = e^{3x}(c_1 \cos 2x + c_2 \sin 2x)$$

Initial-Value and Boundary-Value Problems

An **initial-value problem** for the second-order Equation 1 or 2 consists of finding a solution y of the differential equation that also satisfies initial conditions of the form

$$y(x_0) = y_0 \qquad y'(x_0) = y_1$$

where y_0 and y_1 are given constants. If P, Q, R, and G are continuous on an interval and $P(x) \neq 0$ there, then a theorem found in more advanced books guarantees the existence and uniqueness of a solution to this initial-value problem. Examples 5 and 6 illustrate the technique for solving such a problem.

EXAMPLE 5 Solve the initial-value problem

$$y'' + y' - 6y = 0 \qquad y(0) = 1 \qquad y'(0) = 0$$

SOLUTION From Example 1 we know that the general solution of the differential equation is

$$y(x) = c_1 e^{2x} + c_2 e^{-3x}$$

Differentiating this solution, we get

$$y'(x) = 2c_1 e^{2x} - 3c_2 e^{-3x}$$

Figure 4 shows the graph of the solution of the initial-value problem in Example 5. Compare with Figure 1.

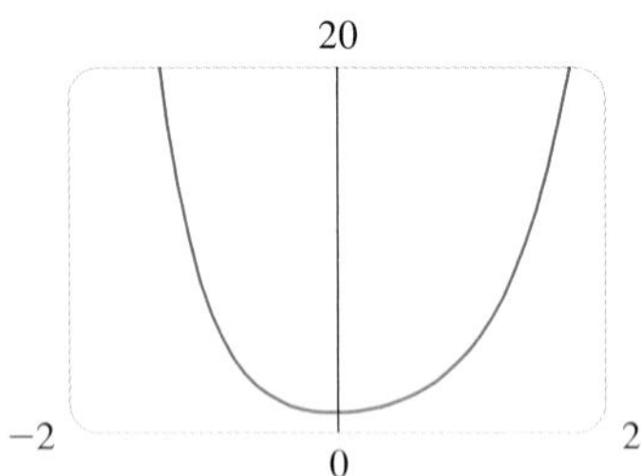

FIGURE 4

To satisfy the initial conditions we require that

[12] $$y(0) = c_1 + c_2 = 1$$

[13] $$y'(0) = 2c_1 - 3c_2 = 0$$

From [13], we have $c_2 = \frac{2}{3}c_1$ and so [12] gives

$$c_1 + \tfrac{2}{3}c_1 = 1 \qquad c_1 = \tfrac{3}{5} \qquad c_2 = \tfrac{2}{5}$$

Thus the required solution of the initial-value problem is

$$y = \tfrac{3}{5}e^{2x} + \tfrac{2}{5}e^{-3x}$$

EXAMPLE 6 Solve the initial-value problem

$$y'' + y = 0 \qquad y(0) = 2 \qquad y'(0) = 3$$

SOLUTION The auxiliary equation is $r^2 + 1 = 0$, or $r^2 = -1$, whose roots are $\pm i$. Thus $\alpha = 0$, $\beta = 1$, and since $e^{0x} = 1$, the general solution is

$$y(x) = c_1 \cos x + c_2 \sin x$$

Since $$y'(x) = -c_1 \sin x + c_2 \cos x$$

The solution to Example 6 is graphed in Figure 5. It appears to be a shifted sine curve and, indeed, you can verify that another way of writing the solution is

$y = \sqrt{13}\sin(x + \phi)$ where $\tan\phi = \frac{2}{3}$

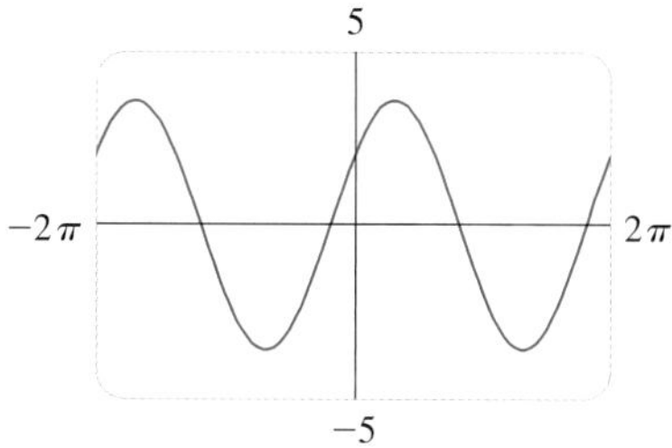

FIGURE 5

the initial conditions become

$$y(0) = c_1 = 2 \qquad y'(0) = c_2 = 3$$

Therefore the solution of the initial-value problem is

$$y(x) = 2\cos x + 3\sin x$$

A **boundary-value problem** for Equation 1 or 2 consists of finding a solution y of the differential equation that also satisfies boundary conditions of the form

$$y(x_0) = y_0 \qquad y(x_1) = y_1$$

In contrast with the situation for initial-value problems, a boundary-value problem does not always have a solution. The method is illustrated in Example 7.

V EXAMPLE 7 Solve the boundary-value problem

$$y'' + 2y' + y = 0 \qquad y(0) = 1 \qquad y(1) = 3$$

SOLUTION The auxiliary equation is

$$r^2 + 2r + 1 = 0 \qquad \text{or} \qquad (r + 1)^2 = 0$$

whose only root is $r = -1$. Therefore the general solution is

$$y(x) = c_1 e^{-x} + c_2 x e^{-x}$$

The boundary conditions are satisfied if

$$y(0) = c_1 = 1$$

$$y(1) = c_1 e^{-1} + c_2 e^{-1} = 3$$

Figure 6 shows the graph of the solution of the boundary-value problem in Example 7.

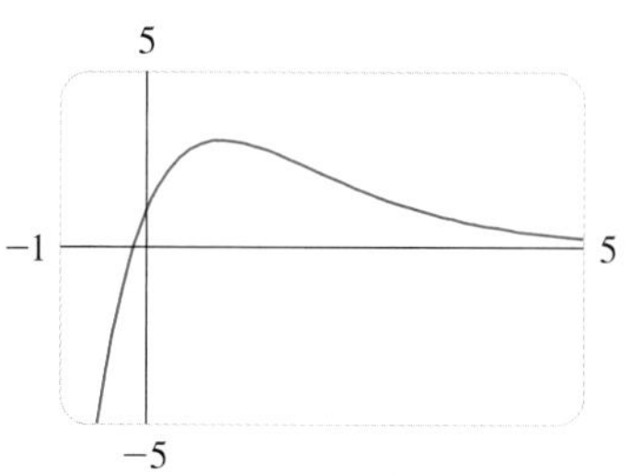

FIGURE 6

The first condition gives $c_1 = 1$, so the second condition becomes

$$e^{-1} + c_2 e^{-1} = 3$$

Solving this equation for c_2 by first multiplying through by e, we get

$$1 + c_2 = 3e \qquad \text{so} \qquad c_2 = 3e - 1$$

Thus the solution of the boundary-value problem is

$$y = e^{-x} + (3e - 1)xe^{-x}$$

Summary: Solutions of $ay'' + by' + c = 0$

Roots of $ar^2 + br + c = 0$	General solution
r_1, r_2 real and distinct	$y = c_1 e^{r_1 x} + c_2 e^{r_2 x}$
$r_1 = r_2 = r$	$y = c_1 e^{rx} + c_2 x e^{rx}$
r_1, r_2 complex: $\alpha \pm i\beta$	$y = e^{\alpha x}(c_1 \cos\beta x + c_2 \sin\beta x)$

17.1 Exercises

1–13 Solve the differential equation.

1. $y'' - y' - 6y = 0$

2. $y'' + 4y' + 4y = 0$

3. $y'' + 16y = 0$

4. $y'' - 8y' + 12y = 0$

5. $9y'' - 12y' + 4y = 0$

6. $25y'' + 9y = 0$

7. $y' = 2y''$

8. $y'' - 4y' + y = 0$

9. $y'' - 4y' + 13y = 0$

10. $y'' + 3y' = 0$

11. $2\dfrac{d^2y}{dt^2} + 2\dfrac{dy}{dt} - y = 0$

12. $8\dfrac{d^2y}{dt^2} + 12\dfrac{dy}{dt} + 5y = 0$

13. $100\dfrac{d^2P}{dt^2} + 200\dfrac{dP}{dt} + 101P = 0$

14–16 Graph the two basic solutions of the differential equation and several other solutions. What features do the solutions have in common?

14. $\dfrac{d^2y}{dx^2} + 4\dfrac{dy}{dx} + 20y = 0$

15. $5\dfrac{d^2y}{dx^2} - 2\dfrac{dy}{dx} - 3y = 0$

16. $9\dfrac{d^2y}{dx^2} + 6\dfrac{dy}{dx} + y = 0$

17–24 Solve the initial-value problem.

17. $y'' - 6y' + 8y = 0, \quad y(0) = 2, \quad y'(0) = 2$

18. $y'' + 4y = 0, \quad y(\pi) = 5, \quad y'(\pi) = -4$

19. $9y'' + 12y' + 4y = 0, \quad y(0) = 1, \quad y'(0) = 0$

20. $2y'' + y' - y = 0, \quad y(0) = 3, \quad y'(0) = 3$

21. $y'' - 6y' + 10y = 0, \quad y(0) = 2, \quad y'(0) = 3$

22. $4y'' - 20y' + 25y = 0, \quad y(0) = 2, \quad y'(0) = -3$

23. $y'' - y' - 12y = 0, \quad y(1) = 0, \quad y'(1) = 1$

24. $4y'' + 4y' + 3y = 0, \quad y(0) = 0, \quad y'(0) = 1$

25–32 Solve the boundary-value problem, if possible.

25. $y'' + 4y = 0, \quad y(0) = 5, \quad y(\pi/4) = 3$

26. $y'' = 4y, \quad y(0) = 1, \quad y(1) = 0$

27. $y'' + 4y' + 4y = 0, \quad y(0) = 2, \quad y(1) = 0$

28. $y'' - 8y' + 17y = 0, \quad y(0) = 3, \quad y(\pi) = 2$

29. $y'' = y', \quad y(0) = 1, \quad y(1) = 2$

30. $4y'' - 4y' + y = 0, \quad y(0) = 4, \quad y(2) = 0$

31. $y'' + 4y' + 20y = 0, \quad y(0) = 1, \quad y(\pi) = 2$

32. $y'' + 4y' + 20y = 0, \quad y(0) = 1, \quad y(\pi) = e^{-2\pi}$

33. Let L be a nonzero real number.
(a) Show that the boundary-value problem $y'' + \lambda y = 0$, $y(0) = 0$, $y(L) = 0$ has only the trivial solution $y = 0$ for the cases $\lambda = 0$ and $\lambda < 0$.
(b) For the case $\lambda > 0$, find the values of λ for which this problem has a nontrivial solution and give the corresponding solution.

34. If a, b, and c are all positive constants and $y(x)$ is a solution of the differential equation $ay'' + by' + cy = 0$, show that $\lim_{x\to\infty} y(x) = 0$.

35. Consider the boundary-value problem $y'' - 2y' + 2y = 0$, $y(a) = c$, $y(b) = d$.
(a) If this problem has a unique solution, how are a and b related?
(b) If this problem has no solution, how are a, b, c, and d related?
(c) If this problem has infinitely many solutions, how are a, b, c, and d related?

Graphing calculator or computer required

1. Homework Hints available at stewartcalculus.com

17.2 Nonhomogeneous Linear Equations

In this section we learn how to solve second-order nonhomogeneous linear differential equations with constant coefficients, that is, equations of the form

$$ay'' + by' + cy = G(x) \tag{1}$$

where a, b, and c are constants and G is a continuous function. The related homogeneous equation

$$ay'' + by' + cy = 0 \tag{2}$$

is called the **complementary equation** and plays an important role in the solution of the original nonhomogeneous equation [1].

> **3 Theorem** The general solution of the nonhomogeneous differential equation [1] can be written as
>
> $$y(x) = y_p(x) + y_c(x)$$
>
> where y_p is a particular solution of Equation 1 and y_c is the general solution of the complementary Equation 2.

PROOF We verify that if y is any solution of Equation 1, then $y - y_p$ is a solution of the complementary Equation 2. Indeed

$$\begin{aligned} a(y - y_p)'' + b(y - y_p)' + c(y - y_p) &= ay'' - ay_p'' + by' - by_p' + cy - cy_p \\ &= (ay'' + by' + cy) - (ay_p'' + by_p' + cy_p) \\ &= G(x) - G(x) = 0 \end{aligned}$$

This shows that every solution is of the form $y(x) = y_p(x) + y_c(x)$. It is easy to check that every function of this form is a solution. ■

We know from Section 17.1 how to solve the complementary equation. (Recall that the solution is $y_c = c_1y_1 + c_2y_2$, where y_1 and y_2 are linearly independent solutions of Equation 2.) Therefore Theorem 3 says that we know the general solution of the nonhomogeneous equation as soon as we know a particular solution y_p. There are two methods for finding a particular solution: The method of undetermined coefficients is straightforward but works only for a restricted class of functions G. The method of variation of parameters works for every function G but is usually more difficult to apply in practice.

The Method of Undetermined Coefficients

We first illustrate the method of undetermined coefficients for the equation

$$ay'' + by' + cy = G(x)$$

where $G(x)$ is a polynomial. It is reasonable to guess that there is a particular solution y_p that is a polynomial of the same degree as G because if y is a polynomial, then $ay'' + by' + cy$ is also a polynomial. We therefore substitute $y_p(x) =$ a polynomial (of the same degree as G) into the differential equation and determine the coefficients.

V EXAMPLE 1 Solve the equation $y'' + y' - 2y = x^2$.

SOLUTION The auxiliary equation of $y'' + y' - 2y = 0$ is

$$r^2 + r - 2 = (r - 1)(r + 2) = 0$$

with roots $r = 1, -2$. So the solution of the complementary equation is

$$y_c = c_1e^x + c_2e^{-2x}$$

Since $G(x) = x^2$ is a polynomial of degree 2, we seek a particular solution of the form

$$y_p(x) = Ax^2 + Bx + C$$

Then $y_p' = 2Ax + B$ and $y_p'' = 2A$ so, substituting into the given differential equation, we have

$$(2A) + (2Ax + B) - 2(Ax^2 + Bx + C) = x^2$$

or
$$-2Ax^2 + (2A - 2B)x + (2A + B - 2C) = x^2$$

Polynomials are equal when their coefficients are equal. Thus

$$-2A = 1 \qquad 2A - 2B = 0 \qquad 2A + B - 2C = 0$$

The solution of this system of equations is

$$A = -\tfrac{1}{2} \qquad B = -\tfrac{1}{2} \qquad C = -\tfrac{3}{4}$$

A particular solution is therefore

$$y_p(x) = -\tfrac{1}{2}x^2 - \tfrac{1}{2}x - \tfrac{3}{4}$$

and, by Theorem 3, the general solution is

$$y = y_c + y_p = c_1e^x + c_2e^{-2x} - \tfrac{1}{2}x^2 - \tfrac{1}{2}x - \tfrac{3}{4}$$

Figure 1 shows four solutions of the differential equation in Example 1 in terms of the particular solution y_p and the functions $f(x) = e^x$ and $g(x) = e^{-2x}$.

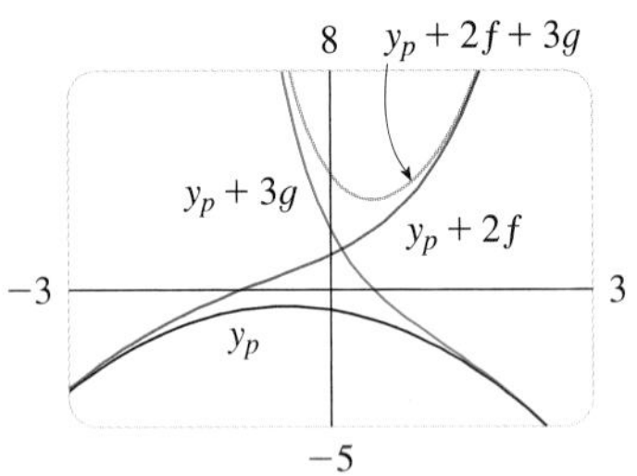

FIGURE 1

If $G(x)$ (the right side of Equation 1) is of the form Ce^{kx}, where C and k are constants, then we take as a trial solution a function of the same form, $y_p(x) = Ae^{kx}$, because the derivatives of e^{kx} are constant multiples of e^{kx}.

EXAMPLE 2 Solve $y'' + 4y = e^{3x}$.

SOLUTION The auxiliary equation is $r^2 + 4 = 0$ with roots $\pm 2i$, so the solution of the complementary equation is

$$y_c(x) = c_1 \cos 2x + c_2 \sin 2x$$

For a particular solution we try $y_p(x) = Ae^{3x}$. Then $y_p' = 3Ae^{3x}$ and $y_p'' = 9Ae^{3x}$. Substituting into the differential equation, we have

$$9Ae^{3x} + 4(Ae^{3x}) = e^{3x}$$

so $13Ae^{3x} = e^{3x}$ and $A = \frac{1}{13}$. Thus a particular solution is

$$y_p(x) = \tfrac{1}{13}e^{3x}$$

and the general solution is

$$y(x) = c_1 \cos 2x + c_2 \sin 2x + \tfrac{1}{13}e^{3x}$$

Figure 2 shows solutions of the differential equation in Example 2 in terms of y_p and the functions $f(x) = \cos 2x$ and $g(x) = \sin 2x$. Notice that all solutions approach ∞ as $x \to \infty$ and all solutions (except y_p) resemble sine functions when x is negative.

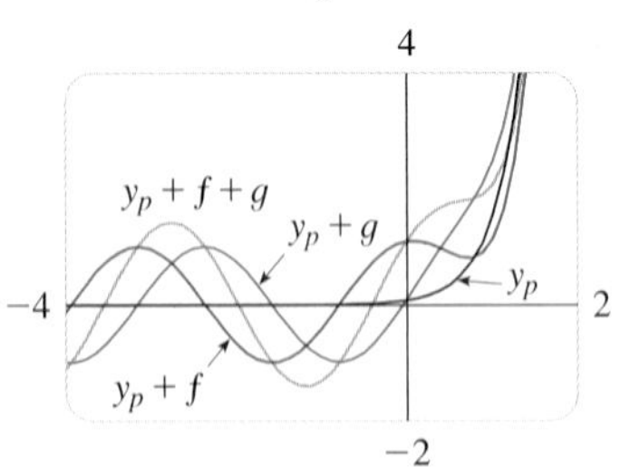

FIGURE 2

If $G(x)$ is either $C \cos kx$ or $C \sin kx$, then, because of the rules for differentiating the sine and cosine functions, we take as a trial particular solution a function of the form

$$y_p(x) = A \cos kx + B \sin kx$$

V **EXAMPLE 3** Solve $y'' + y' - 2y = \sin x$.

SOLUTION We try a particular solution

$$y_p(x) = A \cos x + B \sin x$$

Then $\qquad y_p' = -A\sin x + B\cos x \qquad y_p'' = -A\cos x - B\sin x$

so substitution in the differential equation gives

$$(-A\cos x - B\sin x) + (-A\sin x + B\cos x) - 2(A\cos x + B\sin x) = \sin x$$

or

$$(-3A + B)\cos x + (-A - 3B)\sin x = \sin x$$

This is true if

$$-3A + B = 0 \qquad \text{and} \qquad -A - 3B = 1$$

The solution of this system is

$$A = -\tfrac{1}{10} \qquad B = -\tfrac{3}{10}$$

so a particular solution is

$$y_p(x) = -\tfrac{1}{10}\cos x - \tfrac{3}{10}\sin x$$

In Example 1 we determined that the solution of the complementary equation is $y_c = c_1e^x + c_2e^{-2x}$. Thus the general solution of the given equation is

$$y(x) = c_1e^x + c_2e^{-2x} - \tfrac{1}{10}(\cos x + 3\sin x)$$

If $G(x)$ is a product of functions of the preceding types, then we take the trial solution to be a product of functions of the same type. For instance, in solving the differential equation

$$y'' + 2y' + 4y = x\cos 3x$$

we would try

$$y_p(x) = (Ax + B)\cos 3x + (Cx + D)\sin 3x$$

If $G(x)$ is a sum of functions of these types, we use the easily verified *principle of superposition,* which says that if y_{p_1} and y_{p_2} are solutions of

$$ay'' + by' + cy = G_1(x) \qquad ay'' + by' + cy = G_2(x)$$

respectively, then $y_{p_1} + y_{p_2}$ is a solution of

$$ay'' + by' + cy = G_1(x) + G_2(x)$$

V EXAMPLE 4 Solve $y'' - 4y = xe^x + \cos 2x$.

SOLUTION The auxiliary equation is $r^2 - 4 = 0$ with roots ± 2, so the solution of the complementary equation is $y_c(x) = c_1e^{2x} + c_2e^{-2x}$. For the equation $y'' - 4y = xe^x$ we try

$$y_{p_1}(x) = (Ax + B)e^x$$

Then $y_{p_1}' = (Ax + A + B)e^x$, $y_{p_1}'' = (Ax + 2A + B)e^x$, so substitution in the equation gives

$$(Ax + 2A + B)e^x - 4(Ax + B)e^x = xe^x$$

or

$$(-3Ax + 2A - 3B)e^x = xe^x$$

Thus $-3A = 1$ and $2A - 3B = 0$, so $A = -\frac{1}{3}$, $B = -\frac{2}{9}$, and

$$y_{p_1}(x) = \left(-\tfrac{1}{3}x - \tfrac{2}{9}\right)e^x$$

For the equation $y'' - 4y = \cos 2x$, we try

$$y_{p_2}(x) = C\cos 2x + D\sin 2x$$

In Figure 3 we show the particular solution $y_p = y_{p_1} + y_{p_2}$ of the differential equation in Example 4. The other solutions are given in terms of $f(x) = e^{2x}$ and $g(x) = e^{-2x}$.

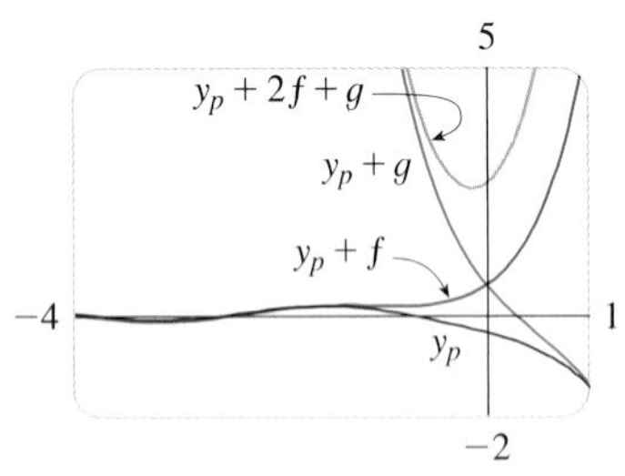

FIGURE 3

Substitution gives

$$-4C\cos 2x - 4D\sin 2x - 4(C\cos 2x + D\sin 2x) = \cos 2x$$

or

$$-8C\cos 2x - 8D\sin 2x = \cos 2x$$

Therefore $-8C = 1$, $-8D = 0$, and

$$y_{p_2}(x) = -\tfrac{1}{8}\cos 2x$$

By the superposition principle, the general solution is

$$y = y_c + y_{p_1} + y_{p_2} = c_1e^{2x} + c_2e^{-2x} - \left(\tfrac{1}{3}x + \tfrac{2}{9}\right)e^x - \tfrac{1}{8}\cos 2x$$

Finally we note that the recommended trial solution y_p sometimes turns out to be a solution of the complementary equation and therefore can't be a solution of the nonhomogeneous equation. In such cases we multiply the recommended trial solution by x (or by x^2 if necessary) so that no term in $y_p(x)$ is a solution of the complementary equation.

EXAMPLE 5 Solve $y'' + y = \sin x$.

SOLUTION The auxiliary equation is $r^2 + 1 = 0$ with roots $\pm i$, so the solution of the complementary equation is

$$y_c(x) = c_1\cos x + c_2\sin x$$

Ordinarily, we would use the trial solution

$$y_p(x) = A\cos x + B\sin x$$

but we observe that it is a solution of the complementary equation, so instead we try

$$y_p(x) = Ax\cos x + Bx\sin x$$

Then

$$y_p'(x) = A\cos x - Ax\sin x + B\sin x + Bx\cos x$$

$$y_p''(x) = -2A\sin x - Ax\cos x + 2B\cos x - Bx\sin x$$

Substitution in the differential equation gives

$$y_p'' + y_p = -2A\sin x + 2B\cos x = \sin x$$

The graphs of four solutions of the differential equation in Example 5 are shown in Figure 4.

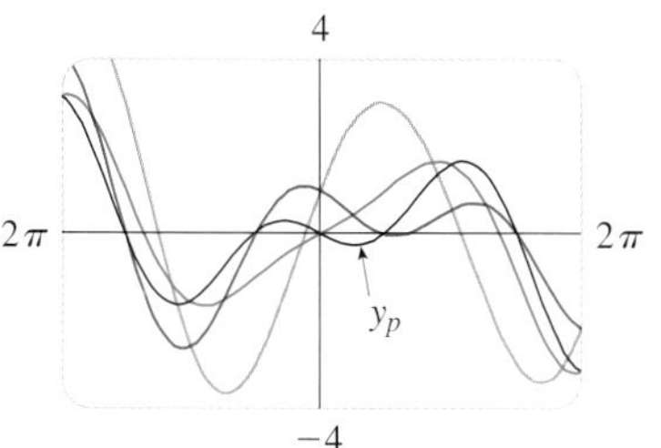

FIGURE 4

so $A = -\frac{1}{2}$, $B = 0$, and

$$y_p(x) = -\tfrac{1}{2}x \cos x$$

The general solution is

$$y(x) = c_1 \cos x + c_2 \sin x - \tfrac{1}{2}x \cos x$$

We summarize the method of undetermined coefficients as follows:

Summary of the Method of Undetermined Coefficients

1. If $G(x) = e^{kx}P(x)$, where P is a polynomial of degree n, then try $y_p(x) = e^{kx}Q(x)$, where $Q(x)$ is an nth-degree polynomial (whose coefficients are determined by substituting in the differential equation).
2. If $G(x) = e^{kx}P(x) \cos mx$ or $G(x) = e^{kx}P(x) \sin mx$, where P is an nth-degree polynomial, then try

$$y_p(x) = e^{kx}Q(x) \cos mx + e^{kx}R(x) \sin mx$$

where Q and R are nth-degree polynomials.

Modification: If any term of y_p is a solution of the complementary equation, multiply y_p by x (or by x^2 if necessary).

EXAMPLE 6 Determine the form of the trial solution for the differential equation $y'' - 4y' + 13y = e^{2x} \cos 3x$.

SOLUTION Here $G(x)$ has the form of part 2 of the summary, where $k = 2$, $m = 3$, and $P(x) = 1$. So, at first glance, the form of the trial solution would be

$$y_p(x) = e^{2x}(A \cos 3x + B \sin 3x)$$

But the auxiliary equation is $r^2 - 4r + 13 = 0$, with roots $r = 2 \pm 3i$, so the solution of the complementary equation is

$$y_c(x) = e^{2x}(c_1 \cos 3x + c_2 \sin 3x)$$

This means that we have to multiply the suggested trial solution by x. So, instead, we use

$$y_p(x) = xe^{2x}(A \cos 3x + B \sin 3x)$$

The Method of Variation of Parameters

Suppose we have already solved the homogeneous equation $ay'' + by' + cy = 0$ and written the solution as

4 $$y(x) = c_1y_1(x) + c_2y_2(x)$$

where y_1 and y_2 are linearly independent solutions. Let's replace the constants (or parameters) c_1 and c_2 in Equation 4 by arbitrary functions $u_1(x)$ and $u_2(x)$. We look for a particular solution of the nonhomogeneous equation $ay'' + by' + cy = G(x)$ of the form

$$\boxed{5} \qquad y_p(x) = u_1(x)\,y_1(x) + u_2(x)\,y_2(x)$$

(This method is called **variation of parameters** because we have varied the parameters c_1 and c_2 to make them functions.) Differentiating Equation 5, we get

$$\boxed{6} \qquad y_p' = (u_1'y_1 + u_2'y_2) + (u_1y_1' + u_2y_2')$$

Since u_1 and u_2 are arbitrary functions, we can impose two conditions on them. One condition is that y_p is a solution of the differential equation; we can choose the other condition so as to simplify our calculations. In view of the expression in Equation 6, let's impose the condition that

$$\boxed{7} \qquad u_1'y_1 + u_2'y_2 = 0$$

Then
$$y_p'' = u_1'y_1' + u_2'y_2' + u_1y_1'' + u_2y_2''$$

Substituting in the differential equation, we get

$$a(u_1'y_1' + u_2'y_2' + u_1y_1'' + u_2y_2'') + b(u_1y_1' + u_2y_2') + c(u_1y_1 + u_2y_2) = G$$

or

$$\boxed{8} \qquad u_1(ay_1'' + by_1' + cy_1) + u_2(ay_2'' + by_2' + cy_2) + a(u_1'y_1' + u_2'y_2') = G$$

But y_1 and y_2 are solutions of the complementary equation, so

$$ay_1'' + by_1' + cy_1 = 0 \qquad \text{and} \qquad ay_2'' + by_2' + cy_2 = 0$$

and Equation 8 simplifies to

$$\boxed{9} \qquad a(u_1'y_1' + u_2'y_2') = G$$

Equations 7 and 9 form a system of two equations in the unknown functions u_1' and u_2'. After solving this system we may be able to integrate to find u_1 and u_2 and then the particular solution is given by Equation 5.

EXAMPLE 7 Solve the equation $y'' + y = \tan x$, $0 < x < \pi/2$.

SOLUTION The auxiliary equation is $r^2 + 1 = 0$ with roots $\pm i$, so the solution of $y'' + y = 0$ is $y(x) = c_1 \sin x + c_2 \cos x$. Using variation of parameters, we seek a solution of the form

$$y_p(x) = u_1(x)\sin x + u_2(x)\cos x$$

Then
$$y_p' = (u_1' \sin x + u_2' \cos x) + (u_1 \cos x - u_2 \sin x)$$

Set

$$\boxed{10} \qquad u_1' \sin x + u_2' \cos x = 0$$

Then $$y_p'' = u_1' \cos x - u_2' \sin x - u_1 \sin x - u_2 \cos x$$

For y_p to be a solution we must have

11 $$y_p'' + y_p = u_1' \cos x - u_2' \sin x = \tan x$$

Solving Equations 10 and 11, we get

$$u_1'(\sin^2 x + \cos^2 x) = \cos x \tan x$$

$$u_1' = \sin x \qquad u_1(x) = -\cos x$$

(We seek a particular solution, so we don't need a constant of integration here.) Then, from Equation 10, we obtain

$$u_2' = -\frac{\sin x}{\cos x} u_1' = -\frac{\sin^2 x}{\cos x} = \frac{\cos^2 x - 1}{\cos x} = \cos x - \sec x$$

Figure 5 shows four solutions of the differential equation in Example 7.

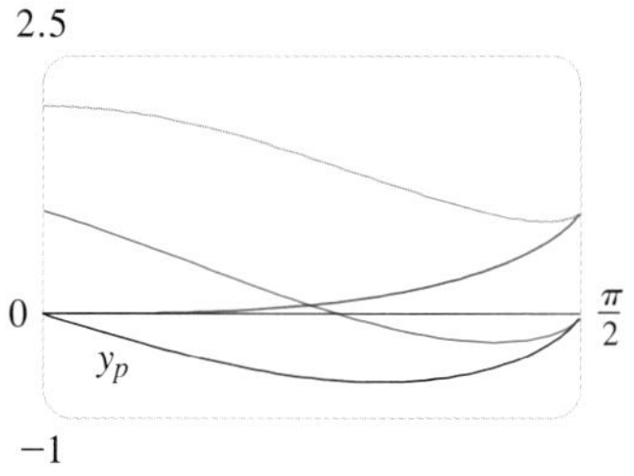

So $$u_2(x) = \sin x - \ln(\sec x + \tan x)$$

(Note that $\sec x + \tan x > 0$ for $0 < x < \pi/2$.) Therefore

$$y_p(x) = -\cos x \sin x + [\sin x - \ln(\sec x + \tan x)] \cos x$$

$$= -\cos x \ln(\sec x + \tan x)$$

and the general solution is

$$y(x) = c_1 \sin x + c_2 \cos x - \cos x \ln(\sec x + \tan x)$$

2.5

0

y_p

$\frac{\pi}{2}$

−1

FIGURE 5

17.2 Exercises

1–10 Solve the differential equation or initial-value problem using the method of undetermined coefficients.

1. $y'' - 2y' - 3y = \cos 2x$

2. $y'' - y = x^3 - x$

3. $y'' + 9y = e^{-2x}$

4. $y'' + 2y' + 5y = 1 + e^x$

5. $y'' - 4y' + 5y = e^{-x}$

6. $y'' - 4y' + 4y = x - \sin x$

7. $y'' + y = e^x + x^3, \quad y(0) = 2, \quad y'(0) = 0$

8. $y'' - 4y = e^x \cos x, \quad y(0) = 1, \quad y'(0) = 2$

9. $y'' - y' = xe^x, \quad y(0) = 2, \quad y'(0) = 1$

10. $y'' + y' - 2y = x + \sin 2x, \quad y(0) = 1, \quad y'(0) = 0$

11–12 Graph the particular solution and several other solutions. What characteristics do these solutions have in common?

11. $y'' + 3y' + 2y = \cos x$

12. $y'' + 4y = e^{-x}$

13–18 Write a trial solution for the method of undetermined coefficients. Do not determine the coefficients.

13. $y'' - y' - 2y = xe^x \cos x$

14. $y'' + 4y = \cos 4x + \cos 2x$

15. $y'' - 3y' + 2y = e^x + \sin x$

16. $y'' + 3y' - 4y = (x^3 + x)e^x$

17. $y'' + 2y' + 10y = x^2 e^{-x} \cos 3x$

18. $y'' + 4y = e^{3x} + x \sin 2x$

Graphing calculator or computer required

1. Homework Hints available at stewartcalculus.com

19–22 Solve the differential equation using (a) undetermined coefficients and (b) variation of parameters.

19. $4y'' + y = \cos x$

20. $y'' - 2y' - 3y = x + 2$

21. $y'' - 2y' + y = e^{2x}$

22. $y'' - y' = e^x$

23–28 Solve the differential equation using the method of variation of parameters.

23. $y'' + y = \sec^2 x,\ 0 < x < \pi/2$

24. $y'' + y = \sec^3 x,\ 0 < x < \pi/2$

25. $y'' - 3y' + 2y = \dfrac{1}{1 + e^{-x}}$

26. $y'' + 3y' + 2y = \sin(e^x)$

27. $y'' - 2y' + y = \dfrac{e^x}{1 + x^2}$

28. $y'' + 4y' + 4y = \dfrac{e^{-2x}}{x^3}$

17.3 Applications of Second-Order Differential Equations

Second-order linear differential equations have a variety of applications in science and engineering. In this section we explore two of them: the vibration of springs and electric circuits.

Vibrating Springs

We consider the motion of an object with mass m at the end of a spring that is either vertical (as in Figure 1) or horizontal on a level surface (as in Figure 2).

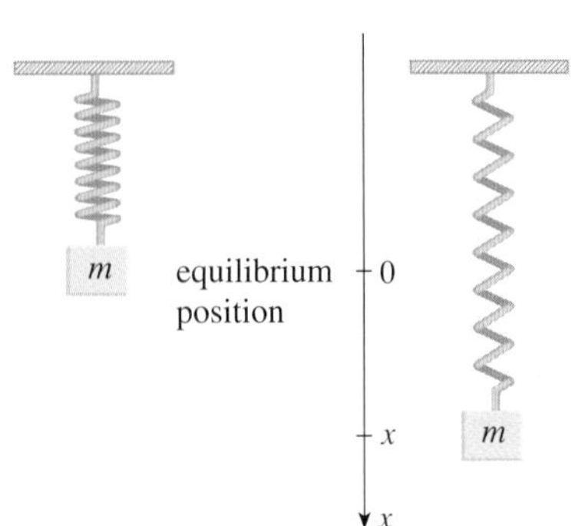

FIGURE 1

In Section 6.4 we discussed Hooke's Law, which says that if the spring is stretched (or compressed) x units from its natural length, then it exerts a force that is proportional to x:

$$\text{restoring force} = -kx$$

where k is a positive constant (called the **spring constant**). If we ignore any external resisting forces (due to air resistance or friction) then, by Newton's Second Law (force equals mass times acceleration), we have

$$\boxed{1} \qquad m\frac{d^2x}{dt^2} = -kx \qquad \text{or} \qquad m\frac{d^2x}{dt^2} + kx = 0$$

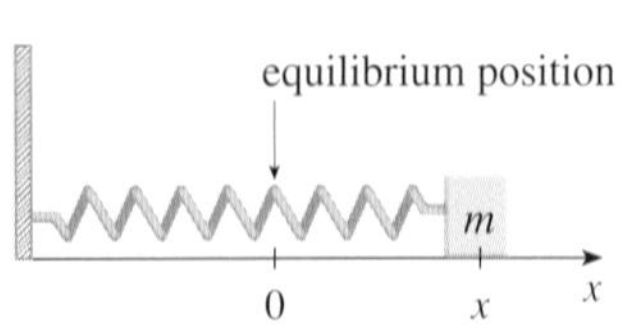

FIGURE 2

This is a second-order linear differential equation. Its auxiliary equation is $mr^2 + k = 0$ with roots $r = \pm\omega i$, where $\omega = \sqrt{k/m}$. Thus the general solution is

$$x(t) = c_1 \cos \omega t + c_2 \sin \omega t$$

which can also be written as

$$x(t) = A\cos(\omega t + \delta)$$

where

$$\omega = \sqrt{k/m} \quad \text{(frequency)}$$

$$A = \sqrt{c_1^2 + c_2^2} \quad \text{(amplitude)}$$

$$\cos \delta = \frac{c_1}{A} \qquad \sin \delta = -\frac{c_2}{A} \qquad (\delta \text{ is the phase angle})$$

(See Exercise 17.) This type of motion is called **simple harmonic motion**.

V EXAMPLE 1 A spring with a mass of 2 kg has natural length 0.5 m. A force of 25.6 N is required to maintain it stretched to a length of 0.7 m. If the spring is stretched to a length of 0.7 m and then released with initial velocity 0, find the position of the mass at any time t.

SOLUTION From Hooke's Law, the force required to stretch the spring is

$$k(0.2) = 25.6$$

so $k = 25.6/0.2 = 128$. Using this value of the spring constant k, together with $m = 2$ in Equation 1, we have

$$2\frac{d^2x}{dt^2} + 128x = 0$$

As in the earlier general discussion, the solution of this equation is

$$x(t) = c_1 \cos 8t + c_2 \sin 8t \tag{2}$$

We are given the initial condition that $x(0) = 0.2$. But, from Equation 2, $x(0) = c_1$. Therefore $c_1 = 0.2$. Differentiating Equation 2, we get

$$x'(t) = -8c_1 \sin 8t + 8c_2 \cos 8t$$

Since the initial velocity is given as $x'(0) = 0$, we have $c_2 = 0$ and so the solution is

$$x(t) = \tfrac{1}{5}\cos 8t$$

Damped Vibrations

FIGURE 3

We next consider the motion of a spring that is subject to a frictional force (in the case of the horizontal spring of Figure 2) or a damping force (in the case where a vertical spring moves through a fluid as in Figure 3). An example is the damping force supplied by a shock absorber in a car or a bicycle.

We assume that the damping force is proportional to the velocity of the mass and acts in the direction opposite to the motion. (This has been confirmed, at least approximately, by some physical experiments.) Thus

$$\text{damping force} = -c\frac{dx}{dt}$$

Schwinn Cycling and Fitness

where c is a positive constant, called the **damping constant**. Thus, in this case, Newton's Second Law gives

$$m\frac{d^2x}{dt^2} = \text{restoring force} + \text{damping force} = -kx - c\frac{dx}{dt}$$

or

$$m\frac{d^2x}{dt^2} + c\frac{dx}{dt} + kx = 0 \tag{3}$$

Equation 3 is a second-order linear differential equation and its auxiliary equation is $mr^2 + cr + k = 0$. The roots are

$$\boxed{4} \qquad r_1 = \frac{-c + \sqrt{c^2 - 4mk}}{2m} \qquad r_2 = \frac{-c - \sqrt{c^2 - 4mk}}{2m}$$

According to Section 17.1 we need to discuss three cases.

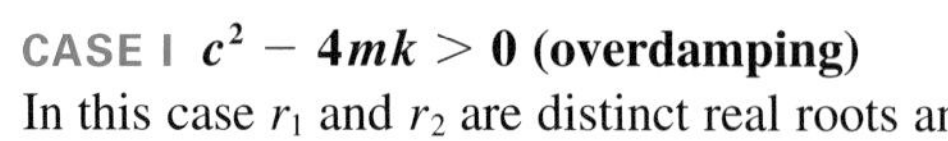

CASE I $c^2 - 4mk > 0$ (overdamping)
In this case r_1 and r_2 are distinct real roots and

$$x = c_1 e^{r_1 t} + c_2 e^{r_2 t}$$

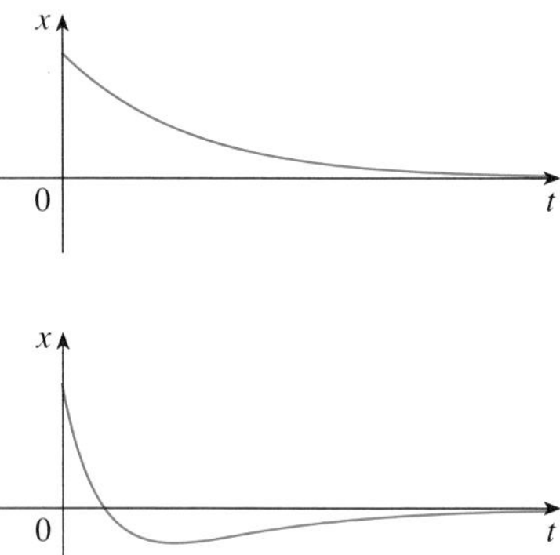

FIGURE 4
Overdamping

Since c, m, and k are all positive, we have $\sqrt{c^2 - 4mk} < c$, so the roots r_1 and r_2 given by Equations 4 must both be negative. This shows that $x \to 0$ as $t \to \infty$. Typical graphs of x as a function of t are shown in Figure 4. Notice that oscillations do not occur. (It's possible for the mass to pass through the equilibrium position once, but only once.) This is because $c^2 > 4mk$ means that there is a strong damping force (high-viscosity oil or grease) compared with a weak spring or small mass.

CASE II $c^2 - 4mk = 0$ (critical damping)
This case corresponds to equal roots

$$r_1 = r_2 = -\frac{c}{2m}$$

and the solution is given by

$$x = (c_1 + c_2 t)e^{-(c/2m)t}$$

It is similar to Case I, and typical graphs resemble those in Figure 4 (see Exercise 12), but the damping is just sufficient to suppress vibrations. Any decrease in the viscosity of the fluid leads to the vibrations of the following case.

CASE III $c^2 - 4mk < 0$ (underdamping)
Here the roots are complex:

$$\left.\begin{matrix} r_1 \\ r_2 \end{matrix}\right\} = -\frac{c}{2m} \pm \omega i$$

where

$$\omega = \frac{\sqrt{4mk - c^2}}{2m}$$

The solution is given by

$$x = e^{-(c/2m)t}(c_1 \cos \omega t + c_2 \sin \omega t)$$

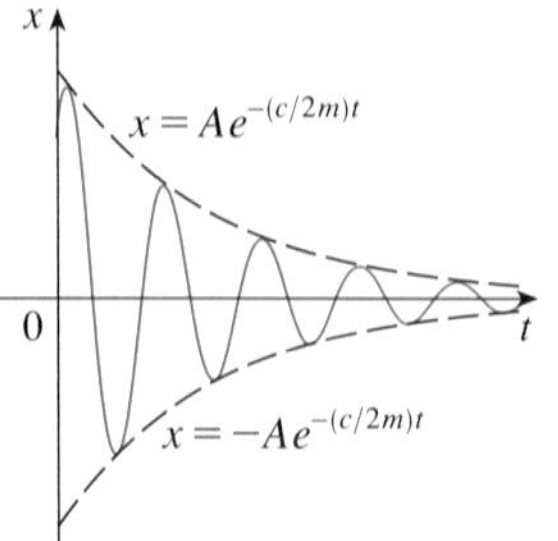

FIGURE 5
Underdamping

We see that there are oscillations that are damped by the factor $e^{-(c/2m)t}$. Since $c > 0$ and $m > 0$, we have $-(c/2m) < 0$ so $e^{-(c/2m)t} \to 0$ as $t \to \infty$. This implies that $x \to 0$ as $t \to \infty$; that is, the motion decays to 0 as time increases. A typical graph is shown in Figure 5.

V EXAMPLE 2 Suppose that the spring of Example 1 is immersed in a fluid with damping constant $c = 40$. Find the position of the mass at any time t if it starts from the equilibrium position and is given a push to start it with an initial velocity of 0.6 m/s.

SOLUTION From Example 1, the mass is $m = 2$ and the spring constant is $k = 128$, so the differential equation [3] becomes

$$2\frac{d^2x}{dt^2} + 40\frac{dx}{dt} + 128x = 0$$

or

$$\frac{d^2x}{dt^2} + 20\frac{dx}{dt} + 64x = 0$$

The auxiliary equation is $r^2 + 20r + 64 = (r + 4)(r + 16) = 0$ with roots -4 and -16, so the motion is overdamped and the solution is

$$x(t) = c_1e^{-4t} + c_2e^{-16t}$$

Figure 6 shows the graph of the position function for the overdamped motion in Example 2.

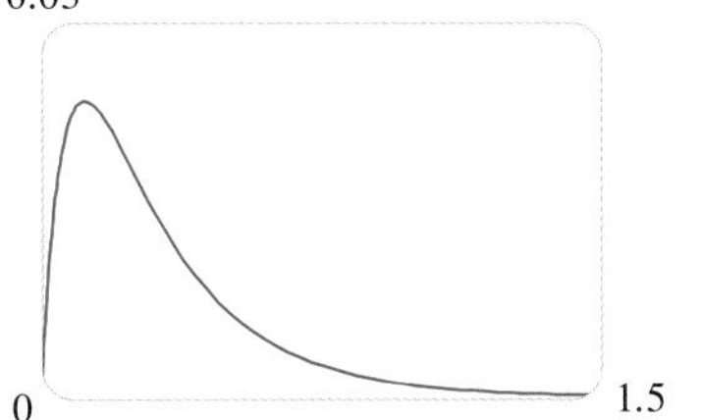

FIGURE 6

We are given that $x(0) = 0$, so $c_1 + c_2 = 0$. Differentiating, we get

$$x'(t) = -4c_1e^{-4t} - 16c_2e^{-16t}$$

so

$$x'(0) = -4c_1 - 16c_2 = 0.6$$

Since $c_2 = -c_1$, this gives $12c_1 = 0.6$ or $c_1 = 0.05$. Therefore

$$x = 0.05(e^{-4t} - e^{-16t})$$

Forced Vibrations

Suppose that, in addition to the restoring force and the damping force, the motion of the spring is affected by an external force $F(t)$. Then Newton's Second Law gives

$$m\frac{d^2x}{dt^2} = \text{restoring force} + \text{damping force} + \text{external force}$$

$$= -kx - c\frac{dx}{dt} + F(t)$$

Thus, instead of the homogeneous equation [3], the motion of the spring is now governed by the following nonhomogeneous differential equation:

5
$$m\frac{d^2x}{dt^2} + c\frac{dx}{dt} + kx = F(t)$$

The motion of the spring can be determined by the methods of Section 17.2.

A commonly occurring type of external force is a periodic force function

$$F(t) = F_0 \cos \omega_0 t \qquad \text{where} \quad \omega_0 \neq \omega = \sqrt{k/m}$$

In this case, and in the absence of a damping force ($c = 0$), you are asked in Exercise 9 to use the method of undetermined coefficients to show that

$$x(t) = c_1 \cos \omega t + c_2 \sin \omega t + \frac{F_0}{m(\omega^2 - \omega_0^2)} \cos \omega_0 t \tag{6}$$

If $\omega_0 = \omega$, then the applied frequency reinforces the natural frequency and the result is vibrations of large amplitude. This is the phenomenon of **resonance** (see Exercise 10).

Electric Circuits

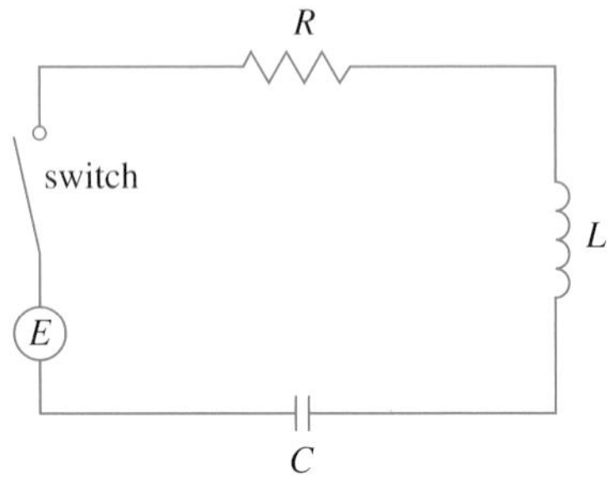

In Sections 9.3 and 9.5 we were able to use first-order separable and linear equations to analyze electric circuits that contain a resistor and inductor (see Figure 5 in Section 9.3 or Figure 4 in Section 9.5) or a resistor and capacitor (see Exercise 29 in Section 9.5). Now that we know how to solve second-order linear equations, we are in a position to analyze the circuit shown in Figure 7. It contains an electromotive force E (supplied by a battery or generator), a resistor R, an inductor L, and a capacitor C, in series. If the charge on the capacitor at time t is $Q = Q(t)$, then the current is the rate of change of Q with respect to t: $I = dQ/dt$. As in Section 9.5, it is known from physics that the voltage drops across the resistor, inductor, and capacitor are

FIGURE 7

$$RI \qquad L\frac{dI}{dt} \qquad \frac{Q}{C}$$

respectively. Kirchhoff's voltage law says that the sum of these voltage drops is equal to the supplied voltage:

$$L\frac{dI}{dt} + RI + \frac{Q}{C} = E(t)$$

Since $I = dQ/dt$, this equation becomes

$$\boxed{L\frac{d^2Q}{dt^2} + R\frac{dQ}{dt} + \frac{1}{C}Q = E(t)} \tag{7}$$

which is a second-order linear differential equation with constant coefficients. If the charge Q_0 and the current I_0 are known at time 0, then we have the initial conditions

$$Q(0) = Q_0 \qquad Q'(0) = I(0) = I_0$$

and the initial-value problem can be solved by the methods of Section 17.2.

A differential equation for the current can be obtained by differentiating Equation 7 with respect to t and remembering that $I = dQ/dt$:

$$L\frac{d^2I}{dt^2} + R\frac{dI}{dt} + \frac{1}{C}I = E'(t)$$

V EXAMPLE 3 Find the charge and current at time t in the circuit of Figure 7 if $R = 40\ \Omega$, $L = 1$ H, $C = 16 \times 10^{-4}$ F, $E(t) = 100\cos 10t$, and the initial charge and current are both 0.

SOLUTION With the given values of L, R, C, and $E(t)$, Equation 7 becomes

8 $$\frac{d^2Q}{dt^2} + 40\frac{dQ}{dt} + 625Q = 100\cos 10t$$

The auxiliary equation is $r^2 + 40r + 625 = 0$ with roots

$$r = \frac{-40 \pm \sqrt{-900}}{2} = -20 \pm 15i$$

so the solution of the complementary equation is

$$Q_c(t) = e^{-20t}(c_1\cos 15t + c_2\sin 15t)$$

For the method of undetermined coefficients we try the particular solution

$$Q_p(t) = A\cos 10t + B\sin 10t$$

Then

$$Q_p'(t) = -10A\sin 10t + 10B\cos 10t$$

$$Q_p''(t) = -100A\cos 10t - 100B\sin 10t$$

Substituting into Equation 8, we have

$$(-100A\cos 10t - 100B\sin 10t) + 40(-10A\sin 10t + 10B\cos 10t)$$
$$+ 625(A\cos 10t + B\sin 10t) = 100\cos 10t$$

or

$$(525A + 400B)\cos 10t + (-400A + 525B)\sin 10t = 100\cos 10t$$

Equating coefficients, we have

$$\begin{aligned} 525A + 400B &= 100 \\ -400A + 525B &= 0 \end{aligned} \qquad \text{or} \qquad \begin{aligned} 21A + 16B &= 4 \\ -16A + 21B &= 0 \end{aligned}$$

The solution of this system is $A = \frac{84}{697}$ and $B = \frac{64}{697}$, so a particular solution is

$$Q_p(t) = \tfrac{1}{697}(84\cos 10t + 64\sin 10t)$$

and the general solution is

$$Q(t) = Q_c(t) + Q_p(t)$$
$$= e^{-20t}(c_1 \cos 15t + c_2 \sin 15t) + \tfrac{4}{697}(21 \cos 10t + 16 \sin 10t)$$

Imposing the initial condition $Q(0) = 0$, we get

$$Q(0) = c_1 + \tfrac{84}{697} = 0 \qquad c_1 = -\tfrac{84}{697}$$

To impose the other initial condition, we first differentiate to find the current:

$$I = \frac{dQ}{dt} = e^{-20t}[(-20c_1 + 15c_2) \cos 15t + (-15c_1 - 20c_2) \sin 15t]$$
$$+ \tfrac{40}{697}(-21 \sin 10t + 16 \cos 10t)$$

$$I(0) = -20c_1 + 15c_2 + \tfrac{640}{697} = 0 \qquad c_2 = -\tfrac{464}{2091}$$

Thus the formula for the charge is

$$Q(t) = \frac{4}{697}\left[\frac{e^{-20t}}{3}(-63 \cos 15t - 116 \sin 15t) + (21 \cos 10t + 16 \sin 10t)\right]$$

and the expression for the current is

$$I(t) = \tfrac{1}{2091}[e^{-20t}(-1920 \cos 15t + 13{,}060 \sin 15t) + 120(-21 \sin 10t + 16 \cos 10t)]$$

NOTE 1 In Example 3 the solution for $Q(t)$ consists of two parts. Since $e^{-20t} \to 0$ as $t \to \infty$ and both $\cos 15t$ and $\sin 15t$ are bounded functions,

$$Q_c(t) = \tfrac{4}{2091} e^{-20t}(-63 \cos 15t - 116 \sin 15t) \to 0 \qquad \text{as } t \to \infty$$

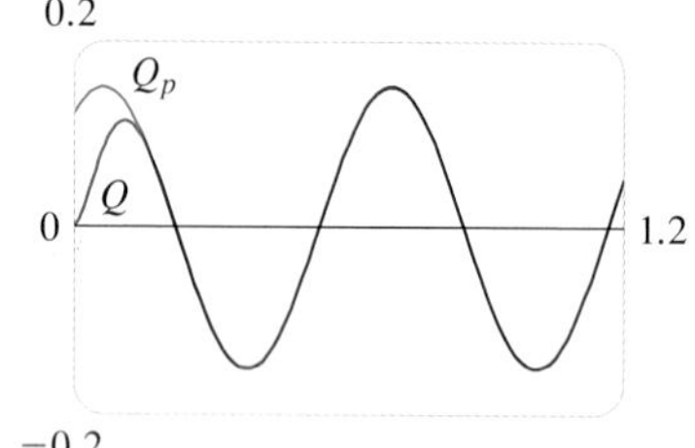

FIGURE 8

So, for large values of t,

$$Q(t) \approx Q_p(t) = \tfrac{4}{697}(21 \cos 10t + 16 \sin 10t)$$

and, for this reason, $Q_p(t)$ is called the **steady state solution**. Figure 8 shows how the graph of the steady state solution compares with the graph of Q in this case.

5 $$m\frac{d^2x}{dt^2} + c\frac{dx}{dt} + kx = F(t)$$

7 $$L\frac{d^2Q}{dt^2} + R\frac{dQ}{dt} + \frac{1}{C}Q = E(t)$$

NOTE 2 Comparing Equations 5 and 7, we see that mathematically they are identical. This suggests the analogies given in the following chart between physical situations that, at first glance, are very different.

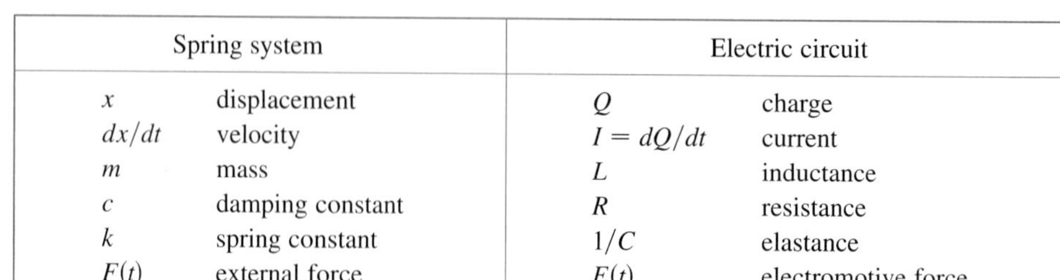

Spring system		Electric circuit	
x	displacement	Q	charge
dx/dt	velocity	$I = dQ/dt$	current
m	mass	L	inductance
c	damping constant	R	resistance
k	spring constant	$1/C$	elastance
$F(t)$	external force	$E(t)$	electromotive force

We can also transfer other ideas from one situation to the other. For instance, the steady state solution discussed in Note 1 makes sense in the spring system. And the phenomenon of resonance in the spring system can be usefully carried over to electric circuits as electrical resonance.

17.3 Exercises

1. A spring has natural length 0.75 m and a 5-kg mass. A force of 25 N is needed to keep the spring stretched to a length of 1 m. If the spring is stretched to a length of 1.1 m and then released with velocity 0, find the position of the mass after t seconds.

2. A spring with an 8-kg mass is kept stretched 0.4 m beyond its natural length by a force of 32 N. The spring starts at its equilibrium position and is given an initial velocity of 1 m/s. Find the position of the mass at any time t.

3. A spring with a mass of 2 kg has damping constant 14, and a force of 6 N is required to keep the spring stretched 0.5 m beyond its natural length. The spring is stretched 1 m beyond its natural length and then released with zero velocity. Find the position of the mass at any time t.

4. A force of 13 N is needed to keep a spring with a 2-kg mass stretched 0.25 m beyond its natural length. The damping constant of the spring is $c = 8$.
   (a) If the mass starts at the equilibrium position with a velocity of 0.5 m/s, find its position at time t.
   (b) Graph the position function of the mass.

5. For the spring in Exercise 3, find the mass that would produce critical damping.

6. For the spring in Exercise 4, find the damping constant that would produce critical damping.

7. A spring has a mass of 1 kg and its spring constant is $k = 100$. The spring is released at a point 0.1 m above its equilibrium position. Graph the position function for the following values of the damping constant c: 10, 15, 20, 25, 30. What type of damping occurs in each case?

8. A spring has a mass of 1 kg and its damping constant is $c = 10$. The spring starts from its equilibrium position with a velocity of 1 m/s. Graph the position function for the following values of the spring constant k: 10, 20, 25, 30, 40. What type of damping occurs in each case?

9. Suppose a spring has mass m and spring constant k and let $\omega = \sqrt{k/m}$. Suppose that the damping constant is so small that the damping force is negligible. If an external force $F(t) = F_0 \cos \omega_0 t$ is applied, where $\omega_0 \neq \omega$, use the method of undetermined coefficients to show that the motion of the mass is described by Equation 6.

10. As in Exercise 9, consider a spring with mass m, spring constant k, and damping constant $c = 0$, and let $\omega = \sqrt{k/m}$. If an external force $F(t) = F_0 \cos \omega t$ is applied (the applied frequency equals the natural frequency), use the method of undetermined coefficients to show that the motion of the mass is given by

$$x(t) = c_1 \cos \omega t + c_2 \sin \omega t + \frac{F_0}{2m\omega} t \sin \omega t$$

11. Show that if $\omega_0 \neq \omega$, but ω/ω_0 is a rational number, then the motion described by Equation 6 is periodic.

12. Consider a spring subject to a frictional or damping force.
   (a) In the critically damped case, the motion is given by $x = c_1 e^{rt} + c_2 t e^{rt}$. Show that the graph of x crosses the t-axis whenever c_1 and c_2 have opposite signs.
   (b) In the overdamped case, the motion is given by $x = c_1 e^{r_1 t} + c_2 e^{r_2 t}$, where $r_1 > r_2$. Determine a condition on the relative magnitudes of c_1 and c_2 under which the graph of x crosses the t-axis at a positive value of t.

13. A series circuit consists of a resistor with $R = 20\ \Omega$, an inductor with $L = 1$ H, a capacitor with $C = 0.002$ F, and a 12-V battery. If the initial charge and current are both 0, find the charge and current at time t.

14. A series circuit contains a resistor with $R = 24\ \Omega$, an inductor with $L = 2$ H, a capacitor with $C = 0.005$ F, and a 12-V battery. The initial charge is $Q = 0.001$ C and the initial current is 0.
   (a) Find the charge and current at time t.
   (b) Graph the charge and current functions.

15. The battery in Exercise 13 is replaced by a generator producing a voltage of $E(t) = 12 \sin 10t$. Find the charge at time t.

16. The battery in Exercise 14 is replaced by a generator producing a voltage of $E(t) = 12 \sin 10t$.
   (a) Find the charge at time t.
   (b) Graph the charge function.

17. Verify that the solution to Equation 1 can be written in the form $x(t) = A \cos(\omega t + \delta)$.

Graphing calculator or computer required   1. Homework Hints available at stewartcalculus.com

18. The figure shows a pendulum with length L and the angle θ from the vertical to the pendulum. It can be shown that θ, as a function of time, satisfies the nonlinear differential equation

$$\frac{d^2\theta}{dt^2} + \frac{g}{L}\sin\theta = 0$$

where g is the acceleration due to gravity. For small values of θ we can use the linear approximation $\sin\theta \approx \theta$ and then the differential equation becomes linear.

(a) Find the equation of motion of a pendulum with length 1 m if θ is initially 0.2 rad and the initial angular velocity is $d\theta/dt = 1$ rad/s.

(b) What is the maximum angle from the vertical?

(c) What is the period of the pendulum (that is, the time to complete one back-and-forth swing)?

(d) When will the pendulum first be vertical?

(e) What is the angular velocity when the pendulum is vertical?

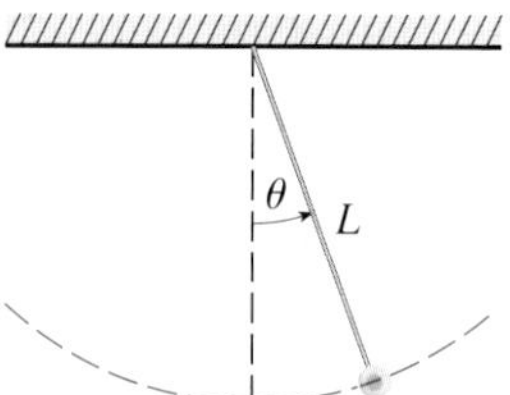

17.4 Series Solutions

Many differential equations can't be solved explicitly in terms of finite combinations of simple familiar functions. This is true even for a simple-looking equation like

$$y'' - 2xy' + y = 0 \tag{1}$$

But it is important to be able to solve equations such as Equation 1 because they arise from physical problems and, in particular, in connection with the Schrödinger equation in quantum mechanics. In such a case we use the method of power series; that is, we look for a solution of the form

$$y = f(x) = \sum_{n=0}^{\infty} c_n x^n = c_0 + c_1 x + c_2 x^2 + c_3 x^3 + \cdots$$

The method is to substitute this expression into the differential equation and determine the values of the coefficients $c_0, c_1, c_2, \ldots$. This technique resembles the method of undetermined coefficients discussed in Section 17.2.

Before using power series to solve Equation 1, we illustrate the method on the simpler equation $y'' + y = 0$ in Example 1. It's true that we already know how to solve this equation by the techniques of Section 17.1, but it's easier to understand the power series method when it is applied to this simpler equation.

V EXAMPLE 1 Use power series to solve the equation $y'' + y = 0$.

SOLUTION We assume there is a solution of the form

$$y = c_0 + c_1 x + c_2 x^2 + c_3 x^3 + \cdots = \sum_{n=0}^{\infty} c_n x^n \tag{2}$$

We can differentiate power series term by term, so

$$y' = c_1 + 2c_2 x + 3c_3 x^2 + \cdots = \sum_{n=1}^{\infty} nc_n x^{n-1}$$

$$y'' = 2c_2 + 2 \cdot 3c_3 x + \cdots = \sum_{n=2}^{\infty} n(n-1)c_n x^{n-2} \tag{3}$$

In order to compare the expressions for y and y'' more easily, we rewrite y'' as follows:

By writing out the first few terms of 4, you can see that it is the same as 3. To obtain 4, we replaced n by $n + 2$ and began the summation at 0 instead of 2.

4
$$y'' = \sum_{n=0}^{\infty} (n + 2)(n + 1)c_{n+2}x^n$$

Substituting the expressions in Equations 2 and 4 into the differential equation, we obtain

$$\sum_{n=0}^{\infty} (n + 2)(n + 1)c_{n+2}x^n + \sum_{n=0}^{\infty} c_n x^n = 0$$

or

5
$$\sum_{n=0}^{\infty} [(n + 2)(n + 1)c_{n+2} + c_n]x^n = 0$$

If two power series are equal, then the corresponding coefficients must be equal. Therefore the coefficients of x^n in Equation 5 must be 0:

$$(n + 2)(n + 1)c_{n+2} + c_n = 0$$

6
$$c_{n+2} = -\frac{c_n}{(n + 1)(n + 2)} \qquad n = 0, 1, 2, 3, \ldots$$

Equation 6 is called a *recursion relation*. If c_0 and c_1 are known, this equation allows us to determine the remaining coefficients recursively by putting $n = 0, 1, 2, 3, \ldots$ in succession.

$$\text{Put } n = 0: \qquad c_2 = -\frac{c_0}{1 \cdot 2}$$

$$\text{Put } n = 1: \qquad c_3 = -\frac{c_1}{2 \cdot 3}$$

$$\text{Put } n = 2: \qquad c_4 = -\frac{c_2}{3 \cdot 4} = \frac{c_0}{1 \cdot 2 \cdot 3 \cdot 4} = \frac{c_0}{4!}$$

$$\text{Put } n = 3: \qquad c_5 = -\frac{c_3}{4 \cdot 5} = \frac{c_1}{2 \cdot 3 \cdot 4 \cdot 5} = \frac{c_1}{5!}$$

$$\text{Put } n = 4: \qquad c_6 = -\frac{c_4}{5 \cdot 6} = -\frac{c_0}{4!\,5 \cdot 6} = -\frac{c_0}{6!}$$

$$\text{Put } n = 5: \qquad c_7 = -\frac{c_5}{6 \cdot 7} = -\frac{c_1}{5!\,6 \cdot 7} = -\frac{c_1}{7!}$$

By now we see the pattern:

$$\text{For the even coefficients, } c_{2n} = (-1)^n \frac{c_0}{(2n)!}$$

$$\text{For the odd coefficients, } c_{2n+1} = (-1)^n \frac{c_1}{(2n + 1)!}$$

Putting these values back into Equation 2, we write the solution as

$$\begin{aligned} y &= c_0 + c_1 x + c_2 x^2 + c_3 x^3 + c_4 x^4 + c_5 x^5 + \cdots \\ &= c_0\left(1 - \frac{x^2}{2!} + \frac{x^4}{4!} - \frac{x^6}{6!} + \cdots + (-1)^n \frac{x^{2n}}{(2n)!} + \cdots\right) \\ &\qquad + c_1\left(x - \frac{x^3}{3!} + \frac{x^5}{5!} - \frac{x^7}{7!} + \cdots + (-1)^n \frac{x^{2n+1}}{(2n+1)!} + \cdots\right) \\ &= c_0 \sum_{n=0}^{\infty} (-1)^n \frac{x^{2n}}{(2n)!} + c_1 \sum_{n=0}^{\infty} (-1)^n \frac{x^{2n+1}}{(2n+1)!} \end{aligned}$$

Notice that there are two arbitrary constants, c_0 and c_1.

NOTE 1 We recognize the series obtained in Example 1 as being the Maclaurin series for $\cos x$ and $\sin x$. (See Equations 11.10.16 and 11.10.15.) Therefore we could write the solution as

$$y(x) = c_0 \cos x + c_1 \sin x$$

But we are not usually able to express power series solutions of differential equations in terms of known functions.

V EXAMPLE 2 Solve $y'' - 2xy' + y = 0$.

SOLUTION We assume there is a solution of the form

$$y = \sum_{n=0}^{\infty} c_n x^n$$

Then
$$y' = \sum_{n=1}^{\infty} n c_n x^{n-1}$$

and
$$y'' = \sum_{n=2}^{\infty} n(n-1) c_n x^{n-2} = \sum_{n=0}^{\infty} (n+2)(n+1) c_{n+2} x^n$$

as in Example 1. Substituting in the differential equation, we get

$$\sum_{n=0}^{\infty} (n+2)(n+1) c_{n+2} x^n - 2x \sum_{n=1}^{\infty} n c_n x^{n-1} + \sum_{n=0}^{\infty} c_n x^n = 0$$

$$\sum_{n=0}^{\infty} (n+2)(n+1) c_{n+2} x^n - \sum_{n=1}^{\infty} 2n c_n x^n + \sum_{n=0}^{\infty} c_n x^n = 0$$

$$\sum_{n=1}^{\infty} 2n c_n x^n = \sum_{n=0}^{\infty} 2n c_n x^n$$

$$\sum_{n=0}^{\infty} [(n+2)(n+1) c_{n+2} - (2n-1) c_n] x^n = 0$$

This equation is true if the coefficient of x^n is 0:

$$(n+2)(n+1) c_{n+2} - (2n-1) c_n = 0$$

7
$$c_{n+2} = \frac{2n-1}{(n+1)(n+2)} c_n \qquad n = 0, 1, 2, 3, \ldots$$

We solve this recursion relation by putting $n = 0, 1, 2, 3, \ldots$ successively in Equation 7:

$$\text{Put } n = 0: \qquad c_2 = \frac{-1}{1 \cdot 2} c_0$$

$$\text{Put } n = 1: \qquad c_3 = \frac{1}{2 \cdot 3} c_1$$

$$\text{Put } n = 2: \qquad c_4 = \frac{3}{3 \cdot 4} c_2 = -\frac{3}{1 \cdot 2 \cdot 3 \cdot 4} c_0 = -\frac{3}{4!} c_0$$

$$\text{Put } n = 3: \qquad c_5 = \frac{5}{4 \cdot 5} c_3 = \frac{1 \cdot 5}{2 \cdot 3 \cdot 4 \cdot 5} c_1 = \frac{1 \cdot 5}{5!} c_1$$

$$\text{Put } n = 4: \qquad c_6 = \frac{7}{5 \cdot 6} c_4 = -\frac{3 \cdot 7}{4!\, 5 \cdot 6} c_0 = -\frac{3 \cdot 7}{6!} c_0$$

$$\text{Put } n = 5: \qquad c_7 = \frac{9}{6 \cdot 7} c_5 = \frac{1 \cdot 5 \cdot 9}{5!\, 6 \cdot 7} c_1 = \frac{1 \cdot 5 \cdot 9}{7!} c_1$$

$$\text{Put } n = 6: \qquad c_8 = \frac{11}{7 \cdot 8} c_6 = -\frac{3 \cdot 7 \cdot 11}{8!} c_0$$

$$\text{Put } n = 7: \qquad c_9 = \frac{13}{8 \cdot 9} c_7 = \frac{1 \cdot 5 \cdot 9 \cdot 13}{9!} c_1$$

In general, the even coefficients are given by

$$c_{2n} = -\frac{3 \cdot 7 \cdot 11 \cdot \cdots \cdot (4n - 5)}{(2n)!} c_0$$

and the odd coefficients are given by

$$c_{2n+1} = \frac{1 \cdot 5 \cdot 9 \cdot \cdots \cdot (4n - 3)}{(2n + 1)!} c_1$$

The solution is

$$\begin{aligned} y &= c_0 + c_1 x + c_2 x^2 + c_3 x^3 + c_4 x^4 + \cdots \\ &= c_0 \left(1 - \frac{1}{2!} x^2 - \frac{3}{4!} x^4 - \frac{3 \cdot 7}{6!} x^6 - \frac{3 \cdot 7 \cdot 11}{8!} x^8 - \cdots \right) \\ &\qquad + c_1 \left(x + \frac{1}{3!} x^3 + \frac{1 \cdot 5}{5!} x^5 + \frac{1 \cdot 5 \cdot 9}{7!} x^7 + \frac{1 \cdot 5 \cdot 9 \cdot 13}{9!} x^9 + \cdots \right) \end{aligned}$$

or

8
$$\begin{aligned} y &= c_0 \left(1 - \frac{1}{2!} x^2 - \sum_{n=2}^{\infty} \frac{3 \cdot 7 \cdot \cdots \cdot (4n - 5)}{(2n)!} x^{2n} \right) \\ &\qquad + c_1 \left(x + \sum_{n=1}^{\infty} \frac{1 \cdot 5 \cdot 9 \cdot \cdots \cdot (4n - 3)}{(2n + 1)!} x^{2n+1} \right) \end{aligned}$$

NOTE 2 In Example 2 we had to *assume* that the differential equation had a series solution. But now we could verify directly that the function given by Equation 8 is indeed a solution.

NOTE 3 Unlike the situation of Example 1, the power series that arise in the solution of Example 2 do not define elementary functions. The functions

$$y_1(x) = 1 - \frac{1}{2!}x^2 - \sum_{n=2}^{\infty} \frac{3 \cdot 7 \cdot \cdots \cdot (4n-5)}{(2n)!} x^{2n}$$

and

$$y_2(x) = x + \sum_{n=1}^{\infty} \frac{1 \cdot 5 \cdot 9 \cdot \cdots \cdot (4n-3)}{(2n+1)!} x^{2n+1}$$

are perfectly good functions but they can't be expressed in terms of familiar functions. We can use these power series expressions for y_1 and y_2 to compute approximate values of the functions and even to graph them. Figure 1 shows the first few partial sums $T_0, T_2, T_4, \dots$ (Taylor polynomials) for $y_1(x)$, and we see how they converge to y_1. In this way we can graph both y_1 and y_2 in Figure 2.

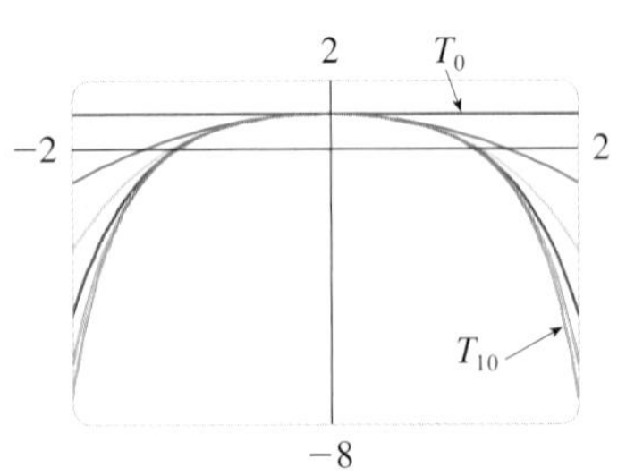

FIGURE 1

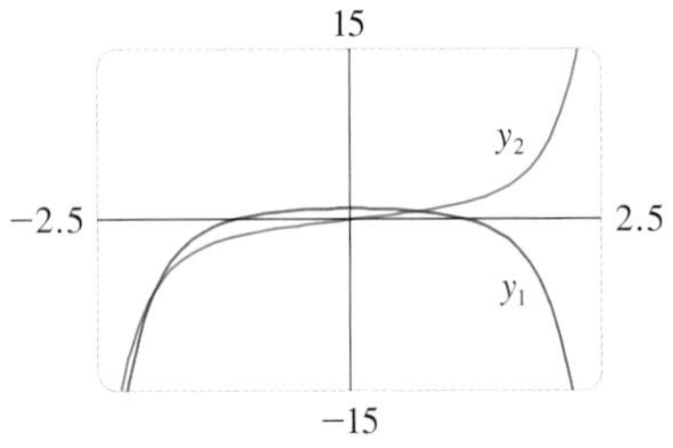

FIGURE 2

NOTE 4 If we were asked to solve the initial-value problem

$$y'' - 2xy' + y = 0 \qquad y(0) = 0 \qquad y'(0) = 1$$

we would observe from Theorem 11.10.5 that

$$c_0 = y(0) = 0 \qquad c_1 = y'(0) = 1$$

This would simplify the calculations in Example 2, since all of the even coefficients would be 0. The solution to the initial-value problem is

$$y(x) = x + \sum_{n=1}^{\infty} \frac{1 \cdot 5 \cdot 9 \cdot \cdots \cdot (4n-3)}{(2n+1)!} x^{2n+1}$$

17.4 Exercises

1–11 Use power series to solve the differential equation.

1. $y' - y = 0$

2. $y' = xy$

3. $y' = x^2 y$

4. $(x - 3)y' + 2y = 0$

5. $y'' + xy' + y = 0$

6. $y'' = y$

7. $(x - 1)y'' + y' = 0$

8. $y'' = xy$

9. $y'' - xy' - y = 0, \quad y(0) = 1, \quad y'(0) = 0$

10. $y'' + x^2 y = 0, \quad y(0) = 1, \quad y'(0) = 0$

11. $y'' + x^2 y' + xy = 0, \quad y(0) = 0, \quad y'(0) = 1$

12. The solution of the initial-value problem

$$x^2 y'' + xy' + x^2 y = 0 \qquad y(0) = 1 \qquad y'(0) = 0$$

is called a Bessel function of order 0.

(a) Solve the initial-value problem to find a power series expansion for the Bessel function.

(b) Graph several Taylor polynomials until you reach one that looks like a good approximation to the Bessel function on the interval $[-5, 5]$.

17 Review

Concept Check

1. (a) Write the general form of a second-order homogeneous linear differential equation with constant coefficients.
   (b) Write the auxiliary equation.
   (c) How do you use the roots of the auxiliary equation to solve the differential equation? Write the form of the solution for each of the three cases that can occur.

2. (a) What is an initial-value problem for a second-order differential equation?
   (b) What is a boundary-value problem for such an equation?

3. (a) Write the general form of a second-order nonhomogeneous linear differential equation with constant coefficients.
   (b) What is the complementary equation? How does it help solve the original differential equation?
   (c) Explain how the method of undetermined coefficients works.
   (d) Explain how the method of variation of parameters works.

4. Discuss two applications of second-order linear differential equations.

5. How do you use power series to solve a differential equation?

True-False Quiz

Determine whether the statement is true or false. If it is true, explain why. If it is false, explain why or give an example that disproves the statement.

1. If y_1 and y_2 are solutions of $y'' + y = 0$, then $y_1 + y_2$ is also a solution of the equation.

2. If y_1 and y_2 are solutions of $y'' + 6y' + 5y = x$, then $c_1 y_1 + c_2 y_2$ is also a solution of the equation.

3. The general solution of $y'' - y = 0$ can be written as

$$y = c_1 \cosh x + c_2 \sinh x$$

4. The equation $y'' - y = e^x$ has a particular solution of the form

$$y_p = Ae^x$$

Exercises

1–10 Solve the differential equation.

1. $4y'' - y = 0$

2. $y'' - 2y' + 10y = 0$

3. $y'' + 3y = 0$

4. $4y'' + 4y' + y = 0$

5. $\dfrac{d^2y}{dx^2} - 4\dfrac{dy}{dx} + 5y = e^{2x}$

6. $\dfrac{d^2y}{dx^2} + \dfrac{dy}{dx} - 2y = x^2$

7. $\dfrac{d^2y}{dx^2} - 2\dfrac{dy}{dx} + y = x \cos x$

8. $\dfrac{d^2y}{dx^2} + 4y = \sin 2x$

9. $\dfrac{d^2y}{dx^2} - \dfrac{dy}{dx} - 6y = 1 + e^{-2x}$

10. $\dfrac{d^2y}{dx^2} + y = \csc x, \quad 0 < x < \pi/2$

11–14 Solve the initial-value problem.

11. $y'' + 6y' = 0, \quad y(1) = 3, \quad y'(1) = 12$

12. $y'' - 6y' + 25y = 0, \quad y(0) = 2, \quad y'(0) = 1$

13. $y'' - 5y' + 4y = 0, \quad y(0) = 0, \quad y'(0) = 1$

14. $9y'' + y = 3x + e^{-x}, \quad y(0) = 1, \quad y'(0) = 2$

15–16 Solve the boundary-value problem, if possible.

15. $y'' + 4y' + 29y = 0, \quad y(0) = 1, \quad y(\pi) = -1$

16. $y'' + 4y' + 29y = 0, \quad y(0) = 1, \quad y(\pi) = -e^{-2\pi}$

17. Use power series to solve the initial-value problem

$$y'' + xy' + y = 0 \qquad y(0) = 0 \qquad y'(0) = 1$$

18. Use power series to solve the equation

$$y'' - xy' - 2y = 0$$

19. A series circuit contains a resistor with $R = 40\ \Omega$, an inductor with $L = 2$ H, a capacitor with $C = 0.0025$ F, and a 12-V battery. The initial charge is $Q = 0.01$ C and the initial current is 0. Find the charge at time t.

20. A spring with a mass of 2 kg has damping constant 16, and a force of 12.8 N keeps the spring stretched 0.2 m beyond its natural length. Find the position of the mass at time t if it starts at the equilibrium position with a velocity of 2.4 m/s.

21. Assume that the earth is a solid sphere of uniform density with mass M and radius $R = 3960$ mi. For a particle of mass m within the earth at a distance r from the earth's center, the gravitational force attracting the particle to the center is

$$F_r = \frac{-GM_r m}{r^2}$$

where G is the gravitational constant and M_r is the mass of the earth within the sphere of radius r.

(a) Show that $F_r = \dfrac{-GMm}{R^3} r$.

(b) Suppose a hole is drilled through the earth along a diameter. Show that if a particle of mass m is dropped from rest at the surface, into the hole, then the distance $y = y(t)$ of the particle from the center of the earth at time t is given by

$$y''(t) = -k^2 y(t)$$

where $k^2 = GM/R^3 = g/R$.

(c) Conclude from part (b) that the particle undergoes simple harmonic motion. Find the period T.

(d) With what speed does the particle pass through the center of the earth?

Appendixes

A Numbers, Inequalities, and Absolute Values

B Coordinate Geometry and Lines

C Graphs of Second-Degree Equations

D Trigonometry

E Sigma Notation

F Proofs of Theorems

G The Logarithm Defined as an Integral

H Complex Numbers

I Answers to Odd-Numbered Exercises

A Numbers, Inequalities, and Absolute Values

Calculus is based on the real number system. We start with the **integers**:

$$\ldots, \quad -3, \quad -2, \quad -1, \quad 0, \quad 1, \quad 2, \quad 3, \quad 4, \quad \ldots$$

Then we construct the **rational numbers**, which are ratios of integers. Thus any rational number r can be expressed as

$$r = \frac{m}{n} \qquad \text{where } m \text{ and } n \text{ are integers and } n \neq 0$$

Examples are

$$\tfrac{1}{2} \qquad -\tfrac{3}{7} \qquad 46 = \tfrac{46}{1} \qquad 0.17 = \tfrac{17}{100}$$

(Recall that division by 0 is always ruled out, so expressions like $\frac{3}{0}$ and $\frac{0}{0}$ are undefined.) Some real numbers, such as $\sqrt{2}$, can't be expressed as a ratio of integers and are therefore called **irrational numbers**. It can be shown, with varying degrees of difficulty, that the following are also irrational numbers:

$$\sqrt{3} \qquad \sqrt{5} \qquad \sqrt[3]{2} \qquad \pi \qquad \sin 1^\circ \qquad \log_{10} 2$$

The set of all real numbers is usually denoted by the symbol $\mathbb{R}$. When we use the word *number* without qualification, we mean "real number."

Every number has a decimal representation. If the number is rational, then the corresponding decimal is repeating. For example,

$$\tfrac{1}{2} = 0.5000\ldots = 0.5\overline{0} \qquad \tfrac{2}{3} = 0.66666\ldots = 0.\overline{6}$$

$$\tfrac{157}{495} = 0.317171717\ldots = 0.3\overline{17} \qquad \tfrac{9}{7} = 1.285714285714\ldots = 1.\overline{285714}$$

(The bar indicates that the sequence of digits repeats forever.) On the other hand, if the number is irrational, the decimal is nonrepeating:

$$\sqrt{2} = 1.414213562373095\ldots \qquad \pi = 3.141592653589793\ldots$$

If we stop the decimal expansion of any number at a certain place, we get an approximation to the number. For instance, we can write

$$\pi \approx 3.14159265$$

where the symbol $\approx$ is read "is approximately equal to." The more decimal places we retain, the better the approximation we get.

The real numbers can be represented by points on a line as in Figure 1. The positive direction (to the right) is indicated by an arrow. We choose an arbitrary reference point O, called the **origin**, which corresponds to the real number 0. Given any convenient unit of measurement, each positive number x is represented by the point on the line a distance of x units to the right of the origin, and each negative number $-x$ is represented by the point x units to the left of the origin. Thus every real number is represented by a point on the line, and every point P on the line corresponds to exactly one real number. The number associated with the point P is called the **coordinate** of P and the line is then called a **coordinate**

line, or a **real number line**, or simply a **real line**. Often we identify the point with its coordinate and think of a number as being a point on the real line.

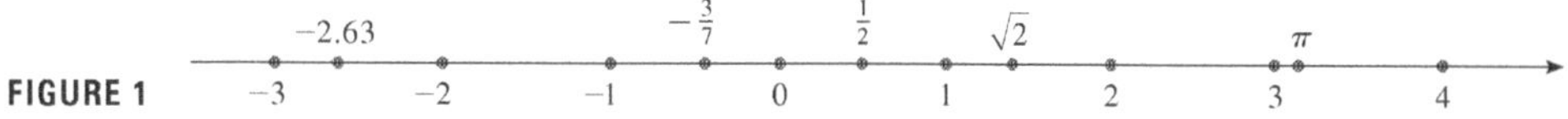

FIGURE 1

The real numbers are ordered. We say a *is less than* b and write $a < b$ if $b - a$ is a positive number. Geometrically this means that a lies to the left of b on the number line. (Equivalently, we say b *is greater than* a and write $b > a$.) The symbol $a \leqslant b$ (or $b \geqslant a$) means that either $a < b$ or $a = b$ and is read "a is less than or equal to b." For instance, the following are true inequalities:

$$7 < 7.4 < 7.5 \qquad -3 > -\pi \qquad \sqrt{2} < 2 \qquad \sqrt{2} \leqslant 2 \qquad 2 \leqslant 2$$

In what follows we need to use *set notation*. A **set** is a collection of objects, and these objects are called the **elements** of the set. If S is a set, the notation $a \in S$ means that a is an element of S, and $a \notin S$ means that a is not an element of S. For example, if Z represents the set of integers, then $-3 \in Z$ but $\pi \notin Z$. If S and T are sets, then their **union** $S \cup T$ is the set consisting of all elements that are in S or T (or in both S and T). The **intersection** of S and T is the set $S \cap T$ consisting of all elements that are in both S and T. In other words, $S \cap T$ is the common part of S and T. The empty set, denoted by $\varnothing$, is the set that contains no element.

Some sets can be described by listing their elements between braces. For instance, the set A consisting of all positive integers less than 7 can be written as

$$A = \{1, 2, 3, 4, 5, 6\}$$

We could also write A in *set-builder notation* as

$$A = \{x \mid x \text{ is an integer and } 0 < x < 7\}$$

which is read "A is the set of x such that x is an integer and $0 < x < 7$."

Intervals

Certain sets of real numbers, called **intervals,** occur frequently in calculus and correspond geometrically to line segments. For example, if $a < b$, the **open interval** from a to b consists of all numbers between a and b and is denoted by the symbol (a, b). Using set-builder notation, we can write

$$(a, b) = \{x \mid a < x < b\}$$

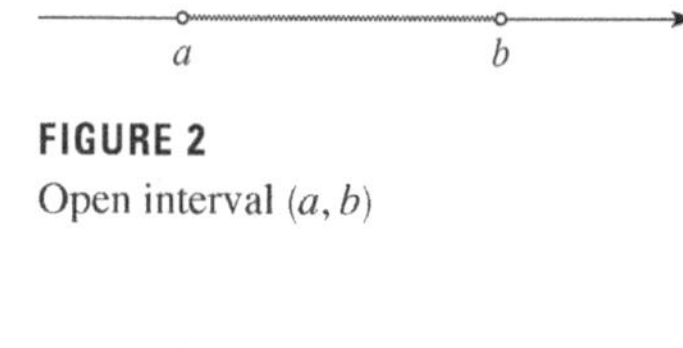

FIGURE 2
Open interval (a, b)

Notice that the endpoints of the interval—namely, a and b—are excluded. This is indicated by the round brackets ( ) and by the open dots in Figure 2. The **closed interval** from a to b is the set

$$[a, b] = \{x \mid a \leqslant x \leqslant b\}$$

FIGURE 3
Closed interval $[a, b]$

Here the endpoints of the interval are included. This is indicated by the square brackets [ ] and by the solid dots in Figure 3. It is also possible to include only one endpoint in an interval, as shown in Table 1.

1 Table of Intervals

Table 1 lists the nine possible types of intervals. When these intervals are discussed, it is always assumed that $a < b$.

Notation	Set description	Picture
(a, b)	$\{x \mid a < x < b\}$	a b
$[a, b]$	$\{x \mid a \leqslant x \leqslant b\}$	a b
$[a, b)$	$\{x \mid a \leqslant x < b\}$	a b
$(a, b]$	$\{x \mid a < x \leqslant b\}$	a b
(a, ∞)	$\{x \mid x > a\}$	a
$[a, \infty)$	$\{x \mid x \geqslant a\}$	a
$(-\infty, b)$	$\{x \mid x < b\}$	b
$(-\infty, b]$	$\{x \mid x \leqslant b\}$	b
$(-\infty, \infty)$	$\mathbb{R}$ (set of all real numbers)	

We also need to consider infinite intervals such as

$$(a, \infty) = \{x \mid x > a\}$$

This does not mean that ∞ ("infinity") is a number. The notation (a, ∞) stands for the set of all numbers that are greater than a, so the symbol ∞ simply indicates that the interval extends indefinitely far in the positive direction.

Inequalities

When working with inequalities, note the following rules.

2 Rules for Inequalities

1. If $a < b$, then $a + c < b + c$.
2. If $a < b$ and $c < d$, then $a + c < b + d$.
3. If $a < b$ and $c > 0$, then $ac < bc$.
4. If $a < b$ and $c < 0$, then $ac > bc$.
5. If $0 < a < b$, then $1/a > 1/b$.

Rule 1 says that we can add any number to both sides of an inequality, and Rule 2 says that two inequalities can be added. However, we have to be careful with multiplication. Rule 3 says that we can multiply both sides of an inequality by a *positive* number, but Rule 4 says that if we multiply both sides of an inequality by a negative number, then we reverse the direction of the inequality. For example, if we take the inequality $3 < 5$ and multiply by 2, we get $6 < 10$, but if we multiply by -2, we get $-6 > -10$. Finally, Rule 5 says that if we take reciprocals, then we reverse the direction of an inequality (provided the numbers are positive).

EXAMPLE 1 Solve the inequality $1 + x < 7x + 5$.

SOLUTION The given inequality is satisfied by some values of x but not by others. To *solve* an inequality means to determine the set of numbers x for which the inequality is true. This is called the *solution set*.

First we subtract 1 from each side of the inequality (using Rule 1 with $c = -1$):

$$x < 7x + 4$$

Then we subtract $7x$ from both sides (Rule 1 with $c = -7x$):

$$-6x < 4$$

Now we divide both sides by -6 (Rule 4 with $c = -\frac{1}{6}$):

$$x > -\tfrac{4}{6} = -\tfrac{2}{3}$$

These steps can all be reversed, so the solution set consists of all numbers greater than $-\frac{2}{3}$. In other words, the solution of the inequality is the interval $\left(-\frac{2}{3}, \infty\right)$.

EXAMPLE 2 Solve the inequalities $4 \leqslant 3x - 2 < 13$.

SOLUTION Here the solution set consists of all values of x that satisfy both inequalities. Using the rules given in [2], we see that the following inequalities are equivalent:

$$4 \leqslant 3x - 2 < 13$$

$$6 \leqslant 3x < 15 \qquad \text{(add 2)}$$

$$2 \leqslant x < 5 \qquad \text{(divide by 3)}$$

Therefore the solution set is $[2, 5)$.

EXAMPLE 3 Solve the inequality $x^2 - 5x + 6 \leqslant 0$.

SOLUTION First we factor the left side:

$$(x - 2)(x - 3) \leqslant 0$$

We know that the corresponding equation $(x - 2)(x - 3) = 0$ has the solutions 2 and 3. The numbers 2 and 3 divide the real line into three intervals:

$$(-\infty, 2) \qquad (2, 3) \qquad (3, \infty)$$

On each of these intervals we determine the signs of the factors. For instance,

$$x \in (-\infty, 2) \quad \Rightarrow \quad x < 2 \quad \Rightarrow \quad x - 2 < 0$$

Then we record these signs in the following chart:

Interval	$x - 2$	$x - 3$	$(x - 2)(x - 3)$
$x < 2$	$-$	$-$	$+$
$2 < x < 3$	$+$	$-$	$-$
$x > 3$	$+$	$+$	$+$

A visual method for solving Example 3 is to use a graphing device to graph the parabola $y = x^2 - 5x + 6$ (as in Figure 4) and observe that the curve lies on or below the x-axis when $2 \leqslant x \leqslant 3$.

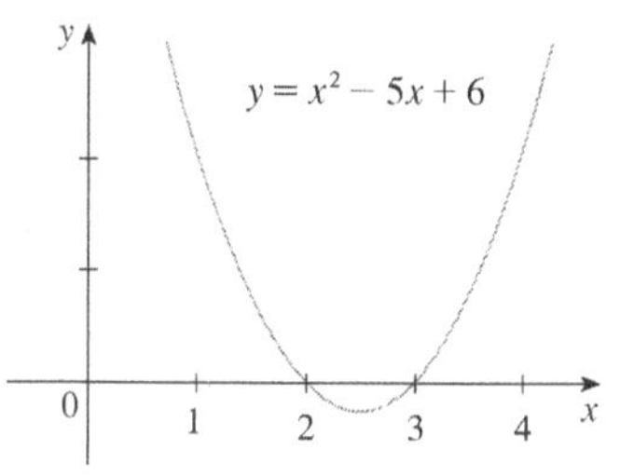

FIGURE 4

Another method for obtaining the information in the chart is to use *test values*. For instance, if we use the test value $x = 1$ for the interval $(-\infty, 2)$, then substitution in $x^2 - 5x + 6$ gives

$$1^2 - 5(1) + 6 = 2$$

The polynomial $x^2 - 5x + 6$ doesn't change sign inside any of the three intervals, so we conclude that it is positive on $(-\infty, 2)$.

Then we read from the chart that $(x - 2)(x - 3)$ is negative when $2 < x < 3$. Thus the solution of the inequality $(x - 2)(x - 3) \leqslant 0$ is

$$\{x \mid 2 \leqslant x \leqslant 3\} = [2, 3]$$

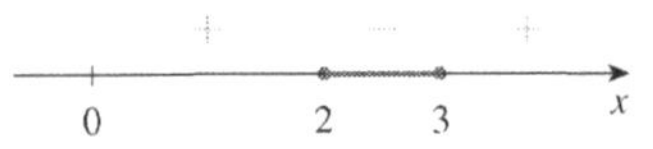

FIGURE 5

Notice that we have included the endpoints 2 and 3 because we are looking for values of x such that the product is either negative or zero. The solution is illustrated in Figure 5.

EXAMPLE 4 Solve $x^3 + 3x^2 > 4x$.

SOLUTION First we take all nonzero terms to one side of the inequality sign and factor the resulting expression:

$$x^3 + 3x^2 - 4x > 0 \qquad \text{or} \qquad x(x - 1)(x + 4) > 0$$

As in Example 3 we solve the corresponding equation $x(x - 1)(x + 4) = 0$ and use the solutions $x = -4$, $x = 0$, and $x = 1$ to divide the real line into four intervals $(-\infty, -4)$, $(-4, 0)$, $(0, 1)$, and $(1, \infty)$. On each interval the product keeps a constant sign as shown in the following chart:

Interval	x	$x - 1$	$x + 4$	$x(x - 1)(x + 4)$
$x < -4$	−	−	−	−
$-4 < x < 0$	−	−	+	+
$0 < x < 1$	+	−	+	−
$x > 1$	+	+	+	+

Then we read from the chart that the solution set is

$$\{x \mid -4 < x < 0 \text{ or } x > 1\} = (-4, 0) \cup (1, \infty)$$

FIGURE 6

The solution is illustrated in Figure 6.

Absolute Value

The **absolute value** of a number a, denoted by $|a|$, is the distance from a to 0 on the real number line. Distances are always positive or 0, so we have

$$|a| \geqslant 0 \qquad \text{for every number } a$$

For example,

$$|3| = 3 \qquad |-3| = 3 \qquad |0| = 0 \qquad |\sqrt{2} - 1| = \sqrt{2} - 1 \qquad |3 - \pi| = \pi - 3$$

In general, we have

3

$$|a| = a \qquad \text{if } a \geqslant 0$$
$$|a| = -a \qquad \text{if } a < 0$$

Remember that if a is negative, then $-a$ is positive.

EXAMPLE 6 Express $|3x - 2|$ without using the absolute-value symbol.

SOLUTION

$$|3x - 2| = \begin{cases} 3x - 2 & \text{if } 3x - 2 \geqslant 0 \\ -(3x - 2) & \text{if } 3x - 2 < 0 \end{cases}$$

$$= \begin{cases} 3x - 2 & \text{if } x \geqslant \frac{2}{3} \\ 2 - 3x & \text{if } x < \frac{2}{3} \end{cases}$$

Recall that the symbol $\sqrt{\ }$ means "the positive square root of." Thus $\sqrt{r} = s$ means $s^2 = r$ and $s \geqslant 0$. Therefore the equation $\sqrt{a^2} = a$ is not always true. It is true only when $a \geqslant 0$. If $a < 0$, then $-a > 0$, so we have $\sqrt{a^2} = -a$. In view of [3], we then have the equation

[4]
$$\sqrt{a^2} = |a|$$

which is true for all values of a.

Hints for the proofs of the following properties are given in the exercises.

[5] **Properties of Absolute Values** Suppose a and b are any real numbers and n is an integer. Then

1. $|ab| = |a||b|$ **2.** $\left|\frac{a}{b}\right| = \frac{|a|}{|b|} \quad (b \neq 0)$ **3.** $|a^n| = |a|^n$

For solving equations or inequalities involving absolute values, it's often very helpful to use the following statements.

[6] Suppose $a > 0$. Then

4. $|x| = a$ if and only if $x = \pm a$

5. $|x| < a$ if and only if $-a < x < a$

6. $|x| > a$ if and only if $x > a$ or $x < -a$

FIGURE 7

For instance, the inequality $|x| < a$ says that the distance from x to the origin is less than a, and you can see from Figure 7 that this is true if and only if x lies between $-a$ and a.

If a and b are any real numbers, then the distance between a and b is the absolute value of the difference, namely, $|a - b|$, which is also equal to $|b - a|$. (See Figure 8.)

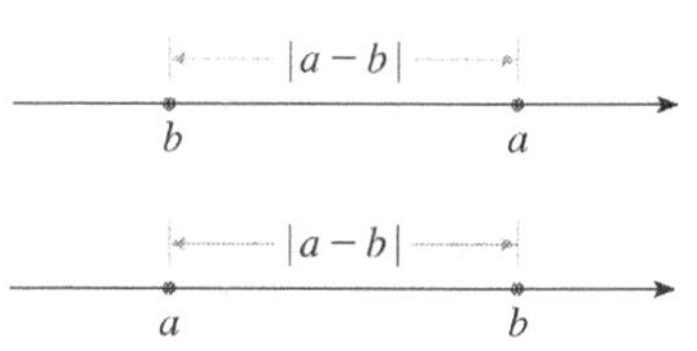

FIGURE 8
Length of a line segment = $|a - b|$

EXAMPLE 6 Solve $|2x - 5| = 3$.

SOLUTION By Property 4 of [6], $|2x - 5| = 3$ is equivalent to

$$2x - 5 = 3 \quad \text{or} \quad 2x - 5 = -3$$

So $2x = 8$ or $2x = 2$. Thus $x = 4$ or $x = 1$.

EXAMPLE 7 Solve $|x - 5| < 2$.

SOLUTION 1 By Property 5 of [6], $|x - 5| < 2$ is equivalent to

$$-2 < x - 5 < 2$$

Therefore, adding 5 to each side, we have

$$3 < x < 7$$

and the solution set is the open interval $(3, 7)$.

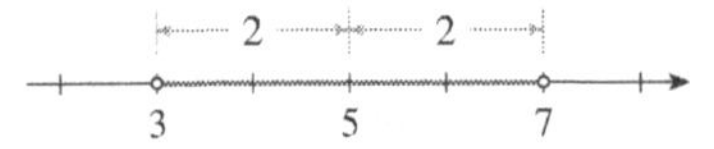

FIGURE 9

SOLUTION 2 Geometrically the solution set consists of all numbers x whose distance from 5 is less than 2. From Figure 9 we see that this is the interval $(3, 7)$.

EXAMPLE 8 Solve $|3x + 2| \geq 4$.

SOLUTION By Properties 4 and 6 of [6], $|3x + 2| \geq 4$ is equivalent to

$$3x + 2 \geq 4 \qquad \text{or} \qquad 3x + 2 \leq -4$$

In the first case $3x \geq 2$, which gives $x \geq \frac{2}{3}$. In the second case $3x \leq -6$, which gives $x \leq -2$. So the solution set is

$$\{x \mid x \leq -2 \text{ or } x \geq \tfrac{2}{3}\} = (-\infty, -2] \cup \left[\tfrac{2}{3}, \infty\right)$$

Another important property of absolute value, called the Triangle Inequality, is used frequently not only in calculus but throughout mathematics in general.

> [7] **The Triangle Inequality** If a and b are any real numbers, then
>
> $$|a + b| \leq |a| + |b|$$

Observe that if the numbers a and b are both positive or both negative, then the two sides in the Triangle Inequality are actually equal. But if a and b have opposite signs, the left side involves a subtraction and the right side does not. This makes the Triangle Inequality seem reasonable, but we can prove it as follows.

Notice that

$$-|a| \leq a \leq |a|$$

is always true because a equals either $|a|$ or $-|a|$. The corresponding statement for b is

$$-|b| \leq b \leq |b|$$

Adding these inequalities, we get

$$-(|a| + |b|) \leq a + b \leq |a| + |b|$$

If we now apply Properties 4 and 5 (with x replaced by $a + b$ and a by $|a| + |b|$), we obtain

$$|a + b| \leq |a| + |b|$$

which is what we wanted to show.

EXAMPLE 9 If $|x - 4| < 0.1$ and $|y - 7| < 0.2$, use the Triangle Inequality to estimate $|(x + y) - 11|$.

SOLUTION In order to use the given information, we use the Triangle Inequality with $a = x - 4$ and $b = y - 7$:

$$\begin{aligned} |(x + y) - 11| &= |(x - 4) + (y - 7)| \\ &\leqslant |x - 4| + |y - 7| \\ &< 0.1 + 0.2 = 0.3 \end{aligned}$$

Thus

$$|(x + y) - 11| < 0.3$$

A Exercises

1–12 Rewrite the expression without using the absolute value symbol.

1. $|5 - 23|$
2. $|5| - |-23|$
3. $|-\pi|$
4. $|\pi - 2|$
5. $|\sqrt{5} - 5|$
6. $||-2| - |-3||$
7. $|x - 2|$ if $x < 2$
8. $|x - 2|$ if $x > 2$
9. $|x + 1|$
10. $|2x - 1|$
11. $|x^2 + 1|$
12. $|1 - 2x^2|$

13–38 Solve the inequality in terms of intervals and illustrate the solution set on the real number line.

13. $2x + 7 > 3$
14. $3x - 11 < 4$
15. $1 - x \leqslant 2$
16. $4 - 3x \geqslant 6$
17. $2x + 1 < 5x - 8$
18. $1 + 5x > 5 - 3x$
19. $-1 < 2x - 5 < 7$
20. $1 < 3x + 4 \leqslant 16$
21. $0 \leqslant 1 - x < 1$
22. $-5 \leqslant 3 - 2x \leqslant 9$
23. $4x < 2x + 1 \leqslant 3x + 2$
24. $2x - 3 < x + 4 < 3x - 2$
25. $(x - 1)(x - 2) > 0$
26. $(2x + 3)(x - 1) \geqslant 0$
27. $2x^2 + x \leqslant 1$
28. $x^2 < 2x + 8$
29. $x^2 + x + 1 > 0$
30. $x^2 + x > 1$
31. $x^2 < 3$
32. $x^2 \geqslant 5$
33. $x^3 - x^2 \leqslant 0$
34. $(x + 1)(x - 2)(x + 3) \geqslant 0$
35. $x^3 > x$
36. $x^3 + 3x < 4x^2$
37. $\dfrac{1}{x} < 4$
38. $-3 < \dfrac{1}{x} \leqslant 1$

39. The relationship between the Celsius and Fahrenheit temperature scales is given by $C = \frac{5}{9}(F - 32)$, where C is the temperature in degrees Celsius and F is the temperature in degrees Fahrenheit. What interval on the Celsius scale corresponds to the temperature range $50 \leqslant F \leqslant 95$?

40. Use the relationship between C and F given in Exercise 39 to find the interval on the Fahrenheit scale corresponding to the temperature range $20 \leqslant C \leqslant 30$.

41. As dry air moves upward, it expands and in so doing cools at a rate of about 1°C for each 100-m rise, up to about 12 km.
    (a) If the ground temperature is 20°C, write a formula for the temperature at height h.
    (b) What range of temperature can be expected if a plane takes off and reaches a maximum height of 5 km?

42. If a ball is thrown upward from the top of a building 128 ft high with an initial velocity of 16 ft/s, then the height h above the ground t seconds later will be

$$h = 128 + 16t - 16t^2$$

During what time interval will the ball be at least 32 ft above the ground?

43–46 Solve the equation for x.

43. $|2x| = 3$
44. $|3x + 5| = 1$
45. $|x + 3| = |2x + 1|$
46. $\left|\dfrac{2x - 1}{x + 1}\right| = 3$

47–56 Solve the inequality.

47. $|x| < 3$
48. $|x| \geqslant 3$
49. $|x - 4| < 1$
50. $|x - 6| < 0.1$
51. $|x + 5| \geqslant 2$
52. $|x + 1| \geqslant 3$
53. $|2x - 3| \leqslant 0.4$
54. $|5x - 2| < 6$
55. $1 \leqslant |x| \leqslant 4$
56. $0 < |x - 5| < \frac{1}{2}$

57–58 Solve for x, assuming a, b, and c are positive constants.

57. $a(bx - c) \geqslant bc$

58. $a \leqslant bx + c < 2a$

59–60 Solve for x, assuming a, b, and c are negative constants.

59. $ax + b < c$

60. $\dfrac{ax + b}{c} \leqslant b$

61. Suppose that $|x - 2| < 0.01$ and $|y - 3| < 0.04$. Use the Triangle Inequality to show that $|(x + y) - 5| < 0.05$.

62. Show that if $|x + 3| < \frac{1}{2}$, then $|4x + 13| < 3$.

63. Show that if $a < b$, then $a < \dfrac{a + b}{2} < b$.

64. Use Rule 3 to prove Rule 5 of [2].

65. Prove that $|ab| = |a||b|$. [*Hint:* Use Equation 4.]

66. Prove that $\left|\dfrac{a}{b}\right| = \dfrac{|a|}{|b|}$.

67. Show that if $0 < a < b$, then $a^2 < b^2$.

68. Prove that $|x - y| \geqslant |x| - |y|$. [*Hint:* Use the Triangle Inequality with $a = x - y$ and $b = y$.]

69. Show that the sum, difference, and product of rational numbers are rational numbers.

70. (a) Is the sum of two irrational numbers always an irrational number?

(b) Is the product of two irrational numbers always an irrational number?

B Coordinate Geometry and Lines

Just as the points on a line can be identified with real numbers by assigning them coordinates, as described in Appendix A, so the points in a plane can be identified with ordered pairs of real numbers. We start by drawing two perpendicular coordinate lines that intersect at the origin O on each line. Usually one line is horizontal with positive direction to the right and is called the x-axis; the other line is vertical with positive direction upward and is called the y-axis.

Any point P in the plane can be located by a unique ordered pair of numbers as follows. Draw lines through P perpendicular to the x- and y-axes. These lines intersect the axes in points with coordinates a and b as shown in Figure 1. Then the point P is assigned the ordered pair (a, b). The first number a is called the **x-coordinate** of P; the second number b is called the **y-coordinate** of P. We say that P is the point with coordinates (a, b), and we denote the point by the symbol $P(a, b)$. Several points are labeled with their coordinates in Figure 2.

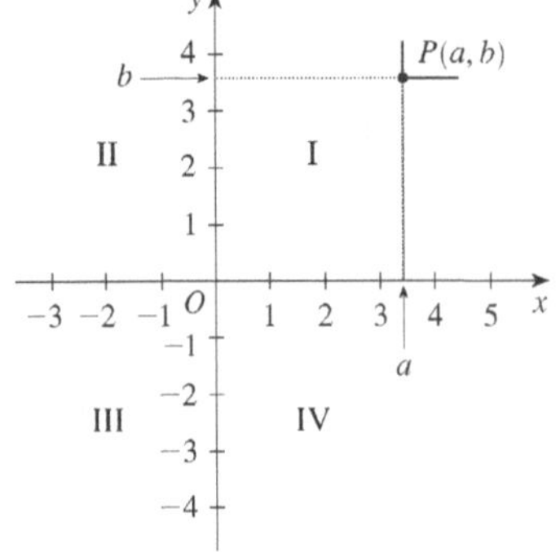

FIGURE 1

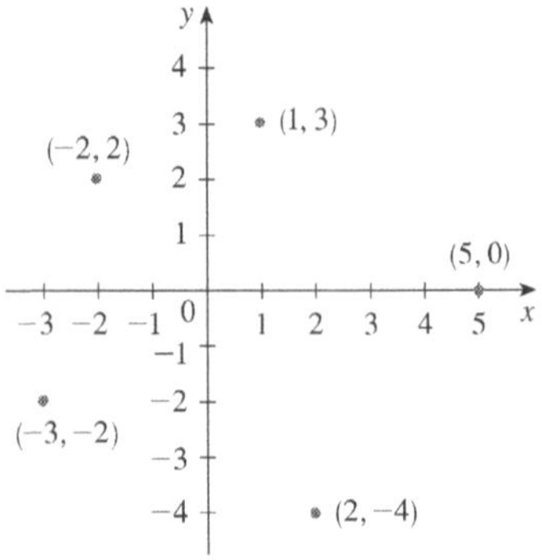

FIGURE 2

By reversing the preceding process we can start with an ordered pair (a, b) and arrive at the corresponding point P. Often we identify the point P with the ordered pair (a, b) and refer to "the point (a, b)." [Although the notation used for an open interval (a, b) is the same as the notation used for a point (a, b), you will be able to tell from the context which meaning is intended.]

This coordinate system is called the **rectangular coordinate system** or the **Cartesian coordinate system** in honor of the French mathematician René Descartes (1596–1650), even though another Frenchman, Pierre Fermat (1601–1665), invented the principles of analytic geometry at about the same time as Descartes. The plane supplied with this coordinate system is called the **coordinate plane** or the **Cartesian plane** and is denoted by $\mathbb{R}^2$.

The x- and y-axes are called the **coordinate axes** and divide the Cartesian plane into four quadrants, which are labeled I, II, III, and IV in Figure 1. Notice that the first quadrant consists of those points whose x- and y-coordinates are both positive.

EXAMPLE 1 Describe and sketch the regions given by the following sets.

(a) $\{(x, y) \mid x \geqslant 0\}$ (b) $\{(x, y) \mid y = 1\}$ (c) $\{(x, y) \mid |y| < 1\}$

SOLUTION

(a) The points whose x-coordinates are 0 or positive lie on the y-axis or to the right of it as indicated by the shaded region in Figure 3(a).

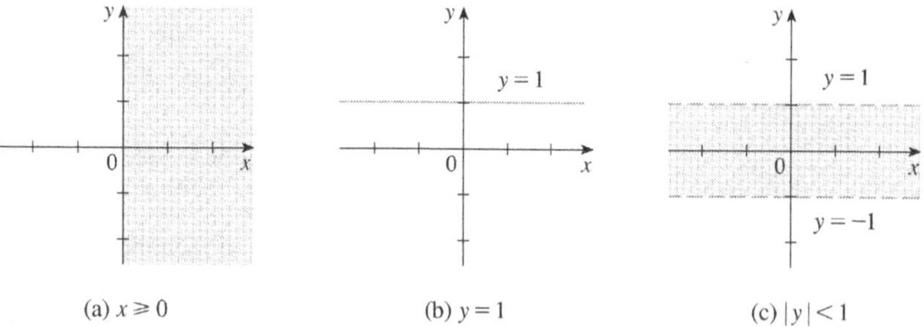

FIGURE 3 (a) $x \geqslant 0$ (b) $y = 1$ (c) $|y| < 1$

(b) The set of all points with y-coordinate 1 is a horizontal line one unit above the x-axis [see Figure 3(b)].

(c) Recall from Appendix A that

$$|y| < 1 \quad \text{if and only if} \quad -1 < y < 1$$

The given region consists of those points in the plane whose y-coordinates lie between -1 and 1. Thus the region consists of all points that lie between (but not on) the horizontal lines $y = 1$ and $y = -1$. [These lines are shown as dashed lines in Figure 3(c) to indicate that the points on these lines don't lie in the set.]

Recall from Appendix A that the distance between points a and b on a number line is $|a - b| = |b - a|$. Thus the distance between points $P_1(x_1, y_1)$ and $P_3(x_2, y_1)$ on a horizontal line must be $|x_2 - x_1|$ and the distance between $P_2(x_2, y_2)$ and $P_3(x_2, y_1)$ on a vertical line must be $|y_2 - y_1|$. (See Figure 4.)

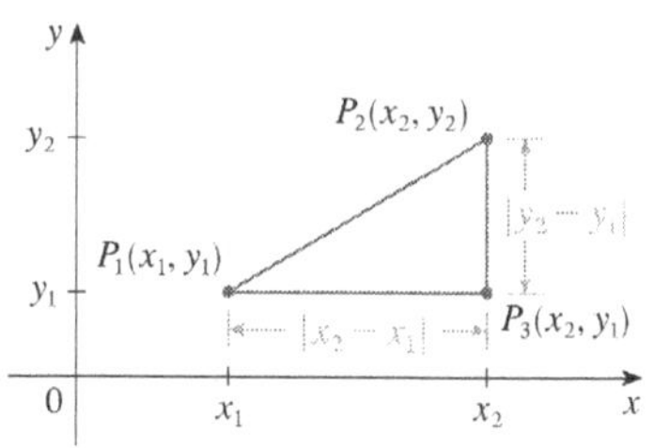

FIGURE 4

To find the distance $|P_1P_2|$ between any two points $P_1(x_1, y_1)$ and $P_2(x_2, y_2)$, we note that triangle $P_1P_2P_3$ in Figure 4 is a right triangle, and so by the Pythagorean Theorem we have

$$|P_1P_2| = \sqrt{|P_1P_3|^2 + |P_2P_3|^2} = \sqrt{|x_2 - x_1|^2 + |y_2 - y_1|^2}$$

$$= \sqrt{(x_2 - x_1)^2 + (y_2 - y_1)^2}$$

1 Distance Formula The distance between the points $P_1(x_1, y_1)$ and $P_2(x_2, y_2)$ is

$$|P_1P_2| = \sqrt{(x_2 - x_1)^2 + (y_2 - y_1)^2}$$

EXAMPLE 2 The distance between $(1, -2)$ and $(5, 3)$ is

$$\sqrt{(5 - 1)^2 + [3 - (-2)]^2} = \sqrt{4^2 + 5^2} = \sqrt{41}$$

Lines

We want to find an equation of a given line L; such an equation is satisfied by the coordinates of the points on L and by no other point. To find the equation of L we use its *slope*, which is a measure of the steepness of the line.

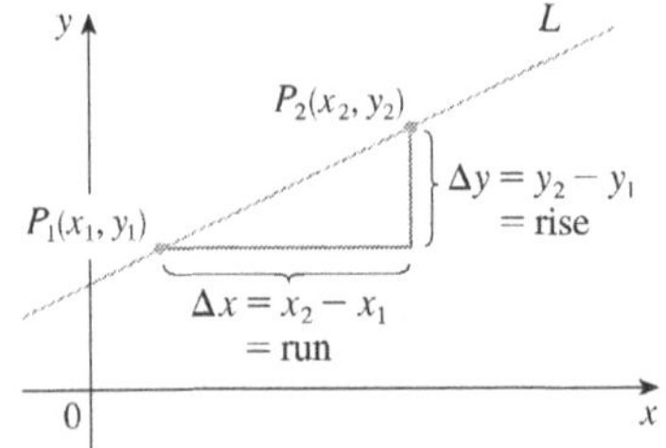

FIGURE 5

2 Definition The **slope** of a nonvertical line that passes through the points $P_1(x_1, y_1)$ and $P_2(x_2, y_2)$ is

$$m = \frac{\Delta y}{\Delta x} = \frac{y_2 - y_1}{x_2 - x_1}$$

The slope of a vertical line is not defined.

Thus the slope of a line is the ratio of the change in y, Δy, to the change in x, Δx. (See Figure 5.) The slope is therefore the rate of change of y with respect to x. The fact that the line is straight means that the rate of change is constant.

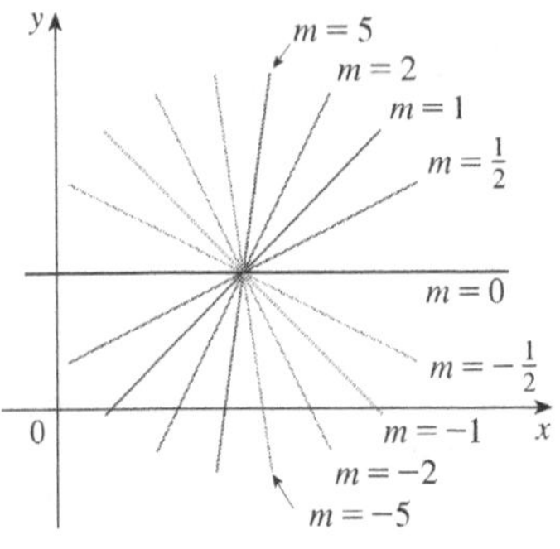

FIGURE 6

Figure 6 shows several lines labeled with their slopes. Notice that lines with positive slope slant upward to the right, whereas lines with negative slope slant downward to the right. Notice also that the steepest lines are the ones for which the absolute value of the slope is largest, and a horizontal line has slope 0.

Now let's find an equation of the line that passes through a given point $P_1(x_1, y_1)$ and has slope m. A point $P(x, y)$ with $x \neq x_1$ lies on this line if and only if the slope of the line through P_1 and P is equal to m; that is,

$$\frac{y - y_1}{x - x_1} = m$$

This equation can be rewritten in the form

$$y - y_1 = m(x - x_1)$$

and we observe that this equation is also satisfied when $x = x_1$ and $y = y_1$. Therefore it is an equation of the given line.

3 Point-Slope Form of the Equation of a Line An equation of the line passing through the point $P_1(x_1, y_1)$ and having slope m is

$$y - y_1 = m(x - x_1)$$

EXAMPLE 3 Find an equation of the line through $(1, -7)$ with slope $-\frac{1}{2}$.

SOLUTION Using [3] with $m = -\frac{1}{2}$, $x_1 = 1$, and $y_1 = -7$, we obtain an equation of the line as

$$y + 7 = -\tfrac{1}{2}(x - 1)$$

which we can rewrite as

$$2y + 14 = -x + 1 \qquad \text{or} \qquad x + 2y + 13 = 0$$

EXAMPLE 4 Find an equation of the line through the points $(-1, 2)$ and $(3, -4)$.

SOLUTION By Definition 2 the slope of the line is

$$m = \frac{-4 - 2}{3 - (-1)} = -\frac{3}{2}$$

Using the point-slope form with $x_1 = -1$ and $y_1 = 2$, we obtain

$$y - 2 = -\tfrac{3}{2}(x + 1)$$

which simplifies to

$$3x + 2y = 1$$

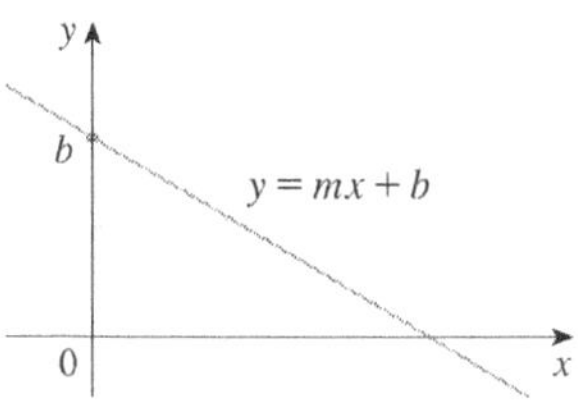

FIGURE 7

Suppose a nonvertical line has slope m and y-intercept b. (See Figure 7.) This means it intersects the y-axis at the point $(0, b)$, so the point-slope form of the equation of the line, with $x_1 = 0$ and $y_1 = b$, becomes

$$y - b = m(x - 0)$$

This simplifies as follows.

> [4] **Slope-Intercept Form of the Equation of a Line** An equation of the line with slope m and y-intercept b is
>
> $$y = mx + b$$

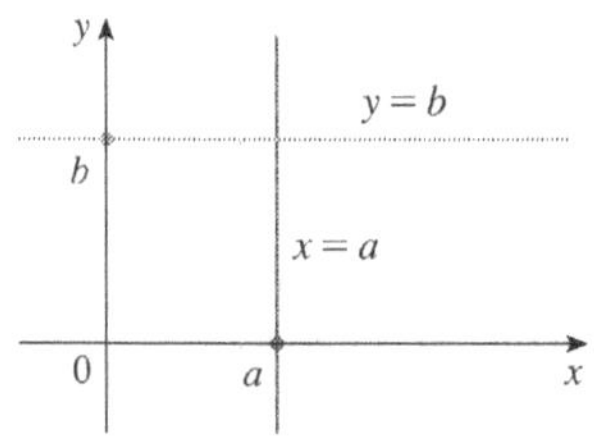

FIGURE 8

In particular, if a line is horizontal, its slope is $m = 0$, so its equation is $y = b$, where b is the y-intercept (see Figure 8). A vertical line does not have a slope, but we can write its equation as $x = a$, where a is the x-intercept, because the x-coordinate of every point on the line is a.

Observe that the equation of every line can be written in the form

[5] $$Ax + By + C = 0$$

because a vertical line has the equation $x = a$ or $x - a = 0$ ($A = 1$, $B = 0$, $C = -a$) and a nonvertical line has the equation $y = mx + b$ or $-mx + y - b = 0$ ($A = -m$, $B = 1$, $C = -b$). Conversely, if we start with a general first-degree equation, that is, an equation of the form [5], where A, B, and C are constants and A and B are not both 0, then we can show that it is the equation of a line. If $B = 0$, the equation becomes $Ax + C = 0$ or $x = -C/A$, which represents a vertical line with x-intercept $-C/A$. If $B \neq 0$, the equation can be rewritten by solving for y:

$$y = -\frac{A}{B}x - \frac{C}{B}$$

and we recognize this as being the slope-intercept form of the equation of a line ($m = -A/B$, $b = -C/B$). Therefore an equation of the form [5] is called a **linear equation** or the **general equation of a line.** For brevity, we often refer to "the line $Ax + By + C = 0$" instead of "the line whose equation is $Ax + By + C = 0$."

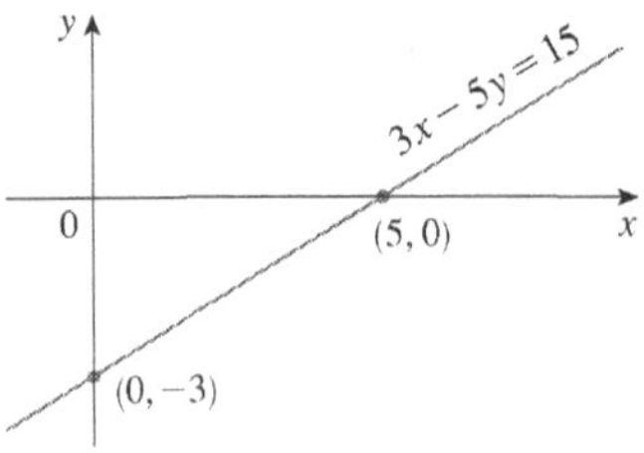

FIGURE 9

EXAMPLE 5 Sketch the graph of the equation $3x - 5y = 15$.

SOLUTION Since the equation is linear, its graph is a line. To draw the graph, we can simply find two points on the line. It's easiest to find the intercepts. Substituting $y = 0$ (the equation of the x-axis) in the given equation, we get $3x = 15$, so $x = 5$ is the x-intercept. Substituting $x = 0$ in the equation, we see that the y-intercept is -3. This allows us to sketch the graph as in Figure 9.

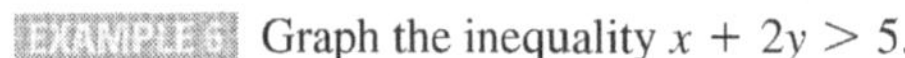

EXAMPLE 6 Graph the inequality $x + 2y > 5$.

SOLUTION We are asked to sketch the graph of the set $\{(x, y) \mid x + 2y > 5\}$ and we begin by solving the inequality for y:

$$\begin{aligned} x + 2y &> 5 \\ 2y &> -x + 5 \\ y &> -\tfrac{1}{2}x + \tfrac{5}{2} \end{aligned}$$

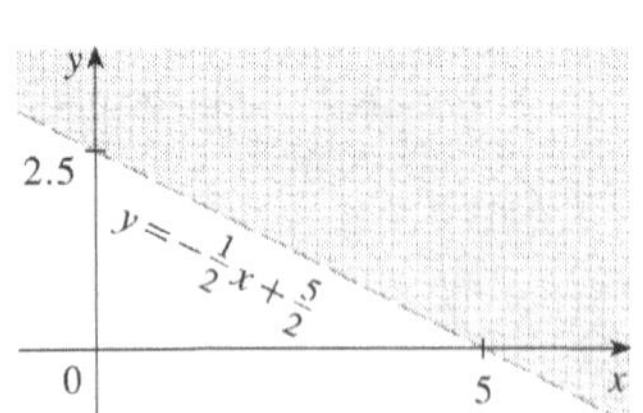

FIGURE 10

Compare this inequality with the equation $y = -\frac{1}{2}x + \frac{5}{2}$, which represents a line with slope $-\frac{1}{2}$ and y-intercept $\frac{5}{2}$. We see that the given graph consists of points whose y-coordinates are *larger* than those on the line $y = -\frac{1}{2}x + \frac{5}{2}$. Thus the graph is the region that lies *above* the line, as illustrated in Figure 10.

Parallel and Perpendicular Lines

Slopes can be used to show that lines are parallel or perpendicular. The following facts are proved, for instance, in *Precalculus: Mathematics for Calculus, Sixth Edition* by Stewart, Redlin, and Watson (Belmont, CA, 2012).

> **[6] Parallel and Perpendicular Lines**
>
> 1. Two nonvertical lines are parallel if and only if they have the same slope.
> 2. Two lines with slopes m_1 and m_2 are perpendicular if and only if $m_1 m_2 = -1$; that is, their slopes are negative reciprocals:
>
> $$m_2 = -\frac{1}{m_1}$$

EXAMPLE 7 Find an equation of the line through the point (5, 2) that is parallel to the line $4x + 6y + 5 = 0$.

SOLUTION The given line can be written in the form

$$y = -\tfrac{2}{3}x - \tfrac{5}{6}$$

which is in slope-intercept form with $m = -\frac{2}{3}$. Parallel lines have the same slope, so the required line has slope $-\frac{2}{3}$ and its equation in point-slope form is

$$y - 2 = -\tfrac{2}{3}(x - 5)$$

We can write this equation as $2x + 3y = 16$.

EXAMPLE 8 Show that the lines $2x + 3y = 1$ and $6x - 4y - 1 = 0$ are perpendicular.

SOLUTION The equations can be written as

$$y = -\tfrac{2}{3}x + \tfrac{1}{3} \qquad \text{and} \qquad y = \tfrac{3}{2}x - \tfrac{1}{4}$$

from which we see that the slopes are

$$m_1 = -\tfrac{2}{3} \qquad \text{and} \qquad m_2 = \tfrac{3}{2}$$

Since $m_1 m_2 = -1$, the lines are perpendicular.

B Exercises

1–6 Find the distance between the points.

1. $(1, 1)$, $(4, 5)$
2. $(1, -3)$, $(5, 7)$
3. $(6, -2)$, $(-1, 3)$
4. $(1, -6)$, $(-1, -3)$
5. $(2, 5)$, $(4, -7)$
6. (a, b), (b, a)

7–10 Find the slope of the line through P and Q.

7. $P(1, 5)$, $Q(4, 11)$
8. $P(-1, 6)$, $Q(4, -3)$
9. $P(-3, 3)$, $Q(-1, -6)$
10. $P(-1, -4)$, $Q(6, 0)$

11. Show that the triangle with vertices $A(0, 2)$, $B(-3, -1)$, and $C(-4, 3)$ is isosceles.

12. (a) Show that the triangle with vertices $A(6, -7)$, $B(11, -3)$, and $C(2, -2)$ is a right triangle using the converse of the Pythagorean Theorem.
    (b) Use slopes to show that ABC is a right triangle.
    (c) Find the area of the triangle.

13. Show that the points $(-2, 9)$, $(4, 6)$, $(1, 0)$, and $(-5, 3)$ are the vertices of a square.

14. (a) Show that the points $A(-1, 3)$, $B(3, 11)$, and $C(5, 15)$ are collinear (lie on the same line) by showing that $|AB| + |BC| = |AC|$.
    (b) Use slopes to show that A, B, and C are collinear.

15. Show that $A(1, 1)$, $B(7, 4)$, $C(5, 10)$, and $D(-1, 7)$ are vertices of a parallelogram.

16. Show that $A(1, 1)$, $B(11, 3)$, $C(10, 8)$, and $D(0, 6)$ are vertices of a rectangle.

17–20 Sketch the graph of the equation.

17. $x = 3$
18. $y = -2$
19. $xy = 0$
20. $|y| = 1$

21–36 Find an equation of the line that satisfies the given conditions.

21. Through $(2, -3)$, slope 6
22. Through $(-1, 4)$, slope -3
23. Through $(1, 7)$, slope $\frac{2}{3}$
24. Through $(-3, -5)$, slope $-\frac{7}{2}$
25. Through $(2, 1)$ and $(1, 6)$
26. Through $(-1, -2)$ and $(4, 3)$
27. Slope 3, y-intercept -2
28. Slope $\frac{2}{5}$, y-intercept 4
29. x-intercept 1, y-intercept -3
30. x-intercept -8, y-intercept 6
31. Through $(4, 5)$, parallel to the x-axis
32. Through $(4, 5)$, parallel to the y-axis
33. Through $(1, -6)$, parallel to the line $x + 2y = 6$
34. y-intercept 6, parallel to the line $2x + 3y + 4 = 0$
35. Through $(-1, -2)$, perpendicular to the line $2x + 5y + 8 = 0$
36. Through $(\frac{1}{2}, -\frac{2}{3})$, perpendicular to the line $4x - 8y = 1$

37–42 Find the slope and y-intercept of the line and draw its graph.

37. $x + 3y = 0$
38. $2x - 5y = 0$

39. $y = -2$

40. $2x - 3y + 6 = 0$

41. $3x - 4y = 12$

42. $4x + 5y = 10$

43–52 Sketch the region in the xy-plane.

43. $\{(x, y) \mid x < 0\}$

44. $\{(x, y) \mid y > 0\}$

45. $\{(x, y) \mid xy < 0\}$

46. $\{(x, y) \mid x \geqslant 1 \text{ and } y < 3\}$

47. $\{(x, y) \mid |x| \leqslant 2\}$

48. $\{(x, y) \mid |x| < 3 \text{ and } |y| < 2\}$

49. $\{(x, y) \mid 0 \leqslant y \leqslant 4 \text{ and } x \leqslant 2\}$

50. $\{(x, y) \mid y > 2x - 1\}$

51. $\{(x, y) \mid 1 + x \leqslant y \leqslant 1 - 2x\}$

52. $\{(x, y) \mid -x \leqslant y < \frac{1}{2}(x + 3)\}$

53. Find a point on the y-axis that is equidistant from $(5, -5)$ and $(1, 1)$.

54. Show that the midpoint of the line segment from $P_1(x_1, y_1)$ to $P_2(x_2, y_2)$ is

$$\left(\frac{x_1 + x_2}{2}, \frac{y_1 + y_2}{2}\right)$$

55. Find the midpoint of the line segment joining the given points.
(a) $(1, 3)$ and $(7, 15)$
(b) $(-1, 6)$ and $(8, -12)$

56. Find the lengths of the medians of the triangle with vertices $A(1, 0)$, $B(3, 6)$, and $C(8, 2)$. (A median is a line segment from a vertex to the midpoint of the opposite side.)

57. Show that the lines $2x - y = 4$ and $6x - 2y = 10$ are not parallel and find their point of intersection.

58. Show that the lines $3x - 5y + 19 = 0$ and $10x + 6y - 50 = 0$ are perpendicular and find their point of intersection.

59. Find an equation of the perpendicular bisector of the line segment joining the points $A(1, 4)$ and $B(7, -2)$.

60. (a) Find equations for the sides of the triangle with vertices $P(1, 0)$, $Q(3, 4)$, and $R(-1, 6)$.
(b) Find equations for the medians of this triangle. Where do they intersect?

61. (a) Show that if the x- and y-intercepts of a line are nonzero numbers a and b, then the equation of the line can be put in the form

$$\frac{x}{a} + \frac{y}{b} = 1$$

This equation is called the **two-intercept form** of an equation of a line.
(b) Use part (a) to find an equation of the line whose x-intercept is 6 and whose y-intercept is -8.

62. A car leaves Detroit at 2:00 PM, traveling at a constant speed west along I-96. It passes Ann Arbor, 40 mi from Detroit, at 2:50 PM.
(a) Express the distance traveled in terms of the time elapsed.
(b) Draw the graph of the equation in part (a).
(c) What is the slope of this line? What does it represent?

C Graphs of Second-Degree Equations

In Appendix B we saw that a first-degree, or linear, equation $Ax + By + C = 0$ represents a line. In this section we discuss second-degree equations such as

$$x^2 + y^2 = 1 \qquad y = x^2 + 1 \qquad \frac{x^2}{9} + \frac{y^2}{4} = 1 \qquad x^2 - y^2 = 1$$

which represent a circle, a parabola, an ellipse, and a hyperbola, respectively.

The graph of such an equation in x and y is the set of all points (x, y) that satisfy the equation; it gives a visual representation of the equation. Conversely, given a curve in the xy-plane, we may have to find an equation that represents it, that is, an equation satisfied by the coordinates of the points on the curve and by no other point. This is the other half of the basic principle of analytic geometry as formulated by Descartes and Fermat. The idea is that if a geometric curve can be represented by an algebraic equation, then the rules of algebra can be used to analyze the geometric problem.

Circles

As an example of this type of problem, let's find an equation of the circle with radius r and center (h, k). By definition, the circle is the set of all points $P(x, y)$ whose distance from

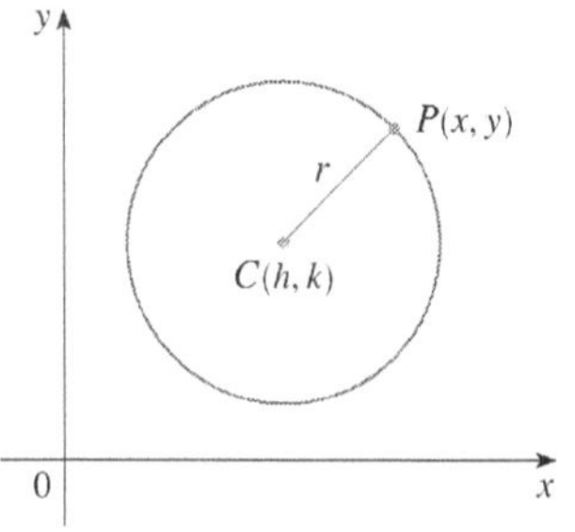

FIGURE 1

the center $C(h, k)$ is r. (See Figure 1.) Thus P is on the circle if and only if $|PC| = r$. From the distance formula, we have

$$\sqrt{(x - h)^2 + (y - k)^2} = r$$

or equivalently, squaring both sides, we get

$$(x - h)^2 + (y - k)^2 = r^2$$

This is the desired equation.

> **1** **Equation of a Circle** An equation of the circle with center (h, k) and radius r is
>
> $$(x - h)^2 + (y - k)^2 = r^2$$
>
> In particular, if the center is the origin $(0, 0)$, the equation is
>
> $$x^2 + y^2 = r^2$$

EXAMPLE 1 Find an equation of the circle with radius 3 and center $(2, -5)$.

SOLUTION From Equation 1 with $r = 3$, $h = 2$, and $k = -5$, we obtain

$$(x - 2)^2 + (y + 5)^2 = 9$$

EXAMPLE 2 Sketch the graph of the equation $x^2 + y^2 + 2x - 6y + 7 = 0$ by first showing that it represents a circle and then finding its center and radius.

SOLUTION We first group the x-terms and y-terms as follows:

$$(x^2 + 2x) + (y^2 - 6y) = -7$$

Then we complete the square within each grouping, adding the appropriate constants (the squares of half the coefficients of x and y) to both sides of the equation:

$$(x^2 + 2x + 1) + (y^2 - 6y + 9) = -7 + 1 + 9$$

or

$$(x + 1)^2 + (y - 3)^2 = 3$$

Comparing this equation with the standard equation of a circle 1, we see that $h = -1$, $k = 3$, and $r = \sqrt{3}$, so the given equation represents a circle with center $(-1, 3)$ and radius $\sqrt{3}$. It is sketched in Figure 2.

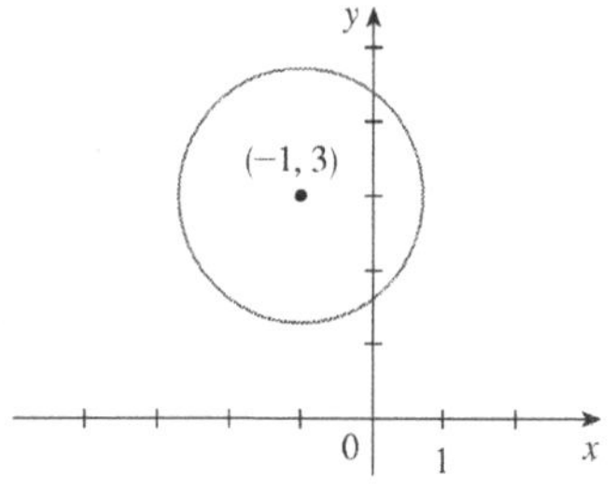

FIGURE 2
$x^2 + y^2 + 2x - 6y + 7 = 0$

Parabolas

The geometric properties of parabolas are reviewed in Section 10.5. Here we regard a parabola as a graph of an equation of the form $y = ax^2 + bx + c$.

EXAMPLE 3 Draw the graph of the parabola $y = x^2$.

SOLUTION We set up a table of values, plot points, and join them by a smooth curve to obtain the graph in Figure 3.

x	$y = x^2$
0	0
$\pm\frac{1}{2}$	$\frac{1}{4}$
± 1	1
± 2	4
± 3	9

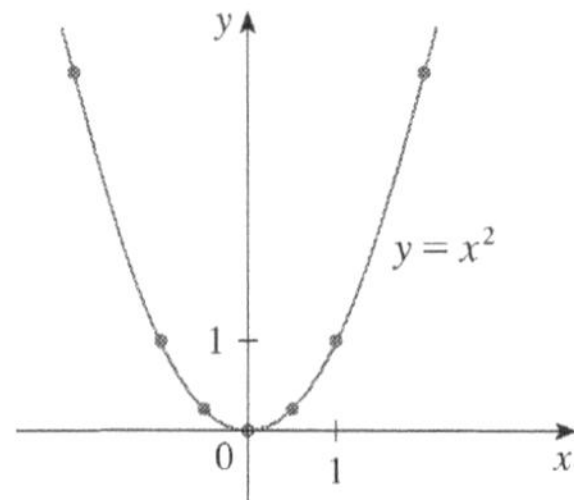

FIGURE 3

Figure 4 shows the graphs of several parabolas with equations of the form $y = ax^2$ for various values of the number a. In each case the *vertex*, the point where the parabola changes direction, is the origin. We see that the parabola $y = ax^2$ opens upward if $a > 0$ and downward if $a < 0$ (as in Figure 5).

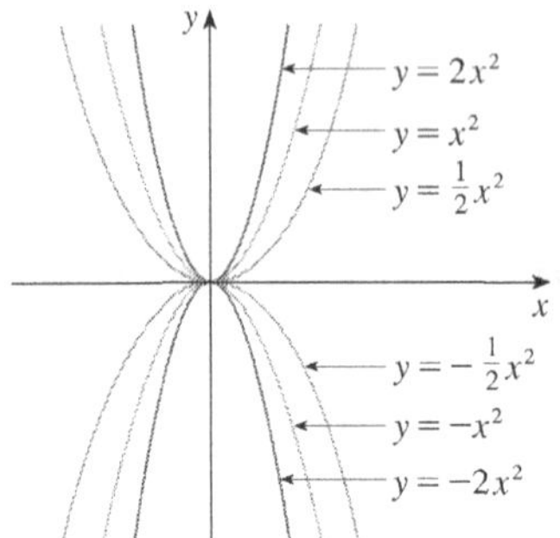

FIGURE 4

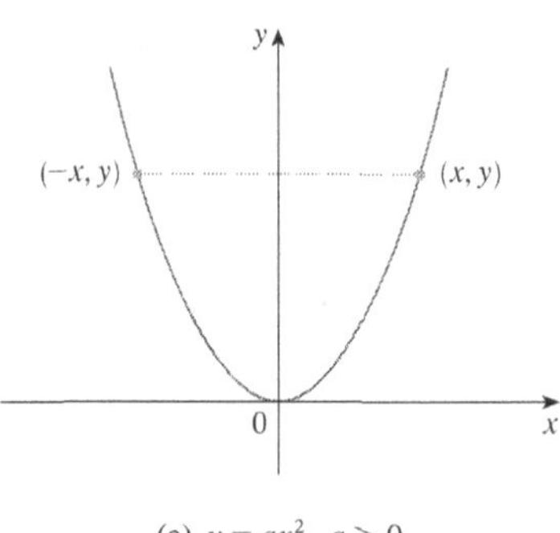

(a) $y = ax^2$, $a > 0$

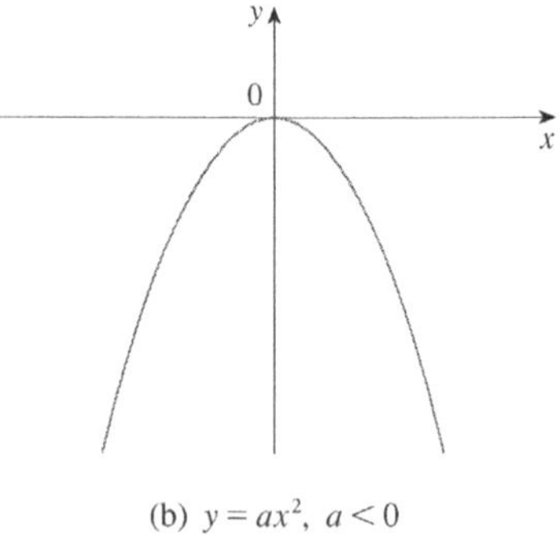

(b) $y = ax^2$, $a < 0$

FIGURE 5

Notice that if (x, y) satisfies $y = ax^2$, then so does $(-x, y)$. This corresponds to the geometric fact that if the right half of the graph is reflected about the y-axis, then the left half of the graph is obtained. We say that the graph is **symmetric with respect to the *y*-axis**.

> The graph of an equation is symmetric with respect to the y-axis if the equation is unchanged when x is replaced by $-x$.

If we interchange x and y in the equation $y = ax^2$, the result is $x = ay^2$, which also represents a parabola. (Interchanging x and y amounts to reflecting about the diagonal line $y = x$.) The parabola $x = ay^2$ opens to the right if $a > 0$ and to the left if $a < 0$. (See

Figure 6.) This time the parabola is symmetric with respect to the x-axis because if (x, y) satisfies $x = ay^2$, then so does $(x, -y)$.

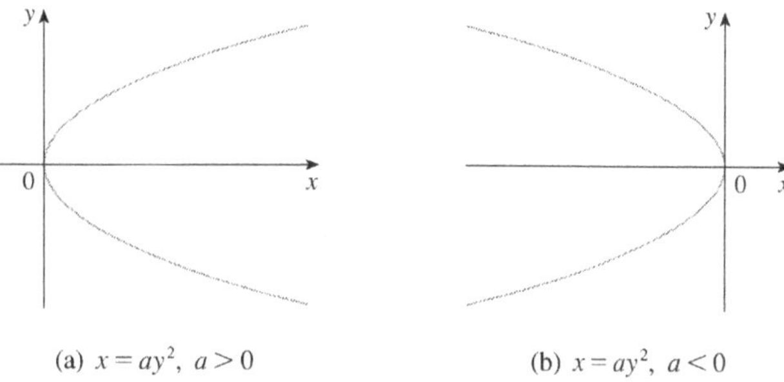

FIGURE 6

(a) $x = ay^2$, $a > 0$ (b) $x = ay^2$, $a < 0$

> The graph of an equation is symmetric with respect to the x-axis if the equation is unchanged when y is replaced by $-y$.

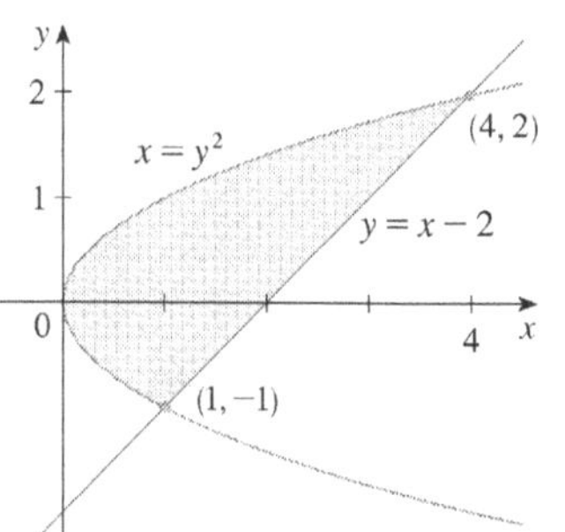

FIGURE 7

EXAMPLE 4 Sketch the region bounded by the parabola $x = y^2$ and the line $y = x - 2$.

SOLUTION First we find the points of intersection by solving the two equations. Substituting $x = y + 2$ into the equation $x = y^2$, we get $y + 2 = y^2$, which gives

$$0 = y^2 - y - 2 = (y - 2)(y + 1)$$

so $y = 2$ or -1. Thus the points of intersection are $(4, 2)$ and $(1, -1)$, and we draw the line $y = x - 2$ passing through these points. We then sketch the parabola $x = y^2$ by referring to Figure 6(a) and having the parabola pass through $(4, 2)$ and $(1, -1)$. The region bounded by $x = y^2$ and $y = x - 2$ means the finite region whose boundaries are these curves. It is sketched in Figure 7.

Ellipses

The curve with equation

$$\boxed{2} \qquad \frac{x^2}{a^2} + \frac{y^2}{b^2} = 1$$

where a and b are positive numbers, is called an **ellipse** in standard position. (Geometric properties of ellipses are discussed in Section 10.5.) Observe that Equation 2 is unchanged if x is replaced by $-x$ or y is replaced by $-y$, so the ellipse is symmetric with respect to both axes. As a further aid to sketching the ellipse, we find its intercepts.

> The ***x*-intercepts** of a graph are the x-coordinates of the points where the graph intersects the x-axis. They are found by setting $y = 0$ in the equation of the graph.
>
> The ***y*-intercepts** are the y-coordinates of the points where the graph intersects the y-axis. They are found by setting $x = 0$ in its equation.

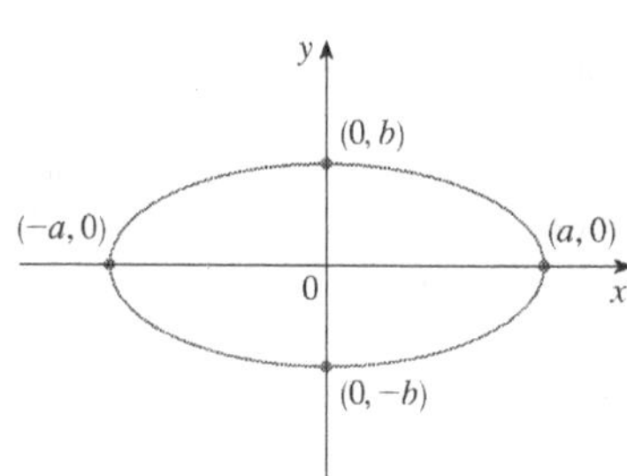

FIGURE 8

$\frac{x^2}{a^2} + \frac{y^2}{b^2} = 1$

If we set $y = 0$ in Equation 2, we get $x^2 = a^2$ and so the x-intercepts are $\pm a$. Setting $x = 0$, we get $y^2 = b^2$, so the y-intercepts are $\pm b$. Using this information, together with symmetry, we sketch the ellipse in Figure 8. If $a = b$, the ellipse is a circle with radius a.

EXAMPLE 3 Sketch the graph of $9x^2 + 16y^2 = 144$.

SOLUTION We divide both sides of the equation by 144:

$$\frac{x^2}{16} + \frac{y^2}{9} = 1$$

The equation is now in the standard form for an ellipse [2], so we have $a^2 = 16$, $b^2 = 9$, $a = 4$, and $b = 3$. The x-intercepts are ± 4; the y-intercepts are ± 3. The graph is sketched in Figure 9.

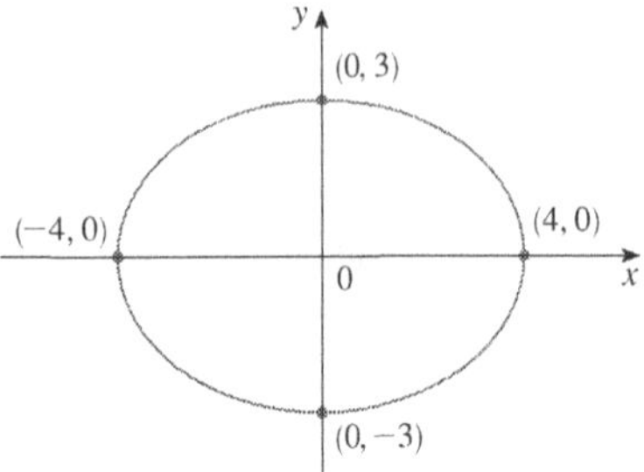

FIGURE 9
$9x^2 + 16y^2 = 144$

Hyperbolas

The curve with equation

$$\boxed{3} \qquad \frac{x^2}{a^2} - \frac{y^2}{b^2} = 1$$

is called a **hyperbola** in standard position. Again, Equation 3 is unchanged when x is replaced by $-x$ or y is replaced by $-y$, so the hyperbola is symmetric with respect to both axes. To find the x-intercepts we set $y = 0$ and obtain $x^2 = a^2$ and $x = \pm a$. However, if we put $x = 0$ in Equation 3, we get $y^2 = -b^2$, which is impossible, so there is no y-intercept. In fact, from Equation 3 we obtain

$$\frac{x^2}{a^2} = 1 + \frac{y^2}{b^2} \geqslant 1$$

which shows that $x^2 \geqslant a^2$ and so $|x| = \sqrt{x^2} \geqslant a$. Therefore we have $x \geqslant a$ or $x \leqslant -a$. This means that the hyperbola consists of two parts, called its *branches.* It is sketched in Figure 10.

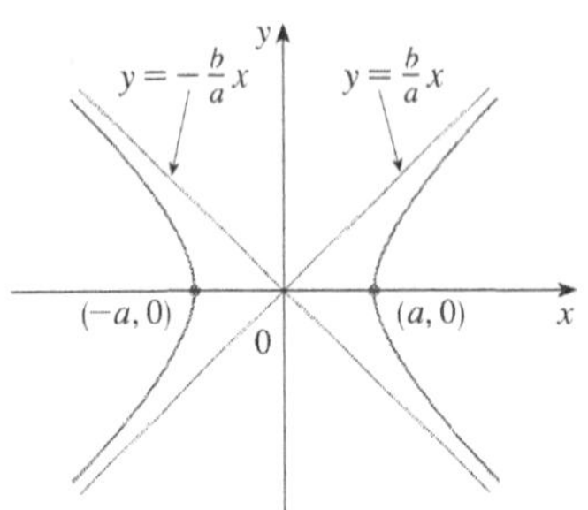

FIGURE 10
The hyperbola $\frac{x^2}{a^2} - \frac{y^2}{b^2} = 1$

In drawing a hyperbola it is useful to draw first its *asymptotes,* which are the lines $y = (b/a)x$ and $y = -(b/a)x$ shown in Figure 10. Both branches of the hyperbola approach the asymptotes; that is, they come arbitrarily close to the asymptotes. This involves the idea of a limit, which is discussed in Chapter 2. (See also Exercise 73 in Section 4.5.)

By interchanging the roles of x and y we get an equation of the form

$$\frac{y^2}{a^2} - \frac{x^2}{b^2} = 1$$

which also represents a hyperbola and is sketched in Figure 11.

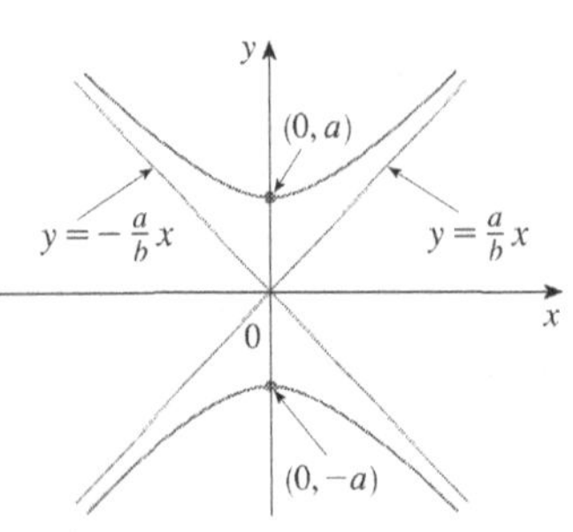

FIGURE 11
The hyperbola $\frac{y^2}{a^2} - \frac{x^2}{b^2} = 1$

EXAMPLE 6 Sketch the curve $9x^2 - 4y^2 = 36$.

SOLUTION Dividing both sides by 36, we obtain

$$\frac{x^2}{4} - \frac{y^2}{9} = 1$$

which is the standard form of the equation of a hyperbola (Equation 3). Since $a^2 = 4$, the x-intercepts are ± 2. Since $b^2 = 9$, we have $b = 3$ and the asymptotes are $y = \pm(\frac{3}{2})x$. The hyperbola is sketched in Figure 12.

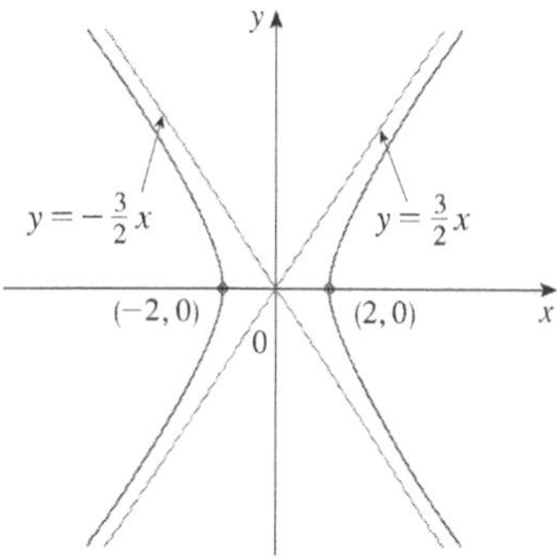

FIGURE 12
The hyperbola $9x^2 - 4y^2 = 36$

If $b = a$, a hyperbola has the equation $x^2 - y^2 = a^2$ (or $y^2 - x^2 = a^2$) and is called an *equilateral hyperbola* [see Figure 13(a)]. Its asymptotes are $y = \pm x$, which are perpendicular. If an equilateral hyperbola is rotated by 45°, the asymptotes become the x- and y-axes, and it can be shown that the new equation of the hyperbola is $xy = k$, where k is a constant [see Figure 13(b)].

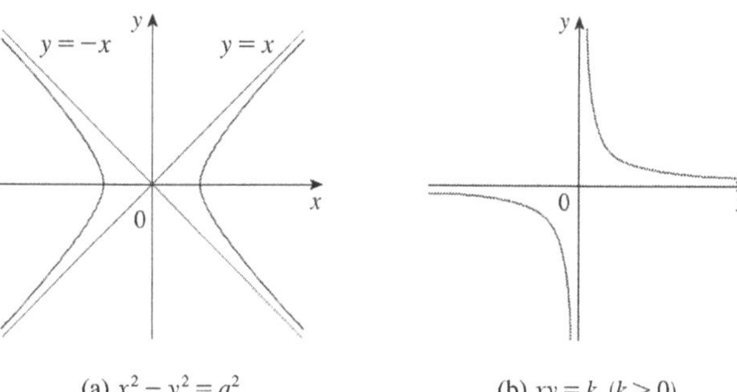

FIGURE 13
Equilateral hyperbolas

Shifted Conics

Recall that an equation of the circle with center the origin and radius r is $x^2 + y^2 = r^2$, but if the center is the point (h, k), then the equation of the circle becomes

$$(x - h)^2 + (y - k)^2 = r^2$$

Similarly, if we take the ellipse with equation

$$\boxed{\frac{x^2}{a^2} + \frac{y^2}{b^2} = 1} \qquad \boxed{4}$$

and translate it (shift it) so that its center is the point (h, k), then its equation becomes

5 $$\frac{(x-h)^2}{a^2} + \frac{(y-k)^2}{b^2} = 1$$

(See Figure 14.)

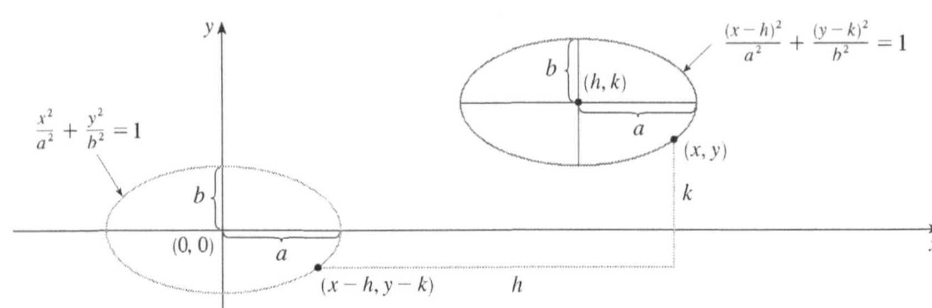

FIGURE 14

Notice that in shifting the ellipse, we replaced x by $x - h$ and y by $y - k$ in Equation 4 to obtain Equation 5. We use the same procedure to shift the parabola $y = ax^2$ so that its vertex (the origin) becomes the point (h, k) as in Figure 15. Replacing x by $x - h$ and y by $y - k$, we see that the new equation is

$$y - k = a(x - h)^2 \qquad \text{or} \qquad y = a(x - h)^2 + k$$

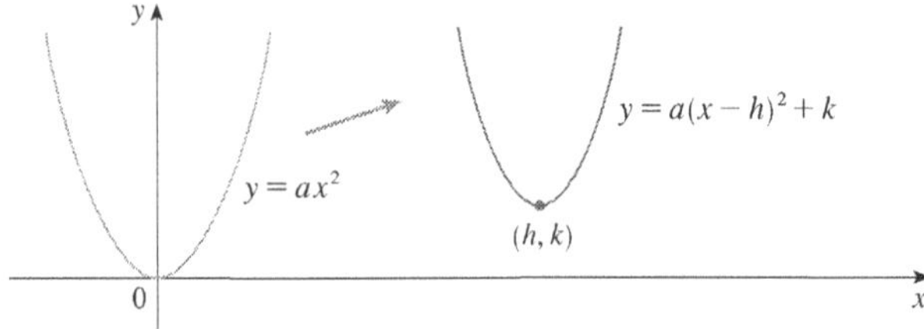

FIGURE 15

EXAMPLE 7 Sketch the graph of the equation $y = 2x^2 - 4x + 1$.

SOLUTION First we complete the square:

$$y = 2(x^2 - 2x) + 1 = 2(x - 1)^2 - 1$$

In this form we see that the equation represents the parabola obtained by shifting $y = 2x^2$ so that its vertex is at the point $(1, -1)$. The graph is sketched in Figure 16.

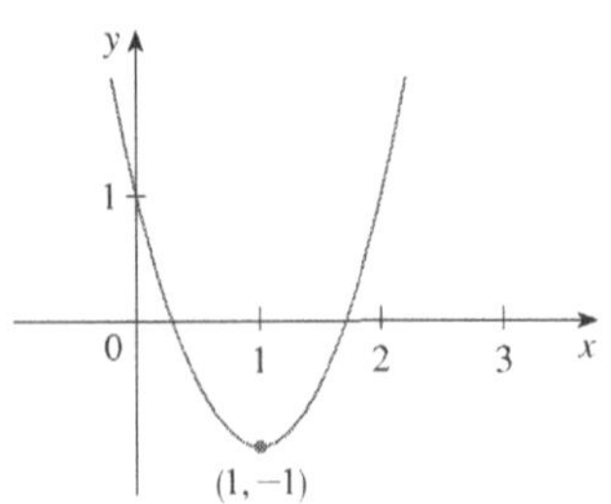

FIGURE 16
$y = 2x^2 - 4x + 1$

EXAMPLE 8 Sketch the curve $x = 1 - y^2$.

SOLUTION This time we start with the parabola $x = -y^2$ (as in Figure 6 with $a = -1$) and shift one unit to the right to get the graph of $x = 1 - y^2$. (See Figure 17.)

FIGURE 17

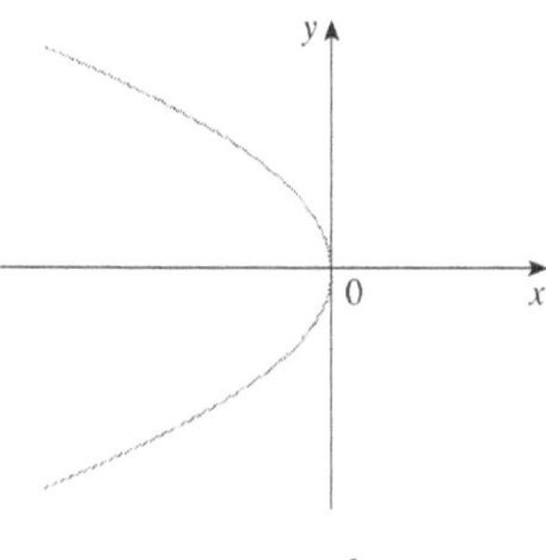

(a) $x = -y^2$

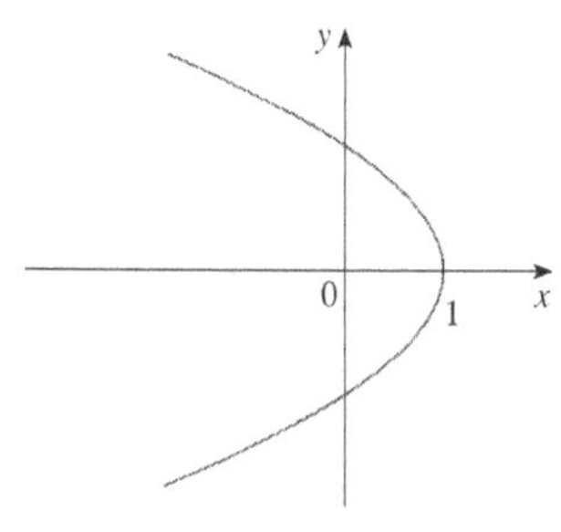

(b) $x = 1 - y^2$

C Exercises

1–4 Find an equation of a circle that satisfies the given conditions.

1. Center $(3, -1)$, radius 5

2. Center $(-2, -8)$, radius 10

3. Center at the origin, passes through $(4, 7)$

4. Center $(-1, 5)$, passes through $(-4, -6)$

5–9 Show that the equation represents a circle and find the center and radius.

5. $x^2 + y^2 - 4x + 10y + 13 = 0$

6. $x^2 + y^2 + 6y + 2 = 0$

7. $x^2 + y^2 + x = 0$

8. $16x^2 + 16y^2 + 8x + 32y + 1 = 0$

9. $2x^2 + 2y^2 - x + y = 1$

10. Under what condition on the coefficients a, b, and c does the equation $x^2 + y^2 + ax + by + c = 0$ represent a circle? When that condition is satisfied, find the center and radius of the circle.

11–32 Identify the type of curve and sketch the graph. Do not plot points. Just use the standard graphs given in Figures 5, 6, 8, 10, and 11 and shift if necessary.

11. $y = -x^2$

12. $y^2 - x^2 = 1$

13. $x^2 + 4y^2 = 16$

14. $x = -2y^2$

15. $16x^2 - 25y^2 = 400$

16. $25x^2 + 4y^2 = 100$

17. $4x^2 + y^2 = 1$

18. $y = x^2 + 2$

19. $x = y^2 - 1$

20. $9x^2 - 25y^2 = 225$

21. $9y^2 - x^2 = 9$

22. $2x^2 + 5y^2 = 10$

23. $xy = 4$

24. $y = x^2 + 2x$

25. $9(x - 1)^2 + 4(y - 2)^2 = 36$

26. $16x^2 + 9y^2 - 36y = 108$

27. $y = x^2 - 6x + 13$

28. $x^2 - y^2 - 4x + 3 = 0$

29. $x = 4 - y^2$

30. $y^2 - 2x + 6y + 5 = 0$

31. $x^2 + 4y^2 - 6x + 5 = 0$

32. $4x^2 + 9y^2 - 16x + 54y + 61 = 0$

33–34 Sketch the region bounded by the curves.

33. $y = 3x, \quad y = x^2$

34. $y = 4 - x^2, \quad x - 2y = 2$

35. Find an equation of the parabola with vertex $(1, -1)$ that passes through the points $(-1, 3)$ and $(3, 3)$.

36. Find an equation of the ellipse with center at the origin that passes through the points $\left(1, -10\sqrt{2}/3\right)$ and $\left(-2, 5\sqrt{5}/3\right)$.

37–40 Sketch the graph of the set.

37. $\{(x, y) \mid x^2 + y^2 \leqslant 1\}$

38. $\{(x, y) \mid x^2 + y^2 > 4\}$

39. $\{(x, y) \mid y \geqslant x^2 - 1\}$

40. $\{(x, y) \mid x^2 + 4y^2 \leqslant 4\}$

D Trigonometry

Angles

Angles can be measured in degrees or in radians (abbreviated as rad). The angle given by a complete revolution contains 360°, which is the same as 2π rad. Therefore

1 $$\pi \text{ rad} = 180°$$

and

2 $$1 \text{ rad} = \left(\frac{180}{\pi}\right)^{\circ} \approx 57.3° \qquad 1° = \frac{\pi}{180} \text{ rad} \approx 0.017 \text{ rad}$$

EXAMPLE 1

(a) Find the radian measure of 60°. (b) Express $5\pi/4$ rad in degrees.

SOLUTION

(a) From Equation 1 or 2 we see that to convert from degrees to radians we multiply by $\pi/180$. Therefore

$$60° = 60\left(\frac{\pi}{180}\right) = \frac{\pi}{3} \text{ rad}$$

(b) To convert from radians to degrees we multiply by $180/\pi$. Thus

$$\frac{5\pi}{4} \text{ rad} = \frac{5\pi}{4}\left(\frac{180}{\pi}\right) = 225°$$

In calculus we use radians to measure angles except when otherwise indicated. The following table gives the correspondence between degree and radian measures of some common angles.

Degrees	0°	30°	45°	60°	90°	120°	135°	150°	180°	270°	360°
Radians	0	$\frac{\pi}{6}$	$\frac{\pi}{4}$	$\frac{\pi}{3}$	$\frac{\pi}{2}$	$\frac{2\pi}{3}$	$\frac{3\pi}{4}$	$\frac{5\pi}{6}$	π	$\frac{3\pi}{2}$	2π

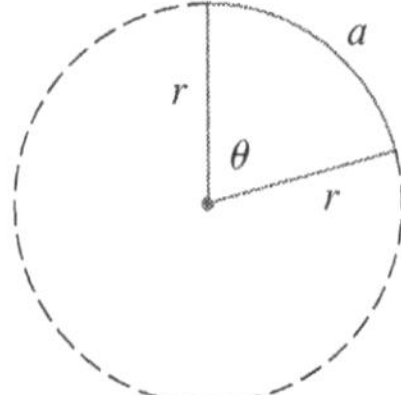

FIGURE 1

Figure 1 shows a sector of a circle with central angle θ and radius r subtending an arc with length a. Since the length of the arc is proportional to the size of the angle, and since the entire circle has circumference $2\pi r$ and central angle 2π, we have

$$\frac{\theta}{2\pi} = \frac{a}{2\pi r}$$

Solving this equation for θ and for a, we obtain

3 $$\theta = \frac{a}{r} \qquad a = r\theta$$

Remember that Equations 3 are valid only when θ is measured in radians.

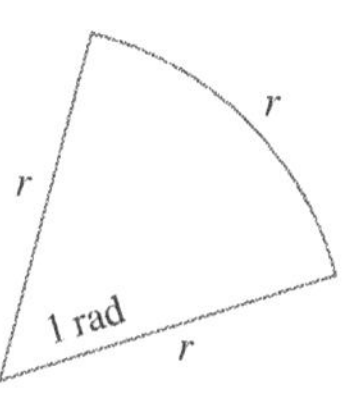

FIGURE 2

In particular, putting $a = r$ in Equation 3, we see that an angle of 1 rad is the angle subtended at the center of a circle by an arc equal in length to the radius of the circle (see Figure 2).

EXAMPLE 2

(a) If the radius of a circle is 5 cm, what angle is subtended by an arc of 6 cm?
(b) If a circle has radius 3 cm, what is the length of an arc subtended by a central angle of $3\pi/8$ rad?

SOLUTION

(a) Using Equation 3 with $a = 6$ and $r = 5$, we see that the angle is

$$\theta = \tfrac{6}{5} = 1.2 \text{ rad}$$

(b) With $r = 3$ cm and $\theta = 3\pi/8$ rad, the arc length is

$$a = r\theta = 3\left(\frac{3\pi}{8}\right) = \frac{9\pi}{8} \text{ cm}$$

The **standard position** of an angle occurs when we place its vertex at the origin of a coordinate system and its initial side on the positive x-axis as in Figure 3. A **positive** angle is obtained by rotating the initial side counterclockwise until it coincides with the terminal side. Likewise, **negative** angles are obtained by clockwise rotation as in Figure 4.

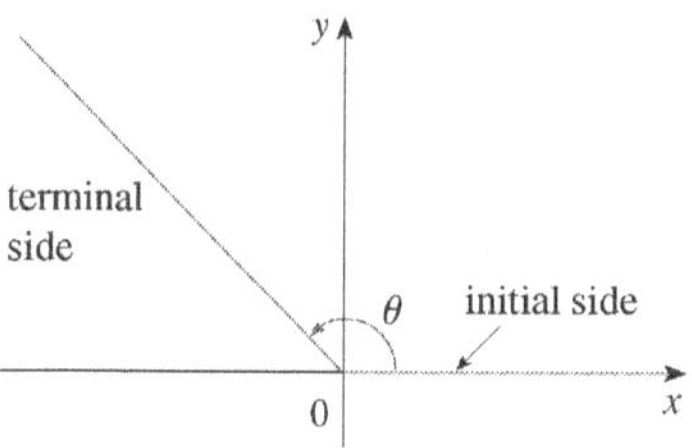

FIGURE 3 $\theta \geqslant 0$

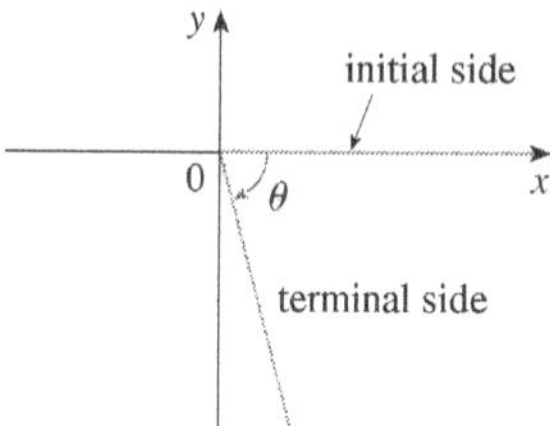

FIGURE 4 $\theta < 0$

Figure 5 shows several examples of angles in standard position. Notice that different angles can have the same terminal side. For instance, the angles $3\pi/4$, $-5\pi/4$, and $11\pi/4$ have the same initial and terminal sides because

$$\frac{3\pi}{4} - 2\pi = -\frac{5\pi}{4} \qquad \frac{3\pi}{4} + 2\pi = \frac{11\pi}{4}$$

and 2π rad represents a complete revolution.

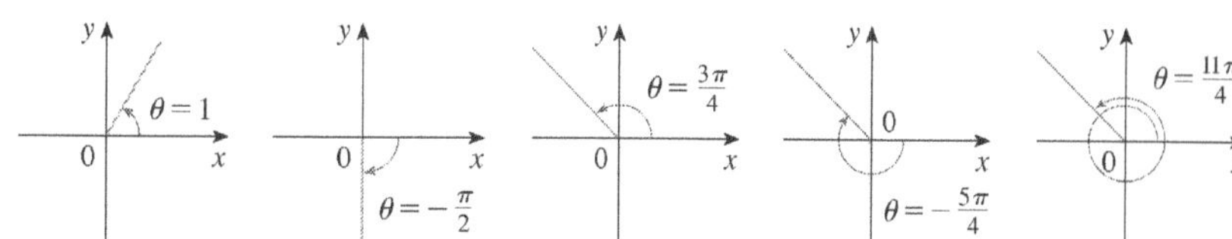

FIGURE 5
Angles in standard position

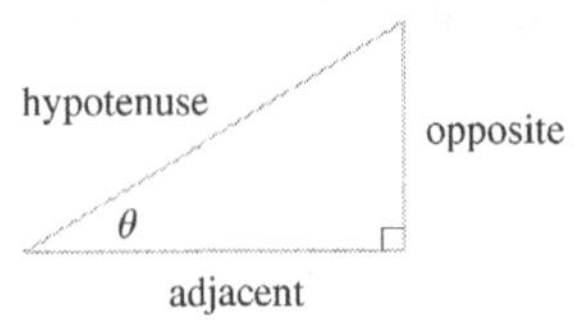

FIGURE 6

The Trigonometric Functions

For an acute angle θ the six trigonometric functions are defined as ratios of lengths of sides of a right triangle as follows (see Figure 6).

4

$$\sin\theta = \frac{\text{opp}}{\text{hyp}} \qquad \csc\theta = \frac{\text{hyp}}{\text{opp}}$$

$$\cos\theta = \frac{\text{adj}}{\text{hyp}} \qquad \sec\theta = \frac{\text{hyp}}{\text{adj}}$$

$$\tan\theta = \frac{\text{opp}}{\text{adj}} \qquad \cot\theta = \frac{\text{adj}}{\text{opp}}$$

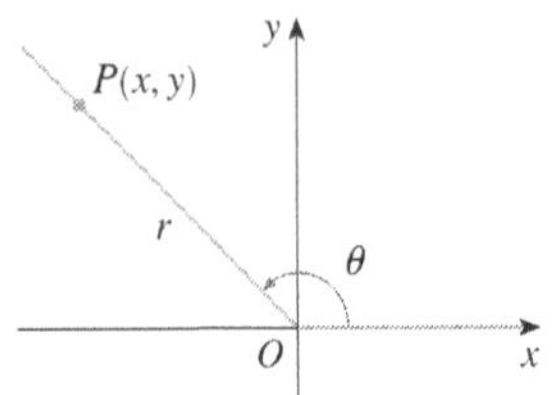

FIGURE 7

This definition doesn't apply to obtuse or negative angles, so for a general angle θ in standard position we let $P(x, y)$ be any point on the terminal side of θ and we let r be the distance $|OP|$ as in Figure 7. Then we define

5

$$\sin\theta = \frac{y}{r} \qquad \csc\theta = \frac{r}{y}$$

$$\cos\theta = \frac{x}{r} \qquad \sec\theta = \frac{r}{x}$$

$$\tan\theta = \frac{y}{x} \qquad \cot\theta = \frac{x}{y}$$

If we put $r = 1$ in Definition 5 and draw a unit circle with center the origin and label θ as in Figure 8, then the coordinates of P are $(\cos\theta, \sin\theta)$.

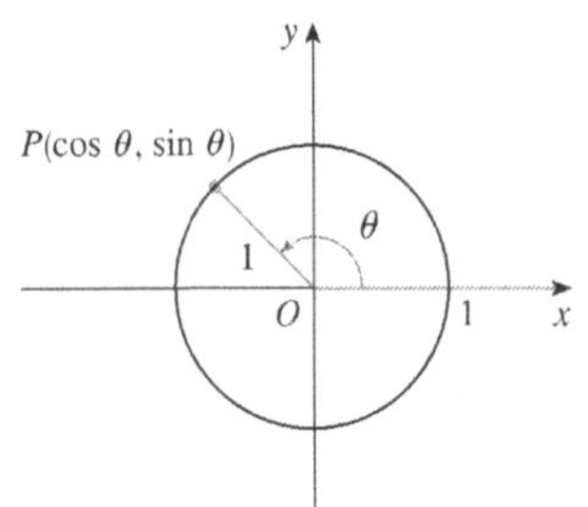

FIGURE 8

Since division by 0 is not defined, $\tan\theta$ and $\sec\theta$ are undefined when $x = 0$ and $\csc\theta$ and $\cot\theta$ are undefined when $y = 0$. Notice that the definitions in **4** and **5** are consistent when θ is an acute angle.

If θ is a number, the convention is that $\sin\theta$ means the sine of the angle whose *radian* measure is θ. For example, the expression $\sin 3$ implies that we are dealing with an angle of 3 rad. When finding a calculator approximation to this number, we must remember to set our calculator in radian mode, and then we obtain

$$\sin 3 \approx 0.14112$$

If we want to know the sine of the angle 3° we would write $\sin 3°$ and, with our calculator in degree mode, we find that

$$\sin 3° \approx 0.05234$$

The exact trigonometric ratios for certain angles can be read from the triangles in Figure 9. For instance,

$$\sin\frac{\pi}{4} = \frac{1}{\sqrt{2}} \qquad \sin\frac{\pi}{6} = \frac{1}{2} \qquad \sin\frac{\pi}{3} = \frac{\sqrt{3}}{2}$$

$$\cos\frac{\pi}{4} = \frac{1}{\sqrt{2}} \qquad \cos\frac{\pi}{6} = \frac{\sqrt{3}}{2} \qquad \cos\frac{\pi}{3} = \frac{1}{2}$$

$$\tan\frac{\pi}{4} = 1 \qquad \tan\frac{\pi}{6} = \frac{1}{\sqrt{3}} \qquad \tan\frac{\pi}{3} = \sqrt{3}$$

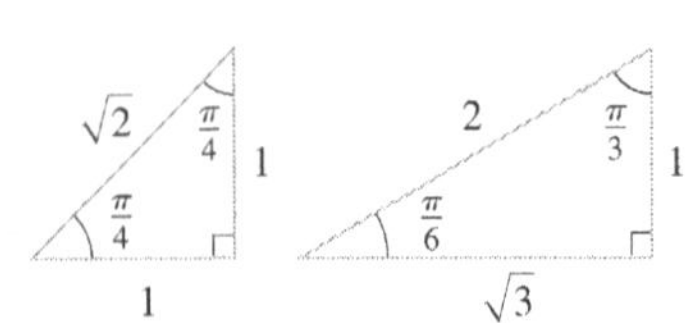

FIGURE 9

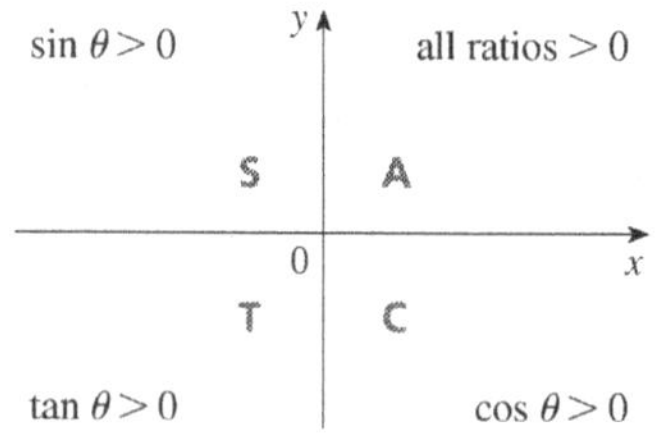

FIGURE 10

The signs of the trigonometric functions for angles in each of the four quadrants can be remembered by means of the rule "All Students Take Calculus" shown in Figure 10.

EXAMPLE 3 Find the exact trigonometric ratios for $\theta = 2\pi/3$.

SOLUTION From Figure 11 we see that a point on the terminal line for $\theta = 2\pi/3$ is $P(-1, \sqrt{3})$. Therefore, taking

$$x = -1 \qquad y = \sqrt{3} \qquad r = 2$$

in the definitions of the trigonometric ratios, we have

$$\sin\frac{2\pi}{3} = \frac{\sqrt{3}}{2} \qquad \cos\frac{2\pi}{3} = -\frac{1}{2} \qquad \tan\frac{2\pi}{3} = -\sqrt{3}$$

$$\csc\frac{2\pi}{3} = \frac{2}{\sqrt{3}} \qquad \sec\frac{2\pi}{3} = -2 \qquad \cot\frac{2\pi}{3} = -\frac{1}{\sqrt{3}}$$

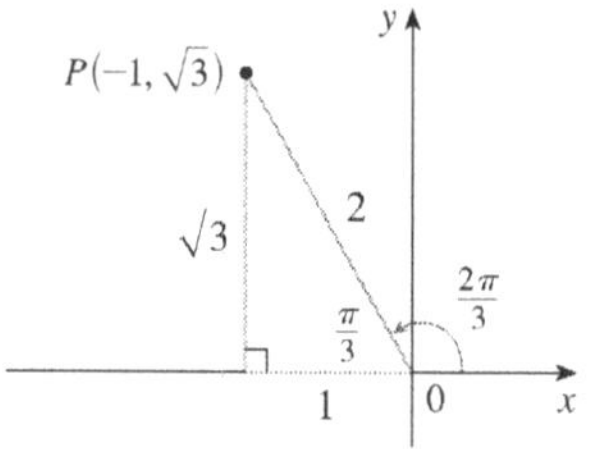

FIGURE 11

The following table gives some values of $\sin\theta$ and $\cos\theta$ found by the method of Example 3.

θ	0	$\frac{\pi}{6}$	$\frac{\pi}{4}$	$\frac{\pi}{3}$	$\frac{\pi}{2}$	$\frac{2\pi}{3}$	$\frac{3\pi}{4}$	$\frac{5\pi}{6}$	π	$\frac{3\pi}{2}$	2π
$\sin\theta$	0	$\frac{1}{2}$	$\frac{1}{\sqrt{2}}$	$\frac{\sqrt{3}}{2}$	1	$\frac{\sqrt{3}}{2}$	$\frac{1}{\sqrt{2}}$	$\frac{1}{2}$	0	-1	0
$\cos\theta$	1	$\frac{\sqrt{3}}{2}$	$\frac{1}{\sqrt{2}}$	$\frac{1}{2}$	0	$-\frac{1}{2}$	$-\frac{1}{\sqrt{2}}$	$-\frac{\sqrt{3}}{2}$	-1	0	1

EXAMPLE 4 If $\cos\theta = \frac{2}{5}$ and $0 < \theta < \pi/2$, find the other five trigonometric functions of θ.

SOLUTION Since $\cos\theta = \frac{2}{5}$, we can label the hypotenuse as having length 5 and the adjacent side as having length 2 in Figure 12. If the opposite side has length x, then the Pythagorean Theorem gives $x^2 + 4 = 25$ and so $x^2 = 21$, $x = \sqrt{21}$. We can now use the diagram to write the other five trigonometric functions:

$$\sin\theta = \frac{\sqrt{21}}{5} \qquad \tan\theta = \frac{\sqrt{21}}{2}$$

$$\csc\theta = \frac{5}{\sqrt{21}} \qquad \sec\theta = \frac{5}{2} \qquad \cot\theta = \frac{2}{\sqrt{21}}$$

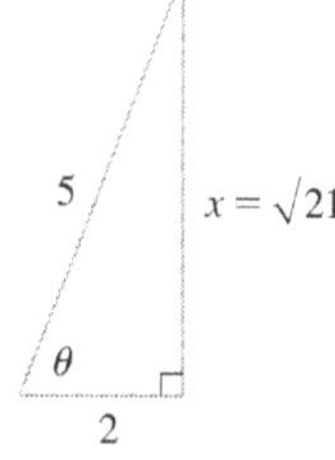

FIGURE 12

EXAMPLE 5 Use a calculator to approximate the value of x in Figure 13.

SOLUTION From the diagram we see that

$$\tan 40° = \frac{16}{x}$$

Therefore

$$x = \frac{16}{\tan 40°} \approx 19.07$$

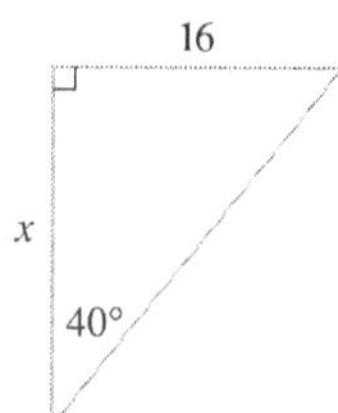

FIGURE 13

Trigonometric Identities

A trigonometric identity is a relationship among the trigonometric functions. The most elementary are the following, which are immediate consequences of the definitions of the trigonometric functions.

6

$$\csc \theta = \frac{1}{\sin \theta} \qquad \sec \theta = \frac{1}{\cos \theta} \qquad \cot \theta = \frac{1}{\tan \theta}$$

$$\tan \theta = \frac{\sin \theta}{\cos \theta} \qquad \cot \theta = \frac{\cos \theta}{\sin \theta}$$

For the next identity we refer back to Figure 7. The distance formula (or, equivalently, the Pythagorean Theorem) tells us that $x^2 + y^2 = r^2$. Therefore

$$\sin^2\theta + \cos^2\theta = \frac{y^2}{r^2} + \frac{x^2}{r^2} = \frac{x^2 + y^2}{r^2} = \frac{r^2}{r^2} = 1$$

We have therefore proved one of the most useful of all trigonometric identities:

7

$$\sin^2\theta + \cos^2\theta = 1$$

If we now divide both sides of Equation 7 by $\cos^2\theta$ and use Equations 6, we get

8

$$\tan^2\theta + 1 = \sec^2\theta$$

Similarly, if we divide both sides of Equation 7 by $\sin^2\theta$, we get

9

$$1 + \cot^2\theta = \csc^2\theta$$

The identities

10a $$\sin(-\theta) = -\sin \theta$$

10b $$\cos(-\theta) = \cos \theta$$

Odd functions and even functions are discussed in Section 1.1.

show that sine is an odd function and cosine is an even function. They are easily proved by drawing a diagram showing θ and $-\theta$ in standard position (see Exercise 39).

Since the angles θ and $\theta + 2\pi$ have the same terminal side, we have

11

$$\sin(\theta + 2\pi) = \sin \theta \qquad \cos(\theta + 2\pi) = \cos \theta$$

These identities show that the sine and cosine functions are periodic with period 2π.

The remaining trigonometric identities are all consequences of two basic identities called the **addition formulas**:

12a $$\sin(x + y) = \sin x \cos y + \cos x \sin y$$

12b $$\cos(x + y) = \cos x \cos y - \sin x \sin y$$

The proofs of these addition formulas are outlined in Exercises 85, 86, and 87.

By substituting $-y$ for y in Equations 12a and 12b and using Equations 10a and 10b, we obtain the following **subtraction formulas**:

13a $$\sin(x - y) = \sin x \cos y - \cos x \sin y$$

13b $$\cos(x - y) = \cos x \cos y + \sin x \sin y$$

Then, by dividing the formulas in Equations 12 or Equations 13, we obtain the corresponding formulas for $\tan(x \pm y)$:

14a $$\tan(x + y) = \frac{\tan x + \tan y}{1 - \tan x \tan y}$$

14b $$\tan(x - y) = \frac{\tan x - \tan y}{1 + \tan x \tan y}$$

If we put $y = x$ in the addition formulas [12], we get the **double-angle formulas**:

15a $$\sin 2x = 2 \sin x \cos x$$

15b $$\cos 2x = \cos^2 x - \sin^2 x$$

Then, by using the identity $\sin^2 x + \cos^2 x = 1$, we obtain the following alternate forms of the double-angle formulas for $\cos 2x$:

16a $$\cos 2x = 2 \cos^2 x - 1$$

16b $$\cos 2x = 1 - 2 \sin^2 x$$

If we now solve these equations for $\cos^2 x$ and $\sin^2 x$, we get the following **half-angle formulas**, which are useful in integral calculus:

17a $$\cos^2 x = \frac{1 + \cos 2x}{2}$$

17b $$\sin^2 x = \frac{1 - \cos 2x}{2}$$

Finally, we state the **product formulas**, which can be deduced from Equations 12 and 13:

18a $$\sin x \cos y = \tfrac{1}{2}[\sin(x + y) + \sin(x - y)]$$

18b $$\cos x \cos y = \tfrac{1}{2}[\cos(x + y) + \cos(x - y)]$$

18c $$\sin x \sin y = \tfrac{1}{2}[\cos(x - y) - \cos(x + y)]$$

There are many other trigonometric identities, but those we have stated are the ones used most often in calculus. If you forget any of the identities 13–18, remember that they can all be deduced from Equations 12a and 12b.

EXAMPLE 6 Find all values of x in the interval $[0, 2\pi]$ such that $\sin x = \sin 2x$.

SOLUTION Using the double-angle formula (15a), we rewrite the given equation as

$$\sin x = 2 \sin x \cos x \qquad \text{or} \qquad \sin x(1 - 2 \cos x) = 0$$

Therefore there are two possibilities:

$$\sin x = 0 \qquad \text{or} \qquad 1 - 2 \cos x = 0$$

$$x = 0, \pi, 2\pi \qquad\qquad \cos x = \tfrac{1}{2}$$

$$x = \frac{\pi}{3}, \frac{5\pi}{3}$$

The given equation has five solutions: 0, $\pi/3$, π, $5\pi/3$, and 2π.

Graphs of the Trigonometric Functions

The graph of the function $f(x) = \sin x$, shown in Figure 14(a), is obtained by plotting points for $0 \leqslant x \leqslant 2\pi$ and then using the periodic nature of the function (from Equation 11) to complete the graph. Notice that the zeros of the sine function occur at the

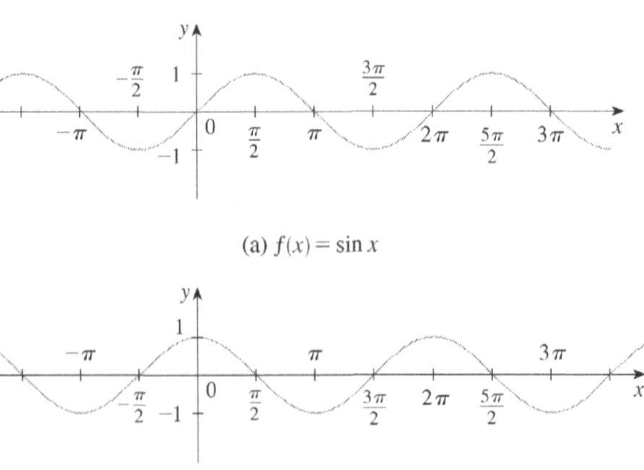

FIGURE 14

integer multiples of π, that is,

$$\sin x = 0 \qquad \text{whenever } x = n\pi, \quad n \text{ an integer}$$

Because of the identity

$$\cos x = \sin\left(x + \frac{\pi}{2}\right)$$

(which can be verified using Equation 12a), the graph of cosine is obtained by shifting the graph of sine by an amount $\pi/2$ to the left [see Figure 14(b)]. Note that for both the sine and cosine functions the domain is $(-\infty, \infty)$ and the range is the closed interval $[-1, 1]$. Thus, for all values of x, we have

$$-1 \leqslant \sin x \leqslant 1 \qquad -1 \leqslant \cos x \leqslant 1$$

The graphs of the remaining four trigonometric functions are shown in Figure 15 and their domains are indicated there. Notice that tangent and cotangent have range $(-\infty, \infty)$, whereas cosecant and secant have range $(-\infty, -1] \cup [1, \infty)$. All four functions are periodic: tangent and cotangent have period π, whereas cosecant and secant have period 2π.

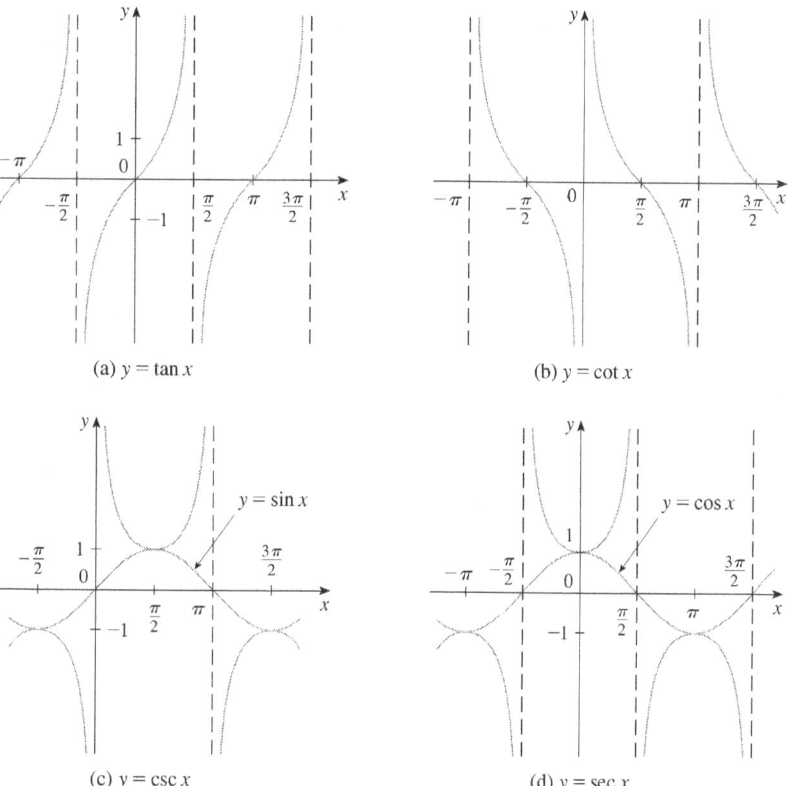

FIGURE 15

D Exercises

1–6 Convert from degrees to radians.

1. 210° **2.** 300° **3.** 9°

4. −315° **5.** 900° **6.** 36°

7–12 Convert from radians to degrees.

7. 4π **8.** $-\frac{7\pi}{2}$ **9.** $\frac{5\pi}{12}$

10. $\frac{8\pi}{3}$ **11.** $-\frac{3\pi}{8}$ **12.** 5

13. Find the length of a circular arc subtended by an angle of $\pi/12$ rad if the radius of the circle is 36 cm.

14. If a circle has radius 10 cm, find the length of the arc subtended by a central angle of 72°.

15. A circle has radius 1.5 m. What angle is subtended at the center of the circle by an arc 1 m long?

16. Find the radius of a circular sector with angle $3\pi/4$ and arc length 6 cm.

17–22 Draw, in standard position, the angle whose measure is given.

17. 315° **18.** −150° **19.** $-\frac{3\pi}{4}$ rad

20. $\frac{7\pi}{3}$ rad **21.** 2 rad **22.** −3 rad

23–28 Find the exact trigonometric ratios for the angle whose radian measure is given.

23. $\frac{3\pi}{4}$ **24.** $\frac{4\pi}{3}$ **25.** $\frac{9\pi}{2}$

26. -5π **27.** $\frac{5\pi}{6}$ **28.** $\frac{11\pi}{4}$

29–34 Find the remaining trigonometric ratios.

29. $\sin\theta = \frac{3}{5}, \quad 0 < \theta < \frac{\pi}{2}$

30. $\tan\alpha = 2, \quad 0 < \alpha < \frac{\pi}{2}$

31. $\sec\phi = -1.5, \quad \frac{\pi}{2} < \phi < \pi$

32. $\cos x = -\frac{1}{3}, \quad \pi < x < \frac{3\pi}{2}$

33. $\cot\beta = 3, \quad \pi < \beta < 2\pi$

34. $\csc\theta = -\frac{4}{3}, \quad \frac{3\pi}{2} < \theta < 2\pi$

35–38 Find, correct to five decimal places, the length of the side labeled x.

35.

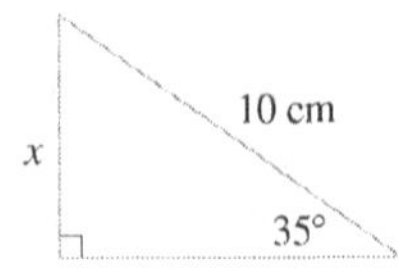

36.

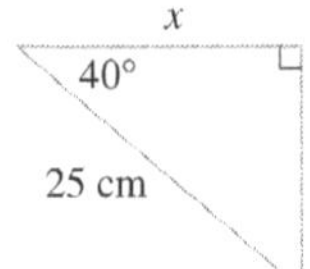

37.

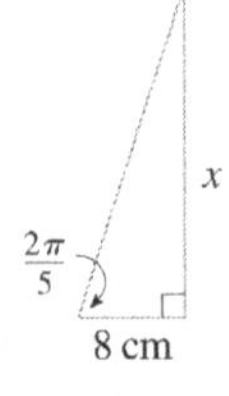

38.

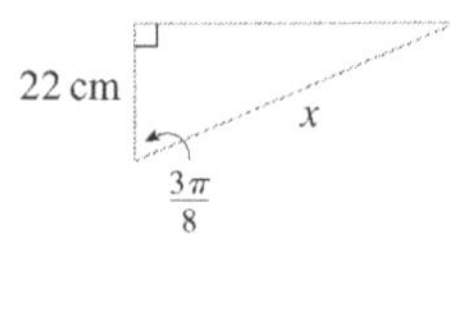

39–41 Prove each equation.

39. (a) Equation 10a (b) Equation 10b

40. (a) Equation 14a (b) Equation 14b

41. (a) Equation 18a (b) Equation 18b
(c) Equation 18c

42–58 Prove the identity.

42. $\cos\left(\frac{\pi}{2} - x\right) = \sin x$

43. $\sin\left(\frac{\pi}{2} + x\right) = \cos x$

44. $\sin(\pi - x) = \sin x$

45. $\sin\theta\cot\theta = \cos\theta$

46. $(\sin x + \cos x)^2 = 1 + \sin 2x$

47. $\sec y - \cos y = \tan y \sin y$

48. $\tan^2\alpha - \sin^2\alpha = \tan^2\alpha \sin^2\alpha$

49. $\cot^2\theta + \sec^2\theta = \tan^2\theta + \csc^2\theta$

50. $2\csc 2t = \sec t \csc t$

51. $\tan 2\theta = \frac{2\tan\theta}{1 - \tan^2\theta}$

52. $\frac{1}{1 - \sin\theta} + \frac{1}{1 + \sin\theta} = 2\sec^2\theta$

53. $\sin x \sin 2x + \cos x \cos 2x = \cos x$

54. $\sin^2 x - \sin^2 y = \sin(x + y)\sin(x - y)$

55. $\frac{\sin\phi}{1 - \cos\phi} = \csc\phi + \cot\phi$

56. $\tan x + \tan y = \dfrac{\sin(x + y)}{\cos x \cos y}$

57. $\sin 3\theta + \sin \theta = 2 \sin 2\theta \cos \theta$

58. $\cos 3\theta = 4 \cos^3\theta - 3 \cos \theta$

59–64 If $\sin x = \frac{1}{3}$ and $\sec y = \frac{5}{4}$, where x and y lie between 0 and $\pi/2$, evaluate the expression.

59. $\sin(x + y)$

60. $\cos(x + y)$

61. $\cos(x - y)$

62. $\sin(x - y)$

63. $\sin 2y$

64. $\cos 2y$

65–72 Find all values of x in the interval $[0, 2\pi]$ that satisfy the equation.

65. $2 \cos x - 1 = 0$

66. $3 \cot^2x = 1$

67. $2 \sin^2x = 1$

68. $|\tan x| = 1$

69. $\sin 2x = \cos x$

70. $2 \cos x + \sin 2x = 0$

71. $\sin x = \tan x$

72. $2 + \cos 2x = 3 \cos x$

73–76 Find all values of x in the interval $[0, 2\pi]$ that satisfy the inequality.

73. $\sin x \leqslant \frac{1}{2}$

74. $2 \cos x + 1 > 0$

75. $-1 < \tan x < 1$

76. $\sin x > \cos x$

77–82 Graph the function by starting with the graphs in Figures 14 and 15 and applying the transformations of Section 1.3 where appropriate.

77. $y = \cos\left(x - \dfrac{\pi}{3}\right)$

78. $y = \tan 2x$

79. $y = \dfrac{1}{3}\tan\left(x - \dfrac{\pi}{2}\right)$

80. $y = 1 + \sec x$

81. $y = |\sin x|$

82. $y = 2 + \sin\left(x + \dfrac{\pi}{4}\right)$

83. Prove the **Law of Cosines**: If a triangle has sides with lengths a, b, and c, and θ is the angle between the sides with lengths a and b, then

$$c^2 = a^2 + b^2 - 2ab \cos \theta$$

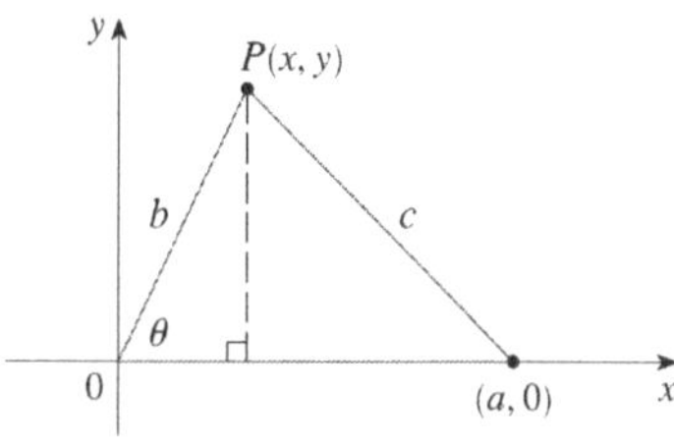

[*Hint:* Introduce a coordinate system so that θ is in standard position, as in the figure. Express x and y in terms of θ and then use the distance formula to compute c.]

84. In order to find the distance $|AB|$ across a small inlet, a point C was located as in the figure and the following measurements were recorded:

$$\angle C = 103° \qquad |AC| = 820 \text{ m} \qquad |BC| = 910 \text{ m}$$

Use the Law of Cosines from Exercise 83 to find the required distance.

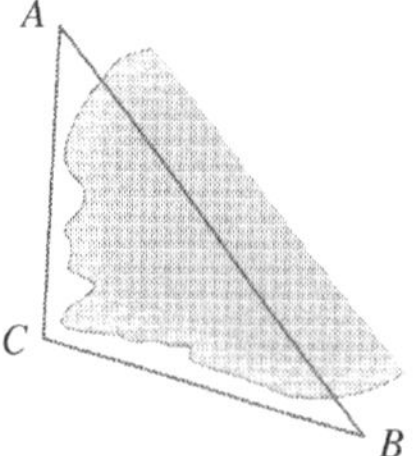

85. Use the figure to prove the subtraction formula

$$\cos(\alpha - \beta) = \cos \alpha \cos \beta + \sin \alpha \sin \beta$$

[*Hint:* Compute c^2 in two ways (using the Law of Cosines from Exercise 83 and also using the distance formula) and compare the two expressions.]

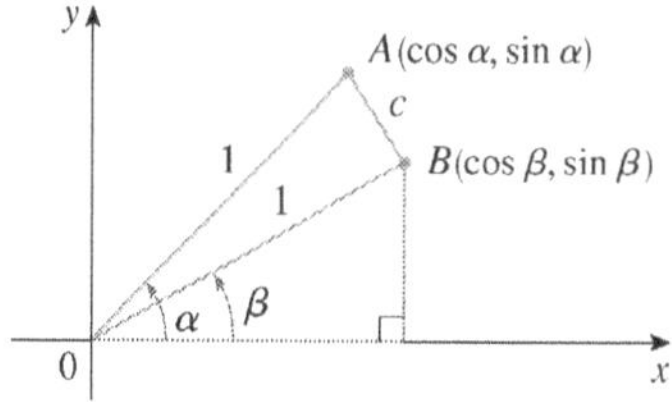

86. Use the formula in Exercise 85 to prove the addition formula for cosine (12b).

87. Use the addition formula for cosine and the identities

$$\cos\left(\frac{\pi}{2} - \theta\right) = \sin \theta \qquad \sin\left(\frac{\pi}{2} - \theta\right) = \cos \theta$$

to prove the subtraction formula (13a) for the sine function.

88. Show that the area of a triangle with sides of lengths a and b and with included angle θ is

$$A = \tfrac{1}{2}ab \sin \theta$$

89. Find the area of triangle ABC, correct to five decimal places, if

$$|AB| = 10 \text{ cm} \qquad |BC| = 3 \text{ cm} \qquad \angle ABC = 107°$$

E Sigma Notation

A convenient way of writing sums uses the Greek letter Σ (capital sigma, corresponding to our letter S) and is called **sigma notation**.

This tells us to end with $i = n$.
This tells us to add.
This tells us to start with $i = m$.

$$\sum_{i=m}^{n} a_i$$

1 Definition If $a_m, a_{m+1}, \ldots, a_n$ are real numbers and m and n are integers such that $m \leqslant n$, then

$$\sum_{i=m}^{n} a_i = a_m + a_{m+1} + a_{m+2} + \cdots + a_{n-1} + a_n$$

With function notation, Definition 1 can be written as

$$\sum_{i=m}^{n} f(i) = f(m) + f(m + 1) + f(m + 2) + \cdots + f(n - 1) + f(n)$$

Thus the symbol $\Sigma_{i=m}^{n}$ indicates a summation in which the letter i (called the **index of summation**) takes on consecutive integer values beginning with m and ending with n, that is, $m, m + 1, \ldots, n$. Other letters can also be used as the index of summation.

EXAMPLE 1

(a) $\displaystyle\sum_{i=1}^{4} i^2 = 1^2 + 2^2 + 3^2 + 4^2 = 30$

(b) $\displaystyle\sum_{i=3}^{n} i = 3 + 4 + 5 + \cdots + (n - 1) + n$

(c) $\displaystyle\sum_{j=0}^{5} 2^j = 2^0 + 2^1 + 2^2 + 2^3 + 2^4 + 2^5 = 63$

(d) $\displaystyle\sum_{k=1}^{n} \frac{1}{k} = 1 + \frac{1}{2} + \frac{1}{3} + \cdots + \frac{1}{n}$

(e) $\displaystyle\sum_{i=1}^{3} \frac{i - 1}{i^2 + 3} = \frac{1 - 1}{1^2 + 3} + \frac{2 - 1}{2^2 + 3} + \frac{3 - 1}{3^2 + 3} = 0 + \frac{1}{7} + \frac{1}{6} = \frac{13}{42}$

(f) $\displaystyle\sum_{i=1}^{4} 2 = 2 + 2 + 2 + 2 = 8$

EXAMPLE 2 Write the sum $2^3 + 3^3 + \cdots + n^3$ in sigma notation.

SOLUTION There is no unique way of writing a sum in sigma notation. We could write

$$2^3 + 3^3 + \cdots + n^3 = \sum_{i=2}^{n} i^3$$

or

$$2^3 + 3^3 + \cdots + n^3 = \sum_{j=1}^{n-1} (j + 1)^3$$

or

$$2^3 + 3^3 + \cdots + n^3 = \sum_{k=0}^{n-2} (k + 2)^3$$

The following theorem gives three simple rules for working with sigma notation.

2 Theorem If c is any constant (that is, it does not depend on i), then

(a) $\sum_{i=m}^{n} ca_i = c \sum_{i=m}^{n} a_i$ (b) $\sum_{i=m}^{n} (a_i + b_i) = \sum_{i=m}^{n} a_i + \sum_{i=m}^{n} b_i$

(c) $\sum_{i=m}^{n} (a_i - b_i) = \sum_{i=m}^{n} a_i - \sum_{i=m}^{n} b_i$

PROOF To see why these rules are true, all we have to do is write both sides in expanded form. Rule (a) is just the distributive property of real numbers:

$$ca_m + ca_{m+1} + \cdots + ca_n = c(a_m + a_{m+1} + \cdots + a_n)$$

Rule (b) follows from the associative and commutative properties:

$$(a_m + b_m) + (a_{m+1} + b_{m+1}) + \cdots + (a_n + b_n)$$
$$= (a_m + a_{m+1} + \cdots + a_n) + (b_m + b_{m+1} + \cdots + b_n)$$

Rule (c) is proved similarly. ■

EXAMPLE 3 Find $\sum_{i=1}^{n} 1$.

SOLUTION
$$\sum_{i=1}^{n} 1 = \underbrace{1 + 1 + \cdots + 1}_{n \text{ terms}} = n$$ ■

EXAMPLE 4 Prove the formula for the sum of the first n positive integers:

$$\sum_{i=1}^{n} i = 1 + 2 + 3 + \cdots + n = \frac{n(n+1)}{2}$$

SOLUTION This formula can be proved by mathematical induction (see page 76) or by the following method used by the German mathematician Karl Friedrich Gauss (1777–1855) when he was ten years old.

Write the sum S twice, once in the usual order and once in reverse order:

$$S = 1 + \quad 2 \quad + \quad 3 \quad + \cdots + (n-1) + n$$
$$S = n + (n-1) + (n-2) + \cdots + \quad 2 \quad + 1$$

Adding all columns vertically, we get

$$2S = (n+1) + (n+1) + (n+1) + \cdots + (n+1) + (n+1)$$

On the right side there are n terms, each of which is $n + 1$, so

$$2S = n(n+1) \qquad \text{or} \qquad S = \frac{n(n+1)}{2}$$ ■

EXAMPLE 5 Prove the formula for the sum of the squares of the first n positive integers:

$$\sum_{i=1}^{n} i^2 = 1^2 + 2^2 + 3^2 + \cdots + n^2 = \frac{n(n+1)(2n+1)}{6}$$

SOLUTION 1 Let S be the desired sum. We start with the *telescoping sum* (or collapsing sum):

Most terms cancel in pairs.

$$\sum_{i=1}^{n} [(1 + i)^3 - i^3] = (2^3 - 1^3) + (3^3 - 2^3) + (4^3 - 3^3) + \cdots + [(n + 1)^3 - n^3]$$

$$= (n + 1)^3 - 1^3 = n^3 + 3n^2 + 3n$$

On the other hand, using Theorem 2 and Examples 3 and 4, we have

$$\sum_{i=1}^{n} [(1 + i)^3 - i^3] = \sum_{i=1}^{n} [3i^2 + 3i + 1] = 3\sum_{i=1}^{n} i^2 + 3\sum_{i=1}^{n} i + \sum_{i=1}^{n} 1$$

$$= 3S + 3\frac{n(n + 1)}{2} + n = 3S + \tfrac{3}{2}n^2 + \tfrac{5}{2}n$$

Thus we have

$$n^3 + 3n^2 + 3n = 3S + \tfrac{3}{2}n^2 + \tfrac{5}{2}n$$

Solving this equation for S, we obtain

$$3S = n^3 + \tfrac{3}{2}n^2 + \tfrac{1}{2}n$$

or

$$S = \frac{2n^3 + 3n^2 + n}{6} = \frac{n(n + 1)(2n + 1)}{6}$$

Principle of Mathematical Induction
Let S_n be a statement involving the positive integer n. Suppose that
1. S_1 is true.
2. If S_k is true, then S_{k+1} is true.

Then S_n is true for all positive integers n.

SOLUTION 2 Let S_n be the given formula.

1. S_1 is true because $\quad 1^2 = \dfrac{1(1 + 1)(2 \cdot 1 + 1)}{6}$

2. Assume that S_k is true; that is,

$$1^2 + 2^2 + 3^2 + \cdots + k^2 = \frac{k(k + 1)(2k + 1)}{6}$$

Then

See pages 76 and 79 for a more thorough discussion of mathematical induction.

$$1^2 + 2^2 + 3^2 + \cdots + (k + 1)^2 = (1^2 + 2^2 + 3^2 + \cdots + k^2) + (k + 1)^2$$

$$= \frac{k(k + 1)(2k + 1)}{6} + (k + 1)^2$$

$$= (k + 1)\frac{k(2k + 1) + 6(k + 1)}{6}$$

$$= (k + 1)\frac{2k^2 + 7k + 6}{6}$$

$$= \frac{(k + 1)(k + 2)(2k + 3)}{6}$$

$$= \frac{(k + 1)[(k + 1) + 1][2(k + 1) + 1]}{6}$$

So S_{k+1} is true.

By the Principle of Mathematical Induction, S_n is true for all n. ■

We list the results of Examples 3, 4, and 5 together with a similar result for cubes (see Exercises 37–40) as Theorem 3. These formulas are needed for finding areas and evaluating integrals in Chapter 5.

3 Theorem Let c be a constant and n a positive integer. Then

(a) $\displaystyle\sum_{i=1}^{n} 1 = n$

(b) $\displaystyle\sum_{i=1}^{n} c = nc$

(c) $\displaystyle\sum_{i=1}^{n} i = \frac{n(n+1)}{2}$

(d) $\displaystyle\sum_{i=1}^{n} i^2 = \frac{n(n+1)(2n+1)}{6}$

(e) $\displaystyle\sum_{i=1}^{n} i^3 = \left[\frac{n(n+1)}{2}\right]^2$

EXAMPLE 6 Evaluate $\displaystyle\sum_{i=1}^{n} i(4i^2 - 3)$.

SOLUTION Using Theorems 2 and 3, we have

$$\begin{aligned}\sum_{i=1}^{n} i(4i^2 - 3) &= \sum_{i=1}^{n} (4i^3 - 3i) = 4\sum_{i=1}^{n} i^3 - 3\sum_{i=1}^{n} i \\ &= 4\left[\frac{n(n+1)}{2}\right]^2 - 3\,\frac{n(n+1)}{2} \\ &= \frac{n(n+1)[2n(n+1) - 3]}{2} \\ &= \frac{n(n+1)(2n^2 + 2n - 3)}{2}\end{aligned}$$

EXAMPLE 7 Find $\displaystyle\lim_{n\to\infty} \sum_{i=1}^{n} \frac{3}{n}\left[\left(\frac{i}{n}\right)^2 + 1\right]$.

The type of calculation in Example 7 arises in Chapter 5 when we compute areas.

SOLUTION

$$\begin{aligned}\lim_{n\to\infty} \sum_{i=1}^{n} \frac{3}{n}\left[\left(\frac{i}{n}\right)^2 + 1\right] &= \lim_{n\to\infty} \sum_{i=1}^{n} \left[\frac{3}{n^3}i^2 + \frac{3}{n}\right] \\ &= \lim_{n\to\infty} \left[\frac{3}{n^3}\sum_{i=1}^{n} i^2 + \frac{3}{n}\sum_{i=1}^{n} 1\right] \\ &= \lim_{n\to\infty} \left[\frac{3}{n^3}\,\frac{n(n+1)(2n+1)}{6} + \frac{3}{n}\cdot n\right] \\ &= \lim_{n\to\infty} \left[\frac{1}{2}\cdot\frac{n}{n}\cdot\left(\frac{n+1}{n}\right)\left(\frac{2n+1}{n}\right) + 3\right] \\ &= \lim_{n\to\infty} \left[\frac{1}{2}\cdot 1\left(1 + \frac{1}{n}\right)\left(2 + \frac{1}{n}\right) + 3\right] \\ &= \tfrac{1}{2}\cdot 1\cdot 1\cdot 2 + 3 = 4\end{aligned}$$

E Exercises

1–10 Write the sum in expanded form.

1. $\sum_{i=1}^{5} \sqrt{i}$

2. $\sum_{i=1}^{6} \frac{1}{i+1}$

3. $\sum_{i=4}^{6} 3^i$

4. $\sum_{i=4}^{6} i^3$

5. $\sum_{k=0}^{4} \frac{2k-1}{2k+1}$

6. $\sum_{k=5}^{8} x^k$

7. $\sum_{i=1}^{n} i^{10}$

8. $\sum_{j=n}^{n+3} j^2$

9. $\sum_{j=0}^{n-1} (-1)^j$

10. $\sum_{i=1}^{n} f(x_i)\,\Delta x_i$

11–20 Write the sum in sigma notation.

11. $1 + 2 + 3 + 4 + \cdots + 10$

12. $\sqrt{3} + \sqrt{4} + \sqrt{5} + \sqrt{6} + \sqrt{7}$

13. $\frac{1}{2} + \frac{2}{3} + \frac{3}{4} + \frac{4}{5} + \cdots + \frac{19}{20}$

14. $\frac{3}{7} + \frac{4}{8} + \frac{5}{9} + \frac{6}{10} + \cdots + \frac{23}{27}$

15. $2 + 4 + 6 + 8 + \cdots + 2n$

16. $1 + 3 + 5 + 7 + \cdots + (2n - 1)$

17. $1 + 2 + 4 + 8 + 16 + 32$

18. $\frac{1}{1} + \frac{1}{4} + \frac{1}{9} + \frac{1}{16} + \frac{1}{25} + \frac{1}{36}$

19. $x + x^2 + x^3 + \cdots + x^n$

20. $1 - x + x^2 - x^3 + \cdots + (-1)^n x^n$

21–35 Find the value of the sum.

21. $\sum_{i=4}^{8} (3i - 2)$

22. $\sum_{i=3}^{6} i(i + 2)$

23. $\sum_{j=1}^{6} 3^{j+1}$

24. $\sum_{k=0}^{8} \cos k\pi$

25. $\sum_{n=1}^{20} (-1)^n$

26. $\sum_{i=1}^{100} 4$

27. $\sum_{i=0}^{4} (2^i + i^2)$

28. $\sum_{i=-2}^{4} 2^{3-i}$

29. $\sum_{i=1}^{n} 2i$

30. $\sum_{i=1}^{n} (2 - 5i)$

31. $\sum_{i=1}^{n} (i^2 + 3i + 4)$

32. $\sum_{i=1}^{n} (3 + 2i)^2$

33. $\sum_{i=1}^{n} (i + 1)(i + 2)$

34. $\sum_{i=1}^{n} i(i + 1)(i + 2)$

35. $\sum_{i=1}^{n} (i^3 - i - 2)$

36. Find the number n such that $\sum_{i=1}^{n} i = 78$.

37. Prove formula (b) of Theorem 3.

38. Prove formula (e) of Theorem 3 using mathematical induction.

39. Prove formula (e) of Theorem 3 using a method similar to that of Example 5, Solution 1 [start with $(1 + i)^4 - i^4$].

40. Prove formula (e) of Theorem 3 using the following method published by Abu Bekr Mohammed ibn Alhusain Alkarchi in about AD 1010. The figure shows a square $ABCD$ in which sides AB and AD have been divided into segments of lengths $1, 2, 3, \ldots, n$. Thus the side of the square has length $n(n + 1)/2$ so the area is $[n(n + 1)/2]^2$. But the area is also the sum of the areas of the n "gnomons" $G_1, G_2, \ldots, G_n$ shown in the figure. Show that the area of G_i is i^3 and conclude that formula (e) is true.

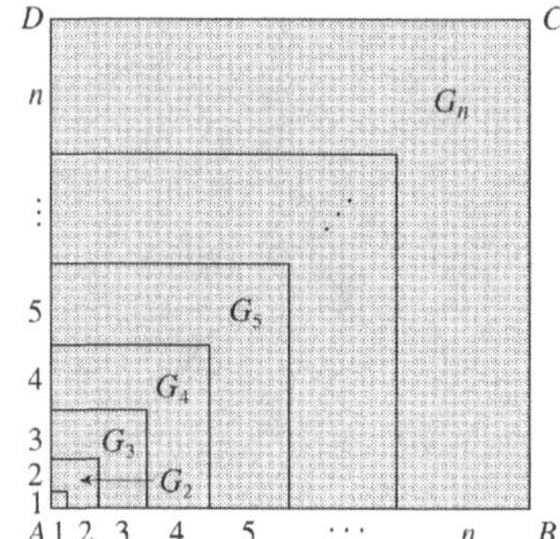

41. Evaluate each telescoping sum.

(a) $\sum_{i=1}^{n} [i^4 - (i - 1)^4]$

(b) $\sum_{i=1}^{100} (5^i - 5^{i-1})$

(c) $\sum_{i=3}^{99} \left(\frac{1}{i} - \frac{1}{i+1}\right)$

(d) $\sum_{i=1}^{n} (a_i - a_{i-1})$

42. Prove the generalized triangle inequality:

$$\left|\sum_{i=1}^{n} a_i\right| \leqslant \sum_{i=1}^{n} |a_i|$$

43–46 Find the limit.

43. $\lim_{n\to\infty} \sum_{i=1}^{n} \frac{1}{n}\left(\frac{i}{n}\right)^2$

44. $\lim_{n\to\infty} \sum_{i=1}^{n} \frac{1}{n}\left[\left(\frac{i}{n}\right)^3 + 1\right]$

45. $\lim_{n\to\infty} \sum_{i=1}^{n} \frac{2}{n}\left[\left(\frac{2i}{n}\right)^3 + 5\left(\frac{2i}{n}\right)\right]$

46. $\lim_{n\to\infty} \sum_{i=1}^{n} \frac{3}{n}\left[\left(1+\frac{3i}{n}\right)^3 - 2\left(1+\frac{3i}{n}\right)\right]$

47. Prove the formula for the sum of a finite geometric series with first term a and common ratio $r \neq 1$:

$$\sum_{i=1}^{n} ar^{i-1} = a + ar + ar^2 + \cdots + ar^{n-1} = \frac{a(r^n - 1)}{r - 1}$$

48. Evaluate $\sum_{i=1}^{n} \frac{3}{2^{i-1}}$.

49. Evaluate $\sum_{i=1}^{n} (2i + 2^i)$.

50. Evaluate $\sum_{i=1}^{m} \left[\sum_{j=1}^{n} (i + j)\right]$.

F Proofs of Theorems

In this appendix we present proofs of several theorems that are stated in the main body of the text. The sections in which they occur are indicated in the margin.

Section 2.3

Limit Laws Suppose that c is a constant and the limits

$$\lim_{x\to a} f(x) = L \quad \text{and} \quad \lim_{x\to a} g(x) = M$$

exist. Then

1. $\lim_{x\to a} [f(x) + g(x)] = L + M$
2. $\lim_{x\to a} [f(x) - g(x)] = L - M$
3. $\lim_{x\to a} [cf(x)] = cL$
4. $\lim_{x\to a} [f(x)g(x)] = LM$
5. $\lim_{x\to a} \frac{f(x)}{g(x)} = \frac{L}{M}$ if $M \neq 0$

PROOF OF LAW 4 Let $\varepsilon > 0$ be given. We want to find $\delta > 0$ such that

$$\text{if} \quad 0 < |x - a| < \delta \quad \text{then} \quad |f(x)g(x) - LM| < \varepsilon$$

In order to get terms that contain $|f(x) - L|$ and $|g(x) - M|$, we add and subtract $Lg(x)$ as follows:

$$\begin{aligned} |f(x)g(x) - LM| &= |f(x)g(x) - Lg(x) + Lg(x) - LM| \\ &= |[f(x) - L]g(x) + L[g(x) - M]| \\ &\leqslant |[f(x) - L]g(x)| + |L[g(x) - M]| \qquad \text{(Triangle Inequality)} \\ &= |f(x) - L||g(x)| + |L||g(x) - M| \end{aligned}$$

We want to make each of these terms less than $\varepsilon/2$.

Since $\lim_{x\to a} g(x) = M$, there is a number $\delta_1 > 0$ such that

$$\text{if} \quad 0 < |x - a| < \delta_1 \quad \text{then} \quad |g(x) - M| < \frac{\varepsilon}{2(1 + |L|)}$$

Also, there is a number $\delta_2 > 0$ such that if $0 < |x - a| < \delta_2$, then

$$|g(x) - M| < 1$$

and therefore

$$|g(x)| = |g(x) - M + M| \leqslant |g(x) - M| + |M| < 1 + |M|$$

Since $\lim_{x \to a} f(x) = L$, there is a number $\delta_3 > 0$ such that

$$\text{if} \quad 0 < |x - a| < \delta_3 \quad \text{then} \quad |f(x) - L| < \frac{\varepsilon}{2(1 + |M|)}$$

Let $\delta = \min\{\delta_1, \delta_2, \delta_3\}$. If $0 < |x - a| < \delta$, then we have $0 < |x - a| < \delta_1$, $0 < |x - a| < \delta_2$, and $0 < |x - a| < \delta_3$, so we can combine the inequalities to obtain

$$\begin{aligned} |f(x)g(x) - LM| &\leqslant |f(x) - L||g(x)| + |L||g(x) - M| \\ &< \frac{\varepsilon}{2(1 + |M|)}(1 + |M|) + |L| \frac{\varepsilon}{2(1 + |L|)} \\ &< \frac{\varepsilon}{2} + \frac{\varepsilon}{2} = \varepsilon \end{aligned}$$

This shows that $\lim_{x \to a} [f(x)g(x)] = LM$. ■

PROOF OF LAW 3 If we take $g(x) = c$ in Law 4, we get

$$\begin{aligned} \lim_{x \to a} [cf(x)] &= \lim_{x \to a} [g(x)f(x)] = \lim_{x \to a} g(x) \cdot \lim_{x \to a} f(x) \\ &= \lim_{x \to a} c \cdot \lim_{x \to a} f(x) \\ &= c \lim_{x \to a} f(x) \qquad \text{(by Law 7)} \end{aligned}$$ ■

PROOF OF LAW 2 Using Law 1 and Law 3 with $c = -1$, we have

$$\begin{aligned} \lim_{x \to a} [f(x) - g(x)] &= \lim_{x \to a} [f(x) + (-1)g(x)] = \lim_{x \to a} f(x) + \lim_{x \to a} (-1)g(x) \\ &= \lim_{x \to a} f(x) + (-1) \lim_{x \to a} g(x) = \lim_{x \to a} f(x) - \lim_{x \to a} g(x) \end{aligned}$$ ■

PROOF OF LAW 5 First let us show that

$$\lim_{x \to a} \frac{1}{g(x)} = \frac{1}{M}$$

To do this we must show that, given $\varepsilon > 0$, there exists $\delta > 0$ such that

$$\text{if} \quad 0 < |x - a| < \delta \quad \text{then} \quad \left| \frac{1}{g(x)} - \frac{1}{M} \right| < \varepsilon$$

Observe that $$\left| \frac{1}{g(x)} - \frac{1}{M} \right| = \frac{|M - g(x)|}{|Mg(x)|}$$

We know that we can make the numerator small. But we also need to know that the denominator is not small when x is near a. Since $\lim_{x \to a} g(x) = M$, there is a number $\delta_1 > 0$ such that, whenever $0 < |x - a| < \delta_1$, we have

$$|g(x) - M| < \frac{|M|}{2}$$

and therefore $$\begin{aligned} |M| &= |M - g(x) + g(x)| \leqslant |M - g(x)| + |g(x)| \\ &< \frac{|M|}{2} + |g(x)| \end{aligned}$$

This shows that

$$\text{if} \quad 0 < |x - a| < \delta_1 \quad \text{then} \quad |g(x)| > \frac{|M|}{2}$$

and so, for these values of x,

$$\frac{1}{|Mg(x)|} = \frac{1}{|M||g(x)|} < \frac{1}{|M|} \cdot \frac{2}{|M|} = \frac{2}{M^2}$$

Also, there exists $\delta_2 > 0$ such that

$$\text{if} \quad 0 < |x - a| < \delta_2 \quad \text{then} \quad |g(x) - M| < \frac{M^2}{2} \varepsilon$$

Let $\delta = \min\{\delta_1, \delta_2\}$. Then, for $0 < |x - a| < \delta$, we have

$$\left| \frac{1}{g(x)} - \frac{1}{M} \right| = \frac{|M - g(x)|}{|Mg(x)|} < \frac{2}{M^2} \frac{M^2}{2} \varepsilon = \varepsilon$$

It follows that $\lim_{x \to a} 1/g(x) = 1/M$. Finally, using Law 4, we obtain

$$\lim_{x \to a} \frac{f(x)}{g(x)} = \lim_{x \to a} f(x) \left(\frac{1}{g(x)} \right) = \lim_{x \to a} f(x) \lim_{x \to a} \frac{1}{g(x)} = L \cdot \frac{1}{M} = \frac{L}{M}$$

2 Theorem If $f(x) \leq g(x)$ for all x in an open interval that contains a (except possibly at a) and

$$\lim_{x \to a} f(x) = L \quad \text{and} \quad \lim_{x \to a} g(x) = M$$

then $L \leq M$.

PROOF We use the method of proof by contradiction. Suppose, if possible, that $L > M$. Law 2 of limits says that

$$\lim_{x \to a} [g(x) - f(x)] = M - L$$

Therefore, for any $\varepsilon > 0$, there exists $\delta > 0$ such that

$$\text{if} \quad 0 < |x - a| < \delta \quad \text{then} \quad |[g(x) - f(x)] - (M - L)| < \varepsilon$$

In particular, taking $\varepsilon = L - M$ (noting that $L - M > 0$ by hypothesis), we have a number $\delta > 0$ such that

$$\text{if} \quad 0 < |x - a| < \delta \quad \text{then} \quad |[g(x) - f(x)] - (M - L)| < L - M$$

Since $a \leq |a|$ for any number a, we have

$$\text{if} \quad 0 < |x - a| < \delta \quad \text{then} \quad [g(x) - f(x)] - (M - L) < L - M$$

which simplifies to

$$\text{if} \quad 0 < |x - a| < \delta \quad \text{then} \quad g(x) < f(x)$$

But this contradicts $f(x) \leq g(x)$. Thus the inequality $L > M$ must be false. Therefore $L \leq M$.

3 **The Squeeze Theorem** If $f(x) \leqslant g(x) \leqslant h(x)$ for all x in an open interval that contains a (except possibly at a) and

$$\lim_{x \to a} f(x) = \lim_{x \to a} h(x) = L$$

then
$$\lim_{x \to a} g(x) = L$$

PROOF Let $\varepsilon > 0$ be given. Since $\lim_{x \to a} f(x) = L$, there is a number $\delta_1 > 0$ such that

$$\text{if} \quad 0 < |x - a| < \delta_1 \quad \text{then} \quad |f(x) - L| < \varepsilon$$

that is,

$$\text{if} \quad 0 < |x - a| < \delta_1 \quad \text{then} \quad L - \varepsilon < f(x) < L + \varepsilon$$

Since $\lim_{x \to a} h(x) = L$, there is a number $\delta_2 > 0$ such that

$$\text{if} \quad 0 < |x - a| < \delta_2 \quad \text{then} \quad |h(x) - L| < \varepsilon$$

that is,

$$\text{if} \quad 0 < |x - a| < \delta_2 \quad \text{then} \quad L - \varepsilon < h(x) < L + \varepsilon$$

Let $\delta = \min\{\delta_1, \delta_2\}$. If $0 < |x - a| < \delta$, then $0 < |x - a| < \delta_1$ and $0 < |x - a| < \delta_2$, so

$$L - \varepsilon < f(x) \leqslant g(x) \leqslant h(x) < L + \varepsilon$$

In particular,
$$L - \varepsilon < g(x) < L + \varepsilon$$

and so $|g(x) - L| < \varepsilon$. Therefore $\lim_{x \to a} g(x) = L$. ■

Section 2.5

Theorem If f is a one-to-one continuous function defined on an interval (a, b), then its inverse function f^{-1} is also continuous.

PROOF First we show that if f is both one-to-one and continuous on (a, b), then it must be either increasing or decreasing on (a, b). If it were neither increasing nor decreasing, then there would exist numbers x_1, x_2, and x_3 in (a, b) with $x_1 < x_2 < x_3$ such that $f(x_2)$ does not lie between $f(x_1)$ and $f(x_3)$. There are two possibilities: either (1) $f(x_3)$ lies between $f(x_1)$ and $f(x_2)$ or (2) $f(x_1)$ lies between $f(x_2)$ and $f(x_3)$. (Draw a picture.) In case (1) we apply the Intermediate Value Theorem to the continuous function f to get a number c between x_1 and x_2 such that $f(c) = f(x_3)$. In case (2) the Intermediate Value Theorem gives a number c between x_2 and x_3 such that $f(c) = f(x_1)$. In either case we have contradicted the fact that f is one-to-one.

Let us assume, for the sake of definiteness, that f is increasing on (a, b). We take any number y_0 in the domain of f^{-1} and we let $f^{-1}(y_0) = x_0$; that is, x_0 is the number in (a, b) such that $f(x_0) = y_0$. To show that f^{-1} is continuous at y_0 we take any $\varepsilon > 0$ such that the interval $(x_0 - \varepsilon, x_0 + \varepsilon)$ is contained in the interval (a, b). Since f is increasing, it maps the numbers in the interval $(x_0 - \varepsilon, x_0 + \varepsilon)$ onto the numbers in the interval $(f(x_0 - \varepsilon), f(x_0 + \varepsilon))$ and f^{-1} reverses the correspondence. If we let δ denote the smaller of the numbers $\delta_1 = y_0 - f(x_0 - \varepsilon)$ and $\delta_2 = f(x_0 + \varepsilon) - y_0$, then the interval $(y_0 - \delta, y_0 + \delta)$ is contained in the interval $(f(x_0 - \varepsilon), f(x_0 + \varepsilon))$ and so is mapped into the interval $(x_0 - \varepsilon, x_0 + \varepsilon)$ by f^{-1}. (See the arrow diagram in Figure 1.) We have therefore found a number $\delta > 0$ such that

$$\text{if} \quad |y - y_0| < \delta \quad \text{then} \quad |f^{-1}(y) - f^{-1}(y_0)| < \varepsilon$$

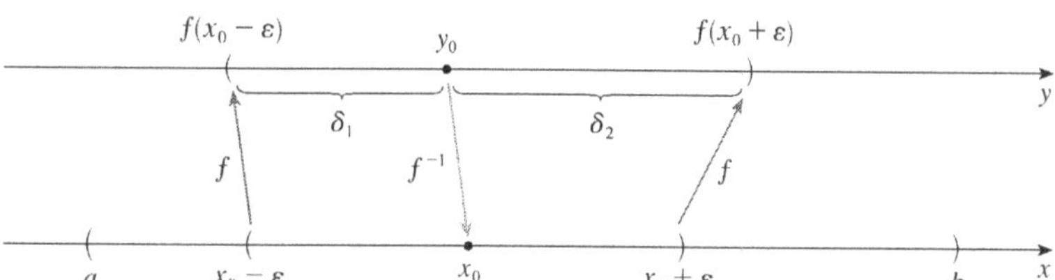

FIGURE 1

This shows that $\lim_{y \to y_0} f^{-1}(y) = f^{-1}(y_0)$ and so f^{-1} is continuous at any number y_0 in its domain. ∎

> **8 Theorem** If f is continuous at b and $\lim_{x \to a} g(x) = b$, then
>
> $$\lim_{x \to a} f(g(x)) = f(b)$$

PROOF Let $\varepsilon > 0$ be given. We want to find a number $\delta > 0$ such that

$$\text{if} \quad 0 < |x - a| < \delta \quad \text{then} \quad |f(g(x)) - f(b)| < \varepsilon$$

Since f is continuous at b, we have

$$\lim_{y \to b} f(y) = f(b)$$

and so there exists $\delta_1 > 0$ such that

$$\text{if} \quad 0 < |y - b| < \delta_1 \quad \text{then} \quad |f(y) - f(b)| < \varepsilon$$

Since $\lim_{x \to a} g(x) = b$, there exists $\delta > 0$ such that

$$\text{if} \quad 0 < |x - a| < \delta \quad \text{then} \quad |g(x) - b| < \delta_1$$

Combining these two statements, we see that whenever $0 < |x - a| < \delta$ we have $|g(x) - b| < \delta_1$, which implies that $|f(g(x)) - f(b)| < \varepsilon$. Therefore we have proved that $\lim_{x \to a} f(g(x)) = f(b)$. ∎

Section 3.3

The proof of the following result was promised when we proved that $\lim_{\theta \to 0} \dfrac{\sin \theta}{\theta} = 1$.

> **Theorem** If $0 < \theta < \pi/2$, then $\theta \leqslant \tan \theta$.

PROOF Figure 2 shows a sector of a circle with center O, central angle θ, and radius 1. Then

$$|AD| = |OA| \tan \theta = \tan \theta$$

We approximate the arc AB by an inscribed polygon consisting of n equal line segments

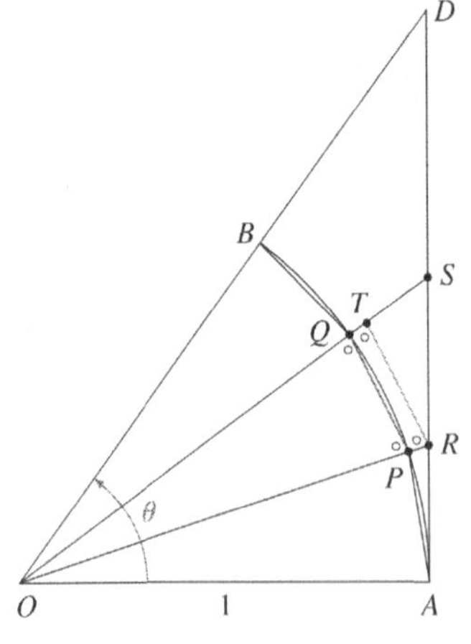

FIGURE 2

and we look at a typical segment PQ. We extend the lines OP and OQ to meet AD in the points R and S. Then we draw $RT \parallel PQ$ as in Figure 2. Observe that

$$\angle RTO = \angle PQO < 90°$$

and so $\angle RTS > 90°$. Therefore we have

$$|PQ| < |RT| < |RS|$$

If we add n such inequalities, we get

$$L_n < |AD| = \tan\theta$$

where L_n is the length of the inscribed polygon. Thus, by Theorem 2.3.2, we have

$$\lim_{n\to\infty} L_n \leqslant \tan\theta$$

But the arc length is defined in Equation 8.1.1 as the limit of the lengths of inscribed polygons, so

$$\theta = \lim_{n\to\infty} L_n \leqslant \tan\theta$$

Section 4.3

> **Concavity Test**
> (a) If $f''(x) > 0$ for all x in I, then the graph of f is concave upward on I.
> (b) If $f''(x) < 0$ for all x in I, then the graph of f is concave downward on I.

PROOF OF (a) Let a be any number in I. We must show that the curve $y = f(x)$ lies above the tangent line at the point $(a, f(a))$. The equation of this tangent is

$$y = f(a) + f'(a)(x - a)$$

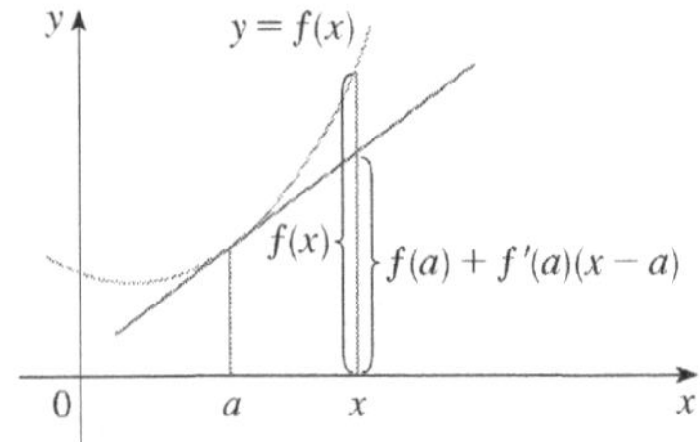

FIGURE 3

So we must show that

$$f(x) > f(a) + f'(a)(x - a)$$

whenever $x \in I$ $(x \neq a)$. (See Figure 3.)

First let us take the case where $x > a$. Applying the Mean Value Theorem to f on the interval $[a, x]$, we get a number c, with $a < c < x$, such that

$$\boxed{1} \qquad f(x) - f(a) = f'(c)(x - a)$$

Since $f'' > 0$ on I, we know from the Increasing/Decreasing Test that f' is increasing on I. Thus, since $a < c$, we have

$$f'(a) < f'(c)$$

and so, multiplying this inequality by the positive number $x - a$, we get

$$\boxed{2} \qquad f'(a)(x - a) < f'(c)(x - a)$$

Now we add $f(a)$ to both sides of this inequality:

$$f(a) + f'(a)(x - a) < f(a) + f'(c)(x - a)$$

But from Equation 1 we have $f(x) = f(a) + f'(c)(x - a)$. So this inequality becomes

$$\boxed{3} \qquad f(x) > f(a) + f'(a)(x - a)$$

which is what we wanted to prove.

For the case where $x < a$ we have $f'(c) < f'(a)$, but multiplication by the negative number $x - a$ reverses the inequality, so we get [2] and [3] as before. ■

Section 4.4

In order to give the promised proof of l'Hospital's Rule, we first need a generalization of the Mean Value Theorem. The following theorem is named after another French mathematician, Augustin-Louis Cauchy (1789–1857).

See the biographical sketch of Cauchy on page 113.

[1] **Cauchy's Mean Value Theorem** Suppose that the functions f and g are continuous on $[a, b]$ and differentiable on (a, b), and $g'(x) \neq 0$ for all x in (a, b). Then there is a number c in (a, b) such that

$$\frac{f'(c)}{g'(c)} = \frac{f(b) - f(a)}{g(b) - g(a)}$$

Notice that if we take the special case in which $g(x) = x$, then $g'(c) = 1$ and Theorem 1 is just the ordinary Mean Value Theorem. Furthermore, Theorem 1 can be proved in a similar manner. You can verify that all we have to do is change the function h given by Equation 4.2.4 to the function

$$h(x) = f(x) - f(a) - \frac{f(b) - f(a)}{g(b) - g(a)}[g(x) - g(a)]$$

and apply Rolle's Theorem as before.

L'Hospital's Rule Suppose f and g are differentiable and $g'(x) \neq 0$ on an open interval I that contains a (except possibly at a). Suppose that

$$\lim_{x \to a} f(x) = 0 \qquad \text{and} \qquad \lim_{x \to a} g(x) = 0$$

or that

$$\lim_{x \to a} f(x) = \pm\infty \qquad \text{and} \qquad \lim_{x \to a} g(x) = \pm\infty$$

(In other words, we have an indeterminate form of type $\frac{0}{0}$ or ∞/∞.) Then

$$\lim_{x \to a} \frac{f(x)}{g(x)} = \lim_{x \to a} \frac{f'(x)}{g'(x)}$$

if the limit on the right side exists (or is ∞ or $-\infty$).

PROOF OF L'HOSPITAL'S RULE We are assuming that $\lim_{x\to a} f(x) = 0$ and $\lim_{x\to a} g(x) = 0$. Let

$$L = \lim_{x\to a} \frac{f'(x)}{g'(x)}$$

We must show that $\lim_{x\to a} f(x)/g(x) = L$. Define

$$F(x) = \begin{cases} f(x) & \text{if } x \neq a \\ 0 & \text{if } x = a \end{cases} \qquad G(x) = \begin{cases} g(x) & \text{if } x \neq a \\ 0 & \text{if } x = a \end{cases}$$

Then F is continuous on I since f is continuous on $\{x \in I \mid x \neq a\}$ and

$$\lim_{x\to a} F(x) = \lim_{x\to a} f(x) = 0 = F(a)$$

Likewise, G is continuous on I. Let $x \in I$ and $x > a$. Then F and G are continuous on $[a, x]$ and differentiable on (a, x) and $G' \neq 0$ there (since $F' = f'$ and $G' = g'$). Therefore, by Cauchy's Mean Value Theorem, there is a number y such that $a < y < x$ and

$$\frac{F'(y)}{G'(y)} = \frac{F(x) - F(a)}{G(x) - G(a)} = \frac{F(x)}{G(x)}$$

Here we have used the fact that, by definition, $F(a) = 0$ and $G(a) = 0$. Now, if we let $x \to a^+$, then $y \to a^+$ (since $a < y < x$), so

$$\lim_{x\to a^+} \frac{f(x)}{g(x)} = \lim_{x\to a^+} \frac{F(x)}{G(x)} = \lim_{y\to a^+} \frac{F'(y)}{G'(y)} = \lim_{y\to a^+} \frac{f'(y)}{g'(y)} = L$$

A similar argument shows that the left-hand limit is also L. Therefore

$$\lim_{x\to a} \frac{f(x)}{g(x)} = L$$

This proves l'Hospital's Rule for the case where a is finite.

If a is infinite, we let $t = 1/x$. Then $t \to 0^+$ as $x \to \infty$, so we have

$$\begin{aligned} \lim_{x\to\infty} \frac{f(x)}{g(x)} &= \lim_{t\to 0^+} \frac{f(1/t)}{g(1/t)} \\ &= \lim_{t\to 0^+} \frac{f'(1/t)(-1/t^2)}{g'(1/t)(-1/t^2)} \qquad \text{(by l'Hospital's Rule for finite } a\text{)} \\ &= \lim_{t\to 0^+} \frac{f'(1/t)}{g'(1/t)} = \lim_{x\to\infty} \frac{f'(x)}{g'(x)} \end{aligned}$$

Section 11.8

In order to prove Theorem 11.8.3, we first need the following results.

Theorem

1. If a power series $\Sigma\, c_n x^n$ converges when $x = b$ (where $b \neq 0$), then it converges whenever $|x| < |b|$.
2. If a power series $\Sigma\, c_n x^n$ diverges when $x = d$ (where $d \neq 0$), then it diverges whenever $|x| > |d|$.

PROOF OF 1 Suppose that $\Sigma\, c_n b^n$ converges. Then, by Theorem 11.2.6, we have $\lim_{n\to\infty} c_n b^n = 0$. According to Definition 11.1.2 with $\varepsilon = 1$, there is a positive integer N such that $|c_n b^n| < 1$ whenever $n \geqslant N$. Thus, for $n \geqslant N$, we have

$$|c_n x^n| = \left|\frac{c_n b^n x^n}{b^n}\right| = |c_n b^n| \left|\frac{x}{b}\right|^n < \left|\frac{x}{b}\right|^n$$

If $|x| < |b|$, then $|x/b| < 1$, so $\Sigma\, |x/b|^n$ is a convergent geometric series. Therefore, by the Comparison Test, the series $\Sigma_{n=N}^{\infty} |c_n x^n|$ is convergent. Thus the series $\Sigma\, c_n x^n$ is absolutely convergent and therefore convergent. ■

PROOF OF 2 Suppose that $\Sigma\, c_n d^n$ diverges. If x is any number such that $|x| > |d|$, then $\Sigma\, c_n x^n$ cannot converge because, by part 1, the convergence of $\Sigma\, c_n x^n$ would imply the convergence of $\Sigma\, c_n d^n$. Therefore $\Sigma\, c_n x^n$ diverges whenever $|x| > |d|$. ■

Theorem For a power series $\Sigma\, c_n x^n$ there are only three possibilities:

1. The series converges only when $x = 0$.
2. The series converges for all x.
3. There is a positive number R such that the series converges if $|x| < R$ and diverges if $|x| > R$.

PROOF Suppose that neither case 1 nor case 2 is true. Then there are nonzero numbers b and d such that $\Sigma\, c_n x^n$ converges for $x = b$ and diverges for $x = d$. Therefore the set $S = \{x \mid \Sigma\, c_n x^n \text{ converges}\}$ is not empty. By the preceding theorem, the series diverges if $|x| > |d|$, so $|x| \leqslant |d|$ for all $x \in S$. This says that $|d|$ is an upper bound for the set S. Thus, by the Completeness Axiom (see Section 11.1), S has a least upper bound R. If $|x| > R$, then $x \notin S$, so $\Sigma\, c_n x^n$ diverges. If $|x| < R$, then $|x|$ is not an upper bound for S and so there exists $b \in S$ such that $b > |x|$. Since $b \in S$, $\Sigma\, c_n b^n$ converges, so by the preceding theorem $\Sigma\, c_n x^n$ converges. ■

3 Theorem For a power series $\Sigma\, c_n(x - a)^n$ there are only three possibilities:

1. The series converges only when $x = a$.
2. The series converges for all x.
3. There is a positive number R such that the series converges if $|x - a| < R$ and diverges if $|x - a| > R$.

PROOF If we make the change of variable $u = x - a$, then the power series becomes $\Sigma\, c_n u^n$ and we can apply the preceding theorem to this series. In case 3 we have convergence for $|u| < R$ and divergence for $|u| > R$. Thus we have convergence for $|x - a| < R$ and divergence for $|x - a| > R$. ■

Section 14.3

Clairaut's Theorem Suppose f is defined on a disk D that contains the point (a, b). If the functions f_{xy} and f_{yx} are both continuous on D, then $f_{xy}(a, b) = f_{yx}(a, b)$.

PROOF For small values of h, $h \neq 0$, consider the difference

$$\Delta(h) = [f(a + h, b + h) - f(a + h, b)] - [f(a, b + h) - f(a, b)]$$

Notice that if we let $g(x) = f(x, b + h) - f(x, b)$, then

$$\Delta(h) = g(a + h) - g(a)$$

By the Mean Value Theorem, there is a number c between a and $a + h$ such that

$$g(a + h) - g(a) = g'(c)h = h[f_x(c, b + h) - f_x(c, b)]$$

Applying the Mean Value Theorem again, this time to f_x, we get a number d between b and $b + h$ such that

$$f_x(c, b + h) - f_x(c, b) = f_{xy}(c, d)h$$

Combining these equations, we obtain

$$\Delta(h) = h^2 f_{xy}(c, d)$$

If $h \to 0$, then $(c, d) \to (a, b)$, so the continuity of f_{xy} at (a, b) gives

$$\lim_{h \to 0} \frac{\Delta(h)}{h^2} = \lim_{(c, d) \to (a, b)} f_{xy}(c, d) = f_{xy}(a, b)$$

Similarly, by writing

$$\Delta(h) = [f(a + h, b + h) - f(a, b + h)] - [f(a + h, b) - f(a, b)]$$

and using the Mean Value Theorem twice and the continuity of f_{yx} at (a, b), we obtain

$$\lim_{h \to 0} \frac{\Delta(h)}{h^2} = f_{yx}(a, b)$$

It follows that $f_{xy}(a, b) = f_{yx}(a, b)$. ■

Section 14.4

8 **Theorem** If the partial derivatives f_x and f_y exist near (a, b) and are continuous at (a, b), then f is differentiable at (a, b).

PROOF Let

$$\Delta z = f(a + \Delta x, b + \Delta y) - f(a, b)$$

According to (14.4.7), to prove that f is differentiable at (a, b) we have to show that we can write Δz in the form

$$\Delta z = f_x(a, b)\, \Delta x + f_y(a, b)\, \Delta y + \varepsilon_1\, \Delta x + \varepsilon_2\, \Delta y$$

where ε_1 and $\varepsilon_2 \to 0$ as $(\Delta x,\ \Delta y) \to (0, 0)$.

Referring to Figure 4, we write

$$\boxed{1} \quad \Delta z = [f(a + \Delta x, b + \Delta y) - f(a, b + \Delta y)] + [f(a, b + \Delta y) - f(a, b)]$$

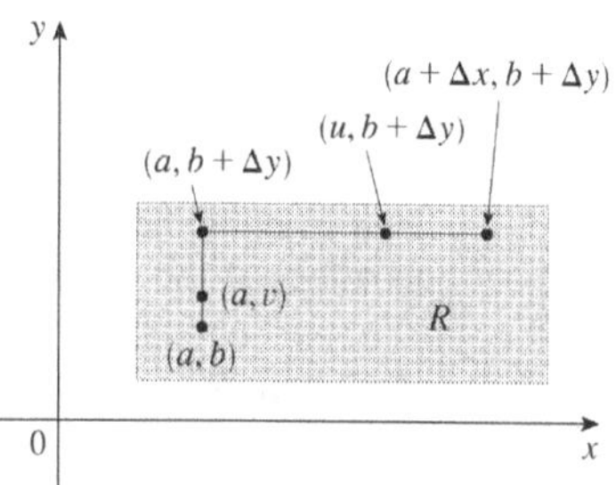

FIGURE 4

Observe that the function of a single variable

$$g(x) = f(x, b + \Delta y)$$

is defined on the interval $[a, a + \Delta x]$ and $g'(x) = f_x(x, b + \Delta y)$. If we apply the Mean Value Theorem to g, we get

$$g(a + \Delta x) - g(a) = g'(u)\,\Delta x$$

where u is some number between a and $a + \Delta x$. In terms of f, this equation becomes

$$f(a + \Delta x, b + \Delta y) - f(a, b + \Delta y) = f_x(u, b + \Delta y)\,\Delta x$$

This gives us an expression for the first part of the right side of Equation 1. For the second part we let $h(y) = f(a, y)$. Then h is a function of a single variable defined on the interval $[b, b + \Delta y]$ and $h'(y) = f_y(a, y)$. A second application of the Mean Value Theorem then gives

$$h(b + \Delta y) - h(b) = h'(v)\,\Delta y$$

where v is some number between b and $b + \Delta y$. In terms of f, this becomes

$$f(a, b + \Delta y) - f(a, b) = f_y(a, v)\,\Delta y$$

We now substitute these expressions into Equation 1 and obtain

$$\begin{aligned} \Delta z &= f_x(u, b + \Delta y)\,\Delta x + f_y(a, v)\,\Delta y \\ &= f_x(a, b)\,\Delta x + [f_x(u, b + \Delta y) - f_x(a, b)]\,\Delta x + f_y(a, b)\,\Delta y \\ &\qquad + [f_y(a, v) - f_y(a, b)]\,\Delta y \\ &= f_x(a, b)\,\Delta x + f_y(a, b)\,\Delta y + \varepsilon_1\,\Delta x + \varepsilon_2\,\Delta y \end{aligned}$$

where
$$\varepsilon_1 = f_x(u, b + \Delta y) - f_x(a, b)$$

$$\varepsilon_2 = f_y(a, v) - f_y(a, b)$$

Since $(u, b + \Delta y) \to (a, b)$ and $(a, v) \to (a, b)$ as $(\Delta x, \Delta y) \to (0, 0)$ and since f_x and f_y are continuous at (a, b), we see that $\varepsilon_1 \to 0$ and $\varepsilon_2 \to 0$ as $(\Delta x, \Delta y) \to (0, 0)$.

Therefore f is differentiable at (a, b). ∎

G The Logarithm Defined as an Integral

Our treatment of exponential and logarithmic functions until now has relied on our intuition, which is based on numerical and visual evidence. (See Sections 1.5, 1.6, and 3.1.) Here we use the Fundamental Theorem of Calculus to give an alternative treatment that provides a surer footing for these functions.

Instead of starting with a^x and defining $\log_a x$ as its inverse, this time we start by defining $\ln x$ as an integral and then define the exponential function as its inverse. You should bear in mind that we do not use any of our previous definitions and results concerning exponential and logarithmic functions.

The Natural Logarithm

We first define $\ln x$ as an integral.

1 Definition The **natural logarithmic function** is the function defined by

$$\ln x = \int_1^x \frac{1}{t}\,dt \qquad x > 0$$

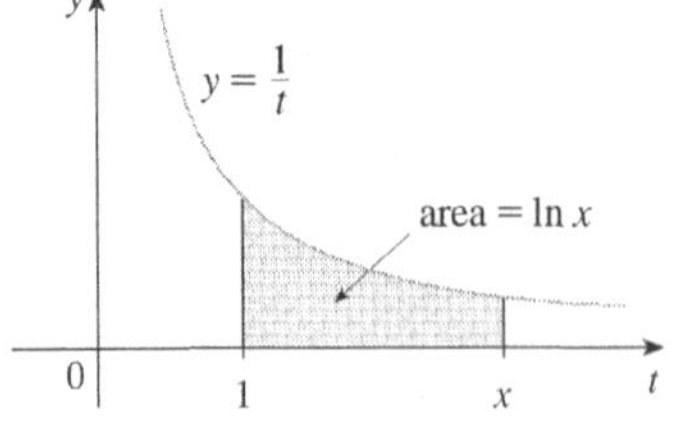

FIGURE 1

The existence of this function depends on the fact that the integral of a continuous function always exists. If $x > 1$, then $\ln x$ can be interpreted geometrically as the area under the hyperbola $y = 1/t$ from $t = 1$ to $t = x$. (See Figure 1.) For $x = 1$, we have

$$\ln 1 = \int_1^1 \frac{1}{t}\,dt = 0$$

For $0 < x < 1$,

$$\ln x = \int_1^x \frac{1}{t}\,dt = -\int_x^1 \frac{1}{t}\,dt < 0$$

and so $\ln x$ is the negative of the area shown in Figure 2.

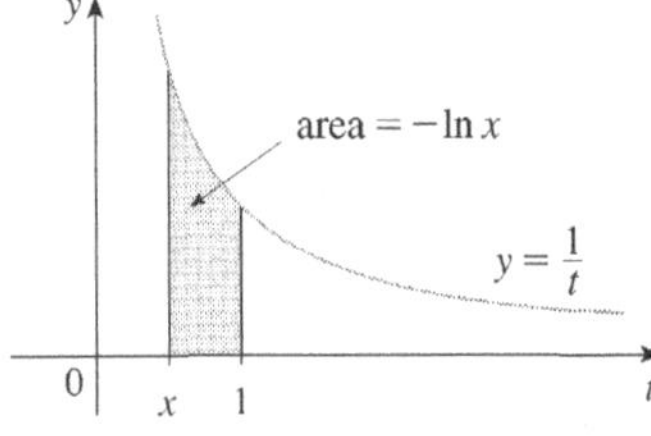

FIGURE 2

V EXAMPLE 1

(a) By comparing areas, show that $\frac{1}{2} < \ln 2 < \frac{3}{4}$.
(b) Use the Midpoint Rule with $n = 10$ to estimate the value of $\ln 2$.

SOLUTION

(a) We can interpret $\ln 2$ as the area under the curve $y = 1/t$ from 1 to 2. From Figure 3 we see that this area is larger than the area of rectangle *BCDE* and smaller than the area of trapezoid *ABCD*. Thus we have

$$\tfrac{1}{2} \cdot 1 < \ln 2 < 1 \cdot \tfrac{1}{2}\left(1 + \tfrac{1}{2}\right)$$

$$\tfrac{1}{2} < \ln 2 < \tfrac{3}{4}$$

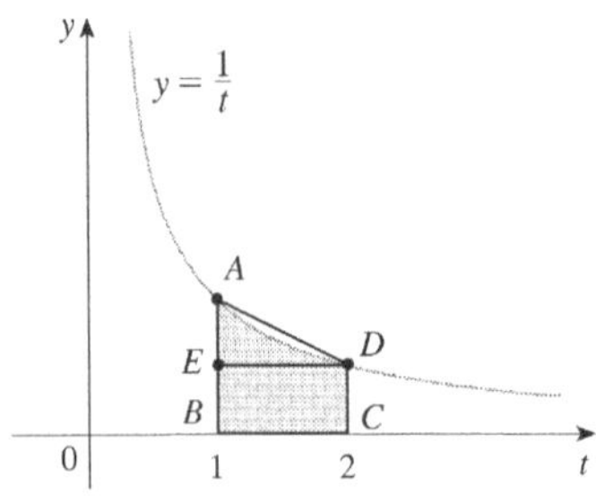

FIGURE 3

(b) If we use the Midpoint Rule with $f(t) = 1/t$, $n = 10$, and $\Delta t = 0.1$, we get

$$\ln 2 = \int_1^2 \frac{1}{t}\,dt \approx (0.1)[f(1.05) + f(1.15) + \cdots + f(1.95)]$$

$$= (0.1)\left(\frac{1}{1.05} + \frac{1}{1.15} + \cdots + \frac{1}{1.95}\right) \approx 0.693$$

Notice that the integral that defines $\ln x$ is exactly the type of integral discussed in Part 1 of the Fundamental Theorem of Calculus (see Section 5.3). In fact, using that theorem, we have

$$\frac{d}{dx}\int_1^x \frac{1}{t}\,dt = \frac{1}{x}$$

and so

2 $$\boxed{\frac{d}{dx}(\ln x) = \frac{1}{x}}$$

We now use this differentiation rule to prove the following properties of the logarithm function.

3 **Laws of Logarithms** If x and y are positive numbers and r is a rational number, then

1. $\ln(xy) = \ln x + \ln y$ **2.** $\ln\left(\frac{x}{y}\right) = \ln x - \ln y$ **3.** $\ln(x^r) = r\ln x$

PROOF

1. Let $f(x) = \ln(ax)$, where a is a positive constant. Then, using Equation 2 and the Chain Rule, we have

$$f'(x) = \frac{1}{ax}\frac{d}{dx}(ax) = \frac{1}{ax}\cdot a = \frac{1}{x}$$

Therefore $f(x)$ and $\ln x$ have the same derivative and so they must differ by a constant:

$$\ln(ax) = \ln x + C$$

Putting $x = 1$ in this equation, we get $\ln a = \ln 1 + C = 0 + C = C$. Thus

$$\ln(ax) = \ln x + \ln a$$

If we now replace the constant a by any number y, we have

$$\ln(xy) = \ln x + \ln y$$

2. Using Law 1 with $x = 1/y$, we have

$$\ln\frac{1}{y} + \ln y = \ln\left(\frac{1}{y}\cdot y\right) = \ln 1 = 0$$

and so $$\ln\frac{1}{y} = -\ln y$$

Using Law 1 again, we have

$$\ln\left(\frac{x}{y}\right) = \ln\left(x\cdot\frac{1}{y}\right) = \ln x + \ln\frac{1}{y} = \ln x - \ln y$$

The proof of Law 3 is left as an exercise. ■

In order to graph $y = \ln x$, we first determine its limits:

4 (a) $\lim_{x \to \infty} \ln x = \infty$ (b) $\lim_{x \to 0^+} \ln x = -\infty$

PROOF

(a) Using Law 3 with $x = 2$ and $r = n$ (where n is any positive integer), we have $\ln(2^n) = n \ln 2$. Now $\ln 2 > 0$, so this shows that $\ln(2^n) \to \infty$ as $n \to \infty$. But $\ln x$ is an increasing function since its derivative $1/x > 0$. Therefore $\ln x \to \infty$ as $x \to \infty$.

(b) If we let $t = 1/x$, then $t \to \infty$ as $x \to 0^+$. Thus, using (a), we have

$$\lim_{x \to 0^+} \ln x = \lim_{t \to \infty} \ln\left(\frac{1}{t}\right) = \lim_{t \to \infty} (-\ln t) = -\infty$$

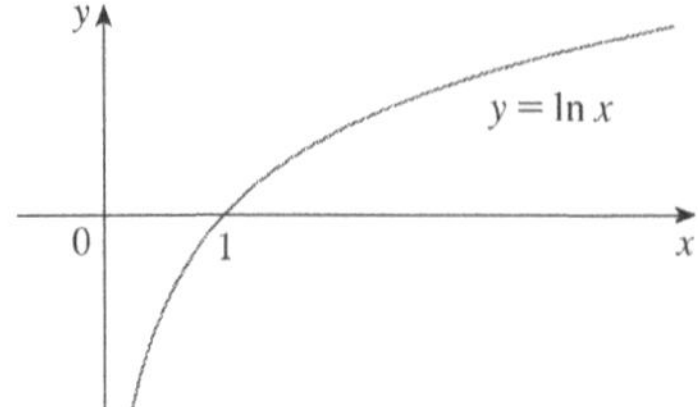

FIGURE 4

If $y = \ln x$, $x > 0$, then

$$\frac{dy}{dx} = \frac{1}{x} > 0 \qquad \text{and} \qquad \frac{d^2y}{dx^2} = -\frac{1}{x^2} < 0$$

which shows that $\ln x$ is increasing and concave downward on $(0, \infty)$. Putting this information together with **4**, we draw the graph of $y = \ln x$ in Figure 4.

Since $\ln 1 = 0$ and $\ln x$ is an increasing continuous function that takes on arbitrarily large values, the Intermediate Value Theorem shows that there is a number where $\ln x$ takes on the value 1. (See Figure 5.) This important number is denoted by e.

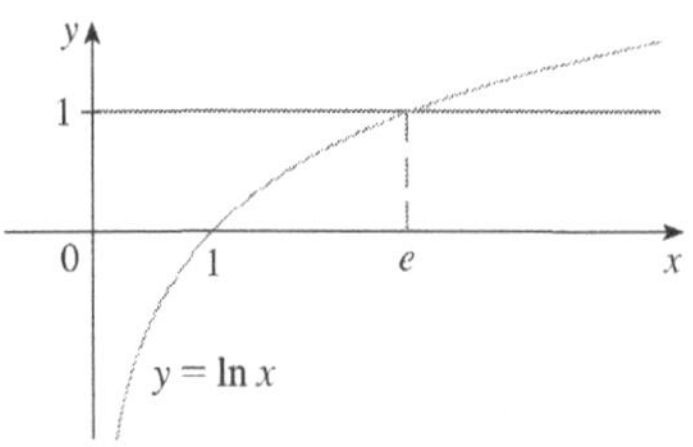

FIGURE 5

5 Definition e is the number such that $\ln e = 1$.

We will show (in Theorem 19) that this definition is consistent with our previous definition of e.

The Natural Exponential Function

Since ln is an increasing function, it is one-to-one and therefore has an inverse function, which we denote by exp. Thus, according to the definition of an inverse function,

$f^{-1}(x) = y \iff f(y) = x$

6 $$\exp(x) = y \iff \ln y = x$$

and the cancellation equations are

$f^{-1}(f(x)) = x$
$f(f^{-1}(x)) = x$

7 $$\exp(\ln x) = x \qquad \text{and} \qquad \ln(\exp x) = x$$

In particular, we have

$$\exp(0) = 1 \quad \text{since} \quad \ln 1 = 0$$

$$\exp(1) = e \quad \text{since} \quad \ln e = 1$$

We obtain the graph of $y = \exp x$ by reflecting the graph of $y = \ln x$ about the line $y = x$.

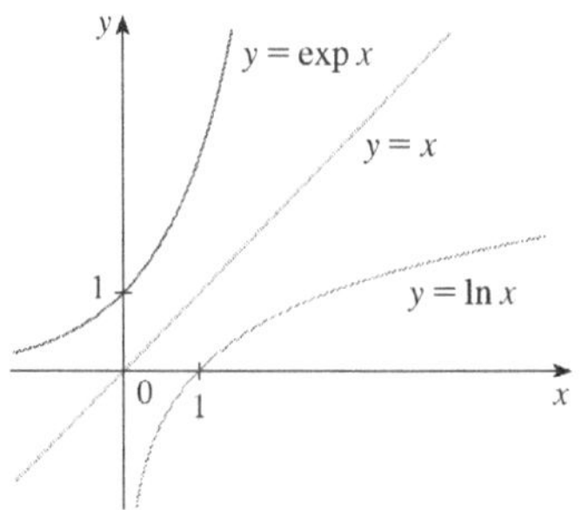

FIGURE 6

(See Figure 6.) The domain of exp is the range of ln, that is, $(-\infty, \infty)$; the range of exp is the domain of ln, that is, $(0, \infty)$.

If r is any rational number, then the third law of logarithms gives

$$\ln(e^r) = r \ln e = r$$

Therefore, by [6], $$\exp(r) = e^r$$

Thus $\exp(x) = e^x$ whenever x is a rational number. This leads us to define e^x, even for irrational values of x, by the equation

$$\boxed{e^x = \exp(x)}$$

In other words, for the reasons given, we define e^x to be the inverse of the function $\ln x$. In this notation [6] becomes

8 $$e^x = y \iff \ln y = x$$

and the cancellation equations [7] become

9 $$e^{\ln x} = x \qquad x > 0$$

10 $$\ln(e^x) = x \qquad \text{for all } x$$

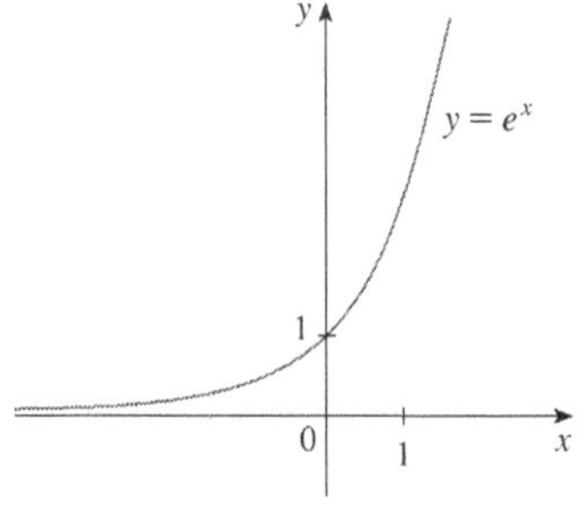

FIGURE 7
The natural exponential function

The natural exponential function $f(x) = e^x$ is one of the most frequently occurring functions in calculus and its applications, so it is important to be familiar with its graph (Figure 7) and its properties (which follow from the fact that it is the inverse of the natural logarithmic function).

Properties of the Exponential Function The exponential function $f(x) = e^x$ is an increasing continuous function with domain $\mathbb{R}$ and range $(0, \infty)$. Thus $e^x > 0$ for all x. Also

$$\lim_{x \to -\infty} e^x = 0 \qquad \lim_{x \to \infty} e^x = \infty$$

So the x-axis is a horizontal asymptote of $f(x) = e^x$.

We now verify that f has the other properties expected of an exponential function.

11 **Laws of Exponents** If x and y are real numbers and r is rational, then

1. $e^{x+y} = e^x e^y$ **2.** $e^{x-y} = \dfrac{e^x}{e^y}$ **3.** $(e^x)^r = e^{rx}$

PROOF OF LAW 1 Using the first law of logarithms and Equation 10, we have

$$\ln(e^x e^y) = \ln(e^x) + \ln(e^y) = x + y = \ln(e^{x+y})$$

Since ln is a one-to-one function, it follows that $e^x e^y = e^{x+y}$.

Laws 2 and 3 are proved similarly (see Exercises 6 and 7). As we will soon see, Law 3 actually holds when r is any real number.

We now prove the differentiation formula for e^x.

12 $$\frac{d}{dx}(e^x) = e^x$$

PROOF The function $y = e^x$ is differentiable because it is the inverse function of $y = \ln x$, which we know is differentiable with nonzero derivative. To find its derivative, we use the inverse function method. Let $y = e^x$. Then $\ln y = x$ and, differentiating this latter equation implicitly with respect to x, we get

$$\frac{1}{y}\frac{dy}{dx} = 1$$

$$\frac{dy}{dx} = y = e^x$$

General Exponential Functions

If $a > 0$ and r is any rational number, then by [9] and [11],

$$a^r = (e^{\ln a})^r = e^{r \ln a}$$

Therefore, even for irrational numbers x, we *define*

13 $$a^x = e^{x \ln a}$$

Thus, for instance,

$$2^{\sqrt{3}} = e^{\sqrt{3}\ln 2} \approx e^{1.20} \approx 3.32$$

The function $f(x) = a^x$ is called the **exponential function with base *a***. Notice that a^x is positive for all x because e^x is positive for all x.

Definition 13 allows us to extend one of the laws of logarithms. We already know that $\ln(a^r) = r \ln a$ when r is rational. But if we now let r be *any* real number we have, from Definition 13,

$$\ln a^r = \ln(e^{r \ln a}) = r \ln a$$

Thus

14 $$\ln a^r = r \ln a \qquad \text{for any real number } r$$

The general laws of exponents follow from Definition 13 together with the laws of exponents for e^x.

15 **Laws of Exponents** If x and y are real numbers and $a, b > 0$, then

1. $a^{x+y} = a^x a^y$ **2.** $a^{x-y} = a^x/a^y$ **3.** $(a^x)^y = a^{xy}$ **4.** $(ab)^x = a^x b^x$

PROOF

1. Using Definition 13 and the laws of exponents for e^x, we have

$$a^{x+y} = e^{(x+y)\ln a} = e^{x\ln a + y\ln a}$$
$$= e^{x\ln a}e^{y\ln a} = a^x a^y$$

3. Using Equation 14 we obtain

$$(a^x)^y = e^{y\ln(a^x)} = e^{yx\ln a} = e^{xy\ln a} = a^{xy}$$

The remaining proofs are left as exercises. ■

The differentiation formula for exponential functions is also a consequence of Definition 13:

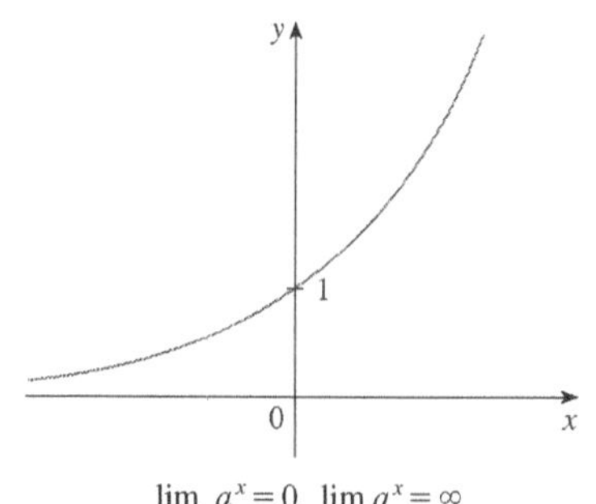

$\lim_{x\to-\infty} a^x = 0, \ \lim_{x\to\infty} a^x = \infty$

FIGURE 8 $y = a^x, \ a > 1$

16
$$\frac{d}{dx}(a^x) = a^x \ln a$$

PROOF

$$\frac{d}{dx}(a^x) = \frac{d}{dx}(e^{x\ln a}) = e^{x\ln a}\frac{d}{dx}(x\ln a) = a^x \ln a$$ ■

If $a > 1$, then $\ln a > 0$, so $(d/dx)\,a^x = a^x \ln a > 0$, which shows that $y = a^x$ is increasing (see Figure 8). If $0 < a < 1$, then $\ln a < 0$ and so $y = a^x$ is decreasing (see Figure 9).

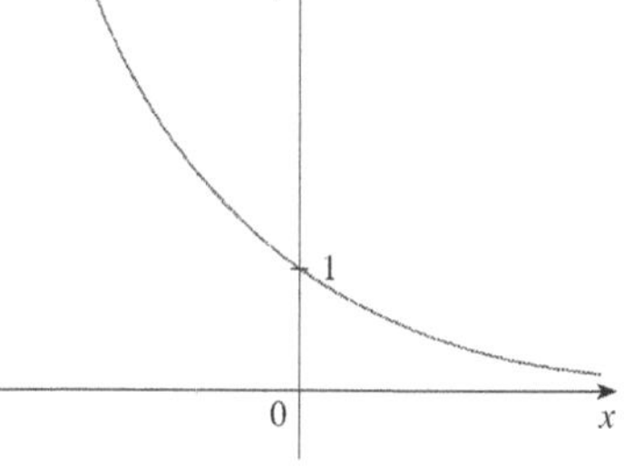

$\lim_{x\to-\infty} a^x = \infty, \ \lim_{x\to\infty} a^x = 0$

FIGURE 9 $y = a^x, \ 0 < a < 1$

General Logarithmic Functions

If $a > 0$ and $a \neq 1$, then $f(x) = a^x$ is a one-to-one function. Its inverse function is called the **logarithmic function with base *a*** and is denoted by $\log_a$. Thus

17
$$\log_a x = y \iff a^y = x$$

In particular, we see that

$$\log_e x = \ln x$$

The laws of logarithms are similar to those for the natural logarithm and can be deduced from the laws of exponents (see Exercise 10).

To differentiate $y = \log_a x$, we write the equation as $a^y = x$. From Equation 14 we have $y \ln a = \ln x$, so

$$\log_a x = y = \frac{\ln x}{\ln a}$$

Since $\ln a$ is a constant, we can differentiate as follows:

$$\frac{d}{dx}(\log_a x) = \frac{d}{dx}\frac{\ln x}{\ln a} = \frac{1}{\ln a}\frac{d}{dx}(\ln x) = \frac{1}{x \ln a}$$

18
$$\frac{d}{dx}(\log_a x) = \frac{1}{x \ln a}$$

The Number e Expressed as a Limit

In this section we defined e as the number such that $\ln e = 1$. The next theorem shows that this is the same as the number e defined in Section 3.1 (see Equation 3.6.5).

19
$$e = \lim_{x \to 0} (1 + x)^{1/x}$$

PROOF Let $f(x) = \ln x$. Then $f'(x) = 1/x$, so $f'(1) = 1$. But, by the definition of derivative,

$$f'(1) = \lim_{h \to 0} \frac{f(1 + h) - f(1)}{h} = \lim_{x \to 0} \frac{f(1 + x) - f(1)}{x}$$

$$= \lim_{x \to 0} \frac{\ln(1 + x) - \ln 1}{x} = \lim_{x \to 0} \frac{1}{x} \ln(1 + x) = \lim_{x \to 0} \ln(1 + x)^{1/x}$$

Because $f'(1) = 1$, we have

$$\lim_{x \to 0} \ln(1 + x)^{1/x} = 1$$

Then, by Theorem 2.5.8 and the continuity of the exponential function, we have

$$e = e^1 = e^{\lim_{x \to 0} \ln(1+x)^{1/x}} = \lim_{x \to 0} e^{\ln(1+x)^{1/x}} = \lim_{x \to 0} (1 + x)^{1/x}$$

G Exercises

1. (a) By comparing areas, show that

$$\tfrac{1}{3} < \ln 1.5 < \tfrac{5}{12}$$

(b) Use the Midpoint Rule with $n = 10$ to estimate $\ln 1.5$.

2. Refer to Example 1.
   (a) Find the equation of the tangent line to the curve $y = 1/t$ that is parallel to the secant line AD.
   (b) Use part (a) to show that $\ln 2 > 0.66$.

3. By comparing areas, show that

$$\frac{1}{2} + \frac{1}{3} + \cdots + \frac{1}{n} < \ln n < 1 + \frac{1}{2} + \frac{1}{3} + \cdots + \frac{1}{n - 1}$$

4. (a) By comparing areas, show that $\ln 2 < 1 < \ln 3$.
   (b) Deduce that $2 < e < 3$.

5. Prove the third law of logarithms. [*Hint:* Start by showing that both sides of the equation have the same derivative.]

6. Prove the second law of exponents for e^x [see [11]].

7. Prove the third law of exponents for e^x [see [11]].

8. Prove the second law of exponents [see [15]].

9. Prove the fourth law of exponents [see [15]].

10. Deduce the following laws of logarithms from [15]:
(a) $\log_a(xy) = \log_a x + \log_a y$
(b) $\log_a(x/y) = \log_a x - \log_a y$
(c) $\log_a(x^y) = y \log_a x$

H Complex Numbers

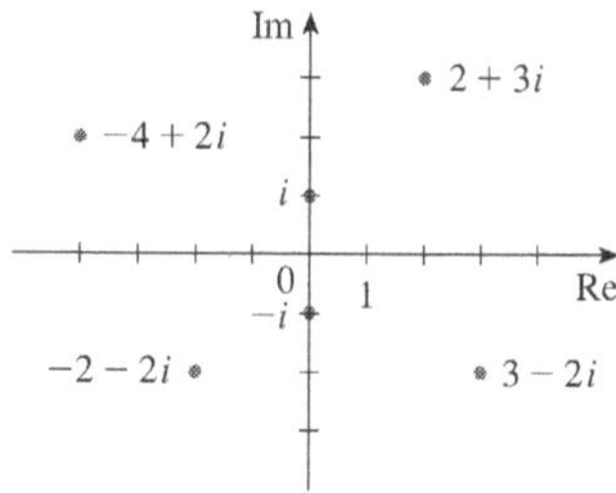

FIGURE 1
Complex numbers as points in the Argand plane

A **complex number** can be represented by an expression of the form $a + bi$, where a and b are real numbers and i is a symbol with the property that $i^2 = -1$. The complex number $a + bi$ can also be represented by the ordered pair (a, b) and plotted as a point in a plane (called the Argand plane) as in Figure 1. Thus the complex number $i = 0 + 1 \cdot i$ is identified with the point $(0, 1)$.

The **real part** of the complex number $a + bi$ is the real number a and the **imaginary part** is the real number b. Thus the real part of $4 - 3i$ is 4 and the imaginary part is -3. Two complex numbers $a + bi$ and $c + di$ are **equal** if $a = c$ and $b = d$, that is, their real parts are equal and their imaginary parts are equal. In the Argand plane the horizontal axis is called the real axis and the vertical axis is called the imaginary axis.

The sum and difference of two complex numbers are defined by adding or subtracting their real parts and their imaginary parts:

$$(a + bi) + (c + di) = (a + c) + (b + d)i$$

$$(a + bi) - (c + di) = (a - c) + (b - d)i$$

For instance,

$$(1 - i) + (4 + 7i) = (1 + 4) + (-1 + 7)i = 5 + 6i$$

The product of complex numbers is defined so that the usual commutative and distributive laws hold:

$$\begin{aligned}(a + bi)(c + di) &= a(c + di) + (bi)(c + di) \\ &= ac + adi + bci + bdi^2\end{aligned}$$

Since $i^2 = -1$, this becomes

$$(a + bi)(c + di) = (ac - bd) + (ad + bc)i$$

EXAMPLE 1

$$\begin{aligned}(-1 + 3i)(2 - 5i) &= (-1)(2 - 5i) + 3i(2 - 5i) \\ &= -2 + 5i + 6i - 15(-1) = 13 + 11i\end{aligned}$$

Division of complex numbers is much like rationalizing the denominator of a rational expression. For the complex number $z = a + bi$, we define its **complex conjugate** to be $\overline{z} = a - bi$. To find the quotient of two complex numbers we multiply numerator and denominator by the complex conjugate of the denominator.

EXAMPLE 2 Express the number $\dfrac{-1 + 3i}{2 + 5i}$ in the form $a + bi$.

SOLUTION We multiply numerator and denominator by the complex conjugate of $2 + 5i$, namely $2 - 5i$, and we take advantage of the result of Example 1:

$$\frac{-1 + 3i}{2 + 5i} = \frac{-1 + 3i}{2 + 5i} \cdot \frac{2 - 5i}{2 - 5i} = \frac{13 + 11i}{2^2 + 5^2} = \frac{13}{29} + \frac{11}{29}i$$

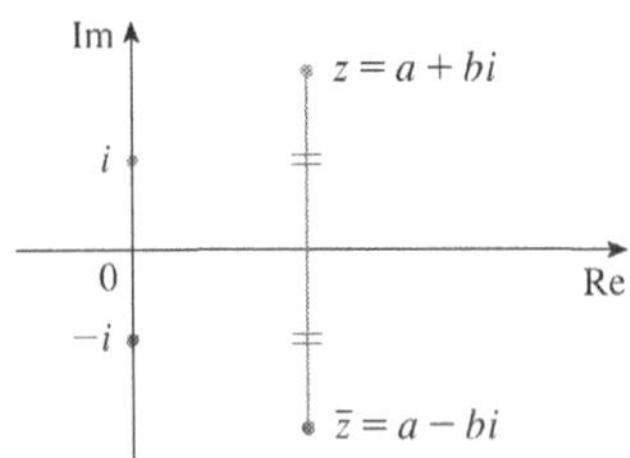

FIGURE 2

The geometric interpretation of the complex conjugate is shown in Figure 2: $\bar{z}$ is the reflection of z in the real axis. We list some of the properties of the complex conjugate in the following box. The proofs follow from the definition and are requested in Exercise 18.

Properties of Conjugates

$$\overline{z + w} = \bar{z} + \bar{w} \qquad \overline{zw} = \bar{z}\,\bar{w} \qquad \overline{z^n} = \bar{z}^n$$

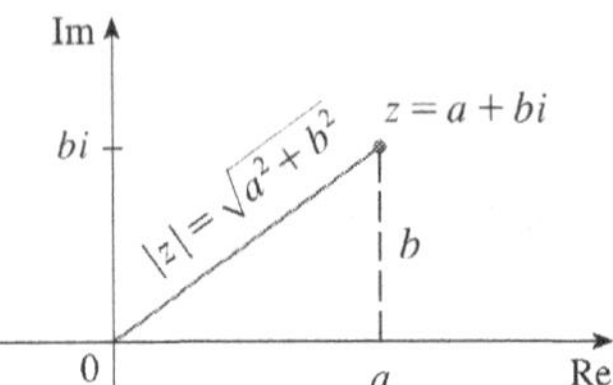

FIGURE 3

The **modulus**, or **absolute value**, $|z|$ of a complex number $z = a + bi$ is its distance from the origin. From Figure 3 we see that if $z = a + bi$, then

$$|z| = \sqrt{a^2 + b^2}$$

Notice that

$$z\bar{z} = (a + bi)(a - bi) = a^2 + abi - abi - b^2i^2 = a^2 + b^2$$

and so

$$z\bar{z} = |z|^2$$

This explains why the division procedure in Example 2 works in general:

$$\frac{z}{w} = \frac{z\bar{w}}{w\bar{w}} = \frac{z\bar{w}}{|w|^2}$$

Since $i^2 = -1$, we can think of i as a square root of -1. But notice that we also have $(-i)^2 = i^2 = -1$ and so $-i$ is also a square root of -1. We say that i is the **principal square root** of -1 and write $\sqrt{-1} = i$. In general, if c is any positive number, we write

$$\sqrt{-c} = \sqrt{c}\, i$$

With this convention, the usual derivation and formula for the roots of the quadratic equation $ax^2 + bx + c = 0$ are valid even when $b^2 - 4ac < 0$:

$$x = \frac{-b \pm \sqrt{b^2 - 4ac}}{2a}$$

EXAMPLE 3 Find the roots of the equation $x^2 + x + 1 = 0$.

SOLUTION Using the quadratic formula, we have

$$x = \frac{-1 \pm \sqrt{1^2 - 4 \cdot 1}}{2} = \frac{-1 \pm \sqrt{-3}}{2} = \frac{-1 \pm \sqrt{3}\, i}{2}$$

We observe that the solutions of the equation in Example 3 are complex conjugates of each other. In general, the solutions of any quadratic equation $ax^2 + bx + c = 0$ with real coefficients a, b, and c are always complex conjugates. (If z is real, $\bar{z} = z$, so z is its own conjugate.)

We have seen that if we allow complex numbers as solutions, then every quadratic equation has a solution. More generally, it is true that every polynomial equation

$$a_n x^n + a_{n-1} x^{n-1} + \cdots + a_1 x + a_0 = 0$$

of degree at least one has a solution among the complex numbers. This fact is known as the Fundamental Theorem of Algebra and was proved by Gauss.

Polar Form

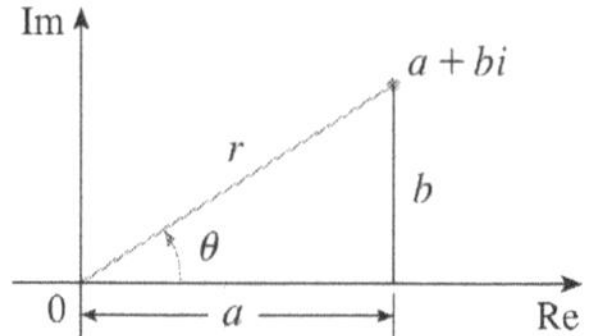

FIGURE 4

We know that any complex number $z = a + bi$ can be considered as a point (a, b) and that any such point can be represented by polar coordinates (r, θ) with $r \geqslant 0$. In fact,

$$a = r\cos\theta \qquad b = r\sin\theta$$

as in Figure 4. Therefore we have

$$z = a + bi = (r\cos\theta) + (r\sin\theta)i$$

Thus we can write any complex number z in the form

$$\boxed{z = r(\cos\theta + i\sin\theta)}$$

where $\qquad r = |z| = \sqrt{a^2 + b^2} \qquad$ and $\qquad \tan\theta = \dfrac{b}{a}$

The angle θ is called the **argument** of z and we write $\theta = \arg(z)$. Note that $\arg(z)$ is not unique; any two arguments of z differ by an integer multiple of 2π.

EXAMPLE 4 Write the following numbers in polar form.
(a) $z = 1 + i$ (b) $w = \sqrt{3} - i$

SOLUTION

(a) We have $r = |z| = \sqrt{1^2 + 1^2} = \sqrt{2}$ and $\tan\theta = 1$, so we can take $\theta = \pi/4$. Therefore the polar form is

$$z = \sqrt{2}\left(\cos\frac{\pi}{4} + i\sin\frac{\pi}{4}\right)$$

(b) Here we have $r = |w| = \sqrt{3 + 1} = 2$ and $\tan\theta = -1/\sqrt{3}$. Since w lies in the fourth quadrant, we take $\theta = -\pi/6$ and

$$w = 2\left[\cos\left(-\frac{\pi}{6}\right) + i\sin\left(-\frac{\pi}{6}\right)\right]$$

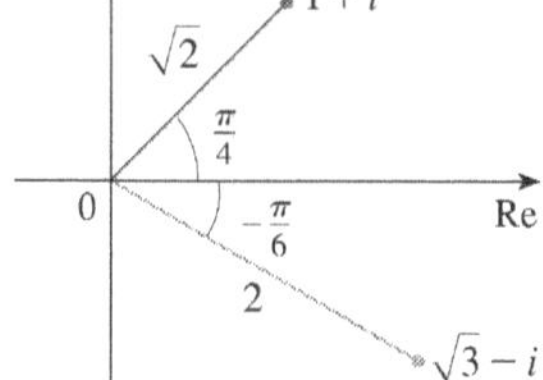

FIGURE 5

The numbers z and w are shown in Figure 5.

The polar form of complex numbers gives insight into multiplication and division. Let

$$z_1 = r_1(\cos\theta_1 + i\sin\theta_1) \qquad z_2 = r_2(\cos\theta_2 + i\sin\theta_2)$$

be two complex numbers written in polar form. Then

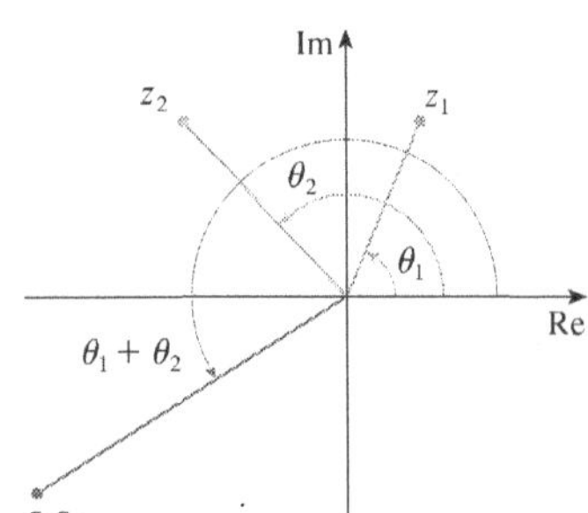

FIGURE 6

$$\begin{aligned} z_1z_2 &= r_1r_2(\cos\theta_1 + i\sin\theta_1)(\cos\theta_2 + i\sin\theta_2) \\ &= r_1r_2[(\cos\theta_1\cos\theta_2 - \sin\theta_1\sin\theta_2) + i(\sin\theta_1\cos\theta_2 + \cos\theta_1\sin\theta_2)] \end{aligned}$$

Therefore, using the addition formulas for cosine and sine, we have

1

$$z_1z_2 = r_1r_2[\cos(\theta_1 + \theta_2) + i\sin(\theta_1 + \theta_2)]$$

This formula says that *to multiply two complex numbers we multiply the moduli and add the arguments.* (See Figure 6.)

A similar argument using the subtraction formulas for sine and cosine shows that *to divide two complex numbers we divide the moduli and subtract the arguments.*

$$\frac{z_1}{z_2} = \frac{r_1}{r_2}[\cos(\theta_1 - \theta_2) + i\sin(\theta_1 - \theta_2)] \qquad z_2 \neq 0$$

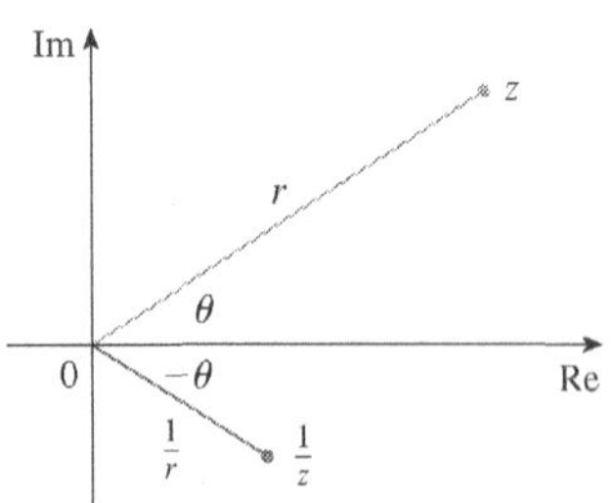

FIGURE 7

In particular, taking $z_1 = 1$ and $z_2 = z$ (and therefore $\theta_1 = 0$ and $\theta_2 = \theta$), we have the following, which is illustrated in Figure 7.

If $z = r(\cos\theta + i\sin\theta)$, then $\dfrac{1}{z} = \dfrac{1}{r}(\cos\theta - i\sin\theta)$.

EXAMPLE 5 Find the product of the complex numbers $1 + i$ and $\sqrt{3} - i$ in polar form.

SOLUTION From Example 4 we have

$$1 + i = \sqrt{2}\left(\cos\frac{\pi}{4} + i\sin\frac{\pi}{4}\right)$$

and

$$\sqrt{3} - i = 2\left[\cos\left(-\frac{\pi}{6}\right) + i\sin\left(-\frac{\pi}{6}\right)\right]$$

So, by Equation 1,

$$\begin{aligned} (1 + i)(\sqrt{3} - i) &= 2\sqrt{2}\left[\cos\left(\frac{\pi}{4} - \frac{\pi}{6}\right) + i\sin\left(\frac{\pi}{4} - \frac{\pi}{6}\right)\right] \\ &= 2\sqrt{2}\left(\cos\frac{\pi}{12} + i\sin\frac{\pi}{12}\right) \end{aligned}$$

This is illustrated in Figure 8.

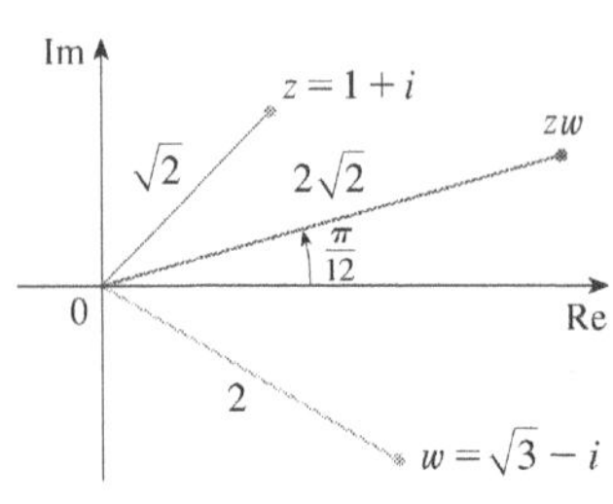

FIGURE 8

Repeated use of Formula 1 shows how to compute powers of a complex number. If

$$z = r(\cos\theta + i\sin\theta)$$

then
$$z^2 = r^2(\cos 2\theta + i\sin 2\theta)$$

and
$$z^3 = zz^2 = r^3(\cos 3\theta + i\sin 3\theta)$$

In general, we obtain the following result, which is named after the French mathematician Abraham De Moivre (1667–1754).

> **2 De Moivre's Theorem** If $z = r(\cos\theta + i\sin\theta)$ and n is a positive integer, then
>
> $$z^n = [r(\cos\theta + i\sin\theta)]^n = r^n(\cos n\theta + i\sin n\theta)$$

This says that *to take the nth power of a complex number we take the nth power of the modulus and multiply the argument by n.*

EXAMPLE 6 Find $(\frac{1}{2} + \frac{1}{2}i)^{10}$.

SOLUTION Since $\frac{1}{2} + \frac{1}{2}i = \frac{1}{2}(1 + i)$, it follows from Example 4(a) that $\frac{1}{2} + \frac{1}{2}i$ has the polar form

$$\frac{1}{2} + \frac{1}{2}i = \frac{\sqrt{2}}{2}\left(\cos\frac{\pi}{4} + i\sin\frac{\pi}{4}\right)$$

So by De Moivre's Theorem,

$$\left(\frac{1}{2} + \frac{1}{2}i\right)^{10} = \left(\frac{\sqrt{2}}{2}\right)^{10}\left(\cos\frac{10\pi}{4} + i\sin\frac{10\pi}{4}\right)$$

$$= \frac{2^5}{2^{10}}\left(\cos\frac{5\pi}{2} + i\sin\frac{5\pi}{2}\right) = \frac{1}{32}i$$

De Moivre's Theorem can also be used to find the nth roots of complex numbers. An nth root of the complex number z is a complex number w such that

$$w^n = z$$

Writing these two numbers in trigonometric form as

$$w = s(\cos\phi + i\sin\phi) \qquad \text{and} \qquad z = r(\cos\theta + i\sin\theta)$$

and using De Moivre's Theorem, we get

$$s^n(\cos n\phi + i\sin n\phi) = r(\cos\theta + i\sin\theta)$$

The equality of these two complex numbers shows that

$$s^n = r \qquad \text{or} \qquad s = r^{1/n}$$

and
$$\cos n\phi = \cos\theta \qquad \text{and} \qquad \sin n\phi = \sin\theta$$

From the fact that sine and cosine have period 2π it follows that

$$n\phi = \theta + 2k\pi \qquad \text{or} \qquad \phi = \frac{\theta + 2k\pi}{n}$$

Thus
$$w = r^{1/n}\left[\cos\left(\frac{\theta + 2k\pi}{n}\right) + i\sin\left(\frac{\theta + 2k\pi}{n}\right)\right]$$

Since this expression gives a different value of w for $k = 0, 1, 2, \ldots, n - 1$, we have the following.

> **3 Roots of a Complex Number** Let $z = r(\cos\theta + i\sin\theta)$ and let n be a positive integer. Then z has the n distinct nth roots
>
> $$w_k = r^{1/n}\left[\cos\left(\frac{\theta + 2k\pi}{n}\right) + i\sin\left(\frac{\theta + 2k\pi}{n}\right)\right]$$
>
> where $k = 0, 1, 2, \ldots, n - 1$.

Notice that each of the nth roots of z has modulus $|w_k| = r^{1/n}$. Thus all the nth roots of z lie on the circle of radius $r^{1/n}$ in the complex plane. Also, since the argument of each successive nth root exceeds the argument of the previous root by $2\pi/n$, we see that the nth roots of z are equally spaced on this circle.

EXAMPLE 7 Find the six sixth roots of $z = -8$ and graph these roots in the complex plane.

SOLUTION In trigonometric form, $z = 8(\cos\pi + i\sin\pi)$. Applying Equation 3 with $n = 6$, we get

$$w_k = 8^{1/6}\left(\cos\frac{\pi + 2k\pi}{6} + i\sin\frac{\pi + 2k\pi}{6}\right)$$

We get the six sixth roots of -8 by taking $k = 0, 1, 2, 3, 4, 5$ in this formula:

$$w_0 = 8^{1/6}\left(\cos\frac{\pi}{6} + i\sin\frac{\pi}{6}\right) = \sqrt{2}\left(\frac{\sqrt{3}}{2} + \frac{1}{2}i\right)$$

$$w_1 = 8^{1/6}\left(\cos\frac{\pi}{2} + i\sin\frac{\pi}{2}\right) = \sqrt{2}\,i$$

$$w_2 = 8^{1/6}\left(\cos\frac{5\pi}{6} + i\sin\frac{5\pi}{6}\right) = \sqrt{2}\left(-\frac{\sqrt{3}}{2} + \frac{1}{2}i\right)$$

$$w_3 = 8^{1/6}\left(\cos\frac{7\pi}{6} + i\sin\frac{7\pi}{6}\right) = \sqrt{2}\left(-\frac{\sqrt{3}}{2} - \frac{1}{2}i\right)$$

$$w_4 = 8^{1/6}\left(\cos\frac{3\pi}{2} + i\sin\frac{3\pi}{2}\right) = -\sqrt{2}\,i$$

$$w_5 = 8^{1/6}\left(\cos\frac{11\pi}{6} + i\sin\frac{11\pi}{6}\right) = \sqrt{2}\left(\frac{\sqrt{3}}{2} - \frac{1}{2}i\right)$$

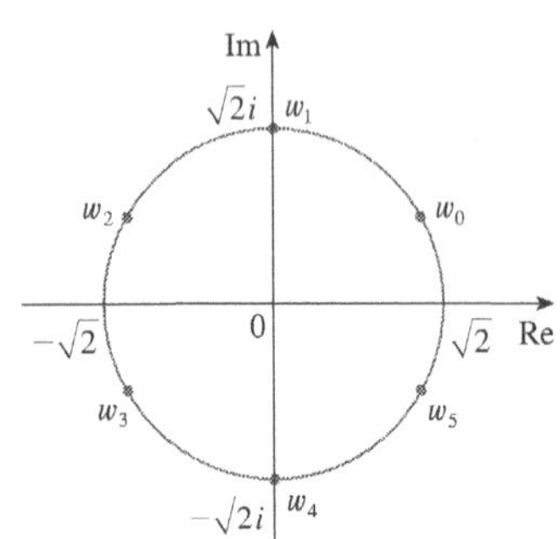

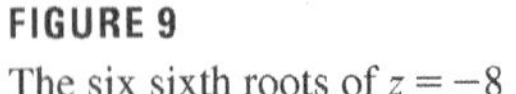

FIGURE 9
The six sixth roots of $z = -8$

All these points lie on the circle of radius $\sqrt{2}$ as shown in Figure 9.

Complex Exponentials

We also need to give a meaning to the expression e^z when $z = x + iy$ is a complex number. The theory of infinite series as developed in Chapter 11 can be extended to the case where the terms are complex numbers. Using the Taylor series for e^x (11.10.11) as our guide, we define

4
$$e^z = \sum_{n=0}^{\infty} \frac{z^n}{n!} = 1 + z + \frac{z^2}{2!} + \frac{z^3}{3!} + \cdots$$

and it turns out that this complex exponential function has the same properties as the real exponential function. In particular, it is true that

5
$$e^{z_1+z_2} = e^{z_1}e^{z_2}$$

If we put $z = iy$, where y is a real number, in Equation 4, and use the facts that

$$i^2 = -1, \quad i^3 = i^2 i = -i, \quad i^4 = 1, \quad i^5 = i, \quad \ldots$$

we get
$$\begin{aligned} e^{iy} &= 1 + iy + \frac{(iy)^2}{2!} + \frac{(iy)^3}{3!} + \frac{(iy)^4}{4!} + \frac{(iy)^5}{5!} + \cdots \\ &= 1 + iy - \frac{y^2}{2!} - i\frac{y^3}{3!} + \frac{y^4}{4!} + i\frac{y^5}{5!} + \cdots \\ &= \left(1 - \frac{y^2}{2!} + \frac{y^4}{4!} - \frac{y^6}{6!} + \cdots\right) + i\left(y - \frac{y^3}{3!} + \frac{y^5}{5!} - \cdots\right) \\ &= \cos y + i \sin y \end{aligned}$$

Here we have used the Taylor series for $\cos y$ and $\sin y$ (Equations 11.10.16 and 11.10.15). The result is a famous formula called **Euler's formula**:

6
$$e^{iy} = \cos y + i \sin y$$

Combining Euler's formula with Equation 5, we get

7
$$e^{x+iy} = e^x e^{iy} = e^x(\cos y + i \sin y)$$

EXAMPLE 8 Evaluate: (a) $e^{i\pi}$ (b) $e^{-1+i\pi/2}$

SOLUTION

(a) From Euler's equation **6** we have

$$e^{i\pi} = \cos \pi + i \sin \pi = -1 + i(0) = -1$$

(b) Using Equation 7 we get

$$e^{-1+i\pi/2} = e^{-1}\left(\cos\frac{\pi}{2} + i \sin\frac{\pi}{2}\right) = \frac{1}{e}[0 + i(1)] = \frac{i}{e}$$

We could write the result of Example 8(a) as

$$e^{i\pi} + 1 = 0$$

This equation relates the five most famous numbers in all of mathematics: 0, 1, e, i, and π.

Finally, we note that Euler's equation provides us with an easier method of proving De Moivre's Theorem:

$$[r(\cos\theta + i \sin\theta)]^n = (re^{i\theta})^n = r^n e^{in\theta} = r^n(\cos n\theta + i \sin n\theta)$$

H Exercises

1–14 Evaluate the expression and write your answer in the form $a + bi$.

1. $(5 - 6i) + (3 + 2i)$

2. $\left(4 - \frac{1}{2}i\right) - \left(9 + \frac{5}{2}i\right)$

3. $(2 + 5i)(4 - i)$

4. $(1 - 2i)(8 - 3i)$

5. $\overline{12 + 7i}$

6. $\overline{2i\left(\frac{1}{2} - i\right)}$

7. $\dfrac{1 + 4i}{3 + 2i}$

8. $\dfrac{3 + 2i}{1 - 4i}$

9. $\dfrac{1}{1 + i}$

10. $\dfrac{3}{4 - 3i}$

11. i^3

12. i^{100}

13. $\sqrt{-25}$

14. $\sqrt{-3}\,\sqrt{-12}$

15–17 Find the complex conjugate and the modulus of the number.

15. $12 - 5i$

16. $-1 + 2\sqrt{2}\,i$

17. $-4i$

18. Prove the following properties of complex numbers.
(a) $\overline{z + w} = \bar{z} + \bar{w}$ (b) $\overline{zw} = \bar{z}\,\bar{w}$
(c) $\overline{z^n} = \bar{z}^n$, where n is a positive integer
[*Hint:* Write $z = a + bi$, $w = c + di$.]

19–24 Find all solutions of the equation.

19. $4x^2 + 9 = 0$

20. $x^4 = 1$

21. $x^2 + 2x + 5 = 0$

22. $2x^2 - 2x + 1 = 0$

23. $z^2 + z + 2 = 0$

24. $z^2 + \frac{1}{2}z + \frac{1}{4} = 0$

25–28 Write the number in polar form with argument between 0 and 2π.

25. $-3 + 3i$

26. $1 - \sqrt{3}\,i$

27. $3 + 4i$

28. $8i$

29–32 Find polar forms for zw, z/w, and $1/z$ by first putting z and w into polar form.

29. $z = \sqrt{3} + i,\quad w = 1 + \sqrt{3}\,i$

30. $z = 4\sqrt{3} - 4i,\quad w = 8i$

31. $z = 2\sqrt{3} - 2i,\quad w = -1 + i$

32. $z = 4\left(\sqrt{3} + i\right),\quad w = -3 - 3i$

33–36 Find the indicated power using De Moivre's Theorem.

33. $(1 + i)^{20}$

34. $\left(1 - \sqrt{3}\,i\right)^5$

35. $\left(2\sqrt{3} + 2i\right)^5$

36. $(1 - i)^8$

37–40 Find the indicated roots. Sketch the roots in the complex plane.

37. The eighth roots of 1

38. The fifth roots of 32

39. The cube roots of i

40. The cube roots of $1 + i$

41–46 Write the number in the form $a + bi$.

41. $e^{i\pi/2}$

42. $e^{2\pi i}$

43. $e^{i\pi/3}$

44. $e^{\ i\pi}$

45. $e^{2+i\pi}$

46. $e^{\pi+i}$

47. Use De Moivre's Theorem with $n = 3$ to express $\cos 3\theta$ and $\sin 3\theta$ in terms of $\cos\theta$ and $\sin\theta$.

48. Use Euler's formula to prove the following formulas for $\cos x$ and $\sin x$:

$$\cos x = \frac{e^{ix} + e^{\ ix}}{2} \qquad \sin x = \frac{e^{ix} - e^{\ ix}}{2i}$$

49. If $u(x) = f(x) + ig(x)$ is a complex-valued function of a real variable x and the real and imaginary parts $f(x)$ and $g(x)$ are differentiable functions of x, then the derivative of u is defined to be $u'(x) = f'(x) + ig'(x)$. Use this together with Equation 7 to prove that if $F(x) = e^{rx}$, then $F'(x) = re^{rx}$ when $r = a + bi$ is a complex number.

50. (a) If u is a complex-valued function of a real variable, its indefinite integral $\int u(x)\,dx$ is an antiderivative of u. Evaluate

$$\int e^{(1+i)x}\,dx$$

(b) By considering the real and imaginary parts of the integral in part (a), evaluate the real integrals

$$\int e^x \cos x\,dx \qquad \text{and} \qquad \int e^x \sin x\,dx$$

(c) Compare with the method used in Example 4 in Section 7.1.

I Answers to Odd-Numbered Exercises

CHAPTER 1

EXERCISES 1.1 ■ PAGE 19

1. Yes

3. (a) 3 (b) -0.2 (c) 0, 3 (d) -0.8
(e) $[-2, 4], [-1, 3]$ (f) $[-2, 1]$

5. $[-85, 115]$ **7.** No

9. Yes, $[-3, 2]$, $[-3, -2) \cup [-1, 3]$

11. Diet, exercise, or illness

13.

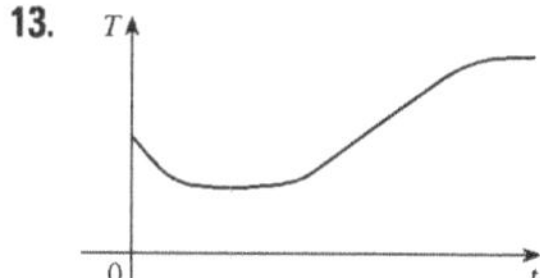

15. (a) 500 MW; 730 MW (b) 4 AM; noon

17.

T
midnight
noon
t

19.

amount
0
price

21.

height of grass
Wed. Wed. Wed. Wed. Wed. t

23. (a)

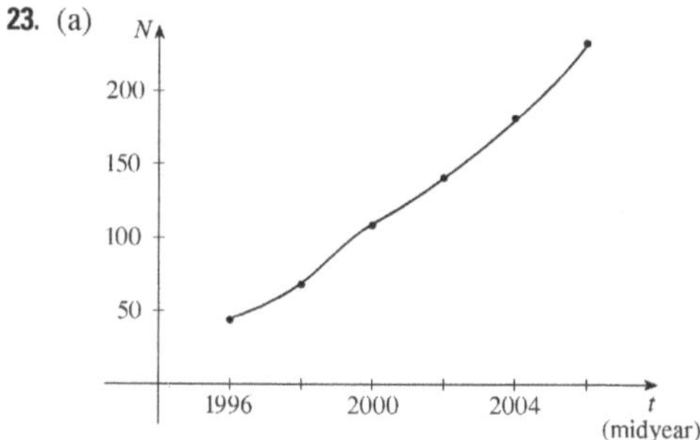

(b) 126 million; 207 million

25. $12, 16, 3a^2 - a + 2, 3a^2 + a + 2, 3a^2 + 5a + 4,$
$6a^2 - 2a + 4, 12a^2 - 2a + 2, 3a^4 - a^2 + 2,$
$9a^4 - 6a^3 + 13a^2 - 4a + 4, 3a^2 + 6ah + 3h^2 - a - h + 2$

27. $-3 - h$ **29.** $-1/(ax)$

31. $(-\infty, -3) \cup (-3, 3) \cup (3, \infty)$ **33.** $(-\infty, \infty)$

35. $(-\infty, 0) \cup (5, \infty)$ **37.** $[0, 4]$

39. $(-\infty, \infty)$

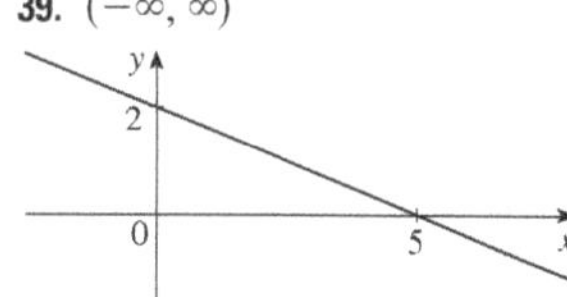

41. $(-\infty, \infty)$

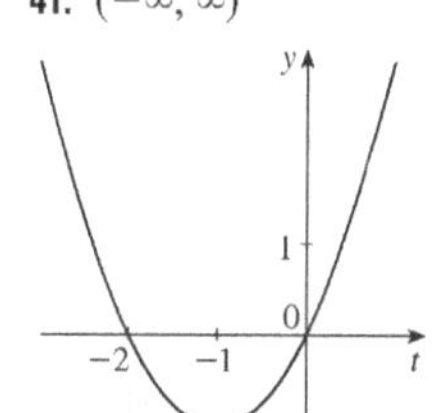

43. $[5, \infty)$

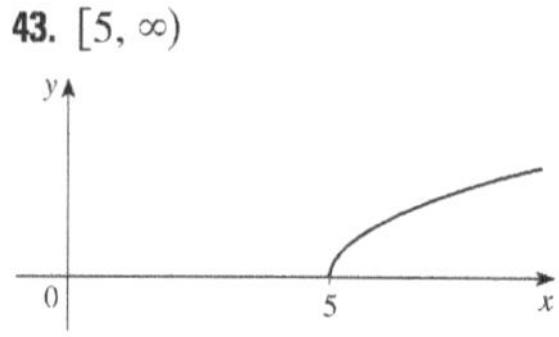

45. $(-\infty, 0) \cup (0, \infty)$

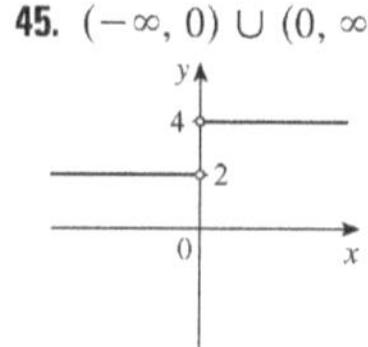

47. $(-\infty, \infty)$

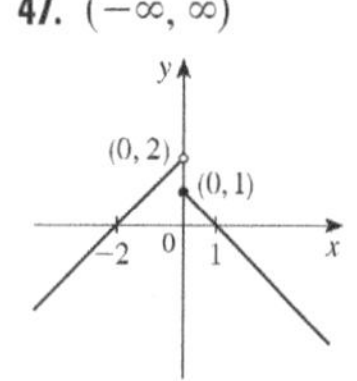

49. $(-\infty, \infty)$

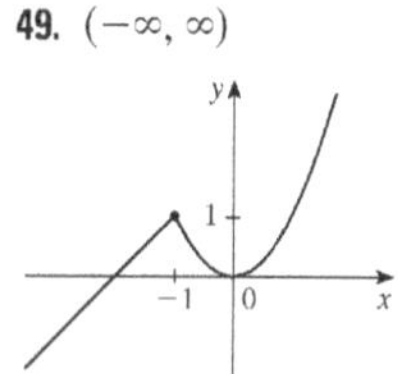

51. $f(x) = \frac{5}{2}x - \frac{11}{2}, 1 \leqslant x \leqslant 5$ **53.** $f(x) = 1 - \sqrt{-x}$

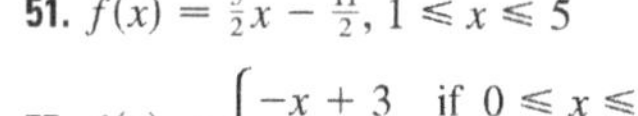

55. $f(x) = \begin{cases} -x + 3 & \text{if } 0 \leqslant x \leqslant 3 \\ 2x - 6 & \text{if } 3 < x \leqslant 5 \end{cases}$

57. $A(L) = 10L - L^2, 0 < L < 10$

59. $A(x) = \sqrt{3}x^2/4, x > 0$ **61.** $S(x) = x^2 + (8/x), x > 0$

63. $V(x) = 4x^3 - 64x^2 + 240x, 0 < x < 6$

65. $F(x) = \begin{cases} 15(40 - x) & \text{if } 0 \leqslant x < 40 \\ 0 & \text{if } 40 \leqslant x \leqslant 65 \\ 15(x - 65) & \text{if } x > 65 \end{cases}$

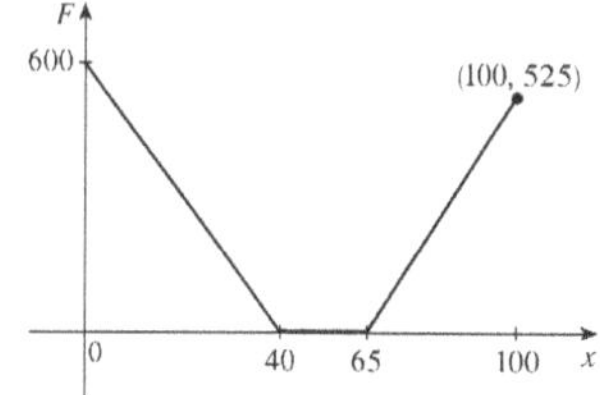

67. (a)

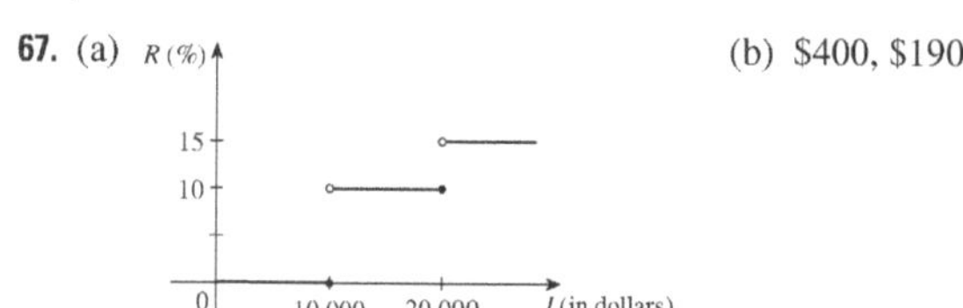

(b) \$400, \$1900

(c)

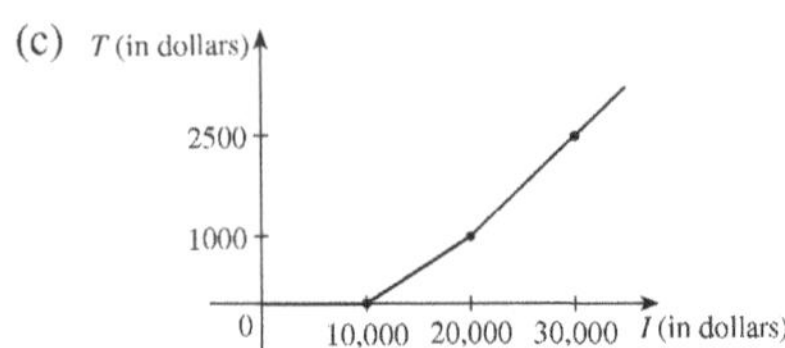

69. f is odd, g is even **71.** (a) $(-5, 3)$ (b) $(-5, -3)$
73. Odd **75.** Neither **77.** Even
79. Even; odd; neither (unless $f = 0$ or $g = 0$)

EXERCISES 1.2 ■ PAGE 33

1. (a) Logarithmic (b) Root (c) Rational
(d) Polynomial, degree 2 (e) Exponential (f) Trigonometric
3. (a) h (b) f (c) g
5. (a) $y = 2x + b$,
where b is the y-intercept.

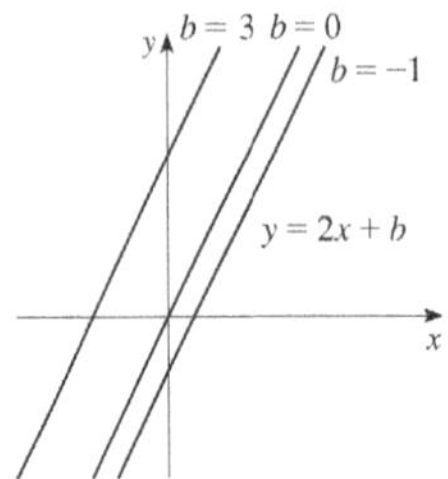

(b) $y = mx + 1 - 2m$,
where m is the slope.
(c) $y = 2x - 3$

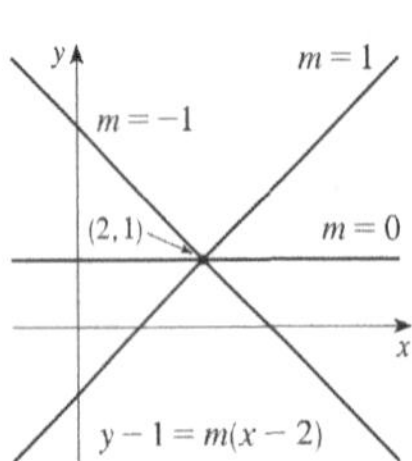

7. Their graphs have slope -1.

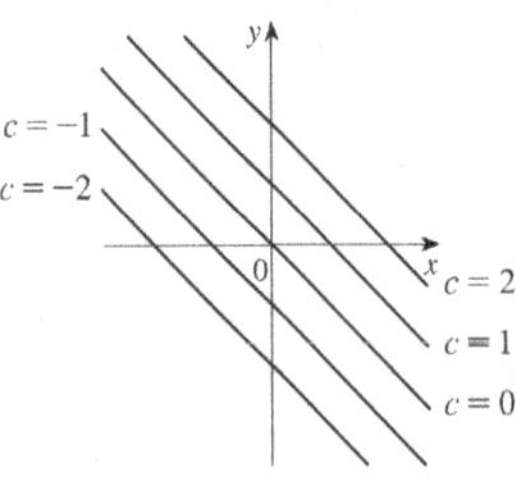

9. $f(x) = -3x(x + 1)(x - 2)$
11. (a) 8.34, change in mg for every 1 year change
(b) 8.34 mg
13. (a)

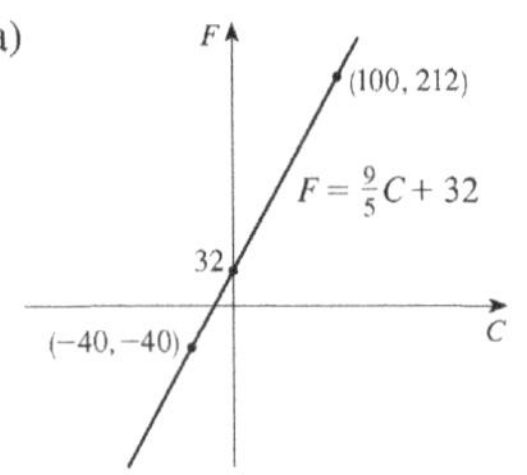

(b) $\frac{9}{5}$, change in °F for every 1°C change; 32, Fahrenheit temperature corresponding to 0°C

15. (a) $T = \frac{1}{6}N + \frac{307}{6}$ (b) $\frac{1}{6}$, change in °F for every chirp per minute change (c) 76°F
17. (a) $P = 0.434d + 15$ (b) 196 ft
19. (a) Cosine (b) Linear
21. (a)

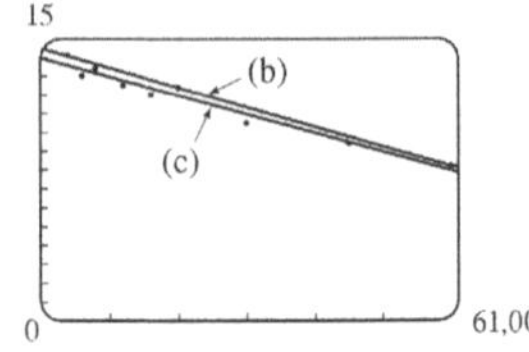

Linear model is appropriate.

(b) $y = -0.000105x + 14.521$
(c) $y = -0.00009979x + 13.951$

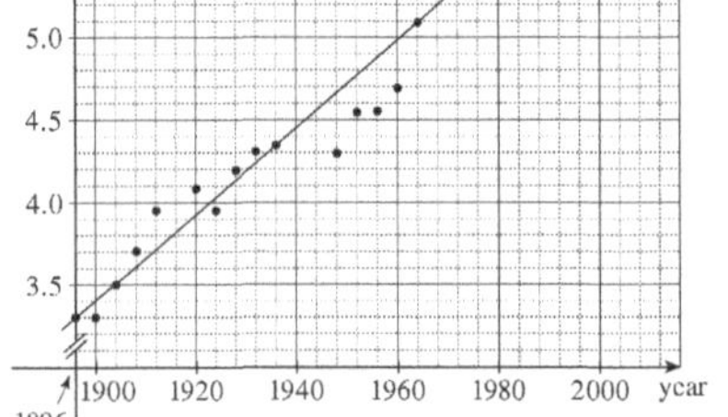

(d) About 11.5 per 100 population (e) About 6% (f) No
23. (a)

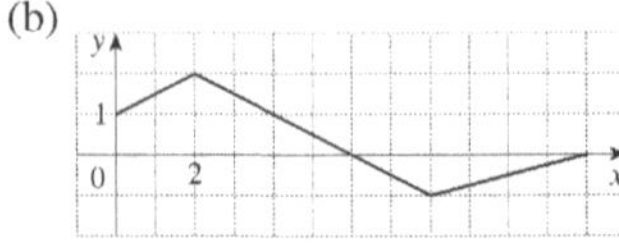

Linear model is appropriate.

(b) $y = 0.0265x - 46.8759$ (c) 6.27 m; higher (d) No
25. Four times as bright
27. (a) $N = 3.1046A^{0.308}$ (b) 18

EXERCISES 1.3 ■ PAGE 42

1. (a) $y = f(x) + 3$ (b) $y = f(x) - 3$ (c) $y = f(x - 3)$
(d) $y = f(x + 3)$ (e) $y = -f(x)$ (f) $y = f(-x)$
(g) $y = 3f(x)$ (h) $y = \frac{1}{3}f(x)$
3. (a) 3 (b) 1 (c) 4 (d) 5 (e) 2
5. (a)

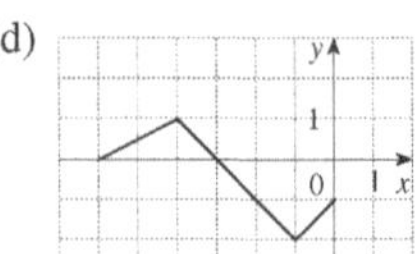

(b) (c) (d)

7. $y = -\sqrt{-x^2 - 5x - 4} - 1$

9.

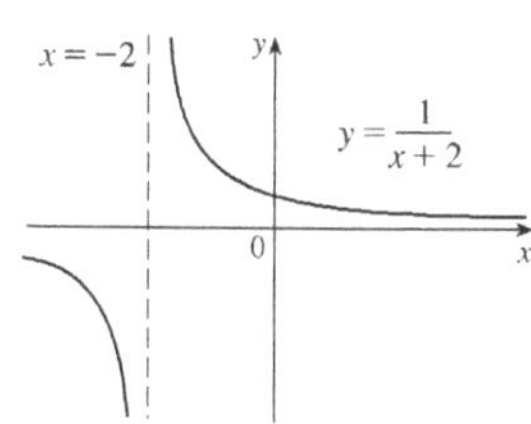

11.

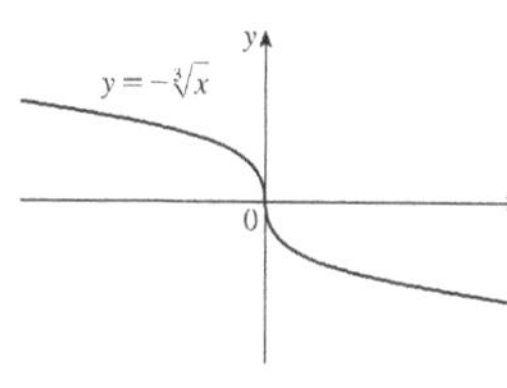

13.

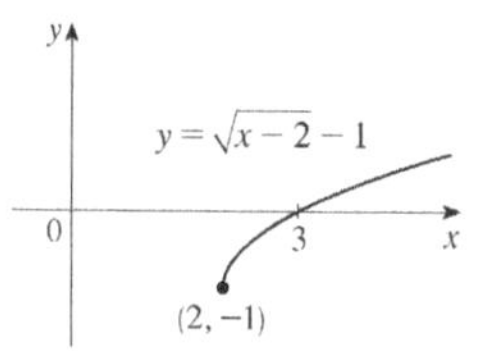

15.

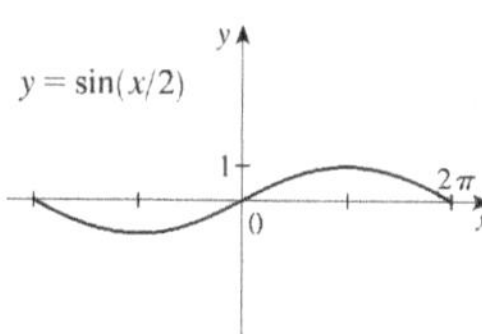

17.

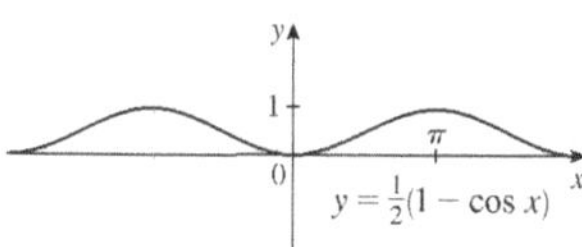

19.

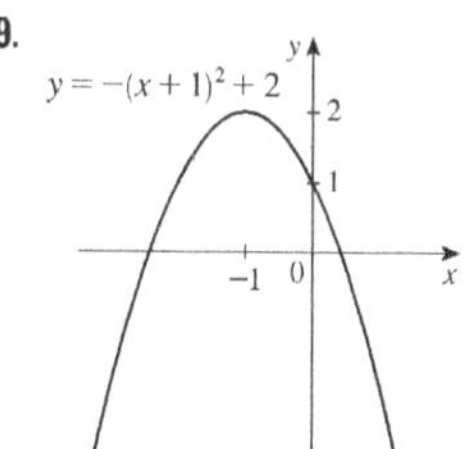

21.

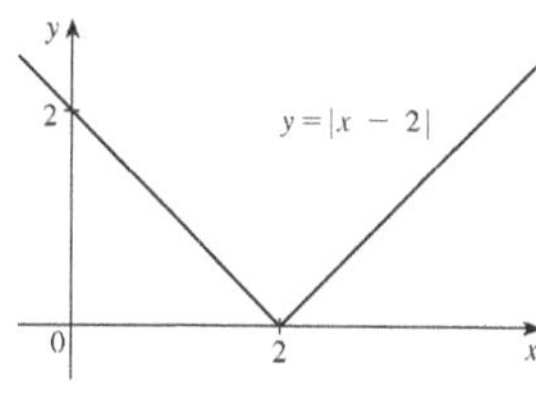

23.

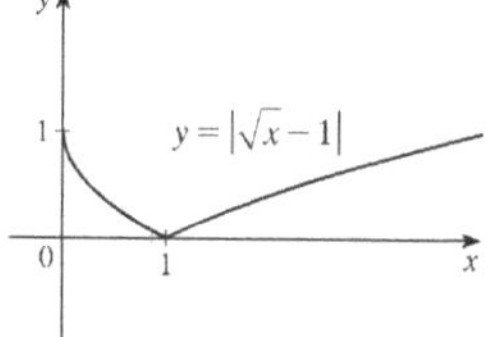

25. $L(t) = 12 + 2\sin\left[\dfrac{2\pi}{365}(t - 80)\right]$

27. (a) The portion of the graph of $y = f(x)$ to the right of the y-axis is reflected about the y-axis.

(b)

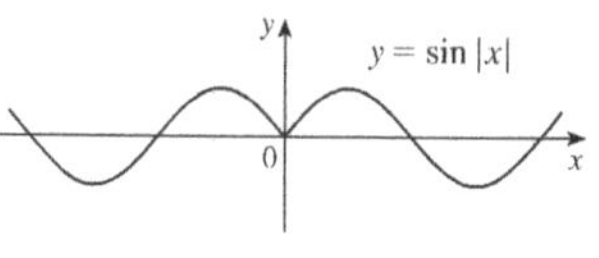

(c)

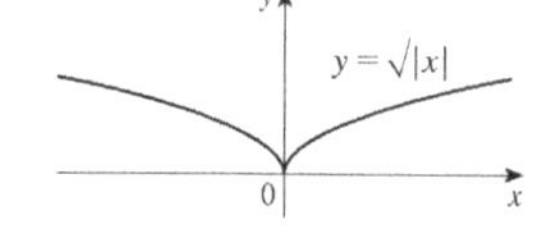

29. (a) $(f + g)(x) = x^3 + 5x^2 - 1$, $(-\infty, \infty)$
(b) $(f - g)(x) = x^3 - x^2 + 1$, $(-\infty, \infty)$
(c) $(fg)(x) = 3x^5 + 6x^4 - x^3 - 2x^2$, $(-\infty, \infty)$
(d) $(f/g)(x) = \dfrac{x^3 + 2x^2}{3x^2 - 1}$, $\{x \mid x \neq \pm 1/\sqrt{3}\}$

31. (a) $(f \circ g)(x) = 4x^2 + 4x$, $(-\infty, \infty)$
(b) $(g \circ f)(x) = 2x^2 - 1$, $(-\infty, \infty)$
(c) $(f \circ f)(x) = x^4 - 2x^2$, $(-\infty, \infty)$
(d) $(g \circ g)(x) = 4x + 3$, $(-\infty, \infty)$

33. (a) $(f \circ g)(x) = 1 - 3\cos x$, $(-\infty, \infty)$
(b) $(g \circ f)(x) = \cos(1 - 3x)$, $(-\infty, \infty)$
(c) $(f \circ f)(x) = 9x - 2$, $(-\infty, \infty)$
(d) $(g \circ g)(x) = \cos(\cos x)$, $(-\infty, \infty)$

35. (a) $(f \circ g)(x) = \dfrac{2x^2 + 6x + 5}{(x + 2)(x + 1)}$, $\{x \mid x \neq -2, -1\}$
(b) $(g \circ f)(x) = \dfrac{x^2 + x + 1}{(x + 1)^2}$, $\{x \mid x \neq -1, 0\}$
(c) $(f \circ f)(x) = \dfrac{x^4 + 3x^2 + 1}{x(x^2 + 1)}$, $\{x \mid x \neq 0\}$
(d) $(g \circ g)(x) = \dfrac{2x + 3}{3x + 5}$, $\{x \mid x \neq -2, -\frac{5}{3}\}$

37. $(f \circ g \circ h)(x) = 3\sin(x^2) - 2$

39. $(f \circ g \circ h)(x) = \sqrt{x^6 + 4x^3 + 1}$

41. $g(x) = 2x + x^2$, $f(x) = x^4$

43. $g(x) = \sqrt[3]{x}$, $f(x) = x/(1 + x)$

45. $g(t) = t^2$, $f(t) = \sec t \tan t$

47. $h(x) = \sqrt{x}$, $g(x) = x - 1$, $f(x) = \sqrt{x}$

49. $h(x) = \sqrt{x}$, $g(x) = \sec x$, $f(x) = x^4$

51. (a) 4 (b) 3 (c) 0 (d) Does not exist; $f(6) = 6$ is not in the domain of g. (e) 4 (f) −2

53. (a) $r(t) = 60t$ (b) $(A \circ r)(t) = 3600\pi t^2$; the area of the circle as a function of time

55. (a) $s = \sqrt{d^2 + 36}$ (b) $d = 30t$
(c) $(f \circ g)(t) = \sqrt{900t^2 + 36}$; the distance between the lighthouse and the ship as a function of the time elapsed since noon

57. (a) (b)

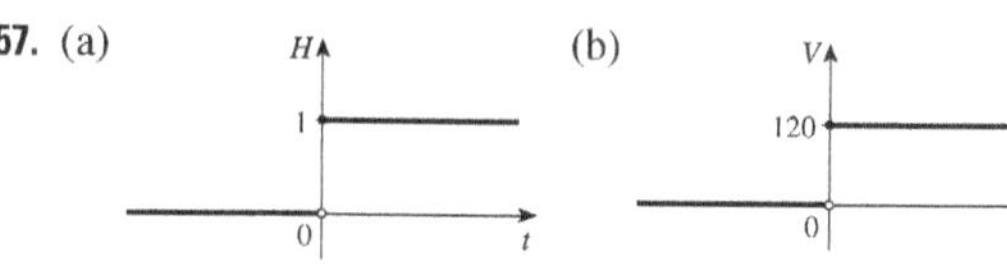

$V(t) = 120H(t)$

(c)

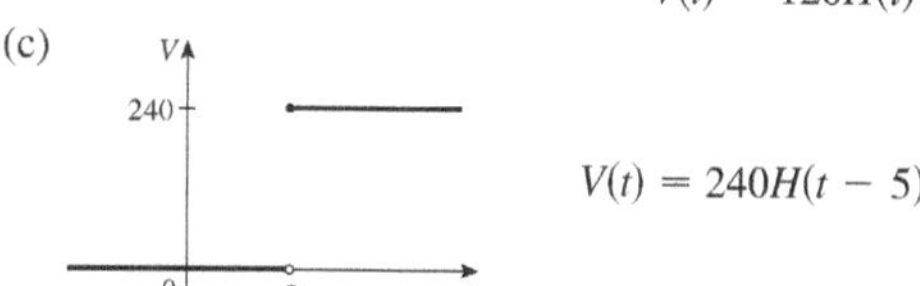

$V(t) = 240H(t - 5)$

59. Yes; m_1m_2

61. (a) $f(x) = x^2 + 6$ (b) $g(x) = x^2 + x - 1$

63. Yes

EXERCISES 1.4 ■ PAGE 50

1. (c)

3.

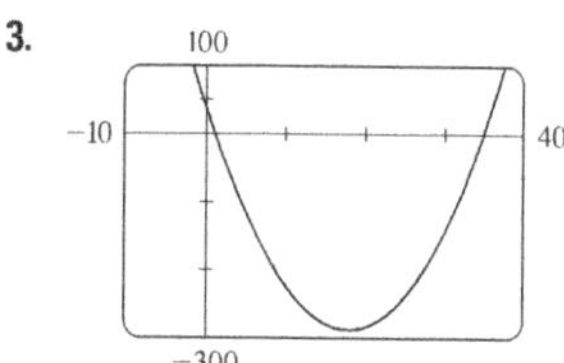

5.

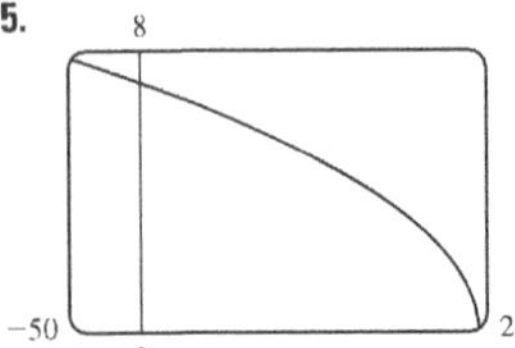

7.

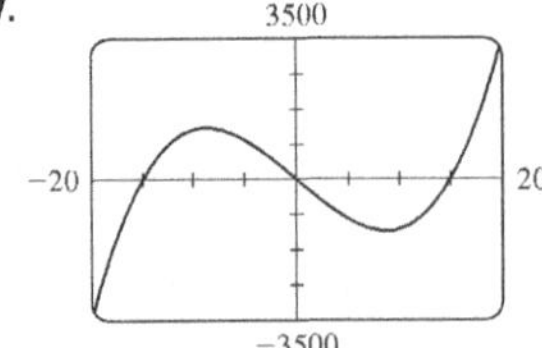

9.

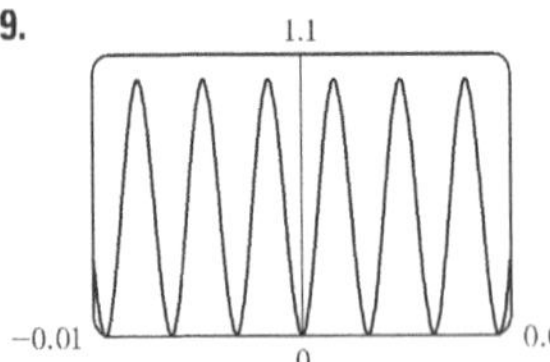

11.

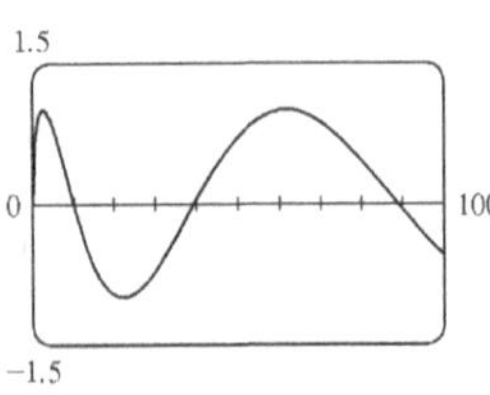

13.

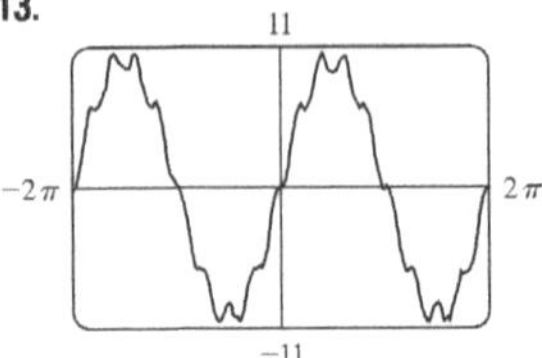

2
$-\frac{\pi}{25}$
$\frac{\pi}{25}$
−2

15. (b) Yes; two are needed

17.

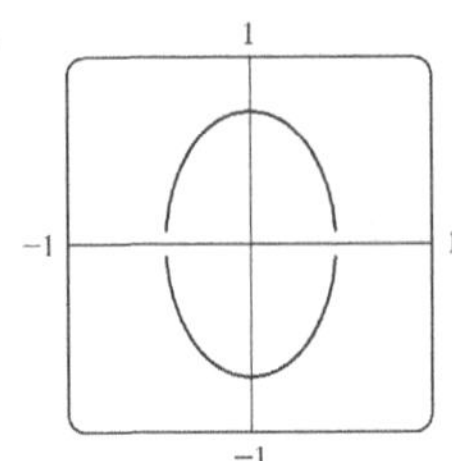

19. No **21.** −0.72, 1.22 **23.** 0.65 **25.** g

27. $-0.31 < x < 0.31$

29. (a)

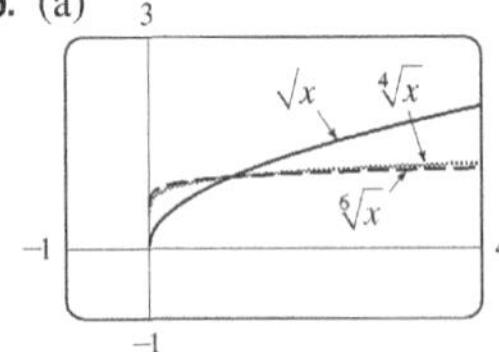

(b)

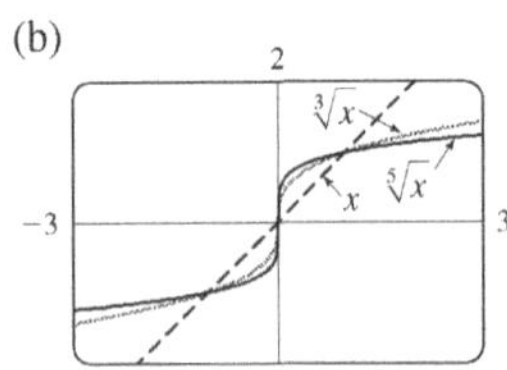

(c)

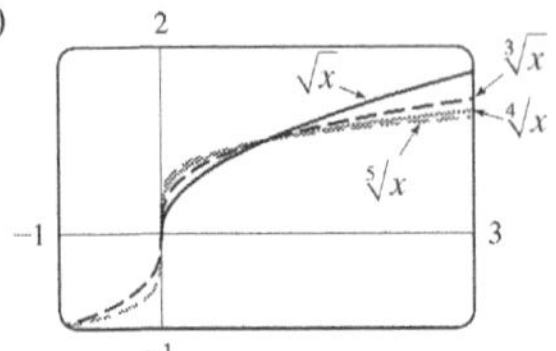

(d) Graphs of even roots are similar to $\sqrt{x}$, graphs of odd roots are similar to $\sqrt[3]{x}$. As n increases, the graph of $y = \sqrt[n]{x}$ becomes steeper near 0 and flatter for $x > 1$.

31.

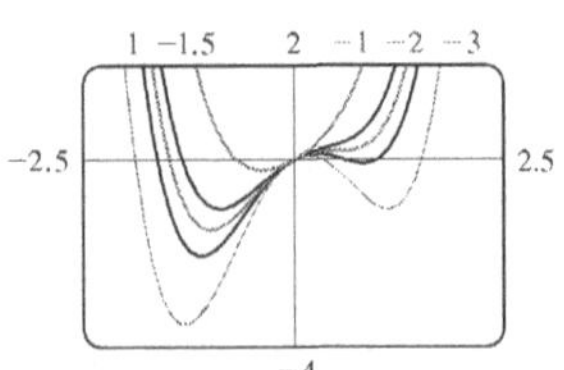

If $c < -1.5$, the graph has three humps: two minimum points and a maximum point. These humps get flatter as c increases until at $c = -1.5$ two of the humps disappear and there is only one minimum point. This single hump then moves to the right and approaches the origin as c increases.

33. The hump gets larger and moves to the right.

35. If $c < 0$, the loop is to the right of the origin; if $c > 0$, the loop is to the left. The closer c is to 0, the larger the loop.

EXERCISES 1.5 ■ PAGE 57

1. (a) 4 (b) $x^{-4/3}$

3. (a) $16b^{12}$ (b) $648y^7$

5. (a) $f(x) = a^x$, $a > 0$ (b) $\mathbb{R}$ (c) $(0, \infty)$

(d) See Figures 4(c), 4(b), and 4(a), respectively.

7. All approach 0 as $x \to -\infty$, all pass through (0, 1), and all are increasing. The larger the base, the faster the rate of increase.

5 $y = 20^x$ $y = 5^x$ $y = e^x$
$y = 2^x$
−1
2
0

9. $y = \left(\frac{1}{3}\right)^x$ $y = \left(\frac{1}{10}\right)^x$ 5 $y = 10^x$ $y = 3^x$

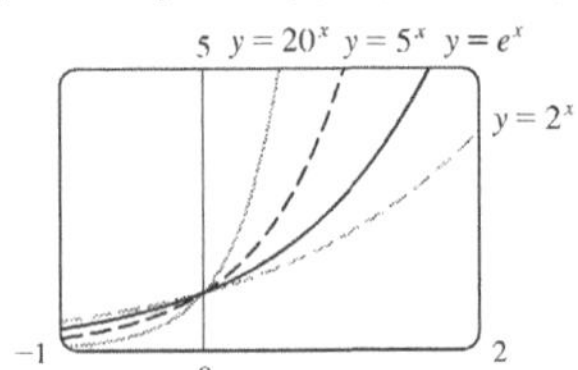

The functions with base greater than 1 are increasing and those with base less than 1 are decreasing. The latter are reflections of the former about the y-axis.

11.

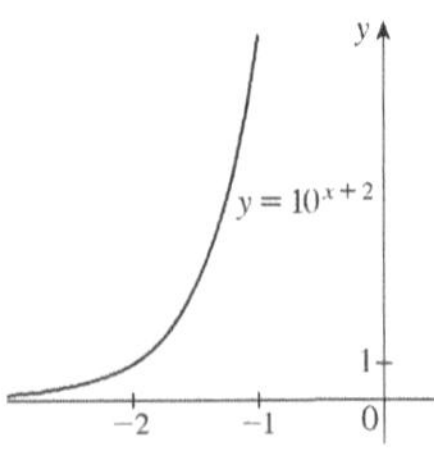

13.

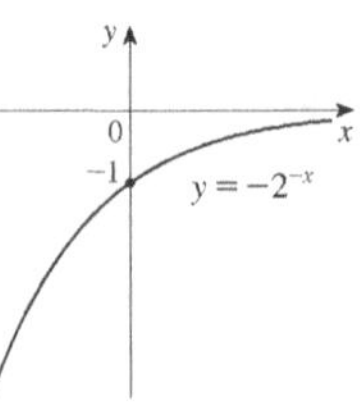

15.

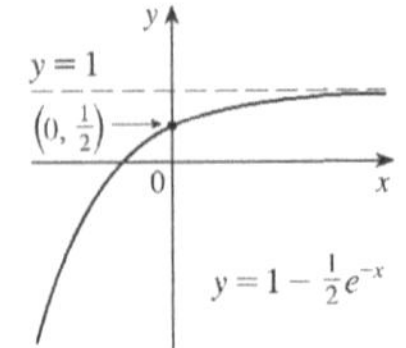

17. (a) $y = e^x - 2$ (b) $y = e^{x-2}$ (c) $y = -e^x$ (d) $y = e^{-x}$ (e) $y = -e^{-x}$

19. (a) $(-\infty, -1) \cup (-1, 1) \cup (1, \infty)$ (b) $(-\infty, \infty)$
21. $f(x) = 3 \cdot 2^x$ **27.** At $x \approx 35.8$
29. (a) 3200 (b) $100 \cdot 2^{t/3}$ (c) 10,159
(d) $t \approx 26.9$ h

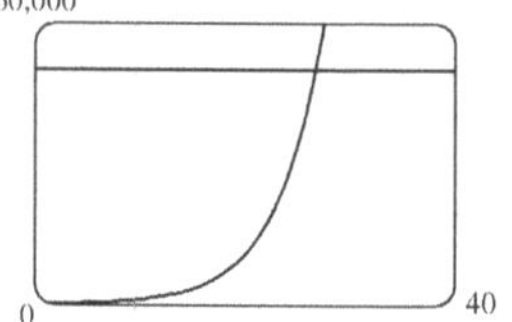

31. $P = 2614.086(1.01693)^t$; 5381 million; 8466 million

EXERCISES 1.6 ■ PAGE 69

1. (a) See Definition 1.
(b) It must pass the Horizontal Line Test.
3. No **5.** No **7.** Yes **9.** No **11.** Yes **13.** No
15. (a) 6 (b) 3 **17.** 0
19. $F = \frac{9}{5}C + 32$; the Fahrenheit temperature as a function of the Celsius temperature; $[-273.15, \infty)$
21. $y = \frac{1}{3}(x-1)^2 - \frac{2}{3}, x \geqslant 1$
23. $y = \frac{1}{2}(1 + \ln x)$ **25.** $y = e^x - 3$
27. $f^{-1}(x) = \sqrt[4]{x-1}$ **29.**

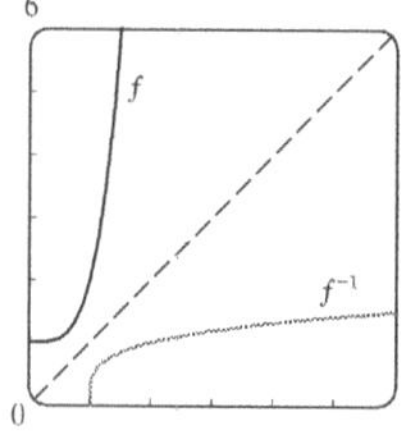

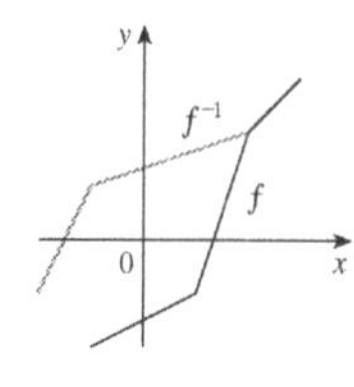

31. (a) $f^{-1}(x) = \sqrt{1 - x^2}, 0 \leqslant x \leqslant 1$; f^{-1} and f are the same function. (b) Quarter-circle in the first quadrant
33. (a) It's defined as the inverse of the exponential function with base a, that is, $\log_a x = y \iff a^y = x$.
(b) $(0, \infty)$ (c) $\mathbb{R}$ (d) See Figure 11.
35. (a) 3 (b) -3 **37.** (a) 3 (b) -2 **39.** ln 1215
41. $\ln \dfrac{\sqrt{x}}{x+1}$
43.

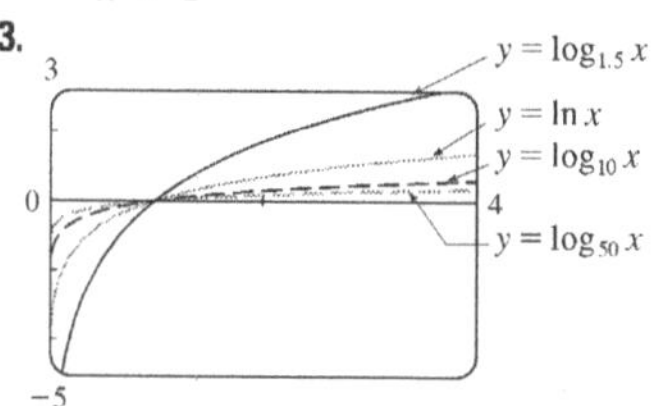

All graphs approach $-\infty$ as $x \to 0^+$, all pass through (1, 0), and all are increasing. The larger the base, the slower the rate of increase.

45. About 1,084,588 mi
47. (a)

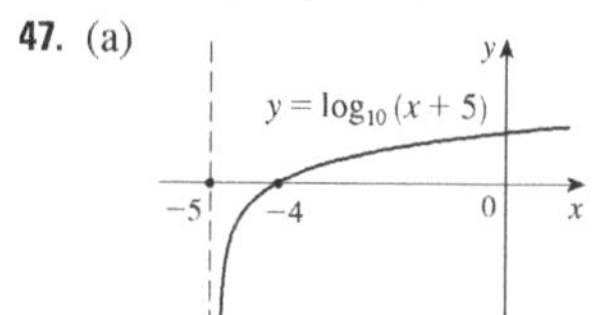

(b)

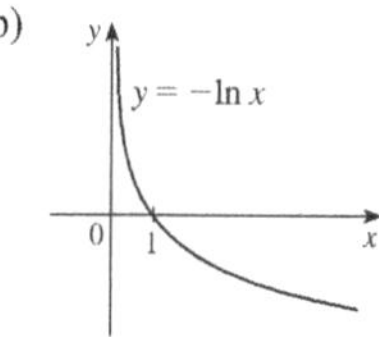

49. (a) $(0, \infty)$; $(-\infty, \infty)$ (b) e^{-2}
(c)

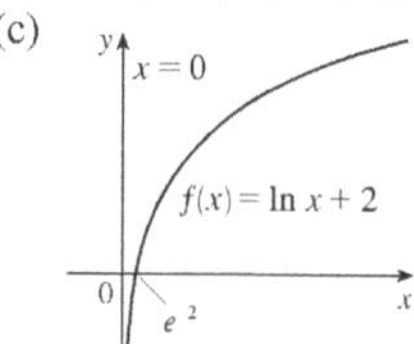

51. (a) $\frac{1}{4}(7 - \ln 6)$ (b) $\frac{1}{3}(e^2 + 10)$
53. (a) $5 + \log_2 3$ or $5 + (\ln 3)/\ln 2$ (b) $\frac{1}{2}(1 + \sqrt{1 + 4e})$
55. (a) $0 < x < 1$ (b) $x > \ln 5$
57. (a) $(\ln 3, \infty)$ (b) $f^{-1}(x) = \ln(e^x + 3)$; $\mathbb{R}$
59.

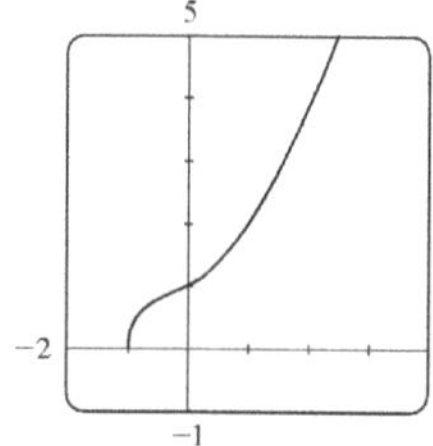

The graph passes the Horizontal Line Test.

$f^{-1}(x) = -\frac{1}{6}\sqrt[3]{4}\left(\sqrt[3]{D - 27x^2 + 20} - \sqrt[3]{D + 27x^2 - 20} + \sqrt[3]{2}\right)$, where $D = 3\sqrt{3}\sqrt{27x^4 - 40x^2 + 16}$; two of the expressions are complex.
61. (a) $f^{-1}(n) = (3/\ln 2)\ln(n/100)$; the time elapsed when there are n bacteria (b) After about 26.9 hours
63. (a) $\pi/3$ (b) π **65.** (a) $\pi/4$ (b) $\pi/4$
67. (a) 10 (b) $\pi/3$
71. $x/\sqrt{1 + x^2}$
73.

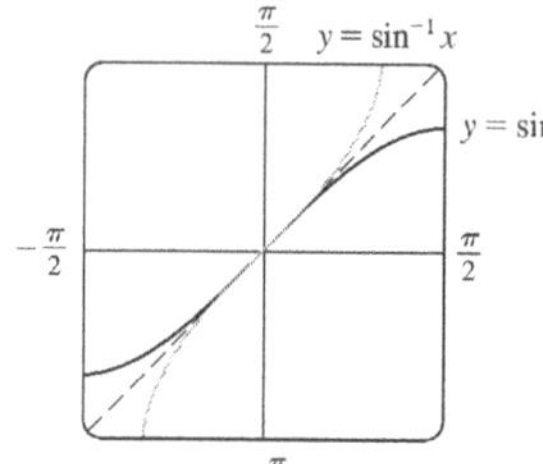

The second graph is the reflection of the first graph about the line $y = x$.

75. (a) $[-\frac{2}{3}, 0]$ (b) $[-\pi/2, \pi/2]$
77. (a) $g^{-1}(x) = f^{-1}(x) - c$ (b) $h^{-1}(x) = (1/c)f^{-1}(x)$

CHAPTER 1 REVIEW ■ PAGE 72

True-False Quiz

1. False **3.** False **5.** True **7.** False **9.** True
11. False **13.** False

Exercises

1. (a) 2.7 (b) 2.3, 5.6 (c) $[-6, 6]$ (d) $[-4, 4]$
(e) $[-4, 4]$ (f) No; it fails the Horizontal Line Test.
(g) Odd; its graph is symmetric about the origin.
3. $2a + h - 2$ **5.** $(-\infty, \frac{1}{3}) \cup (\frac{1}{3}, \infty)$, $(-\infty, 0) \cup (0, \infty)$
7. $(-6, \infty)$, $\mathbb{R}$

9. (a) Shift the graph 8 units upward.
(b) Shift the graph 8 units to the left.
(c) Stretch the graph vertically by a factor of 2, then shift it 1 unit upward.
(d) Shift the graph 2 units to the right and 2 units downward.
(e) Reflect the graph about the x-axis.
(f) Reflect the graph about the line $y = x$ (assuming f is one-to-one).

11.

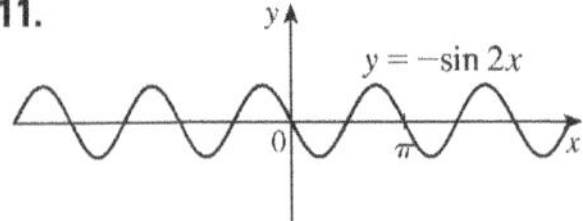

13.

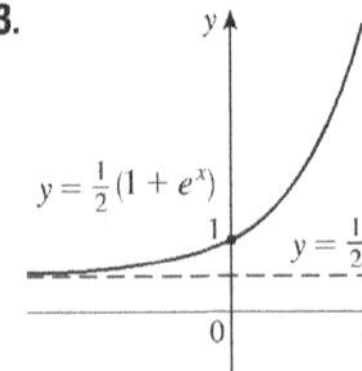

15.

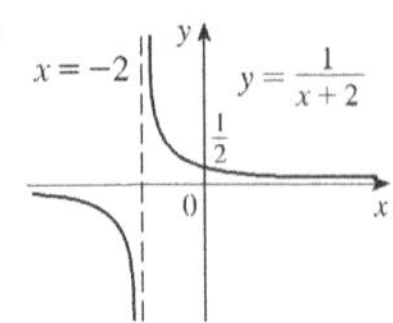

17. (a) Neither (b) Odd (c) Even (d) Neither
19. (a) $(f \circ g)(x) = \ln(x^2 - 9)$, $(-\infty, -3) \cup (3, \infty)$
(b) $(g \circ f)(x) = (\ln x)^2 - 9$, $(0, \infty)$
(c) $(f \circ f)(x) = \ln \ln x$, $(1, \infty)$
(d) $(g \circ g)(x) = (x^2 - 9)^2 - 9$, $(-\infty, \infty)$
21. $y = 0.2493x - 423.4818$; about 77.6 years
23. 1 **25.** (a) 9 (b) 2 (c) $1/\sqrt{3}$ (d) $\frac{3}{5}$
27. (a) ≈ 4.4 years

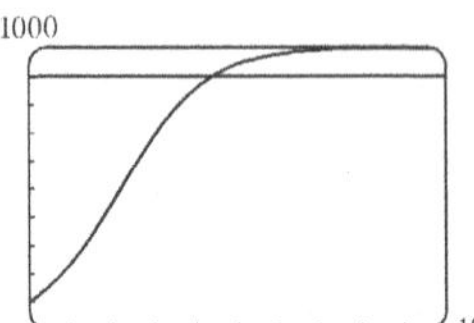

(b) $t = -\ln\left(\dfrac{1000 - P}{9P}\right)$; the time required for the population to reach a given number P.
(c) $\ln 81 \approx 4.4$ years

PRINCIPLES OF PROBLEM SOLVING ■ PAGE 80

1. $a = 4\sqrt{h^2 - 16}/h$, where a is the length of the altitude and h is the length of the hypotenuse
3. $-\frac{7}{3}, 9$
5.

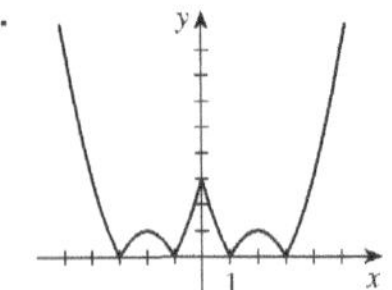

7.

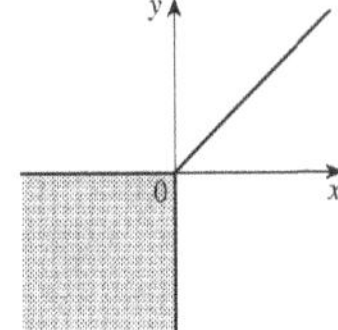

9. (a)

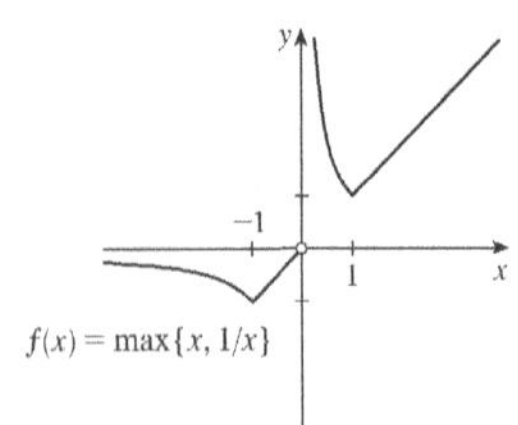

(b)

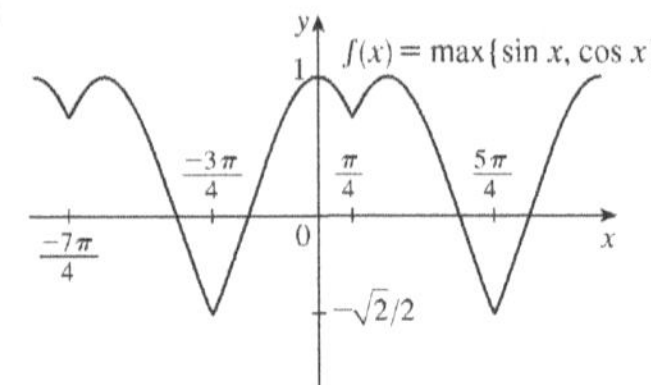

(c)

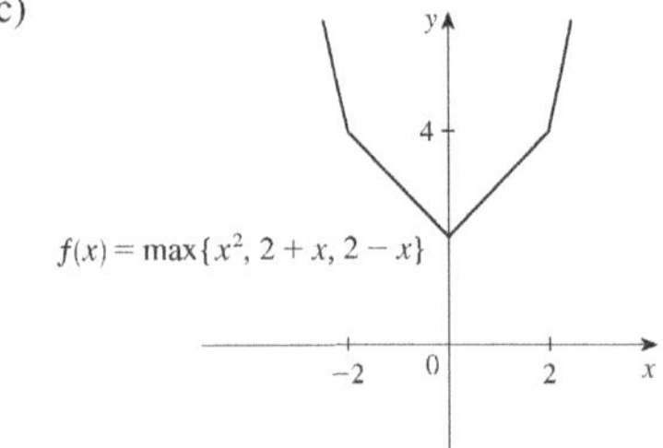

11. 5 **13.** $x \in [-1, 1 - \sqrt{3}) \cup (1 + \sqrt{3}, 3]$
15. 40 mi/h **19.** $f_n(x) = x^{2^{n+1}}$

CHAPTER 2

EXERCISES 2.1 ■ PAGE 86

1. (a) $-44.4, -38.8, -27.8, -22.2, -16.\overline{6}$
(b) -33.3 (c) $-33\frac{1}{3}$
3. (a)(i) 2 (ii) 1.111111 (iii) 1.010101 (iv) 1.001001
(v) 0.666667 (vi) 0.909091 (vii) 0.990099
(viii) 0.999001 (b) 1 (c) $y = x - 3$
5. (a) (i) -32 ft/s (ii) -25.6 ft/s (iii) -24.8 ft/s
(iv) -24.16 ft/s (b) -24 ft/s
7. (a) (i) 4.65 m/s (ii) 5.6 m/s (iii) 7.55 m/s
(iv) 7 m/s (b) 6.3 m/s
9. (a) 0, 1.7321, -1.0847, -2.7433, 4.3301, -2.8173, 0, -2.1651, -2.6061, -5, 3.4202; no (c) -31.4

EXERCISES 2.2 ■ PAGE 96

1. Yes
3. (a) $\lim_{x \to -3} f(x) = \infty$ means that the values of $f(x)$ can be made arbitrarily large (as large as we please) by taking x sufficiently close to -3 (but not equal to -3).

(b) $\lim_{x \to 4^+} f(x) = -\infty$ means that the values of $f(x)$ can be made arbitrarily large negative by taking x sufficiently close to 4 through values larger than 4.

5. (a) 2 (b) 1 (c) 4 (d) Does not exist (e) 3

7. (a) -1 (b) -2 (c) Does not exist (d) 2 (e) 0 (f) Does not exist (g) 1 (h) 3

9. (a) $-\infty$ (b) ∞ (c) ∞ (d) $-\infty$ (e) ∞ (f) $x = -7, x = -3, x = 0, x = 6$

11. $\lim_{x \to a} f(x)$ exists for all a except $a = -1$.

13. (a) 1 (b) 0 (c) Does not exist

15.

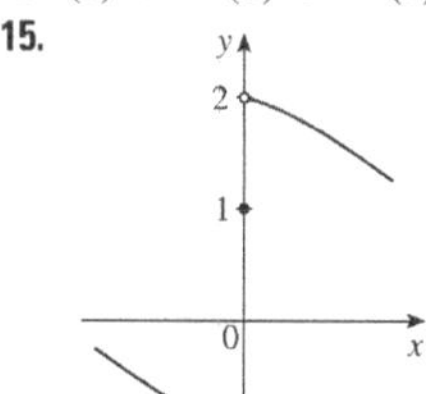

17.

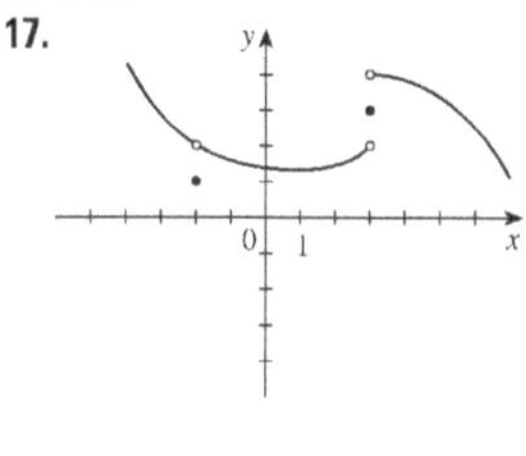

19. $\frac{2}{3}$ **21.** 5 **23.** $\frac{1}{4}$ **25.** $\frac{3}{5}$ **27.** (a) -1.5

29. $-\infty$ **31.** ∞ **33.** $-\infty$ **35.** $-\infty$ **37.** ∞

39. $-\infty$; ∞

41. (a) 2.71828 (b)

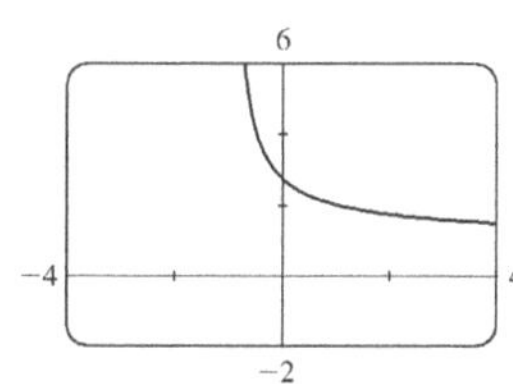

43. (a) 0.998000, 0.638259, 0.358484, 0.158680, 0.038851, 0.008928, 0.001465; 0

(b) 0.000572, -0.000614, -0.000907, -0.000978, -0.000993, -0.001000; -0.001

45. No matter how many times we zoom in toward the origin, the graph appears to consist of almost-vertical lines. This indicates more and more frequent oscillations as $x \to 0$.

47. $x \approx \pm 0.90, \pm 2.24$; $x = \pm \sin^{-1}(\pi/4), \pm(\pi - \sin^{-1}(\pi/4))$

EXERCISES 2.3 ■ PAGE 106

1. (a) -6 (b) -8 (c) 2 (d) -6 (e) Does not exist (f) 0

3. 105 **5.** $\frac{7}{8}$ **7.** 390 **9.** $\frac{3}{2}$ **11.** 4

13. Does not exist **15.** $\frac{6}{5}$ **17.** -10 **19.** $\frac{1}{12}$

21. $\frac{1}{6}$ **23.** $-\frac{1}{16}$ **25.** 1 **27.** $\frac{1}{128}$ **29.** $-\frac{1}{2}$

31. $3x^2$ **33.** (a), (b) $\frac{2}{3}$ **37.** 7 **41.** 6 **43.** -4

45. Does not exist

47. (a)

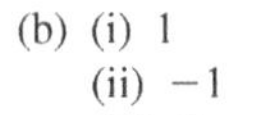

(b) (i) 1
(ii) -1
(iii) Does not exist
(iv) 1

49. (a) (i) 5 (ii) -5 (b) Does not exist
(c)

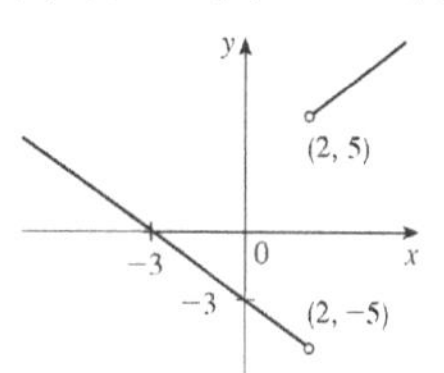

51. (a) (i) -2 (ii) Does not exist (iii) -3
(b) (i) $n - 1$ (ii) n (c) a is not an integer.

57. 8 **63.** 15; -1

EXERCISES 2.4 ■ PAGE 116

1. 0.1 (or any smaller positive number)

3. 1.44 (or any smaller positive number)

5. 0.0906 (or any smaller positive number)

7. 0.011 (or any smaller positive number)

9. (a) 0.031 (b) 0.010

11. (a) $\sqrt{1000/\pi}$ cm (b) Within approximately 0.0445 cm
(c) Radius; area; $\sqrt{1000/\pi}$; 1000; 5; ≈ 0.0445

13. (a) 0.025 (b) 0.0025

35. (a) 0.093 (b) $\delta = (B^{2/3} - 12)/(6B^{1/3}) - 1$, where $B = 216 + 108\varepsilon + 12\sqrt{336 + 324\varepsilon + 81\varepsilon^2}$

41. Within 0.1

EXERCISES 2.5 ■ PAGE 127

1. $\lim_{x \to 4} f(x) = f(4)$

3. (a) $f(-4)$ is not defined and $\lim_{x \to a} f(x)$ [for $a = -2$, 2, and 4] does not exist
(b) -4, neither; -2, left; 2, right; 4, right

5.

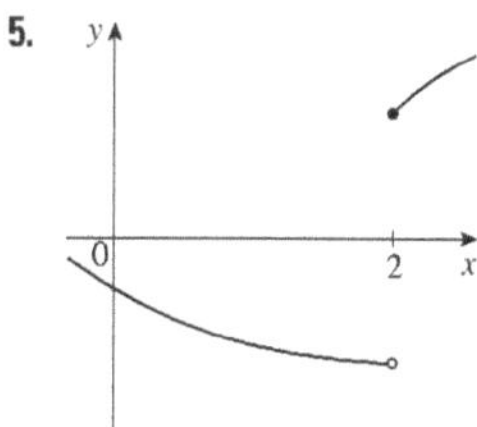

7.

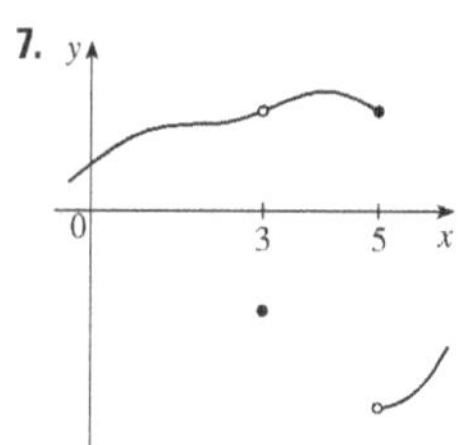

9. (a)

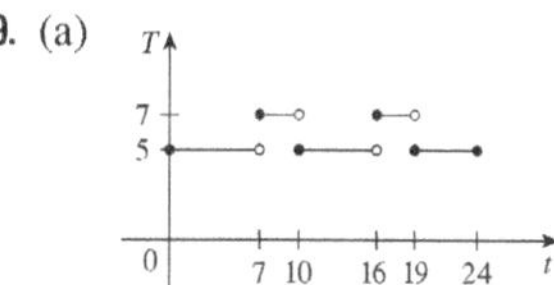

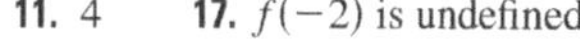

11. 4 **17.** $f(-2)$ is undefined.

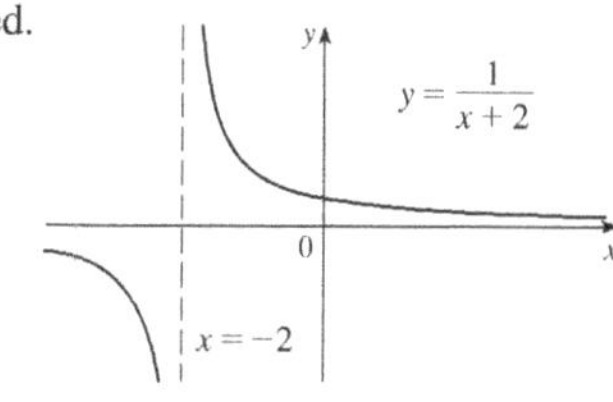

19. $\lim_{x\to 0} f(x)$ does not exist. **21.** $\lim_{x\to 0} f(x) \neq f(0)$

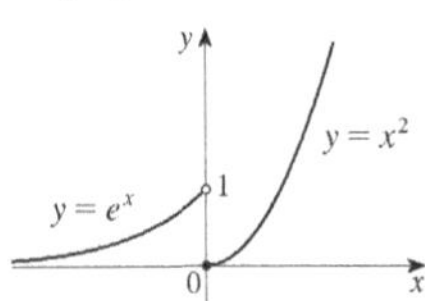

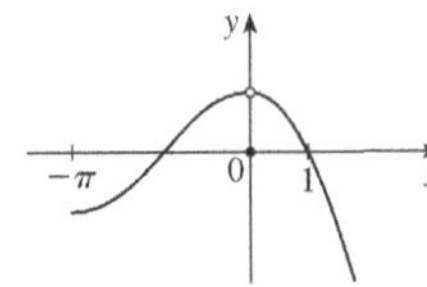

23. Define $f(2) = 3$. **25.** $(-\infty, \infty)$

27. $(-\infty, \sqrt[3]{2}) \cup (\sqrt[3]{2}, \infty)$ **29.** $[-1, 0]$

31. $(-\infty, -1] \cup (0, \infty)$

33. $x = 0$

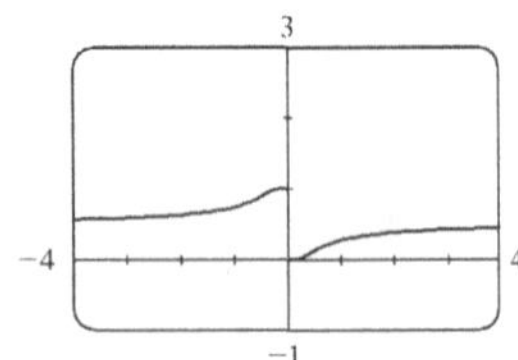

35. $\frac{7}{3}$ **37.** 1

41. 0, left **43.** 0, right; 1, left

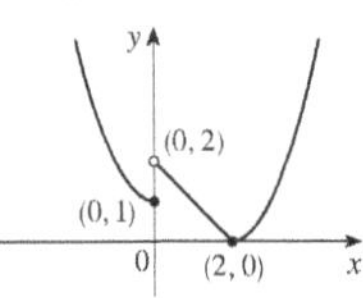

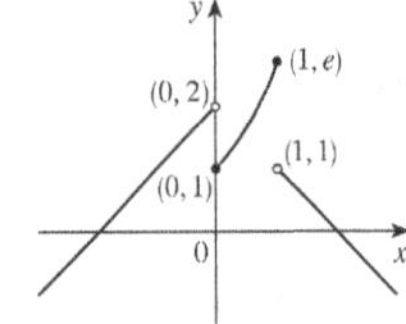

45. $\frac{2}{3}$ **47.** (a) $g(x) = x^3 + x^2 + x + 1$ (b) $g(x) = x^2 + x$

55. (b) (0.86, 0.87) **57.** (b) 70.347 **63.** None

65. Yes

EXERCISES 2.6 ■ PAGE 140

1. (a) As x becomes large, $f(x)$ approaches 5.
(b) As x becomes large negative, $f(x)$ approaches 3.

3. (a) -2 (b) 2 (c) ∞ (d) $-\infty$
(e) $x = 1, x = 3, y = -2, y = 2$

5.

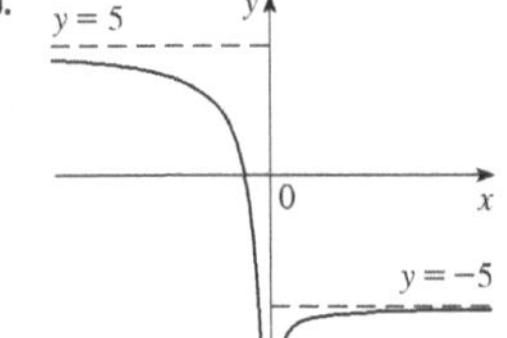

7.

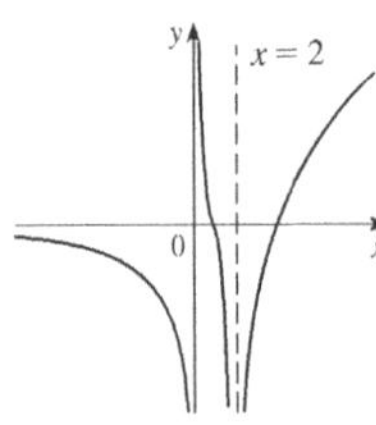

9.

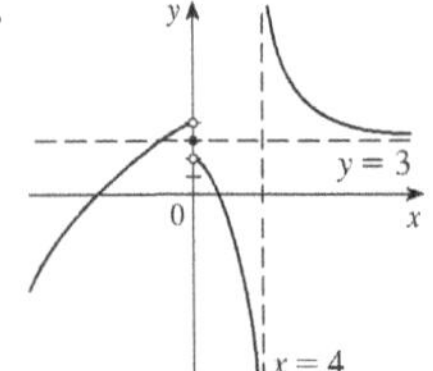

11. 0 **13.** $\frac{3}{2}$ **15.** $\frac{3}{2}$ **17.** 0 **19.** -1 **21.** 4

23. 3 **25.** $\frac{1}{6}$ **27.** $\frac{1}{2}(a - b)$ **29.** ∞

31. $-\infty$ **33.** $\pi/2$ **35.** $-\frac{1}{2}$ **37.** 0

39. (a), (b) $-\frac{1}{2}$ **41.** $y = 2; x = 2$

43. $y = 2; x = -2, x = 1$ **45.** $x = 5$ **47.** $y = 3$

49. $f(x) = \dfrac{2 - x}{x^2(x - 3)}$ **51.** (a) $\frac{5}{4}$ (b) 5

53. $-\infty, -\infty$ **55.** $-\infty, \infty$

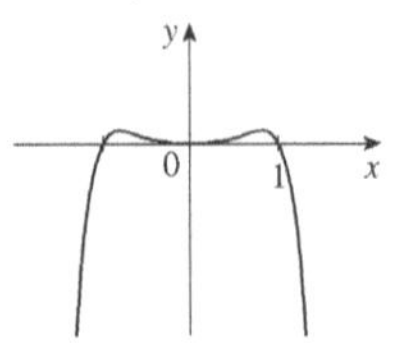

57. (a) 0 (b) An infinite number of times

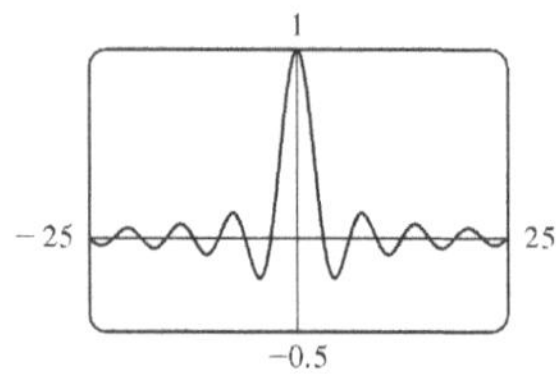

59. (a) 0 (b) $\pm\infty$ **61.** 5

63. (a) v^* (b) ≈ 0.47 s

65. $N \geqslant 15$ **67.** $N \leqslant -6, N \leqslant -22$

69. (a) $x > 100$

EXERCISES 2.7 ■ PAGE 150

1. (a) $\dfrac{f(x) - f(3)}{x - 3}$ (b) $\lim_{x\to 3} \dfrac{f(x) - f(3)}{x - 3}$

3. (a) 2 (b) $y = 2x + 1$ (c)

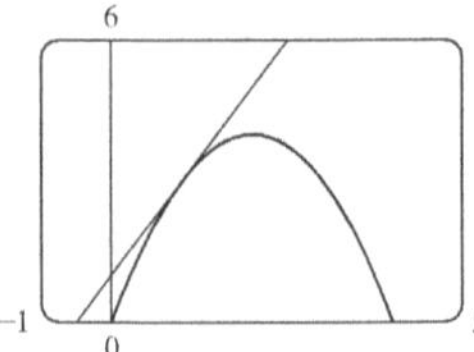

5. $y = -8x + 12$ **7.** $y = \frac{1}{2}x + \frac{1}{2}$

9. (a) $8a - 6a^2$ (b) $y = 2x + 3, y = -8x + 19$
(c)

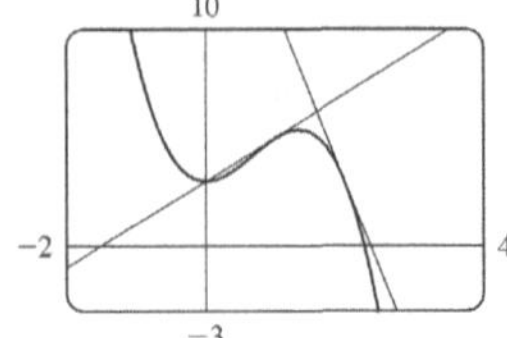

11. (a) Right: $0 < t < 1$ and $4 < t < 6$; left: $2 < t < 3$; standing still: $1 < t < 2$ and $3 < t < 4$

(b)

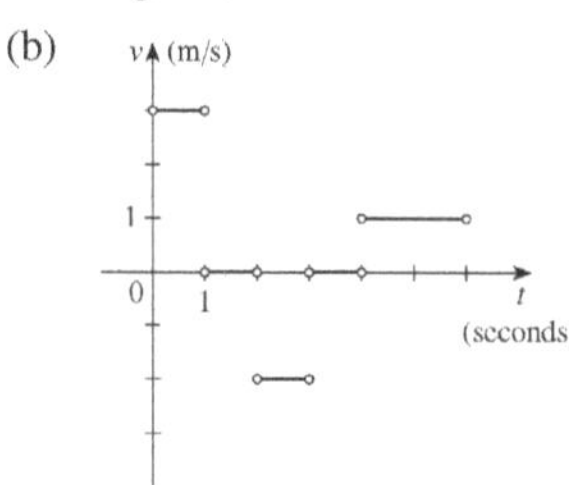

13. -24 ft/s

15. $-2/a^3$ m/s; -2 m/s; $-\frac{1}{4}$ m/s; $-\frac{2}{27}$ m/s

17. $g'(0)$, 0, $g'(4)$, $g'(2)$, $g'(-2)$

19. $f(2) = 3$; $f'(2) = 4$

21.

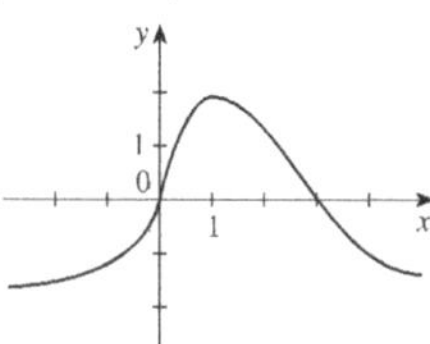

23. $y = 3x - 1$

25. (a) $-\frac{3}{5}$; $y = -\frac{3}{5}x + \frac{16}{5}$ (b)

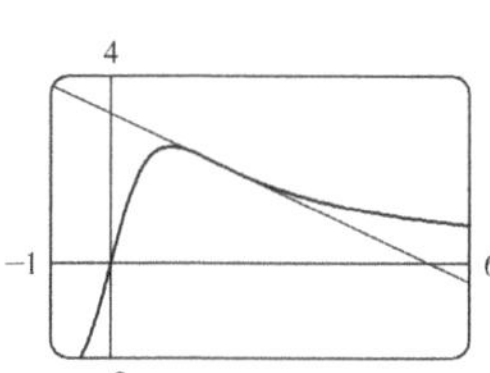

27. $6a - 4$ **29.** $\dfrac{5}{(a+3)^2}$ **31.** $-\dfrac{1}{\sqrt{1-2a}}$

33. $f(x) = x^{10}$, $a = 1$ or $f(x) = (1 + x)^{10}$, $a = 0$

35. $f(x) = 2^x$, $a = 5$

37. $f(x) = \cos x$, $a = \pi$ or $f(x) = \cos(\pi + x)$, $a = 0$

39. 1 m/s; 1 m/s

41.

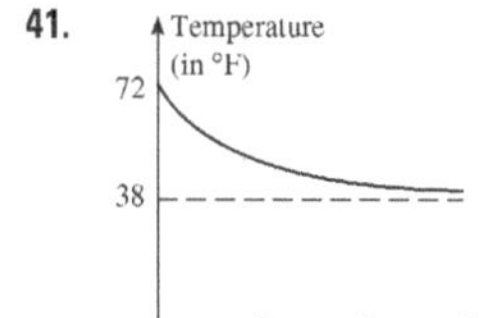

Greater (in magnitude)

43. (a) (i) 23 million/year (ii) 20.5 million/year (iii) 16 million/year
(b) 18.25 million/year (c) 17 million/year

45. (a) (i) \$20.25/unit (ii) \$20.05/unit (b) \$20/unit

47. (a) The rate at which the cost is changing per ounce of gold produced; dollars per ounce
(b) When the 800th ounce of gold is produced, the cost of production is \$17/oz.
(c) Decrease in the short term; increase in the long term

49. The rate at which the temperature is changing at 8:00 AM; 3.75°F/h

51. (a) The rate at which the oxygen solubility changes with respect to the water temperature; (mg/L)/°C
(b) $S'(16) \approx -0.25$; as the temperature increases past 16°C, the oxygen solubility is decreasing at a rate of 0.25 (mg/L)/°C.

53. Does not exist

EXERCISES 2.8 ■ PAGE 162

1. (a) -0.2 (b) 0 (c) 1 (d) 2
(e) 1 (f) 0 (g) -0.2

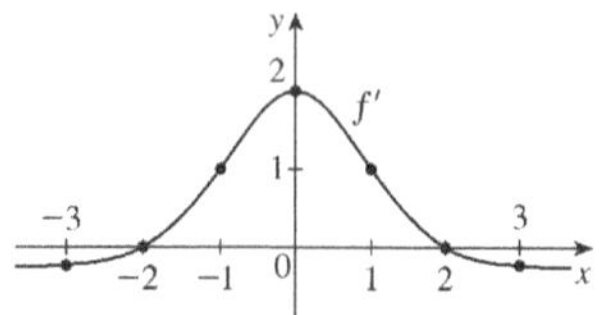

3. (a) II (b) IV (c) I (d) III

5.

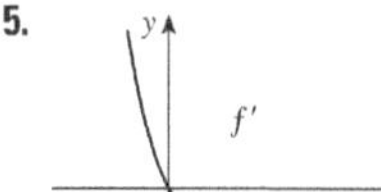

7.

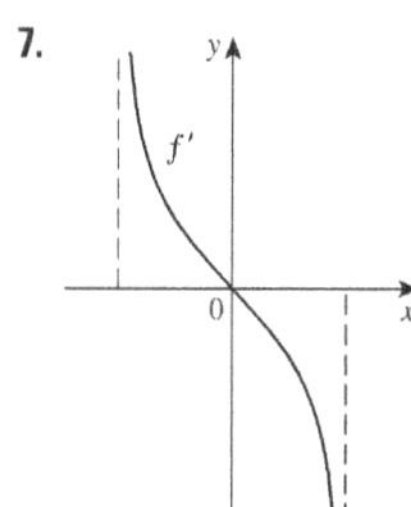

9.

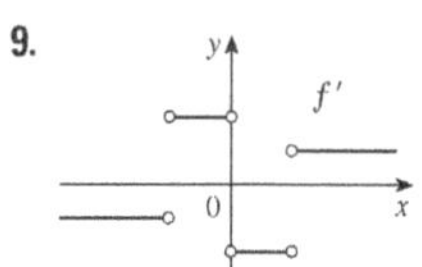

11.

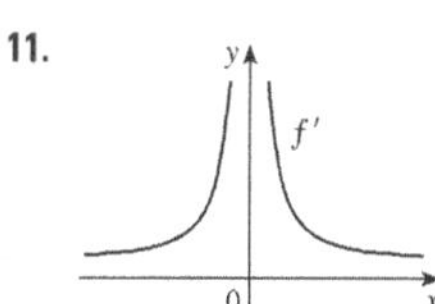

13. (a) The instantaneous rate of change of percentage of full capacity with respect to elapsed time in hours

(b)

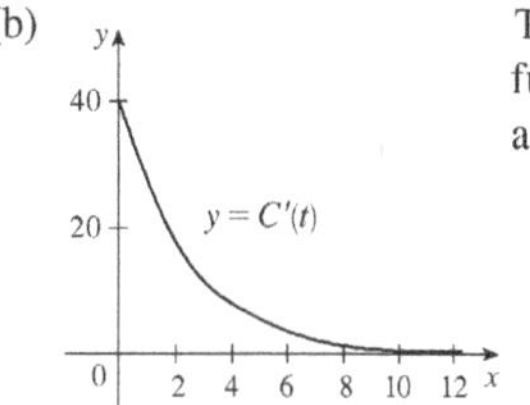

The rate of change of percentage of full capacity is decreasing and approaching 0.

15.

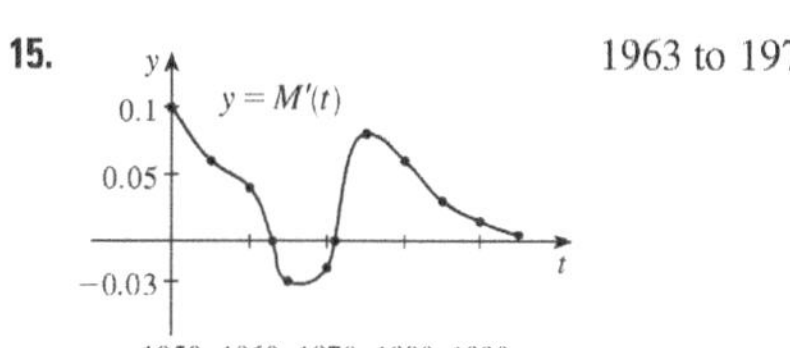

1963 to 1971

17. $f'(x) = e^x$

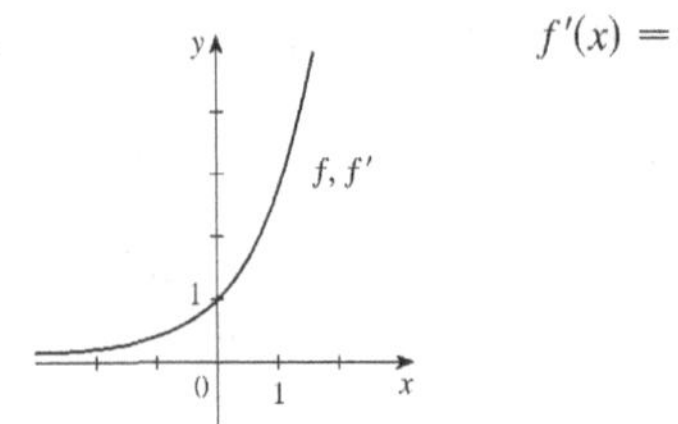

19. (a) 0, 1, 2, 4 (b) $-1, -2, -4$ (c) $f'(x) = 2x$

21. $f'(x) = \frac{1}{2}$, $\mathbb{R}$, $\mathbb{R}$ **23.** $f'(t) = 5 - 18t$, $\mathbb{R}$, $\mathbb{R}$

25. $f'(x) = 2x - 6x^2$, $\mathbb{R}$, $\mathbb{R}$

27. $g'(x) = \dfrac{-1}{2\sqrt{9-x}}$, $(-\infty, 9]$, $(-\infty, 9)$

29. $G'(t) = \dfrac{-7}{(3+t)^2}$, $(-\infty, -3) \cup (-3, \infty)$, $(-\infty, -3) \cup (-3, \infty)$

31. $f'(x) = 4x^3$, $\mathbb{R}$, $\mathbb{R}$ **33.** (a) $f'(x) = 4x^3 + 2$

35. (a) The rate at which the unemployment rate is changing, in percent unemployed per year

(b)

t	$U'(t)$	t	$U'(t)$
1999	−0.2	2004	−0.45
2000	0.25	2005	−0.45
2001	0.9	2006	−0.25
2002	0.65	2007	0.6
2003	−0.15	2008	1.2

37. -4 (corner); 0 (discontinuity)

39. -1 (vertical tangent); 4 (corner)

41.

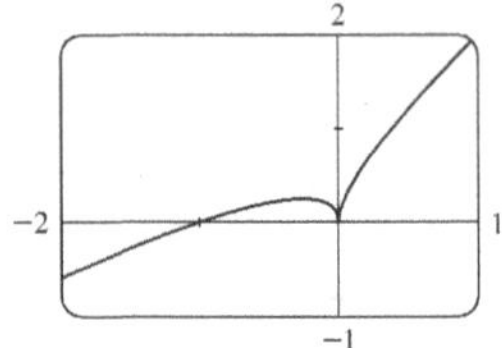

Differentiable at -1; not differentiable at 0

43. $a = f$, $b = f'$, $c = f''$

45. a = acceleration, b = velocity, c = position

47. $6x + 2$; 6

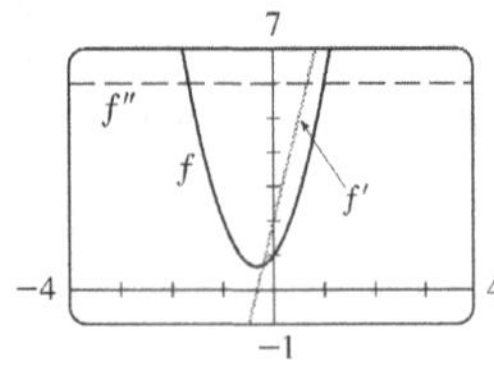

49.

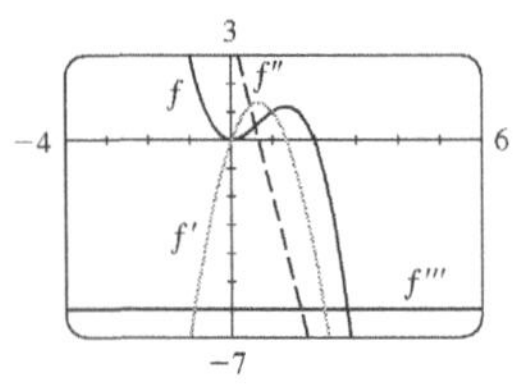

$f'(x) = 4x - 3x^2$,
$f''(x) = 4 - 6x$,
$f'''(x) = -6$,
$f^{(4)}(x) = 0$

51. (a) $\frac{1}{3}a^{-2/3}$

53. $f'(x) = \begin{cases} -1 & \text{if } x < 6 \\ 1 & \text{if } x > 6 \end{cases}$

or $f'(x) = \dfrac{x-6}{|x-6|}$

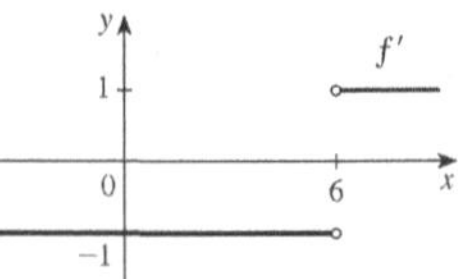

55. (a)

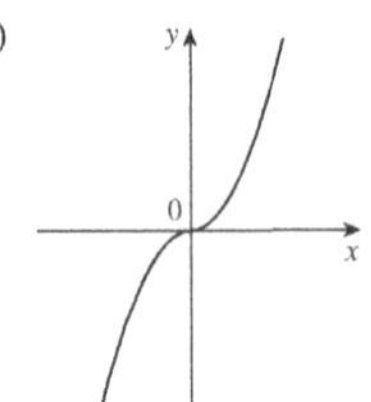

(b) All x
(c) $f'(x) = 2|x|$

59. 63°

CHAPTER 2 REVIEW ■ PAGE 166

True-False Quiz

1. False **3.** True **5.** False **7.** True **9.** True
11. True **13.** False **15.** True **17.** True **19.** False
21. False **23.** True

Exercises

1. (a) (i) 3 (ii) 0 (iii) Does not exist (iv) 2
(v) ∞ (vi) $-\infty$ (vii) 4 (viii) -1
(b) $y = 4$, $y = -1$ (c) $x = 0$, $x = 2$ (d) $-3, 0, 2, 4$

3. 1 **5.** $\frac{3}{2}$ **7.** 3 **9.** ∞ **11.** $\frac{4}{7}$ **13.** $\frac{1}{2}$

15. $-\infty$ **17.** 2 **19.** $\pi/2$ **21.** $x = 0$, $y = 0$ **23.** 1

29. (a) (i) 3 (ii) 0 (iii) Does not exist (iv) 0 (v) 0 (vi) 0
(b) At 0 and 3 (c)

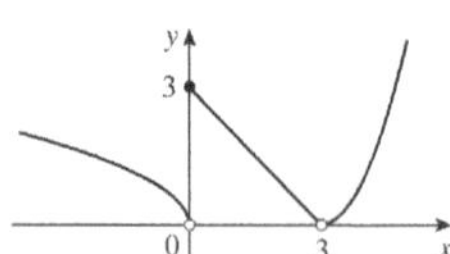

31. $\mathbb{R}$

35. (a) -8 (b) $y = -8x + 17$

37. (a) (i) 3 m/s (ii) 2.75 m/s (iii) 2.625 m/s
(iv) 2.525 m/s (b) 2.5 m/s

39. (a) 10 (b) $y = 10x - 16$
(c)

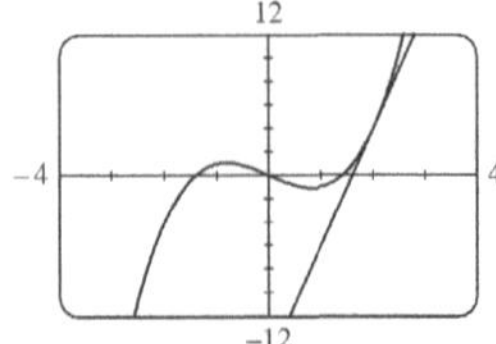

41. (a) The rate at which the cost changes with respect to the interest rate; dollars/(percent per year)
(b) As the interest rate increases past 10%, the cost is increasing at a rate of $1200/(percent per year).
(c) Always positive

43.

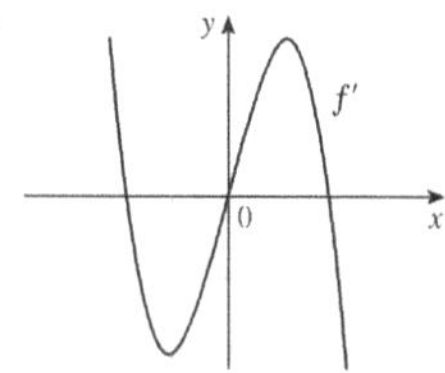

45. (a) $f'(x) = -\frac{5}{2}(3 - 5x)^{-1/2}$ (b) $\left(-\infty, \frac{3}{5}\right], \left(-\infty, \frac{3}{5}\right)$
(c)

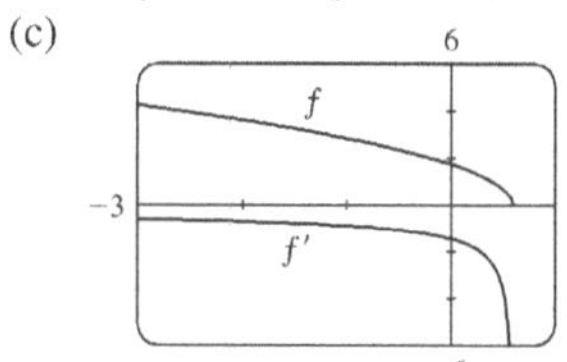

47. -4 (discontinuity), -1 (corner), 2 (discontinuity), 5 (vertical tangent)
49. The rate at which the total value of US currency in circulation is changing in billions of dollars per year; \$22.2 billion/year
51. 0

PROBLEMS PLUS ■ PAGE 170

1. $\frac{2}{3}$ **3.** -4 **5.** (a) Does not exist (b) 1
7. $a = \frac{1}{2} \pm \frac{1}{2}\sqrt{5}$ **9.** $\frac{3}{4}$ **11.** (b) Yes (c) Yes; no
13. (a) 0 (b) 1 (c) $f'(x) = x^2 + 1$

CHAPTER 3

EXERCISES 3.1 ■ PAGE 181

1. (a) See Definition of the Number e (page 180).
(b) 0.99, 1.03; $2.7 < e < 2.8$
3. $f'(x) = 0$ **5.** $f'(t) = -\frac{2}{3}$ **7.** $f'(x) = 3x^2 - 4$
9. $g'(x) = 2x - 6x^2$ **11.** $g'(t) = -\frac{3}{2}t^{\;7/4}$ **13.** $A'(s) = 60/s^6$
15. $R'(a) = 18a + 6$ **17.** $S'(p) = \frac{1}{2}p^{-1/2} - 1$
19. $y' = 3e^x - \frac{4}{3}x^{-4/3}$ **21.** $h'(u) = 3Au^2 + 2Bu + C$
23. $y' = \frac{3}{2}\sqrt{x} + \dfrac{2}{\sqrt{x}} - \dfrac{3}{2x\sqrt{x}}$ **25.** $j'(x) = 2.4x^{1.4}$
27. $H'(x) = 3x^2 + 3 - 3x^{-2} - 3x^{-4}$
29. $u' = \frac{1}{5}t^{-4/5} + 10t^{3/2}$
31. $z' = -10A/y^{11} + Be^y$ **33.** $y = \frac{1}{4}x + \frac{3}{4}$
35. Tangent: $y = 2x + 2$; normal: $y = -\frac{1}{2}x + 2$
37. $y = 3x - 1$ **39.** $f'(x) = 4x^3 - 6x^2 + 2x$
41. (a) (c) $4x^3 - 9x^2 - 12x + 7$

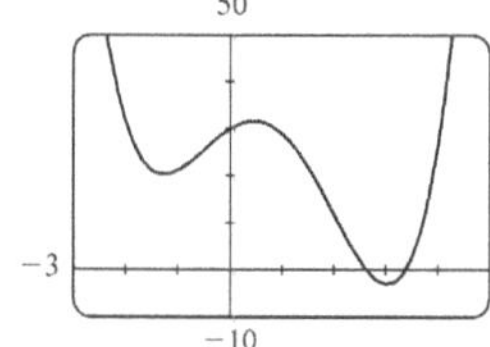

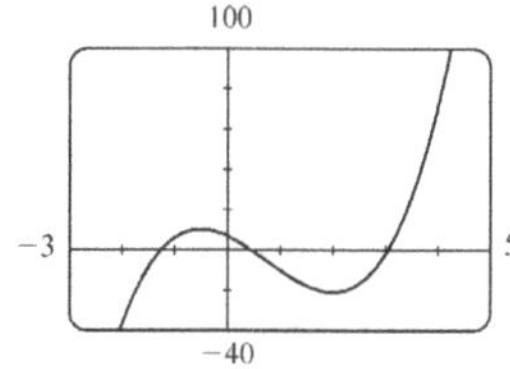

43. $f'(x) = 100x^9 + 25x^4 - 1$; $f''(x) = 900x^8 + 100x^3$
45. $f'(x) = 2 - \frac{15}{4}x^{-1/4}$, $f''(x) = \frac{15}{16}x^{-5/4}$

47. (a) $v(t) = 3t^2 - 3$, $a(t) = 6t$ (b) $12\ \text{m/s}^2$
(c) $a(1) = 6\ \text{m/s}^2$
49. (a) $V = 5.3/P$
(b) -0.00212; instantaneous rate of change of the volume with respect to the pressure at 25°C; m^3/kPa
51. $(-2, 21)$, $(1, -6)$
55. $y = 12x - 15$, $y = 12x + 17$ **57.** $y = \frac{1}{3}x - \frac{1}{3}$
59. $(\pm 2, 4)$ **63.** $P(x) = x^2 - x + 3$ **65.** $y = \frac{3}{16}x^3 - \frac{9}{4}x + 3$
67. No

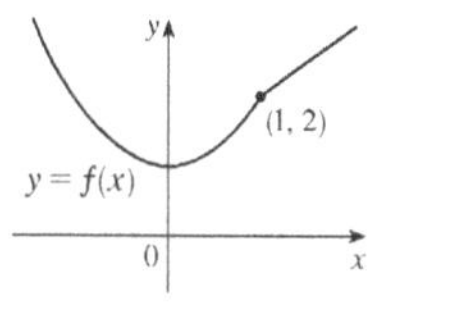

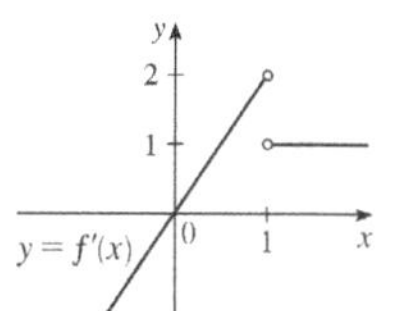

69. (a) Not differentiable at 3 or -3

$$f'(x) = \begin{cases} 2x & \text{if } |x| > 3 \\ -2x & \text{if } |x| < 3 \end{cases}$$

(b)

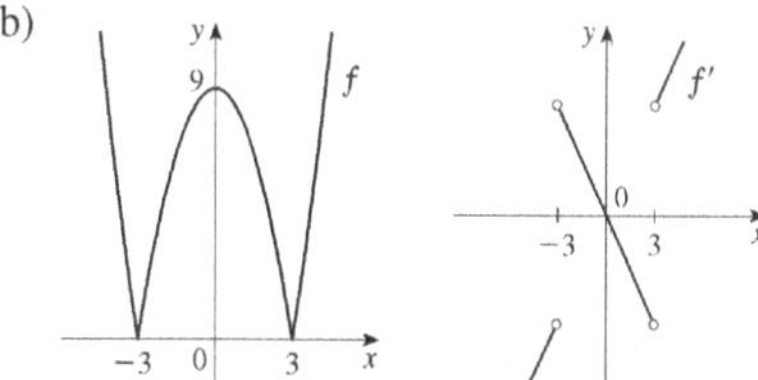

71. $y = 2x^2 - x$ **73.** $a = -\frac{1}{2}$, $b = 2$ **75.** $m = 4$, $b = -4$
77. 1000 **79.** 3; 1

EXERCISES 3.2 ■ PAGE 189

1. $1 - 2x + 6x^2 - 8x^3$ **3.** $f'(x) = e^x(x^3 + 3x^2 + 2x + 2)$
5. $y' = \dfrac{1 - x}{e^x}$ **7.** $g'(x) = \dfrac{10}{(3 - 4x)^2}$ **9.** $H'(u) = 2u - 1$
11. $F'(y) = 5 + \dfrac{14}{y^2} + \dfrac{9}{y^4}$
13. $y' = \dfrac{x^2(3 - x^2)}{(1 - x^2)^2}$ **15.** $y' = \dfrac{2t(-t^4 - 4t^2 + 7)}{(t^4 - 3t^2 + 1)^2}$
17. $y' = e^p\left(1 + \frac{3}{2}\sqrt{p} + p + p\sqrt{p}\right)$ **19.** $y' = 2v - 1/\sqrt{v}$
21. $f'(t) = \dfrac{4 + t^{1/2}}{(2 + \sqrt{t})^2}$ **23.** $f'(x) = \dfrac{-ACe^x}{(B + Ce^x)^2}$
25. $f'(x) = \dfrac{2cx}{(x^2 + c)^2}$
27. $(x^4 + 4x^3)e^x$; $(x^4 + 8x^3 + 12x^2)e^x$
29. $\dfrac{2x^2 + 2x}{(1 + 2x)^2}$; $\dfrac{2}{(1 + 2x)^3}$ **31.** $y = \frac{2}{3}x - \frac{2}{3}$
33. $y = 2x$; $y = -\frac{1}{2}x$
35. (a) $y = \frac{1}{2}x + 1$ (b)

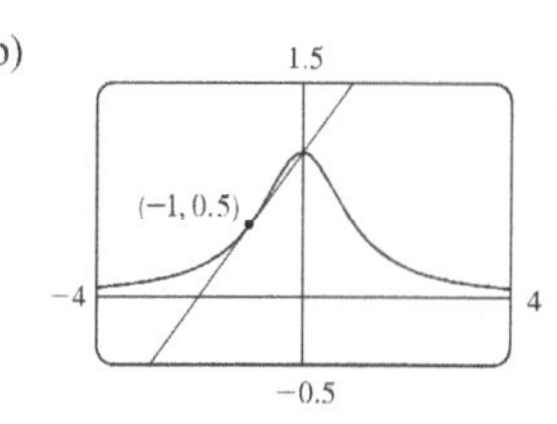

37. (a) $e^x(x^3 + 3x^2 - x - 1)$ (b)

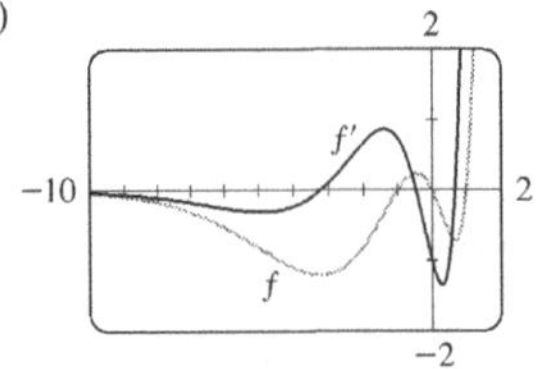

39. (a) $f'(x) = \dfrac{4x}{(x^2 + 1)^2}$; $f''(x) = \dfrac{4(1 - 3x^2)}{(x^2 + 1)^3}$

(b)

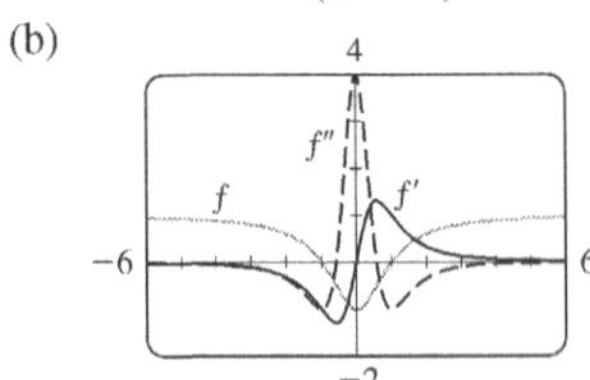

41. $\frac{1}{4}$ **43.** (a) -16 (b) $-\frac{20}{9}$ (c) 20 **45.** 7

47. $y = -2x + 18$

49. (a) 0 (b) $-\frac{2}{3}$

51. (a) $y' = xg'(x) + g(x)$

(b) $y' = \dfrac{g(x) - xg'(x)}{[g(x)]^2}$ (c) $y' = \dfrac{xg'(x) - g(x)}{x^2}$

53. Two, $\left(-2 \pm \sqrt{3}, \frac{1}{2}(1 \mp \sqrt{3})\right)$ **55.** 1

57. \$1.627 billion/year **59.** (c) $3e^{3x}$

61. $f'(x) = (x^2 + 2x)e^x$, $f''(x) = (x^2 + 4x + 2)e^x$, $f'''(x) = (x^2 + 6x + 6)e^x$, $f^{(4)}(x) = (x^2 + 8x + 12)e^x$, $f^{(5)}(x) = (x^2 + 10x + 20)e^x$; $f^{(n)}(x) = [x^2 + 2nx + n(n - 1)]e^x$

EXERCISES 3.3 ■ PAGE 197

1. $f'(x) = 6x + 2\sin x$ **3.** $f'(x) = \cos x - \frac{1}{2}\csc^2 x$

5. $y' = \sec\theta\,(\sec^2\theta + \tan^2\theta)$

7. $y' = -c\sin t + t(t\cos t + 2\sin t)$

9. $y' = \dfrac{2 - \tan x + x\sec^2 x}{(2 - \tan x)^2}$ **11.** $f'(\theta) = \dfrac{\sec\theta\tan\theta}{(1 + \sec\theta)^2}$

13. $y' = \dfrac{(t^2 + t)\cos t + \sin t}{(1 + t)^2}$

15. $f'(x) = e^x\csc x\,(-x\cot x + x + 1)$

21. $y = 2\sqrt{3}x - \frac{2}{3}\sqrt{3}\pi + 2$ **23.** $y = x - \pi - 1$

25. (a) $y = 2x$ (b)

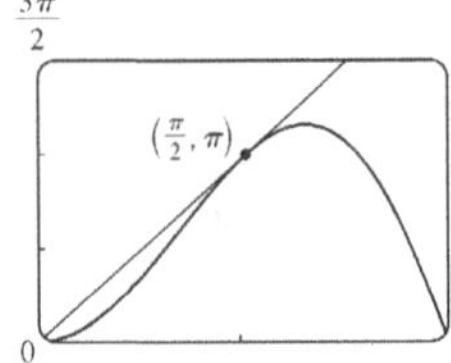

27. (a) $\sec x\tan x - 1$

29. $\theta\cos\theta + \sin\theta$; $2\cos\theta - \theta\sin\theta$

31. (a) $f'(x) = (1 + \tan x)/\sec x$ (b) $f'(x) = \cos x + \sin x$

33. $(2n + 1)\pi \pm \frac{1}{3}\pi$, n an integer

35. (a) $v(t) = 8\cos t$, $a(t) = -8\sin t$

(b) $4\sqrt{3}, -4, -4\sqrt{3}$; to the left

37. 5 ft/rad **39.** 3 **41.** 3 **43.** $-\frac{3}{4}$

45. $\frac{1}{2}$ **47.** $-\sqrt{2}$ **49.** $-\cos x$ **51.** $A = -\frac{3}{10}$, $B = -\frac{1}{10}$

53. (a) $\sec^2 x = \dfrac{1}{\cos^2 x}$ (b) $\sec x\tan x = \dfrac{\sin x}{\cos^2 x}$

(c) $\cos x - \sin x = \dfrac{\cot x - 1}{\csc x}$

55. 1

EXERCISES 3.4 ■ PAGE 205

1. $\dfrac{4}{3\sqrt[3]{(1 + 4x)^2}}$ **3.** $\pi\sec^2\pi x$ **5.** $\dfrac{e^{\sqrt{x}}}{2\sqrt{x}}$

7. $F'(x) = 10x(x^4 + 3x^2 - 2)^4(2x^2 + 3)$

9. $F'(x) = -\dfrac{1}{\sqrt{1 - 2x}}$ **11.** $f'(z) = -\dfrac{2z}{(z^2 + 1)^2}$

13. $y' = -3x^2\sin(a^3 + x^3)$ **15.** $y' = e^{-kx}(-kx + 1)$

17. $f'(x) = (2x - 3)^3(x^2 + x + 1)^4(28x^2 - 12x - 7)$

19. $h'(t) = \frac{2}{3}(t + 1)^{-1/3}(2t^2 - 1)^2(20t^2 + 18t - 1)$

21. $y' = \dfrac{-12x(x^2 + 1)^2}{(x^2 - 1)^4}$ **23.** $y' = \dfrac{3e^{3x}}{\sqrt{1 + 2e^{3x}}}$

25. $y' = 5^{\ 1/x}(\ln 5)/x^2$ **27.** $y' = (r^2 + 1)^{\ 3/2}$

29. $F'(t) = e^{t\sin 2t}(2t\cos 2t + \sin 2t)$

31. $y' = 2\cos(\tan 2x)\sec^2(2x)$ **33.** $y' = 2^{\sin\pi x}(\pi\ln 2)\cos\pi x$

35. $y' = \dfrac{4e^{2x}}{(1 + e^{2x})^2}\sin\dfrac{1 - e^{2x}}{1 + e^{2x}}$

37. $y' = -2\cos\theta\cot(\sin\theta)\csc^2(\sin\theta)$

39. $f'(t) = \sec^2(e^t)e^t + e^{\tan t}\sec^2 t$

41. $f'(t) = 4\sin(e^{\sin^2 t})\cos(e^{\sin^2 t})\,e^{\sin^2 t}\sin t\cos t$

43. $g'(x) = 2r^2p(\ln a)(2ra^{rx} + n)^{p\ \ 1}a^{rx}$

45. $y' = \dfrac{-\pi\cos(\tan\pi x)\sec^2(\pi x)\sin\sqrt{\sin(\tan\pi x)}}{2\sqrt{\sin(\tan\pi x)}}$

47. $y' = -2x\sin(x^2)$; $y'' = -4x^2\cos(x^2) - 2\sin(x^2)$

49. $e^{\alpha x}(\beta\cos\beta x + \alpha\sin\beta x)$; $e^{\alpha x}[(\alpha^2 - \beta^2)\sin\beta x + 2\alpha\beta\cos\beta x]$

51. $y = 20x + 1$ **53.** $y = -x + \pi$

55. (a) $y = \frac{1}{2}x + 1$ (b)

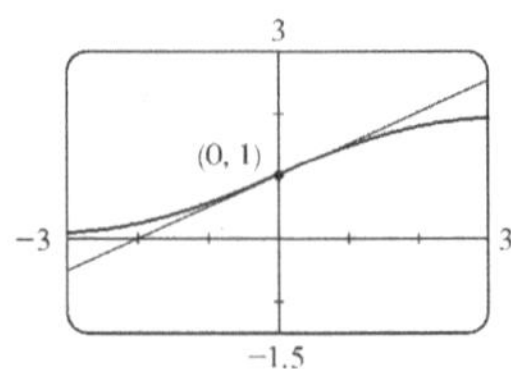

57. (a) $f'(x) = \dfrac{2 - 2x^2}{\sqrt{2 - x^2}}$

59. $((\pi/2) + 2n\pi, 3)$, $((3\pi/2) + 2n\pi, -1)$, n an integer

61. 24 **63.** (a) 30 (b) 36

65. (a) $\frac{3}{4}$ (b) Does not exist (c) -2

67. $-\frac{1}{6}\sqrt{2}$

69. (a) $F'(x) = e^x f'(e^x)$ (b) $G'(x) = e^{f(x)}f'(x)$

71. 120 **73.** 96

77. $-2^{50}\cos 2x$ **79.** $v(t) = \frac{5}{2}\pi\cos(10\pi t)$ cm/s

81. (a) $\dfrac{dB}{dt} = \dfrac{7\pi}{54}\cos\dfrac{2\pi t}{5.4}$ (b) 0.16

83. $v(t) = 2e^{\ 1.5t}(2\pi\cos 2\pi t - 1.5\sin 2\pi t)$

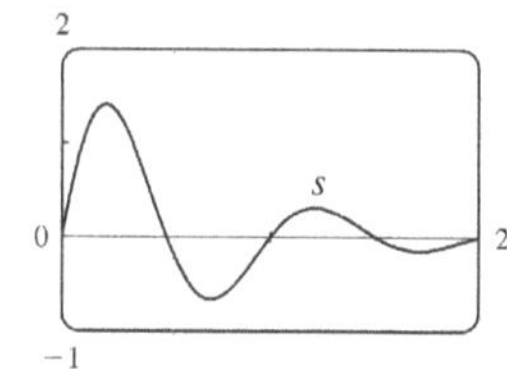

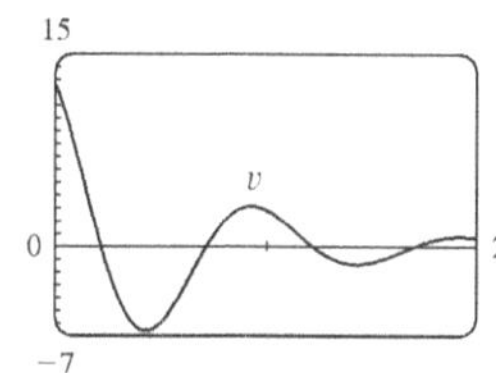

85. dv/dt is the rate of change of velocity with respect to time; dv/ds is the rate of change of velocity with respect to displacement

87. (a) $y = ab^t$ where $a \approx 100.01244$ and $b \approx 0.000045146$
(b) -670.63 μA

89. (b) The factored form **93.** (b) $-n\cos^{n\ 1}x\sin[(n+1)x]$

EXERCISES 3.5 ■ PAGE 215

1. (a) $y' = 9x/y$ (b) $y = \pm\sqrt{9x^2 - 1}$, $y' = \pm 9x/\sqrt{9x^2 - 1}$

3. (a) $y' = -y^2/x^2$ (b) $y = x/(x-1)$, $y' = -1/(x-1)^2$

5. $y' = -\dfrac{x^2}{y^2}$ **7.** $y' = \dfrac{2x + y}{2y - x}$

9. $y' = \dfrac{3y^2 - 5x^4 - 4x^3y}{x^4 + 3y^2 - 6xy}$ **11.** $y' = \dfrac{2x + y\sin x}{\cos x - 2y}$

13. $y' = \tan x\tan y$ **15.** $y' = \dfrac{y(y - e^{x/y})}{y^2 - xe^{x/y}}$

17. $y' = \dfrac{1 + x^4y^2 + y^2 + x^4y^4 - 2xy}{x^2 - 2xy - 2x^5y^3}$

19. $y' = \dfrac{e^y\sin x + y\cos(xy)}{e^y\cos x - x\cos(xy)}$ **21.** $-\frac{16}{13}$

23. $x' = \dfrac{-2x^4y + x^3 - 6xy^2}{4x^3y^2 - 3x^2y + 2y^3}$ **25.** $y = \frac{1}{2}x$

27. $y = -x + 2$ **29.** $y = x + \frac{1}{2}$ **31.** $y = -\frac{9}{13}x + \frac{40}{13}$

33. (a) $y = \frac{9}{2}x - \frac{5}{2}$ (b)

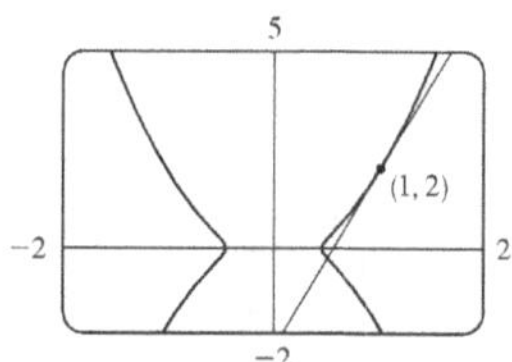

35. $-81/y^3$ **37.** $-2x/y^5$ **39.** $1/e^2$

41. (a)

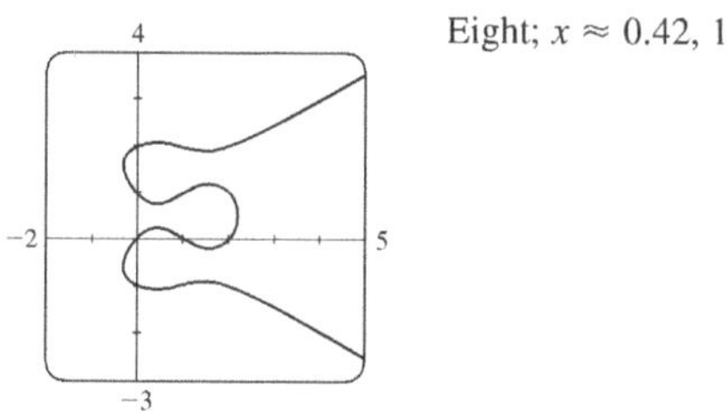

Eight; $x \approx 0.42, 1.58$

(b) $y = -x + 1$, $y = \frac{1}{3}x + 2$ (c) $1 \mp \frac{1}{3}\sqrt{3}$

43. $\left(\pm\frac{5}{4}\sqrt{3}, \pm\frac{5}{4}\right)$ **45.** $(x_0x/a^2) - (y_0y/b^2) = 1$

49. $y' = \dfrac{2\tan^{\ 1}x}{1 + x^2}$ **51.** $y' = \dfrac{1}{\sqrt{-x^2 - x}}$

53. $G'(x) = -1 - \dfrac{x\arccos x}{\sqrt{1 - x^2}}$ **55.** $h'(t) = 0$

57. $y' = \sin^{\ 1}x$ **59.** $y' = \dfrac{\sqrt{a^2 - b^2}}{a + b\cos x}$

61. $1 - \dfrac{x\arcsin x}{\sqrt{1 - x^2}}$

65.

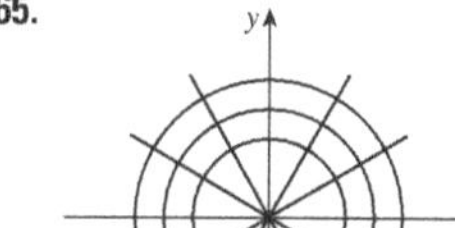

67.

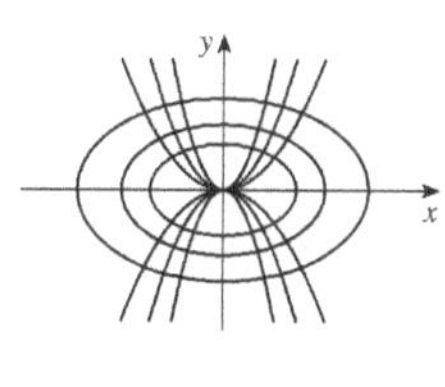

71. (a) $\dfrac{V^3(nb - V)}{PV^3 - n^2aV + 2n^3ab}$ (b) -4.04 L/atm

73. $(\pm\sqrt{3}, 0)$ **75.** $(-1, -1), (1, 1)$ **77.** (b) $\frac{3}{2}$

79. (a) 0 (b) $-\frac{1}{2}$

EXERCISES 3.6 ■ PAGE 223

1. The differentiation formula is simplest.

3. $f'(x) = \dfrac{\cos(\ln x)}{x}$ **5.** $f'(x) = -\dfrac{1}{x}$

7. $f'(x) = \dfrac{3x^2}{(x^3 + 1)\ln 10}$ **9.** $f'(x) = \dfrac{\sin x}{x} + \cos x\ln(5x)$

11. $g'(x) = \dfrac{2x^2 - 1}{x(x^2 - 1)}$ **13.** $G'(y) = \dfrac{10}{2y + 1} - \dfrac{y}{y^2 + 1}$

15. $F'(s) = \dfrac{1}{s\ln s}$ **17.** $y' = \sec^2(\ln(ax + b))\dfrac{a}{ax + b}$

19. $y' = \dfrac{-x}{1 + x}$ **21.** $y' = \dfrac{1}{\ln 10} + \log_{10}x$

23. $y' = x + 2x\ln(2x)$; $y'' = 3 + 2\ln(2x)$

25. $y' = \dfrac{1}{\sqrt{1 + x^2}}$; $y'' = \dfrac{-x}{(1 + x^2)^{3/2}}$

27. $f'(x) = \dfrac{2x - 1 - (x - 1)\ln(x - 1)}{(x - 1)[1 - \ln(x - 1)]^2}$;
$(1, 1 + e) \cup (1 + e, \infty)$

29. $f'(x) = \dfrac{2(x - 1)}{x(x - 2)}$; $(-\infty, 0) \cup (2, \infty)$ **31.** 1

33. $y = 3x - 9$ **35.** $\cos x + 1/x$ **37.** 7

39. $y' = (x^2 + 2)^2(x^4 + 4)^4\left(\dfrac{4x}{x^2 + 2} + \dfrac{16x^3}{x^4 + 4}\right)$

41. $y' = \sqrt{\dfrac{x - 1}{x^4 + 1}}\left(\dfrac{1}{2x - 2} - \dfrac{2x^3}{x^4 + 1}\right)$

43. $y' = x^x(1 + \ln x)$

45. $y' = x^{\sin x}\left(\dfrac{\sin x}{x} + \cos x\ln x\right)$

47. $y' = (\cos x)^x(-x\tan x + \ln\cos x)$

49. $y' = (\tan x)^{1/x}\left(\dfrac{\sec^2 x}{x\tan x} - \dfrac{\ln\tan x}{x^2}\right)$

51. $y' = \dfrac{2x}{x^2 + y^2 - 2y}$ **53.** $f^{(n)}(x) = \dfrac{(-1)^{n\ 1}(n - 1)!}{(x - 1)^n}$

EXERCISES 3.7 ■ PAGE 233

1. (a) $3t^2 - 24t + 36$ (b) -9 ft/s (c) $t = 2, 6$
(d) $0 \leqslant t < 2, t > 6$ (e) 96 ft
(f)

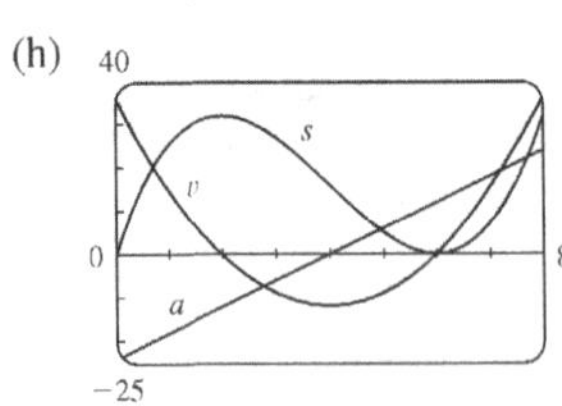

(g) $6t - 24$; -6 ft/s^2

(h)

(i) Speeding up when $2 < t < 4$ or $t > 6$; slowing down when $0 \leqslant t < 2$ or $4 < t < 6$

3. (a) $-\dfrac{\pi}{4}\sin\left(\dfrac{\pi t}{4}\right)$ (b) $-\frac{1}{8}\pi\sqrt{2}$ ft/s (c) $t = 0, 4, 8$
(d) $4 < t < 8$ (e) 4 ft
(f)

(g) $-\frac{1}{16}\pi^2\cos(\pi t/4)$; $\frac{1}{32}\pi^2\sqrt{2}$ ft/s^2
(h)

(i) Speeding up when $0 < t < 2$, $4 < t < 6$, $8 < t < 10$; slowing down when $2 < t < 4$, $6 < t < 8$

5. (a) Speeding up when $0 < t < 1$ or $2 < t < 3$; slowing down when $1 < t < 2$
(b) Speeding up when $1 < t < 2$ or $3 < t < 4$; slowing down when $0 < t < 1$ or $2 < t < 3$

7. (a) 4.9 m/s; -14.7 m/s (b) After 2.5 s (c) $32\frac{5}{16}$ m
(d) ≈ 5.08 s (e) ≈ -25.3 m/s

9. (a) 7.56 m/s (b) 6.24 m/s; -6.24 m/s

11. (a) 30 mm^2/mm; the rate at which the area is increasing with respect to side length as x reaches 15 mm
(b) $\Delta A \approx 2x\,\Delta x$

13. (a) (i) 5π (ii) 4.5π (iii) 4.1π
(b) 4π (c) $\Delta A \approx 2\pi r\,\Delta r$

15. (a) 8π ft^2/ft (b) 16π ft^2/ft (c) 24π ft^2/ft
The rate increases as the radius increases.

17. (a) 6 kg/m (b) 12 kg/m (c) 18 kg/m
At the right end; at the left end

19. (a) 4.75 A (b) 5 A; $t = \frac{2}{3}$ s

23. (a) $dV/dP = -C/P^2$ (b) At the beginning

25. $400(3^t)\ln 3$; ≈ 6850 bacteria/h

27. (a) 16 million/year; 78.5 million/year
(b) $P(t) = at^3 + bt^2 + ct + d$, where $a \approx 0.00129371$, $b \approx -7.061422$, $c \approx 12{,}822.979$, $d \approx -7{,}743{,}770$
(c) $P'(t) = 3at^2 + 2bt + c$
(d) 14.48 million/year; 75.29 million/year (smaller)
(e) 81.62 million/year

29. (a) 0.926 cm/s; 0.694 cm/s; 0
(b) 0; -92.6 (cm/s)/cm; -185.2 (cm/s)/cm
(c) At the center; at the edge

31. (a) $C'(x) = 12 - 0.2x + 0.0015x^2$
(b) \$32/yard; the cost of producing the 201st yard
(c) \$32.20

33. (a) $[xp'(x) - p(x)]/x^2$; the average productivity increases as new workers are added.

35. -0.2436 K/min

37. (a) 0 and 0 (b) $C = 0$
(c) (0, 0), (500, 50); it is possible for the species to coexist.

EXERCISES 3.8 ■ PAGE 242

1. About 235

3. (a) $100(4.2)^t$ (b) ≈ 7409 (c) $\approx 10{,}632$ bacteria/h
(d) $(\ln 100)/(\ln 4.2) \approx 3.2$ h

5. (a) 1508 million, 1871 million (b) 2161 million
(c) 3972 million; wars in the first half of century, increased life expectancy in second half

7. (a) $Ce^{-0.0005t}$ (b) $-2000\ln 0.9 \approx 211$ s

9. (a) $100 \times 2^{-t/30}$ mg (b) ≈ 9.92 mg (c) ≈ 199.3 years

11. ≈ 2500 years **13.** (a) ≈ 137°F (b) ≈ 116 min

15. (a) $13.\overline{3}$°C (b) ≈ 67.74 min

17. (a) ≈ 64.5 kPa (b) ≈ 39.9 kPa

19. (a) (i) \$3828.84 (ii) \$3840.25 (iii) \$3850.08
(iv) \$3851.61 (v) \$3852.01 (vi) \$3852.08
(b) $dA/dt = 0.05A$, $A(0) = 3000$

EXERCISES 3.9 ■ PAGE 248

1. $dV/dt = 3x^2\,dx/dt$ **3.** 48 cm^2/s

5. $3/(25\pi)$ m/min **7.** (a) 1 (b) 25 **9.** -18

11. (a) The plane's altitude is 1 mi and its speed is 500 mi/h.
(b) The rate at which the distance from the plane to the station is increasing when the plane is 2 mi from the station
(c)
(d) $y^2 = x^2 + 1$
(e) $250\sqrt{3}$ mi/h

13. (a) The height of the pole (15 ft), the height of the man (6 ft), and the speed of the man (5 ft/s)
(b) The rate at which the tip of the man's shadow is moving when he is 40 ft from the pole
(c)

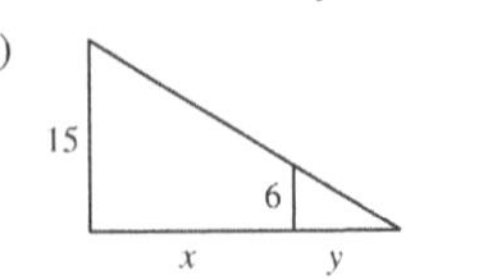

(d) $\dfrac{15}{6} = \dfrac{x + y}{y}$ (e) $\frac{25}{3}$ ft/s

15. 65 mi/h **17.** $837/\sqrt{8674} \approx 8.99$ ft/s

19. -1.6 cm/min **21.** $\frac{720}{13} \approx 55.4$ km/h

23. $(10{,}000 + 800{,}000\pi/9) \approx 2.89 \times 10^5 \text{ cm}^3/\text{min}$
25. $\frac{10}{3}$ cm/min **27.** $6/(5\pi) \approx 0.38$ ft/min **29.** $0.3 \text{ m}^2/\text{s}$
31. 5 m **33.** $80 \text{ cm}^3/\text{min}$ **35.** $\frac{107}{810} \approx 0.132\ \Omega/\text{s}$
37. 0.396 m/min **39.** (a) 360 ft/s (b) 0.096 rad/s
41. $\frac{10}{9}\pi$ km/min **43.** $1650/\sqrt{31} \approx 296$ km/h
45. $\frac{7}{4}\sqrt{15} \approx 6.78$ m/s

EXERCISES 3.10 ■ PAGE 255

1. $L(x) = -10x - 6$ **3.** $L(x) = \frac{1}{4}x + 1$
5. $\sqrt{1-x} \approx 1 - \frac{1}{2}x$;
$\sqrt{0.9} \approx 0.95$,
$\sqrt{0.99} \approx 0.995$

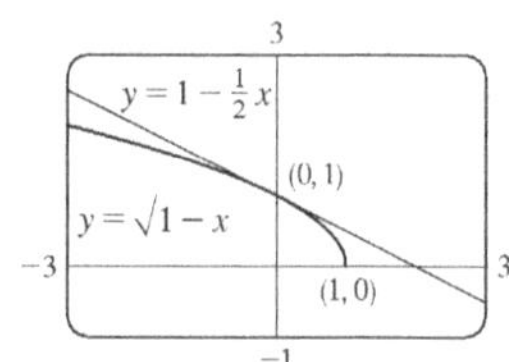

7. $-0.383 < x < 0.516$ **9.** $-0.368 < x < 0.677$
11. (a) $dy = 2x(x\cos 2x + \sin 2x)\,dx$ (b) $dy = \dfrac{t}{1+t^2}\,dt$
13. (a) $dy = \dfrac{\sec^2\sqrt{t}}{2\sqrt{t}}\,dt$ (b) $dy = \dfrac{-4v}{(1+v^2)^2}\,dv$
15. (a) $dy = \frac{1}{10}e^{x/10}\,dx$ (b) 0.01
17. (a) $dy = \dfrac{x}{\sqrt{3+x^2}}\,dx$ (b) -0.05
19. $\Delta y = 0.64$, $dy = 0.8$

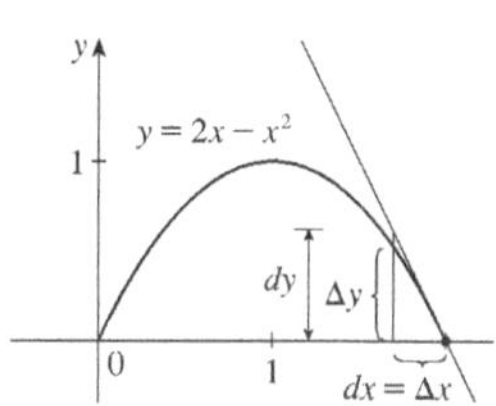

21. $\Delta y = -0.1$, $dy = -0.125$

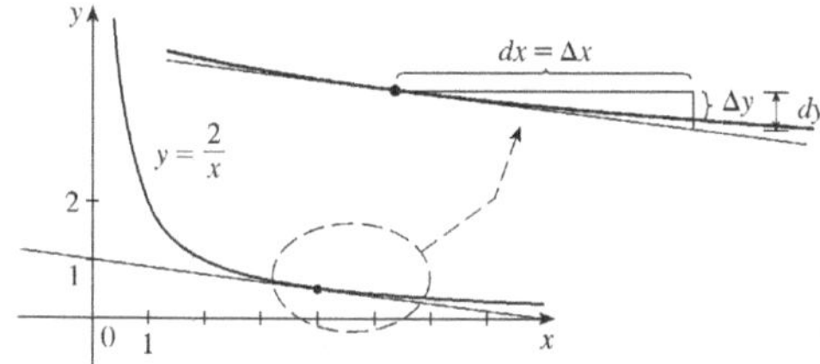

23. 15.968 **25.** $10.00\overline{3}$ **27.** $1 - \pi/90 \approx 0.965$
33. (a) 270 cm^3, 0.01, 1% (b) 36 cm^2, $0.00\overline{6}$, $0.\overline{6}$%
35. (a) $84/\pi \approx 27 \text{ cm}^2$; $\frac{1}{84} \approx 0.012 = 1.2\%$
(b) $1764/\pi^2 \approx 179 \text{ cm}^3$; $\frac{1}{56} \approx 0.018 = 1.8\%$
37. (a) $2\pi rh\,\Delta r$ (b) $\pi(\Delta r)^2 h$
43. (a) 4.8, 5.2 (b) Too large

EXERCISES 3.11 ■ PAGE 262

1. (a) 0 (b) 1 **3.** (a) $\frac{3}{4}$ (b) $\frac{1}{2}(e^2 - e^{-2}) \approx 3.62686$
5. (a) 1 (b) 0
21. $\operatorname{sech} x = \frac{3}{5}$, $\sinh x = \frac{4}{3}$, $\operatorname{csch} x = \frac{3}{4}$, $\tanh x = \frac{4}{5}$, $\coth x = \frac{5}{4}$
23. (a) 1 (b) −1 (c) ∞ (d) −∞ (e) 0 (f) 1
(g) ∞ (h) −∞ (i) 0
31. $f'(x) = x\cosh x$ **33.** $h'(x) = \tanh x$
35. $y' = 3e^{\cosh 3x}\sinh 3x$ **37.** $f'(t) = -2e^t\operatorname{sech}^2(e^t)\tanh(e^t)$
39. $G'(x) = \dfrac{-2\sinh x}{(1+\cosh x)^2}$ **41.** $y' = \dfrac{1}{2\sqrt{x(x-1)}}$
43. $y' = \sinh^{-1}(x/3)$ **45.** $y' = -\csc x$
51. (a) 0.3572 (b) 70.34°
53. (a) 164.50 m (b) 120 m; 164.13 m
55. (b) $y = 2\sinh 3x - 4\cosh 3x$
57. $\left(\ln(1+\sqrt{2}), \sqrt{2}\right)$

CHAPTER 3 REVIEW ■ PAGE 264

True-False Quiz

1. True **3.** True **5.** False **7.** False **9.** True
11. True **13.** True **15.** True

Exercises

1. $4x^7(x+1)^3(3x+2)$ **3.** $\frac{3}{2}\sqrt{x} - \dfrac{1}{2\sqrt{x}} - \dfrac{1}{\sqrt{x^3}}$
5. $x(\pi x\cos\pi x + 2\sin\pi x)$
7. $\dfrac{8t^3}{(t^4+1)^2}$ **9.** $\dfrac{1+\ln x}{x\ln x}$ **11.** $\dfrac{\cos\sqrt{x} - \sqrt{x}\sin\sqrt{x}}{2\sqrt{x}}$
13. $-\dfrac{e^{1/x}(1+2x)}{x^4}$ **15.** $\dfrac{2xy - \cos y}{1 - x\sin y - x^2}$
17. $\dfrac{1}{2\sqrt{\arctan x}\,(1+x^2)}$ **19.** $\dfrac{1-t^2}{(1+t^2)^2}\sec^2\left(\dfrac{t}{1+t^2}\right)$
21. $3^{x\ln x}(\ln 3)(1+\ln x)$ **23.** $-(x-1)^{-2}$
25. $\dfrac{2x - y\cos(xy)}{x\cos(xy)+1}$ **27.** $\dfrac{2}{(1+2x)\ln 5}$
29. $\cot x - \sin x\cos x$ **31.** $\dfrac{4x}{1+16x^2} + \tan^{-1}(4x)$
33. $5\sec 5x$ **35.** $-6x\csc^2(3x^2+5)$
37. $\cos\left(\tan\sqrt{1+x^3}\right)\left(\sec^2\sqrt{1+x^3}\right)\dfrac{3x^2}{2\sqrt{1+x^3}}$
39. $2\cos\theta\tan(\sin\theta)\sec^2(\sin\theta)$
41. $\dfrac{(x-2)^4(3x^2-55x-52)}{2\sqrt{x+1}\,(x+3)^8}$ **43.** $2x^2\cosh(x^2) + \sinh(x^2)$
45. $3\tanh 3x$ **47.** $\dfrac{\cosh x}{\sqrt{\sinh^2 x - 1}}$
49. $\dfrac{-3\sin\left(e^{\sqrt{\tan 3x}}\right)e^{\sqrt{\tan 3x}}\sec^2(3x)}{2\sqrt{\tan 3x}}$ **51.** $-\frac{4}{27}$
53. $-5x^4/y^{11}$ **57.** $y = 2\sqrt{3}x + 1 - \pi\sqrt{3}/3$
59. $y = 2x + 1$ **61.** $y = -x + 2$; $y = x + 2$
63. (a) $\dfrac{10-3x}{2\sqrt{5-x}}$ (b) $y = \frac{7}{4}x + \frac{1}{4}$, $y = -x + 8$
(c)

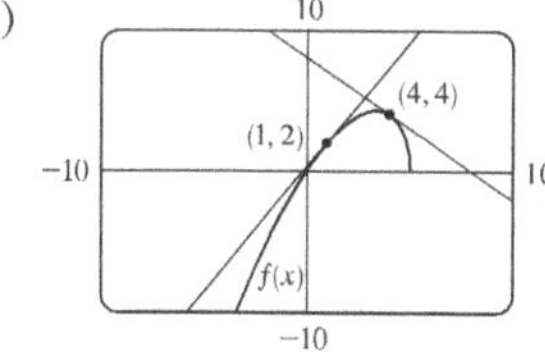

65. $(\pi/4, \sqrt{2}), (5\pi/4, -\sqrt{2})$ **69.** (a) 2 (b) 44
71. $2xg(x) + x^2g'(x)$ **73.** $2g(x)g'(x)$

75. $g'(e^x)e^x$ **77.** $g'(x)/g(x)$ **79.** $\dfrac{f'(x)[g(x)]^2 + g'(x)[f(x)]^2}{[f(x) + g(x)]^2}$

81. $f'(g(\sin 4x))g'(\sin 4x)(\cos 4x)(4)$
83. $(-3, 0)$ **85.** $y = -\frac{2}{3}x^2 + \frac{14}{3}x$
87. $v(t) = -Ae^{-ct}[c\cos(\omega t + \delta) + \omega\sin(\omega t + \delta)]$,
$a(t) = Ae^{-ct}[(c^2 - \omega^2)\cos(\omega t + \delta) + 2c\omega\sin(\omega t + \delta)]$
89. (a) $v(t) = 3t^2 - 12$; $a(t) = 6t$ (b) $t > 2$; $0 \leqslant t < 2$
(c) 23 (d) (e) $t > 2$; $0 < t < 2$

91. 4 kg/m
93. (a) $200(3.24)^t$ (b) $\approx 22{,}040$
(c) $\approx 25{,}910$ bacteria/h (d) $(\ln 50)/(\ln 3.24) \approx 3.33$ h
95. (a) C_0e^{kt} (b) ≈ 100 h **97.** $\frac{4}{3}$ cm^2/min
99. 13 ft/s **101.** 400 ft/h
103. (a) $L(x) = 1 + x$; $\sqrt[3]{1 + 3x} \approx 1 + x$; $\sqrt[3]{1.03} \approx 1.01$
(b) $-0.235 < x < 0.401$
105. $12 + \frac{3}{2}\pi \approx 16.7$ cm^2 **107.** $\frac{1}{32}$ **109.** $\frac{1}{4}$ **111.** $\frac{1}{8}x^2$

PROBLEMS PLUS ■ PAGE 269

1. $(\pm\sqrt{3}/2, \frac{1}{4})$ **5.** $3\sqrt{2}$ **11.** $(0, \frac{5}{4})$
13. (a) $4\pi\sqrt{3}/\sqrt{11}$ rad/s (b) $40(\cos\theta + \sqrt{8 + \cos^2\theta})$ cm
(c) $-480\pi\sin\theta\,(1 + \cos\theta/\sqrt{8 + \cos^2\theta})$ cm/s
17. $x_T \in (3, \infty)$, $y_T \in (2, \infty)$, $x_N \in (0, \frac{5}{3})$, $y_N \in (-\frac{5}{2}, 0)$
19. (b) (i) 53° (or 127°) (ii) 63° (or 117°)
21. R approaches the midpoint of the radius AO.
23. $-\sin a$ **25.** $2\sqrt{e}$ **29.** $(1, -2), (-1, 0)$
31. $\sqrt{29}/58$ **33.** $2 + \frac{375}{128}\pi \approx 11.204$ cm^3/min

CHAPTER 4

EXERCISES 4.1 ■ PAGE 280

Abbreviations: abs, absolute; loc, local; max, maximum; min, minimum

1. Abs min: smallest function value on the entire domain of the function; loc min at c: smallest function value when x is near c
3. Abs max at s, abs min at r, loc max at c, loc min at b and r, neither a max nor a min at a and d
5. Abs max $f(4) = 5$, loc max $f(4) = 5$ and $f(6) = 4$, loc min $f(2) = 2$ and $f(1) = f(5) = 3$

7.
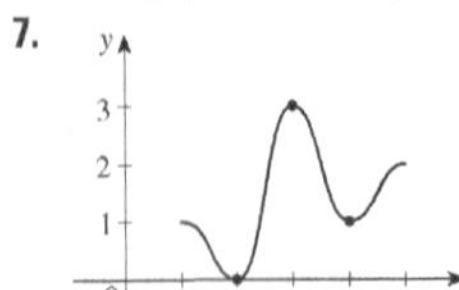

9.
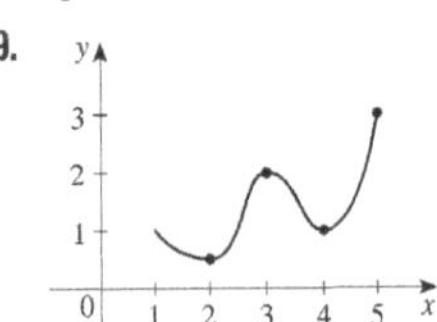

11. (a) (b)
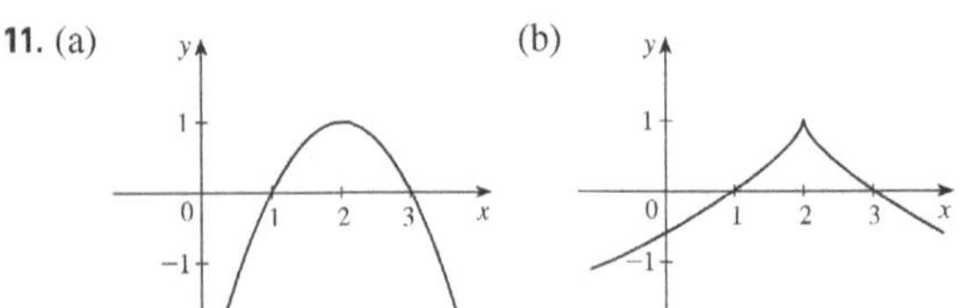

(c)
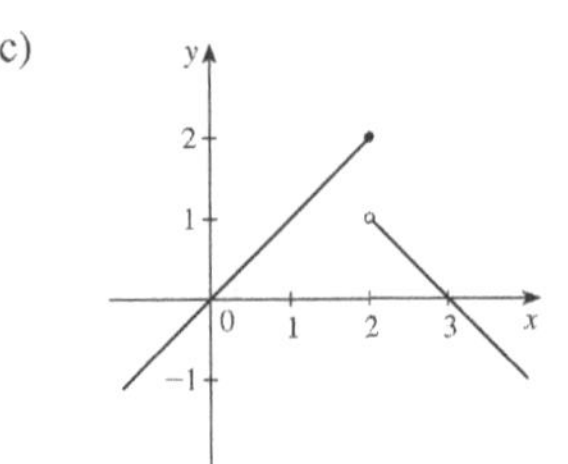

13. (a) (b)
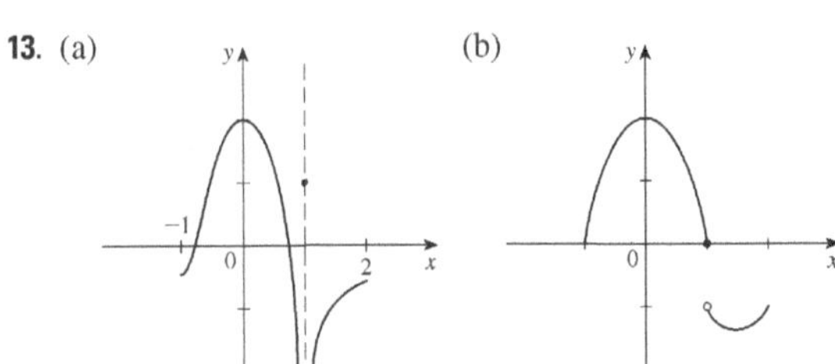

15. Abs max $f(3) = 4$ **17.** Abs max $f(1) = 1$
19. Abs min $f(0) = 0$
21. Abs max $f(\pi/2) = 1$; abs min $f(-\pi/2) = -1$
23. Abs max $f(2) = \ln 2$ **25.** Abs max $f(0) = 1$
27. Abs max $f(3) = 2$ **29.** $\frac{1}{3}$ **31.** $-2, 3$ **33.** 0
35. 0, 2 **37.** $0, \frac{4}{9}$ **39.** $0, \frac{8}{7}, 4$ **41.** $n\pi$ (n an integer)
43. $0, \frac{2}{3}$ **45.** 10 **47.** $f(2) = 16, f(5) = 7$
49. $f(-1) = 8, f(2) = -19$ **51.** $f(-2) = 33, f(2) = -31$
53. $f(0.2) = 5.2, f(1) = 2$ **55.** $f(\sqrt{2}) = 2, f(-1) = -\sqrt{3}$
57. $f(\pi/6) = \frac{3}{2}\sqrt{3}, f(\pi/2) = 0$
59. $f(2) = 2/\sqrt{e}, f(-1) = -1/\sqrt[8]{e}$
61. $f(1) = \ln 3, f(-\frac{1}{2}) = \ln\frac{3}{4}$

63. $f\left(\dfrac{a}{a + b}\right) = \dfrac{a^ab^b}{(a + b)^{a+b}}$

65. (a) 2.19, 1.81 (b) $\frac{6}{25}\sqrt{\frac{3}{5}} + 2, -\frac{6}{25}\sqrt{\frac{3}{5}} + 2$
67. (a) 0.32, 0.00 (b) $\frac{3}{16}\sqrt{3}, 0$ **69.** ≈ 3.9665°C
71. Cheapest, $t \approx 0.855$ (June 1994); most expensive, $t \approx 4.618$ (March 1998)
73. (a) $r = \frac{2}{3}r_0$ (b) $v = \frac{4}{27}kr_0^3$
(c)
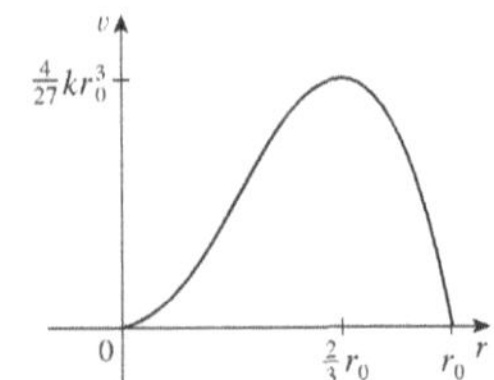

EXERCISES 4.2 ■ PAGE 288

1. 2 **3.** $\frac{9}{4}$ **5.** f is not differentiable on $(-1, 1)$
7. 0.3, 3, 6.3 **9.** 1 **11.** $3/\ln 4$ **13.** 1
15. f is not continous at 3 **23.** 16 **25.** No **31.** No

EXERCISES 4.3 ■ PAGE 297

Abbreviations: inc, increasing; dec, decreasing; CD, concave downward; CU, concave upward; HA, horizontal asymptote; VA, vertical asymptote; IP, inflection point(s)

1. (a) $(1, 3), (4, 6)$ (b) $(0, 1), (3, 4)$ (c) $(0, 2)$
(d) $(2, 4), (4, 6)$ (e) $(2, 3)$

3. (a) I/D Test (b) Concavity Test
(c) Find points at which the concavity changes.

5. (a) Inc on $(1, 5)$; dec on $(0, 1)$ and $(5, 6)$
(b) Loc max at $x = 5$, loc min at $x = 1$

7. (a) 3, 5 (b) 2, 4, 6 (c) 1, 7

9. (a) Inc on $(-\infty, -3), (2, \infty)$; dec on $(-3, 2)$
(b) Loc max $f(-3) = 81$; loc min $f(2) = -44$
(c) CU on $(-\frac{1}{2}, \infty)$; CD on $(-\infty, -\frac{1}{2})$; IP $(-\frac{1}{2}, \frac{37}{2})$

11. (a) Inc on $(-1, 0), (1, \infty)$; dec on $(-\infty, -1), (0, 1)$
(b) Loc max $f(0) = 3$; loc min $f(\pm 1) = 2$
(c) CU on $(-\infty, -\sqrt{3}/3), (\sqrt{3}/3, \infty)$;
CD on $(-\sqrt{3}/3, \sqrt{3}/3)$; IP $(\pm\sqrt{3}/3, \frac{22}{9})$

13. (a) Inc on $(0, \pi/4), (5\pi/4, 2\pi)$; dec on $(\pi/4, 5\pi/4)$
(b) Loc max $f(\pi/4) = \sqrt{2}$; loc min $f(5\pi/4) = -\sqrt{2}$
(c) CU on $(3\pi/4, 7\pi/4)$; CD on $(0, 3\pi/4), (7\pi/4, 2\pi)$;
IP $(3\pi/4, 0), (7\pi/4, 0)$

15. (a) Inc on $(-\frac{1}{3}\ln 2, \infty)$; dec on $(-\infty, -\frac{1}{3}\ln 2)$
(b) Loc min $f(-\frac{1}{3}\ln 2) = 2^{\,2/3} + 2^{1/3}$ (c) CU on $(-\infty, \infty)$

17. (a) Inc on $(1, \infty)$; dec on $(0, 1)$ (b) Loc min $f(1) = 0$
(c) CU on $(0, \infty)$; No IP

19. Loc max $f(1) = 2$; loc min $f(0) = 1$

21. Loc min $f(\frac{1}{16}) = -\frac{1}{4}$

23. (a) f has a local maximum at 2.
(b) f has a horizontal tangent at 6.

25.

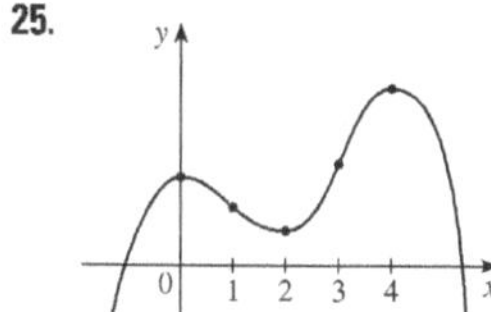

27.

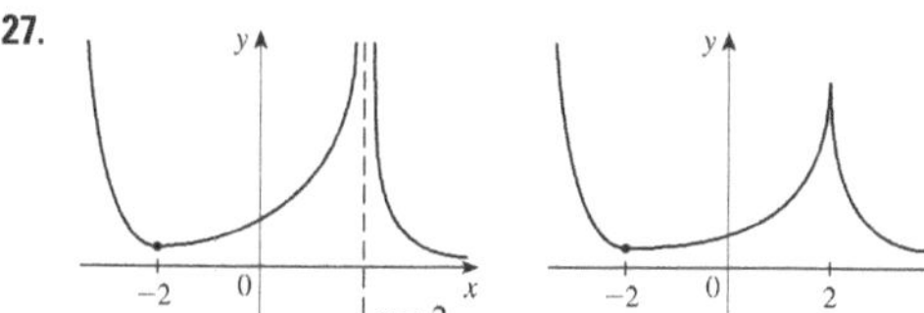

29.

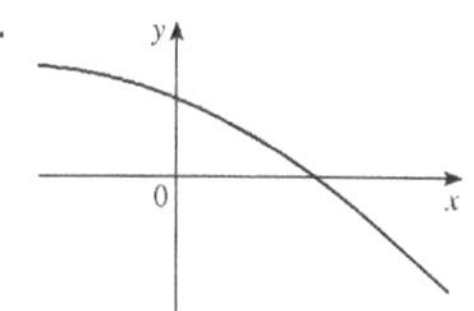

31. (a) Inc on $(0, 2), (4, 6), (8, \infty)$;
dec on $(2, 4), (6, 8)$
(b) Loc max at $x = 2, 6$;
loc min at $x = 4, 8$
(c) CU on $(3, 6), (6, \infty)$;
CD on $(0, 3)$
(d) 3
(e) See graph at right.

33. (a) Inc on $(-\infty, -2), (2, \infty)$; dec on $(-2, 2)$
(b) Loc max $f(-2) = 18$; loc min $f(2) = -14$
(c) CU on $(0, \infty)$, CD on $(-\infty, 0)$; IP $(0, 2)$
(d)

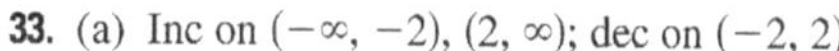

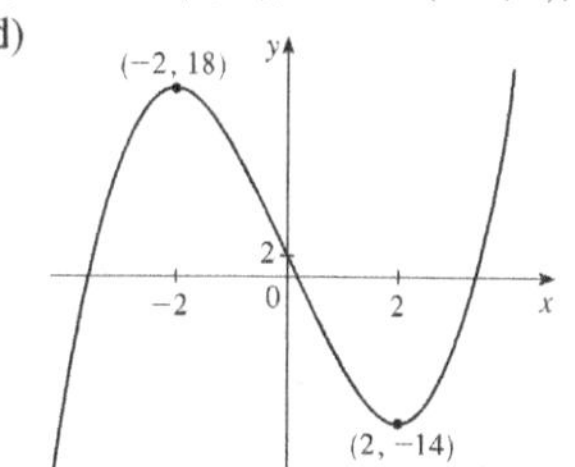

35. (a) Inc on $(-\infty, -1), (0, 1)$;
dec on $(-1, 0), (1, \infty)$
(b) Loc max $f(-1) = 3, f(1) = 3$;
loc min $f(0) = 2$
(c) CU on $(-1/\sqrt{3}, 1/\sqrt{3})$;
CD on $(-\infty, -1/\sqrt{3}), (1/\sqrt{3}, \infty)$;
IP $(\pm 1/\sqrt{3}, \frac{23}{9})$
(d) See graph at right.

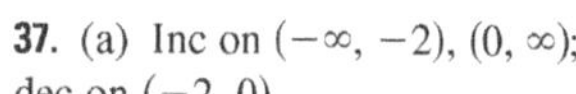

37. (a) Inc on $(-\infty, -2), (0, \infty)$;
dec on $(-2, 0)$
(b) Loc max $h(-2) = 7$;
loc min $h(0) = -1$
(c) CU on $(-1, \infty)$;
CD on $(-\infty, -1)$; IP $(-1, 3)$
(d) See graph at right.

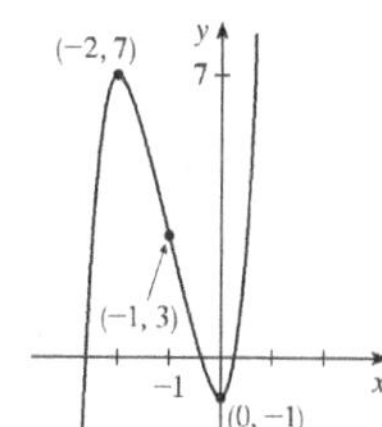

39. (a) Inc on $(-\infty, 4)$; dec on $(4, 6)$
(b) Loc max $f(4) = 4\sqrt{2}$
(c) CD on $(-\infty, 6)$; No IP
(d) See graph at right.

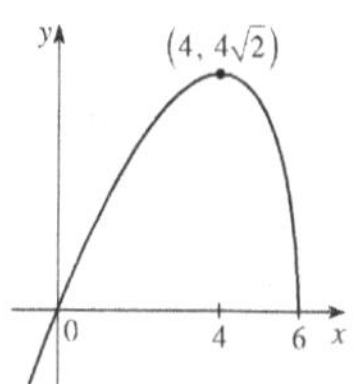

41. (a) Inc on $(-1, \infty)$;
dec on $(-\infty, -1)$
(b) Loc min $C(-1) = -3$
(c) CU on $(-\infty, 0)$, $(2, \infty)$;
CD on $(0, 2)$;
IP $(0, 0)$, $(2, 6\sqrt[3]{2})$
(d) See graph at right.

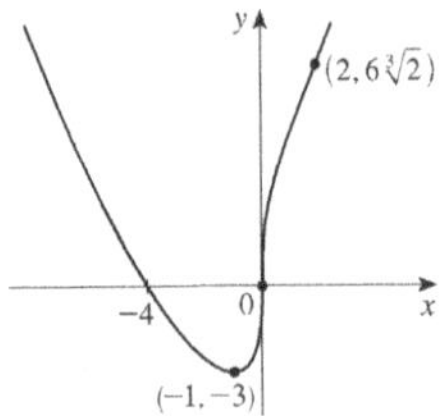

43. (a) Inc on $(\pi, 2\pi)$;
dec on $(0, \pi)$
(b) Loc min $f(\pi) = -1$
(c) CU on $(\pi/3, 5\pi/3)$;
CD on $(0, \pi/3)$, $(5\pi/3, 2\pi)$;
IP $(\pi/3, \frac{5}{4})$, $(5\pi/3, \frac{5}{4})$
(d) See graph at right.

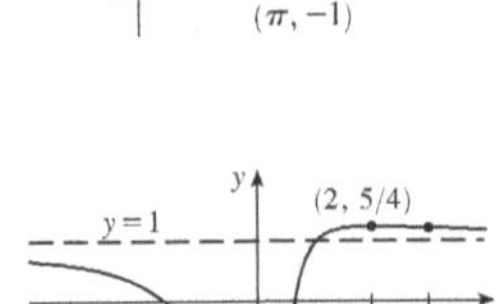

45. (a) VA $x = 0$; HA $y = 1$
(b) Inc on $(0, 2)$;
dec on $(-\infty, 0)$, $(2, \infty)$
(c) Loc max $f(2) = \frac{5}{4}$
(d) CU on $(3, \infty)$;
CD on $(-\infty, 0)$, $(0, 3)$; IP $(3, \frac{11}{9})$
(e) See graph at right.

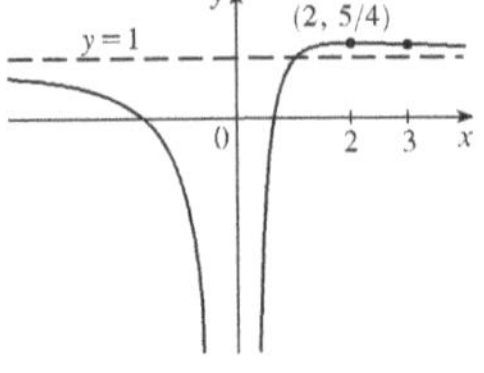

47. (a) HA $y = 0$
(b) Dec on $(-\infty, \infty)$
(c) None
(d) CU on $(-\infty, \infty)$
(e) See graph at right.

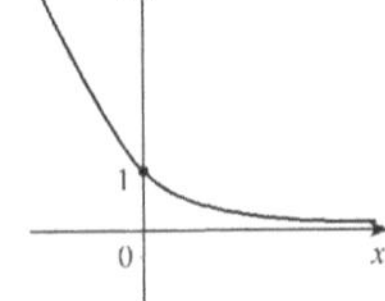

49. (a) HA $y = 0$
(b) Inc on $(-\infty, 0)$, dec on $(0, \infty)$
(c) Loc max $f(0) = 1$
(d) CU on $(-\infty, -1/\sqrt{2})$, $(1/\sqrt{2}, \infty)$;
CD on $(-1/\sqrt{2}, 1/\sqrt{2})$;
IP $(\pm 1/\sqrt{2}, e^{-1/2})$
(e) See graph at right.

51. (a) VA $x = 0$, $x = e$
(b) Dec on $(0, e)$
(c) None
(d) CU on $(0, 1)$; CD on $(1, e)$;
IP $(1, 0)$
(e) See graph at right.

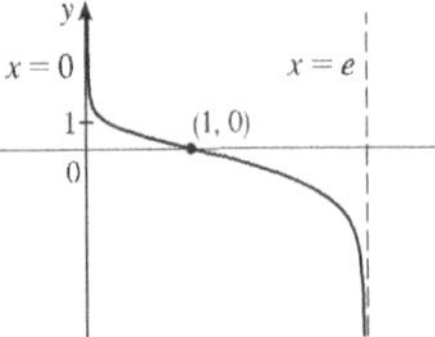

53. $(3, \infty)$
55. (a) Loc and abs max $f(1) = \sqrt{2}$, no min
(b) $\frac{1}{4}(3 - \sqrt{17})$
57. (b) CU on (0.94, 2.57), (3.71, 5.35);
CD on (0, 0.94), (2.57, 3.71), $(5.35, 2\pi)$;
IP (0.94, 0.44), (2.57, −0.63), (3.71, −0.63), (5.35, 0.44)
59. CU on $(-\infty, -0.6)$, $(0.0, \infty)$; CD on $(-0.6, 0.0)$

61. (a) The rate of increase is initially very small, increases to a maximum at $t \approx 8$ h, then decreases toward 0.
(b) When $t = 8$ (c) CU on (0, 8); CD on (8, 18) (d) (8, 350)
63. $K(3) - K(2)$; CD
65. 28.57 min, when the rate of increase of drug level in the bloodstream is greatest; 85.71 min, when rate of decrease is greatest
67. $f(x) = \frac{1}{9}(2x^3 + 3x^2 - 12x + 7)$
69. (a) $a = 0$, $b = -1$ (b) $y = -x$ at (0, 0)

EXERCISES 4.4 ■ PAGE 307

1. (a) Indeterminate (b) 0 (c) 0
(d) ∞, $-\infty$, or does not exist (e) Indeterminate
3. (a) $-\infty$ (b) Indeterminate (c) ∞
5. $\frac{9}{4}$ **7.** 2 **9.** $-\frac{1}{3}$ **11.** $-\infty$ **13.** 2 **15.** $\frac{1}{4}$
17. 0 **19.** $-\infty$ **21.** $\frac{8}{5}$ **23.** 3 **25.** $\frac{1}{2}$ **27.** 1
29. 1 **31.** $1/\ln 3$ **33.** 0 **35.** $-1/\pi^2$ **37.** $\frac{1}{2}a(a - 1)$
39. $\frac{1}{24}$ **41.** π **43.** 3 **45.** 0 **47.** $-2/\pi$ **49.** $\frac{1}{2}$
51. $\frac{1}{2}$ **53.** ∞ **55.** 1 **57.** $e^{\;2}$ **59.** $1/e$
61. 1 **63.** e^4 **65.** $1/\sqrt{e}$ **67.** e^2 **69.** $\frac{1}{4}$ **73.** 1
75. f has an absolute minimum for $c > 0$. As c increases, the minimum points get farther away from the origin.
81. $\frac{16}{9}a$ **83.** $\frac{1}{2}$ **85.** 56 **89.** (a) 0

EXERCISES 4.5 ■ PAGE 317

Abbreviation: int, intercept; SA, slant asymptote

1. A. $\mathbb{R}$ B. y-int 0; x-int 0, 6
C. None D. None
E. Inc on $(-\infty, 2)$, $(6, \infty)$;
dec on $(2, 6)$
F. Loc max $f(2) = 32$;
loc min $f(6) = 0$
G. CU on $(4, \infty)$; CD on $(-\infty, 4)$;
IP $(4, 16)$
H. See graph at right.

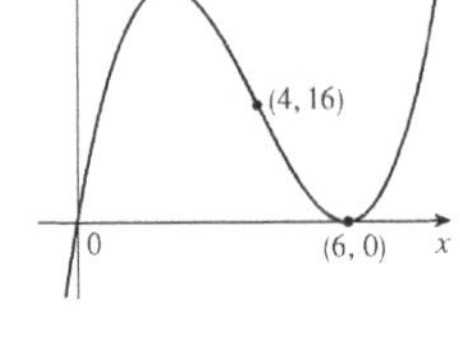

3. A. $\mathbb{R}$ B. y-int 0; x-int 0, $\sqrt[3]{4}$
C. None D. None
E. Inc on $(1, \infty)$; dec on $(-\infty, 1)$
F. Loc min $f(1) = -3$
G. CU on $(-\infty, \infty)$
H. See graph at right.

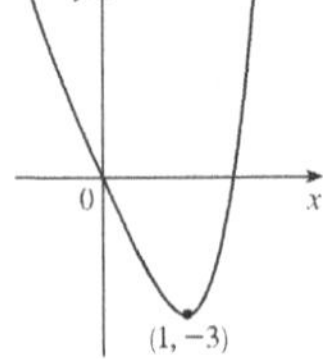

5. A. $\mathbb{R}$ B. y-int 0; x-int 0, 4
C. None D. None
E. Inc on $(1, \infty)$; dec on $(-\infty, 1)$
F. Loc min $f(1) = -27$
G. CU on $(-\infty, 2)$, $(4, \infty)$;
CD on $(2, 4)$;
IP $(2, -16)$, $(4, 0)$
H. See graph at right.

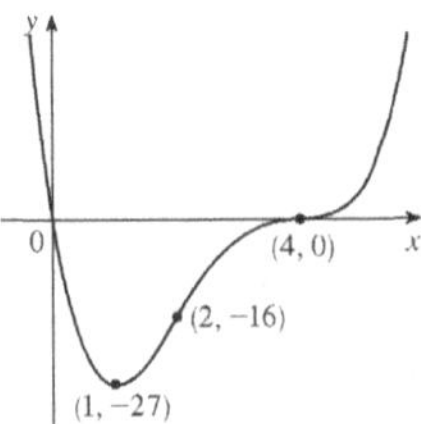

7. A. $\mathbb{R}$ B. y-int 0; x-int 0
C. About (0, 0) D. None
E. Inc on $(-\infty, \infty)$
F. None
G. CU on $(-2, 0)$, $(2, \infty)$;
CD on $(-\infty, -2)$, $(0, 2)$;
IP $\left(-2, -\frac{256}{15}\right)$, $(0, 0)$, $\left(2, \frac{256}{15}\right)$
H. See graph at right.

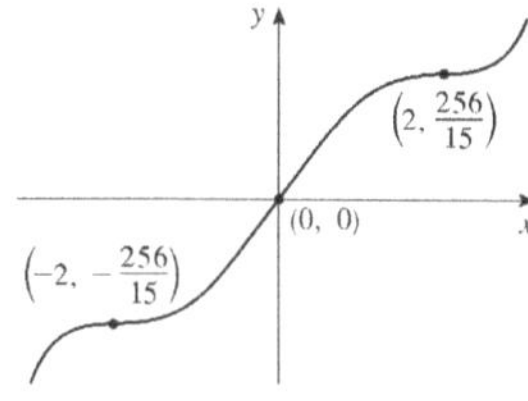

9. A. $\{x \mid x \neq 1\}$ B. y-int 0; x-int 0
C. None D. VA $x = 1$, HA $y = 1$
E. Dec on $(-\infty, 1)$, $(1, \infty)$
F. None
G. CU on $(1, \infty)$; CD on $(-\infty, 1)$
H. See graph at right.

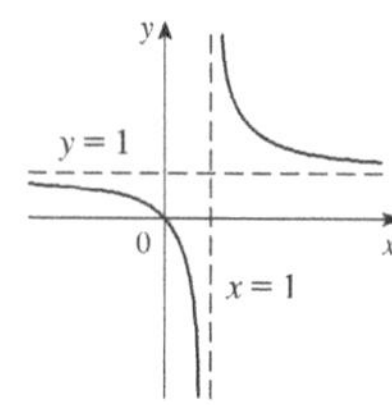

11. A. $(-\infty, 1) \cup (1, 2) \cup (2, \infty)$
B. y-int 0; x-int 0 C. None
D. HA $y = -1$; VA $x = 2$
E. Inc on $(-\infty, 1)$, $(1, 2)$, $(2, \infty)$
F. None
G. CU on $(-\infty, 1)$, $(1, 2)$;
CD on $(2, \infty)$
H. See graph at right.

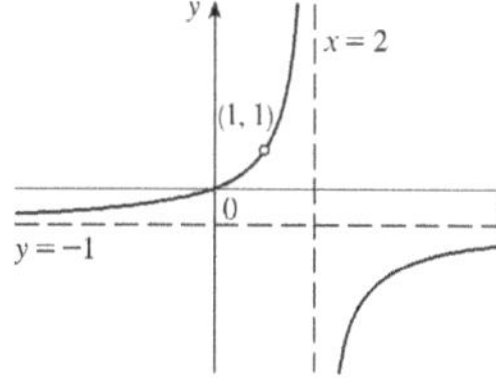

13. A. $\{x \mid x \neq \pm 3\}$ B. y-int $-\frac{1}{9}$
C. About y-axis D. VA $x = \pm 3$, HA $y = 0$
E. Inc on $(-\infty, -3)$, $(-3, 0)$;
dec on $(0, 3)$, $(3, \infty)$
F. Loc max $f(0) = -\frac{1}{9}$
G. CU on $(-\infty, -3)$, $(3, \infty)$;
CD on $(-3, 3)$
H. See graph at right.

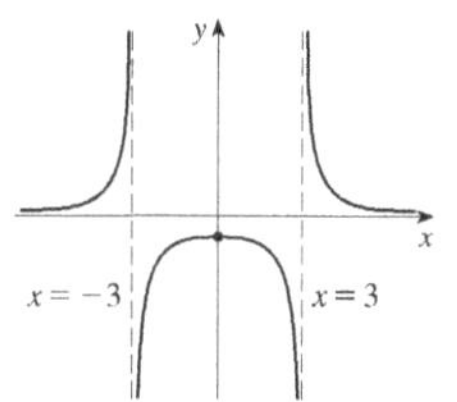

15. A. $\mathbb{R}$ B. y-int 0; x-int 0
C. About (0, 0) D. HA $y = 0$
E. Inc on $(-3, 3)$;
dec on $(-\infty, -3)$, $(3, \infty)$
F. Loc min $f(-3) = -\frac{1}{6}$;
loc max $f(3) = \frac{1}{6}$;
G. CU on $(-3\sqrt{3}, 0)$, $(3\sqrt{3}, \infty)$;
CD on $(-\infty, -3\sqrt{3})$, $(0, 3\sqrt{3})$;
IP $(0, 0)$, $(\pm 3\sqrt{3}, \pm\sqrt{3}/12)$
H. See graph at right.

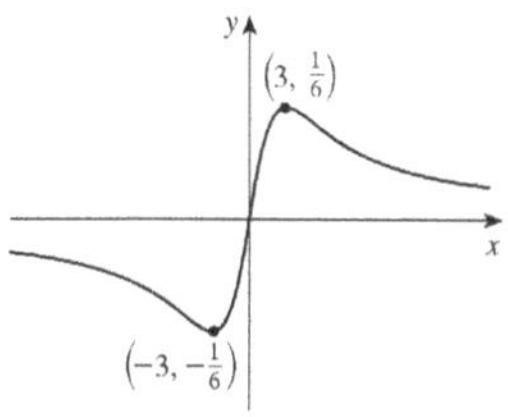

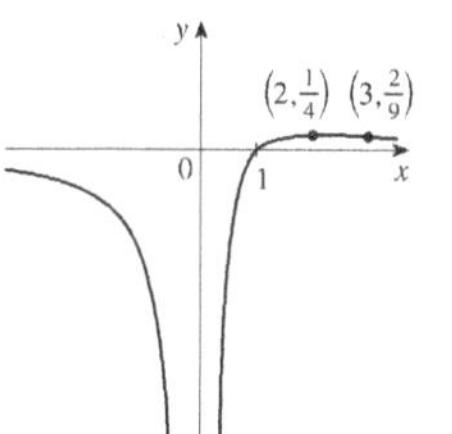

17. A. $(-\infty, 0) \cup (0,\infty)$ B. x-int 1
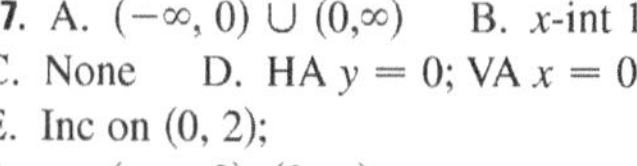
C. None D. HA $y = 0$; VA $x = 0$
E. Inc on $(0, 2)$;
dec on $(-\infty, 0)$, $(2, \infty)$
F. Loc max $f(2) = \frac{1}{4}$
G. CU on $(3, \infty)$;
CD on $(-\infty, 0)$, $(0, 3)$; IP $\left(3, \frac{2}{9}\right)$
H. See graph at right.

19. A. $\mathbb{R}$ B. y-int 0; x-int 0
C. About y-axis D. HA $y = 1$
E. Inc on $(0, \infty)$; dec on $(-\infty, 0)$
F. Loc min $f(0) = 0$
G. CU on $(-1, 1)$;
CD on $(-\infty, -1)$, $(1, \infty)$; IP $\left(\pm 1, \frac{1}{4}\right)$
H. See graph at right.

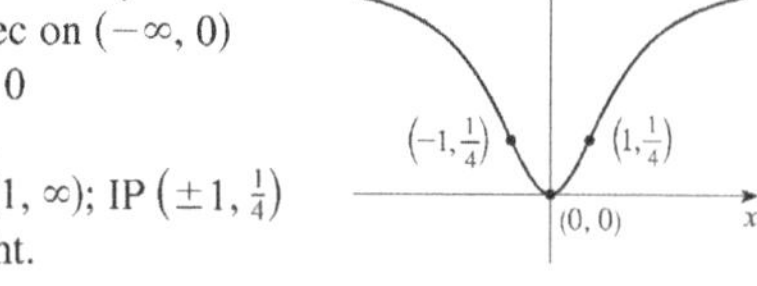

21. A. $[0, \infty)$ B. y-int 0; x-int 0, 3
C. None D. None
E. Inc on $(1, \infty)$; dec on $(0, 1)$
F. Loc min $f(1) = -2$
G. CU on $(0, \infty)$
H. See graph at right.

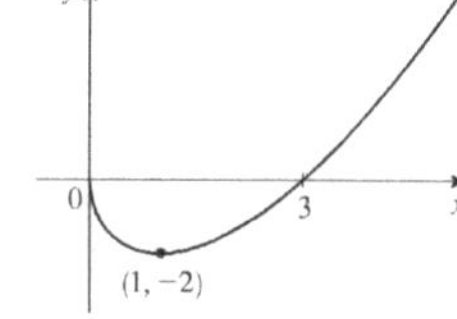

23. A. $(-\infty, -2] \cup [1, \infty)$
B. x-int -2, 1
C. None D. None
E. Inc on $(1, \infty)$; dec on $(-\infty, -2)$
F. None
G. CD on $(-\infty, -2)$, $(1, \infty)$
H. See graph at right.

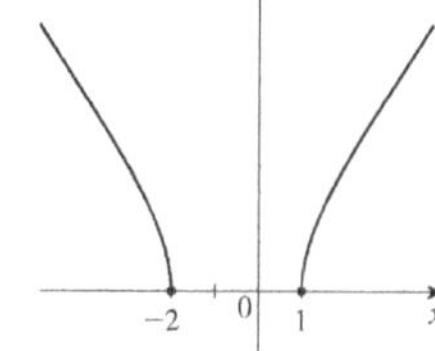

25. A. $\mathbb{R}$ B. y-int 0; x-int 0
C. About the origin
D. HA $y = \pm 1$
E. Inc on $(-\infty, \infty)$ F. None
G. CU on $(-\infty, 0)$;
CD on $(0, \infty)$; IP $(0, 0)$
H. See graph at right.

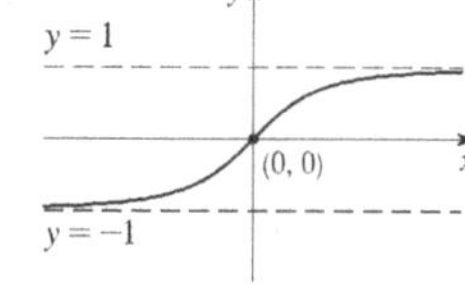

27. A. $\{x \mid |x| \leqslant 1, x \neq 0\} = [-1, 0) \cup (0, 1]$
B. x-int ± 1 C. About (0, 0)
D. VA $x = 0$
E. Dec on $(-1, 0)$, $(0, 1)$
F. None
G. CU on $(-1, -\sqrt{2/3})$, $(0, \sqrt{2/3})$;
CD on $(-\sqrt{2/3}, 0)$, $(\sqrt{2/3}, 1)$;
IP $(\pm\sqrt{2/3}, \pm 1/\sqrt{2})$
H. See graph at right.

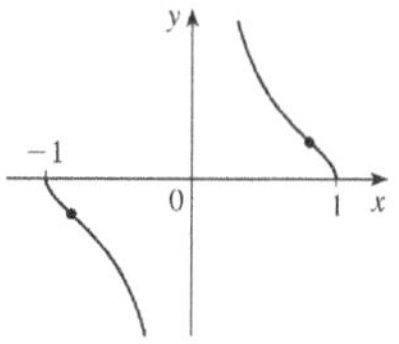

29. A. $\mathbb{R}$ B. y-int 0; x-int 0, $\pm 3\sqrt{3}$ C. About the origin
D. None E. Inc on $(-\infty, -1)$, $(1, \infty)$; dec on $(-1, 1)$
F. Loc max $f(-1) = 2$;
loc min $f(1) = -2$
G. CU on $(0, \infty)$; CD on $(-\infty, 0)$;
IP $(0, 0)$
H. See graph at right.

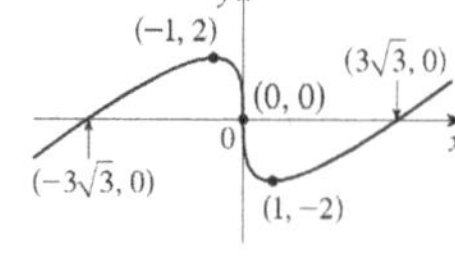

31. A. $\mathbb{R}$ B. y-int -1; x-int ± 1
C. About y-axis D. None
E. Inc on $(0, \infty)$; dec on $(-\infty, 0)$
F. Loc min $f(0) = -1$
G. CU on $(-1, 1)$;
CD on ;
IP
H. See graph at right.

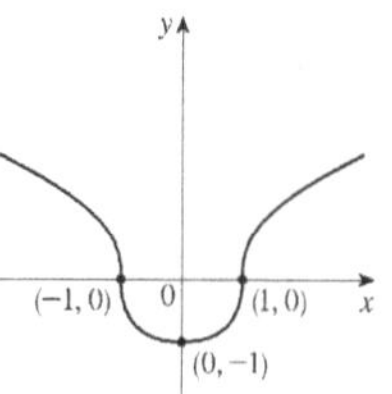

33. A. $\mathbb{R}$ B. y-int 0; x-int $n\pi$ (n an integer)
C. About (0, 0), period 2π D. None
E–G answers for $0 \leqslant x \leqslant \pi$:
E. Inc on $(0, \pi/2)$; dec on $(\pi/2, \pi)$ F. Loc max $f(\pi/2) = 1$
G. Let $\alpha = \sin^{-1}\sqrt{2/3}$; CU on $(0, \alpha)$, $(\pi - \alpha, \pi)$;
CD on $(\alpha, \pi - \alpha)$; IP at $x = 0, \pi, \alpha, \pi - \alpha$
H.

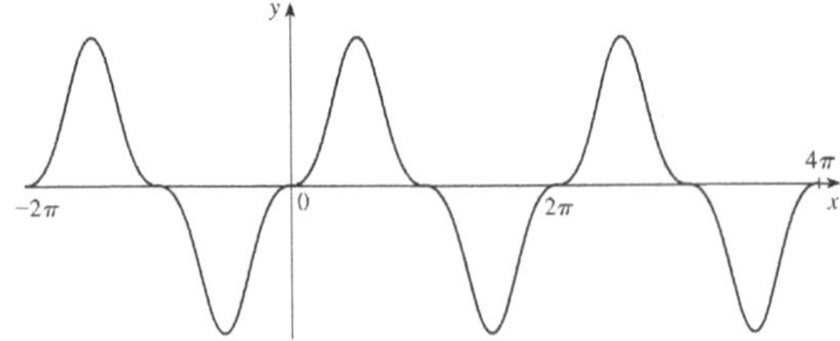

35. A. $(-\pi/2, \pi/2)$ B. y-int 0; x-int 0 C. About y-axis
D. VA $x = \pm\pi/2$
E. Inc on $(0, \pi/2)$;
dec on $(-\pi/2, 0)$
F. Loc min $f(0) = 0$
G. CU on $(-\pi/2, \pi/2)$
H. See graph at right.

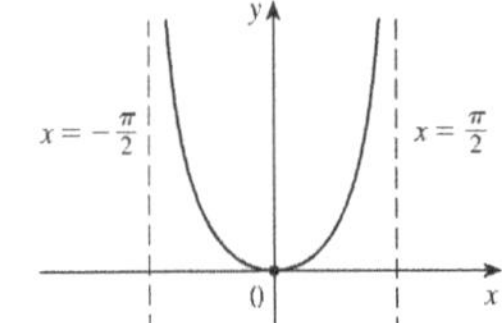

37. A. $(0, 3\pi)$ C. None D. None
E. Inc on $(\pi/3, 5\pi/3)$, $(7\pi/3, 3\pi)$;
dec on $(0, \pi/3)$, $(5\pi/3, 7\pi/3)$
F. Loc min $f(\pi/3) = (\pi/6) - \frac{1}{2}\sqrt{3}$, $f(7\pi/3) = (7\pi/6) - \frac{1}{2}\sqrt{3}$;
loc max $f(5\pi/3) = (5\pi/6) + \frac{1}{2}\sqrt{3}$
G. CU on $(0, \pi)$, $(2\pi, 3\pi)$;
CD on $(\pi, 2\pi)$;
IP $(\pi, \pi/2)$, $(2\pi, \pi)$
H. See graph at right.

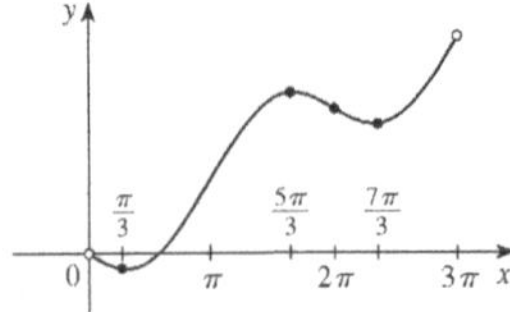

39. A. All reals except $(2n + 1)\pi$ (n an integer)
B. y-int 0; x-int $2n\pi$
C. About the origin, period 2π
D. VA $x = (2n + 1)\pi$
E. Inc on $((2n - 1)\pi, (2n + 1)\pi)$ F. None
G. CU on $(2n\pi, (2n + 1)\pi)$; CD on $((2n - 1)\pi, 2n\pi)$;
IP $(2n\pi, 0)$
H.

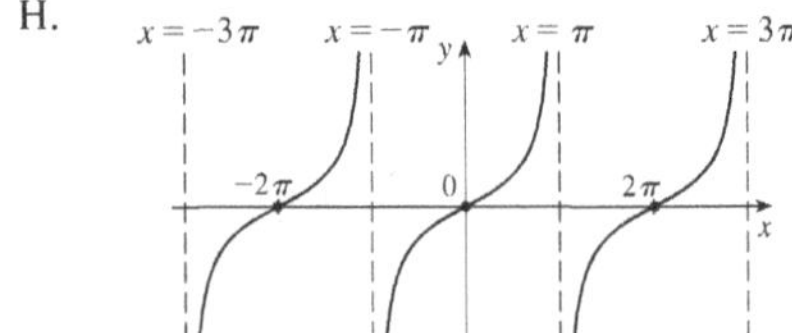

41. A. $\mathbb{R}$ B. y-int $\pi/4$
C. None
D. HA $y = 0$, $y = \pi/2$
E. Inc on $(-\infty, \infty)$ F. None
G. CU on $(-\infty, 0)$; CD on $(0, \infty)$;
IP $(0, \pi/4)$
H. See graph at right.

43. A. $\mathbb{R}$ B. y-int. $\frac{1}{2}$ C. None
D. HA $y = 0$, $y = 1$
E. Inc on $\mathbb{R}$ F. None
G. CU on $(-\infty, 0)$; CD on $(0, \infty)$;
IP $(0, \frac{1}{2})$ H. See graph at right.

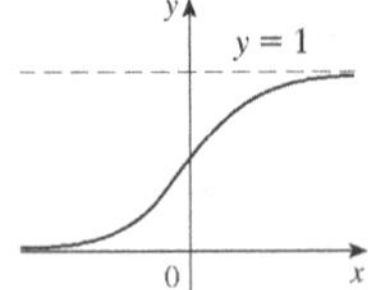

45. A. $(0, \infty)$ B. None
C. None D. VA $x = 0$
E. Inc on $(1, \infty)$; dec on $(0, 1)$
F. Loc min $f(1) = 1$
G. CU on $(0, \infty)$
H. See graph at right.

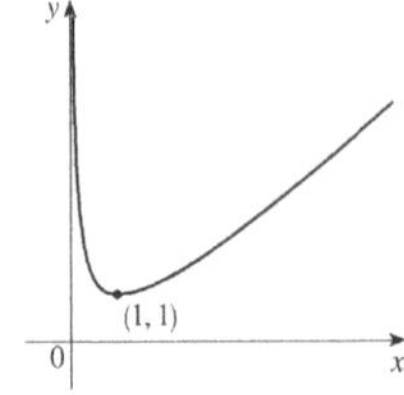

47. A. $\mathbb{R}$ B. y-int $\frac{1}{4}$ C. None
D. HA $y = 0$, $y = 1$
E. Dec on $\mathbb{R}$ F. None
G. CU on ; CD on $(-\infty, \ln\frac{1}{2})$;
IP $(\ln\frac{1}{2}, \frac{4}{9})$
H. See graph at right.

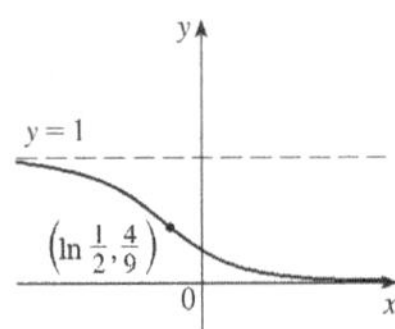

49. A. All x in $(2n\pi, (2n + 1)\pi)$ (n an integer)
B. x-int $\pi/2 + 2n\pi$ C. Period 2π D. VA $x = n\pi$
E. Inc on $(2n\pi, \pi/2 + 2n\pi)$; dec on $(\pi/2 + 2n\pi, (2n + 1)\pi)$
F. Loc max $f(\pi/2 + 2n\pi) = 0$ G. CD on $(2n\pi, (2n + 1)\pi)$
H.

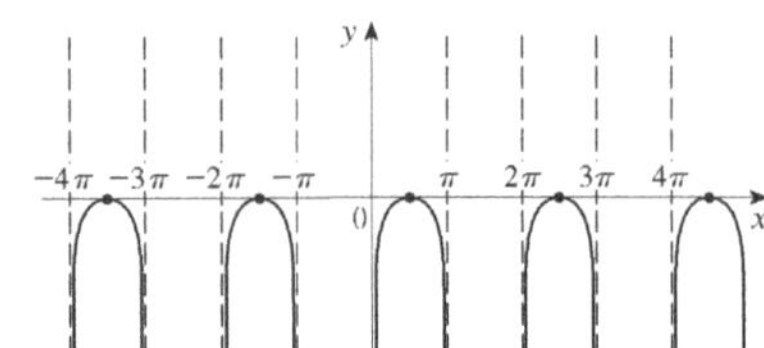

51. A. $(-\infty, 0) \cup (0, \infty)$
B. None C. None D. VA $x = 0$
E. Inc on $(-\infty, -1)$, $(0, \infty)$;
dec on $(-1, 0)$
F. Loc max $f(-1) = -e$
G. CU on $(0, \infty)$; CD on $(-\infty, 0)$
H. See graph at right.

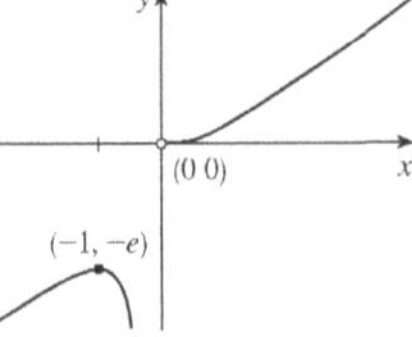

53. A. $\mathbb{R}$ B. y-int 2
C. None D. None
E. Inc on $(\frac{1}{5}\ln\frac{2}{3}, \infty)$; dec on $(-\infty, \frac{1}{5}\ln\frac{2}{3})$
F. Loc min $f(\frac{1}{5}\ln\frac{2}{3}) = (\frac{2}{3})^{3/5} + (\frac{2}{3})^{-2/5}$
G. CU on $(-\infty, \infty)$
H. See graph at right.

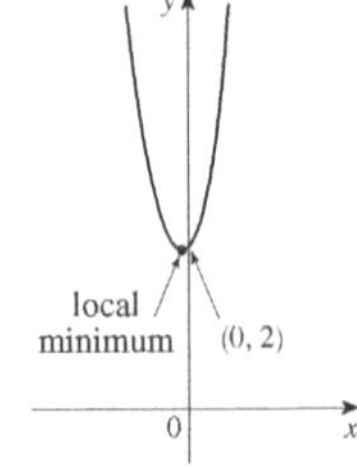

55.

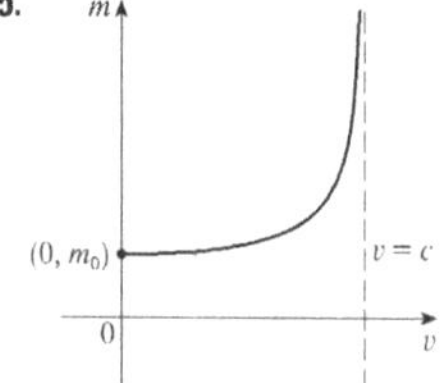

57. (a) When $t = (\ln a)/k$ (b) When $t = (\ln a)/k$
(c)

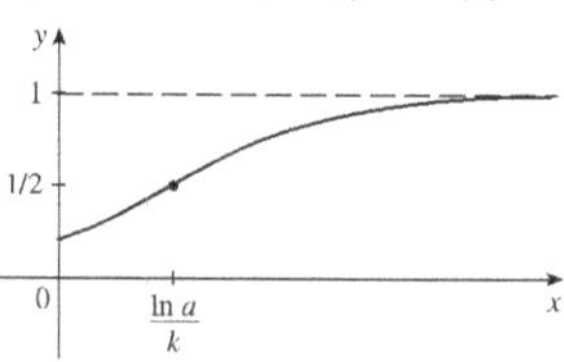

59.

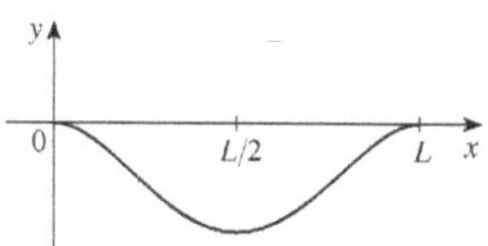

61. $y = x - 1$ **63.** $y = 2x - 2$

65. A. $(-\infty, 1) \cup (1, \infty)$
B. y-int 0; x-int 0 C. None
D. VA $x = 1$; SA $y = x + 1$
E. Inc on $(-\infty, 0)$, $(2, \infty)$;
dec on $(0, 1)$, $(1, 2)$
F. Loc max $f(0) = 0$;
loc min $f(2) = 4$
G. CU on $(1, \infty)$; CD on $(-\infty, 1)$
H. See graph at right.

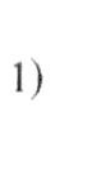

67. A. $(-\infty, 0) \cup (0, \infty)$
B. x-int $-\sqrt[3]{4}$ C. None
D. VA $x = 0$; SA $y = x$
E. Inc. on $(-\infty, 0)$, $(2, \infty)$;
dec on $(0, 2)$
F. Loc min $f(2) = 3$
G. CU on $(-\infty, 0)$, $(0, \infty)$
H. See graph at right.

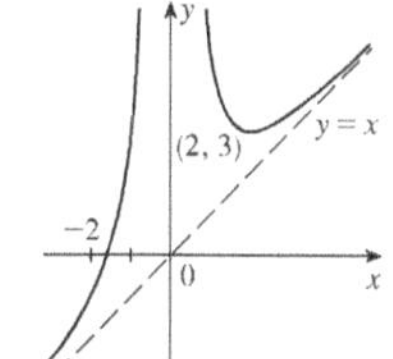

69. A. $\mathbb{R}$ B. y-int 2
C. None
D. SA $y = 1 + \frac{1}{2}x$
E. Inc. on $(\ln 2, \infty)$;
dec on $(-\infty, \ln 2)$
F. Loc min $f(\ln 2) = \frac{3}{2} + \frac{1}{2}\ln 2$
G. CU on $(-\infty, \infty)$
H. See graph at right.

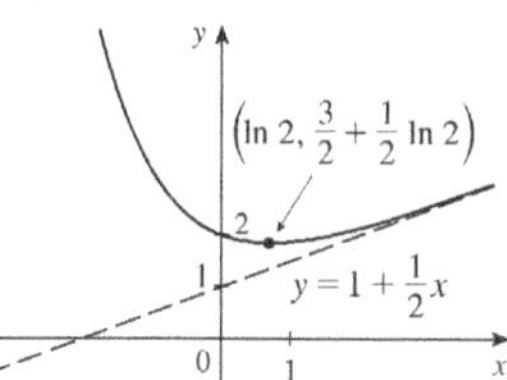

71.

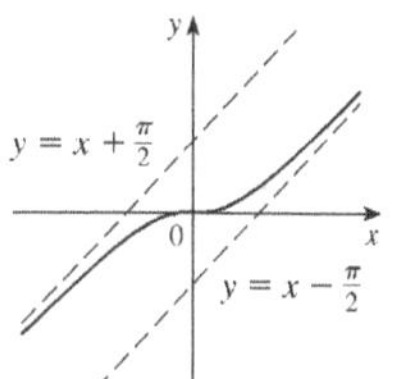

75. VA $x = 0$, asymptotic to $y = x^3$

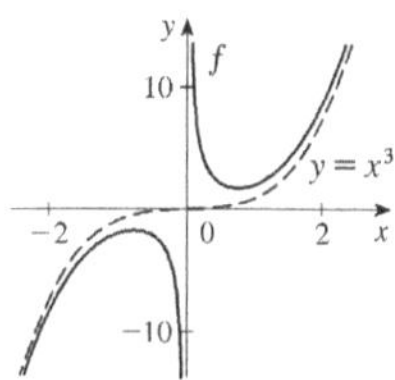

EXERCISES 4.6 ■ PAGE 324

1. Inc on (0.92, 2.5), (2.58, ∞); dec on , (2.5, 2.58);
loc max $f(2.5) = 4$; loc min $f(0.92) \approx -5.12$, $f(2.58) \approx 3.998$;
CU on $(-\infty, 1.46)$, $(2.54, \infty)$;
CD on (1.46, 2.54); IP (1.46, −1.40), (2.54, 3.999)

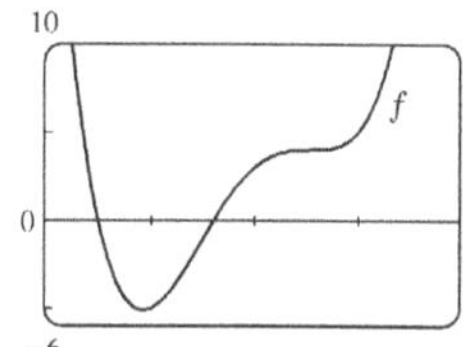

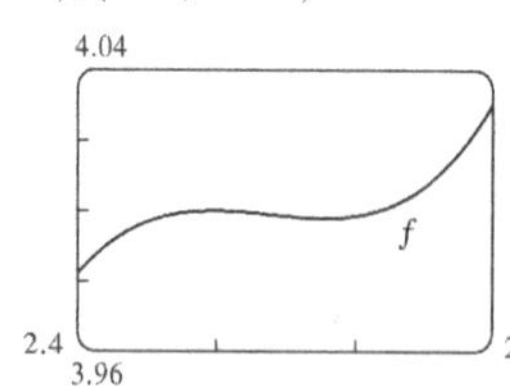

3. Inc on (−15, 4.40), (18.93, ∞);
dec on $(-\infty, -15)$, (4.40, 18.93);
loc max $f(4.40) \approx 53{,}800$; loc min $f(-15) \approx -9{,}700{,}000$,
$f(18.93) \approx -12{,}700{,}000$; CU on $(-\infty, -11.34)$, (0, 2.92),
(15.08, ∞); CD on (−11.34, 0), (2.92, 15.08);
IP (0, 0), $\approx$ (−11.34, −6,250,000), (2.92, 31,800),
(15.08, −8,150,000)

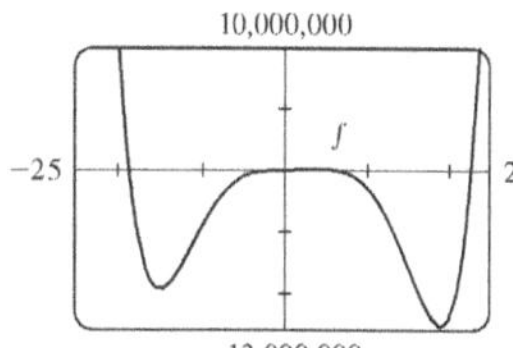

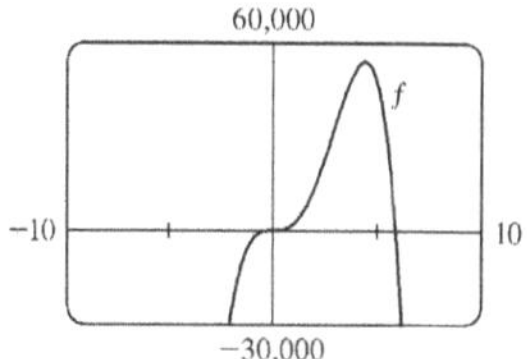

5. Inc on $(-\infty, -1.47)$, (−1.47, 0.66); dec on (0.66, ∞);
loc max $f(0.66) \approx 0.38$; CU on $(-\infty, -1.47)$, (−0.49, 0),
(1.10, ∞); CD on (−1.47, −0.49), (0, 1.10);
IP (−0.49, −0.44), (1.10, 0.31)

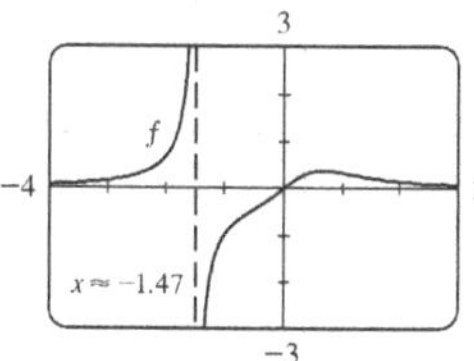

7. Inc on $(-1.40, -0.44)$, $(0.44, 1.40)$; dec on $(-\pi, -1.40)$, $(-0.44, 0)$, $(0, 0.44)$, $(1.40, \pi)$; loc max $f(-0.44) \approx -4.68$, $f(1.40) \approx 6.09$; loc min $f(-1.40) \approx -6.09, f(0.44) \approx 5.22$; CU on $(-\pi, -0.77)$, $(0, 0.77)$; CD on $(-0.77, 0)$, $(0.77, \pi)$; IP $(-0.77, -5.22)$, $(0.77, 5.22)$

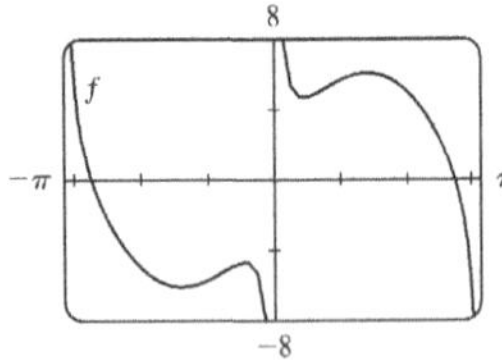

9. Inc on $(-8 - \sqrt{61}, -8 + \sqrt{61})$; dec on $(-\infty, -8 - \sqrt{61})$, $(-8 + \sqrt{61}, 0)$, $(0, \infty)$; CU on $(-12 - \sqrt{138}, -12 + \sqrt{138})$, $(0, \infty)$; CD on $(-\infty, -12 - \sqrt{138})$, $(-12 + \sqrt{138}, 0)$

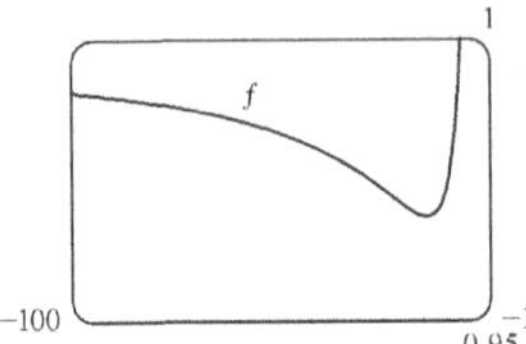

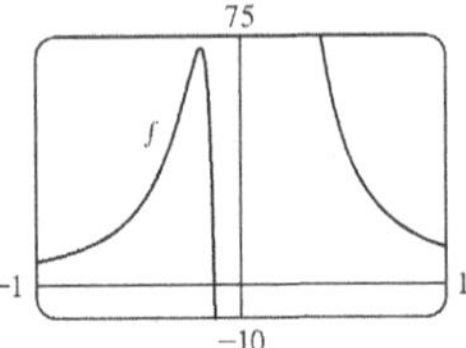

11. (a)

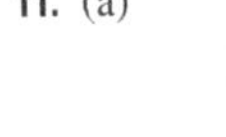

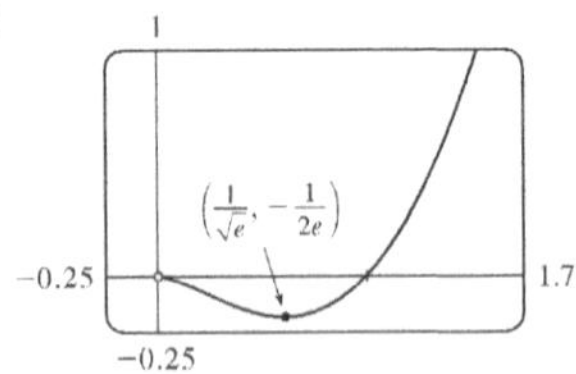

(b) $\lim_{x\to 0^+} f(x) = 0$
(c) Loc min $f(1/\sqrt{e}) = -1/(2e)$; CD on $(0, e^{-3/2})$; CU on $(e^{-3/2}, \infty)$

13. Loc max $f(-5.6) \approx 0.018$, $f(0.82) \approx -281.5$, $f(5.2) \approx 0.0145$; loc min $f(3) = 0$

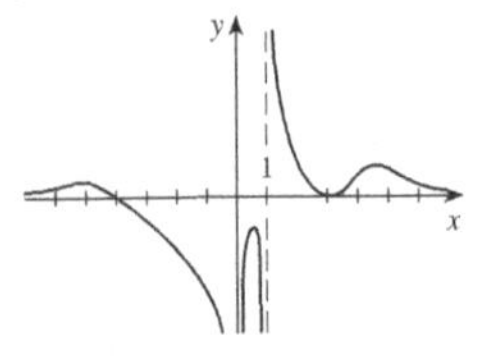

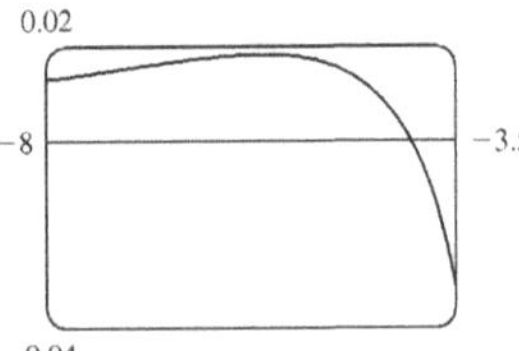

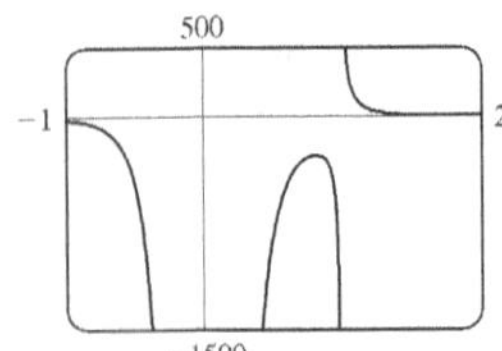

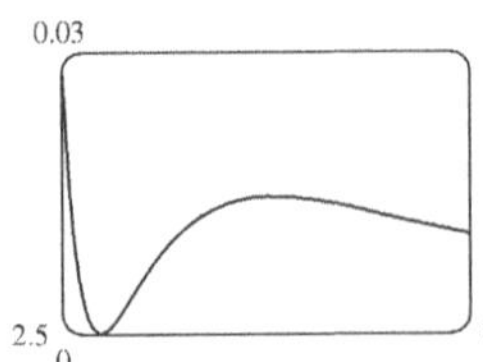

15. $f'(x) = -\dfrac{x(x+1)^2(x^3 + 18x^2 - 44x - 16)}{(x-2)^3(x-4)^5}$

$$f''(x) = 2\frac{(x+1)(x^6 + 36x^5 + 6x^4 - 628x^3 + 684x^2 + 672x + 64)}{(x-2)^4(x-4)^6}$$

CU on $(-35.3, -5.0)$, $(-1, -0.5)$, $(-0.1, 2)$, $(2, 4)$, $(4, \infty)$;
CD on $(-\infty, -35.3)$, $(-5.0, -1)$, $(-0.5, -0.1)$;
IP $(-35.3, -0.015)$, $(-5.0, -0.005)$, $(-1, 0)$, $(-0.5, 0.00001)$, $(-0.1, 0.0000066)$

17. Inc on $(-9.41, -1.29)$, $(0, 1.05)$; dec on $(-\infty, -9.41)$, $(-1.29, 0)$, $(1.05, \infty)$; loc max $f(-1.29) \approx 7.49, f(1.05) \approx 2.35$; loc min $f(-9.41) \approx -0.056, f(0) = 0.5$; CU on $(-13.81, -1.55)$, $(-1.03, 0.60)$, $(1.48, \infty)$; CD on $(-\infty, -13.81)$, $(-1.55, -1.03)$, $(0.60, 1.48)$; IP $(-13.81, -0.05)$, $(-1.55, 5.64)$, $(-1.03, 5.39)$, $(0.60, 1.52)$, $(1.48, 1.93)$

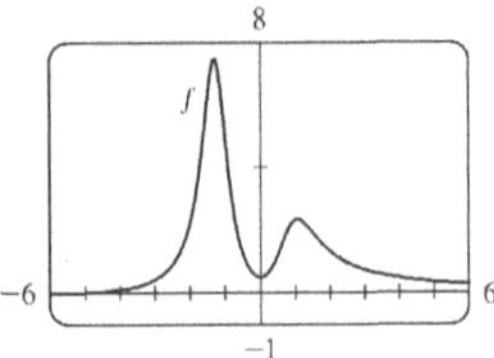

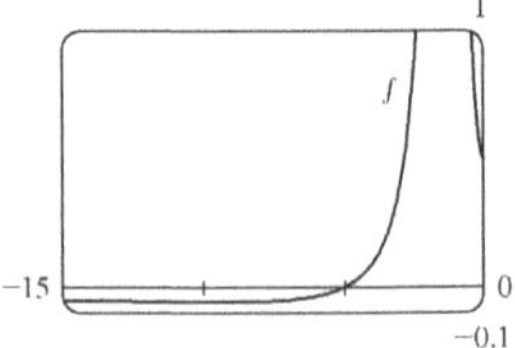

19. Inc on $(-4.91, -4.51)$, $(0, 1.77)$, $(4.91, 8.06)$, $(10.79, 14.34)$, $(17.08, 20)$;
dec on $(-4.51, -4.10)$, $(1.77, 4.10)$, $(8.06, 10.79)$, $(14.34, 17.08)$;
loc max $f(-4.51) \approx 0.62, f(1.77) \approx 2.58, , f(14.34) \approx 4.39$;
loc min $f(10.79) \approx 2.43, f(17.08) \approx 3.49$;
CU on $(9.60, 12.25)$, $(15.81, 18.65)$;
CD on $(-4.91, -4.10)$, $(0, 4.10)$, $(4.91, 9.60)$, $(12.25, 15.81)$, $(18.65, 20)$;
IP at $(9.60, 2.95)$, $(12.25, 3.27)$, $(15.81, 3.91)$, $(18.65, 4.20)$

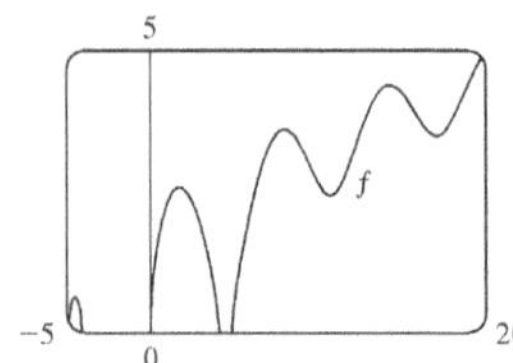

21. Inc on $(-\infty, 0)$, $(0, \infty)$;
CU on $(-\infty, -0.42)$, $(0, 0.42)$;
CD on $(-0.42, 0)$, $(0.42, \infty)$;
IP $(\mp 0.42, \pm 0.83)$

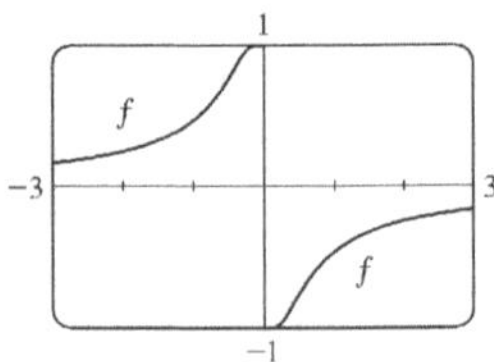

23.

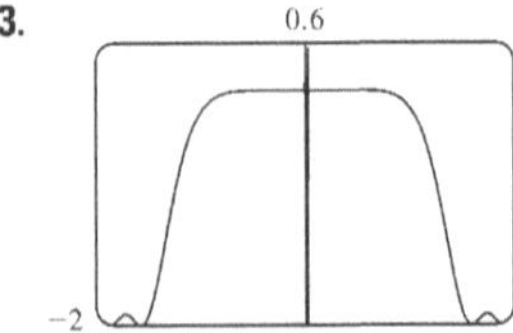

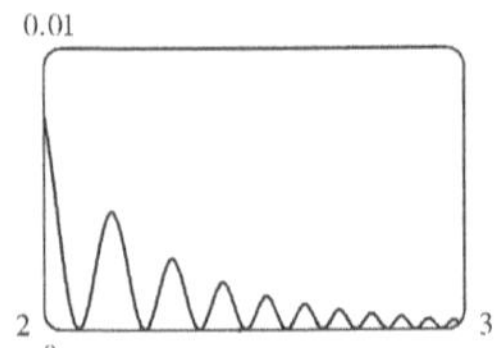

25. (a)

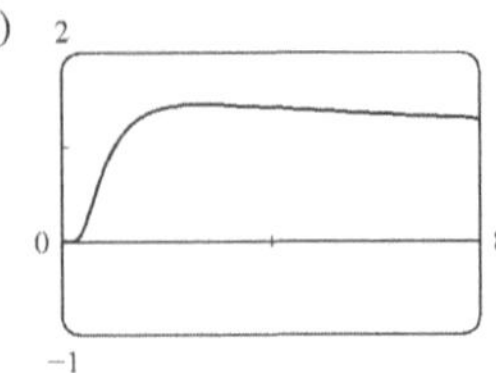

(b) $\lim_{x\to 0^+} x^{1/x} = 0$, $\lim_{x\to\infty} x^{1/x} = 1$
(c) Loc max $f(e) = e^{1/e}$ (d) IP at $x \approx 0.58, 4.37$

27. Max $f(0.59) \approx 1$, $f(0.68) \approx 1$, $f(1.96) \approx 1$;
min $f(0.64) \approx 0.99996$, $f(1.46) \approx 0.49$, $f(2.73) \approx -0.51$;
IP $(0.61, 0.99998)$, $(0.66, 0.99998)$, $(1.17, 0.72)$, $(1.75, 0.77)$, $(2.28, 0.34)$

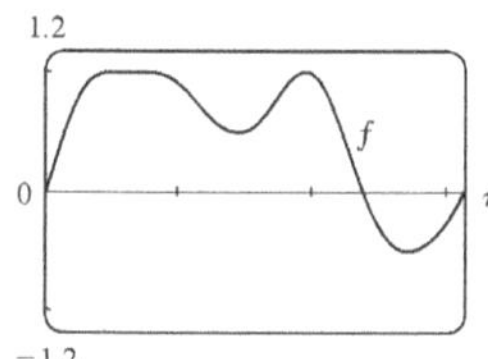

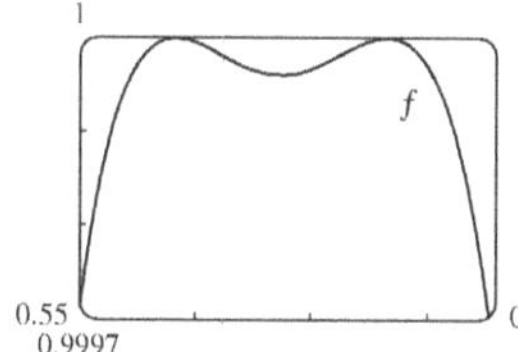

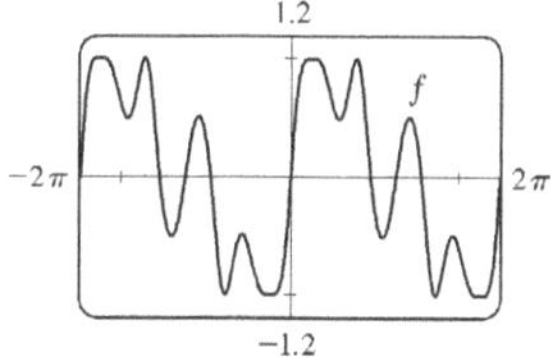

29. For $c \geqslant 0$, there is an absolute minimum at the origin. There are no other maxima or minima. The more negative c becomes, the farther the two IPs move from the origin. $c = 0$ is a transitional value.

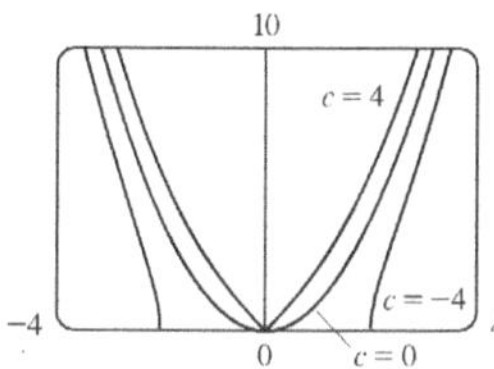

31. For $c < 0$, there is no extreme point and one IP, which decreases along the x-axis. For $c > 0$, there is no IP, and one minimum point.

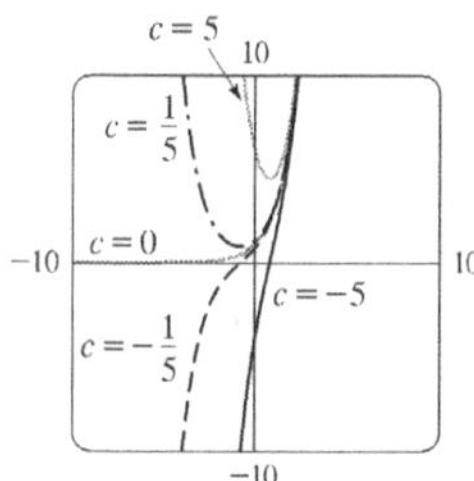

33. For $c > 0$, the maximum and minimum values are always $\pm\frac{1}{2}$, but the extreme points and IPs move closer to the y-axis as c increases. $c = 0$ is a transitional value: when c is replaced by $-c$, the curve is reflected in the x-axis.

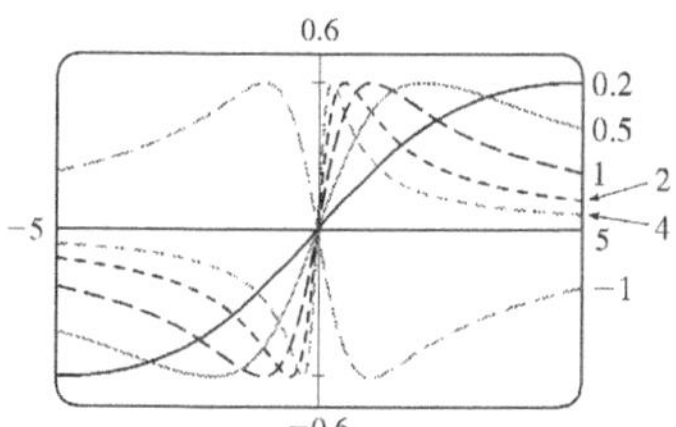

35. For $|c| < 1$, the graph has loc max and min values; for $|c| \geqslant 1$ it does not. The function increases for $c \geqslant 1$ and decreases for $c \leqslant -1$. As c changes, the IPs move vertically but not horizontally.

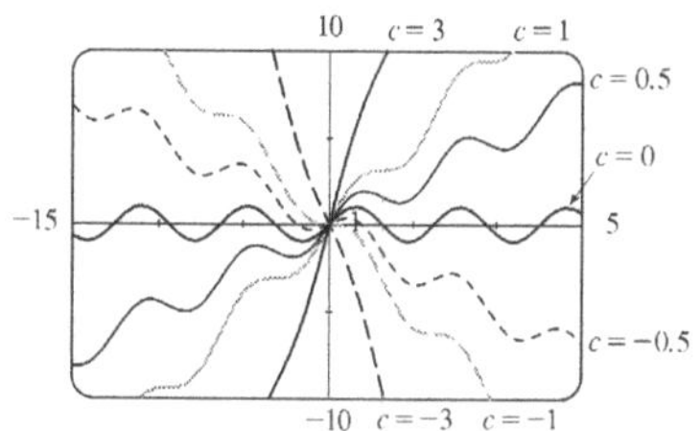

37.

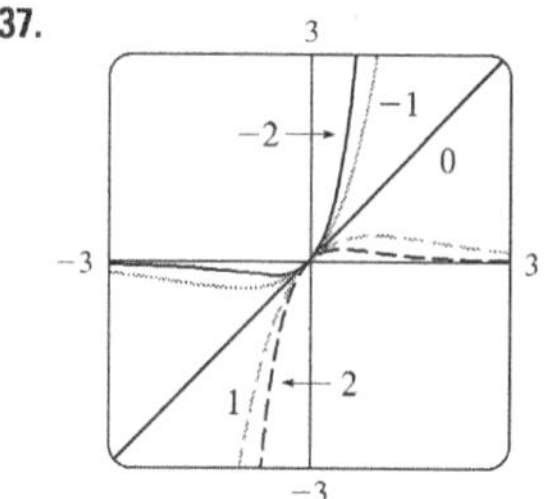

For $c > 0$, $\lim_{x\to\infty} f(x) = 0$ and $\lim_{x\to -\infty} f(x) = -\infty$.
For $c < 0$, $\lim_{x\to\infty} f(x) = \infty$ and $\lim_{x\to-\infty} f(x) = 0$.
As $|c|$ increases, the max and min points and the IPs get closer to the origin.

39. (a) Positive (b)

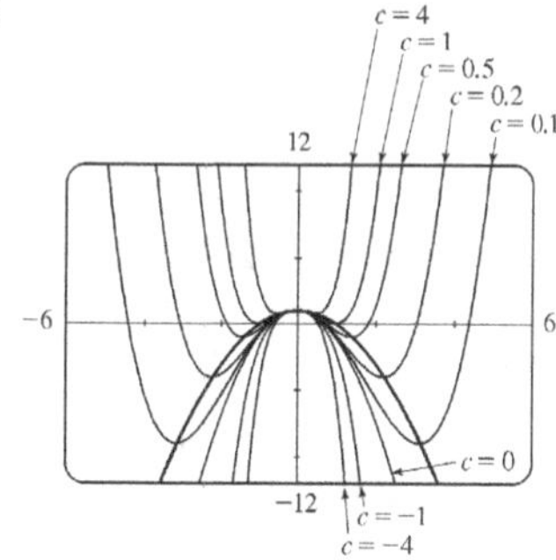

EXERCISES 4.7 ■ PAGE 331

1. (a) 11, 12 (b) 11.5, 11.5 **3.** 10, 10 **5.** $\frac{9}{4}$
7. 25 m by 25 m **9.** $N = 1$
11. (a)

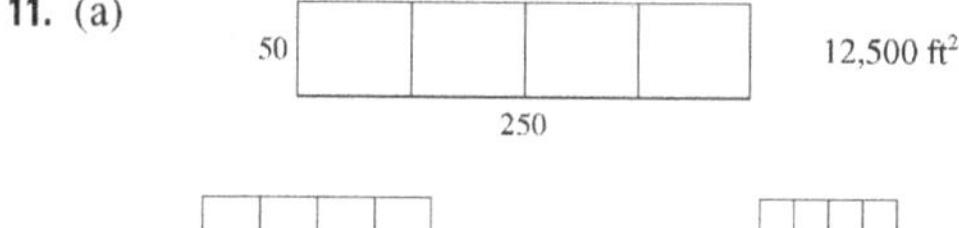

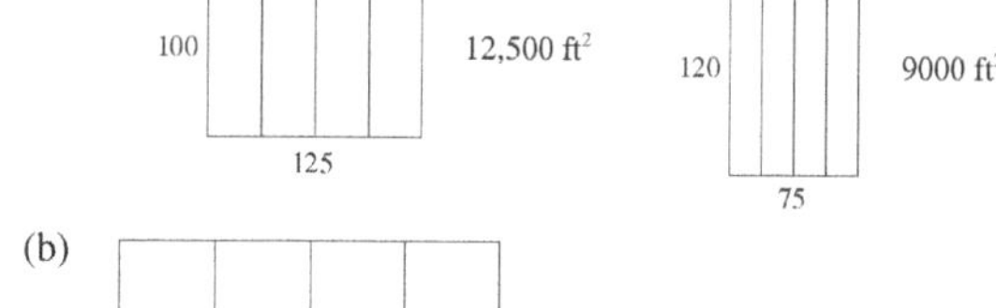

(b)

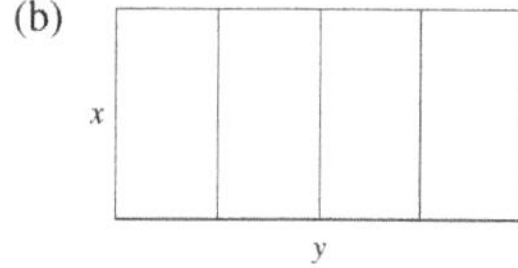

(c) $A = xy$ (d) $5x + 2y = 750$ (e) $A(x) = 375x - \frac{5}{2}x^2$
(f) 14,062.5 ft^2
13. 1000 ft by 1500 ft **15.** 4000 cm^3 **17.** \$191.28
19. $(-\frac{6}{5}, \frac{3}{5})$ **21.** $(-\frac{1}{3}, \pm\frac{4}{3}\sqrt{2})$ **23.** Square, side $\sqrt{2}r$
25. $L/2$, $\sqrt{3}\,L/4$ **27.** Base $\sqrt{3}r$, height $3r/2$
29. $4\pi r^3/(3\sqrt{3})$ **31.** $\pi r^2(1 + \sqrt{5})$ **33.** 24 cm, 36 cm
35. (a) Use all of the wire for the square
(b) $40\sqrt{3}/(9 + 4\sqrt{3})$ m for the square
37. Height = radius = $\sqrt[3]{V/\pi}$ cm
39. $V = 2\pi R^3/(9\sqrt{3})$ **43.** $E^2/(4r)$
45. (a) $\frac{3}{2}s^2 \csc\theta\,(\csc\theta - \sqrt{3}\cot\theta)$ (b) $\cos^{-1}(1/\sqrt{3}) \approx 55°$
(c) $6s[h + s/(2\sqrt{2})]$
47. Row directly to B **49.** ≈ 4.85 km east of the refinery
51. $10\sqrt[3]{3}/(1 + \sqrt[3]{3})$ ft from the stronger source
53. $(a^{2/3} + b^{2/3})^{3/2}$ **55.** $2\sqrt{6}$
57. (b) (i) \$342,491; \$342/unit; \$390/unit (ii) 400
(iii) \$320/unit
59. (a) $p(x) = 19 - \frac{1}{3000}x$ (b) \$9.50
61. (a) $p(x) = 550 - \frac{1}{10}x$ (b) \$175 (c) \$100
65. 9.35 m **69.** $x = 6$ in. **71.** $\pi/6$
73. At a distance $5 - 2\sqrt{5}$ from A **75.** $\frac{1}{2}(L + W)^2$
77. (a) About 5.1 km from B (b) C is close to B; C is close to D; $W/L = \sqrt{25 + x^2}/x$, where $x = |BC|$
(c) ≈ 1.07; no such value (d) $\sqrt{41}/4 \approx 1.6$

EXERCISES 4.8 ■ PAGE 342

1. (a) $x_2 \approx 2.3$, $x_3 \approx 3$ (b) No
3. $\frac{9}{2}$ **5.** a, b, c **7.** 1.1785 **9.** -1.25 **11.** 1.82056420
13. 1.217562 **15.** -1.964636
17. -3.637958, -1.862365, 0.889470
19. 1.412391, 3.057104 **21.** 0, ± 0.902025
23. -1.93822883, -1.21997997, 1.13929375, 2.98984102
25. 0.76682579 **27.** 0.21916368, 1.08422462
29. (b) 31.622777
35. (a) -1.293227, -0.441731, 0.507854 (b) -2.0212
37. (1.520092, 2.306964) **39.** (0.410245, 0.347810)
41. 0.76286%

EXERCISES 4.9 ■ PAGE 348

1. $F(x) = \frac{1}{2}x^2 - 3x + C$ **3.** $F(x) = \frac{1}{2}x + \frac{1}{4}x^3 - \frac{1}{5}x^4 + C$
5. $F(x) = \frac{2}{3}x^3 + \frac{1}{2}x^2 - x + C$ **7.** $F(x) = 5x^{7/5} + 40x^{1/5} + C$
9. $F(x) = \sqrt{2}x + C$ **11.** $F(x) = 2x^{3/2} - \frac{3}{2}x^{4/3} + C$
13. $F(x) = \begin{cases} \frac{1}{5}x - 2\ln|x| + C_1 & \text{if } x < 0 \\ \frac{1}{5}x - 2\ln|x| + C_2 & \text{if } x > 0 \end{cases}$
15. $G(t) = 2t^{1/2} + \frac{2}{3}t^{3/2} + \frac{2}{5}t^{5/2} + C$
17. $H(\theta) = -2\cos\theta - \tan\theta + C_n$ on $(n\pi - \pi/2, n\pi + \pi/2)$, n an integer
19. $F(x) = 5e^x - 3\sinh x + C$
21. $F(x) = \frac{1}{2}x^2 - \ln|x| - 1/x^2 + C$
23. $F(x) = x^5 - \frac{1}{3}x^6 + 4$ **25.** $f(x) = x^5 - x^4 + x^3 + Cx + D$
27. $\frac{3}{20}x^{8/3} + Cx + D$ **29.** $f(t) = -\sin t + Ct^2 + Dt + E$
31. $f(x) = x + 2x^{3/2} + 5$ **33.** $f(t) = 4\arctan t - \pi$
35. $2\sin t + \tan t + 4 - 2\sqrt{3}$
37. $\frac{3}{2}x^{2/3} - \frac{1}{2}$ if $x > 0$; $\frac{3}{2}x^{2/3} - \frac{5}{2}$ if $x < 0$
39. $-x^2 + 2x^3 - x^4 + 12x + 4$ **41.** $-\sin\theta - \cos\theta + 5\theta + 4$
43. $f(x) = 2x^2 + x^3 + 2x^4 + 2x + 3$ **45.** $x^2 - \cos x - \frac{1}{2}\pi x$
47. $-\ln x + (\ln 2)x - \ln 2$ **49.** 10 **51.** b
53.

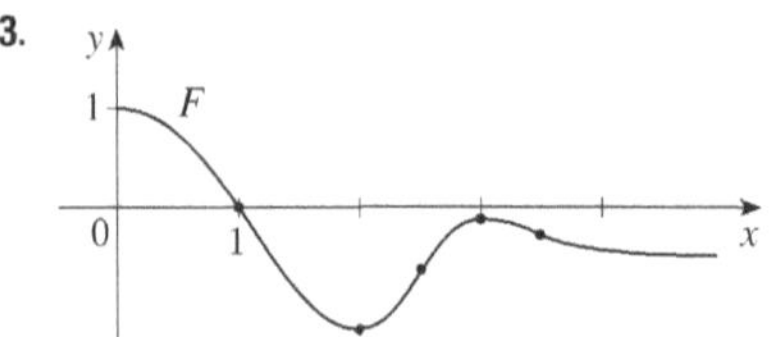

55. **57.**

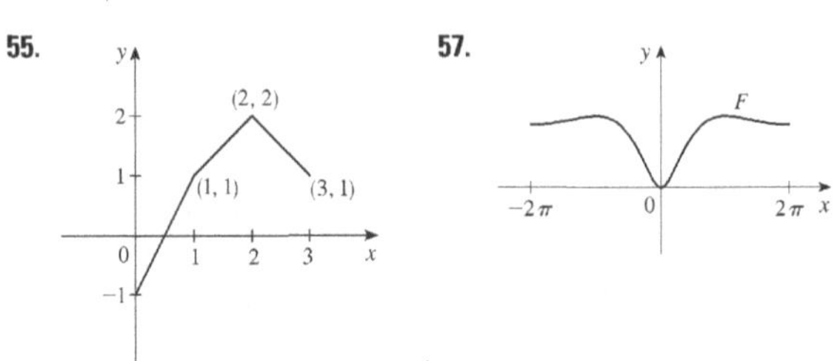

59. $s(t) = 1 - \cos t - \sin t$
61. $s(t) = \frac{1}{3}t^3 + \frac{1}{2}t^2 - 2t + 3$
63. $s(t) = -10\sin t - 3\cos t + (6/\pi)t + 3$
65. (a) $s(t) = 450 - 4.9t^2$ (b) $\sqrt{450/4.9} \approx 9.58$ s
(c) $-9.8\sqrt{450/4.9} \approx -93.9$ m/s (d) About 9.09 s
69. 225 ft **71.** \$742.08 **73.** $\frac{130}{11} \approx 11.8$ s
75. $\frac{88}{15} \approx 5.87$ ft/s^2 **77.** 62,500 km/h^2 ≈ 4.82 m/s^2
79. (a) 22.9125 mi (b) 21.675 mi (c) 30 min 33 s
(d) 55.425 mi

CHAPTER 4 REVIEW ■ PAGE 351

True-False Quiz

1. False **3.** False **5.** True **7.** False **9.** True
11. True **13.** False **15.** True **17.** True **19.** True

Exercises

1. Abs max $f(4) = 5$, abs and loc min $f(3) = 1$
3. Abs max $f(2) = \frac{2}{5}$, abs and loc min $f(-\frac{1}{3}) = -\frac{9}{2}$
5. Abs and loc max $f(\pi/6) = \pi/6 + \sqrt{3}$,
abs min $f(-2) = -\pi - 2$, loc min $f(5\pi/6) = 5\pi/6 - \sqrt{3}$
7. 1 **9.** 8 **11.** 0 **13.** $\frac{1}{2}$
15.

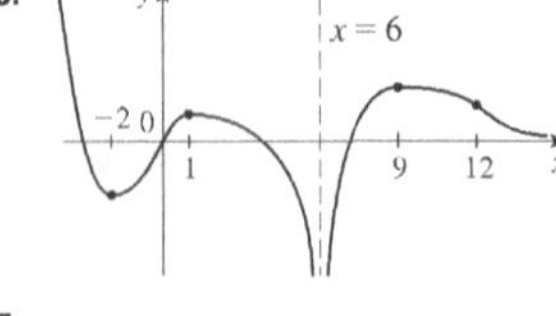

17.

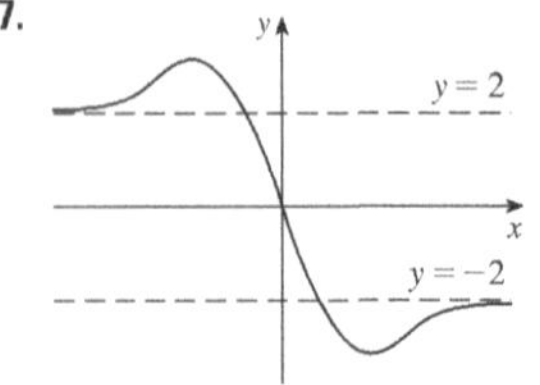

19. A. $\mathbb{R}$ B. y-int 2
C. None D. None
E. Dec on $(-\infty, \infty)$ F. None
G. CU on $(-\infty, 0)$;
CD on $(0, \infty)$; IP $(0, 2)$
H. See graph at right.

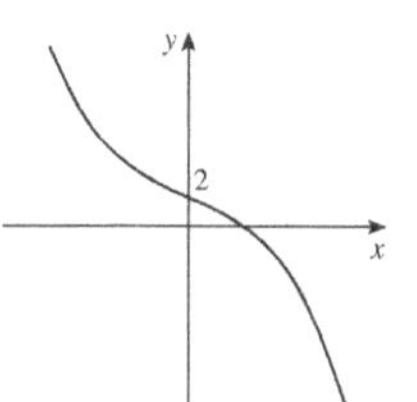

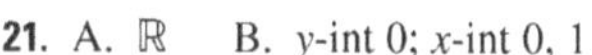

21. A. $\mathbb{R}$ B. y-int 0; x-int 0, 1
C. None D. None
E. Inc on $\left(\frac{1}{4}, \infty\right)$, dec on $\left(-\infty, \frac{1}{4}\right)$
F. Loc min $f\left(\frac{1}{4}\right) = -\frac{27}{256}$
G. CU on $\left(-\infty, \frac{1}{2}\right)$, $(1, \infty)$;
CD on $\left(\frac{1}{2}, 1\right)$; IP $\left(\frac{1}{2}, -\frac{1}{16}\right)$, $(1, 0)$
H. See graph at right.

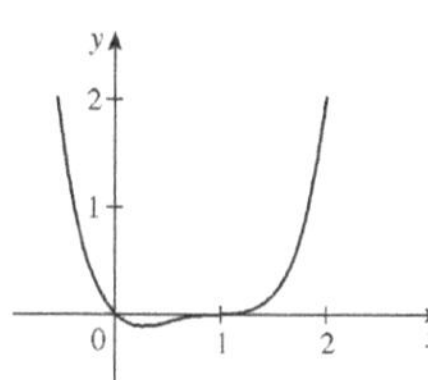

23. A. $\{x \mid x \neq 0, 3\}$
B. None C. None
D. HA $y = 0$; VA $x = 0$, $x = 3$
E. Inc on $(1, 3)$; dec on $(-\infty, 0)$,
$(0, 1)$, $(3, \infty)$
F. Loc min $f(1) = \frac{1}{4}$
G. CU on $(0, 3)$, $(3, \infty)$; CD on $(-\infty, 0)$
H. See graph at right.

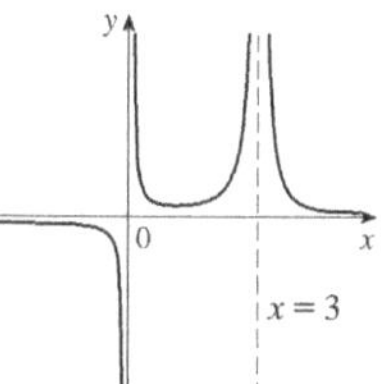

25. A. $\{x \mid x \neq -8\}$
B. y-int 0, x-int 0 C. None
D. VA $x = -8$; SA $y = x - 8$
E. Inc on $(-\infty, -16)$, $(0, \infty)$;
dec on $(-16, -8)$, $(-8, 0)$
F. Loc max $f(-16) = -32$;
loc min $f(0) = 0$
G. CU on $(-8, \infty)$; CD on $(-\infty, -8)$
H. See graph at right.

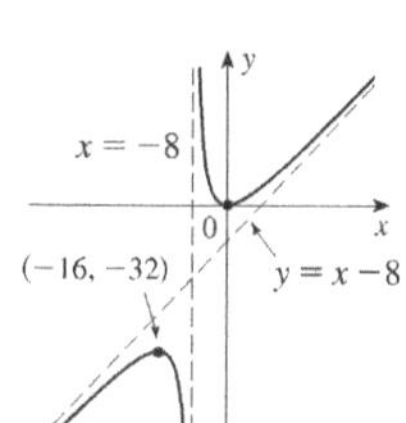

27. A. $[-2, \infty)$
B. y-int 0; x-int -2, 0
C. None D. None
E. Inc on $\left(-\frac{4}{3}, \infty\right)$, dec on $\left(-2, -\frac{4}{3}\right)$
F. Loc min $f\left(-\frac{4}{3}\right) = -\frac{4}{9}\sqrt{6}$
G. CU on $(-2, \infty)$
H. See graph at right.

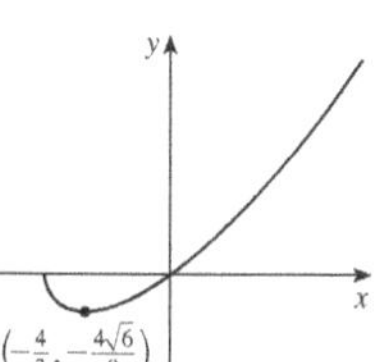
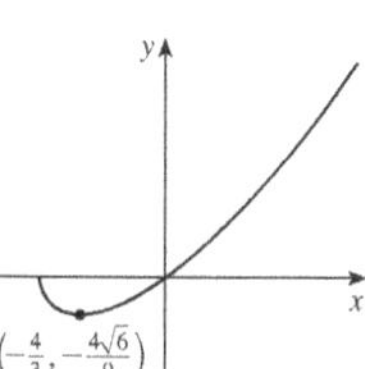
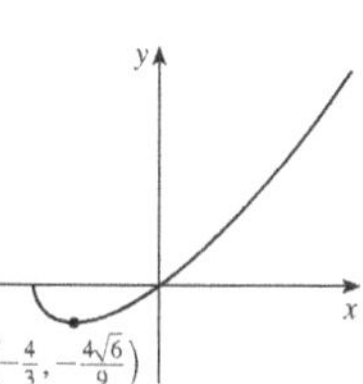
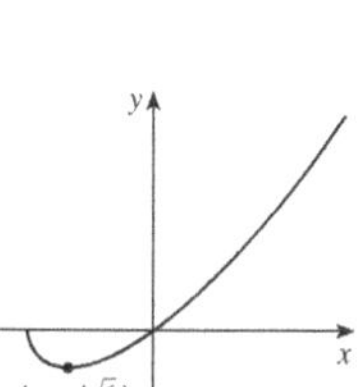

29. A. $[-\pi, \pi]$ B. y-int 0; x-int $-\pi$, 0, π
C. None D. None
E. Inc on $(-\pi/4, 3\pi/4)$; dec on $(-\pi, -\pi/4)$, $(3\pi/4, \pi)$
F. Loc max $f(3\pi/4) = \frac{1}{2}\sqrt{2}e^{3\pi/4}$, loc min $f(-\pi/4) = -\frac{1}{2}\sqrt{2}e^{3\pi/4}$
G. CU on $(-\pi/2, \pi/2)$; CD on $(-\pi, -\pi/2)$, $(\pi/2, \pi)$;
IP $(-\pi/2, -e^{-\pi/2})$, $(\pi/2, e^{\pi/2})$

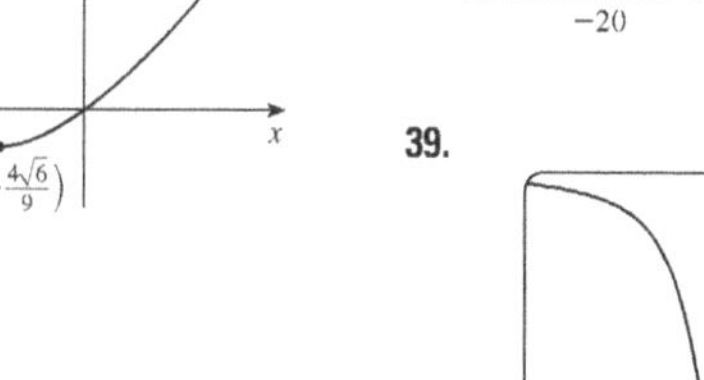

H.

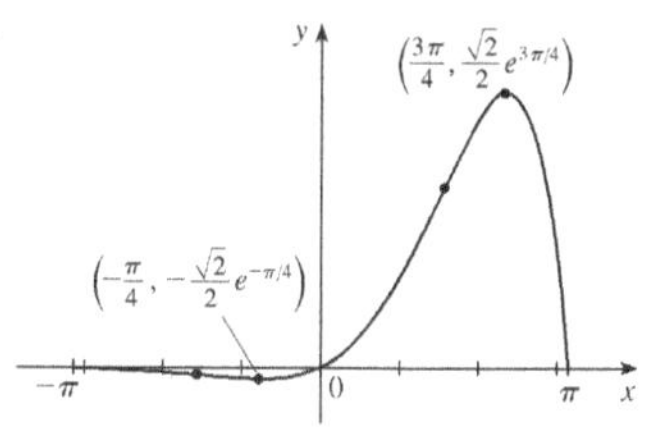

31. A. $\{x \mid |x| \geqslant 1\}$
B. None C. About $(0, 0)$
D. HA $y = 0$
E. Dec on $(-\infty, -1)$, $(1, \infty)$
F. None
G. CU on $(1, \infty)$; CD on $(-\infty, -1)$
H. See graph at right.

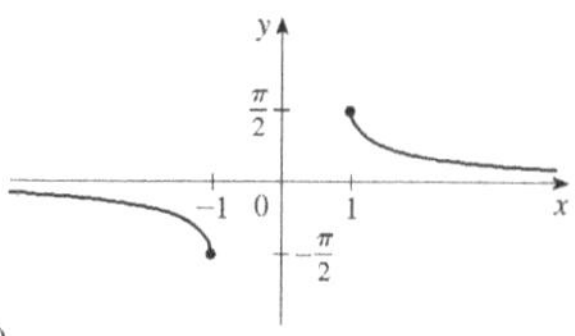

33. A. $\mathbb{R}$
B. y-int -2; x-int 2
C. None D. HA $y = 0$
E. Inc on $(-\infty, 3)$; dec on $(3, \infty)$
F. Loc max $f(3) = e^{-3}$
G. CU on $(4, \infty)$; CD on $(-\infty, 4)$;
IP $(4, 2e^{-4})$
H. See graph at right.

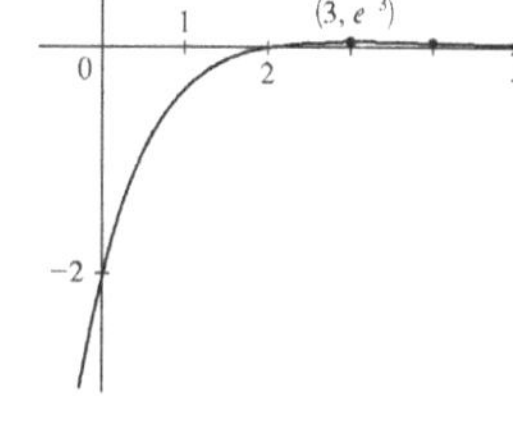

35. Inc on $(-\sqrt{3}, 0)$, $(0, \sqrt{3})$;
dec on $(-\infty, -\sqrt{3})$, $(\sqrt{3}, \infty)$;
loc max $f(\sqrt{3}) = \frac{2}{9}\sqrt{3}$,
loc min $f(-\sqrt{3}) = -\frac{2}{9}\sqrt{3}$;
CU on $(-\sqrt{6}, 0)$, $(\sqrt{6}, \infty)$;
CD on $(-\infty, -\sqrt{6})$, $(0, \sqrt{6})$;
IP $\left(\sqrt{6}, \frac{5}{36}\sqrt{6}\right)$, $\left(-\sqrt{6}, -\frac{5}{36}\sqrt{6}\right)$

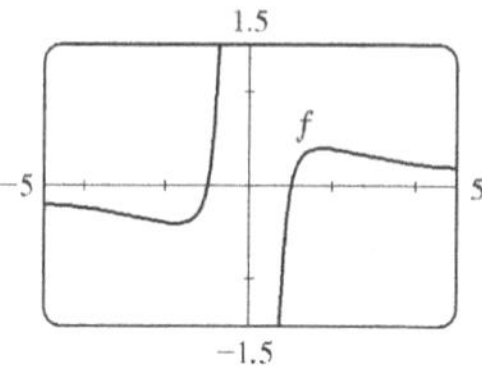

37. Inc on $(-0.23, 0)$, $(1.62, \infty)$; dec on $(-\infty, -0.23)$, $(0, 1.62)$;
loc max $f(0) = 2$; loc min $f(-0.23) \approx 1.96$, $f(1.62) \approx -19.2$;
CU on $(-\infty, -0.12)$, $(1.24, \infty)$;
CD on $(-0.12, 1.24)$; IP $(-0.12, 1.98)$, $(1.24, -12.1)$

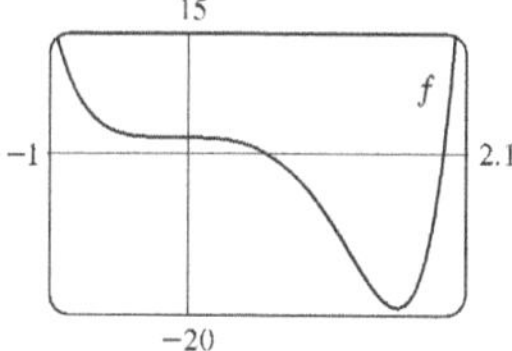

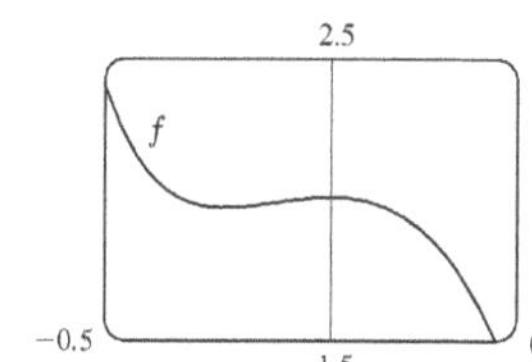

39.

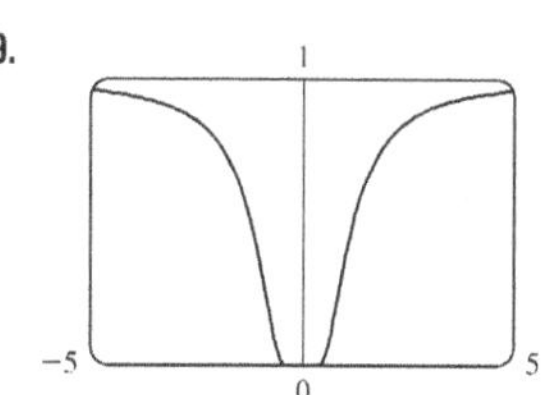

$(\pm 0.82, 0.22)$; $\left(\pm\sqrt{2/3}, e^{-3/2}\right)$

41. -2.96, -0.18, 3.01; -1.57, 1.57; -2.16, -0.75, 0.46, 2.21

43. For $C > -1$, f is periodic with period 2π and has local maxima at $2n\pi + \pi/2$, n an integer. For $C \leqslant -1$, f has no graph. For $-1 < C \leqslant 1$, f has vertical asymptotes. For $C > 1$, f is continuous on $\mathbb{R}$. As C increases, f moves upward and its oscillations become less pronounced.

49. (a) 0 (b) CU on $\mathbb{R}$ **53.** $3\sqrt{3}r^2$

55. $4/\sqrt{3}$ cm from D **57.** $L = C$ **59.** \$11.50

61. 1.297383 **63.** 1.16718557

65. $f(x) = \sin x - \sin^{-1}x + C$

67. $f(x) = \frac{2}{5}x^{5/2} + \frac{3}{5}x^{5/3} + C$

69. $f(t) = t^2 + 3\cos t + 2$

71. $f(x) = \frac{1}{2}x^2 - x^3 + 4x^4 + 2x + 1$

73. $s(t) = t^2 - \tan^{-1}t + 1$

75. (b) $0.1e^x - \cos x + 0.9$

(c)

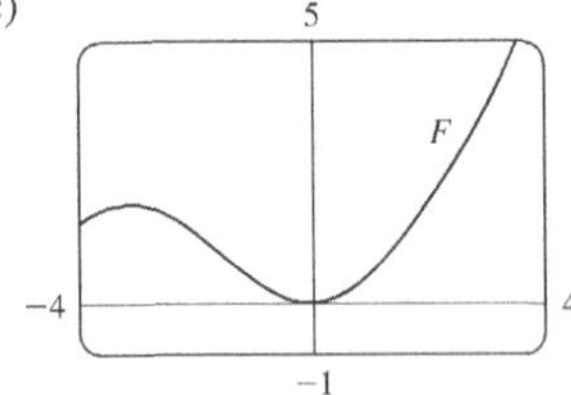

77. No

79. (b) About 8.5 in. by 2 in. (c) $20/\sqrt{3}$ in., $20\sqrt{2/3}$ in.

83. (a) $20\sqrt{2} \approx 28$ ft

(b) $\dfrac{dI}{dt} = \dfrac{-480k(h-4)}{[(h-4)^2 + 1600]^{5/2}}$, where k is the constant of proportionality

PROBLEMS PLUS ■ PAGE 356

3. Abs max $f(-5) = e^{45}$, no abs min **7.** 24

9. $(-2, 4)$, $(2, -4)$ **13.** $(m/2, m^2/4)$ **15.** $a \leqslant e^{1/e}$

19. (a) $T_1 = D/c_1$, $T_2 = (2h\sec\theta)/c_1 + (D - 2h\tan\theta)/c_2$, $T_3 = \sqrt{4h^2 + D^2}/c_1$

(c) $c_1 \approx 3.85$ km/s, $c_2 \approx 7.66$ km/s, $h \approx 0.42$ km

23. $3/(\sqrt[3]{2} - 1) \approx 11\frac{1}{2}$ h

CHAPTER 5

EXERCISES 5.1 ■ PAGE 369

1. (a) $L_4 = 33$, $R_4 = 41$

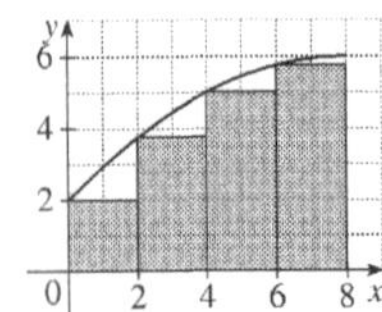

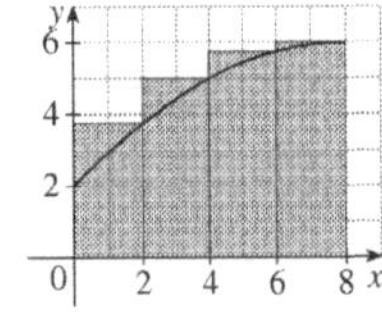

(b) $L_8 \approx 35.2$, $R_8 \approx 39.2$

3. (a) 0.7908, underestimate (b) 1.1835, overestimate

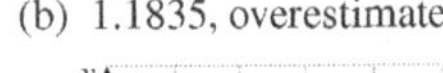

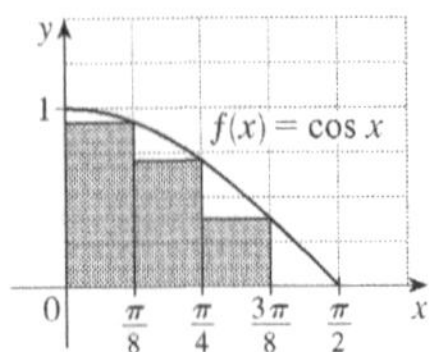

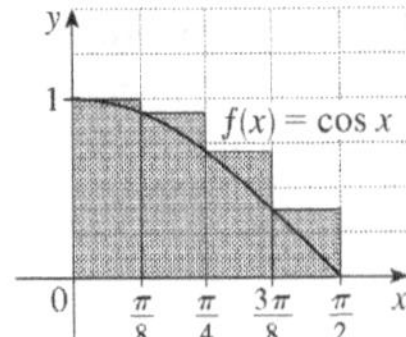

5. (a) 8, 6.875 (b) 5, 5.375

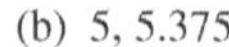

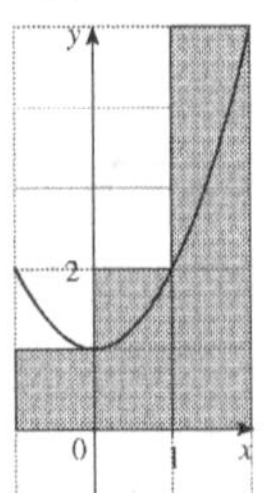

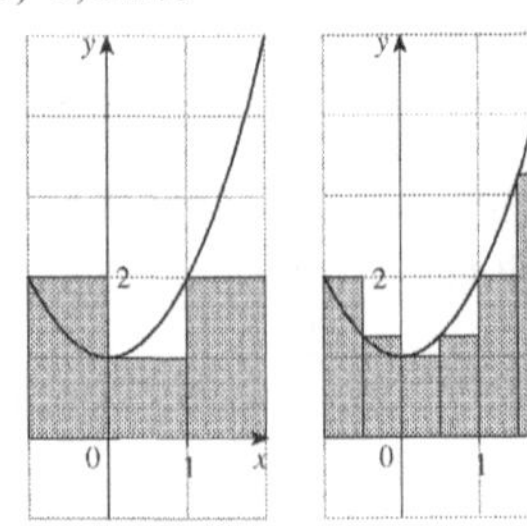

(c) 5.75, 5.9375

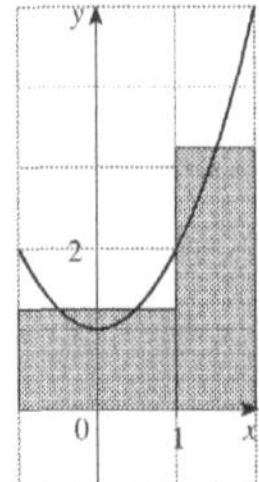

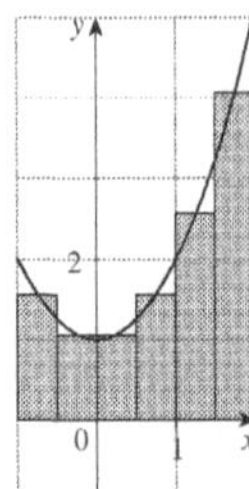

(d) M_6

7. $n = 2$: upper $= 3\pi \approx 9.42$, lower $= 2\pi \approx 6.28$

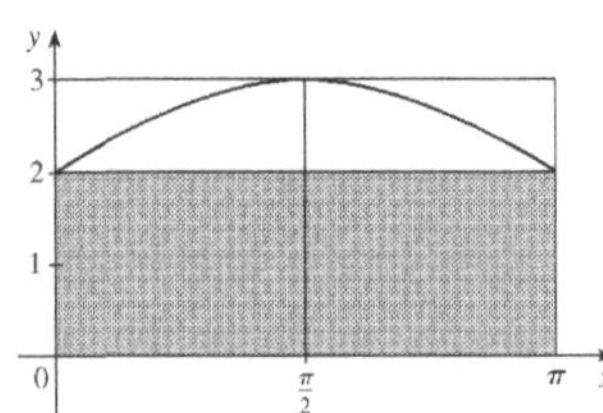

$n = 4$: upper $= (10 + \sqrt{2})(\pi/4) \approx 8.96$,
lower $= (8 + \sqrt{2})(\pi/4) \approx 7.39$

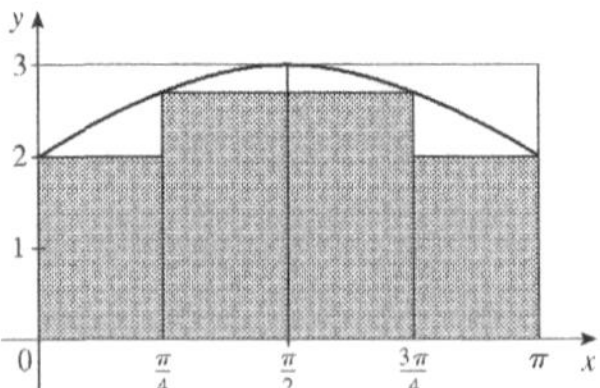

$n = 8$: upper ≈ 8.65, lower ≈ 7.86

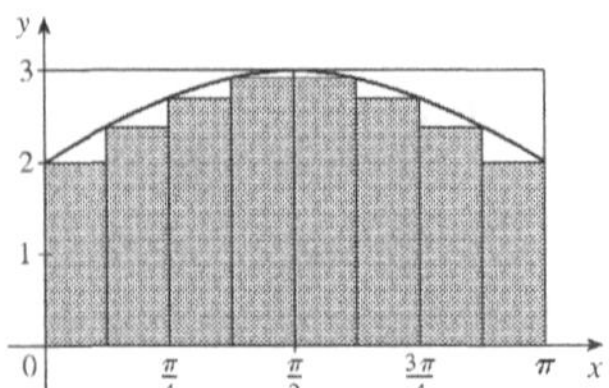

9. 0.2533, 0.2170, 0.2101, 0.2050; 0.2

11. (a) Left: 0.8100, 0.7937, 0.7904; right: 0.7600, 0.7770, 0.7804

13. 34.7 ft, 44.8 ft **15.** 63.2 L, 70 L **17.** 155 ft

19. $\lim_{n\to\infty} \sum_{i=1}^{n} \frac{2(1 + 2i/n)}{(1 + 2i/n)^2 + 1} \cdot \frac{2}{n}$ **21.** $\lim_{n\to\infty} \sum_{i=1}^{n} \sqrt{\sin(\pi i/n)} \cdot \frac{\pi}{n}$

23. The region under the graph of $y = \tan x$ from 0 to $\pi/4$

25. (a) $L_n < A < R_n$

27. (a) $\lim_{n\to\infty} \frac{64}{n^6} \sum_{i=1}^{n} i^5$ (b) $\frac{n^2(n + 1)^2(2n^2 + 2n - 1)}{12}$ (c) $\frac{32}{3}$

29. $\sin b$, 1

EXERCISES 5.2 ■ PAGE 382

1. -6

The Riemann sum represents the sum of the areas of the two rectangles above the x-axis minus the sum of the areas of the three rectangles below the x-axis; that is, the net area of the rectangles with respect to the x-axis.

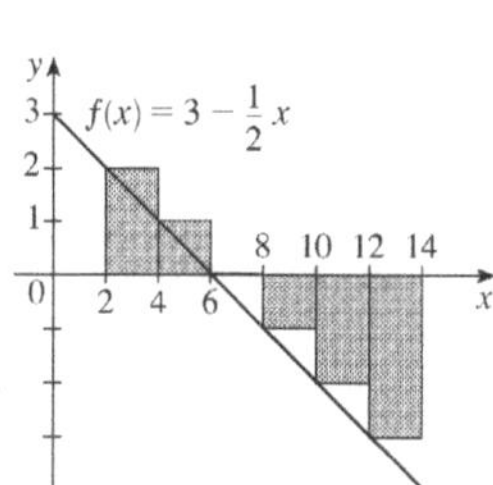

3. 2.322986

The Riemann sum represents the sum of the areas of the three rectangles above the x-axis minus the area of the rectangle below the x-axis.

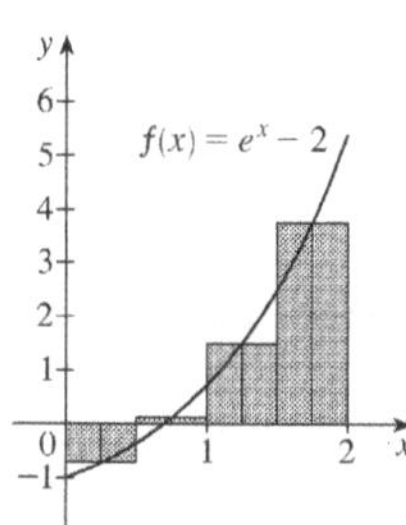

5. (a) 6 (b) 4 (c) 2

7. Lower, $L_5 = -64$; upper, $R_5 = 16$

9. 6.1820 **11.** 0.9071 **13.** 0.9029, 0.9018

15.

n	R_n
5	1.933766
10	1.983524
50	1.999342
100	1.999836

The values of R_n appear to be approaching 2.

17. $\int_2^6 x \ln(1 + x^2)\, dx$ **19.** $\int_2^7 (5x^3 - 4x)\, dx$

21. -9 **23.** $\frac{2}{3}$ **25.** $-\frac{3}{4}$

29. $\lim_{n\to\infty} \sum_{i=1}^{n} \frac{2 + 4i/n}{1 + (2 + 4i/n)^5} \cdot \frac{4}{n}$

31. $\lim_{n\to\infty} \sum_{i=1}^{n} \left(\sin \frac{5\pi i}{n}\right) \frac{\pi}{n} = \frac{2}{5}$

33. (a) 4 (b) 10 (c) -3 (d) 2

35. $\frac{3}{2}$ **37.** $3 + \frac{9}{4}\pi$ **39.** $\frac{5}{2}$ **41.** 0 **43.** 3

45. $e^5 - e^3$ **47.** $\int_{-1}^{5} f(x)\, dx$ **49.** 122

51. B < E < A < D < C **53.** 15

59. $3 \leq \int_1^4 \sqrt{x}\, dx \leq 6$ **61.** $\frac{\pi}{12} \leq \int_{\pi/4}^{\pi/3} \tan x\, dx \leq \frac{\pi}{12}\sqrt{3}$

63. $0 \leq \int_0^2 xe^{-x}\, dx \leq 2/e$ **71.** $\int_0^1 x^4\, dx$ **73.** $\frac{1}{2}$

EXERCISES 5.3 ■ PAGE 394

1. One process undoes what the other one does. See the Fundamental Theorem of Calculus, page 393.

3. (a) 0, 2, 5, 7, 3
(b) (0, 3)
(c) $x = 3$
(d)

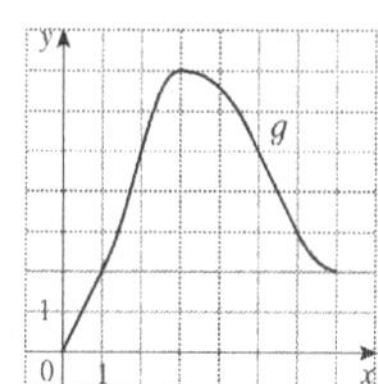

5.

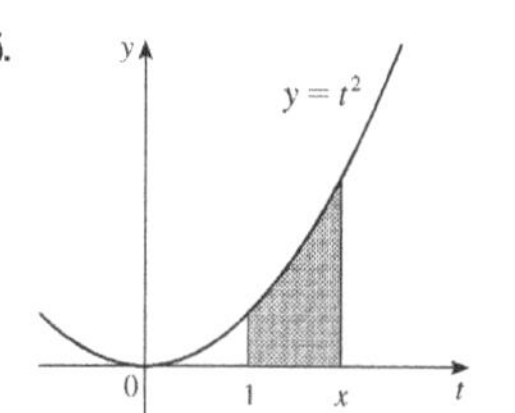

(a), (b) x^2

7. $g'(x) = 1/(x^3 + 1)$ **9.** $g'(s) = (s - s^2)^8$

11. $F'(x) = -\sqrt{1 + \sec x}$ **13.** $h'(x) = xe^x$

15. $y' = \sqrt{\tan x + \sqrt{\tan x}}\, \sec^2 x$

17. $y' = \frac{3(1 - 3x)^3}{1 + (1 - 3x)^2}$ **19.** $\frac{3}{4}$ **21.** 63 **23.** $\frac{52}{3}$

25. $1 + \sqrt{3}/2$ **27.** $-\frac{37}{6}$ **29.** $\frac{40}{3}$ **31.** 1 **33.** $\frac{49}{3}$

35. $\ln 2 + 7$ **37.** $\frac{1}{e + 1} + e - 1$ **39.** $4\pi/3$

41. $e^2 - 1$ **43.** 0

45. The function $f(x) = x^{-4}$ is not continuous on the interval $[-2, 1]$, so FTC2 cannot be applied.

47. The function $f(\theta) = \sec \theta \tan \theta$ is not continuous on the interval $[\pi/3, \pi]$, so FTC2 cannot be applied.

49. $\frac{243}{4}$ **51.** 2

53. 3.75

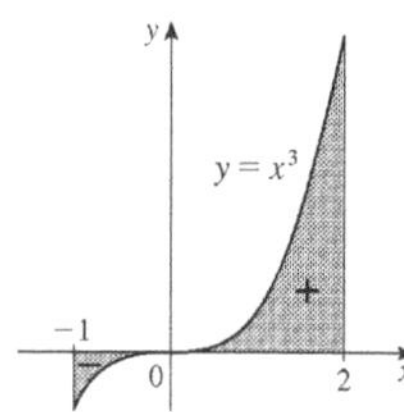

55. $g'(x) = \frac{-2(4x^2 - 1)}{4x^2 + 1} + \frac{3(9x^2 - 1)}{9x^2 + 1}$

57. $F'(x) = 2xe^{x^4} - e^{x^2}$

59. $y' = \sin x \ln(1 + 2\cos x) + \cos x \ln(1 + 2\sin x)$

61. $(-4, 0)$ **63.** 29

65. (a) $-2\sqrt{n}$, $\sqrt{4n - 2}$, n an integer > 0
(b) $(0, 1)$, $(-\sqrt{4n - 1}, -\sqrt{4n - 3})$, and $(\sqrt{4n - 1}, \sqrt{4n + 1})$, n an integer > 0 (c) 0.74

67. (a) Loc max at 1 and 5; loc min at 3 and 7
(b) $x = 9$
(c) $(\frac{1}{2}, 2)$, $(4, 6)$, $(8, 9)$
(d) See graph at right.

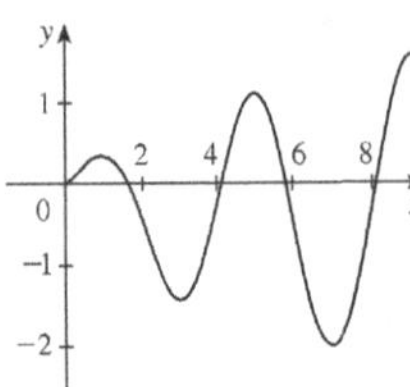

69. $\frac{1}{4}$ **77.** $f(x) = x^{3/2}$, $a = 9$
79. (b) Average expenditure over $[0, t]$; minimize average expenditure

EXERCISES 5.4 ■ PAGE 403

5. $\frac{1}{3}x^3 - (1/x) + C$ **7.** $\frac{1}{5}x^5 - \frac{1}{8}x^4 + \frac{1}{8}x^2 - 2x + C$
9. $\frac{2}{3}u^3 + \frac{9}{2}u^2 + 4u + C$ **11.** $\frac{1}{3}x^3 - 4\sqrt{x} + C$
13. $-\cos x + \cosh x + C$ **15.** $\frac{1}{2}\theta^2 + \csc\theta + C$
17. $\tan\alpha + C$
19. $\sin x + \frac{1}{4}x^2 + C$

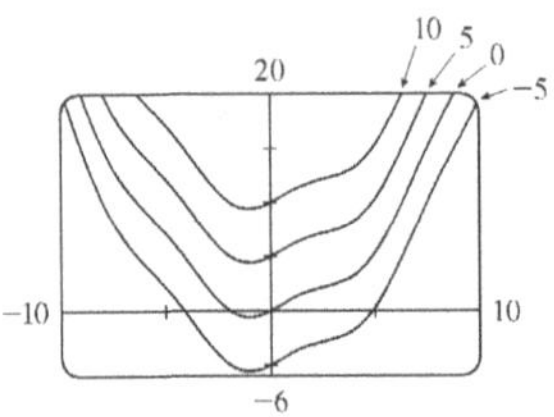

21. $-\frac{10}{3}$ **23.** $\frac{21}{5}$ **25.** -2 **27.** $5e^{\pi} + 1$ **29.** 36
31. $\frac{55}{63}$ **33.** $\frac{3}{4} - 2\ln 2$ **35.** $\dfrac{1}{11} + \dfrac{9}{\ln 10}$ **37.** $1 + \pi/4$
39. $\frac{256}{5}$ **41.** $\pi/3$ **43.** $\pi/6$ **45.** -3.5
47. ≈ 1.36 **49.** $\frac{4}{3}$
51. The increase in the child's weight (in pounds) between the ages of 5 and 10
53. Number of gallons of oil leaked in the first 2 hours
55. Increase in revenue when production is increased from 1000 to 5000 units
57. Newton-meters **59.** (a) $-\frac{3}{2}$ m (b) $\frac{41}{6}$ m
61. (a) $v(t) = \frac{1}{2}t^2 + 4t + 5$ m/s (b) $416\frac{2}{3}$ m
63. $46\frac{2}{3}$ kg **65.** 1.4 mi **67.** \$58,000
69. 5443 bacteria **71.** 4.75×10^5 megawatt-hours

EXERCISES 5.5 ■ PAGE 413

1. $-e^{-x} + C$ **3.** $\frac{2}{9}(x^3 + 1)^{3/2} + C$ **5.** $-\frac{1}{4}\cos^4\theta + C$
7. $-\frac{1}{2}\cos(x^2) + C$ **9.** $-\frac{1}{20}(1 - 2x)^{10} + C$
11. $\frac{1}{3}(2x + x^2)^{3/2} + C$ **13.** $-\frac{1}{3}\ln|5 - 3x| + C$
15. $-(1/\pi)\cos\pi t + C$ **17.** $\dfrac{1}{1 - e^u} + C$
19. $\frac{2}{3}\sqrt{3ax + bx^3} + C$ **21.** $\frac{1}{3}(\ln x)^3 + C$ **23.** $\frac{1}{4}\tan^4\theta + C$
25. $\frac{2}{3}(1 + e^x)^{3/2} + C$ **27.** $\frac{1}{15}(x^3 + 3x)^5 + C$
29. $-\dfrac{1}{\ln 5}\cos(5^t) + C$ **31.** $e^{\tan x} + C$ **33.** $-\dfrac{1}{\sin x} + C$

35. $-\frac{2}{3}(\cot x)^{3/2} + C$ **37.** $\frac{1}{3}\sinh^3 x + C$
39. $-\ln(1 + \cos^2 x) + C$ **41.** $\ln|\sin x| + C$
43. $\ln|\sin^{-1}x| + C$ **45.** $\tan^{-1}x + \frac{1}{2}\ln(1 + x^2) + C$
47. $\frac{1}{40}(2x + 5)^{10} - \frac{5}{36}(2x + 5)^9 + C$
49. $\frac{1}{8}(x^2 - 1)^4 + C$ **51.** $-e^{\cos x} + C$

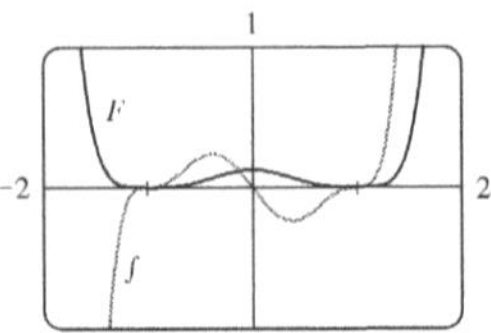

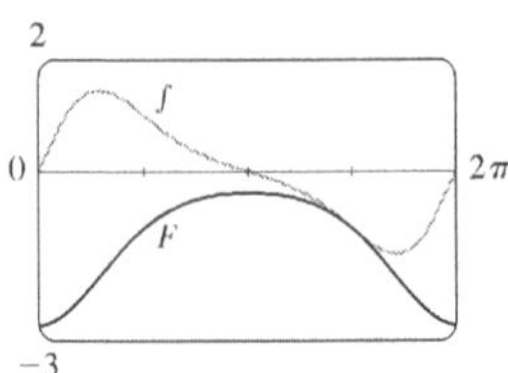

53. $2/\pi$ **55.** $\frac{45}{28}$ **57.** 4 **59.** $e - \sqrt{e}$ **61.** 0
63. 3 **65.** $\frac{1}{3}(2\sqrt{2} - 1)a^3$ **67.** $\frac{16}{15}$ **69.** 2
71. $\ln(e + 1)$ **73.** $\frac{1}{6}$ **75.** $\sqrt{3} - \frac{1}{3}$ **77.** 6π
79. All three areas are equal. **81.** ≈ 4512 L
83. $\dfrac{5}{4\pi}\left(1 - \cos\dfrac{2\pi t}{5}\right)$ L **85.** 5 **91.** $\pi^2/4$

CHAPTER 5 REVIEW ■ PAGE 416

True-False Quiz

1. True **3.** True **5.** False **7.** True **9.** True
11. False **13.** True **15.** False **17.** False

Exercises

1. (a) 8 (b) 5.7

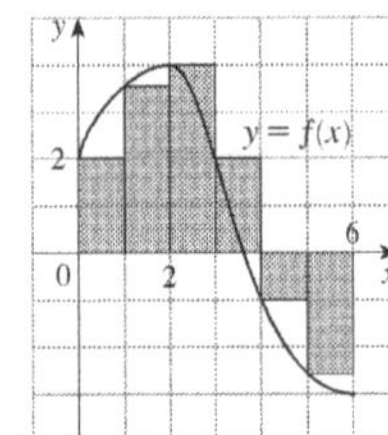

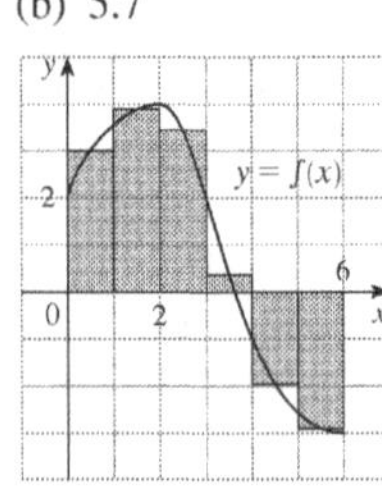

3. $\frac{1}{2} + \pi/4$ **5.** 3 **7.** f is c, f' is b, $\int_0^x f(t)\,dt$ is a
9. 37 **11.** $\frac{9}{10}$ **13.** -76 **15.** $\frac{21}{4}$ **17.** Does not exist
19. $\frac{1}{3}\sin 1$ **21.** 0 **23.** $-(1/x) - 2\ln|x| + x + C$
25. $\sqrt{x^2 + 4x} + C$ **27.** $\dfrac{1}{2\pi}\sin^2\pi t + C$
29. $2e^{\sqrt{x}} + C$ **31.** $-\frac{1}{2}[\ln(\cos x)]^2 + C$
33. $\frac{1}{4}\ln(1 + x^4) + C$ **35.** $\ln|1 + \sec\theta| + C$ **37.** $\frac{23}{3}$
39. $2\sqrt{1 + \sin x} + C$ **41.** $\frac{64}{5}$ **43.** $F'(x) = x^2/(1 + x^3)$
45. $g'(x) = 4x^3\cos(x^8)$ **47.** $y' = (2e^x - e^{\sqrt{x}})/(2x)$
49. $4 \leqslant \int_1^3 \sqrt{x^2 + 3}\,dx \leqslant 4\sqrt{3}$ **55.** 0.280981
57. Number of barrels of oil consumed from Jan. 1, 2000, through Jan. 1, 2008

59. 72,400 **61.** 3 **63.** $c \approx 1.62$
65. $e^{2x}(2x - 1)/(1 - e^{-x})$ **71.** $\frac{2}{3}$

PROBLEMS PLUS ■ PAGE 420

1. $\pi/2$ **3.** $2k$ **5.** -1 **7.** e^{-2} **9.** $[-1, 2]$
11. (a) $\frac{1}{2}(n - 1)n$
(b) $\frac{1}{2}[\![b]\!](2b - [\![b]\!] - 1) - \frac{1}{2}[\![a]\!](2a - [\![a]\!] - 1)$
17. $2(\sqrt{2} - 1)$

CHAPTER 6

EXERCISES 6.1 ■ PAGE 427

1. $\frac{32}{3}$ **3.** $e - (1/e) + \frac{10}{3}$ **5.** $e - (1/e) + \frac{4}{3}$ **7.** $\frac{9}{2}$
9. $\ln 2 - \frac{1}{2}$ **11.** $\frac{8}{3}$ **13.** 72 **15.** $e - 2$ **17.** $\frac{32}{3}$
19. $2/\pi + \frac{2}{3}$ **21.** $2 - 2 \ln 2$ **23.** $\frac{1}{2}$ **25.** $\frac{59}{12}$ **27.** $\ln 2$
29. $\frac{5}{2}$ **31.** $\frac{3}{2}\sqrt{3} - 1$ **33.** 0, 0.90; 0.04
35. -1.11, 1.25, 2.86; 8.38 **37.** 2.80123 **39.** 0.25142
41. $12\sqrt{6} - 9$ **43.** $117\frac{1}{3}$ ft **45.** 4232 cm^2
47. (a) Car A (b) The distance by which A is ahead of B after 1 minute (c) Car A (d) $t \approx 2.2$ min
49. $\frac{24}{5}\sqrt{3}$ **51.** $4^{2/3}$ **53.** ± 6
55. $0 < m < 1$; $m - \ln m - 1$

EXERCISES 6.2 ■ PAGE 438

1. $19\pi/12$

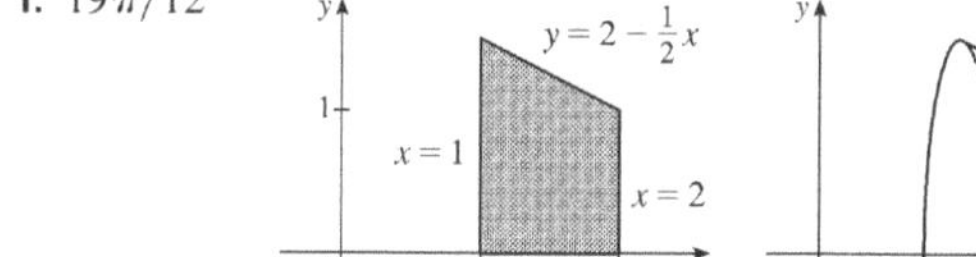

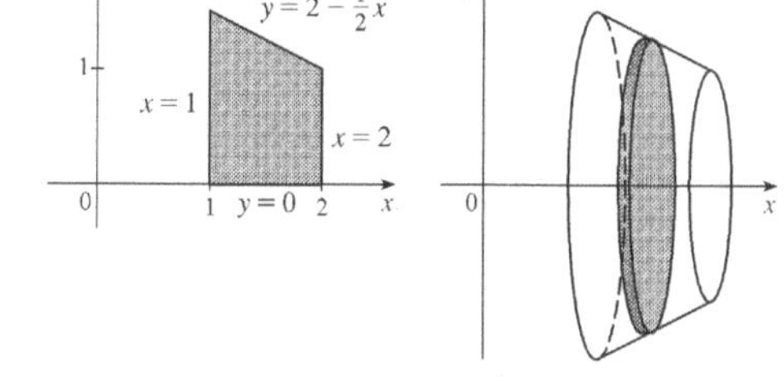

3. 8π

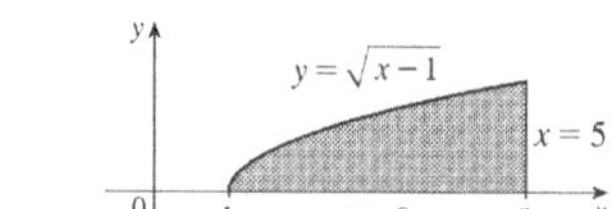

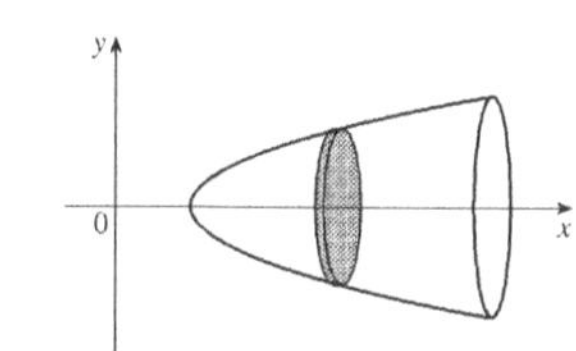

5. 162π

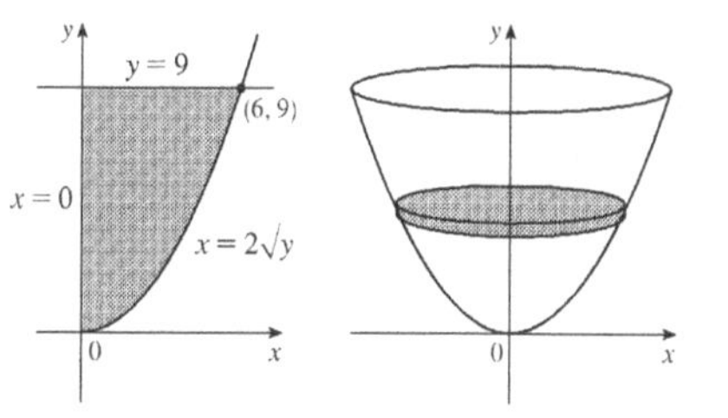

7. $4\pi/21$

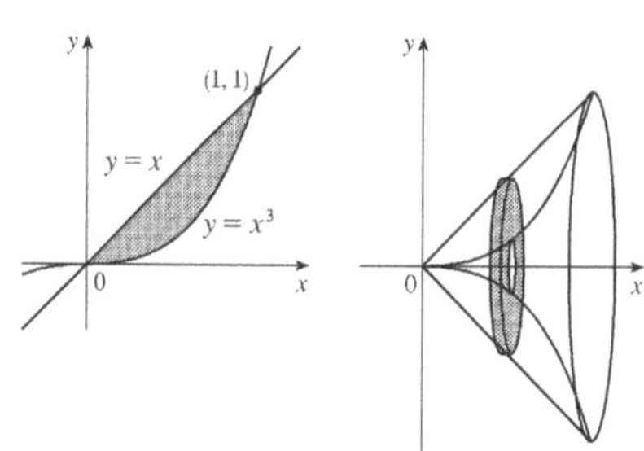

9. $64\pi/15$

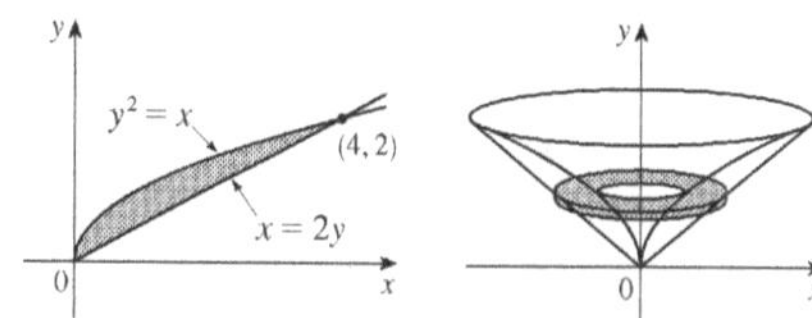

11. $11\pi/30$

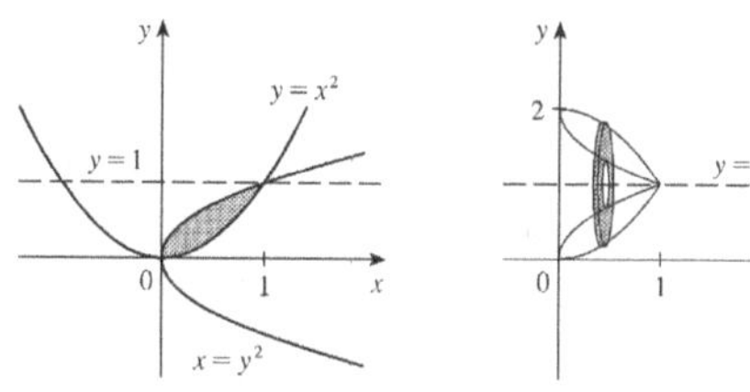

13. $2\pi(\frac{4}{3}\pi - \sqrt{3})$

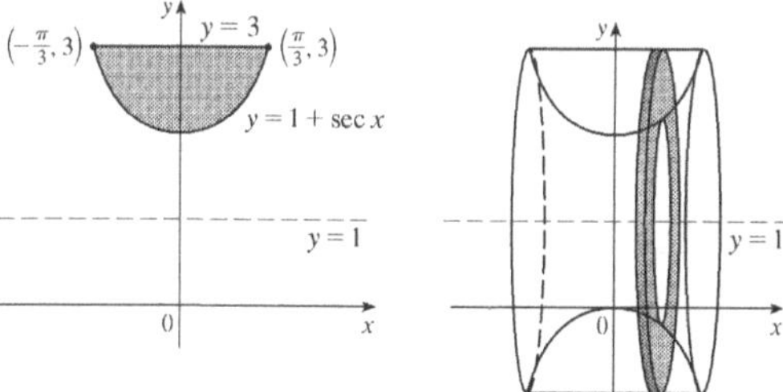

15. $3\pi/5$

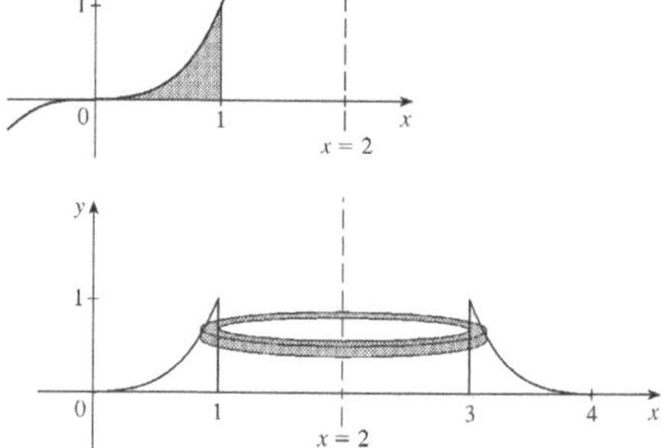

17. $10\sqrt{2}\,\pi/3$

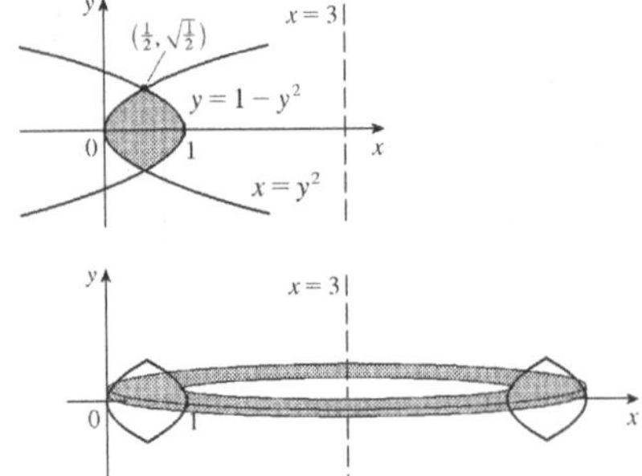

19. $\pi/3$ **21.** $\pi/3$ **23.** $\pi/3$

25. $13\pi/45$ **27.** $\pi/3$ **29.** $17\pi/45$

31. (a) $2\pi\int_0^1 e^{-2x^2}\,dx \approx 3.75825$

(b) $2\pi\int_0^1 \left(e^{-2x^2} + 2e^{-x^2}\right) dx \approx 13.14312$

33. (a) $2\pi\int_0^2 8\sqrt{1 - x^2/4}\;dx \approx 78.95684$

(b) $2\pi\int_0^1 8\sqrt{4 - 4y^2}\;dy \approx 78.95684$

35. $-1.288, 0.884; 23.780$ **37.** $\frac{11}{8}\pi^2$

39. Solid obtained by rotating the region $0 \leqslant x \leqslant \pi$, $0 \leqslant y \leqslant \sqrt{\sin x}$ about the x-axis

41. Solid obtained by rotating the region above the x-axis bounded by $x = y^2$ and $x = y^4$ about the y-axis

43. 1110 cm^3 **45.** (a) 196 (b) 838

47. $\frac{1}{3}\pi r^2 h$ **49.** $\pi h^2\left(r - \frac{1}{3}h\right)$ **51.** $\frac{2}{3}b^2 h$

53. 10 cm^3 **55.** 24 **57.** $\frac{1}{3}$ **59.** $\frac{8}{15}$

61. (a) $8\pi R\int_0^r \sqrt{r^2 - y^2}\,dy$ (b) $2\pi^2 r^2 R$

63. (b) $\pi r^2 h$ **65.** $\frac{5}{12}\pi r^3$ **67.** $8\int_0^r \sqrt{R^2 - y^2}\sqrt{r^2 - y^2}\,dy$

EXERCISES 6.3 ■ PAGE 444

1. Circumference $= 2\pi x$, height $= x(x-1)^2$; $\pi/15$

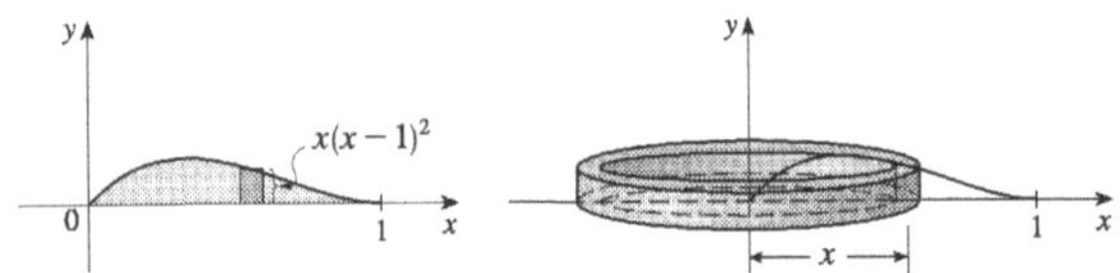

3. $6\pi/7$ **5.** $\pi(1 - 1/e)$ **7.** 8π **9.** 4π **11.** $768\pi/7$

13. $16\pi/3$ **15.** $7\pi/15$ **17.** $8\pi/3$ **19.** $5\pi/14$

21. (a) $2\pi\int_0^2 x^2 e^{-x}\,dx$ (b) 4.06300

23. (a) $4\pi\int_{-\pi/2}^{\pi/2} (\pi - x)\cos^4 x\,dx$ (b) 46.50942

25. (a) $\int_0^\pi 2\pi(4 - y)\sqrt{\sin y}\,dy$ (b) 36.57476

27. 3.68

29. Solid obtained by rotating the region $0 \leqslant y \leqslant x^4$, $0 \leqslant x \leqslant 3$ about the y-axis

31. Solid obtained by rotating the region bounded by (i) $x = 1 - y^2$, $x = 0$, and $y = 0$, or (ii) $x = y^2$, $x = 1$, and $y = 0$ about the line $y = 3$

33. 0.13 **35.** $\frac{1}{32}\pi^3$ **37.** 8π **39.** $4\sqrt{3}\,\pi$ **41.** $4\pi/3$

43. $117\pi/5$ **45.** $\frac{4}{3}\pi r^3$ **47.** $\frac{1}{3}\pi r^2 h$

EXERCISES 6.4 ■ PAGE 449

1. (a) 7200 ft-lb (b) 7200 ft-lb

3. 4.5 ft-lb **5.** 180 J **7.** $\frac{15}{4}$ ft-lb

9. (a) $\frac{25}{24} \approx 1.04$ J (b) 10.8 cm **11.** $W_2 = 3W_1$

13. (a) 625 ft-lb (b) $\frac{1875}{4}$ ft-lb **15.** 650,000 ft-lb

17. 3857 J **19.** 2450 J **21.** $\approx 1.06 \times 10^6$ J

23. $\approx 1.04 \times 10^5$ ft-lb **25.** 2.0 m

29. (a) $Gm_1m_2\left(\dfrac{1}{a} - \dfrac{1}{b}\right)$ (b) $\approx 8.50 \times 10^9$ J

EXERCISES 6.5 ■ PAGE 453

1. $\frac{8}{3}$ **3.** $\frac{45}{28}$ **5.** $(2/\pi)(e - 1)$ **7.** $2/(5\pi)$

9. (a) 1 (b) 2, 4 (c)

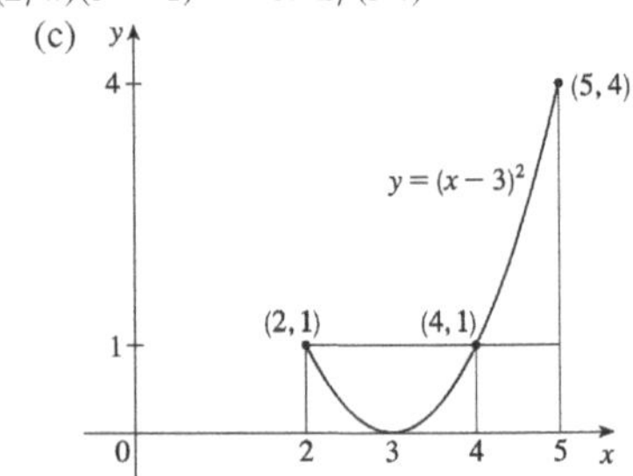

11. (a) $4/\pi$ (b) $\approx 1.24, 2.81$

(c) 3

15. $\frac{9}{8}$ **17.** $(50 + 28/\pi)$°F ≈ 59°F **19.** 6 kg/m

21. About 4056 million (or 4 billion) people

23. $5/(4\pi) \approx 0.4$ L

CHAPTER 6 REVIEW ■ PAGE 457

Exercises

1. $\frac{8}{3}$ **3.** $\frac{7}{12}$ **5.** $\frac{4}{3} + 4/\pi$ **7.** $64\pi/15$ **9.** $1656\pi/5$

11. $\frac{4}{3}\pi(2ah + h^2)^{3/2}$ **13.** $\int_{-\pi/3}^{\pi/3} 2\pi(\pi/2 - x)\left(\cos^2 x - \frac{1}{4}\right) dx$

15. (a) $2\pi/15$ (b) $\pi/6$ (c) $8\pi/15$

17. (a) 0.38 (b) 0.87

19. Solid obtained by rotating the region $0 \leqslant y \leqslant \cos x$, $0 \leqslant x \leqslant \pi/2$ about the y-axis

21. Solid obtained by rotating the region $0 \leqslant x \leqslant \pi$, $0 \leqslant y \leqslant 2 - \sin x$ about the x-axis

23. 36 **25.** $\frac{125}{3}\sqrt{3}\text{ m}^3$ **27.** 3.2 J

29. (a) $8000\pi/3 \approx 8378$ ft-lb (b) 2.1 ft

31. $f(x)$

PROBLEMS PLUS ■ PAGE 459

1. (a) $f(t) = 3t^2$ (b) $f(x) = \sqrt{2x/\pi}$ **3.** $\frac{32}{27}$
5. (b) 0.2261 (c) 0.6736 m
(d) (i) $1/(105\pi) \approx 0.003$ in/s (ii) $370\pi/3$ s ≈ 6.5 min
9. $y = \frac{32}{9}x^2$
11. (a) $V = \int_0^h \pi[f(y)]^2\,dy$
(c) $f(y) = \sqrt{kA/(\pi C)}\,y^{1/4}$. Advantage: the markings on the container are equally spaced.
13. $b = 2a$ **15.** $B = 16A$

CHAPTER 7

EXERCISES 7.1 ■ PAGE 469

1. $\frac{1}{3}x^3\ln x - \frac{1}{9}x^3 + C$ **3.** $\frac{1}{5}x\sin 5x + \frac{1}{25}\cos 5x + C$
5. $-\frac{1}{3}te^{-3t} - \frac{1}{9}e^{-3t} + C$
7. $(x^2 + 2x)\sin x + (2x + 2)\cos x - 2\sin x + C$
9. $x\ln\sqrt[3]{x} - \frac{1}{3}x + C$ **11.** $t\arctan 4t - \frac{1}{8}\ln(1 + 16t^2) + C$
13. $\frac{1}{2}t\tan 2t - \frac{1}{4}\ln|\sec 2t| + C$
15. $x(\ln x)^2 - 2x\ln x + 2x + C$
17. $\frac{1}{13}e^{2\theta}(2\sin 3\theta - 3\cos 3\theta) + C$
19. $z^3e^z - 3z^2e^z + 6ze^z - 6e^z + C$
21. $\dfrac{e^{2x}}{4(2x+1)} + C$ **23.** $\dfrac{\pi - 2}{2\pi^2}$
25. $1 - 1/e$ **27.** $\frac{81}{4}\ln 3 - 5$ **29.** $\frac{1}{4} - \frac{3}{4}e^{-2}$
31. $\frac{1}{6}(\pi + 6 - 3\sqrt{3})$ **33.** $\sin x(\ln\sin x - 1) + C$
35. $\frac{32}{5}(\ln 2)^2 - \frac{64}{25}\ln 2 + \frac{62}{125}$
37. $2\sqrt{x}\sin\sqrt{x} + 2\cos\sqrt{x} + C$ **39.** $-\frac{1}{2} - \pi/4$
41. $\frac{1}{2}(x^2 - 1)\ln(1 + x) - \frac{1}{4}x^2 + \frac{1}{2}x + \frac{3}{4} + C$
43. $-\frac{1}{2}xe^{-2x} - \frac{1}{4}e^{-2x} + C$

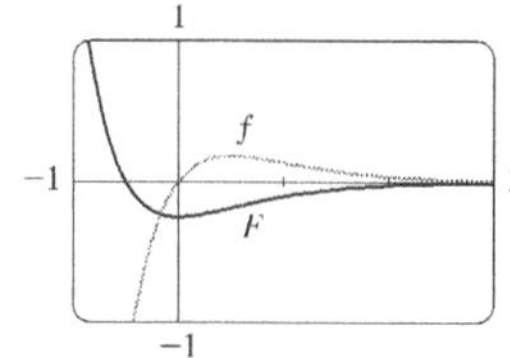

45. $\frac{1}{3}x^2(1 + x^2)^{3/2} - \frac{2}{15}(1 + x^2)^{5/2} + C$

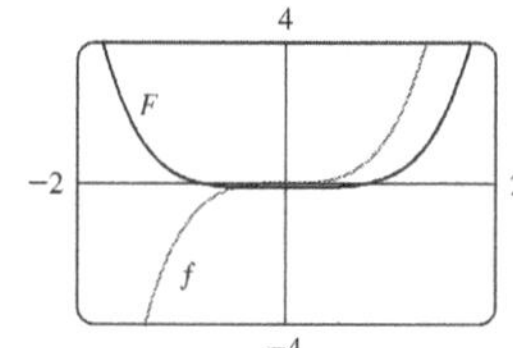

47. (b) $-\frac{1}{4}\cos x\sin^3 x + \frac{3}{8}x - \frac{3}{16}\sin 2x + C$
49. (b) $\frac{2}{3}, \frac{8}{15}$
55. $x[(\ln x)^3 - 3(\ln x)^2 + 6\ln x - 6] + C$
57. $\frac{16}{3}\ln 2 - \frac{29}{9}$ **59.** $-1.75119, 1.17210; 3.99926$
61. $4 - 8/\pi$ **63.** $2\pi e$ **65.** $1 - (2/\pi)\ln 2$
67. $2 - e^{-t}(t^2 + 2t + 2)$ m **69.** 2

EXERCISES 7.2 ■ PAGE 476

1. $\frac{1}{3}\sin^3 x - \frac{1}{5}\sin^5 x + C$ **3.** $\frac{1}{120}$
5. $\dfrac{1}{3\pi}\sin^3(\pi x) - \dfrac{2}{5\pi}\sin^5(\pi x) + \dfrac{1}{7\pi}\sin^7(\pi x) + C$
7. $\pi/4$ **9.** $3\pi/8$ **11.** $\pi/16$
13. $\frac{1}{4}t^2 - \frac{1}{4}t\sin 2t - \frac{1}{8}\cos 2t + C$
15. $\frac{2}{45}\sqrt{\sin\alpha}\,(45 - 18\sin^2\alpha + 5\sin^4\alpha) + C$
17. $\frac{1}{2}\cos^2 x - \ln|\cos x| + C$ **19.** $\ln|\sin x| + 2\sin x + C$
21. $\frac{1}{3}\sec^3 x + C$ **23.** $\tan x - x + C$
25. $\frac{1}{9}\tan^9 x + \frac{2}{7}\tan^7 x + \frac{1}{5}\tan^5 x + C$ **27.** $\frac{117}{8}$
29. $\frac{1}{3}\sec^3 x - \sec x + C$
31. $\frac{1}{4}\sec^4 x - \tan^2 x + \ln|\sec x| + C$
33. $x\sec x - \ln|\sec x + \tan x| + C$ **35.** $\sqrt{3} - \frac{1}{3}\pi$
37. $\frac{22}{105}\sqrt{2} - \frac{8}{105}$ **39.** $\ln|\csc x - \cot x| + C$
41. $-\frac{1}{6}\cos 3x - \frac{1}{26}\cos 13x + C$ **43.** $\frac{1}{8}\sin 4\theta - \frac{1}{12}\sin 6\theta + C$
45. $\frac{1}{2}\sqrt{2}$ **47.** $\frac{1}{2}\sin 2x + C$
49. $x\tan x - \ln|\sec x| - \frac{1}{2}x^2 + C$
51. $\frac{1}{4}x^2 - \frac{1}{4}\sin(x^2)\cos(x^2) + C$

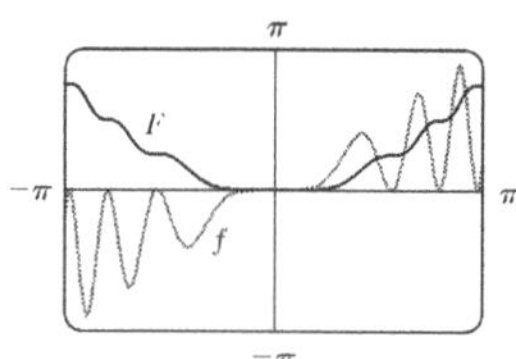

53. $\frac{1}{6}\sin 3x - \frac{1}{18}\sin 9x + C$

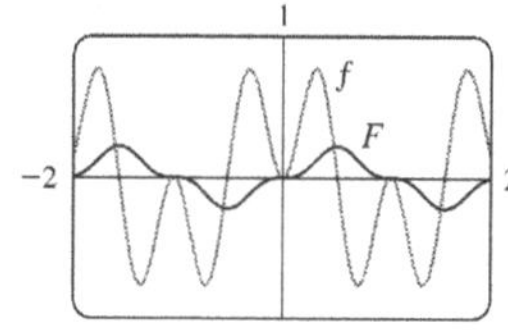

55. 0 **57.** 1 **59.** 0 **61.** $\pi^2/4$ **63.** $\pi(2\sqrt{2} - \frac{5}{2})$
65. $s = (1 - \cos^3\omega t)/(3\omega)$

EXERCISES 7.3 ■ PAGE 483

1. $-\dfrac{\sqrt{4 - x^2}}{4x} + C$ **3.** $\sqrt{x^2 - 4} - 2\sec^{-1}\left(\dfrac{x}{2}\right) + C$
5. $\dfrac{\pi}{24} + \dfrac{\sqrt{3}}{8} - \dfrac{1}{4}$ **7.** $\dfrac{1}{\sqrt{2}a^2}$
9. $\ln(\sqrt{x^2 + 16} + x) + C$ **11.** $\frac{1}{4}\sin^{-1}(2x) + \frac{1}{2}x\sqrt{1 - 4x^2} + C$
13. $\frac{1}{6}\sec^{-1}(x/3) - \sqrt{x^2 - 9}/(2x^2) + C$
15. $\frac{1}{16}\pi a^4$ **17.** $\sqrt{x^2 - 7} + C$
19. $\ln\left|(\sqrt{1 + x^2} - 1)/x\right| + \sqrt{1 + x^2} + C$ **21.** $\frac{9}{500}\pi$
23. $\frac{9}{2}\sin^{-1}((x - 2)/3) + \frac{1}{2}(x - 2)\sqrt{5 + 4x - x^2} + C$
25. $\sqrt{x^2 + x + 1} - \frac{1}{2}\ln(\sqrt{x^2 + x + 1} + x + \frac{1}{2}) + C$
27. $\frac{1}{2}(x + 1)\sqrt{x^2 + 2x} - \frac{1}{2}\ln\left|x + 1 + \sqrt{x^2 + 2x}\right| + C$
29. $\frac{1}{4}\sin^{-1}(x^2) + \frac{1}{4}x^2\sqrt{1 - x^4} + C$
33. $\frac{1}{6}(\sqrt{48} - \sec^{-1}7)$ **37.** $\frac{3}{8}\pi^2 + \frac{3}{4}\pi$
41. $2\pi^2 Rr^2$ **43.** $r\sqrt{R^2 - r^2} + \pi r^2/2 - R^2\arcsin(r/R)$

EXERCISES 7.4 ■ PAGE 492

1. (a) $\dfrac{A}{4x-3}+\dfrac{B}{2x+5}$ (b) $\dfrac{A}{x}+\dfrac{B}{x^2}+\dfrac{C}{5-2x}$

3. (a) $\dfrac{A}{x}+\dfrac{B}{x^2}+\dfrac{C}{x^3}+\dfrac{Dx+E}{x^2+4}$

(b) $\dfrac{A}{x+3}+\dfrac{B}{(x+3)^2}+\dfrac{C}{x-3}+\dfrac{D}{(x-3)^2}$

5. (a) $x^4+4x^2+16+\dfrac{A}{x+2}+\dfrac{B}{x-2}$

(b) $\dfrac{Ax+B}{x^2-x+1}+\dfrac{Cx+D}{x^2+2}+\dfrac{Ex+F}{(x^2+2)^2}$

7. $\frac{1}{4}x^4+\frac{1}{3}x^3+\frac{1}{2}x^2+x+\ln|x-1|+C$

9. $\frac{1}{2}\ln|2x+1|+2\ln|x-1|+C$ **11.** $2\ln\frac{3}{2}$

13. $a\ln|x-b|+C$ **15.** $\frac{7}{6}+\ln\frac{2}{3}$

17. $\frac{27}{5}\ln 2-\frac{9}{5}\ln 3$ (or $\frac{9}{5}\ln\frac{8}{3}$)

19. $10\ln|x-3|-9\ln|x-2|+\dfrac{5}{x-2}+C$

21. $\frac{1}{2}x^2-2\ln(x^2+4)+2\tan^{-1}(x/2)+C$

23. $\ln|x-1|-\frac{1}{2}\ln(x^2+9)-\frac{1}{3}\tan^{-1}(x/3)+C$

25. $-2\ln|x+1|+\ln(x^2+1)+2\tan^{-1}x+C$

27. $\frac{1}{2}\ln(x^2+1)+(1/\sqrt{2})\tan^{-1}(x/\sqrt{2})+C$

29. $\frac{1}{2}\ln(x^2+2x+5)+\frac{3}{2}\tan^{-1}\left(\dfrac{x+1}{2}\right)+C$

31. $\frac{1}{3}\ln|x-1|-\frac{1}{6}\ln(x^2+x+1)-\dfrac{1}{\sqrt{3}}\tan^{-1}\dfrac{2x+1}{\sqrt{3}}+C$

33. $\frac{1}{4}\ln\frac{8}{3}$ **35.** $\frac{1}{16}\ln|x|-\frac{1}{32}\ln(x^2+4)+\dfrac{1}{8(x^2+4)}+C$

37. $\frac{7}{8}\sqrt{2}\tan^{-1}\left(\dfrac{x-2}{\sqrt{2}}\right)+\dfrac{3x-8}{4(x^2-4x+6)}+C$

39. $2\sqrt{x+1}-\ln(\sqrt{x+1}+1)+\ln|\sqrt{x+1}-1|+C$

41. $-2\ln\sqrt{x}-\dfrac{2}{\sqrt{x}}+2\ln(\sqrt{x}+1)+C$

43. $\frac{3}{10}(x^2+1)^{5/3}-\frac{3}{4}(x^2+1)^{2/3}+C$

45. $2\sqrt{x}+3\sqrt[3]{x}+6\sqrt[6]{x}+6\ln|\sqrt[6]{x}-1|+C$

47. $\ln\dfrac{(e^x+2)^2}{e^x+1}+C$

49. $\ln|\tan t+1|-\ln|\tan t+2|+C$

51. $x-\ln(e^x+1)+C$

53. $\left(x-\frac{1}{2}\right)\ln(x^2-x+2)-2x+\sqrt{7}\tan^{-1}\left(\dfrac{2x-1}{\sqrt{7}}\right)+C$

55. $-\frac{1}{2}\ln 3\approx-0.55$

57. $\frac{1}{2}\ln\left|\dfrac{x-2}{x}\right|+C$ **61.** $\frac{1}{5}\ln\left|\dfrac{2\tan(x/2)-1}{\tan(x/2)+2}\right|+C$

63. $4\ln\frac{2}{3}+2$ **65.** $-1+\frac{11}{3}\ln 2$

67. $t=-\ln P-\frac{1}{9}\ln(0.9P+900)+C$, where $C\approx 10.23$

69. (a) $\dfrac{24{,}110}{4879}\dfrac{1}{5x+2}-\dfrac{668}{323}\dfrac{1}{2x+1}-\dfrac{9438}{80{,}155}\dfrac{1}{3x-7}+\dfrac{1}{260{,}015}\dfrac{22{,}098x+48{,}935}{x^2+x+5}$

(b) $\dfrac{4822}{4879}\ln|5x+2|-\dfrac{334}{323}\ln|2x+1|-\dfrac{3146}{80{,}155}\ln|3x-7|+\dfrac{11{,}049}{260{,}015}\ln(x^2+x+5)+\dfrac{75{,}772}{260{,}015\sqrt{19}}\tan^{-1}\dfrac{2x+1}{\sqrt{19}}+C$

The CAS omits the absolute value signs and the constant of integration.

73. $\dfrac{1}{a^n(x-a)}-\dfrac{1}{a^nx}-\dfrac{1}{a^{n-1}x^2}-\cdots-\dfrac{1}{ax^n}$

EXERCISES 7.5 ■ PAGE 499

1. $\sin x+\frac{1}{3}\sin^3x+C$

3. $\sin x+\ln|\csc x-\cot x|+C$

5. $\dfrac{1}{2\sqrt{2}}\tan^{-1}\left(\dfrac{t^2}{\sqrt{2}}\right)+C$ **7.** $e^{\pi/4}-e^{-\pi/4}$

9. $\frac{243}{5}\ln 3-\frac{242}{25}$ **11.** $\frac{1}{2}\ln(x^2-4x+5)+\tan^{-1}(x-2)+C$

13. $-\frac{1}{5}\cos^5t+\frac{2}{7}\cos^7t-\frac{1}{9}\cos^9t+C$ **15.** $x/\sqrt{1-x^2}+C$

17. $\frac{1}{4}\pi^2$ **19.** $e^{e^x}+C$ **21.** $(x+1)\arctan\sqrt{x}-\sqrt{x}+C$

23. $\frac{4097}{45}$ **25.** $3x+\frac{23}{3}\ln|x-4|-\frac{5}{3}\ln|x+2|+C$

27. $x-\ln(1+e^x)+C$

29. $x\ln(x+\sqrt{x^2-1})-\sqrt{x^2-1}+C$

31. $\sin^{-1}x-\sqrt{1-x^2}+C$

33. $2\sin^{-1}\left(\dfrac{x+1}{2}\right)+\dfrac{x+1}{2}\sqrt{3-2x-x^2}+C$

35. $\frac{1}{8}\sin 4x+\frac{1}{16}\sin 8x+C$ **37.** $\frac{1}{4}$

39. $\ln|\sec\theta-1|-\ln|\sec\theta|+C$

41. $\theta\tan\theta-\frac{1}{2}\theta^2-\ln|\sec\theta|+C$ **43.** $\frac{2}{3}\tan^{-1}(x^{3/2})+C$

45. $-\frac{1}{3}(x^3+1)e^{-x^3}+C$

47. $\ln|x-1|-3(x-1)^{-1}-\frac{3}{2}(x-1)^{-2}-\frac{1}{3}(x-1)^{-3}+C$

49. $\ln\left|\dfrac{\sqrt{4x+1}-1}{\sqrt{4x+1}+1}\right|+C$ **51.** $-\ln\left|\dfrac{\sqrt{4x^2+1}+1}{2x}\right|+C$

53. $\dfrac{1}{m}x^2\cosh(mx)-\dfrac{2}{m^2}x\sinh(mx)+\dfrac{2}{m^3}\cosh(mx)+C$

55. $2\ln\sqrt{x}-2\ln(1+\sqrt{x})+C$

57. $\frac{3}{7}(x+c)^{7/3}-\frac{3}{4}c(x+c)^{4/3}+C$

59. $\sin(\sin x)-\frac{1}{3}\sin^3(\sin x)+C$

61. $\csc\theta-\cot\theta+C$ or $\tan(\theta/2)+C$

63. $2(x-2\sqrt{x}+2)e^{\sqrt{x}}+C$

65. $-\tan^{-1}(\cos^2x)+C$ **67.** $\frac{2}{3}[(x+1)^{3/2}-x^{3/2}]+C$

69. $\sqrt{2}-2/\sqrt{3}+\ln(2+\sqrt{3})-\ln(1+\sqrt{2})$

71. $e^x-\ln(1+e^x)+C$

73. $-\sqrt{1-x^2}+\frac{1}{2}(\arcsin x)^2+C$

75. $\frac{1}{8}\ln|x-2|-\frac{1}{16}\ln(x^2+4)-\frac{1}{8}\tan^{-1}(x/2)+C$

77. $2(x-2)\sqrt{1+e^x}+2\ln\dfrac{\sqrt{1+e^x}+1}{\sqrt{1+e^x}-1}+C$

79. $\frac{1}{3}x\sin^3x+\frac{1}{3}\cos x-\frac{1}{9}\cos^3x+C$

81. $2\sqrt{1+\sin x}+C$ **83.** $xe^{x^2}+C$

EXERCISES 7.6 ■ PAGE 504

1. $-\frac{5}{21}$ **3.** $\sqrt{13} - \frac{3}{4}\ln(4 + \sqrt{13}) - \frac{1}{2} + \frac{3}{4}\ln 3$

5. $\frac{\pi}{8} - \frac{1}{4}\ln\left(1 + \frac{1}{16}\pi^2\right)$ **7.** $\frac{1}{6}\ln\left|\frac{\sin x - 3}{\sin x + 3}\right| + C$

9. $-\sqrt{4x^2 + 9}/(9x) + C$ **11.** $e - 2$

13. $-\frac{1}{2}\tan^2(1/z) - \ln|\cos(1/z)| + C$

15. $\frac{1}{2}(e^{2x} + 1)\arctan(e^x) - \frac{1}{2}e^x + C$

17. $\frac{2y - 1}{8}\sqrt{6 + 4y - 4y^2} + \frac{7}{8}\sin^{-1}\left(\frac{2y - 1}{\sqrt{7}}\right)$
$- \frac{1}{12}(6 + 4y - 4y^2)^{3/2} + C$

19. $\frac{1}{9}\sin^3 x\,[3\ln(\sin x) - 1] + C$ **21.** $\frac{1}{2\sqrt{3}}\ln\left|\frac{e^x + \sqrt{3}}{e^x - \sqrt{3}}\right| + C$

23. $\frac{1}{4}\tan x\sec^3 x + \frac{3}{8}\tan x\sec x + \frac{3}{8}\ln|\sec x + \tan x| + C$

25. $\frac{1}{2}(\ln x)\sqrt{4 + (\ln x)^2} + 2\ln\left[\ln x + \sqrt{4 + (\ln x)^2}\right] + C$

27. $-\frac{1}{2}x^{-2}\cos^{-1}(x^{-2}) + \frac{1}{2}\sqrt{1 - x^{-4}} + C$

29. $\sqrt{e^{2x} - 1} - \cos^{-1}(e^{-x}) + C$

31. $\frac{1}{5}\ln|x^5 + \sqrt{x^{10} - 2}| + C$ **33.** $\frac{3}{8}\pi^2$

37. $\frac{1}{3}\tan x\sec^2 x + \frac{2}{3}\tan x + C$

39. $\frac{1}{4}x(x^2 + 2)\sqrt{x^2 + 4} - 2\ln\left(\sqrt{x^2 + 4} + x\right) + C$

41. $\frac{1}{4}\cos^3 x\sin x + \frac{3}{8}x + \frac{3}{8}\sin x\cos x + C$

43. $\frac{1}{4}\tan^4 x - \frac{1}{2}\tan^2 x - \ln|\cos x| + C$

45. (a) $-\ln\left|\frac{1 + \sqrt{1 - x^2}}{x}\right| + C$;
both have domain $(-1, 0) \cup (0, 1)$

EXERCISES 7.7 ■ PAGE 516

1. (a) $L_2 = 6$, $R_2 = 12$, $M_2 \approx 9.6$
(b) L_2 is an underestimate, R_2 and M_2 are overestimates.
(c) $T_2 = 9 < I$ (d) $L_n < T_n < I < M_n < R_n$

3. (a) $T_4 \approx 0.895759$ (underestimate)
(b) $M_4 \approx 0.908907$ (overestimate)
$T_4 < I < M_4$

5. (a) $M_{10} \approx 0.806598$, $E_M \approx -0.001879$
(b) $S_{10} \approx 0.804779$, $E_S \approx -0.000060$

7. (a) 1.506361 (b) 1.518362 (c) 1.511519

9. (a) 2.660833 (b) 2.664377 (c) 2.663244

11. (a) 2.591334 (b) 2.681046 (c) 2.631976

13. (a) 4.513618 (b) 4.748256 (c) 4.675111

15. (a) −0.495333 (b) −0.543321 (c) −0.526123

17. (a) 8.363853 (b) 8.163298 (c) 8.235114

19. (a) $T_8 \approx 0.902333$, $M_8 \approx 0.905620$
(b) $|E_T| \leqslant 0.0078$, $|E_M| \leqslant 0.0039$
(c) $n = 71$ for T_n, $n = 50$ for M_n

21. (a) $T_{10} \approx 1.983524$, $E_T \approx 0.016476$;
$M_{10} \approx 2.008248$, $E_M \approx -0.008248$;
$S_{10} \approx 2.000110$, $E_S \approx -0.000110$
(b) $|E_T| \leqslant 0.025839$, $|E_M| \leqslant 0.012919$, $|E_S| \leqslant 0.000170$
(c) $n = 509$ for T_n, $n = 360$ for M_n, $n = 22$ for S_n

23. (a) 2.8 (b) 7.954926518 (c) 0.2894
(d) 7.954926521 (e) The actual error is much smaller.
(f) 10.9 (g) 7.953789422 (h) 0.0593
(i) The actual error is smaller. (j) $n \geqslant 50$

25.

n	L_n	R_n	T_n	M_n
5	0.742943	1.286599	1.014771	0.992621
10	0.867782	1.139610	1.003696	0.998152
20	0.932967	1.068881	1.000924	0.999538

n	E_L	E_R	E_T	E_M
5	0.257057	−0.286599	−0.014771	0.007379
10	0.132218	−0.139610	−0.003696	0.001848
20	0.067033	−0.068881	−0.000924	0.000462

Observations are the same as after Example 1.

27.

n	T_n	M_n	S_n
6	6.695473	6.252572	6.403292
12	6.474023	6.363008	6.400206

n	E_T	E_M	E_S
6	−0.295473	0.147428	−0.003292
12	−0.074023	0.036992	−0.000206

Observations are the same as after Example 1.

29. (a) 19.8 (b) 20.6 (c) $20.5\overline{3}$

31. (a) 14.4 (b) $\frac{1}{2}$

33. 64.4°F **35.** $37.7\overline{3}$ ft/s **37.** 10,177 megawatt-hours

39. (a) 190 (b) 828

41. 6.0 **43.** 59.4

45.

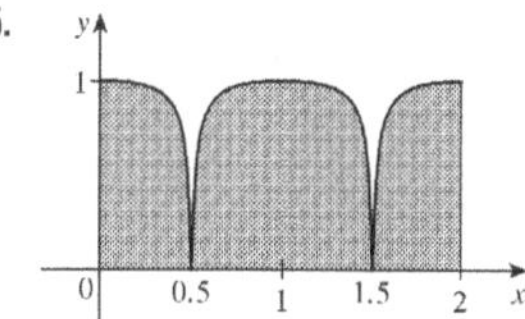

EXERCISES 7.8 ■ PAGE 527

Abbreviations: C, convergent; D, divergent

1. (a), (d) Infinite discontinuity (b), (c) Infinite interval

3. $\frac{1}{2} - 1/(2t^2)$; 0.495, 0.49995, 0.4999995; 0.5

5. 2 **7.** D **9.** $\frac{1}{5}e^{-10}$ **11.** D **13.** 0 **15.** D

17. ln 2 **19.** $-\frac{1}{4}$ **21.** D **23.** $\pi/9$ **25.** $\frac{1}{2}$ **27.** D

29. $\frac{32}{3}$ **31.** D **33.** $\frac{9}{2}$ **35.** D **37.** $-2/e$

39. $\frac{8}{3}\ln 2 - \frac{8}{9}$

41. $1/e$

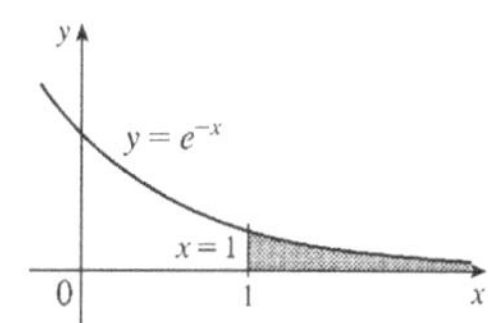

43. $\frac{1}{2}\ln 2$

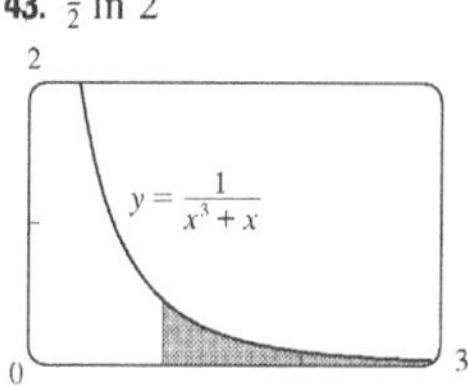

45. Infinite area

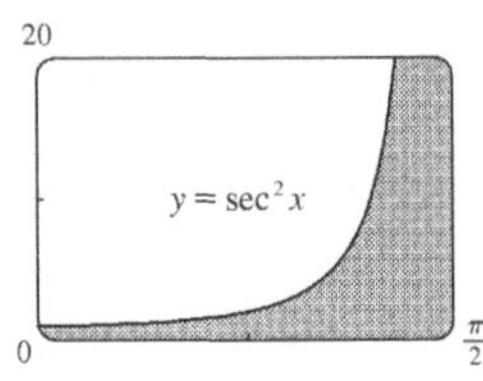

47. (a)

t	$\int_1^t [(\sin^2 x)/x^2]\,dx$
2	0.447453
5	0.577101
10	0.621306
100	0.668479
1,000	0.672957
10,000	0.673407

It appears that the integral is convergent.

(c)

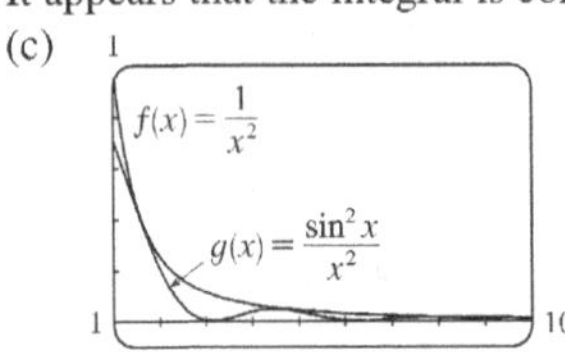

49. C **51.** D **53.** D **55.** π **57.** $p < 1$, $1/(1 - p)$

59. $p > -1$, $-1/(p + 1)^2$ **65.** $\sqrt{2GM/R}$

67. (a)

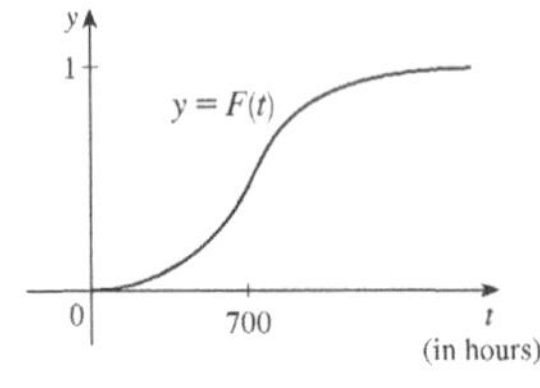

(b) The rate at which the fraction $F(t)$ increases as t increases

(c) 1; all bulbs burn out eventually

69. 1000

71. (a) $F(s) = 1/s$, $s > 0$ (b) $F(s) = 1/(s - 1)$, $s > 1$ (c) $F(s) = 1/s^2$, $s > 0$

77. $C = 1$; $\ln 2$ **79.** No

CHAPTER 7 REVIEW ■ PAGE 530

True-False Quiz

1. False **3.** False **5.** False **7.** False

9. (a) True (b) False **11.** False **13.** False

Exercises

1. $\frac{7}{2} + \ln 2$ **3.** $e - 1$ **5.** $\ln|2t + 1| - \ln|t + 1| + C$

7. $\frac{2}{15}$ **9.** $-\cos(\ln t) + C$ **11.** $\sqrt{3} - \frac{1}{3}\pi$

13. $3e^{\sqrt[3]{x}}(x^{2/3} - 2x^{1/3} + 2) + C$

15. $-\frac{1}{2}\ln|x| + \frac{3}{2}\ln|x + 2| + C$

17. $x \sec x - \ln|\sec x + \tan x| + C$

19. $\frac{1}{18}\ln(9x^2 + 6x + 5) + \frac{1}{9}\tan^{-1}\left[\frac{1}{2}(3x + 1)\right] + C$

21. $\ln\left|x - 2 + \sqrt{x^2 - 4x}\right| + C$

23. $\ln\left|\dfrac{\sqrt{x^2 + 1} - 1}{x}\right| + C$

25. $\frac{3}{2}\ln(x^2 + 1) - 3\tan^{-1}x + \sqrt{2}\tan^{-1}(x/\sqrt{2}) + C$

27. $\frac{2}{5}$ **29.** 0 **31.** $6 - \frac{3}{2}\pi$

33. $\dfrac{x}{\sqrt{4 - x^2}} - \sin^{-1}\left(\dfrac{x}{2}\right) + C$

35. $4\sqrt{1 + \sqrt{x}} + C$ **37.** $\frac{1}{2}\sin 2x - \frac{1}{8}\cos 4x + C$

39. $\frac{1}{8}e - \frac{1}{4}$ **41.** $\frac{1}{36}$ **43.** D

45. $4\ln 4 - 8$ **47.** $-\frac{4}{3}$ **49.** $\pi/4$

51. $(x + 1)\ln(x^2 + 2x + 2) + 2\arctan(x + 1) - 2x + C$

53. 0

55. $\frac{1}{4}(2x - 1)\sqrt{4x^2 - 4x - 3} - \ln\left|2x - 1 + \sqrt{4x^2 - 4x - 3}\right| + C$

57. $\frac{1}{2}\sin x\sqrt{4 + \sin^2 x} + 2\ln\left(\sin x + \sqrt{4 + \sin^2 x}\right) + C$

61. No

63. (a) 1.925444 (b) 1.920915 (c) 1.922470

65. (a) 0.01348, $n \geqslant 368$ (b) 0.00674, $n \geqslant 260$

67. 8.6 mi

69. (a) 3.8 (b) 1.7867, 0.000646 (c) $n \geqslant 30$

71. (a) D (b) C

73. 2 **75.** $\frac{3}{16}\pi^2$

PROBLEMS PLUS ■ PAGE 534

1. About 1.85 inches from the center **3.** 0

7. $f(\pi) = -\pi/2$ **11.** $(b^b a^{-a})^{1/(b-a)}e^{-1}$ **13.** $\frac{1}{8}\pi - \frac{1}{12}$

15. $2 - \sin^{-1}(2/\sqrt{5})$

CHAPTER 8

EXERCISES 8.1 ■ PAGE 543

1. $4\sqrt{5}$ **3.** 3.8202 **5.** 3.6095

7. $\frac{2}{243}(82\sqrt{82} - 1)$ **9.** $\frac{59}{24}$ **11.** $\frac{32}{3}$

13. $\ln(\sqrt{2} + 1)$ **15.** $\frac{3}{4} + \frac{1}{2}\ln 2$ **17.** $\ln 3 - \frac{1}{2}$

19. $\sqrt{2} + \ln(1 + \sqrt{2})$ **21.** 10.0556

23. 15.374568 **25.** 7.118819

27. (a), (b)

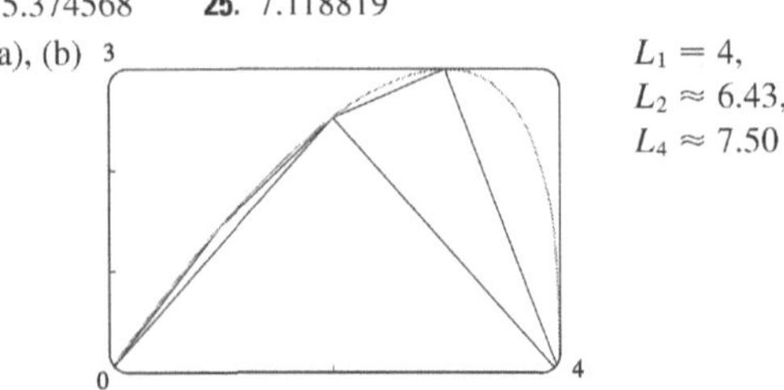

$L_1 = 4$, $L_2 \approx 6.43$, $L_4 \approx 7.50$

(c) $\int_0^4 \sqrt{1 + [4(3 - x)/(3(4 - x)^{2/3})]^2}\,dx$ (d) 7.7988

29. $\sqrt{5} - \ln\left(\frac{1}{2}(1 + \sqrt{5})\right) - \sqrt{2} + \ln(1 + \sqrt{2})$

31. 6

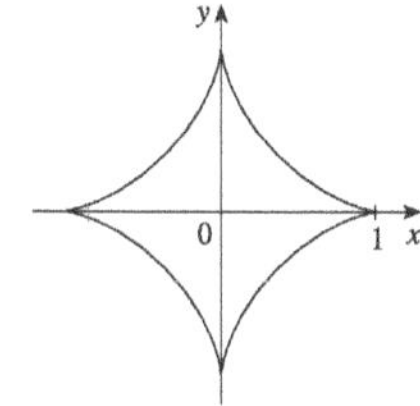

33. $s(x) = \frac{2}{27}\left[(1 + 9x)^{3/2} - 10\sqrt{10}\right]$ **35.** $2\sqrt{2}\left(\sqrt{1 + x} - 1\right)$

37. 209.1 m **39.** 29.36 in. **41.** 12.4

EXERCISES 8.2 ■ PAGE 550

1. (a) (i) $\int_0^{\pi/3} 2\pi \tan x\sqrt{1 + \sec^4 x}\,dx$
(ii) $\int_0^{\pi/3} 2\pi x\sqrt{1 + \sec^4 x}\,dx$ (b) (i) 10.5017 (ii) 7.9353

3. (a) (i) $\int_{-1}^{1} 2\pi e^{-x^2}\sqrt{1 + 4x^2 e^{-2x^2}}\,dx$
(ii) $\int_0^1 2\pi x\sqrt{1 + 4x^2 e^{-2x^2}}\,dx$ (b) (i) 11.0753 (ii) 3.9603

5. $\frac{1}{27}\pi\left(145\sqrt{145} - 1\right)$ **7.** $\frac{98}{3}\pi$

9. $2\sqrt{1 + \pi^2} + (2/\pi)\ln\left(\pi + \sqrt{1 + \pi^2}\right)$ **11.** $\frac{21}{2}\pi$

13. $\frac{1}{27}\pi\left(145\sqrt{145} - 10\sqrt{10}\right)$ **15.** πa^2

17. 1,230,507 **19.** 24.144251

21. $\frac{1}{4}\pi\left[4\ln\left(\sqrt{17} + 4\right) - 4\ln\left(\sqrt{2} + 1\right) - \sqrt{17} + 4\sqrt{2}\right]$

23. $\frac{1}{6}\pi\left[\ln\left(\sqrt{10} + 3\right) + 3\sqrt{10}\right]$

27. (a) $\frac{1}{3}\pi a^2$ (b) $\frac{56}{45}\pi\sqrt{3}a^2$

29. (a) $2\pi\left[b^2 + \dfrac{a^2 b\sin^{-1}\left(\sqrt{a^2 - b^2}/a\right)}{\sqrt{a^2 - b^2}}\right]$

(b) $2\pi\left[a^2 + \dfrac{ab^2\sin^{-1}\left(\sqrt{b^2 - a^2}/b\right)}{\sqrt{b^2 - a^2}}\right]$

31. $\int_a^b 2\pi[c - f(x)]\sqrt{1 + [f'(x)]^2}\,dx$ **33.** $4\pi^2 r^2$

EXERCISES 8.3 ■ PAGE 560

1. (a) 187.5 lb/ft^2 (b) 1875 lb (c) 562.5 lb

3. 6000 lb **5.** 6.7×10^4 N **7.** 9.8×10^3 N

9. 1.2×10^4 lb **11.** $\frac{2}{3}\delta a h^2$

13. 5.27×10^5 N **15.** (a) 314 N (b) 353 N

17. (a) 5.63×10^3 lb (b) 5.06×10^4 lb
(c) 4.88×10^4 lb (d) 3.03×10^5 lb

19. 4148 lb **21.** 330; 22 **23.** 10; 14; (1.4, 1) **25.** $\left(\frac{2}{3}, \frac{2}{3}\right)$

27. $\left(\dfrac{1}{e - 1}, \dfrac{e + 1}{4}\right)$ **29.** $\left(\frac{9}{20}, \frac{9}{20}\right)$

31. $\left(\dfrac{\pi\sqrt{2} - 4}{4\left(\sqrt{2} - 1\right)}, \dfrac{1}{4\left(\sqrt{2} - 1\right)}\right)$ **33.** $\left(\frac{8}{5}, -\frac{1}{2}\right)$

35. 60; 160; $\left(\frac{8}{3}, 1\right)$ **37.** $\left(-\frac{1}{5}, -\frac{12}{35}\right)$ **41.** $\left(0, \frac{1}{12}\right)$ **45.** $\frac{1}{3}\pi r^2 h$

EXERCISES 8.4 ■ PAGE 566

1. \$21,104 **3.** \$140,000; \$60,000 **5.** \$407.25

7. \$12,000 **9.** 3727; \$37,753

11. $\frac{2}{3}\left(16\sqrt{2} - 8\right) \approx$ \$9.75 million **13.** $\dfrac{(1 - k)(b^{2-k} - a^{2-k})}{(2 - k)(b^{1-k} - a^{1-k})}$

15. 1.19×10^{4} cm^3/s **17.** 6.60 L/min **19.** 5.77 L/min

EXERCISES 8.5 ■ PAGE 573

1. (a) The probability that a randomly chosen tire will have a lifetime between 30,000 and 40,000 miles
(b) The probability that a randomly chosen tire will have a lifetime of at least 25,000 miles

3. (a) $f(x) \geqslant 0$ for all x and $\int_{-\infty}^{\infty} f(x)\,dx = 1$ (b) $\frac{17}{81}$

5. (a) $1/\pi$ (b) $\frac{1}{2}$

7. (a) $f(x) \geqslant 0$ for all x and $\int_{-\infty}^{\infty} f(x)\,dx = 1$ (b) 5

11. (a) $e^{4/2.5} \approx 0.20$ (b) $1 - e^{2/2.5} \approx 0.55$ (c) If you aren't served within 10 minutes, you get a free hamburger.

13. $\approx$44% **15.** (a) 0.0668 (b) $\approx$5.21% **17.** $\approx$0.9545

19. (b) 0; a_0 (c) 1×10^{10}

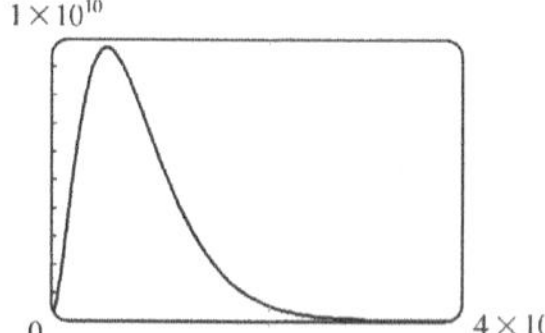

(d) $1 - 41e^{-8} \approx 0.986$ (e) $\frac{3}{2}a_0$

CHAPTER 8 REVIEW ■ PAGE 575

Exercises

1. $\frac{15}{2}$ **3.** (a) $\frac{21}{16}$ (b) $\frac{41}{10}\pi$

5. 3.8202 **7.** $\frac{124}{5}$ **9.** $\approx$458 lb **11.** $\left(\frac{8}{5}, 1\right)$

13. $\left(2, \frac{2}{3}\right)$ **15.** $2\pi^2$ **17.** \$7166.67

19. (a) $f(x) \geqslant 0$ for all x and $\int_{\infty}^{\infty} f(x)\,dx = 1$
(b) $\approx$0.3455 (c) 5, yes

21. (a) $1 - e^{-3/8} \approx 0.31$ (b) $e^{-5/4} \approx 0.29$
(c) $8\ln 2 \approx 5.55$ min

PROBLEMS PLUS ■ PAGE 577

1. $\frac{2}{3}\pi - \frac{1}{2}\sqrt{3}$

3. (a) $2\pi r(r \pm d)$ (b) $\approx 3.36 \times 10^6$ mi^2
(d) $\approx 7.84 \times 10^7$ mi^2

5. (a) $P(z) = P_0 + g\int_0^z \rho(x)\,dx$
(b) $(P_0 - \rho_0 gH)(\pi r^2) + \rho_0 gHe^{L/H}\int_{-r}^{r} e^{x/H} \cdot 2\sqrt{r^2 - x^2}\,dx$

7. Height $\sqrt{2}\,b$, volume $\left(\frac{28}{27}\sqrt{6} - 2\right)\pi b^3$ **9.** 0.14 m

11. $2/\pi$, $1/\pi$ **13.** $(0, -1)$

CHAPTER 9

EXERCISES 9.1 ■ PAGE 584

3. (a) $\frac{1}{2}, -1$ **5.** (d)

7. (a) It must be either 0 or decreasing
(c) $y = 0$ (d) $y = 1/(x + 2)$

9. (a) $0 < P < 4200$ (b) $P > 4200$
(c) $P = 0, P = 4200$

13. (a) III (b) I (c) IV (d) II

15. (a) At the beginning; stays positive, but decreases

(c)

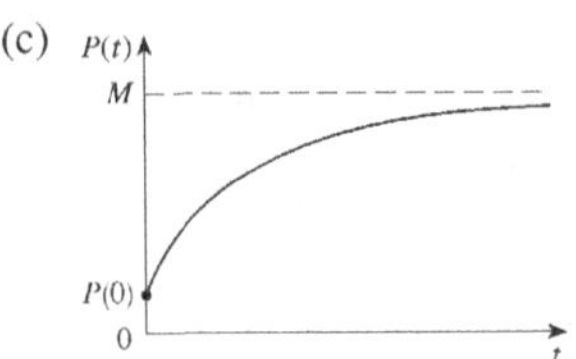

EXERCISES 9.2 ■ PAGE 592

1. (a)

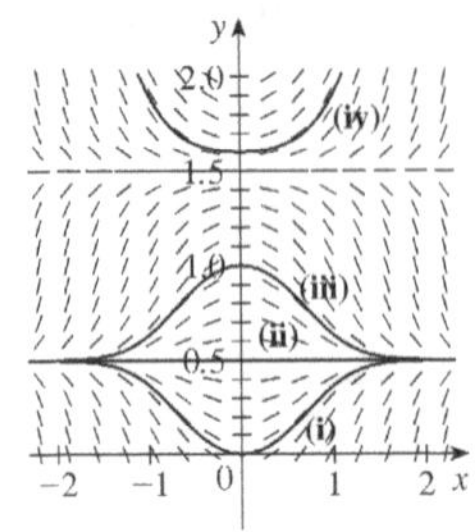

(b) $y = 0.5$, $y = 1.5$

3. III **5.** IV

7.

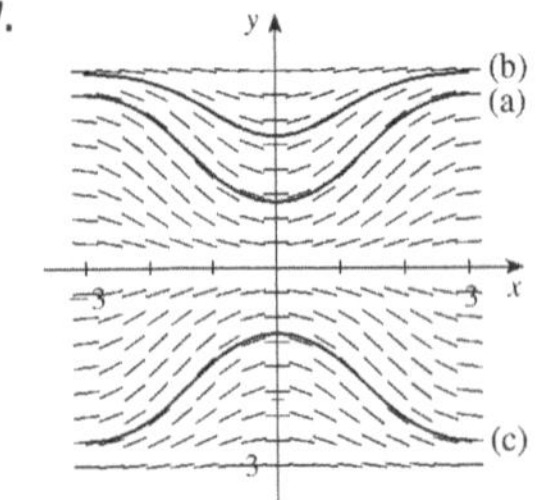

9.

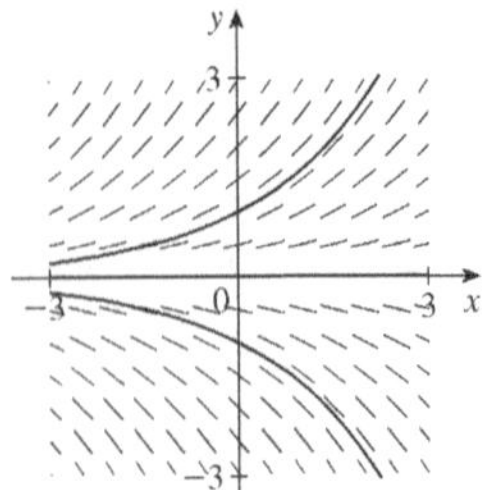

11.

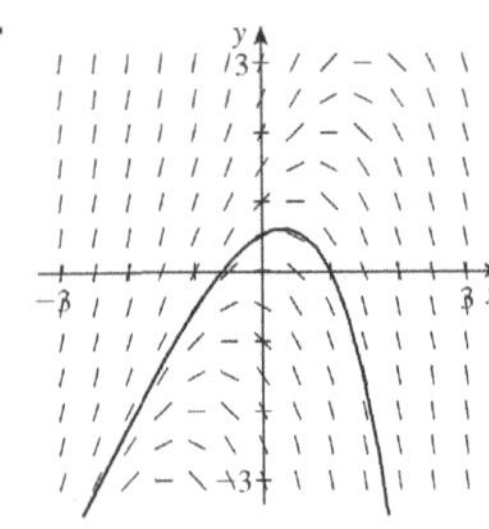

13.

15.

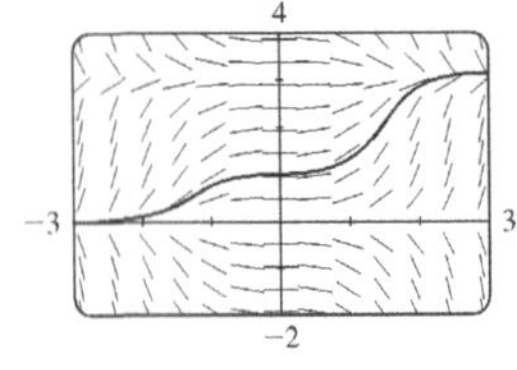

17. $-2 \leqslant c \leqslant 2$; $-2, 0, 2$

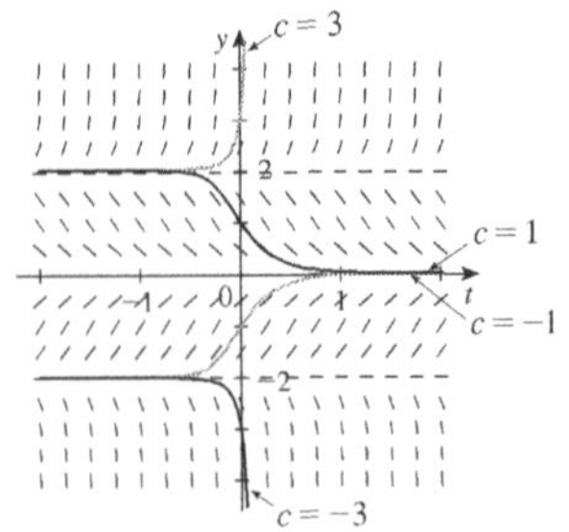

19. (a) (i) 1.4 (ii) 1.44 (iii) 1.4641

(b) Underestimates

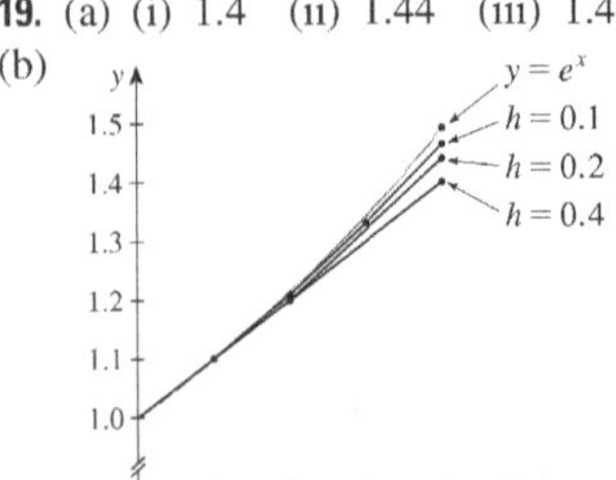

(c) (i) 0.0918 (ii) 0.0518 (iii) 0.0277

It appears that the error is also halved (approximately).

21. $-1, -3, -6.5, -12.25$ **23.** 1.7616

25. (a) (i) 3 (ii) 2.3928 (iii) 2.3701 (iv) 2.3681

(c) (i) -0.6321 (ii) -0.0249 (iii) -0.0022 (iv) -0.0002

It appears that the error is also divided by 10 (approximately).

27. (a), (d)

(b) 3

(c) Yes; $Q = 3$

(e) 2.77 C

EXERCISES 9.3 ■ PAGE 600

1. $y = \dfrac{2}{K - x^2}$, $y = 0$ **3.** $y = \sqrt[3]{3x + 3\ln|x| + K}$

5. $\frac{1}{2}y^2 - \cos y = \frac{1}{2}x^2 + \frac{1}{4}x^4 + C$

7. $e^y(y - 1) = C - \frac{1}{2}e^{-t^2}$ **9.** $p = Ke^{t^3/3 - t} - 1$

11. $y = -\sqrt{x^2 + 9}$ **13.** $u = -\sqrt{t^2 + \tan t + 25}$

15. $\frac{1}{2}y^2 + \frac{1}{3}(3 + y^2)^{3/2} = \frac{1}{2}x^2 \ln x - \frac{1}{4}x^2 + \frac{41}{12}$

17. $y = \dfrac{4a}{\sqrt{3}} \sin x - a$

19. $y = e^{x^2/2}$ **21.** $y = Ke^x - x - 1$

23. (a) $\sin^{-1}y = x^2 + C$

(b) $y = \sin(x^2)$, $-\sqrt{\pi/2} \leqslant x \leqslant \sqrt{\pi/2}$ (c) No

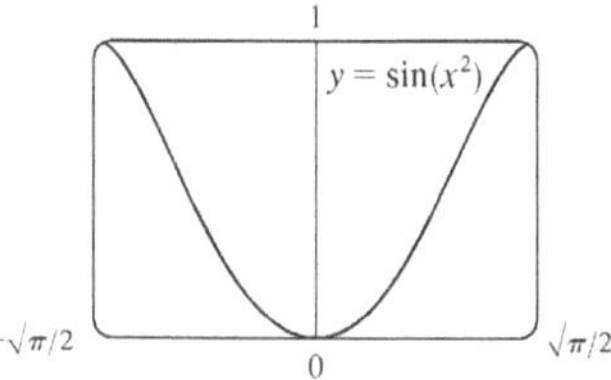

25. $\cos y = \cos x - 1$

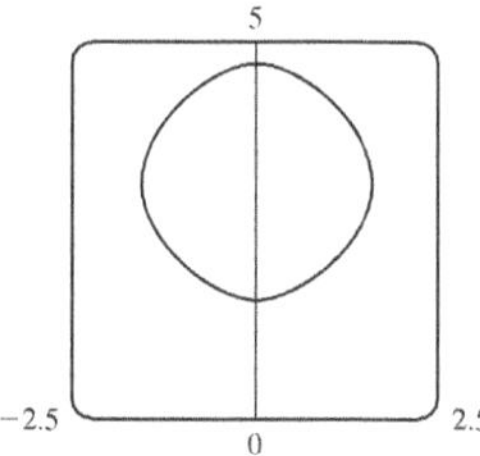

27. (a)

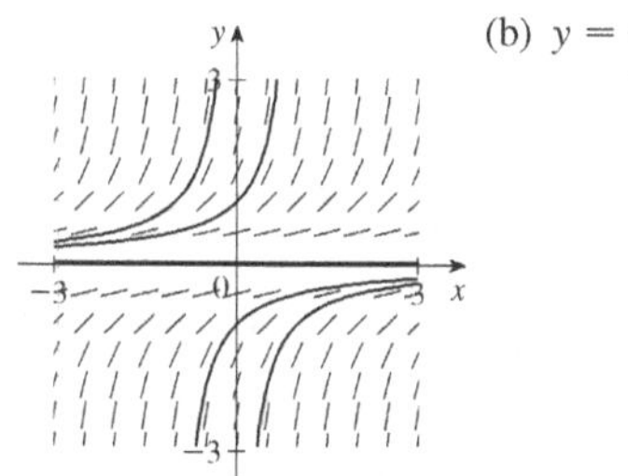

(b) $y = \dfrac{1}{K - x}$

29. $y = Cx^2$

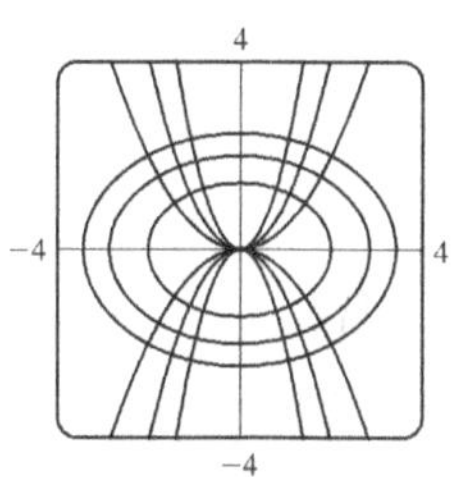

31. $x^2 - y^2 = C$

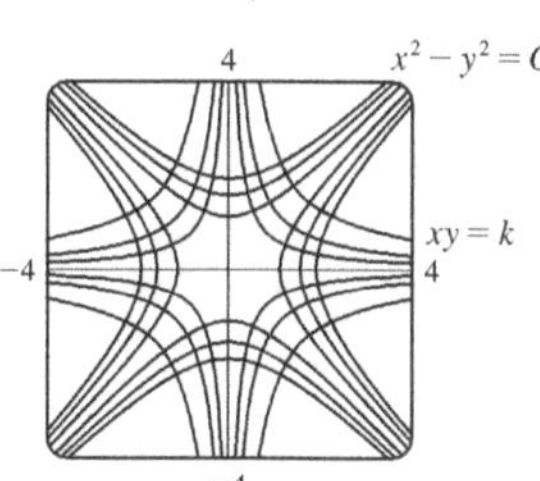

33. $y = 1 + e^{2 - x^2/2}$ **35.** $y = (\frac{1}{2}x^2 + 2)^2$

37. $Q(t) = 3 - 3e^{-4t}$; 3 **39.** $P(t) = M - Me^{-kt}$; M

41. (a) $x = a - \dfrac{4}{(kt + 2/\sqrt{a})^2}$

(b) $t = \dfrac{2}{k\sqrt{a - b}}\left(\tan^{-1}\sqrt{\dfrac{b}{a - b}} - \tan^{-1}\sqrt{\dfrac{b - x}{a - b}}\right)$

43. (a) $C(t) = (C_0 - r/k)e^{-kt} + r/k$

(b) r/k; the concentration approaches r/k regardless of the value of C_0

45. (a) $15e^{-t/100}$ kg (b) $15e^{-0.2} \approx 12.3$ kg

47. About 4.9% **49.** g/k

51. (a) $L_1 = KL_2^k$ (b) $B = KV^{0.0794}$

53. (a) $dA/dt = k\sqrt{A}\,(M - A)$ (b) $A(t) = M\left(\dfrac{Ce^{\sqrt{M}kt} - 1}{Ce^{\sqrt{M}kt} + 1}\right)^2$, where $C = \dfrac{\sqrt{M} + \sqrt{A_0}}{\sqrt{M} - \sqrt{A_0}}$ and $A_0 = A(0)$

EXERCISES 9.4 ■ PAGE 613

1. (a) 100; 0.05 (b) Where P is close to 0 or 100; on the line $P = 50$; $0 < P_0 < 100$; $P_0 > 100$

(c)

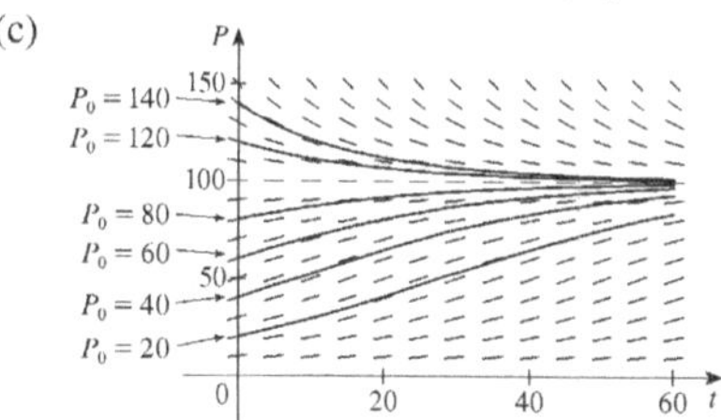

Solutions approach 100; some increase and some decrease, some have an inflection point but others don't; solutions with $P_0 = 20$ and $P_0 = 40$ have inflection points at $P = 50$

(d) $P = 0$, $P = 100$; other solutions move away from $P = 0$ and toward $P = 100$

3. (a) 3.23×10^7 kg (b) ≈ 1.55 years

5. 9000

7. (a) $dP/dt = \frac{1}{265}P(1 - P/100)$, P in billions

(b) 5.49 billion (c) In billions: 7.81, 27.72

(d) In billions: 5.48, 7.61, 22.41

9. (a) $dy/dt = ky(1 - y)$ (b) $y = \dfrac{y_0}{y_0 + (1 - y_0)e^{-kt}}$

(c) 3:36 PM

13. $P_E(t) = 1578.3(1.0933)^t + 94{,}000$;

$P_L(t) = \dfrac{32{,}658.5}{1 + 12.75e^{-0.1706t}} + 94{,}000$

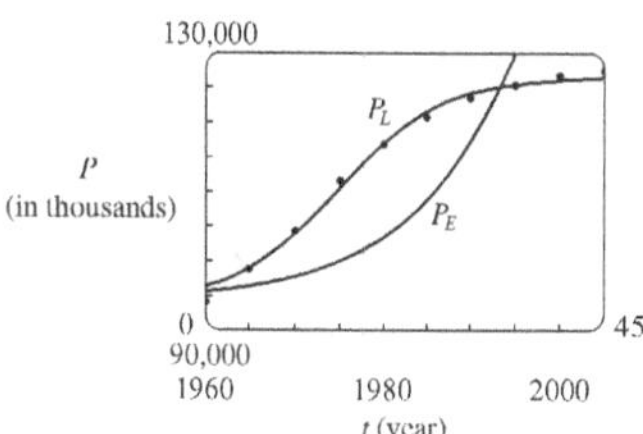

15. (a) $P(t) = \dfrac{m}{k} + \left(P_0 - \dfrac{m}{k}\right)e^{kt}$ (b) $m < kP_0$

(c) $m = kP_0$, $m > kP_0$ (d) Declining

17. (a) Fish are caught at a rate of 15 per week.
(b) See part (d). (c) $P = 250$, $P = 750$
(d)

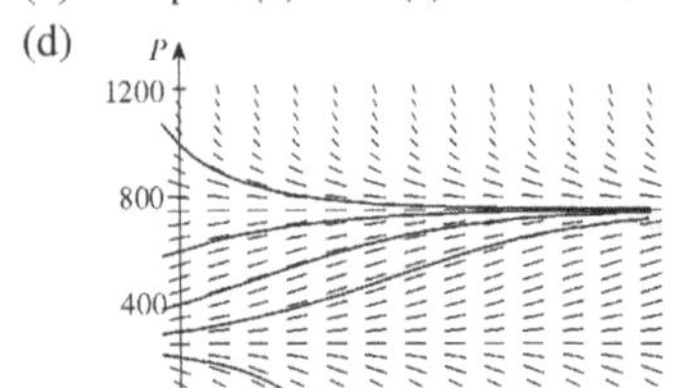

$0 < P_0 < 250$: $P \to 0$;
$P_0 = 250$: $P \to 250$;
$P_0 > 250$: $P \to 750$

(e) $P(t) = \dfrac{250 - 750ke^{t/25}}{1 - ke^{t/25}}$
where $k = \frac{1}{11}, -\frac{1}{9}$

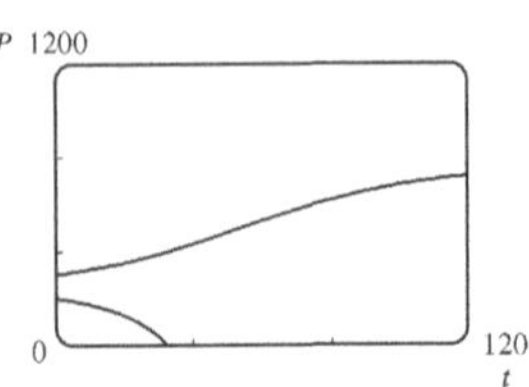

19. (b)

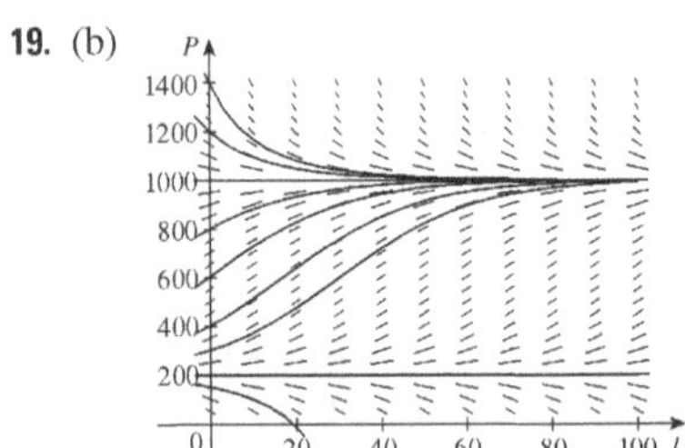

$0 < P_0 < 200$: $P \to 0$;
$P_0 = 200$: $P \to 200$;
$P_0 > 200$: $P \to 1000$

(c) $P(t) = \dfrac{m(M - P_0) + M(P_0 - m)e^{(M-m)(k/M)t}}{M - P_0 + (P_0 - m)e^{(M-m)(k/M)t}}$

21. (a) $P(t) = P_0 e^{(k/r)[\sin(rt - \phi) + \sin\phi]}$ (b) Does not exist

EXERCISES 9.5 ■ PAGE 620

1. Yes **3.** No **5.** $y = 1 + Ce^{-x}$
7. $y = x - 1 + Ce^{-x}$ **9.** $y = \frac{2}{3}\sqrt{x} + C/x$
11. $y = \dfrac{\int \sin(x^2)\,dx + C}{\sin x}$ **13.** $u = \dfrac{t^2 + 2t + 2C}{2(t + 1)}$
15. $y = \dfrac{1}{x}\ln x - \dfrac{1}{x} + \dfrac{3}{x^2}$ **17.** $u = -t^2 + t^3$
19. $y = -x\cos x - x$
21. $y = \dfrac{(x - 1)e^x + C}{x^2}$

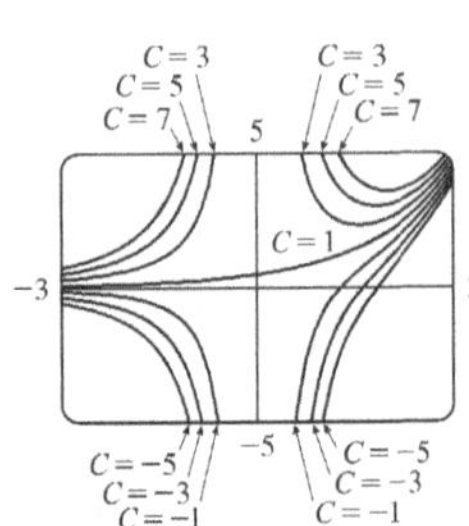

25. $y = \pm\left(Cx^4 + \dfrac{2}{5x}\right)^{1/2}$

27. (a) $I(t) = 4 - 4e^{-5t}$ (b) $4 - 4e^{-1/2} \approx 1.57$ A
29. $Q(t) = 3(1 - e^{-4t})$, $I(t) = 12e^{-4t}$
31. $P(t) = M + Ce^{kt}$

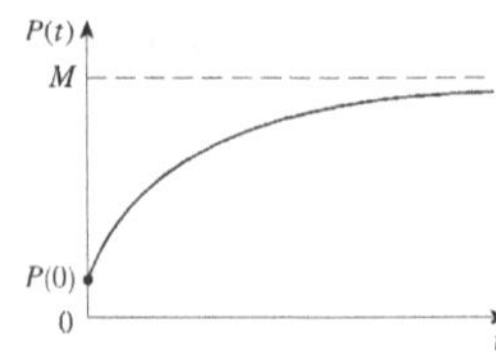

33. $y = \frac{2}{5}(100 + 2t) - 40{,}000(100 + 2t)^{-3/2}$; 0.2275 kg/L
35. (b) mg/c (c) $(mg/c)[t + (m/c)e^{-ct/m}] - m^2g/c^2$
37. (b) $P(t) = \dfrac{M}{1 + MCe^{-kt}}$

EXERCISES 9.6 ■ PAGE 627

1. (a) x = predators, y = prey; growth is restricted only by predators, which feed only on prey.
(b) x = prey, y = predators; growth is restricted by carrying capacity and by predators, which feed only on prey.
3. (a) Competition
(b) (i) $x = 0$, $y = 0$: zero populations
(ii) $x = 0$, $y = 400$: In the absence of an x-population, the y-population stabilizes at 400.
(iii) $x = 125$, $y = 0$: In the absence of a y-population, the x-population stabilizes at 125.
(iv) $x = 50$, $y = 300$: Both populations are stable.
5. (a) The rabbit population starts at about 300, increases to 2400, then decreases back to 300. The fox population starts at 100, decreases to about 20, increases to about 315, decreases to 100, and the cycle starts again.
(b)

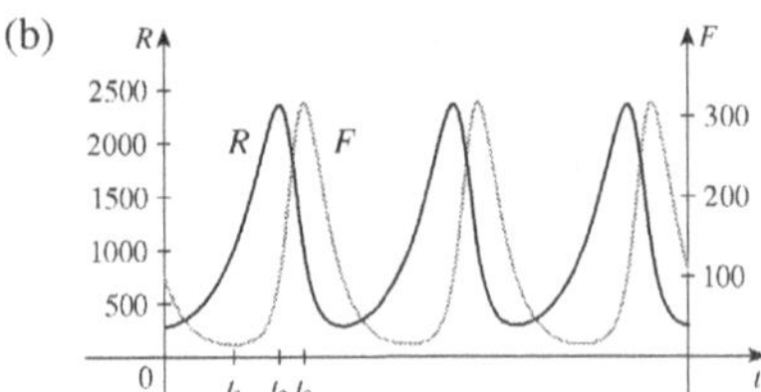

7.

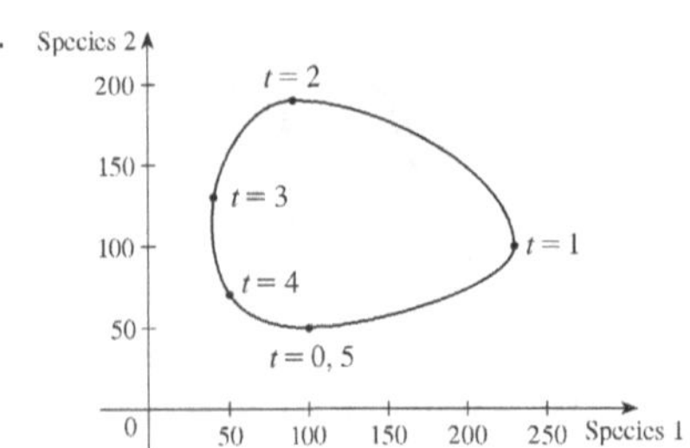

11. (a) Population stabilizes at 5000.
(b) (i) $W = 0$, $R = 0$: Zero populations
(ii) $W = 0$, $R = 5000$: In the absence of wolves, the rabbit population is always 5000.
(iii) $W = 64$, $R = 1000$: Both populations are stable.

(c) The populations stabilize at 1000 rabbits and 64 wolves.

(d)

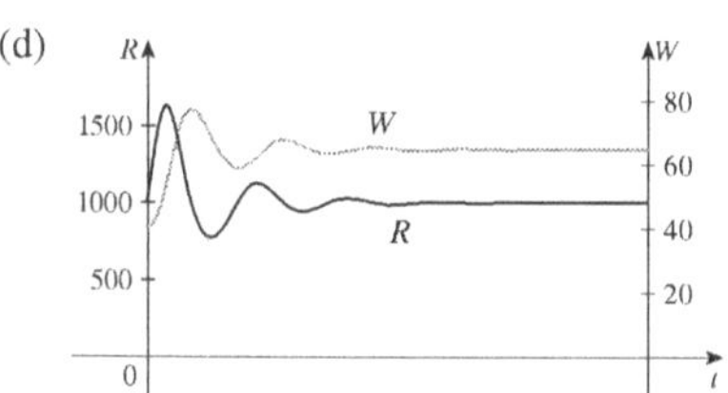

CHAPTER 9 REVIEW ■ PAGE 629

True-False Quiz

1. True **3.** False **5.** True **7.** True

Exercises

1. (a)

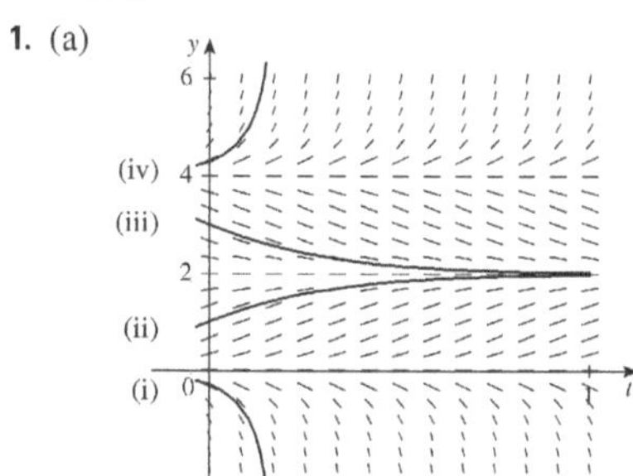

(b) $0 \leqslant c \leqslant 4$; $y = 0$, $y = 2$, $y = 4$

3. (a) $y(0.3) \approx 0.8$

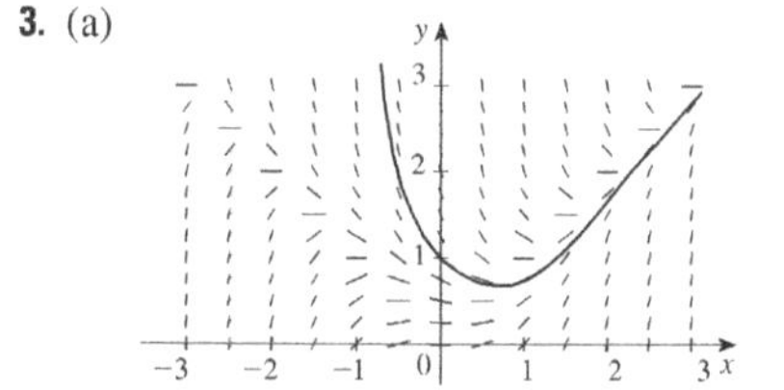

(b) 0.75676

(c) $y = x$ and $y = -x$; there is a loc max or loc min

5. $y = \left(\frac{1}{2}x^2 + C\right)e^{-\sin x}$

7. $y = \pm\sqrt{\ln(x^2 + 2x^{3/2} + C)}$

9. $r(t) = 5e^{t-t^2}$ **11.** $y = \frac{1}{2}x(\ln x)^2 + 2x$ **13.** $x = C - \frac{1}{2}y^2$

15. (a) $P(t) = \dfrac{2000}{1 + 19e^{-0.1t}}$; ≈ 560 (b) $t = -10 \ln \frac{2}{57} \approx 33.5$

17. (a) $L(t) = L_\infty - [L_\infty - L(0)]e^{-kt}$ (b) $L(t) = 53 - 43e^{-0.2t}$

19. 15 days **21.** $k \ln h + h = (-R/V)t + C$

23. (a) Stabilizes at 200,000

(b) (i) $x = 0$, $y = 0$: Zero populations

(ii) $x = 200{,}000$, $y = 0$: In the absence of birds, the insect population is always 200,000.

(iii) $x = 25{,}000$, $y = 175$: Both populations are stable.

(c) The populations stabilize at 25,000 insects and 175 birds.

(d)

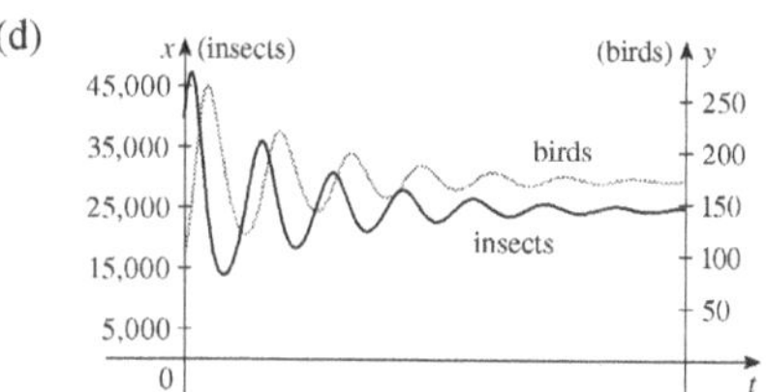

25. (a) $y = (1/k)\cosh kx + a - 1/k$ or $y = (1/k)\cosh kx - (1/k)\cosh kb + h$ (b) $(2/k)\sinh kb$

PROBLEMS PLUS ■ PAGE 633

1. $f(x) = \pm 10e^x$ **5.** $y = x^{1/n}$ **7.** 20°C

9. (b) $f(x) = \dfrac{x^2 - L^2}{4L} - \frac{1}{2}L \ln\left(\dfrac{x}{L}\right)$ (c) No

11. (a) 9.8 h (b) $31{,}900\pi$ ft^2; 2000π ft^2/h

(c) 5.1 h

13. $x^2 + (y - 6)^2 = 25$

15. $y = K/x$, $K \neq 0$

CHAPTER 10

EXERCISES 10.1 ■ PAGE 641

1.

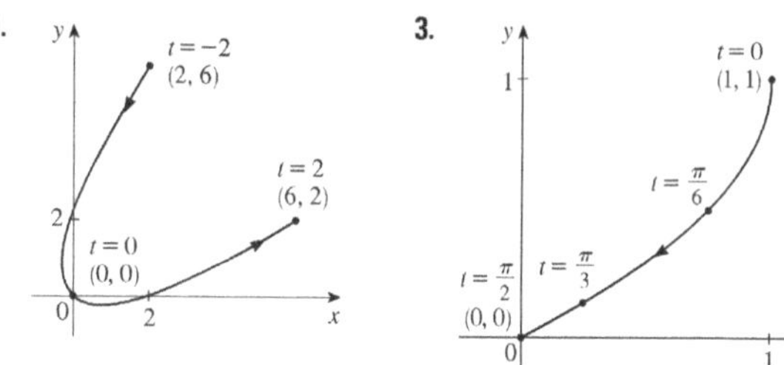

3.

5. (a)

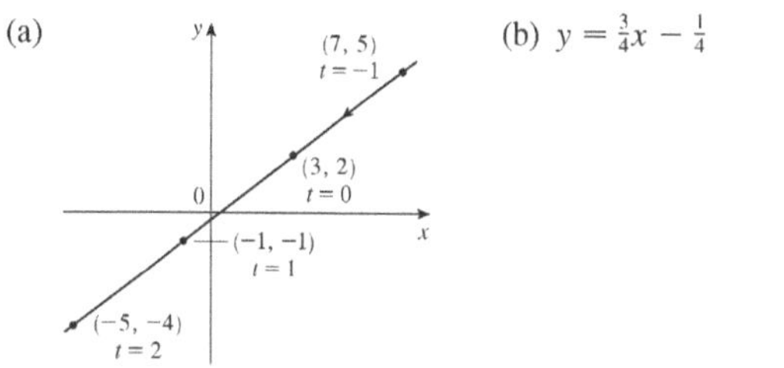

(b) $y = \frac{3}{4}x - \frac{1}{4}$

7. (a)

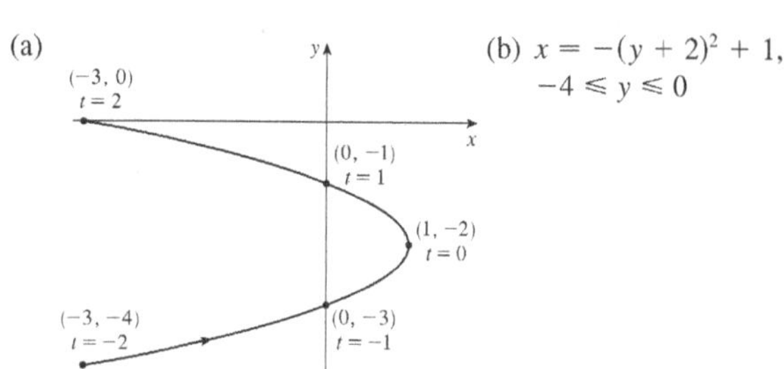

(b) $x = -(y + 2)^2 + 1$, $-4 \leqslant y \leqslant 0$

9. (a)

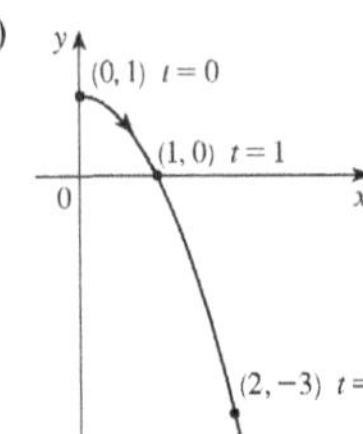

(b) $y = 1 - x^2,\ x \geqslant 0$

11. (a) $x^2 + y^2 = 1, y \geqslant 0$ (b)

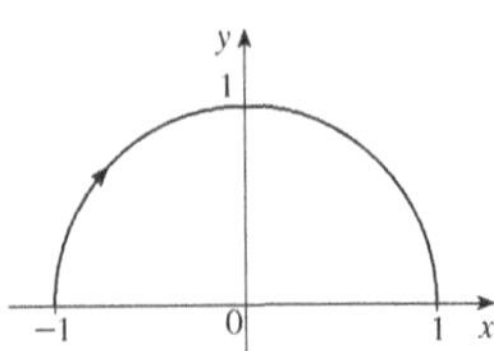

13. (a) $y = 1/x, y > 1$ (b)

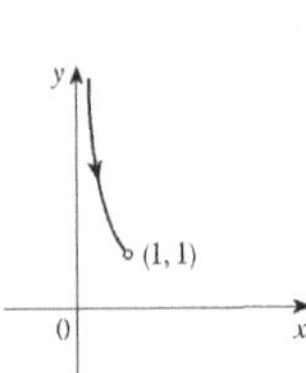

15. (a) $y = \frac{1}{2}\ln x + 1$ (b)

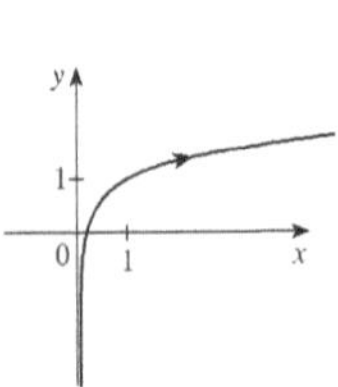

17. (a) $y^2 - x^2 = 1, y \geqslant 1$
(b)

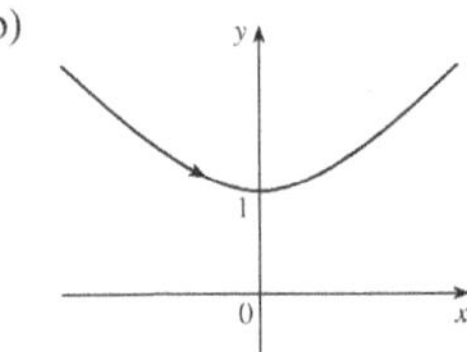

19. Moves counterclockwise along the circle $(x - 3)^2 + (y - 1)^2 = 4$ from (3, 3) to (3, −1)
21. Moves 3 times clockwise around the ellipse $(x^2/25) + (y^2/4) = 1$, starting and ending at (0, −2)
23. It is contained in the rectangle described by $1 \leqslant x \leqslant 4$ and $2 \leqslant y \leqslant 3$.

25.

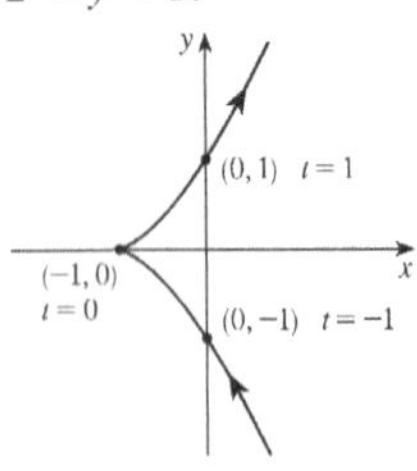

27.

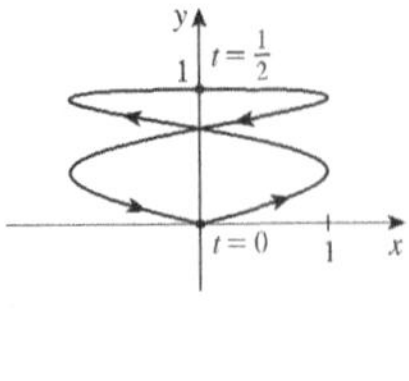

29.

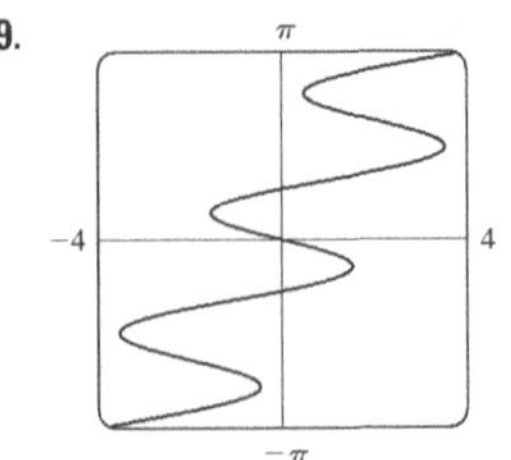

31. (b) $x = -2 + 5t, y = 7 - 8t, 0 \leqslant t \leqslant 1$
33. (a) $x = 2\cos t, y = 1 - 2\sin t, 0 \leqslant t \leqslant 2\pi$
(b) $x = 2\cos t, y = 1 + 2\sin t, 0 \leqslant t \leqslant 6\pi$
(c) $x = 2\cos t, y = 1 + 2\sin t, \pi/2 \leqslant t \leqslant 3\pi/2$
37. The curve $y = x^{2/3}$ is generated in (a). In (b), only the portion with $x \geqslant 0$ is generated, and in (c) we get only the portion with $x > 0$.
41. $x = a\cos\theta, y = b\sin\theta$; $(x^2/a^2) + (y^2/b^2) = 1$, ellipse

43.

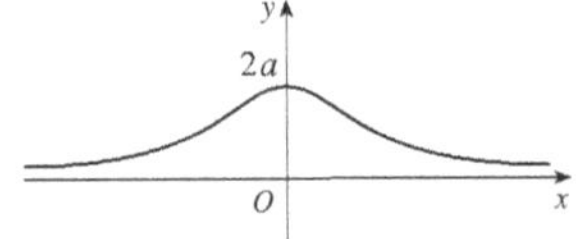

45. (a) Two points of intersection

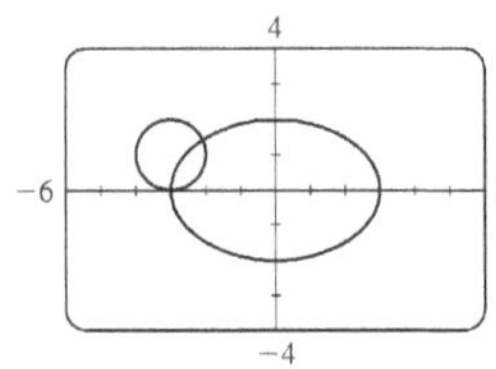

(b) One collision point at (−3, 0) when $t = 3\pi/2$
(c) There are still two intersection points, but no collision point.
47. For $c = 0$, there is a cusp; for $c > 0$, there is a loop whose size increases as c increases.

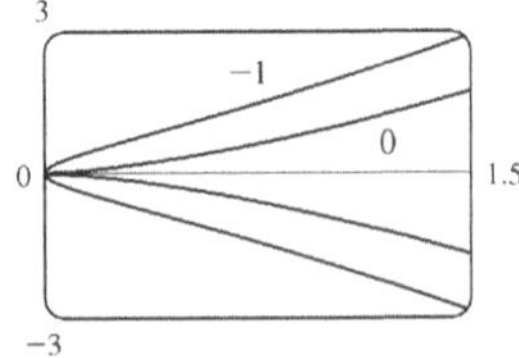

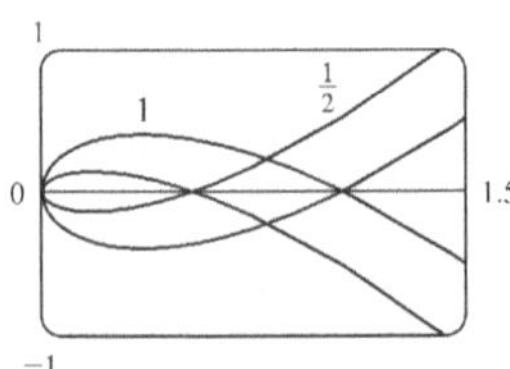

49. The curves roughly follow the line $y = x$, and they start having loops when a is between 1.4 and 1.6 (more precisely, when $a > \sqrt{2}$). The loops increase in size as a increases.
51. As n increases, the number of oscillations increases; a and b determine the width and height.

EXERCISES 10.2 ■ PAGE 651

1. $\dfrac{2t + 1}{t\cos t + \sin t}$ **3.** $y = -\frac{3}{2}x + 7$ **5.** $y = \pi x + \pi^2$
7. $y = 2x + 1$

9. $y = \frac{1}{6}x$

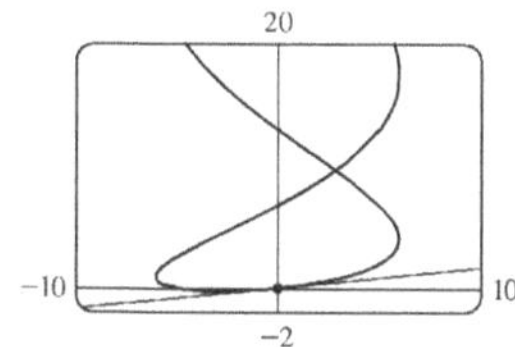

11. $\dfrac{2t+1}{2t}, -\dfrac{1}{4t^3}, t < 0$ **13.** $e^{-2t}(1-t), e^{-3t}(2t-3),\ t > \frac{3}{2}$

15. $-\frac{3}{2}\tan t, -\frac{3}{4}\sec^3 t, \pi/2 < t < 3\pi/2$

17. Horizontal at $(0, -3)$, vertical at $(\pm 2, -2)$

19. Horizontal at $\left(\frac{1}{2}, -1\right)$ and $\left(-\frac{1}{2}, 1\right)$, no vertical

21. $(0.6, 2)$; $\left(5 \cdot 6^{-6/5}, e^{6^{-1/5}}\right)$

23.

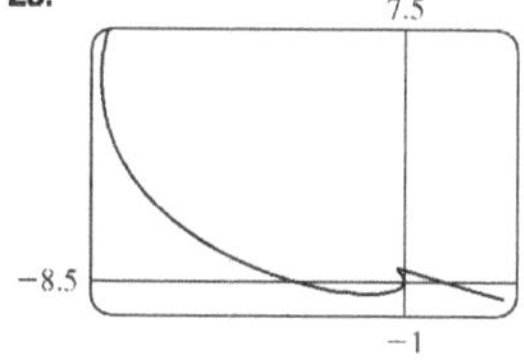

25. $y = x, y = -x$

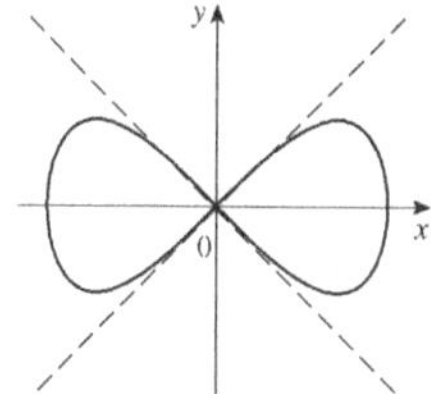

27. (a) $d \sin\theta/(r - d\cos\theta)$ **29.** $\left(\frac{16}{27}, \frac{29}{9}\right), (-2, -4)$

31. πab **33.** $3 - e$ **35.** $2\pi r^2 + \pi d^2$

37. $\int_0^2 \sqrt{2 + 2e^{-2t}}\, dt \approx 3.1416$

39. $\int_0^{4\pi} \sqrt{5 - 4\cos t}\, dt \approx 26.7298$ **41.** $4\sqrt{2} - 2$

43. $\frac{1}{2}\sqrt{2} + \frac{1}{2}\ln(1 + \sqrt{2})$

45. $\sqrt{2}\,(e^{\pi} - 1)$

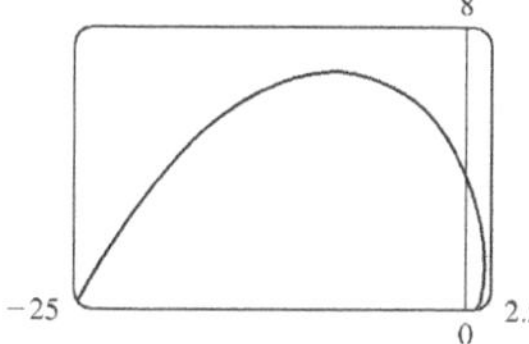

47. 16.7102

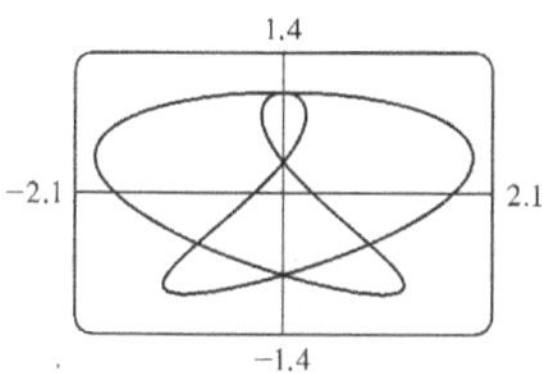

49. 612.3053 **51.** $6\sqrt{2}, \sqrt{2}$

55. (a) $t \in [0, 4\pi]$

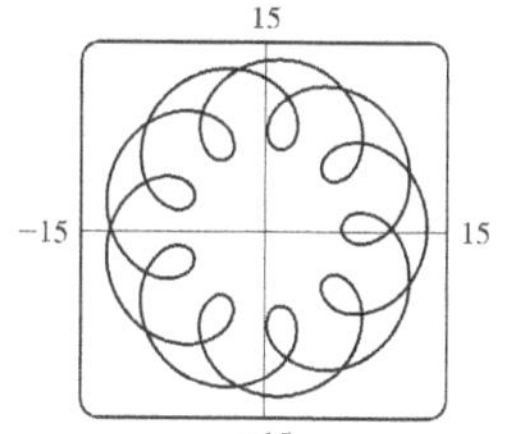

(b) 294

57. $\int_0^{\pi/2} 2\pi t \cos t \sqrt{t^2 + 1}\, dt \approx 4.7394$

59. $\int_0^1 2\pi(t^2 + 1)e^t\sqrt{e^{2t}(t+1)^2(t^2 + 2t + 2)}\, dt \approx 103.5999$

61. $\frac{2}{1215}\pi\left(247\sqrt{13} + 64\right)$ **63.** $\frac{6}{5}\pi a^2$

65. $\frac{24}{5}\pi\left(949\sqrt{26} + 1\right)$ **71.** $\frac{1}{4}$

EXERCISES 10.3 ■ PAGE 662

1. (a)

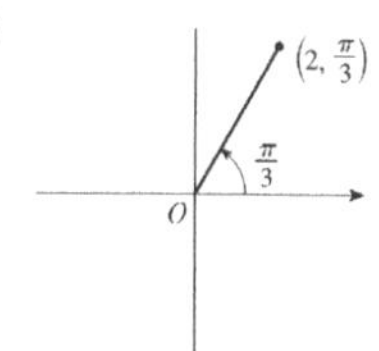

$(2, 7\pi/3), (-2, 4\pi/3)$

(b)

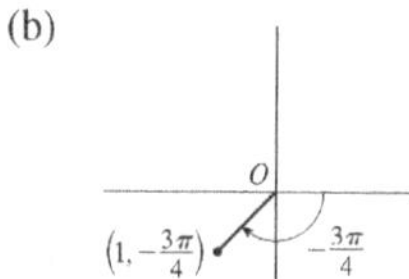

$(1, 5\pi/4), (-1, \pi/4)$

(c)

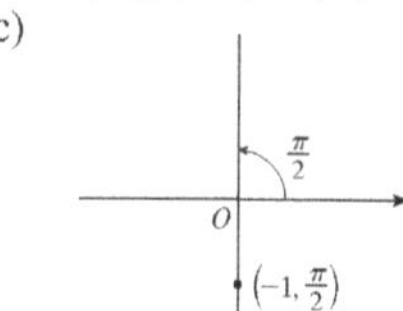

$(1, 3\pi/2), (-1, 5\pi/2)$

3. (a)

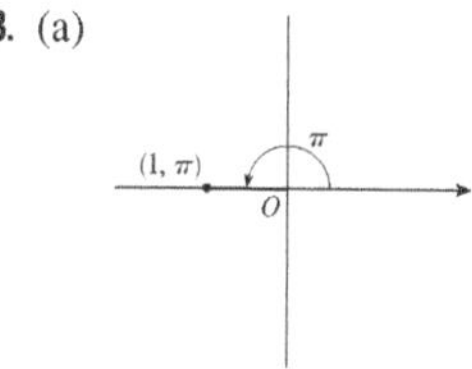

$(-1, 0)$

(b)

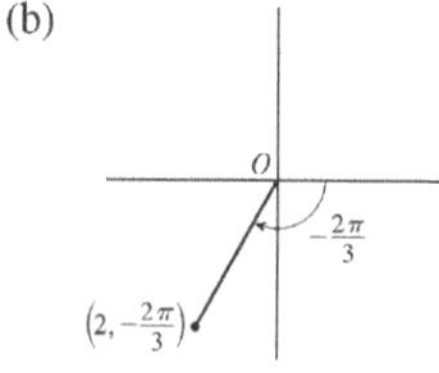

$(-1, -\sqrt{3})$

(c)

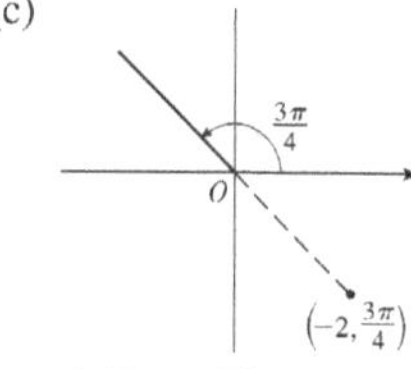

$(\sqrt{2}, -\sqrt{2})$

5. (a) (i) $\left(2\sqrt{2}, 7\pi/4\right)$ (ii) $\left(-2\sqrt{2}, 3\pi/4\right)$

(b) (i) $(2, 2\pi/3)$ (ii) $(-2, 5\pi/3)$

7.

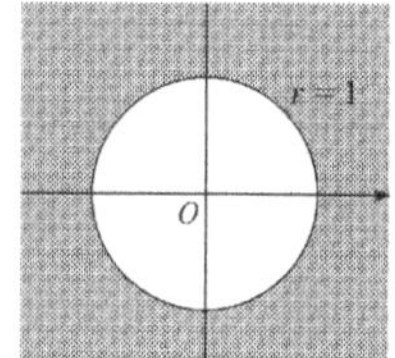

9.

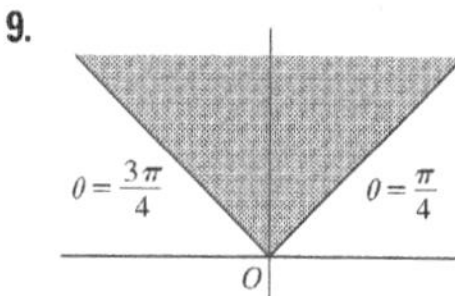

11.

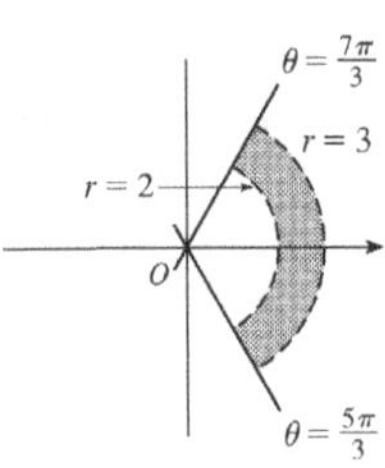

13. $2\sqrt{3}$ **15.** Circle, center O, radius $\sqrt{5}$

17. Circle, center $(1, 0)$, radius 1

19. Hyperbola, center O, foci on x-axis

21. $r = 2 \csc \theta$ **23.** $r = 1/(\sin \theta - 3 \cos \theta)$

25. $r = 2c \cos \theta$ **27.** (a) $\theta = \pi/6$ (b) $x = 3$

29.

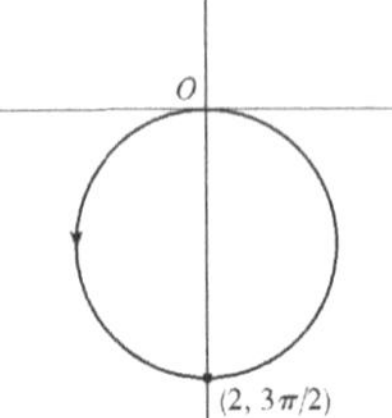

31.

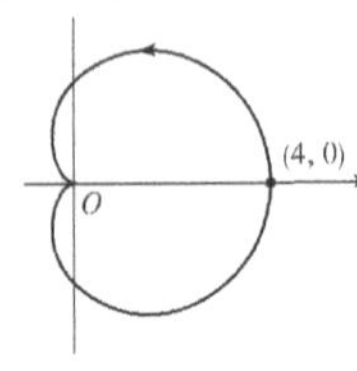

33.

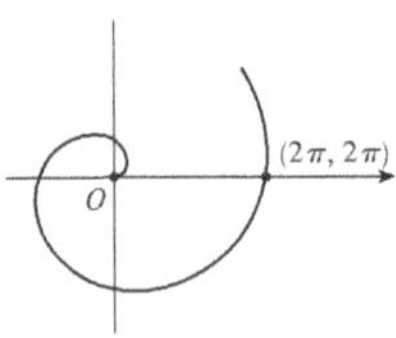

35.

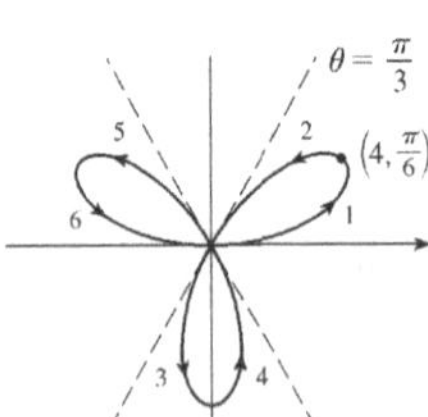

37.

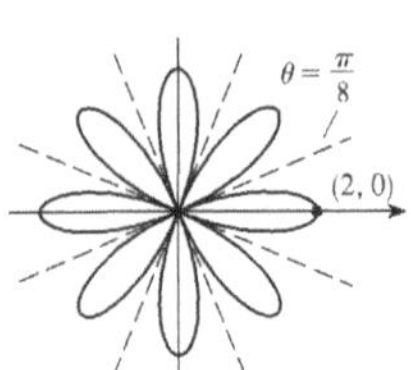

39.

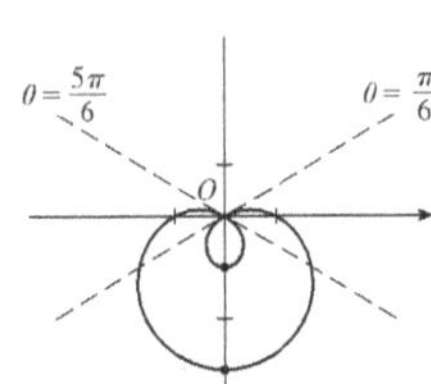

41.

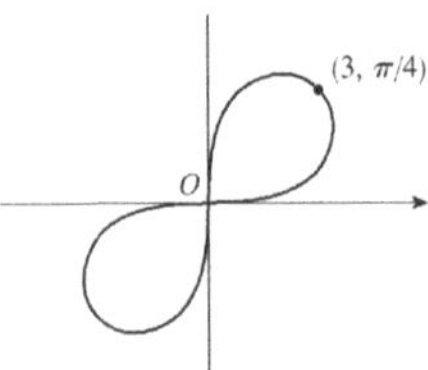

43.

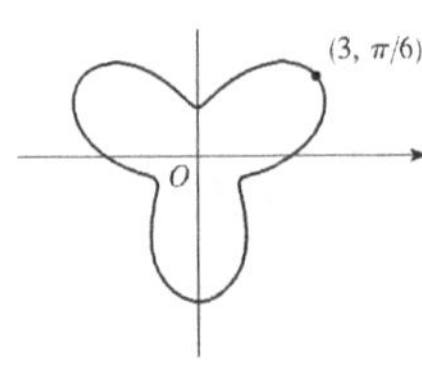

45.

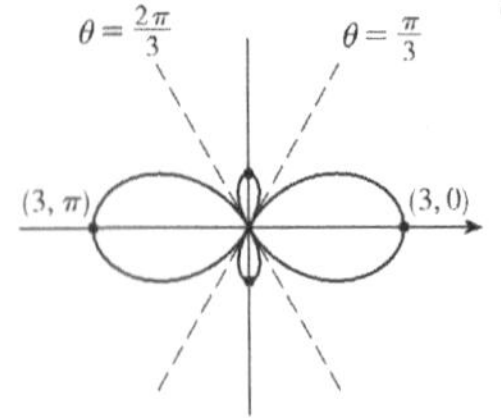

47.

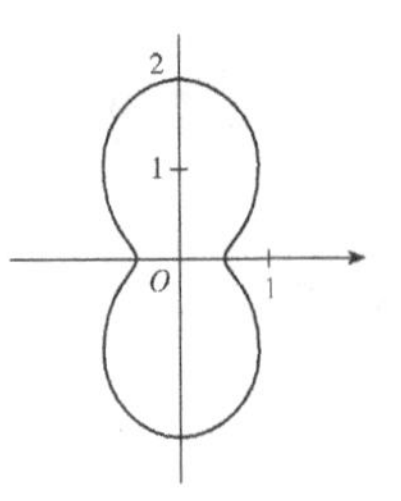

49

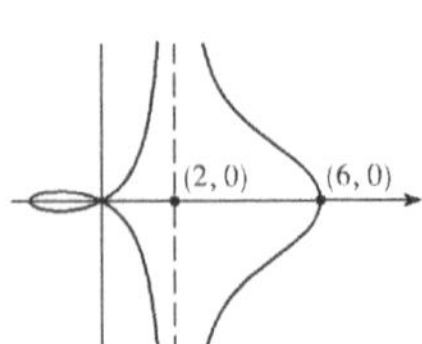

51.

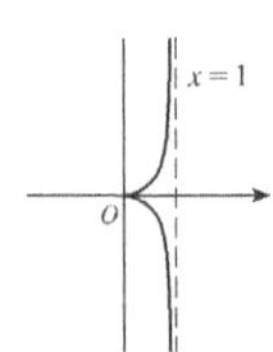

53. (a) For $c < -1$, the inner loop begins at $\theta = \sin^{-1}(-1/c)$ and ends at $\theta = \pi - \sin^{-1}(-1/c)$; for $c > 1$, it begins at $\theta = \pi + \sin^{-1}(1/c)$ and ends at $\theta = 2\pi - \sin^{-1}(1/c)$.

55. $\sqrt{3}$ **57.** $-\pi$ **59.** 1

61. Horizontal at $(3/\sqrt{2}, \pi/4)$, $(-3/\sqrt{2}, 3\pi/4)$; vertical at $(3, 0)$, $(0, \pi/2)$

63. Horizontal at $(\frac{3}{2}, \pi/3)$, $(0, \pi)$ [the pole], and $(\frac{3}{2}, 5\pi/3)$; vertical at $(2, 0)$, $(\frac{1}{2}, 2\pi/3)$, $(\frac{1}{2}, 4\pi/3)$

65. Center $(b/2, a/2)$, radius $\sqrt{a^2 + b^2}/2$

67.

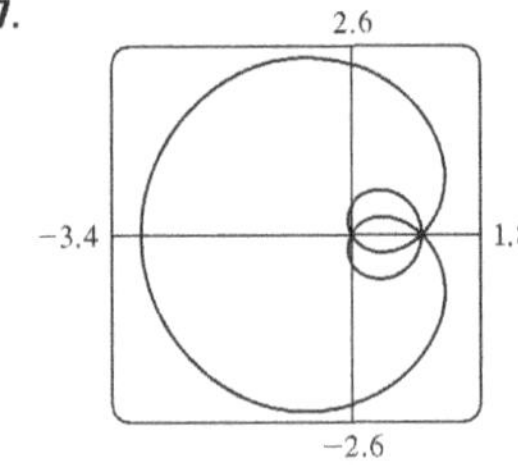

69.

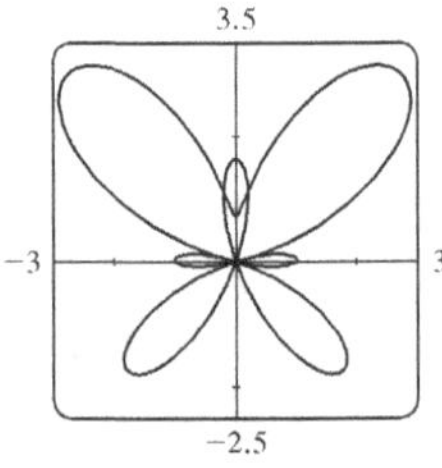

71.

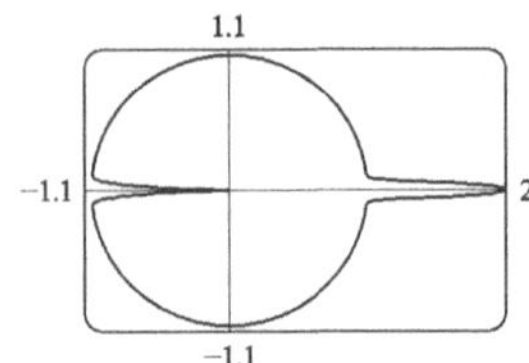

73. By counterclockwise rotation through angle $\pi/6$, $\pi/3$, or α about the origin

75. For $c = 0$, the curve is a circle. As c increases, the left side gets flatter, then has a dimple for $0.5 < c < 1$, a cusp for $c = 1$, and a loop for $c > 1$.

EXERCISES 10.4 ■ PAGE 668

1. $e^{-\pi/4} - e^{-\pi/2}$ **3.** $\frac{9}{2}$ **5.** π^2 **7.** $\frac{41}{4}\pi$

9. π

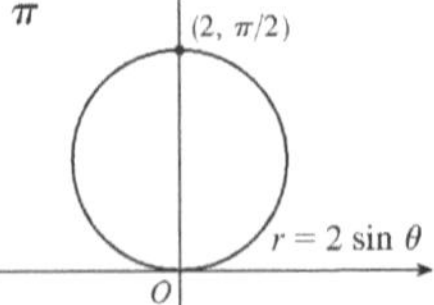

11. 11π

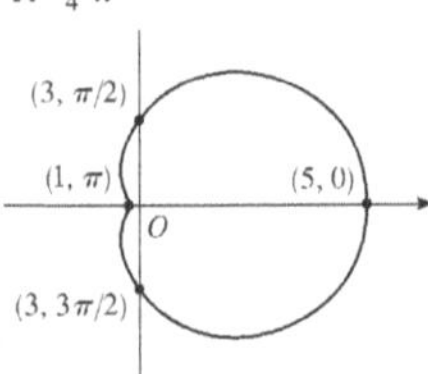

13. $\frac{9}{2}\pi$

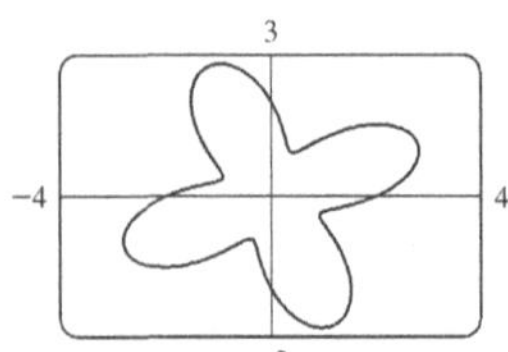

15. $\frac{3}{2}\pi$

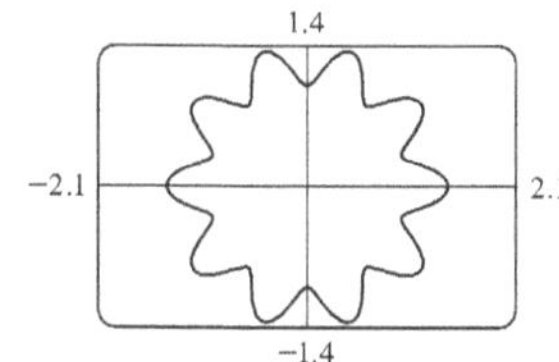

17. $\frac{4}{3}\pi$ **19.** $\frac{1}{16}\pi$ **21.** $\pi - \frac{3}{2}\sqrt{3}$ **23.** $\frac{1}{3}\pi + \frac{1}{2}\sqrt{3}$

25. $4\sqrt{3} - \frac{4}{3}\pi$ **27.** π **29.** $\frac{5}{24}\pi - \frac{1}{4}\sqrt{3}$ **31.** $\frac{1}{2}\pi - 1$

33. $1 - \frac{1}{2}\sqrt{2}$ **35.** $\frac{1}{4}(\pi + 3\sqrt{3})$

37. $(\frac{3}{2}, \pi/6), (\frac{3}{2}, 5\pi/6)$, and the pole

39. $(1, \theta)$ where $\theta = \pi/12, 5\pi/12, 13\pi/12, 17\pi/12$ and $(-1, \theta)$ where $\theta = 7\pi/12, 11\pi/12, 19\pi/12, 23\pi/12$

41. $(\frac{1}{2}\sqrt{3}, \pi/3), (\frac{1}{2}\sqrt{3}, 2\pi/3)$, and the pole

43. Intersection at $\theta \approx 0.89, 2.25$; area ≈ 3.46

45. 2π **47.** $\frac{8}{3}[(\pi^2 + 1)^{3/2} - 1]$

49.  $\frac{16}{3}$

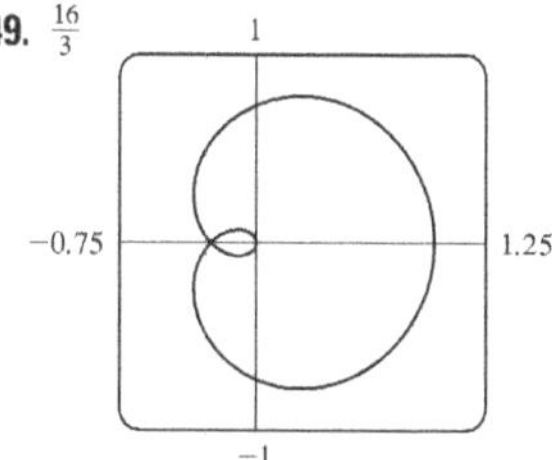

51. 2.4221 **53.** 8.0091

55. (b) $2\pi(2 - \sqrt{2})$

EXERCISES 10.5 ■ PAGE 676

1. $(0, 0), (0, \frac{3}{2}), y = -\frac{3}{2}$

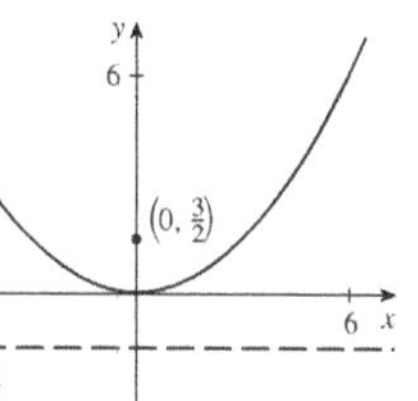

3. $(0, 0), (-\frac{1}{2}, 0), x = \frac{1}{2}$

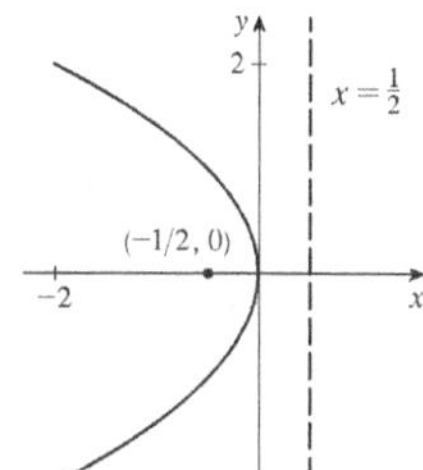

5. $(-2, 3), (-2, 5), y = 1$ **7.** $(-2, -1), (-5, -1), x = 1$

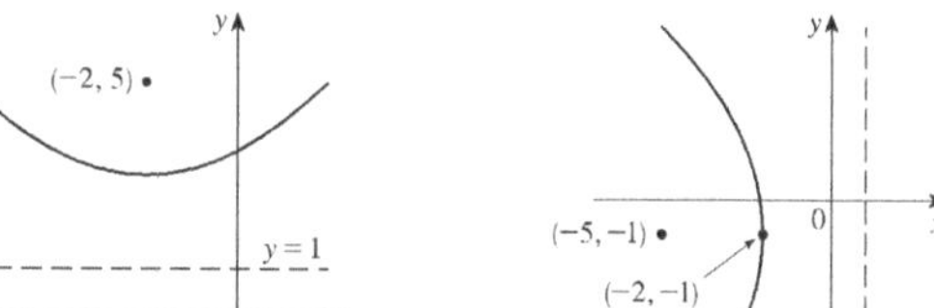

9. $x = -y^2$, focus $(-\frac{1}{4}, 0)$, directrix $x = \frac{1}{4}$

11. $(0, \pm 2), (0, \pm\sqrt{2})$

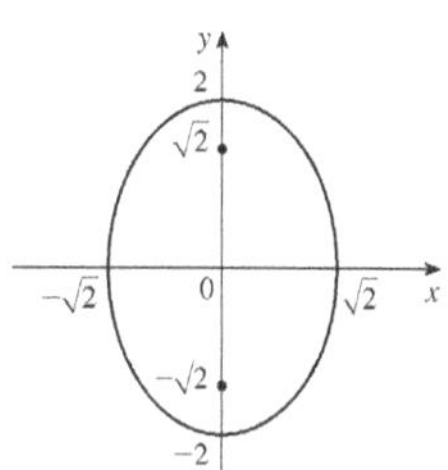

13. $(\pm 3, 0), (\pm 2\sqrt{2}, 0)$

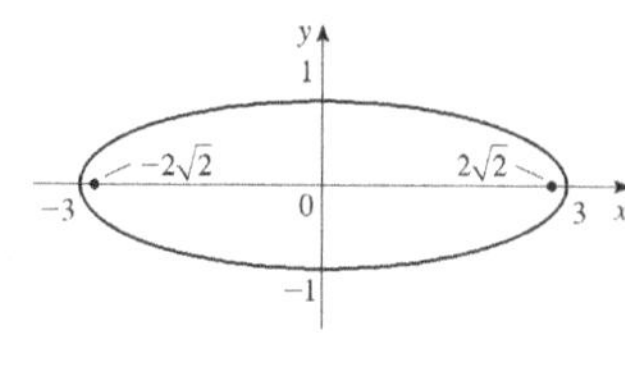

15. $(1, \pm 3), (1, \pm\sqrt{5})$

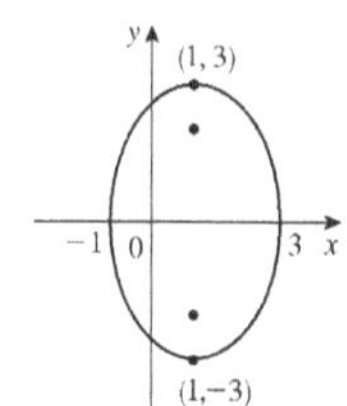

17. $\dfrac{x^2}{4} + \dfrac{y^2}{9} = 1$, foci $(0, \pm\sqrt{5})$

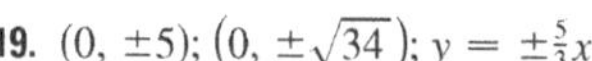

19. $(0, \pm 5); (0, \pm\sqrt{34}); y = \pm\frac{5}{3}x$

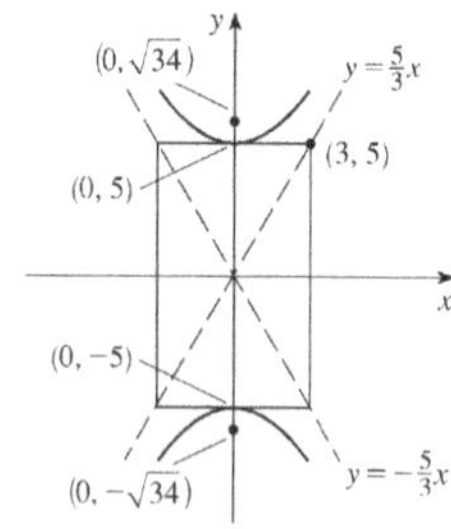

21. $(\pm 10, 0), (\pm 10\sqrt{2}, 0), y = \pm x$

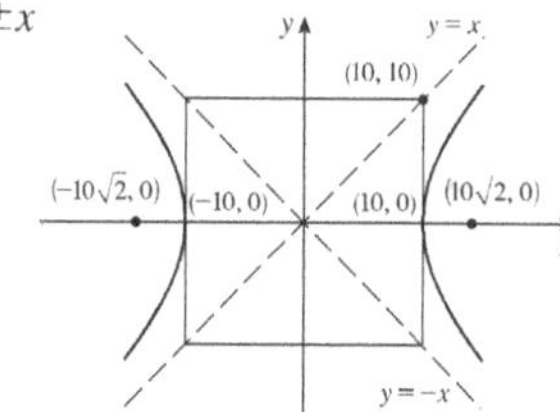

23. $(4, -2), (2, -2)$; $(3 \pm \sqrt{5}, -2)$; $y + 2 = \pm 2(x - 3)$

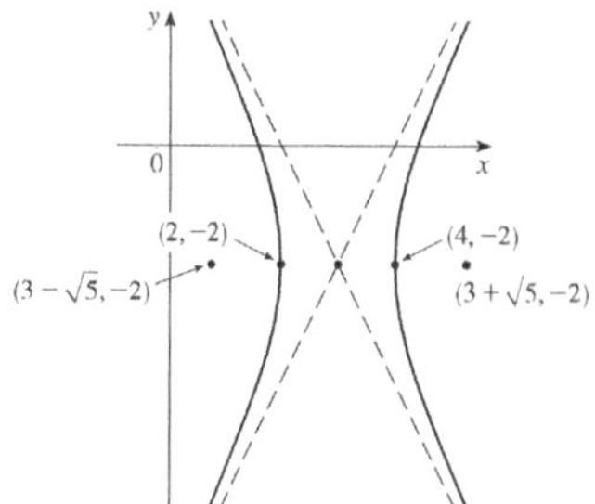

25. Parabola, $(0, -1), (0, -\frac{3}{4})$

27. Ellipse, $(\pm\sqrt{2}, 1), (\pm 1, 1)$

29. Hyperbola, $(0, 1), (0, -3); (0, -1 \pm \sqrt{5})$

31. $y^2 = 4x$ **33.** $y^2 = -12(x + 1)$ **35.** $y - 3 = 2(x - 2)^2$

37. $\dfrac{x^2}{25} + \dfrac{y^2}{21} = 1$ **39.** $\dfrac{x^2}{12} + \dfrac{(y-4)^2}{16} = 1$

41. $\dfrac{(x+1)^2}{12} + \dfrac{(y-4)^2}{16} = 1$ **43.** $\dfrac{x^2}{9} - \dfrac{y^2}{16} = 1$

45. $\dfrac{(y-1)^2}{25} - \dfrac{(x+3)^2}{39} = 1$ **47.** $\dfrac{x^2}{9} - \dfrac{y^2}{36} = 1$

49. $\dfrac{x^2}{3{,}763{,}600} + \dfrac{y^2}{3{,}753{,}196} = 1$

51. (a) $\dfrac{121x^2}{1{,}500{,}625} - \dfrac{121y^2}{3{,}339{,}375} = 1$ (b) $\approx$248 mi

55. (a) Ellipse (b) Hyperbola (c) No curve

59. 15.9

61. $\dfrac{b^2c}{a} + ab \ln\left(\dfrac{a}{b+c}\right)$ where $c^2 = a^2 + b^2$

63. $(0, 4/\pi)$

EXERCISES 10.6 ■ PAGE 684

1. $r = \dfrac{4}{2 + \cos\theta}$ **3.** $r = \dfrac{6}{2 + 3\sin\theta}$

5. $r = \dfrac{8}{1 - \sin\theta}$ **7.** $r = \dfrac{4}{2 + \cos\theta}$

9. (a) $\frac{4}{5}$ (b) Ellipse (c) $y = -1$

(d)

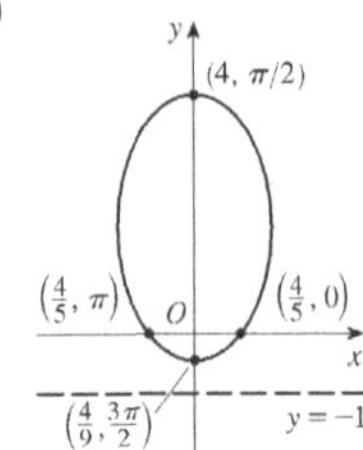

11. (a) 1 (b) Parabola (c) $y = \frac{2}{3}$

(d)

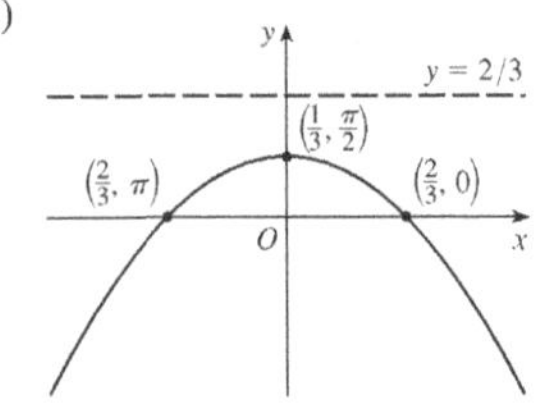

13. (a) $\frac{1}{3}$ (b) Ellipse (c) $x = \frac{9}{2}$

(d)

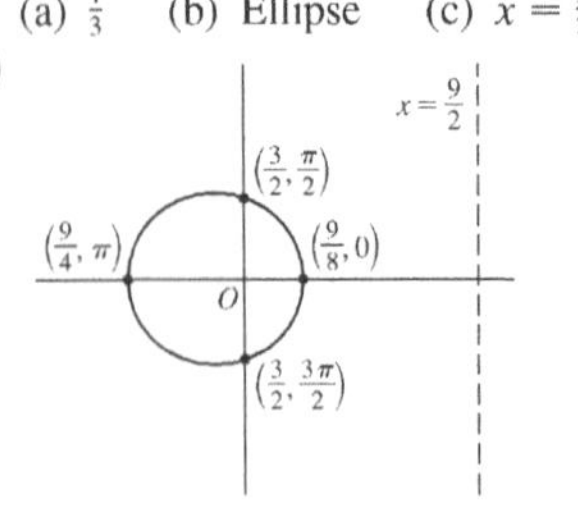

15. (a) 2 (b) Hyperbola (c) $x = -\frac{3}{8}$

(d)

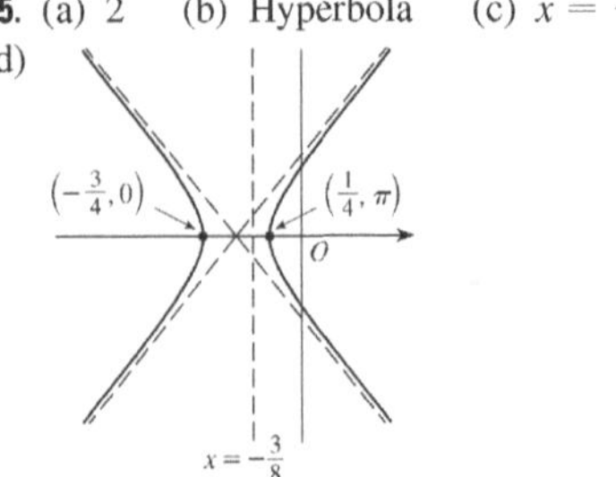

17. (a) $2, y = -\frac{1}{2}$

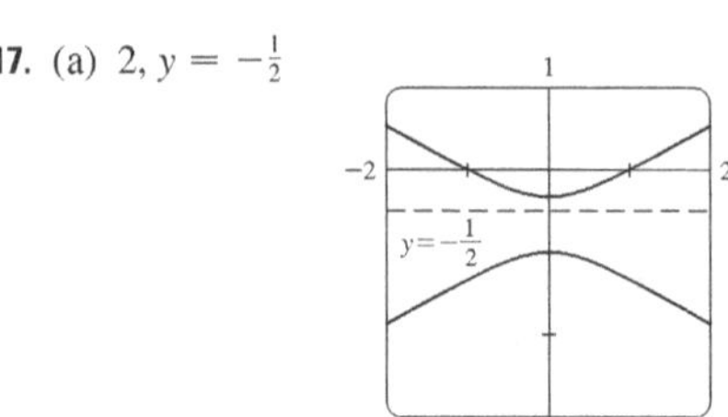

(b) $r = \dfrac{1}{1 - 2\sin(\theta - 3\pi/4)}$

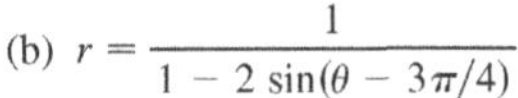

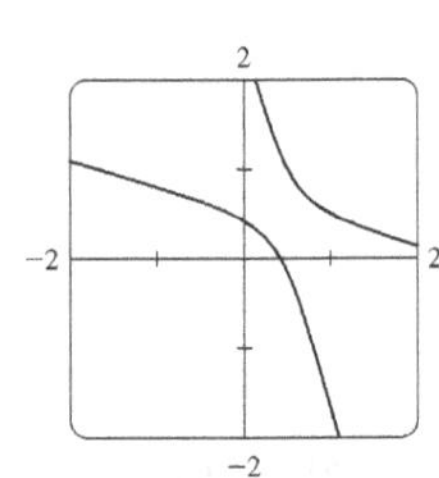

19. The ellipse is nearly circular when e is close to 0 and becomes more elongated as $e \to 1^-$. At $e = 1$, the curve becomes a parabola.

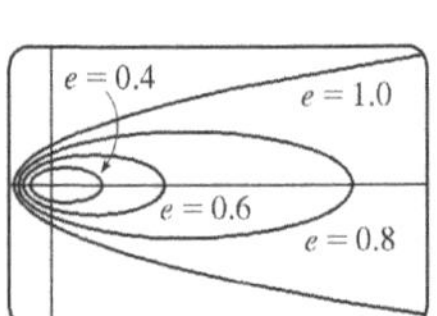

25. $r = \dfrac{2.26 \times 10^8}{1 + 0.093\cos\theta}$

27. 35.64 AU **29.** 7.0×10^7 km **31.** 3.6×10^8 km

CHAPTER 10 REVIEW ■ PAGE 685

True-False Quiz

1. False **3.** False **5.** True **7.** False **9.** True

Exercises

1. $x = y^2 - 8y + 12$ **3.** $y = 1/x$

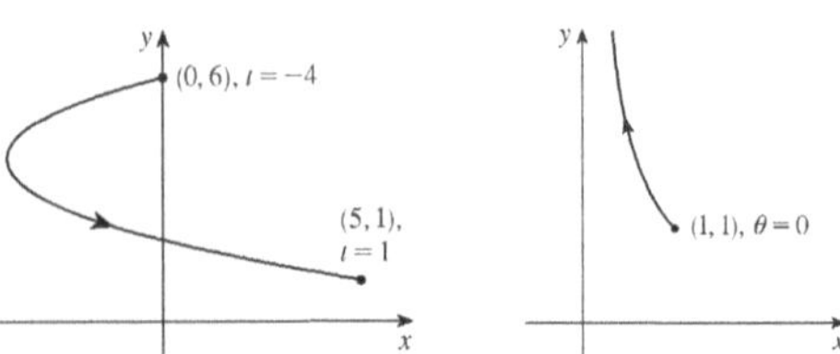

5. $x = t, y = \sqrt{t}$; $x = t^4, y = t^2$; $x = \tan^2 t, y = \tan t, 0 \leq t < \pi/2$

7. (a)

(b) $(3\sqrt{2}, 3\pi/4)$, $(-3\sqrt{2}, 7\pi/4)$

$(-2, 2\sqrt{3})$

9.

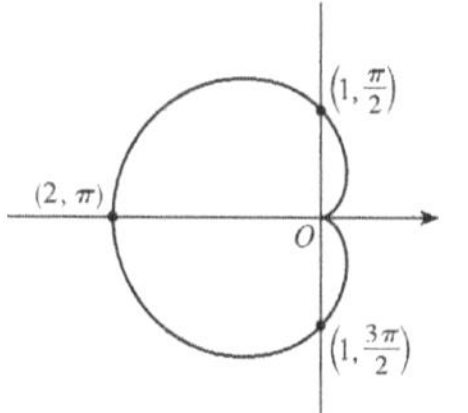

11.

13.

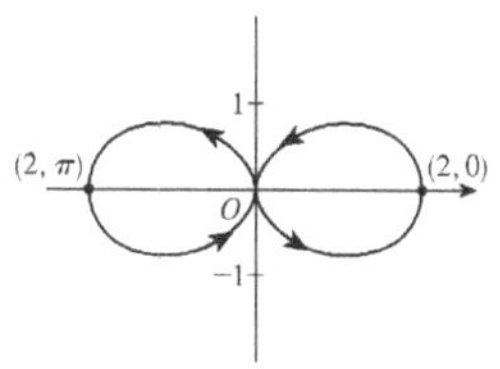

15.

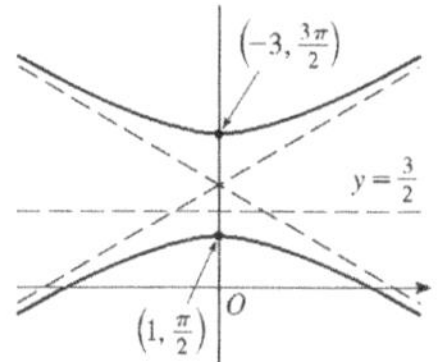

17. $r = \dfrac{2}{\cos\theta + \sin\theta}$

19.

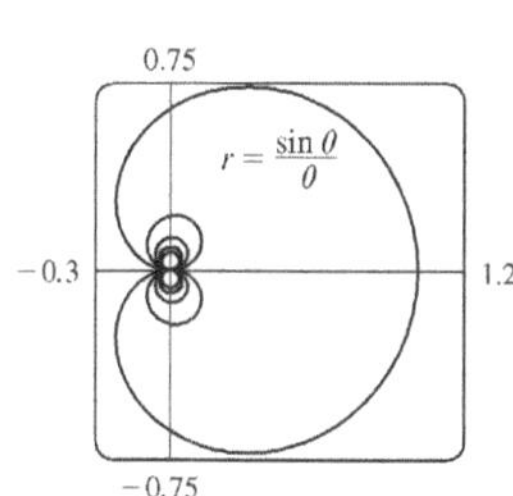

21. 2 **23.** −1

25. $\dfrac{1 + \sin t}{1 + \cos t}, \dfrac{1 + \cos t + \sin t}{(1 + \cos t)^3}$ **27.** $\left(\frac{11}{8}, \frac{3}{4}\right)$

29. Vertical tangent at $\left(\frac{3}{2}a, \pm\frac{1}{2}\sqrt{3}\,a\right), (-3a, 0)$; horizontal tangent at $(a, 0), \left(-\frac{1}{2}a, \pm\frac{3}{2}\sqrt{3}\,a\right)$

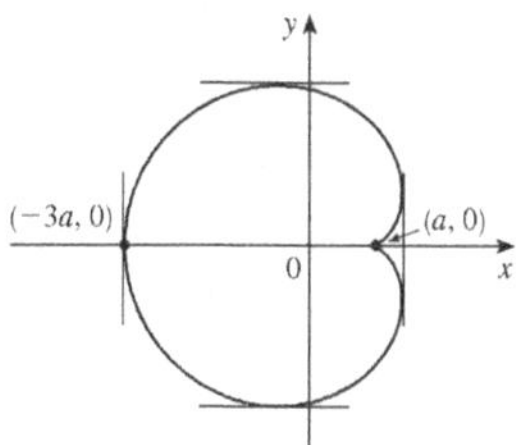

31. 18 **33.** $(2, \pm\pi/3)$ **35.** $\frac{1}{2}(\pi - 1)$

37. $2(5\sqrt{5} - 1)$

39. $\dfrac{2\sqrt{\pi^2 + 1} - \sqrt{4\pi^2 + 1}}{2\pi} + \ln\left(\dfrac{2\pi + \sqrt{4\pi^2 + 1}}{\pi + \sqrt{\pi^2 + 1}}\right)$

41. $471{,}295\pi/1024$

43. All curves have the vertical asymptote $x = 1$. For $c < -1$, the curve bulges to the right. At $c = -1$, the curve is the line $x = 1$. For $-1 < c < 0$, it bulges to the left. At $c = 0$ there is a cusp at $(0, 0)$. For $c > 0$, there is a loop.

45. $(\pm 1, 0), (\pm 3, 0)$

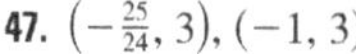

47. $\left(-\frac{25}{24}, 3\right), (-1, 3)$

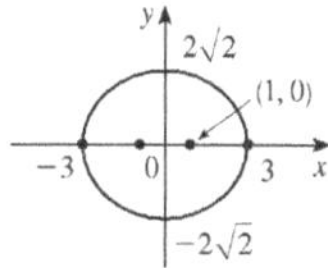

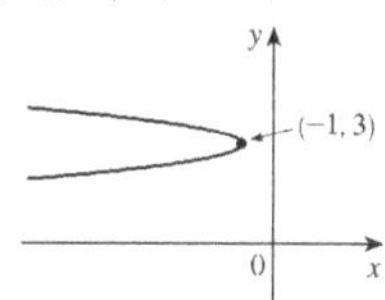

49. $\dfrac{x^2}{25} + \dfrac{y^2}{9} = 1$ **51.** $\dfrac{y^2}{72/5} - \dfrac{x^2}{8/5} = 1$

53. $\dfrac{x^2}{25} + \dfrac{(8y - 399)^2}{160{,}801} = 1$ **55.** $r = \dfrac{4}{3 + \cos\theta}$

57. (a) At $(0, 0)$ and $\left(\frac{3}{2}, \frac{3}{2}\right)$
(b) Horizontal tangents at $(0, 0)$ and $\left(\sqrt[3]{2}, \sqrt[3]{4}\right)$; vertical tangents at $(0, 0)$ and $\left(\sqrt[3]{4}, \sqrt[3]{2}\right)$
(d)

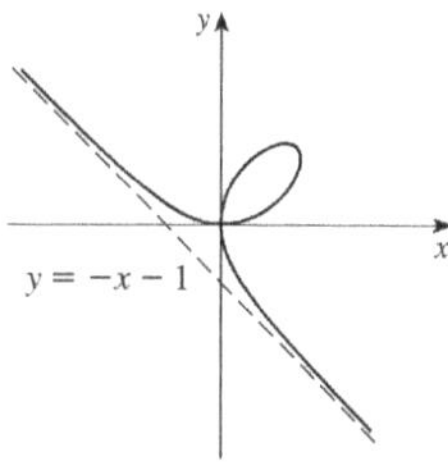

(g) $\frac{3}{2}$

PROBLEMS PLUS ■ PAGE 688

1. $\ln(\pi/2)$ **3.** $\left[-\frac{3}{4}\sqrt{3}, \frac{3}{4}\sqrt{3}\right] \times [-1, 2]$

CHAPTER 11

EXERCISES 11.1 ■ PAGE 700

Abbreviations: C, convergent; D, divergent

1. (a) A sequence is an ordered list of numbers. It can also be defined as a function whose domain is the set of positive integers.
(b) The terms a_n approach 8 as n becomes large.
(c) The terms a_n become large as n becomes large.

3. $1, \frac{4}{5}, \frac{3}{5}, \frac{8}{17}, \frac{5}{13}$ **5.** $\frac{1}{5}, -\frac{1}{25}, \frac{1}{125}, -\frac{1}{625}, \frac{1}{3125}$ **7.** $\frac{1}{2}, \frac{1}{6}, \frac{1}{24}, \frac{1}{120}, \frac{1}{720}$

9. 1, 2, 7, 32, 157 **11.** $2, \frac{2}{3}, \frac{2}{5}, \frac{2}{7}, \frac{2}{9}$ **13.** $a_n = 1/(2n - 1)$

15. $a_n = -3\left(-\frac{2}{3}\right)^{n-1}$ **17.** $a_n = (-1)^{n+1}\dfrac{n^2}{n + 1}$

19. 0.4286, 0.4615, 0.4737, 0.4800, 0.4839, 0.4865, 0.4884, 0.4898, 0.4909, 0.4918; yes; $\frac{1}{2}$

21. 0.5000, 1.2500, 0.8750, 1.0625, 0.9688, 1.0156, 0.9922, 1.0039, 0.9980, 1.0010; yes; 1

23. 1 **25.** 5 **27.** 1 **29.** 1 **31.** D **33.** 0

35. D **37.** 0 **39.** 0 **41.** 0 **43.** 0 **45.** 1

47. e^2 **49.** ln 2 **51.** $\pi/2$ **53.** D **55.** D

57. 1 **59.** $\frac{1}{2}$ **61.** D **63.** 0

65. (a) 1060, 1123.60, 1191.02, 1262.48, 1338.23 (b) D

67. (a) $P_n = 1.08P_{n-1} - 300$ (b) 5734
69. $-1 < r < 1$
71. Convergent by the Monotonic Sequence Theorem; $5 \leqslant L < 8$
73. Decreasing; yes **75.** Not monotonic; no
77. Decreasing; yes
79. 2 **81.** $\frac{1}{2}(3 + \sqrt{5})$ **83.** (b) $\frac{1}{2}(1 + \sqrt{5})$
85. (a) 0 (b) 9, 11

EXERCISES 11.2 ■ PAGE 711

1. (a) A sequence is an ordered list of numbers whereas a series is the *sum* of a list of numbers.
(b) A series is convergent if the sequence of partial sums is a convergent sequence. A series is divergent if it is not convergent.
3. 2
5. 1, 1.125, 1.1620, 1.1777, 1.1857, 1.1903, 1.1932, 1.1952; C
7. 0.5, 1.3284, 2.4265, 3.7598, 5.3049, 7.0443, 8.9644, 11.0540; D

9. −2.40000, −1.92000, −2.01600, −1.99680, −2.00064, −1.99987, −2.00003, −1.99999, −2.00000, −2.00000; convergent, sum = −2

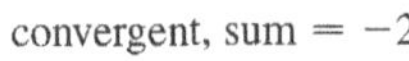

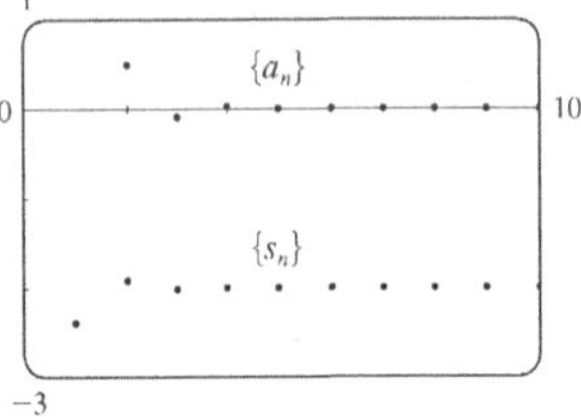

11. 0.44721, 1.15432, 1.98637, 2.88080, 3.80927, 4.75796, 5.71948, 6.68962, 7.66581, 8.64639; divergent

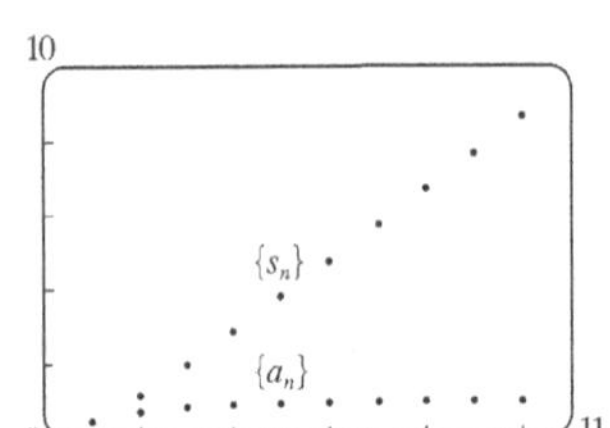

13. 0.29289, 0.42265, 0.50000, 0.55279, 0.59175, 0.62204, 0.64645, 0.66667, 0.68377, 0.69849; convergent, sum = 1

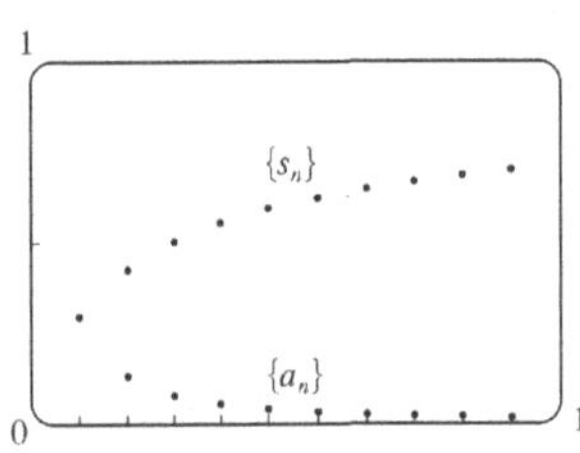

15. (a) C (b) D **17.** D **19.** $\frac{25}{3}$ **21.** 60 **23.** $\frac{1}{7}$
25. D **27.** D **29.** D **31.** $\frac{5}{2}$ **33.** D **35.** D
37. D **39.** D **41.** $e/(e - 1)$ **43.** $\frac{3}{2}$ **45.** $\frac{11}{6}$ **47.** $e - 1$
49. (b) 1 (c) 2 (d) All rational numbers with a terminating decimal representation, except 0.
51. $\frac{8}{9}$ **53.** $\frac{838}{333}$ **55.** 5063/3300
57. $-\frac{1}{5} < x < \frac{1}{5}$; $\frac{-5x}{1 + 5x}$ **59.** $-1 < x < 5$; $\frac{3}{5 - x}$
61. $x > 2$ or $x < -2$; $\frac{x}{x - 2}$ **63.** $x < 0$; $\frac{1}{1 - e^x}$
65. 1 **67.** $a_1 = 0$, $a_n = \frac{2}{n(n + 1)}$ for $n > 1$, sum = 1

69. (a) 157.875 mg; $\frac{3000}{19}(1 - 0.05^n)$ (b) 157.895 mg
71. (a) $S_n = \frac{D(1 - c^n)}{1 - c}$ (b) 5 **73.** $\frac{1}{2}(\sqrt{3} - 1)$
77. $\frac{1}{n(n + 1)}$ **79.** The series is divergent.
85. $\{s_n\}$ is bounded and increasing.
87. (a) $0, \frac{1}{9}, \frac{2}{9}, \frac{1}{3}, \frac{2}{3}, \frac{7}{9}, \frac{8}{9}, 1$
89. (a) $\frac{1}{2}, \frac{5}{6}, \frac{23}{24}, \frac{119}{120}$; $\frac{(n + 1)! - 1}{(n + 1)!}$ (c) 1

EXERCISES 11.3 ■ PAGE 720

1. C

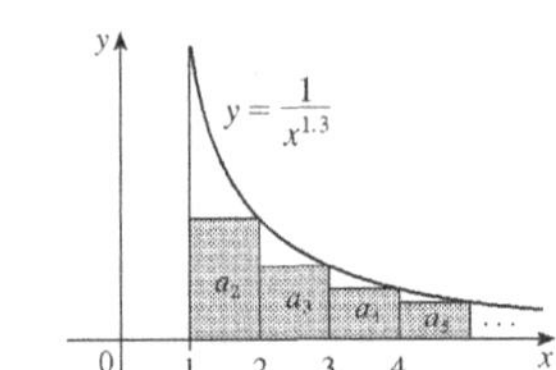

3. D **5.** C **7.** D **9.** C **11.** C **13.** D
15. C **17.** C **19.** C **21.** D **23.** C **25.** C
27. f is neither positive nor decreasing.
29. $p > 1$ **31.** $p < -1$ **33.** $(1, \infty)$
35. (a) $\frac{9}{10}\pi^4$ (b) $\frac{1}{90}\pi^4 - \frac{17}{16}$
37. (a) 1.54977, error $\leqslant$ 0.1 (b) 1.64522, error $\leqslant$ 0.005
(c) 1.64522 compared to 1.64493 (d) $n > 1000$
39. 0.00145 **45.** $b < 1/e$

EXERCISES 11.4 ■ PAGE 726

1. (a) Nothing (b) C **3.** C **5.** D **7.** C **9.** D
11. C **13.** C **15.** D **17.** D **19.** D **21.** C
23. C **25.** D **27.** C **29.** C **31.** D
33. 1.249, error < 0.1 **35.** 0.0739, error $< 6.4 \times 10^{-8}$
45. Yes

EXERCISES 11.5 ■ PAGE 731

1. (a) A series whose terms are alternately positive and negative (b) $0 < b_{n+1} \leqslant b_n$ and $\lim_{n\to\infty} b_n = 0$, where $b_n = |a_n|$ (c) $|R_n| \leqslant b_{n+1}$
3. C **5.** C **7.** D **9.** C **11.** C **13.** D **15.** C
17. C **19.** D **21.** −0.5507 **23.** 5 **25.** 4
27. −0.4597 **29.** 0.0676 **31.** An underestimate
33. p is not a negative integer **35.** $\{b_n\}$ is not decreasing

EXERCISES 11.6 ■ PAGE 737

Abbreviations: AC, absolutely convergent;
CC, conditionally convergent

1. (a) D (b) C (c) May converge or diverge
3. AC **5.** CC **7.** AC **9.** D **11.** AC **13.** AC
15. AC **17.** CC **19.** AC **21.** AC **23.** D **25.** AC
27. AC **29.** D **31.** D **33.** AC

35. (a) and (d)

39. (a) $\frac{661}{960} \approx 0.68854$, error < 0.00521

(b) $n \geq 11$, 0.693109

45. (b) $\sum_{n=2}^{\infty} \frac{(-1)^n}{n \ln n}$; $\sum_{n=1}^{\infty} \frac{(-1)^{n-1}}{n}$

EXERCISES 11.7 ■ PAGE 740

1. C **3.** D **5.** C **7.** D **9.** C **11.** C
13. C **15.** C **17.** C **19.** C **21.** D **23.** D
25. C **27.** C **29.** C **31.** D
33. C **35.** D **37.** C

EXERCISES 11.8 ■ PAGE 745

1. A series of the form $\sum_{n=0}^{\infty} c_n(x-a)^n$, where x is a variable and a and the c_n's are constants

3. 1, $(-1, 1)$ **5.** 1, $[-1, 1)$

7. ∞, $(-\infty, \infty)$ **9.** 2, $(-2, 2)$ **11.** $\frac{1}{3}$, $\left[-\frac{1}{3}, \frac{1}{3}\right]$

13. 4, $(-4, 4]$ **15.** 1, $[1, 3]$ **17.** $\frac{1}{3}$, $\left[-\frac{13}{3}, -\frac{11}{3}\right)$

19. ∞, $(-\infty, \infty)$ **21.** b, $(a-b, a+b)$ **23.** 0, $\left\{\frac{1}{2}\right\}$

25. $\frac{1}{5}$, $\left[\frac{3}{5}, 1\right]$ **27.** ∞, $(-\infty, \infty)$ **29.** (a) Yes (b) No

31. k^k **33.** No

35. (a) $(-\infty, \infty)$

(b), (c)

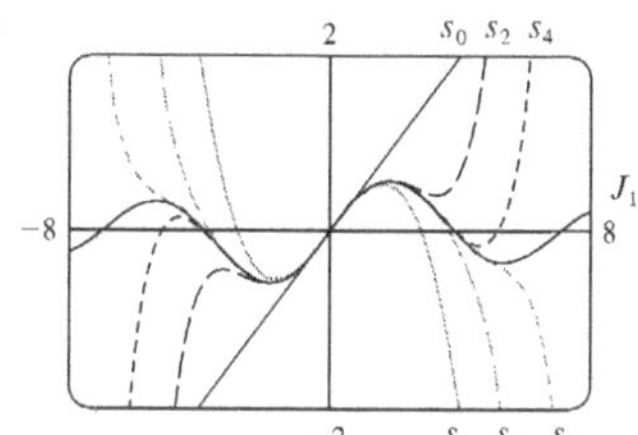

37. $(-1, 1)$, $f(x) = (1+2x)/(1-x^2)$ **41.** 2

EXERCISES 11.9 ■ PAGE 751

1. 10 **3.** $\sum_{n=0}^{\infty} (-1)^n x^n$, $(-1, 1)$ **5.** $2\sum_{n=0}^{\infty} \frac{1}{3^{n+1}} x^n$, $(-3, 3)$

7. $\sum_{n=0}^{\infty} (-1)^n \frac{1}{9^{n+1}} x^{2n+1}$, $(-3, 3)$ **9.** $1 + 2\sum_{n=1}^{\infty} x^n$, $(-1, 1)$

11. $\sum_{n=0}^{\infty} \left[(-1)^{n+1} - \frac{1}{2^{n+1}}\right] x^n$, $(-1, 1)$

13. (a) $\sum_{n=0}^{\infty} (-1)^n (n+1) x^n$, $R = 1$

(b) $\frac{1}{2}\sum_{n=0}^{\infty} (-1)^n (n+2)(n+1) x^n$, $R = 1$

(c) $\frac{1}{2}\sum_{n=2}^{\infty} (-1)^n n(n-1) x^n$, $R = 1$

15. $\ln 5 - \sum_{n=1}^{\infty} \frac{x^n}{n5^n}$, $R = 5$

17. $\sum_{n=0}^{\infty} (-1)^n 4^n (n+1) x^{n+1}$, $R = \frac{1}{4}$

19. $\sum_{n=0}^{\infty} (2n+1) x^n$, $R = 1$

21. $\sum_{n=0}^{\infty} (-1)^n \frac{1}{16^{n+1}} x^{2n+1}$, $R = 4$

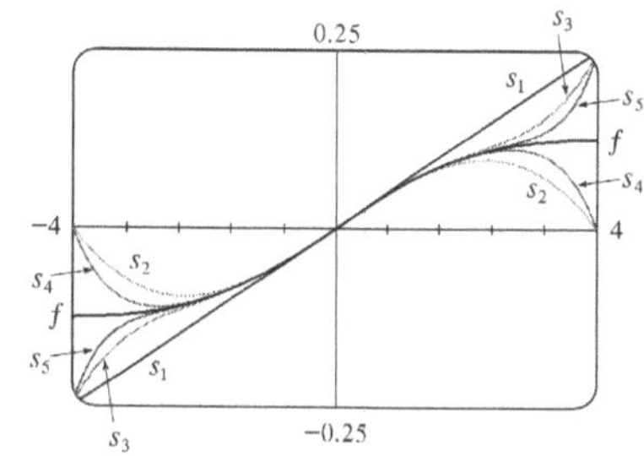

23. $\sum_{n=0}^{\infty} \frac{2x^{2n+1}}{2n+1}$, $R = 1$

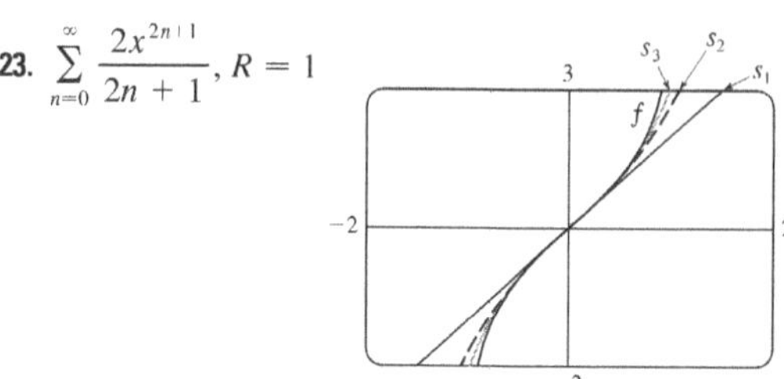

25. $C + \sum_{n=0}^{\infty} \frac{t^{8n+2}}{8n+2}$, $R = 1$

27. $C + \sum_{n=1}^{\infty} (-1)^n \frac{x^{n+3}}{n(n+3)}$, $R = 1$

29. 0.199989 **31.** 0.000983 **33.** 0.19740

35. (b) 0.920 **39.** $[-1, 1]$, $[-1, 1)$, $(-1, 1)$

EXERCISES 11.10 ■ PAGE 765

1. $b_8 = f^{(8)}(5)/8!$ **3.** $\sum_{n=0}^{\infty} (n+1) x^n$, $R = 1$

5. $\sum_{n=0}^{\infty} (n+1) x^n$, $R = 1$

7. $\sum_{n=0}^{\infty} (-1)^n \frac{\pi^{2n+1}}{(2n+1)!} x^{2n+1}$, $R = \infty$

9. $\sum_{n=0}^{\infty} \frac{(\ln 2)^n}{n!} x^n$, $R = \infty$ **11.** $\sum_{n=0}^{\infty} \frac{x^{2n+1}}{(2n+1)!}$, $R = \infty$

13. $-1 - 2(x-1) + 3(x-1)^2 + 4(x-1)^3 + (x-1)^4$, $R = \infty$

15. $\ln 2 + \sum_{n=1}^{\infty} (-1)^{n+1} \frac{1}{n2^n}(x-2)^n$, $R = 2$

17. $\sum_{n=0}^{\infty} \frac{2^n e^6}{n!}(x-3)^n$, $R = \infty$

19. $\sum_{n=0}^{\infty} (-1)^{n+1} \frac{1}{(2n)!}(x-\pi)^{2n}$, $R = \infty$

25. $1 - \frac{1}{4}x - \sum_{n=2}^{\infty} \frac{3 \cdot 7 \cdot \cdots \cdot (4n-5)}{4^n \cdot n!} x^n$, $R = 1$

27. $\sum_{n=0}^{\infty} (-1)^n \frac{(n+1)(n+2)}{2^{n+4}} x^n$, $R = 2$

29. $\sum_{n=0}^{\infty} (-1)^n \frac{\pi^{2n+1}}{(2n+1)!} x^{2n+1}$, $R = \infty$

31. $\sum_{n=0}^{\infty} \frac{2^n + 1}{n!} x^n$, $R = \infty$

33. $\sum_{n=0}^{\infty} (-1)^n \frac{1}{2^{2n}(2n)!} x^{4n+1}$, $R = \infty$

35. $\frac{1}{2}x + \sum_{n=1}^{\infty} (-1)^n \frac{1 \cdot 3 \cdot 5 \cdot \cdots \cdot (2n-1)}{n!\, 2^{3n+1}} x^{2n+1}$, $R = 2$

37. $\sum_{n=1}^{\infty} (-1)^{n+1} \frac{2^{2n-1}}{(2n)!} x^{2n}$, $R = \infty$

39. $\sum_{n=0}^{\infty} (-1)^n \frac{1}{(2n)!} x^{4n}$, $R = \infty$

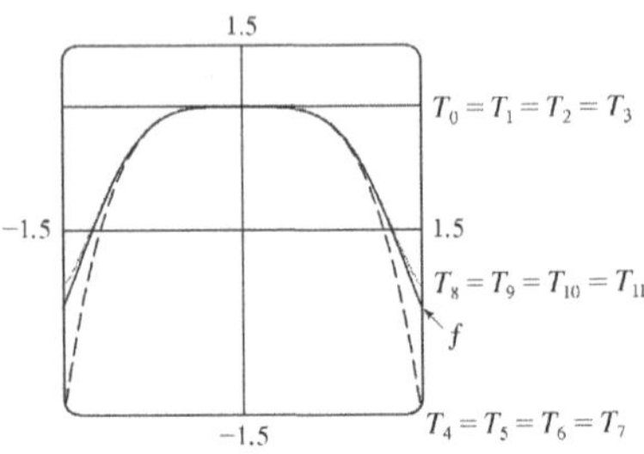

41. $\sum_{n=1}^{\infty} \frac{(-1)^{n-1}}{(n-1)!} x^n$, $R = \infty$

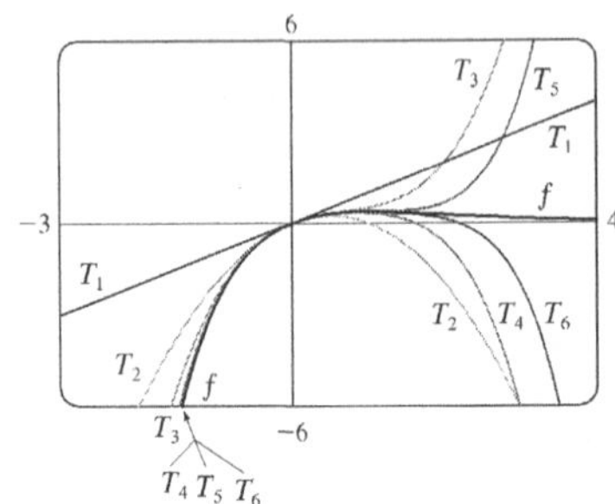

43. 0.99619

45. (a) $1 + \sum_{n=1}^{\infty} \frac{1 \cdot 3 \cdot 5 \cdot \cdots \cdot (2n-1)}{2^n n!} x^{2n}$

(b) $x + \sum_{n=1}^{\infty} \frac{1 \cdot 3 \cdot 5 \cdot \cdots \cdot (2n-1)}{(2n+1)2^n n!} x^{2n+1}$

47. $C + \sum_{n=0}^{\infty} (-1)^n \frac{x^{6n+2}}{(6n+2)(2n)!}$, $R = \infty$

49. $C + \sum_{n=1}^{\infty} (-1)^n \frac{1}{2n\,(2n)!} x^{2n}$, $R = \infty$

51. 0.0059 **53.** 0.40102 **55.** $\frac{1}{2}$ **57.** $\frac{1}{120}$

59. $1 - \frac{3}{2}x^2 + \frac{25}{24}x^4$ **61.** $1 + \frac{1}{6}x^2 + \frac{7}{360}x^4$ **63.** e^{-x^4}

65. $\ln \frac{8}{5}$ **67.** $1/\sqrt{2}$ **69.** $e^3 - 1$

EXERCISES 11.11 ■ PAGE 774

1. (a) $T_0(x) = 1 = T_1(x)$, $T_2(x) = 1 - \frac{1}{2}x^2 = T_3(x)$,
$T_4(x) = 1 - \frac{1}{2}x^2 + \frac{1}{24}x^4 = T_5(x)$,
$T_6(x) = 1 - \frac{1}{2}x^2 + \frac{1}{24}x^4 - \frac{1}{720}x^6$

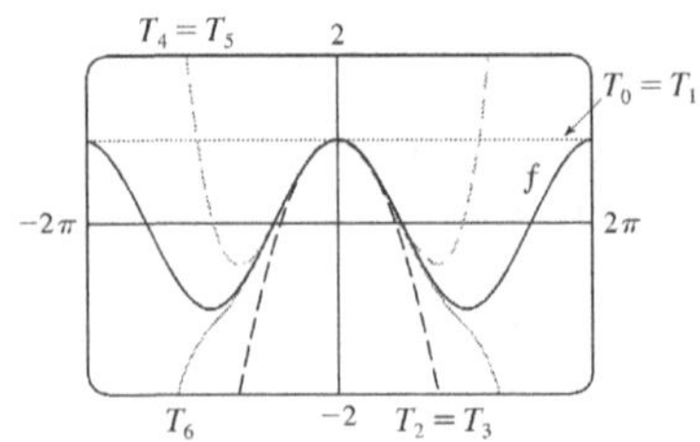

(b)

x	f	$T_0 = T_1$	$T_2 = T_3$	$T_4 = T_5$	T_6
$\frac{\pi}{4}$	0.7071	1	0.6916	0.7074	0.7071
$\frac{\pi}{2}$	0	1	−0.2337	0.0200	−0.0009
π	−1	1	−3.9348	0.1239	−1.2114

(c) As n increases, $T_n(x)$ is a good approximation to $f(x)$ on a larger and larger interval.

3. $\frac{1}{2} - \frac{1}{4}(x-2) + \frac{1}{8}(x-2)^2 - \frac{1}{16}(x-2)^3$

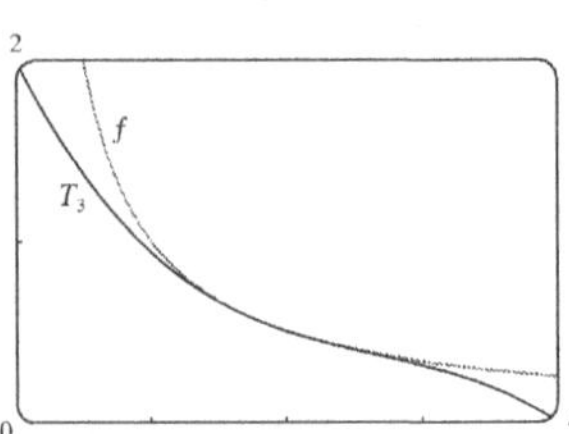

5. $-\left(x - \frac{\pi}{2}\right) + \frac{1}{6}\left(x - \frac{\pi}{2}\right)^3$

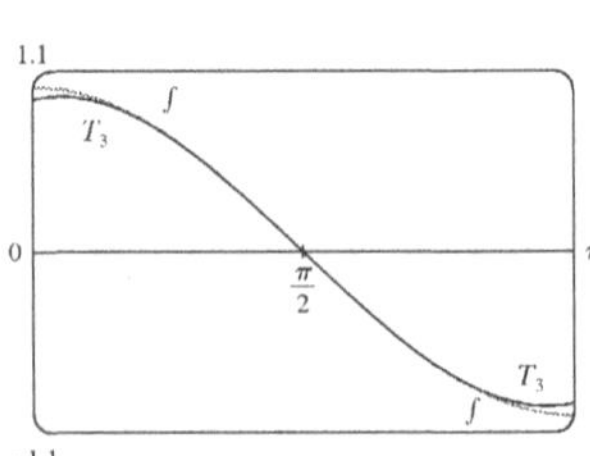

7. $(x-1) - \frac{1}{2}(x-1)^2 + \frac{1}{3}(x-1)^3$

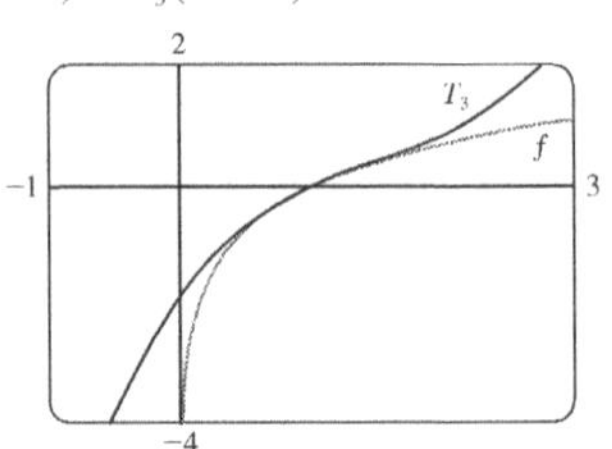

9. $x - 2x^2 + 2x^3$

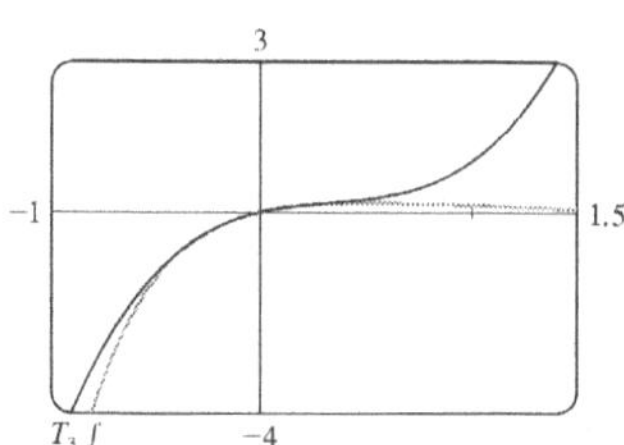

11. $T_5(x) = 1 - 2\left(x - \frac{\pi}{4}\right) + 2\left(x - \frac{\pi}{4}\right)^2 - \frac{8}{3}\left(x - \frac{\pi}{4}\right)^3 + \frac{10}{3}\left(x - \frac{\pi}{4}\right)^4 - \frac{64}{15}\left(x - \frac{\pi}{4}\right)^5$

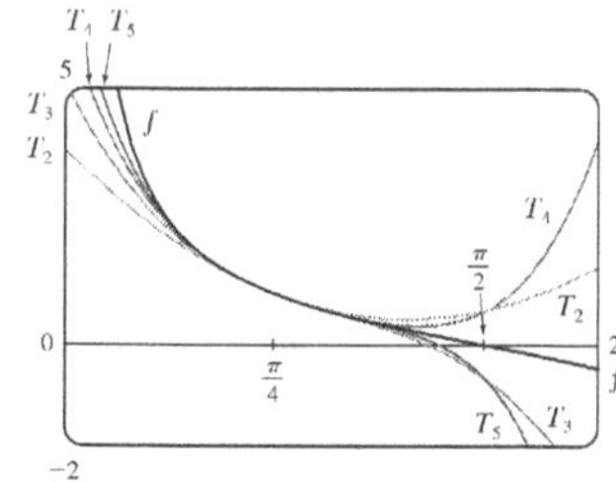

13. (a) $2 + \frac{1}{4}(x - 4) - \frac{1}{64}(x - 4)^2$ (b) 1.5625×10^{-5}

15. (a) $1 + \frac{2}{3}(x - 1) - \frac{1}{9}(x - 1)^2 + \frac{4}{81}(x - 1)^3$ (b) 0.000097

17. (a) $1 + \frac{1}{2}x^2$ (b) 0.0014

19. (a) $1 + x^2$ (b) 0.00006 **21.** (a) $x^2 - \frac{1}{6}x^4$ (b) 0.042

23. 0.17365 **25.** Four **27.** $-1.037 < x < 1.037$

29. $-0.86 < x < 0.86$ **31.** 21 m, no

37. (c) They differ by about 8×10^{-9} km.

CHAPTER 11 REVIEW ■ PAGE 778

True-False Quiz

1. False **3.** True **5.** False **7.** False **9.** False
11. True **13.** True **15.** False **17.** True
19. True **21.** True

Exercises

1. $\frac{1}{2}$ **3.** D **5.** 0 **7.** e^{12} **9.** 2 **11.** C **13.** C
15. D **17.** C **19.** C **21.** C **23.** CC **25.** AC
27. $\frac{1}{11}$ **29.** $\pi/4$ **31.** e^{-e} **35.** 0.9721

37. 0.18976224, error $< 6.4 \times 10^{-7}$

41. 4, $[-6, 2)$ **43.** 0.5, $[2.5, 3.5)$

45. $\frac{1}{2}\sum_{n=0}^{\infty}(-1)^n\left[\frac{1}{(2n)!}\left(x - \frac{\pi}{6}\right)^{2n} + \frac{\sqrt{3}}{(2n+1)!}\left(x - \frac{\pi}{6}\right)^{2n+1}\right]$

47. $\sum_{n=0}^{\infty}(-1)^n x^{n+2}$, $R = 1$ **49.** $\ln 4 - \sum_{n=1}^{\infty}\frac{x^n}{n4^n}$, $R = 4$

51. $\sum_{n=0}^{\infty}(-1)^n\frac{x^{8n+4}}{(2n+1)!}$, $R = \infty$

53. $\frac{1}{2} + \sum_{n=1}^{\infty}\frac{1 \cdot 5 \cdot 9 \cdot \cdots \cdot (4n-3)}{n!\,2^{6n+1}}x^n$, $R = 16$

55. $C + \ln|x| + \sum_{n=1}^{\infty}\frac{x^n}{n \cdot n!}$

57. (a) $1 + \frac{1}{2}(x - 1) - \frac{1}{8}(x - 1)^2 + \frac{1}{16}(x - 1)^3$
(b) (c) 0.000006

59. $-\frac{1}{6}$

PROBLEMS PLUS ■ PAGE 781

1. $15!/5! = 10{,}897{,}286{,}400$
3. (b) 0 if $x = 0$, $(1/x) - \cot x$ if $x \neq k\pi$, k an integer
5. (a) $s_n = 3 \cdot 4^n$, $l_n = 1/3^n$, $p_n = 4^n/3^{n-1}$ (c) $\frac{2}{5}\sqrt{3}$

9. $(-1, 1)$, $\frac{x^3 + 4x^2 + x}{(1 - x)^4}$

11. $\ln\frac{1}{2}$ **13.** (a) $\frac{250}{101}\pi(e^{-(n-1)\pi/5} - e^{-n\pi/5})$ (b) $\frac{250}{101}\pi$

19. $\frac{\pi}{2\sqrt{3}} - 1$

21. $-\left(\frac{\pi}{2} - \pi k\right)^2$ where k is a positive integer

CHAPTER 12

EXERCISES 12.1 ■ PAGE 790

1. $(4, 0, -3)$ **3.** C; A

5. A vertical plane that intersects the xy-plane in the line $y = 2 - x$, $z = 0$

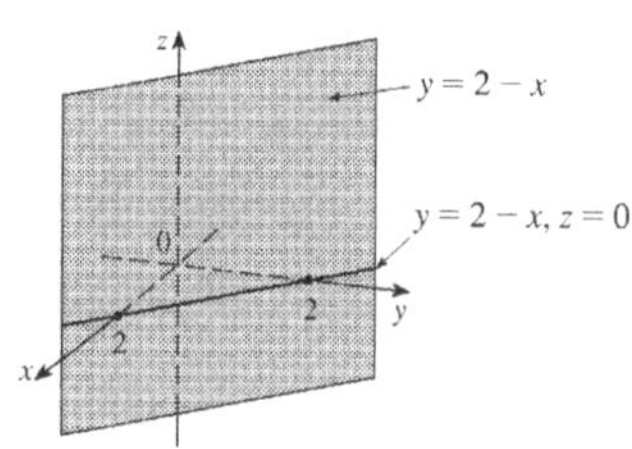

7. (a) $|PQ| = 6$, $|QR| = 2\sqrt{10}$, $|RP| = 6$; isosceles triangle
9. (a) No (b) Yes
11. $(x + 3)^2 + (y - 2)^2 + (z - 5)^2 = 16$;
$(y - 2)^2 + (z - 5)^2 = 7$, $x = 0$ (a circle)
13. $(x - 3)^2 + (y - 8)^2 + (z - 1)^2 = 30$
15. $(1, 2, -4)$, 6 **17.** $(2, 0, -6)$, $9/\sqrt{2}$
19. (b) $\frac{5}{2}$, $\frac{1}{2}\sqrt{94}$, $\frac{1}{2}\sqrt{85}$
21. (a) $(x - 2)^2 + (y + 3)^2 + (z - 6)^2 = 36$
(b) $(x - 2)^2 + (y + 3)^2 + (z - 6)^2 = 4$
(c) $(x - 2)^2 + (y + 3)^2 + (z - 6)^2 = 9$
23. A plane parallel to the yz-plane and 5 units in front of it

25. A half-space consisting of all points to the left of the plane $y = 8$
27. All points on or between the horizontal planes $z = 0$ and $z = 6$
29. All points on a circle with radius 2 with center on the z-axis that is contained in the plane $z = -1$
31. All points on or inside a sphere with radius $\sqrt{3}$ and center O
33. All points on or inside a circular cylinder of radius 3 with axis the y-axis
35. $0 < x < 5$ **37.** $r^2 < x^2 + y^2 + z^2 < R^2$
39. (a) (2, 1, 4) (b)

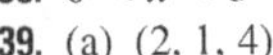

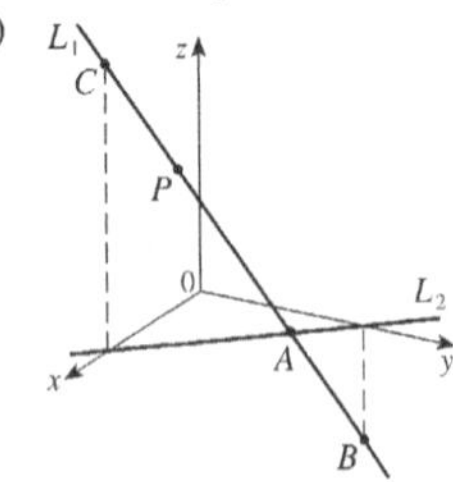

41. $14x - 6y - 10z = 9$, a plane perpendicular to AB
43. $2\sqrt{3} - 3$

EXERCISES 12.2 ■ PAGE 798

1. (a) Scalar (b) Vector (c) Vector (d) Scalar
3. $\overrightarrow{AB} = \overrightarrow{DC}$, $\overrightarrow{DA} = \overrightarrow{CB}$, $\overrightarrow{DE} = \overrightarrow{EB}$, $\overrightarrow{EA} = \overrightarrow{CE}$
5. (a) (b)

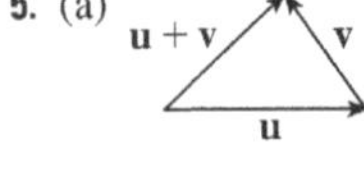

(c) (d)

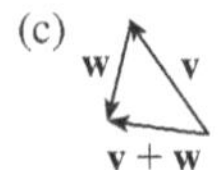

(e) (f)

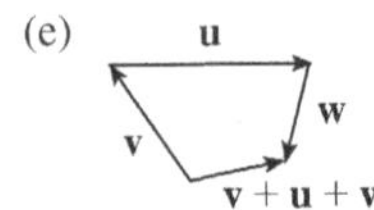

7. $\mathbf{c} = \frac{1}{2}\mathbf{a} + \frac{1}{2}\mathbf{b}$, $\mathbf{d} = \frac{1}{2}\mathbf{b} - \frac{1}{2}\mathbf{a}$
9. $\mathbf{a} = \langle 4, 1 \rangle$ **11.** $\mathbf{a} = \langle 3, -1 \rangle$

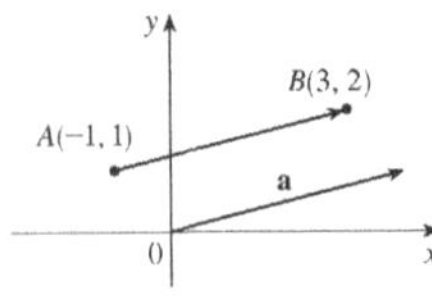

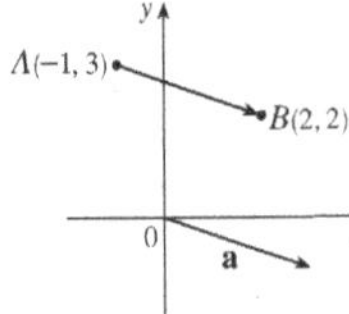

13. $\mathbf{a} = \langle 2, 0, -2 \rangle$ **15.** $\langle 5, 2 \rangle$

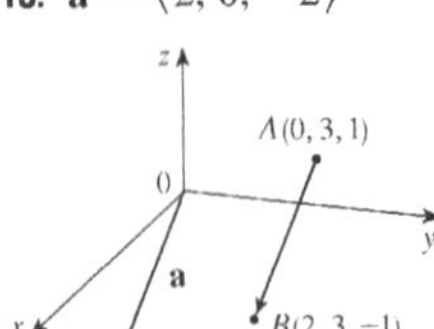

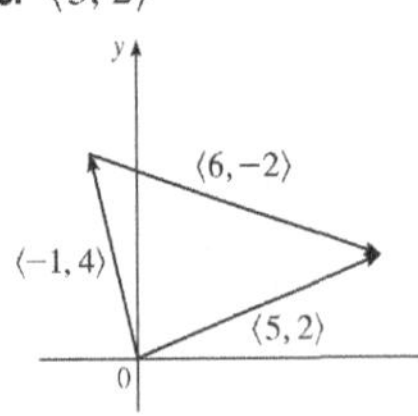

17. $\langle 3, 8, 1 \rangle$

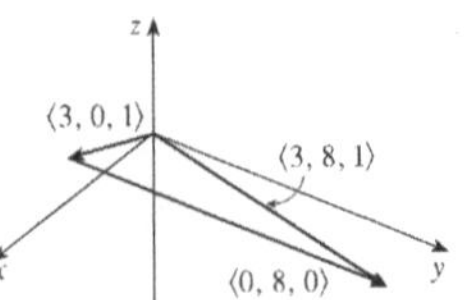

19. $\langle 2, -18 \rangle$, $\langle 1, -42 \rangle$, 13, 10
21. $-\mathbf{i} + \mathbf{j} + 2\mathbf{k}$, $-4\mathbf{i} + \mathbf{j} + 9\mathbf{k}$, $\sqrt{14}$, $\sqrt{82}$
23. $-\dfrac{3}{\sqrt{58}}\mathbf{i} + \dfrac{7}{\sqrt{58}}\mathbf{j}$ **25.** $\frac{8}{9}\mathbf{i} - \frac{1}{9}\mathbf{j} + \frac{4}{9}\mathbf{k}$ **27.** 60°
29. $\langle 2, 2\sqrt{3} \rangle$ **31.** ≈ 45.96 ft/s, ≈ 38.57 ft/s
33. $100\sqrt{7} \approx 264.6$ N, $\approx 139.1°$
35. $\sqrt{493} \approx 22.2$ mi/h, N8°W
37. $\mathbf{T}_1 = -196\,\mathbf{i} + 3.92\,\mathbf{j}$, $\mathbf{T}_2 = 196\,\mathbf{i} + 3.92\,\mathbf{j}$
39. (a) At an angle of 43.4° from the bank, toward upstream (b) 20.2 min
41. $\pm(\mathbf{i} + 4\,\mathbf{j})/\sqrt{17}$ **43.** $\mathbf{0}$
45. (a), (b) (d) $s = \frac{9}{7}$, $t = \frac{11}{7}$

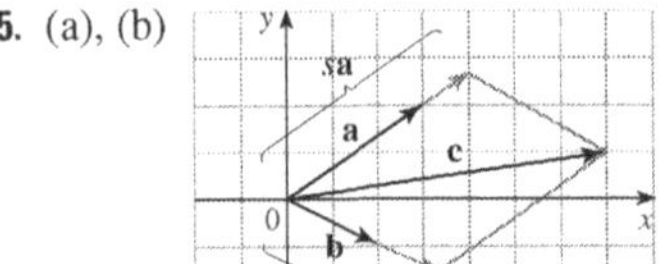

47. A sphere with radius 1, centered at (x_0, y_0, z_0)

EXERCISES 12.3 ■ PAGE 806

1. (b), (c), (d) are meaningful **3.** 14 **5.** 19 **7.** 1
9. −15 **11.** $\mathbf{u} \cdot \mathbf{v} = \frac{1}{2}$, $\mathbf{u} \cdot \mathbf{w} = -\frac{1}{2}$
15. $\cos^{-1}\left(\dfrac{1}{\sqrt{5}}\right) \approx 63°$ **17.** $\cos^{-1}\left(\dfrac{5}{\sqrt{1015}}\right) \approx 81°$
19. $\cos^{-1}\left(\dfrac{7}{\sqrt{130}}\right) \approx 52°$ **21.** 48°, 75°, 57°
23. (a) Neither (b) Orthogonal (c) Orthogonal (d) Parallel
25. Yes **27.** $(\mathbf{i} - \mathbf{j} - \mathbf{k})/\sqrt{3}$ $[\text{or } (-\mathbf{i} + \mathbf{j} + \mathbf{k})/\sqrt{3}]$
29. 45° **31.** 0° at (0, 0), 8.1° at (1, 1)
33. $\frac{2}{3}, \frac{1}{3}, \frac{2}{3}$; 48°, 71°, 48°
35. $1/\sqrt{14}, -2/\sqrt{14}, -3/\sqrt{14}$; 74°, 122°, 143°
37. $1/\sqrt{3}, 1/\sqrt{3}, 1/\sqrt{3}$; 55°, 55°, 55° **39.** 4, $\left\langle -\frac{20}{13}, \frac{48}{13} \right\rangle$
41. $\frac{9}{7}$, $\left\langle \frac{27}{49}, \frac{54}{49}, -\frac{18}{49} \right\rangle$ **43.** $1/\sqrt{21}$, $\frac{2}{21}\mathbf{i} - \frac{1}{21}\mathbf{j} + \frac{4}{21}\mathbf{k}$
47. $\langle 0, 0, -2\sqrt{10} \rangle$ or any vector of the form $\langle s, t, 3s - 2\sqrt{10} \rangle$, $s, t \in \mathbb{R}$
49. 144 J **51.** $2400\cos(40°) \approx 1839$ ft-lb
53. $\frac{13}{5}$ **55.** $\cos^{-1}(1/\sqrt{3}) \approx 55°$

EXERCISES 12.4 ■ PAGE 814

1. $16\,\mathbf{i} + 48\,\mathbf{k}$ **3.** $15\,\mathbf{i} - 3\,\mathbf{j} + 3\,\mathbf{k}$ **5.** $\frac{1}{2}\mathbf{i} - \mathbf{j} + \frac{3}{2}\mathbf{k}$
7. $(1 - t)\,\mathbf{i} + (t^3 - t^2)\,\mathbf{k}$ **9.** $\mathbf{0}$ **11.** $\mathbf{i} + \mathbf{j} + \mathbf{k}$
13. (a) Scalar (b) Meaningless (c) Vector (d) Meaningless (e) Meaningless (f) Scalar

15. $96\sqrt{3}$; into the page **17.** $\langle -7, 10, 8\rangle, \langle 7, -10, -8\rangle$

19. $\left\langle -\frac{1}{3\sqrt{3}}, -\frac{1}{3\sqrt{3}}, \frac{5}{3\sqrt{3}} \right\rangle, \left\langle \frac{1}{3\sqrt{3}}, \frac{1}{3\sqrt{3}}, -\frac{5}{3\sqrt{3}} \right\rangle$

27. 16 **29.** (a) $\langle 0, 18, -9\rangle$ (b) $\frac{9}{2}\sqrt{5}$

31. (a) $\langle 13, -14, 5\rangle$ (b) $\frac{1}{2}\sqrt{390}$

33. 9 **35.** 16 **39.** $10.8 \sin 80° \approx 10.6\ \text{N}\cdot\text{m}$

41. ≈ 417 N **43.** 60°

45. (b) $\sqrt{97/3}$ **53.** (a) No (b) No (c) Yes

EXERCISES 12.5 ■ PAGE 824

1. (a) True (b) False (c) True (d) False (e) False (f) True (g) False (h) True (i) True (j) False (k) True

3. $\mathbf{r} = (2\,\mathbf{i} + 2.4\,\mathbf{j} + 3.5\,\mathbf{k}) + t(3\,\mathbf{i} + 2\,\mathbf{j} - \mathbf{k})$; $x = 2 + 3t, y = 2.4 + 2t, z = 3.5 - t$

5. $\mathbf{r} = (\mathbf{i} + 6\mathbf{k}) + t(\mathbf{i} + 3\,\mathbf{j} + \mathbf{k})$; $x = 1 + t, y = 3t, z = 6 + t$

7. $x = 2 + 2t, y = 1 + \frac{1}{2}t, z = -3 - 4t$; $(x - 2)/2 = 2y - 2 = (z + 3)/(-4)$

9. $x = -8 + 11t, y = 1 - 3t, z = 4; \dfrac{x + 8}{11} = \dfrac{y - 1}{-3}, z = 4$

11. $x = 1 + t, y = -1 + 2t, z = 1 + t$; $x - 1 = (y + 1)/2 = z - 1$

13. Yes

15. (a) $(x - 1)/(-1) = (y + 5)/2 = (z - 6)/(-3)$ (b) $(-1, -1, 0), (-\frac{3}{2}, 0, -\frac{3}{2}), (0, -3, 3)$

17. $\mathbf{r}(t) = (2\mathbf{i} - \mathbf{j} + 4\mathbf{k}) + t(2\mathbf{i} + 7\mathbf{j} - 3\mathbf{k}), 0 \leqslant t \leqslant 1$

19. Skew **21.** $(4, -1, -5)$ **23.** $x - 2y + 5z = 0$

25. $x + 4y + z = 4$ **27.** $5x - y - z = 7$

29. $6x + 6y + 6z = 11$ **31.** $x + y + z = 2$

33. $-13x + 17y + 7z = -42$ **35.** $33x + 10y + 4z = 190$

37. $x - 2y + 4z = -1$ **39.** $3x - 8y - z = -38$

41.

43.

45. $(2, 3, 5)$ **47.** $(2, 3, 1)$ **49.** $1, 0, -1$

51. Perpendicular **53.** Neither, $\cos^{-1}(\frac{1}{3}) \approx 70.5°$

55. Parallel

57. (a) $x = 1, y = -t, z = t$ (b) $\cos^{-1}\left(\dfrac{5}{3\sqrt{3}}\right) \approx 15.8°$

59. $x = 1, y - 2 = -z$ **61.** $x + 2y + z = 5$

63. $(x/a) + (y/b) + (z/c) = 1$

65. $x = 3t, y = 1 - t, z = 2 - 2t$

67. P_2 and P_3 are parallel, P_1 and P_4 are identical

69. $\sqrt{61/14}$ **71.** $\frac{18}{7}$ **73.** $5/(2\sqrt{14})$ **77.** $1/\sqrt{6}$

79. $13/\sqrt{69}$

EXERCISES 12.6 ■ PAGE 832

1. (a) Parabola
(b) Parabolic cylinder with rulings parallel to the z-axis
(c) Parabolic cylinder with rulings parallel to the x-axis

3. Circular cylinder **5.** Parabolic cylinder

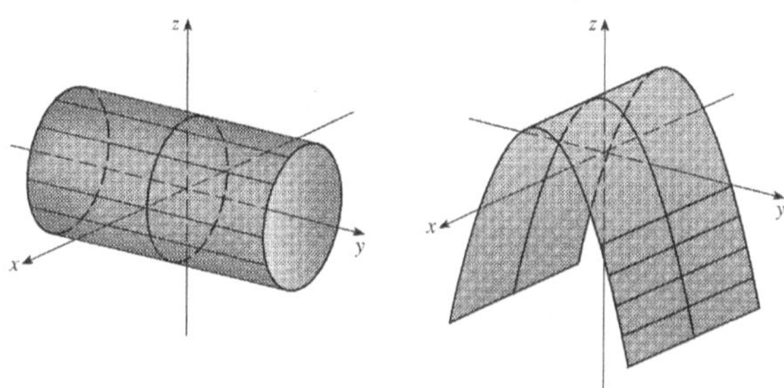

7. Hyperbolic cylinder

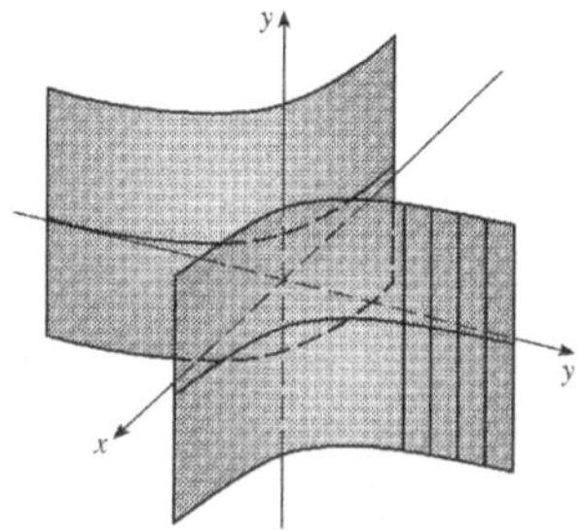

9. (a) $x = k, y^2 - z^2 = 1 - k^2$, hyperbola $(k \neq \pm 1)$;
$y = k, x^2 - z^2 = 1 - k^2$, hyperbola $(k \neq \pm 1)$;
$z = k, x^2 + y^2 = 1 + k^2$, circle
(b) The hyperboloid is rotated so that it has axis the y-axis
(c) The hyperboloid is shifted one unit in the negative y-direction

11. Elliptic paraboloid with axis the x-axis

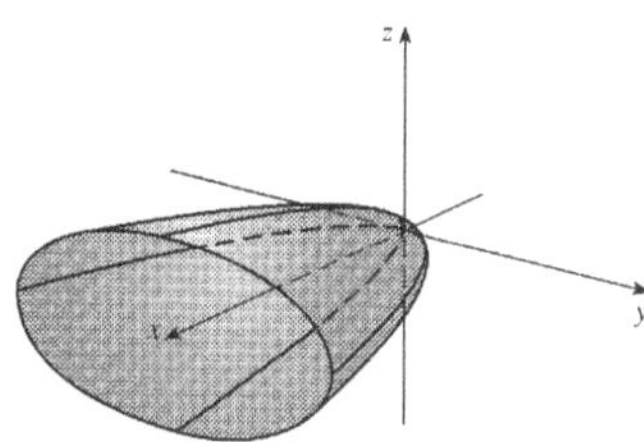

13. Elliptic cone with axis the x-axis

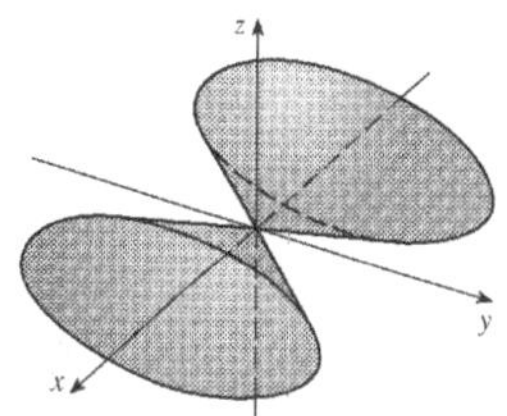

15. Hyperboloid of two sheets

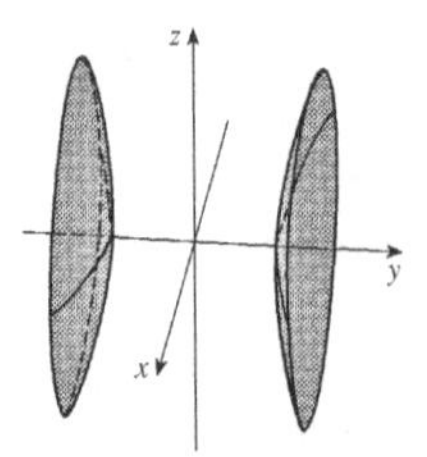

17. Ellipsoid

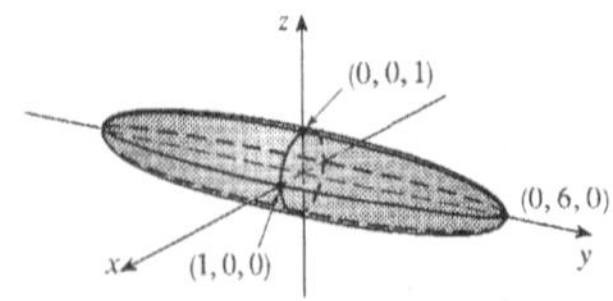

19. Hyperbolic paraboloid

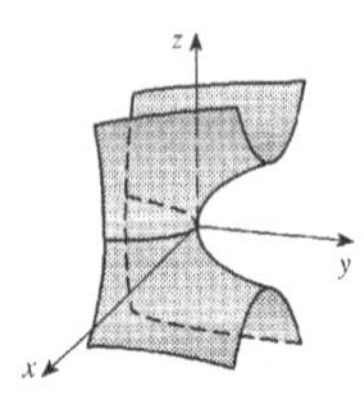

21. VII **23.** II **25.** VI **27.** VIII

29. $y^2 = x^2 + \dfrac{z^2}{9}$

Elliptic cone with axis the y-axis

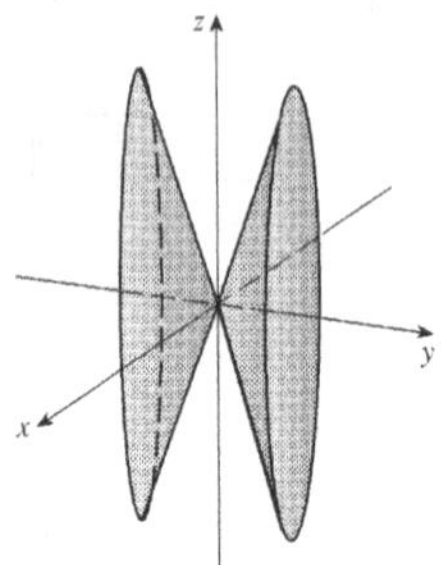

31. $y = z^2 - \dfrac{x^2}{2}$

Hyperbolic paraboloid

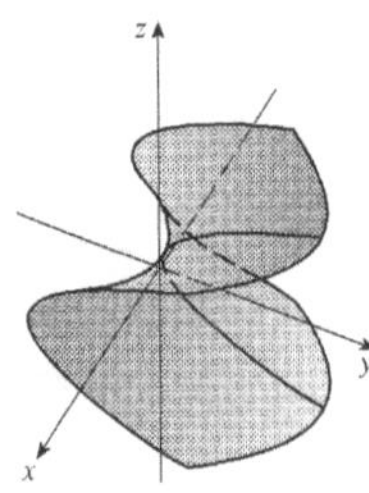

33. $x^2 + \dfrac{(y-2)^2}{4} + (z-3)^2 = 1$

Ellipsoid with center (0, 2, 3)

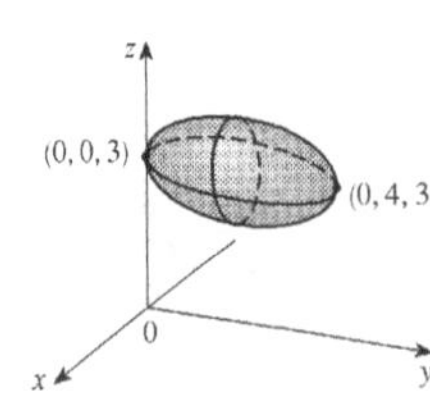

35. $(y+1)^2 = (x-2)^2 + (z-1)^2$
Circular cone with vertex $(2, -1, 1)$ and axis parallel to the y-axis

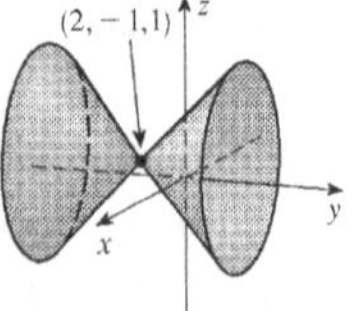

37.

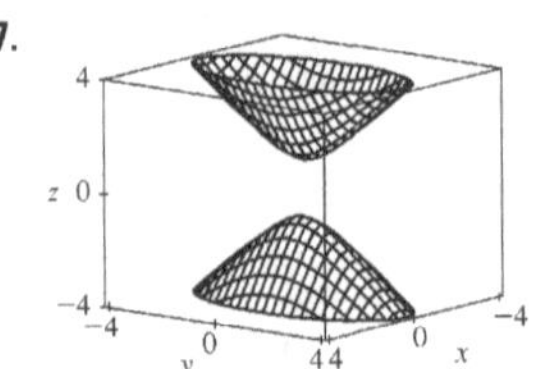

39.

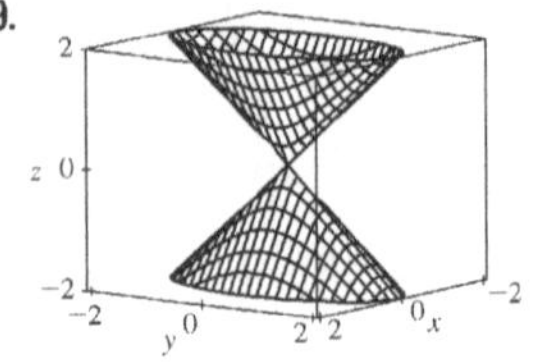

41.

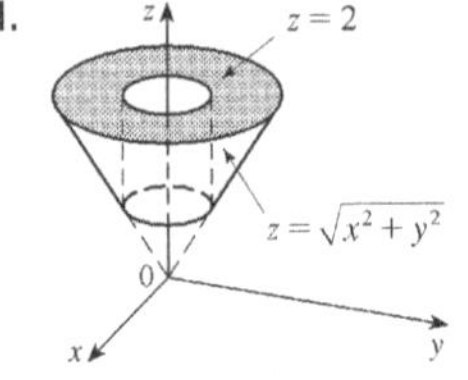

43. $y = x^2 + z^2$ **45.** $-4x = y^2 + z^2$, paraboloid

47. (a) $\dfrac{x^2}{(6378.137)^2} + \dfrac{y^2}{(6378.137)^2} + \dfrac{z^2}{(6356.523)^2} = 1$
(b) Circle (c) Ellipse

51.

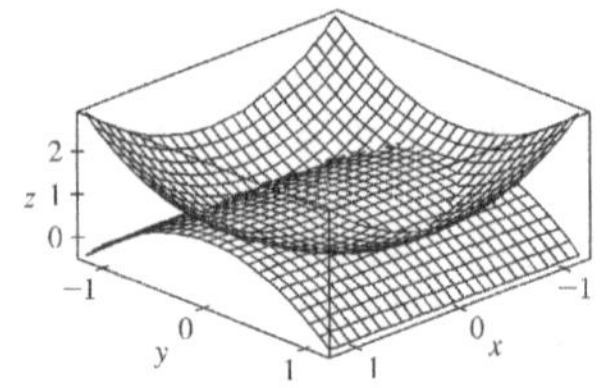

CHAPTER 12 REVIEW ■ PAGE 834

True-False Quiz

1. False **3.** False **5.** True **7.** True **9.** True
11. True **13.** True **15.** False **17.** False
19. False **21.** True

Exercises

1. (a) $(x+1)^2 + (y-2)^2 + (z-1)^2 = 69$
(b) $(y-2)^2 + (z-1)^2 = 68$, $x = 0$
(c) Center $(4, -1, -3)$, radius 5
3. $\mathbf{u} \cdot \mathbf{v} = 3\sqrt{2}$; $|\mathbf{u} \times \mathbf{v}| = 3\sqrt{2}$; out of the page
5. $-2, -4$ **7.** (a) 2 (b) -2 (c) -2 (d) 0
9. $\cos^{-1}(\frac{1}{3}) \approx 71°$ **11.** (a) $\langle 4, -3, 4 \rangle$ (b) $\sqrt{41}/2$
13. 166 N, 114 N
15. $x = 4 - 3t$, $y = -1 + 2t$, $z = 2 + 3t$
17. $x = -2 + 2t$, $y = 2 - t$, $z = 4 + 5t$
19. $-4x + 3y + z = -14$ **21.** $(1, 4, 4)$ **23.** Skew
25. $x + y + z = 4$ **27.** $22/\sqrt{26}$

29. Plane

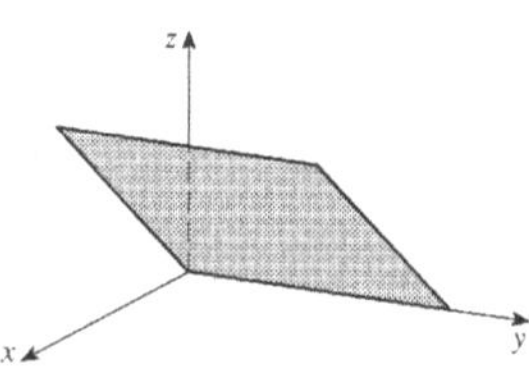

31. Cone

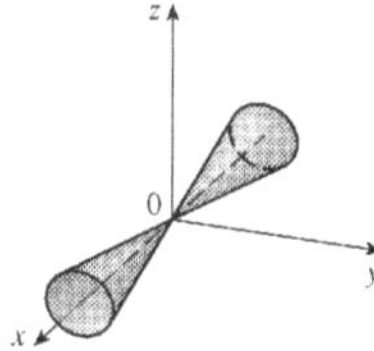

33. Hyperboloid of two sheets

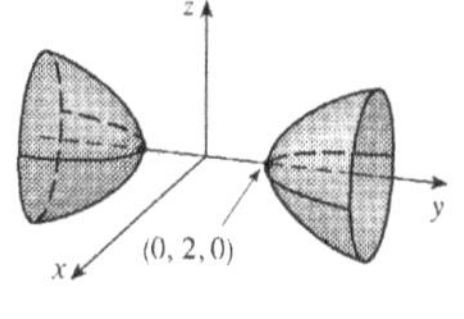

35. Ellipsoid

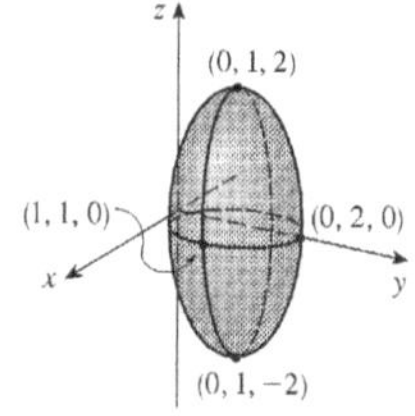

37. $4x^2 + y^2 + z^2 = 16$

PROBLEMS PLUS ■ PAGE 837

1. $(\sqrt{3} - \frac{3}{2})$ m

3. (a) $(x + 1)/(-2c) = (y - c)/(c^2 - 1) = (z - c)/(c^2 + 1)$
(b) $x^2 + y^2 = t^2 + 1, z = t$ (c) $4\pi/3$

5. 20

CHAPTER 13

EXERCISES 13.1 ■ PAGE 845

1. $(-1, 2]$ **3.** $\mathbf{i} + \mathbf{j} + \mathbf{k}$ **5.** $\langle -1, \pi/2, 0 \rangle$

7.

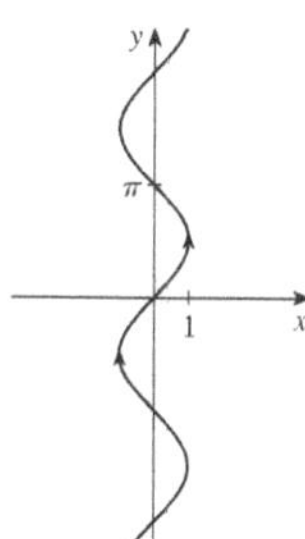

9.

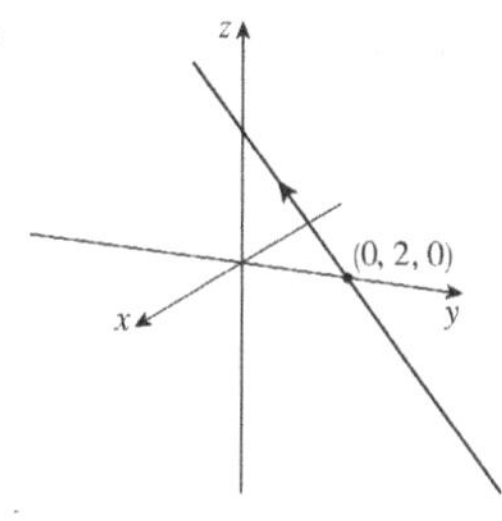

11.

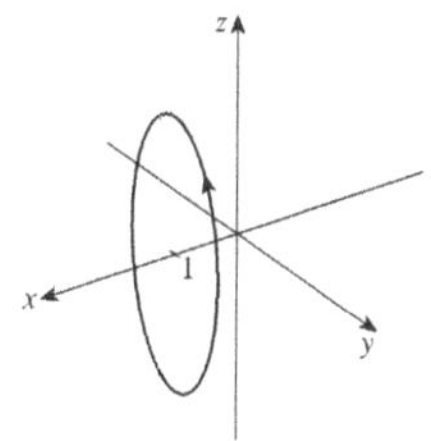

13.

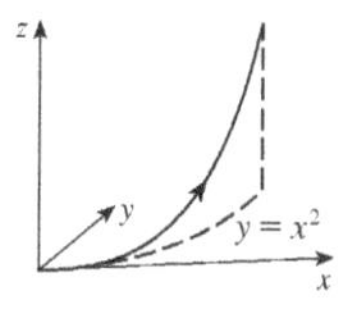

15.

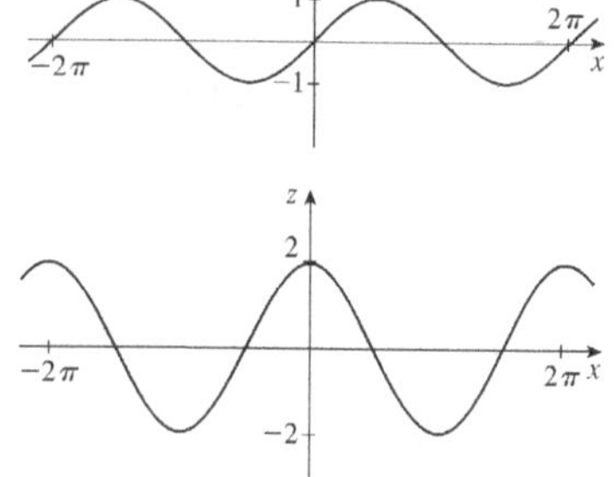

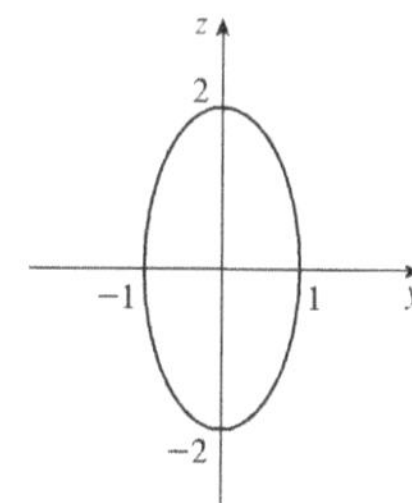

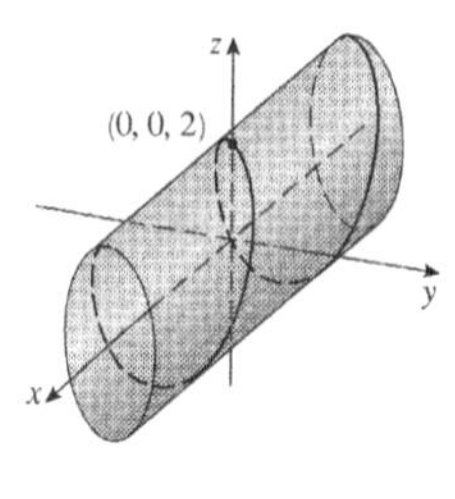

17. $\mathbf{r}(t) = \langle 2 + 4t, 2t, -2t \rangle, 0 \leqslant t \leqslant 1$;
$x = 2 + 4t, y = 2t, z = -2t, 0 \leqslant t \leqslant 1$

19. $\mathbf{r}(t) = \langle \frac{1}{2}t, -1 + \frac{4}{3}t, 1 - \frac{3}{4}t \rangle, 0 \leqslant t \leqslant 1$;
$x = \frac{1}{2}t, y = -1 + \frac{4}{3}t, z = 1 - \frac{3}{4}t, 0 \leqslant t \leqslant 1$

21. II **23.** V **25.** IV

27.

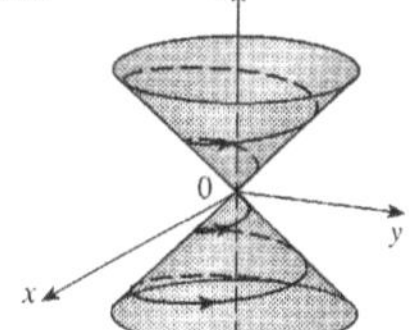

29. (0, 0, 0), (1, 0, 1)

31.

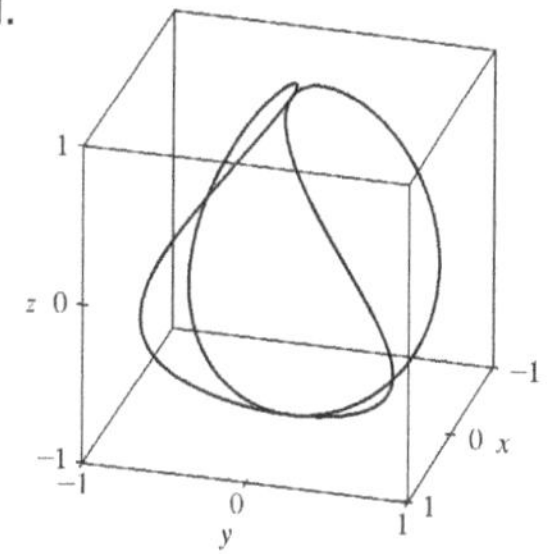

33.

10
z 0
−10
10 0 x −10
10 0 y −10

35.

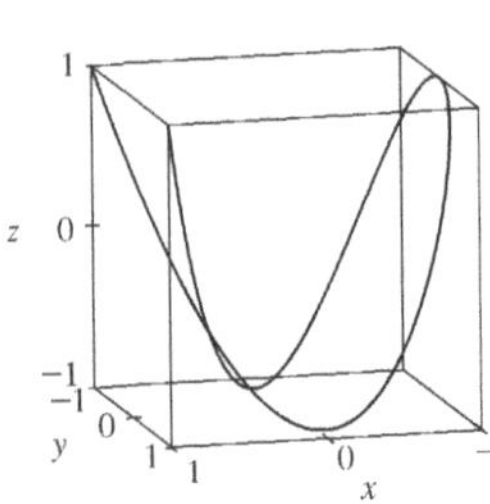

37.

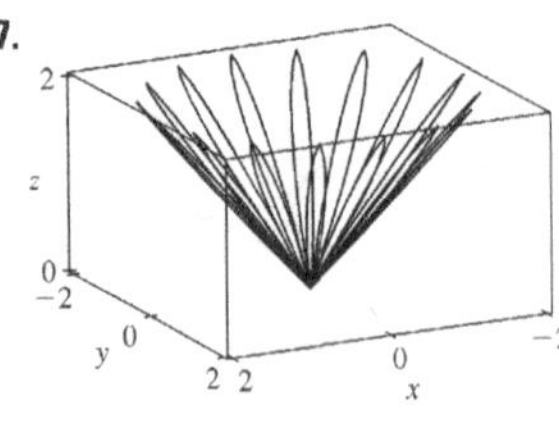

41. $\mathbf{r}(t) = t\,\mathbf{i} + \frac{1}{2}(t^2 - 1)\,\mathbf{j} + \frac{1}{2}(t^2 + 1)\,\mathbf{k}$

43. $\mathbf{r}(t) = \cos t\,\mathbf{i} + \sin t\,\mathbf{j} + \cos 2t\,\mathbf{k},\ 0 \leqslant t \leqslant 2\pi$

45. $x = 2\cos t,\ y = 2\sin t,\ z = 4\cos^2 t$ **47.** Yes

EXERCISES 13.2 ■ PAGE 852

1. (a)

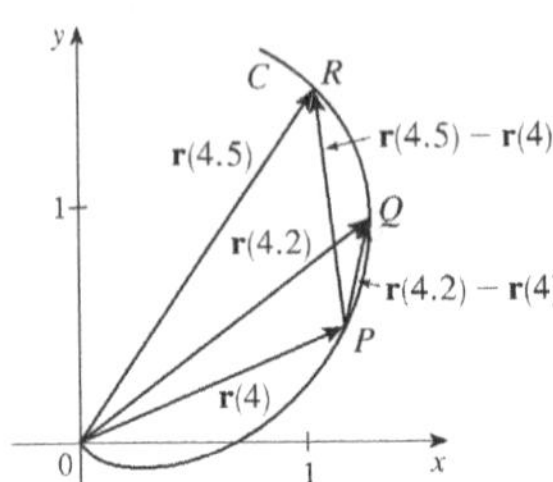

(b), (d)

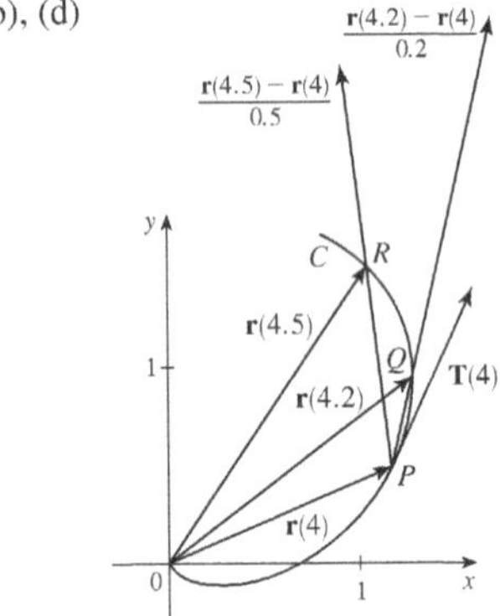

(c) $\mathbf{r}'(4) = \lim_{h \to 0} \dfrac{\mathbf{r}(4 + h) - \mathbf{r}(4)}{h}$; $\mathbf{T}(4) = \dfrac{\mathbf{r}'(4)}{|\mathbf{r}'(4)|}$

3. (a), (c) (b) $\mathbf{r}'(t) = \langle 1, 2t \rangle$

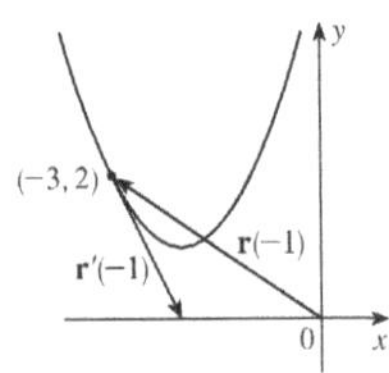

5. (a), (c) (b) $\mathbf{r}'(t) = \cos t\,\mathbf{i} - 2\sin t\,\mathbf{j}$

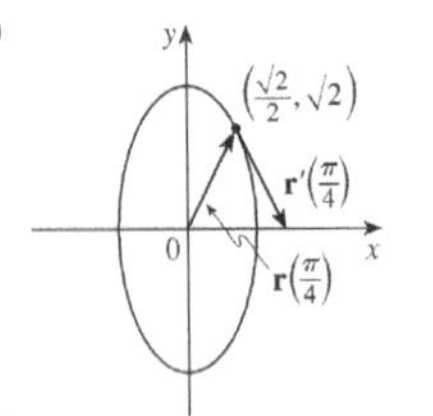

7. (a), (c) (b) $\mathbf{r}'(t) = 2e^{2t}\,\mathbf{i} + e^t\,\mathbf{j}$

9. $\mathbf{r}'(t) = \langle t\cos t + \sin t, 2t, \cos 2t - 2t\sin 2t \rangle$

11. $\mathbf{r}'(t) = \mathbf{i} + (1/\sqrt{t})\,\mathbf{k}$

13. $\mathbf{r}'(t) = 2te^{t^2}\,\mathbf{i} + [3/(1 + 3t)]\,\mathbf{k}$ **15.** $\mathbf{r}'(t) = \mathbf{b} + 2t\mathbf{c}$

17. $\langle \frac{1}{3}, \frac{2}{3}, \frac{2}{3} \rangle$ **19.** $\frac{3}{5}\,\mathbf{j} + \frac{4}{5}\,\mathbf{k}$

21. $\langle 1, 2t, 3t^2 \rangle$, $\langle 1/\sqrt{14}, 2/\sqrt{14}, 3/\sqrt{14} \rangle$, $\langle 0, 2, 6t \rangle$, $\langle 6t^2, -6t, 2 \rangle$

23. $x = 3 + t,\ y = 2t,\ z = 2 + 4t$

25. $x = 1 - t,\ y = t,\ z = 1 - t$

27. $\mathbf{r}(t) = (3 - 4t)\,\mathbf{i} + (4 + 3t)\,\mathbf{j} + (2 - 6t)\,\mathbf{k}$

29. $x = t,\ y = 1 - t,\ z = 2t$

31. $x = -\pi - t,\ y = \pi + t,\ z = -\pi t$

33. $66°$ **35.** $2\,\mathbf{i} - 4\,\mathbf{j} + 32\,\mathbf{k}$ **37.** $\mathbf{i} + \mathbf{j} + \mathbf{k}$

39. $\tan t\,\mathbf{i} + \frac{1}{8}(t^2 + 1)^4\mathbf{j} + (\frac{1}{3}t^3 \ln t - \frac{1}{9}t^3)\mathbf{k} + \mathbf{C}$

41. $t^2\mathbf{i} + t^3\mathbf{j} + (\frac{2}{3}t^{3/2} - \frac{2}{3})\mathbf{k}$

47. $2t\cos t + 2\sin t - 2\cos t\sin t$ **49.** 35

EXERCISES 13.3 ■ PAGE 860

1. $10\sqrt{10}$ **3.** $e - e^{-1}$ **5.** $\frac{1}{27}(13^{3/2} - 8)$ **7.** 18.6833

9. 1.2780 **11.** 42

13. $\mathbf{r}(t(s)) = \dfrac{2}{\sqrt{29}}s\,\mathbf{i} + \left(1 - \dfrac{3}{\sqrt{29}}s\right)\mathbf{j} + \left(5 + \dfrac{4}{\sqrt{29}}s\right)\mathbf{k}$

15. $(3\sin 1, 4, 3\cos 1)$

17. (a) $\langle 1/\sqrt{10}, (-3/\sqrt{10})\sin t, (3/\sqrt{10})\cos t \rangle$, $\langle 0, -\cos t, -\sin t \rangle$ (b) $\frac{3}{10}$

19. (a) $\dfrac{1}{e^{2t} + 1}\langle \sqrt{2}e^t, e^{2t}, -1 \rangle$, $\dfrac{1}{e^{2t} + 1}\langle 1 - e^{2t}, \sqrt{2}e^t, \sqrt{2}e^t \rangle$

(b) $\sqrt{2}e^{2t}/(e^{2t} + 1)^2$

21. $6t^2/(9t^4 + 4t^2)^{3/2}$ **23.** $\frac{4}{25}$ **25.** $\frac{1}{7}\sqrt{\frac{19}{14}}$

27. $12x^2/(1 + 16x^6)^{3/2}$ **29.** $e^x|x + 2|/[1 + (xe^x + e^x)^2]^{3/2}$

31. $(-\frac{1}{2}\ln 2, 1/\sqrt{2})$; approaches 0 **33.** (a) P (b) 1.3, 0.7

35.

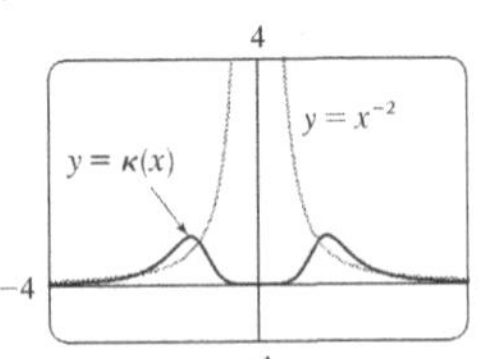

37.

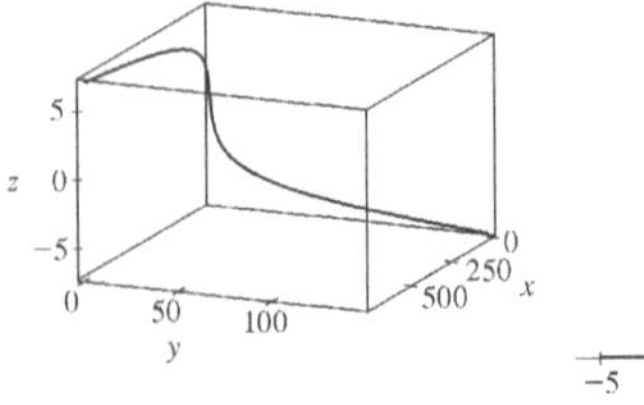

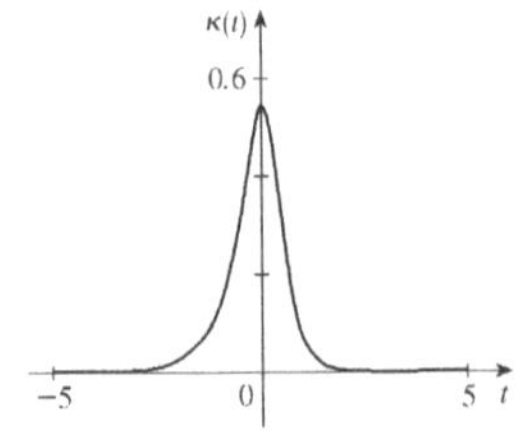

39. a is $y = f(x)$, b is $y = \kappa(x)$

41. $\kappa(t) = \dfrac{6\sqrt{4\cos^2 t - 12\cos t + 13}}{(17 - 12\cos t)^{3/2}}$

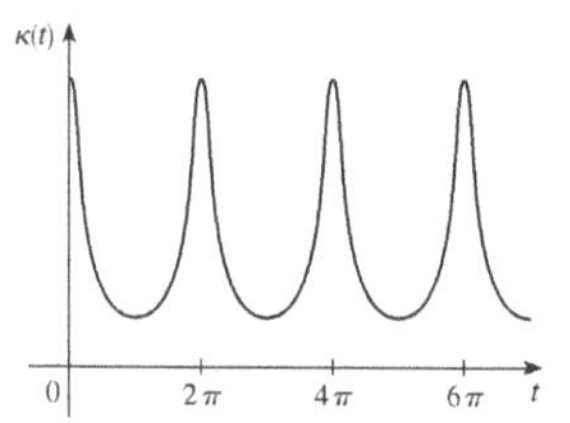

integer multiples of 2π

43. $6t^2/(4t^2 + 9t^4)^{3/2}$

45. $1/(\sqrt{2}e^t)$ **47.** $\langle \frac{2}{3}, \frac{2}{3}, \frac{1}{3} \rangle, \langle -\frac{1}{3}, \frac{2}{3}, -\frac{2}{3} \rangle, \langle -\frac{2}{3}, \frac{1}{3}, \frac{2}{3} \rangle$

49. $y = 6x + \pi, x + 6y = 6\pi$

51. $(x + \frac{5}{2})^2 + y^2 = \frac{81}{4}, x^2 + (y - \frac{5}{3})^2 = \frac{16}{9}$

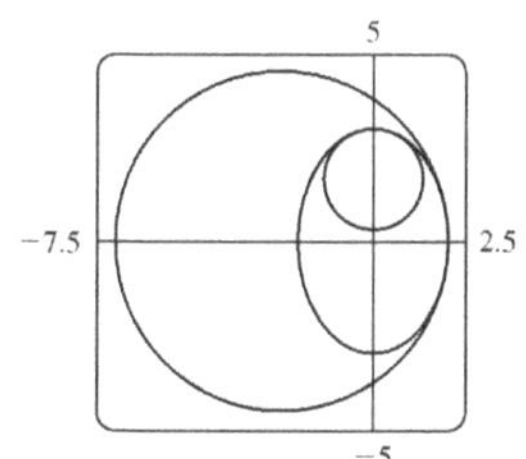

53. $(-1, -3, 1)$

55. $2x + y + 4z = 7, 6x - 8y - z = -3$

63. $2/(t^4 + 4t^2 + 1)$ **65.** 2.07×10^{10} Å ≈ 2 m

EXERCISES 13.4 ■ PAGE 870

1. (a) $1.8\mathbf{i} - 3.8\mathbf{j} - 0.7\mathbf{k}, 2.0\mathbf{i} - 2.4\mathbf{j} - 0.6\mathbf{k}$, $2.8\mathbf{i} + 1.8\mathbf{j} - 0.3\mathbf{k}, 2.8\mathbf{i} + 0.8\mathbf{j} - 0.4\mathbf{k}$
(b) $2.4\mathbf{i} - 0.8\mathbf{j} - 0.5\mathbf{k}$, 2.58

3. $\mathbf{v}(t) = \langle -t, 1 \rangle$
$\mathbf{a}(t) = \langle -1, 0 \rangle$
$|\mathbf{v}(t)| = \sqrt{t^2 + 1}$

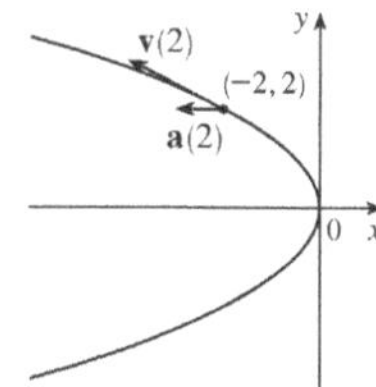

5. $\mathbf{v}(t) = -3 \sin t\, \mathbf{i} + 2 \cos t\, \mathbf{j}$
$\mathbf{a}(t) = -3 \cos t\, \mathbf{i} - 2 \sin t\, \mathbf{j}$
$|\mathbf{v}(t)| = \sqrt{5 \sin^2 t + 4}$

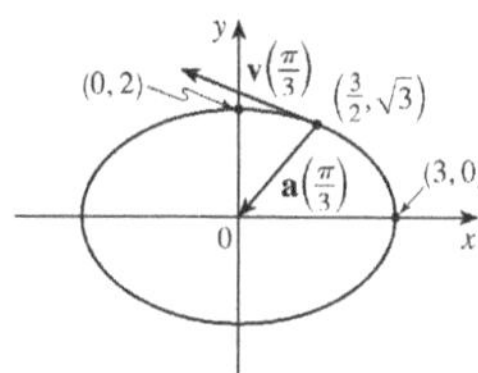

7. $\mathbf{v}(t) = \mathbf{i} + 2t\,\mathbf{j}$
$\mathbf{a}(t) = 2\,\mathbf{j}$
$|\mathbf{v}(t)| = \sqrt{1 + 4t^2}$

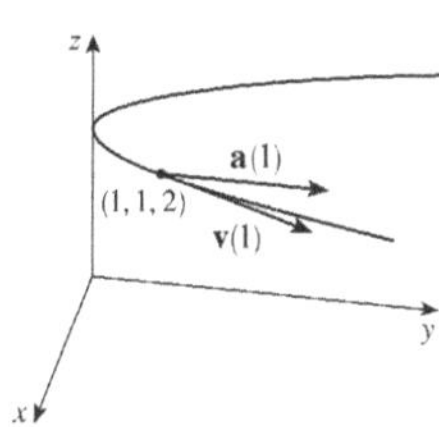

9. $\langle 2t + 1, 2t - 1, 3t^2 \rangle, \langle 2, 2, 6t \rangle, \sqrt{9t^4 + 8t^2 + 2}$

11. $\sqrt{2}\,\mathbf{i} + e^t\,\mathbf{j} - e^{-t}\,\mathbf{k}, e^t\,\mathbf{j} + e^{-t}\,\mathbf{k}, e^t + e^{-t}$

13. $e^t[(\cos t - \sin t)\mathbf{i} + (\sin t + \cos t)\mathbf{j} + (t + 1)\mathbf{k}]$, $e^t[-2 \sin t\,\mathbf{i} + 2 \cos t\,\mathbf{j} + (t + 2)\mathbf{k}], e^t\sqrt{t^2 + 2t + 3}$

15. $\mathbf{v}(t) = t\,\mathbf{i} + 2t\,\mathbf{j} + \mathbf{k}, \mathbf{r}(t) = (\frac{1}{2}t^2 + 1)\,\mathbf{i} + t^2\,\mathbf{j} + t\,\mathbf{k}$

17. (a) $\mathbf{r}(t) = (\frac{1}{3}t^3 + t)\,\mathbf{i} + (t - \sin t + 1)\,\mathbf{j} + (\frac{1}{4} - \frac{1}{4}\cos 2t)\,\mathbf{k}$
(b)

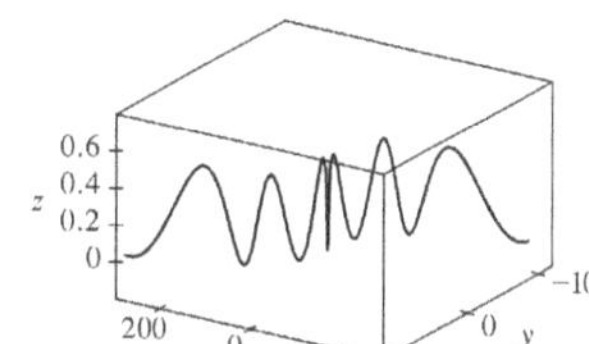

19. $t = 4$

21. $\mathbf{r}(t) = t\,\mathbf{i} - t\,\mathbf{j} + \frac{5}{2}t^2\,\mathbf{k}, |\mathbf{v}(t)| = \sqrt{25t^2 + 2}$

23. (a) ≈ 3535 m (b) ≈ 1531 m (c) 200 m/s

25. 30 m/s **27.** $\approx 10.2°$, $\approx 79.8°$

29. $13.0° < \theta < 36.0°, 55.4° < \theta < 85.5°$

31. $(250, -50, 0)$; $10\sqrt{93} \approx 96.4$ ft/s

33. (a) 16 m (b) $\approx 23.6°$ upstream

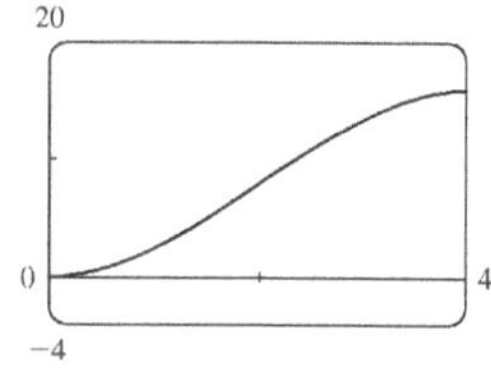

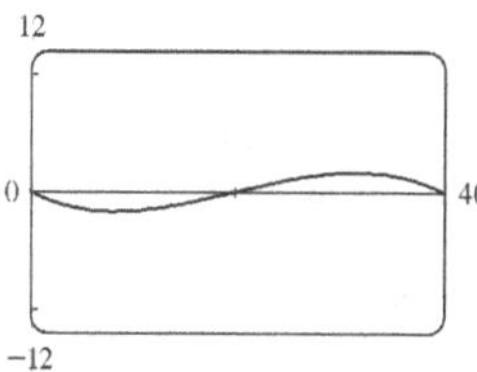

35. The path is contained in a circle that lies in a plane perpendicular to **c** with center on a line through the origin in the direction of **c**.

37. $6t$, 6 **39.** 0, 1 **41.** $e^t - e^{-t}, \sqrt{2}$

43. 4.5 cm/s², 9.0 cm/s² **45.** $t = 1$

CHAPTER 13 REVIEW ■ PAGE 873

True-False Quiz

1. True **3.** False **5.** False **7.** False
9. True **11.** False **13.** True

Exercises

1. (a)

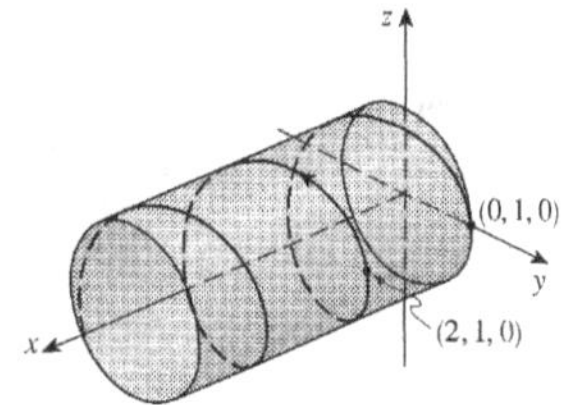

(b) $\mathbf{r}'(t) = \mathbf{i} - \pi \sin \pi t\,\mathbf{j} + \pi \cos \pi t\,\mathbf{k}$,
$\mathbf{r}''(t) = -\pi^2 \cos \pi t\,\mathbf{j} - \pi^2 \sin \pi t\,\mathbf{k}$

3. $\mathbf{r}(t) = 4 \cos t\,\mathbf{i} + 4 \sin t\,\mathbf{j} + (5 - 4 \cos t)\mathbf{k}, 0 \leqslant t \leqslant 2\pi$

5. $\frac{1}{3}\mathbf{i} - (2/\pi^2)\mathbf{j} + (2/\pi)\mathbf{k}$ **7.** 86.631 **9.** $\pi/2$

11. (a) $\langle t^2, t, 1 \rangle / \sqrt{t^4 + t^2 + 1}$
(b) $\langle t^3 + 2t, 1 - t^4, -2t^3 - t \rangle / \sqrt{t^8 + 5t^6 + 6t^4 + 5t^2 + 1}$
(c) $\sqrt{t^8 + 5t^6 + 6t^4 + 5t^2 + 1}/(t^4 + t^2 + 1)^2$

13. $12/17^{3/2}$ **15.** $x - 2y + 2\pi = 0$

17. $\mathbf{v}(t) = (1 + \ln t)\mathbf{i} + \mathbf{j} - e^{-t}\mathbf{k}$,
$|\mathbf{v}(t)| = \sqrt{2 + 2\ln t + (\ln t)^2 + e^{-2t}}$, $\mathbf{a}(t) = (1/t)\mathbf{i} + e^{-t}\mathbf{k}$
19. (a) About 3.8 ft above the ground, 60.8 ft from the athlete
(b) ≈ 21.4 ft (c) ≈ 64.2 ft from the athlete
21. (c) $-2e^{-t}\mathbf{v}_d + e^{-t}\mathbf{R}$
23. (a) $\mathbf{v} = \omega R(-\sin \omega t\,\mathbf{i} + \cos \omega t\,\mathbf{j})$ (c) $\mathbf{a} = -\omega^2\mathbf{r}$

PROBLEMS PLUS ■ PAGE 876

1. (a) 90°, $v_0^2/(2g)$
3. (a) ≈ 0.94 ft to the right of the table's edge, ≈ 15 ft/s
(b) $\approx 7.6°$ (c) ≈ 2.13 ft to the right of the table's edge
5. 56°
7. $\mathbf{r}(u, v) = \mathbf{c} + u\,\mathbf{a} + v\,\mathbf{b}$ where $\mathbf{a} = \langle a_1, a_2, a_3 \rangle$,
$\mathbf{b} = \langle b_1, b_2, b_3 \rangle$, $\mathbf{c} = \langle c_1, c_2, c_3 \rangle$

CHAPTER 14

EXERCISES 14.1 ■ PAGE 888

1. (a) -27; a temperature of -15°C with wind blowing at 40 km/h feels equivalent to about -27°C without wind.
(b) When the temperature is -20°C, what wind speed gives a wind chill of -30°C? 20 km/h
(c) With a wind speed of 20 km/h, what temperature gives a wind chill of -49°C? -35°C
(d) A function of wind speed that gives wind-chill values when the temperature is -5°C
(e) A function of temperature that gives wind-chill values when the wind speed is 50 km/h
3. ≈ 94.2; the manufacturer's yearly production is valued at \$94.2 million when 120,000 labor hours are spent and \$20 million in capital is invested.
5. (a) ≈ 20.5; the surface area of a person 70 inches tall who weighs 160 pounds is approximately 20.5 square feet.
7. (a) 25; a 40-knot wind blowing in the open sea for 15 h will create waves about 25 ft high.
(b) $f(30, t)$ is a function of t giving the wave heights produced by 30-knot winds blowing for t hours.
(c) $f(v, 30)$ is a function of v giving the wave heights produced by winds of speed v blowing for 30 hours.
9. (a) 1 (b) $\mathbb{R}^2$ (c) $[-1, 1]$
11. (a) 3 (b) $\{(x, y, z) \mid x^2 + y^2 + z^2 < 4, x \geq 0, y \geq 0, z \geq 0\}$, interior of a sphere of radius 2, center the origin, in the first octant
13. $\{(x, y) \mid y \leq 2x\}$

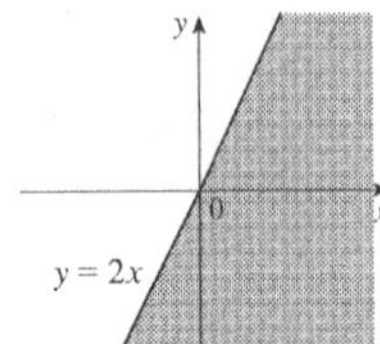

15. $\{(x, y) \mid \frac{1}{9}x^2 + y^2 < 1\}$, $(-\infty, \ln 9]$

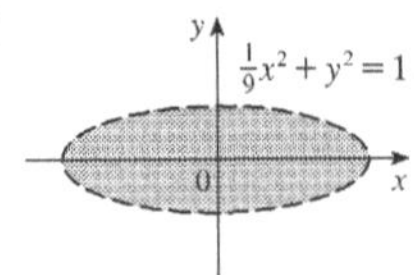

17. $\{(x, y) \mid -1 \leq x \leq 1, -1 \leq y \leq 1\}$

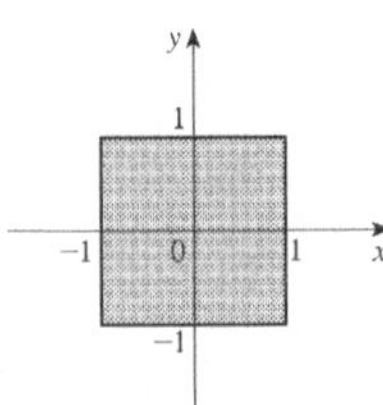

19. $\{(x, y) \mid y \geq x^2, x \neq \pm 1\}$

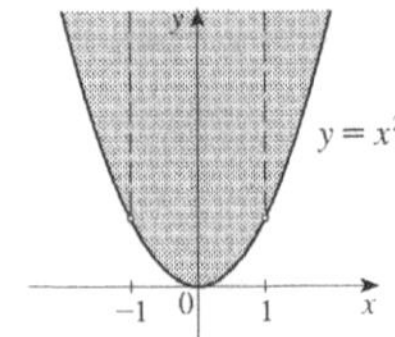

21. $\{(x, y, z) \mid x^2 + y^2 + z^2 \leq 1\}$

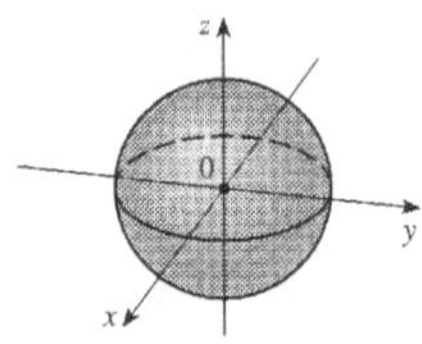

23. $z = 1 + y$, plane parallel to x-axis

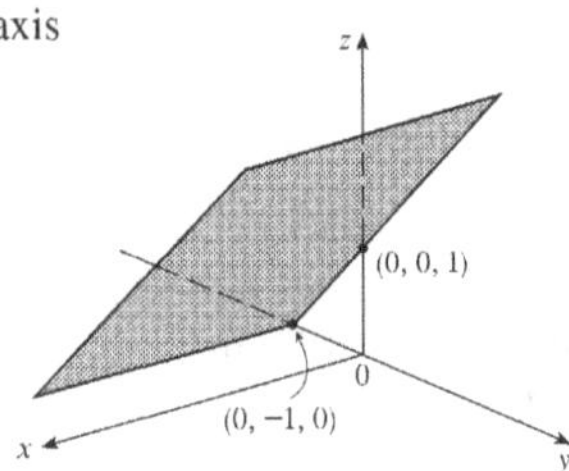

25. $4x + 5y + z = 10$, plane

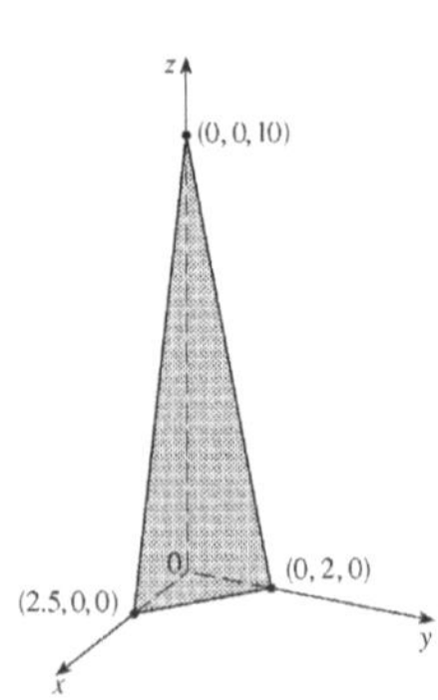

27. $z = y^2 + 1$, parabolic cylinder

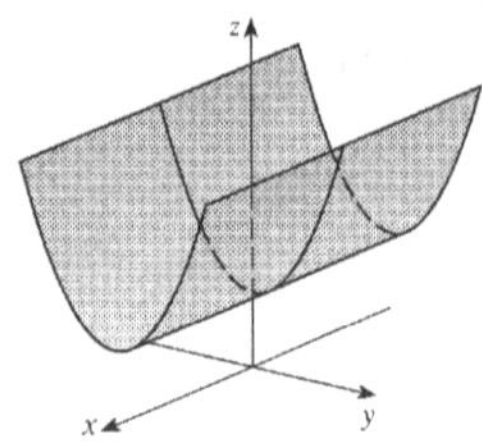

29. $z = 9 - x^2 - 9y^2$, elliptic paraboloid

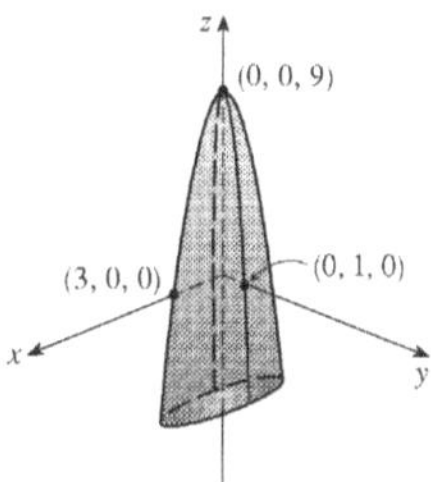

31. $z = \sqrt{4 - 4x^2 - y^2}$, top half of ellipsoid

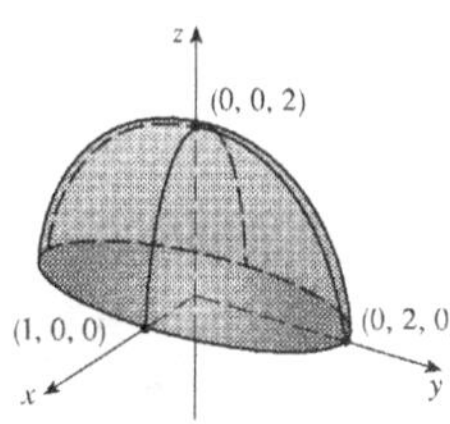

33. ≈56, ≈35 **35.** 11°C, 19.5°C **37.** Steep; nearly flat

39.

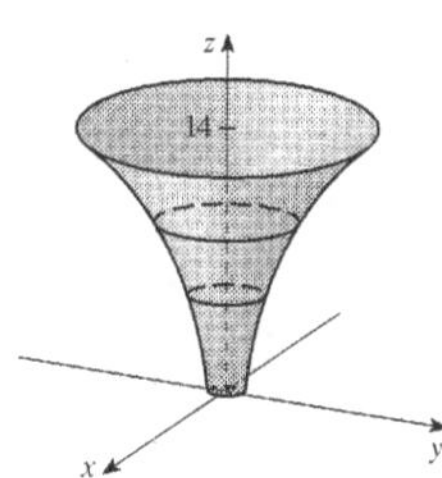

41.

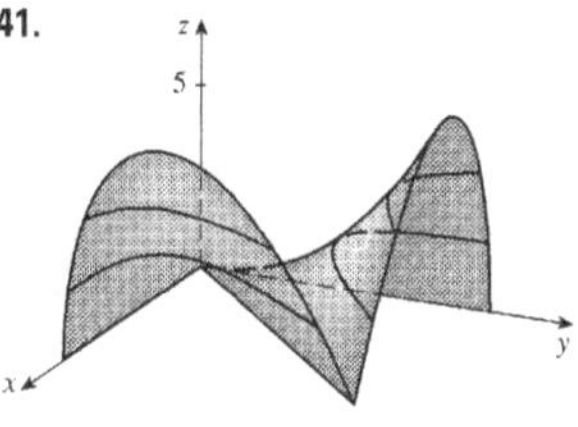

43. $(y - 2x)^2 = k$

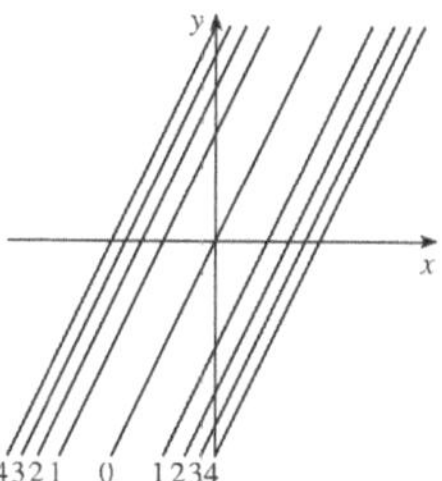

45. $y = -\sqrt{x} + k$

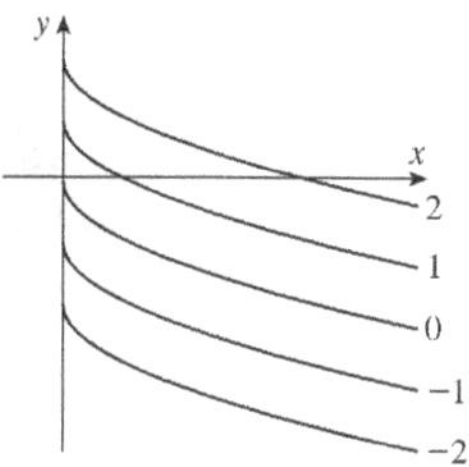

47. $y = ke^{-x}$

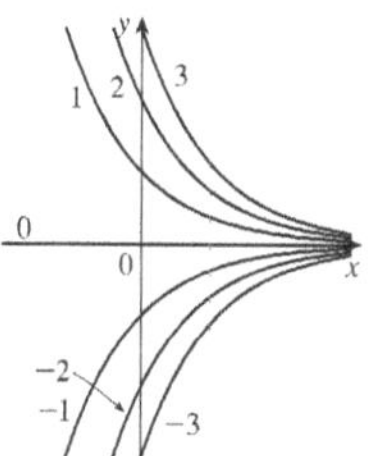

49. $y^2 - x^2 = k^2$

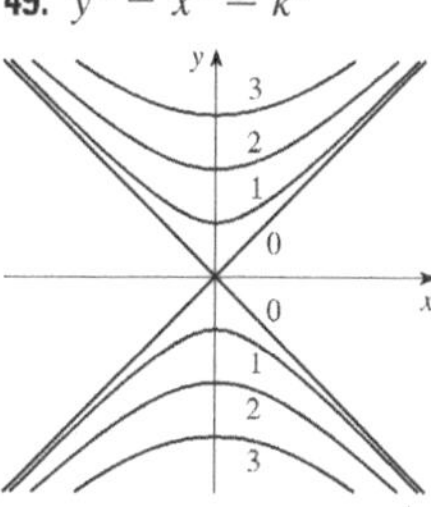

51. $x^2 + 9y^2 = k$

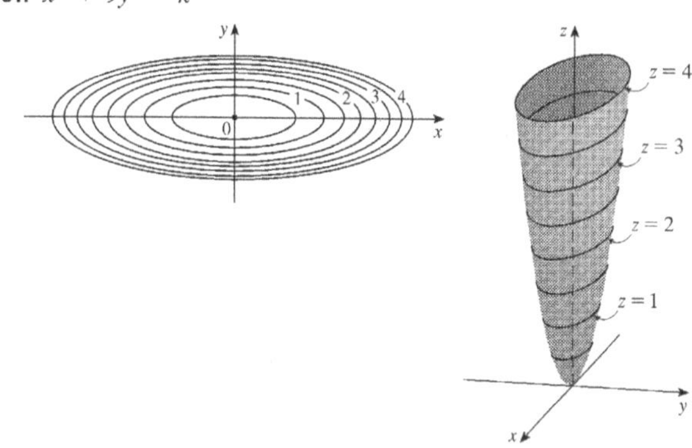

53.

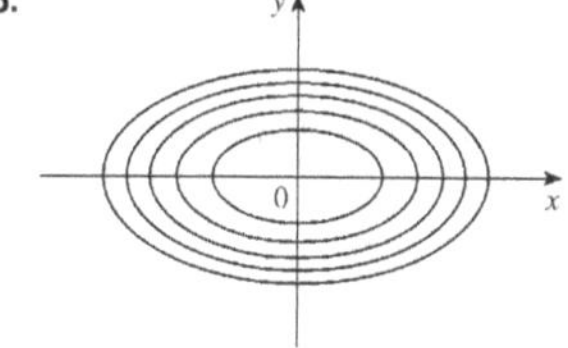

55.

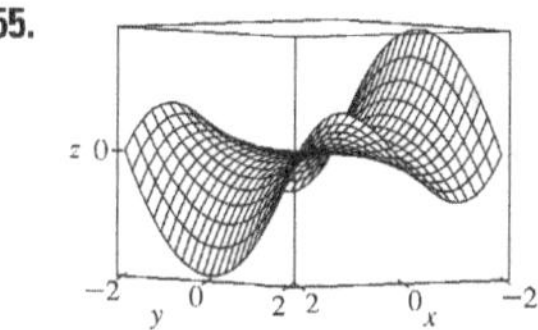

57.

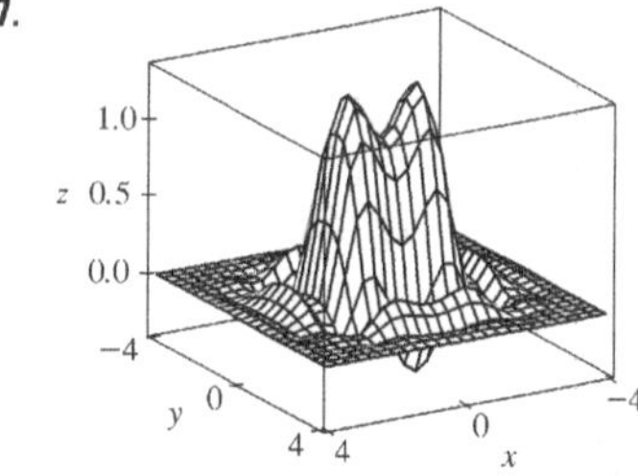

59. (a) C (b) II **61.** (a) F (b) I

63. (a) B (b) VI **65.** Family of parallel planes

67. Family of circular cylinders with axis the x-axis ($k > 0$)

69. (a) Shift the graph of f upward 2 units

(b) Stretch the graph of f vertically by a factor of 2

(c) Reflect the graph of f about the xy-plane

(d) Reflect the graph of f about the xy-plane and then shift it upward 2 units

71.

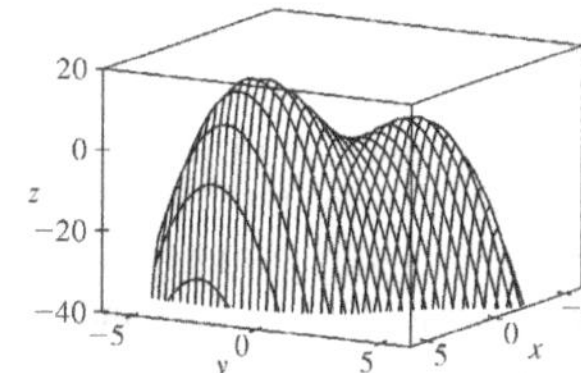

f appears to have a maximum value of about 15. There are two local maximum points but no local minimum point.

73.

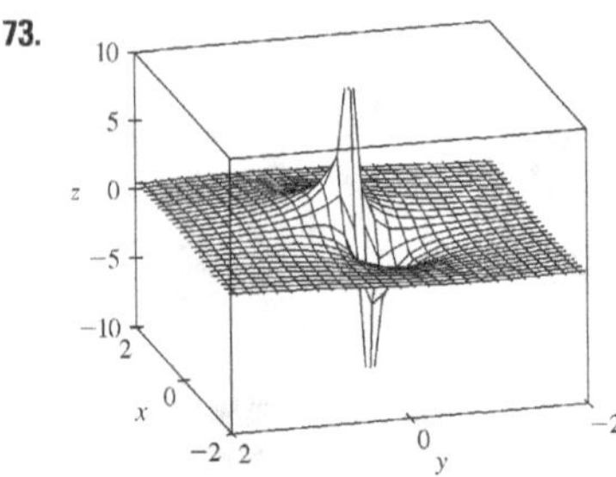

The function values approach 0 as x, y become large; as (x, y) approaches the origin, f approaches $\pm\infty$ or 0, depending on the direction of approach.

75. If $c = 0$, the graph is a cylindrical surface. For $c > 0$, the level curves are ellipses. The graph curves upward as we leave the origin, and the steepness increases as c increases. For $c < 0$, the level curves are hyperbolas. The graph curves upward in the y-direction and downward, approaching the xy-plane, in the x-direction giving a saddle-shaped appearance near $(0, 0, 1)$.

77. $c = -2, 0, 2$ **79.** (b) $y = 0.75x + 0.01$

EXERCISES 14.2 ■ PAGE 899

1. Nothing; if f is continuous, $f(3, 1) = 6$ **3.** $-\frac{5}{2}$

5. 1 **7.** $\frac{2}{7}$ **9.** Does not exist **11.** Does not exist

13. 0 **15.** Does not exist **17.** 2

19. $\sqrt{3}$ **21.** Does not exist

23. The graph shows that the function approaches different numbers along different lines.

25. $h(x, y) = (2x + 3y - 6)^2 + \sqrt{2x + 3y - 6}$; $\{(x, y) \mid 2x + 3y \geqslant 6\}$

27. Along the line $y = x$ **29.** $\mathbb{R}^2$ **31.** $\{(x, y) \mid x^2 + y^2 \neq 1\}$

33. $\{(x, y) \mid x^2 + y^2 > 4\}$ **35.** $\{(x, y, z) \mid x^2 + y^2 + z^2 \leqslant 1\}$

37. $\{(x, y) \mid (x, y) \neq (0, 0)\}$ **39.** 0 **41.** -1

43.

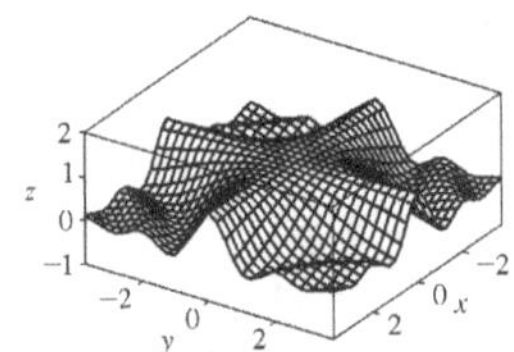

f is continuous on $\mathbb{R}^2$

EXERCISES 14.3 ■ PAGE 911

1. (a) The rate of change of temperature as longitude varies, with latitude and time fixed; the rate of change as only latitude varies; the rate of change as only time varies.
(b) Positive, negative, positive

3. (a) $f_T(-15, 30) \approx 1.3$; for a temperature of -15°C and wind speed of 30 km/h, the wind-chill index rises by 1.3°C for each degree the temperature increases. $f_v(-15, 30) \approx -0.15$; for a temperature of -15°C and wind speed of 30 km/h, the wind-chill index decreases by 0.15°C for each km/h the wind speed increases.
(b) Positive, negative (c) 0

5. (a) Positive (b) Negative

7. (a) Positive (b) Negative

9. $c = f$, $b = f_x$, $a = f_y$

11. $f_x(1, 2) = -8 =$ slope of C_1, $f_y(1, 2) = -4 =$ slope of C_2

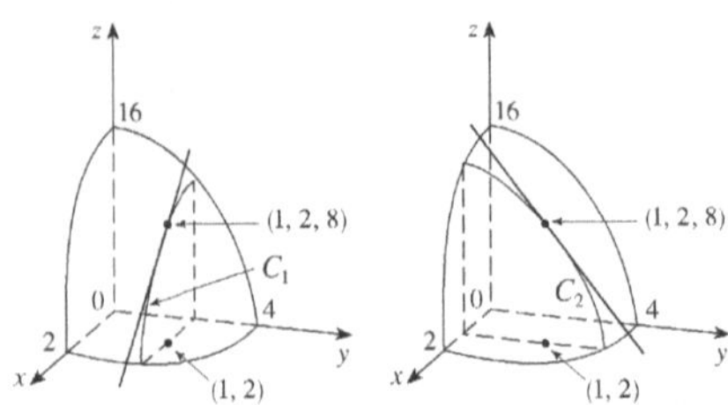

13.

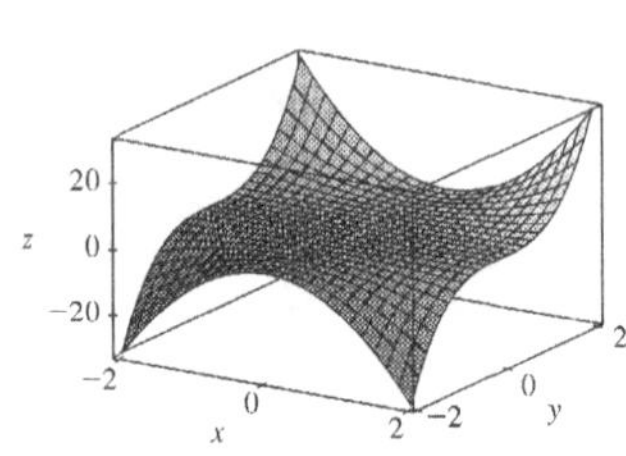

$f(x, y) = x^2y^3$

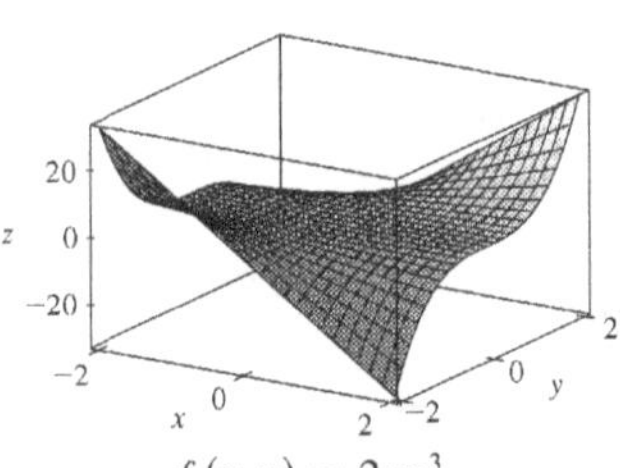

$f_x(x, y) = 2xy^3$

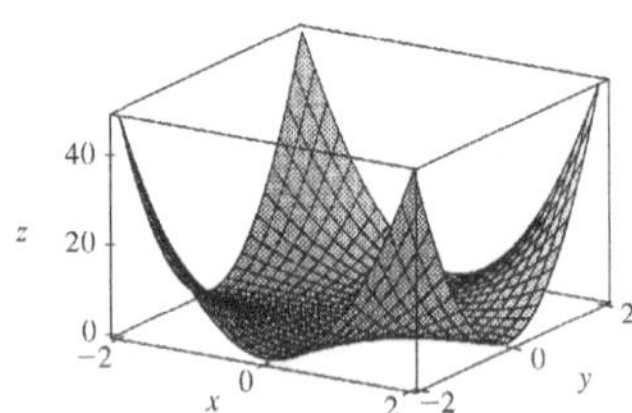

$f_y(x, y) = 3x^2y^2$

15. $f_x(x, y) = -3y$, $f_y(x, y) = 5y^4 - 3x$

17. $f_x(x, t) = -\pi e^{-t}\sin \pi x$, $f_t(x, t) = -e^{-t}\cos \pi x$

19. $\partial z/\partial x = 20(2x + 3y)^9$, $\partial z/\partial y = 30(2x + 3y)^9$

21. $f_x(x, y) = 1/y$, $f_y(x, y) = -x/y^2$

23. $f_x(x, y) = \dfrac{(ad - bc)y}{(cx + dy)^2}$, $f_y(x, y) = \dfrac{(bc - ad)x}{(cx + dy)^2}$

25. $g_u(u, v) = 10uv(u^2v - v^3)^4$, $g_v(u, v) = 5(u^2 - 3v^2)(u^2v - v^3)^4$

27. $R_p(p, q) = \dfrac{q^2}{1 + p^2q^4}$, $R_q(p, q) = \dfrac{2pq}{1 + p^2q^4}$

29. $F_x(x, y) = \cos(e^x)$, $F_y(x, y) = -\cos(e^y)$

31. $f_x = z - 10xy^3z^4$, $f_y = -15x^2y^2z^4$, $f_z = x - 20x^2y^3z^3$

33. $\partial w/\partial x = 1/(x + 2y + 3z)$, $\partial w/\partial y = 2/(x + 2y + 3z)$, $\partial w/\partial z = 3/(x + 2y + 3z)$

35. $\partial u/\partial x = y\sin^{-1}(yz)$, $\partial u/\partial y = x\sin^{-1}(yz) + xyz/\sqrt{1 - y^2z^2}$, $\partial u/\partial z = xy^2/\sqrt{1 - y^2z^2}$

37. $h_x = 2xy\cos(z/t)$, $h_y = x^2\cos(z/t)$, $h_z = (-x^2y/t)\sin(z/t)$, $h_t = (x^2yz/t^2)\sin(z/t)$

39. $\partial u/\partial x_i = x_i/\sqrt{x_1^2 + x_2^2 + \cdots + x_n^2}$

41. $\frac{1}{5}$ **43.** $\frac{1}{4}$ **45.** $f_x(x, y) = y^2 - 3x^2y$, $f_y(x, y) = 2xy - x^3$

47. $\dfrac{\partial z}{\partial x} = -\dfrac{x}{3z}, \dfrac{\partial z}{\partial y} = -\dfrac{2y}{3z}$

49. $\dfrac{\partial z}{\partial x} = \dfrac{yz}{e^z - xy}, \dfrac{\partial z}{\partial y} = \dfrac{xz}{e^z - xy}$

51. (a) $f'(x), g'(y)$ (b) $f'(x + y), f'(x + y)$

53. $f_{xx} = 6xy^5 + 24x^2y$, $f_{xy} = 15x^2y^4 + 8x^3 = f_{yx}$, $f_{yy} = 20x^3y^3$

55. $w_{uu} = v^2/(u^2 + v^2)^{3/2}$, $w_{uv} = -uv/(u^2 + v^2)^{3/2} = w_{vu}$, $w_{vv} = u^2/(u^2 + v^2)^{3/2}$

57. $z_{xx} = -2x/(1 + x^2)^2$, $z_{xy} = 0 = z_{yx}$, $z_{yy} = -2y/(1 + y^2)^2$

63. $24xy^2 - 6y, 24x^2y - 6x$ **65.** $(2x^2y^2z^5 + 6xyz^3 + 2z)e^{xyz^2}$

67. $\theta e^{r\theta}(2\sin\theta + \theta\cos\theta + r\theta\sin\theta)$ **69.** $4/(y + 2z)^3, 0$

71. $6yz^2$ **73.** $\approx 12.2, \approx 16.8, \approx 23.25$ **83.** R^2/R_1^2

87. $\dfrac{\partial T}{\partial P} = \dfrac{V - nb}{nR}, \dfrac{\partial P}{\partial V} = \dfrac{2n^2a}{V^3} - \dfrac{nRT}{(V - nb)^2}$

93. No **95.** $x = 1 + t, y = 2, z = 2 - 2t$ **99.** -2

101. (a)

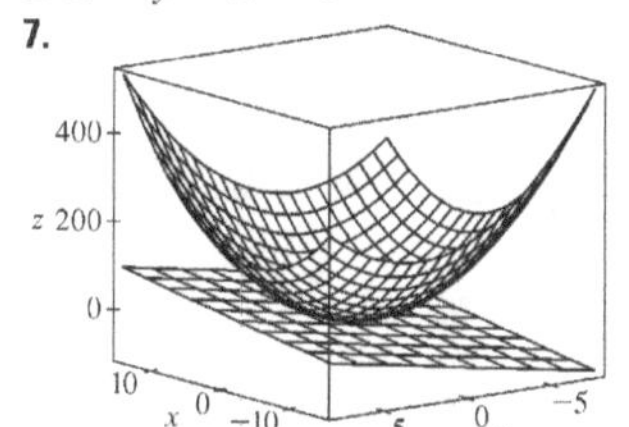

(b) $f_x(x, y) = \dfrac{x^4y + 4x^2y^3 - y^5}{(x^2 + y^2)^2}$, $f_y(x, y) = \dfrac{x^5 - 4x^3y^2 - xy^4}{(x^2 + y^2)^2}$

(c) 0, 0 (e) No, since f_{xy} and f_{yx} are not continuous.

EXERCISES 14.4 ■ PAGE 922

1. $z = -7x - 6y + 5$ **3.** $x + y - 2z = 0$

5. $x + y + z = 0$

7.

9.

11. $6x + 4y - 23$ **13.** $\frac{1}{9}x - \frac{2}{9}y + \frac{2}{3}$ **15.** $1 - \pi y$

19. 6.3 **21.** $\frac{3}{7}x + \frac{2}{7}y + \frac{6}{7}z$; 6.9914

23. $4T + H - 329$; 129°F

25. $dz = -2e^{-2x}\cos 2\pi t\,dx - 2\pi e^{-2x}\sin 2\pi t\,dt$

27. $dm = 5p^4q^3\,dp + 3p^5q^2\,dq$

29. $dR = \beta^2\cos\gamma\,d\alpha + 2\alpha\beta\cos\gamma\,d\beta - \alpha\beta^2\sin\gamma\,d\gamma$

31. $\Delta z = 0.9225, dz = 0.9$ **33.** 5.4 cm² **35.** 16 cm³

37. $\approx -0.0165mg$; decrease

39. $\frac{1}{17} \approx 0.059\ \Omega$ **41.** 2.3% **43.** $\varepsilon_1 = \Delta x, \varepsilon_2 = \Delta y$

EXERCISES 14.5 ■ PAGE 930

1. $(2x + y)\cos t + (2y + x)e^t$

3. $[(x/t) - y\sin t]/\sqrt{1 + x^2 + y^2}$

5. $e^{y/z}[2t - (x/z) - (2xy/z^2)]$

7. $\partial z/\partial s = 2xy^3\cos t + 3x^2y^2\sin t$, $\partial z/\partial t = -2sxy^3\sin t + 3sx^2y^2\cos t$

9. $\partial z/\partial s = t^2\cos\theta\cos\phi - 2st\sin\theta\sin\phi$, $\partial z/\partial t = 2st\cos\theta\cos\phi - s^2\sin\theta\sin\phi$

11. $\dfrac{\partial z}{\partial s} = e^r\left(t\cos\theta - \dfrac{s}{\sqrt{s^2 + t^2}}\sin\theta\right)$, $\dfrac{\partial z}{\partial t} = e^r\left(s\cos\theta - \dfrac{t}{\sqrt{s^2 + t^2}}\sin\theta\right)$

13. 62 **15.** 7, 2

17. $\dfrac{\partial u}{\partial r} = \dfrac{\partial u}{\partial x}\dfrac{\partial x}{\partial r} + \dfrac{\partial u}{\partial y}\dfrac{\partial y}{\partial r}, \dfrac{\partial u}{\partial s} = \dfrac{\partial u}{\partial x}\dfrac{\partial x}{\partial s} + \dfrac{\partial u}{\partial y}\dfrac{\partial y}{\partial s}$, $\dfrac{\partial u}{\partial t} = \dfrac{\partial u}{\partial x}\dfrac{\partial x}{\partial t} + \dfrac{\partial u}{\partial y}\dfrac{\partial y}{\partial t}$

19. $\dfrac{\partial w}{\partial x} = \dfrac{\partial w}{\partial r}\dfrac{\partial r}{\partial x} + \dfrac{\partial w}{\partial s}\dfrac{\partial s}{\partial x} + \dfrac{\partial w}{\partial t}\dfrac{\partial t}{\partial x}$, $\dfrac{\partial w}{\partial y} = \dfrac{\partial w}{\partial r}\dfrac{\partial r}{\partial y} + \dfrac{\partial w}{\partial s}\dfrac{\partial s}{\partial y} + \dfrac{\partial w}{\partial t}\dfrac{\partial t}{\partial y}$

21. 1582, 3164, −700 **23.** $2\pi, -2\pi$

25. $\frac{5}{144}, -\frac{5}{96}, \frac{5}{144}$ **27.** $\dfrac{2x + y\sin x}{\cos x - 2y}$

29. $\dfrac{1 + x^4y^2 + y^2 + x^4y^4 - 2xy}{x^2 - 2xy - 2x^5y^3}$

31. $-\dfrac{x}{3z}, -\dfrac{2y}{3z}$ **33.** $\dfrac{yz}{e^z - xy}, \dfrac{xz}{e^z - xy}$

35. 2°C/s **37.** ≈ -0.33 m/s per minute

39. (a) 6 m³/s (b) 10 m²/s (c) 0 m/s

41. ≈ -0.27 L/s **43.** $-1/(12\sqrt{3})$ rad/s

45. (a) $\partial z/\partial r = (\partial z/\partial x)\cos\theta + (\partial z/\partial y)\sin\theta$, $\partial z/\partial\theta = -(\partial z/\partial x)r\sin\theta + (\partial z/\partial y)r\cos\theta$

51. $4rs\,\partial^2z/\partial x^2 + (4r^2 + 4s^2)\partial^2z/\partial x\,\partial y + 4rs\,\partial^2z/\partial y^2 + 2\,\partial z/\partial y$

EXERCISES 14.6 ■ PAGE 943

1. ≈ -0.08 mb/km **3.** ≈ 0.778 **5.** $2 + \sqrt{3}/2$

7. (a) $\nabla f(x, y) = \langle 2\cos(2x + 3y), 3\cos(2x + 3y)\rangle$ (b) $\langle 2, 3\rangle$ (c) $\sqrt{3} - \frac{3}{2}$

9. (a) $\langle 2xyz - yz^3, x^2z - xz^3, x^2y - 3xyz^2\rangle$ (b) $\langle -3, 2, 2\rangle$ (c) $\frac{2}{5}$

11. $\dfrac{4 - 3\sqrt{3}}{10}$ **13.** $-8/\sqrt{10}$ **15.** $4/\sqrt{30}$

17. $\frac{23}{42}$ **19.** 2/5 **21.** $\sqrt{65}, \langle 1, 8\rangle$

23. $1, \langle 0, 1\rangle$ **25.** $1, \langle 3, 6, -2\rangle$

27. (b) $\langle -12, 92\rangle$

29. All points on the line $y = x + 1$

31. (a) $-40/(3\sqrt{3})$

33. (a) $32/\sqrt{3}$ (b) $\langle 38, 6, 12\rangle$ (c) $2\sqrt{406}$

35. $\frac{327}{13}$ **39.** $\frac{774}{25}$

41. (a) $x + y + z = 11$ (b) $x - 3 = y - 3 = z - 5$

43. (a) $2x + 3y + 12z = 24$ (b) $\dfrac{x-3}{2} = \dfrac{y-2}{3} = \dfrac{z-1}{12}$

45. (a) $x + y + z = 1$ (b) $x = y = z - 1$

47.

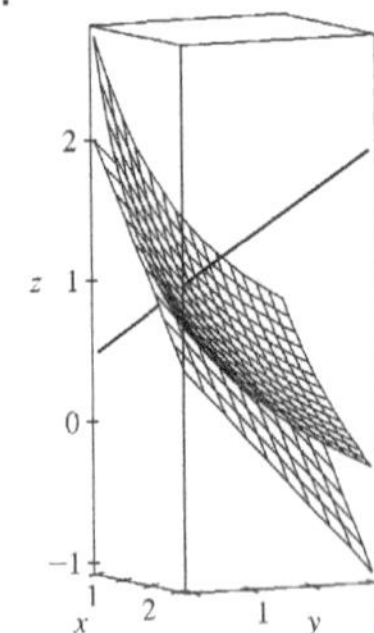

49. $\langle 2, 3\rangle$, $2x + 3y = 12$

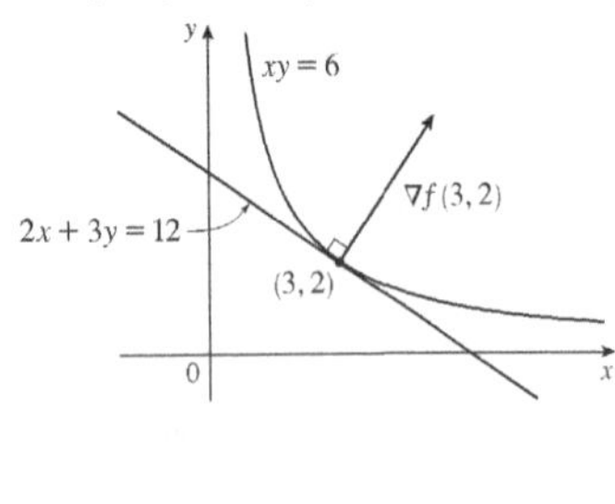

55. No **59.** $\left(-\frac{5}{4}, -\frac{5}{4}, \frac{25}{8}\right)$

63. $x = -1 - 10t$, $y = 1 - 16t$, $z = 2 - 12t$

67. If $\mathbf{u} = \langle a, b\rangle$ and $\mathbf{v} = \langle c, d\rangle$, then $af_x + bf_y$ and $cf_x + df_y$ are known, so we solve linear equations for f_x and f_y.

EXERCISES 14.7 ■ PAGE 953

1. (a) f has a local minimum at (1, 1).
(b) f has a saddle point at (1, 1).

3. Local minimum at (1, 1), saddle point at (0, 0)

5. Minimum $f\left(\frac{1}{3}, -\frac{2}{3}\right) = -\frac{1}{3}$

7. Saddle points at (1, 1), (−1, −1)

9. Maximum $f(0, 0) = 2$, minimum $f(0, 4) = -30$, saddle points at (2, 2), (−2, 2)

11. Minimum $f(2, 1) = -8$, saddle point at (0, 0)

13. None **15.** Minimum $f(0, 0) = 0$, saddle points at $(\pm 1, 0)$

17. Minima $f(0, 1) = f(\pi, -1) = f(2\pi, 1) = -1$, saddle points at $(\pi/2, 0)$, $(3\pi/2, 0)$

21. Minima $f(1, \pm 1) = 3$, $f(-1, \pm 1) = 3$

23. Maximum $f(\pi/3, \pi/3) = 3\sqrt{3}/2$, minimum $f(5\pi/3, 5\pi/3) = -3\sqrt{3}/2$, saddle point at (π, π)

25. Minima $f(0, -0.794) \approx -1.191$, $f(\pm 1.592, 1.267) \approx -1.310$, saddle points $(\pm 0.720, 0.259)$, lowest points $(\pm 1.592, 1.267, -1.310)$

27. Maximum $f(0.170, -1.215) \approx 3.197$, minima $f(-1.301, 0.549) \approx -3.145$, $f(1.131, 0.549) \approx -0.701$, saddle points $(-1.301, -1.215)$, $(0.170, 0.549)$, $(1.131, -1.215)$, no highest or lowest point

29. Maximum $f(0, \pm 2) = 4$, minimum $f(1, 0) = -1$

31. Maximum $f(\pm 1, 1) = 7$, minimum $f(0, 0) = 4$

33. Maximum $f(3, 0) = 83$, minimum $f(1, 1) = 0$

35. Maximum $f(1, 0) = 2$, minimum $f(-1, 0) = -2$

37.

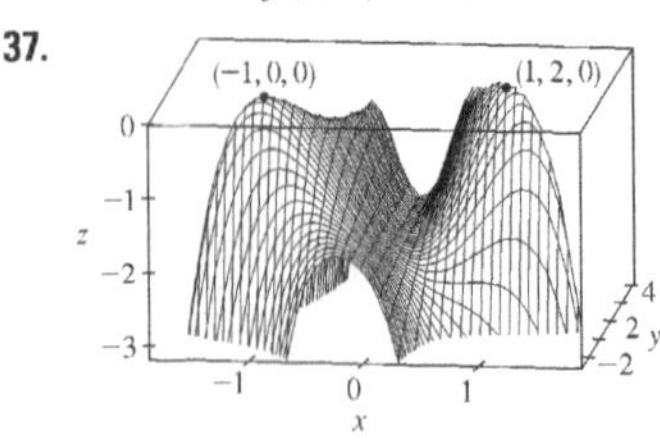

39. $2/\sqrt{3}$ **41.** $(2, 1, \sqrt{5})$, $(2, 1, -\sqrt{5})$ **43.** $\frac{100}{3}, \frac{100}{3}, \frac{100}{3}$

45. $8r^3/(3\sqrt{3})$ **47.** $\frac{4}{3}$ **49.** Cube, edge length $c/12$

51. Square base of side 40 cm, height 20 cm **53.** $L^3/(3\sqrt{3})$

EXERCISES 14.8 ■ PAGE 963

1. ≈59, 30

3. No maximum, minimum $f(1, 1) = f(-1, -1) = 2$

5. Maximum $f(0, \pm 1) = 1$, minimum $f(\pm 2, 0) = -4$

7. Maximum $f(2, 2, 1) = 9$, minimum $f(-2, -2, -1) = -9$

9. Maximum $2/\sqrt{3}$, minimum $-2/\sqrt{3}$

11. Maximum $\sqrt{3}$, minimum 1

13. Maximum $f\left(\frac{1}{2}, \frac{1}{2}, \frac{1}{2}, \frac{1}{2}\right) = 2$, minimum $f\left(-\frac{1}{2}, -\frac{1}{2}, -\frac{1}{2}, -\frac{1}{2}\right) = -2$

15. Maximum $f(1, \sqrt{2}, -\sqrt{2}) = 1 + 2\sqrt{2}$, minimum $f(1, -\sqrt{2}, \sqrt{2}) = 1 - 2\sqrt{2}$

17. Maximum $\frac{3}{2}$, minimum $\frac{1}{2}$

19. Maximum $f(3/\sqrt{2}, -3/\sqrt{2}) = 9 + 12\sqrt{2}$, minimum $f(-2, 2) = -8$

21. Maximum $f(\pm 1/\sqrt{2}, \mp 1/(2\sqrt{2})) = e^{1/4}$, minimum $f(\pm 1/\sqrt{2}, \pm 1/(2\sqrt{2})) = e^{-1/4}$

29–41. See Exercises 39–53 in Section 14.7.

43. Nearest $\left(\frac{1}{2}, \frac{1}{2}, \frac{1}{2}\right)$, farthest $(-1, -1, 2)$

45. Maximum ≈9.7938, minimum ≈−5.3506

47. (a) c/n (b) When $x_1 = x_2 = \cdots = x_n$

CHAPTER 14 REVIEW ■ PAGE 967

True-False Quiz

1. True **3.** False **5.** False **7.** True **9.** False **11.** True

Exercises

1. $\{(x, y) \mid y > -x - 1\}$

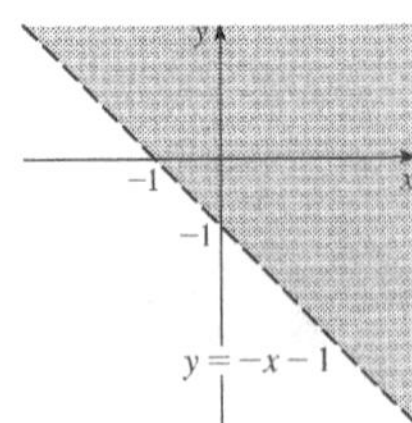

3.

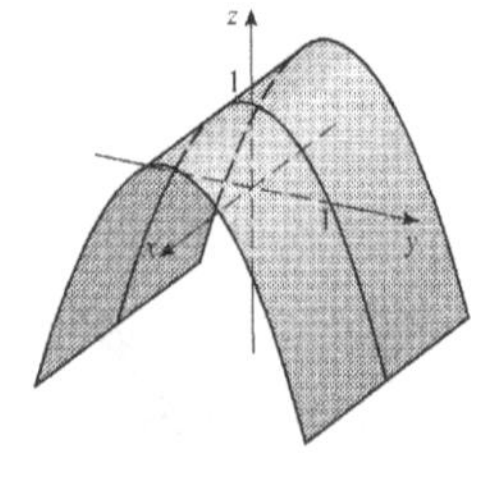

5.

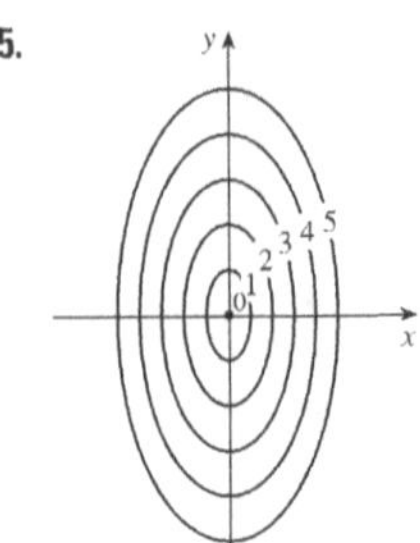

7.

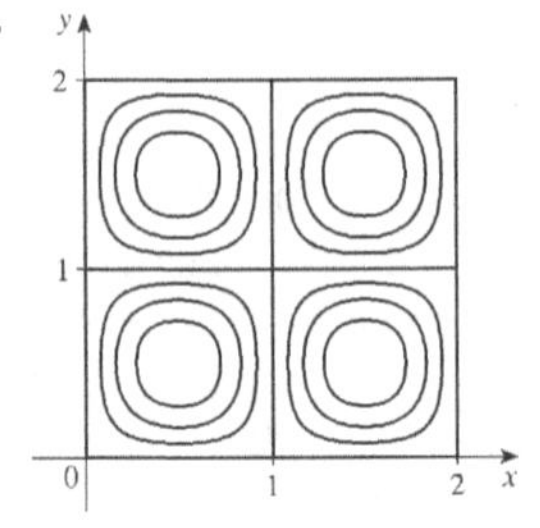

9. $\frac{2}{3}$
11. (a) $\approx 3.5°\text{C/m}, -3.0°\text{C/m}$ (b) $\approx 0.35°\text{C/m}$ by Equation 14.6.9 (Definition 14.6.2 gives $\approx 1.1°\text{C/m}$.)
(c) -0.25
13. $f_x = 32xy(5y^3 + 2x^2y)^7, f_y = (16x^2 + 120y^2)(5y^3 + 2x^2y)^7$
15. $F_\alpha = \dfrac{2\alpha^3}{\alpha^2 + \beta^2} + 2\alpha \ln(\alpha^2 + \beta^2), F_\beta = \dfrac{2\alpha^2\beta}{\alpha^2 + \beta^2}$
17. $S_u = \arctan(v\sqrt{w}), S_v = \dfrac{u\sqrt{w}}{1 + v^2w}, S_w = \dfrac{uv}{2\sqrt{w}(1 + v^2w)}$
19. $f_{xx} = 24x, f_{xy} = -2y = f_{yx}, f_{yy} = -2x$
21. $f_{xx} = k(k-1)x^{k-2}y^lz^m, f_{xy} = klx^{k-1}y^{l-1}z^m = f_{yx}$,
$f_{xz} = kmx^{k-1}y^lz^{m-1} = f_{zx}, f_{yy} = l(l-1)x^ky^{l-2}z^m$,
$f_{yz} = lmx^ky^{l-1}z^{m-1} = f_{zy}, f_{zz} = m(m-1)x^ky^lz^{m-2}$
25. (a) $z = 8x + 4y + 1$ (b) $\dfrac{x-1}{8} = \dfrac{y+2}{4} = \dfrac{z-1}{-1}$
27. (a) $2x - 2y - 3z = 3$ (b) $\dfrac{x-2}{4} = \dfrac{y+1}{-4} = \dfrac{z-1}{-6}$
29. (a) $x + 2y + 5z = 0$
(b) $x = 2 + t, y = -1 + 2t, z = 5t$
31. $(2, \frac{1}{2}, -1), (-2, -\frac{1}{2}, 1)$
33. $60x + \frac{24}{5}y + \frac{32}{5}z - 120$; 38.656
35. $2xy^3(1 + 6p) + 3x^2y^2(pe^p + e^p) + 4z^3(p\cos p + \sin p)$
37. $-47, 108$
43. $\langle 2xe^{yz^2}, x^2z^2e^{yz^2}, 2x^2yze^{yz^2}\rangle$ **45.** $-\frac{4}{5}$
47. $\sqrt{145}/2, \langle 4, \frac{9}{2}\rangle$ **49.** $\approx \frac{5}{8}$ knot/mi
51. Minimum $f(-4, 1) = -11$
53. Maximum $f(1, 1) = 1$; saddle points $(0, 0), (0, 3), (3, 0)$
55. Maximum $f(1, 2) = 4$, minimum $f(2, 4) = -64$
57. Maximum $f(-1, 0) = 2$, minima $f(1, \pm 1) = -3$, saddle points $(-1, \pm 1), (1, 0)$
59. Maximum $f(\pm\sqrt{2/3}, 1/\sqrt{3}) = 2/(3\sqrt{3})$, minimum $f(\pm\sqrt{2/3}, -1/\sqrt{3}) = -2/(3\sqrt{3})$
61. Maximum 1, minimum -1
63. $(\pm 3^{-1/4}, 3^{-1/4}\sqrt{2}, \pm 3^{1/4}), (\pm 3^{-1/4}, -3^{-1/4}\sqrt{2}, \pm 3^{1/4})$
65. $P(2 - \sqrt{3}), P(3 - \sqrt{3})/6, P(2\sqrt{3} - 3)/3$

PROBLEMS PLUS ■ PAGE 971

1. $L^2W^2, \frac{1}{4}L^2W^2$ **3.** (a) $x = w/3$, base $= w/3$ (b) Yes
7. $\sqrt{3/2}, 3/\sqrt{2}$

CHAPTER 15

EXERCISES 15.1 ■ PAGE 981

1. (a) 288 (b) 144 **3.** (a) 0.990 (b) 1.151
5. (a) 4 (b) -8 **7.** $U < V < L$
9. (a) ≈ 248 (b) ≈ 15.5 **11.** 60 **13.** 3
15. 1.141606, 1.143191, 1.143535, 1.143617, 1.143637, 1.143642

EXERCISES 15.2 ■ PAGE 987

1. $500y^3, 3x^2$ **3.** 222 **5.** $32(e^4 - 1)$ **7.** 18
9. $\frac{21}{2}\ln 2$ **11.** $\frac{31}{30}$ **13.** π **15.** 0
17. $9\ln 2$ **19.** $\frac{1}{2}(\sqrt{3} - 1) - \frac{1}{12}\pi$ **21.** $\frac{1}{2}e^{-6} + \frac{5}{2}$
23.

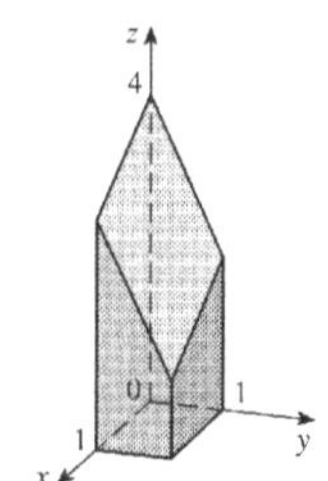

25. 51 **27.** $\frac{166}{27}$ **29.** 2 **31.** $\frac{64}{3}$
33. $21e - 57$

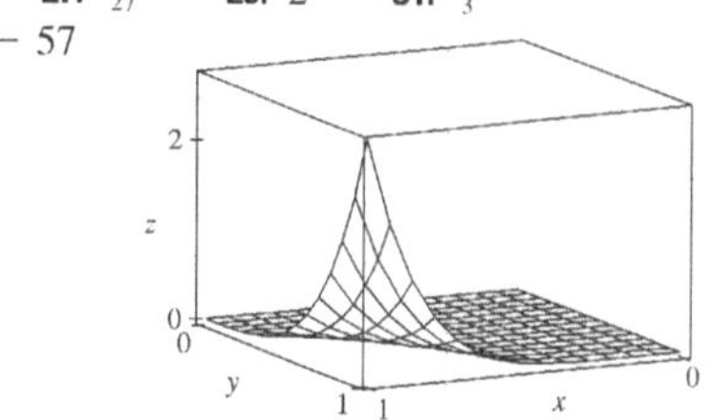

35. $\frac{5}{6}$ **37.** 0
39. Fubini's Theorem does not apply. The integrand has an infinite discontinuity at the origin.

EXERCISES 15.3 ■ PAGE 995

1. 32 **3.** $\frac{3}{10}$ **5.** $\frac{1}{3}\sin 1$ **7.** $\frac{4}{3}$ **9.** π
11. (a) y D 0 x (b)

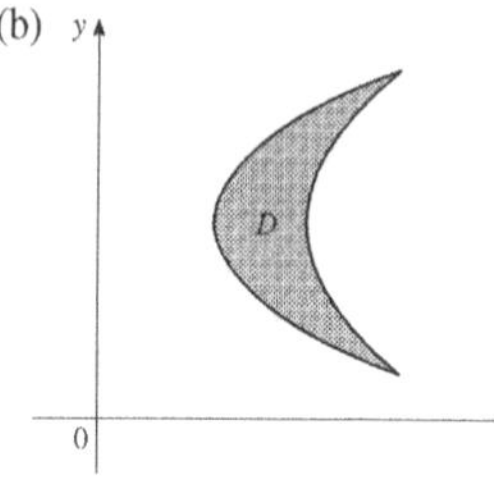

13. Type I: $D = \{(x, y) \mid 0 \leqslant x \leqslant 1, 0 \leqslant y \leqslant x\}$,
type II: $D = \{(x, y) \mid 0 \leqslant y \leqslant 1, y \leqslant x \leqslant 1\}$; $\frac{1}{3}$
15. $\int_0^1 \int_{-\sqrt{x}}^{\sqrt{x}} y\,dy\,dx + \int_1^4 \int_{x-2}^{\sqrt{x}} y\,dy\,dx = \int_{-1}^2 \int_{y^2}^{y+2} y\,dx\,dy = \frac{9}{4}$
17. $\frac{1}{2}(1 - \cos 1)$ **19.** $\frac{11}{3}$ **21.** 0 **23.** $\frac{17}{60}$ **25.** $\frac{31}{8}$
27. 6 **29.** $\frac{128}{15}$ **31.** $\frac{1}{3}$ **33.** 0, 1.213; 0.713 **35.** $\frac{64}{3}$
37.

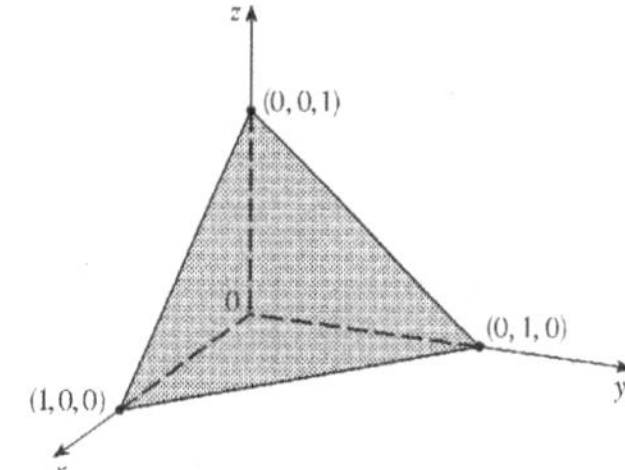

39. 13,984,735,616/14,549,535
41. $\pi/2$

43. $\int_0^1 \int_x^1 f(x, y)\, dy\, dx$

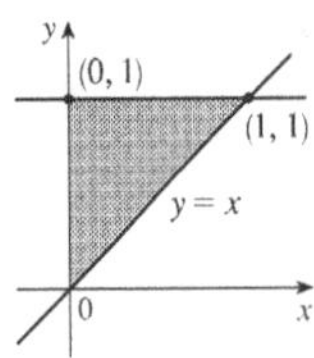

45. $\int_0^1 \int_0^{\cos^{-1} y} f(x, y)\, dx\, dy$

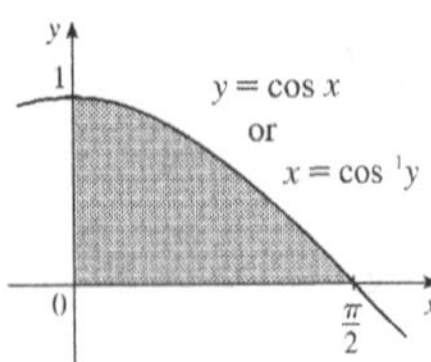

47. $\int_0^{\ln 2} \int_{e^y}^2 f(x, y)\, dx\, dy$

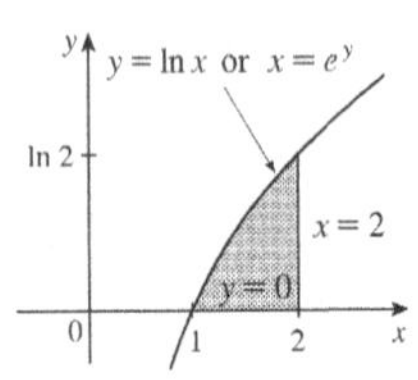

49. $\frac{1}{6}(e^9 - 1)$ **51.** $\frac{1}{3}\ln 9$ **53.** $\frac{1}{3}(2\sqrt{2} - 1)$ **55.** 1
57. $(\pi/16)e^{-1/16} \leqslant \iint_Q e^{-(x^2+y^2)^2}\, dA \leqslant \pi/16$ **59.** $\frac{3}{4}$ **63.** 9π
65. $a^2b + \frac{3}{2}ab^2$ **67.** $\pi a^2 b$

EXERCISES 15.4 ■ PAGE 1002

1. $\int_0^{3\pi/2} \int_0^4 f(r\cos\theta, r\sin\theta)\, r\, dr\, d\theta$ **3.** $\int_{-1}^1 \int_0^{(x+1)/2} f(x, y)\, dy\, dx$

5. $3\pi/4$

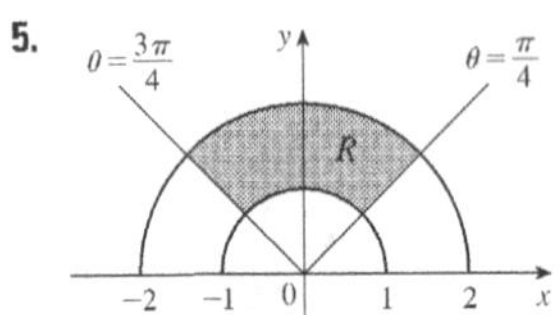

7. $\frac{1250}{3}$ **9.** $(\pi/4)(\cos 1 - \cos 9)$
11. $(\pi/2)(1 - e^{-4})$ **13.** $\frac{3}{64}\pi^2$ **15.** $\pi/12$
17. $\dfrac{\pi}{3} + \dfrac{\sqrt{3}}{2}$ **19.** $\frac{16}{3}\pi$ **21.** $\frac{4}{3}\pi$ **23.** $\frac{4}{3}\pi a^3$
25. $(2\pi/3)[1 - (1/\sqrt{2})]$ **27.** $(8\pi/3)(64 - 24\sqrt{3})$
29. $\frac{1}{2}\pi(1 - \cos 9)$ **31.** $2\sqrt{2}/3$ **33.** 4.5951
35. 1800π ft^3 **37.** $2/(a + b)$ **39.** $\frac{15}{16}$
41. (a) $\sqrt{\pi}/4$ (b) $\sqrt{\pi}/2$

EXERCISES 15.5 ■ PAGE 1012

1. 285 C **3.** $42k$, $(2, \frac{85}{28})$ **5.** 6, $(\frac{3}{4}, \frac{3}{2})$ **7.** $\frac{8}{15}k$, $(0, \frac{4}{7})$
9. $L/4$, $(L/2, 16/(9\pi))$ **11.** $(\frac{3}{8}, 3\pi/16)$ **13.** $(0, 45/(14\pi))$
15. $(2a/5, 2a/5)$ if vertex is $(0, 0)$ and sides are along positive axes
17. $\frac{64}{315}k$, $\frac{8}{105}k$, $\frac{88}{315}k$
19. $7ka^6/180$, $7ka^6/180$, $7ka^6/90$ if vertex is $(0, 0)$ and sides are along positive axes
21. $\rho bh^3/3$, $\rho b^3h/3$; $b/\sqrt{3}$, $h/\sqrt{3}$
23. $\rho a^4\pi/16$, $\rho a^4\pi/16$; $a/2$, $a/2$
25. $m = 3\pi/64$, $(\bar{x}, \bar{y}) = \left(\dfrac{16384\sqrt{2}}{10395\pi}, 0\right)$,
$I_x = \dfrac{5\pi}{384} - \dfrac{4}{105}$, $I_y = \dfrac{5\pi}{384} + \dfrac{4}{105}$, $I_0 = \dfrac{5\pi}{192}$
27. (a) $\frac{1}{2}$ (b) 0.375 (c) $\frac{5}{48} \approx 0.1042$
29. (b) (i) $e^{-0.2} \approx 0.8187$
(ii) $1 + e^{-1.8} - e^{-0.8} - e^{-1} \approx 0.3481$ (c) 2, 5
31. (a) ≈ 0.500 (b) ≈ 0.632
33. (a) $\iint_D k\left[1 - \frac{1}{20}\sqrt{(x - x_0)^2 + (y - y_0)^2}\right] dA$, where D is the disk with radius 10 mi centered at the center of the city
(b) $200\pi k/3 \approx 209k$, $200(\pi/2 - \frac{8}{9})k \approx 136k$, on the edge

EXERCISES 15.6 ■ PAGE 1016

1. $15\sqrt{26}$ **3.** $3\sqrt{14}$ **5.** $12\sin^{-1}(\frac{2}{3})$
7. $(\pi/6)(17\sqrt{17} - 5\sqrt{5})$ **9.** $(2\pi/3)(2\sqrt{2} - 1)$
11. $a^2(\pi - 2)$ **13.** 13.9783 **15.** (a) ≈ 1.83 (b) ≈ 1.8616
17. $\frac{45}{8}\sqrt{14} + \frac{15}{16}\ln[(11\sqrt{5} + 3\sqrt{70})/(3\sqrt{5} + \sqrt{70})]$
19. 3.3213 **23.** $(\pi/6)(101\sqrt{101} - 1)$

EXERCISES 15.7 ■ PAGE 1025

1. $\frac{27}{4}$ **3.** $\frac{16}{15}$ **5.** $\frac{5}{3}$ **7.** $-\frac{1}{3}$ **9.** $\frac{27}{2}$ **11.** $9\pi/8$
13. $\frac{65}{28}$ **15.** $\frac{1}{60}$ **17.** $16\pi/3$ **19.** $\frac{16}{3}$ **21.** $\frac{8}{15}$
23. (a) $\int_0^1 \int_0^x \int_0^{\sqrt{1-y^2}} dz\, dy\, dx$ (b) $\frac{1}{4}\pi - \frac{1}{3}$
25. 0.985
27.

29. $\int_{-2}^2 \int_0^{4-x^2} \int_{-\sqrt{4-x^2-y}/2}^{\sqrt{4-x^2-y}/2} f(x, y, z)\, dz\, dy\, dx$
$= \int_0^4 \int_{-\sqrt{4-y}}^{\sqrt{4-y}} \int_{-\sqrt{4-x^2-y}/2}^{\sqrt{4-x^2-y}/2} f(x, y, z)\, dz\, dx\, dy$
$= \int_{-1}^1 \int_0^{4-4z^2} \int_{-\sqrt{4-y-4z^2}}^{\sqrt{4-y-4z^2}} f(x, y, z)\, dx\, dy\, dz$
$= \int_0^4 \int_{-\sqrt{4-y}/2}^{\sqrt{4-y}/2} \int_{-\sqrt{4-y-4z^2}}^{\sqrt{4-y-4z^2}} f(x, y, z)\, dx\, dz\, dy$
$= \int_{-2}^2 \int_{-\sqrt{4-x^2}/2}^{\sqrt{4-x^2}/2} \int_0^{4-x^2-4z^2} f(x, y, z)\, dy\, dz\, dx$
$= \int_{-1}^1 \int_{-\sqrt{4-4z^2}}^{\sqrt{4-4z^2}} \int_0^{4-x^2-4z^2} f(x, y, z)\, dy\, dx\, dz$

31. $\int_{-2}^2 \int_{x^2}^4 \int_0^{2-y/2} f(x, y, z)\, dz\, dy\, dx$
$= \int_0^4 \int_{-\sqrt{y}}^{\sqrt{y}} \int_0^{2-y/2} f(x, y, z)\, dz\, dx\, dy$
$= \int_0^2 \int_0^{4-2z} \int_{-\sqrt{y}}^{\sqrt{y}} f(x, y, z)\, dx\, dy\, dz$
$= \int_0^4 \int_0^{2-y/2} \int_{-\sqrt{y}}^{\sqrt{y}} f(x, y, z)\, dx\, dz\, dy$
$= \int_{-2}^2 \int_0^{2-x^2/2} \int_{x^2}^{4-2z} f(x, y, z)\, dy\, dz\, dx$
$= \int_0^2 \int_{-\sqrt{4-2z}}^{\sqrt{4-2z}} \int_{x^2}^{4-2z} f(x, y, z)\, dy\, dx\, dz$

33. $\int_0^1 \int_{\sqrt{x}}^1 \int_0^{1-y} f(x, y, z)\, dz\, dy\, dx = \int_0^1 \int_0^{y^2} \int_0^{1-y} f(x, y, z)\, dz\, dx\, dy$
$= \int_0^1 \int_0^{1-z} \int_0^{y^2} f(x, y, z)\, dx\, dy\, dz = \int_0^1 \int_0^{1-y} \int_0^{y^2} f(x, y, z)\, dx\, dz\, dy$
$= \int_0^1 \int_0^{1-\sqrt{x}} \int_{\sqrt{x}}^{1-z} f(x, y, z)\, dy\, dz\, dx = \int_0^1 \int_0^{(1-z)^2} \int_{\sqrt{x}}^{1-z} f(x, y, z)\, dy\, dx\, dz$

35. $\int_0^1 \int_y^1 \int_0^y f(x, y, z)\, dz\, dx\, dy = \int_0^1 \int_0^x \int_0^y f(x, y, z)\, dz\, dy\, dx$
$= \int_0^1 \int_z^1 \int_y^1 f(x, y, z)\, dx\, dy\, dz = \int_0^1 \int_0^y \int_y^1 f(x, y, z)\, dx\, dz\, dy$
$= \int_0^1 \int_0^x \int_z^x f(x, y, z)\, dy\, dz\, dx = \int_0^1 \int_z^1 \int_z^x f(x, y, z)\, dy\, dx\, dz$

37. 64π **39.** $\frac{79}{30}, \left(\frac{358}{553}, \frac{33}{79}, \frac{571}{553}\right)$

41. $a^5, (7a/12, 7a/12, 7a/12)$

43. $I_x = I_y = I_z = \frac{2}{3}kL^5$ **45.** $\frac{1}{2}\pi kha^4$

47. (a) $m = \int_{-1}^1 \int_{x^2}^1 \int_0^{1-y} \sqrt{x^2 + y^2}\, dz\, dy\, dx$
(b) $(\bar{x}, \bar{y}, \bar{z})$, where
$\bar{x} = (1/m) \int_{-1}^1 \int_{x^2}^1 \int_0^{1-y} x\sqrt{x^2 + y^2}\, dz\, dy\, dx$
$\bar{y} = (1/m) \int_{-1}^1 \int_{x^2}^1 \int_0^{1-y} y\sqrt{x^2 + y^2}\, dz\, dy\, dx$
$\bar{z} = (1/m) \int_{-1}^1 \int_{x^2}^1 \int_0^{1-y} z\sqrt{x^2 + y^2}\, dz\, dy\, dx$
(c) $\int_{-1}^1 \int_{x^2}^1 \int_0^{1-y} (x^2 + y^2)^{3/2}\, dz\, dy\, dx$

49. (a) $\frac{3}{32}\pi + \frac{11}{24}$
(b) $\left(\dfrac{28}{9\pi + 44}, \dfrac{30\pi + 128}{45\pi + 220}, \dfrac{45\pi + 208}{135\pi + 660}\right)$
(c) $\frac{1}{240}(68 + 15\pi)$

51. (a) $\frac{1}{8}$ (b) $\frac{1}{64}$ (c) $\frac{1}{5760}$ **53.** $L^3/8$

55. (a) The region bounded by the ellipsoid $x^2 + 2y^2 + 3z^2 = 1$
(b) $4\sqrt{6}\pi/45$

EXERCISES 15.8 ■ PAGE 1031

1. (a)

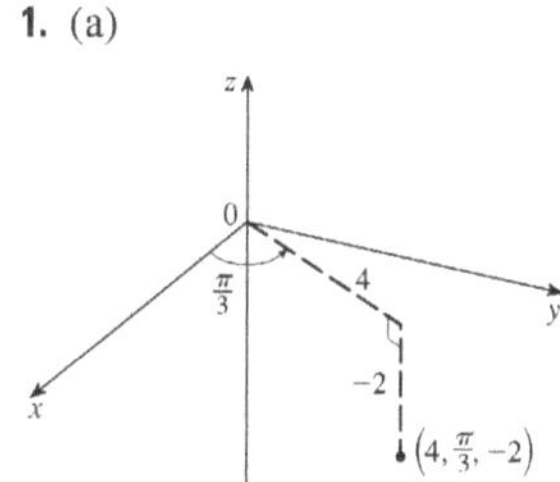

$(2, 2\sqrt{3}, -2)$

(b)

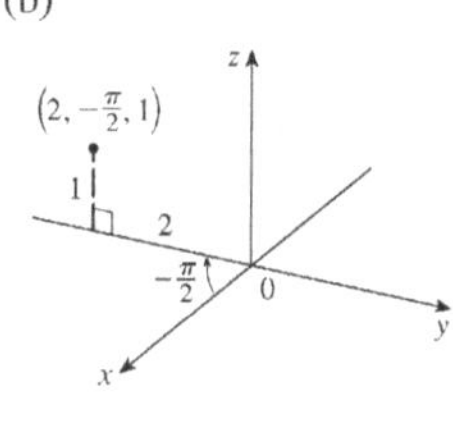

$(0, -2, 1)$

3. (a) $(\sqrt{2}, 3\pi/4, 1)$ (b) $(4, 2\pi/3, 3)$

5. Vertical half-plane through the z-axis

7. Circular paraboloid

9. (a) $z^2 = 1 + r\cos\theta - r^2$ (b) $z = r^2\cos 2\theta$

11.

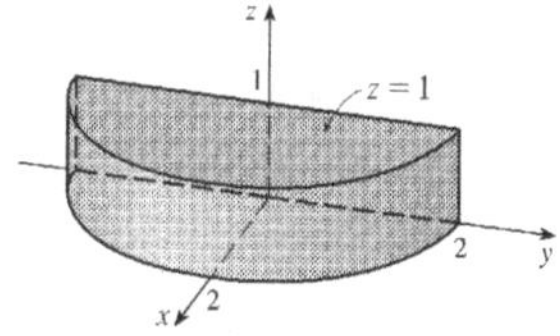

13. Cylindrical coordinates: $6 \leqslant r \leqslant 7, 0 \leqslant \theta \leqslant 2\pi, 0 \leqslant z \leqslant 20$

15.

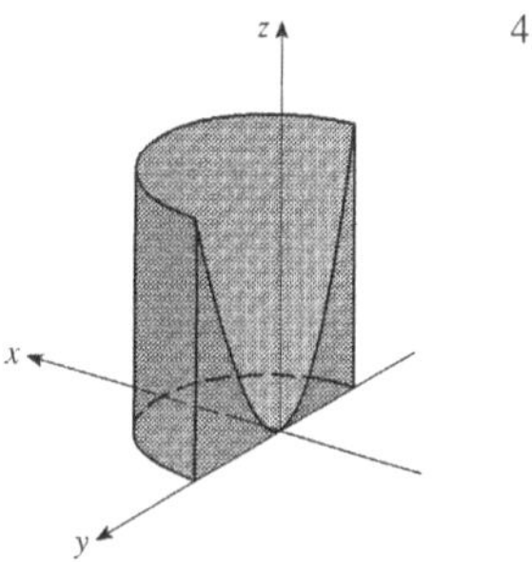

4π

17. 384π **19.** $\frac{8}{3}\pi + \frac{128}{15}$ **21.** $2\pi/5$ **23.** $\frac{4}{3}\pi(\sqrt{2} - 1)$

25. (a) 162π (b) $(0, 0, 15)$

27. $\pi Ka^2/8, (0, 0, 2a/3)$ **29.** 0

31. (a) $\iiint_C h(P)g(P)\, dV$, where C is the cone
(b) $\approx 3.1 \times 10^{19}$ ft-lb

EXERCISES 15.9 ■ PAGE 1037

1. (a)

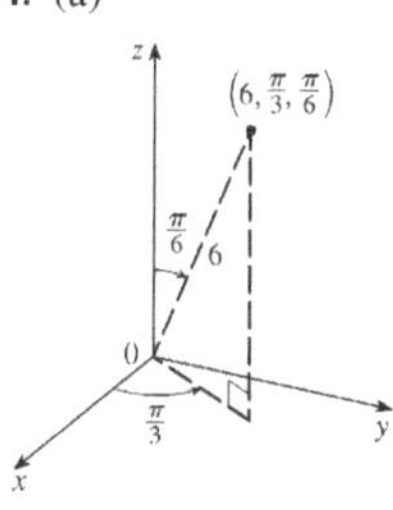

$\left(\dfrac{3}{2}, \dfrac{3\sqrt{3}}{2}, 3\sqrt{3}\right)$

(b)

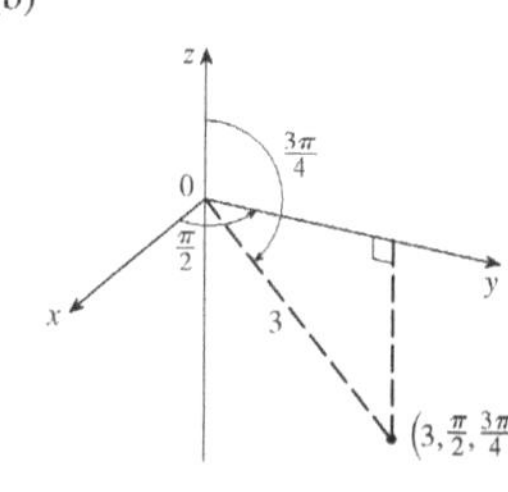

$\left(0, \dfrac{3\sqrt{2}}{2}, -\dfrac{3\sqrt{2}}{2}\right)$

3. (a) $(2, 3\pi/2, \pi/2)$ (b) $(2, 3\pi/4, 3\pi/4)$

5. Half-cone **7.** Sphere, radius $\frac{1}{2}$, center $\left(0, \frac{1}{2}, 0\right)$

9. (a) $\cos^2\phi = \sin^2\phi$ (b) $\rho^2(\sin^2\phi\cos^2\theta + \cos^2\phi) = 9$

11.

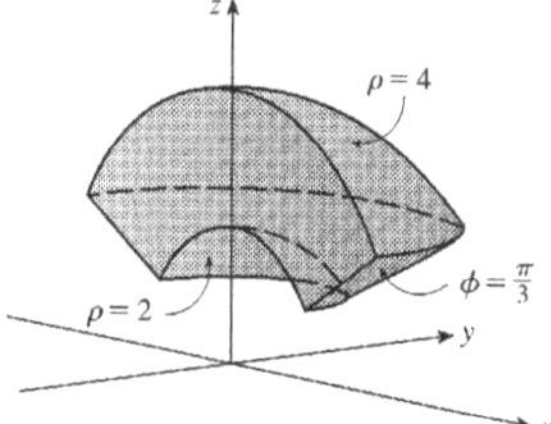

13.

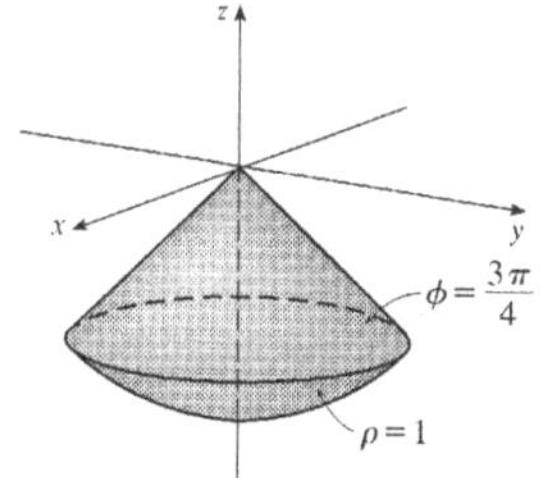

15. $0 \leqslant \phi \leqslant \pi/4, 0 \leqslant \rho \leqslant \cos\phi$

17. 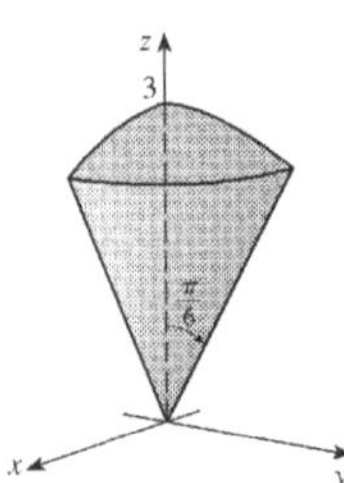

$(9\pi/4)(2 - \sqrt{3})$

19. $\int_0^{\pi/2} \int_0^3 \int_0^2 f(r\cos\theta, r\sin\theta, z)\, r\, dz\, dr\, d\theta$
21. $312{,}500\pi/7$ **23.** $1688\pi/15$ **25.** $\pi/8$
27. $(\sqrt{3} - 1)\pi a^3/3$ **29.** (a) 10π (b) $(0, 0, 2.1)$
31. (a) $(0, 0, \frac{7}{12})$ (b) $11K\pi/960$
33. (a) $(0, 0, \frac{3}{8}a)$ (b) $4K\pi a^5/15$
35. $\frac{1}{3}\pi(2 - \sqrt{2}), (0, 0, 3/[8(2 - \sqrt{2})])$
37. $5\pi/6$ **39.** $(4\sqrt{2} - 5)/15$ **41.** $4096\pi/21$
43. 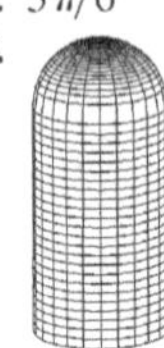 **45.** $136\pi/99$

EXERCISES 15.10 ■ PAGE 1047

1. 16 **3.** $\sin^2\theta - \cos^2\theta$ **5.** 0
7. The parallelogram with vertices $(0, 0)$, $(6, 3)$, $(12, 1)$, $(6, -2)$
9. The region bounded by the line $y = 1$, the y-axis, and $y = \sqrt{x}$
11. $x = \frac{1}{3}(v - u)$, $y = \frac{1}{3}(u + 2v)$ is one possible transformation, where $S = \{(u, v) \mid -1 \leqslant u \leqslant 1, 1 \leqslant v \leqslant 3\}$
13. $x = u\cos v$, $y = u\sin v$ is one possible transformation, where $S = \{(u, v) \mid 1 \leqslant u \leqslant \sqrt{2}, 0 \leqslant v \leqslant \pi/2\}$
15. -3 **17.** 6π **19.** $2\ln 3$
21. (a) $\frac{4}{3}\pi abc$ (b) $1.083 \times 10^{12}\text{ km}^3$ (c) $\frac{4}{15}\pi(a^2 + b^2)abc$
23. $\frac{8}{5}\ln 8$ **25.** $\frac{3}{2}\sin 1$ **27.** $e - e^{-1}$

CHAPTER 15 REVIEW ■ PAGE 1049

True-False Quiz

1. True **3.** True **5.** True **7.** True **9.** False

Exercises

1. ≈ 64.0 **3.** $4e^2 - 4e + 3$ **5.** $\frac{1}{2}\sin 1$ **7.** $\frac{2}{3}$
9. $\int_0^{\pi} \int_2^4 f(r\cos\theta, r\sin\theta)\, r\, dr\, d\theta$
11. The region inside the loop of the four-leaved rose $r = \sin 2\theta$ in the first quadrant
13. $\frac{1}{2}\sin 1$ **15.** $\frac{1}{2}e^6 - \frac{7}{2}$ **17.** $\frac{1}{4}\ln 2$ **19.** 8
21. $81\pi/5$ **23.** $\frac{81}{2}$ **25.** $\pi/96$ **27.** $\frac{64}{15}$
29. 176 **31.** $\frac{2}{3}$ **33.** $2ma^3/9$
35. (a) $\frac{1}{4}$ (b) $(\frac{1}{3}, \frac{8}{15})$
(c) $I_x = \frac{1}{12}$, $I_y = \frac{1}{24}$; $\bar{\bar{y}} = 1/\sqrt{3}$, $\bar{\bar{x}} = 1/\sqrt{6}$
37. (a) $(0, 0, h/4)$ (b) $\pi a^4 h/10$
39. $\ln(\sqrt{2} + \sqrt{3}) + \sqrt{2}/3$ **41.** $\frac{486}{5}$ **43.** 0.0512

45. (a) $\frac{1}{15}$ (b) $\frac{1}{3}$ (c) $\frac{1}{45}$
47. $\int_0^1 \int_0^{1-z} \int_{-\sqrt{y}}^{\sqrt{y}} f(x, y, z)\, dx\, dy\, dz$ **49.** $-\ln 2$ **51.** 0

PROBLEMS PLUS ■ PAGE 1053

1. 30 **3.** $\frac{1}{2}\sin 1$ **7.** (b) 0.90
13. $abc\pi\left(\frac{2}{3} - \frac{8}{9\sqrt{3}}\right)$

CHAPTER 16

EXERCISES 16.1 ■ PAGE 1061

1. 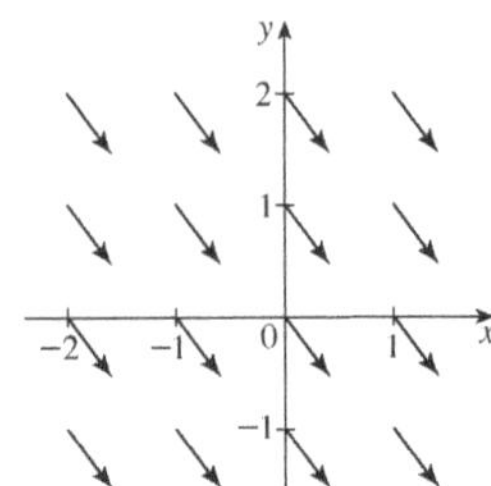

3. 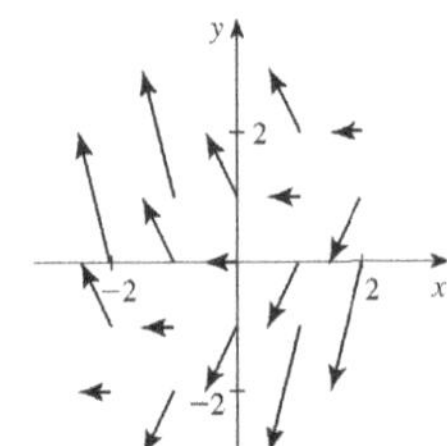

5. 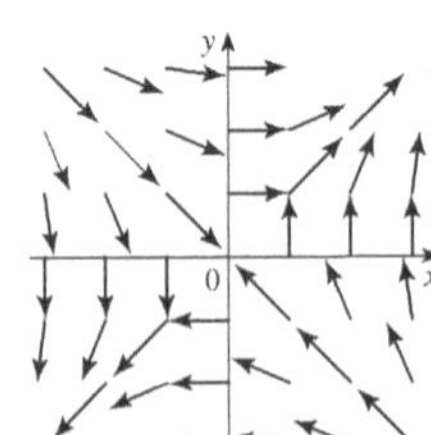

7. 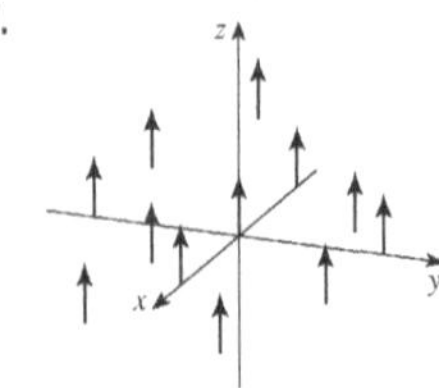

9. 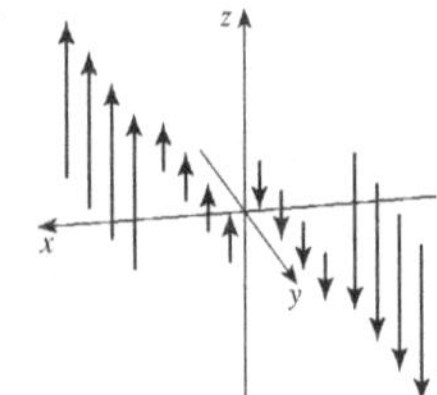

11. IV **13.** I **15.** IV **17.** III

19. The line $y = 2x$

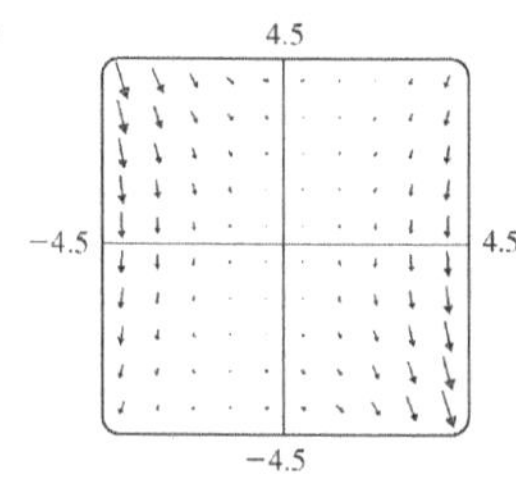

21. $\nabla f(x, y) = (xy + 1)e^{xy}\,\mathbf{i} + x^2e^{xy}\,\mathbf{j}$

23. $\nabla f(x, y, z) = \dfrac{x}{\sqrt{x^2 + y^2 + z^2}}\,\mathbf{i} + \dfrac{y}{\sqrt{x^2 + y^2 + z^2}}\,\mathbf{j} + \dfrac{z}{\sqrt{x^2 + y^2 + z^2}}\,\mathbf{k}$

25. $\nabla f(x, y) = 2x\,\mathbf{i} - \mathbf{j}$

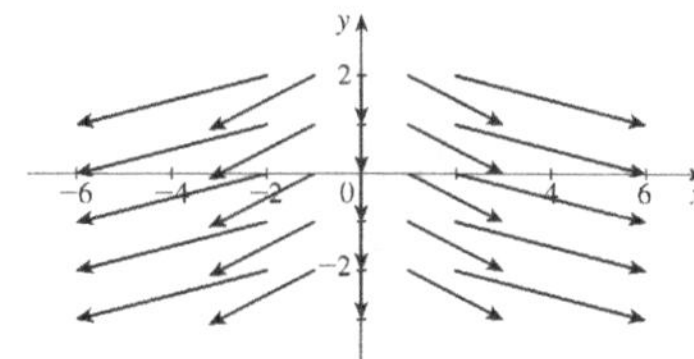

27.

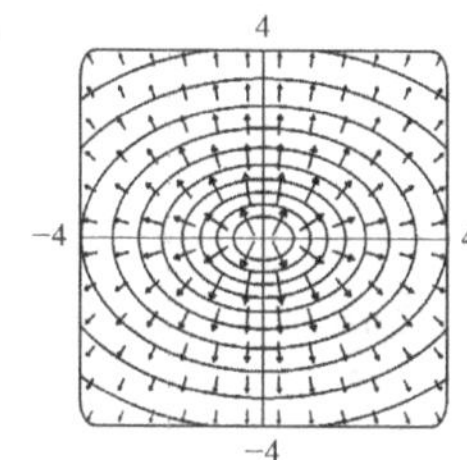

29. III **31.** II **33.** (2.04, 1.03)

35. (a) (b) $y = 1/x,\ x > 0$

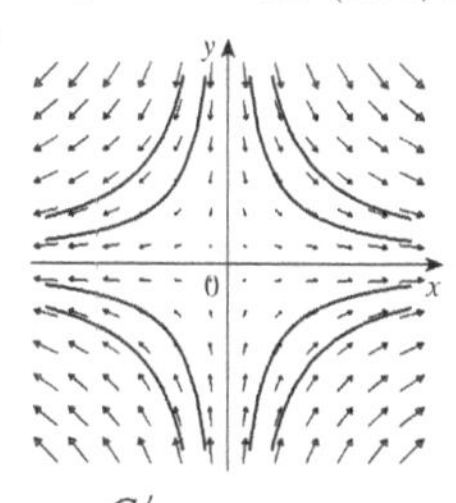

$y = C/x$

EXERCISES 16.2 ■ PAGE 1072

1. $\frac{1}{54}(145^{3/2} - 1)$ **3.** 1638.4 **5.** $\frac{243}{8}$ **7.** $\frac{5}{2}$

9. $\sqrt{5}\pi$ **11.** $\frac{1}{12}\sqrt{14}(e^6 - 1)$ **13.** $\frac{2}{5}(e - 1)$ **15.** $\frac{35}{3}$

17. (a) Positive (b) Negative **19.** 45

21. $\frac{6}{5} - \cos 1 - \sin 1$ **23.** 1.9633 **25.** 15.0074

27. $3\pi + \frac{2}{3}$

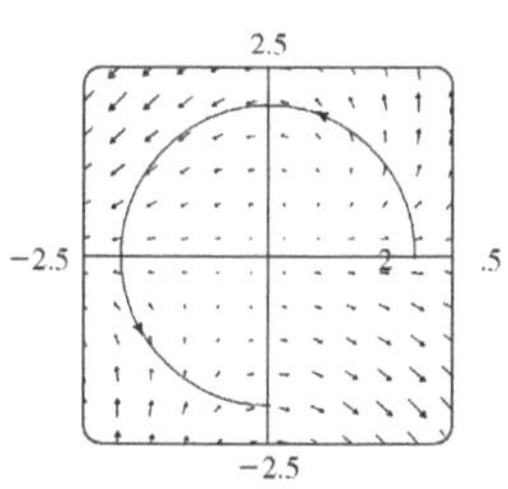

29. (a) $\frac{11}{8} - 1/e$ (b)

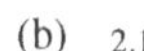

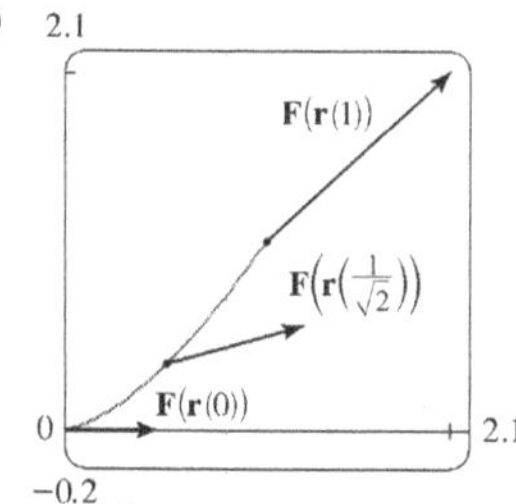

31. $\frac{172{,}704}{5{,}632{,}705}\sqrt{2}(1 - e^{-14\pi})$ **33.** $2\pi k$, $(4/\pi, 0)$

35. (a) $\bar{x} = (1/m)\int_C x\rho(x, y, z)\,ds$,
$\bar{y} = (1/m)\int_C y\rho(x, y, z)\,ds$,
$\bar{z} = (1/m)\int_C z\rho(x, y, z)\,ds$, where $m = \int_C \rho(x, y, z)\,ds$
(b) $(0, 0, 3\pi)$

37. $I_x = k(\frac{1}{2}\pi - \frac{4}{3})$, $I_y = k(\frac{1}{2}\pi - \frac{2}{3})$ **39.** $2\pi^2$ **41.** $\frac{7}{3}$

43. (a) $2ma\,\mathbf{i} + 6mbt\,\mathbf{j}$, $0 \leqslant t \leqslant 1$ (b) $2ma^2 + \frac{9}{2}mb^2$

45. $\approx 1.67 \times 10^4$ ft-lb **47.** (b) Yes **51.** ≈ 22 J

EXERCISES 16.3 ■ PAGE 1082

1. 40 **3.** $f(x, y) = x^2 - 3xy + 2y^2 - 8y + K$

5. Not conservative **7.** $f(x, y) = ye^x + x \sin y + K$

9. $f(x, y) = x \ln y + x^2y^3 + K$

11. (b) 16 **13.** (a) $f(x, y) = \frac{1}{2}x^2y^2$ (b) 2

15. (a) $f(x, y, z) = xyz + z^2$ (b) 77

17. (a) $f(x, y, z) = ye^{xz}$ (b) 4 **19.** $4/e$

21. It doesn't matter which curve is chosen.

23. 30 **25.** No **27.** Conservative

31. (a) Yes (b) Yes (c) Yes

33. (a) No (b) Yes (c) Yes

EXERCISES 16.4 ■ PAGE 1089

1. 8π **3.** $\frac{2}{3}$ **5.** 12 **7.** $\frac{1}{3}$ **9.** -24π **11.** $-\frac{16}{3}$

13. 4π **15.** $-8e + 48e^{-1}$ **17.** $-\frac{1}{12}$ **19.** 3π **21.** (c) $\frac{9}{2}$

23. $(4a/3\pi, 4a/3\pi)$ if the region is the portion of the disk $x^2 + y^2 = a^2$ in the first quadrant

27. 0

EXERCISES 16.5 ■ PAGE 1097

1. (a) $\mathbf{0}$ (b) 3

3. (a) $ze^x\,\mathbf{i} + (xye^z - yze^x)\,\mathbf{j} - xe^z\,\mathbf{k}$ (b) $y(e^z + e^x)$

5. (a) $\mathbf{0}$ (b) $2/\sqrt{x^2 + y^2 + z^2}$

7. (a) $\langle -e^y \cos z, -e^z \cos x, -e^x \cos y \rangle$
(b) $e^x \sin y + e^y \sin z + e^z \sin x$

9. (a) Negative (b) curl $\mathbf{F} = \mathbf{0}$

11. (a) Zero (b) curl $\mathbf{F}$ points in the negative z-direction

13. $f(x, y, z) = xy^2z^3 + K$ **15.** Not conservative

17. $f(x, y, z) = xe^{yz} + K$ **19.** No

EXERCISES 16.6 ■ PAGE 1108

1. P: no; Q: yes
3. Plane through (0, 3, 1) containing vectors $\langle 1, 0, 4\rangle$, $\langle 1, -1, 5\rangle$
5. Hyperbolic paraboloid
7.

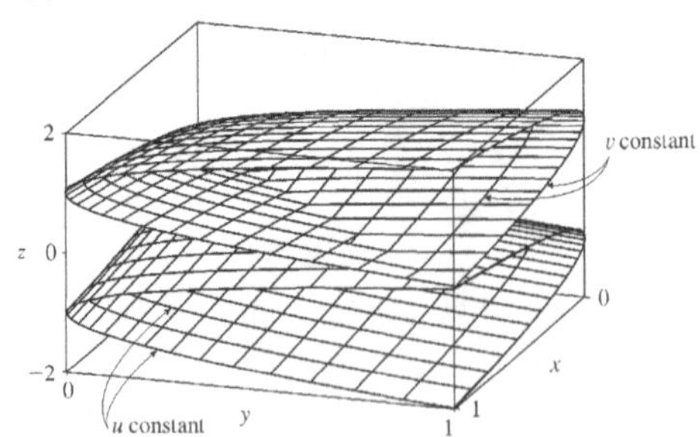

9.

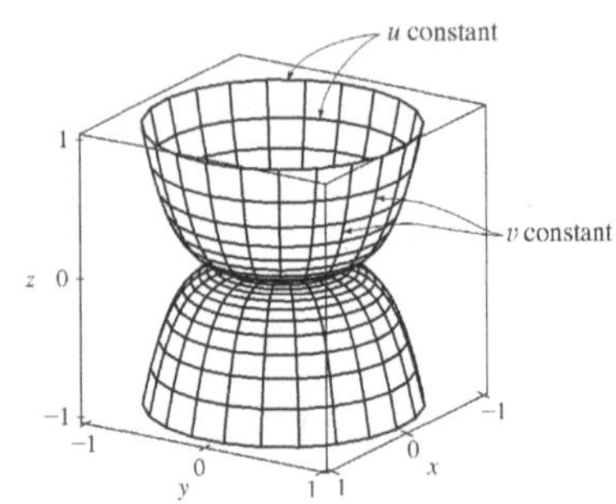

11.

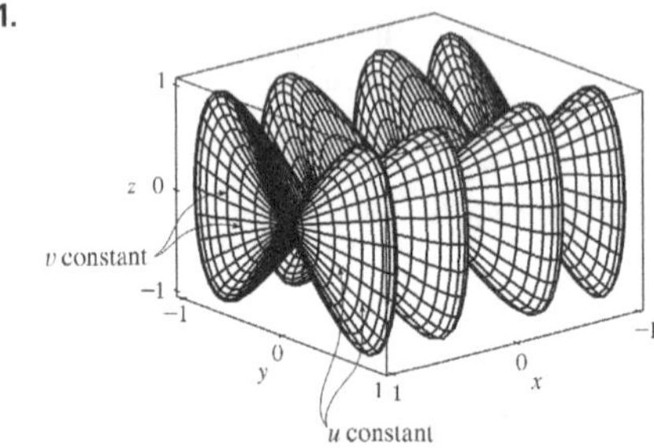

13. IV **15.** II **17.** III
19. $x = u,\ y = v - u,\ z = -v$
21. $y = y,\ z = z,\ x = \sqrt{1 + y^2 + \frac{1}{4}z^2}$
23. $x = 2 \sin\phi \cos\theta,\ y = 2 \sin\phi \sin\theta,$ $z = 2\cos\phi,\ 0 \leqslant \phi \leqslant \pi/4,\ 0 \leqslant \theta \leqslant 2\pi$ $\left[\text{or } x = x,\ y = y,\ z = \sqrt{4 - x^2 - y^2},\ x^2 + y^2 \leqslant 2\right]$
25. $x = x,\ y = 4\cos\theta,\ z = 4\sin\theta,\ 0 \leqslant x \leqslant 5,\ 0 \leqslant \theta \leqslant 2\pi$
29. $x = x,\ y = e^{-x}\cos\theta,$ $z = e^{-x}\sin\theta,\ 0 \leqslant x \leqslant 3,$ $0 \leqslant \theta \leqslant 2\pi$

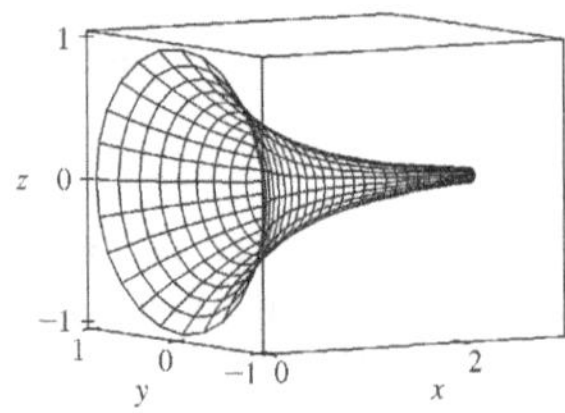

31. (a) Direction reverses (b) Number of coils doubles

33. $3x - y + 3z = 3$ **35.** $\frac{\sqrt{3}}{2}x - \frac{1}{2}y + z = \frac{\pi}{3}$
37. $-x + 2z = 1$ **39.** $3\sqrt{14}$ **41.** $\sqrt{14}\,\pi$
43. $\frac{4}{15}(3^{5/2} - 2^{7/2} + 1)$ **45.** $(2\pi/3)(2\sqrt{2} - 1)$
47. $\frac{1}{2}\sqrt{21} + \frac{17}{4}\left[\ln(2 + \sqrt{21}) - \ln\sqrt{17}\right]$ **49.** 4
51. $A(S) \leqslant \sqrt{3}\,\pi R^2$ **53.** 13.9783
55. (a) 24.2055 (b) 24.2476
57. $\frac{45}{8}\sqrt{14} + \frac{15}{16}\ln\left[(11\sqrt{5} + 3\sqrt{70})/(3\sqrt{5} + \sqrt{70})\right]$
59. (b)

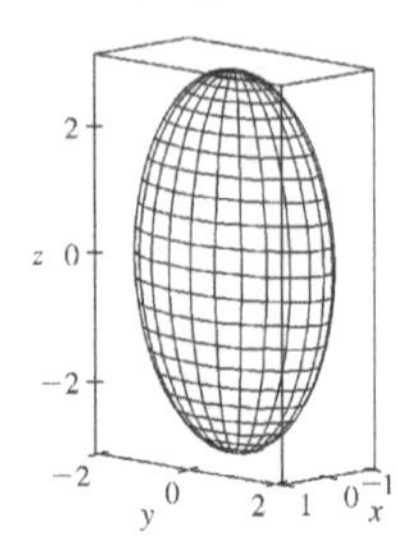

(c) $\int_0^{2\pi}\int_0^{\pi} \sqrt{36\sin^4 u\cos^2 v + 9\sin^4 u\sin^2 v + 4\cos^2 u\sin^2 u}\ du\,dv$

61. 4π **63.** $2a^2(\pi - 2)$

EXERCISES 16.7 ■ PAGE 1120

1. 49.09 **3.** 900π **5.** $11\sqrt{14}$ **7.** $\frac{2}{3}(2\sqrt{2} - 1)$
9. $171\sqrt{14}$ **11.** $\sqrt{21}/3$ **13.** $364\sqrt{2}\,\pi/3$
15. $(\pi/60)(391\sqrt{17} + 1)$ **17.** 16π **19.** 12 **21.** 4
23. $\frac{713}{180}$ **25.** $-\frac{4}{3}\pi$ **27.** 0 **29.** 48 **31.** $2\pi + \frac{8}{3}$
33. 4.5822 **35.** 3.4895
37. $\iint_S \mathbf{F}\cdot d\mathbf{S} = \iint_D \left[P(\partial h/\partial x) - Q + R(\partial h/\partial z)\right] dA$, where D = projection of S on xz-plane
39. $(0, 0, a/2)$
41. (a) $I_z = \iint_S (x^2 + y^2)\rho(x, y, z)\, dS$ (b) $4329\sqrt{2}\,\pi/5$
43. 0 kg/s **45.** $\frac{8}{3}\pi a^3 \varepsilon_0$ **47.** 1248π

EXERCISES 16.8 ■ PAGE 1127

3. 0 **5.** 0 **7.** -1 **9.** 80π
11. (a) $81\pi/2$ (b)

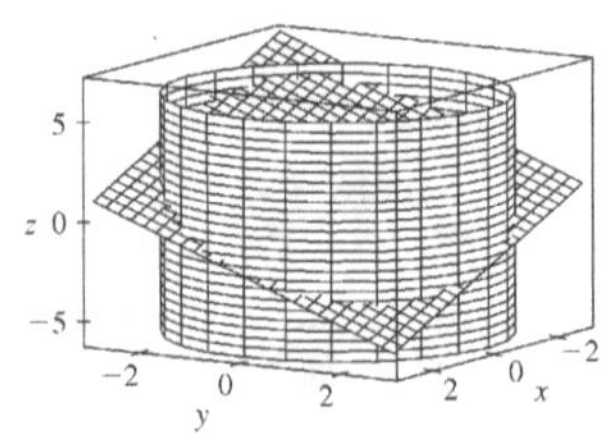

(c) $x = 3\cos t,\ y = 3\sin t,$ $z = 1 - 3(\cos t + \sin t),$ $0 \leqslant t \leqslant 2\pi$

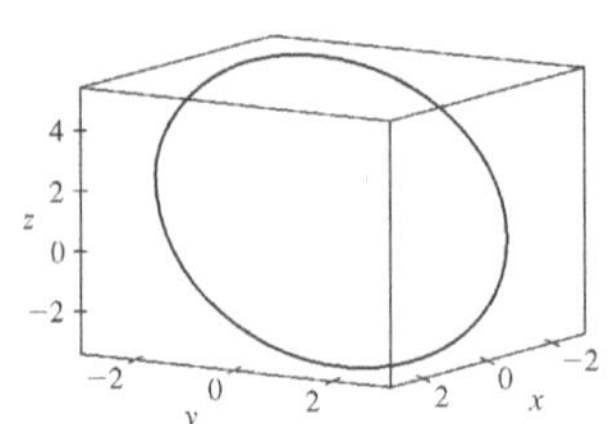

17. 3

EXERCISES 16.9 ■ PAGE 1133

5. $\frac{9}{2}$ **7.** $9\pi/2$ **9.** 0 **11.** $32\pi/3$ **13.** 2π
15. $341\sqrt{2}/60 + \frac{81}{20}\arcsin(\sqrt{3}/3)$
17. $13\pi/20$ **19.** Negative at P_1, positive at P_2
21. $\operatorname{div} \mathbf{F} > 0$ in quadrants I, II; $\operatorname{div} \mathbf{F} < 0$ in quadrants III, IV

CHAPTER 16 REVIEW ■ PAGE 1136

True-False Quiz

1. False **3.** True **5.** False
7. False **9.** True **11.** True

Exercises

1. (a) Negative (b) Positive **3.** $6\sqrt{10}$ **5.** $\frac{4}{15}$
7. $\frac{110}{3}$ **9.** $\frac{11}{12} - 4/e$ **11.** $f(x, y) = e^y + xe^{xy}$ **13.** 0
17. -8π **25.** $\frac{1}{6}(27 - 5\sqrt{5})$ **27.** $(\pi/60)(391\sqrt{17} + 1)$
29. $-64\pi/3$ **33.** $-\frac{1}{2}$ **37.** -4 **39.** 21

CHAPTER 17

EXERCISES 17.1 ■ PAGE 1148

1. $y = c_1e^{3x} + c_2e^{-2x}$ **3.** $y = c_1 \cos 4x + c_2 \sin 4x$
5. $y = c_1e^{2x/3} + c_2xe^{2x/3}$ **7.** $y = c_1 + c_2e^{x/2}$
9. $y = e^{2x}(c_1 \cos 3x + c_2 \sin 3x)$
11. $y = c_1e^{(\sqrt{3}-1)t/2} + c_2e^{-(\sqrt{3}+1)t/2}$
13. $P = e^{-t}\left[c_1 \cos\left(\frac{1}{10}t\right) + c_2 \sin\left(\frac{1}{10}t\right)\right]$
15. All solutions approach either 0 or $\pm\infty$ as $x \to \pm\infty$.

17. $y = 3e^{2x} - e^{4x}$ **19.** $y = e^{-2x/3} + \frac{2}{3}xe^{-2x/3}$
21. $y = e^{3x}(2 \cos x - 3 \sin x)$
23. $y = \frac{1}{7}e^{4x-4} - \frac{1}{7}e^{3-3x}$ **25.** $y = 5 \cos 2x + 3 \sin 2x$
27. $y = 2e^{-2x} - 2xe^{-2x}$ **29.** $y = \dfrac{e - 2}{e - 1} + \dfrac{e^x}{e - 1}$
31. No solution
33. (b) $\lambda = n^2\pi^2/L^2$, n a positive integer; $y = C \sin(n\pi x/L)$
35. (a) $b - a \neq n\pi$, n any integer
(b) $b - a = n\pi$ and $\dfrac{c}{d} \neq e^{a-b}\dfrac{\cos a}{\cos b}$ unless $\cos b = 0$, then $\dfrac{c}{d} \neq e^{a-b}\dfrac{\sin a}{\sin b}$
(c) $b - a = n\pi$ and $\dfrac{c}{d} = e^{a-b}\dfrac{\cos a}{\cos b}$ unless $\cos b = 0$, then $\dfrac{c}{d} = e^{a-b}\dfrac{\sin a}{\sin b}$

EXERCISES 17.2 ■ PAGE 1155

1. $y = c_1e^{3x} + c_2e^{-x} - \frac{7}{65} \cos 2x - \frac{4}{65} \sin 2x$
3. $y = c_1 \cos 3x + c_2 \sin 3x + \frac{1}{13}e^{-2x}$
5. $y = e^{2x}(c_1 \cos x + c_2 \sin x) + \frac{1}{10}e^{-x}$
7. $y = \frac{3}{2} \cos x + \frac{11}{2} \sin x + \frac{1}{2}e^x + x^3 - 6x$
9. $y = e^x\left(\frac{1}{2}x^2 - x + 2\right)$
11.

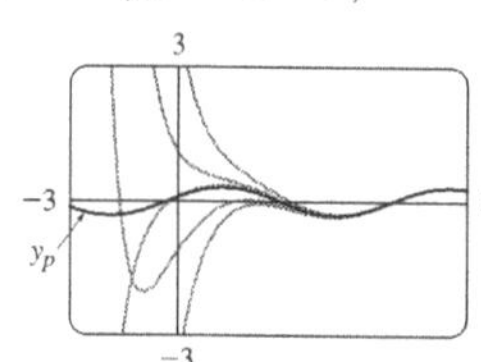

The solutions are all asymptotic to $y_p = \frac{1}{10} \cos x + \frac{3}{10} \sin x$ as $x \to \infty$. Except for y_p, all solutions approach either ∞ or $-\infty$ as $x \to -\infty$.

13. $y_p = (Ax + B)e^x \cos x + (Cx + D)e^x \sin x$
15. $y_p = Axe^x + B \cos x + C \sin x$
17. $y_p = xe^{-x}[(Ax^2 + Bx + C) \cos 3x + (Dx^2 + Ex + F) \sin 3x]$
19. $y = c_1 \cos\left(\frac{1}{2}x\right) + c_2 \sin\left(\frac{1}{2}x\right) - \frac{1}{3} \cos x$
21. $y = c_1e^x + c_2xe^x + e^{2x}$
23. $y = c_1 \sin x + c_2 \cos x + \sin x \ln(\sec x + \tan x) - 1$
25. $y = [c_1 + \ln(1 + e^{-x})]e^x + [c_2 - e^{-x} + \ln(1 + e^{-x})]e^{2x}$
27. $y = e^x\left[c_1 + c_2x - \frac{1}{2}\ln(1 + x^2) + x \tan^{-1}x\right]$

EXERCISES 17.3 ■ PAGE 1163

1. $x = 0.35 \cos(2\sqrt{5}\, t)$ **3.** $x = -\frac{1}{5}e^{-6t} + \frac{6}{5}e^{-t}$ **5.** $\frac{49}{12}$ kg
7.

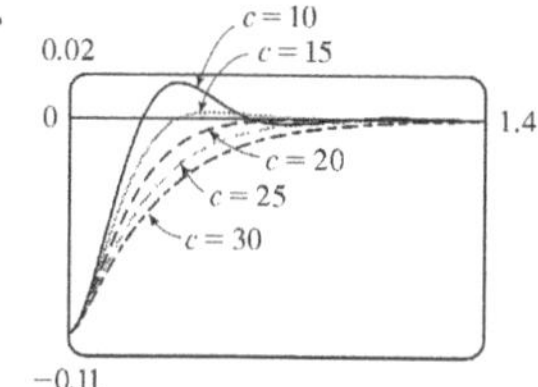

13. $Q(t) = (-e^{-10t}/250)(6 \cos 20t + 3 \sin 20t) + \frac{3}{125}$, $I(t) = \frac{3}{5}e^{-10t} \sin 20t$
15. $Q(t) = e^{-10t}\left[\frac{3}{250} \cos 20t - \frac{3}{500} \sin 20t\right] - \frac{3}{250} \cos 10t + \frac{3}{125} \sin 10t$

EXERCISES 17.4 ■ PAGE 1168

1. $c_0 \sum_{n=0}^{\infty} \dfrac{x^n}{n!} = c_0e^x$ **3.** $c_0 \sum_{n=0}^{\infty} \dfrac{x^{3n}}{3^n n!} = c_0e^{x^3/3}$
5. $c_0 \sum_{n=0}^{\infty} \dfrac{(-1)^n}{2^n n!}x^{2n} + c_1 \sum_{n=0}^{\infty} \dfrac{(-2)^n n!}{(2n + 1)!}x^{2n+1}$
7. $c_0 + c_1 \sum_{n=1}^{\infty} \dfrac{x^n}{n} = c_0 - c_1 \ln(1 - x)$ for $|x| < 1$
9. $\sum_{n=0}^{\infty} \dfrac{x^{2n}}{2^n n!} = e^{x^2/2}$
11. $x + \sum_{n=1}^{\infty} \dfrac{(-1)^n 2^2 5^2 \cdot \cdots \cdot (3n - 1)^2}{(3n + 1)!}x^{3n+1}$

CHAPTER 17 REVIEW ■ PAGE 1169

True-False Quiz

1. True **3.** True

Exercises

1. $y = c_1e^{x/2} + c_2e^{-x/2}$
3. $y = c_1 \cos(\sqrt{3}x) + c_2 \sin(\sqrt{3}x)$

5. $y = e^{2x}(c_1 \cos x + c_2 \sin x + 1)$
7. $y = c_1 e^x + c_2 x e^x - \frac{1}{2}\cos x - \frac{1}{2}(x + 1)\sin x$
9. $y = c_1 e^{3x} + c_2 e^{-2x} - \frac{1}{6} - \frac{1}{5}xe^{-2x}$
11. $y = 5 - 2e^{-6(x-1)}$ **13.** $y = (e^{4x} - e^x)/3$
15. No solution
17. $\displaystyle\sum_{n=0}^{\infty} \frac{(-2)^n n!}{(2n+1)!} x^{2n+1}$
19. $Q(t) = -0.02e^{-10t}(\cos 10t + \sin 10t) + 0.03$
21. (c) $2\pi/k \approx 85$ min (d) $\approx 17{,}600$ mi/h

APPENDIXES

EXERCISES A ■ PAGE A9

1. 18 **3.** π **5.** $5 - \sqrt{5}$ **7.** $2 - x$

9. $|x + 1| = \begin{cases} x + 1 & \text{for } x \geq -1 \\ -x - 1 & \text{for } x < -1 \end{cases}$ **11.** $x^2 + 1$

13. $(-2, \infty)$ **15.** $[-1, \infty)$

17. $(3, \infty)$ **19.** $(2, 6)$

21. $(0, 1]$ **23.** $[-1, \frac{1}{2})$

25. $(-\infty, 1) \cup (2, \infty)$ **27.** $[-1, \frac{1}{2}]$

29. $(-\infty, \infty)$ **31.** $(-\sqrt{3}, \sqrt{3})$

33. $(-\infty, 1]$ **35.** $(-1, 0) \cup (1, \infty)$

37. $(-\infty, 0) \cup (\frac{1}{4}, \infty)$

39. $10 \leq C \leq 35$ **41.** (a) $T = 20 - 10h$, $0 \leq h \leq 12$
(b) $-30°\text{C} \leq T \leq 20°\text{C}$ **43.** $\pm\frac{3}{2}$ **45.** $2, -\frac{4}{3}$
47. $(-3, 3)$ **49.** $(3, 5)$ **51.** $(-\infty, -7] \cup [-3, \infty)$
53. $[1.3, 1.7]$ **55.** $[-4, -1] \cup [1, 4]$
57. $x \geq (a + b)c/(ab)$ **59.** $x > (c - b)/a$

EXERCISES B ■ PAGE A15

1. 5 **3.** $\sqrt{74}$ **5.** $2\sqrt{37}$ **7.** 2 **9.** $-\frac{9}{2}$

17.

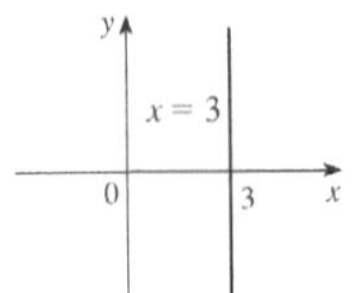

19.

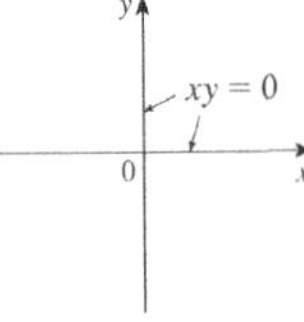

21. $y = 6x - 15$ **23.** $2x - 3y + 19 = 0$
25. $5x + y = 11$ **27.** $y = 3x - 2$ **29.** $y = 3x - 3$
31. $y = 5$ **33.** $x + 2y + 11 = 0$ **35.** $5x - 2y + 1 = 0$
37. $m = -\frac{1}{3}$, $b = 0$ **39.** $m = 0$, $b = -2$ **41.** $m = \frac{3}{4}$, $b = -3$

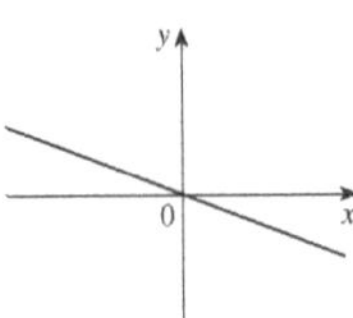

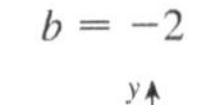

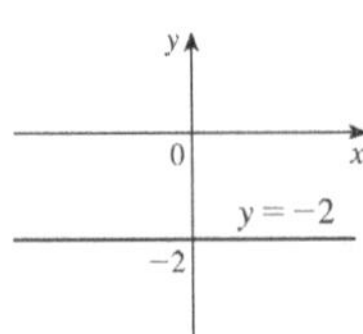

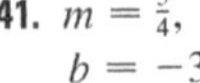

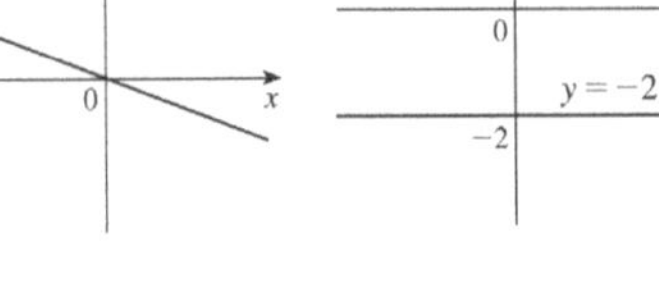

43.

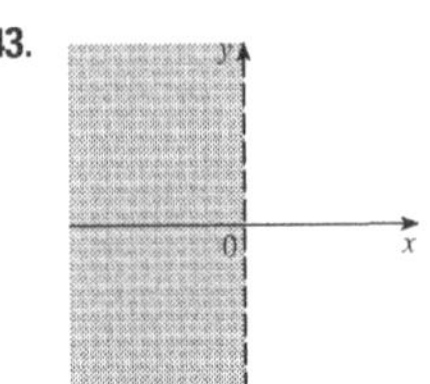

45.

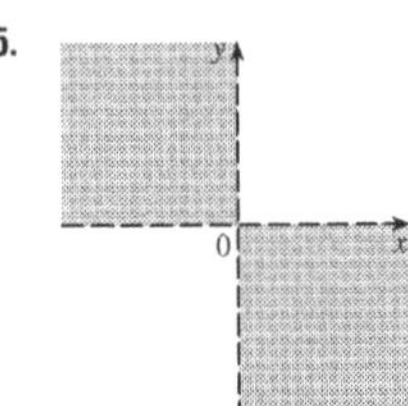

47.

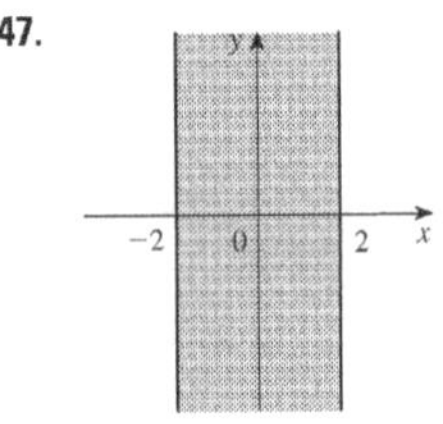

49.

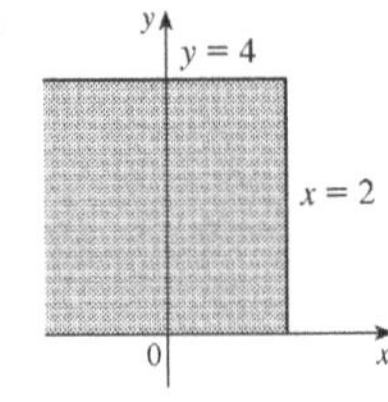

51.

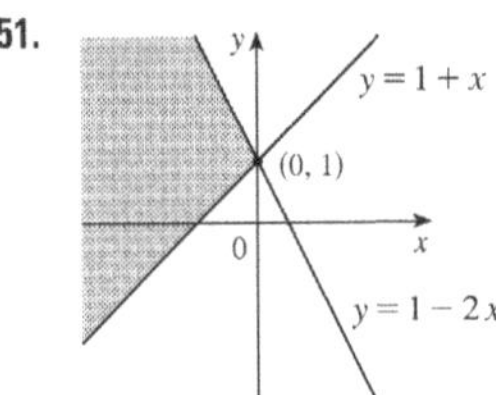

53. $(0, -4)$ **55.** (a) $(4, 9)$ (b) $(3.5, -3)$ **57.** $(1, -2)$
59. $y = x - 3$ **61.** (b) $4x - 3y - 24 = 0$

EXERCISES C ■ PAGE A23

1. $(x - 3)^2 + (y + 1)^2 = 25$ **3.** $x^2 + y^2 = 65$
5. $(2, -5), 4$ **7.** $(-\frac{1}{2}, 0), \frac{1}{2}$ **9.** $(\frac{1}{4}, -\frac{1}{4}), \sqrt{10}/4$

11. Parabola

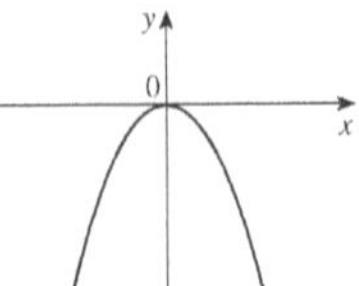

13. Ellipse

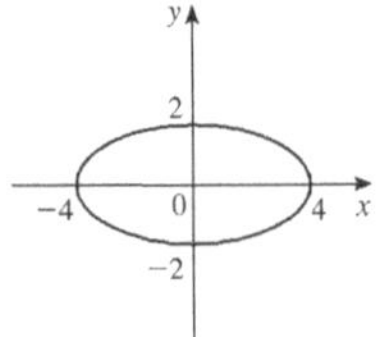

15. Hyperbola

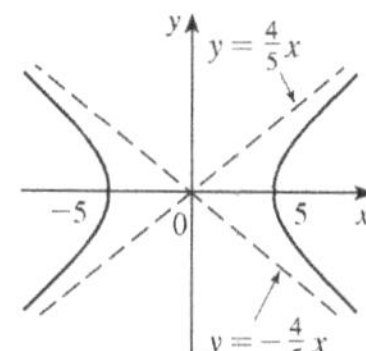

17. Ellipse

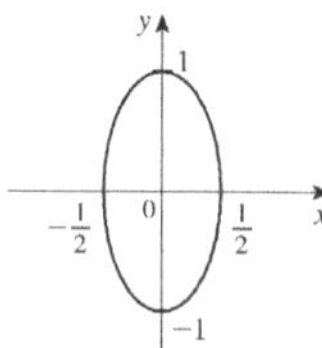

19. Parabola

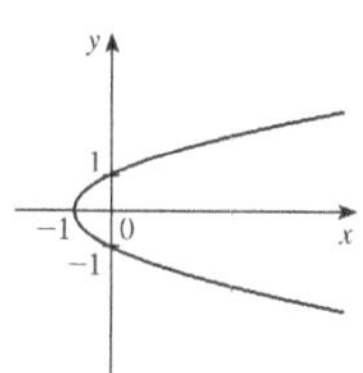

21. Hyperbola

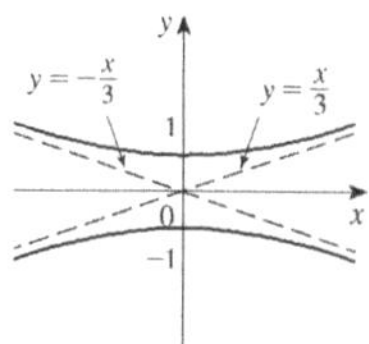

23. Hyperbola

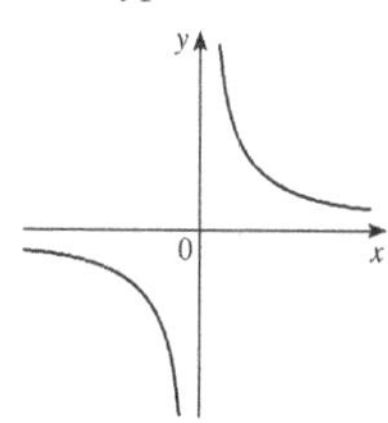

25. Ellipse

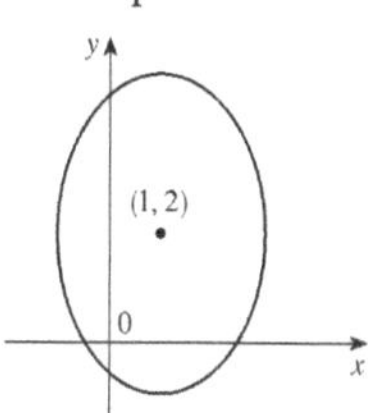

27. Parabola

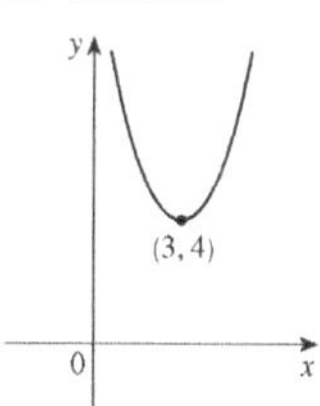

29. Parabola

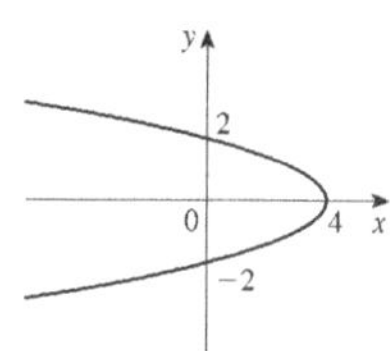

31. Ellipse

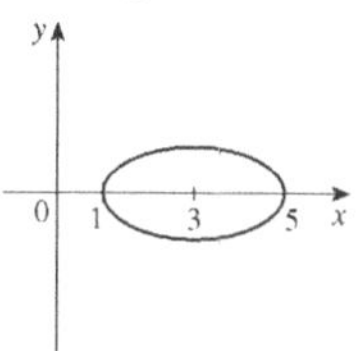

33.

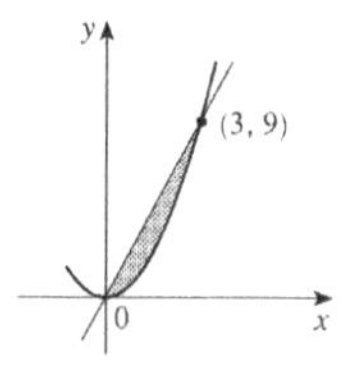

35. $y = x^2 - 2x$

37.

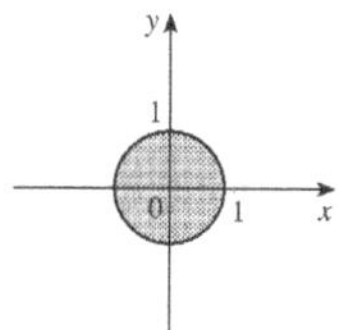

39.

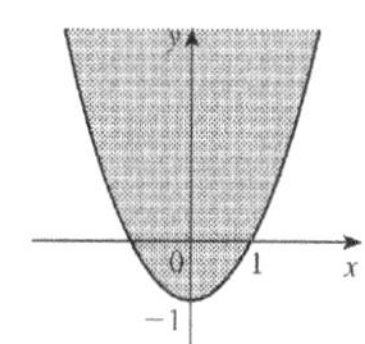

EXERCISES D ■ PAGE A32

1. $7\pi/6$ **3.** $\pi/20$ **5.** 5π **7.** $720°$ **9.** $75°$
11. $-67.5°$ **13.** 3π cm **15.** $\frac{2}{3}$ rad $= (120/\pi)°$

17.

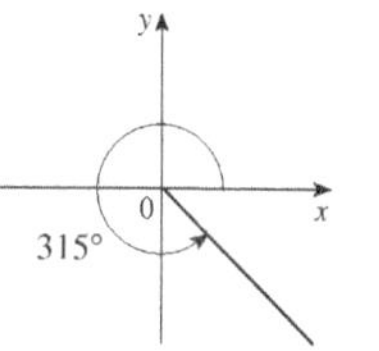

19.

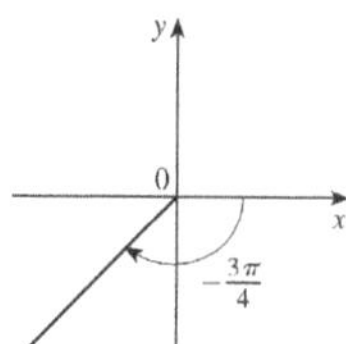

21.

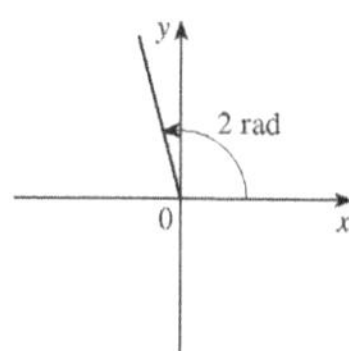

23. $\sin(3\pi/4) = 1/\sqrt{2}$, $\cos(3\pi/4) = -1/\sqrt{2}$, $\tan(3\pi/4) = -1$, $\csc(3\pi/4) = \sqrt{2}$, $\sec(3\pi/4) = -\sqrt{2}$, $\cot(3\pi/4) = -1$

25. $\sin(9\pi/2) = 1$, $\cos(9\pi/2) = 0$, $\csc(9\pi/2) = 1$, $\cot(9\pi/2) = 0$, $\tan(9\pi/2)$ and $\sec(9\pi/2)$ undefined

27. $\sin(5\pi/6) = \frac{1}{2}$, $\cos(5\pi/6) = -\sqrt{3}/2$, $\tan(5\pi/6) = -1/\sqrt{3}$, $\csc(5\pi/6) = 2$, $\sec(5\pi/6) = -2/\sqrt{3}$, $\cot(5\pi/6) = -\sqrt{3}$

29. $\cos\theta = \frac{4}{5}$, $\tan\theta = \frac{3}{4}$, $\csc\theta = \frac{5}{3}$, $\sec\theta = \frac{5}{4}$, $\cot\theta = \frac{4}{3}$

31. $\sin\phi = \sqrt{5}/3$, $\cos\phi = -\frac{2}{3}$, $\tan\phi = -\sqrt{5}/2$, $\csc\phi = 3/\sqrt{5}$, $\cot\phi = -2/\sqrt{5}$

33. $\sin\beta = -1/\sqrt{10}$, $\cos\beta = -3/\sqrt{10}$, $\tan\beta = \frac{1}{3}$, $\csc\beta = -\sqrt{10}$, $\sec\beta = -\sqrt{10}/3$

35. 5.73576 cm **37.** 24.62147 cm **59.** $\frac{1}{15}(4 + 6\sqrt{2})$

61. $\frac{1}{15}(3 + 8\sqrt{2})$ **63.** $\frac{24}{25}$ **65.** $\pi/3, 5\pi/3$

67. $\pi/4, 3\pi/4, 5\pi/4, 7\pi/4$ **69.** $\pi/6, \pi/2, 5\pi/6, 3\pi/2$

71. $0, \pi, 2\pi$ **73.** $0 \leqslant x \leqslant \pi/6$ and $5\pi/6 \leqslant x \leqslant 2\pi$

75. $0 \leqslant x < \pi/4$, $3\pi/4 < x < 5\pi/4$, $7\pi/4 < x \leqslant 2\pi$

77.

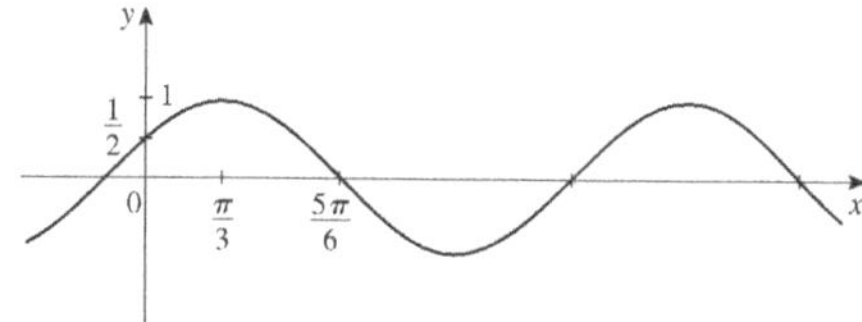

79.

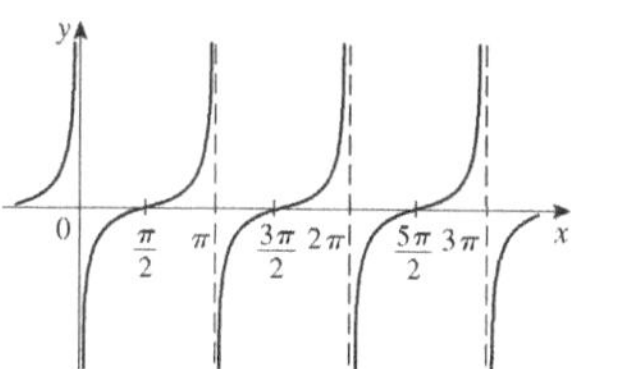

81.

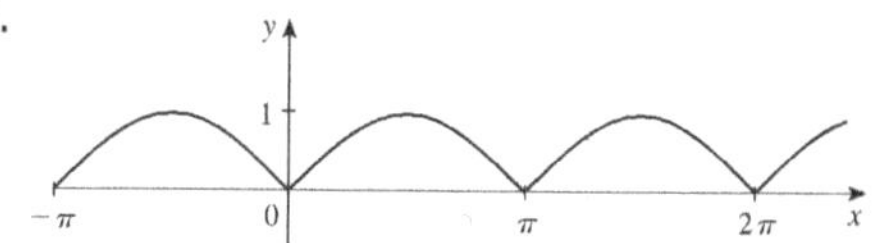

89. 14.34457 cm^2

EXERCISES E ■ PAGE A38

1. $\sqrt{1} + \sqrt{2} + \sqrt{3} + \sqrt{4} + \sqrt{5}$ **3.** $3^4 + 3^5 + 3^6$
5. $-1 + \frac{1}{3} + \frac{3}{5} + \frac{5}{7} + \frac{7}{9}$ **7.** $1^{10} + 2^{10} + 3^{10} + \cdots + n^{10}$
9. $1 - 1 + 1 - 1 + \cdots + (-1)^{n-1}$ **11.** $\sum_{i=1}^{10} i$
13. $\sum_{i=1}^{19} \frac{i}{i+1}$ **15.** $\sum_{i=1}^{n} 2i$ **17.** $\sum_{i=0}^{5} 2^i$ **19.** $\sum_{i=1}^{n} x^i$
21. 80 **23.** 3276 **25.** 0 **27.** 61 **29.** $n(n+1)$
31. $n(n^2 + 6n + 17)/3$ **33.** $n(n^2 + 6n + 11)/3$
35. $n(n^3 + 2n^2 - n - 10)/4$
41. (a) n^4 (b) $5^{100} - 1$ (c) $\frac{97}{300}$ (d) $a_n - a_0$
43. $\frac{1}{3}$ **45.** 14 **49.** $2^{n+1} + n^2 + n - 2$

EXERCISES G ■ PAGE A54

1. (b) 0.405

EXERCISES H ■ PAGE A64

1. $8 - 4i$ **3.** $13 + 18i$ **5.** $12 - 7i$ **7.** $\frac{11}{13} + \frac{10}{13}i$
9. $\frac{1}{2} - \frac{1}{2}i$ **11.** $-i$ **13.** $5i$ **15.** $12 + 5i$, 13
17. $4i$, 4 **19.** $\pm\frac{3}{2}i$ **21.** $-1 \pm 2i$
23. $-\frac{1}{2} \pm (\sqrt{7}/2)i$ **25.** $3\sqrt{2}\,[\cos(3\pi/4) + i\sin(3\pi/4)]$
27. $5\{\cos[\tan^{-1}(\frac{4}{3})] + i\sin[\tan^{-1}(\frac{4}{3})]\}$
29. $4[\cos(\pi/2) + i\sin(\pi/2)]$, $\cos(-\pi/6) + i\sin(-\pi/6)$, $\frac{1}{2}[\cos(-\pi/6) + i\sin(-\pi/6)]$
31. $4\sqrt{2}\,[\cos(7\pi/12) + i\sin(7\pi/12)]$, $(2\sqrt{2})[\cos(13\pi/12) + i\sin(13\pi/12)]$, $\frac{1}{4}[\cos(\pi/6) + i\sin(\pi/6)]$
33. -1024 **35.** $-512\sqrt{3} + 512i$
37. ± 1, $\pm i$, $(1/\sqrt{2})(\pm 1 \pm i)$ **39.** $\pm(\sqrt{3}/2) + \frac{1}{2}i$, $-i$

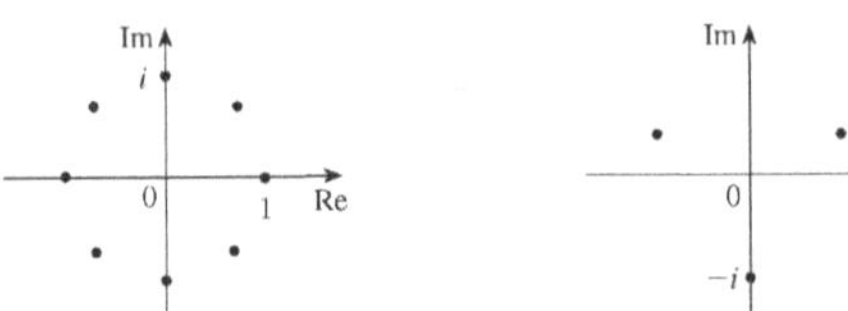

41. i **43.** $\frac{1}{2} + (\sqrt{3}/2)i$ **45.** $-e^2$
47. $\cos 3\theta = \cos^3\theta - 3\cos\theta\sin^2\theta$, $\sin 3\theta = 3\cos^2\theta\sin\theta - \sin^3\theta$

Index

RP denotes Reference Page numbers.

Abel, Niels, 212
absolute maximum and minimum values, 274, 946, 951
absolute value, 16, A6, A58
absolute value function, 16
absolutely convergent series, 732
acceleration as a rate of change, 161, 224
acceleration of a particle, 863
  components of, 866
  as a vector, 863
Achilles and the tortoise, 5
adaptive numerical integration, 515
addition formulas for sine and cosine, A29
addition of vectors, 792, 795
Airy, Sir George, 746
Airy function, 746
algebraic function, 30
alternating harmonic series, 729, 732
alternating series, 727
Alternating Series Estimation Theorem, 730
Alternating Series Test, 727
analytic geometry, A10
angle(s), A24
  between curves, 271
  of deviation, 283
  negative or positive, A25
  between planes, 821
  standard position, A25
  between vectors, 801, 802
angular momentum, 871
angular speed, 864
antiderivative, 344
antidifferentiation formulas, 345
aphelion, 683
apolune, 677
approach path of an aircraft, 208
approximate integration, 506
approximating cylinder, 432
approximating surface, 546
approximation
  by differentials, 253
  to *e*, 180
  linear, 251, 917, 921
  linear, to a tangent plane, 917
  by the Midpoint Rule, 378, 507
  by Newton's method, 339
  by an *n*th-degree Taylor polynomial, 257
  quadratic, 256
  by Riemann sums, 372
  by Simpson's Rule, 511, 513
  tangent line, 251
  by Taylor polynomials, 768
  by Taylor's Inequality, 756, 769
  by the Trapezoidal Rule, 508
Archimedes, 406
Archimedes' Principle, 460, 1134
arc curvature, 853
arc length, 538, 854
  of a parametric curve, 648
  of a polar curve, 667
  of a space curve, 853, 854
arc length contest, 545
arc length formula, 539
arc length function, 541
arcsine function, 67
area, 2, 360
  of a circle, 480
  under a curve, 360, 365, 371
  between curves, 422, 423
  of an ellipse, 479
  by exhaustion, 2, 101
  by Green's Theorem, 1087
  enclosed by a parametric curve, 647
  in polar coordinates, 654, 665
  of a sector of a circle, 665
  surface, 650, 1014, 1104, 1106
  of a surface of a revolution, 545, 551
area function, 385
area problem, 2, 360
argument of a complex number, A59
arithmetic-geometric mean, 702
arrow diagram, 11
astroid, 215, 645
asymptote(s), 311
  in graphing, 311
  horizontal, 131, 311
  of a hyperbola, 674, A20
  slant, 312, 315
  vertical, 94, 311
asymptotic curve, 318
autonomous differential equation, 589
auxiliary equation, 1143
  complex roots of, 1145
  real roots of, 1144
average cost function, 334
average rate of change, 148, 224, 862
average speed of molecules, 528
average value of a function, 451, 452, 570, 979, 1027
average velocity, 4, 84, 145, 224
axes, coordinate, 786, A11
axes of an ellipse, A19
axis of a parabola, 670

bacterial growth, 605, 610
Barrow, Isaac, 3, 101, 153, 386, 406
base of a cylinder, 430
base of a logarithm, 62, A55
  change of, 65
baseball and calculus, 455
basis vectors, 796
Bernoulli, James, 594, 621
Bernoulli, John, 303, 310, 594, 640, 754
Bernoulli differential equation, 621
Bessel, Friedrich, 742
Bessel function, 217, 742, 746
Bézier, Pierre, 653
Bézier curves, 639, 653
binomial coefficients, 760
binomial series, 760
  discovery by Newton, 767
binomial theorem, 175, RP1
binormal vector, 858
blackbody radiation, 777
blood flow, 230, 336, 564
boundary curve, 1122
boundary-value problem, 1147
bounded sequence, 697
bounded set, 951
Boyle's Law, 235
brachistochrone problem, 640
Brahe, Tycho, 867
branches of a hyperbola, 674, A20
Buffon's needle problem, 578
bullet-nose curve, 51, 205

C^1 transformation, 1040
cable (hanging), 258
calculator, graphing, 44, 318, 638, 661. *See also* computer algebra system
calculus, 8
  differential, 3
  integral, 2, 3
  invention of, 8, 406
cancellation equations
  for inverse functions, 61
  for inverse trigonometric functions, 61, 67
  for logarithms, 63
cans, minimizing manufacturing cost of, 337
Cantor, Georg, 713
Cantor set, 713
capital formation, 567
cardiac output, 565
cardioid, 215, 658
carrying capacity, 236, 294, 581, 607
Cartesian coordinate system, A11
Cartesian plane, A11
Cassini, Giovanni, 665
CAS. *See* computer algebra system
catenary, 258
Cauchy, Augustin-Louis, 113, 984, A45
Cauchy-Schwarz Inequality, 807
Cauchy's Mean Value Theorem, A45
Cavalieri, 513
Cavalieri's Principle, 440
center of gravity. *See* center of mass
center of mass, 554, 1004, 1065
  of a lamina, 1005
  of a plate, 557
  of a solid, 1023
  of a surface, 1112
  of a wire, 1065
centripetal force, 875
centroid of a plane region, 556
centroid of a solid, 1023
Chain Rule, 198, 199, 201
  for several variables, 924, 926, 927
change of base, formula for, 65
change of variable(s)
  in a double integral, 999, 1041, 1044
  in integration, 407
  in a triple integral, 1029, 1034, 1046
characteristic equation, 1143
charge, electric, 227, 1003, 1004, 1023, 1160
charge density, 1004, 1023
chemical reaction, 227
circle, area of, 480
circle, equation of, A16
circle of curvature, 859
circular cylinder, 430
circular paraboloid, 832
circulation of a vector field, 1126
cissoid of Diocles, 644, 663
Clairaut, Alexis, 907
Clairaut's Theorem, 907, A48
clipping planes, 826
closed curve, 1077
closed interval, A3
Closed Interval Method, 278
  for a function of two variables, 952
closed set, 951
closed surface, 1116
Cobb, Charles, 879
Cobb-Douglas production function, 880, 910, 963
cochleoid, 686
coefficient(s)
  binomial, 760
  of friction, 198, 281
  of inequality, 429
  of a polynomial, 27
  of a power series, 741
  of static friction, 837
combinations of functions, 39
comets, orbits of, 684
common ratio, 705
comparison properties of the integral, 381
comparison test for improper integrals, 525
Comparison Test for series, 722
Comparison Theorem for integrals, 525
complementary equation, 1149
Completeness Axiom, 698
complex conjugate, A57
complex exponentials, A63
complex number(s), A57
  addition and subtraction of, A57
  argument of, A59
  division of, A57, A60
  equality of, A57
  imaginary part of, A57
  modulus of, A58
  multiplication of, A57, A60
  polar form, A59
  powers of, A61
  principal square root of, A58
  real part of, A57
  roots of, A62
component function, 840, 1057
components of acceleration, 866
components of a vector, 793, 804
composition of functions, 40, 199
  continuity of, 125, 898
  derivative of, 200
compound interest, 241, 309
compressibility, 228
computer algebra system, 90, 502, 638
  for integration, 502, 751
  pitfalls of using, 90
computer algebra system, graphing with, 44
  a curve, 318
  function of two variables, 882
  level curves, 886
  parametric equations, 638
  parametric surface, 1102
  partial derivatives, 907
  polar curve, 661
  sequence, 695
  space curve, 843
  vector field, 1058
concavity, 293
Concavity Test, 293, A44
concentration, 227
conchoid, 641, 663
conditionally convergent series, 733
conductivity (of a substance), 1120
cone, 670, 830
  parametrization of, 1102
conic section, 670, 678
  directrix, 670, 678
  eccentricity, 678
  focus, 670, 672, 678
  polar equation, 680
  shifted, 675, A21
  vertex (vertices), 670
conjugates, properties of, A58
connected region, 1077
conservation of energy, 1081
conservative vector field, 1061, 1082
constant force, 446, 805
constant function, 174
Constant Multiple Law of limits, 99
Constant Multiple Rule, 177
constraint, 957, 961
consumer surplus, 563, 564
continued fraction expansion, 702
continuity
  of a function, 118, 841
  of a function of three variables, 898
  of a function of two variables, 896
  on an interval, 120
  from the left or right, 120
  continuous compounding of interest, 241, 309
continuous random variable, 568
contour curves, 883
contour map, 883, 909
convergence
  absolute, 732
  conditional, 733
  of an improper integral, 520, 523
  interval of, 743
  radius of, 743
  of a sequence, 692
  of a series, 705
convergent improper integral, 520, 523
convergent sequence, 692
convergent series, 705
  properties of, 709
conversion, cylindrical to rectangular coordinates, 1028
cooling tower, hyperbolic, 832
coordinate axes, 786, A11
coordinate planes, 786
coordinate system, A2
  Cartesian, A11
  cylindrical, 1028

polar, 654
retangular, A11
spherical, 1033
three-dimensional rectangular, 786
coplanar vectors, 813
Coriolis acceleration, 874
Cornu's spiral, 652
cosine function, A26
derivative of, 193
graph of, 31, A31
power series for, 758, 760
cost function, 231, 330
critical number, 277
critical point(s), 946, 956
critically damped vibration, 1158
cross product, 808
direction of, 810
geometric characterization of, 811
magnitude of, 811
properties of, 812
cross-section, 430
of a surface, 827
cubic function, 27
curl of a vector field, 1091
current, 227
curvature, 653, 855
curve(s)
asymptotic, 318
Bézier, 639, 653
boundary, 1122
bullet-nose, 51, 205
cissoid of Diocles, 663
closed, 1077
Cornu's spiral, 652
demand, 563
devil's, 215
dog saddle, 891
epicycloid, 645
equipotential, 890
grid, 1100
helix, 841
length of, 538, 853
level, 883
monkey saddle, 891
orientation of, 1068, 1084
orthogonal, 216
ovals of Cassini, 665
parametric, 636, 841
piecewise-smooth,1064
polar, 656
serpentine, 189
simple, 1078
smooth, 538
space, 840, 841
strophoid, 669, 687
swallotail catastrophe, 644
toroidal spiral, 843
trochoid, 643
twisted cubic, 843
witch of Maria Agnesi, 643
curve fitting, 25
curve-sketching procedure, 311
cusp, 641
cycloid, 639
cylinder, 430, 827
parabolic, 827
parametrization of, 1102
cylindrical coordinate system, 1028
conversion equations for, 1028
triple integrals in, 1029
cylindrical coordinates, 1030
cylindrical shell, 441

damped vibration, 1157
damping constant, 1157
decay, law of natural, 237
decay, radioactive, 239
decreasing function, 19
decreasing sequence, 696
definite integral, 372, 974
properties of, 379
Substitution Rule for, 411
of a vector function, 851
definite integration
by parts, 464, 466, 467
by substitution, 411
degree of a polynomial, 27
del (∇), 936
delta (Δ) notation, 147, 148
demand curve, 331, 563
demand function, 330, 563
De Moivre, Abraham, A61
De Moivre's Theorem, A61
density
of a lamina, 1003
linear, 226, 401
liquid, 553
mass vs. weight, 553
of a solid, 1023
dependent variable, 10, 878, 926
derivative(s), 143, 146, 154, 256
of a composite function, 199
of a constant function, 174
directional, 933, 934, 937
domain of, 154
of exponential functions, 180, 203, A54, A55
as a function, 154
higher, 160
higher partial, 906
of hyperbolic functions, 259
of an integral, 387
of an inverse function, 218
of inverse trigonometric functions, 213, 214
left-hand, 165
of logarithmic functions, 218, A51, A54
normal, 1098
notation, 157
notation for partial, 903
partial, 902
of a polynomial, 174
of a power function, 175
of a power series, 748
of a product, 184, 185
of a quotient, 187
as a rate of change, 143
right-hand, 165
second, 160, 850
second directional, 944
second partial, 906
as the slope of a tangent, 143, 148
third, 161
of trigonometric functions, 191, 194
of a vector function, 847
Descartes, René, A11
descent of aircraft, determining start of, 208
determinant, 808
devil's curve, 215
Difference Law of limits, 99
difference quotient, 12
Difference Rule, 178
differentiable function, 157, 918
differential, 253, 919, 921
differential calculus, 3
differential equation, 183, 237, 346, 579, 580, 582
autonomous, 589
Bernoulli, 621
family of solutions, 580, 583
first-order, 582
general solution of, 583
homogeneous, 1142
linear, 616
linearly independent solutions, 1143
logistic, 607, 703
nonhomogeneous, 1142, 1149
order of, 582
partial, 908
second-order, 582, 1142
separable, 594
solution of, 582
differentiation, 157
formulas for, 188, RP5
formulas for vector functions, 850
implicit, 209, 210, 905, 928
logarithmic, 220
partial, 900, 905, 906
of a power series, 748
term-by-term, 748
of a vector function, 850
differentiation operator, 157
Direct Substitution Property, 101
directed line segment, 791
direction field, 585, 586
direction numbers, 818
directional derivative, 933, 934, 937
maximum value of, 938
of a temperature function, 933, 934
second, 944
directrix, 670, 678

discontinuity, 119, 120
discontinuous function, 119
discontinuous integrand, 523
disk method for approximating volume, 432
dispersion, 283
displacement, 145, 401
displacement vector, 791, 805
distance
  between lines, 823
  between planes, 823
  between point and line in space, 815
  between point and plane, 815
  between points in a plane, A11
  between points in space, 788
  between real numbers, A7
distance formula, A12
  in three dimensions, 788
distance problem, 367
divergence
  of an improper integral, 520, 523
  of an infinite series, 705
  of a sequence, 692
  of a vector field, 1094
Divergence, Test for, 709
Divergence Theorem, 1129
divergent improper integral, 520, 523
divergent sequence, 692
divergent series, 705
division of power series, 763
DNA, helical shape of, 842
dog saddle, 891
domain of a function, 10, 878
domain sketching, 878
Doppler effect, 932
dot product, 800
  in component form, 800
  properties of, 801
double-angle formulas, A29
double integral, 974, 976
  change of variable in, 1041, 1044
  over general regions, 988, 989
  Midpoint Rule for, 978
  in polar coordinates, 997, 998, 999
  properties of, 981, 993
  over rectangles, 974
double Riemann sum, 977
Douglas, Paul, 879
Dumpster design, minimizing cost of, 956
dye dilution method, 565

e (the number), 55, 180, A52
  as a limit, 222
  as a sum of an infinite series, 757
eccentricity, 678
electric charge, 1003, 1004, 1023
electric circuit, 593, 596, 619
    analysis of, 1160
electric current to a flash bulb, 83, 207
electric field (force per unit charge), 1060
electric flux, 1119
electric force, 1060
elementary function, integrability of, 498
element of a set, A3
ellipse, 215, 672, 678, A19
  area, 479
  directrix, 678
  eccentricity, 678
  foci, 672, 678
  major axis, 672, 683
  minor axis, 672
  polar equation, 680, 683
  reflection property, 673
  rotated, 217
  vertices, 672
ellipsoid, 828, 830
elliptic paraboloid, 828, 830
empirical model, 25
end behavior of a function, 142
endpoint extreme values, 275
energy
  conservation of, 1081
  kinetic, 455, 1081
  potential, 1081
epicycloid, 645
epitrochoid, 652
equation(s)
  cancellation, 61
  of a circle, A17
  differential (*see* differential equation)
  of an ellipse, 672, 680, A19
  of a graph, A16, A17
  heat conduction, 913
  of a hyperbola, 67, 675, 680, A20
  Laplace's, 908, 1095
  of a line, A12, A13, A14, A16
  of a line in space, 816, 817
  of a line through two points, 818
  linear, 820, A14
  logistic difference, 703
  logistic differential, 581, 615
  Lotka-Volterra, 623
  *n*th-degree, 212
  of a parabola, 670, 680, A18
  parametric, 636, 817, 841, 1099
  of a plane, 819
  of a plane through three points, 821
  point-slope, A12
  polar, 656, 680
  predator-prey, 622, 623
  second-degree, A16
  slope-intercept, A13
  of a space curve, 841
  of a sphere, 789
  symmetric, 818
  two-intercept form, A16
  van der Waals, 914
  vector, 816
  wave, 908
equilateral hyperbola, A21
equilibrium point, 624
equilibrium solution, 581, 623
equipotential curves, 890
equivalent vectors, 792
error
  in approximate integration, 508, 509
  percentage, 254
  relative, 254
  in Taylor approximation, 769
error bounds, 510, 514
error estimate
  for alternating series, 730
  for the Midpoint Rule, 508, 509
  for Simpson's Rule, 514
  for the Trapezoidal Rule, 508, 509
error function, 395
escape velocity, 528
estimate of the sum of a series, 718, 725,
    730, 735
Euclid, 101
Eudoxus, 2, 101, 406
Euler, Leonhard, 55, 589, 715, 721, 757
Euler's formula, A63
Euler's Method, 589, 590
even function, 17, 311
expected values, 1011
exponential decay, 237
exponential function(s), 32, 51, 179, RP4
  with base *a*, A55
  derivative, of 180, 203, A55
  graphs of, 52, 180
  integration of, 377, 408, 762, 763
  limits of, 135, A53
  power series for, 755
  properties of, A53
exponential graph, 52
exponential growth, 237, 610
exponents, laws of, 53, A53, A55
extrapolation, 26
extreme value, 274
Extreme Value Theorem, 275, 951

family
  of epicycloids and hypocycloids, 644
  of exponential functions, 52
  of functions, 49, 322, 323
  of parametric curves, 640
  of solutions, 580, 583
fat circles, 213, 544
Fermat, Pierre, 3, 153, 276, 406, A11
Fermat's Principle, 335
Fermat's Theorem, 276
Fibonacci, 691, 702
Fibonacci sequence, 691, 702
field
  conservative, 1061
  electric, 1060
  force, 1060
  gradient, 942, 1060
  gravitational, 1060
  incompressible, 1095

irrotational, 1094
scalar, 1057
vector, 1056, 1057
velocity, 1056, 1059
First Derivative Test, 291
for Absolute Extreme Values, 328
first octant, 786
first-order linear differential equation, 582, 616
first-order optics, 774
fixed point of a function, 171, 289
flash bulb, current to, 83
flow lines, 1062
fluid flow, 1059, 1095, 1118
flux, 564, 565, 1117, 1119
flux integral, 1117
FM synthesis, 322
foci, 672
focus, 670, 678
of a conic section, 678
of an ellipse, 672, 678
of a hyperbola, 673
of a parabola, 670
folium of Descartes, 209, 687
force, 446
centripetal, 875
constant, 446, 805
exerted by fluid, 552, 553
resultant, 797
torque, 813
force field, 1056, 1060
forced vibrations, 1159
Fourier, Joseph, 233
Fourier series, finite, 478
four-leaved rose, 658
fractions (partial), 484, 485
Frenet-Serret formulas, 862
Fresnel, Augustin, 389
Fresnel function, 389
frustum, 439, 440
Fubini, Guido, 984
Fubini's Theorem, 984, 1017
function(s), 10, 878
absolute value, 16
Airy, 746
algebraic, 30
arc length, 541, 853
arcsine, 67
area, 385
arrow diagram of, 11
average cost, 334
average value of, 452, 570, 979, 1027
Bessel, 217, 742, 746
Cobb-Douglas production, 880, 910, 963
combinations of, 39
component, 840, 1057
composite, 40, 199, 898
constant, 174
continuity of, 118, 896, 898
continuous, 841
cost, 230, 231
cubic, 27
decreasing, 19
demand, 330, 563
derivative of, 146
differentiability of, 157, 918
discontinuous, 119
domain of, 10, 878
elementary, 498
error, 395
even, 17, 311
exponential, 32, 51, 179
extreme values of, 274
family of, 49, 322, 323
fixed point of, 171, 289
Fresnel, 389
Gompertz, 612, 615
gradient of, 936, 938
graph of, 11, 880
greatest integer, 105
harmonic, 908
Heaviside, 44, 91
homogeneous, 932
hyperbolic, 257
implicit, 209
increasing, 19
integrable, 976
inverse, 58, 60
inverse cosine, 68
inverse hyperbolic, 260
inverse sine, 67
inverse tangent, 68
inverse trigonometric, 67, 68
joint density, 1008, 1023
limit of, 87, 109, 893, 898
linear, 23, 881
logarithmic, 32, 62, A50, A55
machine diagram of, 11
marginal cost, 148, 232, 330, 401
marginal profit, 331
marginal revenue, 331
maximum and minimum values of, 274, 946
natural exponential, 56
natural logarithmic, 64
nondifferentiable, 159
of n variables, 887
odd, 18, 311
one-to-one, 59
periodic, 311
piecewise defined, 16
polynomial, 27, 897
position, 145
potential, 1061
power, 28, 174
probability density, 568, 1008
profit, 331
quadratic, 27
ramp, 44
range of, 10, 878
rational, 30, 484, 897
reciprocal, 30
reflected, 36
representation as a power series, 746
representations of, 10, 12
revenue, 331
root, 29
of several variables, 878, 886
shifted, 36
sine integral, 396
smooth, 538
step, 17
stretched, 36
tabular, 13
of three variables, 886
transformation of, 36
translation of, 36
trigonometric, 31, A26
of two variables, 878
value of, 10, 11
vector, 840
Fundamental Theorem of Calculus, 386, 388, 393
higher-dimensional versions, 1135
for line integrals, 1075
for vector functions, 851

G (gravitational constant), 234, 451
Gabriel's horn, 550
Galileo, 640, 647, 670
Galois, Evariste, 212
Gause, G. F., 610
Gauss, Karl Friedrich, 1129, A35
Gaussian optics, 774
Gauss's Law, 1119
Gauss's Theorem, 1129
geometric series, 705
geometry of a tetrahedron, 816
Gibbs, Joseph Willard, 797
Gini, Corrado, 429
Gini coefficient, 429
Gini index, 429
global maximum and minimum, 274
Gompertz function, 612, 615
gradient, 936, 938
gradient vector, 936, 938
interpretations of, 942
gradient vector field, 942, 1060
graph(s)
of an equation, A16, A17
of equations in three dimensions, 787
of exponential functions, 52, 180, RP4
of a function, 11
of a function of two variables, 880
of logarithmic functions, 63, 66
of a parametric curve, 636
of a parametric surface, 1112
polar, 656, 661
of power functions, 29, RP3
of a sequence, 695
of trigonometric functions, 31, A30, RP2

graphing calculator, 44, 318, 638, 661
graphing device. *See* computer algebra system
gravitation law, 234, 451
gravitational acceleration, 446
gravitational field, 1060
great circle, 1039
greatest integer function, 105
Green, George, 1085, 1128
Green's identities, 1098
Green's Theorem, 1084, 1128
  vector forms, 1096
Gregory, James, 199, 475, 513, 750, 754
Gregory's series, 750
grid curves, 1100
growth, law of natural, 237, 606
growth rate, 229, 401
  relative, 237, 606

half-angle formulas, A29
half-life, 239
half-space, 887
hare-lynx system, 626
harmonic function, 908
harmonic series, 708, 717
  alternating, 729
heat conduction equation, 913
heat conductivity, 1120
heat flow, 1119
heat index, 900
Heaviside, Oliver, 91
Heaviside function, 44, 91
Hecht, Eugene, 253, 256, 773
helix, 841
hidden line rendering, 826
higher derivatives, 160
higher partial derivatives, 906
homogeneous differential equation, 1142
homogeneous function, 932
Hooke's Law, 447, 1156
horizontal asymptote, 131, 311
horizontal line, equation of, A13
Horizontal Line Test, 59
horizontal plane, 787
Hubble Space Telescope, 279
Huygens, Christiaan, 640
hydrostatic pressure and force, 552, 553
hydro-turbine optimization, 966
hyperbola, 215, 673, 678, A20
  asymptotes, 674, A20
  branches, 674, A20
  directrix, 678
  eccentricity, 678
  equation, 674, 675, 680, A20
  equilateral, A21
  foci, 673, 678
  polar equation, 680
  reflection property, 678
  vertices, 674
hyperbolic function(s), 257
  derivatives of, 259
  inverse, 260
hyperbolic identities, 258
hyperbolic paraboloid, 829, 830
hyperbolic substitution, 481, 482
hyperboloid, 830
hypersphere, 1027
hypocycloid, 644

i (imaginary number), A57
i (standard basis vector), 796
I/D Test, 290
ideal gas law, 236, 914
image of a point, 1041
image of a region, 1041
implicit differentiation, 209, 210, 905, 928
implicit function, 209, 210
Implicit Function Theorem, 929, 930
improper integral, 519
  convergence or divergence of, 520, 523
impulse of a force, 455
incompressible velocity field, 1095
increasing function, 19
increasing sequence, 696
Increasing/Decreasing Test, 290
increment, 147, 921
indefinite integral(s), 397
  table of, 398
independence of path, 1076
independent random variable, 1010
independent variable, 10, 878, 926
indeterminate difference, 305
indeterminate forms of limits, 301
indeterminate power, 306
indeterminate product, 305
index of summation, A34
inequalities, rules for, A4
inertia (moment of), 1006, 1023, 1074
infinite discontinuity, 120
infinite interval, 519, 520
infinite limit, 93, 115, 136
infinite sequence. *See* sequence
infinite series. *See* series
inflection point, 294
initial condition, 583
initial point
  of a parametric curve, 637
  of a vector, 791, 1146
initial-value problem, 583
inner product, 800
instantaneous rate of change, 85, 148, 224
instantaneous rate of growth, 229
instantaneous rate of reaction, 228
instantaneous velocity, 85, 145, 224
integer, A2
integrable function, 976
integral(s)
  approximations to, 378
  change of variables in, 407, 999, 1040, 1044, 1046
  comparison properties of, 381
  conversion to cylindrical coordinates, 1029
  conversion to polar coordinates, 998
  conversion to spherical coordinates, 1034
  definite, 371, 974
  derivative of, 388
  double (*see* double integral)
  evaluating, 374
  improper, 519
  indefinite, 397
  iterated, 982, 983
  line (*see* line integral)
  patterns in, 505
  properties of, 379
  surface, 1110, 1117
  of symmetric functions, 412
  table of, 463, 495, 500, RP6–10
  triple, 1017, 1018
  units for, 403
integral calculus, 2, 3
Integral Test, 716
integrand, 372
  discontinuous, 523
integration, 372
  approximate, 506
  by computer algebra system, 502
  of exponential functions, 377, 408
  formulas, 463, 495, RP6–10
  indefinite, 397
  limits of, 372
  numerical, 506
  partial, 983
  by partial fractions, 484
  by parts, 464, 465, 466
  of a power series, 748
  of rational functions, 484
  by a rationalizing substitution, 492
  reversing order of, 985, 993
  over a solid, 1030
  substitution in, 407
  tables, use of, 500
  term-by-term, 748
  of a vector function, 847
intercepts, 311, A19
interest compunded continuously, 241
Intermediate Value Theorem, 126
intermediate variable, 926
interpolation, 26
intersection
  of planes, 821
  of polar graphs, area of, 666
  of sets, A3
  of three cylinders, 1032
interval, A3
interval of convergence, 743
inverse cosine function, 68
inverse function(s), 58, 60

inverse sine function, 67
inverse square laws, 35
inverse tangent function, 68
inverse transformation, 1041
inverse trigonometric functions, 67, 68
irrational number, A2
irrotational vector field, 1094
isothermal, 883, 890
isothermal compressibility, 228
iterated integral, 982, 983

j (standard basis vector), 796
Jacobi, Carl, 1043
Jacobian of a transformation, 1043, 1046
jerk, 161
joint density function, 1008, 1023
joule, 446
jump discontinuity, 120

k (standard basis vector), 796
kampyle of Eudoxus, 215
Kepler, Johannes, 682, 867
Kepler's Laws, 682, 867, 868, 872
kinetic energy, 455, 1081
Kirchhoff's Laws, 587, 1160
Kondo, Shigeru, 757

Lagrange, Joseph-Louis, 285, 286, 958
Lagrange multiplier, 957, 958
lamina, 556, 1003, 1005
Laplace, Pierre, 908, 1095
Laplace operator, 1095
Laplace's equation, 908, 1095
lattice point, 272
law of conservation of angular momentum, 871
Law of Conservation of Energy, 1082
law of cosines, A33
law of gravitation, 451
law of laminar flow, 230, 564
law of natural growth or decay, 237
laws of exponents, 53
laws of logarithms, 63
learning curve, 585
least squares method, 26, 955
least upper bound, 698
left-hand derivative, 165
left-hand limit, 92, 113
Leibniz, Gottfried Wilhelm, 3, 157, 386, 406, 594, 767
Leibniz notation, 157
lemniscate, 215
length
  of a curve, 538
  of a line segment, A7, A12
  of a parametric curve, 648
  of a polar curve, 667
  of a space curve, 853
  of a vector, 794
level curve(s), 883, 886
level surface, 887
  tangent plane to, 940
l'Hospital, Marquis de, 303, 310
l'Hospital's Rule, 302, 310, A45
  origins of, 310
libration point, 343
limaçon, 662
limit(s), 2, 87
  calculating, 99
  e (the number) as, 222
  of exponential functions, 135
  of a function, 87, 110
  of a function of three variables, 898
  of a function of two variables, 893
  infinite, 93, 115, 136
  at infinity, 130, 131, 136
  of integration, 372
  left-hand, 92, 113
  of logarithmic functions, 95, A50
  one-sided, 92, 113
  precise definitions, 108, 113, 116, 137, 140
  properties of, 99
  right-hand, 92, 113
  of a sequence, 5, 362, 692
  involving sine and cosine functions, 191, 192, 193
  of a trigonometric function, 193
  of a vector function, 840
Limit Comparison Test, 724
Limit Laws, 99, A39
  for functions of two variables, 896
  for sequences, 693
linear approximation, 251, 917, 921
linear combination, 1142
linear density, 226, 401
linear differential equation, 616, 1142
linear equation, A14
  of a plane, 820
linear function, 23, 881
linearity of an integral, 981
linearization, 251, 917
linearly independent solutions, 1143
linear model, 23
linear regression, 26
line(s) in the plane, 82, A12
  equation of, A12, A13, A14
  equation of, through two points, 818
  horizontal, A13
  normal, 176
  parallel, A14
  perpendicular, A14
  secant, 82, 83
  slope of, A12
  tangent, 82, 83, 144
line(s) in space
  normal, 941
  parametric equations of, 817
  skew, 819
  symmetric equations of, 818
  tangent, 848
  vector equation of, 816, 817
line integral, 1063
  Fundamental Theorem for, 1075
  for a plane curve, 1063
  with respect to arc length, 1066
  for a space curve, 1068
  work defined as, 1070
  of vector fields, 1070, 1071
liquid force, 552, 553
Lissajous figure, 638, 644
lithotripsy, 673
local maximum and minimum values, 274, 946
logarithm(s), 32, 62
  laws of, 63, A51
  natural, 64, A50
  notation for, 64
logarithmic differentiation, 220
logarithmic function(s), 32, 62
  with base *a*, 62, A55
  derivatives of, 218, A55
  graphs of, 63, 66
  limits of, 95, A52
  properties of, 63, 64, A51
logistic difference equation, 703
logistic differential equation, 581, 607
logistic model, 581, 606
logistic sequence, 703
LORAN system, 677
Lorenz curve, 429
Lotka-Volterra equations, 623

machine diagram of a function, 11
Maclaurin, Colin, 754
Maclaurin series, 753, 754
  table of, 761
magnitude of a vector, 794
major axis of ellipse, 672
marginal cost function, 148, 232, 330, 401
marginal productivity, 910
marginal profit function, 331
marginal propensity to consume or save, 712
marginal revenue function, 331
mass
  of a lamina, 1003
  of a solid, 1023
  of a surface, 1112
  of a wire, 1065
mass, center of. *See* center of mass
mathematical induction, 76, 79, 699
  principle of, 76, 79, A36
mathematical model. *See* model(s), mathematical
maximum and minimum values, 274, 946
mean life of an atom, 528
mean of a probability density function, 570
Mean Value Theorem, 284, 285
  for double integrals, 1052
  for integrals, 452

mean waiting time, 570
median of a probability density
function, 572
method of cylindrical shells, 441
method of exhaustion, 2, 101
method of Lagrange multipliers, 957, 958, 961
method of least squares, 26, 955
method of undetermined coefficients, 1149,
1153
midpoint formula, A16
Midpoint Rule, 378, 508
for double integrals, 978
error in using, 508
for triple integrals, 1025
minor axis of ellipse, 672
mixing problems, 598
Möbius, August, 1115
Möbius strip, 1109, 1115
model(s), mathematical, 13, 23
Cobb-Douglas, for production costs, 880,
910, 963
comparison of natural growth vs.
logistic, 610
of electric current, 587
empirical, 25
exponential, 32, 54
Gompertz function, 612, 615
linear, 23
logarithmic, 32
polynomial, 28
for population growth, 237, 580, 612
power function, 28
predator-prey, 622
rational function, 30
seasonal-growth, 615
trigonometric, 31, 32
for vibration of membrane, 742
von Bertalanffy, 631
modeling
with differential equations, 580
motion of a spring, 582
population growth, 54, 237, 580, 606,
612, 630
modulus, A58
moment
about an axis, 555, 1005
of inertia, 1006, 1023, 1074
of a lamina, 556, 1005
of a mass, 555
about a plane, 1023
polar, 1007
second, 1006
of a solid, 1023
of a system of particles, 556
momentum of an object, 455
monkey saddle, 891
monotonic sequence, 696
Monotonic Sequence Theorem, 698
motion of a projectile, 864
motion in space, 862
motion of a spring, force affecting
damping, 1157
resonance, 1160
restoring, 1156
movie theater seating, 456
multiple integrals. *See* double integral;
triple integral(s)
multiplication of power series, 763
multiplier (Lagrange), 957, 958, 961
multiplier effect, 712

natural exponential function, 56, 180, A52
derivative of, 180, A54
graph of, 180
power series for, 754
properties of, A53
natural growth law, 237, 606
natural logarithm function, 64, A50
derivative of, 218, A51
limits of, A51
properties of, A51
n-dimensional vector, 795
negative angle, A25
net area, 373
Net Change Theorem, 401
net investment flow, 567
newton (unit of force), 446
Newton, Sir Isaac, 3, 8, 101, 153, 157, 386,
406, 767, 868, 872
Newton's Law of Cooling, 240, 585
Newton's Law of Gravitation, 234, 451,
868, 1059
Newton's method, 338, 339
Newton's Second Law of Motion, 446, 455,
864, 868, 1156
Nicomedes, 641
nondifferentiable function, 159
nonhomogeneous differential equation,
1142, 1149
nonparallel planes, 821
normal component of acceleration,
866, 867
normal derivative, 1098
normal distribution, 572
normal line, 176, 941
normal plane, 859
normal vector, 820, 858
nth-degree equation, finding roots of, 212
nth-degree Taylor polynomial, 257, 755
number
complex, A57
integer, A2
irrational, A2
rational, A2
real, A2
numerical integration, 506

O (origin), 786
octant, 786
odd function, 18, 311
one-sided limits, 92, 113
one-to-one function, 59
one-to-one transformation, 1041
open interval, A3
open region, 1077
optics
first-order, 774
Gaussian, 774
third-order, 774
optimization problems, 274, 325
orbit of a planet, 868
order of a differential equation, 582
order of integration, reversed, 985, 993
ordered pair, A10
ordered triple, 786
Oresme, Nicole, 708
orientation of a curve, 1068, 1084
orientation of a surface, 1115
oriented surface, 1115
origin, 786, A2, A10
orthogonal curves, 216
orthogonal projection, 807
orthogonal surfaces, 945
orthogonal trajectory, 216, 597
orthogonal vectors, 802
osculating circle, 859
osculating plane, 859
Ostrogradsky, Mikhail, 1129
ovals of Cassini, 665
overdamped vibration, 1158

Pappus, Theorem of, 559
Pappus of Alexandria, 559
parabola, 670, 678, A18
axis, 670
directrix, 670
equation, 670, 671
focus, 670, 678
polar equation, 680
reflection property, 272
vertex, 670
parabolic cylinder, 827
paraboloid, 828, 832
paradoxes of Zeno, 5
parallel lines, A14
parallel planes, 821
parallel vectors, 793
parallelepiped, 430
volume of, 813
Parallelogram Law, 792, 807
parameter, 636, 817, 841
parametric curve, 636, 841
arc length of, 648
area under, 647
slope of tangent line to, 645
parametric equations, 636, 817, 841
of a line in space, 817
of a space curve, 841
of a surface, 1099
of a trajectory, 865

parametric surface, 1099
  graph of, 1112
  surface area of, 1104, 1105
  surface integral over, 1111
  tangent plane to, 1103
parametrization of a space curve, 854
  with respect to arc length, 855
  smooth, 855
paraxial rays, 252
partial derivative(s), 902
  of a function of more than three variables, 905
  interpretations of, 903
  notations for, 903
  as a rate of change, 902
  rules for finding, 903
  second, 906
  as slopes of tangent lines, 903
partial differential equation, 908
partial fractions, 484, 485
partial integration, 464, 465, 466, 983
partial sum of a series, 704
particle, motion of, 862
parts, integration by, 464, 465, 466
pascal (unit of pressure), 553
path, 1076
patterns in integrals, 505
pendulum, approximating the period of, 252, 256
percentage error, 254
perihelion, 683
perilune, 677
period, 311
periodic function, 311
perpendicular lines, A14
perpendicular vectors, 802
phase plane, 624
phase portrait, 624
phase trajectory, 624
piecewise defined function, 16
piecewise-smooth curve, 1064
Planck's Law, 777
plane region of type I, 989
plane region of type II, 990
plane(s)
  angle between, 821
  coordinate, 786
  equation(s) of, 816, 819, 820
  equation of, through three points, 821
  horizontal, 787
  line of intersection, 821
  normal, 859
  osculating, 859
  parallel, 821
  tangent to a surface, 915, 940, 1103
  vertical, 878
planetary motion, 867
  laws of, 682
planimeter, 1087
point of inflection, 294
point(s) in space
  coordinates of, 786
  distance between, 788
  projection of, 787
point-slope equation of a line, A12
Poiseuille, Jean-Louis-Marie, 230
Poiseuille's Laws, 256, 336, 565
polar axis, 654
polar coordinate system, 654
  area in, 665
  conic sections in, 678
  conversion of double integral to, 997, 998
  conversion equations for Cartesian coordinates, 655, 656
polar curve, 656
  arc length of, 667
  graph of, 656
  symmetry in, 659
  tangent line to, 659
polar equation, graph of, 656
polar equation of a conic, 680
polar form of a complex number, A59
polar graph, 656
polar moment of inertia, 1007
polar rectangle, 997
polar region, area of, 665
pole, 654
polynomial, 27
polynomial function, 27
  of two variables, 897
population growth, 54, 237, 605
  of bacteria, 605, 610
  of insects, 494
  models, 580
  world, 54
position function, 145
position vector, 794
positive angle, A25
positive orientation
  of a boundary curve, 1122
  of a closed curve, 1084
  of a surface, 1116
potential, 532
potential energy, 1081
potential function, 1061
pound (unit of force), 446
power, 150
power consumption, approximation of, 403
power function(s), 28
  derviative of, 174
Power Law of limits, 100
Power Rule, 175, 176, 201, 221
power series, 741
  coefficients of, 741
  for cosine and sine, 758
  differentiation of, 748
  division of, 763
  for exponential function, 758
  integration of, 748
  interval of convergence, 743
  multiplication of, 763
  radius of convergence, 743
  representations of functions as, 747
predator-prey model, 236, 622, 623
pressure exerted by a fluid, 552, 553
prime notation, 146, 177
principal square root of a complex number, A58
principal unit normal vector, 858
principle of mathematical induction, 76, 79, A36
principle of superposition, 1151
probability, 568, 1008
probability density function, 568, 1008
problem-solving principles, 75
  uses of, 170, 355, 407, 419
producer surplus, 566
product
  cross, 808 (*see also* cross product)
  dot, 800 (*see also* dot product)
  scalar, 800
  scalar triple, 812
  triple, 812
product formulas, A29
Product Law of limits, 99
Product Rule, 184, 185
profit function, 331
projectile, path of, 644, 864
projection, 787, 804
  orthogonal, 807
p-series, 717

quadrant, A11
quadratic approximation, 256, 956
quadratic function, 27
quadric surface(s), 827
  cone, 830
  cylinder, 827
  ellipsoid, 830
  hyperboloid, 830
  paraboloid, 828, 829, 830
  table of graphs, 830
quaternion, 797
Quotient Law of limits, 99
Quotient Rule, 187

radian measure, 191, A24
radiation from stars, 777
radioactive decay, 239
radiocarbon dating, 243
radius of convergence, 743
radius of gyration, 1008
rainbow, formation and location of, 282
rainbow angle, 283
ramp function, 44
range of a function, 10, 878
rate of change
  average, 148, 224
  derivative as, 148
  instantaneous, 85, 148, 224

rate of growth, 229, 401
rate of reaction, 150, 228, 401
rates, related, 244
rational function, 30, 485, 897
  continuity of, 122
  integration of, 484
rational number, A2
rationalizing substitution for integration, 492
Ratio Test, 734
Rayleigh-Jeans Law, 777
real line, A3
real number, A2
rearrangement of a series, 737
reciprocal function, 30
Reciprocal Rule, 191
rectangular coordinate system, 787, A11
  conversion to cylindrical coordinates, 1028
  conversion to spherical coordinates, 1033
rectilinear motion, 347
recursion relation, 1165
reduction formula, 467
reflecting a function, 36
reflection property
  of conics, 271
  of an ellipse, 673
  of a hyperbola, 678
  of a parabola, 271, 272
region
  connected, 1077
  under a graph, 360, 365
  open, 1077
  plane, of type I or II, 989, 990
  simple plane, 1085
  simple solid, 1129
  simply-connected, 1078
  solid (of type 1, 2, or 3), 1018, 1019, 1020
  between two graphs, 422
regression, linear, 26
related rates, 244
relative error, 254
relative growth rate, 237, 606
relative maximum or minimum, 274
remainder estimates
  for the Alternating Series, 730
  for the Integral Test, 718
remainder of the Taylor series, 755
removable discontinuity, 120
representation(s) of a function, 10, 12, 13
  as a power series, 746
resonance, 1160
restoring force, 1156
resultant force, 797
revenue function, 331
reversing order of integration, 985, 993
revolution, solid of, 435
revolution, surface of, 545
Riemann, Georg Bernhard, 372
Riemann sum(s), 372
  for multiple integrals, 977, 1017
right circular cylinder, 430
right-hand derivative, 165
right-hand limit, 92, 113
right-hand rule, 786, 810
Roberval, Gilles de, 393, 647
rocket science, 964
Rolle, Michel, 284
roller coaster, design of, 184
roller derby, 1039
Rolle's Theorem, 284
root function, 29
Root Law of limits, 101
Root Test, 736
roots of a complex number, A62
roots of an *n*th-degree equation, 212
rubber membrane, vibration of, 742
ruling of a surface, 827
rumors, rate of spread, 233

saddle point, 947
sample point, 365, 372, 975
satellite dish, parabolic, 832
scalar, 793
scalar equation of a plane, 820
scalar field, 1057
scalar multiple of a vector, 793
scalar product, 800
scalar projection, 804
scalar triple product, 812
  geometric characterization of, 813
scatter plot, 13
seasonal-growth model, 615
secant function, A26
  derivative of, 194
  graph of, A31
secant line, 3, 82, 83, 85
secant vector, 848
second derivative, 160, 850
  of a vector function, 850
Second Derivative Test, 295
Second Derivatives Test, 947
second directional derivative, 944
second moment of inertia, 1006
second-order differential equation, 582
  solutions of, 1142, 1147
second partial derivative, 906
sector of a circle, area of, 665
separable differential equation, 594
sequence, 5, 690
  bounded, 697
  convergent, 692
  decreasing, 696
  divergent, 692
  Fibonacci, 691
  graph of, 695
  increasing, 696
  limit of, 5, 362, 692
  logistic, 703
  monotonic, 696
  of partial sums, 704
  term of, 690
series, 6, 704
  absolutely convergent, 732
  alternating, 727
  alternating harmonic, 729, 732, 733
  binomial, 760
  coefficients of, 741
  conditionally convergent, 733
  convergent, 705
  divergent, 705
  geometric, 705
  Gregory's, 750
  harmonic, 708, 717
  infinite, 704
  Maclaurin, 753, 754
  p-, 717
  partial sum of, 704
  power, 741
  rearrangement of, 737
  strategy for testing, 739
  sum of, 6, 705
  Taylor, 753, 754
  term of, 704
  trigonometric, 741
series solution of a differential equation, 1164
set, bounded or closed, 951
set notation, A3
serpentine, 189
shell method for approximating volume, 441
shift of a function, 36
shifted conics, 675, A21
shock absorber, 1157
Sierpinski carpet, 713
sigma notation, 366, A34
simple curve, 1078
simple harmonic motion, 206
simple plane region, 1085
simple solid region, 1129
simply-connected region, 1078
Simpson, Thomas, 512, 513, 972
Simpson's Rule, 511, 513
  error bounds for, 514
sine function, A26
  derivative of, 193, 194
  graph of, 31, A31
  power series for, 758
sine integral function, 396
sink, 1133
skew lines, 819
slant asymptote, 312, 315
slope, A12
  of a curve, 144
slope field, 586
slope-intercept equation of a line, A13
smooth curve, 538, 855
smooth function, 538
smooth parametrization, 855
smooth surface, 1104
Snell's Law, 335
snowflake curve, 782

solid, 430
volume of, 430, 431, 1018, 1019
solid angle, 1139
solid of revolution, 435
rotated on a slant, 551
volume of, 437, 442, 551
solid region, 1129
solution curve, 586
solution of a differential equation, 582
solution of predator-prey equations, 623
source, 1133
space, three-dimensional, 786
space curve, 840, 841, 842, 843
arc length of, 853
speed of a particle, 148, 862
sphere
equation of, 789
flux across, 1117
parametrization of, 1101
surface area of, 1105
spherical coordinate system, 1033
conversion equations for, 1033
triple integrals in, 1034
spherical wedge, 1034
spherical zones, 577
spring constant, 447, 582, 1156
Squeeze Theorem, 105,A42
for sequences, 694
standard basis vectors, 796
standard deviation, 572
standard position of an angle, A25
stationary points, 946
steady state solution, 1162
stellar stereography, 528
step function, 17
Stokes, Sir George, 1123, 1128
Stokes' Theorem, 1122
strategy
for integration, 494, 495
for optimization problems, 325, 326
for problem solving, 75
for related rates, 246
for testing series, 739
for trigonometric integrals, 473, 474
streamlines, 1062
stretching of a function, 36
strophoid, 669, 687
Substitution Rule, 407, 408
for definite integrals, 411
subtraction formulas for sine
and cosine, A29
sum, 365
of a geometric series, 706
of an infinite series, 705
of partial fractions, 485
Riemann, 372
telescoping, 708
of vectors, 792
Sum Law of limits, 99
Sum Rule, 177
summation notation, A34
supply function, 566
surface(s)
closed, 1116
graph of, 1112
level, 887
oriented, 1115
parametric, 1099
positive orientation of, 1116
quadric, 827
smooth, 1104
surface area, 547
of a parametric surface, 650, 1104, 1105
of a sphere, 1105
of a surface $z = f(x, y)$, 1013,
1014, 1106
surface integral, 1110
over a parametric surface, 1111
of a vector field, 1116
surface of revolution, 545
parametric representation of, 1103
surface area of, 547
swallowtail catastrophe curve, 644
symmetric equations of a line, 818
symmetric functions, integrals of, 412
symmetry, 17, 311, 412
in polar graphs, 659
symmetry principle, 556

T and T^{-1} transformations, 1040, 1041
table of differentiation formulas, 188, RP5
tables of integrals, 495, RP6–10
use of, 500
tabular function, 13
tangent function, A26
derivative of, 194
graph of, 32, A31
tangent line(s), 143
to a curve, 3, 82, 143
early methods of finding, 153
to a parametric curve, 645, 646
to a polar curve, 659
to a space curve, 849
vertical, 159
tangent line approximation, 251
tangent plane
to a level surface, 915, 940
to a parametric surface, 1103
to a surface $F(x, y, z) = k$, 916, 940
to a surface $z = f(x, y)$, 915
tangent plane approximation, 917
tangent problem, 2, 3, 82, 143
tangent vector, 848
tangential component of acceleration, 866
tautochrone problem, 640
Taylor, Brook, 754
Taylor polynomial, 257, 755, 956
applications of, 768
Taylor series, 753, 754
Taylor's Inequality, 756
techniques of integration, summary, 495
telescoping sum, 708
temperature-humidity index, 888, 900
term of a sequence, 690
term of a series, 704
term-by-term differentiation and
integration, 748
terminal point of a parametric curve, 637
terminal point of a vector, 791
terminal velocity, 602
Test for Divergence, 709
tests for convergence and divergence
of series
Alternating Series Test, 727
Comparison Test, 722
Integral Test, 716
Limit Comparison Test, 724
Ratio Test, 734
Root Test, 736
summary of tests, 739
tetrahedron, 816
third derivative, 161
third-order optics, 774
Thomson, William (Lord Kelvin), 1085,
1123, 1128
three-dimensional coordinate systems,
786, 787
TNB frame, 858
toroidal spiral, 843
torque, 871
Torricelli, Evangelista, 647
Torricelli's Law, 234
torsion of a space curve, 861
torus, 440, 1110
total differential, 920
total electric charge, 1004, 1023
total fertility rate, 169
trace of a surface, 827
trajectory, parametric equations for, 865
transfer curve, 875
transformation, 1040
of a function, 36
inverse, 1041
Jacobian of, 1043, 1046
one-to-one, 1041
translation of a function, 36
Trapezoidal Rule, 508
error in, 508
tree diagram, 926
trefoil knot, 843
Triangle Inequality, 115, A8
for vectors, 807
Triangle Law, 792
trigonometric functions, 31, A26
derivatives of, 191, 194
graphs of, 31, 32, A30, A31
integrals of, 398, 471
inverse, 67
limits involving, 192, 193
trigonometric identities, A28

trigonometric integrals, 471
 strategy for evaluating, 473, 474
trigonometric series, 741
trigonometric substitutions, 478
 table of, 478
triple integral(s), 1017, 1018
 applications of, 1022
 in cylindrical coordinates, 1029
 over a general bounded region, 1018
 Midpoint Rule for, 1025
 in spherical coordinates, 1034, 1035
triple product, 812
triple Riemann sum, 1017
trochoid, 643
Tschirnhausen cubic, 215, 428
twisted cubic, 843
type I or type II plane region, 989, 990
type 1, 2, or 3 solid region, 1018, 1019, 1020

ultraviolet catastrophe, 777
underdamped vibration, 1158
undetermined coefficients, method of, 1149, 1153
uniform circular motion, 864
union of sets, A3
unit normal vector, 858
unit tangent vector, 848
unit vector, 797

value of a function, 10
van der Waals equation, 216, 914
variable(s)
 change of, 407
 continuous random, 568
 dependent, 10, 878, 926
 independent, 10, 878, 926
 independent random, 1010
 intermediate, 926
variables, change of. *See* change of variable(s)
variation of parameters, method of, 1153, 1154
vascular branching, 336, 337
vector(s), 791
 acceleration, 863
 addition of, 792, 794
 algebraic, 794, 795
 angle between, 801
 basis, 796
 binormal, 858
 combining speed, 799
 components of, 804
 coplanar, 813
 cross product of, 808
 difference, 793
 displacement, 805
 dot product, 801
 equality of, 792
 force, 1059
 geometric representation of, 794
 gradient, 936, 938
 i, **j**, and **k**, 796
 length of, 794
 magnitude of, 794
 multiplication of, 793, 795
 n-dimensional, 795
 normal, 820
 orthogonal, 802
 parallel, 793
 perpendicular, 802
 position, 794
 properties of, 795
 representation of, 794
 scalar mulitple of, 793
 standard basis, 796
 tangent, 848
 three-dimensional, 794
 triple product, 813
 two-dimensional, 794
 unit, 797
 unit normal, 858
 unit tangent, 848
 velocity, 862
 zero, 792
vector equation
 of a line, 816, 817
 of a plane, 820
vector field, 1056, 1057
 conservative, 1061
 curl of, 1091
 divergence of, 1094
 electric flux of, 1119
 flux of, 1117
 force, 1056, 1060
 gradient, 1060
 gravitationsl, 1060
 incompressible, 1095
 irrotational, 1094
 line integral of, 1070, 1071
 potential function, 1080
 surface integral of, 1117
 velocity, 1056, 1059
vector function, 840
 continuity of, 841
 derivative of, 847
 integration of, 851
 limit of, 840
vector product, 808
 properties of, 812
vector projection, 804
vector triple product, 813
vector-valued function. *See* vector function
 continuous, 841
 limit of, 840
velocity, 3, 84, 145, 224, 401
 average, 4, 84, 145, 224
 instantaneous, 85, 145, 224
velocity field, 1059
 airflow, 1056
 ocean currents, 1056
 wind patterns, 1056
velocity gradient, 231
velocity problem, 84, 145
velocity vector, 862
velocity vector field, 1056
Verhulst, Pierre-François, 581
vertex of a parabola, 670
vertical asymptote, 94, 311
vertical line, A13
Vertical Line Test, 15
vertical tangent line, 159
vertical translation of a graph, 36
vertices of an ellipse, 672
vertices of a hyperbola, 674
vibration of a rubber membrane, 742
vibration of a spring, 1156
vibrations, 1156, 1157, 1159
viewing rectangle, 44
visual representations of a function, 10, 12
volume, 431
 by cross-sections, 430, 431, 565
 by cylindrical shells, 441
 by disks, 432, 435
 by double integrals, 974
 of a hypersphere, 1027
 by polar coordinates, 1000
 of a solid, 430, 976
 of a solid of revolution, 435, 551
 of a solid on a slant, 551
 by triple integrals, 1022
 by washers, 434, 435
Volterra, Vito, 623
von Bertalanffy model, 631

Wallis, John, 3
Wallis product, 470
washer method, 434
wave equation, 908
Weierstrass, Karl, 493
weight (force), 446
wind-chill index, 879
wind patterns in San Francisco Bay area, 1056
witch of Maria Agnesi, 189, 643
work (force), 446, 447
 defined as a line integral, 1070
Wren, Sir Christopher, 650

x-axis, 786, A10
x-coordinate, 786, A10
x-intercept, A13, A19
X-mean, 1011

y-axis, 786, A10
y-coordinate, 786, A10
y-intercept, A13, A19
Y-mean, 1011

z-axis, 786
z-coordinate, 786
Zeno, 5
Zeno's paradoxes, 5
zero vectors, 792

SPECIAL FUNCTIONS

Power Functions $f(x) = x^a$

(i) $f(x) = x^n$, n a positive integer

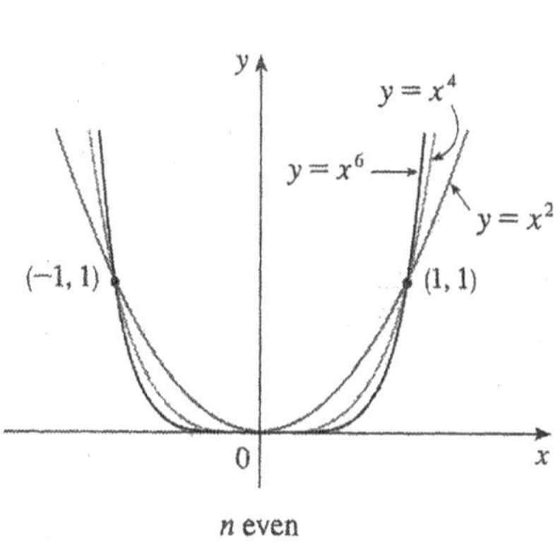

n even

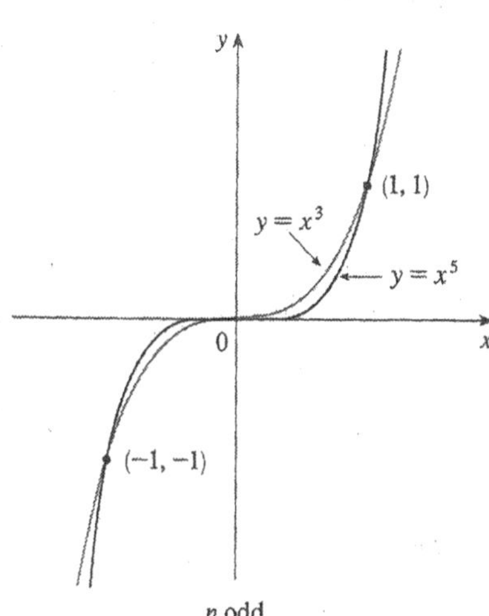

n odd

(ii) $f(x) = x^{1/n} = \sqrt[n]{x}$, n a positive integer

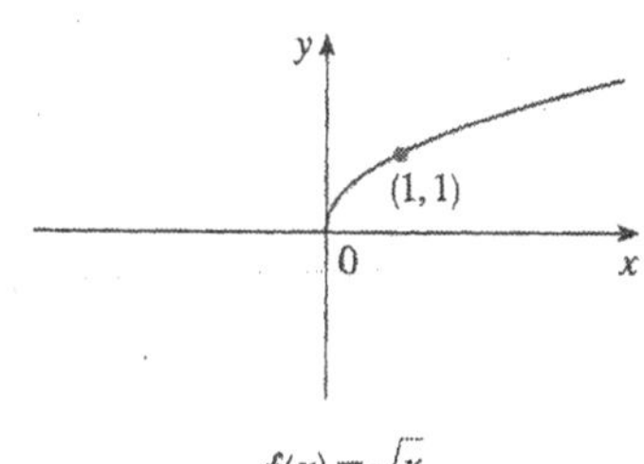

$f(x) = \sqrt{x}$

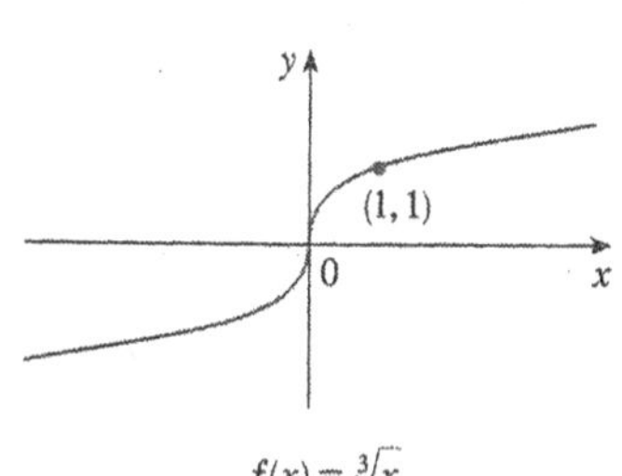

$f(x) = \sqrt[3]{x}$

(iii) $f(x) = x^{-1} = \dfrac{1}{x}$

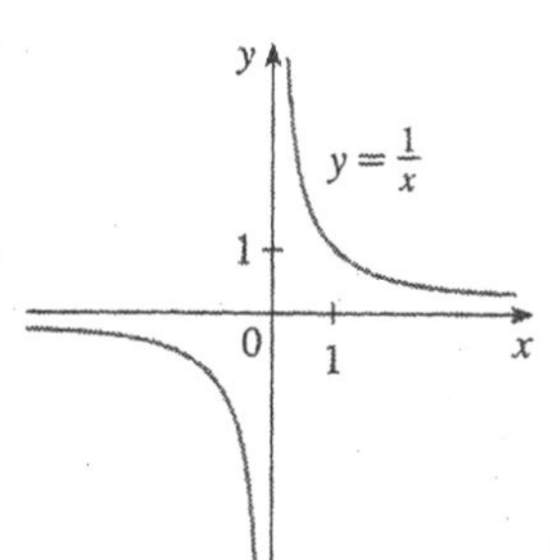

Inverse Trigonometric Functions

$\arcsin x = \sin^{-1}x = y \iff \sin y = x$ and $-\dfrac{\pi}{2} \leqslant y \leqslant \dfrac{\pi}{2}$

$\arccos x = \cos^{-1}x = y \iff \cos y = x$ and $0 \leqslant y \leqslant \pi$

$\arctan x = \tan^{-1}x = y \iff \tan y = x$ and $-\dfrac{\pi}{2} < y < \dfrac{\pi}{2}$

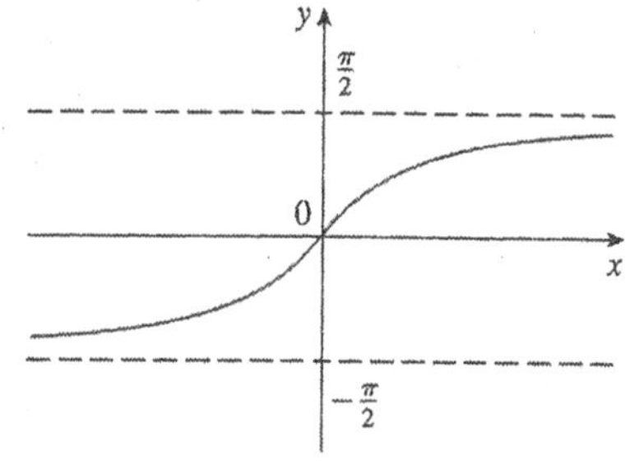

$y = \tan^{-1}x = \arctan x$

$$\lim_{x \to -\infty} \tan^{-1}x = -\frac{\pi}{2}$$

$$\lim_{x \to \infty} \tan^{-1}x = \frac{\pi}{2}$$

SPECIAL FUNCTIONS

Exponential and Logarithmic Functions

$\log_a x = y \iff a^y = x$

$\ln x = \log_e x$, where $\ln e = 1$

$\ln x = y \iff e^y = x$

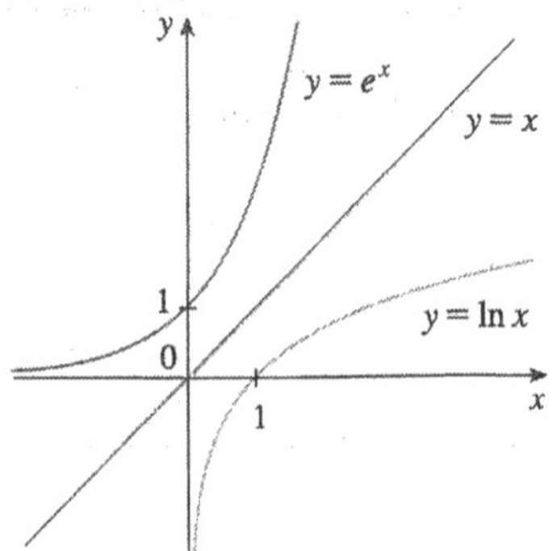

Cancellation Equations

$\log_a(a^x) = x \qquad a^{\log_a x} = x$

$\ln(e^x) = x \qquad e^{\ln x} = x$

Laws of Logarithms

1. $\log_a(xy) = \log_a x + \log_a y$

2. $\log_a\left(\dfrac{x}{y}\right) = \log_a x - \log_a y$

3. $\log_a(x^r) = r \log_a x$

$\lim_{x \to -\infty} e^x = 0 \qquad \lim_{x \to \infty} e^x = \infty$

$\lim_{x \to 0^+} \ln x = -\infty \qquad \lim_{x \to \infty} \ln x = \infty$

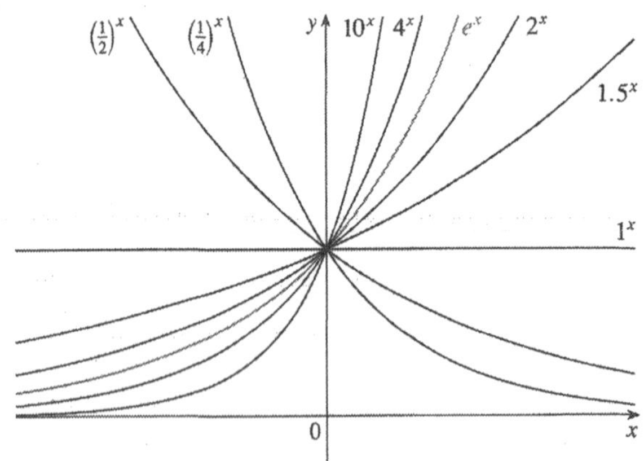

Exponential functions

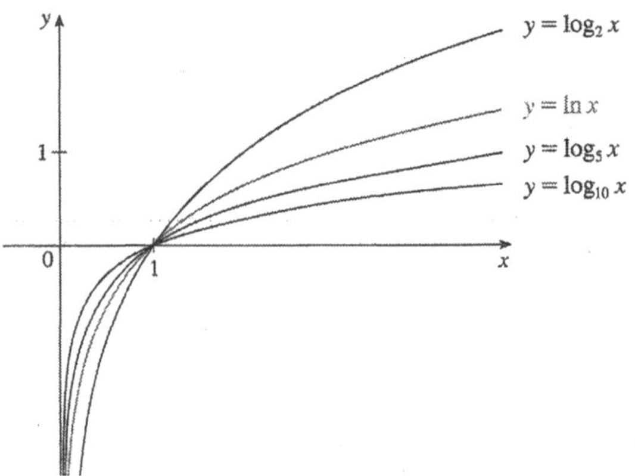

Logarithmic functions

Hyperbolic Functions

$\sinh x = \dfrac{e^x - e^{-x}}{2} \qquad \operatorname{csch} x = \dfrac{1}{\sinh x}$

$\cosh x = \dfrac{e^x + e^{-x}}{2} \qquad \operatorname{sech} x = \dfrac{1}{\cosh x}$

$\tanh x = \dfrac{\sinh x}{\cosh x} \qquad \coth x = \dfrac{\cosh x}{\sinh x}$

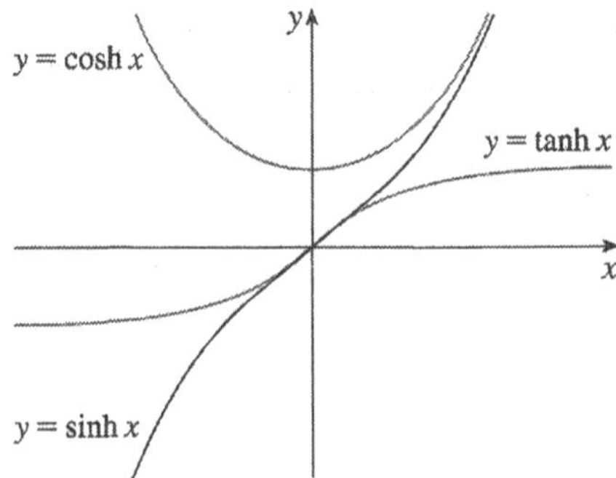

Inverse Hyperbolic Functions

$y = \sinh^{-1}x \iff \sinh y = x \qquad \sinh^{-1}x = \ln\left(x + \sqrt{x^2 + 1}\right)$

$y = \cosh^{-1}x \iff \cosh y = x$ and $y \geq 0 \qquad \cosh^{-1}x = \ln\left(x + \sqrt{x^2 - 1}\right)$

$y = \tanh^{-1}x \iff \tanh y = x \qquad \tanh^{-1}x = \frac{1}{2}\ln\left(\dfrac{1 + x}{1 - x}\right)$

DIFFERENTIATION RULES

General Formulas

1. $\frac{d}{dx}(c) = 0$
2. $\frac{d}{dx}[cf(x)] = cf'(x)$
3. $\frac{d}{dx}[f(x) + g(x)] = f'(x) + g'(x)$
4. $\frac{d}{dx}[f(x) - g(x)] = f'(x) - g'(x)$
5. $\frac{d}{dx}[f(x)g(x)] = f(x)g'(x) + g(x)f'(x)$ (Product Rule)
6. $\frac{d}{dx}\left[\frac{f(x)}{g(x)}\right] = \frac{g(x)f'(x) - f(x)g'(x)}{[g(x)]^2}$ (Quotient Rule)
7. $\frac{d}{dx} f(g(x)) = f'(g(x))g'(x)$ (Chain Rule)
8. $\frac{d}{dx}(x^n) = nx^{n-1}$ (Power Rule)

Exponential and Logarithmic Functions

9. $\frac{d}{dx}(e^x) = e^x$
10. $\frac{d}{dx}(a^x) = a^x \ln a$
11. $\frac{d}{dx} \ln|x| = \frac{1}{x}$
12. $\frac{d}{dx}(\log_a x) = \frac{1}{x \ln a}$

Trigonometric Functions

13. $\frac{d}{dx}(\sin x) = \cos x$
14. $\frac{d}{dx}(\cos x) = -\sin x$
15. $\frac{d}{dx}(\tan x) = \sec^2 x$
16. $\frac{d}{dx}(\csc x) = -\csc x \cot x$
17. $\frac{d}{dx}(\sec x) = \sec x \tan x$
18. $\frac{d}{dx}(\cot x) = -\csc^2 x$

Inverse Trigonometric Functions

19. $\frac{d}{dx}(\sin^{-1}x) = \frac{1}{\sqrt{1 - x^2}}$
20. $\frac{d}{dx}(\cos^{-1}x) = -\frac{1}{\sqrt{1 - x^2}}$
21. $\frac{d}{dx}(\tan^{-1}x) = \frac{1}{1 + x^2}$
22. $\frac{d}{dx}(\csc^{-1}x) = -\frac{1}{x\sqrt{x^2 - 1}}$
23. $\frac{d}{dx}(\sec^{-1}x) = \frac{1}{x\sqrt{x^2 - 1}}$
24. $\frac{d}{dx}(\cot^{-1}x) = -\frac{1}{1 + x^2}$

Hyperbolic Functions

25. $\frac{d}{dx}(\sinh x) = \cosh x$
26. $\frac{d}{dx}(\cosh x) = \sinh x$
27. $\frac{d}{dx}(\tanh x) = \operatorname{sech}^2 x$
28. $\frac{d}{dx}(\operatorname{csch} x) = -\operatorname{csch} x \coth x$
29. $\frac{d}{dx}(\operatorname{sech} x) = -\operatorname{sech} x \tanh x$
30. $\frac{d}{dx}(\coth x) = -\operatorname{csch}^2 x$

Inverse Hyperbolic Functions

31. $\frac{d}{dx}(\sinh^{-1}x) = \frac{1}{\sqrt{1 + x^2}}$
32. $\frac{d}{dx}(\cosh^{-1}x) = \frac{1}{\sqrt{x^2 - 1}}$
33. $\frac{d}{dx}(\tanh^{-1}x) = \frac{1}{1 - x^2}$
34. $\frac{d}{dx}(\operatorname{csch}^{-1}x) = -\frac{1}{|x|\sqrt{x^2 + 1}}$
35. $\frac{d}{dx}(\operatorname{sech}^{-1}x) = -\frac{1}{x\sqrt{1 - x^2}}$
36. $\frac{d}{dx}(\coth^{-1}x) = \frac{1}{1 - x^2}$

TABLE OF INTEGRALS

Basic Forms

1. $\int u\,dv = uv - \int v\,du$

2. $\int u^n\,du = \dfrac{u^{n+1}}{n+1} + C, \quad n \neq -1$

3. $\int \dfrac{du}{u} = \ln|u| + C$

4. $\int e^u\,du = e^u + C$

5. $\int a^u\,du = \dfrac{a^u}{\ln a} + C$

6. $\int \sin u\,du = -\cos u + C$

7. $\int \cos u\,du = \sin u + C$

8. $\int \sec^2 u\,du = \tan u + C$

9. $\int \csc^2 u\,du = -\cot u + C$

10. $\int \sec u \tan u\,du = \sec u + C$

11. $\int \csc u \cot u\,du = -\csc u + C$

12. $\int \tan u\,du = \ln|\sec u| + C$

13. $\int \cot u\,du = \ln|\sin u| + C$

14. $\int \sec u\,du = \ln|\sec u + \tan u| + C$

15. $\int \csc u\,du = \ln|\csc u - \cot u| + C$

16. $\int \dfrac{du}{\sqrt{a^2 - u^2}} = \sin^{-1}\dfrac{u}{a} + C, \quad a > 0$

17. $\int \dfrac{du}{a^2 + u^2} = \dfrac{1}{a}\tan^{-1}\dfrac{u}{a} + C$

18. $\int \dfrac{du}{u\sqrt{u^2 - a^2}} = \dfrac{1}{a}\sec^{-1}\dfrac{u}{a} + C$

19. $\int \dfrac{du}{a^2 - u^2} = \dfrac{1}{2a}\ln\left|\dfrac{u + a}{u - a}\right| + C$

20. $\int \dfrac{du}{u^2 - a^2} = \dfrac{1}{2a}\ln\left|\dfrac{u - a}{u + a}\right| + C$

Forms Involving $\sqrt{a^2 + u^2}$, $a > 0$

21. $\int \sqrt{a^2 + u^2}\,du = \dfrac{u}{2}\sqrt{a^2 + u^2} + \dfrac{a^2}{2}\ln\left(u + \sqrt{a^2 + u^2}\right) + C$

22. $\int u^2\sqrt{a^2 + u^2}\,du = \dfrac{u}{8}(a^2 + 2u^2)\sqrt{a^2 + u^2} - \dfrac{a^4}{8}\ln\left(u + \sqrt{a^2 + u^2}\right) + C$

23. $\int \dfrac{\sqrt{a^2 + u^2}}{u}\,du = \sqrt{a^2 + u^2} - a\ln\left|\dfrac{a + \sqrt{a^2 + u^2}}{u}\right| + C$

24. $\int \dfrac{\sqrt{a^2 + u^2}}{u^2}\,du = -\dfrac{\sqrt{a^2 + u^2}}{u} + \ln\left(u + \sqrt{a^2 + u^2}\right) + C$

25. $\int \dfrac{du}{\sqrt{a^2 + u^2}} = \ln\left(u + \sqrt{a^2 + u^2}\right) + C$

26. $\int \dfrac{u^2\,du}{\sqrt{a^2 + u^2}} = \dfrac{u}{2}\sqrt{a^2 + u^2} - \dfrac{a^2}{2}\ln\left(u + \sqrt{a^2 + u^2}\right) + C$

27. $\int \dfrac{du}{u\sqrt{a^2 + u^2}} = -\dfrac{1}{a}\ln\left|\dfrac{\sqrt{a^2 + u^2} + a}{u}\right| + C$

28. $\int \dfrac{du}{u^2\sqrt{a^2 + u^2}} = -\dfrac{\sqrt{a^2 + u^2}}{a^2 u} + C$

29. $\int \dfrac{du}{(a^2 + u^2)^{3/2}} = \dfrac{u}{a^2\sqrt{a^2 + u^2}} + C$

TABLE OF INTEGRALS

Forms Involving $\sqrt{a^2 - u^2}$, $a > 0$

30. $\displaystyle\int \sqrt{a^2 - u^2}\, du = \frac{u}{2}\sqrt{a^2 - u^2} + \frac{a^2}{2}\sin^{-1}\frac{u}{a} + C$

31. $\displaystyle\int u^2\sqrt{a^2 - u^2}\, du = \frac{u}{8}(2u^2 - a^2)\sqrt{a^2 - u^2} + \frac{a^4}{8}\sin^{-1}\frac{u}{a} + C$

32. $\displaystyle\int \frac{\sqrt{a^2 - u^2}}{u}\, du = \sqrt{a^2 - u^2} - a\ln\left|\frac{a + \sqrt{a^2 - u^2}}{u}\right| + C$

33. $\displaystyle\int \frac{\sqrt{a^2 - u^2}}{u^2}\, du = -\frac{1}{u}\sqrt{a^2 - u^2} - \sin^{-1}\frac{u}{a} + C$

34. $\displaystyle\int \frac{u^2\, du}{\sqrt{a^2 - u^2}} = -\frac{u}{2}\sqrt{a^2 - u^2} + \frac{a^2}{2}\sin^{-1}\frac{u}{a} + C$

35. $\displaystyle\int \frac{du}{u\sqrt{a^2 - u^2}} = -\frac{1}{a}\ln\left|\frac{a + \sqrt{a^2 - u^2}}{u}\right| + C$

36. $\displaystyle\int \frac{du}{u^2\sqrt{a^2 - u^2}} = -\frac{1}{a^2 u}\sqrt{a^2 - u^2} + C$

37. $\displaystyle\int (a^2 - u^2)^{3/2}\, du = -\frac{u}{8}(2u^2 - 5a^2)\sqrt{a^2 - u^2} + \frac{3a^4}{8}\sin^{-1}\frac{u}{a} + C$

38. $\displaystyle\int \frac{du}{(a^2 - u^2)^{3/2}} = \frac{u}{a^2\sqrt{a^2 - u^2}} + C$

Forms Involving $\sqrt{u^2 - a^2}$, $a > 0$

39. $\displaystyle\int \sqrt{u^2 - a^2}\, du = \frac{u}{2}\sqrt{u^2 - a^2} - \frac{a^2}{2}\ln\left|u + \sqrt{u^2 - a^2}\right| + C$

40. $\displaystyle\int u^2\sqrt{u^2 - a^2}\, du = \frac{u}{8}(2u^2 - a^2)\sqrt{u^2 - a^2} - \frac{a^4}{8}\ln\left|u + \sqrt{u^2 - a^2}\right| + C$

41. $\displaystyle\int \frac{\sqrt{u^2 - a^2}}{u}\, du = \sqrt{u^2 - a^2} - a\cos^{-1}\frac{a}{|u|} + C$

42. $\displaystyle\int \frac{\sqrt{u^2 - a^2}}{u^2}\, du = -\frac{\sqrt{u^2 - a^2}}{u} + \ln\left|u + \sqrt{u^2 - a^2}\right| + C$

43. $\displaystyle\int \frac{du}{\sqrt{u^2 - a^2}} = \ln\left|u + \sqrt{u^2 - a^2}\right| + C$

44. $\displaystyle\int \frac{u^2\, du}{\sqrt{u^2 - a^2}} = \frac{u}{2}\sqrt{u^2 - a^2} + \frac{a^2}{2}\ln\left|u + \sqrt{u^2 - a^2}\right| + C$

45. $\displaystyle\int \frac{du}{u^2\sqrt{u^2 - a^2}} = \frac{\sqrt{u^2 - a^2}}{a^2 u} + C$

46. $\displaystyle\int \frac{du}{(u^2 - a^2)^{3/2}} = -\frac{u}{a^2\sqrt{u^2 - a^2}} + C$

TABLE OF INTEGRALS

Forms Involving $a + bu$

47. $\displaystyle\int \frac{u\,du}{a+bu} = \frac{1}{b^2}\left(a + bu - a\ln|a+bu|\right) + C$

48. $\displaystyle\int \frac{u^2\,du}{a+bu} = \frac{1}{2b^3}\left[(a+bu)^2 - 4a(a+bu) + 2a^2\ln|a+bu|\right] + C$

49. $\displaystyle\int \frac{du}{u(a+bu)} = \frac{1}{a}\ln\left|\frac{u}{a+bu}\right| + C$

50. $\displaystyle\int \frac{du}{u^2(a+bu)} = -\frac{1}{au} + \frac{b}{a^2}\ln\left|\frac{a+bu}{u}\right| + C$

51. $\displaystyle\int \frac{u\,du}{(a+bu)^2} = \frac{a}{b^2(a+bu)} + \frac{1}{b^2}\ln|a+bu| + C$

52. $\displaystyle\int \frac{du}{u(a+bu)^2} = \frac{1}{a(a+bu)} - \frac{1}{a^2}\ln\left|\frac{a+bu}{u}\right| + C$

53. $\displaystyle\int \frac{u^2\,du}{(a+bu)^2} = \frac{1}{b^3}\left(a + bu - \frac{a^2}{a+bu} - 2a\ln|a+bu|\right) + C$

54. $\displaystyle\int u\sqrt{a+bu}\,du = \frac{2}{15b^2}(3bu-2a)(a+bu)^{3/2} + C$

55. $\displaystyle\int \frac{u\,du}{\sqrt{a+bu}} = \frac{2}{3b^2}(bu-2a)\sqrt{a+bu} + C$

56. $\displaystyle\int \frac{u^2\,du}{\sqrt{a+bu}} = \frac{2}{15b^3}(8a^2 + 3b^2u^2 - 4abu)\sqrt{a+bu} + C$

57. $\displaystyle\int \frac{du}{u\sqrt{a+bu}} = \frac{1}{\sqrt{a}}\ln\left|\frac{\sqrt{a+bu}-\sqrt{a}}{\sqrt{a+bu}+\sqrt{a}}\right| + C, \quad \text{if } a > 0$

$$= \frac{2}{\sqrt{-a}}\tan^{-1}\sqrt{\frac{a+bu}{-a}} + C, \quad \text{if } a < 0$$

58. $\displaystyle\int \frac{\sqrt{a+bu}}{u}\,du = 2\sqrt{a+bu} + a\int\frac{du}{u\sqrt{a+bu}}$

59. $\displaystyle\int \frac{\sqrt{a+bu}}{u^2}\,du = -\frac{\sqrt{a+bu}}{u} + \frac{b}{2}\int\frac{du}{u\sqrt{a+bu}}$

60. $\displaystyle\int u^n\sqrt{a+bu}\,du = \frac{2}{b(2n+3)}\left[u^n(a+bu)^{3/2} - na\int u^{n-1}\sqrt{a+bu}\,du\right]$

61. $\displaystyle\int \frac{u^n\,du}{\sqrt{a+bu}} = \frac{2u^n\sqrt{a+bu}}{b(2n+1)} - \frac{2na}{b(2n+1)}\int\frac{u^{n-1}\,du}{\sqrt{a+bu}}$

62. $\displaystyle\int \frac{du}{u^n\sqrt{a+bu}} = -\frac{\sqrt{a+bu}}{a(n-1)u^{n-1}} - \frac{b(2n-3)}{2a(n-1)}\int\frac{du}{u^{n-1}\sqrt{a+bu}}$

TABLE OF INTEGRALS

Trigonometric Forms

63. $\int \sin^2 u\, du = \frac{1}{2}u - \frac{1}{4}\sin 2u + C$

64. $\int \cos^2 u\, du = \frac{1}{2}u + \frac{1}{4}\sin 2u + C$

65. $\int \tan^2 u\, du = \tan u - u + C$

66. $\int \cot^2 u\, du = -\cot u - u + C$

67. $\int \sin^3 u\, du = -\frac{1}{3}(2 + \sin^2 u)\cos u + C$

68. $\int \cos^3 u\, du = \frac{1}{3}(2 + \cos^2 u)\sin u + C$

69. $\int \tan^3 u\, du = \frac{1}{2}\tan^2 u + \ln|\cos u| + C$

70. $\int \cot^3 u\, du = -\frac{1}{2}\cot^2 u - \ln|\sin u| + C$

71. $\int \sec^3 u\, du = \frac{1}{2}\sec u \tan u + \frac{1}{2}\ln|\sec u + \tan u| + C$

72. $\int \csc^3 u\, du = -\frac{1}{2}\csc u \cot u + \frac{1}{2}\ln|\csc u - \cot u| + C$

73. $\int \sin^n u\, du = -\frac{1}{n}\sin^{n-1}u \cos u + \frac{n-1}{n}\int \sin^{n-2}u\, du$

74. $\int \cos^n u\, du = \frac{1}{n}\cos^{n-1}u \sin u + \frac{n-1}{n}\int \cos^{n-2}u\, du$

75. $\int \tan^n u\, du = \frac{1}{n-1}\tan^{n-1}u - \int \tan^{n-2}u\, du$

76. $\int \cot^n u\, du = \frac{-1}{n-1}\cot^{n-1}u - \int \cot^{n-2}u\, du$

77. $\int \sec^n u\, du = \frac{1}{n-1}\tan u \sec^{n-2}u + \frac{n-2}{n-1}\int \sec^{n-2}u\, du$

78. $\int \csc^n u\, du = \frac{-1}{n-1}\cot u \csc^{n-2}u + \frac{n-2}{n-1}\int \csc^{n-2}u\, du$

79. $\int \sin au \sin bu\, du = \frac{\sin(a-b)u}{2(a-b)} - \frac{\sin(a+b)u}{2(a+b)} + C$

80. $\int \cos au \cos bu\, du = \frac{\sin(a-b)u}{2(a-b)} + \frac{\sin(a+b)u}{2(a+b)} + C$

81. $\int \sin au \cos bu\, du = -\frac{\cos(a-b)u}{2(a-b)} - \frac{\cos(a+b)u}{2(a+b)} + C$

82. $\int u \sin u\, du = \sin u - u\cos u + C$

83. $\int u \cos u\, du = \cos u + u \sin u + C$

84. $\int u^n \sin u\, du = -u^n \cos u + n\int u^{n-1}\cos u\, du$

85. $\int u^n \cos u\, du = u^n \sin u - n\int u^{n-1}\sin u\, du$

86. $\int \sin^n u \cos^m u\, du = -\frac{\sin^{n-1}u \cos^{m+1}u}{n+m} + \frac{n-1}{n+m}\int \sin^{n-2}u \cos^m u\, du$

$= \frac{\sin^{n+1}u \cos^{m-1}u}{n+m} + \frac{m-1}{n+m}\int \sin^n u \cos^{m-2}u\, du$

Inverse Trigonometric Forms

87. $\int \sin^{-1}u\, du = u\sin^{-1}u + \sqrt{1-u^2} + C$

88. $\int \cos^{-1}u\, du = u\cos^{-1}u - \sqrt{1-u^2} + C$

89. $\int \tan^{-1}u\, du = u\tan^{-1}u - \frac{1}{2}\ln(1+u^2) + C$

90. $\int u\sin^{-1}u\, du = \frac{2u^2-1}{4}\sin^{-1}u + \frac{u\sqrt{1-u^2}}{4} + C$

91. $\int u\cos^{-1}u\, du = \frac{2u^2-1}{4}\cos^{-1}u - \frac{u\sqrt{1-u^2}}{4} + C$

92. $\int u\tan^{-1}u\, du = \frac{u^2+1}{2}\tan^{-1}u - \frac{u}{2} + C$

93. $\int u^n \sin^{-1}u\, du = \frac{1}{n+1}\left[u^{n+1}\sin^{-1}u - \int \frac{u^{n+1}\,du}{\sqrt{1-u^2}}\right], \quad n \neq -1$

94. $\int u^n \cos^{-1}u\, du = \frac{1}{n+1}\left[u^{n+1}\cos^{-1}u + \int \frac{u^{n+1}\,du}{\sqrt{1-u^2}}\right], \quad n \neq -1$

95. $\int u^n \tan^{-1}u\, du = \frac{1}{n+1}\left[u^{n+1}\tan^{-1}u - \int \frac{u^{n+1}\,du}{1+u^2}\right], \quad n \neq -1$

TABLE OF INTEGRALS

Exponential and Logarithmic Forms

96. $\displaystyle\int ue^{au}\,du = \frac{1}{a^2}(au - 1)e^{au} + C$

97. $\displaystyle\int u^n e^{au}\,du = \frac{1}{a}u^n e^{au} - \frac{n}{a}\int u^{n-1}e^{au}\,du$

98. $\displaystyle\int e^{au}\sin bu\,du = \frac{e^{au}}{a^2 + b^2}(a\sin bu - b\cos bu) + C$

99. $\displaystyle\int e^{au}\cos bu\,du = \frac{e^{au}}{a^2 + b^2}(a\cos bu + b\sin bu) + C$

100. $\displaystyle\int \ln u\,du = u\ln u - u + C$

101. $\displaystyle\int u^n \ln u\,du = \frac{u^{n+1}}{(n+1)^2}[(n+1)\ln u - 1] + C$

102. $\displaystyle\int \frac{1}{u\ln u}\,du = \ln|\ln u| + C$

Hyperbolic Forms

103. $\displaystyle\int \sinh u\,du = \cosh u + C$

104. $\displaystyle\int \cosh u\,du = \sinh u + C$

105. $\displaystyle\int \tanh u\,du = \ln\cosh u + C$

106. $\displaystyle\int \coth u\,du = \ln|\sinh u| + C$

107. $\displaystyle\int \operatorname{sech} u\,du = \tan^{-1}|\sinh u| + C$

108. $\displaystyle\int \operatorname{csch} u\,du = \ln|\tanh \tfrac{1}{2}u| + C$

109. $\displaystyle\int \operatorname{sech}^2 u\,du = \tanh u + C$

110. $\displaystyle\int \operatorname{csch}^2 u\,du = -\coth u + C$

111. $\displaystyle\int \operatorname{sech} u\tanh u\,du = -\operatorname{sech} u + C$

112. $\displaystyle\int \operatorname{csch} u\coth u\,du = -\operatorname{csch} u + C$

Forms Involving $\sqrt{2au - u^2}$, $a > 0$

113. $\displaystyle\int \sqrt{2au - u^2}\,du = \frac{u-a}{2}\sqrt{2au - u^2} + \frac{a^2}{2}\cos^{-1}\left(\frac{a-u}{a}\right) + C$

114. $\displaystyle\int u\sqrt{2au - u^2}\,du = \frac{2u^2 - au - 3a^2}{6}\sqrt{2au - u^2} + \frac{a^3}{2}\cos^{-1}\left(\frac{a-u}{a}\right) + C$

115. $\displaystyle\int \frac{\sqrt{2au - u^2}}{u}\,du = \sqrt{2au - u^2} + a\cos^{-1}\left(\frac{a-u}{a}\right) + C$

116. $\displaystyle\int \frac{\sqrt{2au - u^2}}{u^2}\,du = -\frac{2\sqrt{2au - u^2}}{u} - \cos^{-1}\left(\frac{a-u}{a}\right) + C$

117. $\displaystyle\int \frac{du}{\sqrt{2au - u^2}} = \cos^{-1}\left(\frac{a-u}{a}\right) + C$

118. $\displaystyle\int \frac{u\,du}{\sqrt{2au - u^2}} = -\sqrt{2au - u^2} + a\cos^{-1}\left(\frac{a-u}{a}\right) + C$

119. $\displaystyle\int \frac{u^2\,du}{\sqrt{2au - u^2}} = -\frac{(u+3a)}{2}\sqrt{2au - u^2} + \frac{3a^2}{2}\cos^{-1}\left(\frac{a-u}{a}\right) + C$

120. $\displaystyle\int \frac{du}{u\sqrt{2au - u^2}} = -\frac{\sqrt{2au - u^2}}{au} + C$